A TEXTBOOK OF

AUTOMOBILE ENGINEERING

A TEXTBOOK OF
AUTOMOBILE ENGINEERING

[For the Students of B.E./B.Tech.; U.P.S.C. (Engg. Services); Section B of A.M.I.E. (India)]

S.I. UNITS

By

R.K. RAJPUT
M.E. (Heat Power Engg.) Hons.–Gold Medallist ; Grad. (Mech. & Elect. Engg.); M.I.E. (India); M.S.E.S.I.; M.I.S.T.E.; C.E. (India)
Patiala (*Punjab*)

(An ISO 9001:2008 Company)

BENGALURU • CHENNAI • COCHIN • GUWAHATI • HYDERABAD
JALANDHAR • KOLKATA • LUCKNOW • MUMBAI • RANCHI • NEW DELHI
BOSTON (USA) • ACCRA (GHANA) • NAIROBI (KENYA)

A TEXTBOOK OF AUTOMOBILE ENGINEERING

Typeset at Goswami Associates, Delhi
First Edition: 2007; Reprint: 2008, 2009, 2010, 2011, 2012, 2013, Second Edition: 2014, 2015, 2016, 2017 Edition: 2018
ISBN 978-81-7008-991-9

Branches

Bengaluru	080-26 75 69 30	
Chennai	044-24 34 47 26,	24 35 95 07
Cochin	0484-237 70 04,	405 13 03
Guwahati	0361-254 36 69,	251 38 81
Hyderabad	040-27 55 53 83,	27 55 53 93
Jalandhar	0181-222 12 72	
Kolkata	033-22 27 43 84	
Lucknow	0522-220 99 16	
Mumbai	022-24 93 12 61	
Ranchi	0651-220 44 64	

PUBLISHED IN INDIA BY

Laxmi Publications (P) Ltd.
(An ISO 9001:2008 Company)
113, GOLDEN HOUSE, DARYAGANJ,
NEW DELHI - 110002, INDIA
Telephone : 91-11-4353 2500, 4353 2501
Fax : 91-11-2325 2572, 4353 2528
www.laxmipublications.com info@laxmipublications.com

Contents

Chapter ***Pages***

Chapter **Pages**

Chapter **Pages**

PREFACE TO THE SECOND EDITION

I take pleasure in presenting the ***"Second Edition"*** of this book on **"Automobile Engineering"**. The warm reception which the previous edition has received is a matter of great satisfaction to me.

In this edition, the book has been thoroughly revised; a new chapter (No. 13) namely ***"Universities Questions* (Latest) with *Answers/Solutions"*** has been added to make the book more useful to the readers.

Any suggestions for improvement of the book are most welcome.

Er. R.K. Rajput
(***Author***)

PREFACE TO THE FIRST EDITION

The main objective of writing this book on the subject **"Automobile Engineering"** is to present to the student community, a book which contains comprehensive treatment of the subject matter in simple, lucid and direct language and envelopes an ample number of worked examples to illustrate the text.

The book contains twelve chapters in all. All these chapters are saturated with much needed text, supported by simple and self-explanatory figures and worked examples, wherever required. At the end of each chapter **Short Answer Questions, Highlights, Objective Type Questions, Theoretical Questions** and **Unsolved Examples** have been added to make the book a comprehensive and complete unit in all respects.

The book will prove to be a boon to the students preparing for engineering undergraduate, A.M.I.E. section B (India) and competitive examinations.

The author's thanks are due to his wife Ramesh Rajput for extending all cooperation during preparation of the manuscript.

In the end the author wishes to express his gratitude to Shri R.K. Gupta, Chairman, Shri Saurabh Gupta, Managing Director, Laxmi Publications Pvt. Ltd., New Delhi for taking pains in bringing out the book with a very good presentation in a short span of time.

Although every care has been taken to make the book free of errors both in text as well as in worked examples, yet the author shall feel obliged if any errors present are brought to his notice. Constructive criticism of the book will be warmly and thankfully welcomed.

Er. R.K. Rajput
(***Author***)

1

Introduction

1.1. Introduction to an automobile. 1.2. Brief history of automobile. 1.3. Classification of automobiles. 1.4. Parts of an automobile—mechanical portion—chassis and transmission—the engine—electrical system—body or carriage portion. 1.5. Description of an automobile. 1.6. Performance of an automobile—Short Answer Questions—Highlights—Objective Type Questions—Theoretical Questions.

1.1. INTRODUCTION TO AN AUTOMOBILE

Automobile. *An "automobile" is a self-propelled vehicle driven by an internal combustion engine and is used for transportation of passengers and goods on ground.* ***Examples :*** Bus, car, jeep, truck, tractor, scooter, motor cycle.

The modern automobile, in general, is essentially a transportation equipment unit. It consists of a *"frame"* supporting the *"body"* and certain *"power developing and transmitting units" which are further supported by "tyres and wheels" through "springs and axles".*

An *"engine"* supplies the power, which is delivered by the *"transmission system"* to the wheels through the *clutch or fluid coupling.*

Automobile engineering. *Automobile (or automotive) engineering is a branch of engineering in which we study all about the automobiles and have practice to propel them.*

— Mobile or motive means one which can move. Automobile or automotive means one *which itself can move.*

The different names for the automobile are :

- Auto
- Automobile
- Autocar
- Autobuggy
- Car
- Motor
- Motor car
- Motor vehicle
- Motor coach
- Motor wagon
- Horseless coach.

1.2. BRIEF HISTORY OF AUTOMOBILE

The famous years in early automobile history are as follows :

Year	Event
• *1769*	French engineer Captain Nicholas Cugnot of France built the first road vehicle propelled by its own power (Attained a speed of about 2.5. m.p.h. in 15 minutes).
• *1801*	First steam carriage built by Richard Trevithick in England.
• *1804*	Oliver Evans built the finest American self-propelled steam vehicle.
• *1827*	Onesiphare Pacqueur of France invented first differential.
• *1832*	First 3-speed Transmission patented by W.H. James in England.
• *1880*	German and French efforts developed an internal combustion engine vehicle (which was used to carry fruits). The present day automobile is the development of this vehicle.

- *1885* Benz in Germany developed tricycle propelled by an internal combustion engine.
- *1886* One of the first gasoline engine powered automobiles by Gottlieb Daimler of Germany.
- *1894* Panhard and Levassor in France developed a car which incorporated the chief features of the modern automobile.
- *1895* First motor car race held.
- *1897* *First car arrived in India.*
- *1900* The design of the automobile was so improved that it awakened the public to the fact that this new form of transportation was really practical for use.
- *1902* First volume of production car "The Curved Dash Oldsmobile" in America.
- *1906* The production and sale of these vehicles became a business.
- *1908* Ford 'T' model car produced in America by Ford Motor Company (Ford started his model with an initial run of 20000 vehicles, an output unheard at that time).
- *1911* First electric self-starter installed in the automobile.
- *1920* There was a gradual change and refinement in automobile design ....

The design development of recent years has provided the owners with cars that are :

— Safer ;

— Easier to drive ;

— Comfortable ;

— More reliable.

"Germany is the birth place of the automobile. It was invented there, it went through its first paces there and it was developed there to a high level of technical maturity. The list of German automobile pioneers is a long one. Starting with Nicholas Cugnot, August Otto, Carl Benz, Gottlieb Daimler, Wilhelm Maybach and Rudolf Diesel and going all the way upto Ferdinand Posche and Felin Wankel."

... From SCALA INTERNATIONAL

Automobile in India :

— In India the first motor car appeared in 1897. For about 50 years from 1897, India was an importer of automobiles.

— The late Bharat Ratna Sir M. Vivesvaraya made an automobile industry in India, but the government did not approve the plan.

— In 1943 and 1944 two automobile factories namely, Hindustan Motors Limited, Calcutta and premier Automobiles Limited, Bombay were set up in India.

At present, several automobile companies are manufacturing automobiles.

Use of Diesel engines in cars :

Following are the *advantages* of using Diesel engines in cars :

1. Diesel fuel is much cheaper than petrol.

2. The diesel engine uses 50 per cent less fuel in heavy traffic than does the available spark-ignition engine.

— One reason for extra economy is that diesel's *good low-speed torque makes it possible to drive at low speeds in a higher gear.*

3. The high thermal efficiency of the diesel engine, reflected in low specific fuel consumption gives its high advantages in the matter of range.

4. The crankshaft speed of about 5000 r.p.m. enables the diesel engine to propel the cars at high speeds, with acceleration comparable to that of S.I. engine.

5. The use of diesel engine eliminates the complication of spark-ignition with its spark plugs, breaker points, condenser, ignition cables, distributor rotor and other components that may have to be replaced several times during the life of engine.

6. Operating tests are further reduced by having no carburettor to service periodically.

7. Owing to the absence of electrical ignition equipment, non-volatile fuel and lower exhaust temperature, the *diesel engine is a much smaller fire risk* than the petrol engine when working in closed places and in the neighbourhood of inflammable material.

8. The special feature of diesel engines called "*torque back up*" (an inbuilt performance feature) makes it convenient for the vehicle to *climb a steep incline* (such as ghats) *with low consumption at high load.* Under these conditions diesel engines *run comparatively cold* than an equivalent petrol engine under the similar load conditions.

Inspite of all the above mentioned advantages uptill now, diesel engines for cars were rejected from majority of customers, due to their *heavy weight and size, high noise, vibration and smoke levels.* These problems finally led to the development of *compact, light weight and fuel efficient passenger car diesel engines.* In this connection, the reference needs to be mentioned to :

— Over-square bore-stroke ratio ;
— Fast responding value gear trains ;
— Cold starting equipments ;
— Combustion chamber design ;
— Fuel injection equipment ;
— Solution of metallurgical problems to ensure satisfactory working life of aluminium alloy components, crankshaft journals, bearings, cylinder bores and other highly stressed components.

- The sacrifice of German engineer Rudolf Diesel is memorable in Automobile and I.C. engine history. He got very seriously injured when he successfully exploded coal dust as fuel. And in 1913 although he invented Diesel engine using liquid fuel 'Diesel' but he committed suicide as he could not get financial co-operation from his friend for his research and development.

1.3. CLASSIFICATION OF AUTOMOBILES

Automobiles can be *classified with different regards* which are as under :

1. **Purpose :**

(*i*) Passenger carriers — Car, Jeep, Bus etc.
(*ii*) Goods carriers — Truck etc.

2. **Fuel used :**

(*i*) Petrol vehicle — Motor cycles, Scooters, Cars, Station wagons
(*ii*) Diesel vehicle — Trucks, Buses etc.
(*iii*) Gas vehicle — Coal gas turbine
(*iv*) Electric — Using storage batteries
(*v*) Steam — Using steam engine (obsolete).

3. **Capacity :**

(*i*) HTV (Heavy Transport Vehicles) or HMV (Heavy Motor Vehicles) — Trucks, Buses, Dumpers etc.
(*ii*) LTV (Light Transport Vehicles) — Pick up, Station wagon
(*iii*) LMV (Light Motor Vehicles) — Jeep, Cars
(*iv*) Medium vehicle — Station wagon, Tempo, Minibus and small trucks.

4. **Construction :**

(i) Single unit vehicles

(ii) Articulated vehicles and tractors.

5. **Drive :**

(i) Left hand drive — Steering wheel fitted on left hand side.

(ii) Right hand drive — Steering wheel fitted on right hand side.

(iii) Fluid drive — Vehicle employing torque converter, fluid flywheel or hydramatic transmission.

6. **Wheel and axle :**

(i) Two wheeler — Auto cycle, Mopped, Scooter and Motor cycle

(ii) Three wheeler — Three wheeler scooter, Auto rickshaw and Tempo.

(iii) Four wheeler (4 × 2) and (4 × 4) — Cars, Jeep, Station wagon, Pick up, Trucks, Buses.

(iv) Six wheeler (6 × 2) and (6 × 4) — Six wheeler having two gear axles each having four wheels. Vehicles are classified by number of wheels and number of drive wheels where in it is written 4 × 2 or 4 × 4 drive etc. In this case first figure indicates *number of wheels* and the second figure indicates number of *drive wheels*. In case of 4 × 4 drive, the vehicle has 4 wheels and all are drive wheels, however it should be noted that wheels are considered as unit whether they have single tyre or dual tyre.

7. **Suspension system :**

(i) Conventional — Leaf spring

(ii) Independent — Coil spring, Torsion bar, Pneumatic.

8. **Body and number of doors :**

(i) Sedan — Two doors and four doors types

(ii) Convertible — Jeep, Some imported cars

(iii) Station wagon

(iv) Delivery van/Pick ups

(v) Special purpose vehicle — Ambulances, Milk van, Mobile workshop, Mobile hospital etc.

9. **Transmission :**

(i) Conventional — All Indian vehicles

(ii) Semi-automatic — Most of British and Japanese vehicles

(iii) Automatic — Most of the American vehicles.

A *brief classification* of the automobiles is as follows :

1. **Passenger carriers :**

(i) ***Light***

- Scooterets
- Scooters
- Mopeds
- Motor cycles
- Cars
- Pick-ups
- Station wagons.

(*ii*) ***Heavy***

- Buses
  — Single deck
  — Double deck
- Coaches
  — Delux
  — Air-conditioned.

2. **Goods carriers :**

(*i*) ***Light***

- Delivery van
- Light truck

(*ii*) ***Heavy***

- Truck
- Tractor trailer.

1.4. PARTS OF AN AUTOMOBILE

Every automobile consists of the following *two main parts* :

1. Machine portion
2. Carriage portion *i.e.*, Body.

1.4.1. Machine Portion

Refer to Fig. 1.1. (Automobile parts)

Every automobile irrespective of its country of manufacture or model consists of the following *three basic units* :

(*i*) The "*chassis and transmission*".

(*ii*) The "*engine*".

(*iii*) The "*electrical equipment*".

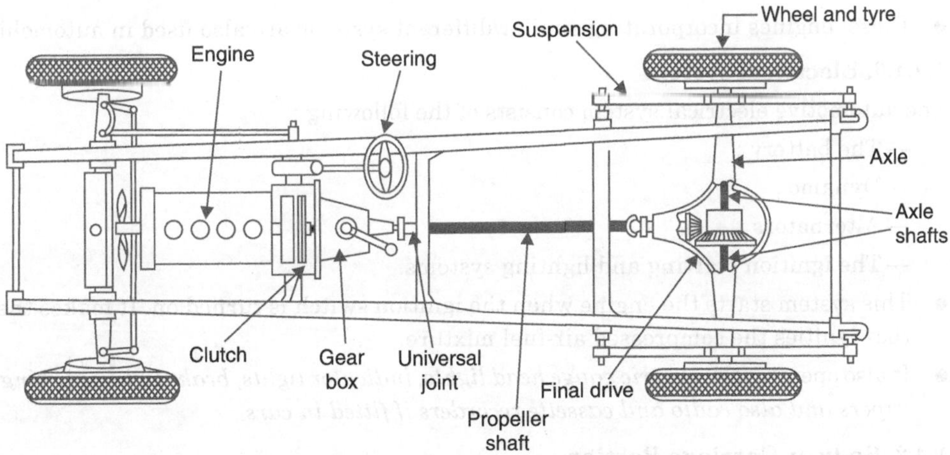

Fig. 1.1. Parts of an automobile.

1.4.1.1. Chassis and transmission

Chassis :

- This part of the automobile supports its body, engine and transmission system.
- The automotive chassis *includes* the following :
  — The frame ;
  — Springs and shock absorbers ;

— Steering system ;

— Brakes ;

— Tyres and wheels.

Transmission :

- This unit *transmits the power from the engine to the wheels.*
- It consists of the following :

— Clutch ;

— Gear box ;

— Universal joint ;

— Final drive ;

— Axles and differential.

1.4.1.2. The engine

An engine in an automobile is the *source of power*. A *petrol engine* consists of the following four basic system, described briefly below :

(*i*) **Full system.** This system mixes petrol with air in the right proportion to give a mixture which when burnt produces pressure. This pressure is then used to move the piston(s).

(*ii*) **Ignition system.** This system of the engine provides regular sparks to set fire to the mixture coming from the fuel system.

(*iii*) **Lubricating system.** This system lubricates the moving parts of the engine so that they can work smoothly.

(*iv*) **Cooling system.** This system with the help of water cools the engine and prevents it from getting hot.

- Diesel engines incorporating similar/different systems are also used in automobiles.

1.4.1.3. Electrical system

The automotive electrical system consists of the following :

— The battery ;

— Dynamo ;

— Alternators ;

— The ignition starting and lighting systems.

- This system starts the engine when the ignition switch is turned on. It makes the spark that ignities the compressed air-fuel mixture.
- It also operates the *electric gauge head lights, indicator lights, brake lights, parking lights, wipers and also radio and cassette recorders if fitted in cars.*

1.4.2. Body or Carriage Portion

- It is that portion of an automobile where passengers have their seats or where the cargo to be carried is placed.
- The body is designed according to the nature of cargo to be carried. The body of a passenger carrier is much different from the body of a load carrier, *i.e.*, truck. Its design depends upon the utility for which the vehicle is meant for. It is made of either wood and steel or steel alone. Modern research has led to the development of plastic body.
- In trucks, tractors and certain other vehicles, a separate cabin know as '*Cab*' is provided for the driver where the instrument panel and controls are housed.

The *layout of an automobile* is shown in Fig. 1.2.

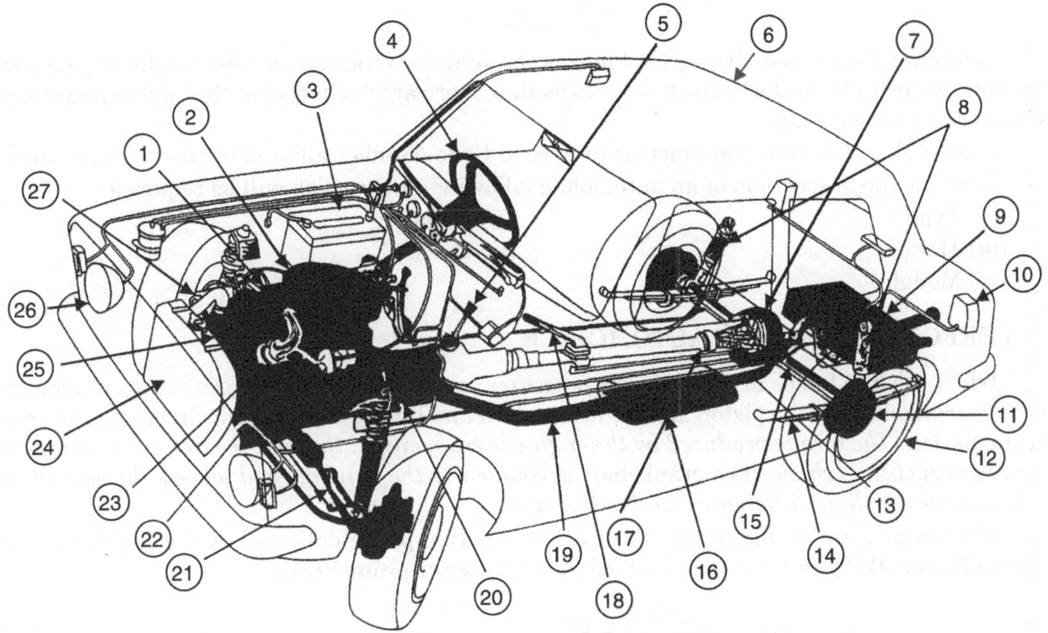

1. Engine 2. Air cleaner 3. Battery 4. Steering wheel 5. Gear lever 6. Car body 7. Differential 8. Shock obsorber 9. Fuel tank 10. Back light 11. Rear brake 12. Wheel and tyres 13. Tail pipe 14. Leaf spring 15. Rear axle 16. Silencer 17. Universal joint 18. Hand brake lever 19. Exhaust pipe 20. Clutch 21. Track rod 22. Carburettor 23. Distributor 24. Radiator 25. Fan 26. Head lamp 27. Alternator.

Fig. 1.2. Layout of an automobile.

1.5. DESCRIPTION OF AN AUTOMOBILE

The following factors should be taken into consideration while writing down the description of an automobile.

1. **Type ....** Bus, truck, car, motor cycle etc.

2. **Capacity .....**Carriage capacity— 5 ton, 3 ton etc. ; 4 seater, 6 seater, 30 seater, 45 seater etc.

3. **Make.** It is the actual name allotted by the manufacturer. In most cases, the make also indicates capacity/H.P. of the engine fitted in the vehicle, such as *Maruti 800.* This means that in Maruti make of car 800 c.c. engine, the total piston displacement is about 800 c.c.

4. **Drive.** The description of an automobile with regard to drive may be given as follows:

(*i*) *Right hand or left hand drive.* It means whether the steering is fitted on the right side or left side.

(*ii*) *Two wheel drive ; 4 wheel drive ; 6 wheel drive.* This means as to how many wheels the engine power flows or how many wheels are directly connected with the engine.

— In *majority of the cars the engine power flows to the rear wheel only and the front wheels are filled on the dead axle.* These types of cars are known as *"two wheel drive vehicles."*

— In certain vehicles, like jeep, all the vehicles are directly in contact with engine and the engine *power could be transmitted to all the four wheels.*

- Drive is usually indicated as under :

Left hand drive ; 4 × 4 (4 wheel drive)

Or

Left hand, Four wheel drive ; **4 × 4** means the vehicle contains 4 wheels and the engine power could flow towards all the 4 wheels, **6 × 4** means that there are 6 wheels but the engine power could flow towards 4 wheels only.

5. **Model.** ......Year of manufacture or special Code Number allotted by the manufacturer.

Thus for the description of an automobile following *information* will be required :

(*i*) Type (*ii*) Capacity
(*iii*) Make (*iv*) Drive
(*v*) Model.

1.6. PERFORMANCE OF AN AUTOMOBILE

When the fuel burns in the cylinder, *pressures* are developed. These pressures are transmitted to the crankshaft by the piston and connecting rod and torque is produced which sets the crankshaft in motion. *The torque produced by the engine is transmitted through the drive line to the road wheels to propel the vehicle* (The crankshaft is *coupled to the driving road wheels through clutch, gearbox, propeller shaft, differential and axle shafts*).

The torque is measured in Nm (SI units) ; the actual power delivered by the engine is known as Brake Power (B.P.) and is measured by dynamometer or prony brake.

$$\text{B.P.} = \frac{2\pi NT}{60 \times 1000} \text{ kW} \quad ...(1.1)$$

where, T = Torque, Nm, and

N = Speed in r.p.m. (revolutions per minute).

— The torque increases with the increase in engine speed upto a certain point after which it starts to fall down even though the engine speed continues to increase. The number of r.p.m. at which the torque begins to decrease, depends upon *engine design*. At higher speeds, engine vacuum falls down and less fuel enters the cylinders resulting in lesser force available at the piston and hence the fall in torque as shown in Fig. 1.3.

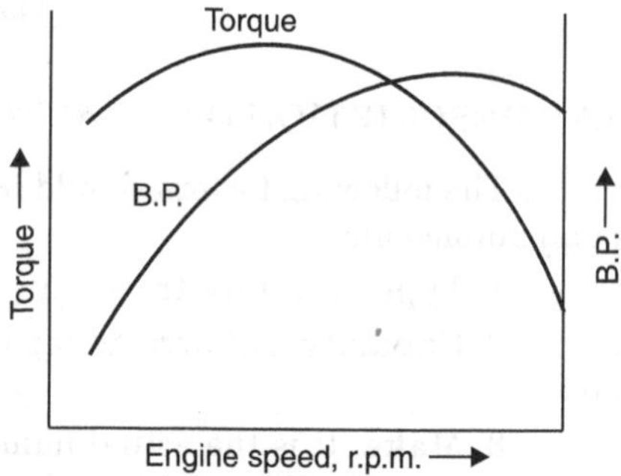

Fig. 1.3. Typical curves of torque and B.P. with speed of an engine.

The torque available at the contact between driving wheels and road is referred to as *tractive effort. Gear box and final drive at differential act as leverage to multiply torque which is inversely proportional to speed.* If the gear speed is lowered, the torque shall be increased in the same ratio and vice versa.

Let, T_w = Torque at driving wheels,

G = Gear box ratio,

η_t = Overall transmission efficiency,

T_E = Engine torque (Nm), and

N = r.p.m. of the crankshaft.

Then, $$T_w = G \times \eta_t \times T_E \quad ...(1.2)$$

Engine torque, $$T_E = \frac{B.P. \times 60 \times 1000}{2\pi N} \text{ Nm}, \quad ...(1.3)$$

where B.P. is in kW.

Tractive effort, $$F = \frac{T_w}{R_w} \quad \text{...(1.4)}$$

where, R_w = Radius of the driving wheel.

The ratio between engine r.p.m. and vehicle speed depends upon overall gear ratio. A vehicle having four speed gear box shall have four different speeds and ratio between engine r.p.m. and vehicle speed shall be different.

$$\text{R.P.M. of driving wheel} = \frac{V}{2\pi R_w}$$

where, V = Vehicle speed in metres/min., and

R_w = Radius of wheel in metres.

$$\text{Vehicle speed} = \text{Wheel circumference} \times \frac{N}{G}$$

or,
$$V = \frac{2\pi R_w N}{G} \text{ m/min} \quad \text{...(1.5)}$$

$$\therefore \quad \text{Engine r.p.m.,} \quad N = \frac{V \times G}{2\pi R_w} \quad \text{...[1.5(a)]}$$

$$\text{Vehicle speed,} \quad \frac{V \times 1000}{60} = \frac{2\pi R_w N}{G}$$

where V is in km/h.

∴ The *ratio between engine r.p.m.* (N) and vehicle speed (V)

or,
$$\frac{N}{V} = \frac{1000 \times G}{2\pi R_w \times 60} = 2.65 \left(\frac{G}{R_w}\right) \quad \text{...(1.6)}$$

Also,
$$V = \frac{2\pi R_w N}{G} \text{ m/min} \quad \text{[Eqn. (1.5) as above]}$$

$$\therefore \quad G = \frac{2\pi R_w N}{V} \quad \text{...(1.7)}$$

- The engine torque (T_E) can be *increased by reduction gearing*. The torque transmitted by the engine through gearbox and propeller shaft to the final drive is *increased in every gear speed except in top (direct) and overdrive.*
- The torque transmitted by propeller shaft is further increased by means of gear reduction of final drive (drive piston and ring gear at differential). The *torque of final drive, provided a differential is fitted, is always equally divided between each axle shaft irrespective of speed of road vehicles although it does not apply to limited-slip type of differential.*
- *The speed of propeller shaft is always less than the engine speed except in top gear or when overdrive is engaged.*
- *The speeds of axle shafts are always less than the speed of propeller shaft owing to final drive gear reduction.*

Power at Driving Wheels :

The power available at the driving wheels to drive the vehicle ranges from about 60 to 75%. The various power losses which take place between engine and the driving wheels are :

(*i*) Power loss due to friction of piston, bearings and gears in the engine (The power available at engine flywheel is about 85%).

(*ii*) Power loss from clutch to drive wheels due to friction in clutch, gearbox, universal joints, final drive, differential and between tyres and ground.

(*iii*) Transmission line losses.

The power lost in transmission of power from engine to road wheels reflects the transmission efficiency (η_t) which is taken into account while calculating power available at road wheels.

The *thrust* known as *tractive effort* provided by the engine at the driving road wheels varies at different engine speeds and gear positions.

A moving vehicle is opposed by various forces known as *resistances.* For keeping the vehicle moving, a driving force or tractive effort (*F*) equal to the sum of all the resistances has to be applied to it. When *F* exceeds the sum of the resistances, the excess value *F* will accelerate the vehicle, where as, when *F* is less than the sum of the resistances, the vehicle, will decelerate.

The main forces which *oppose* the motion of a vehicle are :

1. Rolling resistance.
2. Wind or air resistance
3. Gradient resistance.

1. **Rolling resistance (R_r) :**

- It is *mainly due to the friction between wheel tyres and road surface.* It depends upon the following *factors* :
  — Load on each road wheel ;
  — Type of tyre tread ;
  — Wheel inflation pressure ;
  — Nature of road surface.
- It is measured in kg or N and is expressed as kg/tonne or N/tonne of vehicle weight or as a *percentage of the vehicle weight.* Rolling resistance on average type of road surface is between **1 to 2% of vehicle weight.**

Rolling resistance, $R_r = k_r W$

where, k_r = Constant of rolling resistance, and
W = Total weight of vehicle.

2. **Wind or air resistance (R_a) :**

- This type of resistance depends upon the following **factors** :
  — The shape and size of vehicle body ;
  — Air velocity ;
  — Speed of the vehicle.
- It *increases as the square of the vehicle speed* owing to which much importance is given to streamlining and frontal area of modern automobiles. In calculating air resistance, air velocity is usually neglected.

Air resistance, $R_a = k_a A V^2$

where, k_a = Coefficient of air resistance,
A = Projected frontal area, m^2, and
V = Vehicle speed, km/h.

The values of k_a for "*best streamlined cars,*" "*average cars*" and "*buses and trucks*" are 0.00235, 0.0032 and 0.0046 respectively.

3. **Gradient resistance (R_g) :**

- This resistance is due to *steepness of road gradient.* It is subject to vehicle weight and road gradient. It *does not depend upon vehicle speed.*
- Gradient resistance, $R_g = \frac{W}{G}$, or, $W \sin \theta$

(if gradient is expressed in angular dimensions)

where, W = Total weight of vehicle,

G = Gradient, and

θ = Inclination. (For small values, $\tan \theta = \sin \theta$)

When the vehicle is moving along a level road, total resistance,

$$R_{total} = R_r + R_a, \text{ and}$$

while moving up a gradient,

$$R_{total} = R_r + R_a + R_g.$$

Power-to-weight Ratio :

- An automobile's performance much depends upon its power-to-weight ratio. The *best performance can be achieved by keeping the weight of the vehicle down to a minimum and installing engines of high B.P.* The higher the effective B.P. of the engine and lower the total weight of the vehicle, the better will be its hill climbing abilities, the higher the maximum speed and better its acceleration.
- A *well designed streamlined car having a high power-to-weight ratio registers a low fuel consumption at any given speed.*
- Power-to-weight ratio (B.H.P. per ton) in small and medium cars ranges from 30 to 90 ; special high performance cars have the ratios upto 230.
- Considering maximum speed, there should be *minimum body resistance* in addition to high power-to-weight ratio because the *air resistance* of an automobile body and chassis *increases as the square of the speed whereas the power varies as cube of the speed.*

Although performance of an automobile is governed by the power-to-weight ratio and the resistances affecting its movement yet the *road performance is usually assessed by its rate of acceleration from rest and the top speed attainable on level roads.* The acceleration is usually measured in terms of time the car takes to reach a given speed from rest.

Note. The progressive increase in power-to-weight ratio has brought about improvement in the performance of an automobile. The *total weight* of the automobile has been *reduced* considerably by the *use of lighter materials and improved methods of chassis and body construction.* The *power output* of the engine has been *increased* by using aluminium and magnesium alloys, *improved piston, piston ring, cylinder and combustion chamber design, increased compression ratios due to use of aluminium alloy pistons and cylinder heads, better engine balancing, lubrication and cooling systems, fuels of higher octane values, balanced fuel supply and correct ignition.*

SHORT ANSWER QUESTIONS

Q. 1. What is an automobile ?

Ans. An automobile is a *self-propelled vehicle* which is used for the transportation of passengers and goods on the ground.

Q. 2. What is a self-propelled vehicle ?

Ans. A self-propelled vehicle is one is which the power required for the propulsion is produced *in the vehicle itself.*

Q. 3. Give six examples of automobiles.

Ans. The six examples of automobiles are :

(*i*) Car (*ii*) Bus (*iii*) Truck
(*iv*) Tractor (*v*) Motor Cycle (*vi*) Scooter.

Q. 4. State the basis on which automobiles are classified.

Ans. Following are the basis on which automobiles are classified :

Purpose ; Fuel used ; Capacity ; Construction ; Drive ; Wheel and axel ; Suspension system ; Body and number of doors ; Transmission.

Q. 5. Enumerate the specifications which should be considered while purchasing an automobile.

Ans. Following specifications should be considered while purchasing an automobile :

Type ; Capacity ; Make ; Drive ; Model.

Q. 6. Who built the first road vehicle propelled by its own power ?

Ans. French engineer Captain Nicholas Cugnol (Year-1769).

Q. 7. Give examples of Indian manufactures of scooters.

Ans. Andhra Pradesh scooters ; Bajaj Auto ; Lohia Machines ; Scooter India ; Girnar Scooters.

Q. 8. Give examples of Indian manufacturers of passenger cars.

Ans. Premier Automobiles Ltd. Bombay ; Hindustan Motors Ltd., Calcutta ; Maruti Udyog Ltd., Gurgaon ; Standards Motor Products of India Ltd., Madras ; Sipani Automobiles ; Mohindra and Mohindra Ltd., Bombay ; Telco, Poona ; Hyundai Motor India Ltd.

Q. 9. Give examples of Indian manufacturers of tractors and three wheelers.

Ans. Eicher Tractors Ltd. Faridabad ; Escorts Ltd., Faridabad ; Escorts Tractors Ltd., New Delhi ; Gujrat Tractor Corpn. Ltd., Baroda ; HMT Pinjore ; Mohindra and Mohindra, Bombay ; Punjab Tractors, Mohali ; Tractors and Farm Equipment Ltd. ; Bajaj Auto Ltd., Pune ; Automobile Products of India.

HIGHLIGHTS

1. An "*automobile*" is a self-propelled vehicle driven by an internal combustion engine and is used for transportation of goods and passangers on ground.
2. Every automobile irrespective of its country of manufacture or model consists of the following *three basic units* :
   (*i*) The chassis and transmission ;
   (*ii*) The engine ;
   (*iii*) The electrical equipment.
3. The following factors should be taken into consideration while writing down the description of an automobile :
   (*i*) Type (*ii*) Capacity
   (*iii*) Make (*iv*) Drive
   (*v*) Model.
4. The engine torque can be increased by reduction gearing.
5. The main forces which oppose the motion of a vehicle are :
   (*i*) Rolling resistance (*ii*) Wind or air resistance
   (*iii*) Gradient resistance.
6. A well designed streamlined car having a high power-to-weight ratio register a low fuel consumption at any given speed.

OBJECTIVE TYPE QUESTIONS

Fill in the Blanks or Say "Yes" or "No"

1. An ...... is a self-propelled vehicle driven by an internal combustion engine and is used for transportation of passengers and goods on ground.
2. Mobile or motive means one which cannot move.
3. First motor car race was held in 1895.
4. First car arrived an India in ....
5. ...... is the birth place of the automobile.
6. Left hand drive means steering wheel fitted on left hand side.
7. Every automobile consists of machine portion and carriage portion.
8. Chassis of the automobile supports its body, engine and ...... system.
9. The ...... unit transmits the power from the engine to the wheels.
10. An ...... in an automobile is the source of power.

Answers

1. automobile	**2.** No	**3.** Yes	**4.** 1897	**5.** Germany
6. Yes	**7.** Yes	**8.** transmission	**9.** transmission	**10.** engine.

THEORETICAL QUESTIONS

1. Define the following terms :
   (*i*) Vehicle (*ii*) Motor vehicle
   (*iii*) Automobile (*iv*) Automobile engineering.
2. What do you understand by Automobile engineering ? What is the importance of automobile engineering over the other branches of engineering ?
3. Give a brief history of automobile.
4. What are the advantages of diesel engines in cars ?
5. Give the classification of automobiles.
6. What are the different types of automobile vehicles.
7. Explain briefly various parts of an automobile with the help of simple sketch.
8. Describe briefly the following three basic units of an automobile :
   (*i*) The chassis and transmission.
   (*ii*) The engine.
   (*iii*) The electrical equipment.
9. Explain briefly the factors which should be taken into consideration while writing down the description of an automobile.
10. Explain briefly the main forces which oppose the motion of a vehicle.
11. Explain the effect of "Power-to-weight" ratio on an automobile's performance.

2

Power Unit—Automobile Engines

I. Introduction to internal combustion engines : 2.1. Heat engines. 2.2. Development of I.C. engines. 2.3. Classification of I.C. engines. 2.4. Applications of I.C. engines. 2.5. Engine cycle-Energy balance. 2.6. Basic idea of I.C. engines. 2.7. Different parts of I.C. engines. 2.8. Terms connected with I.C. engines. 2.9. Working cycles. 2.10. Indicator diagram. 2.11. Four-stroke cycle engines. 2.12. Two stroke cycle engines. 2.13. Intake for compression ignition engines. 2.14. Comparison of four stroke and two stroke cycle engines. 2.15. Comparison of spark ignition (S.I.) and compression ignition (C.I.) engines. 2.16. Comparison between a petrol engine and a diesel engine. 2.17. How to tell a two stroke cycle engine from a four stroke cycle engine ?

II. Combustion in S.I. engines : 2.18. Introduction—Definition of combustion—Ignition limits. 2.19. Combustion phenomenon—Normal combustion—Abnormal combustion. 2.20. Effect of engine variables on ignition lag. 2.21. Spark advance and factors affecting ignition timing. 2.22. Pre-ignition. 2.23. Detonation—Introduction—Process of detonation or knocking—Theories of detonation—Effects of detonation—Factors affecting detonation/knock. 2.24. Performance number (PN). 2.25. Highest useful compression ratio (HUCR). 2.26. Combustion chamber design—S.I. engines—Induction swirl—Squish and tumble—Quench area—Turbulence—Flame propagation—Swirl ratio—Surface-to-volume ratio—Stroke-to-bore ratio—Compression ratio (C.R.). 2.27. Some types of combustion chambers—Divided combustion chambers.

III. Combustion in C.I. engines : 2.28. Introduction. 2.29. Combustion phenomenon in C.I. engines. 2.30. Fundamentals of the combustion process in diesel engines. 2.31. Delay period (or ignition lag) in C.I. Engines. 2.32. Diesel knock. 2.33. C.I. engine combustion chambers—Primary considerations in the design of combustion chambers for C.I. engines—Basic methods of generating air swirl in C.I. engines combustion chambers—Types of combustion chambers. 2.34. Cold starting of C.I engines.

IV. Two stroke engines : 2.35. General aspects—Construction and working—Comparison between two stroke cycle and four stroke cycle engines—Disadvantages of two stroke S.I. engine compared to two stroke C.I. engine—Reasons for use of two stroke engines for marine propulsion—Reasons for the use of two stroke S.I. engines for low horse power two wheelers. 2.36. Intake for two stroke cycle engines. 2.37. Scavenging process. 2.38. Scavenging parameters. 2.39. Scavenging systems. 2.40. Crankcase scavenging. 2.41. Scavenging pumps and blowers.

V. Fuels for I.C. engines : 2.42. Conventional fuels (for I.C. engines)—Introduction—Desirable properties of good I.C. engine fuels—Gaseous fuels—Liquid fuels—Structure of petroleum—Petroleum and composition of crude oil—Fuels for spark-ignition engines—Knock rating of S.I. engines fuels—Miscellaneous properties of S.I. engines fuels—Diesel fuel. 2.43. Alternative fuels for I.C. engines—General aspects—Advantages and disadvantages of using alternative fuels—Alcohol—Alcohol-gasoline fuel blends—Hydrogen—Natural gas (methane)—LPG and LNG—Biogas.

VI. Supercharging of I.C. engines : 2.44. Purpose of supercharging. 2.45. Supercharging of S.I. engines—Naturally aspirated cycle of operation—Supercharged cycle of operation—Comparison of actual naturally aspirated and supercharged engine pressure–volume diagrams—Boost pressure and pressure ratio—The effect of pressure ratio on air charge temperature—Thermodynamic cycle and supercharging power—Supercharging limits of S.I. engines. 2.46. Supercharging of C.I.

I. Introduction to Internal Combustion Engines

2.1. HEAT ENGINES

Any type of engine or machine which derives heat energy from the combustion of fuel or any other source and converts this energy into mechanical work is termed as a ***heat engine.***

Heat engines may be *classified* into two main classes as follows :

1. External Combustion Engines.
2. Internal Combustion Engines.

1. **External combustion engines (E.C. engines) :**

In this case, *combustion of fuel takes place outside the cylinder as in case* of *steam engines* where the heat of combustion is employed to generate steam which is used to move a piston in a cylinder. Other examples of external combustion engines are *hot air engines, steam turbine* and *closed cycle gas turbine*. These engines are generally used for driving locomotives, ships, generation of electric power etc.

2. **Internal combustion engines (I.C. engines) :**

In this case, *combustion of the fuel with oxygen of the air occurs within the cylinder of the engine.* The internal combustion engines group includes engines employing mixtures of combustible gases and air, known as *gas engines*, those *using lighter liquid fuel* or *spirit* known as *petrol engines* and those using heavier liquid fuels, known as *oil compression ignition* or *diesel engines*.

- Now a days, **I.C. engines** *are most-commonly used in* **automobiles.**

The detailed classification of heat engines is given in Fig. 2.1.

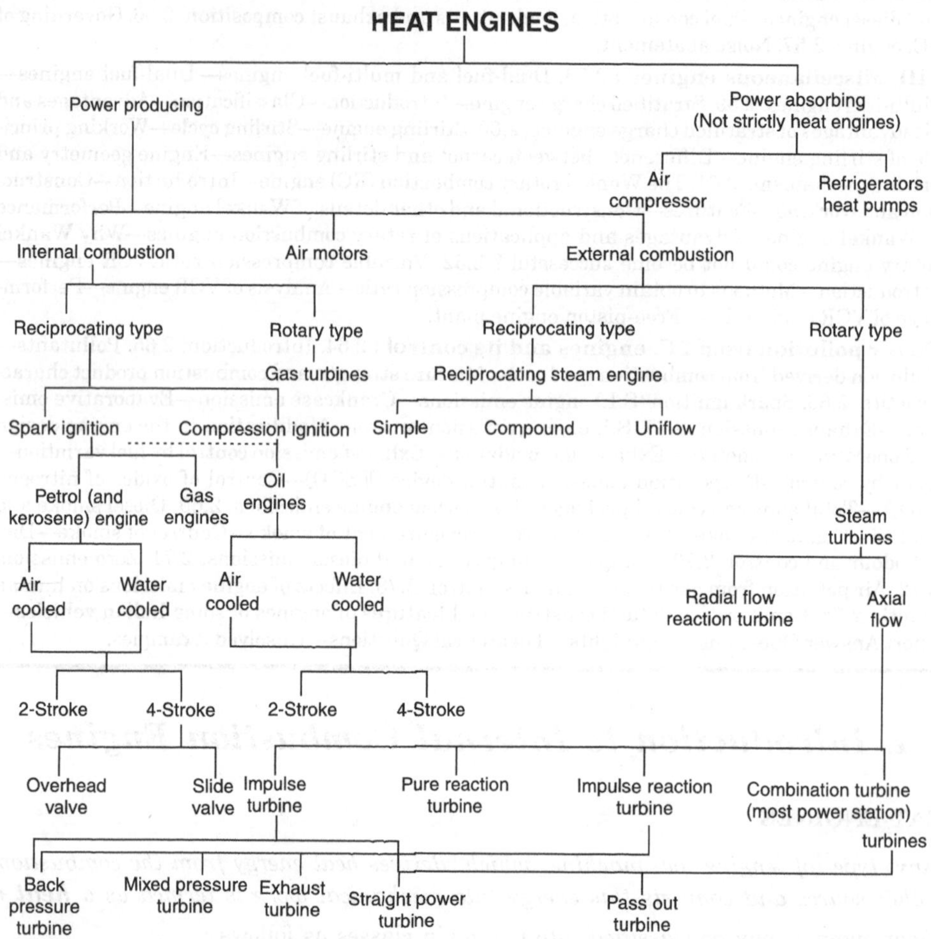

Fig. 2.1. Classification of heat engines.

Advantages of reciprocating internal combustion engines over external combustion engines :

Reciprocating internal combustion engines offer the following *advantages* over external combustion engines :

1. Overall efficiency is high.
2. Greater mechanical simplicity.
3. Weight-to-power ratio is generally low.
4. Generally lower initial cost.
5. Easy starting from cold conditions.
6. These units are compact and thus require less space.

Advantages of the external combustion engines over internal combustion engines :

The external combustion engines claim the following *advantages* over internal combustion engines :

1. Starting torque is generally high.
2. Because of external combustion of fuel, cheaper fuels can be used. Even solid fuels can be used advantageously.
3. Due to external combustion of fuel it is possible to have flexibility in arrangement.
4. These units are self-starting with the working fluid whereas in case of internal combustion engines, some additional equipment or device is used for starting the engines.

2.2. DEVELOPMENT OF I.C. ENGINES

Brief early history of development of I.C. engines is as follows :

- Many different styles of internal combustion engines were built and tested during the seond half of the 19th century.
- The first fairly practical engine was invented by J.J.E. Lenoir which appeared on the scene about 1860. During the next decade, several hundred of these engines were built with power upto about 4.5 kW and mechanical efficiency upto 5%.
- The Otto-Langen engine with efficiency improved to about 11% was first introduced in 1867 and several thousands of these were produced during the next decade. This was a type of atmospheric engine with the power stroke propelled by atmospheric pressure acting against a vacuum.
- Although many people were working on four-stroke cycle design, Otto was given credit when his prototype engine was built in 1876.
- In the 1880s, the internal combustion engines first appeared in automobiles. Also in this decade the two-stroke cycle engine became practical and was manufactured in large number.
- Rudolf Diesel, by 1892, had perfected his compression ignition engine into basically the same diesel engine known today. This was after years of development work which included the use of solid fuel in his early experimental engines.
- *Early compression engines were noisy, large, slow, single cylinder engines. They were, however, generally more efficient than spark ignition engines.*
- It wasn't until the 1920s that multicylinder compression ignition engines were made small enough to be used with automobile and trucks.
- *Wakdle's first rotary engine* was tested at NSV, Germany in 1957.
- The practical *stirling engines* in small number are being produced since 1965.
  - These engines require costly material and advanced technology for manufacture.
  - Thermal efficiencies higher than 30% have been obtained.
  - The *advantages* of stirling engine *are low exhaust emission and multi-fuel capability.*

2.3. CLASSIFICATION OF I.C. ENGINES

Internal combustion engines may be classified as given below :

1. **According to cycle of operation :**

(*i*) Two stroke cycle engines

(*ii*) Four stroke cycle engines.

2. **According to cycle of combustion :**

(*i*) Otto cycle engine (combustion at constant volume)

(*ii*) Diesel cycle engine (combustion at constant pressure)

(*iii*) Dual-combustion or Semi-Diesel cycle engine (combustion partly at constant volume and partly at constant pressure).

3. **According to arrangement of cylinder :** Refer Fig. 2.2.

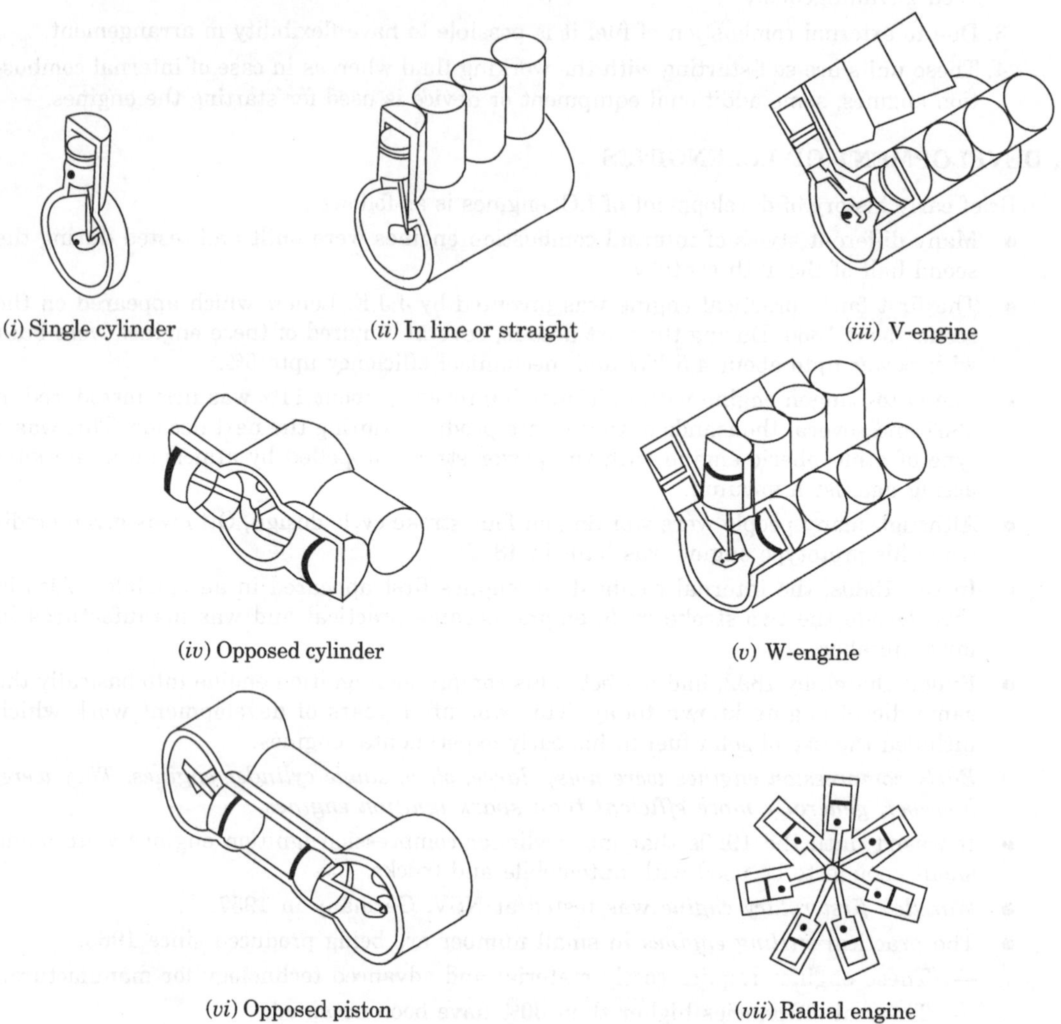

Fig. 2.2. Engine classification by cylinder arrangement.

(*i*) **Single cylinder engine.** Engine has one cylinder and piston connected to the crankshaft.

(*ii*) **In-line or straight engines.** Cylinders are positioned in a straight line one behind the other along the length of the crankshaft.

(*iii*) **V-engine :**

- An engine with two cylinder banks (*i.e.,* two-in-line engines) inclined at an angle to each other and with one crankshaft.

- Most of the bigger automobiles use the 8-cylinder V-engine (4-cylinder in-line on each side of V).

(*iv*) **Opposed cylinder engine :**

- Two banks of cylinders opposite to each other on a single crankshaft (a V-engine with 180° V).
- These are common on small aircraft and some automobiles with even number of cylinders from two to eight or more.

(*v*) **W-engine :**

- Same as V-engine except with three banks of cylinders on the same crankshaft.
- Not common, but some have been developed for racing automobiles.

(*vi*) **Opposed piston engine :**

- In this type of engine there are two pistons in each cylinder with the combustion chamber in the centre between the pistons.
- A single combustion process causes two power strokes, at the same time, with each piston being pushed away from the centre and delivering power to a separate crankshaft at each end of this cylinder.

(*vii*) **Radial engine :**

- It is an engine with pistons positioned in a circular plane around the central crankshaft. The connecting rods of the pistons are connected to a master rod which, in turn, is connected to the crankshaft.
- In a radial engine the bank of cylinders always has an *odd* number of cylinders ranging from 3 to 13 or more.
- Operating on a four-stroke cycle, every other cylinder fires and has a power stroke as the crankshaft rotates, giving a smooth operation.
- Many medium and large size propeller-driven aircraft use radial engines. For large aircraft two or more banks of cylinders are mounted together, one behind the other on a single crankshaft, making one powerful smooth engine.
- Very large ship engines exist with upto 54 cylinders, six banks of 9 cylinders each.

4. **According to their uses :**

(*i*) Stationary engine (*ii*) Portable engine
(*iii*) Marine engine (*iv*) Automobile engine
(*v*) Aero-engine etc.

5. **According to the speed of the engine :**

(*i*) Low speed engine (*ii*) Medium speed engine
(*iii*) High speed engine.

6. **According to method of ignition :**

(*i*) Spark-ignition engine (*ii*) Compression-ignition engine.

7. **According to method of cooling the cylinder :**

(*i*) Air-cooled engine (*ii*) Water-cooled engine.

8. **According to method of governing :**

(*i*) Hit and miss governed engine (*ii*) Quality governed engine
(*iii*) Quantity governed engine.

9. **According to valve arrangement :**
(*i*) Over head valve engine
(*ii*) *L*-head type engine
(*iii*) *T*-head type engine
(*iv*) *F*-head type engine.

10. **According to number of cylinders :**
(*i*) Single-cylinder engine
(*ii*) Multi-cylinder engine.

11. **According to air intake process :**
(*i*) *Naturally aspirated.* No intake air pressure boost system.

(*ii*) *Supercharged.* Intake air pressure increased with the compressor driven off the engine crankshaft.

(*iii*) *Turbocharged.* Intake air pressure increased with the turbine-compressor driven by the engine exhaust gases.

(*iv*) *Crankcase compressed.* Two stroke-cycle engine which uses the crankcase as the intake air compressor. Limited development work has also been done on the design and construction of four-stroke cycle engines with crankcase compression.

12. **According to fuel employed :**
(*i*) Oil engine
(*ii*) Petrol engine
(*iii*) Gas engine
(*iv*) Kerosene engine
(*v*) LPG engine
(*vi*) Alcohol-ethyl, methyl engine
(*vii*) Dual-fuel engine
(*viii*) Gasohol (90% gasoline and 10% alcohol).

13. **Method of fuel input for S.I. engines :**
(*i*) *Carburetted.*
(*ii*) *Multipoint port fuel injection.* One or more injectors at each cylinder intake.
(*iii*) *Throttle body fuel injection.* Injectors upstream in intake manifold.

2.4. APPLICATION OF I.C. ENGINES

The I.C. engines are generally used for :
(*i*) Road vehicles (*e.g.*, scooter, motorcycle, buses etc.)
(*ii*) Aircraft
(*iii*) Locomotives
(*iv*) Construction in civil engineering equipment such as bull-dozer, scraper, power shovels etc.
(*v*) Pumping sets
(*vi*) Cinemas
(*vii*) Hospital
(*viii*) Several industrial applications.

The *applications of various engines separately* are listed below :

1. **Small two-stroke petrol engines :**

- These engines are employed where *simplicity and the low cost of the prime mover are primary considerations.*
- The 50 c.c. engines develops maximum brake power (B.P.) of 1.5 kW at 5000 r.p.m. and is *used in mopeds.*
- The 100 c.c. engine developing maximum brake power of about 3 kW at 5000 r.p.m. is *used in scooters.* The 150 c.c. engine develops maximum brake power of about 5 kW at 5000 r.p.m.

- The 250 c.c. engine developing a maximum brake power of about 9 kW at 4500 r.p.m. is generally *used in motor cycles.*
- These engines also find applications in very small electric generating sets, pumping sets etc.

2. **Small four-stroke petrol engines :**

- These engines are primarily used in automobiles.
- These are also used in pumping sets and mobile electric generating sets.

These days diesel engines are taking them over, in the above mentioned applications.

3. **Four stroke diesel engines :**

- The four-stroke diesel engine (a versatile prime mover) is manufactured in diameter ranging from 50 mm to 600 mm with speeds ranging from 100 to 4400 r.p.m., the power delivered per cylinder varying from 1 to 1000 kW.
- Diesel engine is employed for the following :
  - — Pumping sets
  - — Construction machinery
  - — Air compressors and drilling jigs
  - — Tractors
  - — Jeeps, cars and taxies
  - — Mobile and stationary electric generating plant
  - — Diesel-electric locomotive
  - — Boats and ships.

4. **Two stroke diesel engines :**

- These engines having very high power are usually employed for *ship propulsion* and generally have bores above 60 cm, uniflow with exhaust valves or loop scavenged.

Example. Nordberg, *2 stroke, 12-cylinder 80 cm bore and 155 cm stroke, diesel engine engine develops 20000 kW at 120 r.p.m.*

5. **Radial piston engine in small aircraft propulsion :**

- Radial four stroke petrol engines having power range from 300 kW to 4000 kW have been used in small aircrafts.
- In modern large aircrafts, instead of these engines, gas turbine plant as turboprop engine or turbojet engine and gas turbine engines are used.

2.5. ENGINE CYCLE-ENERGY BALANCE

Refer Fig. 2.3. It shows the energy flow through the reciprocating engine. The analysis is based on the first law of thermodynamics which states that energy can neither be created nor destroyed, it can be converted from one form to other.

- In an I.C. engine fuel is fed to the combustion chamber where it burns in the presence of air and its chemical energy is converted into heat. All this energy is not available for driving the piston since a portion of this energy is lost through exhaust, coolant and radiation. The remaining energy is converted to power and is called indicated energy or *indicated power* (I.P.). The ratio of this energy to the input fuel energy is called indicated thermal efficiency $[\eta_{th.(I)}]$.

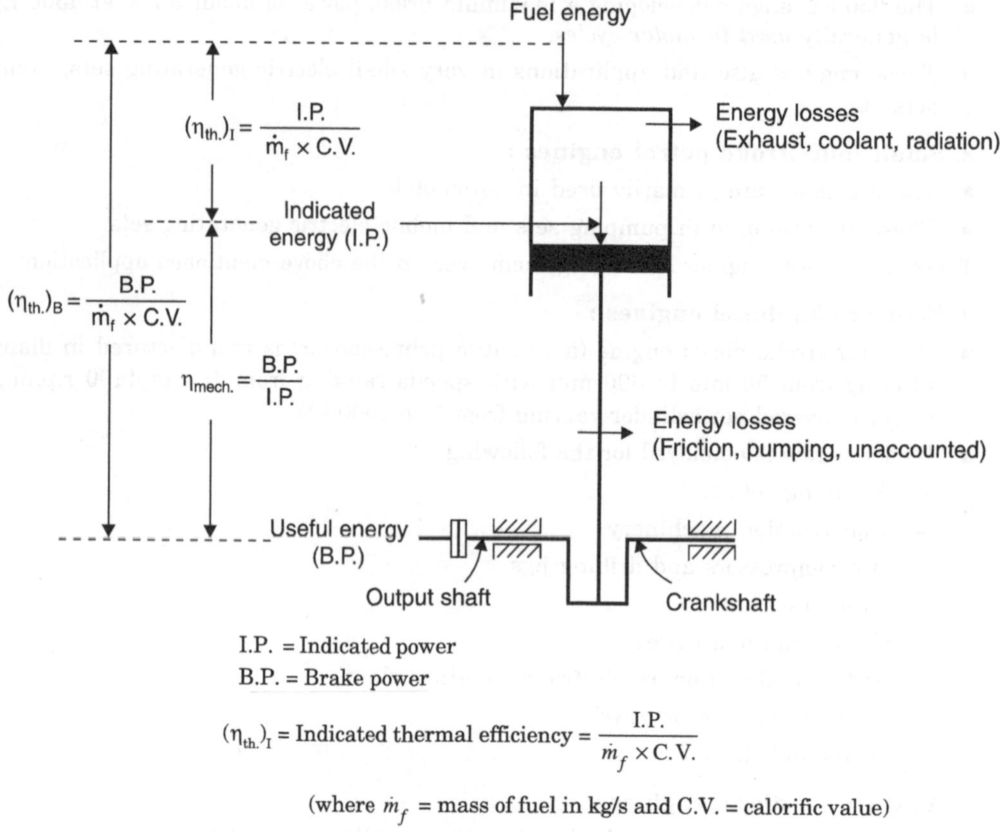

I.P. = Indicated power
B.P. = Brake power

$$(\eta_{th.})_I = \text{Indicated thermal efficiency} = \frac{I.P.}{\dot{m}_f \times C.V.}$$

(where $\dot{m}_f$ = mass of fuel in kg/s and C.V. = calorific value)

$(\eta_{th.})_B$ = Brake thermal efficiency.

Fig. 2.3. The energy flow through the reciprocating engine.

- The energy available at the piston passes through the connecting rod to the crankshaft. In this transmission of energy/power there are losses due to friction, pumping, etc. The sum of all these losses, converted to power, is termed as **Friction Power (F.P.).** The remaining energy is the *useful mechanical energy and is termed as* **shaft energy** or **Brake Power (B.P.).** The *ratio of energy at shaft to fuel input energy is called* **brake thermal efficiency** $[\eta_{th.(B)}]$.
- The *ratio of shaft energy to the energy available at the piston is called* **mechanical efficiency** $(\eta_{mech.})$.

2.6. BASIC IDEA OF I.C. ENGINES

The basic idea of internal combustion engine is shown in Fig. 2.4. The cylinder which is closed at one end is filled with a mixture of fuel and air. As the crankshaft turns it pushes piston. The piston is forced up and compresses the mixture in the top of the cylinder. The mixture is set alight and, as it burns, it creates a gas pressure on the piston, forcing it down the cylinder. This motion is shown by arrow '1'. The piston pushes on the rod which pushes on the crank. The crank is given rotary (turning) motion as shown by the arrow '2'. The fly wheel fitted on the end of the crankshaft stores energy and keeps the crank turning steadily.

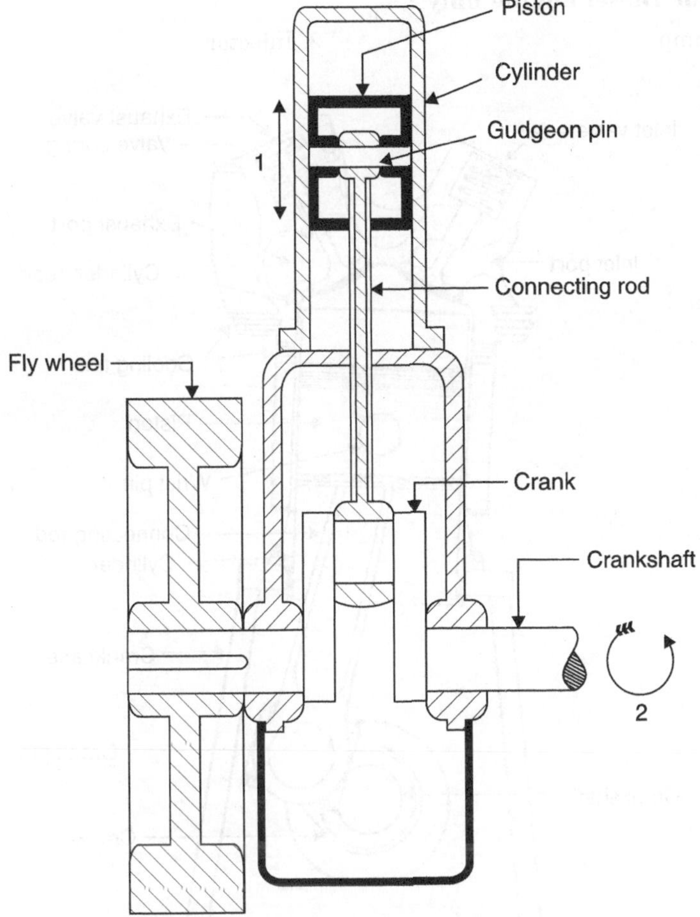

Fig. 2.4. Basic idea of I.C. engine.

2.7. DIFFERENT PARTS OF I.C. ENGINES

Here follows the detail of the various parts of an internal combustion engine.

A cross-section of an air-cooled I.C. engine with principal parts is shown in Fig. 2.5.

A. Parts common to both petrol and diesel engine :

1. Cylinder
2. Cylinder head
3. Piston
4. Piston rings
5. Gudgeon pin
6. Connecting rod
7. Crankshaft
8. Crank
9. Engine bearing
10. Crankcase
11. Flywheel
12. Governor
13. Valves and valve operating mechanisms.

B. Parts for petrol engines only :

1. Spark plugs
2. Carburettor
3. Fuel pump.

C. Parts for Diesel engine only :

1. Fuel pump. 2. Injector.

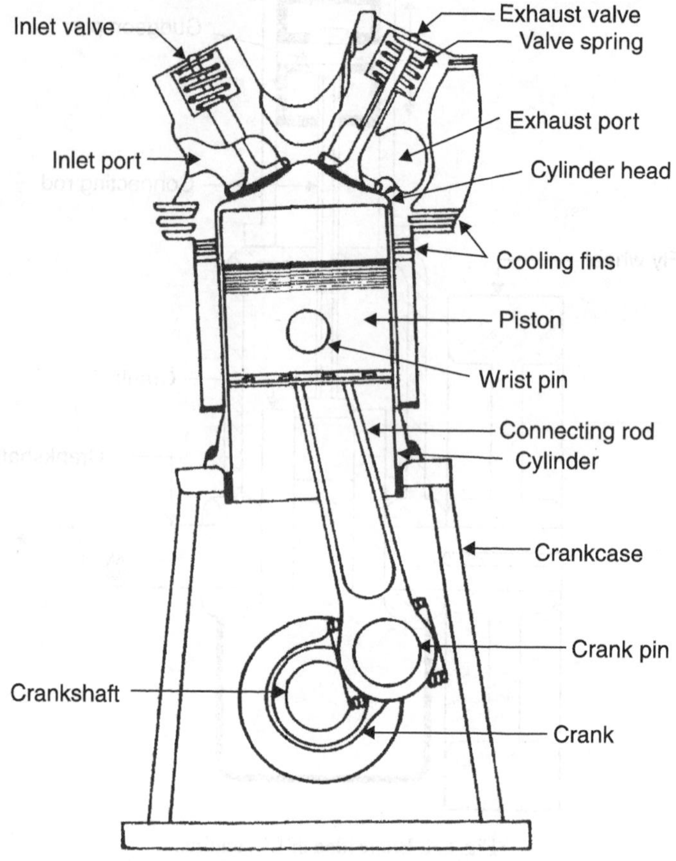

Fig. 2.5. Air-cooled I.C. engine.

A. Parts Common to Both Petrol and Diesel Engines :

1. **Cylinder :**

The cylinder contains gas under pressure and guides the piston. It is in direct contact with the products of combustion and it must be cooled. The ideal form consists of a plain cylindrical barrel in which the piston slides. The movement of the piston or stroke being in most cases, longer than the bore. This is known as the *"stroke-bore ratio"*. The upper end consists of a combustion or clearance space in which the ignition and combustion of the charge takes place. In practice, it is necessary to depart from the ideal hemispherical slope in order to accommodate the valves, sparking plugs etc. and to control the combustion. Sections of an air-cooled cylinder and a water-cooled cylinder are shown in Figs. 2.6 and 2.7, respectively. *The cylinder is made of hard grade cast-iron and is usually, cast in one piece.*

2. Cylinder head :

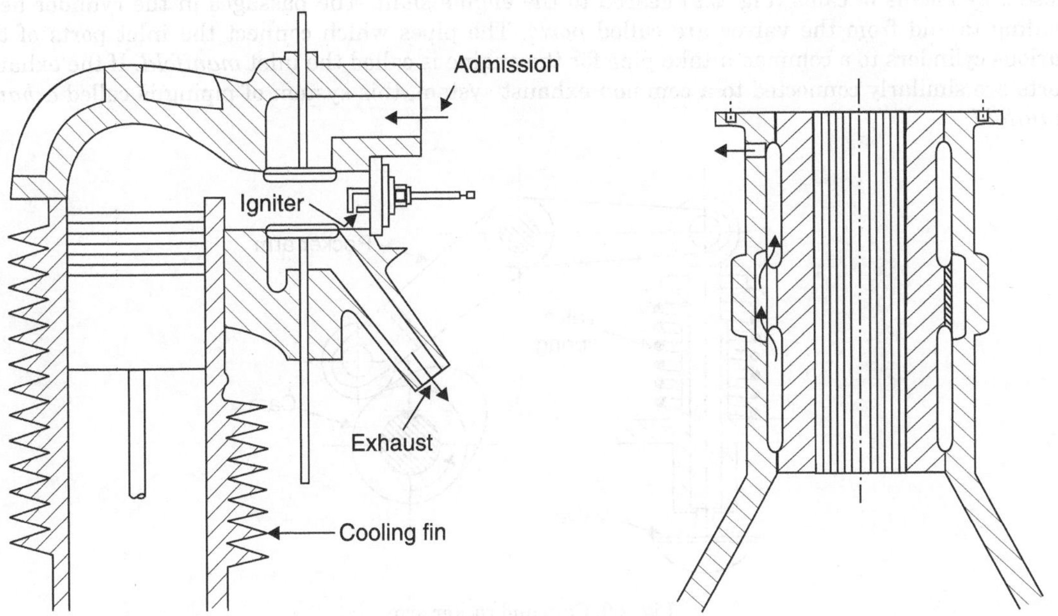

Fig. 2.6. Air-cooled cylinder.

Fig. 2.7. Water-cooled cylinder.

One end of the cylinder is closed by means of a *removable cylinder head* (Fig. 2.6) which usually contains the inlet or admission valve [Fig. 2.8 (*a*)] for admitting the mixture of air and

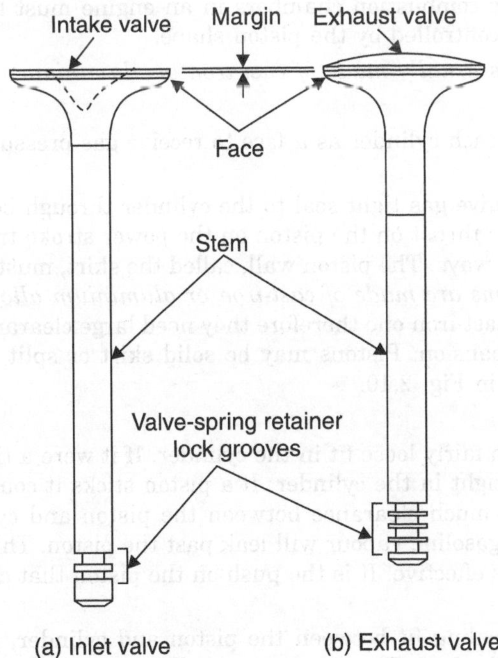

Fig. 2.8

fuel and exhaust valve [Fig. 2.8 (*b*)] for discharging the product of combustion. Two valves are kept closed, by means of cams (Fig. 2.9) geared to the engine shaft. The passages in the cylinder head leading to and from the valves are called *ports*. The pipes which connect the inlet ports of the various cylinders to a common intake pipe for the engine is called the inlet *manifold*. If the exhaust ports are similarly connected to a common exhaust system, this system of piping is called *exhaust manifold*.

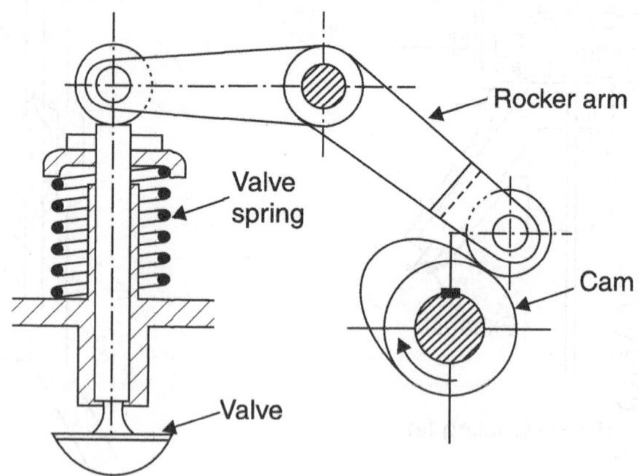

Fig. 2.9. Cam and rocker arm.

The main purpose of the cylinder head is to seal the working ends of the cylinders and not to permit entry and exit of gases on cover head valve engines. The inside cavity of head is called the *combustion chamber*, into which the mixture is compressed for firing. Its *shape controls the direction and rate of combustion*. Heads are drilled and tapped with correct thread to take the ignition spark plug. All the combustion chambers in an engine must be of same shape and size. The shape may be in part controlled by the piston shape.

The cylinder head is usually made of cast-iron or aluminium.

3. **Piston :**

A piston is fitted to each cylinder as a face to receive gas pressure and transmit the thrust to the connecting rod.

The piston must (*i*) give gas tight seal to the cylinder through bore, (*ii*) slide freely, (*iii*) be light and (*iv*) be strong. The thrust on the piston on the power stroke tries to tilt the piston as the connecting rod swings, side ways. The piston wall, called the skirt, must be strong enough to stand upto this side thrust. *Pistons are made of cast-iron or aluminium alloy for lightness.* Light alloy pistons expand more than cast-iron one therefore they need large clearances to the bore, when cold, or special provision for expansion. Pistons may be solid skirt or split skirt. A section through a split skirt piston is shown in Fig. 2.10.

4. **Piston rings :**

The piston must be a fairly loose fit in the cylinder. If it were a tight fit, it would expand as it got hot and might stick tight in the cylinder. If a piston sticks it could ruin the engine. On the other hand, if there is too much clearance between the piston and cylinder walls, much of the pressure from the burning gasoline vapour will leak past the piston. This means, that the push on the piston will be much less effective. It is the push on the piston that delivers the power from the engines.

To provide a good sealing fit between the piston and cylinder, pistons are equipped with piston rings, as shown in Fig. 2.10. The rings are usually made of cast-iron of fine grain and high elasticity which is not affected by the working heat. Some rings are of alloy spring steel. They are split at one point so that they can be expanded and slipped over the end of the piston and into ring grooves which have been cut in the piston. When the piston is installed in the cylinder the rings

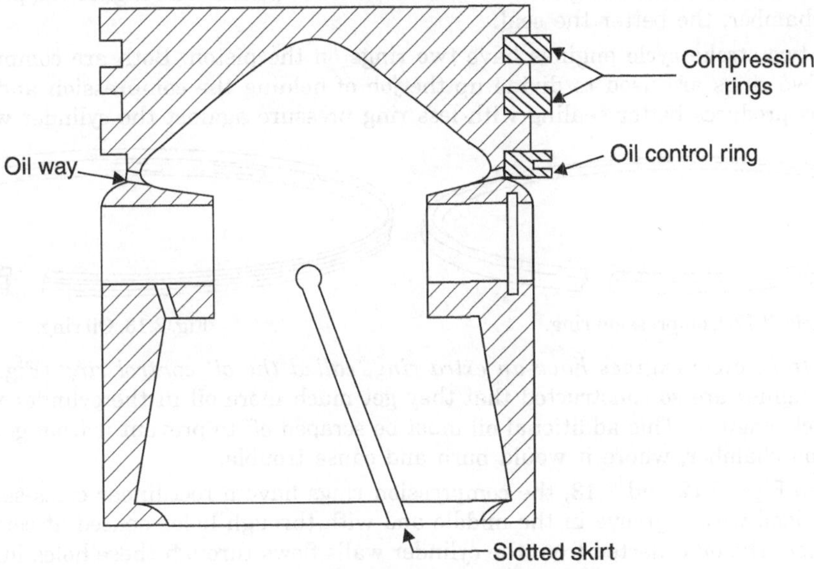

Fig. 2.10. Section through a split skirt piston.

are compressed into ring grooves which have been cut in the piston. When the piston is installed in the cylinder, the rings are compressed into the ring grooves so that the split ends come almost togethor. The rings fit tightly against the cylinder wall and against the sides of the ring grooves in the piston. Thus, *they form a good seal between the piston and the cylinder wall.* The rings can expand or contract as they heat and cool and still make a good deal. Thus they are free to slide up and down the cylinder wall.

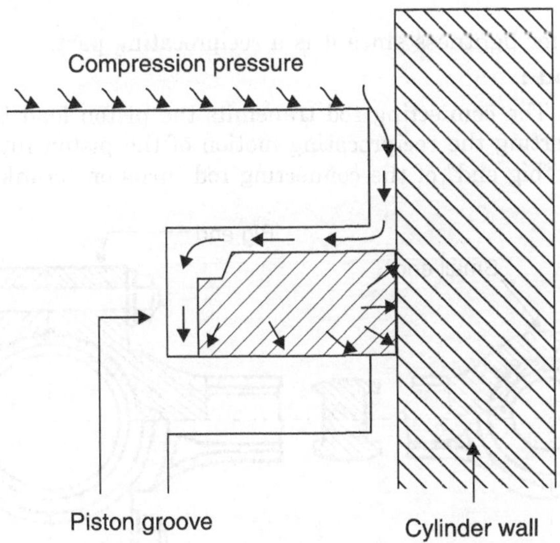

Fig. 2.11. Working of a piston ring.

Fig. 2.11 shows how the piston ring works to hold in the compression and combustion pressure. The arrows show the pressure above the piston passing through clearance between the piston and the cylinder wall. It presses down against the top and against the back of the piston rings as shown by the arrows. This pushes the piston ring firmly against the bottom of the piston ring groove. As a result there are good seals at both of these points. The higher the pressure in the combustion chamber, the better the seal.

Small two stroke cycle engines have two rings on the piston. Both are compression rings (Fig. 2.12). Two rings are used to divide up the job of holding the compression and combustion pressure. This produces better sealing with less ring pressure against the cylinder wall.

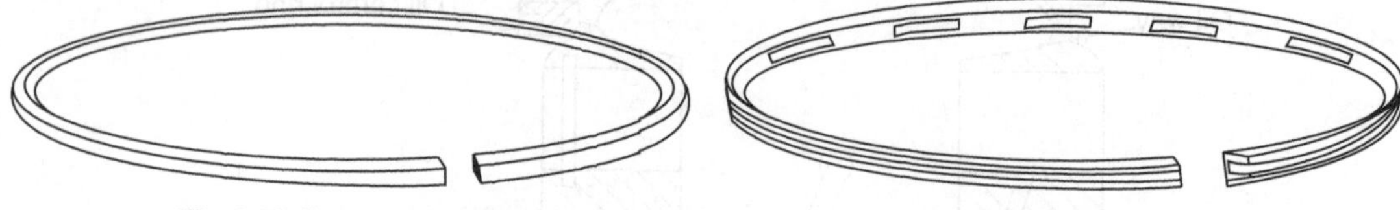

Fig. 2.12. Compression ring. Fig. 2.13. Oil ring.

Four stroke cycle engines have an extra ring, called the oil control ring (Fig. 2.13). Four stroke cycle engines are so constructed that they get much more oil in the cylinder wall than do two stroke cycle engines. This additional oil must be scraped off to prevent it from getting up into the combustion chamber, where it would burn and cause trouble.

Refer to Figs. 2.12 and 2.13, the compression rings have a rectilinear cross-section and oil rings are provided with a groove in the middle and with through holes spaced at certain interval from each other. The oil collected from the cylinder walls flows through these holes into the piston groove whence through the holes in the body of the piston and down its inner walls into the engine crankcase.

5. **Gudgeon pin** (or wrist pin or piston pin) :

These are *hardened steel parallel spindles* fitted through the piston bosses and the small end bushes or eyes to allow the connecting rods to swivel. Gudgeon pins are a press fit in the piston bosses of light alloy pistons when cold. For removal or fitting, the piston should be dipped in hot water or hot oil, this expands the bosses and the pins can be removed or fitted freely without damage.

It is made hollow for lightness since it is a reciprocating part.

6. **Connecting rod :**

Refer to Fig. 2.14. The connecting rod transmits the piston load to the crank, causing the latter to turn, thus converting the reciprocating motion of the piston into a rotary motion of the crankshaft. The lower or "big end" of the connecting rod turns on "crank pins".

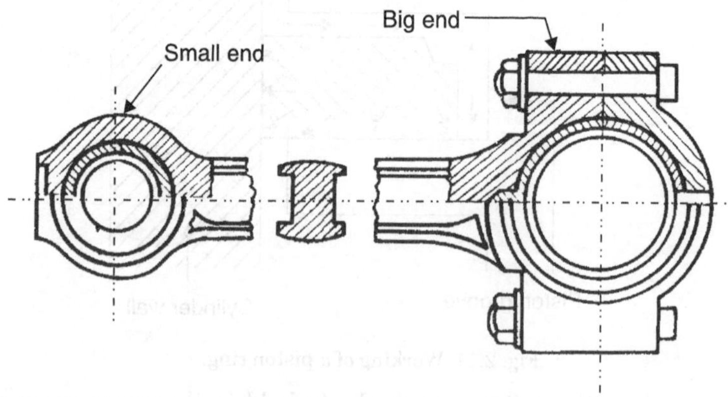

Fig. 2.14. Connecting rod.

The connecting rods are made of nickle, chrome and chrome vandium steels. For small engines the material may be aluminium.

7. Crank :

The piston moves up and down in the cylinder. This up and down motion is called *reciprocating motion*. The piston moves in a straight line. The straight line motion must be changed to rotary, or turning motion, in most machines, before it can do any good. That is rotary motion is required to make wheels turn, a cutting blade spin or a pulley rotate. To change the reciprocating motion to rotary motion a crank and connecting rod are used. (Figs. 2.15 and 2.16). The connecting rod connects the piston to the crank.

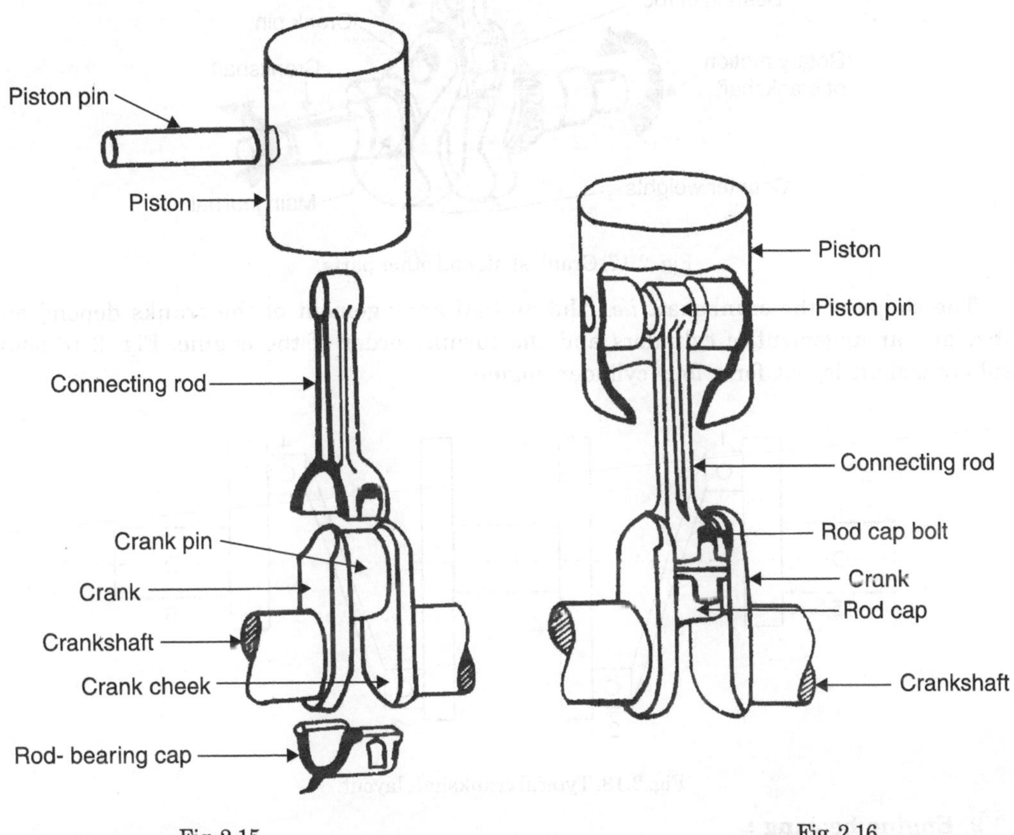

Fig. 2.15 Fig. 2.16

Note. The crank end of the connecting rod is called rod *"big end"*. The piston end of the connecting rod is called the rod *"small end"*.

8. Crankshaft :

The crank is part of the crankshaft. The crankshaft of an internal combustion engine receives via its cranks the efforts supplied by the pistons to the connecting rods. All the engines auxiliary mechanisms with mechanical transmission are geared in one way or the another to the crankshaft. *It is usually a steel forging, but some makers use special types of cast-iron such as spheroidal graphitic or nickel alloy castings* which are cheaper to produce and have good service life. Refer to Fig. 2.17. The crankshaft converts the reciprocating motion to rotary motion. The crankshaft mounts in bearings which, encircle the journals so it can rotate freely.

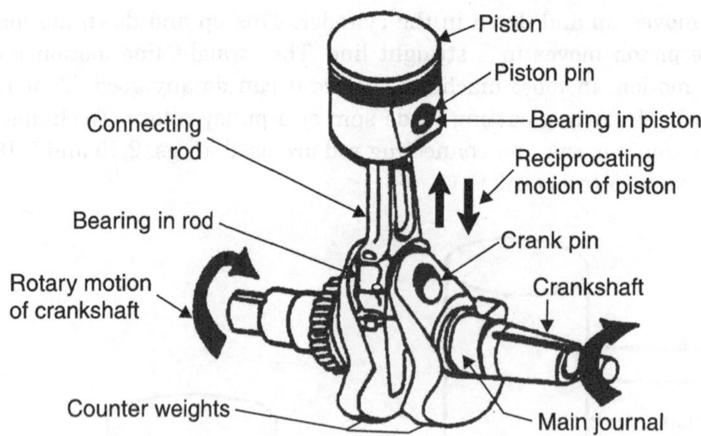

Fig. 2.17. Crank shaft and other parts.

The shape of the crankshaft *i.e.*, the mutual arrangement of the cranks depend on the number and arrangement of cylinders and the turning order of the engine. Fig. 2.18 shows a typical crankshaft layout for a four cylinder engine.

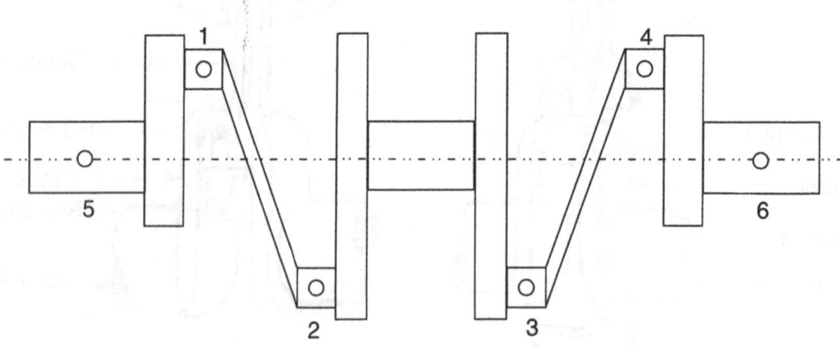

Fig. 2.18. Typical crankshaft layout.

9. **Engine bearing :**

The crankshaft is supported by bearing. The connecting rod big end is attached to the crank pin on the crank of the crankshaft by a bearing. A piston pin at the rod small end is used to attach the rod to the piston. The piston pin rides in bearings. Everywhere there is rotary action in the engine, bearings are used to support the moving parts. The purpose of bearing is to reduce the friction and allow the parts to move easily. Bearings are lubricated with oil to make the relative motion easier.

Bearings used in engines are of two types : *sliding* or *rolling* (Fig. 2.19).

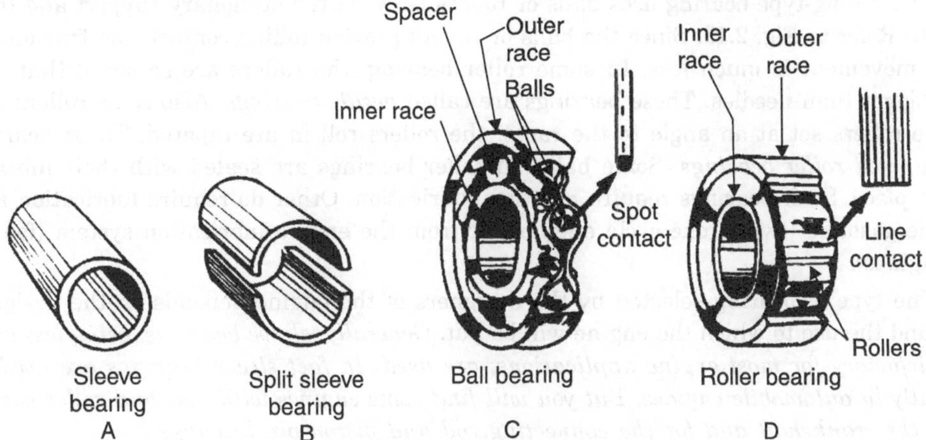

Fig. 2.19. Bearings.

The sliding type of bearings are sometimes called *bushings or sleeve bearings* because they are in the shape of a sleeve that fits around the rotating journal or shaft. The sleeve-type connecting rod big end bearings usually called simply rod bearings and the crankshaft supporting bearings called the main bearings are of the split sleeve type. They must be split in order to permit their assembly into the engine. In the rod bearing, the upper half of the bearing is installed in the rod, the lower half is installed in the rod bearing cap. When the rod cap is fastened to the rod shown in Fig. 2.16 a complete sleeve bearing is formed. Likewise, the upper halves of the main bearings are assembled in the engine and then the main bearing caps, with the lower bearing halves are attached to the engine to complete the sleeve bearings supporting the crankshaft.

The typical bearing half is made of steel or bronze back to which a lining of relatively soft bearing material is applied. Refer to Fig. 2.20. This relatively soft bearing material, which is made of several materials such as copper, lead, tin and other metals, has the ability to conform to slight irregularities of the shaft rotating against it. If wear does take placo, it is the bearing that wears and the bearing can be replacod instead of much more expansive crankshaft or other engine part.

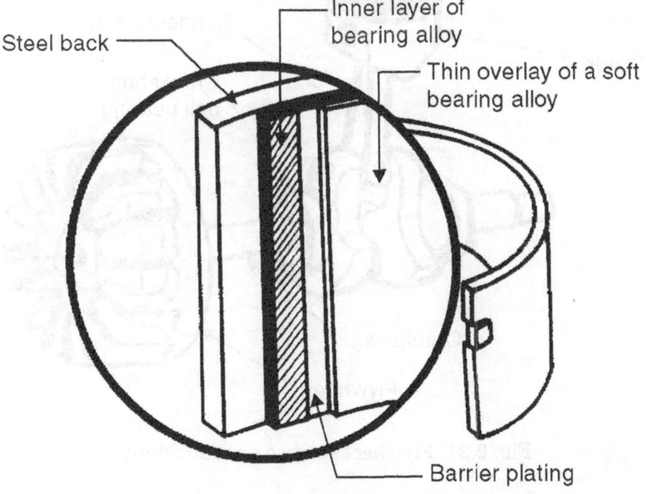

Fig. 2.20. Bearing half (details).

The rolling-type bearing uses balls or rollers between the stationary support and the rotating shaft. Refer to Fig. 2.19. Since the balls or rollers provide rolling contact, the frictional resistance to movement is much less. In some roller bearing, the rollers are so small that they are hardly bigger than needles. These bearings are called *needle bearings*. Also some rollers bearings have the rollers set at an angle to the races, the rollers roll in are tapered. These bearings are called *tapered roller bearings*. Some ball and roller bearings are sealed with their lubricant already in place. Such bearings require no other lubrication. Other do require lubrication from the oil in the gasoline (two stroke cycle engines) or from the engine lubrication system (four stroke cycle engines).

The type of bearing selected by the designers of the engine depends on the design of the engine and the use to which the engine will be put. *Generally, sleeve bearings, being less expensive and satisfactory for most engine applications, are used. In fact sleeve bearings are used almost universally in automobile engines. But you will find some engines with ball and roller bearings to support the crankshaft and for the connecting rod and piston-pin bearings.*

10. **Crankcase :**

The main body of the engine to which the cylinders are attached and which contains the crankshaft and crankshaft bearing is called *crankcase*. This member also holds other parts in alignment and resists the explosion and inertia forces. It also protects the parts from dirt etc. and serves as a part of lubricating system.

11. **Flywheel :**

Refer to Figs. 2.4 and 2.21. A flywheel (steel or cast-iron disc) secured on the crankshaft performs the following *functions* :

(*a*) Brings the mechanism out of dead centres.

(*b*) Stores energy required to rotate the shaft during preparatory strokes.

(*c*) Makes crankshaft rotation more uniform.

(*d*) Facilitates the starting of the engine and overcoming of short time over loads as, for example, when the machine is started from rest.

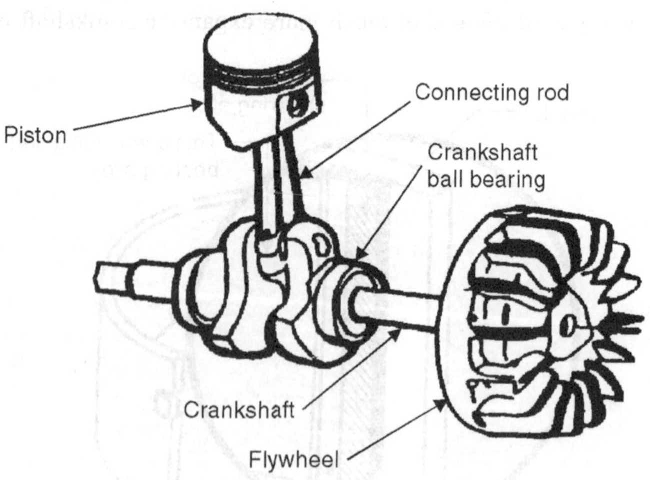

Fig. 2.21. Flywheel secured on crankshaft.

The weight of the flywheel depends upon the nature of variation of the pressure. The flywheel for a double-acting steam engine is lighter than that of a single-acting one. Similarly, the flywheel for a two-stroke cycle engine is lighter than a flywheel used for a four-stroke cycle engine. *Lighter flywheels are used for multi-cylinder engines.*

12. **Governor :**

A governor may be defined *as a device for regulating automatically output of a machine by regulating the supply of working fluid.* When the speed decreases due to increase in load the supply valve is opened by mechanism operated by the governor and the engine therefore speeds up again to its original speed. If the speed increases due to a decrease of load the governor mechanism closes the supply valve sufficiently to slow the engine to its original speed. *Thus the function of a governor is to control the fluctuations of engine speed due to changes of load.*

Comparison between a Flywheel and a Governor

	Flywheel	Governor
1.	It is provided on engines and fabricating machines *viz.*, rolling mills, punching machines ; shear machines, presses etc.	It is provided on prime movers such as engines and turbines.
2.	Its function is to store the available mechanical energy when it is in excess of the load requirement and to part with the same when the available energy is less than that required by the load.	Its function is to regulate the supply of driving fluid producing energy, according to the load requirement so that at different loads almost a constant speed is maintained.
3.	It works continuously from cycle to cycle.	It works intermittently *i.e.* only when there is change in load.
4.	In engines it takes care of fluctuations of speed during thermodynamic cycle.	It takes care of fluctuations of speed due to variation of load over long range of working engines and turbines.
5.	In fabrication machines it is very economical to use it in that it reduces capital investment on prime movers and their running expenses.	But for governor, there would have been unnecessarily more consumption of driving fluid. Thus it economies its consumption.

Types of governor :

Governors are classified as follows :

1. **Centrifugal governor :**

(*i*) *Gravity controlled*, in which the centrifugal force due to the revolving masses is largely balanced by gravity.

(*ii*) *Spring controlled*, in which the centrifugal force is largely balanced by springs.

2. **Inertia and flywheel governors :**

(*i*) Centrifugal type, in which centrifugal forces play the major part in the regulating action.

(*ii*) Inertia governor, in which the inertia effect predominates.

The *inertia type* governors are fitted to the crankshaft or flywheel of an engine and so differ radically in appearance from the centrifugal governors. The balls are so arranged that the inertia force caused by an angular acceleration or retardation of the shaft tends to alter their positions. The amount of displacement of governor balls is controlled by suitable springs and through the governor mechanism, alters the fuel supply to the engine. The inertia governor is more sensitive than centrifugal but it becomes very difficult to balance the revolving parts. For this reason *centrifugal governors are more frequently used.* We shall discuss centrifugal governors only.

Important *centrifugal governors* are :

1. Watt governor
2. Porter governor
3. Proell governor
4. Hartnell governor.

1. **Watt governor :**

It is the primitive governor as used by Watt on some of his early steam engines. It is used for a very slow speed engine and this is why it has now become obsolete.

Refer Fig. 2.22. Two arms are hinged at the top of the spindle and two revolving balls are fitted on the other ends of the arms. One end of each of the links are hinged with the arms, while the other ends are hinged with the sleeve, which may slide over the spindle. The speed of the crankshaft is transmitted to the spindle through a pair of bevel gears by means of a suitable arrangement. So the rotation of the spindle of the governor causes the weights to move away from the centre due to the centrifugal force. This makes the sleeve to move in the upward direction. This movement of the sleeve is transmitted by the lever to the throttle valve which partially closes or opens the steam pipe and reduces or increases the supply of steam to the engine. So the engine speed may be adjusted to a normal limit.

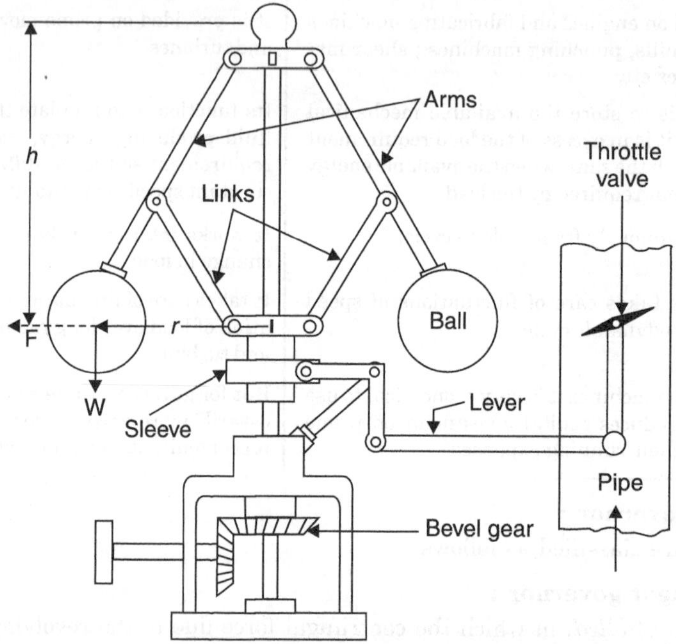

Fig. 2.22. Watt governor.

2. **Porter governor :**

Fig. 2.23 shows diagrammatically a Porter governor where two or more masses called the governor balls rotate about the axis of the governor shaft which is driven through suitable gearing from the engine crankshaft. The governor balls are attached to the arms. The lower arms are attached to the *sleeve which acts as a central weight*. If the speed of the rotation of the balls increases owing to a decrease of load on the engine, the governor balls fly outwards and the sleeve moves upwards thus closing the fuel passage till the engine speed comes back to its designed speed. If the engine speed decreases owing to an increase of load, the governor balls fly inwards and the sleeve moves downwards thus opening the fuel passage more for oil till the engine speed comes back to its designed speed. The engine is said to be running at its designed speed when the outward inertia or centrifugal force is just balanced by the inward controlling force.

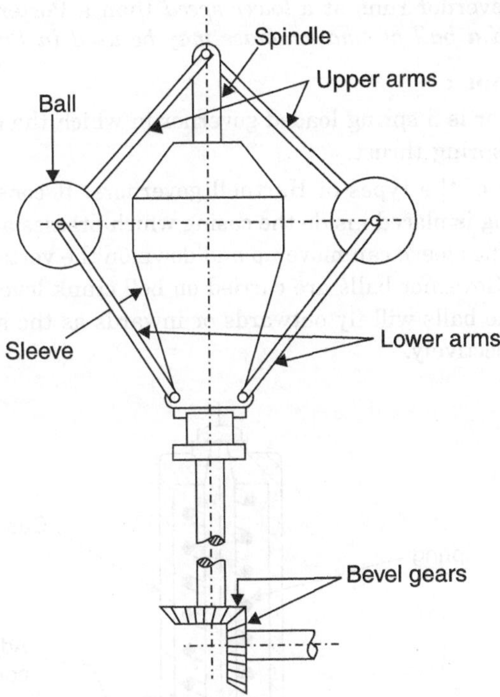

Fig. 2.23. Porter governor.

3. Proell governor :

Refer Fig. 2.24. It is a modification of porter governor. The governor balls are carried on an *extension of the lower arms*. For given values of weight of the ball, weight of the sleeve and height

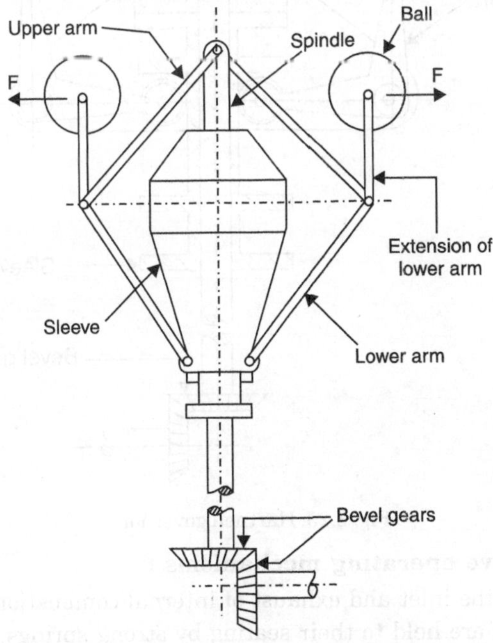

Fig. 2.24. Proell governor.

of the governor, a Proell governor runs at a *lower speed* than a Porter governor. *In order to give the same equilibrium speed a ball of smaller mass may be used in Proell governor.*

4. **Hartnell governor :**

The Hartnell governor is a spring loaded governor in which the controlling force, to a great extent, is provided by the spring thrust.

Fig. 2.25 shows one of the types of Hartnell governors. It consists of casing fixed to the spindle. A compressed spring is placed inside the casing which presses against the top of the casing and on adjustable collars. The sleeve can move up and down on the vertical spindle depending upon the speed of the governor. Governor balls are carried on bell crank lever which are pivoted on the lower end of the casing. The balls will fly outwards or inwards as the speed of the governor shaft increases or decreases respectively.

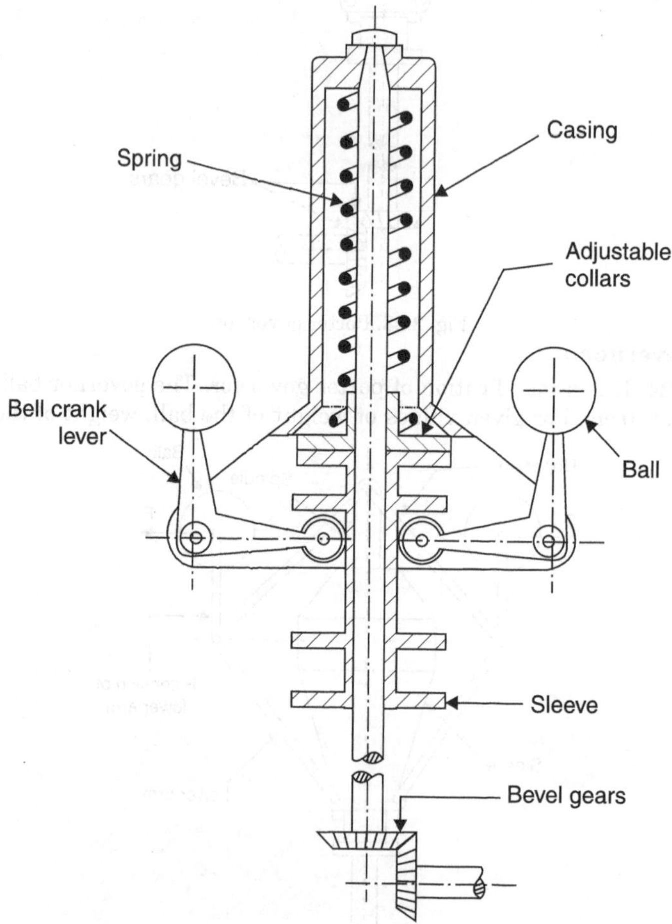

Fig. 2.25. Hartnell governor.

13. **Valves and valve operating mechanisms :**

With few exceptions the inlet and exhaust of internal combustion engines are controlled by poppet valves. These valves are held to their seating by strong springs, and as the valves usually open inwards, the pressure in the cylinder helps to keep them closed. The valves are lifted from their seats and the ports opened either by cams having projecting portion designed to give the period of opening required or by eccentrics operating through link-work. Of these two methods the cam gear is more commonly used, but in either case it is necessary that the valve gear shaft of an engine should rotate but once from beginning to end of a complete cycle, however many strokes may be involved in the completion of that cycle. This is necessary to secure a continuous regulation of the valve gear as required. For this purpose the cams or eccentrics of four-stroke engines are mounted on shafts driven by gearing at half the speed of the crankshaft. The curves used for the acting faces of the cams depend on the speed of the engine and rapidity of valve opening desired.

Fig. 2.26 shows a valve gear for I.C. engine. It consists of poppet valve, the steam bushing or guide, valve spring, spring retainer, lifter or push rod, camshaft and half speed gear for a four-

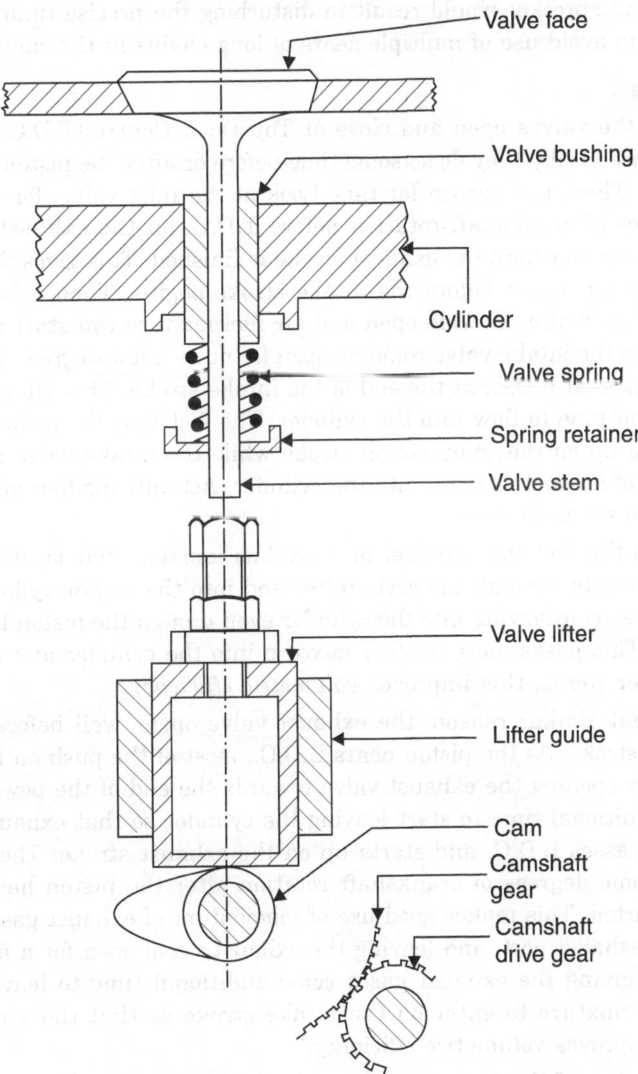

Fig. 2.26. Valve gear for I.C. engine.

stroke engine. The poppet valve, in spite of its shortcomings of noise and difficulties of cooling is commonly used due to its simplicity and capacity for effective sealing under all operating conditions. The valve is subjected to very heavy duty. It holds in combustion chamber and is exposed to high temperatures of burning gases. Exhaust valve itself may attain a high temperature while external cooling is not available. Special heat resisting alloys are therefore used in the construction of the exhaust valve and it may sometimes have a hollow construction filled with mineral salts to provide for heat dissipation. The salts become liquid when valve is working and transfer heat from the head to the stem from which it is carried through the stem guide to the cylinder block.

The timing of the valves *i.e.*, their opening and closing with respect to the travel of the piston is very important thing for efficient working of the engine. The drive of the camshaft is arranged through gears or chain and sprocket (called timing gear or timing chain). Any wearing of the gears or chain and sprocket would result in disturbing the precise timing of the valves. It is desirable, therefore, to avoid use of multiple gears of long chains in the camshaft drive.

Valve timing :

Theoretically the valves open and close at Top Dead Centre (T.D.C.) or at Bottom Dead Centre (B.D.C.) but practically they do so some time before or after the piston reaches the upper or lower limit of travel. There is a reason for this. Look at the inlet valve, for example. It normally opens several degrees of crankshaft-rotation before T.D.C. on the exhaust stroke. That is the intake valve begins to open before the exhaust stroke is finished. This gives the valve enough time to reach the fully open position before the intake stroke begins. Then, when the intake stroke starts, the intake valve is already wide open and air fuel mixture can start to enter the cylinder, immediately. Likewise the intake valve remains open for quite a few degrees of crankshaft rotation after the piston has passed B.D.C. at the end of the intake stroke. This allows additional time for air fuel mixture to continue to flow into the cylinder. The fact that the piston has already passed B.D.C. and is moving up on the compression stroke while the intake valve is still open does not effect the movement of air-fuel mixture into the cylinder. Actually air-fuel mixture is still flowing in as the intake valve starts to close.

This is due to the fact that air-fuel mixture has inertia. That is, it attempts to keep on flowing after it once starts through the carburettor and into the engine cylinder. The momentum of the mixture then keeps it flowing into the cylinder even though the piston has started up on the compression stroke. This packs more air-fuel mixture into the cylinder and results in a stronger power stroke. In other words, this improves *volumetric efficiency*.

For a somewhat similar reason, the exhaust valve opens well before the piston reaches B.D.C. on the power stroke. As the piston nears B.D.C., most of the push on the piston has ended and nothing is lost by opening the exhaust valve towards the end of the power stroke. This gives the exhaust gases additional time to start leaving the cylinder so that exhaust is well started by the time the piston passes B.D.C. and starts up on the exhaust stroke. The exhaust valve then starts opening for some degrees of crankshaft rotation after the piston has passed T.D.C. and intake stroke has started. This makes good use of momentum of exhaust gases. They are moving rapidly towards the exhaust port, and leaving the exhaust valve open for a few degrees after the intake stroke starts giving the exhaust gases some additional time to leave the cylinder. This allows more air-fuel mixture to enter on the intake stroke so that the stronger power stroke results. That is, it improves volumetric efficiency.

The actual timing of the valves varies with different four stroke cycle engines, but the typical example for an engine is shown in Fig. 2.27. Note that the inlet valve opens 15° of crankshaft rotation before T.D.C. on the exhaust stroke and stays open until 50° of crankshaft rotation after B.D.C. on the compression stroke. The exhaust valve opens 50° before B.D.C. on the power stroke and stays open 15° after T.D.C. on the inlet stroke. This gives the two valves an overlap of 30° at the end of exhaust stroke and beginning of the *compression stroke.*

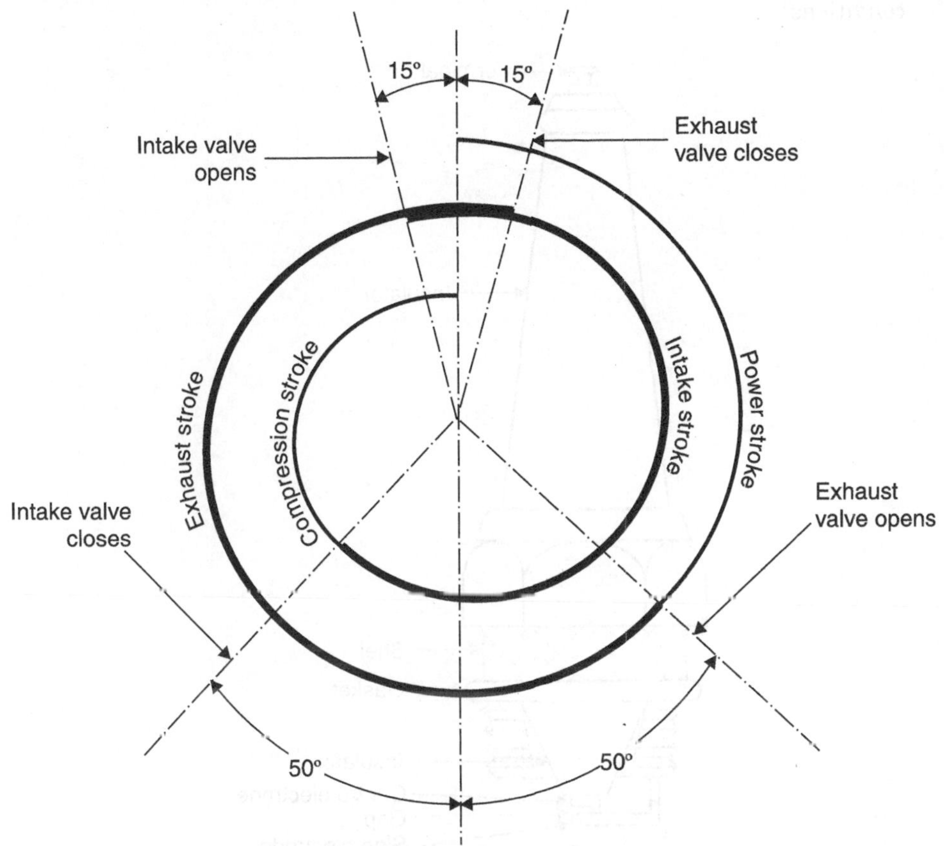

Fig. 2.27. Typical valve timing diagram.

B. Parts for Petrol Engines Only :

1. Spark plugs :

The main function of a spark-plug is to conduct the high potential from the ignition system into the combustion chamber. It provides the proper gap across which spark is produced by applying high voltage, to ignite the mixture in combustion chamber.

A spark-plug entails the following requirements :

(*i*) It must withstand peak pressures up to atleast 55 bar.

(*ii*) It must provide suitable insulation between two electrodes to prevent short circuiting.

(*iii*) It must be capable of withstanding high temperatures to the tune of 2000°C to 2500°C over long periods of operation.

(*iv*) It must offer maximum resistance to erosion burning away of the spark points irrespective of the nature of fuel used.

(*v*) It must possess a high heat resistance so that the electrodes do not become sufficiently hot to cause the preignition of the charge within the engine cylinder.

(*vi*) The insulating material must withstand satisfactorily the chemical reaction effects of the fuel and hot products of combustion.

(*vii*) Gas tight joints between the insulator and metal parts are essential under all operating conditions.

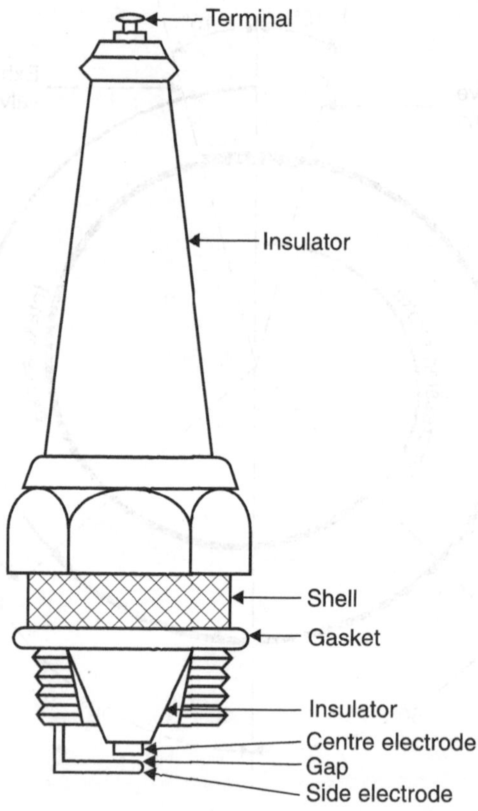

Fig. 2.28. Spark-plug.

Refer Fig. 2.28. The spark-plug consists of a metal shell having two electrodes which are insulated from each other with an air gap. High tension current jumping from the supply electrode produces the necessary spark. Plugs are sometimes identified by the heat range or the relative temperature obtained during operation. The correct type of plug with correct width of gap between the electrodes are important factors. The spark-plug gap can be easily checked by means of a feeler gauge and set as per manufacturer's specifications. It is most important that while adjusting the spark plug it is the outer earthed electrode *i.e.*, tip which is moved in or out gradually for proper setting of the gap. No bending force should be applied on the centre-electrode for adjusting the gap as this can cause crack and fracture of insulation and the plug may become absolutely useless.

Porcelain is commonly used as insulating material in spark-plugs, as it is cheap and easy to manufacture. Mica can also be used as insulating material for spark-plugs. Mica, however, cannot withstand high temperatures successfully.

- **Operating Heat Range :**
  - — A spark-plug heat range is a measure of the plug's ability to transfer heat from the central electrode and insulator nose to the cylinder-head and cooling system.
  - — When the heat absorbed by the plug's central electrode and insulator nose exceeds the capability of the plug to dissipate this heat in the same time, then the plug will *overheat* and the central electrode temperature will rise above its safe operating limit of about 900°C to 950°C. *Above the plug's upper working temperature-limit, the central electrode will glow and ignite the air-fuel mixture before the timed spark actually occurs.* This condition is *known as* **auto-ignition** as it *automatically starts the combustion process independently of the controlled ignition spark.* The danger of this occuring is in the fact that it may take place relatively early in the compression stroke. Consequently, the pressure generated in the particular cylinder suffering from auto-ignition will oppose the upward movement of the piston. Excessive mechanical stresses will be produced in the reciprocating and rotating components and an abnormal rise in the cylinder temperature would, if allowed to continue, damage the engine.
  - — If the plug's ability to transfer heat away from the central electrode and insulator tip *exceeds* that of the input heat from combustion, over the same time span, then the plug's central electrode and insulator nose *would operate at such a low temperature as to permit the formation of carbon deposits around the central nose of the plug.* This *critical lower temperature region is usually between 350°C and 400°C* and, at temperatures below this, carbon or oil deposits will *foul the insulation,* creating conducting shunts to the inside of the metal casing of the plug. Consequently, if *deposits are permitted to form, a proportion of the ignition spark energy will bypass the plug gap so that there will be insufficient energy left to ionize the electrode with the result that misfiring will result.* Establishing a heat balance between the plug's input and output heat flow, so that the *plug's temperature remains just in excess of 400°C, provides a self cleaning action on both the surfaces of the electrodes and insulator.*
  - — A good spark-plug design tries to match the heat flowing from the plug to the heat flowing into it, caused by combustion under all working conditions, *so that the plug operates below the upper temperature limit at full load, but never drops below the lower limit when idling or running under light-load conditions.*

- **Firing Voltage :**

  A certain *minimum voltage* is necessary to make the spark jump the electrode air gap, the actual magnitude of the voltage required will depend upon the following *factors* :

  (*i*) Compression pressure (*ii*) Mixture strength

  (*iii*) Electrode gap (*iv*) Electrode tip temperature.

- **Tightness of Spark-plug :**
  - — The seat joint tightness is essential for good heat dissipation.
  - — Spark-plugs should not be over tightened otherwise the plug metal casing may become distorted, causing the central electrode insulator to break its seal and become loose. *Combustion gas may then escape through the plug with the result that it overheats.*

— An under-tightened plug may work itself loose and cause combustion gas to escape between the plug and cylinder-head plug hole threads to the atmosphere, again this will *result in overheating and rapid deterioration of the electrode tips.*

It is best to torque the plug to a definite degree of tightness.

2. Carburettor

The function of a carburettor is to atomise and metre the liquid fuel and mix it with the air as it enters the induction system of the engine, maintaining under all conditions of operation fuel-air proportions appropriate to those conditions.

All modern carburettors are based upon Bernoulli's theorem,

$$C^2 = 2gh$$

where, C is the velocity in metres/sec, g is the acceleration due to gravity in metre/sec^2 and h is the head causing the flow expressed in metres of height of a column of the fluid.

The equation of mass rate of flow is given by,

$$m = \rho A\sqrt{2gh}$$

where, ρ is the density of the fluid and A is the cross-sectional area of fluid stream.

In Fig. 2.29 is shown simple carburettor. L is the float chamber for the storage of fuel. The fuel supplied under gravity action or by fuel pump enters the float chamber through the filter F.

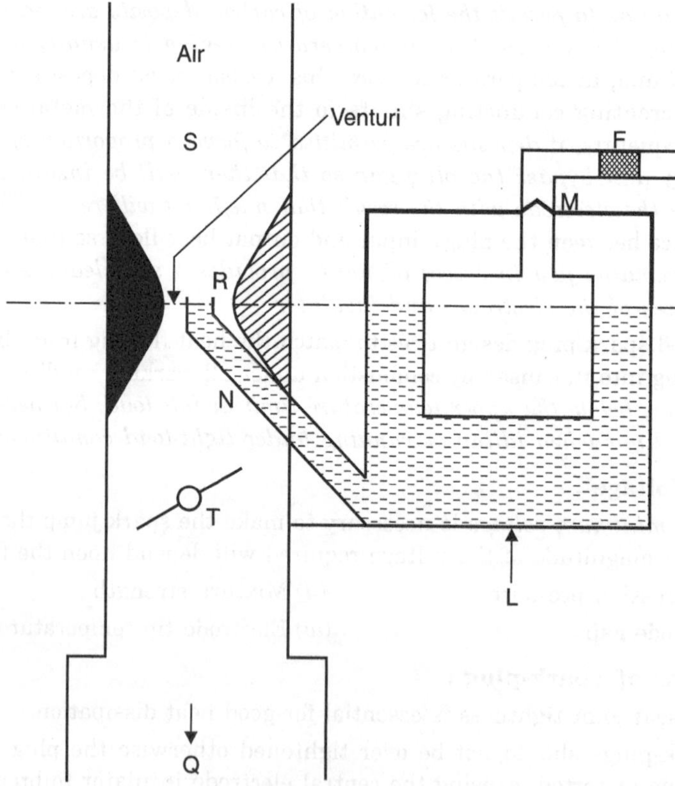

Fig. 2.29. Simple carburettor.

The arrangement is such that when the oil reaches a particular level the float valve *M* blocks the inlet passage and thus cuts off the fuel oil supply. On the fall of oil level, the float descends down, consequently intake passage opens and again the chamber is filled with oil. Then the float and the float valve maintains a constant fuel oil level in the float chamber. *N* is the jet from which the fuel is sprayed into the air stream as it enters the carburettor at the inlet *S* and passes through the throat or venturi *R*. The fuel level is slightly below the outlet of the jet when the carburettor is inoperative.

As the piston moves down in the engine cylinder, suction is produced in the cylinder as well as in the induction manifold *Q* as a result of which air flows through the carburettor. The velocity of air increases as it passes through the constriction at the venturi *R* and pressure decreases due to conversion of a portion of pressure head into kinetic energy. Due to decreased pressure at the venturi and hence by virtue of difference in pressure (between the float chamber and the venturi) the jet issues fuel oil into air stream. Since the jet has a very fine bore, the oil issuing from the jet is in the form of fine spray ; it vapourises quickly and mixes with the air. This air-fuel mixture enters the engine cylinder ; its quantity being controlled by varying the position of the throttle valve *T*.

Limitations :

(*i*) Although theoretically the air fuel ratio supplied by a simple (single jet) carburettor should remain constant as the throttle goes on opening, actually it provides increasingly richer mixture as the throttle is opened. This is because of the reason that the density of air tends to decrease as the rate of flow increases.

(*ii*) During idling, however, the nearly closed throttle causes a reduction in the mass of air flowing through the venturi. At such low rates of air flow, the pressure difference between the float chamber and the fuel discharge nozzle becomes very small. It is not sufficient to cause fuel to flow through the jet.

(*iii*) Carburettor does not have arrangement for providing rich mixture during starting and warm up.

In order to *correct for faults* :

(*i*) number of compensating devices are used for (*ii*) an idling jet is used which helps in running the engine during idling. For (*iii*) choke arrangement is used.

3. **Fuel pump** (for carburettor-petrol engine).

Refer Fig. 2.30. This type of pump is used in petrol engine for supply of fuel to the carburettor. Due to rotation of the crankshaft the cam pushes the lever in the upward direction. One end of the lever is hinged while the other end pulls the diaphragm rod with the *diaphragm*. So the diaphragm comes in the downward direction against the compression of the spring and thus a vacuum is produced in the pump chamber. This causes the fuel to enter into the pump chamber from the *glass bowl* through the *strainer* and the inlet valve, the impurities of the fuel ; if there is any, deposit at the bottom of the glass bowl. On the return stroke the spring pushes the diaphragm in the upward direction forcing the fuel from the pump chamber into the carburettor through the *outlet valve*.

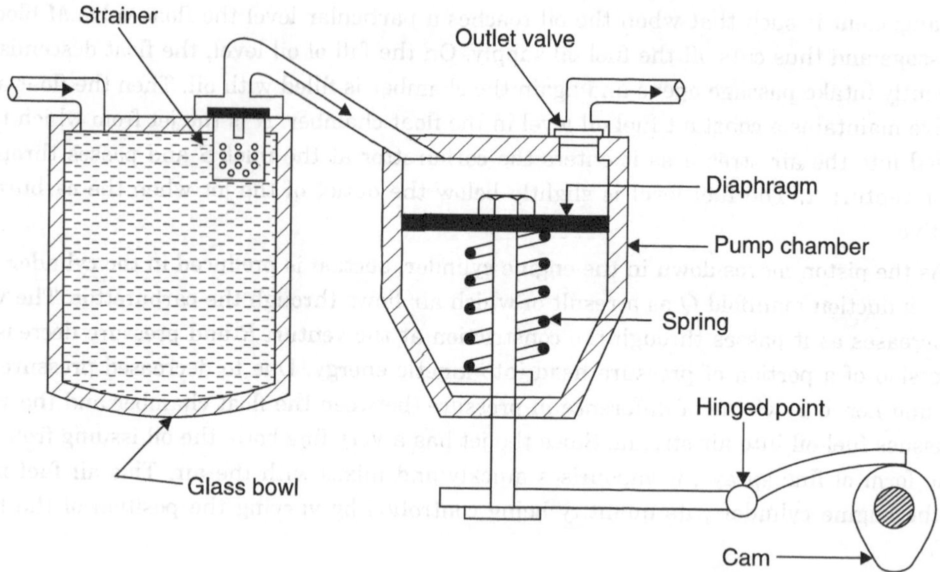

Fig. 2.30. Fuel pump for carburettor.

C. Parts for Diesel Engine Only :

1. **Fuel pump :**

Refer to Fig. 2.31. *L* is the plunger which is driven by a cam and tappet mechanism at the bottom (not shown in the figure), *B* is the barrel in which the plunger reciprocates. There is the rectangular vertical groove in the plunger which extends from top to another helical groove. *V* is the delivery valve which lifts off its seat under the liquid fuel pressure and against the spring force (*S*). The fuel pump is connected to fuel atomiser through the passage *P*, *SP* and *Y* are the spill and supply ports respectively. When the plunger is at its bottom stroke the ports *SP* and *Y* are uncovered (as shown in the Fig. 2.31) and oil from low pressure pump (not shown) after being filtered is forced into the barrel. When the plunger moves up due to cam and tappet mechanism, a stage reaches when both the ports *SP* and *Y* are closed and with the further upward movement of the plunger the fuel gets compressed. The high pressure thus developed lifts the delivery valve off its seat and fuel flows to atomiser through the passage *P*. With further rise of the plunger, at a certain moment, the port *SP* is connected to the fuel in the upper part of the plunger through the rectangular vertical groove by the helical groove ; as a result of which a sudden drop in pressure occurs and the delivery valve falls back and occupies its seat against the spring force. The plunger is rotated by the rack *R* which is moved in or out by the governor. *By changing the angular position of the helical groove (by rotating the plunger) of the plunger relative to the supply port, the length of stroke during which the oil is delivered can be varied and thereby quantity of fuel delivered to the engine is also varied accordingly.*

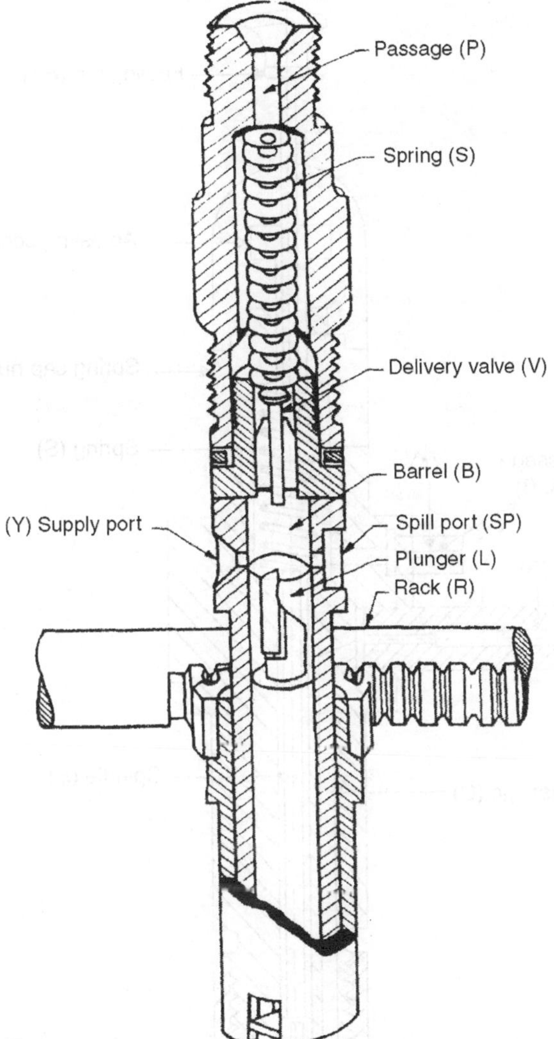

Fig. 2.31. Fuel pump.

2. **Fuel atomiser or injector :**

Refer Fig. 2.32. It consists of a nozzle valve (*NV*) fitted in the nozzle body (*NB*). The nozzle valve is held on its seat by a spring '*S*' which exerts pressure through the spindle *E*. '*AS*' is the adjusting screw by which the nozzle valve lift can be adjusted. Usually the nozzle valve is set to lift at 135 to 170 bar pressure. *FP* is the feeling pin which indicates whether valve is working properly or not. The oil under pressure from the fuel pump enters the injector through the passages *B* and *C* and lifts the nozzle valve. The fuel travels down the nozzle *N* and injected into the engine cylinder in the form of fine sprays. When the pressure of the oil falls, the nozzle valve occupies its seat under the spring force and fuel supply is cut-off. Any leakage of fuel accumulated above the valve is led to the fuel tank through the passage *A*. The leakage occurs when the nozzle valve is *worn out*.

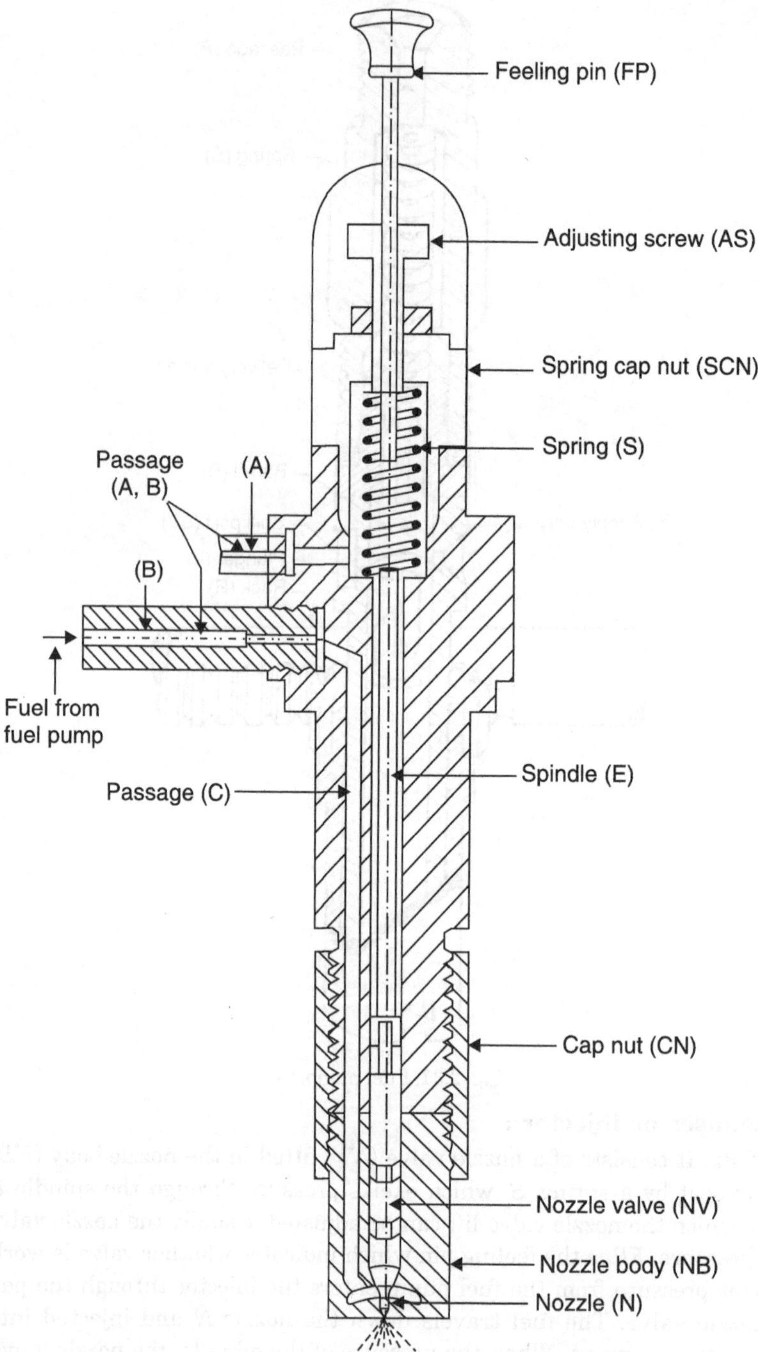

Fig. 2.32. Fuel atomiser or injector.

List of engine parts, materials, method of manufacture and functions :

	Name of the part	Material	Function	Method of manufacture
1.	*Cylinder*	Hard grade cast-iron	Contains gas under pressure and guides the piston.	Casting
2.	*Cylinder head*	Cast-iron or aluminium	Main function is to seal the working end of the cylinder and not to permit entry and exit of gases on overhead valve engines.	Casting, forging
3.	*Piston*	Cast-iron or aluminium alloy	It acts as a face to receive gas pressure and transmits the thrust to the connecting rod.	Casting, forging
4.	*Piston rings*	Cast-iron	Their main function is to provide a good sealing fit between the piston and cylinder.	Casting
5.	*Gudgeon pin*	Hardened steel	It supports and allows the connecting rod to swivel.	Forging
6.	*Connecting rod*	Alloy steel ; for small engines the material may be aluminium	It transmits the piston load to the crank, causing the latter to turn, thus converting the reciprocating motion of the piston into rotary motion of the crankshaft.	Forging
7.	*Crankshaft*	In general the crankshaft is made from a high tensile forging, but special cast-irons are sometimes used to produce a light weight crankshaft that does not require a lot of machining.	It converts the reciprocating motion of the piston into the rotary motion.	Forging
8.	*Main bearings*	The typical bearing half is made of steel or bronze back to which a lining of relatively soft bearing material is applied.	The function of bearing is to reduce the friction and allow the parts to move easily.	Casting
9.	*Flywheel*	Steel or cast-iron.	In engines it takes care of fluctuations of speed during thermodynamic cycle.	Casting
10.	*Inlet valve*	Silicon chrome steel with about 3% carbon.	Admits the air or mixture of air and fuel into engine cylinder.	Forging
11.	*Exhaust valve*	Austenitic steel	Discharges the product of combustion.	Forging

2.8. TERMS CONNECTED WITH I.C. ENGINES

Refer to Fig. 2.33.

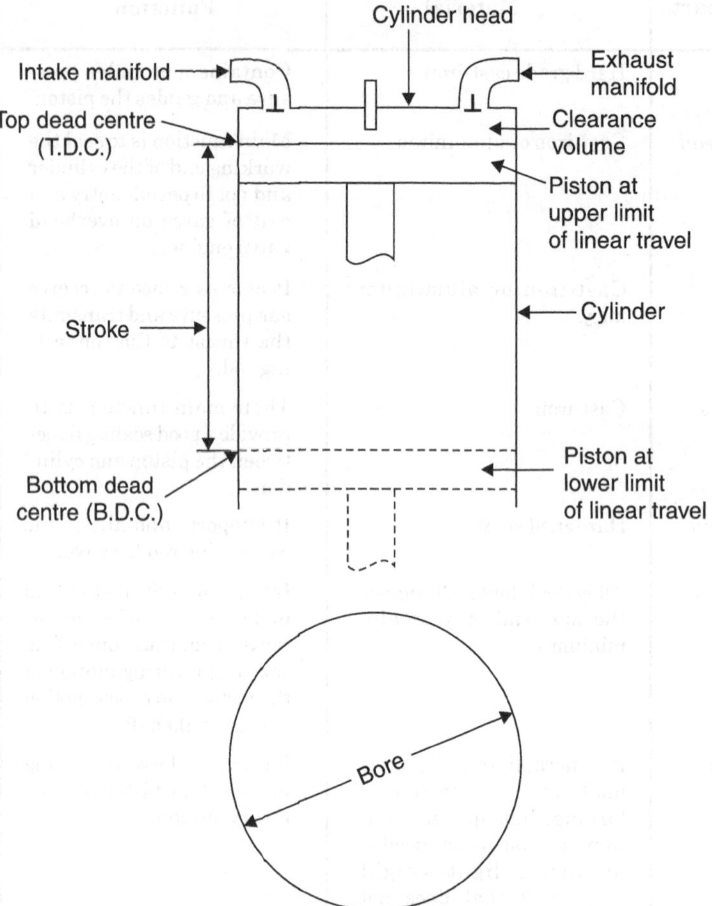

Fig. 2.33. Terms relating to I.C. engines.

Bore. *The inside diameter of the cylinder is called "bore".*

Stroke. As the piston reciprocates inside the engine cylinder, it has got limiting upper and lower positions beyond which it cannot move and reversal of motion takes place at these limiting positions.

The linear distance along the cylinder axis between two limiting positions, is called "stroke".

Top Dead Centre (T.D.C.). *The top most position of the piston towards cover end side of the cylinder is called "top dead centre".* In case of horizontal engines, this is known as *inner dead centre.*

Bottom Dead Centre (B.D.C.). *The lowest position of the piston towards the crank end side of the cylinder is called "bottom dead centre".* In case of horizontal engines it is called *outer dead centre.*

Clearance volume. *The volume contained in the cylinder above the top of the piston, when the piston is at top dead centre, is called the "clearance volume".*

- Bore sizes of engines range from 0.5 m down to 0.5 cm. The ratio of bore of stroke D/L, for small engines is usually from 0.8 to 1.2.

— An engine with $L = D$ is often called a **square** *engine* ;

— If $L > D$ the engine is **under square ;**

— If $L < D$ the engine is **over square.**

Very large engines are *always under square,* with *stroke lengths up to four times bore diameter.*

Swept volume. *The volume swept through by the piston in moving between top dead centre and bottom dead centre is called "swept volume or piston displacement".* Thus, when piston is at bottom dead centre, total volume = swept volume + clearance volume.

- Typical values for engine displacement range from 0.1 cm^3 for small model airplanes to about 8 litres for large automobiles to much large number for large ship engines. The displacement of a modern average automobile engine is about two to three litres.
- For a given displacement volume, a longer stroke allows for a smaller bore (under square), resulting in less surface area in the combustion chamber and correspondingly less heat loss. This increases thermal efficiency within the combustion chamber. However, the longer stroke results in higher piston speed and higher friction losses that reduce the output power which can be obtained off the crankshaft. If the stroke is shortened, the bore must be increased and the engine will be over square. This decreases friction losses but increases heat transfer losses. *Most modern automobile engines are near square, with some slightly over square and some slightly under square.*

Compression ratio. *It is ratio of total cylinder volume to clearance volume.*

Refer to Fig. 2.33. Compression ratio (r) is given by

$$r = \frac{V_s + V_c}{V_c}$$

where, V_s = Swept volume, and V_c = Clearance volume.

The compression ratio varies from 5 : 1 to 11 : 1 (average value 7 : 1 to 9 : 1) in *S.I. engines* and from 12 : 1 to 24 : 1 (average value 15 : 1 to 18 : 1) in *C.I. engines.*

- Modern spark ignition (S.I.) engines have compression ratios of 8 to 11, while compression ignition (C.I.) engines have compression ratios in the range 12 to 24. *Engines with superchargers or turbochargers usually have lower compression ratios than naturally aspirated engines.*
- Various attempts have been made to develop engines with a *variable compression ratio.* One such system uses a *split piston that expands due to changing hydraulic pressure caused by engine speed and load.* Some two-stroke cycle engines have been built which have a sleeve-type valve that changes the slot opening on the exhaust port. The piston where the exhaust port is fully closed can be adjusted by several degrees of engine rotation. This changes the *effective compression ratio* of the engine.

Piston speed. *The average speed of the piston is called "piston speed".*

Piston speed = $2\,LN$

where, L = Length of the stroke, and

N = Speed of the engine in r.p.m.

- Average engine speed for all engines will normally be in the range of 5 to 15 m/s with large diesel engines on the low end and high performance automobile engines on the high end. There are following *two reasons* why engines operate in this range :

  — *First,* this is about the *safe limit which can be tolerated by material strength of the engine components.*

  — The *second* reason why maximum average piston speed is limited is *because of the gas flow into and out of cylinders.* Piston speed determines the instantaneous flow rate of air-fuel into the cylinder during intake and exhaust flow out of the cylinder during the exhaust stroke. *Higher piston speeds would require larger valves to allow for higher flow rates.* In most engines, valves are at a maximum size with no room for enlargement.

Some Other Terms :

Direct Injection (D.I.)—*Fuel injection into the main combustion chamber of an engine.* Engines have either one main combustion chamber (open chamber) or a divided combustion chamber made up of a main chamber and a smaller connected secondary chamber.

Indirect Injection (I.D.I.)—*Fuel injection into the secondary chamber of an engine with a divided combustion chamber.*

Smart Engine—*Engine with computer controls that regulate operating characteristics* such as *air-fuel ratio, ignition timing, valve timing, exhaust control, intake tuning* etc.

Engine Management System (E.M.S.)—Computer and electronics used to control smart engines.

Wide Open Throttle (W.O.T.)—Engine operated with *throttle valve fully open when maximum power and/or speed is desired.*

Ignition Delay (I.D.). It is the *time interval between ignition initiation and the actual start of combustion.*

Air-Fuel Ratio (A/F). It is the *ratio of the air to mass of fuel input into engine.*

Fuel-Air Ratio (F/A). It is the *ratio of fuel to mass of air input into engine.*

2.9. WORKING CYCLES

An internal combustion engine can work on any one of the following cycles :

(*a*) Constant volume or Otto cycle.

(*b*) Constant pressure or Diesel cycle.

(*c*) Dual combustion cycle.

These may be either *four stroke cycle or two stroke cycle engines.*

(*a*) **Constant volume or Otto cycle.** The cycle is so called because heat is supplied at constant volume. Petrol, gas, light oil engines work on this cycle. In the case of a petrol engine the proper mixing of petrol and air takes place in the carburettor which is situated outside the engine cylinder. The proportionate mixture is drawn into the cylinder during the suction stroke. In a gas engine also, air and gas is mixed outside the engine cylinder and this mixture enters the cylinder during the suction stroke. In light oil engines the fuel is converted to vapours by a vapouriser which receives heat from the exhaust gases of the engine and their mixture flows towards engine cylinder during suction stroke.

(*b*) **Constant pressure or diesel cycle.** In this cycle only air is drawn in the engine cylinder during the suction stroke, this air gets compressed during the compression stroke and its pressure and temperature increase by a considerable amount. Just before the end of the stroke a metered quantity of fuel under pressure adequately more than that developed in the engine cylinder is injected in the form of fine sprays by means of a fuel injector. Due to very high pressure and temperature of the air the fuel ignites and hot gases thus produced throw the piston downwards and work is obtained. *Heavy oil engines make use of this cycle.*

(*c*) **Dual combustion cycle.** This cycle is also called *semi-diesel cycle.* It is so named because heat is added *partly at constant volume and partly at constant pressure.* In this cycle only air is drawn in the engine cylinder during suction stroke. The air is then compressed in hot combustion chamber at the end of the cylinder during the compression stroke to a pressure of about 26 bar. The heat of compressed air together with heat of combustion chamber ignites the fuel. The fuel is injected into the cylinder just before the end of compression stroke where it ignites immediately. The fuel injection is continued until the point of cut-off is reached. The burning of fuel at first takes place at constant volume and continues to burn at constant pressure during the first part of expansion or working stroke. The field of application of this cycle is heavy oil engines.

2.10. INDICATOR DIAGRAM

An indicator diagram is a graph between pressure and volume ; the former being taken on vertical axis and the latter on the horizontal axis. This is obtained by an instrument known as *indicator*. The indicator diagrams are of two types : (*a*) Theoretical or hypothetical, (*b*) Actual. The theoretical or hypothetical indicator diagram is always longer in size as compared to the actual one, since in the former losses are neglected. The ratio of the area of the actual indicator diagram to the theoretical one is called *diagram factor*.

2.11. FOUR STROKE CYCLE ENGINES

Here follows the description of the four-stroke otto and diesel-cycle engines.

Otto engines. The Otto four stroke cycle refers to its use in petrol engines, gas engines, light oil engines in which the mixture of air and fuel are drawn in the engine cylinder. Since ignition in these engines is due to a spark, therefore they are also called *spark ignition engines*.

The various strokes of a four stroke (Otto) cycle engine are detailed below.

Refer Fig. 2.34.

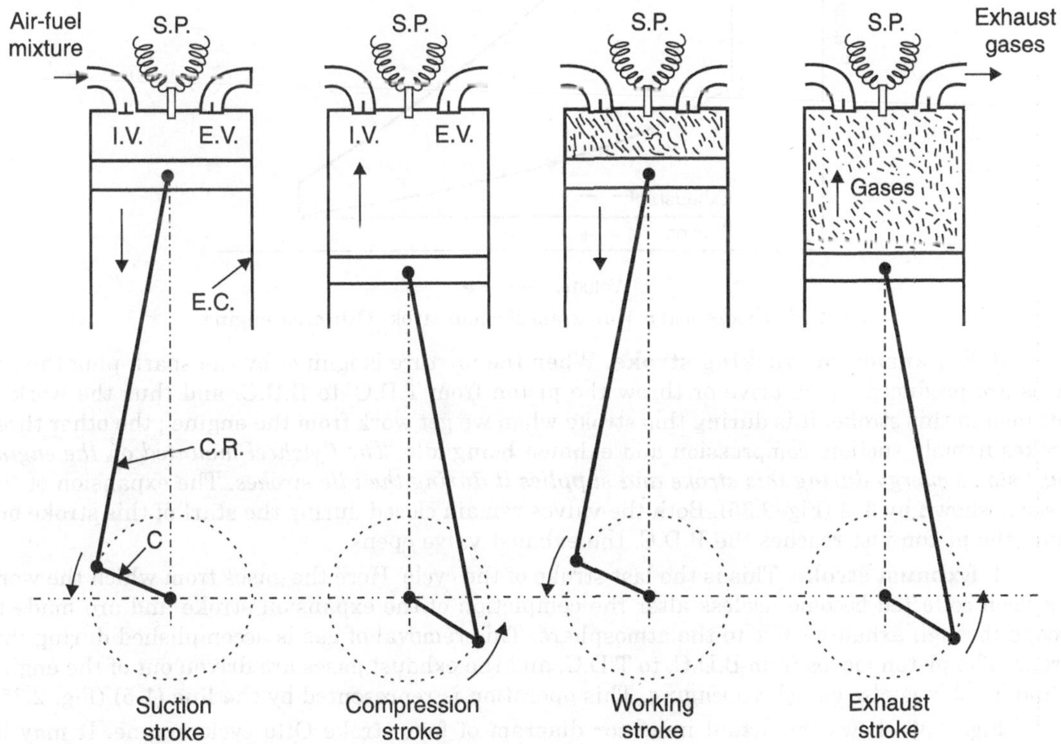

Fig. 2.34. Four stroke Otto cycle engine.

1. **Suction stroke.** During this stroke (also known as *induction stroke*) the piston moves from Top Dead Centre (T.D.C.) to Bottom Dead Centre (B.D.C.) ; the inlet valve opens and proportionate air-fuel mixture is sucked in the engine cylinder. This operation is represented by the line 5–1 (Fig. 2.32). The exhaust valve remains closed throughout the stroke.

2. **Compression stroke.** In this stroke, the piston moves (1–2) towards T.D.C. and compresses the enclosed air-fuel mixture drawn in the engine cylinder during suction stroke. The pressure of the mixture rises in the cylinder to a value of about 8 bar. Just before the end of this stroke the spark-plug initiates a spark which ignites the mixture and combustion takes place at constant volume (line 2–3) (Fig. 2.35). Both the inlet and exhaust valves remain closed during the stroke.

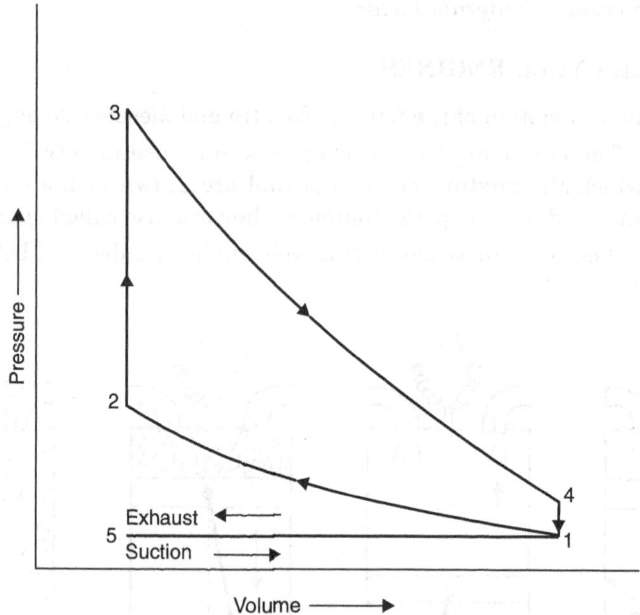

Fig. 2.35. Theoretical *p-V* diagram of a four stroke Otto cycle engine.

3. **Expansion or working stroke.** When the mixture is ignited by the spark-plug the hot gases are produced which drive or throw the piston from T.D.C. to B.D.C. and thus the work is obtained in this stroke. It is during this stroke when we get work from the engine ; the other three strokes namely suction, compression and exhaust being idle. *The flywheel mounted on the engine shaft stores energy during this stroke and supplies it during the idle strokes.* The expansion of the gases is shown by 3–4 (Fig. 2.35). Both the valves remain closed during the start of this stroke but when the piston just reaches the B.D.C. the exhaust valve opens.

4. **Exhaust stroke.** This is the last stroke of the cycle. Here the gases from which the work has been collected become useless after the completion of the expansion stroke and are made to escape through exhaust valve to the atmosphere. This removal of gas is accomplished during this stroke. The piston moves from B.D.C. to T.D.C. and the exhaust gases are driven out of the engine cylinder ; this is also called *scavenging*. This operation is represented by the line (1-5) (Fig. 2.35).

Fig. 2.36 shows the actual indicator diagram of four stroke Otto cycle engine. It may be noted that line 5-1 is below the atmospheric pressure line. This is due to the fact that owing to restricted area of the inlet passages the entering fuel air mixture cannot cope with the speed of the piston. The exhaust line 4-5 is slightly above the atmospheric pressure line. This is due to restricted exhaust passages which do not allow the exhaust gases to leave the engine-cylinder quickly.

The loop which has area 4-5-1 is called *negative loop* ; it gives the pumping loss due to admission of air-fuel mixture and removal of exhaust gases. The area 1-2-3-4 is the total or gross work obtained from the piston and net work can be obtained by subtracting area 451 from the area 1-2-3-4.

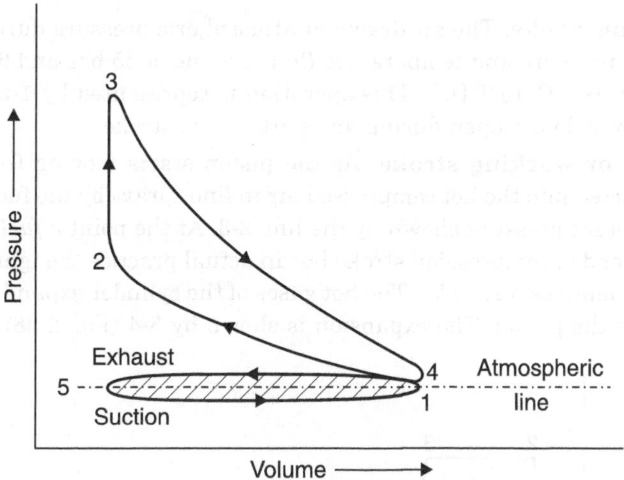

Fig. 2.36. Actual p-V diagram of a four stroke Otto cycle engine.

Diesel engines (four stroke cycle). As is the case of Otto four stroke ; this cycle too is completed in four strokes as follows. (Refer to Fig. 2.37).

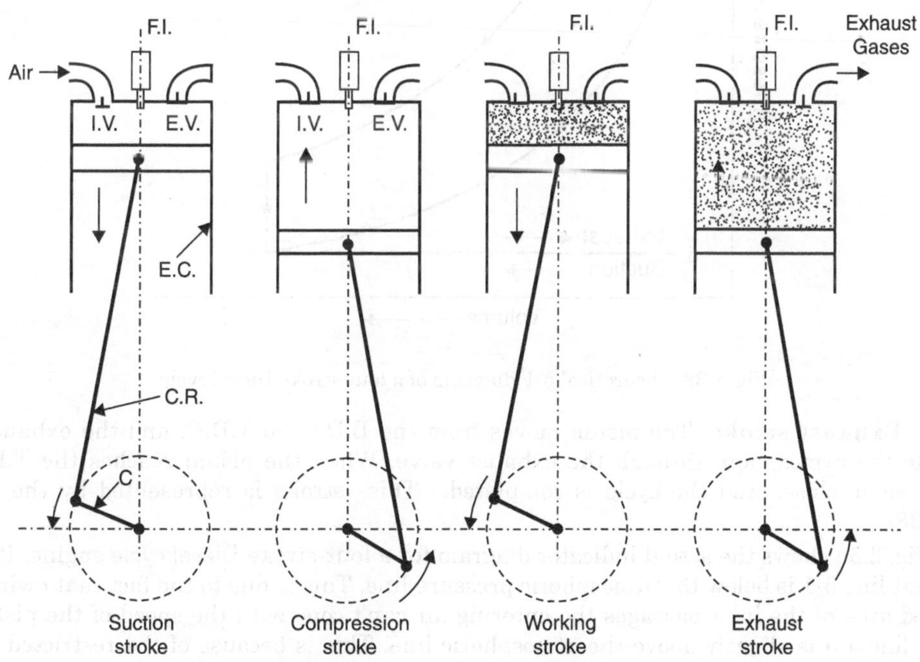

Fig. 2.37. Four stroke Diesel cycle engine.

1. **Suction stroke.** With the movement of the piston from T.D.C. to B.D.C. during this stroke, the inlet valve opens and the air at atmospheric pressure is drawn inside the engine cylinder ; the exhaust valve however remains closed. This operation is represented by the line 5-1 (Fig. 2.38).

2. **Compression stroke.** The air drawn at atmospheric pressure during the suction stroke is compressed to high pressure and temperature (to the value of 35 bar and 600°C respectively) as the piston moves from B.D.C. to T.D.C. This operation is represented by 1-2 (Fig. 2.38). Both the inlet and exhaust valves do not open during any part of this stroke.

3. **Expansion or working stroke.** As the piston starts moving from T.D.C. a metered quantity of fuel is injected into the hot compressed air in fine sprays by the fuel injector and it (fuel) starts burning at constant pressure shown by the line 2-3. At the point 3 *fuel supply is cut off.* The fuel is injected at the end of compression stroke but in actual practice the ignition of the fuel starts before the end of the compression stroke. The hot gases of the cylinder expand adiabatically to point 4, thus doing work on the piston. The expansion is shown by 3-4 (Fig. 2.38).

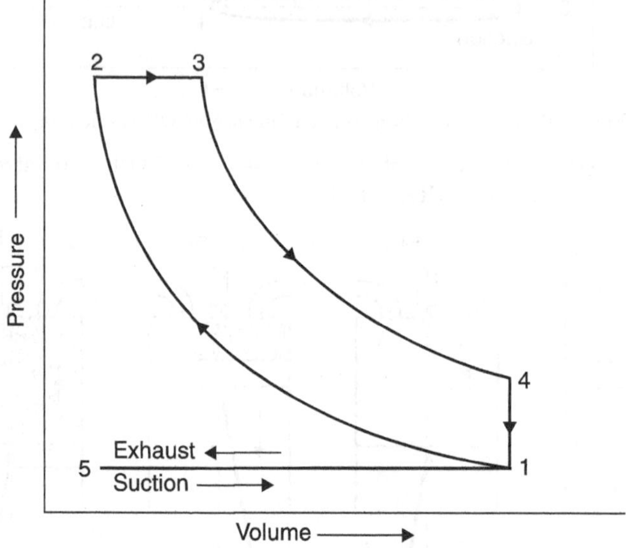

Fig. 2.38. Theoretical *p-V* diagram of a four stroke Diesel cycle.

4. **Exhaust stroke.** The piston moves from the B.D.C. to T.D.C. and the exhaust gases escape to the atmosphere through the exhaust valve. When the piston reaches the T.D.C. the exhaust valve closes and the cycle is completed. This stroke is represented by the line 1-5 (Fig. 2.38).

Fig. 2.39 shows the actual indicator diagram for a four-stroke Diesel cycle engine. It may be noted that line 5-1 is below the atmospheric pressure line. This is due to the fact that owing to the restricted area of the inlet passages the entering air can't cope with the speed of the piston. The exhaust line 4-5 is slightly above the atmospheric line. This is because of the restricted exhaust passages which do not allow the exhaust gases to leave the engine cylinder quickly.

The loop of area 4-5-1 is called negative loop ; it gives the pumping loss due to admission of air and removal of exhaust gases. The area 1-2-3-4 is the total or gross work obtained from the piston and net work can be obtained by subtracting area 4-5-1 from area 1-2-3-4.

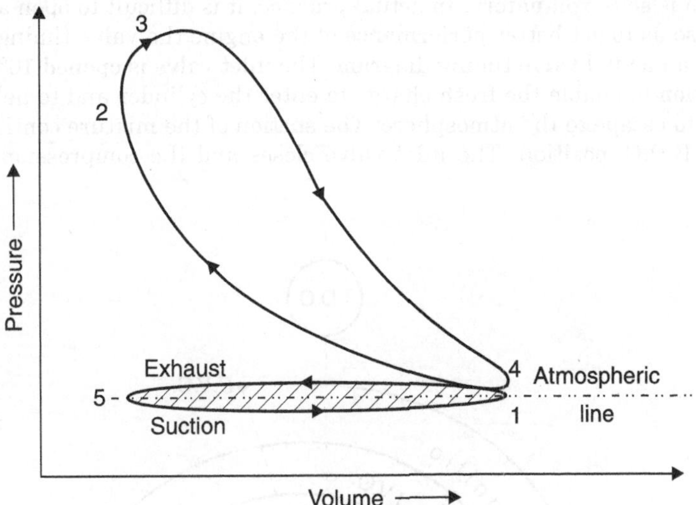

Fig. 2.39. Actual p-V diagram of four stroke Diesel cycle.

Valve Timing Diagrams (Otto and Diesel engines) :

1. **Otto engines.** Fig. 2.40 shows a theoretical valve timing diagram for *four stroke "Otto*

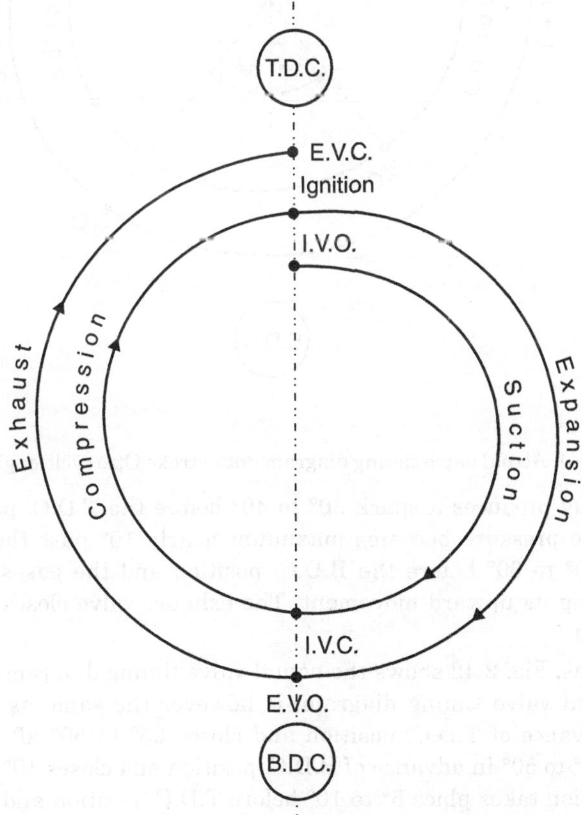

Fig. 2.40. Theoretical valve timing diagram (four stroke Otto cycle engine).

cycle" engine which is self-explanatory. In actual practice, it is difficult to open and close the valve instantaneously ; so as to get better performance of the engine the valve timings are modified. In Fig. 2.41 is shown an actual valve timing diagram. The inlet valve is opened 10° to 30° in advance of the T.D.C. position to enable the fresh charge to enter the cylinder and to help the burnt gases at the same time, to escape to the atmosphere. The suction of the mixture continues up to 30°-40° or even 60° after B.D.C. position. The inlet valve closes and the compression of the entrapped mixture starts.

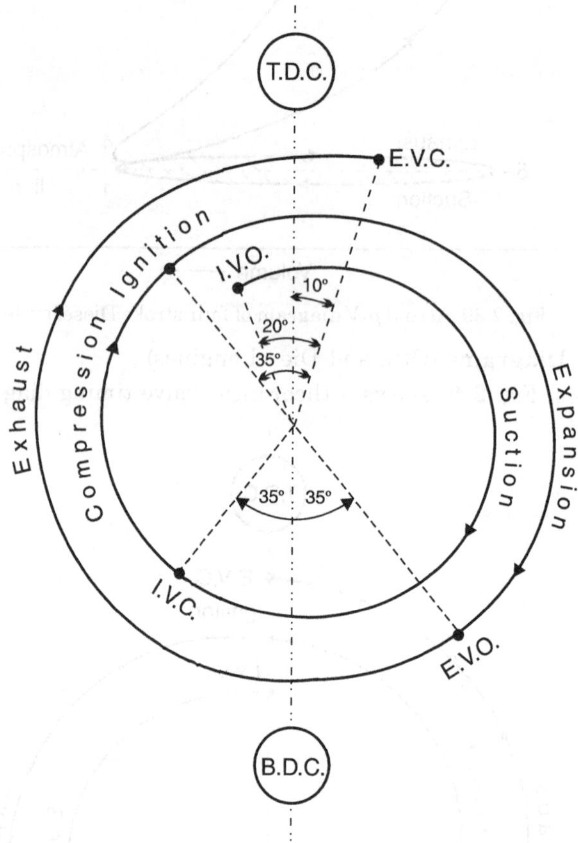

Fig. 2.41. Actual valve timing diagram (four stroke Otto cycle engines).

The sparking plug produces a spark 30° to 40° before the T.D.C. position ; thus fuel gets more time to burn. The pressure becomes maximum nearly 10° past the T.D.C. position. The exhaust valve opens 30° to 60° before the B.D.C. position and the gases are driven out of the cylinder by piston during its upward movement. The exhaust valve closes when piston is nearly 10° past T.D.C. position.

2. **Diesel engines.** Fig. 2.42 shows the actual valve timing diagram of a *four stroke "Diesel cycle" engine* (theoretical valve timing diagram, is however the same as Fig. 2.40). Inlet valve opens 10° to 25° in advance of T.D.C. position and closes 25° to 50° after the B.D.C. position. Exhaust valve opens 30° to 50° in advance of B.D.C. position and closes 10° to 15° after the T.D.C. position. The fuel injection takes place 5° to 10° before T.D.C. position and continues up to 15° to 25° near T.D.C. position.

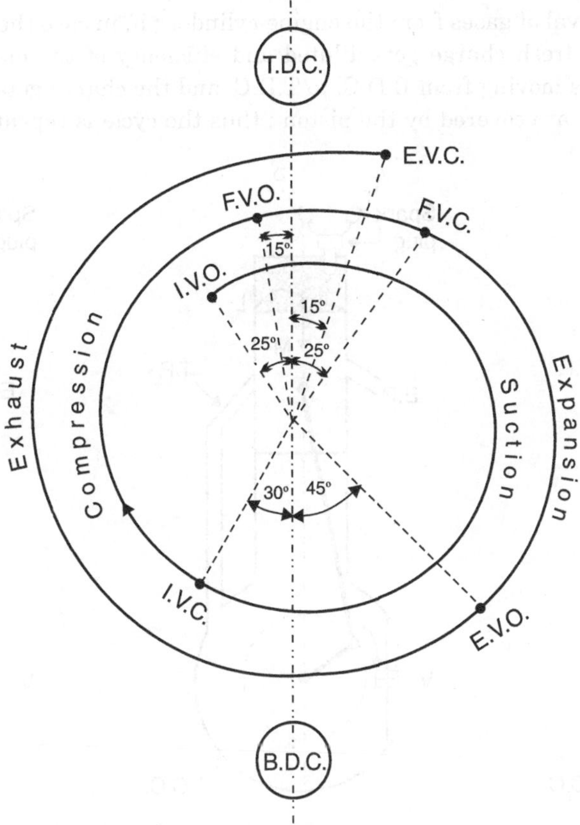

Fig. 2.42. Actual valve timing diagram (four stroke Diesel cycle engines).

2.12. TWO STROKE CYCLE ENGINES

In 1878, Dugald-clerk, a British engineer introduced a cycle which could be completed in two *strokes of piston rather than four strokes* as is the case with the four stroke cycle engines. The engines using this cycle were called two stroke cycle engines. In this engine suction and exhaust strokes are eliminated. Here *instead of valves, ports are used. The exhaust gases are driven out from engine cylinder by the fresh charge of fuel entering the cylinder nearly at the end of the working stroke.*

Fig. 2.43 shows a two stroke petrol engine (used in scooters, motor cycles etc.). The cylinder L is connected to a closed crank chamber C.C. During the upward stroke of the piston M, the gases in L are compressed and at the same time fresh air and fuel (petrol) mixture enters the crank chamber through the valve V. When the piston moves downwards, V closes and the mixture in the crank chamber is compressed. Refer to Fig. 2.43 (*i*), the piston is moving upwards and is compressing an explosive charge which has previously been supplied to L. Ignition takes place at the end of the stroke. The piston then travels downwards due to expansion of the gases (Fig. 2.43 (*ii*)) and near the end of this stroke the piston uncovers the exhaust port (E.P.) and the burnt exhaust gases escape through this port (Fig. 2.43 (*iii*)). The transfer port (T.P.) then is uncovered immediately, and the compressed charge from the crank chamber flows into the cylinder and is deflected upwards by the hump provided on the head of the piston. It may be noted that the incoming air-petrol mixture helps the removal of gases from the engine-cylinder ; if, in case these exhaust gases do not leave the cylinder, the fresh charge gets diluted and efficiency of the engine will decrease. The piston then again starts moving from B.D.C. to T.D.C. and the charge gets compressed when E.P. (exhaust port) and T.P. are covered by the piston ; thus the cycle is repeated.

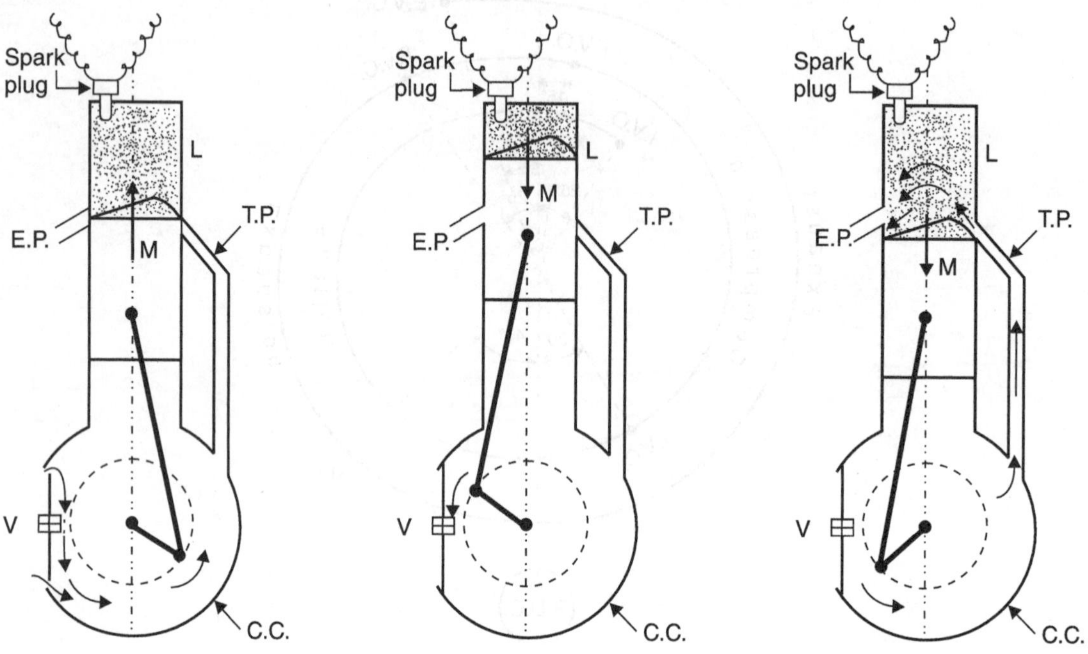

Fig. 2.43. Two stroke cycle engine.

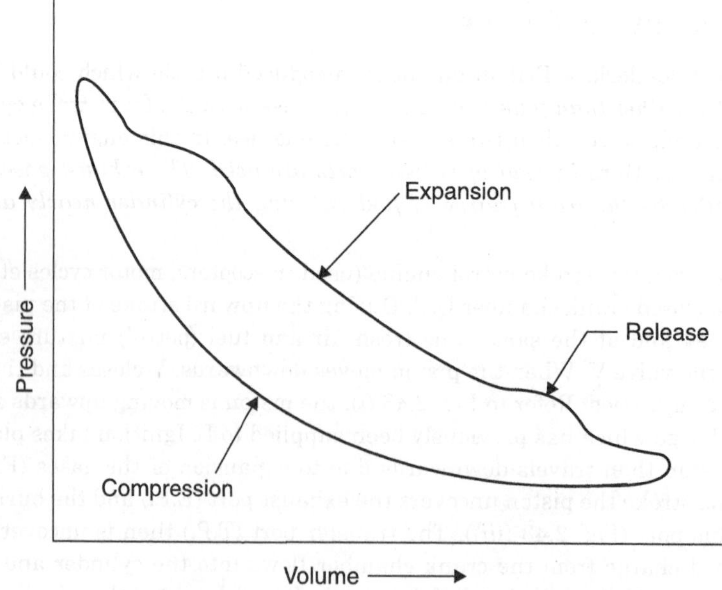

Fig. 2.44. *p-V* diagram for a two stroke cycle engine.

Fig. 2.44 shows the *p-V* diagram for a two stroke cycle engine. It is only for the main cylinder or the top side of the piston. Fig. 2.45 shows self-explanatory port timing diagram for a two stroke cycle engine.

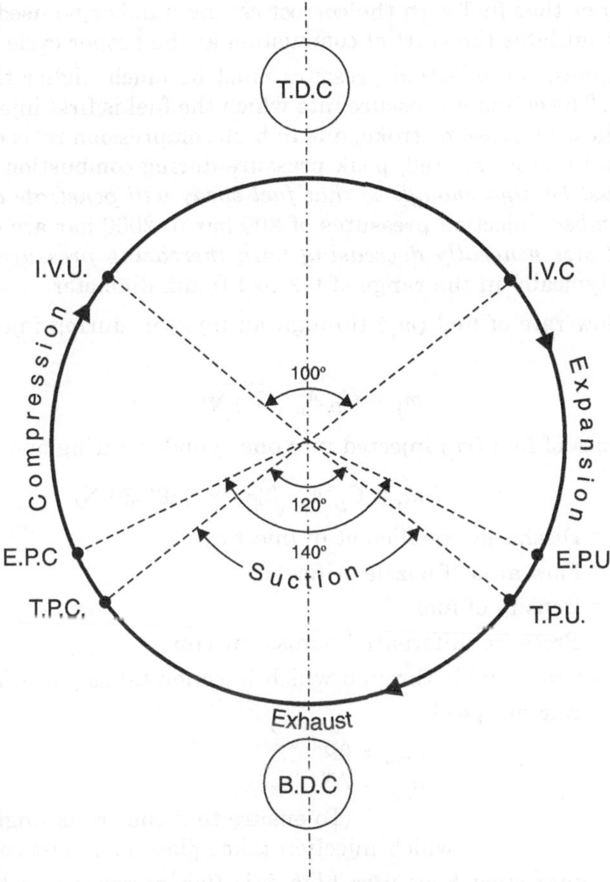

Fig. 2.45. Port timing diagram.

In a two stroke Diesel cycle engine all the operations are the same as in the spark ignition (Otto cycle) engine with the differences ; firstly in this case, only air is admitted into cylinder instead of air-fuel mixture and secondly fuel injector is fitted to supply the fuel instead of a sparking plug.

2.13. INTAKE FOR COMPRESSION IGNITION ENGINES

- *The Compression Ignition (C.I.) engines are operated unthrottled with engine speed and power controlled by the amount of fuel injected during each cycle.* This allows for high volumetric efficiency at all speeds, *with the intake system designed for very little flow restriction of the incoming air.* Further raising the volumetric efficiency is the fact that no fuel is added until late in compression stroke, after air intake is fully completed. In addition many C.I. engines are *turbocharged, which enhances air intake even more.*

- The addition of fuel is made late in the compression stroke, starting somewhere around 20° before T.D.C. Injectors mounted in the cylinder head inject directly into the combustion chamber, where self ignition occurs due to the high temperature of the air caused by compression heating.
- It is important that fuel with the correct cetane number be used in an engine so that self-ignition initiates the start of combustion at the proper cycle position.
- For C.I. engines, the injection pressure must be much higher than that required for S.I. engines. The cylinder pressure into which the fuel is first injected is very high near the end of the compression stroke, due to high compression ratio of C.I. engines. By the time the final fuel is injected, peak pressure during combustion is being experienced. *Pressure must be high enough so that fuel spray will penetrate across the entire combustion chamber.* Injection pressures of 200 bar to 2000 bar are common *with average fuel droplet size generally decreasing with increasing pressure.* Orifice hole size of injectors is typically in the range of 0.2 to 1.0 mm diameter.
- The mass flow rate of fuel ($\dot{m}_f$) through an injector, during injection, is given by the relation :

$$\dot{m}_f = C_D A_n \sqrt{2\rho_f \Delta p} \quad ...(2.1)$$

The total mass of fuel (m_f) injected into one cylinder during one cycle is given as :

$$m_f = C_D A_n \sqrt{2\rho_f \Delta p} \; (\Delta\theta/360\ N) \quad ...(2.2)$$

where, C_D = Discharge co-efficient of injector,
A_n = Flow area of nozzle orifice(s),
ρ_f = Density of fuel,
Δp = Pressure differential across injector,
$\Delta\theta$ = Crank angle through which injection takes place (in degrees), and
N = Engine speed.

Again, $$p_{inj.} \simeq \Delta p \quad ...(2.3)$$

and, $$p_{inj.} \simeq N^2 \quad ...(2.4)$$

(To ensure that the crank angle of rotation through which injection takes place is almost constant for all speeds)

— *Large engines must have very high injection pressure and high spray velocity.*

— *For optimum fuel viscosity and spary penetration, it is important to have fuel at the correct temperature.*

Often engines are equipped with temperature sensors and means of heating or cooling the incoming fuel. *Many large truck engines are equipped with heated fuel filters. This allows the use of cheaper fuel that has less viscosity control.*

— *In small engines more costly, lower viscosity fuel is required.*

2.14. COMPARISON OF FOUR STROKE AND TWO STROKE CYCLE ENGINES

S.No.	Aspects	Four stroke cycle engines	Two stroke cycle engines
1.	*Completion of cycle*	The cycle is completed in *four strokes of the piston* or in *two revolutions of the crankshaft.* Thus one power stroke is obtained in every two revolutions of the crankshaft.	The cycle is completed in *two strokes of the piston* or in *one revolution of the crankshaft.* Thus one power stroke is obtained in each revolution of the crankshaft.

2.	*Flywheel required heavier or lighter*	Because the turning-movement is not so uniform and hence *heavier* flywheel is needed.	More uniform turning movement and hence *lighter* flywheel is needed.
3.	*Power produced for same size of engine*	Again because of one power stroke for two revolutions, power produced for same size of engine is *small* or for the same power the engine is heavy and bulky.	Because of one power stroke for one revolution, power produced for same size of engine is *more* (theoretically twice, actually about 1.8 times) or for the same power the engine is light and compact.
4.	*Cooling and lubrication requirements*	Because of one power stroke in two revolutions *lesser* cooling and lubrication requirements. Lesser rate of wear and tear.	Because of one power stroke in one revolution *greater* cooling and lubrication requirement. Great rate of wear and tear.
5.	*Value and valve mechanism*	The four stroke engine *contains* valve and valve mechanism.	Two stroke engines have *no* valves but only ports (some two stroke engines are fitted with conventional exhaust valves).
6.	*Initial cost*	Because of the heavy weight and complication of valve mechanism, *higher* is the initial cost.	Because of light weight and simplicity due to absence of valve mechanism, *cheaper* in initial cost.
7.	*Volumetric efficiency*	Volumetric efficiency *more* due to more time of induction.	Volumetric efficiency *less* due to lesser time for induction.
8.	*Thermal and part-load efficiencies*	Thermal efficiency higher, part load efficiency better than two stroke cycle engine.	Thermal efficiency lower, part load efficiency lesser than four stroke cycle engine.
9.	*Applications*	Used where efficiency is important ; in *cars, buses, trucks, tractors, industrial engines, aeroplane, power generators etc.*	In two stroke petrol engine some fuel is exhausted during scavenging. Used where (*a*) *low cost*, and (*b*) *compactness and light weight important.* Two stroke (air cooled) petrol engines used in very small sizes only, *lawn movers, scooters motor cycles* (lubricating oil mixed with petrol). Two stroke diesel engines used in *very large sizes* more than 60 cm bore, for *ship propulsion* because of *low weight and compactness.*

2.15. COMPARISON OF SPARK IGNITION (S.I.) AND COMPRESSION IGNITION (C.I.) ENGINES

S.No.	Aspects	S.I. engines	C.I. engines
1.	*Thermodynamic cycle*	Otto cycle	Diesel cycle ...... For slow speed engines. Dual cycle ...... For high speed engines.
2.	*Fuel used*	Petrol	Diesel.

3.	*Air-fuel ratio*	10 : 1 to 20 : 1	18 : 1 to 100 : 1.
4.	*Compression ratio*	upto 11 ; Average value 7 to 9 ; Upper limit of compression ratio fixed by *anti-knock quality of fuel.*	12 to 24 ; Average value 15 to 18 ; Upper limit of compression ratio is limited by *thermal and mechanical stresses.*
5.	*Combustion*	Spark ignition	Compression ignition.
6.	*Fuel supply*	By carburettor ...... cheap method	By injection ...... expensive method.
7.	*Operating pressure*		
	(*i*) Compression pressure	7 bar to 15 bar	30 bar to 50 bar
	(*ii*) Maximum pressure	45 bar to 60 bar	60 bar to 120 bar.
8.	*Operating speed*	High speed : 2000 to 6000 r.p.m.	Low speed : 400 r.p.m. Medium speed : 400 to 1200 r.p.m. High speed : 1200 to 3500 r.p.m.
9.	*Control of power*	Quantity governing ...... by throttle	Quality governing ...... by rack.
10.	*Calorific value*	44 MJ/kg	42 MJ/kg.
11.	*Cost of running*	High	Low.
12.	*Maintenance cost*	Minor maintenance required.	Major overall required but less frequently.
13.	*Supercharging*	Limited by *detonation.* Used only in *aircraft engines.*	Limited by *blower power and mechanical and thermal stresses. Widely used.*
14.	*Two stroke operation*	*Less suitable*, fuel loss in scavenging. But small two stroke engines are used in mopeds, scooters and motorcycles due to their *simplicity and low cost.*	No fuel loss in scavenging. *More suitable.*
15.	*High powers*	No	Yes.
16.	*Distribution of fuel*	A/F ratio is not optimum in multi-cylinder engines.	Excellent distribution of fuel in multi-cylinder engines.
17.	*Starting*	Easy, low cranking effort.	Difficult, high cranking effort.
18.	*Exhaust gas temperature*	High, due to low thermal efficiency.	Low, due to high thermal efficiency.
19.	*Weight per unit power*	Low (0.5 to 4.5 kg/kW).	High (3.3 to 13.5 kg/kW).
20.	*Initial capital cost*	Low	High due to heavy weight and sturdy construction, costly construction, 1.25-1.5 times.
21.	*Noise and vibration*	Less	More idle noise problem.
22.	*Uses*	Mopeds, scooters, motorcycles, simple engine passenger cars, aircrafts etc.	Buses, trucks locomotives, tractors, earth moving machinery and stationary generating plants.

2.16. COMPARISON BETWEEN A PETROL ENGINE AND A DIESEL ENGINE

S.No.	Petrol engine	Diesel engine
1.	Air petrol mixture is sucked in the engine cylinder during suction stroke.	Only air is sucked during suction stroke.
2.	Spark-plug is used.	Employs an injector.
3.	Power is produced by spark ignition.	Power is produced by compression ignition.
4.	Thermal efficiency up to 25%.	Thermal efficiency up to 40%.
5.	Occupies less space.	Occupies more space.
6.	More running cost.	Less running cost.
7.	Light in weight.	Heavy in weight.
8.	Fuel (Petrol) costlier.	Fuel (Diesel) cheaper.
9.	Petrol being volatile is dangerous.	Diesel is non-dangerous as it is non-volatile.
10.	Pre-ignition possible.	Pre-ignition not possible.
11.	Works on Otto cycle.	Works on Diesel cycle.
12.	Less dependable.	More dependable.
13.	Used in *cars* and *motor cycles*.	Used in heavy duty vehicles like *trucks, buses* and *heavy machinery*.

2.17. HOW TO TELL A TWO STROKE CYCLE ENGINE FROM A FOUR STROKE CYCLE ENGINE ?

S.No.	Distinguishing features	Four stroke cycle engine	Two stroke cycle engine
1.	*Oil sump and oil-filter plug*	It has an oil sump and oil-filter plug.	It does not have oil sump and oil-filter plug.
2.	*Oil drains etc.*	It requires oil drains and refills periodically, just an automobile do.	In this type of engine, the oil is added to the gasoline so that a mixture of gasoline and oil passes through the carburettor and enters the crankcase with the air.
3.	*Location of muffler (exhaust silencer)*	It is installed at the head end of the cylinder at the exhaust valve location.	It is installed towards the middle of the cylinder, at the exhaust port location.
4.	*Name plate*	If the name plate mentions the type of oil and the crankcase capacity, or similar data, it is a four stroke cycle engine.	If the name plate tells to mix oil with the gasoline, it is a two stroke cycle engine.

II. Combustion in S.I. Engines

2.18. INTRODUCTION TO COMBUSTION IN S.I. ENGINES

2.18.1. Definition of Combustion

Combustion *may be defined as a relatively rapid chemical combination of hydrogen and carbon in the fuel with the oxygen in the air, resulting in liberation of energy in the form of heat.*

Following conditions are necessary for combustion to take place :

1. A combustible mixture.
2. Some means to initiate combustion.
3. Stabilization and propagation of flame in the combustion chamber.

In spark ignition (S.I.) engines, a carburettor generally supplies a combustible mixture and the electric spark from a spark-plug initiates the combustion.

2.18.2. Ignition Limits

- It has been observed through experiments that ignition of charge is only possible within certain limits of fuel-air ratio.
  The 'ignition limits' correspond approximately to those mixture ratios, at lean and rich ends of the scale, where the heat released by the spark is no longer sufficient to initiate combustion in the neighbouring unburnt mixture.
- Fig. 2.46 shows the ignition limits for hydrocarbons.

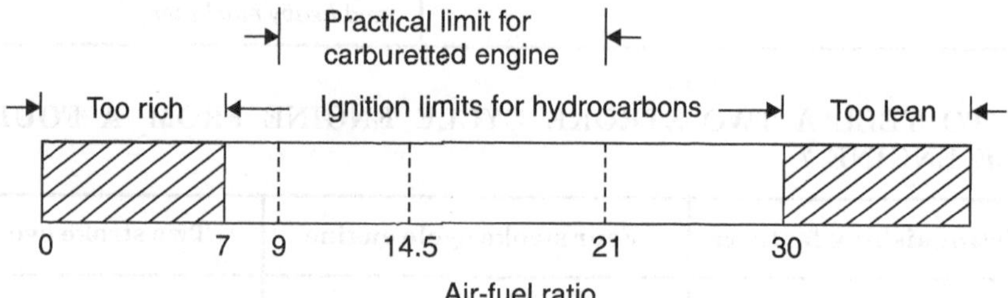

Fig. 2.46. Ignition limits for hydrocarbons.

— The ignition limits are wider at increased temperatures because of *higher rates of reaction and higher thermal diffusivity coefficients of the mixture.*

— The lower and upper limits of ignition of the mixture depend upon the *temperature and mixture ratio.*

— In case of hydrocarbon fuel the stoichiometric air-fuel ratio is about 15 : 1 and the air-fuel ratio lies between about 30 : 1 and 7 : 1.

2.19. COMBUSTION PHENOMENON

2.19.1. Normal Combustion

In a S.I. engine a single intensely high temperature spark passes across the electrodes, leaving behind a thin thread of flame. From this thin thread, combustion spreads to the envelope of mixture immediately surrounding it at a rate which *depends primarily upon the temperature of the flame front itself and to a secondary degree, upon both the temperature and the density of the surrounding envelope.* In the actual engine cylinder, the mixture is not at rest but is in highly turbulent condition. The *turbulence breaks the filament of a flame into a ragged front, thus presenting a far greater area of surface from which heat is being radiated ; hence its advance is speeded up enormously.*

According to Ricardo, the combustion process can be imagined as if developing in the following *two stages* :

(*i*) The growth and development of a self-propagating nucleus of flame (*ignition lag*). This is a *chemical process* and depends upon the following :

— The nature of fuel ;

— The temperature and pressure ;

— The proportion of the exhaust gas ;

— The temperature co-efficient of the fuel *i.e.*, the relationship between temperature and rate of acceleration of oxidation or burning.

(*ii*) The spread of the flame throughout the combustion chamber.

Fig. 2.47 shows the *p*-θ diagram of a petrol engine :

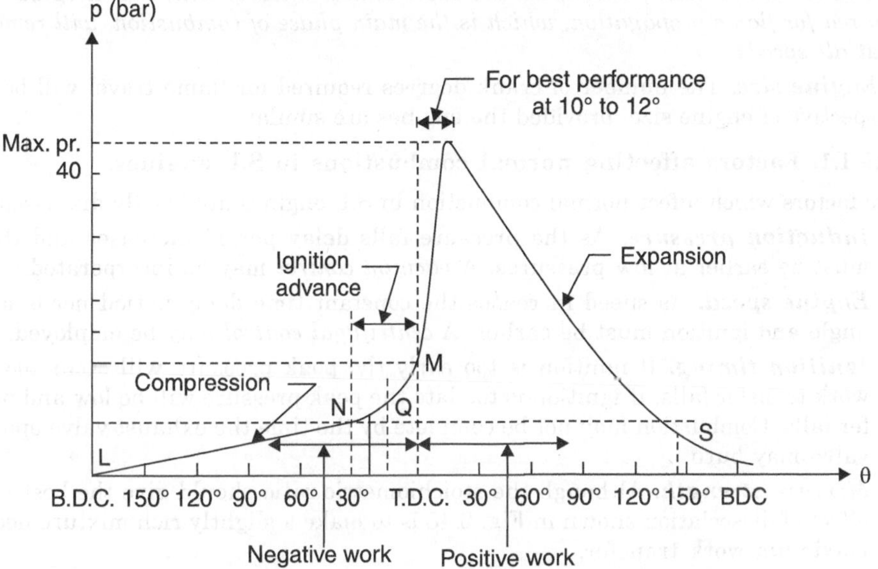

Fig. 2.47. Pressure-crank angle diagram of a petrol engine.

LNQM assumes *compression curve* having *no ignition.*

- First stage of combustion, the *ignition lag,* starts from this point and *no pressure rise is noticeable.*
- *Q* is the point where the *pressure rise can be detected.* From this point it deviates from the simple compression (motoring) curve.
- The time lag between first igniting of fuel and the commencement of the main phase of combustion is called the *period of incubation* or is also known as **ignition lag**. The time is normally about 0.0015 seconds. The maximum pressure is reached at about 12° after top dead centre point. Although the point of maximum pressure marks the completion of flame travel, it does not mean that at this point the whole of the heat of fuel has been liberated, for even after the passage of the flame, some further chemical adjustments due to reassociation, etc., will continue to a greater or less degree throughout the expansion stroke. This is known as ***after burning***.

Effect of engine variables on flame propagation :

1. ***Fuel-air ratio.*** When the *mixture is made leaner or is enriched and still more, the velocity of flame diminishes.*

2. ***Compression ratio.*** The *speed of combustion increases with increase of compression ratio.* The increase in compression ratio results in increase in temperature which increases the tendency of the engine to detonate.

3. ***Intake temperature and pressure.*** *Increase in intake temperature and pressure increases the flame speed.*

4. ***Engine load.*** As the *load on the engine increases,* the cycle pressures increase and hence the flame *speed increases.*

5. ***Turbulence.*** The flame speed is very low in non-turbulent mixture. A turbulent motion of the mixture intensifies the processes of heat transfer and mixing of the burned and unburned portions in the flame front. These two factors cause the *velocity of turbulent flame to increase practically in proportion to the turbulent velocity.*

6. ***Engine speed.*** The *flame speed increases almost linearly with engine speed.* The *crank angle required for flame propagation, which is the main phase of combustion, will remain almost constant at all speeds.*

7. ***Engine size.*** The number of crank degrees required for flame travel will be about the same irrespective of engine size, provided the engines are similar.

2.19.1.1. Factors affecting normal combustions in S.I. engines.

The factors which affect normal combustion in S.I. engines are briefly discussed below :

1. ***Induction pressure.*** As the pressure falls delay period increases and the ignition must be earlier at low pressures. A *vacuum control* may be incorporated.
2. ***Engine speed.*** As speed increases the constant time delay period needs more crank angle and ignition must be earlier. A *centrifugal control* may be employed.
3. ***Ignition timing.*** If ignition is too early the peak pressure will occur too early and work transfer falls. If ignition is too late the peak pressure will be low and work transfer falls. Combustion may not be complete by the time the exhaust valve opens and the valve may burn.
4. ***Mixture strength.*** Although the stoichiometric ratio should give the best results, the effect of dissociation shown in Fig. 2.48 is to make a slightly rich mixture necessary for maximum work transfer.
5. ***Compression ratio.*** An increase in compression ratio increases the maximum pressure and the work transfer.

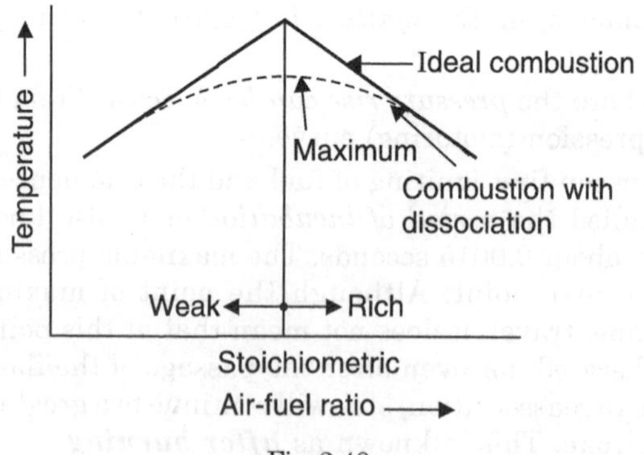

Fig. 2.48

6. **Combustion chamber.** The combustion chamber should be designed to *give a short flame path to avoid knock and it should promote optimum turbulence.*

7. **Fuel choice.**
   - The induction period of the fuel will affect the delay period.
   - The calorific value and the enthalpy of vaporisation will affect the temperatures achieved.

2.19.2. Abnormal Combustion

Due to excessively weak mixtures combustion may be slow or may be mis-timed. This is however obvious.

There are *two combustion abnormalities,* which are less obvious :

- The *first* of these is **pre** or **post ignition** of the mixture by *incandescent carbon particles in the chamber.* This will have the *effect of reducing the work transfer.*
- The *second* abnormality is generally known as *knock* and is a complex condition with many facets. A simple explanation shows that knock occurs when the unburnt portion of the gas in the combustion chamber is heated by combustion and radiation so that *its temperature becomes greater than the self ignition temperature. If normal progressive combustion is not completed before the end of the induction period then a simultaneous explosion of the unburnt gas will occur.* This explosion is accompanied by a detonation (pressure) wave which will be repeatedly reflected from the cylinder walls setting up a high frequency resonance which gives an audible noise. The detonation wave causes excessive stress and also destroys the thermal boundary layer at the cylinder walls causing overheating.

[**Note.** Refer to articles 2.22 and 2.23 for details of pre-ignition and detonation respectively.]

2.20. EFFECT OF ENGINE VARIABLES ON IGNITION LAG

Ignition lag (the time lag between first igniting of fuel and the commencement of the main phase of combustion) is not a period of inactivity but is a chemical process. The ignition lag in terms of crank angles is 10° to 20° and in terms of time, 0.0015 second or so.

The duration of ignition lag depends on the following factors :

1. **Fuel.** Ignition lag depends on chemical nature of fuel. The *higher the Self Ignition Temperature (S.I.T.) of fuel, longer the ignition lag.*
2. **Mixture ratio.** Ignition lag is the *smallest for the mixture ratio which gives the maximum temperature.* This mixture ratio is *somewhat richer than stoichiometric ratio.*
3. **Initial temperature and pressure.** Ignition lag is *reduced if the initial temperature and pressure are increased* (and these can be increased by increasing the compression ratio).
4. **Turbulence.** *Ignition lag is not much affected by the turbulence.*

2.21. SPARK ADVANCE AND FACTORS AFFECTING IGNITION TIMING

Spark advance. In order to obtain maximum power from an engine, the compressed mixture must deliver its maximum pressure at a time when the piston is about to commence its outward stroke and is nearest to T.D.C. Since there is a time lag between the occurrence of spark and the burning of the mixture, the *spark must take place before the piston reaches T.D.C. on its compression stroke, i.e.,* the *spark timing is advanced.* Usually the spark should occur at about 15° before T.D.C.

The correct instant for the introduction of spark is mainly determined by the **"ignition lag"**. The *factors affecting the ignition timings* are discussed as follows :

1. **Engine speed.** Suppose an engine has an ignition advance of θ degrees and operating speed in n r.p.s. Then time available for initiation of combustion is $\frac{\theta}{360n}$ seconds. Now if the engine speed is increased to $2n$ r.p.s. then in order to have the same time available for combustion, an ignition advance for 2θ degrees is required. Thus as the engine speed is *increased,* it will be necessary to advance the ignition progressively.

2. **Mixture strength.** In general *rich mixtures burn faster.* Hence, if the engine is operating with rich mixtures the optimum spark timings must be *retarded, i.e.,* the number of crank angle before T.D.C. at the time of ignition is *decreased* and the spark occurs *closer to T.D.C.*

3. **Part-load operation.** Part-load operation of a spark-ignition engine is affected by throttling the incoming charge. Due to throttling a small amount of charge enters the cylinder, and the dilution due to residual gases is also greater. In order to overcome the problem of exhaust gas dilution and the low charge density, at part-load operation the *spark advance must be increased.*

4. **Type of fuel.** Ignition delay will depend upon the type of fuel used in the engine. For *maximum power and economy a slow burning fuel needs a higher spark advance than a fast burning fuel.*

2.22. PRE-IGNITION

Refer to Fig. 2.49.

- **Pre-ignition** is *the ignition of the homogeneous mixture in the cylinder, before the timed ignition spark occurs, caused by the local overheating of the combustible mixture.* For premature ignition of any local hot-spot to occur in advance of the timed spark on the combustion stroke it must attain a minimum temperature of something like 700–800°C.
- Pre-ignition is *initiated by some overheated projecting part such as the sparking plug electrodes, exhaust valve head, metal corners in the combustion chamber, carbon deposits or protruding cylinder head gasket rim etc.*
  — However, *pre-ignition is also caused by persistent detonating pressure shockwaves* scoring away the stagnant gases which normally protect the combustion chamber walls. The resulting increased heat flow through the walls, raises the surface temperature of any protruding poorly cooled part of the chamber, and this therefore provides a focal point for pre-ignition.
- The initiation of ignition and the propagation of the flame front from the heated hot-spot is similar to that produced by the spark-plug when it fires, the only difference between the hot-spot and spark-plug is their respective *instant of ignition.* Thus, the sparking-plugs provides a timed and controlled moment of ignition whereas the heated surface forming the hot-spot builds upto the ignition temperature during each compression stroke and therefore the *actual instant of ignition is unpredicable.*
- The early ignition created by pre-ignition extends the total time and the burnt gases remain in the cylinder and therefore *increases the heat transfer on the chamber walls, as a result, the self-ignition temperature will occur earlier and earlier on each successive compression stroke.* Consequently, the peak cylinder pressure (which normally occurs at its optimum position of 10°–15° after T.D.C.) will progressively *advance* its position to T.D.C. where the cylinder pressure and temperature will be maximised.

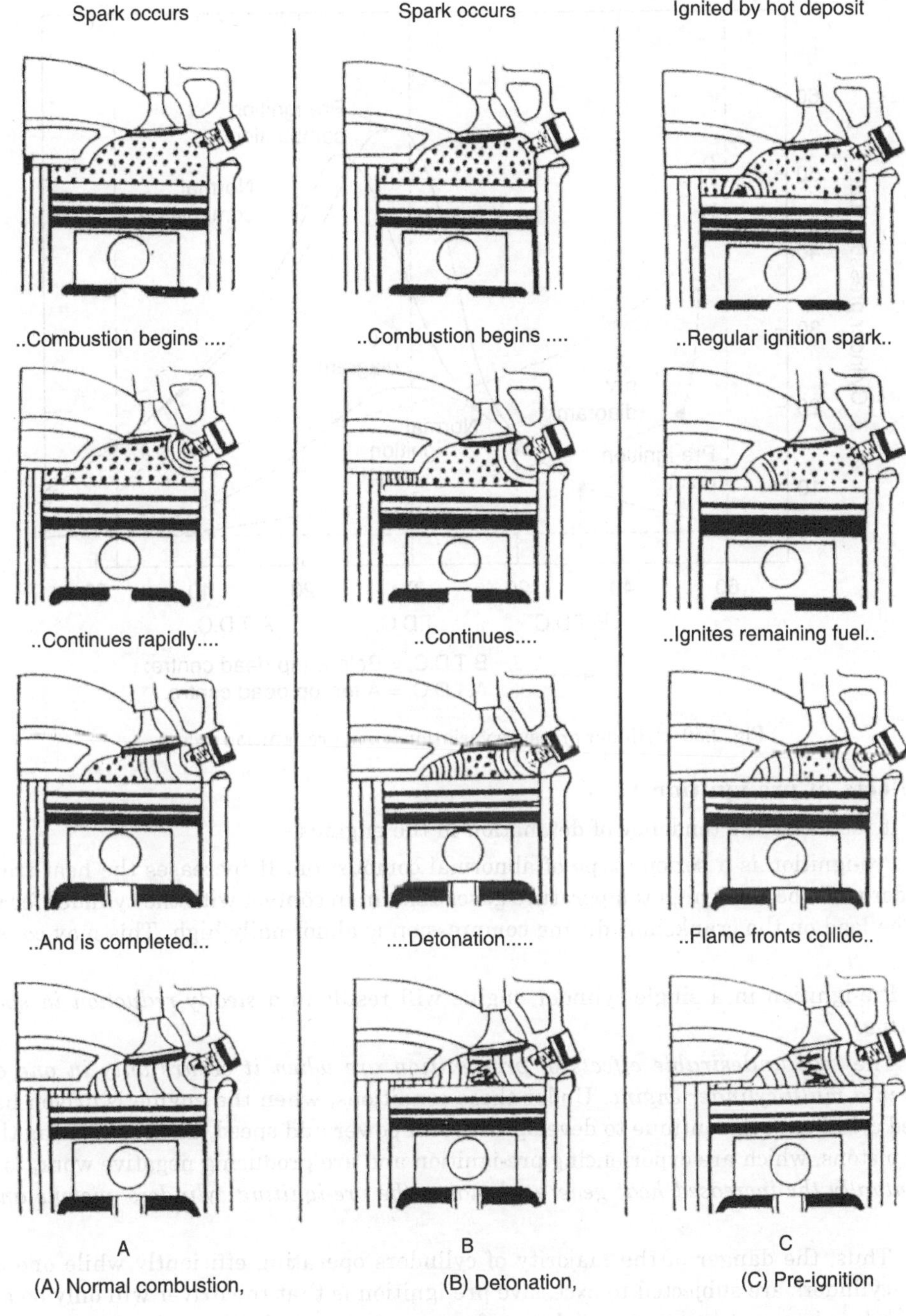

(A) Normal combustion, (B) Detonation, (C) Pre-ignition

Fig. 2.49

— **The accumulated effects of an extended combustion time and rising peak cylinder pressure and temperature cause the self-ignition temperature to creep further and further ahead of T.D.C., and with it, peak cylinder pressure, which will now take place before T.D.C. so that negative work** ***will be done in compressing the combustion products*** **(Fig. 2.50).**

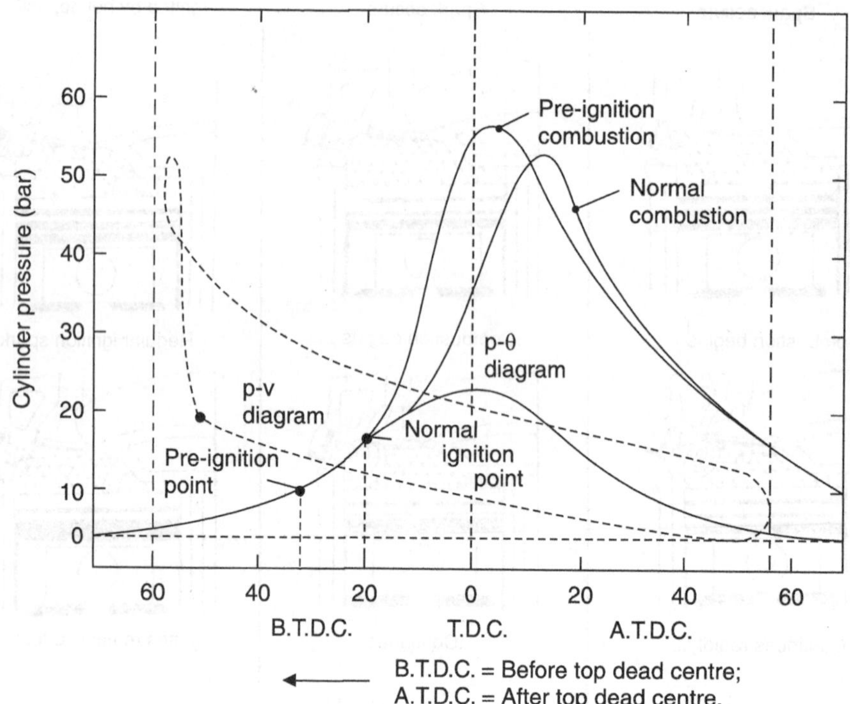

Fig. 2.50. Cylinder pressure variation when pre-ignition occurs.

Effects of pre-ignition :

1. It increases the tendency of detonation in the engines.

2. Pre-ignition is a serious type of abnormal combustion. It increases the heat transfer to the cylinder walls because high temperature gases remain in contact with the cylinder for a longer period. The load on the crankshaft during compression is abnormally high. This may *cause crank failure.*

3. Pre-ignition in a single-cylinder engine will result in a *steady reduction in speed and power output.*

4. The *real undesirable effects of pre-ignition are when it occurs only in one or more cylinders in a multi-cylinder engine.* Under these conditions, when the engine is driven hard, the unaffected cylinders will continue to develop their full power and speed, and so will drag the other piston or pistons, which are experiencing pre-ignition and are producing negative work, to and fro until *eventually the increased heat generated makes the pre-igniting cylinders' pistons and rings sieze.*

— Thus, the danger of the majority of cylinders operating efficiently while one or more cylinders are subjected to excessive pre-ignition is that the driver will only be aware of a loss in speed and power and therefore may try to work the engine harder to compensate for this, which only *intensifies the pre-ignition situation until siezure occurs.*

The following points are *worth noting* :

1. Pre-ignition is not responsible for abnormally high cylinder pressure, but there can be a slight pressure rise above the normal due to the ignition point and, therefore, the peak pressure creeping forward to the T.D.C. position where maximum pressure occurs.

2. If pre-ignition occurs at the same time as the timed sparking plug fires, combustion will appear as normal. Therefore, if ignition is switched-off, the engine would continue to operate at the same speed as if it were controlled by the conventional timed spark, provided the self-ignition temperature continues to occur at the same point.

3. *Over-heated spark-plugs and exhaust valves which are the main causes of pre-ignition should be carefully avoided in the engines.*

Tests for pre-ignition

- *The standard test for pre-ignition is to shut-off the ignition. If the engine still fires, it is assumed that pre-ignition was taking place when the ignition was on.* Experience shows that this assumption is not always valid. *Sudden loss of power with no evidence of mechanical malfunctioning is fairly good evidence of pre-ignition.*
- The *best proof of pre-ignition is the appearance of an indicator card taken with a high speed indicator of the balanced-pressure type.*

2.23. DETONATION

2.23.1. Introduction

At present the amount of power that can be developed in the cylinder of a petrol engine is *fixed by the liability of a fuel to detonate, i.e.*, just before the flame has completed its course across the combustion chamber and remaining unburnt charge fires throughout its mass spontaneously without external assistance.

The result is a tremendously rapid and local increase in pressure which sets up pressure waves that hit the cylinder walls with such violence that the walls emit a sound like a '*ping*'. It is the ping that manifests detonation. *Thus a very sudden rise of pressure during combustion accompanied by metallic hammer like sound is called* **detonation.**

The region in which detonation occurs is farthest removed from the sparking plug, and is named the "detonation zone" and even with severe detonation this zone is rarely more than one quarter the clearance volume.

2.23.2. Process of Detonation or Knocking

- The process/phenomenon of detonation or knocking may be explained by referring to the Fig. 2.51, which shows the cross-section of the combustion chamber with flame advancing from the spark plug location *A*. The advancing flame front compresses the end charge *BB′D* farthest from the spark plug, thus raising its temperature. The temperature of the end charge also increases due to heat transfer from the hot advancing flame front. Also some preflame oxidation may take place in end charge leading to further increase in its temperature.

  If the end charge *BB′D* reaches its auto-ignition temperature and remains for some time to complete the preflame reactions, the charge will autoignite leading to *knocking combustion*. During the preflame reaction period the flame front could move from *BB′* to *CC′*, and the knock occurs due to auto-ignition of the charge ahead of *CC′*. Here we have combustion unaccompanied by flame, producing a very high rate of pressure rise.

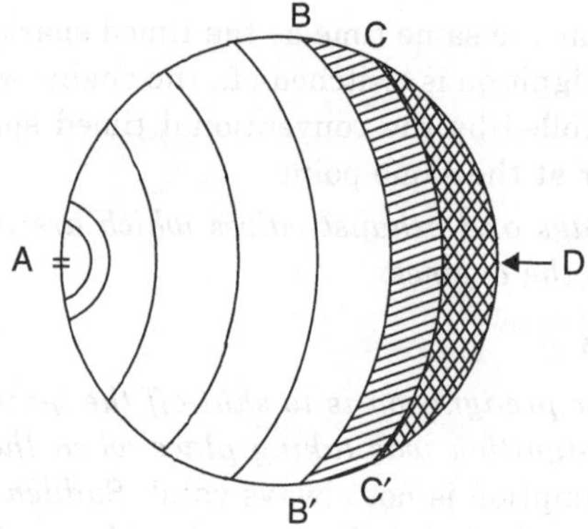

Fig. 2.51. Combustion with knocking.

- The pressure-time diagram of detonating combustion in S.I. engines is drawn and labelled below :

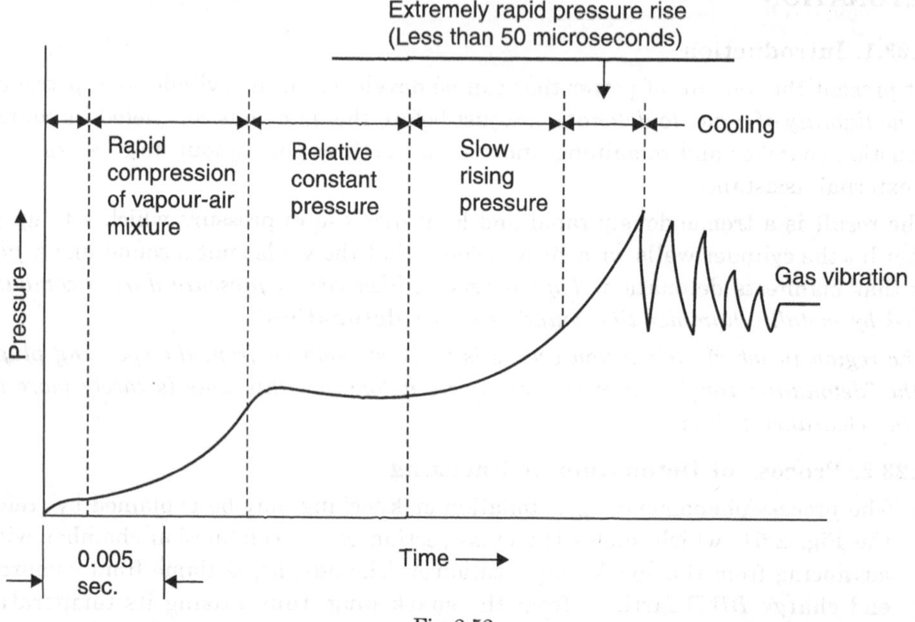

Fig. 2.52

- The *'intensity of detonation'* will depend mainly upon the *amount of energy contained in the 'end-mixture' and the rate of chemical reaction which releases it in the form of heat and a high intensity pressure-wave.* Thus, the earlier in the combustion process the detonation commences, the more unburnt end-mixture will be available to intensify the detonation. As little as 5 per cent of the total mixture charge when spontaneously ignited will be sufficient to produce a very violent knock.

2.23.3. Theories of Detonation :

There are two general theories of knocking/detonation :

(*i*) *The auto-ignition theory.*

(*ii*) *The detonation theory.*

(*i*) **Auto-ignition theory.** Auto-ignition refers to initiation of combustion without the necessity of a flame. The auto-ignition theory of knock assumes that the flame velocity is *normal* before the onset-of auto-ignition and that gas vibrations are created by a number of end-gas elements auto-igniting almost simultaneously.

(*ii*) **Detonation theory.** In the auto-ignition theory, it is assumed that the flame velocity is normal before the onset of auto-ignition whereas in detonation theory a true *detonating wave* formed by preflame reactions has been proposed as the mechanism for explosive auto-ignition. Such a shock wave would travel through the chamber at about twice the sonic velocity and would compress the gases to pressures and temperatures where the reaction should be practically instantaneous.

In fact knocking or detonation is a complex phenomenon and no single explanation may be sufficient to explain it fully.

2.23.4. Effects of Detonation

1. Noise and roughness.
2. Mechanical damage.
3. Carbon deposits.
4. Increase in heat transfer.
5. Decrease in power output and efficiency.
6. Pre-ignition.

Control of detonation :

The detonation can be controlled or even stopped by the following *methods* :

1. Increasing engine r.p.m.
2. Retarding spark.
3. Reducing pressure in the inlet manifold by throttling.
4. Making the ratio too lean or too rich, preferably latter.
5. ***Water injection.*** Water injection increases the delay period as well as reduces the flame temperature.
6. Use of high octane fuel can eliminate detonation. High octane fuels are obtained by adding additives known as dopes (such as tetra-ethyl lead, benzol, xylene etc.), to petrol.

Fig. 2.49 shows normal combustion, detonation and pre-ignition.

2.23.5. Factors Affecting Detonation/Knocks :

The likelihood of knock is increased by any reduction in the induction period of combustion and any reduction in the progressive explosion flame velocity. Particular factors are listed below :

1. **Fuel choice.** A low self-ignition temperature promotes knock.

2. **Induction pressure.** Increase of pressure decreases the self-ignition temperature and the induction period. Knock will tend to occur *at full throttle*.

3. **Engine speed.** Low engine speeds will give low turbulence and low flame velocities (combustion period is constant in angle) and knock may occur at low speed.

4. **Ignition timing.** *Advanced ignition timing* increases peak pressures and *promotes* knock.

5. **Mixture strength.** *Optimum mixture strength* gives high pressures and *promotes* knock.

6. **Compression ratio.** *High compression ratios* increase the cylinder pressures and *promotes* knock.

7. **Combustion chamber design.** Poor design gives long flame paths, poor turbulence and insufficient cooling all of which *promote* knock.

8. **Cylinder cooling.** *Poor cooling* raises the mixture temperature and *promotes* knock.

2.24. PERFORMANCE NUMBER (PN)

Performance number is a *useful measure of detonation tendency.* It has been developed from the conception of knock limited indicated mean effective pressure (*klimep*), when inlet pressure is used as the dependent variable.

$$\text{Performance number (PN)} = \frac{\textit{klimep} \text{ of test fuel}}{\textit{klimep} \text{ of iso-octane}}$$

The performance number is *obtained on specified engine, under specified set of conditions by varying the inlet pressure.*

2.25. HIGHEST USEFUL COMPRESSION RATIO (HUCR)

The **highest useful compression ratio** *is the highest compression ratio employed at which a fuel can be used in a specified engine under specified set of operating conditions, at which detonation first becomes audible with both the ignition and mixture strength adjusted to give the highest efficiency.*

2.26. COMBUSTION CHAMBER DESIGN—S.I. ENGINES

Engine torque, power output and fuel consumption are profoundly influenced by the following :

(*i*) Engine compression ratio.

(*ii*) Combustion chamber and piston crown shape.

(*iii*) The number and size of the inlet and exhaust valves.

(*iv*) The position of the sparking-plug.

The following are the objects of good combustion chamber design :

1. *To optimize the filling and emptying of the cylinder with fresh unburnt charge respectively over the engine's operating speed range.*

2. *To create the condition in the cylinder for the air and fuel to be thoroughly mixed and then excited into a highly turbulent state so that the burning of the charge will be completed in the shortest possible time.*

3. *To prevent the possibility of detonation at all times, as far as possible.*

In order to achieve these fundamental requirements it is imperative to be aware of the *factors that contribute towards inducing the charge to enter the cylinder, to mix intimately, to burn both rapidly and smoothly and to expel the burnt gases.*

2.26.1. Induction Swirl

Refer to Fig. 2.53.

- ***Swirl*** *is the rotational flow of charge within the cylinder about* its axis.

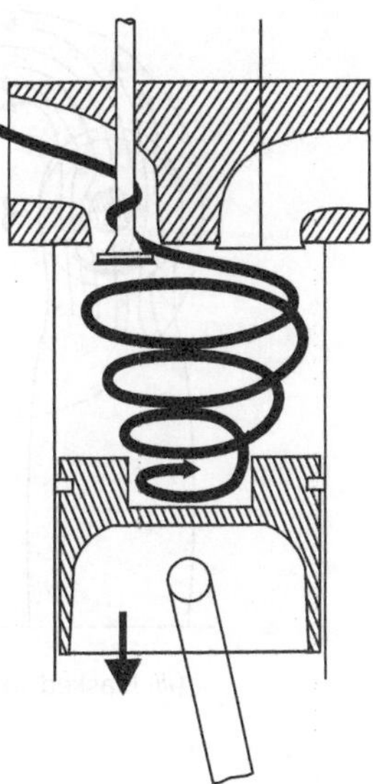

Fig. 2.53. Induction swirl.

- *Swirl* is generated by constructing the intake system to give a tangential component to the intake flow as it enters the cylinder. This is done by *shaping and contouring the intake manifold, valve ports and even the piston face.*
- *Swirl greatly enhances the mixing of air and fuel* to give a homogeneous mixture in the very short time available for this in modern high speed engines. It is also a main mechanism for spreading of the flame front during the combustion process.

The induction ports are classified as follows : Refer to Fig. 2.54.

1. *Direct straight port.*
2. *Deflector wall port.*
3. *Masked valve port.*
4. *Helical port.* The intensity of swirl is influenced by the steepness of the port helix and the mean diameter of the spiral flow path about the valve axis.
   — Helical ports usually provide *higher flow discharges for equivalvent levels of swirl compared with directed ports because* the whole periphery of the valve opening area can be fully utilized, and, as a result, higher volumetric efficiencies can be obtained in the low-to-mid speed range of the engine.
   — These ports are *less sensitive to their position relative to the cylinder axis* since the swirl generated depends mainly on the port geometry above the valve and not how it enters the cylinder. Generally, the magnitude of swirl rises with increased valve lift.
   — These ports, however, *suffer from a loss of volumetric efficiency in the upper speed range of the order 5 to 10%.*

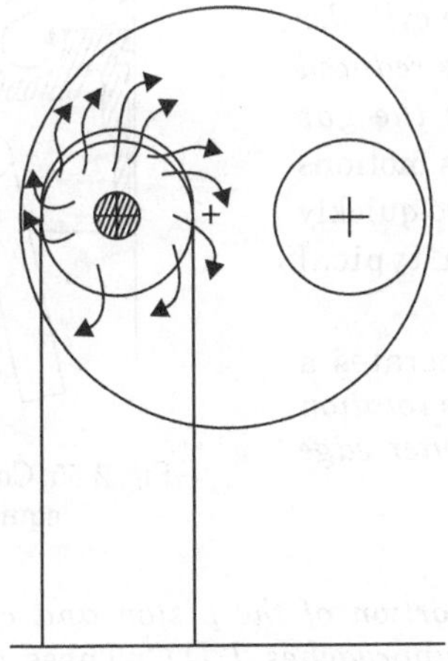

(*i*) Directed straight port

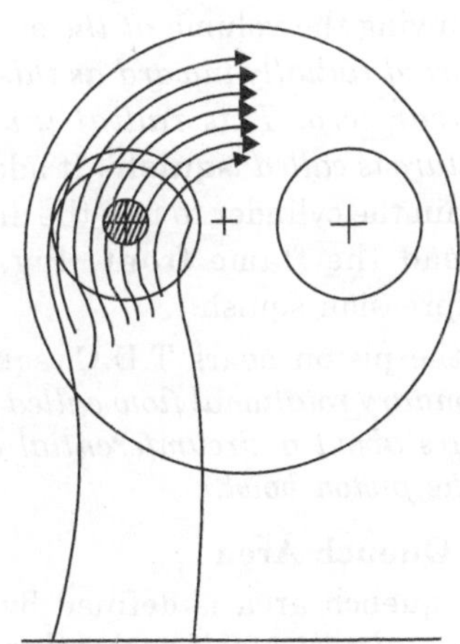

(*ii*) Deflector wall port

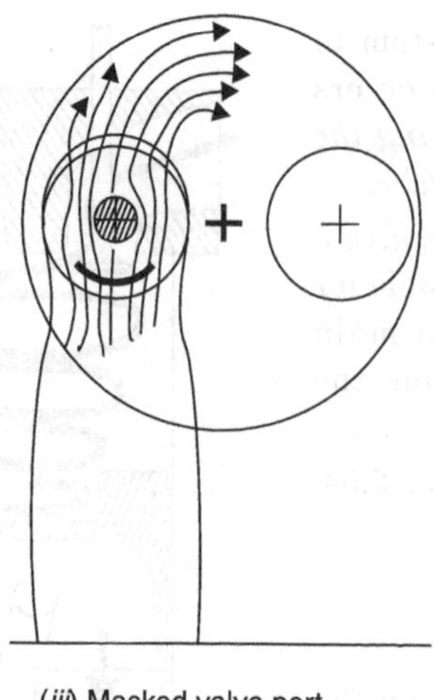
(iii) Masked valve port

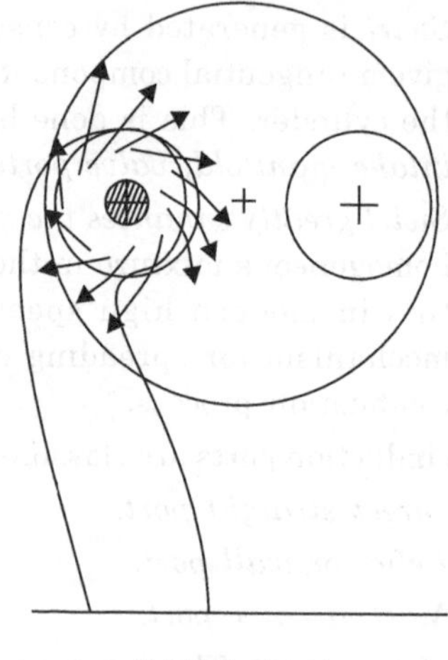
(iv) Helical port

Fig. 2.54. Induction ports.

- In **chamber wall deflected induction swirl,** the downward and circular movement of the mixture *generates an expanding, and then a contracting special swirl about the cylinder axis during both the induction and compression strokes, respectively.*

Methods of Intensifying the Rate of Burning :

2.26.2. Squish and Tumble

- As the piston approaches T.D.C. at the end of compression stroke, the volume around the outlet edges of the combustion chamber is suddenly reduced to a very small value. Many modern combustion chamber designs have most of the clearance volume near the centreline of the cylinder. As the piston approaches T.D.C. the gas mixture occupying the volume *at the outer radius of the cylinder is forced radially inward as this outer volume is reduced to near zero. This radial inward motion of the gas mixture is called* **'squish'.** It adds to other mass motions within the cylinder to mix the air and fuel and to quickly spread the flame front. Fig. 2.55 shows a typical compression squish.
- As the piston nears T.D.C. squish motion generates a *secondary rotational flow called* **"tumble"**. *This rotation occurs about a circumferential axis near the outer edge of the piston bowl.*

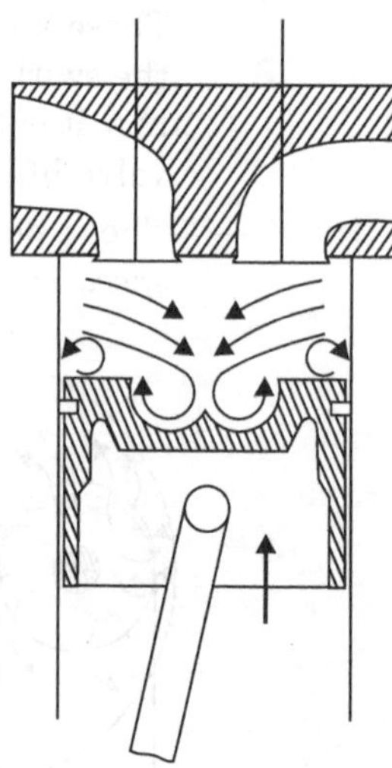
Fig. 2.55. Compression squish.

2.26.3. Quench Area

- The quench area is defined by the *parallel portion of the piston and cylinder head which almost touch each other as the piston approaches T.D.C.* These opposing flat surfaces sandwiching a thin lamina of charge between them, *have a large surface relative to the small volume trapped between them.* Consequently there will be a large amount of heat transferred from this thin lamina of hot charge through the metal walls. The result is a rapid cooling or quenching effect, by these parallel surfaces.

- The quench area is defined as *percentage of opposing flat area relative to the piston crown area.*

2.26.4. Turbulence

- *"Turbulence" consists of randomly dispersed vortices of different sizes which become superimposed into the air, or air and petrol mixture flow stream* (Fig. 2.56). These vortices, which are carried along with the flow stream, represent small irregular breakways that take on a concentric spiral motion (Fig. 2.57).

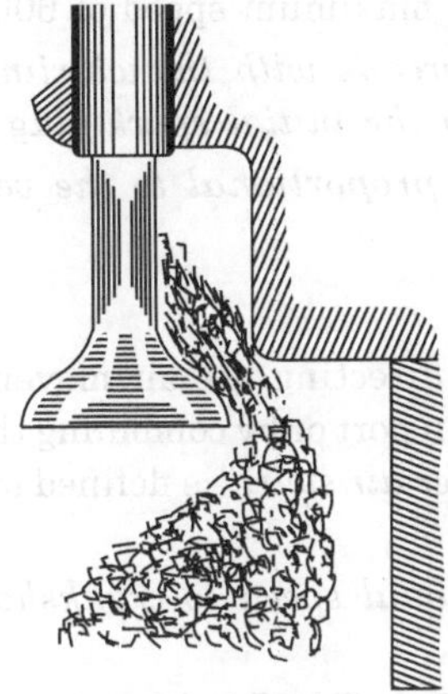

Fig. 2.56. Intake turbulent mixture flow.

Fig. 2.57. Illustration of turbulence superimposed on mixture swirl.

- As the vortices whirl they will contact adjacent vortices causing viscous shear interaction. This rapidly speeds up the rate of heat transfer and fuel mixing.
- The amount of vortex activity, that is the formation of new vortices and the disintegration of others, *increases the turbulent flow with rising engine speed.*
  — Turbulence plays a very important role in combustion phenomenon in S.I. (as well C.I.) engines. The flame speed is very low in non-turbulent mixtures. *A turbulent motion of the mixture intensifies the processes of heat transfer and mixing of the burned and unburned portions in the flame front* (diffusion). These two factors cause the velocity of turbulent flame to increase practically in proportion to the turbulent velocity. The turbulence of the mixture is due to admission of fuel-air mixture through comparatively narrow sections of the intake pipe, valves etc., in the suction stroke. The turbulence can be increased at the end of the compression stroke by suitable design of combustion chamber which involves the geometry of cylinder head and piston crown.
  — The degree of turbulence increases *directly* with the piston speed.

The *effects of turbulence* can be summed up as follows :

1. Turbulence accelerates chemical action by intimate mixing of fuel and oxygen. Thus *weak mixtures can be burnt.*
2. The increase of flame speed due to turbulence reduces the combustion time and hence *minimises the tendency to detonate.*

3. Turbulence increase the heat flow to the cylinder wall and in the limit excessive turbulence may *extinguish the flame.*
4. Excessive turbulence results in the more rapid pressure rise (though maximum pressure may be lowered) and the high pressure rise causes the crankshaft to spring and rest of the engine to vibrate with high periodicity, resulting in *rough and noisy running of the engine.*

2.26.5. Flame Propagation

- Typical flame propagation velocities range from something like 15 to 70 m/s. This would relate to the combuston flame velocity increasing from about 15 m/s at an idle speed of about 1000 r.p.m. to roughly 70 m/s at a maximum speed of 6000 r.p.m.
- *When ignition occurs the nucleus of the flame spreads with the whirling or rotating vortices in the form of ragged burning crust from the initial spark-plug ignition site.*
- *The speed of the flame propagation is roughly proportional to the velocity at the periphery of the vortices.*

2.26.6. Swirl Ratio

- Induction swirl can be generated by tangentially directing the air movement into the cylinder either by creating a preswirl in the induction port or by combining the tangential-directed flows with a preswirl helical port. *"Cylinder air swirl"* is defined *as the angular rotational speed about the cylinder axis.*
- ***Swirl ratio*** is defined as the *ratio of air rotational speed to crankshaft rotational speed.*
  — Helical ports can achieve swirl ratio of 3 to 5 at T.D.C. with a flat piston crown. However, if a bowl in the piston chamber is used, the swirl ratio can be increased to about 15 at T.D.C.

2.26.7. Surface-to-Volume Ratio

- In order to *minimise* the heat losses and formation of hydrocarbons within the combustion

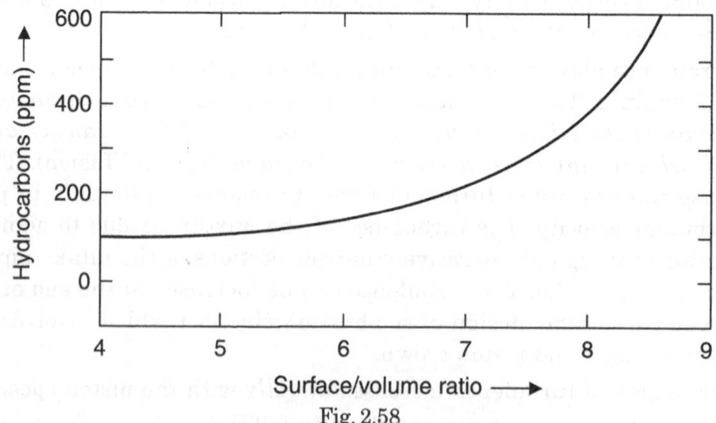

Fig. 2.58

chamber, the chamber volume should be maximised relative to its surface area, that is, the *chamber's surface area should be as small as possible relative to the volume occupied by the combustion chamber* (Fig. 2.58). The surface-to-volume ratio is the *ratio of the combustion surface area to that of its volume.*

- The surface-to-volume ratio *increases linearly with rising compression ratio.*

2.26.8. Stroke-to-Bore Ratio

- For various engines the stroke-to-bore ($L : D$) ratio can range from 0.6 : 1 to 1.4 : 1.
  - — When $L = D$, the $L : D$ ratio is said to be *square* ;
  - — When $L < D$, the $L : D$ ratio is said to be *oversquare* ;
  - — When $L > D$, the engine is said to be *undersquare.*
- *"Oversquare" engines are more suitable for saloon car petrol engines,* whereas *"undersquare" engines are better utilised for large diesel engines.*

2.26.9. Compression Ratio (C.R.)

- When compression ratio increases from 5 : 1 to 10 : 1 the cylinder's compression pressure increases from 8.0 bar to 19.0 bar respectively (Fig. 2.59). Correspondingly, the maximum cylinder pressure increases from 32 bar to 82 bar and b.m.e.p. generated also increases from 9.4 bar to 11.8 bar over the same compression range respectively.
- The effect of higher cylinder pressure is to cause a corresponding rise in cylinder temperature from 360°C to 520°C over the same compression ratio rise. Raising the cylinder temperature reduces the ignition delay period for one set engine speed (Fig. 2.60). Thus, for an engine running in its mid-speed range, the ignition timing would be reduced from 37.5° to 12.5° before T.D.C. if its compression ratio is increased from 5 : 1 to 10 : 1.

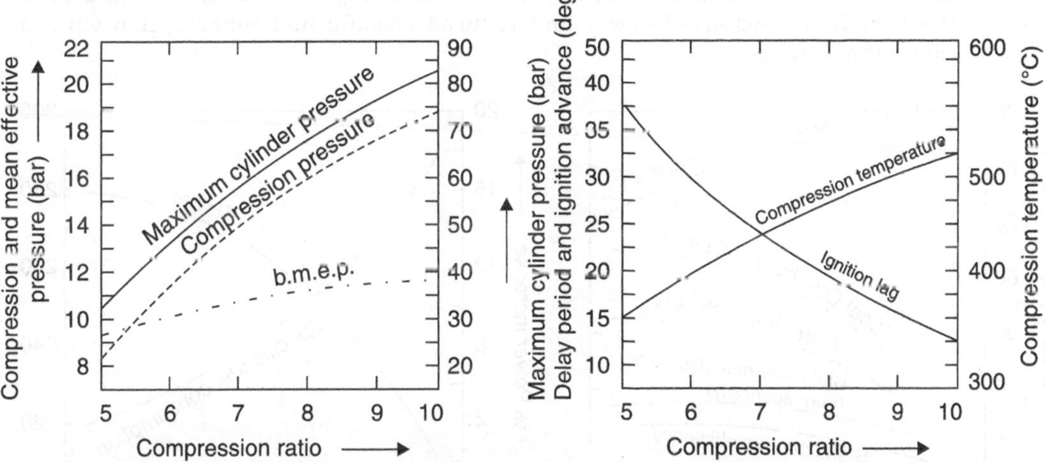

Fig. 2.59. Effect of compression ratio on the b.m.e.p. compression and maximum cylinder pressures.

Fig. 2.60. Effect of compression ratio on the air temperature and ignition lag.

- The effects of compression ratio on the characteristic pressure-volume diagram and the characteristic pressure-crank angle diagram for a petrol engine are shown in Fig. 2.61 and 2.62, respectively.
- *The main reason for raising the engine compression ratio is due to the increased density of the air-fuel mixture at the point of ignition, so that when the energy is released it is better utilized. It therefore, raises both the engine thermal efficiency and the developed power.*
- Out of the *major unwanted side effects* of raising the compression ratio is that there will be a corresponding *increase in cylinder pressure which,* in turn, *increases the piston-ring to cylinder-wall friction and compression and expansion heat losses.* Consequently, the *higher compression ratio produces a reduction in the mechanical efficiency.* Subsequently, increasing the compression ratio produces an *increase in thermal efficiency but at the expense of a falling mechanical efficiency.*

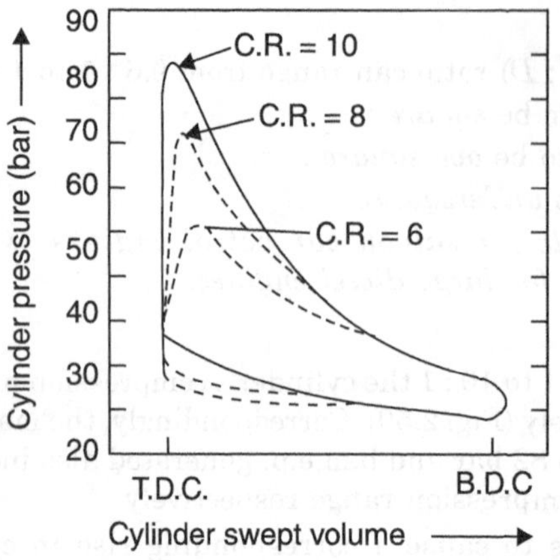

Fig. 2.61. Effect of compression ratio on the characteristic pressure-volume diagram for a petrol engine.

Fig. 2.62. Effect of compression ratio on the characteristic pressure-crank-angle diagram for a petrol engine.

- The *merits* and *limitations* of raising the compression ratio with regards to thermal efficiency and mechanical efficiency are shown in Fig. 2.63, whereas Fig. 2.64 shows the *benefits* of increased power and reduced specific fuel consumption with rising compression ratio.

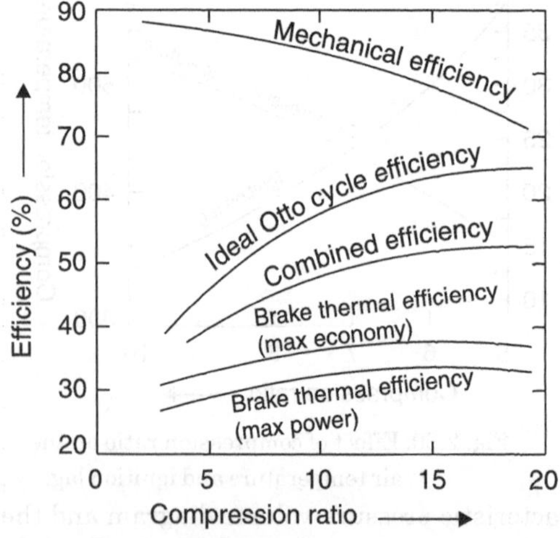

Fig. 2.63. Effect of compression ratio on an engine's thermal and mechanical efficiencies.

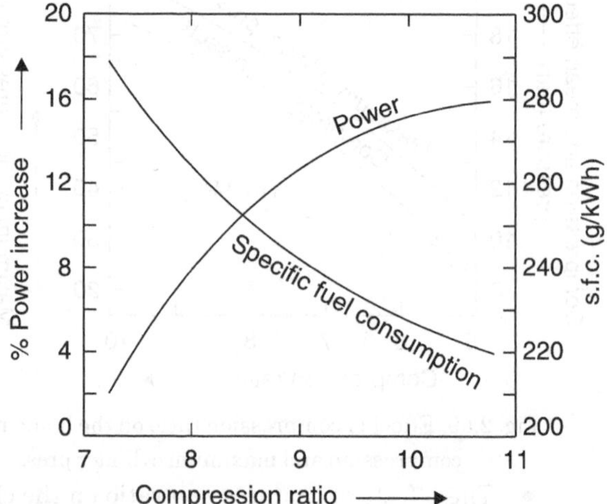

Fig. 2.64. Effect of compression on engine power and specific fuel consumption.

For S.I. engines' combustion design practice, summarily, the following are required :

1. The smallest ratio of chamber surface-area to chamber volume as possible ... *to minimise heat losses to the cooling system.*
2. The shortest flame-front travel distance as possible ... *to minimise the combustion period.*
3. The provision for quenching the mixture farthest from the sparking plug ... *to prevent the end-gas overheating.* However, it must not be excessive as this would prevent the end-gases burning and, therefore, it would cause a high level of hydrocarbons to be expelled to the exhaust.

4. The most central sparking plug position possible ... *to minimise the flame spread path* (or, alternately, *twin plugs* can be used to achieve the same objective).
5. The location of the sparking plug should be as close as to the exhaust valve as possible ... *to maximise the temperature of the mixture surrounding the sparking plug electrodes.*
6. The incoming mixture must have adequate swirl (but not too much as this could lead to excessive heat losses) ... *to mix the air and fuel rapidly and intimately.*
7. The provision for squish zones ... *to excite the mixture into a state of turbulence just before the combustion occurs.*
8. The provision for cooling of the exhaust valve ... *to prevent overheating, distorting, and burning occurring.*
9. The provision for incoming fresh charge to sweep past and cool the sparking plug electrodes ... *to avoid pre-ignition under wide throttle opening.*
10. The utilisation of the highest possible compression ratio ... *to maximise the engine's thermal efficiency without promoting detonation.*
11. The inlet and exhaust valve sizes and numbers should be adequate ... *to expel the exhaust-gases and to fill the cylinder with the maximum mass of fresh charge in the upper speed range.*
12. The degree of turbulence created should be controlled ... *to prevent excessively high rates of burning and, correspondingly, limit very high rates of pressure rise which would cause rough and noisy running.*

2.27. SOME TYPES OF COMBUSTION CHAMBERS

A few representative types of combustion chambers of which there are many more variations are enumerated and discussed below :

1. *T*-head combustion chamber.
2. *L*-head combustion chamber.
3. *I*-head (or overhead valve) combustion chamber.
4. *F*-head combustion chamber.

It may be noted that these chambers are *designed to obtain the objectives namely* :

- A *high combustion rate at the start.*
- A *high surface-to-volume ratio near the end of burning.*
- A *rather centrally located spark plug.*

1. **T-head combustion chamber.** Refer to Fig. 2.65.

This type of combustion chamber (earliest type) was used by Ford-motor corporation in 1908 in its famous model '*T*'.

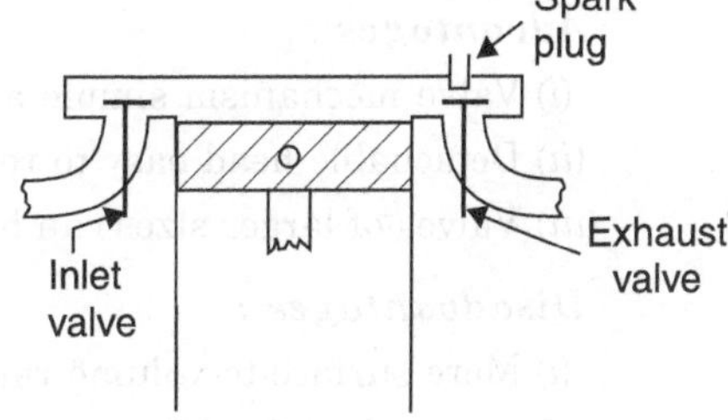

Fig. 2.65. *T*-head combustion chamber.

The *T*-head design has the following *disadvantages* :

(*i*) Requires two cam shafts (for actuating the inlet valve and exhaust valve separately) by two cams mounted on the two cam shafts.

(*ii*) Very prone to detonation. There was violent detonation even at a compression ratio of 4 (with a fuel of octane number of 50).

2. **L-head combustion chamber.** Refer to Fig. 2.66.

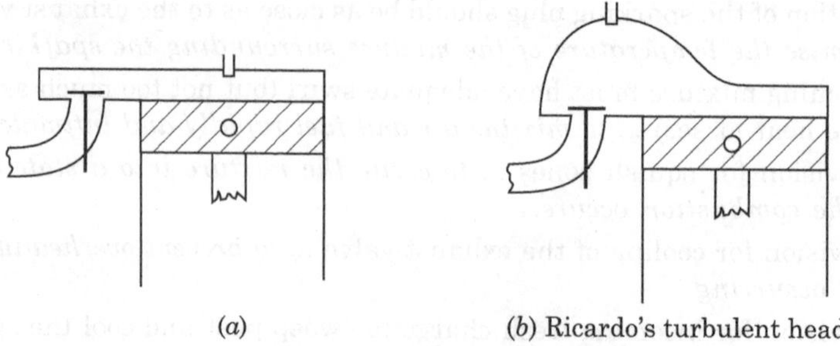

Fig. 2.66. *L*-head combustion chamber.

- It is a modification of the *T*-head type of combustion chamber. It provides the *two valves on the same side of the cylinder,* and the valves are operated through tappet by a *single camshaft.*
- Fig. 2.66 (*a*) and (*b*) shows two types of this side-valve engine. In these types *it is easy to lubricate the valve mechanism* with the detachable head, it may be noticed that *the cylinder head can be removed for cleaning or decarburising without disturbing valve-gear etc.*
  - In Fig. 2.66 (*a*), the air flow has to take two right-angled turns to enter the cylinder. This causes a loss of velocity head, and a loss in turbulence level, *resulting in slow combustion process.*
  - Fig. 2.66 (*b*) is the ***Ricardo's turbulent head design.*** The main body of the combustion chamber is concentrated over the valves leaving a slightly restricted passage communicating with the cylinder, thereby *creating additional turbulence during the compression stroke.* This design *reduces the knocking tendency by shortening the effective flame travel length* by bringing that portion of the head which lay over the further side of the piston into as close a contact as possible with the piston crown, forming a *quench space.* The thin layer of mixture (entrapped between the relatively cool piston and also cooled head) loses its heat rapidly, thereby avoiding knock. By placing the spark plug in the centre of the effective combustion space but with slight bias towards the exhaust valve, the flame travel length is reduced.

Advantages :

(*i*) Valve mechanism simple and easy to lubricate.
(*ii*) Detachable head easy to remove for cleaning and decarburising.
(*iii*) Valves of larger sizes can be provided.

Disadvantages :

(*i*) More surface-to-volume ratio and therefore *more heat loss.*
(*ii*) Longer length of flame travel.
(*iii*) Valve size restricted.
(*iv*) Thermal failure in cylinder block also. In *L*-head engine the thermal failure is confined to cylinder head only.

3. **I-head (or overhead valve) combustion chamber.** Refer to Fig. 2.67.

This type of combustion chamber has both the inlet valve and the exhaust valve located in the cylinder head. An overhead engine is superior to side valve engine at high compression ratios.

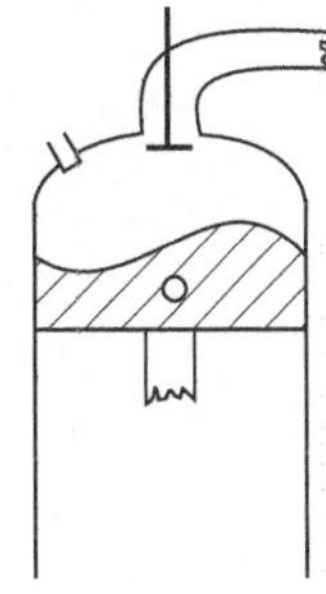

Fig. 2.67. *I*-head combustion chamber.

Advantages :

(*i*) Reduced pumping losses.

(*ii*) Higher volumetric efficiency (since the larger valves and larger lifts can be accommodated).

(*iii*) Less prone to detonation (since the path of flame travel is reduced).

(*iv*) Less force on the head bolts and therefore less possibility of leakage of compression gases or jacket water.

(*v*) Lower surface-volume ratio and, therefore, less heat loss and less air pollution.

(*vi*) Easier to cast.

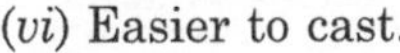

4. **F-head combustion chamber :**

- In such a combustion chamber one valve is in head and other in the block. This design is a compromise between *L*-head and *I*-head combustion chambers.
- One of the most perfect *F*-head engines (wedge type) is the one used by the Rover company for several years. Its *advantages* are :

(*i*) High volumetric efficiency.

(*ii*) Maximum compression ratio for fuel of given octane rating.

(*iii*) High thermal efficiency.

(*iv*) It can operate on leaner air-fuel ratios without misfiring.

The *drawback* of this design is the *complex mechanism for operation of valves and expensive special shaped piston.*

— Another successful design of this type of chamber is that used in Willys jeep.

Note 1. Some modern engines have divided chamber. These offer *high volumetric efficiency, good fuel economy,* and *cycle operation flexibility.* Their main *disadvantages* are *greater heat loss,* due to high surface area, and *high cost* and *difficulty in manufacturing.*

2. Very large engines are almost always C.I. engines. Because of their large combustion chambers and corresponding long flame travel distance, combined with slow engine speed, they would require very high octane fuel and very low compression ratio if operated as an S.I. engine. With the very long real time of combustion in the cylinder, it would be impossible to avoid serious knock problems.

2.27.1. Divided Combustion Chambers

- Some engines have divided combustion chambers, usually with about *80 percent* of the clearance volume in the **main chamber** above the piston and about *20 percent* of the volume as a **secondary chamber** *connected through a small orifice* (Fig. 2.68). *Combustion is started in the small secondary chamber, and the flame then passes through the orifice, where it ignites the main chamber.* Intake swirl is not as important in the main chamber of this type of engine, so the intake system can be designed for greater volumetric efficiency. It is desirable to have very high swirl in the secondary chamber, and the orifice between the chambers is shaped to supply this often, the *secondary chamber is called a "swirl chamber".* As the gases in the secondary chamber are consumed by combustion, the pressure rises and flaming gas expands back through the orifice and acts as a *torch ignition* for the main chamber. The expanding gas rushing back through the orifice creates a large secondary swirl in the main chamber,

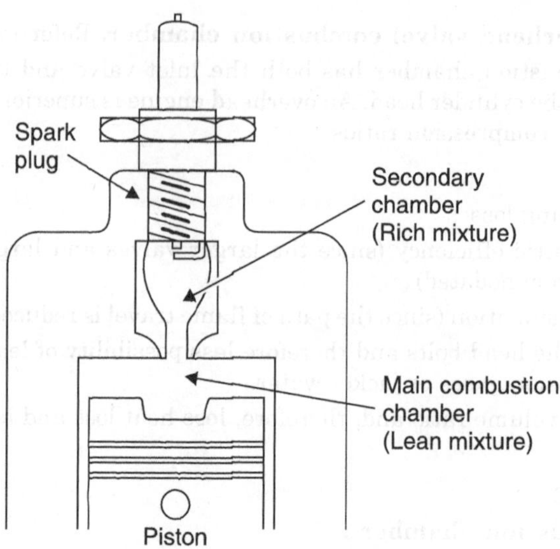

Fig. 2.68. Divided combustion chamber.

which enhances the combustion there. Creating an orifice that can do all this is a major design challenge.

- A divided chamber engine, oftenly, will also be a stratified charge engine. *The intake system is designed to supply a rich mixture in the secondary chamber and a lean mixture in the main chamber.* The rich mixture with very high swirl in the secondary chamber will ignite readily and combust very quickly. The flaming gases expanding back through the orifice will then ignite the lean mixture in the main chamber, a mixture often so lean that it would be difficult to ignite with a spark-plug alone. The net result is *an engine that has good ignition and combustion, yet operates mostly lean to give good fuel economy*. Placement and timing of intake valves injectors to supply the proper air and fuel to all parts of this engine are extremely important.

Note. A variation of this type of combustion chamber on some C.I. engines is one with a totally passive secondary chamber, with all valves and injectors located in the main chamber. When combustion occurs in the main chamber, high pressure forces gas through the very small orifice and raises the pressure in the secondary chamber also. *When the pressure in the main chamber is reduced during the power stroke, the high pressure gases in the secondary chamber flow back into the main chamber. This holds the pressure in the main chamber to a higher level for a short time and gives a smooth, slightly greater force off the piston during the power stroke.* This kind of secondary chamber usually consists of about 5-10 percent of the clearance volume.

WORKED EXAMPLES

Example 2.1. *A S.I. engine operating at 1200 r.p.m. has a 10.2 cm bore with spark-plug offset by 6 mm from the centre. The spark plug is fired at 20°C before T.D.C. It takes 6.5° of engine rotation for combustion to develop and get into flame propagation mode, where the average flame speed is 15.8 m/s. Calculate :*

(i) Time of one combustion process (i.e., time for flame front to reach the farthest cylinder wall) in sec. ;

(ii) Crank angle position at the end of combustion. **(Madras University)**

Solution. Maximum distance of flame travel

$$= \frac{1}{2} \text{ bore + spark-plug offset}$$

$$= \frac{1}{2} \times 10.2 + \frac{6}{10} = 5.7 \text{ cm}$$

Time of flame travel $= \dfrac{5.7 \times 10^{-2}}{15.8} = 3.6076 \times 10^{-3}$ s

1200 r.p.m. $= \dfrac{1200}{60} \times 360 = 7200$ deg./s

$\therefore$ Crank angle for flame travel

$$= 3.6076 \times 10^{-3} \times 7200 = 25.975 \text{ deg.}$$

Time for combustion to develop = 6.5 crank degrees

$$= \frac{6.5}{7200} \quad \text{or} \quad 0.9028 \times 10^{-3} \text{ s}$$

(*i*) **Time for one combustion process**

= Time to develop + Time for propagation

$= 0.9028 \times 10^{-3}$ s $+ 3.6076 \times 10^{-3}$

$=$ **4.5104×10^{-3}. (Ans.)**

(*ii*) **Total crank rotation**

= 6.5 + 26.975 = 33.48 degrees of crank rotation.

Since spark is fired at 20° before T.D.C, the crank position will be (33.48 – 20)

or, **13.48 degrees after T.D.C. (Ans.)**

Example 2.2. *In a trial on S.I. engine at full speed full power (i.e., fully open throttle) the spark occurred 26° bT.D.C (before Top Dead Centre) and delay ended 4° bT.D.C. Assuming that the combustion period should finish 13° aT.D.C. (after top dead centre) for maximum power and that the effect of half closing the throttle at constant speed is to increase the delay period by 14% of the valve at full throttle, estimate the optimum spark timing for maximum power under following conditions :*

(*i*) *Under full throttle conditions when the engine is operated at half the maximum speed ;*

(*ii*) *When the engine is operated at conditions of half the maximum speed and the throttle half open.*

State how these alterations in optimum spark timing may be achieved in practice.

(Bombay University)

Solution.

• *The delay period, at constant throttle, is constant in time and thus increases in angle with the speed.*

• The *combustion period is constant in crank angle.*

The delay period = From 26° T.D.C. to 4° *b*T.D.C. *i.e.*, 22°

The combustion period = From 4° T.D.C. to 13° *a*T.D.C. *i.e.*, 17°.

(*i*) Full throttle half speed will result in delay angle being reduced to $\frac{22}{2}$ = 11° for the same time thus ignition timing should be arranged so that the total of 11 + 17 = 28°, ends 13° *a*T.D.C.

$\therefore$ **Time of spark** = 28 – 13 = **15° *b*T.D.C. (Ans.)**

A *centrifugal device* is used to accomplish this task.

(*ii*) Half throttle half speed will result in an increase of 14% in delay time over that at full throttle half speed *i.e.*, by

$$\frac{14}{100} \times 11 = 1.54°$$

∴ Delay angle = 11 + 1.54 = 12.54°

Combustion period remains same as 17°

∴ Total period = 12.54 + 17 = 29.54° ; end is 13° *a*T.D.C.

∴ **Time of spark** = 29.54 – 13 = **16.54° *b*T.D.C. (Ans.)**

This is accomplished by *a vacuum device connected to the inlet manifold.*

III. Combustion in C.I. Engines

2.28. INTRODUCTION TO COMBUSTION IN C.I. ENGINES

The compression ignition (C.I.) engine was developed by Dr. Rudolf Diesel, he got a patent of his engine in 1892.

It is a very important prime mover these days and is finding wide applications in :

— Buses trucks, tractors ;

— Locomotives ;

— Pumping sets ;

— Stationary industrial applications ;

— Small and medium electric power generation ;

— Marine propulsion.

The following points are worthnoting about C.I. engines :

- *Its thermal efficiency is higher than S.I. engines.*
- C.I. engine fuels (diesel oils) are less expensive than S.I. engine fuels (petrol or gasoline). Furthermore, since C.I. engines fuels have a higher specific gravity than petrol, and since fuel is sold on the volume basis (litres) and not on mass basis (kg), more kg of fuel per litre are obtained in purchasing C.I. engine fuel.

Due to the above mentioned factors the *running cost of C.I. engines is much less than S.I. engines* and as a consequence these engines find wide application in industrial, transport and other applications.

- A *C.I. engine is not much favoured in passenger cars due to the following reasons :*

  (*i*) Heavier weight ;

  (*ii*) Noise and vibration ;

  (*iii*) Smoke ;

  (*iv*) Odour.

  — In view of the utilisation of heavier compression ratios (12 : 1 to 22 : 1 compared to 6 : 1 to 11 : 1 of S.I. engines) the heavy forces act on the parts of the engine and therefore heavier parts are required.

  — Also, because of heterogeneous mixture, lean mixture is used.

These factors make the engine heavier.

— The *incomplete combustion of heterogeneous mixture, and droplet combustion result in the smoke and odour.*

- C.I. engines are manufactured in the following range of speeds, speeds and power outputs :

Particulars	*Range*
1. Piston diameters	50 mm to 900 mm
2. Speeds	100 r.p.m. to 4400 r.p.m
3. Power output	2 B.P. to 40000 B.P.

2.29. COMBUSTION PHENOMENON IN C.I. ENGINES

- The process of combustion in the Compression Ignition (C.I.) engine is fundamentally different from that in a spark-ignition engine. In C.I. engine combustion occurs by the high temperature produced by the compression of the air, *i.e.*, it is an *auto-ignition.* For this a minimum compression ratio of 12 is required. The efficiency of the cycle increases with higher values of compression ratio but the maximum pressure reached in the cylinder also increases. This requires heavier construction. The upper limit of compression ratio in a C.I. engine is due to mechanical factor and is a compromise between high efficiency and low weight and cost. The normal compression ratios are in the range of 14 to 17, but may be upto 23. The air-fuel ratios used in the C.I. engine lie between 18 and 25 as against about 14 in the S.I. engine, and *hence C.I. engines are bigger and heavier for the same power than S.I. engines.*
- In the C.I. engine, the intake is air alone and the fuel is injected at high pressure in the form of fine droplets near the end of compression. This leads to delay period in the C.I. engine, is greater than that in the S.I. engine. The *exact phenomenon of combustion in the C.I. engine* is explained below.

— Each minute droplet of fuel as it enters the highly heated air of engine cylinder is quickly surrounded by an envelope of its own vapour and this, in turn and at an appreciable interval is inflamed at the surface of the envelope. To evaporate the liquid, latent heat is abstracted from the surrounding air which reduces the temperature of the thin layer of air surrounding the droplet, and some time must elapse before this temperature can be raised again by abstracting heat from the main bulk of air in this vicinity. As soon as this vapour and the air in actual contact with it reach a certain temperature, ignition will take place. Once ignition has been started and a flame established the heat required for further evaporation will be supplied from that released by combustion. The vapour would be burning as fast as it can find fresh oxygen, *i.e., it will depend upon the rate at which it is moving through the air or the air is moving past it.*

— In the C.I. engine, the fuel is not fed in at once but is spread over a definite period. The first arrivals meet air whose temperature is only a little above their self-ignition temperature and the delay is more or less prolonged. The later arrivals find air already heated to a far higher temperature by the burning of their predecessors and therefore light up much more quickly, almost as they issue from the injector nozzle, but their subsequent progress is handicapped for there is less oxygen to find.

— If the air within the cylinder were motionless, only a small proportion of the fuel would find sufficient oxygen, for it is impossible to distribute the droplets uniformly throughout the combustion space. Therefore some air movement is absolutely essential, as in the S.I. engine. But there is a fundamental difference between the air movements in the two types of engines. In the S.I. engine we call it *turbulence* and mean a confusion of whirls and eddies with no general direction of flow, (to break up the surface of the flame front, and to distribute the shreds of flame throughout an externally prepared combustible mixture). In the C.I. engine we call it *air swirl* and mean an orderly movement of the whole body of the air, with or without some eddying or turbulence, so as to bring a continuous supply of fresh air to each burning droplet and sweep away the products of combustion which otherwise tend to suffocate it.

Three phases of C.I. engine combustion :

In the C.I. engine, combustion may be considered in *three distinct stages* as shown in Fig. 2.69.

1. Ignition delay period.
2. Period of rapid or uncontrolled combustion.
3. Period of controlled combustion.

The *third phase is followed by* ***after burning*** (or burning on the expansion stroke), which may be called the *fourth phase of combustion.*

1. Ignition delay period

- The delay period is counted from the start of injection to the point where the p-θ combustion curve departs from air compression (or no ignition or motoring) curve.
- The delay period can be roughly sub-divided into ***physical delay*** and ***chemical delay.*** The period of *"physical delay" is the time between the beginning of injection and the attainment of chemical reaction conditions.* In the physical delay period, the fuel is atomized, vaporized, mixed with air, and raised in temperature. In the *"chemical delay" period reaction starts slowly and then accelerates until inflammation or ignition takes place* (it may be noted that the ignition delay in the S.I. engine is essentially equivalent to the chemical delay in the C.I. engine).

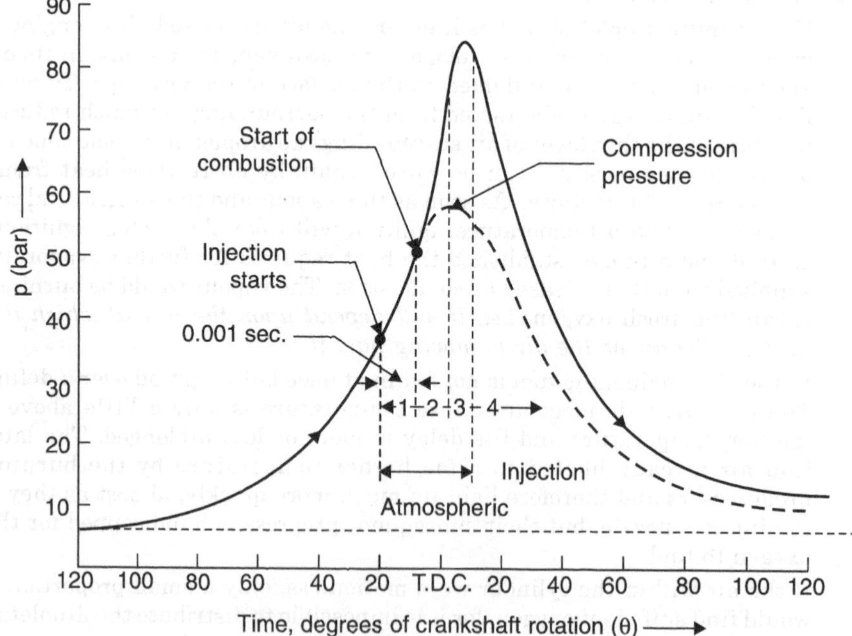

Fig. 2.69. Combustion phenomenon of C.I. engine.

- The delay period exerts a great influence in the C.I. engine combustion phenomenon. It is clear that the pressure reached during the second stage will depend upon the duration of the delay period ; *the longer the delay, the more rapid and higher the pressure rise,* since more fuel will be present in the cylinder before the rate of burning comes under control. This causes rough running and may cause *diesel knock.* Therefore we must *aim to keep the delay period as short as possible, both for the sake of smooth running and in order to maintain control over the pressure changes. But some delay period is necessary otherwise the droplet would not be dispersed in the air for complete combustion.* However, the delay period imposed upon is greater than what is needed and the designer's efforts are to shorten it as much as possible.

2. **Period of rapid or uncontrolled combustion.** The second stage of combustion in C.I. engines, after the delay period, is the *period of rapid or uncontrolled combustion.* This period is *counted from the end of the delay period to the point of maximum pressure on the indicator diagram.* In this second stage of combustion, the rise of pressure is rapid because during the delay period the droplets of fuel have had time to spread themselves out over a wide area and they have fresh air all around them. About one-third of heat is evolved during this process.

The rate of pressure rise depends on the amount of fuel present at the end of delay period, degree of turbulence, fineness of atomization and spray pattern.

3. **Period of controlled combustion.** At the end of second stage of combustion, the temperature and pressure, are so high that the fuel droplets injected in the third stage burn almost as they enter and any further pressure rise can be controlled by purely mechanical means, *i.e.,* by the injection rate. The period of controlled combustion is assumed to end at maximum cycle temperature.The heat evolved by the end of controlled combustion is about 70 to 80 per cent.

4. **After burning :**

- The combustion continues even after the fuel injection is over, because of poor distribution of fuel particles. This burning may continue in the expansion stroke upto 70° to 80° of crank travel from T.D.C. This continued burning, called the *after burning,* may be considered as the *fourth stage of the combustion.* The total heat evolved by the end of entire combustion process is 95 to 97% ; 3 to 5% of heat goes as unburned fuel in exhaust.
- In the *p-V* diagram, the stages of combustion are not seen because of little movement of piston with crank angle at the end and reversal of stroke. So for studying the combustion stages, therefore, a pressure-crank angle or time, *p-θ* or *p-t* diagram is invariably used. In the actual diagram, the various stages of combustion look merged, yet the individual stage is distinguishable.

Factors affecting combustion in C.I. engine :

The factors affecting combustion in C.I. engine are as follows :

(1) Ignition quality of fuel (cetane number).

(2) Injection pressure of droplet size.

(3) Injection advance angle.

(4) Compression ratio.

(5) Intake temperature.

(6) Jacket water temperature.

(7) Intake pressure, supercharging.

(8) Engine speed.

(9) Load and Air to fuel ratio.

(10) Engine size.

(11) Type of combustion chamber.

2.30. FUNDAMENTALS OF THE COMBUSTION PROCESS IN DIESEL ENGINES

Effect of Compression Ratio and Engine Speed on Cylinder Pressure and Temperature :

- The power output of a *diesel engine* is controlled by *varying the amount of fuel spray injected into a cylinder filled with compressed and heated air* whereas the *petrol engine* is controlled by *throttling the pre-mixed charge entering the cylinder.*
- The *pressure and temperature reached at the end of the compression stroke will depend primarily upon the compression ratio, intake temperature and speed of the engine.*
  — It has been observed that injection usually commences 15° to 20° before T.D.C. when both cylinder pressures and temperatures are much lower. As an *example*, a 15 : 1 compression ratio engine would have something like 600°C maximum temperature at T.D.C. but at 15° before T.D.C. this would only amount to 530°C.
  — Further it can be seen that the pressure and temperature rise in the cylinder with increased speed is largely *due to the reduced time available for compressed air to escape past the piston rings and heat to be lost through the cylinder walls and head.*

Diesel Engine Heterogeneous Charge Mixing :

The air-fuel mixture formation, in the diesel engine, is of a *heterogeneous* natures, that is, it is locally concentrated at various sites and is therefore *unevenly distributed throughout the cylinder and combustion chamber.*

— Injected fuel spray penetrates the highly compressed and heated air mass where it is pulverised into many very small droplets in a localised formation. The mixing of the localised spray of fuel droplets in the hot air charge causes stoichiometeric (14.7 : 1 by weight) air-fuel ratio combustion zones to be established which are completely surrounded by pure air only. Thus the overall (averaged out) air-fuel mixture ratio range may vary from a rich, full load, 20 : 1, to a weak no-load, 100 : 1, air-fuel ratio.

— Most engines *operate with at least* **20% excess air** *due to difficulty of introducing sufficient exposed oxygen to the fuel vapour in the given time available so that the combustion process can be completed before the exhaust valve opens.* If the oxygen supply is *partially prevented from getting to the fuel vapour early enough during the power stroke then incomplete combustion, polluted exhaust gas and dark smoke will result.*

Diesel Engine Injected Spray Combustion Process :

- Towards the end of the compression stroke when injection of the fuel into the combustion chamber commences, the quantity of fuel discharged is spread out over a predetermined period.
- The fuel spray enters the hot combustion chamber but does not immediately ignite, instead it breaks up into very small droplets (Fig. 2.70) and once these liquid droplets are formed, *their outer surfaces will immediately start to evaporate so there will be a liquid core surrounded with a layer of vapour.* At this point it should be explained that the burning of a hydrocarbon fuel in air is purely an *oxidation process.* Thus, initially, heat liberated from the oxidation of the fuel vapour is *less than the rate at which heat is extracted by convection and conduction,* but eventually a *critical temperature* is reached when the *rate of heat generated by oxidation exceeds the heat being dissipated by convection and radiation.* As a result, the temperature rises which, in turn, speeds up the oxidation process thus further increasing the heat released until a *flame site or sites*

are established, this being known as the **ignition** and the temperature at which it occurs is called the **self-ignition temperature** of the fuel under these condition. *The heat required for further evaporation of the fuel droplets will thus be provided from heat released by the oxidation process, which is referred to as* **combustion.**

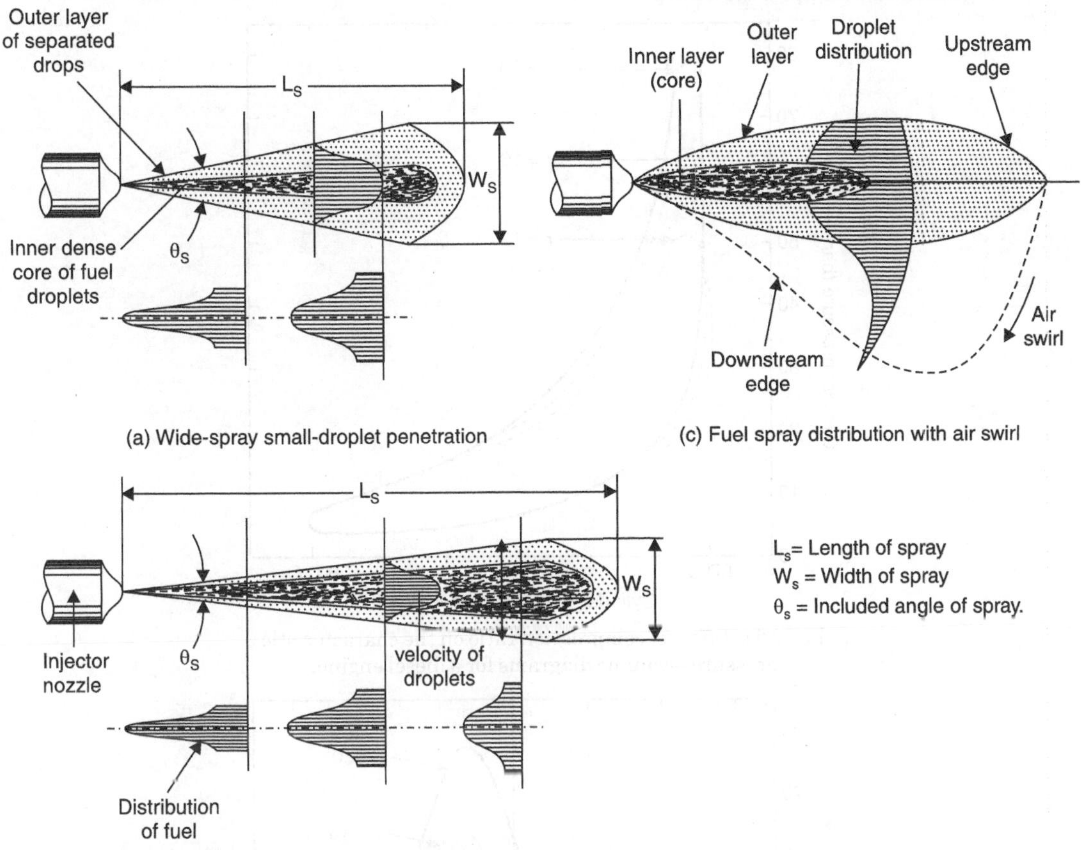

Fig. 2.70. Injected fuel spray characteristics.

- The liquid core, now surrounded by layers of heated vapour, oxidises burns as fast as it can ; that is it finds fresh oxygen to keep the chemical reaction going on.
- When the ***physical delay*** to convert the fuel spray into tiny droplets and the ***chemical reaction delay*** to establish ignition from the initial oxidation process are over, *the rate of burning is dependent on the speed at which the droplets are moving through the air or the air is moving past the droplets.*

Compression Ratio (r) :

Increase in compression ratio exercises the following *effects* :

- The cylinder compression pressure and temperature *increase* ; the ignition time lag between the point of injection to the instant when ignition first commences *reduces.*
- The density and turbulence of the charge *increase*, and this *increases the rate of burning and, accordingly the rate of pressure rise and the magnitude of the peak cylinder pressure reached.* The characteristics of the pressure rise relative to the piston stroke or crank-angle movement is illustrated in Fig. 2.71 and Fig. 2.72.

- Thermal efficiency and the specific fuel consumption are *improved* (Fig. 2.73)
- Raising compression ratio *results in reduction in the mechanical efficiency* as shown in Fig. 2.74 (since the higher cylinder pressures increase the pumping losses, friction losses and compression and expansion losses as more work is done in squeezing together the trapped air charge).

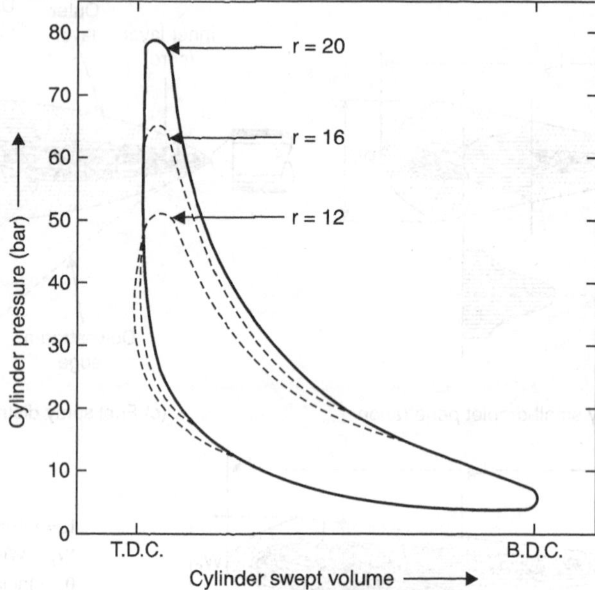

Fig. 2.71. Effect of compression ratio on the characteristic pressure—volume diagrams for a diesel engine.

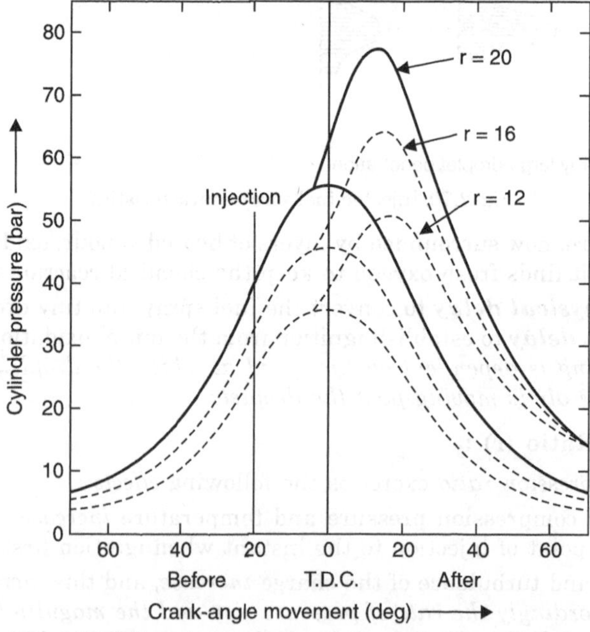

Fig. 2.72. Effect of compression ratio on the characteristic pressure –crank-angle movement diagrams for a diesel engine.

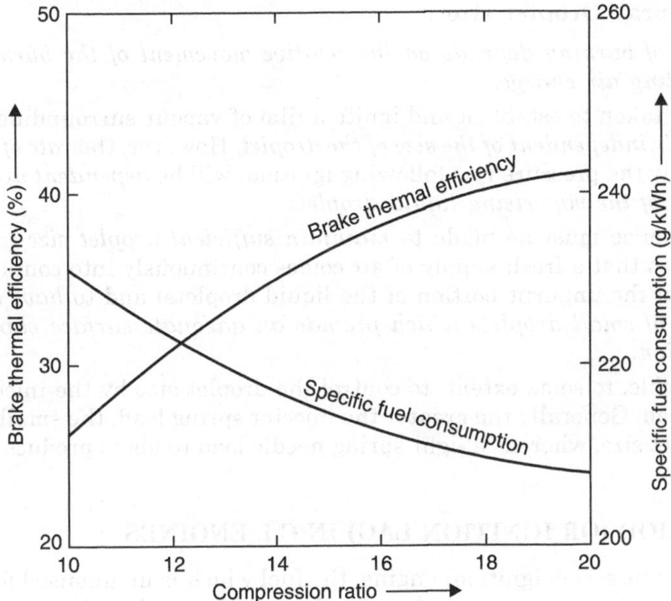

Fig. 2.73. Effect of compression ratio on the thermal efficiency and specific fuel consumption.

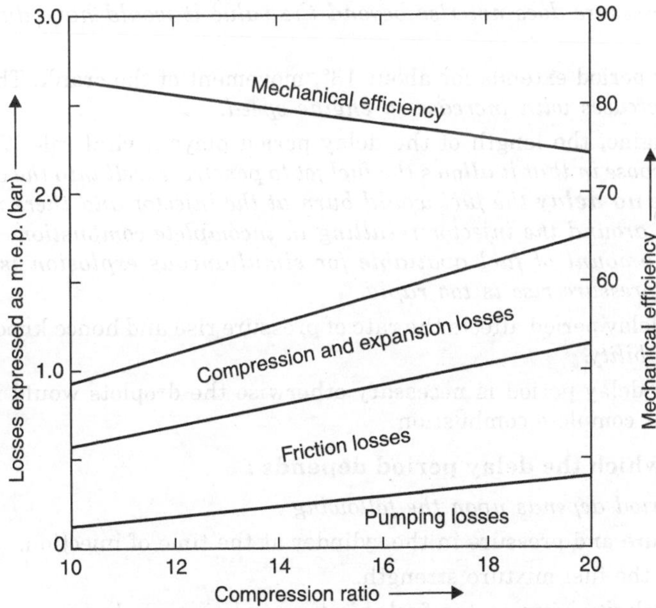

Fig. 2.74. Effect of compression ratio on the pumping, friction, compression and expansion losses and the resultant mechanical efficiency.

Injection Spray Droplet size :

- *The rate of burning depends on the relative movement of the burning droplets to the surrounding air charge.*
- The time taken to establish and ignite a film of vapour surrounding a liquid droplet is *practically independent of the size of the droplet.* However, the *rate of burning* and correspondingly the pressure rise following ignition, will be *dependent upon the exposed surface area of the vaporising liquid droplets.*
- A compromise must be made to *maintain sufficient droplet size* (and, therefore, momentum so that a fresh supply of air comes continuously into contact with the shrinking size of the unburnt portion of the liquid droplets) and to *have available sufficient numbers of small droplets which provide an adequate surface vapour area for rapid combustion.*
- It is possible, to some extent, to control the droplet size by the injection needle spring closing load. Generally the greater the injector spring load, the smaller and finer will be the droplet size, whereas a light spring needle load tends to produce coarse liquid droplets.

2.31. DELAY PERIOD (OR IGNITION LAG) IN C.I. ENGINES

- In C.I. (compression ignition) engine, the fuel which is in atomised form is considerably colder than the hot compressed air in the cylinder. Although the actual ignition is almost instantaneous, *an appreciable time elapses before the combustion is in full progress. This time occupied is called the* **delay period** or **ignition lag**. *It is the time immediately following injection of the fuel during which the ignition process is being initiated and the pressure does not rise beyond the value it would have due to compression of air.*
- The delay period extends for about 13°, movement of the crank. The time for which it occurs *decreases with increase in engine speed.*
- In C.I. engine, the length of the delay period plays a vital role. *This period serves a useful purpose in that it allows the fuel jet to penetrate well into the combustion space. If there were* ***no delay*** *the fuel would burn at the injector and there would be an oxygen deficiency around the injector resulting in incomplete combustion. If the* ***delay is too long*** *the amount of fuel available for simultaneous explosion is too great and the resulting pressure rise is too rapid.*
  - — The delay period affects the rate of pressure rise and hence knocking. It also affects *startability.*
  - — Some delay period is necessary otherwise the droplets would not be dispersed in air for complete combustion.

Factors on which the delay period depends :

The *delay period depends upon the following :*

(*i*) Temperature and pressure in the cylinder at the time of injection.

(*ii*) Nature of the fuel mixture strength.

(*iii*) Relative velocity between the fuel injection and air turbulence.

(*iv*) Presence of residual gases.

(*v*) Rate of fuel injection.

(*vi*) To small extent the finess of the fuel spray.

- *The delay period increases with load but is not much affected by injection pressure.*

- *The delay period should be as short as possible since a long delay period gives a more rapid rise in pressure and thus causes knocking.*

Effects of Various Factors on Delay Period :

Effects of various factors such as fuel properties, intake temperature, compression ratio, engine speed, type of combustion chamber, and injection advance are discussed below :

1. **Fuel properties :**

- The *Self-Ignition Temperature (S.I.T.)* is the most important property of the fuel which affects the delay period.
  — *A lower S.I.T. means a wide margin between it and the temperature of compressed air and hence lower delay period.*
  — *Higher cetane number means a lower delay period and smoother engine operation.* Cetane number depends on the *chemical composition of fuel.* The *more paraffinic hydrocarbons are contained in fuel, higher will be the cetane number.*
- The other fuel properties which affect delay period are :
  (*i*) Volatility ;
  (*ii*) Latent heat ;
  (*iii*) Viscosity ;
  (*iv*) Surface tension.
  — *Volatility and latent* heat affect the *time taken to form an envelope of vapour.*
  — The *viscosity and surface tension* influence the *fineness of atomisation.*

2. **Intake temperature :**

- *Increase in intake temperature* would result in increase in compressed air temperature which would *reduce the delay period.*

3. **Compression ratio :**

- *Increase in compression ratio reduces delay period as it raises both temperature and density.*
  — With increase in compression ratio, temperature of air increases. At the same time the minimum auto-ignition temperature decreases due to increased density of compressed air resulting in closer contact of molecules which thereby reduces the time of reaction when fuel is injected.
  — As *the difference between compressed air temperature and minimum auto-ignition temperature increases, the delay period decreases.*

4. **Engine speed :**

- Delay period can be given either in terms of *absolute time* (in milliseconds) *or crank angle rotation.*
  — At *constant speed,* delay period *is proportional to the delay angle.*
  — In *variable speed operation,* delay period may *decrease in terms of milliseconds but increase in terms of crank angles.*

5. **Type of combustion chamber :**

- *A pre-combustion chamber gives shorter delay compared to an open type of combustion chamber.*

6. **Injection advance :**

- *Delay period increases with increase in injection advance angle.* The reason for increase in delay period with increase in injection advance angle is that pressures and temperatures are lower when injection begins.

- When injection advance angles are small, delay period reduces and operation of engine is smoother but power is reduced because large amount of fuel burns during expansion.

Abnormal Combustion in C.I. engines :

In C.I. engines, abnormal combustion is not a great problem as in S.I. engines. The only abnormality is ***"diesel knock"***. *This occurs when the delay period is excessively long so that there is a large amount of fuel in the cylinder for the simultaneous explosion phase.* The rate of pressure rise per degree of crank angle is then so great that an *audible knocking sound occurs.* Running is rough and if allowed to become extreme the *increase in mechanical and thermal stresses may damage the engine.* ***Knock*** *is thus a function of the fuel chosen and may be avoided by choosing a fuel with characteristics that do not give too long a delay period.*

2.32. DIESEL KNOCK

- *Diesel knock is the sound produced by the very rapid rate of pressure rise during the early part of the uncontrolled second phase of combustion.* The primary cause of an excessively high pressure rise is due to a *prolonged delay period* (Fig. 2.75). An extensive delay period can be due to the following *factors* :
  - (*i*) A low design compression ratio permitting only a marginal self-ignition temperature to be reached.
  - (*ii*) A low combustion pressure due to worn piston rings or badly seating valves.
  - (*iii*) Poor fuel ignition quality ; that is a low cetane number fuel.
  - (*iv*) A poorly atomized fuel spray preventing early ignition to be established.
  - (*v*) An inadequate injector needle spring load producing coarse droplet formation.
  - (*vi*) A very low air intake temperature in cold wintry weather and during cold starting.

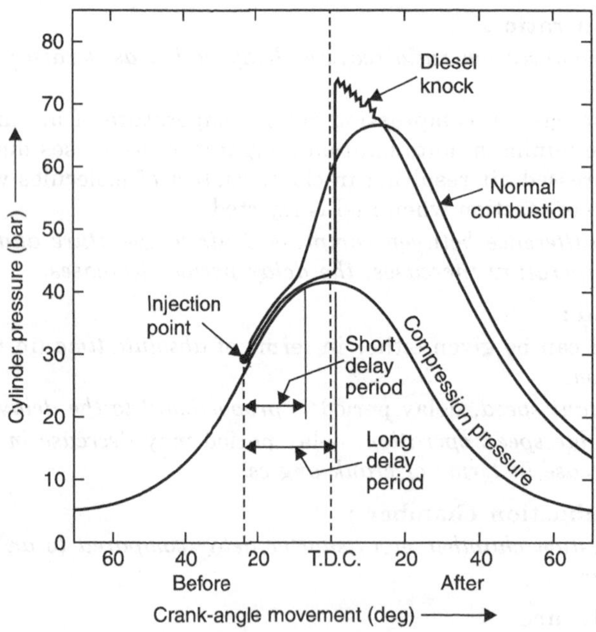

Fig. 2.75. Effect of short and long delay period on the characteristic p-θ diagram.

- A *very long ignition lag after injection causes a large proportion of the fuel discharge to enter the cylinder and to atomise before ignition and the propagation of burning actually occurs. Accordingly, when combustion does commence a relative amount of heat energy will be released almost immediately, this correspondingly produces the abnormally high rate of pressure rise, which is mainly responsible for rough and noisy combustion process under these condition (Fig. 2.75).*
- It has been observed generally, that provided the rate of pressure increase does not exceed 3 bar *per degree of crank-angle movement,* combustion will be relatively smooth, whereas *between a 3 and 4 bar pressure rise there is a tendency to knock and, above this rate of pressure rise, diesel knock will be prominent.*

Differences in the knocking phenomenon of the S.I. and C.I. Engines :

The following are the *differences in the knocking phenomena of the S.I. and C.I. engines* :

1. In the *S.I. engine*, the detonation occurs *near the end of combustion* whereas in the *C.I.* engine detonation occurs *near the beginning of combustion.*
2. The detonation in the S.I. engine is of a homogeneous charge causing very high rate of pressure rise and very high maximum pressure. In the C.I. engine, the fuel and air are imperfectly mixed and hence the rate of pressure rise is *normally lower than that in the detonating part of the charge in the S.I. engine.*
3. In the C.I. engine the fuel is injected into the cylinder only at the end of the compression stroke, *there is no question of pre-ignition as in S.I. engine.*
4. *In the S.I. engine, it is relatively easy to distinguish between knocking and non-knocking operation as the human ear easily finds the distinction.*
5. *Factors that tend to reduce detonation in the S.I. engine increase knocking in the C.I. engine.*

Methods of controlling diesel knock (Reducing delay period) :

The *diesel knock can be controlled by reducing delay period.* The delay is reduced by the following :

(*i*) High charge temperature.
(*ii*) High fuel temperature.
(*iii*) Good turbulence.
(*iv*) A fuel with a short induction period.

2.33. C.I. ENGINE COMBUSTION CHAMBERS

2.33.1. Primary Considerations in the Design of Combustion Chambers for C.I. Engines

In C.I. engines fuel is injected into the combustion chamber at about 15°C before T.D.C. during the compression stroke. For the best efficiency the combustion must complete within 15° to 20° of crank rotation after T.D.C. in the working stroke. Thus it is clear that injection and combustion both must complete in the short time. For best combustion mixing should be completed in the short time.

- In S.I. engine mixing takes place in carburettor, however in C.I. engines this has to be done in the combustion chamber. To achieve this requirement in a short period is an extremely difficult job particularly in high speed C.I. engines.
- From combustion phenomenon of C.I. engines it is evident that fuel-air contact must be limited during the delay period in order to limit $\frac{dp}{dt}$, the rate of pressure rise in the

second stage of combustion. This result can be obtained by *shortening the delay time. To achieve high efficiency and power the combustion must be completed when the piston is nearer to T.D.C., it is necessary to have rapid mixing of fuel and air during the third stage of combustion.*

- The design of combustion chamber for C.I. engines *must also take consideration of fuel injection system and nozzles to be used.*

The considerations can be summarized as follows :

1. High thermal efficiency.
2. Ability to use less expensive fuel (multi-fuel).
3. Ease of starting.
4. Ability to handle variations in speed.
5. Smoothness of operation *i.e.*, avoidance of diesel knock and noise.
6. Low exhaust emission.
7. Nozzle design.
8. High volumetric efficiency.
9. High brake mean effective pressure.

2.33.2. Basic Methods of generating Air Swirl in C.I. Engines Combustion Chambers

There are *there basic methods of generating swirl in a C.I. engine combustion chamber,* which are mentioned below :

1. By *directing the flow of air during its entry to the cylinder, known as* ***induction swirl.*** This method is *used in open combustion chamber.*

2. By *forcing the air through a tangential passage into a separate swirl chamber during compression stroke,* known as ***compression swirl.*** This method is used in *swirl chambers.*

3. By *use of initial pressure rise due to partial combustion to create swirl turbulence,* known as ***combustion induced swirl.*** This method is used in *pre-combustion chambers and air-cell chambers.*

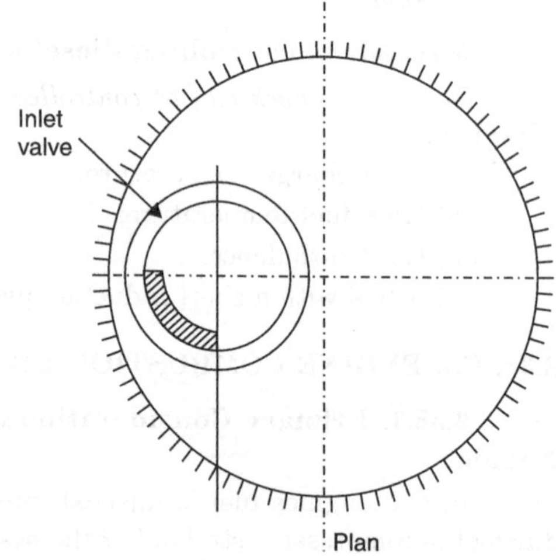

Fig. 2.76. Induction swirl by masking the inlet valve.

Induction swirl : Refer Fig. 2.76.

- In a *four stroke engine* induction swirl can be obtained either by careful formation of air intake passages or masking or shrouding a portion of circumference of inlet valve. The angle of mask is from 90° to 140° of the circumference.
- In *two stroke engine,* induction swirl is *created by suitable inlet port forms.*
- The induction swirl generated by air intake passages is very weak. If a masked inlet valve is used, it provides an obstruction in the passage which reduces volumetric efficiency. Therefore *swirl generated is weak* even with this method. With a weak swirl, a single orifice injection cannot provide the desired air fuel mixing. Therefore, with induction swirl, *we have to use a multiple-orifice injector.*

Advantages of induction swirl :

1. Easier starting (due to low intensity of swirl).
2. High excess air (low temperature), low turbulence (less heat loss), therefore indicated thermal efficiency is high.
3. Production of swirl requires no additional work.
4. Used with low speeds, therefore low quality of fuel can be used.

Disadvantages :

1. Shrouded valves, smaller valves, low volumetric efficiency.
2. Weak swirl, low air utilisation (60%), lower m.e.p. and large size (costly) engine.
3. Weak swirl, multi-orifice nozzle, high induction pressure, clogging of holes, high maintenance.
4. Swirl not proportional to speed ; efficiency not maintained at variable speed engine.
5. Influence minimum quantity of fuel. Complication at high loads and idling.

Compresssion swirl : Refer to Fig. 2.77.

- The second method of generating swirl is by compression swirl in what is known as *swirl chamber*. A swirl chamber is a *divided chamber*. A divided combustion chamber is defined as one in which combustion space is divided into two or more distinct compartments, between which there are restrictions or throats small enough so that considerable pressure differences occur between them during combustion process.
- This swirl is maximum at about 15° before T.D.C. *i.e.*, close to the time of injection. The fuel is injected into the swirl chamber and ignition and bulk of combustion takes place therein. A considerable amount of heat is lost when products of combustion pass back through the same throat and this loss of heat is reduced by employing a heat insulated chamber. Thus, it serves as a thermal regenerator receiving heat during combustion and expansion and returning the heat to air during compression stroke. However the loss of heat to surface of combustion chamber is greater than induction swirl.

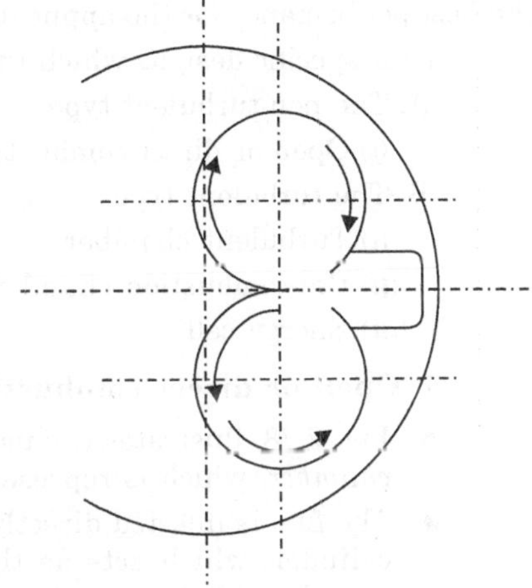

Fig. 2.77. Compression swirl.

- In *combustion swirl, a very strong swirl which increases with speed is generated.*

Advantage of compression swirl :

1. Large valves, high volumetric efficiency.
2. Single injector, pintle type (self cleaning), less maintenance.
3. Smooth engine operation.
4. Greater air utilization due to strong swirl. Smaller (cheaper) engine.
5. Swirl proportional to speed, suitable for variable speed operation.

Disadvantages :

1. Cold starting trouble due to high loss due to strong swirl, mechanical efficiency lower.
2. Less excess air ; lower indicated efficiency ; 5 to 8% more fuel consumption ; decreased exhaust valve life.
3. Cylinder more expensive in construction.
4. Work absorbed in producing swirl, mechanical efficiency lower.

Combustion induced swirl :

- This type of *swirl is induced by use of initial pressure rise due to partial combustion.*
- The combustion chambers which use this type of *swirl are not much favoured these days.*

2.33.3. Types of Combustion Chambers

In C.I. engines several types of combustion chambers are used. Each of these has its own peculiarities, and desirable, as well as undesirable features. Any one of these combustion chambers may produce good results in one field of application, but less desirable, or even poor results in another. No one combustion chamber design has yet been developed which will produce the best result is all types of engines. The particular design chosen, then, must be that which accomplishes the best performance for the application desired.

Four specific designs which find wide use in C.I. engines are discussed below :

A. The non-turbulent type
   (*i*) Open or direct combustion chamber.

B. The turbulent type
   (*i*) Turbulent chamber
   (*ii*) Pre-combustion chamber
   (*iii*) Energy cell.

1. **Open or direct combustion chamber :**

- Fig. 2.78 illustrates the usual design of *open combustion chamber*, which is representative of non-turbulent type.
- The fuel is injected directly into the upper portion of the cylinder, which acts as the combustion chamber. This type depends little on turbulence to perform the mixing. Consequently, the heat loss to the chamber walls is relatively low, and *easier starting results.* In order to obtain proper penetration and dispersal of the fuel necessary for mixing with the air, however, high injection pressures and multi-orifice nozzles are required. This necessitates small nozzle openings and results in more frequent clogging or diversion of the fuel spray by accumulated carbon particles, with consequent higher maintenance costs.

Fig. 2.78. Open or direct combustion chamber.

- This type of chamber is ordinarily used on *low speed engines,* where injection is spread through a greater period of time and thus ignition delay is a relatively less important factor. Consequently, *less costly fuels with longer ignition delay may be used.*
- Many attempts were made to improve the air motion in open chambers, the important are :
  (*a*) *by shrouding the inlet valve,* Refer Fig. 2.79 (*a*)
  (*b*) *by providing squish,* Refer Fig. 2.79 (*b*)

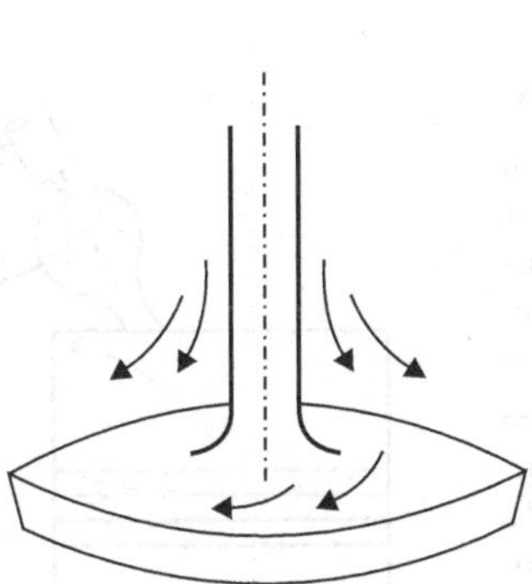

Fig. 2.79. (*a*) Air motion by shrouding the inlet valve.

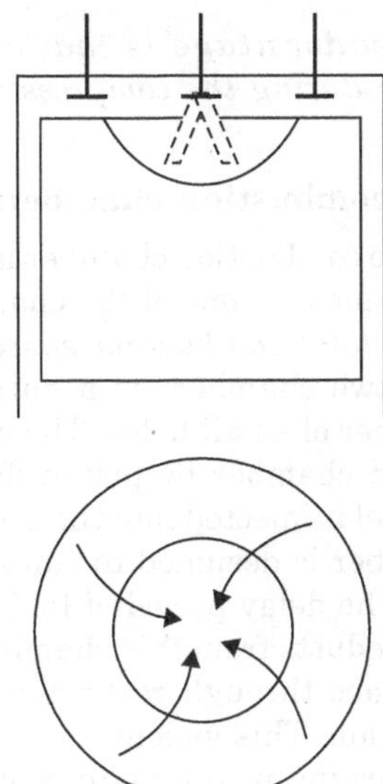

Fig. 2.79. (*b*) Squish air motion inside cylinder.

— By shrouding the inlet valve swirl motion is given to the air entering the cylinder which is believed to persist during compression stroke and the time of injection. This system gives better performance at low speeds, however volumetric efficiency reduces on account of reduction in inlet area due to shroud.

— Squish is provided by pushing the air at the end of the compression stroke in the space whose diameter is smaller than the cylinder bore. Because of the small clearance between the head and piston top when at T.D.C. air is pushed into combustion space providing air movement known as *squish*. The squish helps in mixing of fuel and air.

2. **Turbulent chamber :**

In the '*turbulent chamber*' (Fig. 2.80) the upward moving piston forces all the air (or 70–80% of all air) at a greater velocity into a small antechamber, thus imparting a rotary motion to the air passing the pintle type nozzle. As the fuel is injected into the rotating air, it is partially mixed with this air, and commences to burn. The pressure built up in the antechamber by the expanding burning gases force the burning and unburned fuel and air mixtures back into the main chamber, again imparting high turbulence and further assisting combustion.

Advantages :

(*i*) The insulated or hot running combustion chamber shortens the delay period and limits the rate of pressure rise, resulting in smoother running.

(*ii*) The turbulence is responsible for rapid mixing and burning of fuel during the third stage of combustion.

(*iii*) Suitable for high speeds as the amount of turbulence is proportional to piston or engine speed. The burning in the third stage will be completed early without resulting in late burning.

(*iv*) The demands on the fuel injection system are not severe as it is not to be depended upon for mixing, distribution, etc.

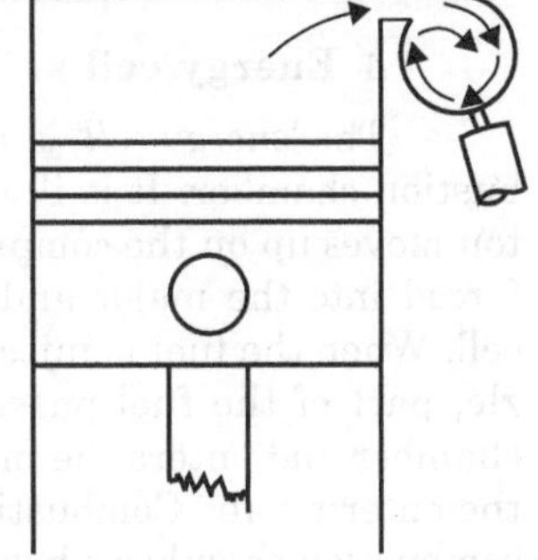

Fig. 2.80. Turbulent chamber.

The ***disadvantage*** is *that cold starting is difficult since air lose heat to combustion chamber walls during the compression stroke. The combustion chamber is relatively cool at the starting.*

3. **Pre-combustion chamber.** Refer to Fig. 2.81.

Here the combustion chamber is separated into *two chambers.* The smaller one of the chambers occupy about *30 per cent of total combustion space.* The communication between two chambers is a narrow restricted passage or a number of small holes. The air is forced into the pre-combustion chamber by piston during the compression stroke. Fuel is injected into the pre-combustion chamber. The chamber is designed to run hot and this results in shortening the delay period of fuel which is highly desirable. The products from this chamber rushes into main combustion space through restricted passages, creating violent air motion. This violent air motion helps in rapid mixing and burning in the main combustion space. The fuel reaching the main combustion space has practically no delay period as the temperature is already high due to combustion in pre-combustion chamber and combustion in main chamber is rapid and complete (*i.e.*, third stage of combustion) due to violent air motion.

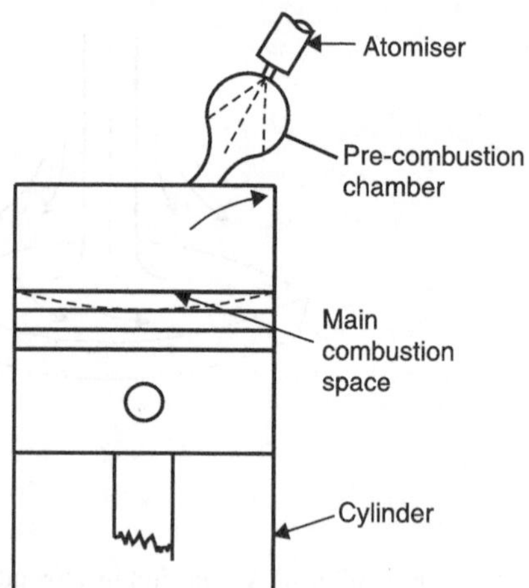

Fig. 2.81. Pre-combustion chamber.

Advantages :

(*i*) Due to short or practically no delay period for the fuel entering the main combustion space, *tendency to knock is minimum,* and as *such running is smooth.*

(*ii*) The combustion in the third stage is *rapid.*

(*iii*) As the mixing of fuel and air is thorough due to vialent projection of combustion products from pre-chamber, the *fuel injection system design need not be critical.*

Disadvantages :

(*i*) The velocity of burning mixture is too high during the passage from prechambers, so the heat loss is very high. This causes reduction in the thermal efficiency, which can be offset by increasing the compression ratio.

(*ii*) Cold starting will be difficult as the air loses heat to chamber walls during compression.

4. **Energy cell :**

The '*energy cell*' is more complex than the pre-combustion chamber. It is illustrated in Fig. 2.82. As the piston moves up on the compression stroke, some of the air is forced into the major and minor chambers of the energy cell. When the fuel is injected through the pintle type nozzle, part of the fuel passes across the main combustion chamber and enters the minor cell, where it is mixed with the entering air. Combustion first commences in the main combustion chamber where the temperature is higher, but the rate of burning is slower in this location, due to insufficient mixing of the fuel and air. The burning in the minor cell is slower at the start, but due to better mixing,

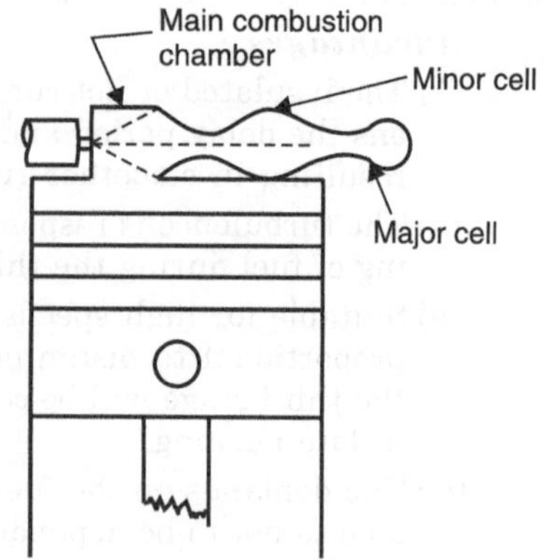

Fig. 2.82. Energy cell.

progresses at a more rapid rate. The pressures built up in the minor cell, therefore, force the burning gases out into the main combustion chamber, thereby creating added turbulence and producing better combustion in this chamber. In the mean time, pressure is built up in the major cell, which then prolongs the action of the jet stream entering the main chamber, thus continuing to induce turbulence in the main chamber.

5. **M. Combustion chamber :**

- After twenty years of research in 1954, Dr. Meuner of M.A.N., Germany developed M-process engine which ran without typical diesel combustion noise and hence it was named '*whisper engine*'.
- Fig. 2.83 shows a combustion chamber developed for small high speed engines. It differs from the other open combustion chamber engines in the respect that *fuel spray impinges tangentially on, and spreads over, the surface of a spherical space in the piston.* There is always some impingement of spray on the combustion chamber walls in all successful diesel engine designs. This impingement was not considered desirable till M.A.N. combustion system was experimented.
- The M.A.N. system's theory is that *enough of spray will ignite before impingement so that delay period will be normal while most of the fuel spray will evaporate from the hemispherical combustion space in piston prior to combustion.* Thus the *second stage of combustion is slowed down avoiding excessive rate of pressure rise.* Shrouded inlet valve is used to give air swirl in direction of arrow.

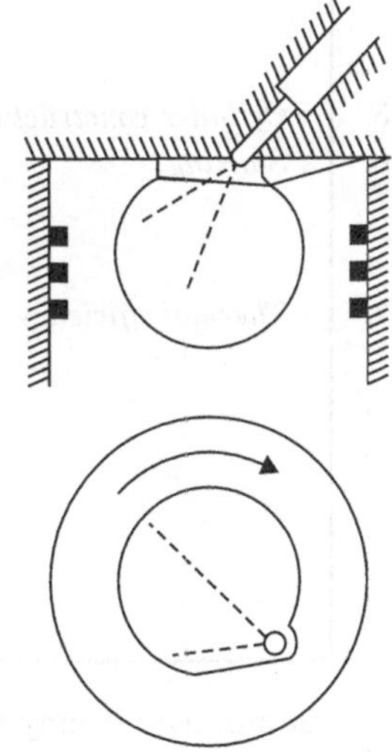

Fig. 2.83. M.A.N. 'M' combustion chamber.

Advantages :

'M-chamber' claims the following *advantages :*

(*i*) Low peak pressure.

(*ii*) Low rate of pressure rise.

(*iii*) Low smoke level.

(*iv*) Ability to operate on a wide range of liquid fuels (multi-fuel capability).

Disadvantages :

(*i*) Low volumetric efficiency.

(*ii*) Since fuel vaporisation depends upon the surface temperature of the combustion chamber, cold starting requires certain aids.

(*iii*) At starting and idling conditions hydrocarbon emissions may occur.

Table below gives *comparison between open combustion chambers and divided combustion chambers.*

S. No.	Aspects	Open combustion chamber	Divided combustion chamber
1.	*Fuel used*	Can consume fuels of good ignition quality, *i.e.*, of shorter ignition delay or higher cetane number.	Can consume fuels of poor ignition quality *i.e.*, larger ignition delay. or lower cetane number.
2.	*Type of injection nozzle used*	Requires multiple hole injection nozzles for proper mixing of fuel	It is able to use single hole injection nozzles and moderate injection

		and air, and also higher injection pressures.	pressures. It can tolerate greater degree of nozzle fouling.
3.	*Sensitivity to fuel spray characteristic*	Sensitive.	Insensitive.
4.	*Mixing of fuel and air*	Mixing of fuel and air is not so efficient and thus high fuel/air ratios are not feasible without smoke.	Ability to use higher fuel/air ratios without smoke, due to proper mixing and consequent high air utilization factor.
5.	*Cylinder construction*	Cylinder construction is simple.	More expensive cylinder construction.
6.	*Starting*	Easy cold starting.	Difficult cold starting because of greater heat loss through the throat.
7.	*Thermal efficiency*	Open combustion chambers are thermally more efficient.	Divided combustion chambers suffer from irreversibilities like throttling through the throat during the compression and expansion ; thus leading to pressure losses and available heat losses. Therefore, these engines are thermally less efficient comparatively.

Summarily it may be said that a particular combustion chamber design must be chosen to perform a given job. No one combustion chamber can produce an ultimate of performance in all tasks. In most engineering work, the design of the chamber must be based on a compromise, after full considerations of the following factors :

(*i*) Heat lost to combustion chamber walls.
(*ii*) Injection pressure.
(*iii*) Nozzle design.
(*iv*) Maintenance.
(*v*) Ease of starting.
(*vi*) Fuel requirement.
(*vii*) Utilisation of air.
(*viii*) Weight relation of engine to power output.
(*ix*) Capacity for variable speed operation.

2.34. COLD STARTING OF C.I. ENGINES

- The important requirement of a C.I. engine is its easy starting from cold. To fulfil this requirement frequently compression ratios higher than necessary are used. Cold work, even so, may become difficult under the following conditions :
  — When the cylinder liner is heavily worn ;
  — When the valves are leaky ;
  — Extreme cold climate (like Himalyan region).

Therefore, sometimes, it is necessary to provide some electrical aid for cold starting.

- *Open chamber direct injection engines are easiest to cold start because of the following reasons :*
  - (*i*) They have smallest surface to volume (S/V) ratio, as a consequence heat loss is minimum.
  - (*ii*) They have lowest intensity of swirl, due to which stagnant gas film remains on the cylinder walls which reduces heat transfer.

Cold starting aids for C.I. engines :

Several methods have been used in the past to *achieve easy cold starting*. Few of them are listed below :

1. *Preheating the engine cylinder by warm water.*
2. *Injection of a small quantity of lubricating oil or fuel oil.* This method temporarily raises the compression ratio, and seals the piston rings and valves.
3. *Provision of cartridges.*
4. *Modifying valve timings for starting.*
5. *Starting as petrol engine* by providing a carburettor and a spark-plug. At starting compression ratio is reduced by providing an auxiliary chamber.

Modern starting aids of high speed engines :

The following basic three types of starting aids are used on modern high speed diesel engines :

1. *Electric glow plugs* (in the combustion chamber)
2. *Manifold heaters* (which ignite a small feed of fuel)
3. Injection of ether.

IV. Two Stroke Engines

2.35. GENERAL ASPECTS

2.35.1. Construction and Working

- In 1878, Dugald-clerk, a British engineer introduced a cycle which could be completed in two *strokes of piston rather than four strokes* as is the case with the four stroke cycle engines. The engines using this cycle were called *two stroke cycle engines*. In this engine suction and exhaust strokes are eliminated. Here *instead of valves, ports are used. The exhaust gases are driven out from engine cylinder by the fresh charge of fuel entering the cylinder nearly at the end of the working stroke.*
- Fig. 2.84 shows a two-stroke petrol engine (used in scooters, motor cycle etc.) Refer to Art. 2.12 also.
  - — The cyclinder L is connected to a closed crankcase C.C.
  - — During the upward stroke of the piston M, the gases in L are compressed and at the same time fresh air and fuel (petrol) mixture enters the crank chamber through the valve V.
  - — When the piston moves downwards, V closes and the mixture in the crank chamber is compressed.
  - — Refer Fig. 2.84 (*i*), the piston is moving upwards and is compressing an explosive charge which has previously been supplied to L. Ignition takes place at the end of the stroke. The piston then travels downwards due to expansion of the gases

(Fig. 2.84 (*ii*)) and near the end of this stroke the piston uncovers the exhaust port (E.P.) and the burnt exhaust gases escape through this port [Fig. 2.84 (*iii*)].

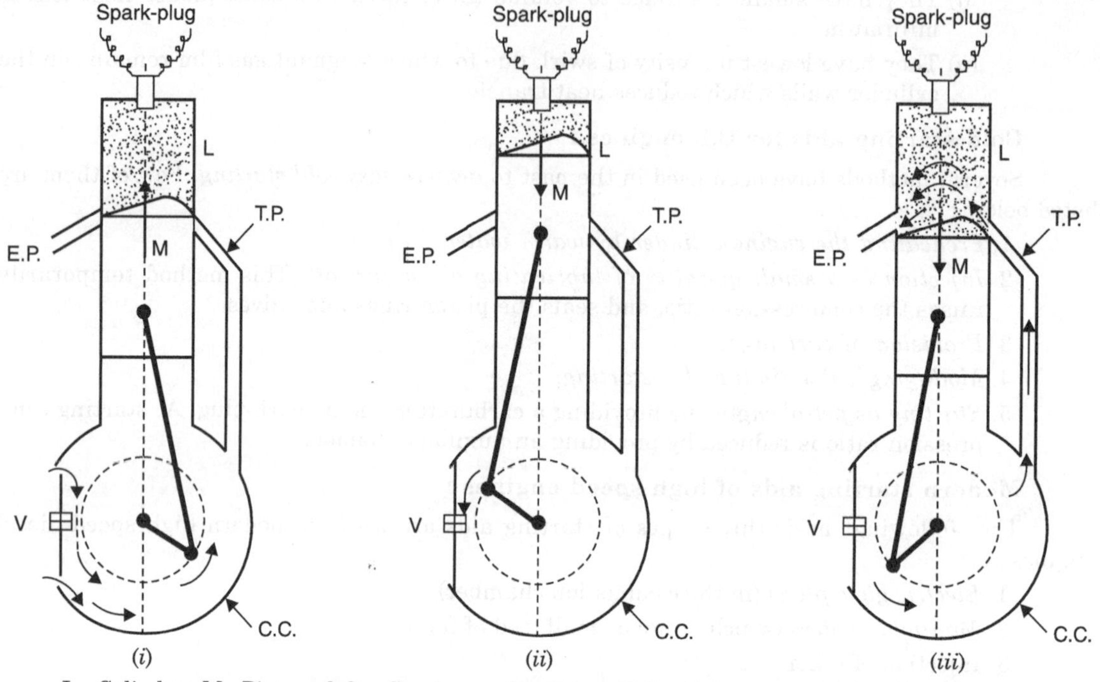

L = Cylinder ; *M* = Piston ; *C.C.* = Crankcase ; *V* = Valve ; *E.P.* = Exhaust port ; *T.P.* = Transfer port.

Fig. 2.84. Two stroke cycle engine (crankcase scavenged).

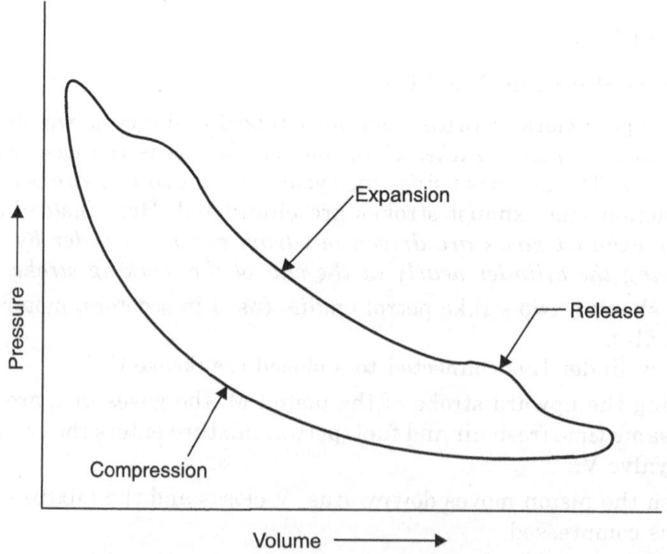

Fig. 2.85. *p-V* diagram for a two stroke cycle engine.

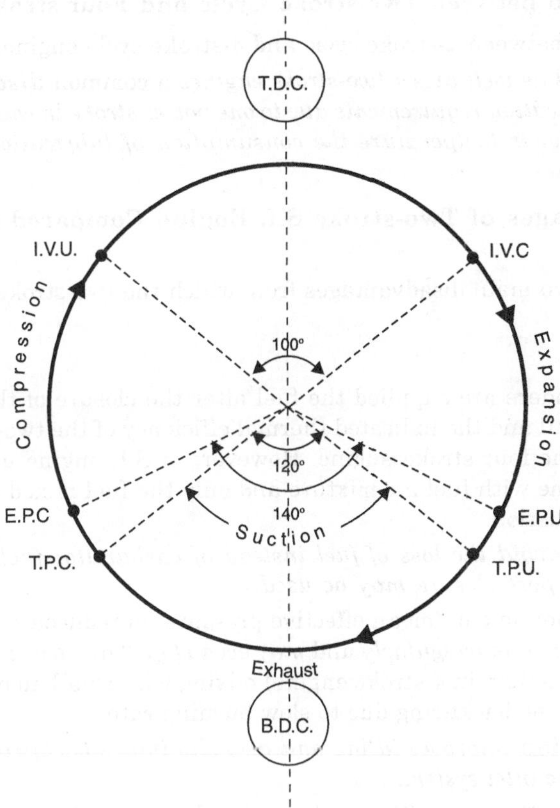

Fig. 2.86. Port timing diagram.

— The transfer port (T.P.) then is uncovered immediately, and the compressed charge from the crank chamber flows into the cylinder and is deflected upwards by the hump provided on the head of the piston. It may be noted that the *incoming air petrol mixture helps the removal of gases from the engine-cylinder ; if, in case these exhaust gases do not leave the cylinder, the fresh charge gets diluted and efficiency of the engine will decrease.*

— The piston then again starts moving from B.D.C. to T.D.C. and the charge gets compressed when E.P. (exhaust port) and T.P. are covered by the piston ; thus the *cycle is repeated.*

- Fig. 2.85 show the *p-V* diagram for a two stroke cycle engine. It is *only for the main cylinder or the top side of the piston.*
- Fig. 2.86 shows self-explanatory port timing diagram for a two stroke cycle engine.

In a **two stroke Diesel cycle** *engine all the operations are the same as in the spark ignition (Otto cycle) engine with the differences ; firstly in this case,* **only air** *is admitted into cylinder instead of air fuel mixture and secondly* **fuel injector** *is fitted to supply the fuel instead of a sparking plug.*

Note. The top of the piston usually has a projection/hump to deflect the fresh air to sweep up to the top of the cylinder before flowing to the exhaust ports. This serves the following two purposes :

(*i*) To scavenge the upper part of the cylinder of combustion products.

(*ii*) To prevent the fresh charge from flowing directly to the exhaust ports.

The same objective can be achieved *without piston deflector by proper shaping of the transfer port.*

2.35.2. Comparison between Two-stroke Cycle and Four-stroke Cycle Engines

- For comparison between 2-stroke cycle and 4-stroke cycle engines refer to Art. 2.14.
- *For all the petrol as well diesel two-stroke engines* a common **disadvantage** is *greater cooling and lubrication requirements due to one power stroke in each revolution of crankshaft. Due to higher temperature the consumption of lubrication oil is also high in two-stroke engines.*

2.35.3. Disadvantages of Two-stroke S.I. Engine Compared to Two-stroke C.I. Engine

Following are the two main disadvantages from which the two-stroke S.I. engines suffer :

1. Loss of fuel
2. Idling difficulty.

- In case two cylinders are supplied the fuel after the closure of the exhaust ports, the fuel loss will be nil and the indicated thermal efficiency of the two-stroke engine will be comparable as the four-stroke engine. However, in S.I. engine using carburettor, the scavenging is done with fuel-air mixture and only the fuel mixed with the retained air is used for combustion.
  — *In order to avoid the loss of fuel instead of carburettor fuel injection just before the exhaust port closure may be used.*
- At *low speeds* when m.e.p. (mean effective pressure) is reduced to about 2 bar, the two stroke S.I. engine *runs irregularly* and *may even stop*. This is owing to large amount of residual gas (more than in 4-stroke engine) mixing with small amount of charge. At low speeds there may be backfiring due to slow burning rate.
  — **Fuel injection** *improves idling and also eliminates backfiring as there is no fuel present in the inlet system.*

In case of C.I. engine there is neither fuel loss (as the charge is only air) nor difficulty in idling since there is no reduction in fresh charge (air).

2.35.4. Reasons for Use of Two-stroke C.I. Engines for Marine Propulsion

Two-stroke C.I. engines find wide use in marine propulsion for the following *reasons* :

1. More uniform torque, the ideal requirement for the propeller.
2. More cooling is required in two stroke engines, plenty of sea water is available for cooling.
3. In C.I. engines there in no loss of fuel in scavenging. Hence they have higher thermal efficiency.
4. *Propeller imposes the condition that maximum power must be developed at about 100 r.p.m. Two stroke engines may be made of slow speed, and with large displacement volume (over 60 cm bore) and of capacity 5000 kW and above. These slow speed engines can be coupled directly to the propeller of the ship, without the necessity of gear reduction.*

- For marine propulsion, *two-stroke C.I. opposed engine (cross-head type) is mainly used.*

2.35.5. Reasons for the Use of Two-stroke S.I. Engines for Low Horse Power Two Wheelers

- When applied to S.I. engines, the Two-stroke cycle engine has certain disadvantages which have restricted its use to small low horse power engines.
  — In S.I. engines the charge consists of a mixture of air and fuel. *During scavenging both, inlet and exhaust ports are open simultaneously for sometime. Some part of the fresh charge escapes with exhaust which results in higher fuel consumption and lower thermal efficiency.*

- For *small two-wheeler engines the* **fuel economy is not a vital factor.** Here ***light-weight and low initial cost are the main considerations,*** *which are the main characteristics of two-stroke S.I. engines.*

2.36. INTAKE FOR TWO STROKE CYCLE ENGINES

- *In two stroke cycle engines inlet air must be input at a pressure greater than atmospheric.* At the start of the intake process, following blowdown, the cylinder is still filled with exhaust gas at atmospheric pressure. There is no exhaust stroke. Air under pressure enters the cylinder and pushes most of the remaining exhaust residual out of the still-open exhaust port. This is called **scavenging.** When most of the exhaust gas is out, the exhaust port closes and the cylinder is filled with air.
  — *At part throttle inlet pressure is low, and this results in poorer scavenging.*
- Generally following two methods are used for putting air into the cylinders :
  (*i*) Through normal intake valves ;
  (*ii*) Through intake slots in the cylinder walls.
  — *The intake air is pressurised using a supercharger, turbocharger, or crankcase compression.*
- There are open combustion chambers in the two stroke cycle engines. It would be extremely difficult to get proper scavenging in a cylinder with a divided chamber.
- In some automobile engines *standard-type superchargers are used and the air is input through intake valves with no fuel added.* The compressed air scavenges the cylinder and leaves it filled with air and a small amount of exhaust residual. After the intake valve is closed, fuel is injected directly into the combustion chamber by injectors mounted in the cylinder head. This is done to avoid HC pollution from fuel passing into the exhaust system, when both exhaust and intake valves are open. *In some automobile engines, air is injected with the fuel. This speeds evaporation and mixing, which is required because of the very short time of the compression stroke.*
  — Fuel injection pressure is of order of 500 to 600 kPa, while air injection pressure is slightly less at about 500 kPa.
  — For *"S.I. engine" fuel injection occurs early in the compression stroke, immediately after the exhaust valve closes.* In *"C.I. engines" the injection occurs late in the compression stroke, a short time before combustion starts.*
- In just about all two stroke cycle engines, *due to cost, crankcase compression is used to force air into and scavenge the cylinders.*
  — In these engines, air is introduced at atmospheric pressure into the cylinder below the piston through a one-way valve when the piston is near T.D.C. The power stroke pushes the piston down and compresses the air in the crankcase, which has been designed for this dual purpose. The compressed air then passes through an input channel into the combustion chambers. In *modern automobiles engines* the fuel is then added with injectors, as with supercharged engines the fuel is then added with injectors, as with supercharged engines. In *small engines the fuel is usually added with a carburettor to the air as it enters the crankcase. This is done to keep the cost down on* small engines, simple carburettors being cheap to build. The fuel injectors will probably become more common as pollution laws become more stringent.
- In case of two stroke cycle engines *using crankcase compression, lubricating oil must be added to the inlet air.* The crankcase in these engines cannot be used as the oil reservoir as with most other engines. Instead, the surfaces of the engine components

are lubricated by *oil vapour carried by the intake air.* In some engines, lubricating oil is mixed directly with the fuel and is vaporised in the carburettor along with the fuel. Other engines have a separate oil reservoir and feed lubricant directly into the intake air flow. *Two negative results* occur beacuse of this method of lubrication : (*i*) Some oil vapour gets into the exhaust flow during valve overlap and contributes directly to HC exhaust emissions ; (*ii*) Combustion is less efficient due to the poorer fuel quality of the oil.

— *Engines which use superchargers or turbochargers generally use standard pressurised lubrication systems, with crankcase serving as the oil reservior.*

- *In order to avoid an excess of exhaust residual no pockets of stagnant flow or dead zones can be allowed in the scavenging process. This is controlled by :*
  (*i*) The size and position of the intake and exhaust slots or valves ;
  (*ii*) The geometry of the slots in the wall ;
  (*iii*) The contoured flow deflectors on the piston face.

2.37. SCAVENGING PROCESS

- In a two stroke engine because of non-availability of an exhaust stroke (unlike four-stroke engine) at the end of an expression stroke, its combustion chamber is left full of combustion products. *The process of clearing the cylinder after the expansion stroke is called* **scavenging process.** Scavenging process is the *replacement of combustion products in the cylinder from previous power stroke with fresh air charge to be burned in the next-cycle.* This process must be completed in a very short duration available between the end of the expansion stroke and start of the process of charging.
- The efficiency of a two stroke engine greatly depends on the effectiveness of the scavenging process.

  Inadequate/poor/bad scavenging leads to the following :

  (*i*) Low mean indicated pressure (m.e.p.) which *results in high weight and high cost per kW (shaft output) for the engine.*

  (*ii*) Low amount of oxygen availability which results in incomplete combustion leading to *higher specific fuel combustion.*

  (*iii*) Contamination of lubricating oil to a greater extent which *reduces the lubricating qualities* and eventually results in *increased wear of piston and cylinder liners.*

  (*iv*) *Higher mean temperature and greater heat stresses on the cylinder walls.*
- Scavenging process can be divided into the following *four distinct stages :*

  (*i*) **Pre-blowdown.** On the opening of inlet port, the gases expanding in the main cylinder tend to escape from it and to pre-discharge into the scavenge air mainfold. This pre-blowdown process ends with the opening of exhaust port.

  (*ii*) **Blowdown.** With the opening of exhaust ports, the gases existing in the cylinder at the end of expansion stroke discharge *simultaneously* into exhaust manifold ; consequently the pressure of main cylinder drops to a value lower than existing in scavenge air manifold. This blowdown process terminates at the moment the gas pressure inside the cylinder attains a value slightly less than air pressure inside scavenge manifold.

  (*iii*) **Scavenging.** This phase of the scavenging process starts at the moment the spontaneous exhaust of gases from cylinder terminates and ends at the moment the exhaust ports are closed. The *scavenge air sweeps out all residual gases remaining in the main cylinder at the end of spontaneous exhaust and replaces them as completely as possible with fresh charge.*

(*iv*) **Additional charging.** After the completion of scavenging phase, the fresh charge continues to flow till the scavenge ports are open and pressure in the cylinder rises. This last phase results in better filling of the cylinder.

- Fig. 2.87 shows a scavenging process on pressure-crank angle (p-θ) for a loop-scavenged cylinder of a typical two stroke engine.
  - It shows the adiabatic compression curve from B.D.C. when the exhaust pressure is 1.013 bar (atmospheric).
  - After the exhaust port opens at 70° before T.D.C., the pressure in the cylinder falls rapidly in the blowdown process (The ***blowdown*** is defined as *the crank angle from the point exhaust port opens to the point the cylinder pressure reaches exhaust pressure*). In practice due to inertia effect of the gases after the blowdown the pressure in the cylinder *normally falls below the exhaust pressure for a few degrees.* Immediately after the exhaust port opens, the inlet port also begins to open. The interval may be of the order of 20° of the crank angle. With the fall in cylinder pressure below scavenging pressure, the fresh charge gets introduced in the cylinder and continues as long as the inlet port is open, and the total inlet pressure becomes more than the pressure in the cylinder.

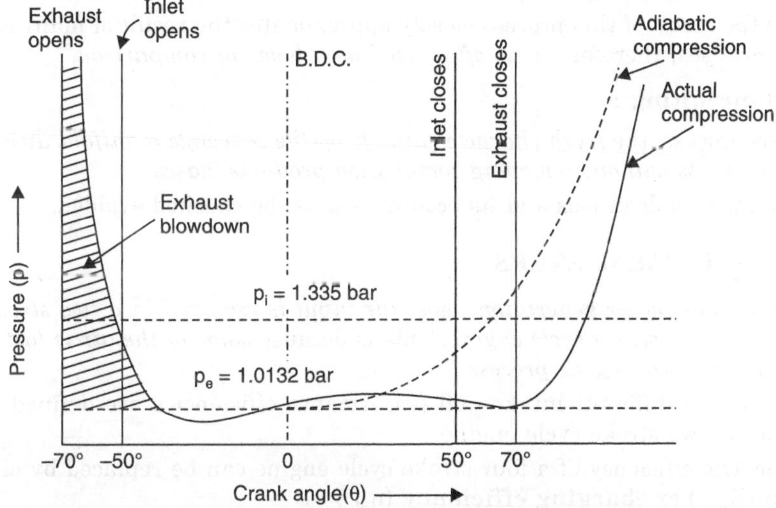

Fig. 2.87. Scavenging process on p-θ for a loop-scavenged cylinder of a typical two stroke engine.

  - Whereas the gases flow into the inlet ports, the exhaust gases continue to flow out of exhaust port, due to the fact that these started in this direction at high velocity during blow-down. Also due to the fresh mixture entering through inlet port there is building up of pressure in the cylinder higher than the exhaust system pressure.
  - **Scavenging angle.** It is defined as the *crank angle during which both inlet and exhaust ports are open.*
  - **Scavenging period.** It is the *time period taken for scavenging angle.*
- After the closure of the two ports, cycle is completed by compression, combustion and expansion in the cylinder as in four-stroke engine.

Theoretical Scavenging Processes :

Following are the three theoretical scavenging processes :

1. Perfect scavenging.
2. Perfect mixing.
3. Short circuiting.

1. **Perfect scavenging :**

- In this type of scavenging, fresh air pumped into the cylinder by the blower through the inlet ports at the lower end of cylinder pushes the combustion products ahead of itself and of the cylinder through the exhaust valve at the other end.
- The air and combustion products *do not mix together.*
- So long as any products remain in the cylinder, the flow through the exhaust valves consists of products only.

2. **Perfect mixing :**

- In this process, the *incoming fresh charge mixes completely and instantaneously with the cylinder contents and a portion of this mixture passes out of exhaust ports at a rate equal to that entering the cylinder.*
- The outgoing (homogeneous) mixture consists initially of combustion products only but then gradually changes to pure air.
- Since the result of this process closely approximates the result of many actual scavenging processes, therefore, it is *often used as a basis of comparison.*

3. **Short-circuiting :**

- In this process, the *fresh charge coming from the scavenge manifold directly goes out of exhaust ports without removing combustion products/gases.*
- It results in a dead loss and its occurence must be checked/avoided.

2.38. SCAVENGING PARAMETERS

- *For the same power generation, more air input is required in a two stroke cycle engine than in a four stroke cycle engine.* This is because *some of the air is lost in the overlap period of the scavenging process.*
- A number of different intake and performance efficiencies are defined for the intake process of two stroke cycle engine.
- Volumetric efficiency of a four-stroke cycle engine can be replaced by either **delivery ratio ($R_{del.}$)** or **charging efficiency ($\eta_{ch.}$) :**

$$\textbf{Delivery ratio} = R_{del.} = \frac{m_{mi}}{V_s \rho_a} \qquad \text{...(2.5)}$$

$$\textbf{Charging efficiency} = \eta_{ch.} = \frac{m_{mt}}{V_s \rho_a} \qquad \text{...(2.6)}$$

where, m_{mi} = Mass of air-fuel mixture *ingested* into the cylinder,

m_{mt} = Mass of air-fuel mixture *trapped* in cylinder after all valves are closed,

V_s = Swept volume, and

ρ_a = Density of air at ambient conditions.

Typical values : $0.65 < R_{del.} < 0.95$

$0.50 < \eta_{ch.} < 0.75.$

— *Delivery ratio is greater than charging efficiency* because some of the air-fuel mixture ingested into the cylinder is lost out of the exhaust port before it is closed.

— In case of those engines that inject fuel after the valves are closed, the *mass of mixture in these equations* (2.5 and 2.6) *should be replaced with the mass of ingested air. Sometimes, ambient air density is replaced by the density of air in the inlet runner downstream of the supercharger.*

Other efficiencies :

$$\textbf{Trapping efficiency} = \eta_{trap.} = \frac{m_{mt}}{m_{mi}} = \frac{\eta_{ch.}}{R_{del.}} \quad ...(2.7)$$

$$\textbf{Scavenging efficiency} = \eta_{sc.} = \frac{m_{mt}}{m_{tc}} \quad ...(2.8)$$

$$\textbf{Relative charge} = C_{rel.} = \frac{m_{tc}}{V_s \rho_a} = \frac{\eta_{ch.}}{\eta_{sc.}} \quad ...(2.9)$$

where, m_{tc} = Mass of total charge trapped in the cylinder, including exhaust residual.

Typical values : $0.65 < \eta_{trap.} < 0.80$

$0.75 < \eta_{sc.} < 0.90$

$0.60 < C_{rel.} < 0.90$.

Pressure loss coefficient. *It is defined as the ratio between the main upstream and downstream pressures during the scavenging period and represents the pressure loss to which the scavenge air is subjected when it crosses the cylinder.*

Excess air factor (λ). The value ($R_{del.}$–1) is called the excess air factor. Thus if the R_{del} (delivery ratio) is 1.3 the excess air factor is 0.3.

Measurement of Scavenging Efficiency :

The following procedure is adopted in *diesel engines for measuring the scavenging efficiency :*

- A small sample of the combustion products is drawn just before the exhaust valve opens or during the earlier part of blowdown.
- The sample is analysed.
- The results obtained are compared with standard curves of exhaust products *vs. F/A* ratio. This determines the *F/A* ratio that must have existed in the cylinder before combustion.
- Knowing the quantity of fuel injected per cycle, the quantity of fresh air retained in the cylinder per cycle is determined. Air present in the residual gas is not considered as it represent a constant quantity which does not participate in combustion process.

2.39. SCAVENGING SYSTEMS

Different scavenging systems/arrangements *based on charge flow* are enumerated and described below :

1. Uniflow scavenging
2. Loop or reverse scavening
3. Cross scavenging.

1. **Uniflow scavenging :**

It is the *most perfect method of scavenging.*

- The fresh charge is admitted at one end of the cylinder and the exhaust escapes at the other end. The air flow is from end to end, and little short-circuiting between the intake and exhaust openings is possible.
- The three available arrangements for uniflow scavenging are shown in Fig. 2.88.

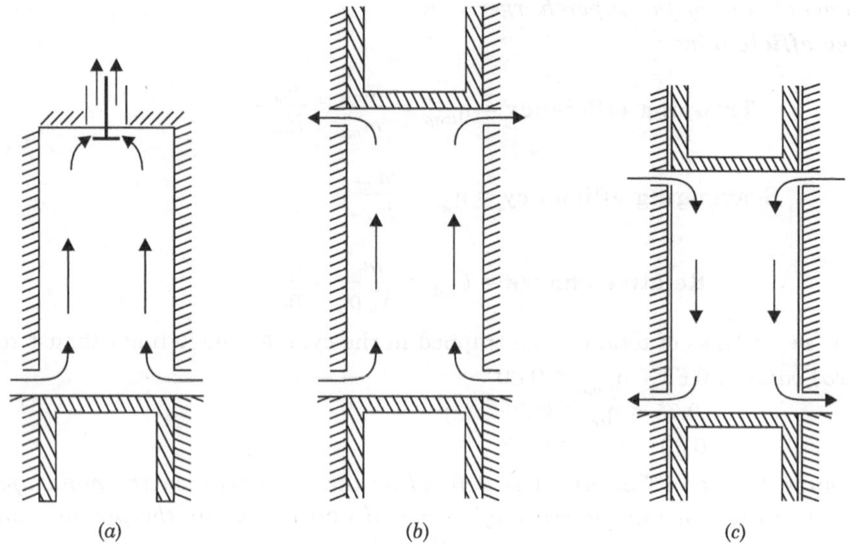

Fig. 2.88. Uniflow scavenging.

— A *poppet valve* is used [Fig. 2.88 (*a*)] to admit the inlet charge or for the exhaust, as the case may be.

— In Fig. 2.88 (*b*) the inlet and exhaust ports are both *controlled by separate pistons* that move in opposite directions (opposed piston engines).

— In Fig. 2.88 (*c*) the inlet and exhaust ports are *controlled by the combined motion of piston and sleeve.*

- *All uniflow systems permit unsymmetrical port timings and supercharging.*
- Due to absence of any eddies or turbulence (at least theoretically) *it is easier in a uniflow scavenging system to push the combustion products out of the cylinder without mixing with it and short circuiting.* Thus this system ***has the highest scavenging efficiency.***
- Since this systems requires either opposed systems, poppet valves or sleeve valve (all of which increase the complication) its *construction is not simple.*

2. **Loop or reverse scavenging :**

In *loop or reverse scavenging, the fresh air first sweeps across the piston top, moves up and then down and finally out through the exhaust.* The system *avoids the short-circuiting of the cross-scavenged engine and thus improves upon its scavenging efficiency.*

- In the MAN type of loop scavenge, Fig. 2.89. (*a*), the exhaust and inlet ports are on the same side, the *exhaust above the inlet.*
- In the Schnuerle type, Fig. 2.89. (*b*), the ports are *side by side.*

  The Curtis type of scavenging, Fig. 2.89 (*c*), is similar to the Schnuerle type, except that *upwardly directed inlet ports are placed also opposite the exhaust ports.*

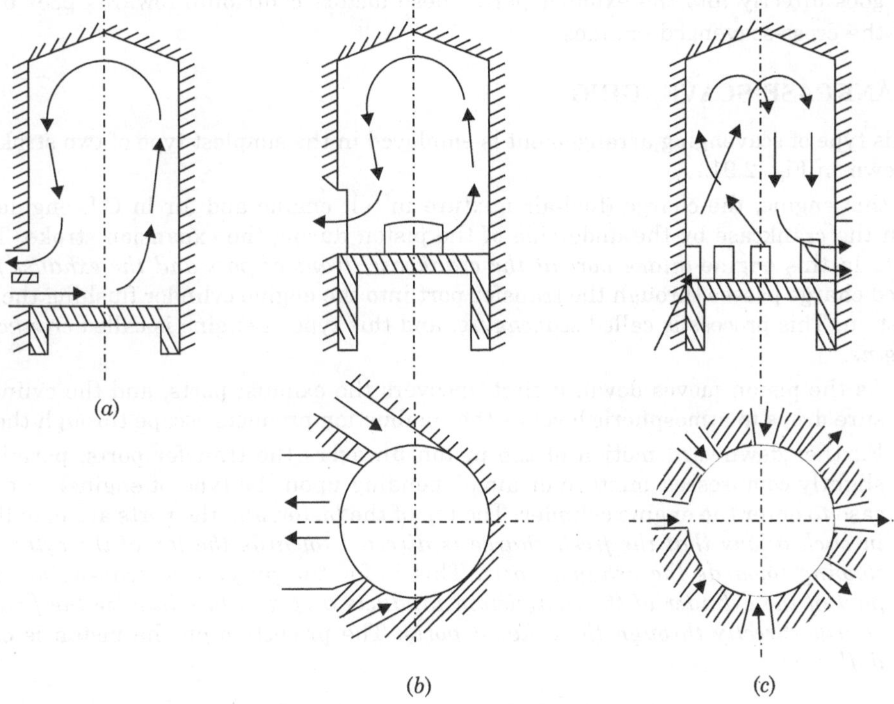

Fig. 2.89. Methods of loop scavenging.

- Owing to the absence of cams, valves and valve gear, loop or reverse scavenged *engines are simple and sturdy*. They have a *high resistance to thermal stresses and are thus much suited to higher supercharge.*
- In a loop scavenged two stroke engine, the *major mechanical problem* is that of obtaining an adequate oil supply to the cylinder wall consistent with reasonable lubricating oil consumption and cylinder wear.

3. **Cross-scavenging**

In this system the *inlet and exhaust ports are located on opposite sides of the cylinder (Fig. 2.90).*

The incoming flow is directed upwards by the deflector on the piston, and the cylinder head reverses the direction of flow, so that exhaust gases are forced through the exhaust port.

- In this type of arrangement the engine is structurally simpler than that with the uniflow scavenging (due to the absence of valves, distributors, and relative drive devices).
- The main *demerit* of this system is that *scavenging air is not able to get rid of the layer of exhaust gas near the wall resulting in poor scavenging.* A small portion of fresh charge

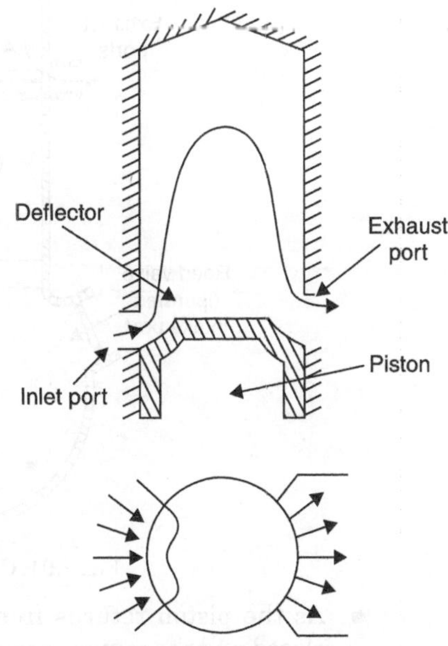

Fig. 2.90. Cross scavenging.

goes directly into the exhaust port. These factors contribute towards poor b.m.e.p. of the cross-scavenged engines.

2.40. CRANKCASE SCAVENGING

This type of scavenging arrangement is employed in the simplest type of two stroke engine, and is shown in Fig. 2.91.

In this engine, the charge (fuel-air mixture in S.I. engine and air in C.I. engine) is compressed in the crankcase by the underside of the piston during the expansion stroke. There are three ports in this engine-*intake port at the crankcase, transfer port and the exhaust port.* The compressed charge passes through the transfer port into the engine cylinder flushing the products of combustion. This process is called *scavenging*, and this type of engine is called *crankcase scavenged engine.*

- As the piston moves down, it first uncovers the exhaust ports, and the cylinder pressure drops to atmospheric level as the combustion products escape through these ports.
- Further downward motion of the piston uncovers the transfer ports, permitting the slightly compressed mixture or air (depending upon the type of engine) in the crankcase to enter the engine cylinder. The top of the piston and the ports are usually *shaped in such a way that the fresh charge is directed towards the top of the cylinder before flowing towards the exhaust ports.* This is for the *purpose of scavenging the upper part of the cylinder of the combustion products and also to minimize the flow of fresh charge directly through the exhaust ports.* The projection on the piston is called the *deflector.*

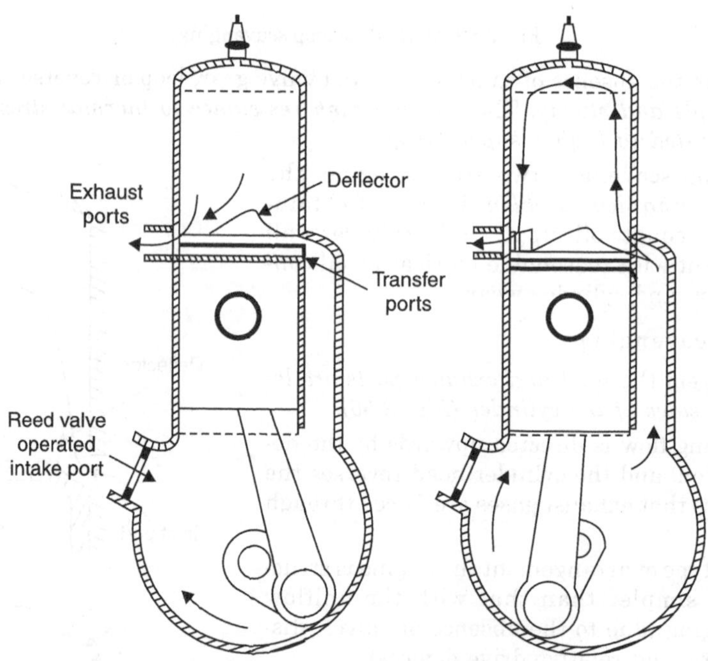

Fig. 2.91. Crankcase-scavenged two stroke engine.

- As the piston returns from B.D.C. the transfer ports and then the exhaust ports are closed and the compression of the charge begins. Motion of the piston during compression lowers the pressure in the crankcase so that the fresh charge is drawn into the crankcase through the inlet reed valve.

- Ignition and expansion take place in the usual way, and the *cycle is repeated.*

Demerits :

1. This system is *very uneconomical and inefficient in operation.* This is owing to the fact that amount of air which can be used for scavenging is less than the swept volume of the cylinder due to *low volumetric efficiency of the crankcase which contains a large dead space.* Thus the delivery ratio ($R_{del.}$) is always *less than unity* and as such it is not possible to scavenge the cylinder completely of the combustion products and some residual gases always remain in the cylinder. Consequently the crankcase-scavenged engine has a lower m.e.p., typical valves being 3 to 4 bar. Since the charge transferred through the transfer port is only 40-50 percent of the cylinder volume, the engine output is strictly limited.
2. Due to mixing of the oil vapours from the crankcase with the scavenging air, oil consumption is increased.

In view of the above demerits the *crankcase scavenging is not preferred and a scavenging pump is essential for a high output two stroke engine.*

2.41. SCAVENGING PUMPS AND BLOWERS

Since the piston of a two stroke engine cannot carry out the pumping action, therefore, a separate pumping mechanism, called the *scavenging pump,* is needed to supply scavenging air to the cylinder.

Following types of pumps are used : Crankcase compression (Refer to Art. 2.40), piston, roots and centrifugal blowers.

- **Piston type pump** shown in Fig. 2.92 is used for *low speed and single or two cylinder engines.*

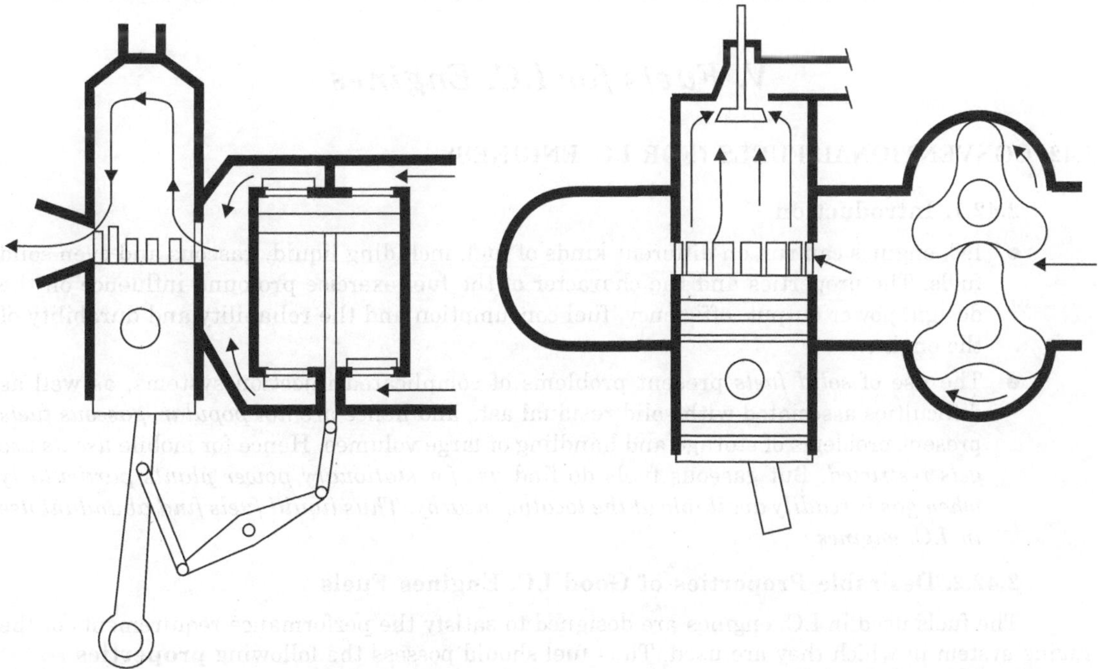

Fig. 2.92. Piston type pump. Fig. 2.93. Roots blower.

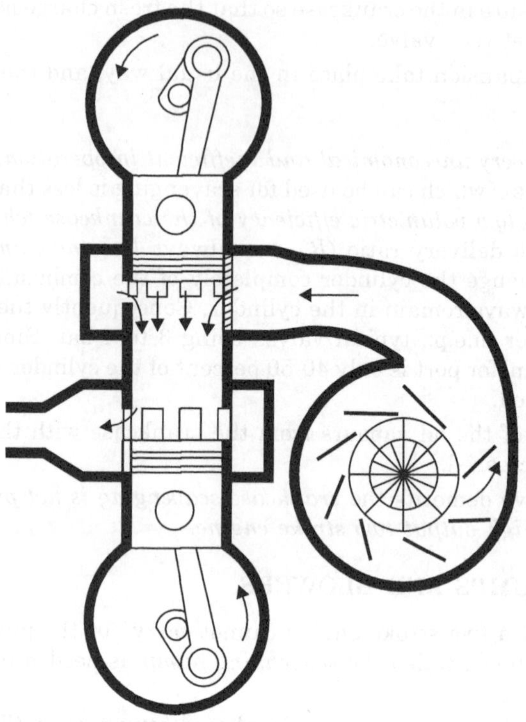

Fig. 2.94. Centrifugal blower.

- **Roots blower** shown in Fig. 2.93 is used for *small and medium engines.*
- **Centrifugal blower** shown in Fig. 2.94 is employed for *large and high output engines.*

V. Fuels for I.C. Engines

2.42. CONVENTIONAL FUELS (FOR I.C. ENGINES)

2.42.1. Introduction

- I.C. engines can run on different kinds of fuel, including liquid, gaseous and even solid fuels. The properties and the character of the fuel exercise profound influence on the design, power output, efficiency, fuel consumption and the reliability and durability of the engine.
- The use of *solid fuels* present problems of complicated injection systems, as well as difficulties associated with solid residual ash, and hence are *not popular, gaseous fuels* present problems of storage and handling of large volumes. Hence for mobile *use its use gets restricted.* But gaseous fuels do find *use for stationary power plants particularly when gas is readily available at the location nearby. Thus liquid fuels find abundant use in I.C. engines.*

2.42.2. Desirable Properties of Good I.C. Engines Fuels

The fuels used in I.C. engines are designed to satisfy the performance requirements of the engine system in which they are used. Thus fuel should possess the following **properties :**

1. High energy density (kJ/kg).
2. Good combustion qualities.
3. High thermal stability.
4. Low toxicity.
5. Low pollution.
6. Easy transportation/transferability and storage.
7. Compatibility with the engine hardware.
8. Good fire safety.
9. Low deposit forming tendency.
10. Economically viable in very large quantities.
11. Easy mixing with air and low latent heat of evaporation (h_{fg}).
12. No chemical reaction with engine components through which it flows.

2.42.3. Gaseous Fuels

These fuels are used in S.I. engines. The different gaseous fuels are enumerated and discussed below :

1. Natural gas
2. Manufactured gases
3. By-product gases
4. Sewage sludge gas
5. Biogas.

1. **Natural gas :**

- Its composition varies with source but mainly it contains CH_4 (75 to 95 percent) and remaining C_2H_6 and N_2. From some areas, the natural gas obtained contains H_2S which is much harmful to the engines.
- It is available with oil wells and is *colourless* and *odourless*.
- It is found in several parts of the world but particularly in U.S.A. It is also carried from the place of availability to the place of use through hundreds of kilometres pipeline.

2. **Manufactured gases :**

The gases are manufactured by various methods, discussed briefly below :

- ***Coal gas*** is manufactured by *heating soft coal in closed vessel*. The contents of the gas depend upon the type of coal and method of operation used in manufacturing.
  — A clean coal gas contains : 33% H_2 and 66% CH_4.
  — Its energy content is *50 percent of natural gas.*
- ***Water gas*** is *formed by using steam*. For its manufacture, the *water and air are passed alternately through a bed of hot carbon.*
  — It contains H_2, CO and N_2.

3. **By-product gases :**

- The gases produced during manufacture of other substances are known as *by-product gases*.
- ***Blast furnace gas*** is a by-product of steel plants. It contains CO and N_2. It contains large amount of dust particles ; therefore, it should be cleaned by an effective method before its use in the engine.

4. **Sewage sludge gas :**

- It contains CH_4 and CO_2 with very small percentage of H_2S.
- This gas is made available from present well developed sewage disposal plants.

5. **Biogas :**

- This gas is produced from the cow dung which is available in large quantities in India.
- It is easy to produce (with appropriate chemical reaction) and use locally.

Except natural gas, all other gases mentioned above are generally employed for running I.C. engines whose power is used locally to run different types of equipments like *small electric generators, pumps etc.*

Advantages of gaseous fuels :

(*i*) Easily compressed and stored.
(*ii*) Easily carried through pipes.
(*iii*) Easy starting of engines.
(*iv*) Easy to maintain A/F ratio in multi-cylinder engines, as compared to liquid fuels.

Disadvantages :

(*i*) High cost (on the basis of energy content)
(*ii*) High purifying cost.
(*iii*) Storage volume per unit energy very large.
(*iv*) As compared to engines using liquid fuels, the size and weight of the engine (kg/kW) is considerably large.
(*v*) The cost (capital and running) of the plants manufacturing gases is considerably high.

2.42.4. Liquid Fuels

Following are the three principal commercial types of liquid fuels :

1. Benzol ; 2. Alcohol ; 3. Refined products of petroleum.

1. **Benzol :**

- It consists of benzene (C_6H_6) and toulene (C_7H_8) and is obtained as a by-product of high temperature coal carbonization.
- It possesses anti-knock quality. As compared to gasoline, its heating value is low.

2. **Alcohol :**

- It has good anti-knock qualities.
- Its heating value is low as compared to gasoline.
- It is more expensive to produce.
- It is used as fuel blended with gasoline.
- It can be manufactured from grain, sugarcane and waste products.

3. **Refined products of petroleum :**

- It is the main source of liquid fuels for I.C. engines.
- It is used in the form of gasoline, kerosene, and diesel oil.

The liquid fuels are ***classified*** in two ***groups*** :

(*i*) *Liquid fuels which are vaporised easily : "Petrol"* and *"Alcohol"* . These are commonly used in S.I. engines.
(*ii*) *Liquid fuel which is directly injected in the combustion chamber :* "Diesel or fuel oil".

2.42.5. Structure of Petroleum

In I.C. engines the fuels which are usually used are *complex mixtures of hydrocarbons.* These fuels are obtained by *refining petroleum.*

Basically, petroleum is a mixture of hydrocarbons, compounds *made up exclusively of carbon and hydrogen atoms ;* it may also contain small quantities of other compounds having sulphur, oxygen and nitrogen (In some petroleum, metallic compounds such as derivatives of vanadium, iron, nickel, arsentic etc., are also found).

The constituents of petroleum are *classified* into the following four main groups :

	Constituents	***General formula***	
1.	*Paraffins*	C_nH_{2n+2}	(where n is the number of carbon atoms)
2.	*Olefins*	C_nH_{2n}	
3.	*Naphthenes*	C_nH_{2n}	
4.	*Aromatics*	C_nH_{2n-6}	

— Within each group also, the physical properties of individual compound differ according to the number of carbon and hydrogen atoms in the molecule.

— The physical differences between compounds, even in any group, influence the way fuel evaporates and hence the formation of combustible mixture.

— The difference in chemical properties of hydrocarbon from different groups affect the combustion process and hence the proportions of fuel and air requirements.

1. **Paraffins :**

(*i*) ***Straight chain or normal paraffins :***

- It consists of a straight chain molecular structure as shown in Fig. 2.95.
- The names of hydrocarbon in this series end with *ane* as in methane, propane, hexane etc.
- The straight chain paraffins are saturated compounds as valency of carbon is fully utilised and therefore, they are very *stable.*

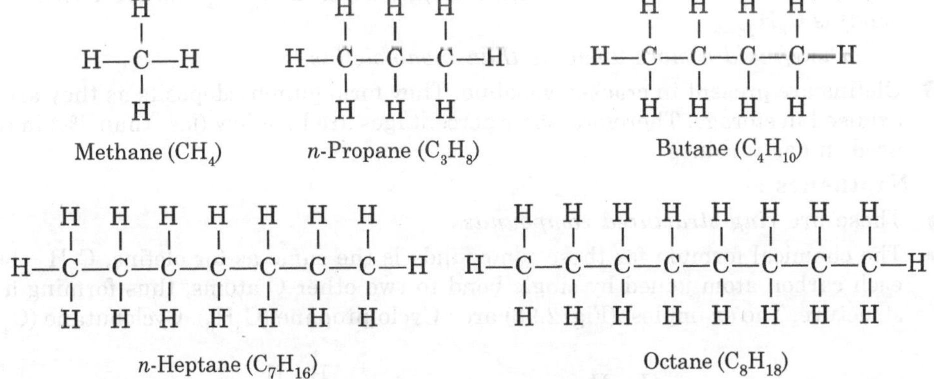

Fig. 2.95. Straight chain paraffin.

(*ii*) ***Branched chain or iso-paraffins :***

- The *carbon atoms are branched in these compounds.*
- Branched chain or iso-paraffins have an open structure which is branched as shown in Fig. 2.96. Iso-octane, triptane etc. are examples of this type.

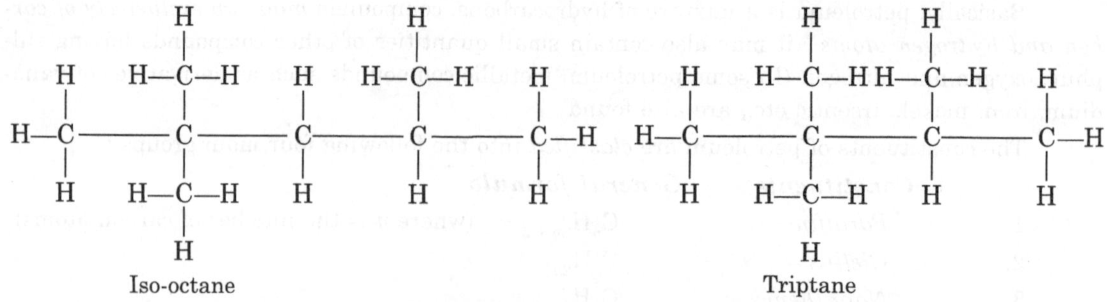

Fig. 2.96. Branched chain or iso-paraffins.

- Iso-paraffins are also *stable compounds* and *highly knock-resistant when used as S.I. engine fuels.*

2. **Olefins :**

- These are *compounds with one or more double bonded carbon atoms in straight chain.* The names end with *ene* for one double bond and *adiene* for two double bonds. The examples are : Hexene and Butadiene (Fig. 2.97)

```
    H   H   H   H    H H              H   H   H   H
    |   |   |   |    | |              |   |   |   |
H—C—C—C—C— C=C—H     H—C = C — C = C —H
    |   |   |   |
    H   H   H   H
```

Hexene, C_6H_{12} (mono-olefin) Butadiene, C_4H_6 (diolefin)

Fig. 2.97. Olefins.

- The general formula for *mono-olefins* (single bond) is C_nH_{2n} and for *diolefins* (double bond) is C_nH_{2n-2}.
  — *Diolefins are more unstable than mono-olefins*
- Olefins are present in cracked gasoline. They form gummy deposits as they are readily oxidised in storage. Therefore, their percentages are kept low (less than 3%) in the fuels used in gas turbines.

3. **Napthenes :**

- These are *ring structured compounds.*
- The chemical formula for these compounds is the same as for olefins, C_nH_{2n} but have each carbon atom joined by single bond to two other C atoms, thus forming a ring of structure. The examples, (Fig. 2.98) are : Cyclo-propane (C_3H_6), Cyclobutane (C_4H_8) etc.

```
     H  H                  H  H
      \/                   |  |
      C                 H—C—C—H
     /\                    |  |
  H—C—C—H               H—C—C—H
    |  |                   |  |
    H  H                   H  H
```

Cyclo-propane (C_3H_6) Cyclobutane (C_4H_8)

Fig. 2.98

- Although napthenes have the same formula as for olefins, the properties are *radically different.*
- The napthenes are *saturated* compounds whereas olefins are unsaturated.

4. **Aromatics :**

- These compounds (C_nH_{2n-6}) have a ring type structure for all or most of the carbon atoms, to which are attached H or group of C and H atoms ; the examples are shown in Fig. 2.99.
- In all aromatics, a *benzene, (C_6H_6) molecule exists as central structure* and other aromatics are formed by replacing one or more of the hydrogen atoms of the benzene molecule with an organic radical such as paraffins, napthenes and olefins. By adding a methyl group (CH_3) ; benzene is converted to toluene ($C_6H_5CH_3$). The three double bonds make aromatics very active and therefore they are *highly unsaturated compounds.*

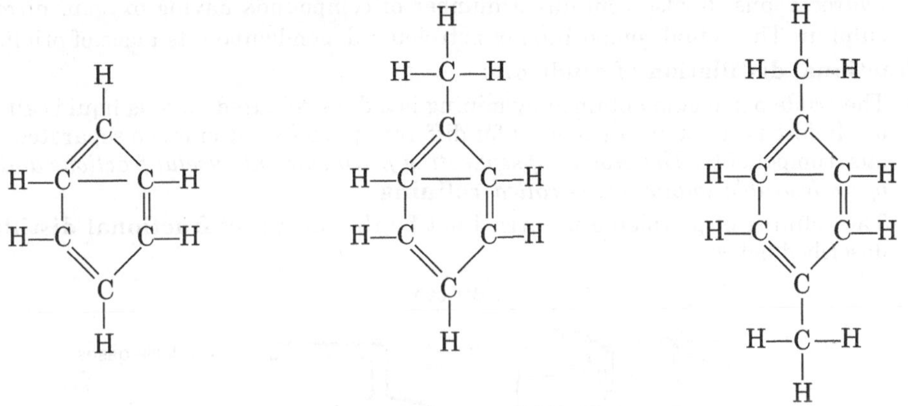

(*a*) Benzene (C_6H_6)

(*b*) Toluene (C_7H_8) or $C_6H_5CH_3$

(*c*) Xylene

Fig. 2.99. Structures of aromatics.

— In most unaltered gasolines, both benzene and toluene are present to a modest extent.

Following are a few *special properties of aromatics :*

(*i*) Offer highest resistance to knocking in S.I. engines.

(*ii*) Suitability of these fuels for C.I. engines is just reverse of their suitability for S.I. engines.

— Therefore, *"paraffins" are most suitable fuels for C.I. engines* and *"aromatics" are most suitable fuels for S.I. engines.*

(*iii*) With the increase in the number of atoms in the molecular structure, the boiling temperature of fuel generally increases.

(*iv*) As the proportion of H_2-atoms to C-atoms in the molecule increases the calorific value of fuel increases. Thus, paraffins have lower calorific value whereas *aromatics have highest calorific value.*

2.42.6. Petroleum and Composition of Crude Oil

- Petroleum is a dark viscous oily liquid known as *rock oil* (In Greek, *petra*—rock, *oleum*—oil). It is formed from the bacterial decomposition of the remains of animals and plants which got buried under the sea millions of years ago. When these organisms died, they sank to the bottom and got covered by sand and clay. Over a period of millions of years, these remains got converted into hydrocarbons by heat, pressure and catalytic action. The hydrocarbons formed rose through porous rocks until they were trapped by impervious rocks forming an oil trap.
- Natural gas is found above the petroleum oil trapped under rocks. The crude petroleum is obtained by drilling a hole into the earth's crust and sinking pipes into it. When the pipe reaches the oil deposit, natural gas comes out with a great pressure. After the pressure has subsided, the crude oil is pumped out of the oil well. *This process of obtaining crude oil from its sources* is called **mining.**
- The *crude oil is a mixture of hydrocarbons* such as alkanes, cycloalkanes and aromatic hydrocarbons. It also contains a number of compounds having oxygen, nitrogen and sulphur. The actual composition of petroleum depends upon its place of origin.

Fractional distillation of crude oil :

- The crude petroleum obtained by mining is a dark coloured viscous liquid called **crude** oil. Before petroleum can be used for different purposes, it must be separated into various components. *The process of separating petroleum into useful fractions and removal of undesirable impurities is called* **refining.**
- The refining of petroleum is carried out by the process of **fractional distillation** *as* described below :

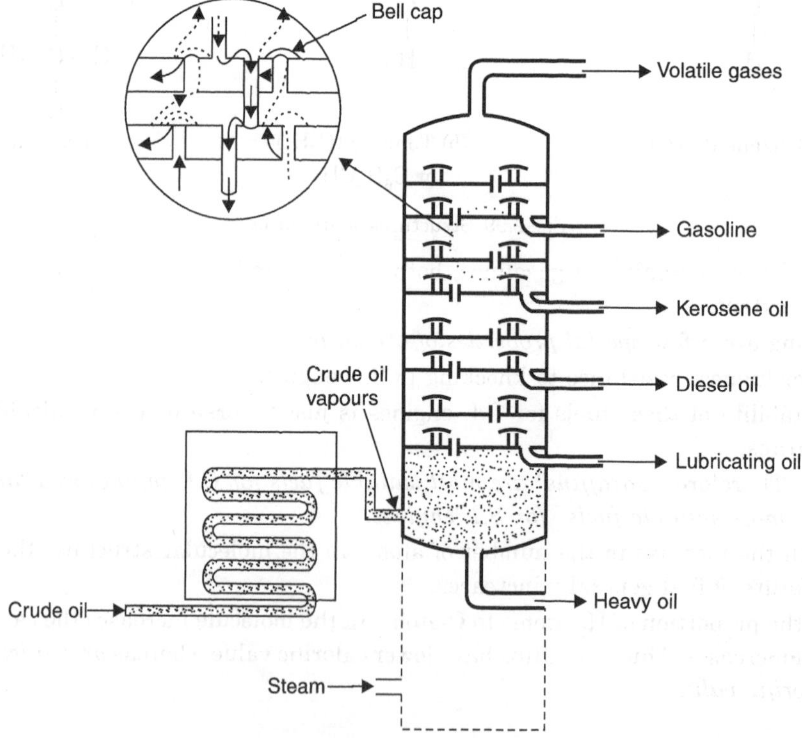

Fig. 2.100. Refining of petroleum.

— The refining of petroleum is done in big refineries. The first step in the refining process is neutralisation of crude oil by washing with acidic or basic solution as needed.

— Then the oil is heated in a furnace to about 675 K and the vapour thus obtained are introduced into a fractionating tower. The tower is divided into a number of compartments by means of shelves (trays) having holes (Fig. 2.100). The holes are covered with *bubble caps* which allow the lighter more volatile components to pass up the column while the heavier, less volatile components condense and flow into trays below. Each shelf is provided with an overflow pipe which keeps the liquid to a certain level and allows the rest to flow down to the lower shelf. Therefore, *during fractional distillation, the fractions with lower boiling points rise up the tower and condense at different levels depending upon the boiling points.* For example, the crude oil is fed at the base at about 675 K. At this temperature all the components except asphalt vapourise. As the mixture of hot vapour rises in the column, it cools. Therefore, the component with the highest boiling point liquefies first and is collected. Then a little, higher in the column, the component having slightly lower boiling point liquefies and so on. The residual gases which do not condense escape from the upper part of the tower. *The fractions are separated at different boiling points and are thus collected at different heights in the column.* The important fractions of petroleum refining are given in the table below.

Table. Different fractions of petroleum refining

Fraction (K)	Boiling range (K)	Approximate composition	Uses
1. Gaseous	113 to 303	C_1—C_5	As gaseous fuel, for producing carbon black and is also used for preparing ammonia, methyl alcohol and gasoline
2. Petroleum ether or ligroin	303 to 363	C_5—C_7	As a solvent for oils, fats, rubber and also in dry cleaning.
3. Gasoline or petrol (Sp. gravity = 0.7 to 0.8)	343 to 473	C_7—C_{12}	Mainly as a *motor fuel.*
4. Kerosene (Sp. gravity = 0.8 to 0.85)	448 to 548	C_{12}—C_{15}	As an illuminant fuel and for preparing *petrol gas.*
5. Gas oil, fuel oil, and diesel oil	523 to 673	C_{15}—C_{18}	In furnace oil, as a fuel for *diesel engines* and also in cracking.
6. Lubricating oils, greases and petroleum jelly	623 and higher	C_{16} and higher	Used mainly as *lubricants.*
7. Paraffin wax	melts between 325 to 330	C_{20} and higher	Used for manufacturing candles, waxed papers and for water proofing.
8. Petroleum coke	residue	C_{30} and higher	Used as artificial asphalt, fuel and also in making electrodes.

2.42.7. Fuels for Spark-Ignition Engines

Gasoline, a mixture of various hydrocarbons (such as paraffins, olefins, napthenes, and aromatics) is the major fuel used for S.I. engines. The composition depends upon the source of crude oil and the nature of refining process.

The following are the *requirements of an ideal gasoline* :

1. High calorific value.
2. Knock-resistance.
3. Easy to handle.
4. Easy availability at reasonable cost.
5. Quick evaporation (when injected by carburettor in the current of air).
6. Clean burning and no deposition of the residue.
7. No pre-ignition.
8. No tendency to decrease the volumetric efficiency of engine.

2.42.7.1. Volatility

- *"Volatility"* is commonly defined as *the evaporating tendency of a liquid fuel.*
- This quality of the fuel has great significance for carburetted engines. This will decide the fuel vapour to air ratio in the cylinder at the time of ignition. As $F_R = 0.6$ is the lowest limit for satisfactory ignition and flame propagation, therefore, volatility of fuel must ensure to give at least this *fuel vapour to air ratio* at the time of ignition under all conditions of operation including starting from cold.
- The volatility of gasoline is generally characterised by the following two laboratory tests.

  (*i*) ASTM distillation test (*ii*) Reid vapour test.

ASTM distillation test :

Fig. 2.101 shows the apparatus used for ASTM distillation test :

- 100 cubic centimetres of gasoline fuel is taken in the flask and heated. The flask is fitted with a thermometer to record the temperature of vapour being formed and collected in a graduated measuring cylinder.

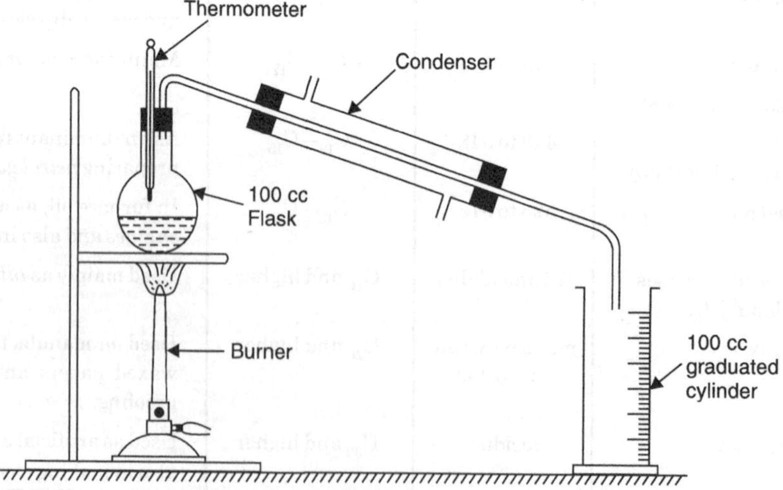

Fig. 2.101. ASTM distillation apparatus.

- When the first drop of condensed vapour drops from condenser, the temperature is recorded. This temperature is called ***initial boiling point.***
- The vapour temperature is recorded at each successive 10 percent of condensed vapour collected. When 95 percent has been distilled the burner flame is increased and the maximum temperature is recorded as the *'end point'*. The mass of the residue in the flask is also recorded.

Fig. 2.102 shows distillation curves (ASTM) for various products of petroleum refining.

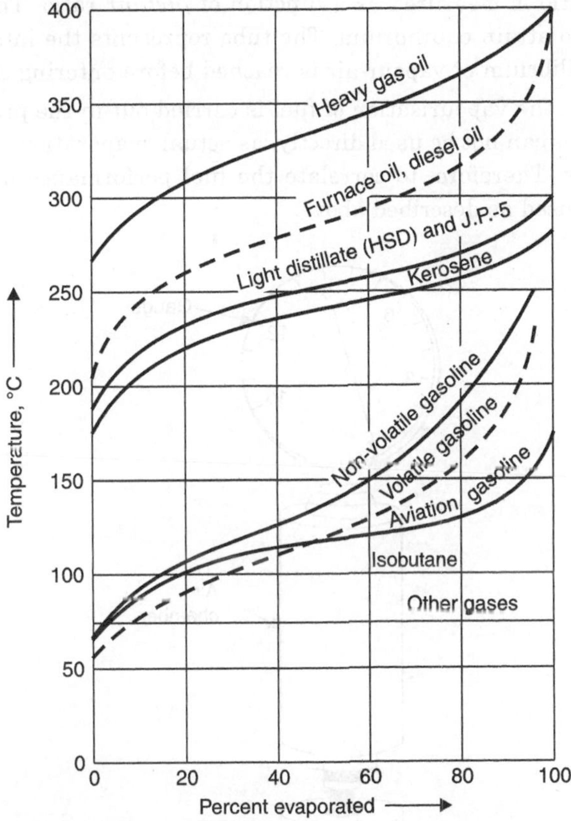

Fig. 2.102. Distillation curves (ASTM).

Reid vapour pressure :

- The *volatility* of petrol is also defined in *terms of Reid vapour pressure.* This is a measure of the vapour pressure of oil at 38°C expressed as millimetres of mercury or in pounds per square inch pressure and indicates initial tendency of a fuel to vapour-lock.
- The apparatus used for determination of Reid vapour pressure is shown in Fig. 2.103.
  - A chilled fuel sample is placed in the Reid bomb and then immersed in water bath held at 38°C ; the *air chamber contains an air volume equal to four times the volume of fuel.* The pressure indicated by the pressure gauge will indicate the *pressure rise due to vaporisation of fuel and increase in volume due to increase*

in temperature. If the latter is subtracted from the total pressure rise we get the Reid vapour pressure (the increase in pressure due to *vaporisation of a given quantity of fuel under a given quantity of fuel under given condition of temperature*).

Equilibrium air distillation (EAD) :

- The ASTM distillation curve is not a true boiling point curve of the fuel. Therefore, it cannot directly relate to fuel performance in the engine. In this case, the fuel is allowed to evaporate into an air stream moving through a long tube with low velocity. The exit *vapour-air* ratio is measured as a function of *fuel-air* ratio. The tube should be sufficiently long to attain equilibrium. The tube represents the intake manifold of the engine and equilibrium of vapour-air is reached before entering the engine.
- In ASTM-test, the vapourisation of fuel is carried out in the presence of vapour of fuel so these curves can not be used directly, as actual evaporation of fuel takes place in the current of air. Therefore, to correlate the fuel performance in the engine, EAD test apparatus is used as described above.

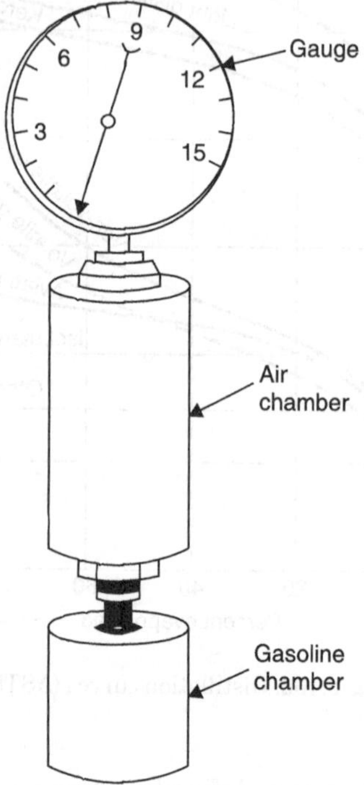

Fig. 2.103. Apparatus for determining Reid vapour pressure.

Fig. 2.104 shows the EAD test curves for a typical gasoline and also A/F ratio volatility curves.

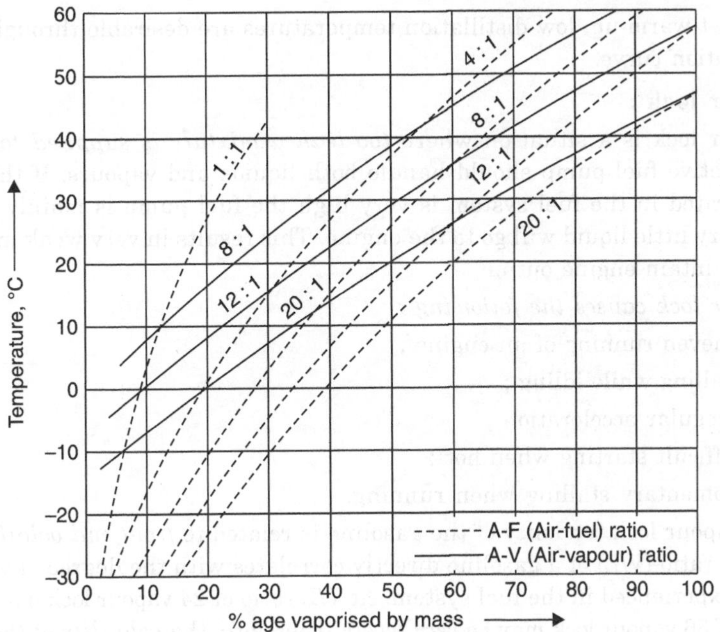

Fig. 2.104. EAD curves for a typical gasoline.

2.42.7.2. Effects of Volatility on S.I. Engine Performance

Volatility of a liquid is its tendency to evaporate under a given set of conditions. It is an extremely important characteristic of S.I. engine which affects engine performance and fuel economy characteristics.

— Cold starting of S.I. engine is improved if front end volatility is higher but it may lead to increased problems of hot starting and vapour lock.

— The *mid range* (20 to 80%) *portion* should be volatile enough to give satisfactory air-fuel ratios under a variety of operating conditions.

— *Low tail end volatility will help in good mixture distribution and hence good fuel economy.*

A few important effects of volatility on S.I. engine performance are enumerated and described below :

1. Starting and warm up.
2. Vapour lock.
3. Evaporation loss.
4. Crankcase dilution.
5. Operating range performance.
6. Spark-plug fouling.
7. Formation of sludge deposits.

1. **Starting and warm up :**

- For easy starting of the engine a certain part of the gasoline should vaporise at room temperature. To fulfil this condition 1 to 10 per cent of the distillation curve must have a low boiling temperature. With the warming up of the engine, the temperature gradually increases to the operating temperature.

- For best warm-up, low distillation temperatures are desirable throughout the range of distillation curve.

2. **Vapour lock :**

- Vapour lock is a situation where *too lean a mixture is supplied to the engine.* The automotive fuel pump should handle both liquids and vapours. If the amount of fuel evaporated in the fuel system is very high the fuel pump is mainly pumping vapour and very little liquid will go to the engine. This results in very weak mixture which can not maintain engine output.

  Vapour lock causes the following :

  (*i*) Uneven running of an engine ;

  (*ii*) Stalling while idling ;

  (*iii*) Irregular acceleration ;

  (*iv*) Difficult starting when hot ;

  (*v*) Momentary stalling when running.

- The vapour lock tendency of the gasoline is related to *front end volatility.* The vapour-liquid ratio (V/L) of a gasoline directly correlates with the degree of vapour lock likely to be experienced in the fuel system. At V/L ratio of 24 vapour lock may *start,* and at V/L ratio of 36 vapour lock may be *very severe.* Therefore, the *volatility of the gasoline should be maintained as low as practicable to prevent this type of difficulty.*

3. **Evaporation loss :**

- The evaporation loss (from carburettors and storage tanks) *depends on the vapour pressure which is a function of fraction components and initial temperature.* These losses can be as high as 10 to 15 per cent.

- The evaporation loss *not only decreases the fuel economy, but also decreases its anti-knock quality as the lighter fraction have higher anti-knock properties.*

  The evaporation loss is a *reason for restricting the low end volatility of the fuels.*

4. **Crankcase dilution :**

- If very frequent starting of the engine with low engine temperature is necessary, very rich mixtures have to be supplied and some of un-evaporated fuel leaks past the piston rings and goes to crankcase. Consequently lubricating oil gets diluted. *This dilution decreases the viscosity of the lubricating oil and also washes away the lubricating oil film on engine cylinder walls.* It is found that the tendency of fuels to dilute the lubricating oils lies in the order of 90% ASTM temperature. Thus control of 90% ASTM temperature combined with proper ventilation of cranckcase reduces the dilution of crank-case oil.

- In the engines using *heavier fuels* like kerosene and other distillates, the problem of dilution and poor lubrication of pistons and rings may be *severe.*

5. **Operating range performance :**

- The acceleration of an engine depends upon its ability to deliver suddenly to intake an extra supply of fuel air mixture in a sufficiently vaporised form to burn quickly. Good acceleration occurs when air-fuel vapour ratio of 12 : 1 is supplied.

- The ability to accelerate falls off as available mixture becomes lean.

6. **Spark-plug fouling :**

Spark plug fouling is caused due to deposition of some high boiling hydrocarbons. Lower the tail-end volatility less are the chances of spark-plug fouling.

7. **Formation of sludge deposits :**

The sludge deposition inside an engine is caused by certain types of high boiling hydrocarbons. These deposits *can cause piston ring plugging and sticking and valve sticking resulting in poor operation and poor fuel economy.*

2.42.8. Knock Rating of S.I. Engines Fuels

2.42.8.1. Self-ignition characteristics of fuels

- When the temperature of an air-fuel mixture is raised high enough, the mixture will *self-ignite* without the need of a spark-plug or other external ignites. *The temperature above which this occurs is called the "Self-Ignition Temperature (S.I.T.)".* This is the basic principle of ignition in a compression ignition engine. The compression ratio is high enough so that the temperature rises above S.I.T. during the compression stroke. Self-ignition then occurs when the fuel is injected into the combustion chamber. On the other hand, self-ignition (or pre-ignition, or auto-ignition) is not desirable in an S.I. engine, where a spark-plug is used to ignite the air-fuel at the proper time in the cycle. *When self-ignition does occur in an S.I. engine higher than desirable, pressure pulses are generated. These high pressure pulses can cause damage to the engine and quite often are in the audible frequency range.* This phenomenon is often called **knock** or **ping.**
- Fig. 2.105 shows the basic process of what happens when self-ignition occurs.

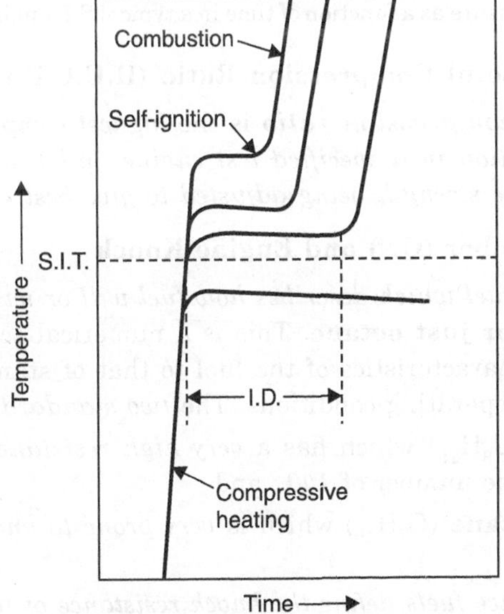

S.I.T. → Self-ignition temperature

I.D. → Ignition delay

Fig. 2.105. Self-ignition characteristics of fuels.

— If the temperature of a fuel is raised above the Self-Ignition Temperature (S.I.T.), the fuel will spontaneously ignite after a short Ignition Delay (I.D.). *The higher above S.I.T. which the fuel is heated, the shorter will be I.D.*

— **Ignition delay** is generally a very small fraction of a second (generally of the order of *thousandths of a second*). During this time, pre-ignition reaction occurs, including oxidation of some fuel components and even cracking of some large hydrocarbon components into smaller hydrocarbon molecules. These pre-ignition reactions raise the temperature at local spots, which then promotes additional reactions until, finally, the actual combustion reaction occurs.

- Fig. 2.106 shows the cylinder pressure as a function of time in a typical S.I. engine combustion chamber showing (*i*) normal combustion, (*ii*) combustion with light knock and (*iii*) combustion with heavy knock.

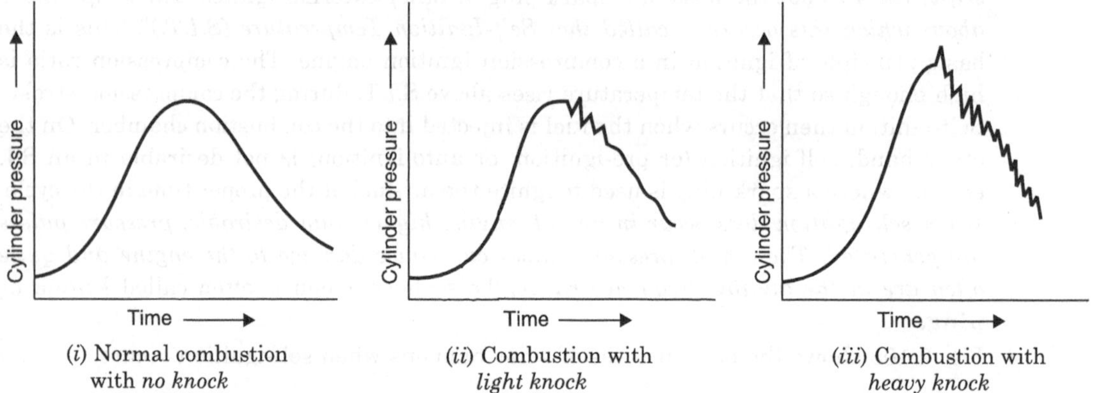

(*i*) Normal combustion with *no knock* (*ii*) Combustion with *light knock* (*iii*) Combustion with *heavy knock*

Fig. 2.106. Cylinder pressure as a function of time in a typical S.I. engine combustion chamber.

2.42.8.2. Highest Useful Compression Ratio (H.U.C.R.)

The ***highest useful compression ratio*** is the *highest compression ratio at which a fuel can be used without detonation in a specified test engine under specified operating conditions and the ignition and mixture strength being adjusted to give best efficiency.*

2.42.8.3. Octane number (ON) and Engine Knock

- *The property of a fuel which describes how fuel will or will not self-ignite is called the* **octane number or just octane**. This is a numerical scale generated by comparing the self-ignition characteristics of the fuel to that of standard fuels in a specific test engine at specific operating conditions. The *two standard reference fuels* are :

— **Iso-octane** (C_8H_{18}) which has a *very high resistance to knock* and therefore is given an octane number of *100* ; and

— **Normal heptane** (C_7H_{16}) which is *very prone to knock* and is therefore given a *zero* value.

Blends of these reference fuels define the knock resistance of intermediate octane numbers, and thus a blend of 10% *n*-heptane and 90% iso-octane by volume has an octane number of 90.

- *The higher the octane number of fuel, the less likely it will self-ignite. Engines with low compression ratios can use fuels with lower octane numbers, but high-compression engines must use high-octane fuel to avoid self-ignition and knock.*

Test procedure for finding octane number (ON) of fuel :

To find ON of a fuel, the following *test procedure* is used :

— The test-engine is run at specified conditions using the fuel being tested. (The *specified/fixed conditions* to give maximum knock response are : Air inlet temperature, coolant temperature, engine speed, ignition advance setting, mixture strength etc.)

— *Compression ratio is adjusted until a standard level of knock is experienced.* (Intensity of knock is measured *with a magnetostriction knock detector*).

— The test fuel is then replaced with a mixture of the two standard fuels. The intake system of the engine is designed such that the blend of the two standard fuels can be varied to any percent from all iso-octane to all n-heptane.

— The blend of fuels is varied until the same knock characteristics are observed as with the test fuel.

The percent of iso-octane in the fuel blend is the ON given to the test fuel. For instance, a fuel that has the same knock characteristics as a blend of 85% iso-octane and 15% n-heptane would have an ON of 85.

The relationship between octane number and compression ratio is approximately shown in Fig. 2.107.

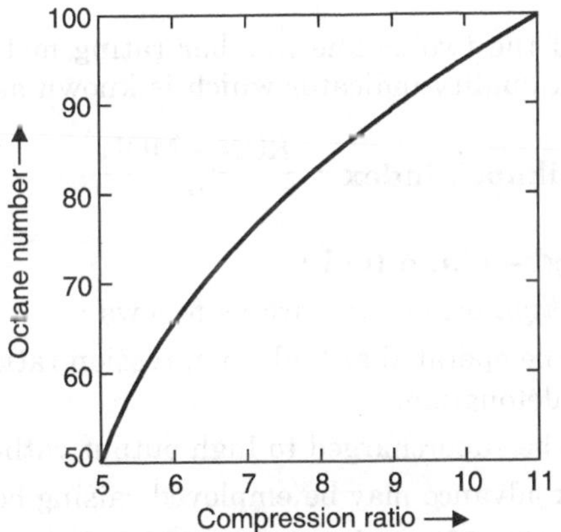

Fig. 2.107. Relationship between octane number and the highest useful compression ratio.

- There are several different tests used for rating octane numbers, each of which will give a slightly different ON value. The *two common methods* of rating gasoline and other automobile S.I. fuels are the **Motor method** and the **Research method.** These give the *Motor Octane Number* (MON) and Research Octane Number (RON). Another *less common method* is the **Aviation method** which is used for aircraft fuel and gives an Aviation Octane Number (AON).

  — The engine used to measure MON and RON was developed in the 1930s. It is a single-cylinder, overhead valve engine that operates on the 4-stroke Otto cycle. It has a variable compression ratio which can be adjusted from 3 to 30.

Research Octave Number (RON) method :

— This method measures anti-knock performance under *relatively mild operating conditions.*

— It is considered to be similar to the detonation tendency of a fuel when the engine is *accelerating from 'low speed' in top gear with a wide open throttle under 'medium load'*. Inlet air temperature = 52°C ; coolant temperature = 100°C ; engine speed = 600 r.p.m. ; ignition advance setting = 13°C BTDC.

Motor Octane Number (MON) method :

— This method measures anti-knock performance under *relatively severe operating conditions.*

— It is considered to be similar to the detonation tendency of a fuel when the engine is driven at *'medium speed'* in top gear with a wide-open throttle under *'heavy load'*. Inlet air temperature = 150°C ; coolant temperature = 100°C ; engine speed = 900 r.p.m., ignition advance setting = 19–26° BTDC.

Fuel sensitivity (FS)

- *The difference in octane number between the Research method and the Motor method octane numbers is known as the fuel* ***sensitivity*** *;* thus

Fuel sensitivity = RON – MON

- Fuel sensitivity is a *good measure of how sensitive knock characteristics of a fuel will be to engine geometry.* A low FS number will usually mean that knock characteristics of that fuel are insensitive to engine geometry. FS numbers generally range from 0 to 10.

Antiknock index

- The average of the two octane number rating methods, RON and MON, is very good antiknock quality indicator which is known as the **antiknock index** ; thus

$$\textbf{Antiknock index} = \frac{\text{RON} + \text{MON}}{2}$$

Advantages of high-octane fuel :

The *advantages of high-octane fuel* are as follows :

1. The engine can be operated at high compression ratio and therefore, with high efficiency without detonation.
2. The engine can be supercharged to high output without detonation.
3. Optimum spark advance may be employed raising both power and efficiency.

- High octane fuels (upto 100) can be produced by *refining techniques*, but it is done more cheaply, and more frequently, by the use of anti-knock *additives,* such *as tetraethyl lead.* (An addition of 1.1 cm^3 of tetraethyl lead to one litre of 80 octane petrol increases the octane number to 90). Fuels have been developed which have a higher anti-knock rating than iso-octane and this has lead to an extension of the octane scale.

2.42.9. Miscellaneous Properties of S.I. Engine Fuels

Miscellaneous properties of S.I. fuels are described below :

1. **Gum content :**

- There is a tendency in some gasolines to deposit gum, a solid oxidation product, in fuel systems and on valve guides. Excessive gum formation often causes sticking of valves and plugging of fuel passages.
- The gum formation is reduced by mixing *inhibitors* (special chemicals) with gasoline.
- The *oxidised gasoline shows a loss of anti-knock quality.*

2. **Sulphur content :**

The presence of sulphur content in gasolines is objectionable since it may lead to the formation of sulphuric acid in the presence of moisture. The sulphuric acid has corrosive effect on engine parts.

3. **Tetra-ethyl lead :**

- It causes deposits on cylinder walls, spark-plug and valves etc. which lead to the corrosion of spark-plug and exhaust valves. These troubles are minimised by adding ethylene-dibromide ($C_2H_4Br_2$).
- It is a very dangerous poison acting on the skin and in vapour form, the lungs.

2.42.10. Diesel Fuel :

- Diesel fuel (diesel oil, fuel oil) is obtainable over a large range of molecular weights and physical properties. It is classified by various methods, some using *numerical scales* and some designating it for *various uses*. Generally speaking, the *greater the refining done on a sample of fuel, the lower is its molecular weight, the lower is its viscosity, and the greater is its cost.*
  - — *"Numerical scales"* usually range from 1 to 5 or 6, with sub categories using alphabetical letters (*e.g.*, A1, 2D, etc.). *The lowest numbers have the lowest molecular weights and lowest viscosity. These are the fuels typically used in C.I. engines. Fuels with the largest numbers are very viscous and can only be used in large, massive heating units.* Each classification has acceptable limits set on various physical properties, such as viscosity, flask point, power point, cetane number, sulphur content etc.
  - — *Another method of classifying diesel fuel* to used in I.C. engines is to *designate it for its intended use.* These designations include, bus, truck, railroad, marine and stationary fuel, going from lowest molecular weight to highest.
- For convenience, diesel fuels for I.C. engines can be divided into two *extreme categories* :
  (*i*) Light diesel fuel (molecular weight 170 appr.)
  (*ii*) Heavy diesel fuel (molecular weight 200 appr.)

  Most diesel fuel used in engines will fit in this range.
  - — *Light diesel fuel* will be less viscous and easier to pump, will generally inject into smaller droplets, and will be more costly.
  - — *Heavy diesel* fuel can generally be used in larger engines with higher injection pressures and heated intake systems.

Often an automobile or light truck can use a less costly heavier fuel in the summer, but must change to a lighter, less viscous fuel in cold weather because of cold starting and fuel line pumping problems.

2.42.10.1. Cetane Number (CN)

- *The cetane number of a diesel fuel is a measure of its ignition quality.* When a fuel is injected into the hot compressed air in the cylinder, it must first be raised to a temperature high enough to ignite the air-fuel mixture. This requires a certain amount of time, known as *ignition delay.*
- Though ignition delay is affected by several engine design parameters such as compression ratio, injection rate, injection time, inlet air temperature etc., it is also dependent on hydrocarbon composition of the fuel and to some extent on its volatility characteristic.
- *The cetane number is a numerical measure of the influence the diesel fuel has in determining the ignition delay.*

- *Higher the cetane rating of the fuel lesser is the propensity for diesel knock.*
- Ignition quality is usually determined by an **engine bench test** which *measures the ignition time delay under standard carefully controlled conditions.*
  - — In such a test, the unknown fuel is rated on a scale between 0 and 100 against a pair of pure hydrocarbon reference fuels. **Cetane** ($C_{16}H_{34}$) (*n*-hexadecane) a straight chain paraffin which has a *very high ignition quality (short delay)* and *does not readily knock*, is assigned to the top of the scale by a *cetane number of 100*, whereas **heptamethylnonane** (HMN) which has a very low *ignition quality (long delay)* and *readily knocks*, is represented at the bottom end of the scale by a cetane number of 15. *Originally*, the low ignition quality reference fuel was *alpha methyl napthalane* ($C_{11}H_{10}$) which was given a cetane number of zero. However, *heptamethylnonane, a, more stable compound but with a slightly better ignition quality (CN = 15), now replaces it.*

    Hence, the cetane number (CN) is shown by,

    $$\text{Cetane number (CN)} = \text{Percent cetane} \quad 0.15 \times \text{percent heptamethylnonane}$$
  - — A standard single-cylinder pre-chamber variable compression ratio engine is used operating under fixed conditions : Inlet temperature = 65.5°C ; Jacket temperature = 100°C ; Speed = 900 r.p.m. ; Injection timing = 13° BTDC ; Injection pressure = 103.5 bar. The *engine is run on a supply of commercial fuel of unknown cetane number under standard operating conditions.* With the injection timing fixed to 13° BTDC, the compression ratio is varied until combustion commences at TDC (by observing the rapid rise in cylinder pressure) thereby producing a 13° delay period of 0.0024s at 900 r.p.m.. A *selection of reference fuel blends are then tested, where again the compression ratio is adjusted for each blend to obtain the standard 13° delay period. The percentage of cetane is one of the blends of reference fuels which gives exactly the same ignition delay (ignition quality) when subjected to the same compression ratio is called the* **cetane number** *of the fuel.* Thus, a commercial 40 cetane fuel would have an ignition delay performance equivalent to that of a blend of 40% cetane and 60% heptamethylnonane (HMN) by volume.
- *For higher speed engines the cetane number required is 50, for medium speed engine about 40, and for slow speed engines about 30.*
- Cetane number is the most important single fuel property which affects the exhaust emissions, noise and startability of a diesel engine. In general, *lower the cetane number higher are the hydrocarbon emissions and noise levels.* Low cetane fuels increase ignition delay so that start of combustion is near to top dead centre. This is similar to retarding of injection timing which is also known to result in higher hydrocarbon levels.
- In general, a *high octane value implies a low cetane value.*
- The relation between cetane number and delay period for a particular engine at a particular set of running conditions is illustrated in Fig. 2.108.

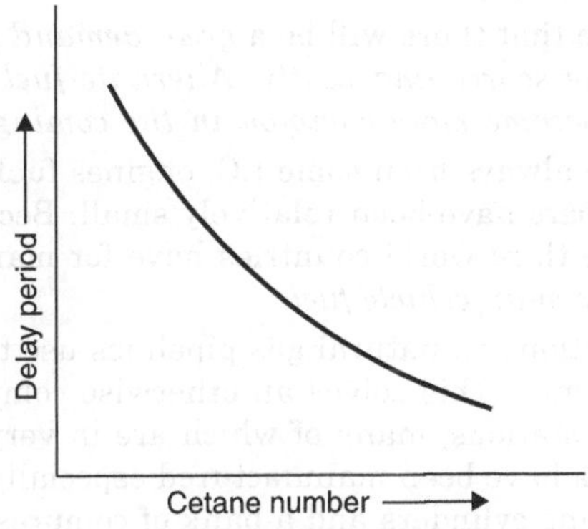

Fig. 2.108

- An approximate inverse relationship between cetane (CN) and octane (ON) number is given as :

$$CN = 60 - \frac{ON}{2}$$

(accurate within ± 5%)

2.42.10.2. Diesel Index (DI)

- The diesel index is a *cheap method of predicting ignition quality.*
- This scale is made possible because ignition quality is sensitive to hydrocarbon compositions ; that is, *paraffins have high ignition quality and aromatic and compounds have low ignition quality.*
- The *diesel index gives an indication of the ignition quality obtained from certain physical characteristics of the fuel as opposed to an actual determination in a test engine.* The index derived from the knowledge of aniline point American Petroleum Institute (API) gravity, which is put together as follows :

$$\text{Diesel index (DI)} = \text{Aniline point (°F)} \times \frac{\text{API gravity (deg)}}{100}$$

— The **aniline point of the fuel** is *the temperature at which equal parts of fuel and pure aniline dissolve in each other.* It therefore gives an indication of the chemical composition of the fuel since the more paraffinic the fuel the higher the solution temperature. Likewise, a higher API gravity reflects a low specific gravity and indicates a high paraffinic content, which corresponds to a good ignition quality.

Note. The correlation between the diesel-index and cetane number is *not exact* and *with certain fuel consumption it is not reliable but, nevertheless, it can be a useful indicator for estimating ignition quality.*

2.43. ALTERNATIVE FUELS FOR I.C. ENGINES

2.43.1. General Aspects

- The *crude oil and petroleum products,* sometimes during the 21st century will *become very scarce and costly to find and produce.* At the same time, there will likely be an increase in the number of automobiles and other I.C. engines. Although fuel economy of engines is greatly improved from the past and will probably continue to be improved, numbers only dictate that there will be a *great demand for fuel in the coming decades. Gasoline will become scarce and costly. Alternate fuel technology, availability, and use must and will become more common in the coming decades.*

- Although there have always been some I.C. engines fueled with non-gasoline or diesel oil fuels, there numbers have been relatively small. Because of the high cost of petroleum products, some third-world countries have for many years been using *manufactured alcohol as their main vehicle fuel.*
- Several pumping stations on natural gas pipelines use the pipeline gas to fuel the engines driving the pumps. This solves an otherwise complicated problem of delivering fuel to the pumping stations, many of which are in very isolated regions. Some large displacement engines have been manufactured especially for pipeline work. These consist of a bank of engine cylinders and a bank of compressor cylinders connected to the same crankshaft and contained in a single engine block similar to a V-style engine.
- *Another reason motivating the development of alternative fuels for the I.C. engine is concern over the emission problems of gasoline engines.* Combined with other air polluting systems, the large number of automobiles is a major contributor to the air quality problem of the world. Vast improvements have been made in reducing emissions given off by an automobile engine.
- Still another reason for alternative fuel development in India and other industrialised countries is the fact that a large percentage of crude oil must be imported from other countries which control the large oil fields.

Some **alternative fuels** *which can replace conventional fuels in I.C. engines are :*

(*i*) Alcohol (Methyl and ethyl)
(*ii*) Hydrogen
(*iii*) Natural gas
(*iv*) LPG (Liquified petroleum gas) and LNG (Liquified natural gas)
(*v*) Biogas.

2.43.2. Advantages and Disadvantages of using Alternative Fuels

Advantages of using alternative fuels :

1. Alcohols can be produced from highly reliable and long-lasting raw material sources like sugarcane, starchy materials, corns, potatoes etc. Thus they are renewable energy sources.
2. Biogas plants can be conveniently and economically installed in village and farms.
3. Manufacture of biogas from cow-dung will give, as a byproduct, manure of very high quality for use in the-farms.
4. Natural gas is available in plenty.
5. Coal gas is produced by coal gasification, and coal is abundant to last much longer than the liquid petroleums.
   — *Alcohols, biogas and natural gas have much higher octane number, and are suitable for use in S.I. engines with little modification. There is less problem of knocking or detonation. Higher compression ratios can be used to give more power and increased thermal efficiency.*
6. The exhaust from engines using gaseous fuels contains less pollutants. Use of hydrogen gives absolutely clean exhaust.

Disadvantages :

Use of alcohols as an alternative fuels have the following *disadvantages :*

1. Social problems due to prohibition, as it can be consumed as liquor by human beings.

2. The carburettor would *need modification*, as the stoichiometric *air-fuel ratio with alcohols is quite low* (of the order of 10 : 1)
3. The calorific value of alcohol fuels is *low* as compared to that of diesel or petrol.
   — The *m.e.p.* and power output from a given size engine will be low with coal *gas and biogas.*
   — The handling and transportation of natural gas is costly.
   — High compressor power is required to compress it for storage otherwise it will need large storage space.
   — Hydrogen is highly explosive, and its handling is risky.

2.43.3. Alcohol

Alcohol is an attractive alternative fuel because it can be obtained from a number of sources, both natural and manufactured. *Methanol* (methyl alcohol) and *ethanol* (ethyl alcohol) are two kinds of alcohol that seem most promising and have had the most development as engine fuel.

Advantages :

The advantages of alcohol as a fuel are :

1. It is high octane fuel with anti-knock index numbers (octane number on fuel pump) of over 100. High octane numbers result, at least in part, from the high flame speed of alcohol.
   — Engines using high-octane fuel can run more efficiently by using higher compression ratios.
2. It can be obtained from a number of sources, both natural and manufactured.
3. It has high evaporative cooling (h_{fg}) which results in *cooler intake process and compression stroke.* This *raises the volumetric efficiency of the engine and reduces the required work input in the compression stroke.*
4. Generally *less overall emissions* when compared with gasoline.
5. Low sulphur content in the fuel.
6. When burned, it forms *more moles of exhaust which gives higher pressure and more power in the expansion stroke.*
7. The contamination of matter in alcohols is less dangerous than petrol or diesel because alcohols are less toxic to humans and has a recognizable taste.

Disadvantages :

1. *Low energy content* of the fuel (Almost twice as much alcohol as gasoline must be burned to give the same energy input to the engine).
2. The exhaust contains more aldehydes. If as much alcohol fuel was consumed as gasoline, *aldehyde emissions would be a serious exhaust pollutions problem.*
3. As compared to gasoline, alcohol is much more corrosive on copper, brass, aluminium, rubber and many plastics. This puts some restriction on the design and manufacturing of engines to be used with this fuel.
   — Methanol is very corrosive on metals.
4. In general, the ignition characteristics are poor.
5. Vapour lock in fuel delivery system.
6. Owing to low vapour pressure and evaporation, the cold weathering starting characteristics are poor.
7. Due to low vapour pressure, there is a danger of storage tank flammability. Air can leak into storage tanks and create a combustible mixture.

8. Alcohols have almost invisible flames, which is considered dangerous when handling fuel. Again, a small amount of gasoline remove this danger.
9. Low flame temperatures generate less NO_x, but the resulting lower exhaust temperatures take longer to heat the catalytic converter to an efficient operating temperature.
10. When refuelling an automobile, headaches and dizziness have been experienced (due to the strong odour of alcohol).

Note :

- Alcohols are considered as clean burning renewable alternative fuels which can come to our rescue to meet the challenge of vehicular fuel oil scarcity and fouling of environment by exhaust emissions.
- Alcohols make very poor diesel engine fuels as their 'cetane number' is considerably lower.
  — Alcohols can be used in dual fuel engines or with assisted ignition in diesel engine. In a dual fuel mode, alcohol is inducted along with the air, compressed and then ignited by a pilot spray of diesel oil.

2.43.3.1. Methanol

- Of all the fuels being considered as an alternate to gasoline, methanol is one of the more promising and has experienced major research and development.
- Methanol can be obtained from many sources, both *fossil* and *renewable.* These include *coal, petroleum, natural gas, biomass, wood, landfills, and even the ocean.* However, any source that requires extensive manufacturing or processing raises the price of the fuel and requires an energy input back into the overall environmental picture, both unattractive.
- Methanol behaves much like petroleum and so, it can be stored and shifted in the same manner.
  — It is more flexible fuel than hydrocarbon fuels permitting wider variation from ideal A/F ratios.
  — It has relatively good lean combustion characteristics compared to hydrocarbon fuels. *Its wider inflammability limits and higher flame speeds have showed higher thermal efficiency and lesser exhaust emissions compared with petrol engines.*
- Depending on gasoline-methanol mixture, some changes in fuel supply are essential. Simple modifications to the carburettor or fuel injection can allow methanol to replace petrol easily.

Some important features of methanol as fuel :

1. The specific heat consumption with methanol as fuel is 50 percent less than petrol engine.
2. Exhaust CO and HC are decreased continuously with blends containing higher and higher percentage of methanol. But exhaust aldehyde concentration shows a reversed trend.
3. Methanol can be used as supplementary fuel in heavy vehicles powered by C.I. engines with consequent saving in diesel oil and reduced exhaust pollution.

Advantages of methanol :

1. Owing to its *excellent anti-knock characteristics, it is much suitable for S.I. engines.*
2. Methanol use maintains good air quality (Methanol emits less amount of CO_2 and other polluting gases as compared to gasoline fueled vehicles).

3. Tertiary butial alcohol is used as an octane improving agent.
4. One percent methanol in petrol is used to prevent freezing of fuel in winter.
5. Iso-propyl alcohol is used as anti-icing agent in carburettor.
6. Addition of methanol causes methanol gasoline blend to evaporate at much faster rate than pure gasoline below its boiling point.

Performance of methanol as I.C. engine fuel :

The performance characteristics are considered as :

— Thermal efficiency verses A/F ratio ;
— Effect of speed on power output and specific heat consumption with petrol engine ;
— Effect of A/F ratio on exhaust emission.

- Fig. 2.109 (*a*) and (*b*) shows the effect of A/F ratio and speed on brake power.

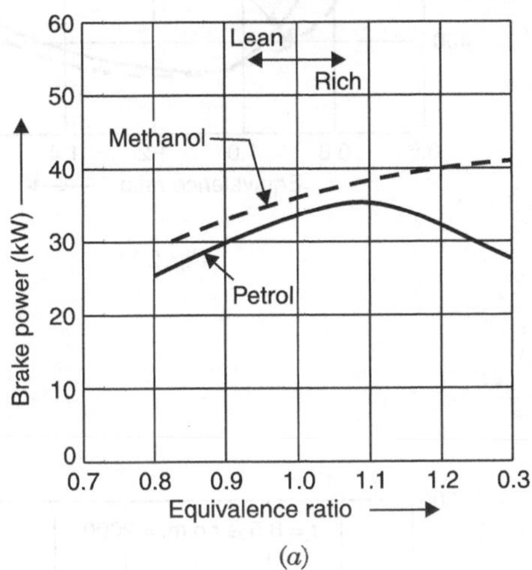

(*a*)

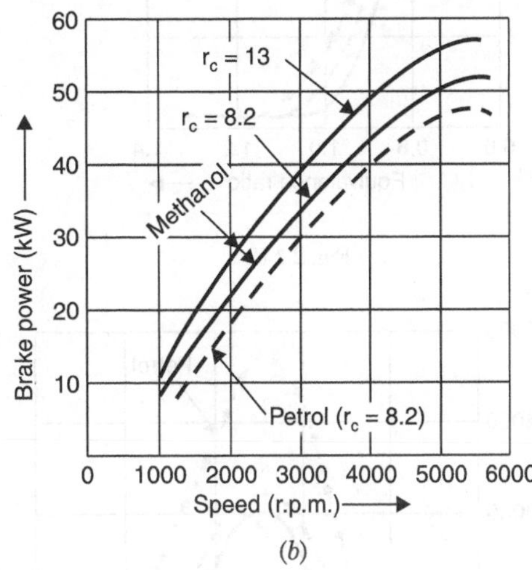

(*b*)

Fig. 2.109. Effect of A/F ratio and speed on brake power.

- Fig. 2.110 shows the effect of load on specific heat consumption.
- Fig. 2.111 shows the effect of speed on volumetric efficiency.

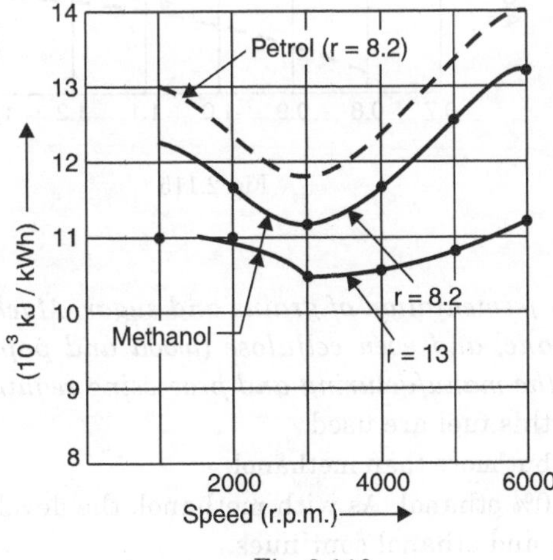

Fig. 2.110

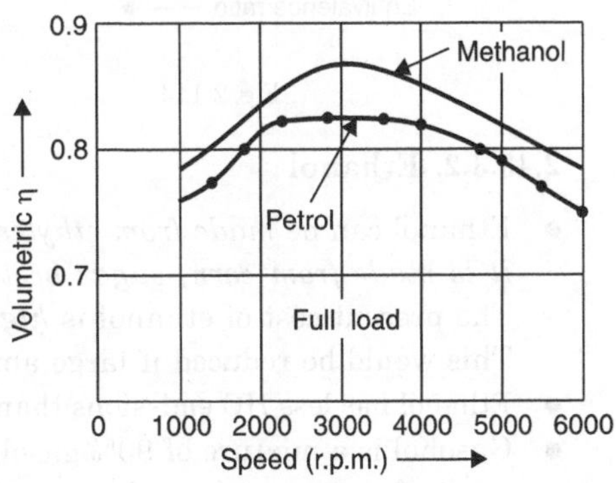

Fig. 2.111

- **Fig. 2.112 to 2.113 show the effect of equivalence ratio on all important objectionable emissions, CO, HC, NO_x and aldehydes respectively, for Petrol and Methanol.**

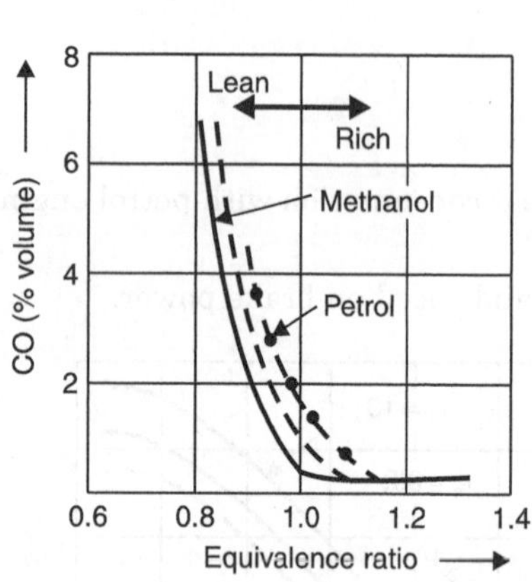

Fig. 2.112

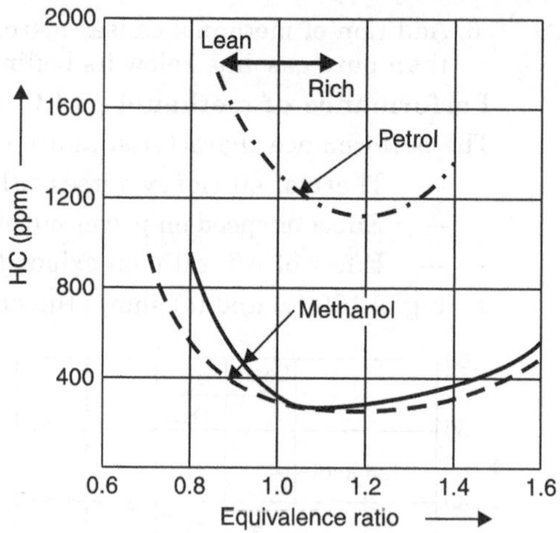

Fig. 2.113

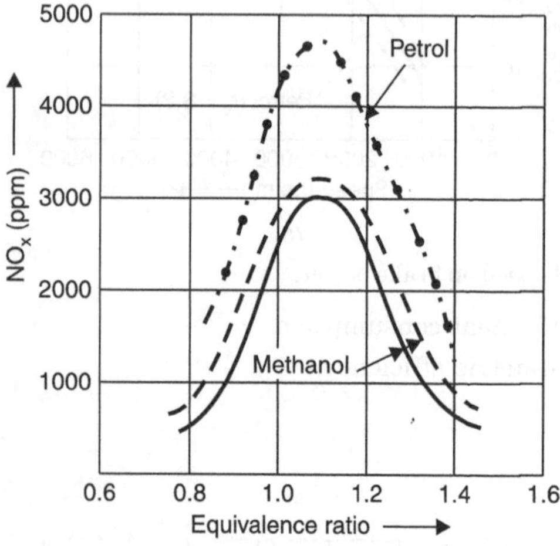

Fig. 2.114

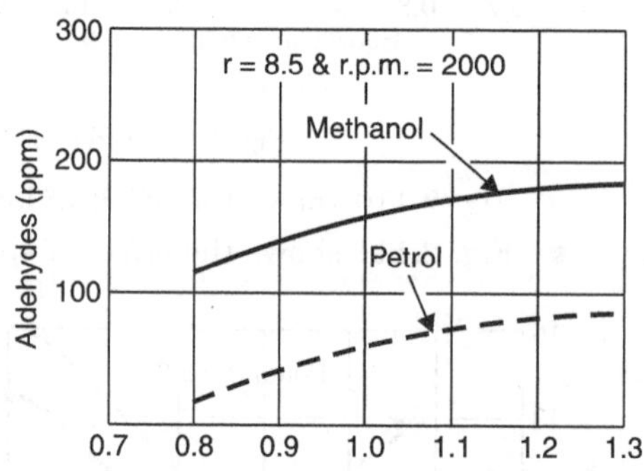

Fig. 2.115

2.43.3.2. Ethanol

- Ethanol can be *made from ethylene or from fermentation of grains and sugar. Much of it is made from corn, sugar beets, sugarcane, and even cellulose (wood and paper).* The present cost of ethanol is *high due to the manufacturing and processing required.* This would be reduced if large amounts of this fuel are used.
- Ethanol has less HC emissions than gasoline but more than methanol.
- Gasohol is a mixture of 90% gasoline and 10% ethanol. As with methanol, the development of systems using mixtures of gasoline and ethanol continues.

— Two mixture combinations that are important are E 85 (85% ethanol) and E10 (gasohol). E 85 is basically an alcohol fuel with 15% gasoline added to eliminate some of the problems of pure alcohol (*i.e.*, cold starting, tank flammability, etc.). E 10 reduces the use of gasoline with no modification needed to the automobile engine. Flexible-fuel engines are being tested which can operate on any ratio of ethanol-gasoline.

Performance of engine using ethanol :

The effect of speed on power output, brake specific heat consumption and thermal efficiency of an engine using ethanol is compared with gasoline engine as shown in Fig. 2.116 to 2.118.

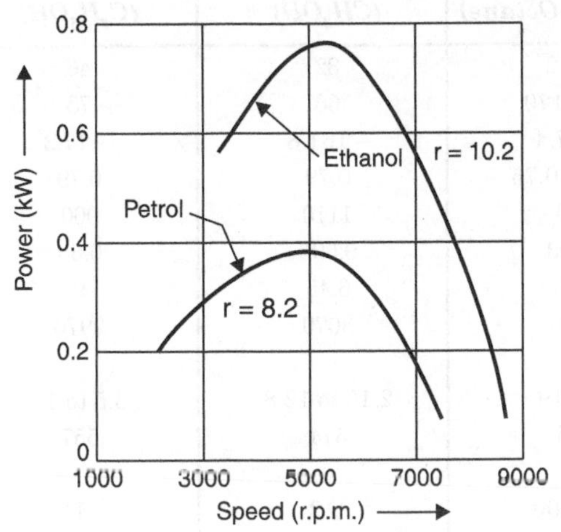

Fig. 2.116

Fig. 2.117

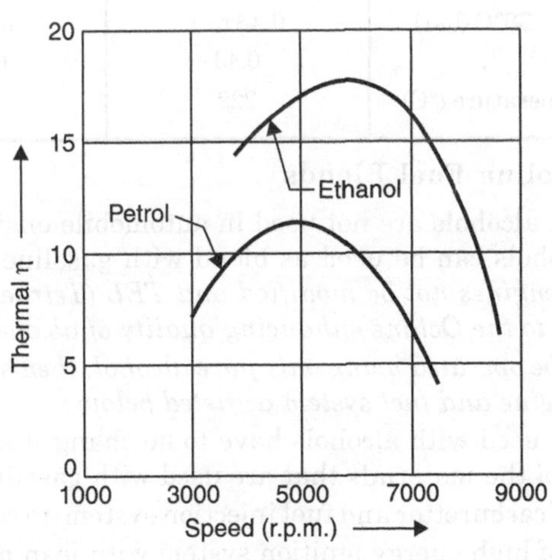

Fig. 2.118

- The *power output* of ethanol engine is *higher* compared to gasoline engine at all speeds.
- The *brake specific heat consumption* is *improved* with ethanol engine compared with petrol engine.

- The *maximum thermal efficiency* of ethanol engine is *higher* than petrol engine. The efficiency curve of ethanol engine is flat for a wide range of speed which indicates that the part load efficiency is much compared with petrol engine.
- The *engine torque* is *considerably higher* for ethanol as compared to petrol engine.

2.43.3.3. Properties of methanol and ethanol

The physical and chemical properties of gasoline, methanol and ethanol are listed in the table below.

Table. Important properties of gasoline, methanol, ethanol

S. No.	*Property*	*Gasoline nearly C_8H_{18} (Iso-Octane)*	*Methyl alcohol (CH_3OH)*	*Ethyl-alcohol (C_2H_5OH)*
1.	Molecular weight	114.2	32	46
2.	Boiling point at 1 bar °C	43 to 170	66	78
3.	Freezing point °C	– 107.4	– 161.8	–117.2
4.	Specific gravity (150°C)	0.72 to 0.75	0.79	0.79
5.	Latent heat (kJ/kg)	400	1110	900
6.	Viscosity (centipoise)	0.503	0.596	0.60
7.	Stoichiometric A : F (ratio)	14.6	6.45	9
8.	Mixture heating value (kJ/kg) (for stoic-mixture)	2930	3070	2970
9.	Ignition Limits (A/F)	8 to 19	2.15 to 12.8	3.5 to 17
10.	Self ignition Temp.	335	574	557
11.	Octane Number			
	(*a*) Research	80 to 90	112	111
	(*b*) Motor	85	91	92
12.	Cetane Number	15	3	8
13.	Lower C.V. (kJ/kg)	44100	19740	26880
14.	Vapour pressure at 38°C (bar)	0.48 to 1	0.313	0.17
15.	Flame speed (m/sec)	0.43	0.76	—
16.	Auto-ignition temperature (°C)	222	467	—

2.43.4. Alcohol-Gasoline Fuel Blends

- Normally straight alcohols are not used in automobile engines except methanol in racing cars. The alcohols can be used as blend with gasoline as this has the advantages that the *existing engines not be modified and TEL (Tetraethyl lead) can be eliminated from gasoline due to the Octane enhancing quality of alcohol.*
- If *the engine is to be operated using only pure alcohol, then some major modifications are required in the engine and fuel system as listed below :*
  - (*i*) The materials used with alcohols have to be changed since both alcohols are corrosive to many of the materials that are used with gasoline.
  - (*ii*) Adjustment of carburettor and fuel injection system to compensate for leaning effect.
  - (*iii*) Introduction of high energy ignition system with lean mixture.
  - (*iv*) Alteration in fuel pump and circulation system to avoid vapour lock as methanol vapourisation rate is very high.
  - (*v*) Increase in compression ratio to make better antiknock properties of the fuel.
  - (*vi*) Addition of detergent and volatile primers to reduce engine deposits and assit in cold starting.

(*vii*) Use of cooler running spark-plugs for avoiding pre-ignition.

- The *blends* have the following important *advantages* over pure ethanol :
  (*i*) The engine can be started easily.
  (*ii*) No abnormal corrosion compared with pure ethanol.
  (*iii*) Lubrication in petrol and alcohol blend is more or less same.
  (*iv*) Some benzene is added to prevent the separation of layers of petrol and alcohol.

When blends are used, the following minor modifications in engine are required :

(*i*) The carburettor jet needs to be increased to increase the flow 1.56 times that of petrol.

(*ii*) The float is to be weighed down, to correct level due to higher specific gravity.

(*iii*) Modification of air inlet to get less air as blend requires less air for complete combustion than petrol.

(*iv*) Provision of a specific arrangement of heating the carburettor and intake manifold as lower vapour pressure of alcohol makes the starting difficult below 70°C.

- Fig. 2.119 shows the change in distillation by admixture of methanol and ethanol.
- Fig. 2.120 shows the leaning effect of alcohol admixture to petrol.
- Fig. 2.121 shows the antiknock quality of methanol-gasoline blends.

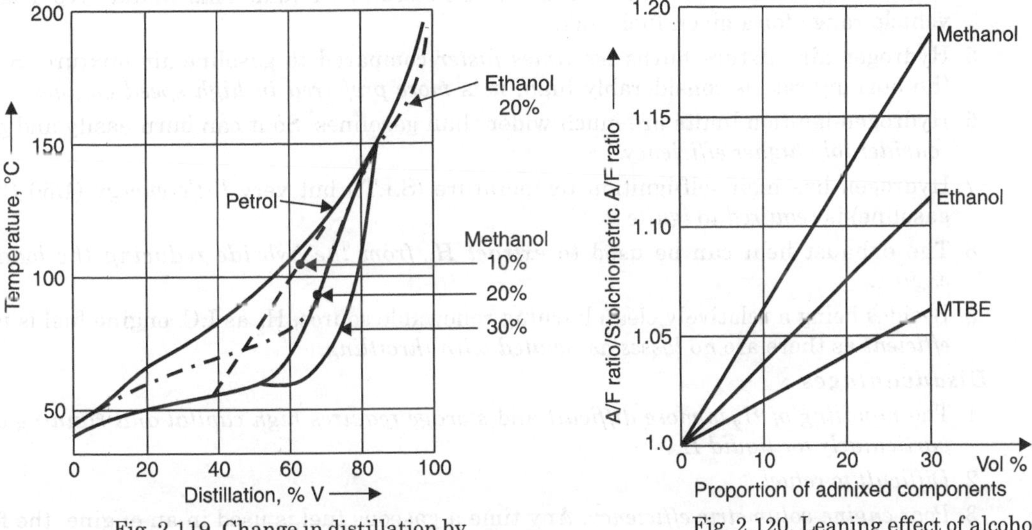

Fig. 2.119. Change in distillation by admixture of methanol and ethanol.

Fig. 2.120. Leaning effect of alcohol admixture to pertol.

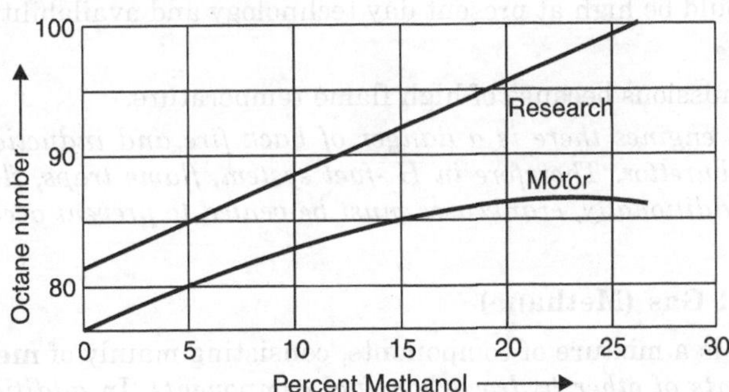

Fig. 2.121. Antiknock quality of methanol-gasoline blends.

2.43.5. Hydrogen

A number of companies have built automobiles with prototype or modified engines which operate on hydrogen.

— A H_2-powered car being developed in Melbourne University Department of Mechanical Engineering is achieving 40 percent energy saving over conventional petrol engine. The car is a converted Ford Cortina Wagon which carries enough fuel in 4-cylinders to travel upto 50 km at a speed of 130 km/h.

— German cars are so developed that they can be converted for driving either gasoline or H_2 by making a few simple adjustments. Presently a few vehicles are running on road as there are very few public filling stations for liquid H_2. Many more are planned for future.

Advantages of hydrogen as I.C. engine fuel :

1. *Low emissions.* Essentially no CO or HC in the exhaust as there is no carbon in the fuel. Most exhaust would be H_2O and N_2.
2. *Fuel availability.* There are a number of different ways of making hydrogen, including electrolysis of water.
3. Fuel leakage to environment is *not a pollutant.*
4. High energy content per unit volume when stored as liquid. This would give a large vehicle range for a given fuel tank.
5. Hydrogen-air mixture burns *ten-times faster* compared to gasoline-air mixture. Since the burning rate is considerably high, it is *more preferred in high speed engines.*
6. Hydrogen-ignition limits are much wider than gasolines. So it can burn easily and *give considerably higher efficiency.*
7. Hydrogen has high self-ignition temperature (S.I.T.) but very *little energy* (1/50 th of gasoline) is *required to ignite it.*
8. The exhaust heat can be used to *extract H_2 from the hybride reducing the load of engine.*
9. Besides being a relatively clean burning renewable source, H_2 as I.C. engine fuel is *very efficient* as there are *no losses associated with throttling.*

Disadvantages :

1. The *handling of H_2 is more difficult* and *storage requires high capital and running cost particularly for liquid H_2.*
2. *Difficult to refuel.*
3. *Poor engine volumetric efficiency.* Any time a gaseous fuel is used in an engine, the fuel will displace some of the inlet air and poorer volumetric efficiency will result.
4. Fuel cost would be high at present day technology and availability.
5. Can detonate.
6. High NO_x emissions because of high flame temperature.
7. *In hydrogen engines there is a danger of back fire and induction ignition which can melt the carburettor. Therefore in H_2-fuel system, flame traps, flask back arresters are necessary. Additionally, crankcases must be vented to prevent accumulation of explosive mixtures.*

2.43.6. Natural Gas (Methane)

- *Natural gas* is a mixture of components, consisting mainly of **methane** (60–98%) with *small amounts of other hydrocarbon fuel components.* In *addition it contains various* *amounts of N_2, CO_2, He and traces of other gases.* Its sulphur content ranges from very little (sweet) to larger amounts (sour).

An *ideal composition of CNG* as an automative fuel is as follows :

Methane = 90% (minimum) ; *Ethane content* = 4% (maximum) ; *Propane content* = 1.7% (maximum) ; *C_4 and higher* = 0.7% (maximum) ; *C_6 and higher* = 0.2% (maximum) ; *(CO_2 + N_2)* = 0.2% (maximum) ; *Hydrogen* = 0.1% (maximum) ; *Carbon monoxide* = 0.1% (maximum) ; *Oxygen* = 0.5% (maximum) ; *Sulphur* = 10% ppm (maximum).

- It is stored as **Compressed Natural Gas (CNG)** at pressures of 7 to 21 bar and a temperature around –160°C.
- *As a fuel it works best in an engine system with a single-throttle body fuel injector. This gives a longer mixing time, which is needed by this fuel.*
- Tests using CNG in various sized vehicles continue to be conducted by government agencies and private industry.

Properties of CNG :

The properties of CNG are almost similar to that of methane :

- Methane has very good antiknock qualities which means it *does not ignite readily. Anti-knock Octane number of CNG is nearly 130, so it burns at much higher temperature compared with petrol unleaded (Octane No = 95) and diesel which have low octane number.*
- Owing to better antiknock quality of CNG it can be safely used in engines with a compression ratio as high as 12 : 1 compared with petrol (maximum 10 : 1).
  — *The CNG fuel used engines have higher thermal efficiencies than those fuelled by gasoline.* In addition to this, the reduction in the pollutants emitted by CNG engine is noticeable.
- CNG is *non-toxic* and *lighter than air so when leakage occurs it quickly disappears* unlike gasoline which paddles and evaporates.
- The presence of ethane and propane even in small percentages (5% and 2%, respectively) affect the burning properties of CNG. Both the gases try to lower the Octane characteristics and causes pre-ignition and reduced fuel efficiency.

Advantages of CNG :

(*i*) *High octane number* makes it a *very good S.I. engine fuel.*

(*ii*) *Low engine emissions.* Less aldehydes than with methanols.

(*iii*) It is cheap (It costs about 25 to 50% less than gasoline and more than 50% less than other alternative fuels, such as methanol and ethanol.

(*iv*) It is engine friendly.

(*v*) It is safe in operation.

(*vi*) Fuel fairly abundant world-wide. Natural gas is the second most abundant fuel available in India after coal.

(*vii*) Easy to tap.

(*viii*) It is odourless.

(*ix*) It is clean.

Disadvantages of CNG :

(*i*) Low energy density resulting in low engine performance.

(*ii*) Low engine volumetric efficiency because it is a gaseous fuel.

(*iii*) Need for large pressurised fuel storage tank.

(*iv*) Inconsistent fuel properties.

(*v*) Refueling is a slow process.

(*vi*) The storage cylinder takes a lot of space as the gas once filled has to travel at least of 400 km. But now a days there are byfuel and dual-fuel engines which can run on CNG and other fuel.

2.43.7. LPG and LNG

- LPG (*Liquified Petroleum Gas*) is mainly *propane* but may also contain a small proportion of butane and possibly, some ethane and a little pentane in heavier vapour form. The heavier fractions tend to occur in LPG produced by distillation of crude oil.
  - — Propane has a higher octane number, burns more clearly and saves on maintenance costs.
  - — Propane is gaining as a gasoline substitute because it costs 60% of petrol and gives 90% mileage of its fellow gasoline.
- LNG (*Liquified Natural Gas*) comes from dry natural reservoirs *mainly* CH_4 with very small percentages of ethane and propane.
  - — The major difficulty encountered in the use of this gas is its boiling temperature – 161.5°C.

2.43.8. Biogas

2.43.8.1. Introduction

- The biogas is generally produced from by dung from different beasts as *cow, buffalo, goat, sheep, horse, donkey* and *elephant*. Some other sources are :

(*i*) Sewage (*ii*) Crop residue
(*iii*) Vegetable wastes (*iv*) Water hyacinth
(*v*) Alga (*vi*) Poultry droppings
(*vii*) Pig manure (*viii*) Ocean kelp.

- Biogas is produced by *digestion, pyrolysis* or *hydrogasification*. **Digestion** is a biological process that occurs in absence of oxygen and in the presence of anaerobic organism at ambient pressures and temperatures of 35–70°C. The container in which digestion takes place is known as the *digester*. Biogas plants have been built in various designs.

2.43.8.2. Composition and Properties of Biogas

- Its main combustible component is CH_4 and another major component is CO_2 *which reduces its octane number*. The components of biogas with composition are given below :

Component	*Composition (% volume)*
CH_4	50—60
CO_2	30—45
H_2 and N_2	5—10
H_2S and O_2	Traces

Octane rating : 110 with CO_2
130 without CO_2

- Biogas *possesses excellent antiknock properties* with an equivalent Octane number in excess of 120 compared with 87 for regular petrol.

- Its auto-ignition temperature is higher than petrol which makes it a *safer fuel*.
- Being a gas it mixes readily with air even at low temperature, therefore, there is no need to provide rich mixture during starting or idling.
- Although its calorific value is lesser than petrol, it is possible to use higher compression ratio for the same size engine thus making it possible to generate the same amount of power.

Use of biogas in S.I. engines :

S.I. engines can be operated on biogas after starting the engine by using petrol. Biogas can be used in these engines in *two forms* as :

(*i*) To run the engine entirely on biogas.

(*ii*) Dual fuel engine where engine can run on both fuels. (This arrangement is preferred these days).

2.43.8.3. Advantages of using biogas as fuel in C.I. engine

The biogas can be used in C.I. engine as a dual fuel and improves engine performance.

The following are the *advantages* of using biogas as fuel in C.I. engine :

(*i*) A uniform gas-air mixture is available in multi-cylinder engine at all times.

(*ii*) Due to clean operation of the engine there is virtually no CO-emission in exhaust.

(*iii*) When biogas is used as a fuel, NO_x emissions are reduced by about 60 per cent.

(*iv*) Soot is virtually eliminated and exhaust is found to have less pungent odour than that obtained while operating the engine with diesel oil.

VI. Supercharging of I.C. Engines

2.44. PURPOSE OF SUPERCHARGING

The purpose of supercharging is *to raise the volumetric efficiency above that value which can be obtained by normal aspiration.*

- The engine is an air pump. Increasing the air consumption permits greater quantities of fuel to be added, and results in a greater potential output. *The indicated power produced is almost directly proportional to the engine air consumption.* While brake power is not so closely related to air consumption, it is nevertheless, dependent upon the mass of air consumed. It is desirable, then, that the *engine takes in greatest possible mass of air.*
- Three possible methods which might be utilized to increase the air consumption of an engine are :
  1. *Increasing the piston displacement,* but this increases the size and weight of the engine, and *introduces additional cooling problems.*
  2. *Running the engine at higher speeds,* which results in *increased fluid and mechanical friction losses, and imposes greater inertia stresses on engine parts.*
  3. ***Increasing the density of the charge,*** *such that a greater mass of charge is introduced into the same volume or same total piston displacement.*

The *last method* of increasing the air capacity of an engine is *widely used,* and is termed **supercharging.**

Supercharger :

- *The apparatus used to increase the air density is known as a* ***supercharger.*** It is merely a *compressor* which provides a denser charge to the engine, thereby enabling the consumption of a greater mass of charge with the same total piston displacement. During the process of compressing the charge, *the supercharger produces the following effects :*
  (*i*) *Provides better mixing of the air-fuel mixture.* The turbulent effect created by the supercharger assists in additional mixing of the fuel and air particles. The arrangement of certain types of superchargers, particularly the *centrifugal type,* also encourages more even distribution of the charge to the cylinders.
  (*ii*) *The temperature of the charge is raised as it is compressed, resulting in a higher temperature within the cylinders.* This is partially beneficial in that it helps to produce better vaporisation of fuel (in case of S.I. engines) but deterimental in that it tends to lessen the density of the charge. The increase in temperature of the charge also effects the detonation of the fuel.
  Supercharging tends to increase the possibility of detonation in a S.I. engine and lessen the possibility in a C.I. engine.
  (*iii*) Power is required to drive the supercharger. This is usually taken from the engine and thereby removes, from over-all engine output, some of the gain in power obtained through supercharging.

Compressors used are of the following three types :

(*i*) ***Positive displacement type*** used with many reciprocating engines in stationary plants, vehicles and marine installations.

(*ii*) ***Axial flow type*** seldom used to supercharge reciprocating engines, it is widely used as the compressor unit of the *gas turbines.*

(*iii*) ***Centrifugal type*** *widely used* as the *supercharger for reciprocating engines, as well as compressor for gas turbines.* It is almost exclusively used as the supercharger with reciprocating power plants for *aircraft because it is relatively light and compact, and produces continuous flow rather than pulsating flow as in some positive displacement types.*

- A *correctly matched supercharger* will raise the cylinder's brake mean effective pressure (b.m.e.p.) *to well above that of a naturally aspirated engine without creating excessively high peak cylinder pressures ; the actual increase in the brake mean effective pressure is basically determined by the level of boost pressure the supercharged system is designed to deliver.*
- *Large commercial vehicle diesel engines are frequently turbocharged, with the objectives of raising b.m.e.p. (and therefore torque and power output) and at the same time reducing the engine's maximum speed.* The *other benefits of raising the cylinder mean pressure and decreasing the engine's limiting speed is that the engine mechanical losses and noise are reduced and there is an improvement in fuel consumption, normally, an added bonus of prolonged engine life expectancy.*

Object of supercharging. The objects of supercharging include one or more of the following :

1. *To increase the power output for a given weight and bulk of the engine* .... relates to *aircraft, marine and automotive engines.*
2. *To compensate for the loss of power due to altitude* .... Relates to *aircraft and other engines which are used at high altitudes.*
3. *To obtain more power from an existing engine.*

Effects of supercharging on performance of the engine :

1. The '*power output*' of a supercharged engine is *higher* than its naturally aspirated counterpart.
2. The '*mechanical efficiencies*' of supercharged engines are *slightly better* than the naturally aspirated engines.
3. Inspite of better mixing and combustion due to reduced delay a mechanically supercharged otto engine almost always have '*specific fuel consumption*' *higher than a naturally aspirated engine.*

2.45. SUPERCHARGING OF S.I. ENGINES

The schematic arrangement for supercharging S.I. engine is shown in Fig. 2.122.

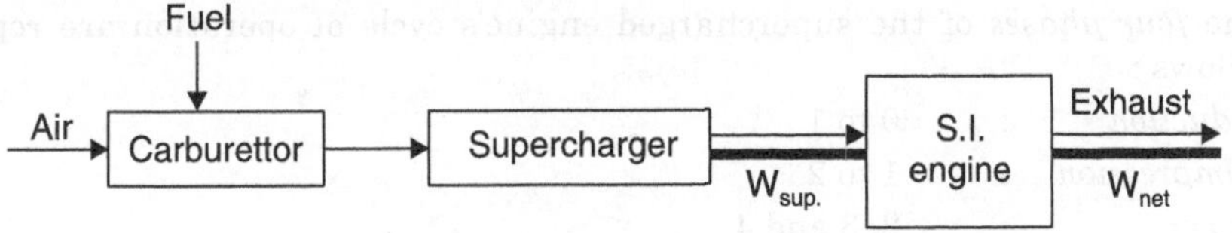

Fig. 2.122

The theoretical operating cycles for both the natural aspirated and supercharged petrol engines can be compared on pressure-volume (p-v) diagrams as shown in Fig. 2.123 and 2.124, respectively. *The larger upper loop is a measure of the positive power developed in the cylinder while the lower loop represents the negative power needed to fill the cylinder with fresh charge.*

2.45.1. Naturally Aspirated Cycle of Operation. Refer Fig. 2.123.

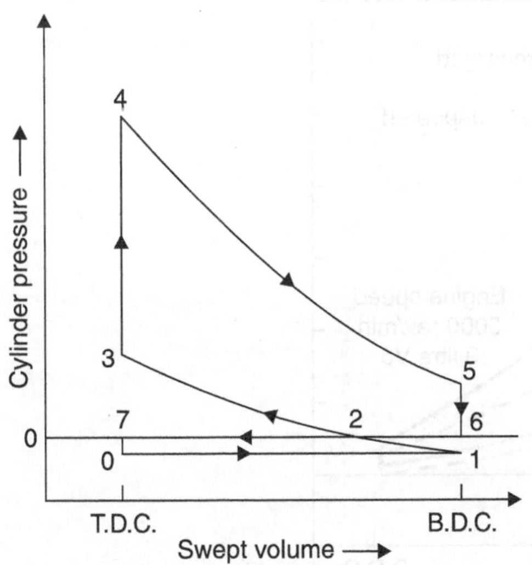

Fig. 2.123. Theoretical naturally aspirated petrol engine (constant volume) pressure-volume diagram.

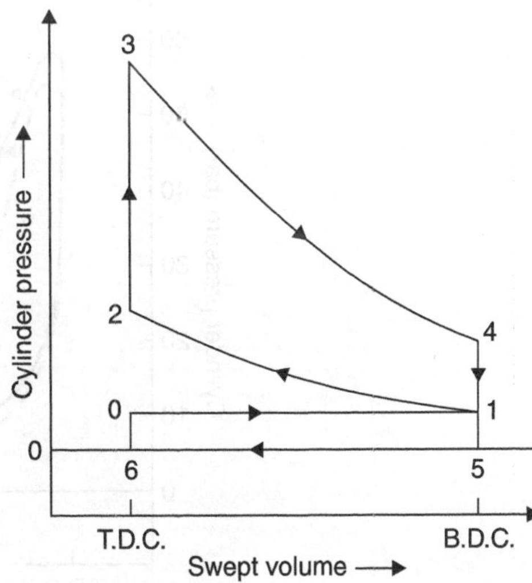

Fig. 2.124. Theoretical supercharged petrol engine (constant volume) pressure-volume diagram.

- The *large area* enclosed in the upper loop (2-3-4-5-6-2) is *proportional to the useful work* performed by combustion on the piston whereas the *small loop area* (0-1-2-7-0), which is below the atmospheric line, is a *measure of the work done in inducing the fresh charge into the cylinder.*

- The *four phases* of the naturally aspirated engine's cycle are represented as follows :

Induction	:	0 to 1
Compression	:	1 to 3
Power	:	3, 4 and 5
Exhaust	:	5, 6 and 7

2.45.2. Supercharged Cycle of Operation. Refer to Fig. 2.124.

- With this pressurised charged system the *large upper loop area* (1-2-3-4-1) represents a measure of the work done in moving the piston to and fro so that the crankshaft rotates. Conversely, the *small loop area* (0-1-5-6-0) *which is above the atmospheric line represents the work done in pumping the fresh charge into the cylinder.*
- The *four phases* of the supercharged engine's cycle of operation are represented as follows :

Induction	:	0 to 1
Compression	:	1 to 2
Power	:	2, 3 and 4
Exhaust	:	4, 5 and 6.

2.45.3. Comparison of Actual Naturally Aspirated and Supercharged Engine Pressure-Volume Diagrams

Fig. 2.125 shows a direct comparison of how the actual cylinder pressure varies relative to the cylinder's swept volume for both a naturally aspirated and a *supercharging petrol engine with wide-open throttle.*

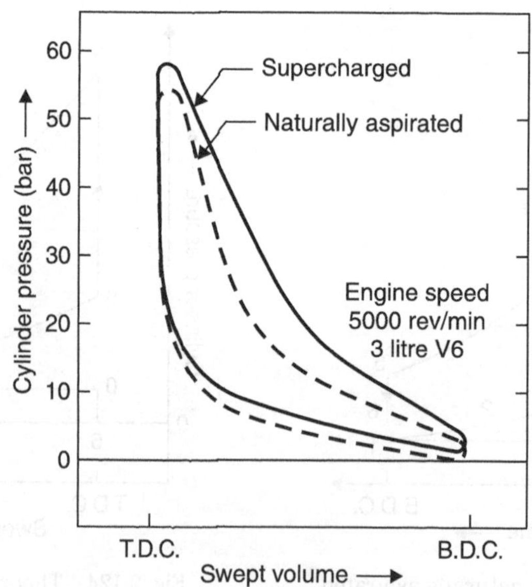

Fig. 2.125. Comparison of actual naturally aspirated and supercharged petrol engine pressure-volume diagrams.

— The two enclosed loops show that the lower compression stroke part of the curves for the supercharged engine is higher than for the naturally aspirated engine, whereas the peak cylinder pressures for the supercharged engine are only marginally higher.

— However, the *main advantage gained by supercharging the cylinders is that the vertical distance between the upper and lower curves for the supercharged engine is greater throughout the cylinder swept volume, which indicates that the mean effective pressure is greater.*

— Finally, the loop area enclosed by the supercharged engine is much larger, the *proportional difference being a measure of the increased power in the supercharged engine.*

2.45.4. Boost Pressure and Pressure Ratio

- **Boost pressure** *refer to the gauge pressure recorded when the air or mixture supply has passed through the supercharger.*
- **Pressure ratio** is the *ratio of absolute pressure to that of the atmospheric pressure.*
  (Here, absolute pressure = boost pressure + atmospheric pressure)

The *intensity of supercharging* can be broadly classified *as follows :*

S. No.	Particulars	Degree of charging			
		Naturally aspirated	Low	Medium	High
1.	*Boost pressure (bar)*	0.0 and below	0.0—0.5	0.5—1.0	1.0 and above
2.	*Pressure ratio*	1.0 and below	1.0—1.5 : 1	1.5—2.0 : 1	2.0 and above

2.45.5. The Effect of Pressure Ratio on Air Charge Temperature

The relationship between supercharger pressure ratio increase and air charge temperature is shown in Fig. 2.126, where it can be seen that as *the boost pressure increases, so does the discharge air temperature.* The lower curve shows the theoretical temperature rise with increased pressure ratio, whereas the band between the lower and upper curves is the *working, temperature variation likely to be encountered due to the turbulence and friction resistance generated by the compression process.*

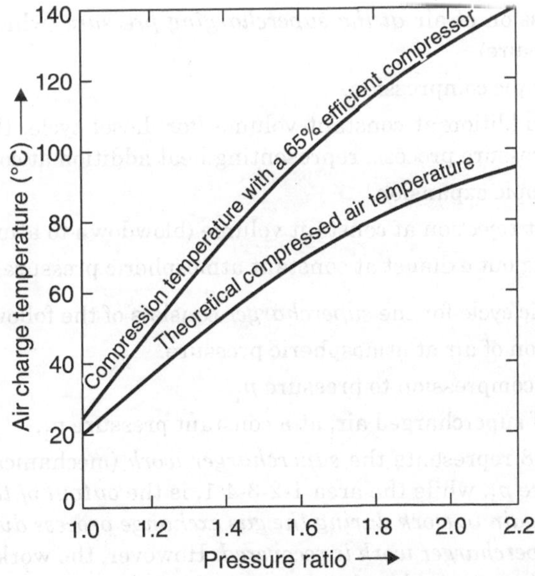

Fig. 2.126. Relationship between charge temperature and boost pressure ratio if heat is not dissipated.

Thus the minimum compression air temperatures with an intake temperature of 20°C for pressure ratios 1.2, 1.4, 1.6, 1.8, 2.0 and 2.2 are 35.7°C, 49.5°C, 62°C, 73.6°C, 84°C and 94°C respectively. In practice, with compressor efficiencies of the order of 60–75% and the churning, turbulence, and frictional factors, the actual output air charge temperature can be considerably higher, particularly at the higher boost levels.

2.45.6. Thermodynamic Cycle and Supercharging Power

Fig. 2.127 shows the thermodynamic cycle of a supercharged I.C. engine on the *p-v* diagram for an *ideal otto cycle.*

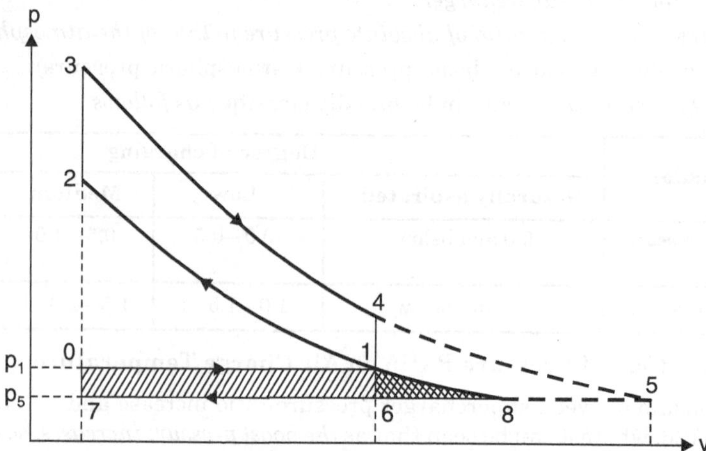

Fig. 2.127. Thermodynamic cycle of supercharged engine on *p-v* diagram for an ideal otto cycle.

- The pressure p_1 represents the *supercharging pressure* and p_5 is the *exhaust pressure.*
- The thermodynamic cycle, consists of the following *processes :*
  - *(i)* **0–1.** Admission of air *at the supercharging pressure* (which is *greater than* atmospheric pressure).
  - *(ii)* **1–2.** Isentropic compression.
  - *(iii)* **2–3.** Heat addition at constant volume (for diesel cycle, this will be replaced by a constant pressure process, representing heat addition at constant pressure.
  - *(iv)* **3–4.** Isentropic expansion.

    4–1–6. Heat rejection at constant volume (blowdown to atmospheric pressure).

    6–7. Driving out exhaust at constant atmospheric pressure.

The thermodynamic cycle for the *supercharger* consists of the following *processes* :

(i) **7–6–8.** Admission of air at atmospheric pressure.

(ii) **8–1.** Isentropic compression to pressure p_1.

(iii) **1–0.** Delivery of supercharged air, at a constant pressure p_1.

- Area 8-6-7-0-1-8 represents the *supercharger work* (mechanically driven) in supplying air at a pressure p_1, while the area 1-2-3-4-1, is the *output of the engine.* Area 0-1-6-7-0 represents the *gain in work during the gas exchange process due to supercharging. Thus a part of the supercharger work is recovered.* However, the work represented by the area 1-6-8-1 *cannot be recovered and represents a loss of work.*

Supercharging power :

Refer to Fig. 2.128.

p_1, v_1, T_1 = Initial conditions of air at entry to the supercharger ;

p_2, v_2, T_2 = Final conditions of air at exit from the supercharger.

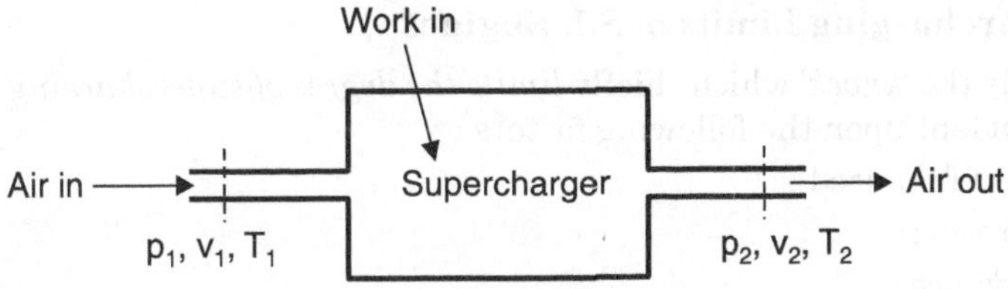

Fig. 2.128. Steady flow process.

Assuming adiabatic compression of air, the work done on the supercharger *per kg of air* is given by

$$W = -\int v\,dp = h_2 - h_1$$

$$= c_p (T_2 - T_1) = c_p T_1 \left(\frac{T_2}{T_1} - 1\right) = c_p T_1 \left[\left(\frac{p_2}{p_1}\right)^{\frac{\gamma-1}{\gamma}} - 1\right] \times \frac{1}{\eta_{adi.}},$$

where, $\eta_{adi.}$ is the adiabatic efficiency of the supercharger/compressor.

∴ *Power required to drive the supercharger (P)* is then given by,

$$P = \frac{\dot{m}_a c_p T_1}{\eta_{adi.}} \left[\left(\frac{p_2}{p_1}\right)^{\frac{\gamma-1}{\gamma}} - 1\right] \text{kW}$$

where, $\dot{m}_a$ = Amount of air in kg/s supplied by the supercharger, and

c_p = Specific heat of air in kJ/kg K.

This power may be supplied by :

— Gas turbine driven by the exhaust gas energy of the engine.

— Separate drive by motor or any other prime mover driving the supercharger.

— Connecting the supercharger to the engine output shaft.

In all the above cases the *gain in the power output of the engine would be many times the power required to drive the supercharger/compressor.*

Important points worth noting regarding supercharging of S.I. engines :

- Supercharging (of S.I. engines) is employed *only for aircraft and racing car engines,* due to the fact that the increase in supercharging pressure increases the *tendency to detonate and pre-ignite.*
- The supercharged petrol engines have a greater fuel consumption than naturally aspirated engines.
- Increased intake pressure and temperature *reduces ignition delay and increase flame speed.* The increased flame speeds make the petrol engine *more sensitive to fuel-air ratio and engine cannot run on weak mixtures without knock.* Rich mixtures are used to control detonation, *which further increases the specific fuel consumption of the engine.*
- Supercharging of petrol engines, because of poor fuel economy, is used in the following cases :

— When a *large amount of power is needed,* or

— When *more power is needed to compensate altitude loss.*

Note. In highly supercharged engine, knocking can be controlled by *injection of water in the combustion chamber* (however use of large amount of liquid becomes prohibitive). Alternatively the *charge may be cooled before it is fed to the engine.*

2.45.7. Supercharging Limits of S.I. Engines

It is primarily the '*knock*' which chiefly *limits the degree of supercharging in S.I. engines.* The knock limit is dependent upon the following factors :

(*i*) The type of fuel used ;

(*ii*) Mixture ratio ;

(*iii*) Spark advance ;

(*iv*) Design features of the engine (important being cooling systems and valve timing)

— It has been observed that for *volatile petroleum fuels* of high octane number the knocking and pre-ignition tendency is reduced at very rich and very lean mixture, and that the fuels of same octane value have different response to supercharging. In the case of alcoholic fuels the knock is reduced at rich mixtures (because of the cooling effect of high latent heat of these fuels).

— Very *lean and very rich mixtures give non-knocking operations. The strongest knocking,* however, occurs *near chemically correct mixtures.* The use of rich mixture results in a higher specific fuel consumption for a supercharged engine though knocking is controlled). A slight reduction in lean mixture makes the engine operation irregular and intermittent.

— The *ignition timings and thermal load on the engine affect the knock limit of S.I. engine.* The *ignition must be retarded at high intake pressure and temperature.*

- In general, supercharger pressure of 1.3 to 1.5 bar is used, which corresponds to about 30 to 50 percent supercharging.

2.46. SUPERCHARGING OF C.I. ENGINES

Fig. 2.129 shows the schematic arrangement and *p–v* diagram for a supercharged constant pressure (diesel) cycle.

- Unlike S.I. engines supercharging does not result in any combustion problem, *rather it improves combustion in diesel engine.* Increase in pressure and temperature of the intake air reduces significantly delay and hence the rate of pressure rise resulting in a *better,*

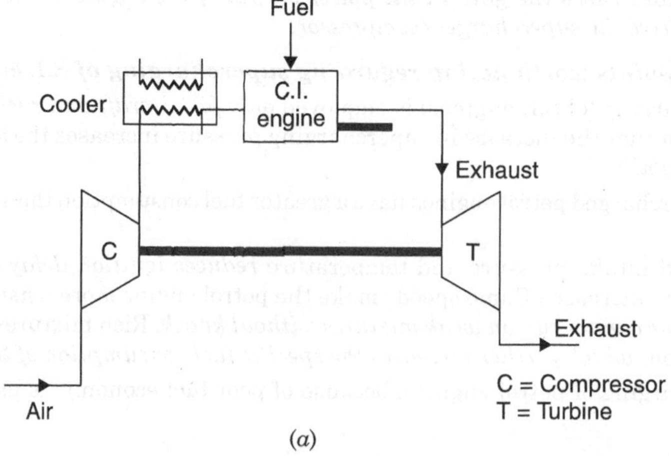

(*a*)

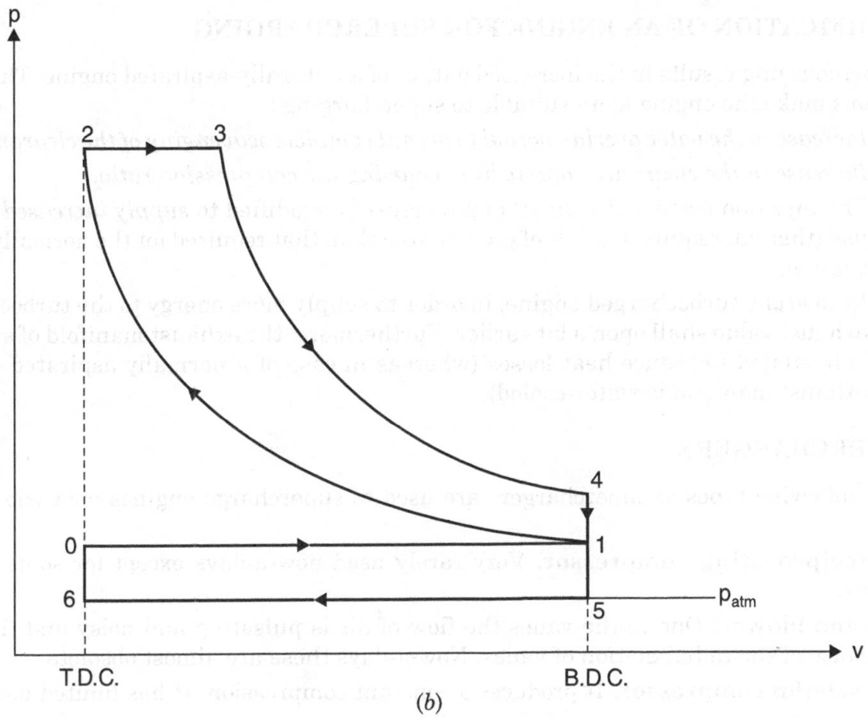

Fig. 2.129. Supercharging C.I. engine.

quieter and smoother combustion. This improvement in combustion allows a poor quality fuel to be used in a diesel engine and it is also not sensitive to the type of fuel used. *The increase in intake temperature reduces volumetric and thermal efficiency but increase in density due to pressure compensates for this and intercooling is not necessary except for highly supercharged engines.*

- *If an unsupercharged engine is supercharged it will increase the reliability and durability of the engine due to smoother combustion and lower exhaust temperatures. The degree of supercharging is limited by thermal and mechanical load on the engine and strongly depends on the type of supercharger used and design of the engine.*

2.46.1. Supercharging Limits of C.I. Engines

The supercharging limits for a C.I. engine (unlike S.I. engine where the limits of supercharging are due to combustion) is reached by ***thermal loading.*** The very high temperature of the piston and cylinder causes scuffing of piston rings and heavy liner wear.

It has been observed that load on bearing is increased due to increased pressure in the cylinder.

The *main considerations* in limiting the degree of supercharging of a C.I. engine are :

(*i*) Durability ;
(*ii*) Reliability ;
(*iii*) Fuel economy.

— The reliability of the engine decreases with the increase in maximum pressure in the cylinder. This also increases the thermal load on the engine due to the increase in the rate of heat release.

2.47. MODIFICATION OF AN ENGINE FOR SUPERCHARGING

Supercharging results in the increased output of a naturally-aspirated engine. The following *modifications* make the engine more suitable to supercharging :

- *Increase in the valve overlap period to permit complete scavenging of the clearance volume.*
- *Increase in the clearance volume by decreasing the compression ratio.*
- The injection system of a diesel engine must be modified to *supply increased amount of fuel* (this will require a nozzle of greater area than that required for the normally aspirated engine).
- In case of a turbocharged engine, in order to supply more energy to the turbocharger the exhaust value shall open a bit earlier. Furthermore, the exhaust manifold of such engine is insulated to reduce heat losses (whereas in case of a normally aspirated engine the exhaust manifold is water-cooled).

2.48. SUPERCHARGERS

The following types of superchargers are used to supercharge engines for various applications :

1. **Reciprocating compressor.** Very rarely used now-a-days except for some stationary installations.

2. **Vane blower.** Due to the vanes the flow of air is pulsating and noisy and the speed is limited because of the radial motion of vanes. Now-a-days these are almost obsolete.

3. **Lysholm compressor.** It produces a constant compression. It has limited use due to its mechanical complexity.

4. **Roots blower.** It is *suitable for low and medium speed engines for stationary and marine installations.*

5. **Centrifugal compressor.** It is simple, small, cheap and has a good efficiency in the range of pressure ratio of 1.5 to 3.0 and is *commonly used for supercharging.*

— The limited speed range of this compressor makes it suitable for constant speed type engines, such as aircraft engines. However, the exhaust driven centrifugal compressor is very popular because its operational speed matches with the exhaust turbine speed satisfactorily.

— *Disadvantage.* Occurrence of surge.

2.49. SUPERCHARGING ARRANGEMENTS

Fig. 2.130 shows the different supercharging arrangements.

- Fig. 2.130 (*i*) shows a compressor coupled to the engine with step-up gearing for increasing the speed of the compressor. Here, a portion of the engine output is used to drive the compressor. The net output increase due to supercharging is obtained by subtracting this power from the engine gross output. The purpose of aftercooler is to send cool air to the engine for further increasing the density of the intake air.
- Fig. 2.130 (*ii*) shows an arrangement in which compressor is driven by a turbine which is run by the exhaust of the engine. The compressor and the turbine are not mechanically coupled to the engine.
- Fig. 2.130 (*iii*) refers to an arrangement in which engine, turbine and compressor all are geared together. The wright Turbo-compound air plane engine is an example of this type of arrangement. Here if the turbine output is insufficient to run the compressor

particularly at part loads, then the remaining load of the compressor is taken care of by the engine.

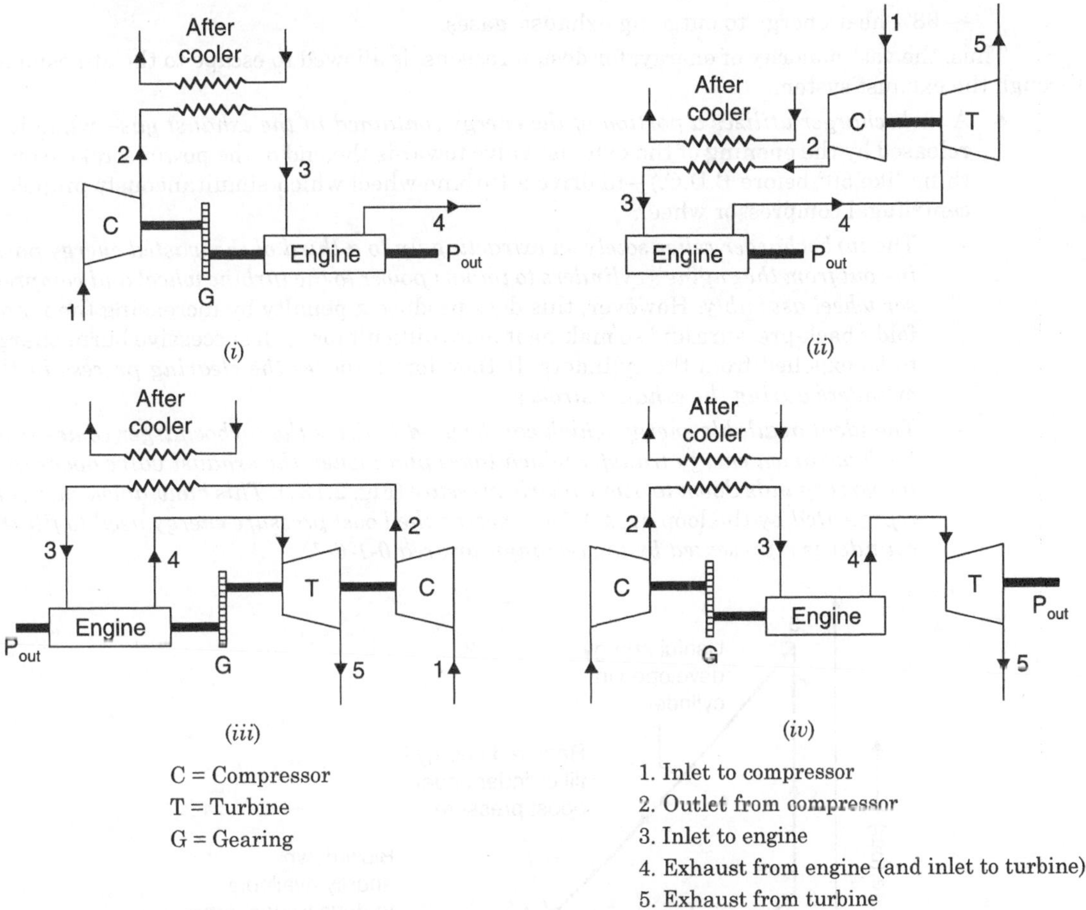

Fig. 2.130. Supercharging arrangements.

- Fig. 2.130 (*iv*) shows an arrangement of supercharging in which engine supplies its total power to the compressor and the exhaust gases from the engine run the turbine giving the power output. Such plants are called *fuel-piston engines.*

2.50. TURBOCHARGERS

2.50.1. Introduction

- ***Turbochargers*** *are centrifugal compressors driven by the exhaust gas turbines.* By utilising the exhaust energy of the engine it recovers a substantial part of energy which would otherwise go waste ; thus the *turbocharger will not draw upon the engine power.* These are *now-a-days extensively used to supercharging almost all types of two stroke engines.*
- A *typical petrol engine may harness up to 30% of the energy contained in the fuel supplied to do useful work under optimum conditions but the remaining 70% of this energy is lost in the following way :*

— 7% heat energy to friction, pumping and dynamic movement ;
— 9% heat energy to surrounding air ;
— 16% heat energy to engine's coolant system ;
— 38% heat energy to outgoing exhaust gases.

Thus, the vast majority of energy, for design reasons, is allowed to escape to the atmosphere through the exhaust system.

- A *turbocharger utilizes a portion of the energy contained in the exhaust gas*—when it is released by the opening of the exhaust valve towards the end of the power stroke (something like 50° before B.D.C.)—to drive a turbine wheel which simultaneously propels a centrifugal compressor wheel.
  — The *turbocharger relies solely on extracting up to a third of the wasted energy passing out from the engine's cylinders to impart power to the turbine wheel and compressor wheel assembly*. However, this does produce a penalty by increasing the manifold's back-pressure and so making it more difficult for each successive burnt charge to be expelled from the cylinders. It therefore *impedes the clearing process in the cylinders during the exhaust strokes.*
  — The *ideal available energy which can be used to drive the turbocharger comes from the blow-down energy transfer which takes place when the exhaust valve opens and the gas expands down to atmospheric pressure* (Fig. 2.131). *This blow-down energy is represented* by the loop area 4-5-6 *whereas the boost pressure energy used to fill the cylinder is represented by the rectangular area 0-1-6-7.*

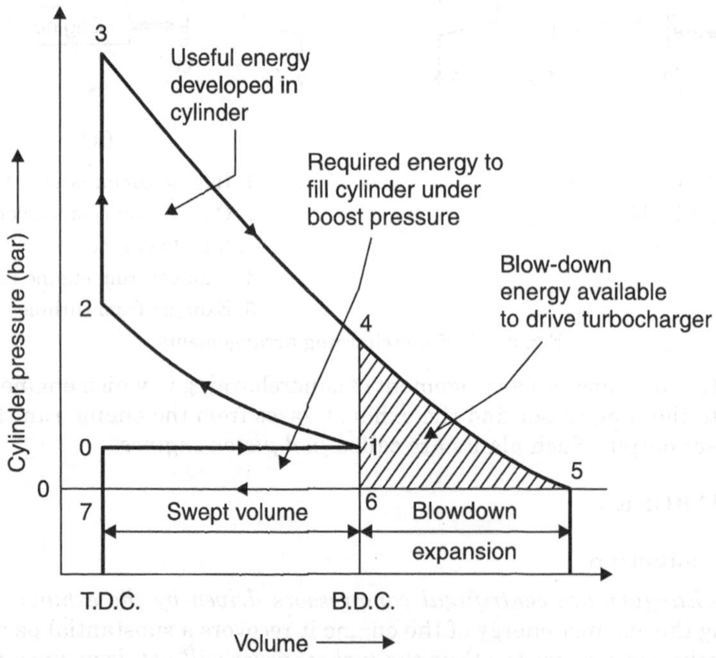

Fig. 2.131. Petrol engine cycle pressure-volume diagram showing available exhaust gas energy.

- *Turbocharged engines produce higher cylinder volumetric efficiencies compared with the normally aspirated induction systems. Therefore, there will be higher peak cylinder pressures which increase the mechanical loading of the engine components and could cause detonation in petrol engines. Therefore, it is usual to reduce the engine's compression ratio by a factor of one or two. Thus, a compression ratio of 10 : 1 for a normally aspirated engine would be derated to 9 : 1 for a low boost pressure or even reduced to 8 : 1 if a medium to high boost pressure is to be introduced. Similarly, for a direct injection diesel engine which might normally have a compression ratio of 16 : 1, when lightly turbocharged the compression ratio may be lowered to 15 : 1 and, if much higher supercharged inlet pressures are to be used, the compression ratio may have to be brought down to something like 14 : 1.*

- The *compression of the charge entering the cells of the impeller depends upon the centrifugal force effect which increases with the square of the rotational speed of the impellor wheel. Consequently, under light load and low engine speed conditions the energy released with the exhaust gases will be relatively small and is therefore insufficient to drive the turbine assembly at very high speeds.* Correspondingly, there will be very little extra boost pressure to make any marked improvement to the engine's torque and power output in the low-speed range of the engine. Thus, in effect, the turbocharged engine will operate with almost no boost pressure and with a reduced compression ratio compared with the equivalent naturally aspirated engine. Hence, *in the very low speed range, the turbocharged engine may have torque and power outputs and fuel consumption values which are inferior to the unsupercharged engine.*
- *Another inherent undesirable characteristic of turbochargers is that when the engine is suddenly accelerated there will be a small time delay before the extra energy discharged into the turbine housing volute can speed up the turbine wheel. Thus, during this transition period, there will be very little improvement in the cylinder filling process, and hence the rise in cylinder brake mean effective pressure will be rather sluggish.*

2.50.2. Altitude Compensation

Engine power outputs are tested and rated at sea-level where the atmospheric air is most dense ; however, as a vehicle climbs, its altitude is increased and the air becomes thinner, that is, less dense. The consequence is a decrease in volumetric efficiency as less air will be drawn into the cylinders per cycle, with a corresponding reduction in engine power since power is directly related to

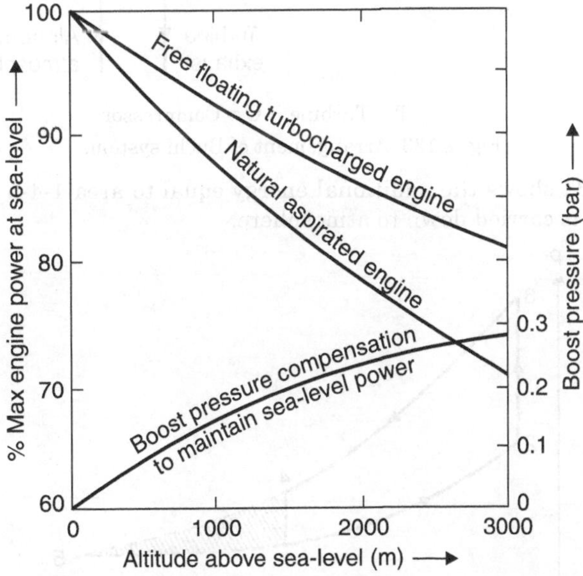

Fig. 2.132. Effect of altitude on rated engine power at sea level for both naturally-aspirated and turbocharged engines.

the actual mass of charge burnt in the cylinder's every power stroke. A *naturally aspirated engine will have its power output reduced by approximately 13% if it is operated approximately 1000 m above sea-level. Supercharging the cylinders enables the engine's rated power above sea-level to be maintained or even exceeded.*

With a turbocharged engine there will still be some power loss with the engine operating at high altitudes, but the *loss will be far less than if the engine breathing depended only on natural aspiration. As can be seen (Fig. 2.132) at 1000 m the power loss is only 8% compared with the naturally aspirated engine where the power decrease is roughly 13%.*

2.50.3. Turbocharging-Buchi System

- Exhaust gases coming out from the engine, when the exhaust valve opens, are at a pressure well above atmospheric pressure, and their temperature is also high. A part of the heat energy contained in the gases can be utilized by further expansion of the gases in an exhaust turbine down to atmospheric pressure. The extra power available from the turbine is used to drive the compressor. The compressor, in turn, will supply more air to the engine. Such utilization of the exhaust energy to drive the supercharger is called *Buchi system of turbocharging.*

— Fig. 2.133 shows the arrangement of Buchi system.

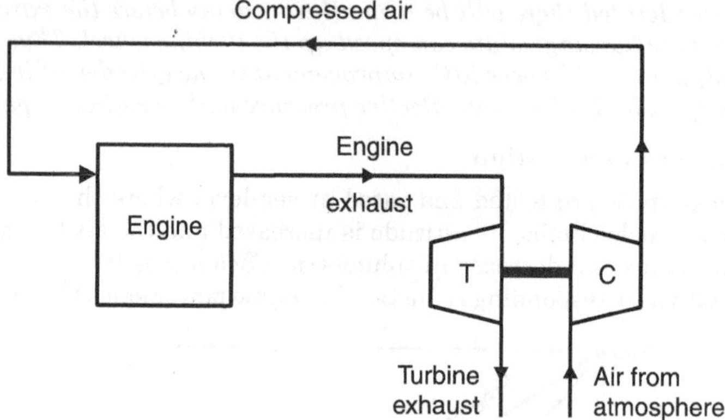

T = Turbine, C = Compressor

Fig. 2.133. Arrangement of Buchi system.

— Fig. 2.134 shows the additional energy equal to area 1-4-5 available when the expansion is carried down to atmosphere.

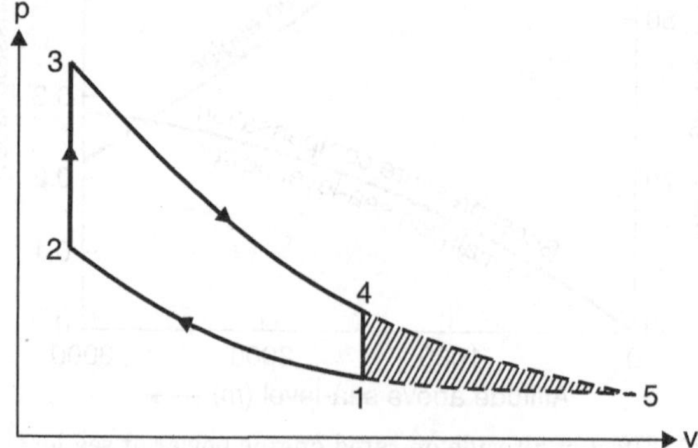

Fig. 2.134. Diagram shows additional energy available when expansion is carried down to atmosphere.

- The *power developed by the turbocharger is sufficient to drive the compressor, and overcome its mechanical friction.* The turbocharger is independent of the engine, and it is only connected to it by a simple exhaust pipe. The speed range of the turbocharger is from 20000 to 30000 r.p.m.
- For supplying adequate energy to the turbocharger, the *exhaust valve is opened much before the B.D.C.* in contrast to a naturally aspirated engine. *This permits the exhaust gases to escape at a higher pressure and temperature, giving turbocharger enough energy to drive the compressor. The loss in piston work due to early opening of the exhaust valve is more than offset by better charging and scavenging of the engine.*

2.50.4. Methods of Turbocharging

The following are the main types of turbocharging methods :

1. **Constant pressure turbocharging :**

- The various cylinders discharge their exhaust into a common manifold at pressures higher than the atmospheric.

The exhaust gases (from all the cylinders) undergo expansion in the exhaust valves, without doing any work, to an approximately constant pressure in the common manifold and then enter the turbine. Thus the blow-down energy in the form of internal energy is converted into work in the turbine. The higher the pressure ratio of the turbine, the higher is the recovery of blow-down energy. During the whole of the cycle the exhaust gases are maintained at constant pressure to make use of a pure reaction turbine.

2. **Pulse turbocharging :**

- In this method of supercharging, as soon as the exhaust valve opens a considerable part of the blow-down energy is converted into *exhaust pulses.* These pulses enter the turbine (through narrow exhaust pipes by the shortest possible route) where a major proportion of the energy is recovered.
- In order that exhaust process of various cylinders do not interfere with one another, separate exhaust pipes are used.

3. **Pulse converter :**

This turbocharging method permits the advantages of the pulse and constant-pressure tubocharging methods simultaneously. The combination of these two systems is done by *connecting the different branches of exhaust manifolds together in a specially designed venturi junction, called "pulse converter", before the turbine.*

4. **Two-stage turbocharger :**

Two-stage turbocharging is defined as *the use of two turbochargers of different sizes in series, e.g.*, a high pressure stage operating on pulse system and low-pressure stage on constant pressure operation. This type of arrangement is employed for diesel engines requiring very high degree of supercharging, b.m.e.p. ranging from 25 to 30 bar, which can not be obtained in a single-stage turbocharger.

5. **Miller turbocharging :**

- The system of turbocharging is based upon the idea of *increasing the expansion ratio relative to compression ratio by means of early closure of inlet valve as the boost pressure is increased.*
- The Miller turbocharging is not very attractive unless two-stage turbocharging is necessary because of other reasons such as need to reduce exhaust valve failures.

6. **Hyperbar turbocharging :**

Fig. 2.135 shows a hyperbar turbocharged engine. It consists of the following components :

1. A low compression (7 : 1) diesel engine ;
2. A high pressure ratio (upto 5.1) turbine ;
3. A by-pass control;
4. An auxiliary combustion chamber/ located between the direct exhaust valve and the turbocharger turbine.

— Firstly, using an electrical starter, the turbocharger is started ; it is kept running by bypassing the air and injecting first into the auxiliary combustion chamber while the engine in inoperative.

— Then after sometime when appropriate pressures and temperatures are reached the diesel engine is started.

— The amount of air bypassed and the fuel injected into the auxiliary combustion chamber are controlled in accordance with operating conditions.

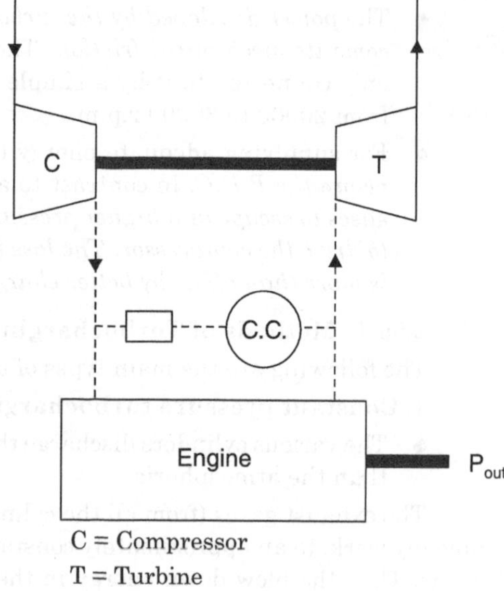

Fig. 2.135. Hyperbar turbocharging.

Advantages :

1. High power-to-weight ratio.
2. High brake mean effective pressure to the tune of 30 bar can be obtained (peak pressure limited to 140 bar).
3. Thermal loading-moderate.
4. Surge pre-operation (due to by pass control).
5. Good torque and acceleration available.

Disadvantages :

1. Due to low compression ratio, there is high fuel consumption over the entire range of operation.
2. System is complex.

2.50.5. Limitations of Turbocharging

Turbocharging entails the following *limitations :*

1. Special exhaust manifolds are required for the turbocharging system.
2. In order to inject more fuel per unit time fuel injection needs modification.
3. In contrast to a naturally aspirated engine which can digest solid particles in the inlet air without undue stress, a turbocharged engine *can pass only the most minute material particles without damage.*
4. It is difficult to obtain good efficiency over a wide range of operations since the efficiency of the turbine blades is very sensitive to gas velocity.

Example 2.3. *An unsupercharged petrol engine develops 735 kW with air fuel ratio 12.8. The bsfc in 0.350 kg/kWh and mechanical efficiency 86%. The inlet pressure is 730 mm of mercury absolute and the mixture temperature is 325 K. The engine is supercharged to a pressure ratio of 1.6 by a supercharger of adiabatic efficiency 0.7 and mechanical efficiency 0.9. Assuming that air-fuel ratio remains unchanged and I.P. is proportional to inlet density, calculate the power required to run the supercharger. Assume that volumetric efficiency does not change due to supercharging.*

(Madras University)

Solution. *Given :* $\frac{p_2}{p_1} = 1.6$, $p_1 = 730$ mm of Hg abs., $T_1 = 325$ K

$$\eta_{adi.} = 0.7,\ \eta_{mech.} = 0.9$$

$$\frac{T_2}{T_1} = \left(\frac{p_2}{p_1}\right)^{\frac{\gamma-1}{\gamma}} = (1.6)^{\frac{1.4-1}{1.4}} = 1.1437$$

$\therefore$ $$T_2 = 325 \times 1.1437 = 371.7 \text{ K}$$

Also, $$\eta_{adi.} = \frac{T_2 - T_1}{T_2' - T_1}$$

or, $$T_2' = T_1 + \frac{T_2 - T_1}{\eta_{adi.}}$$

Fig. 2.136

$\therefore$ $$T_2' = 325 + \frac{371.7 - 325}{0.7} = 391.7 \text{ K}$$

Indicated work of supercharger $= c_p\,(T_2 - T_1)$ kJ/kg of air.

Work required to drive the supercharger

$$W_{sup.} = \frac{c_p(T'_2 - T_1)}{\eta_{mech.}} = \frac{1.005(391.7 - 325)}{0.9} = 74.48 \text{ kJ/kg of air}$$

When unsupercharged :

Inlet pressure, $$p_1 = \frac{730}{1000} \times \frac{9.81 \times (13.6 \times 1000)}{1000} = 97.39 \text{ kN/m}^2$$

Density, $$\rho_{un\,sup.} = \frac{p_1}{RT_1} = \frac{97.39}{0.287 \times 325} = 1.044 \text{ kg/m}^3$$

Air consumption, $$\dot{m}_{a(unsup).} = \frac{0.35 \times 735 \times 12.8}{3600} = 0.9047 \text{ kg/s}$$

When supercharged :

Density, $$\rho_{sup.} = \frac{p_2}{RT_2} = \frac{1.6 \times 97.39}{0.287 \times 391.7} = 1.386 \text{ kg/m}^3$$

Air consumption, $$\dot{m}_{a(sup.)} = m_{a(unsup.)} \times \frac{\rho_{sup}}{\rho_{unsup.}}$$

$$= 0.9147 \times \frac{1.386}{1.044} = 1.2143 \text{ kg/s}$$

∴ Power required to run the supercharger,

$$\mathbf{P_{sup}} = \dot{m}_{a(sup)} \times W_{sup}$$

$$= 1.2143 \times 74.48 = \mathbf{90.44\ kW.} \quad \textbf{(Ans.)}$$

Example 2.4. *A diesel engine operating on four-stroke cycle is to be designed to operate with following characteristics at sea level, where the mean conditions are 1.0132 bar and 10°C.*

B.P. = 260 kW, volumetric efficiency = 78% (at sea level free air conditions), specific fuel consumption = 0.247 kg/B.P.h. ; A/F ratio = 17 ; speed = 1500 r.p.m.

Calculate the required engine capacity and the anticipated brake mean effective pressure.

The engine is fitted with a supercharger so that it may be operated at an altitude of 2700 m where the atmospheric pressure is 0.72 bar. The power taken by a supercharger is 8 per cent of the total power produced by the engine and the temperature of the air leaving the supercharger is 32°C. The air-fuel ratio and thermal efficiency remain the same for the supercharged engine as when running unsupercharged at sea level, as does the volumetric efficiency. Calculate the increase of air pressure required at the supercharger to maintain the same net output of 260 kW. Take R = 0.287 kJ/kgK.

Solution. *Given :* $p_1 = 1.0132$ bar, $T_1 = 10 + 273 = 283$ K, B.P. = 260 kW,

$\eta_{vol.} = 78\%$, s.f.c. = 0.247 kg / B.P. *h*, A / F ratio = 1.7 : 1,

$N = 1500$ r.p.m.

Engine capacity :

Fuel consumption, $$m_f = \frac{s.f.c. \times \text{B.P.}}{60} \text{ kg/min.} = \frac{0.247 \times 260}{60} = 1.07 \text{ kg/min.}$$

Air consumption = Fuel consumption × A / F ratio

= 1.07 × 17 = 18.19 kg / min.

Air consumption per stroke $$= \frac{\text{Air consumption in kg / min.}}{\text{No. of cycles / min.}}$$

$$= \frac{\text{Air consumption in kg/min.}}{N/2} = \frac{18.19}{1500/2} = 0.0242 \text{ kg}$$

Let V_s be the swept volume, then mass of free air corresponding to swept volume $= \frac{p_1 V_s}{RT_1}$

$$= \frac{(1.0132 \times 10^5) \times V_s}{287 \times 283} = 1.247\ V_s \text{ kg}$$

Volumetric efficiency, $$\eta_{vol.} = \frac{\text{Mass of air taken in per stroke}}{\text{Mass of free air corresponding to swept volume}}$$

∴ $$0.78 = \frac{0.0242}{1.247\ V_s} \quad \text{or} \quad V_s = \frac{0.0242}{0.78 \times 1.247} = 0.02488 \text{ m}^3$$

i.e., *Engine capacity* $= \mathbf{0.02488\ m^3}$. **(Ans.)**

Brake mean effective pressure, p_{mb} (bar) :

We know that $$\text{B.P.} = \frac{p_{mb}\ LANk \times 10}{6} \text{ kW}$$

$$260 = \frac{p_{mb} \times V_s \times Nk \times 10}{6} \text{ kW} \qquad (\because \; L \times A = V_s)$$

$$= \frac{p_{mb} \times 0.02488 \times 1500 \times \frac{1}{2} \times 10}{6}$$

$\therefore$ $\quad \mathbf{p_{mb} = 8.36\ bar.}$ **(Ans.)**

Supercharged engine :

Increase of pressure required :

Gross power produced by the engine $= 260 + 0.08 \times$ gross power

$\therefore$ Gross power produced by the engine $= \dfrac{260}{(1 - 0.08)} = 282.6$ kW

$\therefore$ Mass of air required $= \dfrac{18.19 \times 282.6}{260} = 19.77$ kg

$$\eta_{vol.} = \frac{\text{Mass of air taken in per cycle}}{\text{Mass of air corresponding to swept volume measured at outlet conditions of supercharger}}$$

$$0.78 = \frac{\dfrac{19.77}{1500/2}}{\dfrac{(p_2 \times 10^5) \times 0.02488}{287 \times (32 + 273)}} = \frac{0.02636}{0.0284 p_2}$$

or
$$p_2 = \frac{0.02636}{0.78 \times 0.0284} \text{ bar} = 1.19 \text{ bar}$$

$\therefore$ **Increase of pressure required** = 1.19 – 0.72 = **0.47 bar. (Ans.)**

VII. Testing and Performance of I.C. Engines

2.51. INTRODUCTION TO TESTING AND PERFORMANCE OF I.C. ENGINES

- The primary task of the development engineer is to reduce the capital and running cost of the engine. This involves trial of various design concepts. The parameters are so enormous and different in nature that it is almost physically impossible to take care of all of them during the design of the engine. Therefore, it is necessary to conduct the test on the engine and determine the measures which should be taken to improve the engine performance. The nature and the type of the test to be conducted will depend upon a great number of factors such as, the degree of development of the particular design, the accuracy required, the funds available, the nature of the manufacturing company etc.
- The testing of the engine is necessary to verify the performance of the engine as per the specification of the manufacturer.

2.52. PERFORMANCE PARAMETERS

Engine performance is an indication of the degree of success with which it does its assigned job *i.e., conversion of chemical energy contained in the fuel into the useful mechanical work.*

In evaluation of engine performance certain basic parameters are chosen and the effect of various operating conditions, design concepts and modifications on these parameters are studied. The *basic performance parameters* are numerated and discussed below :

1. Power and mechanical efficiency
2. Mean effective pressure and torque
3. Specific output
4. Volumetric efficiency
5. Fuel-air ratio
6. Specific fuel consumption
7. Thermal efficiency and heat balance
8. Exhaust smoke and other emissions
9. Specific weight.

1. **Power and mechanical efficiency :**

(*i*) ***Indicated power.*** *The total power developed by combustion of fuel in the combustion chamber is called indicated power.*

$$\text{I.P.} = \frac{n p_{mi} LANk \times 10}{6} \text{ kW} \quad \text{...(2.10)}$$

where , n = Number of cylinders,

p_{mi} = Indicated mean effective pressure, bar,

L = Length of stroke, m,

A = Area of piston, m^2, and

$k = \frac{1}{2}$ for 4-stroke engine

= 1 for 2-stroke engine.

(*ii*) ***Brake power (B.P.).*** *The power developed by an engine at the output shaft is called the brake power.*

$$\text{B.P.} = \frac{2\pi NT}{60 \times 1000} \text{ kW} \quad \text{...(2.11)}$$

where, N = Speed in r.p.m., and

T = Torque in N-m.

The difference between I.P. and B.P. is called *frictional power, F.P.*

i.e., $$\text{F.P.} = \text{I.P.} - \text{B.P.} \quad \text{...(2.12)}$$

The ratio of B.P. to I.P. is called *mechanical efficiency*

i.e., *Mechanical efficiency,* $$\eta_{mech.} = \frac{\text{B.P.}}{\text{I.P.}} \quad \text{...(2.13)}$$

2. **Mean effective pressure and torque :**

"Mean effective pressure" is defined as hypothetical pressure which is thought to be acting on the piston throughout the power stroke. If it is based on I.P. it is called *indicated mean effective pressure* ($I_{m.e.p.}$ or p_{mi}) and if based on B.P. it is called brake mean effective pressure ($B_{m.e.p.}$ or p_{mb}). Similarly, frictional mean effective pressure ($F_{m.e.p.}$ or p_{mf}) can be defined as :

$$F_{m.e.p.} = I_{m.e.p.} - B_{m.e.p.} \quad \text{...(2.14)}$$

The torque and mean effective pressure are related by the engine size.

Since the power (P) of an engine is dependent on its size and speed, therefore it is not possible to compare engine on the basis of either power or torque. *Mean effective pressure is the true indication of the relative performance of different engines.*

3. Specific output :

It is defined as the *brake output per unit of piston displacement* and is given by :

$$\text{Specific output} = \frac{\text{B.P.}}{A \times L}$$

$$= \text{Constant} \times p_{mb} \times \text{r.p.m.} \quad ...(2.15)$$

For the same piston displacement and brake mean effective pressure (p_{mb}) an engine running at higher speed will give more output.

4. Volumetric efficiency :

It is defined as the ratio of actual volume (reduced to N.T.P.) of the charge drawn in during the suction stroke to the swept volume of the piston.

The average value of this efficiency is from 70 to 80 per cent but in case of *supercharged engine* it may be *more* than 100 per cent, if air at about atmospheric pressure is forced into the cylinder at a pressure greater than that of air surrounding the engine.

5. Fuel-air ratio :

It is the ratio of the mass of fuel to the mass of air in the fuel-air mixture.

Relative fuel air ratio is defined as the ratio of the actual fuel-air ratio to that of stoichiometric fuel-air ratio required to burn the fuel supplied.

6. Specific fuel consumption (s.f.c.) :

It is the mass of fuel consumed per kW developed per hour, and is a criterion of economical power production.

i.e., $$\text{s.f.c.} = \frac{\dot{m}_f}{\text{B.P.}} \text{ kg/kWh.}$$

7. Thermal efficiency and heat balance :

Thermal efficiency. *It is the ratio of indicated work done to energy supplied by the fuel.*

If $\dot{m}_f$ = Mass of fuel used in kg/s, and

C = Calorific value of fuel (lower),

Then *indicated thermal efficiency* (based on I.P.),

$$\eta_{th.}\,(I) = \frac{\text{I.P.}}{\dot{m}_f \times C} \quad ...(2.16)$$

and, *brake thermal efficiency* (based on B.P.)

$$\eta_{th.}\,(B) = \frac{\text{B.P.}}{\dot{m}_f \times C} \quad ...(2.17)$$

Heat balance sheet :

The performance of an engine is generally given by heat balance sheet.

To draw a heat balance sheet for I.C. engine, it is run at constant load. Indicator diagram is obtained with the help of an indicator. The quantity of fuel used in a given time and its calorific value, the amount, inlet and outlet temperature of cooling water and the weight of exhaust gases are recorded. After calculating I.P. and B.P. the heat in different items is found as follows :

Heat supplied by fuel :

For *petrol and oil engines,* heat supplied = $m_f \times C$, where m_f and C are mass used per minute (kg) and lower calorific value (kJ or kcal) of the fuel respectively.

For *gas engines*, heat supplied = $V \times C$, where V and C is volume at N.T.P. (m^3/min.) and lower calorific value of gas respectively.

(*i*) ***Heat absorbed in I.P. :***

Heat equivalent of I.P. (per minute) = I.P. × 60 kJ ...(2.18)

(*ii*) ***Heat taken away by cooling water :***

If m_w = Mass of cooling water used per minute,

t_1 = Initial temperature of cooling water, and

t_2 = Final temperature of cooling water,

Then, heat taken away by water = $m_w \times c_w \times (t_2 - t_1)$ ...(2.19)

where, c_w = specific heat of water.

(*iii*) ***Heat taken away by exhaust gases :***

If m_e = Mass of exhaust gases (kg/min),

c_{pg} = Mean specific heat at constant pressure,

t_e = Temperature of exhaust gases, and

t_r = Room (or boiler house) temperature,

Then heat carried away by exhaust gases = $m_e \times c_{pg}(t_e - t_r)$ ...(2.20)

Note. The mass of exhaust gases can be obtained by adding together mass of fuel supplied and mass of air supplied.

The heat balance sheet from the above data can be drawn as follows :

Item	*kJ*	*Per cent*
Heat supplied by fuel	...	...
(*i*) Heat absorbed in I.P.	...	...
(*ii*) Heat taken away by cooling water	...	...
(*iii*) Heat carried away by exhaust gases	...	...
(*iv*) Heat unaccounted for (by difference)	...	...
Total	...	...

8. **Exhaust smoke and other emissions :**

Smoke is an indication of incomplete combustion. It limits the output of an engine if air pollution control is the consideration. *Exhaust emissions* have of late become a matter of grave concern and with the enforcement of legislation on air pollution in many countries, it has become necessary to view them as performance parameters.

9. **Specific weight :**

It is defined as the weight of the engine in kg for each B.P. developed. It is an indication of the engine bulk.

2.53. BASIC MEASUREMENTS

To evaluate the performance of an engine following *basic measurements* are usually undertaken :

1. Speed
2. Fuel consumption
3. Air consumption
4. Smoke density

5. Exhaust gas analysis
6. Brake power
7. Indicated power and friction power
8. Heat going to cooling water
9. Heat going to exhaust.

1. Measurement of speed :

The speed may be measured by :

(*i*) Revolution counters
(*ii*) Mechanical tachometer
(*iii*) Electrical tachometer.

2. Fuel measurement :

The fuel consumed by an engine can be measured by the following methods :

(*i*) Fuel flow method
(*ii*) Gravemetric method
(*iii*) Continuous flow meters.

3. Measurement of air consumption :

The air consumption can be measured by the following methods :

(*i*) Air box method
(*ii*) Viscous-flow air meter.

(*i*) ***Air box method :***

Fig. 2.137 shows the arrangement of the system. It consists of air-tight chamber fitted with a sharp edged orifice of known coefficient of discharge. The orifice is located away from the suction connection to the engine. Due to the suction of engine, there is a pressure depression in the air box or chambor which causes the flow through the orifice. For obtaining a steady flow, the volume of chamber should be sufficiently large compared with the swept volume of the cylinder , generally

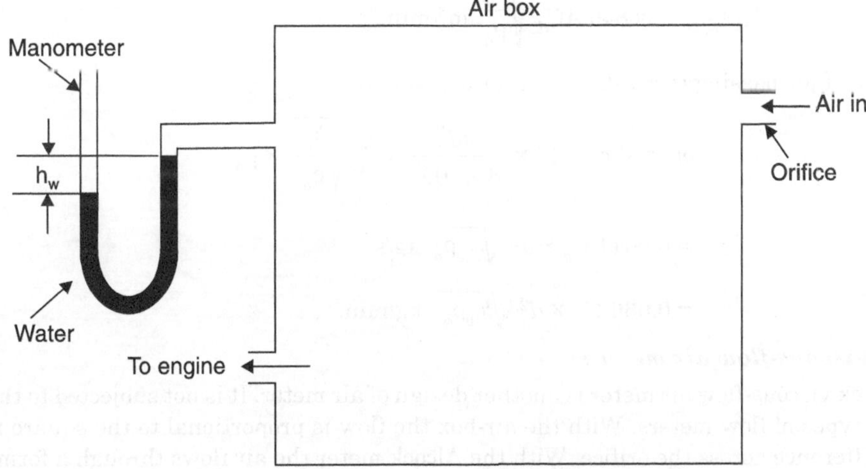

Fig. 2.137. Air-box method for measuring air.

500 to 600 times the swept volume. It is assumed that the intermittent suction of the engine will not effect the air pressure in the air box as volume of the box is sufficiently large, and pressure in the box remains same.

A water manometer is used to measure the pressure difference causing the flow through the orifice. The depression across the orifice *should not exceed 100 to 150 mm of water.*

Let A = Area of orifice, m^2,

d = Diameter of orifice, cm,

h_w = Head of water in cm causing the flow,

C_d = Coefficient of discharge for orifice,

ρ_a = Density of air in kg/m^3 under atmospheric conditions, and

ρ_w = Density of water in kg/m^3.

Head in metres of air (H) is given by :

$$H.\rho_a = \frac{h_w}{100}\,\rho_w$$

$$\therefore \qquad H = \frac{h_w}{100} \times \frac{\rho_w}{\rho_a} = \frac{h_w}{100} \times \frac{1000}{\rho_a} = \frac{10h_w}{\rho_a} \text{ m of air}$$

The velocity of air passing through the orifice is given by,

$$C_a = \sqrt{2gH} \text{ m/s} = \sqrt{2g\,\frac{10h_w}{\rho_a}} \text{ m/s}$$

The volume of air passing through the orifice,

$$V_a = C_d \times A \times C_a = C_d A \sqrt{2g\,\frac{10h_w}{\rho_a}} = 14\,AC_d \sqrt{\frac{h_w}{\rho_a}} \text{ m}^3\text{/s}$$

$$= 840\,AC_d \sqrt{\frac{h_w}{\rho_a}} \text{ m}^3\text{/min.}$$

Mass of air passing through the orifice is given by

$$m_a = V_a\rho_a = 14 \times \frac{\pi d^2}{4 \times 100^2} \times C_d \sqrt{\frac{h_w}{\rho_a}} \times \rho_a$$

$$= 0.0011\, C_d \times d^2 \sqrt{h_w \rho_a} \text{ kg/s}$$

$$= 0.066\, C_d \times d^2 \sqrt{h_w \rho_a} \text{ kg/min.} \qquad ...(2.21)$$

(*ii*) ***Viscous-flow air meter :***

Alcock viscous-flow air meter is another design of air meter. It is not subjected to the errors of the simple types of flow meters. With the air-box the flow is proportional to the square root of the pressure difference across the orifice. With the Alcock meter the air flows through a form of honey-comb so that flow is viscous. *The resistance of the element is directly proportional to the air velocity and is measured by means of an inclined manometer.* Felt pads are fitted in the manometer connections to damp out fluctuations. The meter is shown in Fig. 2.138.

The accuracy is improved by fitting a damping vessel between the meter and the engine to reduce the effect of pulsations.

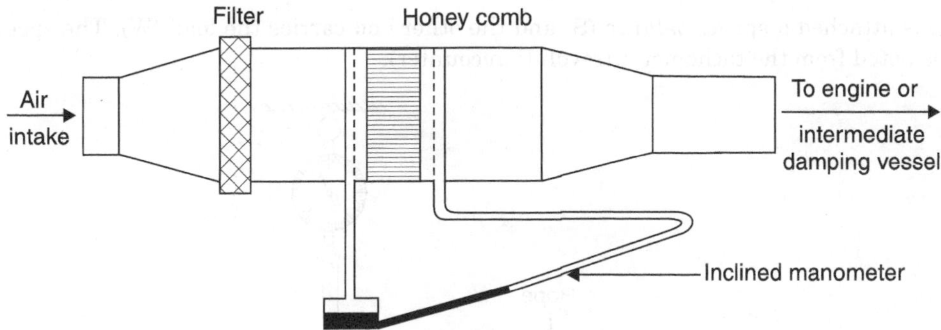

Fig. 2.138. Alcock viscous-flow air meter.

4. **Measurement of exhaust smoke :**

The following smoke meters are used :

(*i*) Bosch smoke meter (*ii*) Hatridge smoke meter

(*iii*) PHS smoke meter.

5. **Measurement of exhaust emission :**

Substances which are emitted to the atmosphere from any opening down stream of the exhaust part of the engine are termed as "exhaust emissions". Some of the more commonly used instruments for measuring exhaust components are given below :

(*i*) Flame ionisation detector (*ii*) Spectroscopic analysers

(*iii*) Gas chromatography.

6. **Measurement of B.P. :**

The B.P. of an engine can be determined by a brake of some kind applied to the brake pulley of the engine. The arrangement for determination of B.P. of the engine is known as *dynamometer*. The dynamometers are classified into following two classes :

(*i*) Absorption dynamometers (*ii*) Transmission dynamometers.

(*i*) ***Absorption dynamometers.*** *Absorption dynamometers are those that absorb the power to be measured by friction.* The power absorbed in friction is finally dissipated in the form of heat energy.

Common forms of absorption dynamometers are :

—Prony brake —Rope brake

—Hydraulic brake —Fan brake

—Electrical brake dynamometers

(*a*) Eddy current dynamometer

(*b*) Swinging field d.c. dynamometer.

(*ii*) ***Transmission dynamometers.*** These are also called *torquemeters*. These are very accurate and are used where continuous transmission of load is necessary. These are used mainly in *automatic units*.

Here we shall discuss *Rope brake dynamometer* only :

Rope brake dynamometer :

Refer Fig. 2.139. A rope is wound round the circumference of the brake wheel. To prevent the rope from slipping small wooden blocks (not shown in the Fig. 2.139) are laced to rope. To one end of the rope is attached a spring balance (S) and the other end carries the load (W). The speed of the engine is noted from the tachometer (revolution counter).

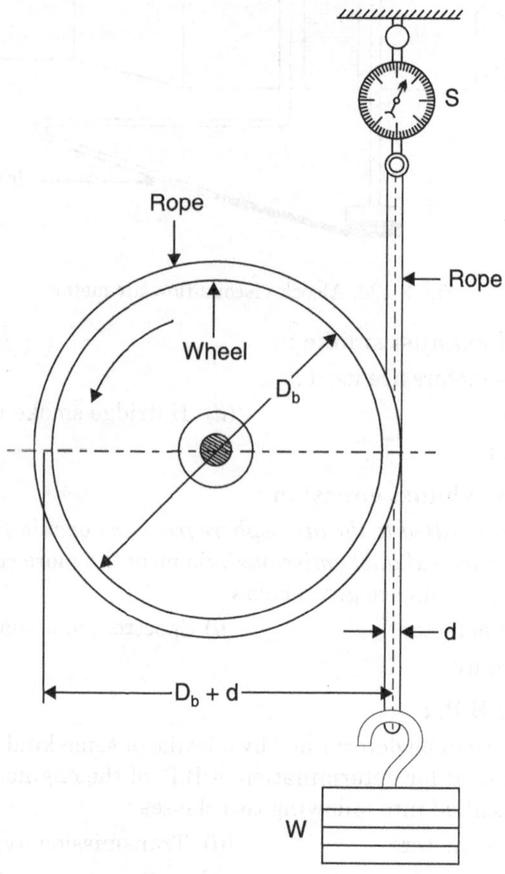

Fig. 2.139. Rope brake dynamometer.

If, W = Weight at the end of the rope, N,

S = Spring balance reading, N,

N = Engine speed, r.p.m.,

D_b = Diameter of the brake wheel, m,

d = Diameter of the rope, m, and

($D_b + d$) = Effective diameter of the brake wheel,

Then work/revolution = Torque × angle turned per revolution

$$= (W - S) \times \left(\frac{D_b + d}{2}\right) \times 2\pi = (W - S)(D_b + d) \times \pi$$

Work done/min. $= (W - S)\,\pi\,(D_b + d)\,N$

Work done/sec. $= \dfrac{(W - S)\,\pi\,(D_b + d)\,N}{60}$

$$\therefore \quad \text{B.P.} = \frac{(W - S)\,\pi\,(D_b + d)\,N}{60 \times 1000} \text{ kW} \qquad \text{...(2.22)}$$

$$= \frac{(W - S)\,\pi\,D_b N}{60 \times 1000} \text{ ...... if } d \text{ is neglected}$$

$$\text{or,} \quad \left(= \frac{T \times 2\pi N}{60 \times 1000} \text{ kW} \right) \qquad \text{...(2.23)}$$

Rope brake is cheap and easily constructed but not very accurate because of changes in friction coefficient of the rope with temperature.

Measurement of Indicated power (I.P.) :

The power developed in the engine cylinder or at the piston is necessarily greater than that at the crankshaft due to engine losses. Thus,

I.P. = B.P. + engine losses.

Indicated power is usually determined with the help of a p-V diagram taken with the help of an indicator. In case indicated power cannot be measured directly, it is made possible by measuring the brake power and also the engine losses. If the *indicator diagram* is available, the indicated power may be *computed by measuring the area of* diagram, either with a *planimeter* or by *ordinate method,* and dividing by the stroke measurement in order to obtain the mean effective pressure (m.e.p.).

i.e.,

$$p_{mi} = \frac{\text{Net area of diagram in mm}^2}{\text{Length of diagram in mm}} \times \text{Spring constant}$$

where p_{mi} is in bar. ...(2.24)

(The spring constant is given in *bar per mm of vertical movement of the indicator stylus.*)

Engine indicators :

The main types of engine indicators are :

1. Piston indicator.
2. Balanced diaphragm type indicator :
   (*i*) The Farnborough balanced engine indicator
   (*ii*) Dickinson-Newell indicator
   (*iii*) MIT balanced pressure indicator
   (*iv*) Capacitance-type balance pressure indicator.
3. Electrical indicators.

In addition to this, *optical indicators* are also used.

Calculation of indicated power (I.P.) :

If p_{mi} = Indicated mean effective pressure, bar,
A = Area of piston, m^2,
L = Length of stroke, m,
N = Speed of the engine, r.p.m., and
$k = \frac{1}{2}$ for 4-stroke engine
$= 1$ for 2-stroke engine,

Then, force on the piston = $p_{mi} \times A \times 10^5$ N

Work done per working stroke = Force × length of stroke

$= p_{mi} \times A \times 10^5 \times L$ N-m

Work done per second = Work done per stroke × number of working stroke per second

$$= p_{mi} \times L \times A \times 10^5 \times \frac{N}{60} \times k \text{ N-m/s or J/s}$$

$$= \frac{p_{mi} \times LANk \times 10^5}{60 \times 1000} \text{ kW}$$

i.e., Indicated power, $\text{I.P.} = \dfrac{p_{mi}\, LANk \times 10}{6}$ kW

If n is the number of cylinders, then

$$\text{I.P.} = \frac{np_{mi}\, LANk \times 10}{6} \text{ kW} \quad ...(2.25)$$

Morse test :

This test is only applicable to *multi-cylinder engines.*

The engine is run at the required speed and the torque is measured. One cylinder is cut out, by shorting the plug if an S.I. engine is under test, or by disconnecting an injector if a C.I. engine is under test. *The speed falls because of the loss of power with one cylinder cut out, but is restored by reducing the load.* The torque is measured again when the speed has reached its original value. If the values of I.P. of the cylinders are denoted by I_1, I_2, I_3 and I_4 (considering a four-cylinder engine), and the power losses in each cylinder are denoted by L_1, L_2, L_3 and L_4, then the value of B.P., B at the test speed with all cylinders firing is given by

$$B = (I_1 - L_1) + (I_2 - L_2) + (I_3 - L_3) + (I_4 - L_4) \quad ...(i)$$

If *number 1 cylinder is cut out,* then the contribution I_1 is lost ; and if the losses due to that cylinder remain the same as when it is firing, then the B.P., B_1, now obtained at the same speed is

$$B_1 = (0 - L_1) + (I_2 - L_2) + (I_3 - L_3) + (I_4 - L_4) \quad ...(ii)$$

Subtracting equation (*ii*) from (*i*), we get

$$B - B_1 = I_1 \quad ...(2.26)$$

Similarly, $B - B_2 = I_2$ when cylinder number 2 is cut out

and, $B - B_3 = I_3$ when cylinder number 3 is cut out

and, $B - B_4 = I_4$ when cylinder number 4 is cut out.

Then, for the engine,

$$I = I_1 + I_2 + I_3 + I_4 \quad ...(2.27)$$

Assumptions :

1. Speed of engine and throttle opening or fuel injection setting has to remain same in both the cases of four cylinders working and three cylinders working with one cylinder cut out.
2. Frictional power and pumping power are functions of speed and must be kept constant.
3. Same throttle opening is expected to supply same quantity of fuel.
4. Cylinder individually will develop same power whether all four are working or only three cylinders are working with one cylinder consuming power.

Limitations :

1. This is applicable to multi-cylinder engines only (three or more cylinders).
2. Results are liable to errors due to change in mixture distribution and other conditions by cutting out one cylinder.

Cutting out of one cylinder may greatly affect the pulsations in exhaust system which may significantly change the engine performance by imposing different back pressures.

Engine Friction :

Frictional losses in an engine

The total engine friction can be divided into *five main components*. These are :

1. Crankcase mechanical friction.
2. Blow-by losses (compression-expansion pumping loss).
3. Exhaust and inlet system throttling losses.
4. Combustion chamber pumping loop losses.
5. Piston mechanical friction.

1. ***Crankcase mechanical friction :***

It can be further sub-divided into :

(*i*) Bearing friction

(*ii*) Valve gear friction, and

(*iii*) Pump and miscellaneous friction.

The bearing friction includes the friction due to main bearing connecting rod bearing and other bearings. Bearing friction is viscous in nature and depends upon the oil viscosity, the speed, size and geometry of the journal.

The valve gear friction losses vary with the engine design variables and no general equation is available predicting them.

2. ***Blow-by losses :***

It is the phenomenon of *leakage of combustion products past the piston and piston rings from the cylinder to the crankcase.* These losses depend on the inlet pressure and compression ratio. *These losses vary as the square root of inlet pressure, and increase as the compression ratio is increased. Blow-by losses are reduced as the engine speed is increased.*

3. ***Exhaust and inlet throttling loss :***

The standard practice of sizing the exhaust valves is to make them a certain percentage *smaller* than the inlet valves. This usually results in an insufficiently sized exhaust valve and hence results in *exhaust pumping loss.*

4. ***Combustion chamber pumping loop losses :***

In the case of pre-combustion chamber engines an additional loss occurs. This is the loss occurring *due to the pumping work required to pump gases into and out of the pre-combustion chamber.* The exact value of this would depend upon the orifice size connecting the pre-combustion chamber and the main chamber, and the speed. *Higher the speed greater is the loss and smaller the orifice size, greater is the loss.*

5. ***Piston mechanical friction :***

It can be sub-divided into :

(*i*) Viscous friction.

(*ii*) Non-viscous-friction—(*a*) friction due to ring tension.

(*b*) friction due to gas pressure forces behind the ring.

- The *viscous friction depends upon the viscosity of the oil and the temperature of the various parts of the piston.* The degree to which the upper part of the piston can be lubricated also effects the viscous friction.
- The oil film thickness between piston and the cylinder is also affected by the piston side thrust and the resulting vibration.

Effect of engine variables on engine friction :

1. ***Effect of stroke/bore ratio.*** The effect of stroke / bore ratio on engine friction and economy is very small. High stroke / bore ratio engines have equally good friction mep values as that for low stroke / bore ratio engine. Indications are that at high speeds the higher stroke / bore ratio engine may be at some disadvantage.

2. ***Effect of cylinder size and number of cylinders.*** The friction and economy improves as a smaller number of larger cylinders are used. This is because the proportion between the working piston area and its friction producing area, *i.e.*, circumference is reduced.

3. ***Effect of piston rings.*** The effect of number of piston ring is not very critical and this number is usually chosen on the basis of cost, size and other requirements rather than on the basis of their effect on friction.

4. ***Effect of compression ratio.*** Friction means effective pressure increases as the compression ratio is increased. But the mechanical efficiency either remains the constant or improves as the compression ratio is increased. If the displacement is varied to keep the maximum engine torque constant, this results in better part load friction characteristics.

5. ***Effect of engine speed.*** Engine friction *increases rapidly as the speed increases.* The best way to improve mechanical efficiency at high speed is to *increase the number of cylinders.*

6. ***Effect of oil viscosity.*** *Higher the oil viscosity greater is the friction loss.* The temperature of the oil in the crankcase significantly affects the friction losses, wear and service life of an engine. *As the oil temperature increases the viscosity decreases and friction losses are reduced during a certain temperature range.*

7. ***Effect of cooling water temperature.*** A rise in cooling water temperature *reduces* engine friction through its effect on oil viscosity.

During starting operation the temperature of both the oil and the water is low, hence, the viscosity is high. This results in high starting friction losses and rapid engine wear.

8. ***Effect of engine load.*** As the load increases the maximum pressure in the cylinder has a *tendency to increase slightly.* This results in *slightly higher friction values.*

Measurement of frictional power (F.P.) :

The frictional power of an engine can be determined by the following methods :

1. Willan's line method (used for C.I. engines only)
2. Morse test.
3. Motoring test.
4. Difference between I.P. and B.P.

1. ***Willan's line method :***

At a constant engine speed the load is reduced in increments and the corresponding B.P. and gross fuel consumption readings are taken. A graph is then drawn of fuel consumption against B.P. as in Fig. 2.140. The graph drawn in called the Willan's line (analogous to Willan's line for a steam engine), and is extrapolated back to cut the B.P. axis at the point *L*. The reading *OL* is taken as the power loss of the engine at that speed. The fuel consumption at zero B.P. is given by *OM* ; and if the relationship between fuel consumption and B.P. is assumed to be linear, then a fuel consumption *OM* is equivalent to a power loss of *OL*.

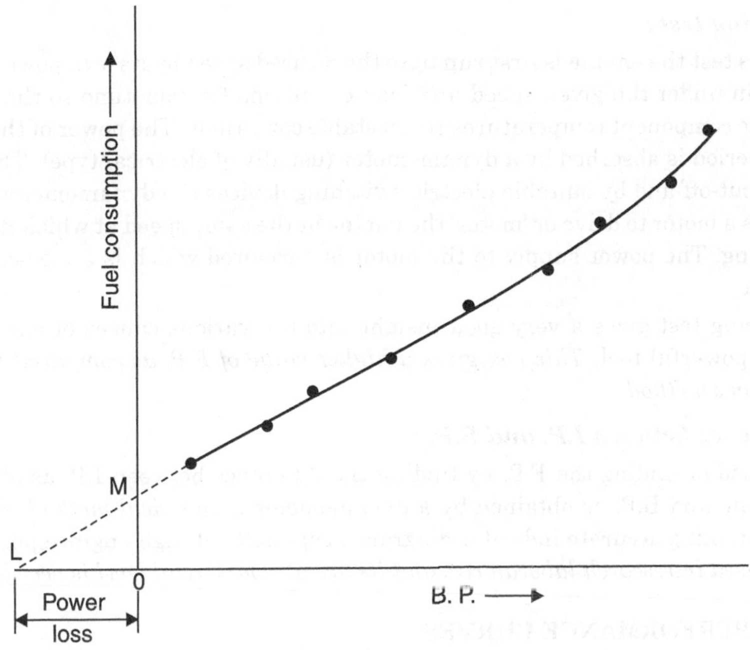

Fig. 2.140. Willan's line method.

- This method is used *only in case of unthrottled engines* as discussed below :
  — The Willians line is plotted for fuel consumption versus load (from no load to full load) *at constant speed.* The intersection of this line on the negative side of X-axis gives the friction power of the engine at that speed. The friction power (F.P.) is assumed constant from no load to full load at that constant speed. The F.P. includes not only the mechanical friction, but also pumping power.

For a throttled engine if such a trial is carried out, the throttle position has to be varied from almost closed at not load to full open at maximum load, to keep the engine speed constant. Therefore the pumping load will be bigger at no load, and reduce gradually as the load is increased. In other words, *the pumping power and therefore the F.P. will not remain constant, as is the assumption in Willian's line method.*

Limitations :

The following are the *limitations* of Willan's line method.

1. Applicable mainly to C.I. engines only.

2. The fuel-consumption— brake power line is not straight line, but turns up slightly at low loads and considerably near full load. *Unless sufficient data are taken to accurately plot the straight line portion of the curve, the results will be significantly in error.*

2. ***Morse test :***

In 'Morse test' (already discussed), frictional power can be found by subtracting $(B.P.)_n$ from $(I.P.)_n$

i.e.,

$$F.P. = (I.P.)_n - (B.P.)_n$$

where n is the number of cylinders.

3. ***Motoring test :***

- In this test the engine is first run upto the desired speed by its own power and allowed to remain under the given speed and load conditions for sometime so that oil, water and engine component temperatures reach stable conditions. The power of the engine during this period is absorbed by a dynamometer (usually of electrical type). The fuel supply is then cut-off and by suitable electric switching devices the dynamometer is converted to run as a motor to drive or 'motor' the engine at the same speed at which it was previously running. The power supply to the motor is measured which is a measure of F.P of the engine.
- Motoring test gives a very good insight into the various causes of losses and is much more powerful tool. *This test gives a higher value of F.P. as compared to that given by Willian's method.*

4. ***Difference between I.P. and B.P. :***

The method of finding the F.P. by finding the difference between I.P. as obtained from an indicator diagram, and B.P. as obtained by a dynamometer is the *ideal method.* However, due to difficulties in obtaining accurate indicator diagrams, especially at high engine speeds, this method is *usually only used in research laboratories and its use at commercial level is very limited.*

2.54. ENGINE PERFORMANCE CURVES

- The following parameters are of importance
  — Torque
  — Power
  — Specific fuel consumption and its inverse
  — I.C. engine efficiency
  — *Brake mean effective pressure* is defined as brake power/swept volume × cyclic frequency. In this relation swept volume is constant. Also brake power = torque × angular speed, thus at constant speed, the b.e.m.p. is directly proportional to torque and either may be used. The ratio of brake mean effective pressure to the indicated mean effective pressure (from indicator diagram) may seem to be equal to the ratio of brake power to the indicated power which is defined as mechanical efficiency. Thus indicator diagram and the output torque may be connected with a suitable allowance for mechanical efficiency.
- The torque-speed relations (Fig. 2.141) exhibit a curve even though it would be seen reasonable to expect the mean effective pressure to be constant at all speeds. However, at low speeds leakage through valves etc. becomes of greater significance so that the m.e.p. falls and at high speeds the volumetic efficiency causes the induced mass to fall with a parallel fall in m.e.p. *The power curves are the product of the torque curves with speed.*

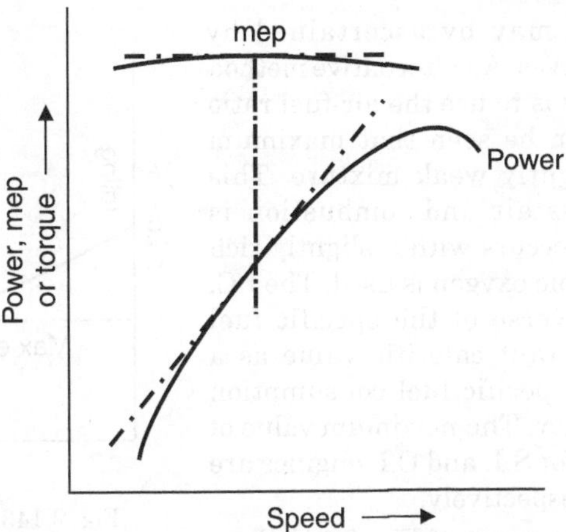

Fig. 2.141. Power-speed and torque-speed curves for the I.C. engine.

Specific fuel consumption relations :

S.I. Engines : Refer to Fig. 2.142. The curves are plotted for *constant throttle opening, constant speed and constant ignition setting.* The *only variable is the air-fuel ratio.* The effect of

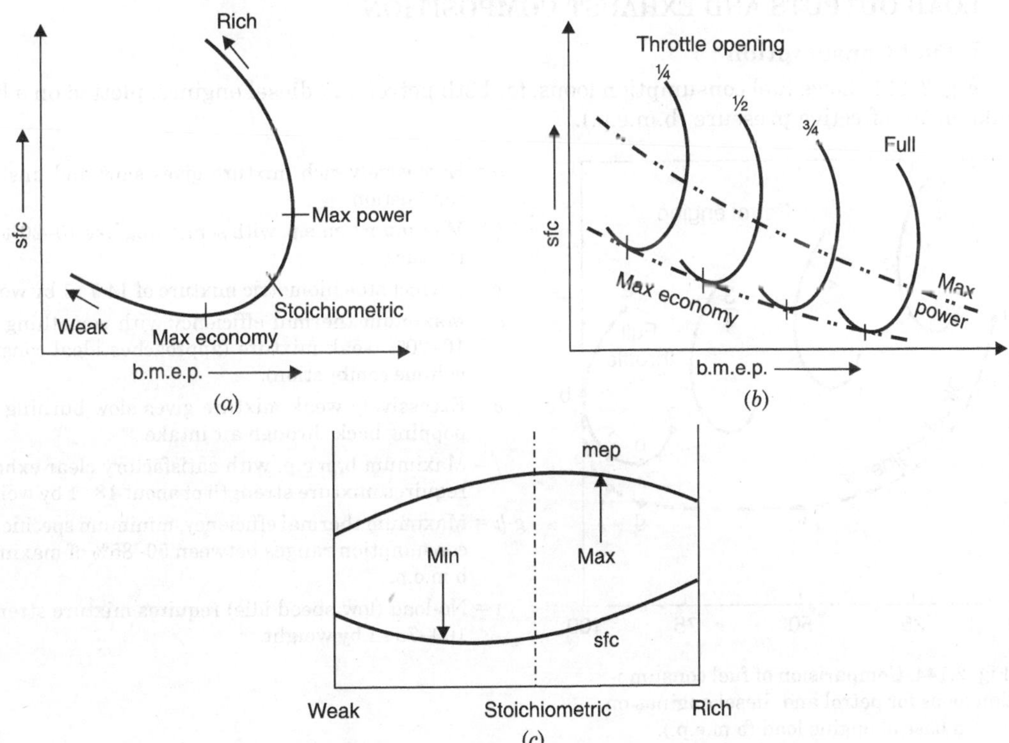

Fig. 2.142. Specific fuel consumption-brake mean effective pressure curves for the S.I. engine.

ignition setting or speed may by ascertained by producing a new family of curves. An alternative method of plotting these parameters is to use the air-fuel ratio as the abscissa. Here it can be seen that maximum economy occurs with a slightly weak mixture. This means that there is excess air and combustion is complete. Maximum power occurs with a slightly rich mixture when all the available oxygen is used. The I.C. engine efficiency is the inverse of the specific fuel consumption with the constant calorific value as a factor. Thus the curves of specific fuel consumption (s.f.c.) also represent efficiency. The maximum value of brake I.C. engine efficiency for S.I. and C.I. engines are of the order 35% and 40% respectively.

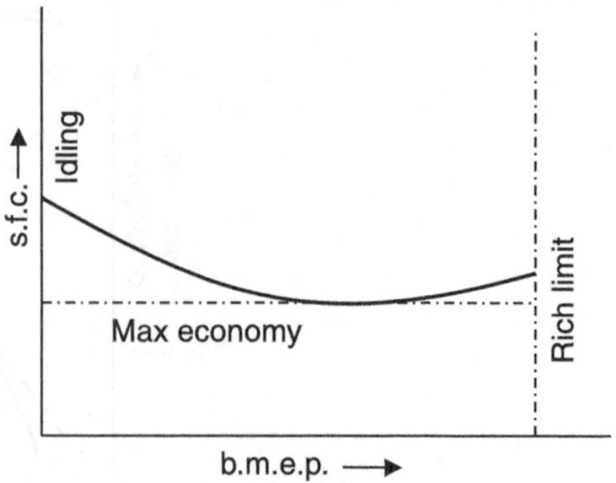

Fig. 2.143. Specific fuel consumption—brake mean effective pressure curve for the C.I. engine.

C.I. engines. The flat curve of Fig. 2.143 illustrates that at *part load the compression ignition engine is more economical than the spark ignition engine.* This is the benefit of quality control rather than quantity control of power.

2.55. COMPARISON OF PETROL AND DIESEL ENGINES—FUEL CONSUMPTION, LOAD OUTPUTS AND EXHAUST COMPOSITION

I. Fuel Consumption :

Fig. 2.144 shows fuel consumption loops, for both petrol and diesel engines, plotted on a base of brake mean effective pressure (b.m.e.p.).

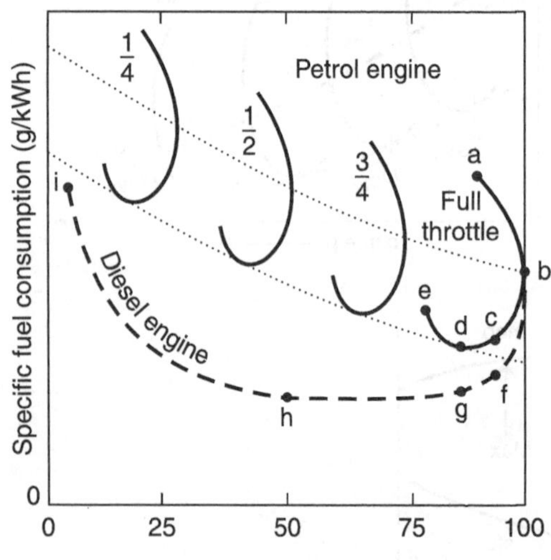

Fig. 2.144. Comparision of fuel consumption loops for petrol and diesel engines on a base of engine load (b.m.e.p.).

a = Excessively rich mixture gives slow and unstable combustion.

b = Maximum b.m.e.p. with something like 10–20% rich mixture.

c = Correct stoichiometric mixture of 14.7 : 1 by weight

d = Maximum thermal efficiency with something like 10–20% weak mixture (approaches ideal constant volume combustion).

e = Excessively weak mixture gives slow burning and popping back through air intake.

f = Maximum b.m.e.p. with satisfactory clear exhaust requires mixture strength of about 18 : 1 by weight.

g-h = Maximum thermal efficiency, minimum specific fuel consumption ranges between 50–85% of maximum b.m.e.p.

i = No-load (low speed idle) requires mixture strength 100–75 : 1 by weight.

- In case of a diesel *engine, load and speed output is controlled entirely by varying the quantity of fuel injected into the cylinder without misfiring occurring,* that is, from 0–100% of the maximum b.m.e.p. developed.
- With the petrol engines, however, if there was no throttle (full throttle position) the effects of varying the mixture strength from the richest position (*a*) to the weakest position

(*e*) produces variations of b.m.e.p. (load) on only 25% of the maximum possible b.m.e.p. that is, from 75–100% of maximum b.m.e.p. (Fig. 2.144). Therefore the *petrol engine's output control can not be achieved alone by varying the mixture strength* and *therefore throttling the mixture coming into the cylinder becomes essential.*

The *diesel engine is thus* **"quality"** *controlled and the petrol engine is* **"quantity"** *controlled.*

- The different points on the consumption loops are listed on the right side of the Fig. 2.144.

II. Load Outputs :

- In case of a *diesel engine* the *pumping losses remain reasonably constant from full load to no load* whereas the pumping losses for a *petrol engine progressively rise as the b.m.e.p. is reduced due to the increased throttling.* This is reflected in the petrol engine by the mechanical efficiency reducing as b.m.e.p. decreases which causes the consumption loops to take up higher and higher levels of consumption as the intake becomes more throttled (Fig. 2.145). In contrast with the diesel engine, as the load is reduced the single consumption loop only begins to rise when the b.m.e.p. drops below about half the engine's maximum b.m.e.p. (Fig. 2.145).

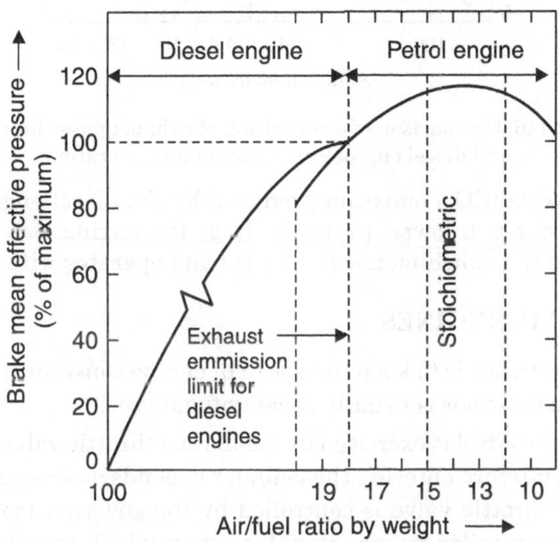

Fig. 2.145. Comparison of load (b.m.e.p.) for petrol and diesel engine on base of air-fuel ratio.

- A *petrol engine* can *operate effectively* under steady conditions over a *mixture strength ranging from 20 : 1 to 10 : 1,* whereas a *diesel engine* which can *normally only utilize 80% of its air charge with a reasonably clear exhaust, will operate from 18 : 1, that is 20% weak at full load to 100 : 1 at no load.* Consequently, the petrol engine will always have a 15 to 20% higher maximum b.m.e.p. than a similar diesel engine (Fig. 2.145).

III. Exhaust Composition

Fig. 2.146 shows the *comparison of composition of exhaust gases for petrol and diesel engines on a base of air-fuel ratio ;* this figure supplies the following information :

— The petrol engine has considerable amounts of carbon monoxide (CO) formed as the mixture moves from the stoichiometric (14.7 : 1 by weight) to the rich region of upto 10 : 1. This being within the petrol engine's mixture range when operating between idle and full power conditions. In contrast the diesel engine under full load never operates with mixture strengths greater than 20% rich, that is 18 : 1 by weight, so that no carbon monoxide (CO) is present in the exhaust (Fig. 2.146).

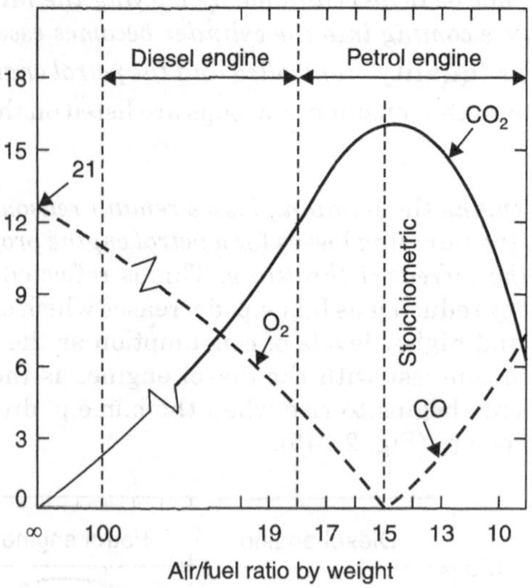

Fig. 2.146. Comparison of composition of exhaust gases for petrol and diesel engines on a base of air-fuel ratio.

— The carbon dioxide (CO_2) emission produced by the diesel engine relative to the petrol engine is always much lower, particularly as the engine load is reduced, whereas the petrol engine in the stoichimetric (14.7 : 1) band operates with the highest level of CO_2.

2.56. GOVERNING OF I.C. ENGINES

The function of a governor is to keep the speed of engine constant irrespective of the changes in load on the engine. The governor is usually of centrifugal type.

In petrol engine, the control is exercised by means of a throttle valve which is placed in intake manifold. The quantity of mixture entering the cylinder depends on the amount of opening of throttle valve. The position of throttle valve is controlled by the governor (centrifugal type). In diesel engines, the flow of fuel is controlled by centrifugal governor which actuates link rods which in turn operate some device on the fuel pump and consequently portion of the fuel bypasses. The governor in plunger type injection pump alters the relative angular position of the plunger.

Following are the *methods of governing I.C. engines* :

(*i*) Hit and miss method.

(*ii*) Quality governing.

(*iii*) Quantity governing.

(*i*) **Hit and miss method.** Refer to Fig. 2.147. When the speed increases the permissible value the governor sleeve *S* gets lifted up, as a result of which the lever *A* lifts the distant piece *B*, so that the pecker *K* misses it. Thus the gas inlet valve *V* does not open and the usual charge does not enter the cylinder. This continues until the speed is reduced and *B* occupies its initial position. Explosions are thus missed intermittently but every charge is of normal strength. *This method is commonly used in gas engines.*

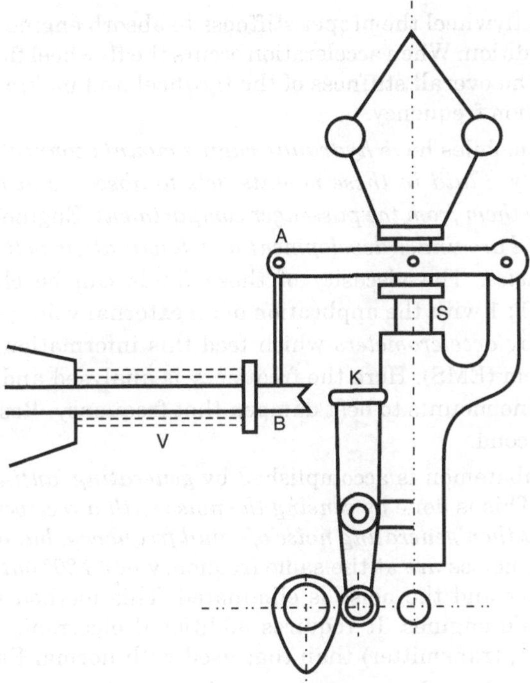

Fig. 2.147. Hit and miss governing.

(*ii*) **Quality governing.** In this method of governing, the *mixture strength is altered.* In gas engine it is effected by reducing the amount of gas supplied to the engine. This is accomplished by varying the lift of the gas valve. In oil engines, quality governing is carried out by varying the quality of fuel oil entering the cylinder per cycle, it is done by changing the angular position of the helical groove of the pump plunger.

In this type of governing, the ignition is not always satisfactory and thermal efficiency is reduced.

(*iii*) **Quantity governing.** Here *mixture strength remaining the same, the quantity of mixture entering the cylinder is altered.* When the speed is too high a lesser amount of charge is admitted into the cylinder. The compression ratio and air standard efficiency remain unchanged. The pressure after compression and during working stroke is lower, the less work thus obtained during the cycle reduces the speed. *This method is preferred for large engines.*

2.57. NOISE ABATEMENT

A lot of research and developement is in progress towards *reducing engine and exhause noise.* This can be accomplished by the following three ways :

1. Passive 2. Semi-active 3. Active.

- Noise reduction is accomplished passively by *correct design and the use of proper materials.*
  - — The use of *ribs and stiffners, composite materials, and sandwich construction is now routine.* This type of construction *reduces noise vibrations in the various engine components.*
- In **semi-active** *noise abatement systems,* **hydraulics** *are often used.*
  - — Some engines are equipped with flywheels which have hydraulic passages through which fluid flows. At idle and other constant-speed operation, the system is designed

to give the flywheel the proper stiffness to absorb engine vibrations for frequencies at that condition. When acceleration occurs the flywheel fluid flows to other locations, changing the overall stiffness of the flywheel and making it more absorbent to the new vibration frequency.

— Some automobiles have *hydraulic engine mounts connecting the engine to the automobile body. Fluid in these mounts acts to absorb and dampen engine vibrations and isolate them from the passenger compartment.* Engine mounts using *electrorheological fluid are under development which will allow better vibration dampening at all frequencies.* The viscosity of these fluids can be changed by as much as a factor of 50 : 1 with the application of an external voltage. Engine noise (vibration) is sensed by *accelerometers* which feed this information into the engine management system (EMS). Here the frequency is analysed and proper voltage is applied to the engine mounts to best dampen that frequency. Response time is of the order of 0.005 second.

- **Active noise** abatemen is accomplished by *generating 'antinoise'* to cancel out engine exhaust noise. This is done by *sensing the noise with a receiver, analyzing the frequency of the noise, and then generating noise of equal frequency, but out of phase with the original noise.* If the noises are at the same frequency but *180° out of phase,* the wave fronts *cancel each other* and the noise is eliminated. This method works well with constant speed automobile engines. It requires additional electronic equipment (receiver, frequency analyzer, transmitter) than that used with normal EMS computers.

  — Some automobiles have receivers and transmitters mounted under seats in the passenger compartment as an active engine noise abatement system. Similar systems are used near the end of the tail pipe, a major source of engine-related noise.

- Noise reduction has been so successful that some automobiles are now equipped with a safety switch on the starter. At idle speed, the engine is so quiet that the safety switch is required to keep drivers from trying to start the engine when it is already running.

Example 2.5. *A two-stroke cycle internal combustion engine has a mean effective pressure of 6 bar. The speed of the engine is 1000 r.p.m. If the diameter of piston and stroke are 110 mm and 140 mm respectively, find the indicated power developed.*

Solution. Mean effective pressure (indicated), p_{mi} = 6 bar

Engine speed, $N = 1000$ r.p.m.

Diameter of the piston, $D = 110$ mm = 0.11 m

Stroke length, $L = 140$ mm = 0.14 m

Indicated power developed, P :

Indicated power, $\text{I.P.} = \dfrac{np_{mi}\, LANk \times 10}{6}$ kW

Here, n = No. of cylinders = 1,

and, $k = 1$ ..... for 2-stroke cycle engine.

$$\therefore \text{ I.P.} = \frac{1 \times 6 \times 0.14 \times \frac{\pi}{4} \times (0.11)^2 \times 1000 \times 1 \times 10}{6} = \textbf{13.3 kW.} \quad \textbf{(Ans.)}$$

Example 2.6. *A four-cylinder four-stroke petrol engine develops 14.7 kW at 1000 r.p.m. The mean effective pressure is 5.5 bar. Calculate the bore and stroke of the engine, if the length of stroke is 1.5 times the bore.*

Solution. Number of cylinder, $n = 4$

Power developed, $P = 14.7$ kW

Engine speed, $N = 1000$ r.p.m.

Indicated mean effective pressure, $p_{mi} = 5.5$ bar

Length of stroke, $L = 1.5\ D$ (bore)

For four-stroke cycle, $k = \frac{1}{2}$

L :, D :

Indicated power developed,
$$\text{I.P.} = \frac{np_{mi}\ LANk \times 10}{6}\ \text{kW}$$

$$14.7 = \frac{4 \times 5.5 \times 1.5D \times \left(\frac{\pi}{4} \times D^2\right) \times 1000 \times \frac{1}{2} \times 10}{6}$$

$$\therefore \quad D^3 = \frac{14.7 \times 6 \times 4 \times 2}{4 \times 5.5 \times 1.5 \times \pi \times 1000 \times 10} = 0.0006806$$

$\therefore$ $D =$ **0.0879 or 87.9 mm. (Ans.)**

and $L =$ **1.5 × 87.9 = 131.8 mm. (Ans.)**

Example 2.7. *A single-cylinder, four-stroke cycle oil engine is fitted with a rope brake. The diameter of the brake wheel is 600 mm and the rope diameter is 26 mm. The dead load on the brake is 200 N and the spring balance reads 30 N. If the engine runs at 450 r.p.m., what will be the brake power of the engine ?*

Solution. Diameter of the brake wheel, $D_b = 600$ mm $= 0.6$ m

Rope diameter, $d = 26$ mm $= 0.026$ m

Dead load on the brake, $W = 200$ N

Spring balance reading, $S = 30$ N

Engine speed, $N = 450$ r.p.m.

Brake power, B.P. :

Brake power is given by,
$$B.P. = \frac{(W - S)\,\pi\,(D_b + d)N}{60 \times 1000}\ \text{kW}$$

$$= \frac{(200 - 30)\pi(0.6 + 0.026) \times 450}{60 \times 1000} = \mathbf{2.5\ kW.}\quad \textbf{(Ans.)}$$

Example 2.8. *A four-cylinder four-stroke S.I. engine has a compression ratio of 8 and bore of 100 mm, with stroke equal to the bore. The volumetric efficiency of each cylinder is equal to 75%. The engine operates at a speed of 4800 r.p.m. with an air-fuel ratio 15.*

Given that the calorific value of fuel = 42 MJ/kg, atmospheric density = 1.12 kg/m³, mean effective pressure in the cylinder = 10 bar and mechanical efficiency of the engine = 80%, determine the indicated thermal efficiency and the brake power. **(GATE)**

Solution. *Given* : Number of cylinders, $n = 4$, $k = \frac{1}{2}$ (engine being four-stroke),

$r = 8$, D = 100 mm = 0.1 m ; $L = D = 0.1\ m$; $\eta_{vol} = 75\%$; $N = 4800$ r.p.m.,

Air-fuel ratio = 15 ; $C = 42$ MJ/kg ; $\rho = 1.12$ kg/m^3 ; $p_{mi} = 10$ bar, $\eta_{mech} = 80\%$.

Indicated thermal efficiency, $\eta_{th.\,(I)}$:

Indicated power, $\quad \text{I.P.} = \dfrac{n\, p_{mi}\, LANk \times 10}{6}$

$$= \frac{4 \times 10 \times 0.1 \times \left(\frac{\pi}{4} \times 0.1^2\right) \times 4800 \times \frac{1}{2} \times 10}{6} \text{ kW} = 125.66 \text{ kW}$$

Air consumption $\quad = n \times \dfrac{\pi}{4} D^2 \times L \times \dfrac{N}{2} \times \eta_{vol.}$

$$= 4 \times \frac{\pi}{4} \times 0.1^2 \times 0.1 \times \frac{4800}{2} \times 0.75$$

$$= 5.655 \text{ m}^3\text{/min} = 0.09425 \text{ m}^3\text{/s}$$

Mass flow of air, $\quad \dot{m}_a = 0.09425 \times 1.12 = 0.1056$ kg/s

Fuel consumption, $\quad \dot{m}_f = \dfrac{0.1056}{\text{Air fuel ratio}} = \dfrac{0.1056}{15} = 0.00704$ kg/s

$\therefore \quad \eta_{th(I)} = \dfrac{\text{I.P.}}{\dot{m}_f \times C}$

$$= \frac{125.66 \times 10^3}{0.00704 \times 42 \times 10^6} = 0.425 \text{ or } \mathbf{42.5\%}. \quad \textbf{(Ans.)}$$

Brake power, B.P.

Brake power $\quad = \text{Indicated power} \times \eta_{mech}$

$= 125.66 \times 0.8 =$ **100.53 kW. (Ans.)**

Example 2.9. *The following particulars were obtained in a trial on a four-stroke gas engine :*

Duration of trial	*= 1 hour*
Revolutions	*= 14000*
Number of missed cycle	*= 500*
Net brake load	*= 1470 N*
Mean effective pressure	*= 7.5 bar*
Gas consumption	*= 20000 litres*
L.C.V. of gas at supply condition	*= 21 kJ/litre*
Cylinder diameter	*= 250 mm*
Stroke	*= 400 mm*
Effective brake circumference	*= 4 m*
Compression ratio	*= 6.5 : 1*

Calculate : (i) Indicated power (ii) Brake power
(iii) Mechanical efficiency (iv) Indicated thermal efficiency
(v) Relative efficiency.

Solution. $\quad N = \dfrac{14000}{60} = \dfrac{700}{3}$ r.p.m., $W - S = 1470$ N,

$$p_{mi} = 7.5 \text{ bar} ; \; V_g = \frac{20000}{3600} = 5.55 \text{ litres/s},$$

$$D = 250 \text{ mm} = 0.25 \text{ m}, \; L = 400 \text{ mm} = 0.4 \text{ m}$$

$$\pi D_b = 4 \text{ m}, \; r = 6.5, \; n = 1.$$

(*i*) **Indicated power, I.P. :**

$$\text{I.P.} = \frac{np_{mi}LANk \times 10}{6}$$

$$Nk = \left(\frac{14000}{2} - 500\right) \Big/ 3600 = \frac{6500}{60} \text{ working cycles / min.}$$

$$\therefore \quad \text{I.P.} = \frac{1 \times 7.5 \times 0.4 \times \pi / 4 \times 0.25^2 \times (6500/60) \times 10}{6}$$

$$= \mathbf{26.59\ kW.} \quad \textbf{(Ans.)}$$

(*ii*) **Brake power, B.P. :**

$$\text{B.P.} = \frac{(W - S)\, \pi\, D_b N}{60 \times 1000} = \frac{1470 \times 4 \times (700/3)}{60 \times 1000} = \mathbf{22.86\ kW.} \quad \textbf{(Ans.)}$$

(*iii*) **Mechanical efficiency, $\eta_{mech.}$:**

$$\eta_{mech.} = \frac{\text{B.P.}}{\text{I.P.}} = \frac{22.86}{26.59} = \mathbf{0.859\ or\ 85.9\%.} \quad \textbf{(Ans.)}$$

(*iv*) **Indicated thermal efficiency, $\eta_{th.(I)}$:**

$$\eta_{th(I)} = \frac{\text{I.P.}}{V_g \times C} = \frac{26.59}{5.5 \times 21} = \mathbf{0.23\ or\ 23\%.} \quad \textbf{(Ans.)}$$

(*v*) **Relative efficiency, $\eta_{relative}$:**

$$\eta_{relative} = \frac{\eta_{thermal}}{\eta_{air\text{-}standard}}$$

$$\eta_{air\text{-}standard} = 1 - \frac{1}{(r)^{\gamma-1}} = 1 - \frac{1}{(6.5)^{1.4-1}} = 0.527 \text{ or } 52.7\%$$

$$\therefore \quad \eta_{relative} = \frac{0.23}{0.527} = \mathbf{0.436\ or\ 43.6\%.} \quad \textbf{(Ans.)}$$

Example 2.10. *Air consumption for a four-stroke petrol engine is measured by means of a circular orifice of diameter 3.2 cm. The coefficient of discharge for the orifice is 0.62 and the pressure across the orifice is 150 mm of water. The barometer reads 760 mm of Hg. Temperature of air in the room is 20°C. The piston displacement volume is 0.00178 m³. The compression ratio is 6.5. The fuel consumption is 0.135 kg/min of calorific value 43900 kJ/kg. The brake power developed at 2500 r.p.m. is 28 kW. Determine :*

(*i*) *The volumetric efficiency on the basis of air alone.*

(*ii*) *The air-fuel ratio.*

(*iii*) *The brake mean effective pressure.*

(*iv*) *The relative efficiency on the brake thermal efficiency basis.*

Solution. Diameter of circular orifice, $d = 3.2 \text{ cm} = 0.032 \text{ m}$

Coefficient of discharge, $C_d = 0.62$

Pressure across orifice, $h_w = 150$ mm of water

Temperature of air in the room $= 20°\text{C}$

Piston displacement $= 0.00178 \text{ m}^3$

Compression ratio, $r = 6.5$

Fuel consumption $= 0.135$ kg/min

Calorific value of fuel, $C = 43900$ kJ/kg

Brake power, B.P. = 28 kW

Speed = 2500 r.p.m.

$k = \frac{1}{2}$ .... for four-stroke cycle, engine.

(*i*) **Volumetric efficiency on the basis of air alone :**

Characteristic gas equation is written as

$$pV = mRT$$

or,

$$\frac{m}{V} = \frac{p}{RT} = \frac{1.0132 \times 10^5}{287 \times (20 + 273)} = 1.2 \text{ kg/m}^3$$

Also, 150 mm of $H_2O = \frac{150}{1000} \times 1000 = 150$ kg/m^2

Thus head of air column causing flow,

$$H = \frac{150}{1.2} = 125 \text{ m}$$

Thus air flow through the orifice

$$= \text{Air consumption} = C_d \times A \times \sqrt{2gH}$$

$$= 0.62 \times \frac{\pi}{4} \times (0.032)^2 \times \sqrt{2 \times 9.81 \times 125} = 0.0247 \text{ m}^3/\text{s}$$

Therefore, air consumption per stroke

$$= \frac{0.0247 \times 60}{\left(\frac{2500}{2}\right)} = 0.001185 \text{ m}^3$$

∴ *Volumetric efficiency,* $\eta_{vol.} = \frac{\text{Air consumption of stroke}}{\text{Piston displacement}}$

$$= \frac{0.001185}{0.00178} = \mathbf{0.665} \quad \textbf{or} \quad \mathbf{66.5\%}. \quad \textbf{(Ans.)}$$

(*ii*) **Air-fuel ratio :**

Mass of air drawn into the cylinder per min.

$$= 0.0247 \times 60 \times 1.2 = 1.778 \text{ kg/min}$$

∴ *Air-fuel ratio* $= \frac{1.778}{0.135} = \mathbf{13.67 : 1.}$ **(Ans.)**

(*iii*) **Brake mean effective pressure, p_{mb} :**

$$\text{B.P.} = \frac{n \times p_{mb} LANk \times 10}{6}$$

$$28 = \frac{1 \times p_{mb} \times 0.00178 \times 2500 \times \frac{1}{2} \times 10}{6} \qquad [\because \; LA = 0.00178 \text{ m}^3]$$

$$\therefore \quad p_{mb} = \frac{28 \times 6 \times 2}{0.00178 \times 2500 \times 10} = \mathbf{7.55 \text{ bar.}} \quad \textbf{(Ans.)}$$

(*iv*) **Relative efficiency :**

$$\eta_{\text{air-standard}} = 1 - \frac{1}{(r)^{\gamma-1}} = 1 - \frac{1}{(6.5)^{1.4-1}} = 0.527 \text{ or } 52.7\%$$

Brake thermal efficiency,

$$\eta_{\text{th.(B)}} = \frac{\text{B.P.}}{\dot{m}_f \times C} = \frac{28}{\frac{0.135}{60} \times 43900} = 0.2835 \text{ or } 28.35\%$$

$$\therefore \quad \eta_{\text{relative}} = \frac{\eta_{\text{thermal (B)}}}{\eta_{\text{air-standard}}} = \frac{0.2835}{0.527} = \mathbf{0.5379 \text{ or } 53.79\%.} \quad \textbf{(Ans.)}$$

Example 2.11. *A single-cylinder four-stroke diesel engine gave the following results while running on full load :*

Area of indicator card	*= 300 mm²*
Length of diagram	*= 40 mm*
Spring constant	*= 1 bar/mm*
Speed of the engine	*= 400 r.p.m.*
Load on the brake	*= 370 N*
Spring balance reading	*= 50 N*
Diameter of brake drum	*= 1.2 m*
Fuel consumption	*= 2.8 kg/h*
Calorific value of fuel	*= 41800 kJ/kg*
Diameter of the cylinder	*= 160 mm*
Stroke of the piston	*= 200 mm*

Calculate : (i) Indicated mean effective pressure. (ii) Brake power and brake mean effective pressure. (iii) Brake specific fuel consumption, brake thermal and indicated thermal efficiencies.

Solution. $N = 400$ r.p.m., $W = 370$ N, $S = 50$ N, $D_b = 1.2$ m,

$m_f = 2.8$ kg / h, $C = 41800$ kJ / kg, $D = 0.16$ m, $L = 0.2$ m,

$k = \frac{1}{2}$ ...... for 4-stroke cycle engine.

(*i*) **Indicated mean effective pressure, p_{mi} :**

$$p_{mi} = \frac{\text{Area of indicator diagram or card} \times \text{spring costant}}{\text{Length of diagram}}$$

$$= \frac{300 \times 1}{40} = \mathbf{7.5 \text{ bar.}} \quad \textbf{(Ans.)}$$

Indicated power, $$\text{I.P.} = \frac{np_{mi}LANk \times 10}{6} = \frac{1 \times 7.5 \times 0.2 \times \pi/4 \times 0.16^2 \times 400 \times \frac{1}{2} \times 10}{6}$$

$$= 10.05 \text{ kW.}$$

(*ii*) **B.P. , p_{mb} :**

Brake power, $$\text{B.P.} = \frac{(W - S)\,\pi\,D_b N}{60 \times 1000} = \frac{(370 - 50)\pi \times 1.2 \times 400}{60 \times 1000} = \mathbf{8.04 \text{ kW.}} \quad \textbf{(Ans.)}$$

Also, $$\text{B.P.} = \frac{np_{mb} \times LANk \times 10}{6}$$

$$8.04 = \frac{1 \times p_{mb} \times 0.2 \times \pi / 4 \times 0.16^2 \times 400 \times \frac{1}{2} \times 10}{6}$$

$$\therefore \quad p_{mb} = \frac{8.04 \times 6 \times 4 \times 2}{0.2 \times \pi \times 0.16^2 \times 400 \times 10} = \textbf{6 bar.} \quad \textbf{(Ans.)}$$

(*iii*) **b.s.f.c., $\eta_{th(B)}$, $\eta_{th.(I)}$:**

Brake specific fuel consumption,

$$b.s.f.c = \text{Fuel consumption per B.P. hour}$$

$$= \frac{2.8}{8.04} = \textbf{0.348 kg / B.P. hour.} \quad \textbf{(Ans.)}$$

Brake thermal efficiency,

$$\eta_{th.(B)} = \frac{\text{B.P.}}{\dot{m}_f \times C} = \frac{8.04}{\frac{2.8}{3600} \times 41800} = \textbf{0.2473 or 24.73\%.} \quad \textbf{(Ans.)}$$

Indicated thermal efficiency,

$$\eta_{th.(I)} = \frac{\text{B.P.}}{\dot{m}_f \times C} = \frac{10.05}{\frac{2.8}{3600} \times 41800} = \textbf{0.3091 or 30.91\%.} \quad \textbf{(Ans.)}$$

Example 2.12. *A four-cylinder, four-stroke cycle engine, 82.5 mm bore × 130 mm stroke develops 28 kW while running at 1500 r.p.m. and using a 20 per cent rich mixture. If the volume of the air in the cylinder when measured at 15.5°C and 762 mm of mercury is 70 per cent of the swept volume, the theoretical air-fuel ratio is 14.8, heating value of petrol used is 45980 kJ/kg and the mechanical efficiency of the engine is 90%, find :*

(*i*) *The indicated thermal efficiency.*

(*ii*) *The brake mean effective pressure.*

Take R = 287 N-m/kg K.

Solution. Number of cylinders, $n = 4$

Engine bore, $D = 0.0825$ m

Stroke length, $L = 0.13$ m

Brake power, B.P. = 28 kW

Engine speed, $N = 1500$ r.p.m.

Theoretical air-fuel ratio = 14.8

Calorific value of fuel, $C = 45980$ kJ/kg

Mechanical efficiency, $\eta_{mech.} = 90\%$.

(*i*) **Indicated thermal efficiency, $\eta_{th. (I)}$:**

Swept volume, $V_s = \pi/4\ D^2L = \pi/4 \times 0.0825^2 \times 0.13 = 0.000695\ \text{m}^3$

$$\text{Volume of air drawn in} = \frac{70}{100} \times 0.000695 = 0.0004865\ \text{m}^3$$

Given : $p = \frac{762}{760} \times 1.0132 = 1.015$ bar

$V = 0.0004865\ \text{m}^3$ (calculated above)

$R = 287$ N-m / kg K

$$T = 15.5 + 273 = 288.5 \text{ K}$$

$\therefore$ m = Mass of air/stroke/cylinder

$$= \frac{pV}{RT} = \frac{1.015 \times 10^5 \times 0.0004865}{287 \times 288.5} = 0.000596 \text{ kg}$$

Theoretical mass of air used per minute

$$= 0.000596 \times \frac{1500}{2} \times 4 = 1.788 \text{ kg}$$

Theoretical air-fuel ratio = 14.8

$\therefore$ Theoretical mass of fuel used / min

$$= \frac{1.788}{14.8} = 0.1208 \text{ kg/min}$$

When using 20% rich mixture, then

$$\dot{m}_f = \text{Mass of fuel burnt / sec}$$

$$= \frac{0.1208}{60} \times \frac{120}{100} = 0.002416 \text{ kg/s}$$

Now, $$\eta_{mech.} = \frac{\text{B.P.}}{\text{I.P.}} = \frac{28}{\text{I.P.}}$$

$$0.9 = \frac{28}{\text{I.P.}}$$

$\therefore$ $$\text{I.P.} = \frac{28}{0.9} = 31.11 \text{ kW.}$$

Indicated thermal efficiency,

$$\eta_{th.(I)} = \frac{\text{I.P.}}{\dot{m}_f \times C} = \frac{31.11}{0.002416 \times 45980} = \mathbf{0.28 \text{ or } 28\%.} \quad \textbf{(Ans.)}$$

(*ii*) **Brake mean effective pressure, p_{mb} :**

$$\text{B.P.} = \frac{n \times p_{mb} LANk \times 10}{6}$$

$$28 = \frac{4 \times p_{mb} \times 0.13 \times \pi / 4 \times 0.0825^2 \times 1500 \times \frac{1}{2} \times 10}{6}$$

$\therefore$ $$p_{mb} = \frac{28 \times 6 \times 4 \times 2}{4 \times 0.13 \times \pi \times 0.0825^2 \times 1500 \times 10} = \mathbf{8.06 \text{ bar.}} \quad \textbf{(Ans.)}$$

HEAT BALANCE SHEET

Example 2.13. *In a trial of a single-cylinder oil engine working on dual cycle, the following observations were made :*

Compression ratio = *15*

Oil consumption = *10.2 kg/h*

Calorific value of fuel = *43890 kJ/kg*
Air consumption = *3.8 kg/min*
Speed = *1900 r.p.m.*
Torque on the brake drum = *186 N-m*
Quantity of cooling water used = *15.5 kg/min*
Temperature rise = *36°C*
Exhaust gas temperature = *410°C*
Room temperature = *20°C*
c_p *for exhaust gases* = *1.17 kJ/kg K*

Calculate : (i) Brake power,
(ii) Brake specific fuel consumption, and
(iii) Brake thermal efficiency.

Draw heat balance sheet on minute basis.

Solution. $n = 1$, $r = 15$, $m_f = 10.2$ kg/h, $C = 43890$ kJ/kg, $m_a = 3.8$ kg/min.,
$N = 1900$ r.p.m., $T = 186$ N-m, $m_w = 15.5$ kg/min, $t_{w_2} - t_{w_1} = 36°C$,
$t_g = 410°C$, $t_r = 20°C$, $c_p = 1.17$.

(*i*) **Brake power, B.P. :**

$$\text{B.P.} = \frac{2\pi NT}{60 \times 1000} = \frac{2\pi \times 1900 \times 186}{60 \times 1000} = \mathbf{37\ kW.} \quad \textbf{(Ans.)}$$

(*ii*) **Brake specific fuel consumption, b.s.f.c. :**

$$\text{b.s.f.c.} = \frac{10.2}{37} = \mathbf{0.2756\ kg/kWh.} \quad \textbf{(Ans.)}$$

(*iii*) **Brake thermal efficiency,**

$$\eta_{th(B)} = \frac{\text{B.P.}}{\dot{m}_f \times C} = \frac{37}{\frac{10.2}{3600} \times 43890} = \mathbf{0.2975\ or\ 29.75\%.} \quad \textbf{(Ans.)}$$

Heat supplied by the fuel per minute

$$= \frac{10.2}{60} \times 43890 = 7461 \text{ kJ/min}$$

(*i*) Heat equivalent of B.P.

$$= \text{B.P.} \times 60 = 37 \times 60 = 2220 \text{ kJ/min.}$$

(*ii*) Heat carried away by cooling water

$$= m_w \times c_{pw} (t_{w_2} - t_{w_1}) = 15.5 \times 4.18 \times 36 = 2332 \text{ kJ/min.}$$

(*iii*) Heat carried away by exhaust gases

$$= m_g \times c_{pg} \times (t_g - t_r)$$

$$= \left(\frac{10.2}{60} + 3.8\right) \times 1.17 \times (410 - 20) = 1811 \text{ kJ/min.}$$

Heat balance sheet (minute basis)

Item	*kJ*	*Per cent*
Heat supplied by fuel	***7461***	***100***
(*i*) Heat absorbed in B.P.	2220	29.8
(*ii*) Heat taken away by cooling water	2332	31.2
(*iii*) Heat carried away by exhaust gases	1811	24.3
(*iv*) Heat unaccounted for (by difference)	1098	14.7
Total	7461	100

MORSE TEST

Example 2.14. *In a test of a four-cylinder, four-stroke engine 75 mm bore and 100 mm stroke, the following results were obtained at full throttle at a particular constant speed and with fixed setting of fuel supply of 6.0 kg/h.*

B.P. with all cylinder working = *15.6 kW*
B.P. with cylinder no. 1 cut-out = *11.1 kW*
B.P. with cylinder no. 2 cut-out = *11.03 kW*
B.P. with cylinder no. 3 cut-out = *10.88 kW*
B.P. with cylinder no. 4 cut-out = *10.66 kW*

If the calorific value of the fuel is 83600 kJ/kg and clearance volume is 0.0001 m³, calculate :

(*i*) *Mechanical efficiency,* (*ii*) *Indicated thermal efficiency, and*

(*iii*) *Air standard efficiency.*

Solution. B.P. = I.P. – F.P.

Assuming that the engine is running at constant speed the *frictional and pumping losses remain constant.* Now if one cylinder is cut-out it will not produce any power but the frictional loss and power lost in operating the valves will remain the same as the speed of the engine is constant. The B.P. reduction at the crankshaft due to one cylinder cut out will be *exactly equal to the I.P. produced by that cylinder.*

Therefore,

I.P. produced in cylinder 1, $IP_1 = BP - BP_1 = 15.6 - 11.1 = 4.5$ kW
I.P. produced in cylinder 2, $IP_2 = BP - BP_2 = 15.6 - 11.03 = 4.57$ kW
I.P. produced in cylinder 3, $IP_3 = BP - BP_3 = 15.6 - 10.88 = 4.72$ kW
I.P. produced in cylinder 4, $IP_4 = BP - BP_4 = 15.6 - 10.66 = 4.94$ kW
Total I.P. produced $= IP_1 + IP_2 + IP_3 + IP_4$

$$\text{I.P.} = 4.5 + 4.57 + 4.72 + 4.94 = 18.73 \text{ kW}$$

(*i*) **Mechanical efficiency, $\eta_{mech.}$:**

$$\eta_{mech.} = \frac{\text{B.P.}}{\text{I.P.}} = \frac{15.6}{18.73} = \mathbf{0.833 \text{ or } 83.3\%.} \quad \textbf{(Ans.)}$$

(*ii*) **Indicated thermal efficiency, $\eta_{th(I)}$:**

$$\eta_{th.(I)} = \frac{\text{I.P.}}{\dot{m}_f \times C} = \frac{18.73}{\frac{6}{3600} \times 83600} = \mathbf{0.1344 \text{ or } 13.44\%.} \quad \textbf{(Ans.)}$$

(iii) **Air-standard efficiency, $\eta_{air\text{-}standard}$:**

Stroke volume, $V_s = \frac{\pi}{4} D^2 L = \frac{\pi}{4} \times 0.075^2 \times 0.1 = 0.0004417 \text{ m}^3$

Clearance volume, $V_c = 0.0001 \text{ m}^3$

Compression ratio, $r = \frac{V_s + V_c}{V_c} = \frac{0.0004417 + 0.0001}{0.0001} = 5.4$

$$\therefore \quad \eta_{air\text{-}standard} = 1 - \frac{1}{(r)^{\gamma - 1}} = 1 - \frac{1}{(5.4)^{1.4-1}} = \mathbf{0.49} \quad \textbf{or} \quad \mathbf{49\%}. \quad \textbf{(Ans.)}$$

VIII. Miscellaneous Engines

2.58. DUAL-FUEL AND MULTI-FUEL ENGINES

2.58.1. Dual-Fuel Engines

2.58.1.1. Introduction

- Owing to various technical and financial reasons, some engines are designed to operate, *using a combination of two fuels.* For instance, in some third-world countries *dual-fuel engines are used because of high cost of diesel fuel.*
- Large C.I. engines are run on a *combination of methane and diesel oil.* Methane is the main fuel because it is more cheaply available. However, methane is not a good C.I. fuel by itself because it *does not readily self-ignite* (due to its *high octane number*). *A small amount of diesel oil is injected at the proper cycle time. This ignites in a normal manner and initiates combustion in the methane-air mixture filling the cylinder.*
- On these type of engines, *combination of fuel input systems are needed.*
- *Dual-fuel operation combines in a simple manner the possibility of operating a diesel engine on liquid fuels such as diesel oil or gas oil and on gaseous fuels such as natural gas, sewage gas and cook oven gas etc.*
  - The engine can be *switched from dual-fuel operation almost instantaneously in case of emergency.*

2.58.1.2. Working of dual-fuel engine

Fig. 2.148 shows a diesel engine with necessary modifications, which it requires when it *works on dual-fuel.*

The working of a dual-fuel engine, which works on a diesel cycle, is given below :

- The mixture of air and gaseous fuel is compressed in the cylinder similar to air compression in diesel engine.
- A small amount of liquid fuel, (diesel fuel), called *pilot fuel* or the *secondary fuel* (5 to 7% of the fuel at full load) is injected near T.D.C., irrespective of load on the engine. This pilot injection acts as a *source of ignition.* The gas-air mixture in the vicinity of injected spray ignites at number of places establishing a number of flame-front and combustion continues rapidly and smoothly. (In a four-stroke cycle engine the gas is supplied in the intake manifold where it mixes with the incoming air to form a homogeneous mixture. In two-stroke cycle engines the gas may be injected at low pressures, from 1.5 to 3.5 bar, directly into the combustion chamber after the closure of the exhaust ports). It is

interesting to note that in the dual-fuel engine, the combustion starts similar to C.I. engine but it propagates by flame similar to S.I. engine.

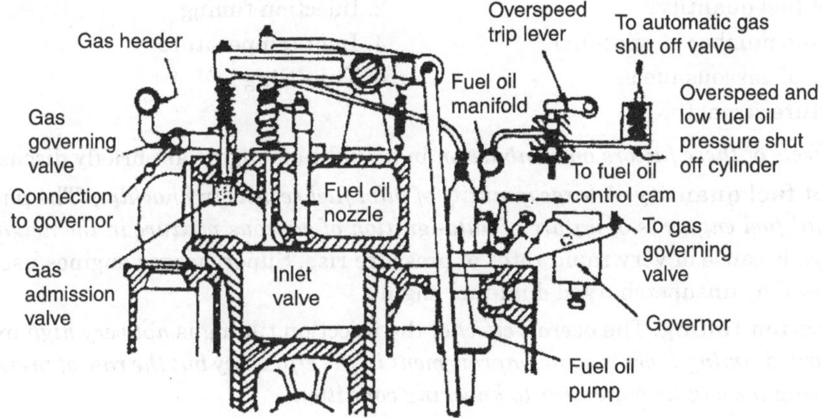

Fig. 2.148. Dual-fuel engine.

- The engine output (power) is generally controlled by *altering the amount* of primary gaseous fuel added to inlet manifold (the pilot oil quantity is usually kept constant).
- This type of engine is *capable of running on either gas or diesel oil or a combination of these two over a wide range of temperature ratios.*
- Fig. 2.149 shows the p-θ diagram for a dual-fuel engine working on different fuels (change from one fuel to another being quite smooth).

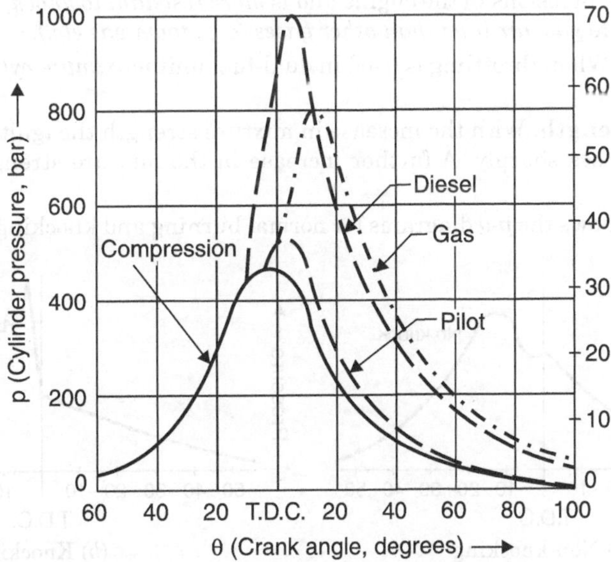

Fig. 2.149. Typical p-θ diagram for a dual-fuel engine.

2.58.1.3. Factors affecting combustion in a dual-fuel engine :

The following factors affect combustion in dual-fuel engine :

1. Pilot fuel quantity
2. Injection timing
3. Cetane number of pilot-fuel
4. Inlet temperature
5. Type of gaseous fuels
6. Throttling
7. Mixture strength.

The *effects of these factors on combustion* in a dual-fuel engine are briefly discussed below :

1. **Pilot fuel quantity.** A *large quantity of pilot fuel results in knocking* (The *phenomenon of knock in a dual-fuel engine is of nature of auto-ignition of gaseous mixture in the neighbourhood of injected spray*), because of very rapid rates of pressure rise. Supercharger engines use about same pilot fuel as used by unsupercharged dual-fuel engine.

2. **Injection timing.** The *overall effect* of the injection timing is *not very high* except that at a *slightly retarded timing there is some improvement in the efficiency but the rate of pressure rise also increases making the engine more near to knocking condition.*

3. **Centane number of pilot fuel :**

- The *use of low cetane number fuels results in poor performance of the engine* and *greatly affects the combustion.*
- The ignition quality of fuel, in general, exercises little effect on combustion in dual-fuel engines as compared to the ignition quality of the primary fuel.

4. **Inlet temperature.** The *temperature of inlet charge greatly affects power output, ignition limits, the rate of pressure rise* and hence the *knocking limits of a particular fuel-air mixture.*

5. **Type of gaseous fuels.** The type of gaseous fuel used decides amply the knock-limited output of the dual-fuel engine. Methane (main constituent of natural gas) does not undergo decomposition during the compressions in the engine and is *more resistant to knock, pre-ignition and back-firing from cylinder into gas/air inlet than other gases (e.g., town gas etc.).*

6. **Throttling.** When throttling is used in dual-fuel unit *maximum cylinder pressure reduces greatly.*

7. **Mixture strength.** With the increase in mixture strength the ignition delay reduces rapidly and then again rises sharply. A further increase in the mixture strength leads to failure of ignition.

- Fig. 2.150 shows the p-θ diagrams for normal burning and knocking for dual-fuel engines.

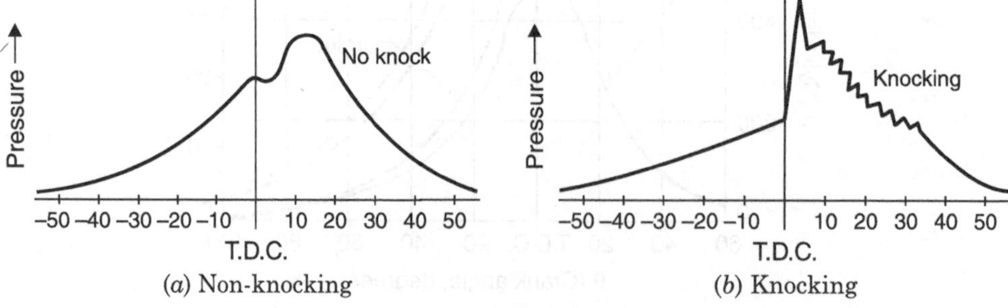

Fig. 2.150. Typical p-θ diagrams for a dual-fuel engine.

— The pressure oscillations during expansion are high and similar to S.I. engine knocking condition. *A small increase in gaseous fuel beyond a limit can cause a very severe knocking which may damage the engine.*

— The *knock tendency of the engine can be reduced by supplying excess air with low initial temperature and by using the antiknock additive* (*e.g.*, tetramethyl lead).

2.58.1.4. Performance of dual-fuel engines

• Fig. 2.151 shows the relative performance of duel-fuel engines.

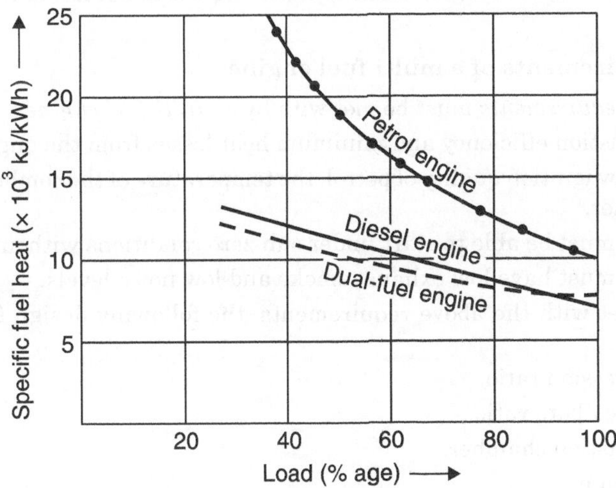

Fig. 2.151. Comparison of performance of dual-fuel engine with other engines.

— At *part load conditions,* in case of *dual-fuel engine, the thermal efficiency is low (38.5% against 40% of diesel engine) and specific fuel consumption high* because of increased delay periods at weak mixture of air and gas. This requires *either adjustment of mixture strength or injecting more pilot fuel at part loads.*

— *At full load, dual-fuel engine is superior to diesel engine since the gaseous fuel fills all parts of the combustion chamber and allows more air to take part in the combustion.*

2.58.1.5. Advantages and applications of dual-fuel engines

Advantages :

1. A dual-fuel engine can run on either of the fuel and diesel requirement is hardly five percent if it runs on gas.
2. Reduced pollution.
3. It is preferred when cheap gas is easily available.
4. Owing to clean combustion in the engine the wear and tear of the engine as well as consumption of lubricating oil are reduced.
5. The engine's utility is considerably increased due to the possibility of instantaneous change over from gas to diesel and *vice-versa.*
6. Best suited for low pressure liquified gas (LPG) which evaporates very easily.
7. Very attractive power generation system because of its greater flexibility of operation compared with conventional diesel engines.

Applications :

1. A typical use of dual-fuel engine is to produce synthetic gas (mixture of CO and H_2) by burning CH_4 and simultaneously developing power.
2. A lot of conventional fuel can be saved by using dual-fuel engines particularly for irrigation purposes.

2.58.2. Multi-fuel Engines

2.58.2.1. Introduction

A **multi-fuel engine** is *one which can operate satisfactorily (with substantially unchanged performance and efficiency) on a wide variety of fuels ranging from diesel oil, crude oil, IP-4 to lighter fuel like gasoline, and even normal lubricating oil.* The use of low ignition quality fuels require higher compression ratio for burning, therefore, diesel engines are prefered to use multi-fuels in the engine.

2.58.2.2. Requirements of a multi-fuel engine

The following *requirements* must be met with by a *multi-fuel engine* :

1. Good combustion efficiency and minimum heat losses from the engine.
2. In view of low ignition quality of petrol, the temperature of the combustion chamber should be comparatively higher.
3. The engine must be able to start under sub-zero conditions without any external aid.
4. The engine must have low exhaust smoke and low noise levels.

In order to meet with the above requirements, the following design features of the engine need be considered :

(*i*) High compression ratio.
(*ii*) Large stroke / bore ratio.
(*iii*) Open combustion chamber.
(*iv*) Injection pump.

2.58.2.3. Difficulties associated with multi-fuel operation

The multi-fuel operation entails the following *difficulties* :

1. Tendency of vapour lock in the fuel pump while using lighter fuels.
2. Tendency of increased wear in the fuel pump due to lower lubricity of gasoline.
3. In view of differences in heating values and compressibility of fuels, different volumes (of fuels) need be injected.

2.58.2.4. Performance of multi-fuel engine

- Fig. 2.152 shows the performance of multi-fuel engine.

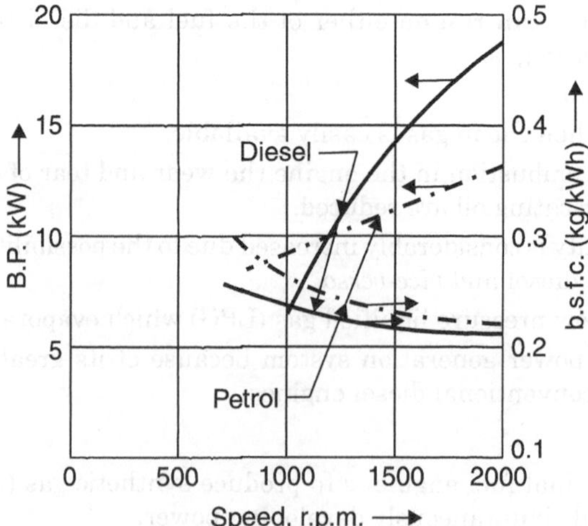

Fig. 2.152

— The engine output is 15 to 20 percent less with petrol as fuel compared with diesel oil at *higher speed.* The difference is less in lower speeds.

— The b.s.f.c. (brake specific fuel consumption) for petrol engine is also higher at lower speed compared with diesel fuel. The difference narrows down at higher speeds.

2.59. STRATIFIED CHARGE ENGINE

2.59.1. Introduction

- The **stratified charge engine** is *usually defined as a S.I. engine* (stratified diesel engine has also been developed) *in which the mixture in the zone of spark-plug is very much richer than that in the rest of the combustion chamber i.e., one which burns leaner overall fuel-air mixtures.*
  - — **Charge stratification** *means providing different fuel-air mixture strengths at various places in the combustion chamber.*
- Whereas several S.I. engines are designed to have a homogeneous air-fuel mixture throughout the combustion chamber, some modern stratified charge engines are designed to have a *different air-fuel ratio at different locations within the combustion chamber.* A *rich mixture that ignites readily is desired around the spark plug, while the major volume of the combustion chamber is filled with a very lean mixture that gives good fuel economy. Special intake systems are necessary to supply this non-homogeneous mixture, combination of multiple valves and multiple fuel injectors, alongwith flexible valve and injection timing are used to accomplish the desired results.*
- Some stratified charge S.I. engines are operated with no throttle, which raises the volumetric efficiency. Speed is controlled by proper timing and quantity of fuel input.
- The *stratified charge engine combines the advantages of both petrol engines* (very good full load power characteristics *e.g.*, high degree of air utilisation, high speed and flexibility) *and diesel engines* (good part-load characteristics) and *at the same time avoids as far as possible their disadvantages.*

The following are the *advantages of burning overall fuel-air mixtures :*

(*i*) Higher thermodynamic efficiency.

(*ii*) Reduced air pollution.

2.59.2. Classification

According to the method of formation of the heterogeneous mixture in the combustion chamber, the stratified charge engines can be classified as follows :

I. ***Those using fuel injection and positive ignition (including swirl stratified charge engines)***

(*a*) *Stratification by fuel injection and positive ignition :*

1. Ricardo system.
2. Pre-chamber stratified charge engine.
3. Volkswagon PCI stratified charge engine.
4. Broderson method of stratification.

(*b*) *Swirl stratified charge engine :*

1. Witzby swirl stratification process.
2. Texaco combustion process (TCP).
3. Ford combustion process (FCP).
4. Ford PROCO.
5. Deutz combustion process (AD-process).

II. ***Stratification by carburetion alone***

1. Russian stratified charge concept.
2. Institute Francias Du Petrols (IFP) process.
3. Honda CVCC (Compound Vertex Controlled Combustion) engine.

Here we shall discuss in detail only Texaco Combustion Process (TCP).

Texaco combustion process :

The carburetted S.I. engine has several problems and areas of potential weaknesses, these are given below :

(*i*) The end charge has a long residence time. It may become highly reactive, and it may cause knock. Therefore a fuel of high octane rating is required.

(*ii*) Homogeneous mixtures with the air-fuel limits of propagating a flame yield relatively low enthalpy efficiencies and also relatively high HC and NO_x emissions.

(*iii*) Throttling as a means of controlling output induces a pumping loss.

(*iv*) Flame quenching at the walls adds to air pollution.

To overcome these weaknesses (and hopefully without adding others) stratified charged engine has been developed.

- The Texaco Combustion Process (TCP) is illustrated in Fig. 2.153. This system was patented by Texas company (USA) in 1949.

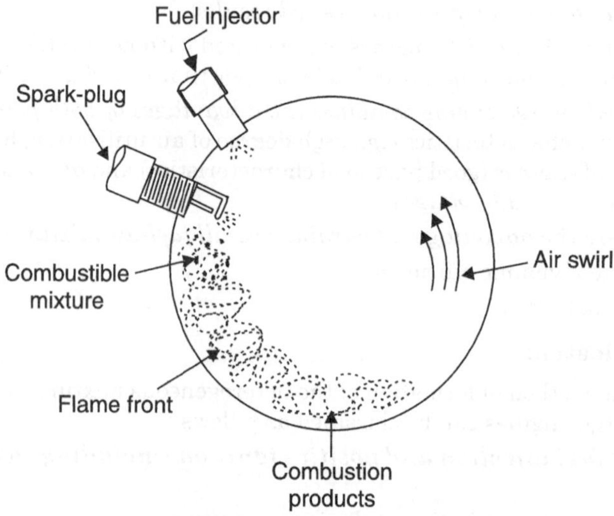

Fig. 2.153. Barber's Texaco combustion process.

— In this system, *the fuel spray is injected at the end of the compression stroke toward spark-plug which is entrained by swirl air and forms almost stoichiometric mixture near spark plug.*

— The *charge from the injected region starts burning successively but the flame front located between the spark-plug and the injector remains almost stationary.*

- This system entails the following *advantages :*
  - (*i*) Gives good performance over whole range of load and speed with a wide range of fuels ranging from premium gasoline to high cetane diesel fuel.
  - (*ii*) Besides giving better part load efficiency the starting and warm-up characteristics of TCP are very good due to employment of cup combustion chamber.

(*iii*) The inherent knock resistance of TCP allows the use of higher compression ratio or turbocharging without imposing any octane requirement.

(*iv*) The lean overall mixture can be used.

(*v*) Low exhaust smoke.

- TCP system has proved very versatile as regards multi-fuel requirements and are *used in military vehicle engines.*

2.59.3. Advantages and Disadvantages of Stratified Charge Engines

The *advantages* and *disadvantages* of stratified charged engines are listed below :

Advantages :

1. Compact, lightweight design and good fuel economy.
2. Good part-load efficiency.
3. Exhibit multi-fuel capability.
4. The rich mixture near the spark-plug and lean mixture near the piston surface *provides cushioning to the explosive combustion.*
5. *Resist the knocking* and provide smooth combustion resulting in smooth and quiet engine operation over the entire speed and load range.
6. Low level of exhaust emissions ; NO_x is *reduced* considerably.
7. Usually *no starting problem.*
8. Can *tolerate wide quality of fuels.*
9. Can be manufactured by the existing technology.

Disadvantages :

1. For a given engine size, charge stratification results in *reduced power.*
2. These engines create *high noise level at low load conditions.*
3. *More complex design* to supply rich and lean mixture, and quantity is varied with load on the engine.
4. *Higher weight* than that of a conventional engine.
5. Unthrottled stratified charge engines *emit high percentage of HC* (due to either incomplete combustion of lean charge or occasional misfire of the charge at part load conditions).
6. Reliability is yet to be well established.
7. Higher manufacturing cost.

2.60. STIRLING ENGINE

2.60.1. Stirling Cycle

The Stirling cycle is ***superior** to the Carnot cycle because of the following reasons :*

1. The Stirling cycle is *practicable,* whereas the *Carnot cycle cannot be realised in practice* due to wide variation in speed during a cycle (alternately very high for adiabatic and very slow for the isothermal part of the cycle).

2. The work output per cycle and *m.e.p. of Stirling cycle are high. The Carnot cycle needs a very long stroke and produces a very narrow strip of work giving a low m.e.p.*

2.60.2. Working Principle of Stirling Engine

The basic principle of working of stirling engine is the same as that of conventional engine. *The alternate compression at low temperature and expansion at high temperature of a working fluid is the basis for the stirling engine.* However, the working fluid is heated in a radically different manner. *It burns fuel outside the engine itself, and continuously.*

The following features *distinguish* the stirling engine from other heat engines :

1. The air and fuel are *externally burned* and the heat generated is transferred to the working fluid (may be air or any other suitable gas) of stirling engine where the *working fluid works in a closed cycle.*

2. The cyclic flow of working fluid within the engine is achieved solely through geometric volume changes, and without the use of intermittently closed valves or ports.

3. An intermittent flow heat exchanger stores a large portion of the heat of the working fluid after expansion and subsequently returns it to the working fluid after compression, thereby accomplishing thermal regeneration.

Fuel used for stirling engine :

Since the stirling engine is an *external combustion engine, it possesses multi-fuel capacity.* It can use any petroleum fraction such as *gasoline, diesel, methanol-gasoline blends, etc., with no octane or cetane requirements.* Thus the stirling engine has the *desirable characteristic of adaptability to changing fuel availability.*

2.60.3. Differences between Carnot and Stirling Engines

The differences between Carnot and Stirling engines are given below :

S. No.	Carnot engine	Stirling engine
1.	Its cycle consists of *two isentropic and two isothermal processes.*	Its cycle consists of *two isothermal and two constant volume processes.*
2.	No heat is supplied during isentropic process.	Regenerative heat exchange takes place during constant volume processes. This is an ideal cycle and the net heat flow to the working fluid is zero in the constant volume processes, or the net effect of these processes is adiabatic.
3.	For its operation only one piston-cylinder is required.	For its operation two cylinder pistons are required (the displacer piston and the power piston).
4.	The cycle is *not practicable* as alternately the isothermal and isentropic processes are to be slow and fast respectively. The cycle is of *academic importance only.*	The cycle is *practicable,* as all the processes are performed at the same speed. Engines working on Stirling cycle have been in existence.
5.	It requires a very long stroke, and produces, *a very low m.e.p.,* due to a narrow indicator diagram.	It does not require a long stroke, and hence the *m.e.p. is relatively quite high.*

2.60.4. Engine Geometry and Driving Mechanism

The following are the three basic configurations/arrangements of engine layout :

Type 1 : *Two separate cylinders, cold and hot sides of the engine, each with working piston*—**Two-piston mechanism.**

Type 2 : A *Single cylinder with two pistons, one is the working piston and other is a displacer*—**Two -piston engine.**

Type 3 : Two *cylinders with a working piston in the cooled cylinder and displacer piston with the heated cylinder*—**Piston displacer engine.**

2.60.4.1. Two-piston mechanism

Fig. 2.154 shows the arrangement of the system. The whole mechanism can be divided into two systems :

(*i*) The displacer system ; (*ii*) The power system.

— *The function of the* ***displacer system*** *is to heat and cool the working fluid.*

— *The function of the* ***power system*** *is to obtain the power during expansion and supply power during compression.*

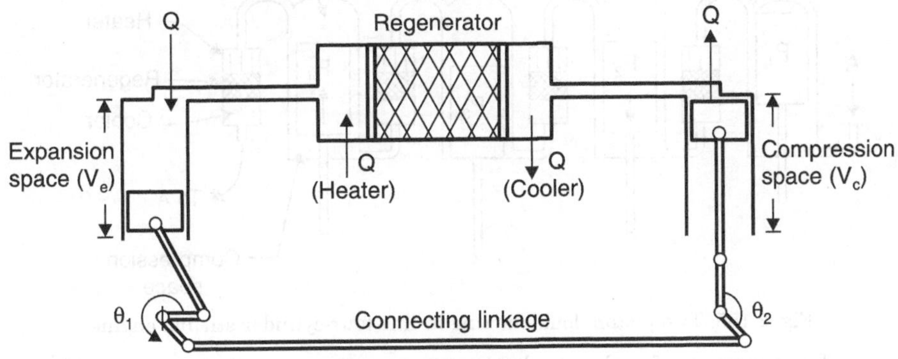

Fig. 2.154. Two-piston mechanism.

The total working volume (V_t) is given by :

$$V_t = V_e + V_c + V_d$$

where, V_e = Volume of expansion space,

V_c = Volume of compression space,

V_d = Volume of regenerator, heater, cooler and connecting tubes.

This total volume undergoes changes due to the motion of the two pistons.

When the crank rotates through one revolution, the following processes take place :

1. The working fluid is passed from hot space to the cold space via *heater, regenerator* and *cooler.* This is accomplished by phasing the two pistons such that there will be small change in the total working volume. Thus, the *heat associated with the working fluid is passed on to the regenerator at almost constant volume.*

2. The working fluid is compressed when it is mainly within the cold space and the cooler.

3. After the working fluid undergoes compression, it is passed from the cold space to hot space through the *cooler, regenerator* and *heater.* During this operation the heat of the regenerator is transferred back to the working fluid at *constant volume.*

4. The working fluid being located mainly within hot space and the heater, an expansion takes place thereby producing work output.

— In this system, these four processes are *not distinct and separate* but *overlap* each other, whereas in an ideal cycle the heat is rejected and received at constant volume and the compression and expansion occur isothermally.

— So long as the compression and expansion are isothermal and regeneration is effective, the thermal efficiency of the engine is least affected with the variation of total volume (V_t).

2.60.4.2. Two-piston engine

- Fig. 2.155 shows a new arrangement (the arrangement discussed earlier in Art. 2.60.4.1 is unsuitable for higher capacity engines) of a double acting suitable for *high capacity engine.* Here, compression of the working fluid occurs on one side and expansion on the other side of the engine.

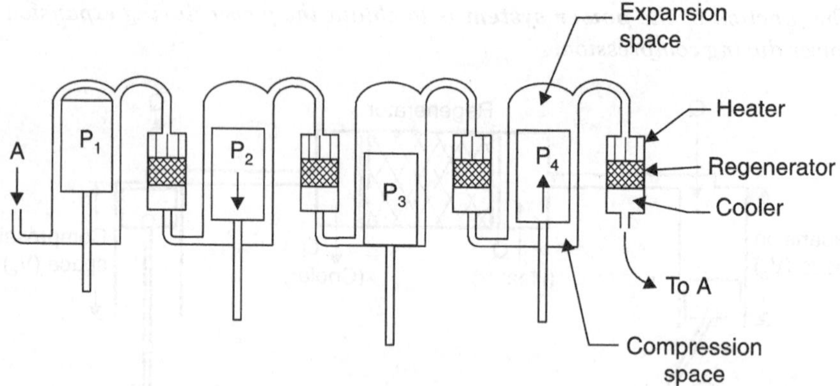

Fig. 2.155. Two-piston double-acting in-line four-cylinder stirling engine.

— The top space of each cylinder is an expansion space (hot space) and bottom space is compression space (cold space).

— The hot space of the cylinder is connected to the cold space of the other cylinder through heater, regenerator and cooler.

— The pistons move with a suitable phase shift (90° for four-cylinder engine).

- The *major drawback* of such an engine *is greater difficulty in obtaining a uniform flow through the cooler, regenerator and heater.* In view of greater aerodynamic and thermodynamic complexities involved this engine is more *difficult to be optimised for high efficiency.*

2.60.4.3. Piston displacer engine

- The piston displacer engine consists of a cylinder into which two pistons–one is known as *displacer piston* and other *power piston.* The cylinder of displacer piston is divided into hot space (expansion) and cold space (compression). The displacer piston reciprocates within the hot portion of the cylinder (open one) while the power piston reciprocates within the cold portion of the cylinder (lower one).
- *The main function of the displacer piston is to heat and cool the working fluid while the main function of the power-piston is to compress and expand the gas.*

Thus, the working of stirling engine is completely governed by relative motion of these two pistons.

2.60.4.4. Advantages and disadvantages

The following are the *advantages* and *disadvantages* of a stirling engines vis-a-vis conventional I.C. engine.

Advantages :

1. Since it is an external combustion engine, it possesses *multi-fuel capability.*
2. Owing to rhombic drive (this mechanism consists of two crankshafts rotating in opposite directions, each having a crank on it ; two synchronising gears are used to time these crankshafts which in turn drive the output shaft), it is a *perfectly balanced engine.*

3. *Reduced exhaust emission* (due to overall lean mixtures and exhaust gas circulation).
4. *Part load efficiency considerably higher.*
5. The engine can be *safely overloaded for a brief period.*
6. *Higher thermal efficiency* (Typical value is 45% at 100 bar and 300°C).
7. *Noise-free engine* (because combustion takes place outside the engine).
8. The *starting of the engine is reliable* (because it depends only on the ignition of fuel in the burner).
9. *Suitable for a variety of applications* (*e.g.*, automobiles, ship propulsion or power generation) since a stirling engine has almost *constant torque characteristics over the entire range of speed.*
10. *No lubricating oil required* (the use of rollsock seal has eliminated the use of lubricating oil).

Disadvantages :

1. *Complex engine design* (due to rhombic drive, regenerator, heater and cooler).
2. *Requires large quantity of cooling water and bigger radiator.*
3. The engines *requires a blower* to force the air through the preheater and the combustion chamber. This reduces the engine efficiency, and increases the noise.
4. The *major disadvantage* of a stirling engine is its ***high cost.***

2.60.4.5. Comparison between stirling engine and I.C. engine

The comparison between stirling engine and I.C. engine is given (in tabular form) below :

S. No.	Aspects	I.C. engine	Stirling engine
1.	*Fuel addition*	Atomised fuel is added in air, either before or after compression.	Continuous combustion takes place in combustion chamber and energy is transferred into or out of engine through an heat exchanger
2.	*Valve or ports*	Valves or ports required.	Valves or ports (for induction or exhaust) not required since the gas charge remains permanently inside the engine only.
3.	*Speed variation*	Achieved by controlling the amount of fuel injected or the quantity of mixture.	Achieved by changing the mass of working gas within the engine.
4.	*Heat rejection*	Achieved by high temperature exhaust.	Achieved through heat exchanger and hence it requires large cooling system.
5.	*Noise*	Noisy (due to presence of valves and periodic explosion).	Noiseless, smooth working.
6.	*Cooling system*	Sealing of cooling of system not absolutely essential.	Should be perfectly sealed.
7.	*Sealing*	The sealing can be done by the use of oil.	This engine does not use oils since it will cause contamination of the working fluid.

2.61. THE WANKEL ROTARY COMBUSTION (RC) ENGINE

2.61.1. Introduction

- Scheffel in 1952, to get a patent for *rotary engine*, utilised the *principle that oval or elliptical rotors can be designed to maintain contact, while turning about fixed centres, and that three or more rotors can be run to enclose between them a continuously varying volume.* The four volumes between the rotors, with the suitable arrangement of ports, ignition system and adequate compression ratio, could be made to execute a *four-phase otto cycle.* However, this design failed due to its complexity and great difficulties/problems involved in its manufacture.
- Felia Wankel (German inventor), in 1954, got a patent for design of four-phase rotary engine working on the otto cycle principle.
- Later Dr. Froede made certain modifications and an engine was developed, called as KKM (Kreiskolben Motor) ; now popularly known as Wankel Rotating Combustion (RC) engine.

2.61.2. Construction and Working

Construction. Refer Fig. 2.156. It consists of the following parts :

(*i*) **Rotor** (Three lobed).

(*ii*) **Eccentric or output shaft** with its integral eccentric. No connecting rod is required as the rotor rotates directly on the eccentric shaft. The *output torque is transmitted to the shaft through eccentric.*

(*iii*) **Internal and external timing gears.** They maintain the phase relationship between the rotation of the rotor and the eccentric shaft and eventually *control the orbital motion of the rotor.*

Working :

- The Wankel engine works on the *four-phase principle.*

  (The word *phase* corresponds to *stroke* of the reciprocating engine)

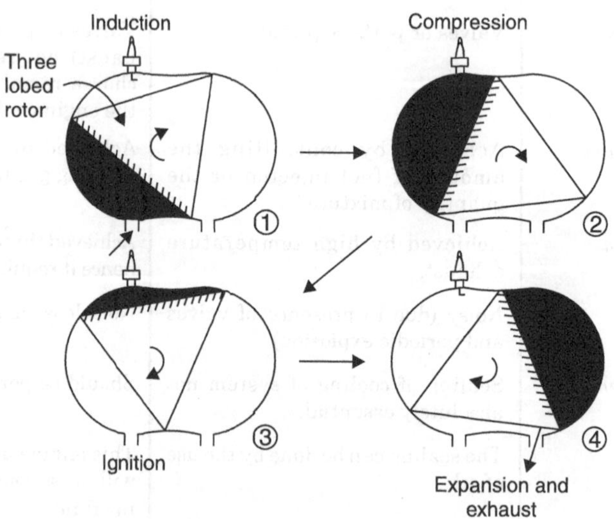

Fig. 2.156. The Wankel engine.

- The engine having three lobed rotor is driven eccentrically in a casing in such a way that there are *three separate volumes trapped between the rotor and the casing. These three volumes perform "induction", "compression", "combustion", "expansion" and "exhaust" processes in sequences.* There are three power impulses for each revolution of the rotor, and since the eccentric or output shaft rotates at three times the speed of the rotor, *there is only one power impulse for each revolution of the output shaft of a single bank rotary engine.*

One complete *thermodynamic cycle* is completed over 360° rotation of the rotor ; the suction phase takes 90° of rotor movement and so the also other three phases. One *thermodynamic phase* is completed every 270° rotation of the output shaft, since the output shaft makes three revolutions for every single rotation of the rotor.

2.61.3. Features

1. *Simple construction, less mechanical loss, smooth motion and does not require a cranking mechanism.*

2. *Good power volume ratio.*

3. *No reciprocating parts* and hence no balancing problem and complicated engine vibrations eliminated.

4. Due to the absence of intake-exhaust valve mechanism, the *correct timings* for opening and closing (the ports) *can be maintained even at high speeds.*

5. *Low torque fluctuation.*

There are problems in the design, notabty of sealing and of heat transfer but these have been overcome sufficiently well for spark ignition engine to be marketed.

2.61.4. Constructional and Other Details of Wankel Engine

Refer Fig. 2.157.

1. **Rotor housing and housing materials :**

- **Rotor** are generally made from *high-grade malleable spheroidal graphite iron.*
- The **rotor housing** is an *aluminium silicon alloy,* bonded to the cylinder-bore walls in thin sheet metal, the outer surfaces of which have a saw-tooth finish to improve adhesion and thermal conductivity. This lining is then given a hard chromium vanadium plating, which in turn is plated with more chrome but in a thin, porous and oil retaining layer. End and intermediate rotor housings are made from implated high silicon aluminium alloy.
- **Apex and side seal blades** can be made from *cast-iron* but the more popular types are made from *hard carbon material.*
- Both the **leaf and washer springs** can be made from *beryllium* copper, which has the ability to retain its elasticity when operating under working temperatures.

2. **Rotor seals :**

- The *planetary motion of the rotor within the epitrochoid bore of the rotor housing is designed to maintain a contact between the triangular corner of the rotor and the cylinder walls.* Peripheral radial corner blades, known as the *apex seals,* are necessary to prevent gas leakage between the three cylinder spaces created by the three-sided rotor. Similarly, *side seals* between the flat rotor sides and the end and intermediate housing side walls are essential to stop engine oil reaching the cylinders, and gas from combustion escaping into the eccentric output shaft region.
- The gas-tight sealing between the rotor and housing may be considered in terms of primary and secondary sealing areas.

— The *primary sealing areas* are those between the sealing elements (apex and side blades) and the cylinder housing bore and side walls.

— The *secondary sealing areas* are those between the sealing elements and grooves, plots and trunnion counter bores.

— The total sealed system around the rotor is known as a *sealed grid.*

- In order to ensure *minimum leakage at each rotor apex,* slotted cylindrical trunnion blocks are counter-sunk on both sides of the rotor at each rotor corner, the object being to provide a much longer leakage path at different meeting points where the apex and side blades overlap.

3. **Combustion and spark-plug location :**

- The improvement in combustion process can generally be made by installing *two spark-plugs :* one in the *lead combustion space and the other in the trailing region of the combustion-chamber.*

A tilting or rocking action takes place when the rotor is near the T.D.C. position and combustion commences. This tends to expand the gases in the leading portion of the combustion chamber but to compress them in the trailing region.

— *Positioning a plug in the high-pressure trailing region of the combustion chamber tends to produce slightly more power and use less fuel,* whereas if the plug *is located in the leading portion of the chamber, better starting and idle running is obtained.*

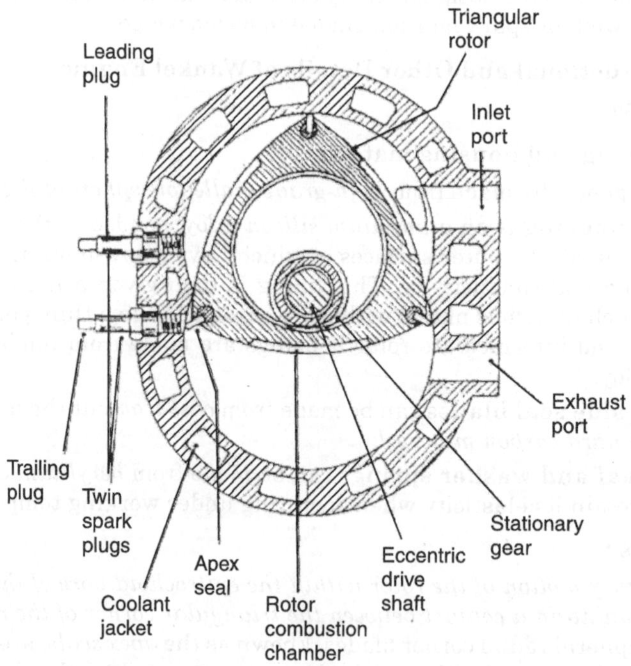

Fig. 2.157. Wankel engine-sectioned view.

— For achieving high efficiency it is necessary to have rapid combustion and pressure rise, but the backward flame propagation of the burning gases in the chamber is opposed by the forward movement of the gases as they are swept around with the rotor, which, in effect, slow down the speed of combustion and therefore does not give optimum performance.

- *Efficient combustion is produced by having a very advanced timing,* this *can be partially achieved by positioning a spark-plug in the trailing region of the bore wall as far as possible from the waist or minor axis of the epitrochoid wall profile.*

4. **Cooling system :**

- The engine block is made up of the following five separate section housings :
  - The front and rear end housings ;
  - Two rotor bore housings ;
  - Intermediate housing.

All of the above housings are clamped together by bolts.

- ***Coolant circulation*** *is provided around the bore walls by axial passages* cast *into each housing section, while the side walls of the twin cylinders are cooled by extended coolant passages adjacent to the side walls cast in both the end covers and the intermediate housing.*

Clearance and Swept Volume, Compression Ratio, Engine Displacement Capacity :

1. **Clearance and swept volume :**

The **clearance volume** (V_c) is *defined as the volume created between the curved chamber bore, flat side walls and one of the rotor flanks when in the nearest-approach position (T.D.C.)* adjacent to the waisted portion of the epitrochoid bore.

The **swept volume** (V_s) is the difference *of the maximum volume to which the chamber expands (B.D.C.) as the rotor revolves and the minimum volume of the combustion chamber at the point of greatest compression (T.D.C.)* It may be given by the following :

$$V_s = 3\sqrt{3}\,RWe$$

where, R = Radius of rotor from its centre to one of the apex tips (cm),
W = Width of the rotor (cm), and
e = Distance between the axes of the output shaft and eccentric lobe, known as the *eccentricity* (cm).

2. **Compression ratio :**

The Compression ratio (r), will be the conventional formula, as follows :

$$r = \frac{V_s + V_c}{V_c}$$

(where V_s and V_c are the swept volume and clearance volume respectively, in cm)

$$\therefore \quad r = \frac{(3\sqrt{3}RWe) + V_c}{V_c}.$$

3. **Engine displacement capacity :**

In a Wankel engine there are *three power strokes per rotor revolution,* and the output shaft completes three revolutions for every one revolution of the rotor. Therefore, there is *one power stroke per revolution of the output shaft.* This, therefore, can be seen to be equivalent to a two-cylinder four-stroke engine which also has a power stroke every crankshaft revolution. Because of this reasoning a *two-chamber swept volume equivalence is generally recognized.*

The displacement capacity (V_d) is therefore *twice the swept volume of a single chamber multiplied by the number of rotors* ; that is

$$V_d = 2\, V_s\, n$$

where, V_s = Swept volume, and

n = Number of rotors.

2.61.5. Performance of Wankel Engine

1. **Power and specific fuel combustion :**

Fig. 2.158 shows the effect of the speed on the power output and specific fuel consumption for I.C. engine (conventional) and Wankel engine.

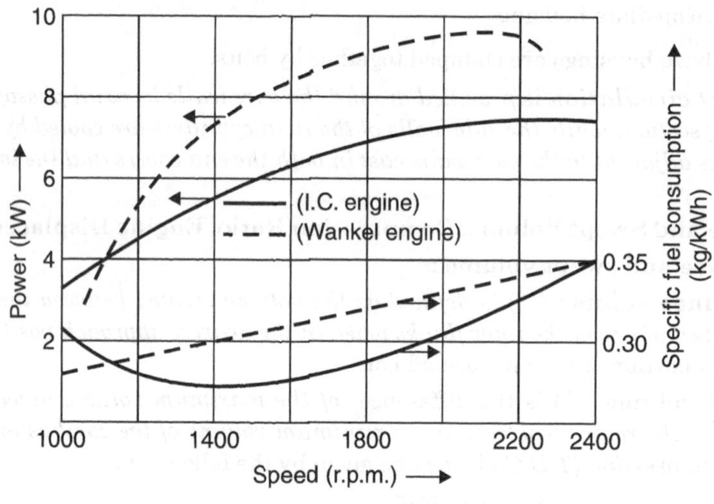

Fig. 2.158

— The *power of Wankel engine,* as compared with conventional I.C. engine, *increases rapidly.*

— The *frictional losses of Wankel engine,* with increase in speed, are *less* than I.C. engines (conventional).

— The *specific fuel consumption of wankel engine* is *higher* than I.C. engines (conventional).

2. **Brake thermal efficiency :**

- The thermal efficiency of Wankel engine is *less* than the conventional I.C. engine since the *cooling loss is more.*
- The thermal efficiency *can be increased by creating squish flow by providing recess in the rotor.*

3. **Exhaust emission.** Fig. 2.159 shows the emission of HC, CO and NO_x from Wankel engine (engine speed being 1450 r.p.m.) taking A/F ratio as variable.

— The presence of unburnt HC in the exhaust gases is due to quenching of the flame by the surface of the combustion chamber. The HC presence in this layer depends upon A/F ratio, temperature and turbulence in the layer.

— The variation of CO (percentage) is also shown in the Fig. 2.159. The emission of CO is higher at rich A/F ratio and decreases with increasing A/F ratio as sufficient air is available to burn CO as mixture is lean.

— The formation of NO_x depends upon O_2 content and peak cycle temperature. Since the maximum temperature in Wankel engine is *lower* than conventional I.C. engine, therefore *NO_x formation is low.*

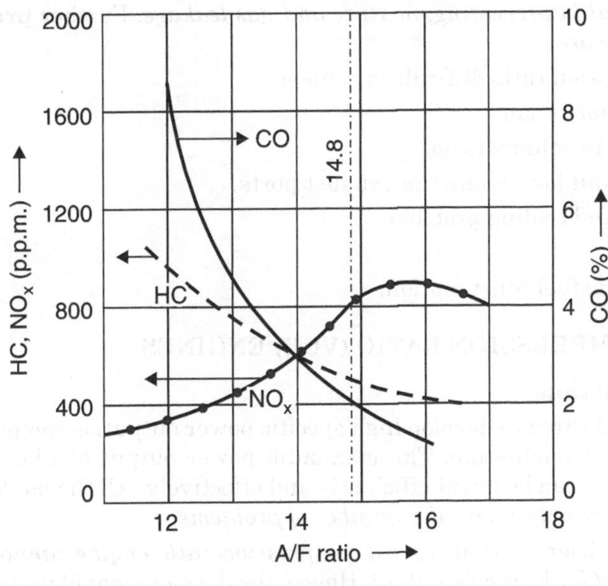

Fig. 2.159

2.61.6. Advantages and Applications of Rotary Combustion Engines

Advantages :

1. Lighter in weight, High power-to-weight ratio.
2. Compact, occupies less space.
3. Engine completely balanced, hence free from vibration and inertia forces.
4. Less noise.
5. Low grade petrol can be used, as it is less knock prone.
6. Low maintenance requirements.
7. The size of the engine is appreciably smaller than conventional I.C. engines for the same output.
8. It generates power after every rotation whereas 4-stroke engine develops power after every two revolutions.

Applications :

1. High performance motor cycle (Comotor, Luxemberg).
2. Snow mobile (Outboard marine in USA).
3. Outboard engine (Yanmar).
4. Helicopter engine (Curtis Wright).
5. General purpose engines.
6. Industrial engines (Ingersoll-Rand, USA).
7. Passenger Cars (Several models in Japan and Germany).

2.61.7. Why Wankel Rotary Engine could not become Successful ?

Wankel rotary engine could not become successful mainly *due to the failure of apex seal on the straight line contact between the rotor apex and the casing.* Also the side *gas seals and corner seals have not been successful in preventing, mixture and gas leakage.* Further **problems / limitations** *with the Wankel engine are :*

1. High compression ratio difficult to achieve.
2. Inefficient combustion.
3. High surface to volume ratio.
4. Scraping of wall layers into the exhaust ports.
5. Lubrication and cooling problem.
6. Shorter life.
7. Higher specific fuel consumption.

2.62. VARIABLE COMPRESSION RATIO (VCR) ENGINES

2.62.1. Introduction

- It is always desired to develop high specific power output, accompanied by good reliability and longer engine life. The maximum power output can be obtained if maximum amount of fuel can be burnt efficiently and effectively. All the *methods which are used to increase power output entail a number of problems.*
- It has been observed that *a fixed compression ratio engine cannot meet the various requirements of high specific output.* Hence, the development of variable compression ratio (VCR) engine seems to be a necessity.
  - In *the VCR engine a "high compression ratio" is employed for good stability and low load operation and a "low compression ratio" is used at full-load to allow the turbocharger to boost the intake pressures without increasing the peak cycle pressure.* It is worthnoting that the greater volume of clearance volume at lower compression ratio would result in increased air intake for the same peak compression pressure and consequently more output will be obtained.
- A S.I. engine can be used as a VCR engine. The concept of variable compression ratio, however, can be more suitably used with the turbocharged diesel engine because of the following *reasons :*

  (*i*) The *VCR concept is beneficial only at part load* and the part-load efficiency of the diesel engine is higher than that of the gasoline engine.

  (*ii*) The diesel engine has better multi-fuel capability.

2.62.2. Methods to Obtain Variable Compression Ratio

Variable compression ratio can be obtained :

(*i*) ***By changing the clearance volume.*** It *requires raising or lowering of either entire crankshaft or piston crown.*

(*ii*) ***By changing both the clearance and swept volume of the engine.*** This method requires a variable throw crankshaft for changing the stroke.

The following mechanisms have been proposed for the VCR engine.

- Fig. 2.160, shows the mechanism by which the displacement can be varied by altering the stroke length. It is not favoured since it mechanism is quite complex.
- Fig. 2.161, shows the mechanism of Varimax engine developed by Tecquiment Ltd. (U.K.) which uses the movement of crankshaft for altering the compression ratio. This mechanism changes the compression ratio from 4.5 : 1 to 20 : 1.

- The most promising VCR mechanism is the use of special piston to lower or raise the piston skirt as developed by British International Combustion Engine, Research Institute (BICERI).

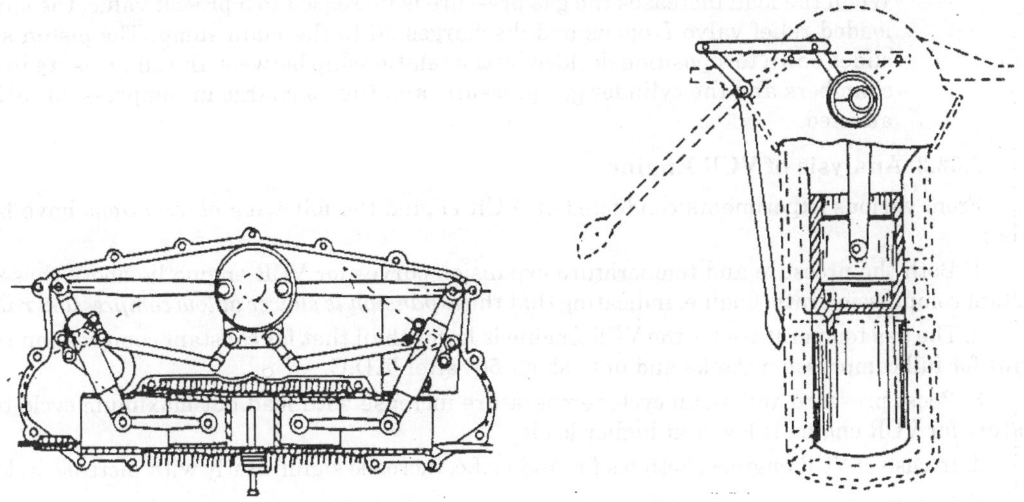

Fig. 2.160. VCR engine by changing the stroke length.

Fig. 2.161. Varimax VCR engine.

Fig. 2.162 shows the VCR system, as developed by BICERI.

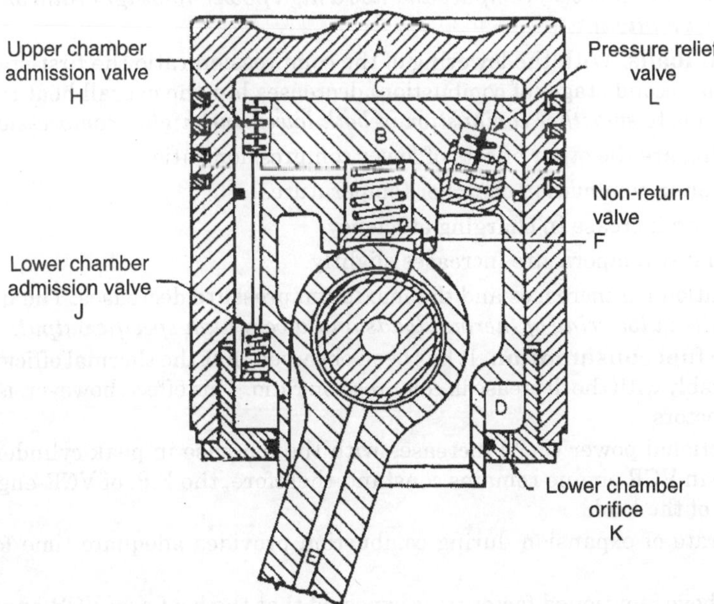

Fig. 2.162. BICERI variable compression ratio engine piston.

— It consists of two main parts A and B as shown in the figure. The part B, known as *carrier* is mounted on the gudgeon pin and part A, known as *shell*, slides on carrier B to vary the clearance volume. These two parts of the piston are so arranged that two chambers C and D are formed between them. The chambers are kept full of lubricating oil via a hole in the connecting rod and non-return valve F from the lubricating system.

— The gas load is carried by the oil in the upper chamber C.

— When the load increases the gas pressure is increased to a pre-set value, the spring loaded relief valve *L* opens and discharges oil to the main sump. The piston shell slides down to a position decided by the relationship between the oil pressure in two chambers and the cylinder gas pressure, and thus a charge in compression ratio is affected.

2.62.3. Analysis of VCR Engine

From various experiments conducted on VCR engine the following *observations* have been made :

1. Both the pressure and temperature expansion curves for VCR-engine lie above those for constant compression ratio engine, indicating that the *expansion is slower at low compression ratios.*

2. The gas temperature for the VCR-engine is lower than that for constant compression ratio engine for full compression stroke and upto about 50° after T.D.C.

3. Boost pressure and mean cycle temperature increase with load but maximum cycle temperature for VCR engine is lower at higher loads.

4. In case of VCR engines, both b.s.f.c. and i.s.f.c. increase significantly with increase in load.

2.62.4. Performance of VCR Engine

The performance aspects of VCR-engine are discussed below :

1. **Power output.** It has been observed that for the same engine dimension a *VCR-engine develops more power* (*i.e.*, it is *very compact and has a high power-to-weight ratio without any penalty on specific fuel consumption*).

2. **Thermal loads.** With the lowering of the compression ratio the first-stage of combustion increases while the second stage (of combustion) decreases but the overall heat release duration is shortened. This leads to *smoother combustion of both lower and higher compression ratios.*

The following are the other effects of lower compression ratio :

(*i*) Reduction in combustion chamber temperature.

(*ii*) There is an increase in charging efficiency.

(*iii*) The exhaust temperature increases slightly.

(*iv*) The ignition lag increases and the maximum pressure decreases. The use of VCR principle *results in lowering of thermal loads and a very high specific output.*

3. **Specific fuel consumption.** It has been observed that the thermal efficiency of the engine decreases considerably with the decrease in compression ratio. This effect, however, is counterbalanced by the following factors :

(*i*) The frictional power (F.P.) increases with the increase in peak cylinder pressure. Since the peak pressure in VCR engine remains constant, therefore, the F.P. of VCR-engine remains constant irrespective of the load.

(*ii*) Lower rate of expansion during combustion provides adequate time for combustion to complete.

Owing to above mentioned factors it is expected that the b.s.f.c. of VCR-engine should not be much higher than the conventional engine.

4. **Engine noise.** *The noise emanating from the engine depends upon the peak pressure in the cylinder and rate of pressure rise.* The peak pressure affects the lower frequency noise while the rate of pressure rise affects the high frequency noise. *As the peak pressure in VCR-engine remains same irrespective of load on engine, the low frequency noise is reduced. However, as the compression ratio is reduced (i.e., high rates of pressure rise) the noise from the VCR-engine is mainly high frequency noise.*

5. **Starting.** As the VCR-engine uses higher compression ratio at low loads, it has a *good cold starting and idling performance at low ambient temperatures.*

6. **Multi-fuel capability.** The VCR-engine has a good multi-fuel capability due to higher compression ratio at starting and part-load operation.

— The opposed piston type engine is especially suitable for multi-fuel operation.

2.63. FREE-PISTON ENGINE PLANT

Free-piston engine plants are the conventional gas turbine plants *with the difference that the air compressor and combustion chamber are replaced by a free piston engine.* The working of such a plant is explained in Fig. 2.163.

Refer Fig. 2.163. A free-piston engine unit comprises of five cylinders with two assemblies of pistons that move opposite to each other. The pistons are powered by a diesel cylinder located in the centre. Each assembly has a diesel piston at the centre of the unit. These are rigidly connected to large-diameter pistons reciprocating in the large air-compressor cylinders. Each assembly has air-cushion cylinders at the end.

Refer Fig. 2.163 (*a*) :

— The pistons are at their innermost position.

— The diesel cylinder has charge of compressed air at about 100 bar ready for firing.

— The air-compression cylinders are filled with air at atmospheric pressure ; the inlet valves IV_1 and IV_2 have just closed. The air trapped in the *air-cushion cylinders,* also called *bounce cylinders,* is at its lowest pressure.

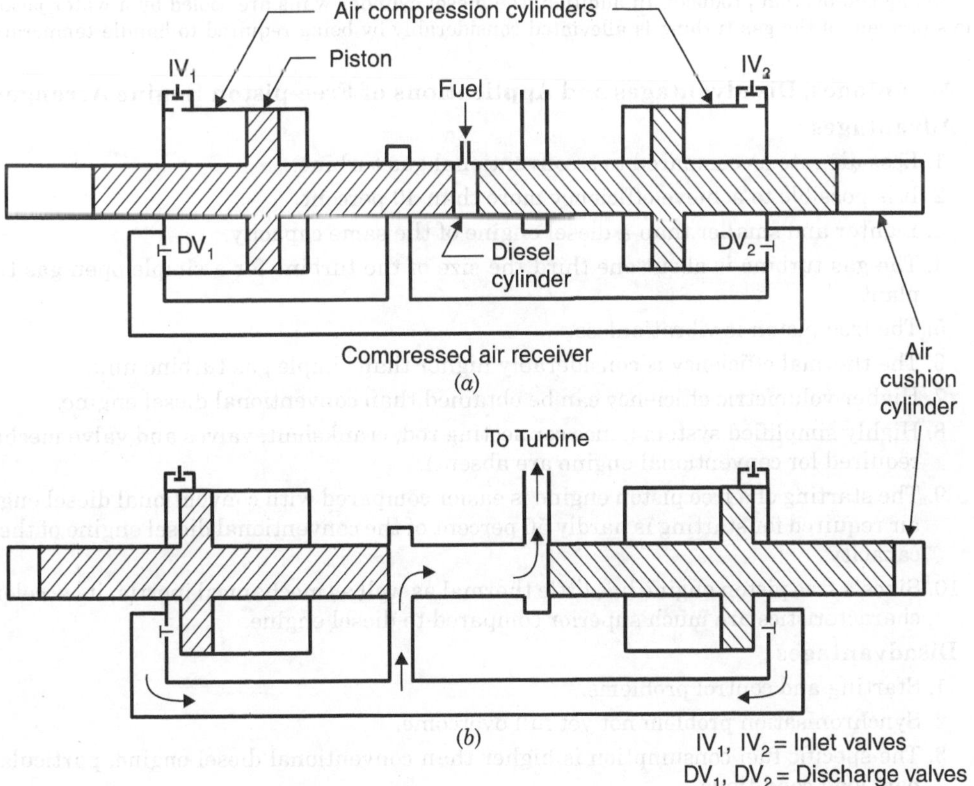

Fig. 2.163. Free-piston engine plant.

— The discharge valves DV_1 and DV_2 in the *air compression cylinders* are held shut by the high-pressure air in the compressed-air receiver that connects the air compression cylinders and the diesel cylinders.

— The compressed air-charge in the diesel cylinder is at high temperature from work of compression, so that a charge of fuel injected immediately ignites and burns. The resulting sudden rise of pressure in the diesel cylinder forces the pistons apart. As they move apart, the exhaust port *leading to the turbine* uncovers first. The combustion products at a pressure of 3 to 6 bar and temperature of about 550°C rush out *through the exhaust to the turbine.* As the piston continue to move to their outermost positions, the inlet port from the compressed-air receiver is uncovered and compressed air from the receiver enters the diesel cylinder to scavenge out the combustion products and fill the cylinder with a fresh charge of air [Fig. 2.163 (*b*)]. During this outward stroke, the air in the cushion cylinders has been compressed. This air now expands and pushes on the end pistons to return the assemblies back to the innermost position, in turn compressing the fresh air charge trapped in the diesel cylinder by the opposed piston closing on each other.

— *Thermal efficiencies of 35%* and more can be developed by these plants. This can be achieved *because the 1900°C combustion gas in the diesel cylinder does work directly on the piston in compressing air. The higher the temperature of gases doing work, the more efficient the process.*

Note. The temperature to the tune of 1900°C can be tolerated because *it appears only instantaneously at ignition and cyclically.* During most of the cycle the cylinder is being cooled by relatively cooler air charges and expanding combustion products. In addition, the diesel cylinder walls are cooled by a water jacket. The materials problem of the gas turbine is alleviated considerably by being required to handle temperatures of 550°C.

Advantages, Disadvantages and Applications of Free-piston Engine Arrangement

Advantages :

1. Less air rate as compared to a conventional gas turbine.
2. It is possible to achieve efficiency more than 40 percent.
3. Lighter and smaller than a diesel engine of the same capacity.
4. The gas turbine is about one third the size of the turbine for a simple open gas turbine plant.
5. The free-piston is vibrationless.
6. The thermal efficiency is considerably higher than simple gas turbine unit.
7. Higher volumetric efficiency can be obtained than conventional diesel engine.
8. Highly simplified system (since connecting rod, crankshaft, valves and valve mechanism required for conventional engine are absent).
9. The starting of a free piston engine is easier compared with conventional diesel engine as air required for starting is hardly 50 percent of the conventional diesel engine of the same capacity.
10. Since a free piston engine has little thermal as well as mechanical inertia, its acceleration characteristics are much superior compared to diesel engine.

Disadvantages :

1. Starting and control problems.
2. Synchronisation problem not yet full overcome.
3. The specific fuel consumption is higher than conventional diesel engine, particularly at part load conditions.
4. The fuel supply can be varied only in a limited range. Any fluctuation in the fuel supply will make the operation unstable.

Applications :

The following are the *applications of free-piston engines* :

1. Widely used as a submarine air compressor units.
2. Suitable for power generation in medium power range. Below 300 kW diesel engines are indispensable as free piston engine of comparable sizes are not being built on commercial scale.
3. Free piston engines are specially suitable for pumping oil ; also the same oil can be used as fuel.

IX. Air Pollution from I.C. Engines and its Control

2.64. INTRODUCTION TO AIR POLLUTION FROM I.C. ENGINES

- **Air pollution** *can be defined as an addition to our atmosphere of any material which will have a deleterious effect on life upon our planet.* Besides I.C. engines other sources such as electric power stations, industrial and domestic fuel consumers also add pollution.
- There has been a great concern, in recent years, that the *internal combustion engines* is responsible for too much atmospheric pollution, which is detrimental to human health and the environment. Thus concerted efforts are being made to *reduce the responsible pollutants emitted from the exhaust system without sacrificing power and fuel consumption.*

2.65. POLLUTANTS

2.65.1. Pollution Derived from Combustion Products

Pollutants are produced by the *incomplete burning* of the air-fuel mixture in the combustion chamber. The *major pollutants* emitted from the exhaust due to *incomplete combustion* are :

1. Carbon monoxide (CO)
2. Hydrocarbons (HC)
3. Oxides of nitrogen (NO_x).

Other products produced are *acetylene, aldehydes* etc. If, however, combustion is complete the only products being expelled from the exhaust would be *water vapour* which is harmless, and *carbon dioxide,* which is an inert gas and, as such it is not directly harmful to humans.

1. **Carbon monoxide (CO) :**

- It is a *colourless gas* of about the same density as air.
- It is a *poisonous gas* which, when inhaled, replaces the oxygen in the blood stream so that the body's metabolism can not function correctly.
- Small amounts of CO concentrations, when breathed in, slow down physical and mental activity and produces headaches, while large concentration will kill.

Mechanism of formation of CO :

CO is generally formed when the mixture is rich in fuel. The amount of CO formed increases as the mixture becomes more and more rich in fuel. A small amount of CO will come out of the exhaust even when the mixture is slightly lean in fuel. This is due to the fact that equilibrium is not established when the products pass to the exhaust. At the high temperature developed during the combustion, the products formed are unstable, and the following reactions take place before the equilibrium is established.

$$2H_2O + O_2 \longrightarrow 2(1-y)\,H_2O + 2yH_2 + yO_2$$

where, y is the fraction of H_2O dissociated.

$$C + O_2 \longrightarrow CO_2 \longrightarrow (1-x)CO_2 + xCO + \frac{x}{2}\,O_2$$

As the products cool down to exhaust temperature, major part of CO reacts with oxygen to form CO_2. However, a relatively small amount of CO will remain in exhaust, its concentration increasing with rich mixtures.

2. **Hydrocarbons (HC) :**

- Hydrocarbons, derived from unburnt fuel emitted by exhausts, engine crankcase fumes and vapour escaping from the carburettor are also harmful to health.

Mechanism of formation of HC :

— Due to existence of local very rich mixture pockets at much lower temperatures than the combustion chambers, unburnt hydrocarbons may appear in the exhaust.

— The hydrocarbons also appear due to flame quenching near the metallic walls.

A significant portion of this unburnt hydrocarbon may burn during expansion and exhaust strokes if the oxygen concentration and exhaust temperature is suitable for complete oxidation. Otherwise a large amount of hydrocarbon will go out with the exhaust gases.

3. **Oxides of nitrogen (NO_x) :**

- Oxides of nitrogen and other obnoxious substances are produced in very small quantities and, in certain environments, can cause pollution ; while *prolonged exposure is dangerous to health.*

Mechanism of formation of nitric oxide (NO) :

— At high combustion temperatures, the following chemical reactions take place behind the flame :

$$N_2 + O_2 \longrightarrow 2NO$$
$$N_2 + 2H_2O \longrightarrow 2NO + 2H_2$$

Chemical equilibrium calculations show that a significant amount of NO will be formed at the *end of combustion.* The majority of NO formed will however decompose at the low temperatures of exhaust. But due to very low reaction rate at the exhaust temperature a part of NO formed remains in exhaust. It is far in excess of the equilibrium composition at that temperature as the formation of NO freezes at low exhaust temperatures.

— *The NO formation will be less in rich mixtures than in lean mixtures.*

Smoke or particulate :

- Solid particles are usually formed by *dehydrogenation, polymerisation* and *agglomeration.*
- In the combustion process of different hydrocarbons, acetylene (C_2H_2) is formed as an intermediate product. These acetylene molecules after simultaneous polymerisation and dehydration produce carbon particles, which are the main constituent of the particulate.

Aldehydes :

- Due to very slow chemical reaction during delay period in the diesel engines, *aldehydes are formed as intermediate products.* In some parts of the spray the aldehydes will be left after the initial reactions. These aldehydes may be oxidised in the later part of the cycle, if the mixture temperature is high, and if there is sufficient oxygen.

- *At heavy loads, due to lack of oxygen, an increase in aldehyde emission in the exhaust is observed.*

Following points are worth noting :

1. If the *air-fuel mixture is **too rich*** there is insufficient air for complete combustion and some of the fuel will not be burnt or at least only partly burnt. Since hydrogen has a greater affinity for oxygen, hydrogen will take all the oxygen it needs leaving the carbon with a deficiency of oxygen. *As a result of the shortage of oxygen a percentage of the carbon will be converted to carbon monoxide as well as carbon dioxide, and, with very rich mixtures, particles of pure unburnt carbon may be expelled from the exhaust as "black smoke".*

Incomplete combustion due to partial oxidation of the hydrocarbon fuel also produces other products such as *acetylene* and *aldehyde.* These products, when expelled from the exhaust, leave an unpleasant smell and are particularly noticeable during engine warm-up when a rich mixture is provided.

2. If the *mixture is made **too weak*** it is unlikely that the atomised liquid fuel will be thoroughly mixed throughout the combustion chamber so that *slow burning, incomplete combustion and misfiring may result.*

A further characteristic of weak mixtures is that the excess oxygen (which has not taken part in the combustion process) at very high temperature is able to combine with some of the nitrogen that constitutes about three-quarters of air, to form oxides of nitrogen such as *nitrogen peroxide* (NO_2). *The amount of nitrogen peroxide produced will increase as the mixture weakens until it peaks at just over an air-fuel ratio of 15.5 : 1, beyond this point the temperature of combustion begins to fall below that necessary for formation of nitrogen peroxide so that with further reduction in mixture strength the amount of nitrogen peroxide progressively decreases.*

2.65.2. Mixture Strength and Combustion Product Characteristics

- The *chemically correct air-fuel ratio by mass for complete combustion is known as* **stoichiometric ratio.**
- Refer Fig. 2.164. It shows how the three main exhaust pollutants products (CO, HC, NO_x) vary from different air-fuel ratio operating on either side of the stoichiometric ratio for a very rich mixture (11 : 1) to a very lean mixture (18 : 1).

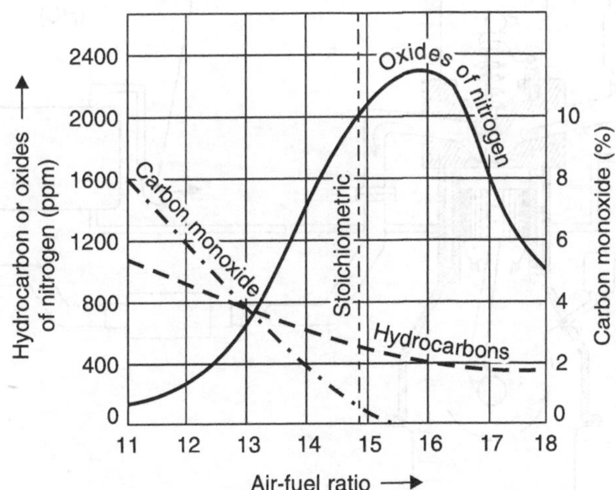

Fig. 2.164. Effects of mixture strength on exhaust composition of a petrol engine.

— The amount of CO produced in the exhaust is about 8% for an 11 : 1 air-fuel, ratio, but this percentage steadily decreases to zero as the mixture is reduced to just beyond the stoichiometric ratio (on the lean side).

— HC produced in the exhaust gases amounts to about 1100 parts per million (ppm) with a rich 11 : 1 air-fuel ratio and, as the mixture strength approaches the stoichiometric ratio, it progressively falls to around 500 ppm. A further weakening of the mixture to 18 : 1 air-fuel ratio only reduces HC content to approximately 350 ppm.

— *Oxides of nitrogen* products formed during combustion are very low at 100 ppm with a rich air-fuel ratio of 11 : 1. As the mixture strength approaches the stoichiometric ratio it rises fairly rapidly to 2000 ppm, and a further reduction of the mixture strength to 15 : 1 peaks the oxides of nitrogen to something like 2300 ppm, weakening the mixture beyond this point rapidly reduces it until, at an 18 : 1 air-fuel ratio, it is 1000 ppm.

2.66. SPARK IGNITION (S.I.) ENGINE EMISSIONS

The following are the three main sources from which pollutants are emitted from the S.I. engine (Refer to Fig. 2.165) :

1. **The crankcase.** *Where piston blow-by fumes and oil mist are vented to the atmosphere.*

2. **The fuel system.** *Where evaporative emissions from the carburettor or petrol injection air intake and fuel tank are vented to the atmosphere.*

3. **The exhaust system.** *Where the products of incomplete combustion are expelled from the tail pipe into the atmosphere.*

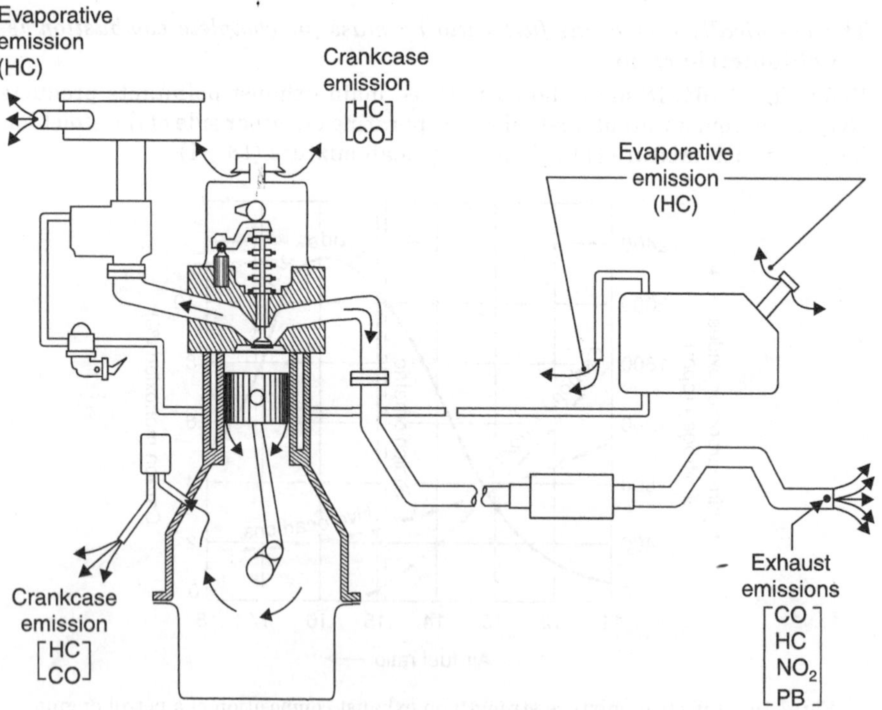

Fig. 2.165. Spark ignition engine emissions.

2.66.1. Crankcase Emission

Piston ring blow-by :

- The piston and its rings are designed to form a gas-tight seal between the sliding piston spirt and the cylinder walls. However, in *practice there will always be some compressed charge and burnt fumes which manage to escape past the compression and oil control piston rings and therefore enter the crankcase* (Fig. 2.166). These gases which find their way past the piston ring belt may be unburnt air-fuel mixture hydrocarbons, or burnt (or partially burnt) products of combustion, CO_2, H_2O (steam) or CO.

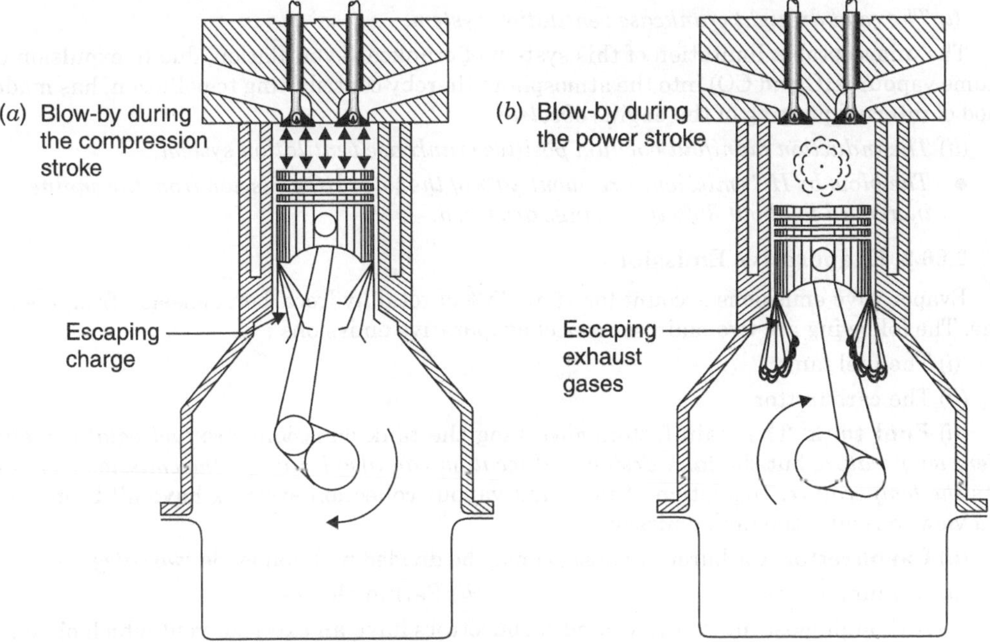

Fig. 2.166. Piston ring blow-by.

- Piston blow-by increases with engine speed and, in particular, as the piston rings and cylinder bore wears (Fig. 2.167), the blow-by becomes more noticeable in the upper speed range.
- Blow-by takes place between the piston ring gap, piston-ring to piston-groove clearance and in the T.D.C. region where the piston ring circumferential shape can not accurately follow the contour of an oval or bell mouth cylinder wall.
- Besides effectively reducing the engine compression ratio and the power developed, the *effects of piston blow-by* are two-fold :

  (*i*) *It can lead to a high concentration of combustible air-fuel mixture which could cause an explosion in the crankcase.*

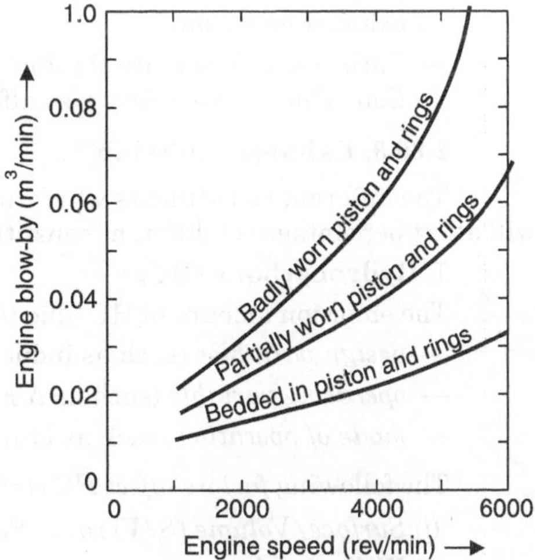

Fig. 2.167. Effect of engine speed on piston and ring blow-by.

(*ii*) The air-fuel mixture, partially burnt and fully burnt vapour fumes, will *condense and contaminate the engine's lubricating oil.*

Since it is impossible to eliminate piston blow-by completely, *an organised connection is deliberately created which circulates the crankcase and rocker or camshaft cover spaces and consequently carries the unwanted fumes out with it.* The removal of blow-by gases and vapour fumes from the crankcase is obtained by *creating a partial depression at the outlet location so* that blow-by gases under pressure (escaping between the piston and cylinder wall) are attracted towards the lower pressure region of the crankcase, at which point they are expelled.

Following are the *two methods of creating the extraction depression :*

(*i*) *The road draught crankcase ventilation system.*

The unacceptable limitation of this system of crankcase ventilation, due to expulsion of gas and fume vapour (HC and CO) into the atmosphere thereby contributing to pollution, has made this method of internal purging of the engine *obsolete.*

(*ii*) *The induction manifold vacuum positive crankcase ventilation system.*

- *The blow-by HC emissions are about 20% of the total HC emission from the engine ; this is increased to about 30% if the rings are worn.*

2.66.2. Evaporative Emission

Evaporative emissions account for 15 to 25% of total hydrocarbon emission from a gasoline engine. The following are two main sources of evaporative emissions :

(*i*) The fuel tank.

(*ii*) The carburettor.

(*i*) **Fuel tank.** The main factors governing the tank emissions are *fuel volatility* and the *ambient temperature* but the *tank design and location can also influence the emissions as location* affects *the temperature.* Insulation of tank and vapour collection systems have all been explored with a view to reduce the tank emission.

(*ii*) **Carburettor.** Carburettor emission may be divided with following two *categories* :

(*i*) Running losses ; (*ii*) Parking losses.

— Although most internally vented carburettors have an external vent which open at idle throttle position, the existing pressure forces prevent outflow of vapours to the atmosphere. Internally vented carburettor may enrich the mixture which in turn increases exhaust emission.

— Carburettor losses are significant only during hot condition when the vehicle is in operation. The fuel volatility also affects the carburettor emissions.

2.66.3. Exhaust Emission

The different constituents which are exhausted from S.I. engine and different factors which will affect percentages of different constituents are discussed below :

1. **Hydrocarbons (HC) :**

The emission amount of HC (due to incomplete combustion) is closely related to :

— *design variables* (such as induction system and combustion chamber design) ;

— *operating variable* (such as A/F ratio, speed, load) ;

— *mode of operation* (such as idling, running or accelerating).

The following *factors affect HC emission :*

(*i*) *Surface/Volume (S/V) ratio.* Fig. 2.168 shows the effect of S/V ratio on HC emission.

(*ii*) *Wall quenching.*

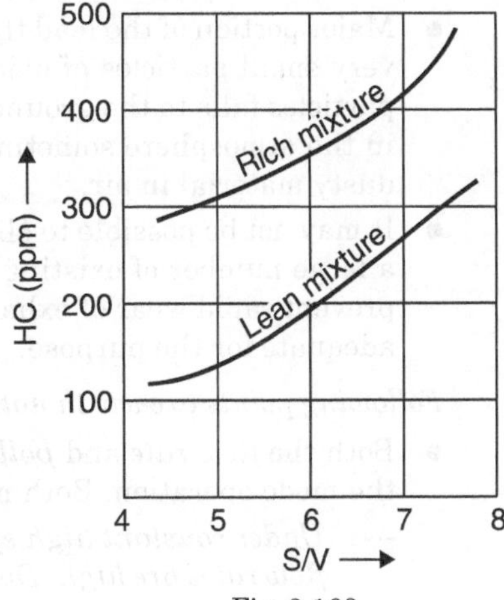

Fig. 2.168

(*iii*) *Incomplete combustion.* When the mixture supplied is rich or lean ; the flame propagation becomes weak which causes incomplete combustion and results in HC emission. The incomplete flame propagation is caused by the following factors :

(*a*) Low charge temperature.

(*b*) Too rich or too lean mixture.

(*c*) Poor condition of the ignition system.

(*d*) Non-uniform fuel in the mixture supplied to the engine.

(*e*) Large exhaust residual gases left in the cylinder.

(*iv*) *Spark-plug timing.* It is observed that HC emission is reduced by retarding the spark-plug timing during low speed (lower than 40 km/h) but has no effect when running at 40 km/h.

(*v*) *Compression ratio.* It has been observed through experiments that *emission of HC in exhaust is decreased with an increase in compression ratio.*

2. **Carbon monoxide (CO) :**

- If the oxidation of CO to CO_2 is not complete, CO remains in the exhaust.
- It can be said theoretically that, the petrol engine exhaust can be made free from CO by operating it at A/F ratio = 15. However, some CO is always present in the exhaust even at lean mixture and can be as high as 1 per cent.
- The *percentage of CO increases during engine idling but decreases with speed.*
  — Whatever may be condition of running at any load or speed, and A/F ratio, it *is not possible to completely eliminate CO and 0.5 percent is considered a reasonable goal.*
- CO emissions are *lowest during acceleration and at steady* speeds. They are, however, *high during idling and reach maximum during deceleration.*

3. **Oxides of nitrogen (NO_x) :**

- Oxides of nitrogen occur mainly in the form of NO and NO_2 and are generally formed at high temperature.
- The maximum NO_x levels are observed with A/F ratios of about 10 percent above stoichiometric. More air than this reduces peak temperature and therefore NO_x concentration falls, even free O_2 is available.
- The following factors affect the formation of NO_x :

  (*i*) *A/F ratio ;* (*ii*) *r.p.m. ;*

  (*iii*) *Angle of advance.* The decreasing angle of advance decreases appreciably the formation of NO_x.

  — It has also been observed that *NO_x increases with increasing manifold pressure, engine load and compression ratio.* This characteristic is different from HC and CO emissions which are nearly independent of engine load except for idling and deceleration.

Lead emission :

- *Lead emissions come only from S.I. engines.*
- In the fuel, lead is present as lead tetroethyl or tetramethyl, to control the self ignition tendency of fuel-air mixtures that is responsible for knock (to improve the octane rating of the fuel).

- Major portion of the lead that enters the engine is emitted from the exhaust which forms very small particles of oxides and oxyhalides in the atmosphere. A portion of the lead particles falls to the ground very quickly, others are small enough to remain suspended in the atmosphere sometime, before they fall out, usually after coagulation with other dusty material in air.
- It may not be possible to eliminate lead completely from all petrols immediately because a large number of existing engines rely upon the lubrication provided by a lead film to prevent rapid wear of exhaust valve seats. However, a very small lead content would be adequate for the purpose.

Following points are worth noting :

- Both the *flow rate* and *pollutant concentration*, for exhaust emissions, can change with the mode operation. Both must be considered in determining emissions.
  - *Under constant high speed conditions, exhaust* HC *concentrations are low while the flow rates are high. During accelerations the flow rate is low but* HC *concentration is high.*
  - The *concentration* of HC in the crankcase and evaporative losses is virtually *independent* of operating conditions, but the *flow rates* from each of these sources *change* during various operations.

  Thus, on km basis CO and HC emissions decrease with increasing driving speed while NO_x emissions are relatively not affected.
- In *a poorly maintained engine the exhaust pollution is more.*
  - An automatic choke sticking in the closed position or a very dirty air cleaner element can reduce air-fuel ratio, generally increasing HC and CO emissions.
  - A misfire allows an entire air-fuel charge to be exhausted without combustion.

2.67. S.I. ENGINE EMISSION CONTROL

The main methods, among various methods, for S.I. engine emission control are :

1. Modification in the engine design and operating parameters.
2. Treatment of exhaust products of combustion.
3. Modification of the fuels.

2.67.1. Modification in the Engine Design and Operating Parameters

Engine design modification improves upon the emission quality. A few parameters which improve an emission are discussed below :

1. **Combustion chamber configuration :**

Modification of combustion chamber involves avoiding flame quenching zones where combustion might otherwise be incomplete and resulting in high HC emission. This includes :

- — Reduced surface to volume (S/V) ratio ;
- — Reduced squish area ;
- — Reduced space around piston ring ;
- — Reduced distance of the top piston ring from the top of the piston.

2. **Lower compression ratio :**

- Lower compression ratio reduces the quenching effect by reducing the quenching area, thus *reducing HC.*
- Lower compression ratio also *reduces NO_x* emissions due to lower maximum temperature.
- Lower compression, however, *reduces thermal efficiency and increases fuel consumption.*

The decrease in compression ratio is becoming very important design parameter, because it will result in decrease in pollutants from the new engine. Besides, the octane *number requirements* will *decrease* and *high leaded gasolines will not be necessary, or the lead compound will be phased out to be replaced by un-leaded gasoline.*

3. **Modified induction system :**

In a multi-cylinder engine it is always difficult to supply designed A/F ratio under all conditions of load and power. This can be achieved by proper design of induction system or using high velocity or multi-choke carburettors.

4. **Ignition timing :**

- The ignition timing control is so adjusted as to provide normal required spark advance during cruising and retard the same for idle running. *NO_x emissions are reduced due to lowering of maximum combustion temperatures. Also HC emission gets reduced due to high exhaust temperatures. However, cooling requirements increase. The fuel economy also suffers to some extent accompanied by some power loss.* Thus a judicious balance needs to be struck between fuel economy, power loss and pollutants.

5. **Reduced valve overlap :**

- Increased overlap allows some fresh charge to escape directly and increase emission level. This can be controlled by *reducing valve overlap.*
- A new variable valve timing (VVT) allows for controlled scheduling of valve timing events ; improves engine performance. It is also claimed VVT system will work best with petrol injection. This system is also applicable to petrol as well as to diesel engines.

2.67.2. Exhaust Gas Oxidation

The *exhaust gas coming out of exhaust manifold is treated to reduce HC and CO emissions.*

The devices used to accomplish it are discussed below :

1. **After-burner :**

- An *"after-burner"* (Fig. 2.169) is *a burner where air is supplied to the exhaust gases and mixture is burnt with the help of ignition system.* The HC and CO which are formed in the engine combustion because of inadequate O_2 and inadequate time to burn are further burnt by providing air in a separate box, known as *after-burner.* The after-burner is located very near to the exhaust manifold with an intention that there is no fall in the temperature of exhaust. The oxidation of HC in the after-burner depends upon the temperature of exhaust and mixing provided in the after-burner. NO_x emission is *not* affected by the air injection.

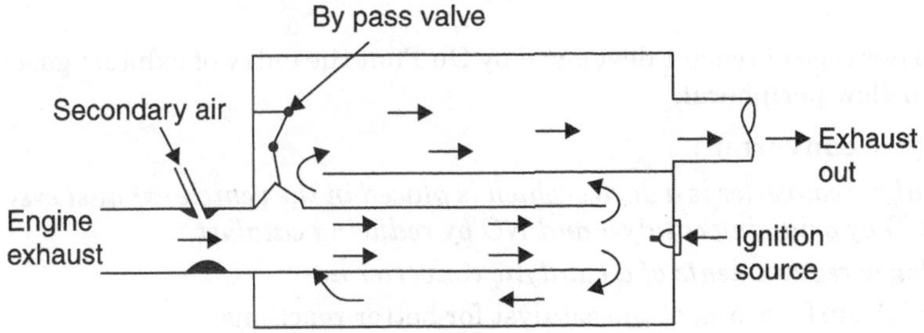

Fig. 2.169. Direct flame after-burner.

- This arrangement/system is not successful in reducing the emission owing to the difficulty in sustaining the combustion during low HC emissions due to high heat losses over a large area.

2. **Exhaust manifold reactor :**

- The exhaust manifold reactor is a further development of after-burner where the design is changed so as to minimise the heat loss and to provide sufficient time for mixing of exhaust and secondary air.
- Fig. 2.170 shows a General Motors, air injection system.

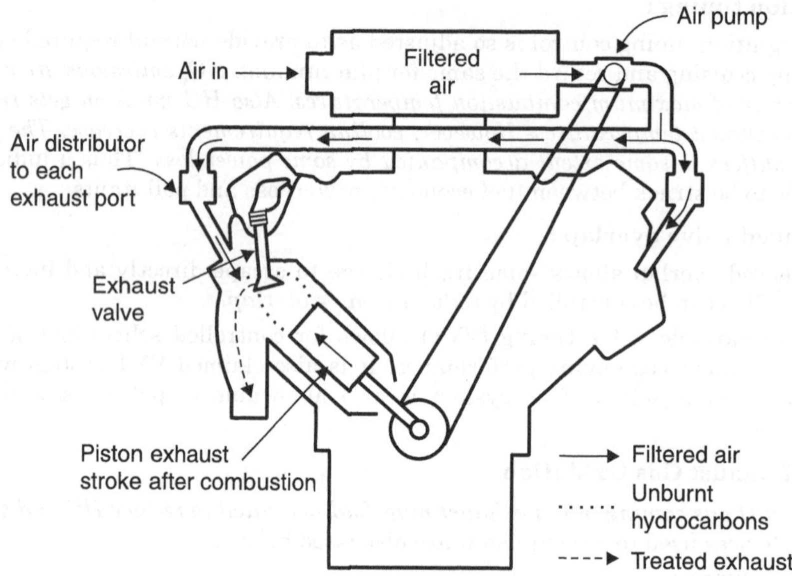

Fig. 2.170. General motors air injection system.

— Here a positive displacement vane pump, driven by the engine, inducts air from the air cleaner or from separate air filter.

— The air passes into an internal or external distributing manifold, with tubes feeding a metered amount into the exhaust port of each cylinder and close to the exhaust valve.

— Since the exhaust gases are at high temperature, the injected air reacts with HC, CO and aldehydes to reduce greatly the concentration of such emissions.

The injected air is closely metered otherwise it can decrease the temperature of the exhaust gas.

- In earlier type of reactor developed by Du Pont the entry of exhaust gases was radial and the air flow peripheral.

3. **Catalytic converter :**

- A *catalytic converter is a device which is placed in the vehicle exhaust system to reduce HC and CO by oxidising catalyst and NO by reducing catalyst.*
- The *basic requirements of a catalytic converter are :*

  (*i*) High surface area of the catalyst for better reactions.

  (*ii*) Good chemical stability to prevent any deterioration in performance.

  (*iii*) Low volume heat capacity to reach the operating temperatures.

(*iv*) Physical durability with attrition resistance.

(*v*) Minimum pressure drop during the flow of exhaust gases through the catalyst bed ; this will not increase back pressure of the engine.

Fig. 2.171 shows a catalytic converter, developed by the Ford Company. It consists of two separate elements, one for NO_x and the other for HC/CO emissions. The secondary air is injected ahead of the first element. The flow in the converter is axial.

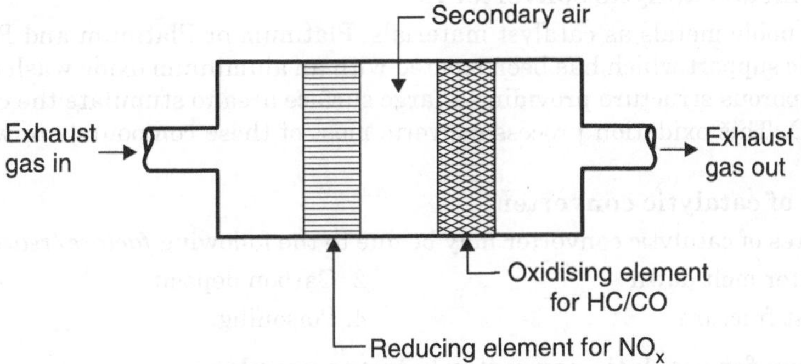

Fig. 2.171. Catalytic converter.

— ***Oxidation catalytic reactions.*** CO, HC and O_2 from air are catalytically converted to CO_2 and H_2O and number of catalysts are known to be effective noble metals like platinum and plutonium, copper, vanadium, iron, cobalt, nickel, chromium etc.

— ***Reduction catalytic reactions.*** The primary concept is to offer the NO molecule an activation site, say nickel or copper grids in the presence of CO but not O_2 which will cause oxidation, to form N_2 and CO_2. The NO may react with a metal molecule to form an oxide which then in turn, may react with CO to restore the metal molecule.

Rhodium is best catalyst to control NO_x but A/F ratio must be within a narrow range of 14.6 : 1 to 14.7 : 1.

Major drawbacks of catalytic converter are as under :

(*i*) Owing to the exothermic reactions in the catalyst bed the exhaust systems are *hotter than normal.*

(*ii*) Cars equipped with such converter *should not use leaded fuel as lead destroys complete catalytic activity.*

(*iii*) If the fuel contains sulphur (as diesel oil) emission of SO_3 is increased.

Three-way, Two-way and noble metal catalytic converters :

1. **Three-way catalytic converter :**

If an engine is operated at all times with an air-fuel ratio close to stoichiometric, then both NO reduction and CO and HC oxidation can be done in a single catalyst bed. The catalyst effectively brings the exhaust gas composition to a near equilibrium state at these exhaust conditions, *i.e.*, a composition of CO_2, H_2O and N_2. Enough reducing gas will be present to reduce NO, and enough O_2 to oxidize the CO and hydrocarbons (HC). Such a converter is called **three-way catalytic converter,** *since it removes all the three pollutants.* There is a narrow band of air-fuel ratios near stoichiometric in which high conversion efficiencies for all three pollutants are available. *Commercial three-way catalysts contain platinum, rhodium with some A_2O_3, NiO and CeO_2.* Alumina is the preferred support material.

2. **Two-way catalytic converter :**

In two-way catalytic converter, the exhaust line has two catalytic converters in it. A reduction catalyst is required to reduce NO_2. An oxidation catalyst is placed downstream of the reduction catalyst to convert the excess HC and CO. The advantage of two-way converter is that it *allows a partial decoupling of emission control from engine operations.* Therefore the conversion efficiencies for HC and CO are very high at normal exhaust temperatures.

3. **Noble metal catalytic converter :**

They use noble metals as catalyst materials. Platinum or Platinum and Palladium are applied to a ceramic support which has been treated with an aluminium oxide wash coat. This results in an extremely porous structure providing a large surface area to stimulate the combination of O_2 with HC and CO. This oxidation process converts most of these compounds to water vapour and CO_2.

Failures of catalytic converter :

The failures of catalytic converter may be due to the following *factors/reasons* :

1. Converter melt down
2. Carbon deposit
3. Catalyst fracture
4. Poisoning.

The reason for catalytic converter being so popular :

The most common range of exhaust gas temperatures for S.I. engines is 400°C–600°C, and the exhaust gas may contain modest amount of O_2 (when lean) and more substantial amount of CO (when rich). In contrast diesel engines always operate lean. The exhaust therefore contains substantial amount of oxygen and is at low temperatures (200°C–500°C) removal of gaseous pollutants from the exhaust gases after they leave the engine may be either thermal or catalytic.

In order to oxidize the hydrocarbons in the gas phase without catalyst (thermal), a resident time of the order of 50 minutes, and a temperature in excess of 700°C are required. To oxidize CO temperatures in excess of 700°C are required. Temperatures high enough for this purpose can be obtained by spark retard (with loss in efficiency) and insulation of exhaust parts and manifold. The residence time can be increased by increasing the exhaust manifold volume to form a thermal reactor. However this approach has limited application.

Catalytic oxidation of CO and HC in the exhaust can be achieved at temperatures as low as 250°C. Thus effective removal of the pollutants occur over a much wider range of exhaust temperatures than can be achieved with thermal oxidation.

The only satisfactory method known for the removal of NO from exhaust gases involve catalytic processes. Removal of NO by catalytic oxidation to NO_2 requires *temperatures less than 400°C* (from equilibrium conditions), and subsequent removal of NO_2 produced. Catalytic reaction of NO with added NH_3 is not practicable because of the transient variation in NO produced in the engine.

Reduction of NO by CO, HC or H_2 in the exhaust to produce N_2 is the preferred catalytic process. *It is only feasible in S.I. engine exhaust.*

Therefore, use of catalytic convertor for CO, HC and NO_2 removal in IC engines has become wide-spread.

2.67.3. Exhaust Emission Control by Fuel Variation

- The ability of a fuel to burn in mixtures leaner than stoichiometric ratio is a rough indication of its potential emission reducing characteristics and reduced fuel consumption.
- If gasoline is changed to propane as engine fuel CO emission can substantially be reduced with reduced HC and NO_x ; and in changing from propane to methane the CO as well HC emission touch zero level and only the NO_x remains as a significant factor.
- From pollution point of view, both methane and steam reformed hexane are very attractive fuels but we are unable to use at present for want of technological progress.

2.67.4. Blow-by Control

- *The basic principle of crankcase blow-by control system is recirculation of vapours back to the inlet manifold.*
- Fig. 2.172 shows a positive crankcase ventilation (PCV) system.
  - — The filtered air is drawn from the air cleaner and passes on to the crankcase. The *air and the blow-by gases passes through flow calibrated PCV valve before being drawn into the intake manifold in order to restrict the air flow while idling when the vacuum is high at the manifold and the blowby is less.* PCV valve is spring loaded. At wide open throttle, the air flow gets unrestricted but flow rate is metered by the valve opening.

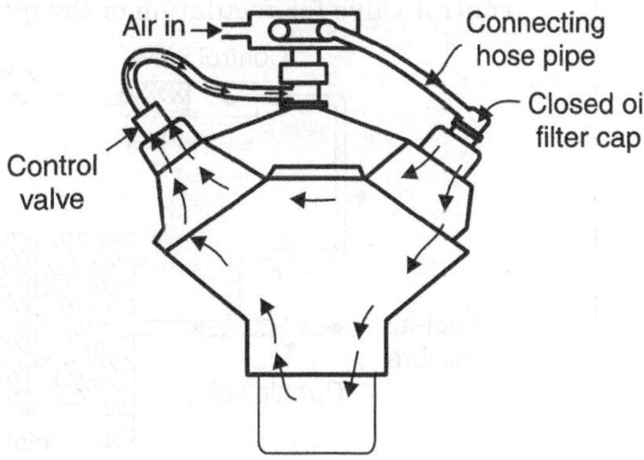

Fig. 2.172. PCV system.

  - — In view of the added blowby flow, the carburettor has to be calibrated. Since blow-by gas and air by-pass the carburettor in entering the manifold, the failure of PCV will give improper fuel ratios. Due to improper seating of the PCV valve at high vacuum, large quantity of air and also lubricating oil mist will enter the manifold. This will lead to lean mixtures. Also, the valve getting stuck to due to contaminants, will lead to rich mixtures.

    In case of valve failure, in this system, blow-by is unable to go to atmosphere.

2.67.5. Evaporation Emission Control Device (EECD)

In one of general motors system the filter pipe is sealed with a pressure cap. It comprises a built in vacuum relief to allow air to enter as the fuel gets consumed. An inverted bowl is placed inside the tank. It traps air as tank is filled to allow for liquid fuel volume expansion. The air gets released from the bowl by two small orifices in the top. The tank is ventilated to a cannister which has activated carbon particles. This can hold 0.2 kg of fuel vapour upon shut down. But when the engine is running, filtered air is drawn through the bottom of cannister, removing the absorbed vapour. This mixture is sent to air cleaner or to the intake manifold, in proportion to air flow rate.

2.67.6. Control of Oxides of Nitrogen (NO_x)

The concentration of oxides of nitrogen in the exhaust is closely related to the peak cycle temperature. The following are the three methods (investigated so far) for reducing peak cycle temperature and thereby reducing NO_x emission.

1. Exhaust Gas Recirculation (EGR).
2. Catalyst.
3. Water injection.

1. **Exhaust Gas Recirculation (EGR) :**

- This method is *commonly used to reduce NO_x* in petrol as well as diesel engines. In S.I engines, about 10 percent recirculation reduces NO_x emission by 50 percent. Unfortunately, the consequently poorer combustion directly increases HC emission and calls for mixture enrichment to restore combustion regularity which gives a further indirect increase of both HC and CO.

- Fig. 2.173 shows the arrangement of Exhaust Gas Recirculation (EGR) system. A portion (about 10% to 15%) of the exhaust gases is recirculated to cylinder intake charge, and this reduces the quantity of O_2 available for combustion. The exhaust gas for recirculation is taken (as shown in Fig. 2.173) through an orifice and passed through control valve for regulation of the quantity of recirculation.

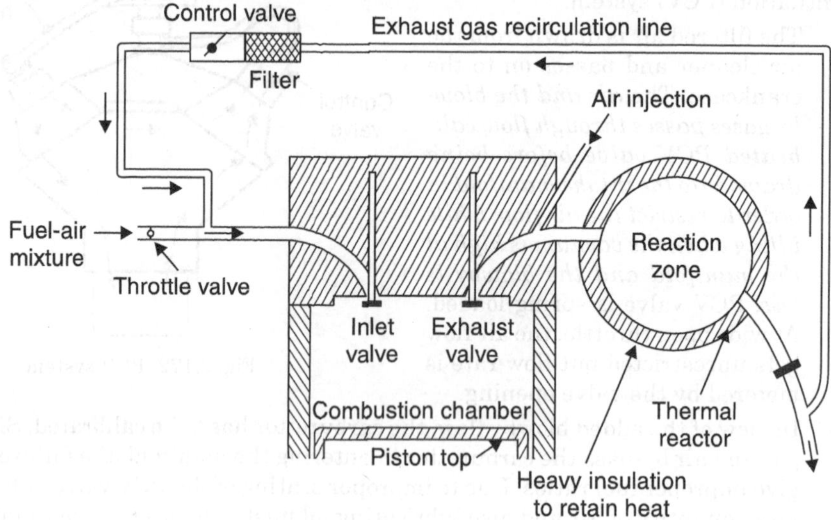

Fig. 2.173. EGR system.

- The effect of A/F ratio of NO_x emission taking EGR parameter is shown in Fig. 2.174.
  — It may be observed from the figure that, maximum emission of NO_x takes place during lean mixture limits when gas recirculation is least effective. Whereas, for emission of hydrocarbon (HC) and carbon monoxide (CO) lean mixture is preferred, 15 percent recycling reduces NO_x by 80 percent but increases HC and CO by 50 to 80%. These are two conflicting requirements of this emission control system and this problem has been solved by adopting package system which have both NO_x and HC/CO control devices.

2. **Catalyst.** A few types of catalysts have been tested to reduce the emission of NO_x; a copper catalyst has been used in the presence of CO for this purpose. The research is going on to develop a good catalyst.

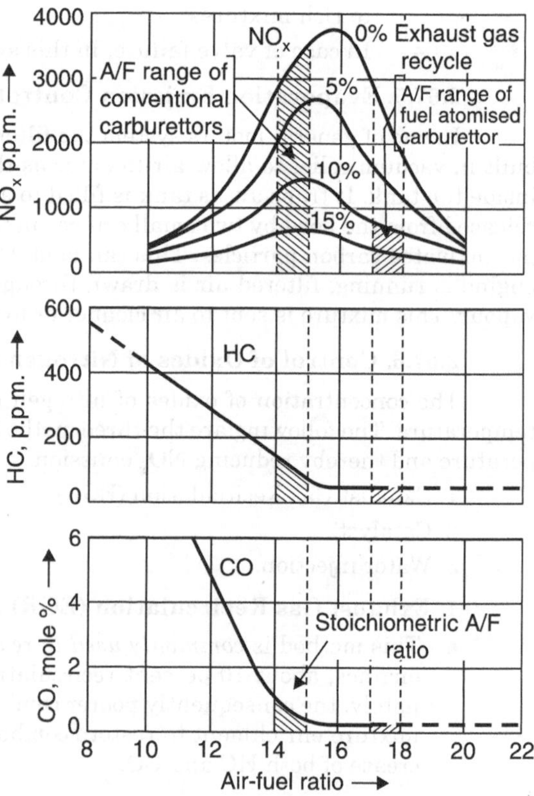

Fig. 2.174. Effect of recycling of gas on NO_x concentration.

3. **Water injection.** It has been observed that the specific fuel consumption decreases a few percent at medium water injection rate. Attempts have been made to use water as a device for controlling the NO_x. This method, because of its complexity, is rarely used.

2.67.7. Total Emission Control Packages

Earlier (previous articles) we have seen that any method which is used to decrease NO_x tries to increase HC and CO and vice-versa. Thus it is of paramount importance to develop a method/system which should reduce emissions of NO_x, HC, CO to a desired level simultaneously. After a long and detailed experimental study of various possible systems, the following two systems/packages have been developed to achieve the required results :

1. Thermal reactor package.
2. Catalytic converter package.

Using this approach, the following are the *three basic methods of emission control :*

(*i*) Thermal reactors, which rely on homogeneous oxidation to control CO and HC ;

(*ii*) Oxidation catalyst for CO and HC ;

(*iii*) Dual catalyst system (here a reduction catalyst for NO_x and an oxidation catalyst for CO and HC are connected in series).

— Where control of NO_x is required with the first two methods, EGR is added to the system.

1. **Thermal reactor package :** Refer to Fig. 2.175.

- A *thermal reactor* is a chamber which is designed to provide adequate residence time for allowing appreciable oxidation of CO and HC to take place. For enhancing the conversion of CO to CO_2 the exhaust temperature is increased by retarding the spark.

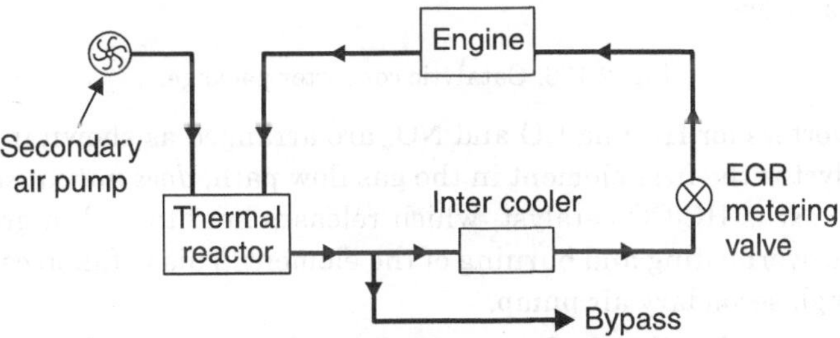

Fig. 2.175. Thermal reactor package (Ford).

- Actual thermal reactor (made of high nickel steel) that is used on a car consists of two enlarged exhaust manifolds which allow greater residence time for burning HC and CO with oxygen in the pumped in air. For keeping a flame constantly burning (and there by assuming complete combustion) a secondary air pump injects fresh air into the reactor ; this reduces HC and CO. About 10 to 75 percent of the gas is recirculated after cooling in the intercooler to reduce the formation of NO_x.

— In this packing system are also included the following :

(*i*) Enriched and stage carburettor temperature controls ;

(*ii*) Crankcase valve to control blow-by gases ;

(*iii*) Special evaporation control valves.

— In this package emission of NO_x, HC and CO are reduced to a required level but at the cost of 20 per cent less power and 10 per cent more fuel consumption.

— This converter can be employed for a run of 1.5 lacs km.

2. **Catalytic converter package :**

The *working principle of this package is to control the emission levels of various pollutants by changing the chemical characteristics of the exhaust gases.* The catalytic converter package as compared to thermal reactor package requires non-leaded fuel as lead reduces the catalytic action.

The *major advantage* of this converter (as compared to thermal reactor) is that it allows a partial decoupling of emission control from engine operation in that the conversion efficiencies for HC and CO are very high at normal exhaust temperatures.

- Fig. 2.176 shows the arrangement of *catalytic converter package.*

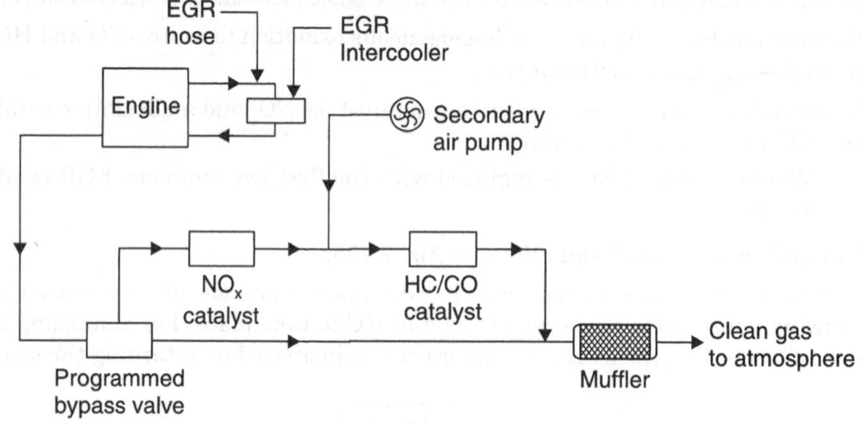

Fig. 2.176. Catalytic converter package.

— Converters for HC and CO and NO_x are arranged as shown in the figure. The NO_x catalyst is the first element in the gas flow path, *does not cause release of any heat.* The next is HC/CO catalyst, which releases heat to such a great extent that may cause overheating and burning of the element. This is taken care of by injecting air through secondary air pump.

— A bypass valve ahead of converter is used to increase the converter life (to about 0.8 lacs km).

— For better control of NO_x, exhaust gas is circulated via an intercooler back to air cleaner.

- For this system, the power loss is about 30% and the fuel consumption is about 10% more than normal.

2.68. DIESEL ENGINE EMISSIONS

- Emissions from diesel engines can be classified in the same categories as those for the gasoline engines but the level of emission in these categories vary considerably. Typical level of the constituents of the exhaust products of combustion in four-stroke cycle and two-stroke cycle are given in the Table (next page) at idling, accelerating, partial load and full load.

Table : Diesel Engine Emissions

Engine exhaust constituents	Concentration as measured in exhaust products			
	Idling	Accelerating	Partial load operation	Fuel load operation
Two-stroke cycle engine				
1. CO%	0.01	0.25	0.01	0.35
2. CO_2%	0.85	5.5	3.8	5.30
3. HC ppm	250	500	350	550
4. NO_x ppm	200	1200	1100	1250
5. RCH ppm	17.0	9.5	1.0	5.5
6. Smoke Hartrige Unit	4	44	4	10
Four-stroke cycle engine (Un-supercharged)				
1. CO%	0.02	0.08	0.04	0.25
2. CO_2%	2.5	3.5	5.5	6.7
3. HC ppm	180	330	210	150
4. NO_x ppm	330	920	590	780
5. RCHO ppm	8.0	7.5	5.0	1.5
6. Smoke, Hartrige Unit	4	44	4	10

- As shown in Fig. 2.177, it is worth noting that NO_x concentration in unsupercharged 4-stroke-cycle diesel engines vary linearly with power (output). The design and operating parameters to control NO_2 are the same as for S.I. engines already discussed.

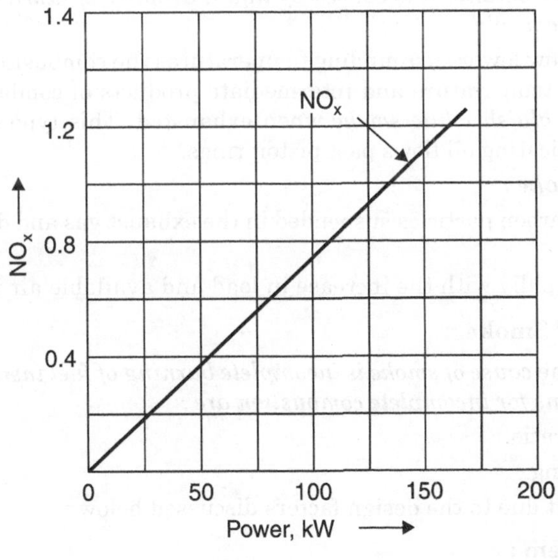

Fig. 2.177

2.69. DIESEL SMOKE AND CONTROL

2.69.1. Exhaust Smoke

In C.I. engines, any volume of the combustion chamber in which the fuel is burnt at a relatively F/A ratio greater than 1.5 ($F_R > 1.5$) at pressures developed in these engines produces **soot.** The quantity of soot formed depends on the following *factors* :

(*i*) The local F/A ratios ;

(*ii*) The type of fuel ;

(*iii*) The pressure.

If this soot is able to find *adequate air* (O_2) which on the whole is much in excess of the requirements of perfect combustion, it will *burn completely.* If it is unable to find air (O_2) in the combustion cycle, it will *pass as exhaust.* And if the *quantity is sufficient, it will be visible and that is called* **smoke.** The *colour of the smoke depends on the size of soot particles.*

— Formation of smoke is basically a process of conversion of molecules of hydrocarbon fuels into particles of soot. It should be noted that soot is not *carbon* but simply an aggomeration of very large polybenzenoid free radicals. It is also observed that soot formation during the early part of the actual combustion process is common to all diesel engines but it is consumed during later part of combustion.

— Pyrolysis of fuel molecules themselves is thought to be responsible for soot formation. Fuel heated with insufficient O_2 will give carbonaceous deposits. It is believed that the "heavy ends" of diesel fuel may pyrolyze to yield the type of smoke that is observed from the diesel engine. This is believed to be the path of formation of polycyclic aromatic hydrocarbons (benzo-pyrene) found in soot.

— Many theories have been put forward for the formation of smoke but the basic reactions leading to the formation of smoke are not fully known.

- The smoke of a diesel engine is, in general, of two basic types :

  (*i*) ***Blue-white smoke.*** *It is caused by liquid droplets of lubricating oil or fuel oil while starting from cold.*

  Owing to low lower surrounding temperatures the combustion products are at a relatively low temperature and intermediate products of combustion *do not burn.* This results in *bluish white smoke* when exhausted. This type of smoke is also formed when lubricating oil flows past piston rings.

  (*ii*) ***Black smoke :***

- It consists of carbon particles suspended in the exhaust gas and depends largely upon A/F ratio.
- It increases rapidly with the increase in load and available air is depleted.

2.69.2. Causes of Smoke

It is known that the *cause of smoke is incomplete burning of fuel inside the combustion chamber.* The *two major reasons for incomplete combustion are :*

(*i*) Incorrect A/F ratio.

(*ii*) Improper mixing.

These might result due to the design factors discussed below :

1. **Injection system :**

The following injection characteristics substantially *increase the smoke levels.*

- Unsuitable droplet size.
- Inadequate or excess penetration.

- Excessive duration of injection.
- Secondary injection.
- Improper dispersion atomisation.

2. **Rating :**

- It has been observed that smoke limited power is reached much before the thermal load limited power ; and the inevitability of smoke in exhaust is inherent in less loads. Thus it seems, smoke is unavoidable in CI engine, only its level can be kept as low as possible.

3. **Load :**

- Smoke may be defined as visible products of combustion. Rich mixtures (higher loads) results in higher smoke because of *non-availability of oxygen for combustion.*
- The smoke level rises from no-load to full-load (Refer to Fig. 2.178). During the first part (no-load, say, about half load), the smoke level is more or less constant, as there is always excess air present. However in higher load range there is an *abrupt* rise in smoke level due to less available oxygen.

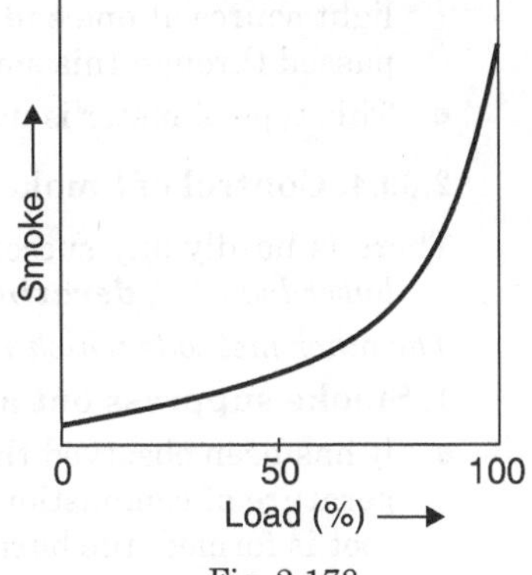

Fig. 2.178

4. **Fuel :**

- The white smoke produced in an engine depends upon the quality of fuel.
- Generally, more volatile fuels give less smoke than heavier fuels of similar cetane number.
- The cetane number exercises no effect on production of black smoke.

5. **Fuel-air ratio :**

- The smoke increases with the increasing fuel-air ratio. This increase in smoke occurs with as much as 25% excess air in the cylinder, clearly indicates that, even in the presence of excess oxygen, diesel engine has a mixing problem.

6. **Engine type and speed :**

- The smoke levels at higher loads are higher in naturally aspirated engines than turbocharged engines because the latter have adequate oxygen even at full-load.
- At low as well as high speed the smoke is worse.

7. **Maintenance :**

- The smoke levels greatly depend upon the condition of the engine. Good maintenance is a must for lower smoke levels (the maintenance affects the injection characteristics and the quantity of lubricating oil which passes across the piston rings and thus exercises a significant effect on engine tendency to generate smoke).

2.69.3. Measurement of Smoke

Two basic types of smoke meters for measuring smoke density are :

(*i*) *Filter darkening type.* (Examples : Bosch smoke meter ; Vem Brand smoke meter).

(*ii*) *Light extinction type.* (Examples : Hartridge smoke meter ; UTAC smoke meter).

The *'light extinction meter'* can be used for *continuous measurements* while the *'filter type'* can be *used only under steady state conditions.*

1. **Bosch smoke meter :**

- It is a filter darkening type of smoke meter.

- In this meter, a measured volume of exhaust gas is drawn through a filter paper which is blackened to various degrees depending upon the carbon present in the exhaust gas. The *density of "soot" is measured by timing out amount of light reflected from the sooted paper.*
- The specifications of sample volume, filter paper size etc. are well defined.

2. **Hatridge smoke meter :**

- It works on the principle of high extinction.
- In this meter exhaust sample is passed through tubes of about 0.46 m length which has light source at one end and photocell or solar cell at the other end. The amount of light passed through this smoke column is used as indication of level of smoke.
- This type of meter is useful for continuous testing and can be employed in vehicles.

2.69.4. Control of Smoke

There is hardly any successful method to control the soot except the *engine has to run at lower load i.e., **derating** and **maintain** the engine at best possible condition.*

The other methods which may be used for the control of smoke are :

1. **Smoke suppress out additives :**

- It has been observed that some barium compounds when added in fuel reduce the temperature of combustion and avoid the soot formation. It is further observed that if the soot is formed, the barium compounds break them in very fine particles and reduce the smoke.
- The use *of barium salts,* however, *enhances the deposit formation tendencies of engine and reduces the life of the fuel filter.*

2. **Fumigation :**

- It is a method of *introducing small amount of fuel with the intake manifold.* This *initiates pre-combustion reactions* before and during the compression stroke resulting in reduced chemical delay (because the intermediate products such as peroxides and aldehydes react more rapidly with oxygen than original hydrocarbons. The shortening *in chemical delay period curbs thermal cracking which is responsible for soot formation.*
- Fumigation rate of about 11 to 15% gives the best smoke improvement.

3. **Catalytic mufflers :**

- The use of catalytic mufflers, unlike petrol engine, are *not very effective.*
- Much development is needed in such-devices before they can be put to use.

2.69.5. Diesel Odour and Control

It has been observed through some experiments that *the products of partial oxidation are the main cause of odour in diesel exhaust.* This partial oxidation may be because of either very lean mixtures such as during idling or due to quenching effect.

The *following factors affect odour production :*

1. Fuel-air ratio.
2. Engine operation mode.
3. Engine type.
4. Fuel consumption.
5. Odour suppressant additives.

Control of Odour :

- Several manufacturers chaim that odour additive compounds can reduce the intensity of odour, but it has been found in practice that these additives hardly have any effect on odour formation etc.

- The control of odours by using catalyst is under development. It has been found experimentally that a few oxidation catalysts reduce intensity of odour.

2.70. COMPARISON OF GASOLINE AND DIESEL EMISSIONS

Fig. 2.179 shows the comparison of emissions and odour from gasoline and diesel engines.

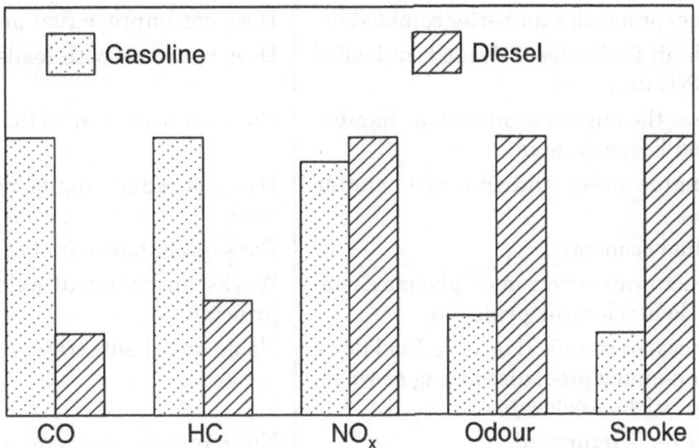

Fig. 2.179. Comparison of emissions and odour from gasoline and diesel engines.

- It is evident from the figure that there is a marked difference between the products of combustion of gasoline and diesel engines.
- Whereas gasoline petrol engines have a somewhat similar emission pattern, all diesel engines have different emission characteristics.

The combustion in a diesel engine occurs over a wide range of fuel-air ratios ranging from very lean mixture to very rich mixture while the gasoline combustion is of relatively homogeneous nature.

In order to account for various differences in the design and operation principles of gasoline and diesel engine the following *correction factor* is applied for comparing these two types of engines :

Correction factor = 15/(%CO_2 + %CO + %C)

- If the diesel engines are maintained properly, they have very little CO in their exhaust and a small quantity of smoke. On the other hand gasoline engines exhaust significant amount of CO and Unburnt hydrocarbon (UBHC). Thus *diesel engine is cleaner as compared to petrol engine.*

2.71. ZERO EMISSION

- For "*zero emission*" the Electronic-Catalytic Converter (ECC) assumes significance. In the recent development India-based Hydrodrive Systems and Controls has launched common Electronic-Catalytic Converter for automobile fuels *viz* ; the diesel, leaded and unleaded gasoline.
- *ECC excites the fuel molecules with microwave electronics and does molecular engineering of fuel for complete combustion.*
- ECC can be installed as a pre-engine device.
- ECC is the *first catalytic converter to offer fuel savings due to improved combustion.*

The comparison between ECC and conventional CC is given in the chart below :

S. No.	Electronic catalytic converter	Conventional exhaust catalytic converter
1.	Used as a pre-engine fuel converter.	Used as an exhaust converter.
2.	Cleans the process of combustion.	Cleans the exhaust gases.
3.	Improves fuel properties for better combustion.	Does not improve fuel properties.
4.	Works with all fuels-diesel, leaded, unleaded gasoline, CNG and LPG.	Does not work with leaded petrol.
5.	Decarbonizes the engine combustion chamber with use and keeps it clean.	Does not decarbonize the combustion chamber.
6.	Reduces engine noise and improves engine smoothness.	Does not reduce engine noise or engine smoothness.
7.	Enhances fuel economy.	Does not enhance fuel economy.
8.	Works on microwave electronics, plasma chemistry, particle accelerator principle.	Works on chemical oxidation and reduction principle.
9.	No Nobel metal or chemical is used. Fuel flows through open-ended pipe, also acting as waveguide and particle accelerator.	Noble metal substrates or chemicals are used.
10.	Engine pick up is improved.	No change in engine pick up.
11.	Same size for all engines, all fuels and cubic capacity.	Different sizes for different engines cubic capacity and different fuels.
12.	No warming up time required for cold start emission control.	Warm up time required for cold start emission control unless provided with electrical heating for the catalyst.

2.72. AIR POLLUTION FROM GAS TURBINES AND ITS CONTROL

- The pollution from gas turbines is *low* as compared to pollution from convention piston engines, due to the fact that *former operate at much leaner mixture* used in the latter, and still further, the *exhaust is diluted due to the presence of excess air.*
- As compared to S.I. engines, a gas turbine exhaust about 1 to 20% HC, 1 to 10% CO and 30 to 80% NO_x. Besides this exhaust also contains oxides of sulphur, aerosol , smoke particles and odourants.

Control of pollutants emanated from gas turbines :

- **Reduction of CO concentration** can be accomplished by using following methods :

(*i*) By improving atomisation of fuel.

(*ii*) By redistributing the air flow to bring the primary zone equivalence ratio to an optimum value of around 0.85.

(*iii*) By increasing the residence time.

(*iv*) By reducing the film cooling air.

(*v*) By fuel staging (In this technique supply of fuel is cut-off to some nozzles and diverted to the remainder ; it reduces emissions at low power conditions by improving the quality of automisation).

- **Reduction of UBHC (Unburnt hydrocarbon)** requires the same treatment as employed for reducing CO.
- **Reduction of NO_x.** The NO_x emission increases exponentially as : $NO_x \; \alpha (e)^{0.09T}$, where T is reaction temperature in K.

Thus, for reducing NO_x concentration, it will be required to lower the reaction temperature T. The reduction in both flame temperature and residence time are readily accomplished by increasing the air flow in the primary zone, but increases the production of HC and CO. Thus any attempt made to reduce HC and CO increases NO_x and vice-versa.

— The NO_x level as well as the smoke levels can be reduced by *providing lean combustion zone.*

2.73. EFFECTS OF ENGINE EMISSIONS ON HUMAN HEALTH

The *effects of different engine emissions* on *human health are discussed below :*

1. **Sulphur dioxide (SO_2) :**

- It is an irritant gas and affects the mucous membrane when inhaled. In the presence of water vapour it forms sulphurous and sulphuric acids. These acids cause severe bronchospasma at very low levels of concentration.
- Diseases like bronchitis and asthama are aggrevated by a high concentration of SO_2.

2. **Carbon-monoxide (CO) :**

- It has a strong affinity (200 times) for combining with the haemoglobin of the blood to form carboxyhaemoglobin. This reduces the ability of the haemoglobin to carry oxygen to the blood tissues.
- *CO affects the central nervous system.*
- It is also responsible for heart attacks and a high mortality rate.

3. **Oxides of nitrogen (NO_x) :**

- These are known to cause occupational diseases. It is estimated that eye and nasal irritation will be observed after exposure to about 15 p.p.m. of nitrogen oxide, and pulmonary discomfort after brief exposure to 25 p.p.m. of nitrogen oxide.
- It also aggrevates diseases like bronchitis and asthama.

4. **Hydrocarbon vapours :**

- They are primarily irritating.
- They are major contributors to eye and respiratory irritation caused by photochemical smog.

5. **Compounds of Incomplete combustion :**

Exhaust discharge from IC engines carry compounds of incomplete combustion (polycyclic organic compounds and aliphatic hydrocarbons), which act as carcinogenic agents and are responsible for lungs cancer.

6. **Lead :**

- Inorganic lead compounds (discharged from vehicles using leaded petrol) cause a variety of human health disorders.
- The effects include gastrointestinal damage, liver and kidney damage, abnormality in fertility and pregnancy etc.

7. **Smoke :**

It is visible carbon particles.

- It causes irritation in eyes and lungs, and visibility reduction. It also causes other respiratory diseases.

Generally speaking, susceptibility to the effects of exhaust emissions is greatest amongst infants and the elderly. Those with chronic diseases of lungs or heart are thought to be at great risk.

2.74. COMPARATIVE DATA OF CONSTRUCTIONAL FEATURES OF ENGINES OF SOME INDIA VEHICLES

S. No.	Item	Maruti 800	Jeep Gypsy	Hindustan Ambassador	Premier Padmini	Jeep (Mahindra) Universal	Santro Epsilon Engine	Maruti Esteem	Maruti Baleno	Maruti Zen
1.	*No of cylinders and arrangements*	3 in line	4 in line	4 in line	4 in line	4 in line	4 in line	4 in line	4 in line	4 in line
2.	*Firing order*	1,3,2	1,3,4,2	1,3,4,2	1,3,4,2	1,3,4,2	1,3,4,2	1,3,4,2	1,3,4,2	1,3,4,2
3.	*Bore*	68.5 mm	65.5 mm	73.02 mm	68 mm	79.3 mm	66 mm	74 mm	75 mm	72 mm
4.	*Stroke*	72 mm	72 mm	88.9 mm	75 mm	111.12 mm	73 mm	75.5 mm	90 mm	61 mm
5.	*Cubic capacity*	796 cc	970 cc	1489 cc	1089 cc	2199 cc	999 cc	1298 cc	1590 cc	993 cc
6.	*Compression ratio*	8.7 : 1	8.8 : 1	7.2 : 1	7.3 : 1	7.4 : 1	8.9 : 1	8.8 : 1	8.8 : 1	8.8 : 1
7.	*Cylinder liner*	No separate cylinder liner	No cylinder liner	No cylinder liner, originally	Originally, no separate cylinder liner	No cylinder liner	No cylinder liner	No cylinder liner	No cylinder liner	No cylinder liner
8.	*Piston*	Cast aluminium alloy cum concave skirt	Aluminium alloy concave crown split skirt	Aluminium alloy oval tapered split skirt	Aluminium alloy ground T-slot	Aluminium alloy cam grounded	Cast alluminium alloy	Cast aluminium alloy	Cast aluminium alloy	Cast aluminium alloy
9.	*Compression rings*	2	2	3	1	2	2	2	2	2
10.	*Oil control rings*	1	1	1	2	1	1	1	1	1
11.	*Gudgeon pin connection*	Fully floating	Fully floating	Clamped in connecting rod	Fully floating	Clamped in connecting rod	Semi-floating	Fully floating	Fully floating	Fully floating

12.	*No. of crank-shaft main bearings*	4	5	3	3	3	5	5	5	5
13.	*No of cam-shaft bearings*	4	5	3	3	3	5	5	5	5
14.	*Camshaft drive*	Grooved rubber belt drive	Grooved rubber belt drive	Chain	Chain	Single helical gear	Cogged belt drive	Grooved belt drive	Grooved belt drive	Grooved belt drive
15.	*Valve type*	Poppet	Poppet	Poppet	Poppet	Poppet	Poppet	Poppet	Poppet	Poppet
16.	*Valve mechanism*	Overhead	Overhead	Overhead	Overhead	Overhead inlet and side exhaust	Single over-head cam (SOHC)	Single overhead cam (SOHC)	Single overhead cam (SOHC)	Single overhead cam (SOHC)
17.	*Valve seat*	Originally no separate valve seat insert is fitted.	Originally no sepa-rate valve seat insert is fitted.	Originally no separate valve seat insert is fitted.	Cadmium steel insert.	Eatonite E.M.S. 58, insert for exhaust is fitted.	No sepa-rate valve seat	No sepa-rate valve seat	No sepa-rate valve seat	No separate valve seat
18.	*Valve seat angle*	45°	45°	45°	45°	45°				

SHORT ANSWER QUESTIONS

Q. 1. What is the main difference between I.C. engine and steam engine ?

Ans. In I.C. engine, the combustion of fuel takes place inside the engine cylinder, while in steam engine, the combustion of fuel taken place outside the engine cylinder.

Q. 2. What does p-V diagram of a 4-stroke cycle represent ?

Ans. It represents the variation of pressure and volume of the engine during every stroke of the piston.

Q. 3. State any one difference between diesel engine and petrol engine.

Ans. In diesel engine, the fuel (diesel) is fed into the cylinder by a fuel injector and is mixed with air inside the cylinder. In petrol engine, the air and petrol are mixed in the carburettor before they enter the cylinder.

Q. 4. Name the largest part in the engine.

Ans. Cylinder block.

Q. 5. What is a monoblock cylinder ?

Ans. It is a cylinder block in which the crankcase is cast integrally.

Q. 6. What causes distortion of engine block ?

Ans. Internal stresses causing warpage, incorrect placing of masses around the cylinder, warpage of head, bore in several directions cause distortion of engine block.

Q. 7. What material is used in piston construction ?

Ans. Close-grained cast-iron and some aluminium alloys.

Q. 8. What is meant by cylinder or piston scoring ?

Ans. It refers to grooves in the cylinder wall or piston or in both. It is caused by the piston scraping the cylinder wall in its movement without proper lubrication.

Q. 9. What is meant by piston seizure ?

Ans. It refers to binding of the piston and cylinder wall, as a result of the lubrication having been destroyed by excessive temperature and friction.

Q. 10. Why are piston pins made hollow ?

Ans. To reduce the inertia load due to reciprocating parts.

Q. 11. What is a "heat dam" in a piston ?

Ans. A narrow cut into the piston above the top ring is called a heat dam. This is to reduce the amount of heat reaching the top ring.

Q. 12. What is meant by "blow by" in piston arrangement ?

Ans. 'Blow by' is the escape of an unburnt air-fuel mixture and burnt gases from the combustion chamber, past the pistons and into the crankcase or in other words the gas blown by the pistons.

Q. 13. What is "squish" in piston ?

Ans. In some combustion chambers, the piston squishes or squeezes some part of the mixture at the end of the compression stroke. This squeezing of the mixture is termed as piston squish.

Q. 14. Why are T-slots provided in a piston ?

Ans. The object of a T-slot is to enable the cold clearance between the skirt and the cylinder bore, to be reduced without risk of seizure when hot.

Q. 15. What are the functions of compression rings in the piston ?

Ans. They act as a seal between the piston and the cylinder and prevent the leakage of high pressure gas from blowing into the crankcase.

Q. 16. How many piston rings are usually provided on diesel engines ?

Ans. Six rings are provided on each piston in diesel engines. The four top rings are conventional compression rings while the ring immediately above the wrist pin and the one in the piston skirt are oil control rings.

Q. 17. How are crankshafts usually made ?

Ans. Crankshafts are single-piece high carbon steel drop forgings which have been carefully machined and heat treated.

Q. 18. Why is crankcase ventilation used in an automobile engine ?

Ans. Crankcase ventilation is provided to reduce dilution due to the presence of water particles, sludge and blow by gases.

Q. 19. How does the flywheel work ?

Ans. The flywheel stores excess energy in power stroke, and helps back in compression stroke, maintaining uniform speed.

Q. 20. Where is single-cylinder engine used ?

Ans. Mopeds, scooters and motor cycles.

Q. 21. Where is four-cylinder engine used ?

Ans. Cars and jeeps.

Q. 22. Where is six-cylinder engine used ?

Ans. Buses and trucks.

Q. 23. What is the main disadvantage of a Wankel rotary engine ?

Ans. Higher oil consumption per B.P.

Q. 24. Define volumetric efficiency.

Ans. The volume of air-fuel mixture drawn into the cylinder at atmospheric pressure during the intake stroke compared to the volume of the cylinder is known as the *volumetric efficiency.*

Q. 25. What is 'air standard efficiency' ?

Ans. The thermal efficiency of an ideal air standard cycle is called the *air standard efficiency* of a cycle.

Q. 26. What is Morse test ?

Ans. Morse test enables us to calculate I.P. of each cylinder of a multi-cylinder engine.

Q. 27. What is volatility of a fuel ?

Ans. It is the property of the fuel which states the ease with which a liquid vaporizes.

Q. 28. What are the effects of low volatility of gasoline ?

Ans. Low volatility of gasoline results in poor vaporization and condensation in intake manifold.

Q. 29. What is meant by unleaded gasoline ?

Ans. Unleaded gasoline refers to petrol to which no lead compounds have been intentionally added. Even gasoline which contains 0.05 g or less of lead per gallon may be referred to as unleaded gasoline.

Q. 30. What is a 'vapour lock' ?

Ans. A vapour lock is the condition in the fuel system when fuel vaporizes, causing the presence of bubbles in the gasoline vapour disturbing fuel delivery to the carburettor.

Q. 31. What is 'octane rating' of fuel ?

Ans. It is measure of fuel ability to resist ignition knock.

Q. 32. What is tetraethyl lead ?

Ans. It is a common additive to gasoline to increase its knock resistance.

Q. 33. What are the factors causing higher oil consumption ?

Ans. (*i*) Worn out piston rings, (*ii*) Leakage through gaskets, and (*iii*) Worn out valve guides.

Q. 34. What does the cetane number of diesel fuel indicate ?

Ans. Cetane number indicates its ability to ensure quiet operation or ignition quality.

Q. 35. How does turbulence aid combustion ?

Ans. By causing an effective thorough mixing of the hot compressed air with the injected fuel oil, thereby bringing about more complete combustion and uniform cylinder pressure.

Q. 36. How is an engine supercharged ?

Ans. An engine is supercharged by an air compressor or blower which supplies the excess air.

Q. 37. What is purpose of supercharging ?

Ans. To increase the power which may be required at intervals and give more power where space does not permit larger engines.

HIGHLIGHTS

I. Introduction to I.C. engines

1. Any type of engine or machine which derives heat energy from the combustion of fuel or any other source and converts this energy into mechanical work is termed as a **heat engine.**
2. The function of a carburettor is to atomise and meter the liquid fuel and mix it with air as it enters the injection system of the engine maintaining under all conditions of operation fuel air proportion approximate to those conditions.
3. The two basic ignition systems in current use are :
   (*i*) Battery or coil ignition system
   (*ii*) Magneto ignition system.
4. Following are the methods of governing I.C. engines :
   (*i*) Hit and miss method (*ii*) Quality governing
   (*iii*) Quantity governing.
5. Pre-ignition is the premature combustion which starts before the application of spark. Overheated spark plugs and exhaust valves which are the main causes of pre-ignition should be carefully avoided in engines.
6. A very sudden rise to pressure during combustion accompanied by metallic hammer like sound is called **detonation.** The region in which detonation occurs is farthest removed from the sparking plug, and is named the 'detonation zone' and even with severe detonation this zone is rarely more than that one quarter the clearance volume.
7. The **octane number** is the percentage of octane in the mixture [of iso-octane (high rating) and normal heptane (low rating), by volume] which knocks under the same conditions as the fuel.
8. **Delay period or ignition lag** is the time immediately following injection of fuel during which the ignition process is being initiated and the pressure does not rise beyond the value it would have due to compression of air.
9. Higher the cetane rating of the fuel lesser is the propensity for diesel knock. In general a high octane value implies a low cetane value.
10. The purpose of **supercharging** is to raise the volumetric efficiency above that value that which can be obtained by normal aspiration. Supercharging of petrol engines, because of its poor fuel economy, is not very popular and is used only when a large amount of power is needed or when more power is needed to compensate altitude loss.
11. **Dissociation** refers to disintegration of burnt gases at high temperatures. It is a reversible process and increases with temperature. Dissociation, in general, causes a loss of power and efficiency.

12. **Performance of I.C. engines.** Some important relations :

(*i*) Indicated power (I.P.) = $\frac{np_{mi}LANk \times 10}{6}$ kW

(*ii*) Brake Power (B.P.) = $\frac{(W-S)\,\pi\,(D_b+d)N}{60 \times 1000}$ kW or $\left(= \frac{2\pi NT}{60 \times 1000} \text{kW}\right)$

(*iii*) Mechanical efficiency, $\eta_{mech.} = \frac{\text{B.P.}}{\text{I.P.}}$

(*iv*) Thermal efficiency (indicated), $\eta_{th.(I)} = \frac{\text{I.P.}}{\dot{m}_f \times C}$

and thermal efficiency (brake), $\eta_{th.(B)} = \frac{\text{B.P.}}{\dot{m}_f \times C}$

where $\dot{m}_f$ = mass of fuel used in kg/sec.

(*v*) $\eta_{relative} = \frac{\eta_{thermal}}{\eta_{air\text{-}standard}}$

(*vi*) Measurement of air consumption by **air box method** :

Volume of air passing through the orifice, $V_a = 840\, AC_d \sqrt{\frac{h_w}{\rho_a}}$

and, mass of air passing through the orifice,

$$m_a = 0.066\, C_d \times d^2 \sqrt{h_w \rho_a} \text{ kg/min}$$

where, A = Area of orifice, m²

d = Diameter of orifice, cm

h_w = Head of water in 'cm' causing the flow

ρ_a = Density of air in kg/m³ under atmospheric conditions.

II. Combustion in S.I. engines

1. **Combustion** may be defined as a relatively rapid chemical combination of hydrogen and carbon in the fuel with the oxygen in the air, resulting in liberation of energy in the form of heat.
2. **Ignition lag** is the time lag between the first igniting of fuel and commencement of the main phase of combustion.
3. **Pre-ignition** is the ignition of the homogeneous mixture in the cylinder, before the timed ignition spark occurs, caused by the local overheating of the combustible mixture. The standard test for pre-ignition is to shut-off the ignition. If the engine still fires, it is assumed that pre-ignition was taking place when the ignition was on.
4. A very sudden rise of pressure during combustion accompanied by metallic hammer like sound is called **detonation.**
5. **Performance number** (PN) is a useful measure of detonation tendency,

$$\text{PN} = \frac{\text{klimep of test fuel}}{\text{klimep of iso-octane}}.$$

6. The *Highest Useful Compression Ratio* (HUCR) is the highest compression ratio employed at which a fuel can be used in a specified engine under a specified set of operating conditions, at which detonation first becomes audible with both the ignition and mixture strength adjusted to give the highest efficiency.
7. **Swirl** is rotational flow of charge within the cylinder about its axis. It is generated by constructing the intake system to give a tangential component to the intake flow as it enters the cylinder.

8. **Squish** is the radial inward motion of the gas mixture. As the piston near T.D.C. squish motion generates a secondary rotational flow called "*tumble*". This rotation occurs about a circumferential axis near the outer edge of the piston bowl.
9. **Quench area** is defined by the parallel portion of the piston and cylinder head which almost touch each other as the piston approaches T.D.C. It is defined as percentage of opposing flat area relative to the piston crown area.
10. **Turbulence** consists of randomly dispersed vortices of different sizes which become superimposed into the air, or air and petrol mixture flow stream.
11. The **speed of the flame propagation** is roughly proportional to the velocity at the periphery of the vortices.
12. **Cylinder air swirl** is defined as the ratio of angular rotational speed about the cylinder axis.
13. **Swirl ratio** is defined as the ratio of air rotational speed to crankshaft rotational speed.
14. The **surface to volume ratio is** the ratio of the combustion surface area to that of its volume. It increases linearly with rising compression ratio.
15. **Over-square** ($L < D$) engines are more suitable for saloon car petrol engines, whereas *under square* ($L > D$) engines are better utilised for large diesel engines.

III. Combustion in C.I. engines

1. The three phases of C.I. engine combustion are :

   (*i*) Ignition delay period (*ii*) Period of rapid or uncontrolled combustion

   (*iii*) Period of controlled combustion.

   The third phase is followed by after burning, which may be called the fourth phase of combustion.
2. The *period of physical delay* is the time between the beginning of injection and attainment of chemical reaction conditions.

   In the *chemical delay period* reaction starts slowly and then accelerates until inflammation or ignition takes place.
3. The *delay period* is the time immediately following injection of the fuel during which the ignition process is being initiated and the pressure does not rise beyond the value it would have due to compression of air.
4. The delay period should be as short as possible since a long delay period gives a more rapid rise in pressure and thus causes knocking.
5. *Diesel knock* is the sound produced by the very rapid rate of pressure rise during the early part of the uncontrolled second phase of combustion.
6. Four specific designs which find wide use in C.I. engines are :

   A. The non-turbulent type :

   (*i*) Open combustion chamber

   B. The turbulent type:

   (*i*) Turbulent chamber

   (*ii*) Pre-combustion chamber

   (*iii*) Energy cell.

IV. Two stroke cycle engines

1. In a two stroke engines the cycle is completed in two strokes, *i.e.*, one revolution of the crankshaft as against two revolutions of four-stroke cycle.
2. The two main disadvantages from which the two stroke S.I. engines suffer are :

   (*i*) Loss of fuel (*ii*) Idling difficulty.
3. The two stroke C.I. oppose engine cross-head type is mainly used for marine propulsion.
4. The process of clearing the cylinder after the expansion stroke is called *scavenging process.*

5. Scavenging process consists of the following four distinct stages :
   (*i*) Pre-blowdown (*ii*) Blowdown
   (*iii*) Scavenging (*iv*) Additional charging.
6. *Scavenging angle* is defined as the crank angle during which both inlet and exhaust ports are open.
7. *Scavenging period* is the time period taken for scavenging angle.
8. *Theoretical scavenging processes are :*
   (*i*) Perfect scavenging ; (*ii*) Perfect mixing ;
   (*iii*) Short-circuiting.
9. Scavenging systems, based on charge flow, are classified as follows :
   (*i*) Uniflow scavenging (*ii*) Loop or reverse scavenging
   (*iii*) Cross scavenging.

V. Conventional fuels

1. The constituents petroleum are classified into the following four groups :
   (*i*) Paraffins (*ii*) Olefins
   (*iii*) Naphthenes (*iv*) Aromatics.
2. The process of separating petroleum into useful fractions and removal of undesirable impurities is called *refining.*
3. *Volatility* is commonly defined as the *evaporating tendency of a liquid fuel.* The volatility of gasoline is generally characterised by the following two laboratory tests :
   (*i*) ASTM distillation test (*ii*) Reid vapour test.
4. *Vapour lock* is a situation where too lean a mixture is supplied to the engine. The vapour lock tendency of the gasoline is related to front end volatility.
5. When self-ignition does occur in S.I. engine higher than desirable, pressure pulses are generated. The higher above self-ignition temperature which the fuel is heated, the shorter will be ignition delay
6. The *Highest Useful Compression Ratio* (HUCR) is the highest compression ratio at which a fuel can be used without detonation in a specified test-engine under specified operating conditions and the ignition and mixture strength being adjusted to give best efficiency.
7. The property of a fuel which describes how fuel will or will not self-ignite is called the *Octane number* or just *Octane.* Engines with low compression ratios can use fuels with lower octane number, but high-compression engines must use high-octane fuel to avoid self-ignition and knock.
8. The difference in octane number between reseach method and motor method octane numbers is known as the *fuel sensitivity.*
9. Antiknock index = $\frac{\text{RON + MON}}{2}$.
10. High octane fuels (upto 100) can be produced by *refining techniques,* but it is done more cheaply, and more frequently, by the use of antiknock *additives* such as *tetraethyl lead.*
11. Cetane number of diesel fuel is a measure of its ignition quality.
    In general, lower the cetane number higher are the hydrocarbon emissions and noise levels. In general, a high octane value implies a low cetane value.
12. Diesel index (D.I.) = Aniline point (°F) × $\frac{\text{API gravity (deg)}}{100}$

VI. Alternative fuels

1. Some alternative fuels which can replace conventional fuels in I.C. engine are :
   (*i*) Alcohol (methyl and ethyl) (*ii*) Hydrogen
   (*iii*) Natural gas (*iv*) LPG and LNG
   (*v*) Biogas.

2. The power output of ethanol engine is higher compared to gasoline engine at all speeds.
3. The CNG fuel used engines have higher thermal efficiencies than those fuelled by gasoline.
4. Biogas possesses excellent antiknock properties with an equivalent octane number in excess of 120 compared with 87 for regular petrol.

VII. Supercharging of I.C. engines

1. The purpose of *supercharging* is to raise the volumetric efficiency above that value which can be obtained by normal aspiration.
2. *Boost pressure* refers to the gauge pressure recorded when the air or mixture supply has passed through the supercharger.
3. *Pressure ratio* is the ratio of absolute pressure to that of the atmospheric pressure.
   (Here, absolute-pressure = boost pressure + atmospheric pressure)
4. The supercharging limits for a C.I. engine (unlike S.I. engines where the limits of supercharging are due to combustion) is reached by *thermal loading*.
5. The following types of superchargers are used to supercharge engines for various applications :
   Reciprocating compressors ; Vane blower ; lysholm compressor ;
   Roots blower ; centrifugal compressor.
6. *Turbochargers* are centrifugal compressors driven by the exhaust gas turbines.
7. *Methods of turbocharging :*
   (*i*) Constant pressure turbocharging (*ii*) Pulse turbocharging
   (*iii*) Pulse converter (*iv*) Two-stage turbocharger
   (*v*) Miller turbocharging (*vi*) Hyperbar turbocharging.

VIII. Miscellaneous engines

1. *Dual-fuel operation,* combines in a simple manner the possibility of operating a diesel engine on liquid fuels such as diesel oil or gas oil and on gaseous fuels such as natural gas, sewage gas and cook oven gas etc.
2. At full-load, dual-fuel engine is superior to diesel engine.
3. A *multi-fuel engine* is one which can operate satisfactorily on a wide variety of fuels ranging from diesel oil, crude oil, JP-4 to lighter fuels like gasoline, and even normal lubricating oil.
4. The *stratified charge engine* is usually defined as a S.I. engine (stratified diesel engine has also been developed) in which the mixture in the zone of spark-plug is very much richer than that in the rest of the combustion chamber *i.e.*, one which burns leaner overall fuel-air mixtures. *Charge stratification* means providing different fuel-air mixture strengths at various places in the combustion chamber.
5. The basic principle of working of *stirling engine* is the same as that of conventional engine. *The alternate compression at low temperature and expansion at high temperature of a working fluid in the basis for the stirling engine.*
6. In the *VCR-engine* a high compression is used for good stability and low load operation and a low compression is used at full-load to allow the turbocharger to boost the intake pressures without increasing the peak cycle pressure.
7. *Free-piston engine plants* are the conventional gas turbine plants with the *difference that the air compressor and combustion chamber are replaced by free-piston engine.*

IX. Air pollution from I.C. engines and its control

1. *Air pollution* can be defined as an addition to our atmosphere of any material which will have a deleterious effect on life upon our planet.
2. The major pollutants emitted from the exhaust due to incomplete combustion are :
   CO, HC, NO_x and other products *e.g.*, acetylene, aldehydes etc.
3. CO is generally formed when mixture is rich in fuel.

4. NO formation will be less in rich mixtures than in lean mixtures.
5. The chemical correct air-fuel ratio by mass for complete combustion is known as *stoichiometric ratio.*
6. The blowby HC emissions are about 20% of the total HC emission from the engine ; this is increased to about 30% if the rings are worn.
7. The following factors affect the formation of NO_x :
   (*i*) A/F ratio (*ii*) r.p.m. (*iii*) Angle of advance.
8. In a poorly maintained engine the exhaust pollution is more.
9. A *catalytic converter* is a device which is placed in the vehicle exhaust system to reduce HC and CO by oxidising catalyst and NO by reducing catalyst.
10. Rhodium is the best catalyst to control NO_x but A/F ratio must be within a narrow range of 14.6 : 1 to 14.7 : 1.
11. The basic principle of crankcase blowby control system is recirculation of vapours back to the intake manifold.
12. Oxides of nitrogen can be controlled by :
    (*i*) Exhaust gas recirculation (*ii*) Catalyst (*iii*) Water injection.
13. The working principle of *catalytic converter package* is to control the emission levels of various pollutants by changing the chemical characteristics of the exhaust gases.

OBJECTIVE TYPE QUESTIONS

Choose the Correct Answer

1. In a four-stroke cycle engine, the four operations namely suction, compression, expansion and exhaust are completed in the number of revolutions of crankshaft equal to
   (*a*) four (*b*) three (*c*) two (*d*) one.
2. In a two-stroke cycle engine, the operations namely suction, compression, expansion and exhaust are completed in the number of revolutions of crankshaft equal to
   (*a*) four (*b*) three (*c*) two (*d*) one.
3. In a four-stroke cycle S.I. engine the camshaft runs
   (*a*) at the same speed as crankshaft (*b*) at half the speed of crankshaft
   (*c*) at twice the speed of crankshaft (*d*) at any speed irrespective of crankshaft speed.
4. The following is an S.I. engine
   (*a*) Diesel engine (*b*) Petrol engine (*c*) Gas engine (*d*) none of the above.
5. The following is C.I. engine
   (*a*) Diesel engine (*b*) Petrol engine (*c*) Gas engine (*d*) none of the above.
6. In a four-stroke cycle petrol engine, during suction stroke
   (*a*) only air is sucked in (*b*) only petrol is sucked in
   (*c*) mixture of petrol and air is sucked in (*d*) none of the above.
7. In a four-stroke cycle diesel engine, during suction stroke
   (*a*) only air is sucked in (*b*) only fuel is sucked in
   (*c*) mixture of fuel and air is sucked in (*d*) none on the above.
8. The two-stroke cycle engine has
   (*a*) one suction valve and one exhaust valve operated by one cam
   (*b*) one suction valve and one exhaust valve operated by two cams
   (*c*) only ports covered and uncovered by piston to effect charging and exhausting
   (*d*) none of the above.
9. For same output, same speed and same compression ratio the thermal efficiency of a two-stroke cycle petrol engine as compared to that for four-stroke cycle petrol engine is
   (*a*) more (*b*) less
   (*c*) same as long as compression ratio is same (*d*) same as long as output is same.

10. The ratio of brake power to indicated power of an I.C. engine is called
(*a*) mechanical efficiency (*b*) thermal efficiency
(*c*) volumetric efficiency (*d*) relative efficiency.

11. The specific fuel consumption of a diesel engine as compared to that for petrol engines is
(*a*) lower (*b*) higher
(*c*) same for same output (*d*) none of the above.

12. The thermal efficiency of petrol engine as compared to diesel engine is
(*a*) lower (*b*) higher
(*c*) Same for same output (*d*) same for same speed.

13. Compression ratio of petrol engines is in the range of
(*a*) 2 to 3 (*b*) 7 to 10 (*c*) 16 to 20 (*d*) none of the above.

14. Compression ratio of diesel engines may have a range
(*a*) 8 to 10 (*b*) 10 to 15 (*c*) 16 to 20 (*d*) none of the above.

15. The thermal efficiency of good I.C. engine at the rated load is in the range of
(*a*) 80 to 90% (*b*) 60 to 70% (*c*) 30 to 35% (*d*) 10 to 20%.

16. In case of S.I. engine, to have best thermal efficiency the fuel air mixture ratio should be
(*a*) lean (*b*) rich
(*c*) may be lean or rich (*d*) chemically correct.

17. The fuel air ratio, for maximum power of S.I. engines, should be
(*a*) lean (*b*) rich
(*c*) may be lean or rich (*d*) chemically correct.

18. In case of petrol engine, at starting
(*a*) rich fuel-air ratio is needed (*b*) weak fuel-air ratio is needed
(*c*) chemically correct fuel-air ratio is needed (*d*) any fuel-air ratio will do.

19. Carburettor is used for
(*a*) S.I. engines (*b*) Gas engines (*c*) C.I. engines (*d*) none of the above.

20. Fuel injector is used in
(*a*) S.I. engines (*b*) Gas engines (*c*) C.I. engines (*d*) none of the above.

21. Very high speed engines are generally
(*a*) Gas engines (*b*) S.I. engines (*c*) C.I. engines (*d*) Steam engines.

22. In S.I. engine, to develop high voltage for spark-plug
(*a*) battery is installed (*b*) distributor is installed
(*c*) carburettor is installed (*d*) ignition coil is installed.

23. In S.I. engine, to obtain required firing order
(*a*) battery is installed (*b*) distributor is installed
(*c*) carburettor is installed (*d*) ignition coil is installed.

24. For petrol engines, the method of governing employed is
(*a*) quantity governing (*b*) quality governing
(*c*) hit and miss governing (*d*) none of the above.

25. For diesel engines, the method of governing employed is
(*a*) quantity governing (*b*) quality governing
(*c*) hit and miss governing (*d*) none of the above.

26. Voltage developed to strike spark in the spark-plug is in the range
(*a*) 6 to 12 volts (*b*) 1000 to 2000 volts
(*c*) 20000 to 25000 volts (*d*) none of the above.

27. In a four-cylinder petrol engine the standard firing order is
(*a*) 1-2-3-4 (*b*) 1-4-3-2 (*c*) 1-3-2-4 (*d*) 1-3-4-2.

28. The torque developed by the engine is maximum
(*a*) at minimum speed of engine
(*b*) at maximum speed of engine
(*c*) at maximum volumetric efficiency speed of engine
(*d*) at maximum power speed of engine.

29. Iso-octane content in a fuel for S.I. engines
(*a*) retards auto-ignition
(*b*) accelerates auto-ignition
(*c*) does not affect auto-ignition
(*d*) none of the above.

30. Normal heptane content in a fuel for S.I. engines
(*a*) retards auto-ignition
(*b*) accelerates auto-ignition
(*c*) does not affect auto-ignition
(*d*) none of the above.

31. The knocking in S.I. engines increases with
(*a*) increase in inlet air temperature
(*b*) increase in compression ratio
(*c*) increase in cooling water temperature
(*d*) all of the above.

32. The knocking in S.I. engines gets reduced
(*a*) by increasing the compression ratio
(*b*) by retarding the spark advance
(*c*) by increasing inlet air temperature
(*d*) by increasing the cooling water temperature.

33. Increasing the compression ratio in S.I. engines
(*a*) increases the tendency for knocking
(*b*) decreases tendency for knocking
(*c*) does not affect knocking
(*d*) none of the above.

34. The knocking tendency in petrol engines will increase when
(*a*) speed is decreased
(*b*) speed is increased
(*c*) fuel-air ratio is made rich
(*d*) fuel-air ratio is made lean.

35. The ignition quality of fuels for S.I. engines is determine by
(*a*) cetane number rating
(*b*) octane number rating
(*c*) calorific value rating
(*d*) volatility of the fuel.

36. Petrol commercially available in India for Indian passenger cars has octane number in the range
(*a*) 40 to 50
(*b*) 60 to 70
(*c*) 80 to 85
(*d*) 95 to 100.

37. Centane number of the fuel used commercially for diesel engines in India is in the range
(*a*) 80 to 90
(*b*) 60 to 80
(*c*) 60 to 70
(*d*) 40 to 45

38. The knocking tendency in C.I. engines increases with
(*a*) decrease of compression ratio
(*b*) increase of compression ratio
(*c*) increasing the temperature of inlet air
(*d*) increasing cooling water temperature.

39. Desirable characteristic of combustion chamber for S.I. engines to avoid knock is
(*a*) small bore
(*b*) short ratio of flame path to bore
(*c*) absence of hot surfaces in the end region of gas
(*d*) all of the above.

40. Accumulation of carbon deposits on the cylinder head of an I.C. engine leads to increase in ...
(*a*) piston displacement
(*b*) clearance volume
(*c*) compression ratio
(*d*) swept volume.

41. Which of the following motor cycles has more than one cylinder ?
(*a*) Bullet
(*b*) Yezdi
(*c*) Rajdoot
(*d*) Yamah.

42. A diesel engine is generally more efficient than a petrol engine because of
(*a*) proper air-fuel mixing and combustion
(*b*) high calorific value of diesel fuel
(*c*) knock-free operation
(*d*) high compression ratio.

43. Vapour lock refers to which of the following ?
(*a*) excess supply of fuel to engine
(*b*) blocking of carburettor jets
(*c*) complete or partial stoppage of fuel supply due to vaporisation of fuel in the supply line
(*d*) supply of air-fuel mixture containing liquid particles.

44. White deposits on the face of a spark-plug indicates that
(*a*) the engine is excessively advanced (*b*) mixture is too rich
(*c*) gap between the electrodes is too large (*d*) the rating of the spark plug is too high.

45. The injection presence in diesel engines is of the order of
(*a*) 30-40 bar (*b*) 100-150 bar (*c*) 170-220 bar (*d*) 400-600 bar.

46. The ignition temperature of diesel fuel is about
(*a*) 200°C (*b*) 400°C (*c*) 550°C (*d*) 700°C.

47. In a petrol engine the delay period is of the order of
(*a*) 0.001 s (*b*) 0.002 s (*c*) 0.015 s (*d*) 0.06 s

48. ... is not the effect of detonation
(*a*) Loud and pulsating noise (*b*) High local stresses
(*c*) High operating temperature (*d*) Loss in efficiency and power output.

49. The ignition quality of a petrol engine fuel is expressed as
(*a*) octane number (*b*) cetane number (*c*) API gravity (*d*) SAE rating.

50. The use of tetraethyl lead in gasoline is being gradually discontinued since its presence
(*a*) decreases the engine speed (*b*) blocks the catalytic converter
(*c*) makes the fuel costly (*d*) gives bad odour.

51. Which of the following fuels has a centane number of 100 ?
(*a*) Normal heptane (*b*) Ethyl fluid (*c*) Cetane (*d*) α-methyl napthalene.

52. During idling a petrol engine requires
(*a*) lean mixture (*b*) rich mixture
(*c*) variable mixture (*d*) chemically correct mixture.

53. Stoichiometric ratio is
(*a*) actual ratio of air to fuel for maximum efficiency
(*b*) chemically correct air-fuel ratio by weight
(*c*) chemically correct air-fuel ratio by volume
(*d*) none of the above.

54. The capacity of most of the mopads in India is
(*a*) 50 CC (*b*) 150 CC (*c*) 200 CC (*d*) 250 CC.

55. Due to which of the following reasons diesel engines are preferred for road transport ?
(*a*) Complete combustion of charge (*b*) Low operating cost
(*c*) Low specific fuel consumption over a large range of load
(*d*) Easy starting.

56. ... is used for the insulating body of a spark-plug
(*a*) Dolomite (*b*) Alumina (*c*) Glass (*d*) Silica.

57. The spark advance is usually specified in terms of
(*a*) degrees of crank rotation (*b*) time in seconds
(*c*) engine speed in rev./sec. (*d*) none of the above.

58. In an I.C. engine if intake air temperature increases, its efficiency will
(*a*) decrease (*b*) increase
(*c*) remain same (*d*) cannot be predicted.

59. Which of the following statements is *incorrect* ?
(*a*) Petrol engines work on Otto cycle
(*b*) For same power output petrol engines occupy more space than diesel engines
(*c*) In a four-stroke engine a power stroke is obtained in four strokes
(*d*) Thermal efficiency of four-stroke engine is more due to positive scavenging.

60. In a diesel engine if one of the cylinders receives more fuel than the others then which of the following will happen for that cylinder ?
(*a*) Exhaust temperature will be high (*b*) Exhaust will be smoky
(*c*) Piston rings would stick into piston grooves (*d*) Engine will start overheating
(*e*) All of the above.

61. The carbon accumulation in an engine cylinder results in
(*a*) increase of effective compression ratio (*b*) increase of volumetric efficiency
(*c*) increase of clearance volume (*d*) increase of ignition time.

62. The compression ratio in diesel engine is ... in comparison to expansion ratio
(*a*) less (*b*) more (*c*) same (*d*) variable.

63. In an automobile the magneto is basically
(*a*) d.c. generator (*b*) a.c. generator (*c*) transformer (*d*) capacitor.

64. Scavenging is usually done to increase which of the following ?
(*a*) Power output (*b*) Fuel consumption (*c*) Thermal efficiency (*d*) Speed.

65. For a petrol engine for vehicles the air-fuel ratio for maximum power generation is of the order of
(*a*) 8 : 1 (*b*) 12 : 1 (*c*) 18 : 1 (*d*) 20 : 1.

66. In loop scavenging the top of the piston is
(*a*) convex shaped (*b*) depressed (*c*) slanted (*d*) contoured.

67. Which of the following statements is *correct* regarding normal heptane ?
(*a*) It retards auto-ignition (*b*) It accelerates auto-ignition
(*c*) It helps to resist auto-ignition (*d*) It does not affect auto-ignition.

68. Due to which of the following reasons the piston rings are plated with chromium, cadmium or phosphate ?
(*a*) To prevent clogging (*b*) To improve heat transfer
(*c*) To reduce wear and eliminate scuffing (*d*) To improve surface finish.

69. The specific gravity of diesel oil is
(*a*) 0.6 (*b*) 0.75 (*c*) 0.85 (*d*) 1.2.

70. Detonation can be controlled by
(*a*) reducing the r.p.m. (*b*) retarding the spark timing
(*c*) varying compression ratio (*d*) any of the above.

71. Due to which of the following the tendency of a diesel engine to knock increases ?
(*a*) Increase in engine speed (*b*) Increase in compression ratio
(*c*) Increase in octane value of fuel (*d*) Increase in engine power.

72. The part load efficiency of a carburettor is
(*a*) constant (*b*) maximum (*c*) option (*d*) poor.

73. ... can work on very lean mixture.
(*a*) C.I. engine (*b*) S.I. engine (*c*) Two-stroke engine (*d*) Four-stroke engine.

74. Thermal efficiency of I.C. engines on weak mixture is
(*a*) lower (*b*) higher (*c*) unaffected (*d*) unpredictable

75. Cetane number is the measure of
(*a*) viscosity of fuel (*b*) ignition quality
(*c*) calorific value of fuel (*d*) auto-ignition temperature.

76. In a S.I. engine an ignition coil performs which of the following functions ?
(*a*) Regulates battery voltage (*b*) Avoids sparking
(*c*) Controls spark (*d*) Supplies high voltage to the spark-plug.

77. ... does not assist in getting higher output from diesel engines.
(*a*) High fuel-air ratio (*b*) High compression ratio
(*c*) High excess air (*d*) Fine atomisation of fuel.

78. Which of the following factors does not promote detonation in S.I. engines ?
(*a*) High self ignition temperature of fuel
(*b*) Increase in inlet pressure and temperature of charge
(*c*) Higher compression ratio
(*d*) Advanced spark timing.

79. Which of the following statements is *correct* ? The phenomenon of pre-ignition
(*a*) always occurs in petrol engines (*b*) always occurs in diesel engines
(*c*) never occurs in diesel engines (*d*) increases the power output of an engine.

80. The octane rating of the commercially available petrol in India is
(*a*) 15-35 (*b*) 45-55 (*c*) 60-70 (*d*) 85-90.

81. ... lubrication technique is used for lubrication of the cylinder of a scooter engine.
(*a*) Petrol (*b*) Splash (*c*) Gravity feed (*d*) Forced feed.

82. In four-stroke engines the camshaft rotates at ... the crankshaft speed
(*a*) half (*b*) three-fourth (*c*) equal (*d*) double.

83. Which of the following is the distinctive features of an I.C. engine ?
(*a*) Easy and instantaneous starting (*b*) High overall efficiency
(*c*) Low weight to power ratio
(*d*) Combustion and conversion of heat energy into mechanical work occur inside a cylinder
(*e*) All of the above.

84. The minimum number of rings in a piston are
(*a*) two (*b*) three (*c*) four (*d*) six.

85. ... process is not associated with diesel cycle.
(*a*) Constant pressure (*b*) Constant volume (*c*) Adiabatic (*d*) Isothermal.

86. Highest useful compression ratio is the compression ratio at which
(*a*) the engine consumes minimum fuel for a particular power output
(*b*) the engine gives maximum power output
(*c*) The engine maintains operating pressures and temperatures within prescribed limits
(*d*) the engine can operate without detonation.

87. A two-stroke cycle engine as compared to four-stroke cycle engine
(*a*) has lower fuel consumption (*b*) can be easily started
(*c*) is smaller in size for the same output (*d*) has lesser shocks and vibrations.

88. For which of the following engines a prony brake is used to measure brake power ?
(*a*) Single cylinder engine (*b*) Low speed engine
(*c*) Low power engine (*d*) Variable speed engine.

89. What does scavenging air mean ?
(*a*) Burnt air containing combustion products (*b*) Air sent under compression
(*c*) Forced air for cooling the engine cyclinder
(*d*) Air used for forcing the burnt gases out of cylinder during the exhaust period.

90. ...... is the basic requirement of a good combustion chamber.
(*a*) Low volumetric efficiency (*b*) High compression ratio
(*c*) Low compression ratio (*d*) High power output and high thermal efficiency.

91. ...... is the method of governing used in petrol engine.
(*a*) Quality governing (*b*) Hit and miss governing
(*c*) Quantity governing (*d*) Partial governing.

92. ...... is the method of governing used in diesel engine.
(*a*) Quality governing (*b*) Hit and miss governing
(*c*) Quantity governing (*d*) Any of the above.

93. Hunting occurs due to which of the following ?
(*a*) Faulty governor
(*b*) Poor-control by the governor
(*c*) Over-control by the governor
(*d*) Bad engine design.

94. Maximum torque is generated by an engine when
(*a*) it runs at lowest speed
(*b*) it develops maximum power
(*c*) it consumes maximum fuel
(*d*) it runs at maximum speed.

95. With an increase of the number of cylinders in a multicylinder engine the power to weight ratio
(*a*) decreases
(*b*) increases
(*c*) remains unaffected
(*d*) none of the above.

96. What will happen if petrol is used in diesel engine ?
(*a*) Black smoke will be produced.
(*b*) Low power will be produced.
(*c*) Higher knocking will occur.
(*d*) Efficiency will be low.

97. What will happen if diesel is fed by mistake in the oil tank of a petrol engine ?
(*a*) The engine will not run.
(*b*) The engine will knock.
(*c*) The engine will detonate.
(*d*) The engine will give lot of smoke.

98. With which of the following tendency of detonation is S.I. engines increases ?
(*a*) Increase of compression ratio
(*b*) Decrease of compression ratio
(*c*) Increase of engine speed
(*d*) Decrease of engine speed.

99. Performance numbers are ...
(*a*) indicative of the fuels having anti-knock qualities superior to iso-octane
(*b*) indicative of the fuels having anti-knock qualities superior to cetane
(*c*) indices of efficiency of petrol engines
(*d*) indices of efficiency of diesel engines.

100. The bi-fuel engine uses which of the following ?
(*a*) Gas fuel during start up and liquid fuel as the basic fuel.
(*b*) Liquid fuel during start up and gas as the basic fuel.
(*c*) Two fuels used in two combustion chambers.
(*d*) None of the above.

101. Lean air-fuel mixture is required for
(*a*) idling
(*b*) acceleration
(*c*) starting
(*d*) cruising.

102. ...... is not a part of petrol engine.
(*a*) Air filter
(*b*) Induction coil
(*c*) Value mechanism
(*d*) Fuel injector.

103. Regarding two-stroke engines which of the following statements is *correct* ?
(*a*) There is only one valve for inlet and exhaust.
(*b*) Charge enters the engine cylinder through ports only.
(*c*) A diesel engine cannot operate on two-stroke cycle.
(*d*) Compression ratio is always lower than that of four-stroke cycle engine.

104. Which of the following statements is *incorrect* ?
(*a*) Choke is kept open when cranking a cold engine for starting.
(*b*) A carburettor prepares a homogenous air-fuel mixture by atomising and vaporising the fuel.
(*c*) Throttle valve controls the supply of air-fuel mixture.
(*d*) Vacuum at the throat of venturi sucks the fuel through the fuel-jet.

105. ...... of heat supplied in the form of fuel in a four-stroke engine is carried away by exhaust gases.
(*a*) 3-7 percent
(*b*) 8-12 percent
(*c*) 20-35 percent
(*d*) 45-55 percent

106. Petrol engines are adjusted to give minimum brake specific fuel consumption at
(*a*) no load
(*b*) 20-30 percent of full load
(*c*) about 70 percent of full load
(*d*) near full load.

107. Regarding contact breaker which of the following statements is Incorrect ?
(*a*) Spark takes place when the points are open (*b*) Excessive gas results in rapid burning of points
(*c*) contact points are generally made of tungsten
(*d*) Points are opened by the cam and closed by the spring tension.

108. ... regulates the pressure strokes in the fuel injection pump of a diesel engine.
(*a*) Pump shaft (*b*) Control rack (*c*) Lift of plunger (*d*) Needle valve.

109. Which of the following could be the probable reason of power loss in a diesel engine ?
(*a*) Low injection pressure (*b*) Restricted exhaust
(*c*) Ineffective cooling (*d*) Clogging of air cleaner.

110. ... is not a part of magneto-ignition system.
(*a*) Condenser (*b*) Induction coil (*c*) Battery (*d*) Circuit breaker.

111. Due to which of the following reasons the petrol engines are usually not supercharged ?
(*a*) Drop in volumetric efficiency. (*d*) Increased knocking.
(*c*) Increased specific fuel consumption. (*d*) Power loss.

112. ... identifies the anti-knock quality of diesel fuel.
(*a*) Octane number (*b*) cetane number (*c*) either of the above (*d*) none of the above.

113. Volumetric efficiency of a well designed engine may be in the range
(*a*) below 20 percent (*b*) 30-40 percent (*c*) 50-70 percent (*d*) 75-90 percent.

114. The advancing of spark timing in a S.I. engine will
(*a*) reduce knocking tendency (*b*) increase knocking tendency
(*c*) not have any effect (*d*) none of the above.

115. Free piston engines find application in
(*a*) gas turbines (*b*) mining installations
(*c*) compressed air supply (*d*) supercharging of diesel engines.

116. In a C.I. engine higher combustion chamber wall temperature will
(*a*) reduce exhaust temperature (*b*) reduce knocking tendency
(*c*) increase knocking tendency (*d*) have no effect.

117. ... acts as ignition accelerator for C.I. engines fuel.
(*a*) Hydrogen peroxide (*b*) Acetone peroxide (*c*) *n*-heptane (*d*) none of the above.

118. Why are fuel ignition accelerators added in C.I. engine ?
(*a*) To reduce combustion chamber temperature (*b*) To reduce combustion knock
(*c*) To accelerate combustion knock (*d*) To increase delay period.

119. In which of the following engines crankcase explosion occurs ?
(*a*) S.I. engines (*b*) four-stroke S.I. engines
(*c*) two-stroke C.I. engines (*d*) four-stroke C.I. engines.

120. Highest useful compression ratio is the compression ratio at which
(*a*) an engine operates smoothly (*b*) detonation first becomes audible
(*c*) an engine can be safely operated (*d*) an engine gives maximum thermal efficiency.

121. Crankshafts are generally
(*a*) die cast (*b*) sand cast
(*c*) forged (*d*) turned from bar stock.

122. ... has maximum resistance to detonation.
(*a*) Alcohol (*b*) Benzene (*c*) Toluene (*d*) Iso-octane.

123. Why are oil rings slotted ?
(*a*) To minimise friction. (*b*) To remove oil from cylinder.
(*c*) To reduce the bulk.
(*d*) To provide an escape for the oil that the slot edges cut from the cylinder wall.

124. In isochronous governors the speed drop is
(*a*) zero (*b*) 5 percent (*c*) 30 percent (*d*) 50 percent.

125. The top ring nearest to the piston crown is known as
(*a*) compression ring (*b*) oil ring (*c*) scrapper ring (*d*) groove ring.

126. A diesel engine as compared to petrol engine (both running at full load) is
(*a*) less efficient (*b*) more efficient (*c*) equally efficient (*d*) none of the above.

127. The level of oil in engine cylinder should be checked when the engine is
(*a*) running (*b*) not running (*c*) during starting (*d*) during cranking.

128. Endurance for I.C. engines is conducted for
(*a*) 200 hours (*b*) 300 hours (*c*) 400 hours (*d*) 500 hours.

129. Movement of air inside engine cylinder does not help in
(*a*) reducing noise (*b*) mixing of fuel with air
(*c*) distribution of fuel (*d*) reduction of after burning.

130. What is swirl in C.I. engines ?
(*a*) Circular motion imparted to suction air. (*b*) Radial motion imparted to fuel-air mixture.
(*c*) Directional movement of fuel spray.
(*d*) Circular motion imparted to gases after combustion.

131. In a C.I. engine squish is created
(*a*) towards the end of compression stroke (*b*) at the end of suction stroke
(*c*) at the beginning of suction stroke (*d*) during combustion.

132. An increase in the mean effective pressure of a diesel engine with fixed compression ratio can be obtained with increase in
(*a*) cut-off ratio (*b*) engine speed (*c*) back pressure (*d*) charge density.

133. By which of the following a two-stroke engine is usually identified ?
(*a*) Absence of valves. (*b*) Size of flywheel.
(*c*) Location of fuel tank. (*d*) Weight of engine.

134. The ... knocking tendency in S.I. engines can be decreased by
(*a*) adding benzole (*b*) decreasing compression ratio
(*c*) controlling ignition timing (*d*) adding dopes (like tetraethyle lead etc.)

135. Due to which of the following injection lag in diesel engines is caused ?
(*a*) Leakage past the fuel-oil plunger. (*b*) Compressibility of fuel.
(*c*) Expansion of fuel-oil discharge lines under high pressure.
(*d*) All of the above.

136. For reducing wear and eliminate scuffing, the piston rings are
(*a*) lubricated (*b*) made of cast iron
(*c*) provided with stepped groove (*d*) plated with chromium or cadmium.

137. Due to which of the following reasons a diesel engine gives a smoky exhaust ?
(*a*) Water in the fuel. (*b*) Fuel injection is late.
(*c*) Fuel is not distributed equally to all the cylinders.
(*d*) Exhaust valve receives too much lube oil. (*e*) All of the above.

138. Compared to a diesel engine the compression ratio in petrol engine is kept low because
(*a*) engine design becomes simple (*b*) it provides fuel economy
(*c*) petrol engine is a light engine
(*d*) higher compression ratio (in petrol engine) would lead to pre-ignition of fuel
(*e*) all of the above.

139. Compared to petrol engines, diesel engines require
(*a*) smaller flywheel (*b*) bigger flywheel (*c*) same size flywheel (*d*) no flywheel.

140. Free acids in diesel oil for diesel engine lead to which of the following ?
(*a*) Excessive fuel consumption (*b*) Deposition on engine parts
(*c*) Damaging of both the storage tank and the engine
(*d*) Excessive engine wear.

141. Which type of cleaner in case of diesel engines is most effective ?
(*a*) Wet type (*b*) Dry type (*c*) Oil bath type (*d*) Whirl type.

142. What happens when cooling water temperature in petrol engine is increased ?
(*a*) The knocking tendency increases. (*b*) The knocking tendency decreases.
(*c*) The knocking tendency remains unaffected. (*d*) Unpredictable.

143. In a petrol engine ... gas gets exhausted out without burning and without transformation
(*a*) CO (*b*) CO_2 (*c*) Nitrogen (*d*) O_2.

144. ...... smooths out power impulses from an I.C. engine
(*a*) Flywheel (*b*) Gear box (*c*) Governor (*d*) Crankshaft.

145. What is the advantage of reversing the flow of air in an air cleaner ?
(*a*) The velocity of air is increased. (*b*) The velocity of air is reduced.
(*c*) The air flow is increased.
(*d*) A large percentage of foreign matter is thrown out.

146. In scooters fins are provided over engine cylinder for
(*a*) good appearance (*b*) higher efficiency
(*c*) higher strength of cylinder (*d*) better cooling.

147. Which of the following systems of lubrication is used for motor cycles and scooters ?
(*a*) Wet sump method. (*b*) Splash lubrication.
(*c*) Forced lubrication system.
(*d*) Mixing about 5 percent lubricating oil with petrol.

148. How can the ignition timing of a multicylinder petrol engine be adjusted ?
(*a*) By adjusting ignition coil postilion. (*b*) By rotating the crank.
(*c*) By rotating the distributor. (*d*) By adjusting the spark-plug gap.

149. In I.C. engines leakage past the piston rings and valves ...... with increase in speed.
(*a*) decreases (*b*) increases (*c*) remains same (*d*) none of the above.

150. Why are the exhaust pipes of engines covered with insulating material ?
(*a*) To conserve heat (*b*) To keep the exhaust pipes warm
(*c*) To reduce heat transfer to the engine room (*d*) To increase the engine efficiency.

151. A fuel will denote less if
(*a*) it has constant self-ignition temperature (*b*) it has lower self-ignition temperature
(*c*) it has higher self-ignition temperature (*d*) none of the above.

152. Why are the pistons usually given a coating (*e.g.*, tin coating) ?
(*a*) To increase lubrication effect. (*b*) To conduct heat efficiently.
(*c*) To reduce weight. (*d*) To reduce possibility of scoring.

153. For preventing knock in S.I. engines, the charge away from spark plug should have
(*a*) rich mixture (*b*) low temperature (*c*) long ignition delay (*d*) low density.
(*e*) all of the above.

154. Due to which of the following violent sound pulsations within the cylinder of an I.C. engine are caused ?
(*a*) Pre-ignition (*b*) Heavy supercharging
(*c*) Detonation (*d*) Heavy turbulence.

155. By which of the following air is supplied in a naturally aspirated diesel engine ?
(*a*) A centrifugal blower (*b*) A supercharger
(*c*) Forced chamber (*d*) A vacuum chamber.

156. ...... is the lightest and most volatile liquid fuel.
(*a*) Kerosene (*b*) Gasoline (*c*) Diesel (*d*) Fuel oil.

157. By installing a supercharger on a 4-stroke cycle diesel engine upto ...... percent increase in power can result in.
(*a*) 20 (*b*) 40 (*c*) 70 (*d*) 100.

158. In ... engines supercharging is essential.
(*a*) marine (*b*) petrol (*c*) aircraft (*d*) diesel.

159. By which of the following methods diesel smoke can be reduced ?
(*a*) Adherence to proper fuel specification (*b*) Using additives in the fuel
(*c*) Avoidance of overloading (*d*) Reducing maximum flow of fuel
(*e*) All of the above.

160. The exhaust valve of an engine is ...... in size in comparison on inlet valve.
(*a*) smaller (*b*) bigger
(*c*) same (*d*) varies from design to design.

161. On which of the following cycles are most high speed engines experated ?
(*a*) Two-stroke cycle (*b*) Carnot cycle (*c*) Diesel cycle (*d*) Dual Cycle.

162. Aluminium alloy is usually used to make engines pistons because it
(*a*) wears less (*b*) is lighter (*c*) is stronger (*d*) absorbs shocks.

163. With which of the following engines is the term scavenging associated ?
(*a*) Aero engines (*b*) Diesel engines
(*c*) High efficiency engines (*d*) None of the above.

164. The rating of a diesel engine will ...... with increase in air-inlet temperature
(*a*) increase parabolically (*b*) increase linearly
(*c*) decrease parabolically (*d*) decrease linearly.

165. Which of the following statements is *correct* ? For the same compression ratio
(*a*) Diesel cycle is more efficient than Otto cycle (*b*) Otto cycle is more efficient than diesel cycle
(*c*) Both diesel and auto cycles are equally efficient
(*d*) Compression ratio has no relation with efficiency.

Answers

1. (*c*)	**2.** (*d*)	**3.** (*b*)	**4.** (*b*)	**5.** (*a*)	**6.** (*c*)	**7.** (*a*)
8. (*c*)	**9.** (*b*)	**10.** (*a*)	**11.** (*a*)	**12.** (*a*)	**13.** (*b*)	**14.** (*c*)
15. (*c*)	**16.** (*a*)	**17.** (*b*)	**18.** (*a*)	**19.** (*a*)	**20.** (*c*)	**21.** (*b*)
22. (*b*)	**23.** (*b*)	**24.** (*a*)	**25.** (*b*)	**26.** (*c*)	**27.** (*d*)	**28.** (*c*)
29. (*a*)	**30.** (*b*)	**31.** (*d*)	**32.** (*b*)	**33.** (*a*)	**34.** (*a*)	**35.** (*b*)
36. (*c*)	**37.** (*d*)	**38.** (*a*)	**39.** (*d*)	**40.** (*c*)	**41.** (*d*)	**42.** (*d*)
43. (*c*)	**44.** (*a*)	**45.** (*b*)	**46.** (*b*)	**47.** (*b*)	**48.** (*d*)	**49.** (*a*)
50. (*b*)	**51.** (*c*)	**52.** (*b*)	**53.** (*b*)	**54.** (*a*)	**55.** (*c*)	**56.** (*b*)
57. (*a*)	**58.** (*a*)	**59.** (*b*)	**60.** (*e*)	**61.** (*a*)	**62.** (*b*)	**63.** (*a*)
64. (*a*)	**65.** (*b*)	**66.** (*d*)	**67.** (*b*)	**68.** (*c*)	**69.** (*c*)	**70.** (*b*)
71. (*b*)	**72.** (*d*)	**73.** (*a*)	**74.** (*b*)	**75.** (*b*)	**76.** (*d*)	**77.** (*c*)
78. (*a*)	**79.** (*c*)	**80.** (*d*)	**81.** (*a*)	**82.** (*a*)	**83.** (*e*)	**84.** (*a*)
85. (*d*)	**86.** (*d*)	**87.** (*c*)	**88.** (*c*)	**89.** (*d*)	**90.** (*d*)	**91.** (*c*)
92. (*a*)	**93.** (*c*)	**94.** (*a*)	**95.** (*a*)	**96.** (*c*)	**97.** (*a*)	**98.** (*a*)
99. (*a*)	**100.** (*b*)	**101.** (*d*)	**102.** (*d*)	**103.** (*b*)	**104.** (*a*)	**105.** (*c*)
106. (*d*)	**107.** (*b*)	**108.** (*b*)	**109.** (*a*)	**110.** (*c*)	**111.** (*b*)	**112.** (*b*)
113. (*d*)	**114.** (*b*)	**115.** (*a*)	**116.** (*b*)	**117.** (*b*)	**118.** (*b*)	**119.** (*a*)
120. (*b*)	**121.** (*c*)	**122.** (*d*)	**123.** (*d*)	**124.** (*a*)	**125.** (*a*)	**126.** (*b*)

127. (*b*)	**128.** (*d*)	**129.** (*a*)	**130.** (*a*)	**131.** (*a*)	**132.** (*a*)	**133.** (*a*)
134. (*d*)	**135.** (*d*)	**136.** (*d*)	**137.** (*c*)	**138.** (*e*)	**139.** (*b*)	**140.** (*c*)
141. (*c*)	**142.** (*a*)	**143.** (*c*)	**144.** (*a*)	**145.** (*d*)	**146.** (*d*)	**147.** (*d*)
148. (*c*)	**149.** (*a*)	**150.** (*c*)	**151.** (*c*)	**152.** (*d*)	**153.** (*e*)	**154.** (*c*)
155. (*d*)	**156.** (*b*)	**157.** (*d*)	**158.** (*c*)	**159.** (*e*)	**160.** (*b*)	**161.** (*d*)
162. (*b*)	**163.** (*c*)	**164.** (*d*)	**165.** (*b*).			

THEORETICAL QUESTIONS

I. Introduction to I.C. engines

1. Name the two general classes of combustion engines and state how do they basically differ in principle ?
2. Discuss the relative advantages and disadvantages of internal combustion and external combustion engines.
3. What are the two basic types of internal combustion engines ? What are the fundamental differences between the two ?
4. What is the function of a governor ? Enumerate the types of governors and discuss with a neat sketch the Porter governor.
5. Differentiate between a flywheel and a governor.
6. (*a*) State the function of a carburettor in a petrol engine.
   (*b*) Describe a simple carburettor with a neat sketch and also state its limitations.
7. Explain with neat sketches the construction and working of the following :
   (*i*) Fuel pump (*ii*) Injector.
8. Explain the following terms as applied to I.C. engines :
   Bore, stroke, T.D.C., B.D.C., clearance volume, swept volume, compression ratio and piston speed.
9. Explain with suitable sketches the working of a four-stroke otto engine.
10. Discuss the difference between ideal and actual valve timing diagrams of a petrol engine.
11. In what respects four-stroke diesel cycle (compression ignition) engine differs from four-stroke cycle spark ignition engine ?
12. Discuss the difference between theoretical and actual valve timing diagrams of a diesel engine ?
13. What promotes the development of two-stroke engines ? What are the two main types of two-stroke engines ?
14. Describe with a suitable sketch the two-stroke cycle spark ignition (SI) engine. How its indicator diagram differs from that of four-stroke cycle engine ?
15. Compare the relative advantages and disadvantages of four-stroke and two stroke cycle engines.

II. Combustion in S.I. engines

1. Define 'combustion'. State the general conditions necessary for combustion.
2. Discuss the ignition limits of hydrocarbon fuels.
3. Explain briefly combustion phenomenon in S.I. engines.
4. What do you mean by pre-ignition ? How can it be detected ?
5. Explain the difference between (*i*) pre-ignition, (*ii*) auto-ignition, and (*iii*) detonation.
6. Explain the phenomenon of auto-ignition. Explain how auto-ignition is responsible for knocking in S.I. engines.
7. Explain the phenomenon of knocking in S.I. engines. What are the different factors which influence the knocking ? Describe the methods used to suppress it.
8. Explain the main factors that influence the flame speed.
9. What is performance number ?

10. What are the factors that limit the compression ratio that can be used in petrol engines ?
11. "Abnormal combustion knock produced by surface ignition in S.I. engines is more harmful than normal combustion knock". Justify the statement.
12. How Tetraethyl Lead (T.E.L) improves the quality of fuel for S.I. engine ?
13. What do you mean by octane number of 85 and octane number of 75 ? What is H.U.C.R. ?
14. Space of the clearance volume controls the detonation in case of S.I. engine. Comment.
15. What is ignition lag ? Discuss the effect of engine variables on ignition lag.
16. Discuss the effects of the following variables on engine heat transfer :
    (*i*) Spark advance (*ii*) Engine output
    (*iii*) Pre-ignition and knocking.
17. "The highest compression ratio that can be used in a S.I. engine is limited by the detonation characteristics of the available fuel". Justify the statement.
18. "The retarding of spark timing in a S.I. engine will reduce detonation". Justify the statement.
19. What action can be taken with regard to the following variables, in order to reduce the possibility of detonation in a S.I. engine ? Justify your answers by reasons.
    (*i*) Compression ratio (*ii*) Mass of charge induced
    (*iii*) Mixture inlet temperature (*iv*) Engine speed
    (*v*) Distance of flame travel.
20. Discuss the effect of the following engine variables on flame propagation :
    (*i*) Compression ratio (*ii*) Fuel-air ratio
    (*iii*) Turbulence (*iv*) Engine load
    (*v*) Engine speed.
21. "Auto-ignition is the cause of detonation". Justify the statement.
22. "Compressed natural gas (CNG) is preferable in S.I. engine than C.I. engine" ? Justify the statement.
23. Why is spark advance required ? Discuss the factor that affect ignition timing.
24. On what basis are S.I. engines fuels compared when they are better than iso-octane in anti-knock characteristics ?
25. Discuss the three basic requirements of a good S.I. engine combustion chamber.
26. Discuss the general principles of S.I. engine combustion chamber design.
27. What are the advantages of overhead valve combustion chamber over side valve combustion chamber ?

III. Combustion in C.I. engines

1. When was C.I. engine developed and by whom ?
2. State the applications of C.I. engines.
3. Enlist the reasons for which C.I. engine is not much favoured in passenger cars.
4. Explain briefly the combustion phenomenon in C.I. engine.
5. Describe briefly various phases of C-I. engine combustion.
6. State the various factors which affect combustion in C.I. engine.
7. Explain briefly diesel engine injected spray combustion process.
8. What is delay period in C.I. engines ?
9. What is the difference between physical delay and chemical delay ?
10. State the factors on which delay period depends.
11. Explain the effect of the following factors on delay period :
    (*i*) Fuel properties (*ii*) Intake temperature
    (*iii*) Compression ratio (*iv*) Engine speed
    (*v*) Type of combustion chamber (*vi*) Injection advance.
12. Explain briefly the phenomenon of "Diesel knock".
13. State the differences in the knocking phenomena of S.I. and C.I. engines.

14. Enlist various methods of controlling diesel knock.
15. What should be the primary considerations in the design of combustion chambers for C.I. engines ?
16. Explain briefly basic methods of generating air swirl in C.I. engines combustion chambers.
17. Enlist the advantages and disadvantages of induction swirl.
18. State the advantages and disadvantages of compression swirl.
19. Explain briefly any two of the following combustion chambers :
    (*i*) Open or direct combustion chamber (*ii*) Turbulent chamber
    (*iii*) Pre-combustion chamber (*iv*) Energy cell.
20. Give the comparison between open combustion chambers and divided combustion chambers.
21. Write short note on cold starting of C.I. engines.
22. Explain briefly cold starting aids for C.I. engines.
23. Explain the phenomenon of knock in C.I. engines and compare it with S.I. engine knock.
24. How does the mixture composition in combustion chamber of a C.I. engine differ from that of a S.I. engine ?
25. "The factors that tend to increase detonation in S.I. engine tend to reduce knocking in C.I. engine". Discuss the above statement with reference to the following influencing factors :
    (*i*) Compression ratio (*ii*) Inlet temperature
    (*iii*) Inlet pressure (*iv*) Self-ignition temperature of fuel
    (*v*) Time lag of ignition of fuel (*vi*) r.p.m.
    (*vii*) Combustion chamber wall temperature.
26. Why does rate of pressure rise during combustion is limited to a certain value ?
27. Discuss the influence of ignition delay on combustion processes in S.I. and C.I. engines. Explain how the presence of a knock inhibitor in fuel oil helps to change the ignition delay in C.I. engines.
28. "The requirement of air motion and swirl in a C.I. engine combustion chamber is more strigent than in a S.I. engine". Justify the statement.
29. "The induction swirl in a C.I. engine helps in increasing indicated thermal efficiency". Justify the statement.
30. How are C.I. engine combustion classified ? What type of swirl is used in these chambers ?
31. "In agriculture field, it is better to use C.I. engine than S.I. engine". Justify the statement.
32. How can a diesel engine be converted to CNG engine ?
33. "The maximum substitution of diesel engine by CNG in a C.I. engine is limited by the cetane characteristics of the available fuel". Justify the statement.
34. Write a short note on aids for starting C.I. engines under extreme cold climate.
35. Describe the M-combustion system and discuss its relative merits with respect to D.I. chambers.

IV. Two stroke cycle engines

1. "In a two-stroke engine it is better to have deflector top type piston". Justify the statement.
2. Discuss the two main disadvantages of two-stroke cycle S.I. engine. How are these disadvantages avoided in the two-stroke cycle C.I. engine ?
3. Discuss briefly 'Mist lubrication system'.
4. Why do the two-stroke C.I. engine find wide use in marine propulsions ?
5. Why are two-stroke S.I. engines more commonly used in low horse power two wheelers ?
6. Why is crankcase scavenging used only for low power engines ?
7. Why are two-stroke diesel engines, for large power, more common than two-stroke S.I. engines ?
8. What is the reason that two-stroke engine is not used in car even though it develops theoretically twice power than that of four-stroke engine ?
9. Explain with suitable sketches the following scavenging processes :
    (*i*) Uniflow scavenging (*ii*) Loop scavenging.

10. Explain briefly crankcase scavenging.
11. Explain the scavenging process in two-stroke engine. Disucss three scavenging processes used in two-stroke engine.
12. Define scavenging and scavenging efficiency. Explain with sketches different scavenging arrangements based on charge flow.
13. How the valve timings of a two-stroke engine differ from that of four-stroke cycle engine ?
14. What is the difference between the valve timing of a crankcase-scavenged and supercharged two-stroke engine ?
15. Compare the relative merits and demerits of different scavenging systems.
16. How is the supercharging of two-stroke engines done ?

V. Conventional fuels

1. What are the desirable properties of good I.C. engines fuels ?
2. Enumerate and describe briefly the gaseous fuels.
3. How are constituents of petroleum classified ?
4. Explain briefly the chemical structure of petroleum.
5. What are five primary hydrocarbon families found in petroleum ? Which are chain types ? Which are ring types ? Which of primary families tends to be better S.I. engine fuel and C.I. engine fuel ?
6. What are different kinds of fuels used in an I.C. engine ?
7. What are the important properties which S.I. engine fuel possess ?
8. What are requirements of an ideal gasoline fuel ?
9. What is volatility ?
10. Discuss the significance of distillation curve.
11. Why volatility is an important quality of S.I. engine fuels ?
12. Explain briefly the following in regard to a fuel :
    (*i*) Vapour lock characteristics.
    (*ii*) Crankcase dilution.
13. "While volatility of the fuel is a determining factor in the selection of fuels for S.I. engines, ignition quality of the fuel is the primary deciding factor for C.I. engines". Discuss briefly the statements.
14. Distinguish clearly between 'Octane Numbers' and 'Cetane Number'. What is their significance in rating of fuels for S.I. and C.I. engines ?
15. What are the reference fuels for 'Octane Number' ?
16. What are the reference fuels for 'Cetane Number' ?
17. What is performance number (PN) ?
18. What is the significance of ASTM distillation curve ?
19. Explain the effect of fuel viscosity on diesel engine performance.
20. What qualities are desired in fuels to inhibit detonation ?

VI. Supercharging of I.C. engines

1. What is supercharging ?
2. Enumerate the main objects of supercharging.
3. What is a supercharger ?
4. Explain briefly supercharging of S.I. engines.
5. Give the comparison of 'Actual naturally aspirated' and 'supercharged engine' pressure-volume diagrams.
6. Define the terms 'Boost pressure' and 'Pressure ratio'.
7. Explain briefly the effect of pressure ratio on air charge temperature.
8. Explain briefly the thermodynamic cycle of supercharged engine on p-v diagram for an ideal otto cycle.

9. Derive an expression for the power required for an I.C. engine supercharger.
10. What are supercharging limits of S.I. engines ?
11. Explain briefly supercharging of C.I. engines.
12. What are supercharging limits of C.I. engines ?
13. Enumerate the main considerations in limiting the degree of supercharging of a C.I. engine.
14. What modifications are necessary for a supercharged engine ?
15. List the types of superchargers used to supercharge engines for various applications.
16. Explain briefly with neat sketches various types of supercharging arrangements.
17. What are turbochargers ?
18. Describe the Buchi system of turbocharging.
19. Explain briefly the main types of supercharging methods.
20. What is Miller turbocharging ?
21. Explain briefly with a neat sketch Hyperbar turbocharging. State its advantages and disadvantages as well.
22. What is pulse converter ?
23. What are the limitations of supercharging ?
24. "S.I. engines are generally not supercharged". Justify this statement.
25. "Supercharging is preferred in diesel engine than petrol engine". Justify the statement.
26. "Supercharging is essential for an aircraft engine". Justify this statement.
27. Explain the factors that limit the extent of supercharging of S.I. and C.I. engines.
28. What do you mean by supercharging of I.C. engines ? Explain why supercharging is essential for the aircraft engines ?

VII. Testing and performance of I.C. engines

1. (*a*) What do you mean by performance of I.C. engine ?
   (*b*) Discuss briefly the basic performance parameters.
   (*c*) Discuss with suitable sketch the brake rope dynamometer.
2. Describe how the I.P. of a multicylinder engine is measured ?
3. Describe the method commonly used in laboratory for measuring the air supplied to an I.C. engine.
4. Derive the formula used for finding the mass of air supplied to an engine using an orifice tank.

VIII. Miscellaneous engines

1. What is a dual-fuel engine ?
2. Explain with a neat diagram the working of a dual-fuel engine.
3. Enumerate and explain briefly the various factors which affects combustion in a dual-fuel engine.
4. Discuss briefly the performance characteristics of dual-fuel engines.
5. State the advantages and applications of dual-fuel engines.
6. What is a multi-fuel engine ?
7. What are the requirements of a multi-fuel engine ?
8. List the difficulties associated with multi-fuel operation.
9. Discuss briefly the performance characteristics of multi-fuel engine.
10. What is a stratified charge engine ?
11. Give the classification of stratified charge engines.
12. Explain with a neat sketch Texaco combustion process.
13. State the advantages and disadvantages of stratified charge engines.
14. What is the working principle of stirling engine ?

15. List the features which distinguish the stirling engine from other heat engines.
16. The stirling cycle is superior to the carnot cycle, how ?
17. List the fuels which are used in a stirling engine.
18. Give the differences (in a tabular form) between carnot and stirling engines.
19. Explain briefly the following with regard to a stirling engine :
    (*i*) Two piston mechanism
    (*ii*) Two piston engine.
20. State the advantages and disadvantages of a stirling engine vis-a-vis conventional I.C. engine.
21. Give the comparison between stirling engine and I.C. engine.
22. Describe with sketches the working principle of Wankel rotary combustion engine.
23. List the features of the Wankel engine.
24. Explain briefly with regard to Wankel engine :
    (*i*) Clearance and swept volume ; (*ii*) Compression ratio ;
    (*iii*) Engine displacement capacity.
25. Describe briefly the performance characteristics of Wankel engine.
26. State the advantages and applications of rotary combustion engines.
27. Wankel rotary engine could not become successful, why ?
28. What are the advantages of variable compression ratio engine ?
29. Discuss the important designs of variable compression ratio engines and comment on their salient points.
30. Discuss the performance aspects of VCR-engine.
31. Describe with a neat sketch a free-piston engine plant.
32. State the advantages, disadvantages and applications of free-piston engines.

IX. Air pollution from I.C. engines and its control

1. What do you mean by "Air pollution" ?
2. What are the main sources of pollutants from gasoline/petrol engine ?
3. What are the main pollutants emitted by petrol engine ?
4. State the mechanism of formation of CO.
5. How are hydrocarbons (HC) formed ?
6. Explain briefly the mechanism of formation of nitric oxide (NO).
7. What are the effects of the following factors on the exhaust emission ?
    (*i*) Air fuel ratio ; (*ii*) Surface volume ratio ;
    (*iii*) Engine speed.
8. Explain briefly the various sources from which pollutants are emitted from S.I. engine.
9. Define the crankcase blowby and explain how it can be controlled.
10. Discuss briefly the following with regard to S.I. Engines :
    (*i*) Crankcase emission (*ii*) Evaporative emission
    (*iii*) Exhaust emission.
11. What are the sources of HC formation in petrol engine ? Explain different factors which affect HC formation.
12. Explain briefly the factors which affect the formation of NO_x.
13. What are the sources of HC formation in petrol engine ? Explain various factors which affect the HC formation.
14. What are the sources of evaporative emission in petrol engines ? How can it be controlled ?
15. What do you mean by crankcase blowby ? How can it be controlled ?
16. Explain briefly various methods by which S.I. engine emission can be controlled.

17. Explain in detail any two of the following methods of S.I. engine emission control :
    (*i*) Modification in the engine design and operating parameters.
    (*ii*) Treatment of exhaust products of combustion.
    (*iii*) Modification of the fuels.
18. Describe with sketches the following methods of petrol exhaust emission control :
    (*i*) After-burner (*ii*) Exhaust manifold reactor
    (*iii*) Catalytic converter system.
19. Explain briefly the Exhaust Gas Recirculation (EGR) device for the control of NO_x.
20. What do you mean by "Total emission control packages" ? Describe with neat sketches two types of total emission control packages.
21. Explain briefly the following :
    (*i*) Three-way catalytic converter (*ii*) Two-way catalytic converter
    (*iii*) Noble metal catalytic converter.
22. Explain briefly the "Evaporation Emission Control Device (EECD)".
23. Discuss the emissions from diesel engines. On what factors these emissions depend ?
24. Define 'smoke' and discuss different factors which affect smoke formation in C.I. engines.
25. What is the cause of diesel smoke ? Discuss the ways by which diesel smoke can be controlled.
26. What is the mechanism of smoke formation ?
27. How can the smoke intensity be measured ?
28. Describe two important types of smoke meters.
29. Explain the effect of engine load on diesel engine smoke.
30. What is diesel odour ? Explain the effect of different factors on the formation of odour in C.I. engines.
31. How odour can be controlled ?
32. Compare the diesel engine and gasoline engine emissions.
33. Discuss the air pollution from gas turbines and compare its with emissions from petrol engines.
34. Give the comparison between "electronic catalytic converter" and "conventional exhaust catalytic converter".
35. Discuss the effects of emissions on human health.

UNSOLVED EXAMPLES

1. A single-cylinder petrol engine working on two-stroke cycle develops indicated power of 5 kW. If the mean effective pressure is 7.0 bar and the piston diameter is 100 mm, calculate the average speed of the piston.
   [**Hint.** Average piston speed = 2 LN.] [**Ans.** 109.1 m/s]
2. A four-cylinder petrol engine works on a mean effective pressure of 5 bar and engine speed of 1250 r.p.m. Find the indicated power developed by the engine if the bore is 100 mm and stroke 150 mm.
   [**Ans.** 6.11 kW]
3. A four-cylinder four-stroke S.I. engine is designed to develop 44 kW indicated power at a speed of 3000 r.p.m. The compression ratio used is 6. The law of compression and expansion is $pV^{1.3}$ = constant and heat addition and rejection takes place at constant volume. The pressure and temperature at the beginning of compression stroke are 1 bar and 50°C. The maximum pressure of the cycle is limited to 30 bar. Calculate the diameter and stroke of each cylinder assuming all cylinders have equal dimensions. Assume diagram factor = 0.8 and ratio of stroke/bore = 15. [**Ans.** D = 95 mm, L = 142.5 mm]
4. During the trial of a four-stroke diesel engine, the following observations were recorded :
   Area of indicator diagram = 475 mm², length of indicator diagram = 62 mm, spring number = 1.1 bar/mm, diameter of piston = 100 mm, length of stroke = 150 mm, engine r.p.m. = 375.

Determine : (*i*) Indicated mean effective pressure

(*ii*) Indicated power. **[Ans.** (*i*) 8.43 bar ; (*ii*) 3.1 kW]

5. A four-cylinder, four-stroke diesel engine runs at 1000 r.p.m. The bore and stroke of each cylinder are 100 mm and 160 mm respectively. The cut-off is 6.62% of the stroke. Assuming that the initial condition of air inside the cylinder are 1 bar and 20°C, mechanical efficiency of 75%, calculate the air-standard efficiency and brake power developed by the engine.

6. A V-8 four-stroke petrol engine is required to give 186.5 kW at 440 r.p.m. The brake thermal efficiency can be assumed to be 32% at the compression ratio of 9 : 1. The air/fuel ratio is 12 : 1 and the volumetric efficiency at this speed is 69%. If the stroke to bore ratio is 0.8, determine the engine displacement required and the dimensions of the bore and stroke. The calorific value of the fuel is 44200 kJ/kg, and the free air conditions are 1.013 bar and 15°C. **[Ans.** 5.12 litres ; 100.6 mm ; 80.5 mm]

7. During the trial (60 minutes) on a single-cylinder oil engine having cylinder diameter 300 mm, stroke 450 mm and working on the four-stroke cycle, the following observations were made :

Total fuel used = 9.6 litres, calorific value of fuel = 45000 kJ/kg, total number of revolutions = 12624, gross indicated mean effective pressure = 7.24 bar, pumping i.m.e.p. = 0.34 bar, net load on the brake = 3150 N, diameter of brake wheel drum = 1.78 m, diameter of the rope = 40 mm, cooling water circulated = 545 litres, cooling water temperature rise = 25°C, specific gravity of oil = 0.8.

Determine : (*i*) Indicated power. (*ii*) Brake power.

(*iii*) Mechanical efficiency.

Draw up the heat balance sheet on minute basis. **[Ans.** (*i*) 77 kW ; (*ii*) 61.77 kW ; (*iii*) 80.22%]

8. Air enters an air compressor (run by a supercharged diesel engine) at 27°C and is passed on to a cooler where 1160 kJ/min. are rejected. The air leaves the cooler at 67°C and 1.7 bar. Some air from the compressor is bled after the cooler to supercharge the engine. The volumetric efficiency of the engine is 75 percent based on the intake manifold temperature of 67°C and pressure of 1.7 bar.

The engine has the following specifications :

B.P. = 30 kW ; Speed = 2000 r.p.m. ; Number of cylinders = 4 ; Bore = 100 mm ;
Stroke = 110 mm ; Mechanical efficiency = 80 percent.

Assuming the internal efficiency of the compressor as 90 percent, determine the following :

(*i*) The indicated mean effective pressure of the engine ;

(*ii*) The air consumption rate of the engine ;

(*iii*) The air handling capacity of the compressor. **[Ans.** (*i*) 6.5 bar ; (*ii*) 4.53 kg/min. ; 11.57 kg/min.]

9. The average indicated power developed in I.C. engine is 12.9 kW/m³ of free air induced per minute. The engine is three-litre four-stroke engine running at 3500 r.p.m., and has a volumetric efficiency of 80%, referred to free conditions of 1.013 bar and 15°C. It is proposed to fit a blower, driven mechanically from the engine. The blower has an isentropic efficiency of 75% and works through a pressure ratio of 1.7, at the pressure and temperature of the delivery from the blower. Calculate the increase in B.P. to be expected from the engine. Take all the mechanical efficiencies as 80%. **[Ans.** 25.3 kW]

3

Fuel Supply System

3.1. AIR, FUEL AND EXHAUST GASES CIRCUITS OF PETROL AND DIESEL ENGINES

3.1.1. Air, Fuel and Exhaust Gases Circuits of a Petrol Engine

The air, fuel and exhaust gases circuits of a petrol engine is shown in Fig. 3.1.

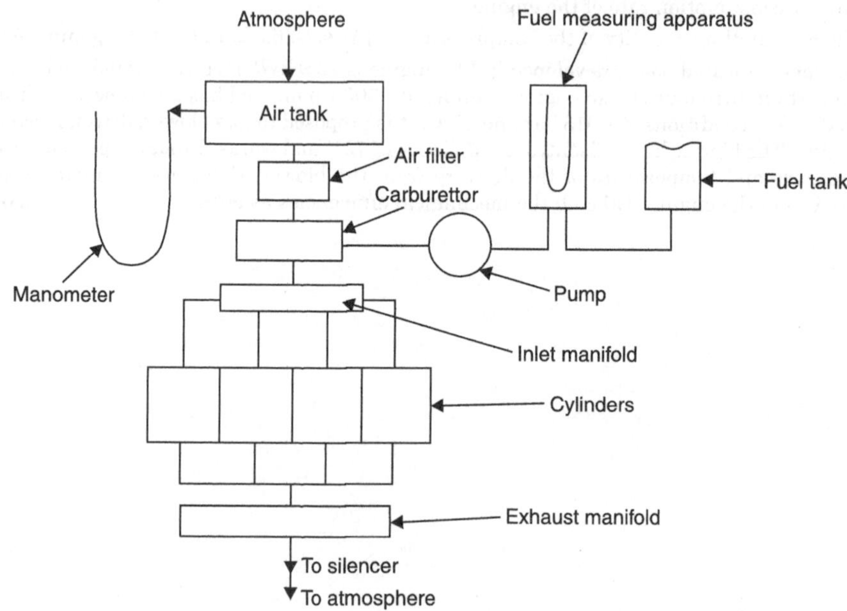

Fig. 3.1. Air, fuel and exhaust gases circuit of a petrol engine.

- During suction or induction stroke the *"air"* is sucked in the air tank from the atmosphere. The manometer fitted on the side measures the pressure. The air then passes through the air filter or cleaner where it gets cleaned.
- On the other side *"fuel"* from the fuel tank reaches the carburettor after passing through the filter and mixes with the clean air. The mixture then enters the engine cylinder where during the compression stroke it is compressed and work is obtained during the working or expansion stroke.
- The working stroke is followed by the exhaust stroke which forces the *"exhaust gases"* to exhaust manifold and finally to the *atmosphere via exhaust silencer.*

Air cleaner or filter. The induction air cleaner has the function of cleaning air before it enters the cylinders of the engine and assists in preventing the fouling of carburettor parts and spark-plugs and reducing abrasive wear on piston rings, cylinder walls, valves bearings and any relative moving parts.

Fig. 3.2 shows an oil bath air cleaner. It is very efficient and the operation is carried out in two stages. In the first stage the reversal of air flow causes dust particles to impinge on the surface of the oil, and during the second stage the partially clean oil laden air passes up into the steel mesh element where the oil retains the dust. The cleaner may be designed so that air to other air accesses to carburettor and float chamber are also filtered.

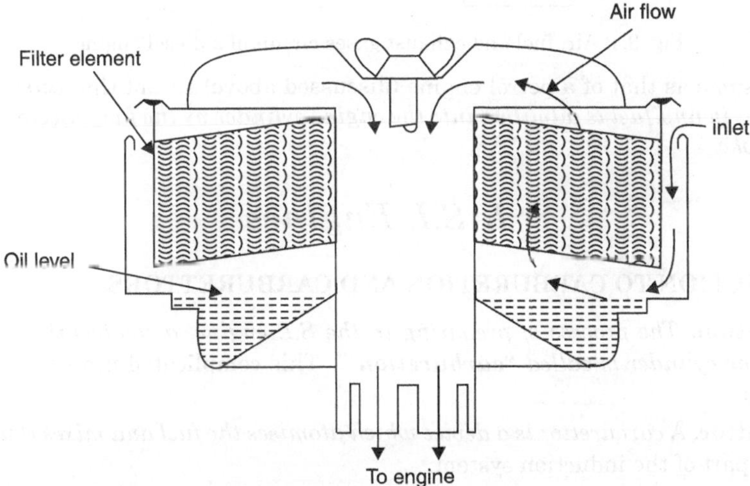

Fig. 3.2. Air cleaner or filter.

Fig. 3.3 shows an *exhaust silencer.*

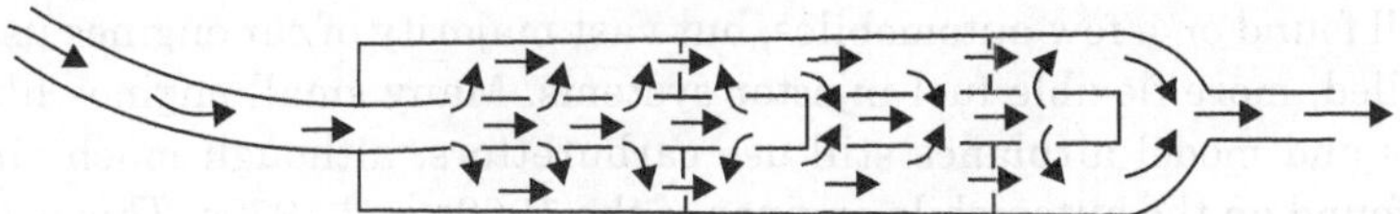

Fig. 3.3. Exhaust silencer.

3.1.2. Air, Fuel and Exhaust Gases Circuit of a Diesel Engine

The air, fuel and exhaust gases circuit is shown in Fig. 3.4.

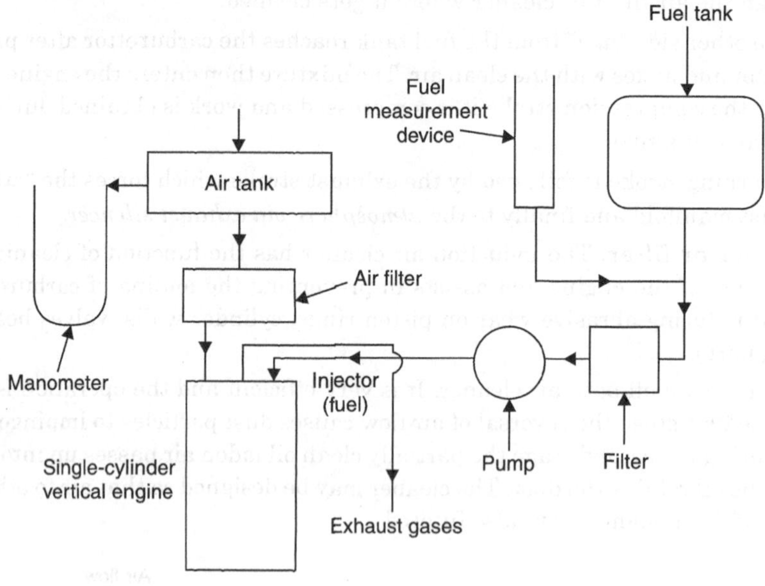

Fig. 3.4. Air, fuel and exhaust gases circuit of a diesel engine.

It is the same as that of a petrol engine (discussed above) except that *during suction stroke only air is sucked in and fuel is admitted into the engine cylinder by the fuel injector at the end of the compression stroke.*

I. S.I. Engines

3.2. INTRODUCTION TO CARBURETION AND CARBURETTORS

Carburetion. *The process of preparing in the S.I. engine, a combustible fuel-air mixture outside the engine cylinder is called* ***"carburetion".*** This complicated process is achieved in the *induction system.*

Carburettor. A *carburettor is a device which atomises the fuel and mixes it with air.* It is the *most* important part of the induction system.

— For several decades, carburettors were used on most S.I. engines as the means of adding fuel to the intake air. The basic principle on which the carburettor works is extremely simple, but by the 1980s, when fuel injectors finally replaced it as the main fuel input system, it had evolved into a complicated, sophisticated, expensive system. Carburettors are still found on a few automobiles, but vast majority of car engines use simpler, better controlled, more flexible fuel injector systems. Many small engines like those on lawn movers and model airplanes still use carburettors, although much simpler ones than those found on the automobile engines of the 1960s and 1970s. *This is to keep the cost of these engines down, simple carburettor being cheap to manufacture while fuel injectors require more costly control systems. Even on some of these small engines, carburettors are being replaced with fuel injectors as pollution laws become more stringent.*

3.3. INDUCTION SYSTEM

The schematic arrangement of induction system is shown in Fig. 3.5.

- The pipe that carries the prepared mixture to the engine cylinders is called the *intake manifold*.
- The carburettor is the focal point of the induction system.
- The fuel system, comprising the fuel supply tank and necessary fuel pumps, lines and filters supply liquid fuel to the carburettor.
- During the motion stroke vacuum is created in the cylinder which causes the air to flow through the carburettor and the fuel to be sprayed from the fuel jets. Due to the volatility of the fuel, most of the fuel vaporises and forms a combustible fuel-air mixture. However, some of the larger drops may reach the cylinder in the liquid form and must be vaporised and mixed with air during the compression stroke before the electric spark ignites the mixture.

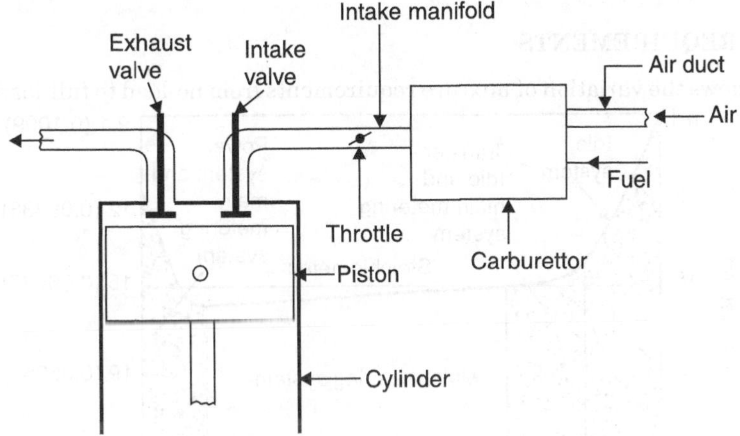

Fig. 3.5. Schematic arrangement of the 'Induction System'.

- The throttle located in the carburettor, regulates the quantity of the mixture.
- There is a limited range of A / F ratio in a homogenous mixture within which combustion in S.I. engines is sustainable. Outside this range, the ratio is too rich or too lean to sustain flame propagation. This range of useful A /F ratio is from approx. 20 : 1 (lean) to 8: 1 (rich).

3.4. FACTORS INFLUENCING CARBURETION

The *various factors which influence the process of carburetion are as follows :*

1. The engine speed ; the time available for the preparation of the mixture.
2. The vaporisation characteristics of fuel.
3. The temperature of the incoming air.
4. The design of the carburettor.

- In case of modern high speed engines, the time duration available for the formation of mixture is very small and limited. The time duration for mixture formation and induction may be of the order of 10 to 5 milliseconds.
- Atomisation, mixing and vaporisation are the processes which require a finite time to occur. The time available for mixture formation is very small in high speed engines (For

example, in an engine running at 3000 r.p.m., the induction process lasts for less than 0.02 second). For completion of these processes in such a small period a great ingenuity is required in designing the carburettor system.

- In order to achieve high quality carburetion within such a short time requires good vaporisation characteristics of the fuel which are ensured by presence of high volatile hydrocarbons in the fuel.
- The temperature is a factor which effectively controls vaporisation process of the fuel. If the temperature of the incoming air is high, it results in higher rates of vaporisation. The mixture temperature can be increased by heating the induction manifold but it will result in reducing power due to reduction in mass flow rate.
- For a S.I. engine, the design of carburetion system is very complicated owing to the fact that the air-fuel ratio required by it varies widely over its range of operation, particularly for an automotive engine. For idling as well as for maximum power rich mixture is required.

3.5. MIXTURE REQUIREMENTS

Fig. 3.6 shows the variation of mixture requirements from no-load to full-load in a S.I. engine.

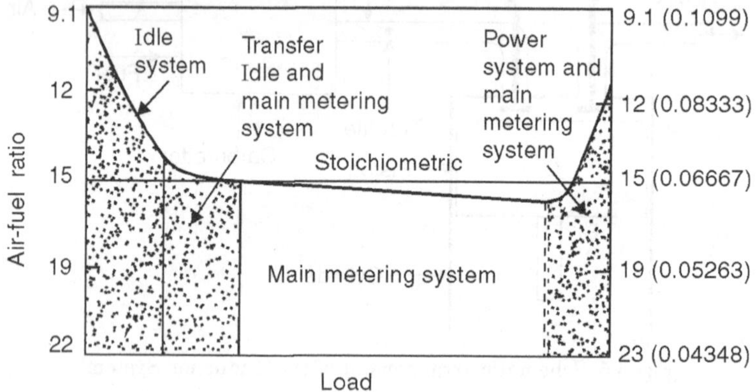

Fig. 3.6. Mixture requirements of automotive S.I. engine.

1. **Idling and low speed** (*From no-load to about 20% of rated power*) :

Idling refers to *no power demand.* During idling air supply is throttled and residual gases make up a large fraction of the charge at the end of the suction period. In addition, during valve overlap period some exhaust gases are drawn back into the cylinder. The result is that a chemically correct (stoichiometric) mixture of air and fuel (≈ 15 : 1) would be so diluted by residual gases that combustion would be erratic or impossible. *A rich mixture, therefore, must be supplied during idling (say A/F ratio* 11 : 1 or 12 : 1). *The richness should gradually change to slightly lean for the second range as* shown in Fig. 3.6.

2. **Cruising or normal power** (*from about 25% to about 75% of rated power*) :

In the *normal power range the main consideration is fuel economy.* Because mixture of fuel and air is never completely homogeneous the stoichiometric mixture of fuel and air will not burn completely and some fuel will be wasted. For this reason an *excess of air, say 10% above theoretically correct (≈ 16.5 : 1), is supplied in order to ensure complete burning of the fuel.*

3. **Maximum power** (*From 75% to 100% of rated power*):

Maximum power is obtained when all the air supplied is fully utilized. As the mixture is not completely homogeneous a rich mixture must be supplied to assure utilization of air (though this would mean wasting some fuel, which would pass in exhaust in unburned state). The *air-fuel ratio for maximum power is about 13 : 1.*

Running on the weakest mixture. *This results in high efficiency and there is fuel economy. On normal loads engines work on weak mixture.*

Running on richest mixture. *Engines run on rich mixture during idling and during the overload. The effect is lowering of efficiency and pollution problems.*

- Automobiles engines generally operate well below full power and a complicated system for making the mixture rich is neither called for nor economically advisable, although some means are employed to enrich the mixture. A more representative curve for an automobile engine is shown in Fig. 3.7.

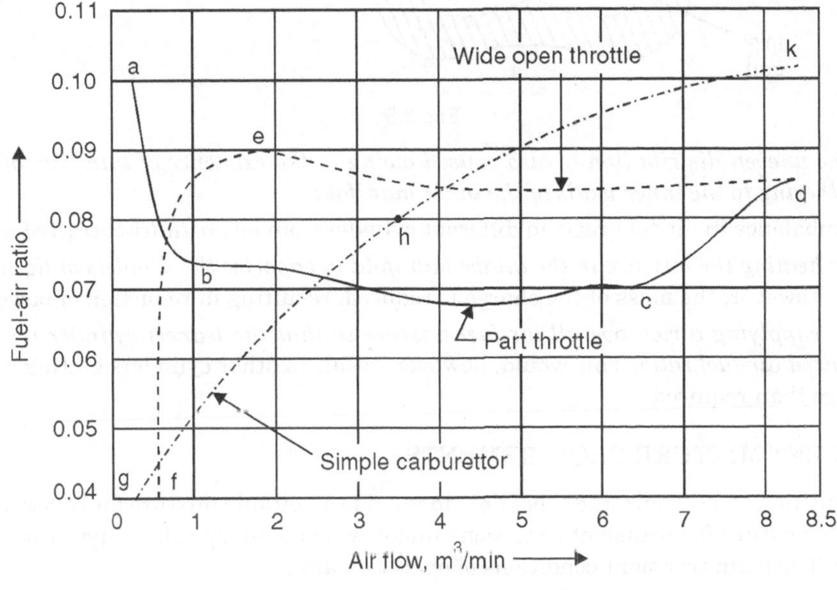

Fig. 3.7

— The portion of the curve from *d* to *e* shows the fuel-air requirements for wide open throttle (W.O.T.) and the load is further increased.

— The Fig. 3.7 exhibits that a simple elementary carburettor is incapable to provide the F/A ratio as desired at part throttle as shown by *a-b-c-d* or W.O.T. given by *d.e.f.* The simple carburettor gives the curve as shown by *g-h-k*.

3.6. DISTRIBUTION

Ideally, a carburettor should supply mixture of the same fuel-air ratio to each cylinder of a multi-cylinder engine ; this condition is very difficult to achieve practically.

- Since in a carburettor complete atomisation and vaporisation of the fuel is not achieved, therefore, the mixture passing through the intake manifold generally contains a certain amount of petrol in the droplet form. These droplets possess greater inertia than the gaseous mixture. Thus when there is an abrupt change in the direction of flow the droplets continue to move in their original direction (Refer Fig. 3.8) and consequently variation of A / F ratio between cylinders takes place, *the outer cylinders getting richer mixture than the inner cylinders.*

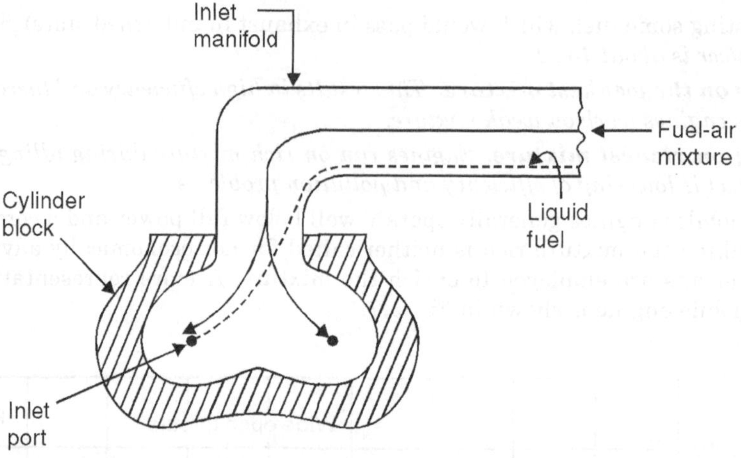

Fig. 3.8

- *The uneven distribution is also caused owing to the existence of thin film of liquid film adhering to the inner walls of the inner manifold.*

The imbalance in air-fuel ratio in different cylinders can be *partially* corrected as follows :

(*i*) *By heating the mixture in the intake manifold to vaporise the droplets of liquid.* By doing so, however, the mass of the charge is reduced, resulting in reduction of power output.

(*ii*) *By supplying a rich overall air-fuel mixture so that the leanest cylinder receives the required air-fuel ratio.* This would, however, result in other cylinders getting a richer mixture than required.

3.7. TRANSIENT MIXTURE REQUIREMENTS

The function of a carburettor is not only to provide a suitable mixture for *steady-running* but also to supply mixture for transient conditions under which load, speed, temperature or pressure alter rapidly. The main transient conditions of operation are :

— Starting and warming up ;

— Acceleration, and deceleration.

The mixture requirements under transient conditions are different from those of steady running due to the following *reasons* :

(*i*) The evaporation may be incomplete ;

(*ii*) The quantity of liquid fuel in the inlet manifold may be increasing or decreasing ;

(*iii*) The distribution of fuel to various cylinders may be different.

1. **Starting and warming up requirements :**

When the engine is started from cold, its speed and temperature are low and as such much of 'heavy ends' (The hydrocarbons with high vapour pressures and low boiling points are called *'light ends'* and those which are less volatile are called 'heavy ends') supplied by the carburettor do not vaporise and remain in liquid form. Further vaporised fuel may recondense on coming in contact with cold cylinder walls and piston head. Thus, even when the F/A ratio at the carburettor is well within the normal combustion limits or petrol-air mixtures, the ratio of the *'evaporated fuel' to air in the cylinder may be too lean to ignite.* Consequently it is *necessary to supply a rich mixture during starting*, as much as 5 to 10 times the normal amount of petrol (A / F ratio 3 : 1 to 15 : 1 or F/A ratio 0.3 to 0.7), in order that 'light ends' are available for proper ignition. With the warming up of the engine there is an increase in the amount of evaporated fuel and hence the mixture ratio should be progressively made leaner, too rich *evaporated* F/A ratio is avoided.

2. **Acceleration requirements :**

- With regard to engines, the terms acceleration generally refers to an increase in engine speed resulting from opening the throttle. The main object of opening the throttle, however, is *to provide an increase in torque* and whether or not an increase in speed follows depends on the nature of load.
- Under steady running conditions, the fuel evaporated in the intake manifold moving much faster than the liquid film formed on the induction system walls, does not cause any problem. But when the throttle is *suddenly* opened *e.g.*, during acceleration, the liquid fuel lags behind and, the cylinder receives temporarily a lean mixture whilst actually, to produce more instantaneous power for acceleration, a rich mixture is needed. Hence, a suitable mechanism (acceleration pump) is required to provide rich mixture during the acceleration period.

Note. The petrol to be used should be carefully made to suit the engine and the climate of the place since too high volatility or too low volatility, both create difficulties in operation.

— *Too high volatility* may form bubbles in the carburettor and fuel lines particularly when the engine temperature are high, which interfere with the supply and metering of the fuel and may disturb the F/A ratio so seriously that engine may stop working.

— *Too low volatility* may cause petrol to condense on the cylinder walls, diluting and removing the lubricating oil film ; ultimately the petrol may reach the cranckcase past the piston rings and dilute the engine oil. Condensation of petrol on cylinder walls also causes carbon deposits.

3.8. A SIMPLE OR ELEMENTARY CARBURETTOR

In order to understand a modern carburettor (a very complex device) it helps first to study a *simple carburettor* which supplies fuel-air mixture for cruising or normal range of speed and then to add other devices or attachments to take care of other function like *starting, idling, accelerating,* decelerating and other variable load and speed operations.

Fig. 3.9, shows a schematic diagram of a simple or elementary carburettor.

- It consists of a *float chamber, nozzle with metering orifice, venturi* and *throttle valve.*
  — The **float chamber** is meant for storage of fuel. The fuel supplied under gravity action or by fuel pump enters the float chamber through a filter. The arrangement is such that when oil reaches a particular level the *needle/float valve* blocks the inlet passage and thus cuts off the fuel oil supply. On the fall of oil level, the *float* descends down, consequently intake passage open and again the chamber is filled with oil. Then the float and the needle/float valve maintains a constant fuel oil level in the float chamber. There is a *nozzle (discharge jet)* from which the fuel is sprayed into the air stream as it enters the inlet and passes through the venturi or throat. *The fuel level is slightly below the outlet of the jet when the carburettor is inoperative.*
- As the piston moves down in the engine cylinder, suction is produced in the cylinder as well as in the induction manifold as a result of which air flows through the carburettor. *The velocity of air increases as it passes through the constriction at the venturi and the pressure decreases due to conversion of a portion of pressure head into kinetic energy.* Due to decreased pressure at the venturi and hence by virtue of difference in pressure (between the float chamber and the venturi) the jet issues fuel oil into air stream. Since the jet has a fine bore, the oil issuing from the jet is in the form of *fine spray* ; it vaporises quickly

and mixes with air. The air-fuel mixture enters the engine cylinder ; *it quantity being controlled by the position of the "throttle valve".*

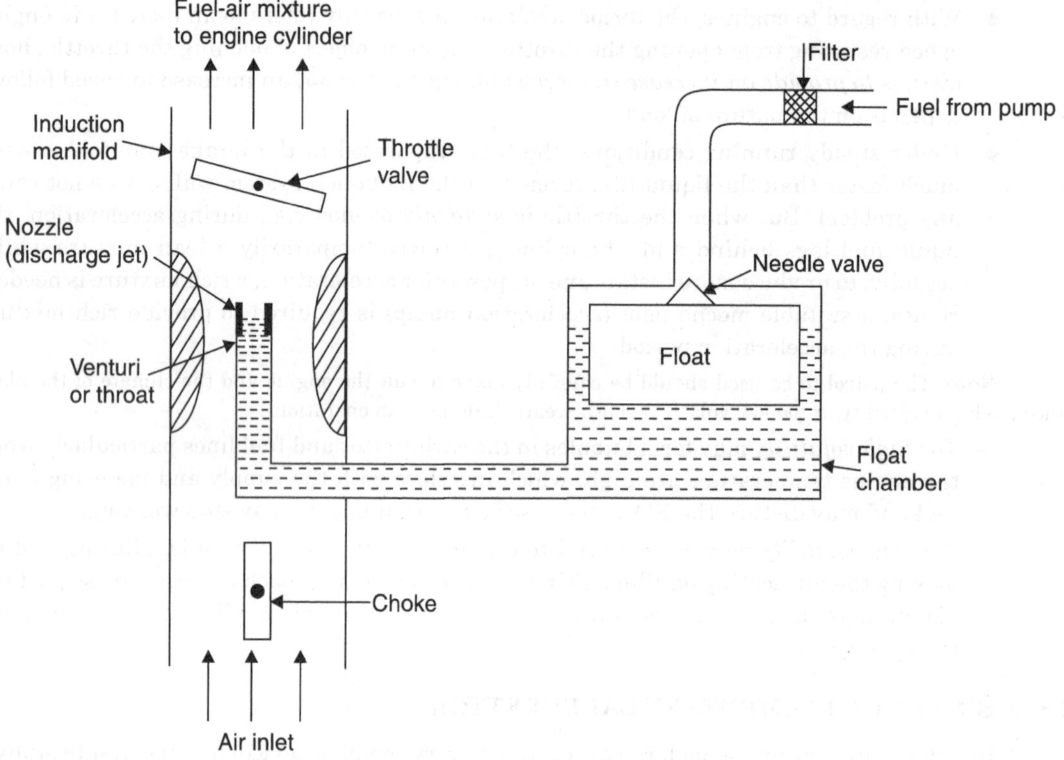

Fig. 3.9. A simple or elementary carburettor.

Limitations :

(*i*) Although theoretically the air-fuel ratio supplied by a simple (single jet) carburettor should remain constant as the throttle goes on opening, actually *it provides increasingly richer mixture as the throttle is opened.* This is because of the reason that the density of air tends to decrease as the rate of flow increases.

— This fault is corrected by *using a number of compensating devices.*

(*ii*) During idling, however, the nearly closed throttle causes a reduction in the mass of air flowing through the venturi. At such low rates of air flow, the pressure difference between the float chamber and the fuel discharge nozzle becomes very small. It is not adequate enough to cause fuel to flow through the jet.

— This fault may be corrected by *using an idling jet which helps, in running the engine during idling.*

(*iii*) Carburettor does not have arrangement for providing rich mixture during starting and warm up.

— This limitation is taken of by using a *choke arrangement.*

3.9. COMPLETE CARBURETTOR

For meeting the demand of the engine under all conditions of operation, the following additional devices/systems are added to the simple carburettor :

1. Main metering system
2. Idling system
3. Power enrichment or economiser system
4. Acceleration pump system
5. Choke.

1. **Main metering system :**

The main metering system of a carburettor should be so designed as to supply a *nearly constant fuel-air ratio over a wide range of operation.* This F/A ratio is approximately equal to 0.064 (A/F ratio ≈ 15.6) for best economy at full throttle. In order to correct the tendency of the simple carburettor to give progressively richer mixtures with load speed, the following *automatic compensating devices* are incorporated in the main metering system :

(*i*) Compensating jet device.

(*ii*) Emulsion tube or air bleeding device.

(*iii*) Back suction control or pressure reduction method.

(*iv*) Auxiliary valve carburettor.

(*v*) Auxiliary port carburettor.

These devices are explained below :

(*i*) ***Compensating jet device :***

A schematic diagram of a compensating jet device is shown in Fig. 3.10.

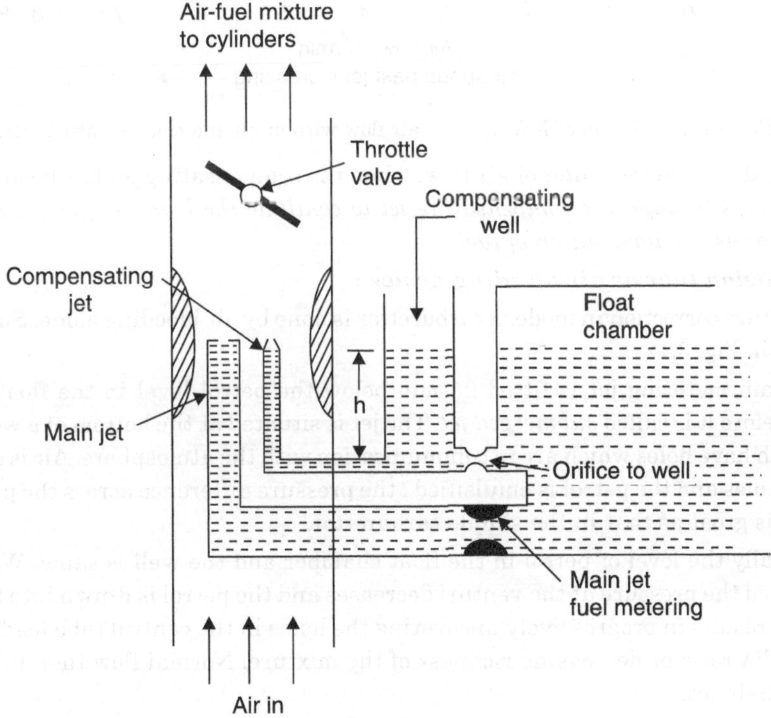

Fig. 3.10. A compensating jet device.

- In this device, in addition to the main jet, a compensating jet is provided which is in communication with a compensating well. The compensating well is also vented to atmosphere (like the main float chamber) ; it is supplied with fuel from the main float chamber through a restricting orifice.
- As the air flow increases, the level of fuel in the compensating well decreases, thus reducing the fuel supply through the compensating jet. The compensating jet thus progressively makes the mixture leaner as the main jet progressively makes the mixture richer, the sum of the two remaining constant as shown in Fig. 3.11. The main jet and compensating curves are more less reciprocals of each other.

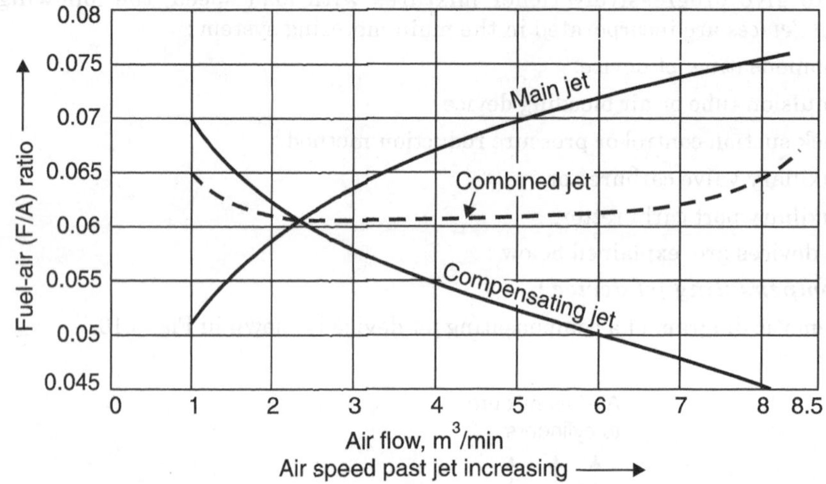

Fig. 3.11. Variation of F/A ratio vs. air flow with main and compensating jets.

— At even higher *rates* of air flow, when the compensating jet has been emptied, *air is bled through the compensating jet to continue the leaness effect,* and incidently to assist in *atomisation of fuel.*

(*ii*) ***Emulsion tube or air bleeding device*** :

The mixture correction in modern carburettor is done by air bleeding alone. Such an arrangement is shown in Fig. 3.12.

- A main metering jet is fitted 25 mm below the petrol level in the float chamber and therefore it is called *submerged jet.* The jet is situated at the bottom of a well, the sides of which have holes which are in communication with the atmosphere. Air is drawn through the holes and the petrol is emulsified ; the pressure difference across the petrol column is not as great as that in the simple carburettor.
- Initially the level of petrol in the float chamber and the well is same. When throttle is opened the pressure at the venturi decreases and the petrol is drawn into the air stream. This results in progressively uncovering the holes in the central tube leading to decreasing F/A ratio or decreasing richness of the mixture. Normal flow then takes place from the main jet.

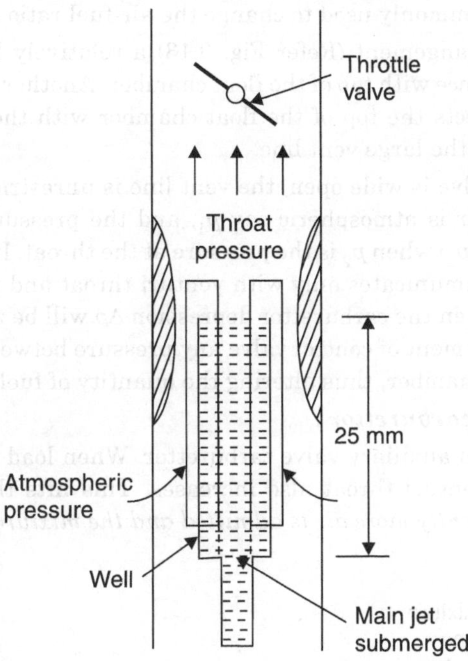

Fig. 3.12. Correction in modern carburettors by air bleeding.

(iii) ***Back suction control or pressure reduction method :***

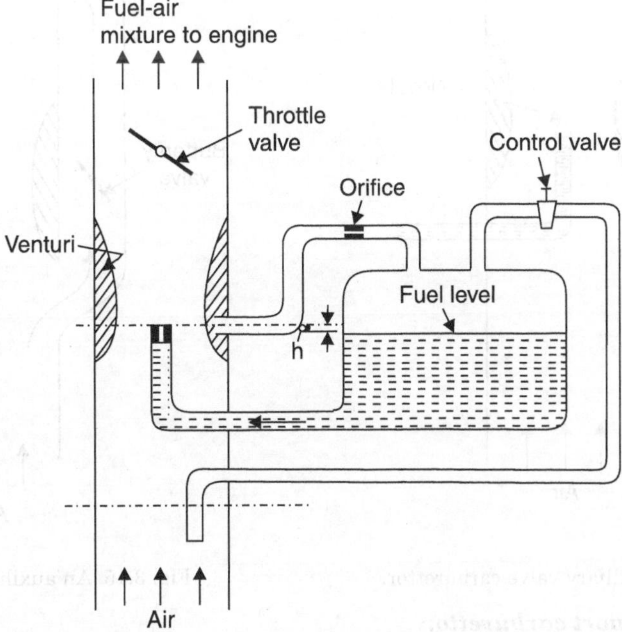

Fig. 3.13. Back-suction control or pressure reduction method.

- This method is commonly used to change the air-fuel ratio in large carburettors.
- In this device/arrangement (Refer Fig. 3.13) a relatively large vent line connects the carburettor entrance with top of the float chamber. Another line, containing a very small orifice line, connects the top of the float chamber with the venturi / throat. A control valve is placed in the large vent line.
  — When the valve is wide open, the vent line is unrestricted and the pressure in the float chamber is atmospheric say p_1, and the pressure differential acting on the orifice is (p_1-p_2) when p_2 is the pressure at the throat. If the valve is closed, the float chamber communicates only with venturi throat and pressure on the fuel surface will be p_2. Then the carburettor depression Δp will be zero and no fuel can flow. By proper adjustment of control valve any pressure between p_1 and p_2 can be obtained in the float chamber, thus altering the quantity of fuel discharged by the nozzle.

(*iv*) ***Auxiliary valve carburettor :***

Fig. 3.14 shows an auxiliary valve carburettor. When load on the engine increases, the vacuum at the venturi throat also increases. This lifts the valve against the spring force and *consequently more air is admitted and the mixture is prevented from becoming over-rich.*

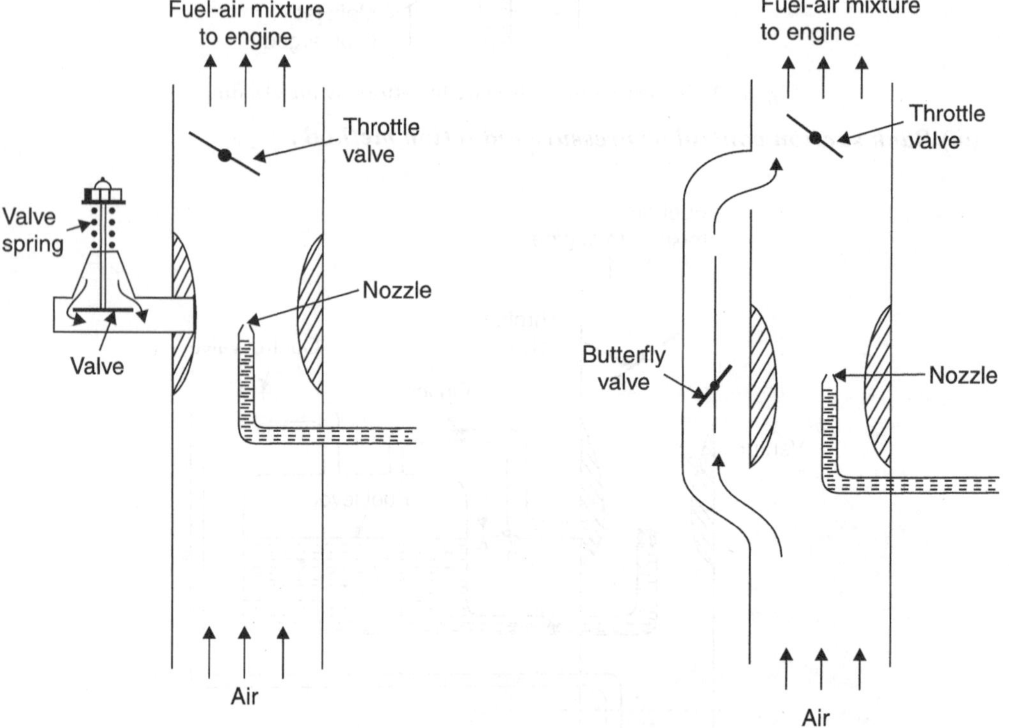

Fig. 3.14. An auxiliary valve carburettor.

Fig. 3.15. An auxiliary port carburettor.

(*v*) ***Auxiliary port carburettor.***

- This method is used in *aircraft carburettors for altitude compensation.*

- Fig. 3.15 shows an auxiliary port carburettor. When the butterfly valve is opened, additional air is admitted and at the same time the depression at the venturi throat is reduced ; this results in decreasing the quantity of fuel drawn in.

2. **Idling system :**

- As earlier discussed that at idling and low load an engine requires a rich mixture having about air-fuel ratio 12 : 1. The main metering system not only fails to supply the enrich mixture at low air flows but also cannot supply any fuel during idling operation. It is due to this reason that a separate idling jet must be incorporated in the basic carburettor.
- Fig. 3.16 shows an idling jet. It consists of a small fuel line from the float chamber to a point on the engine side of the throttle ; this line contains a fixed fuel orifice.
  - When *throttle is practically closed,* the full manifold suction operates on the outlet to this jet. Besides local suction is increased due to very high velocity past the throttle valve. Fuel therefore can be lifted by the additional height upto the discharge point, but this *occurs only at very low rates of air-flow.*

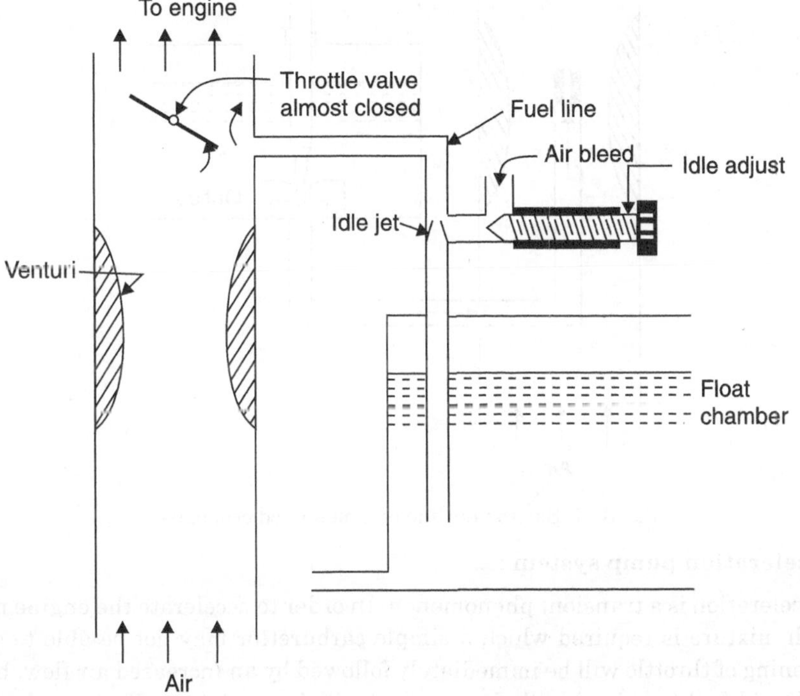

Fig. 3.16. Idling jet.

  - When the *throttle is opened,* the main jet gradually takes over and the idle jet eventually becomes ineffective.
  - The *idle adjust* (a needle valve controlling the air bleed, which is manually operated) *regulates the desired A/F ratio for the idling jet.*

3. **Power enrichment or economiser system :**

At the maximum power range of operation from 75% to 100% load, a device should be available to allow richer mixture (F/A about 0.08) to be supplied despite the compensating leanness.

Meter rod economiser shown in Fig. 3.17 is such a device. It simply provides a large orifice opening to the main jet when the throttle is opened beyond specified limit. The rod may be tapered or stepped.

— An economiser is a valve which remains closed at normal cruise operation and gets opened to supply enriched mixture at full throttle operation. It regulates the additional fuel supply for the above operation. (The term economiser is rather *misleading*. It stems from the fact that such a device *provides a rich uneconomical mixture at high load demand without interfering with economical operation in the normal power range*).

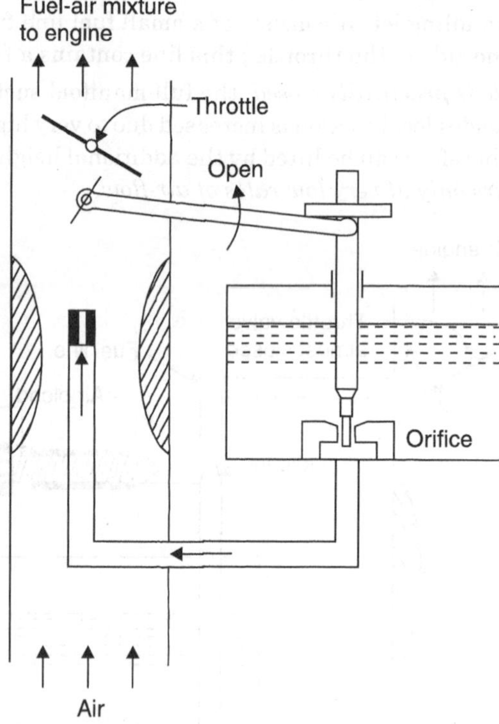

Fig. 3.17. Skelton outline of a meter rod economiser.

4. **Acceleration pump system :**

- Acceleration is a transient phenomenon. In order to accelerate the engine rapidly, a very rich mixture is required which a simple carburettor may not be able to supply. Rapid opening of throttle will be immediately followed by an increased air flow, but the inertia of liquid fuel (gasoline) will give momentarily lean mixture. Thus acceleration mixture required may not be met with in practice. To overcome this difficult situation an *acceleration pump is incorporated.*
- Fig. 3.18 shows an *acceleration pump.* It consists of a spring-load plunger. Also is provided a linkage mechanism so that when throttle is rapidly opened the plunger moves into the cylinder and forces an additional jet of fuel into the venturi. An arrangement is also provided which ensures that when throttle is opened slowly, the fuel in the pump cylinder is not forced into the venturi but leaks past plunger or some holes into the float chamber.

— In some carburettors, instead of providing mechanical linkages, an arrangement is made so *that the pump plunger is held up by manifold vacuum*. Whenever this vacuum is reduced by rapid opening of throttle a spring forces the plunger down pumping the fuel through the jet.

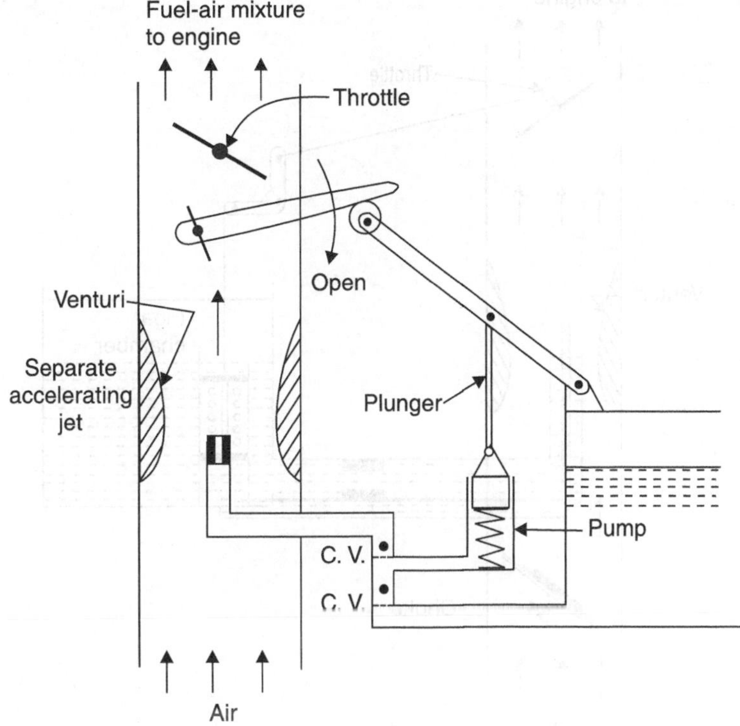

Fig. 3.18. Acceleration pump.

5. **Choke :**

- Starting of a vehicle which is kept stationary for a long period (may be overnight.) during cool winter seasons, is often more difficult. At low cranking speeds and intake temperatures a very rich mixture is required to initiate combustion. Sometimes as high as *five to ten times* more fuel (than usual mixtures) is required. The main reason is that very large fraction of fuel may remain liquid suspended in air even in the cylinder, and only the vapour fraction can provide a combustible mixture with air. The most popular method of providing such mixture is by the use of *choke*.
- A *choke* is simply butterfly valve located between the entrance to the carburettor and the venturi throat as shown in Fig. 3.19.
  — When the choke is partly closed, large pressure drop occurs at the venturi throat, would normally result from the amount of air passing through the venturi throat. The very large carburettor depression at the throat inducts large amount of fuel from the main nozzle and provides a very rich mixture so that the ratio of the evaporated fuel to air in the cylinder is within the combustible limits.
- Sometimes the choke valves are made with spring-loaded by-pass to ensure that large carburettor depression and excessive choking does not persist after the engine has started,

and reached a desired speed. The choke can be made to operate automatically by means of a thermostat so that choke is closed when engine is cold and goes out of operation when the engine warms up after starting.

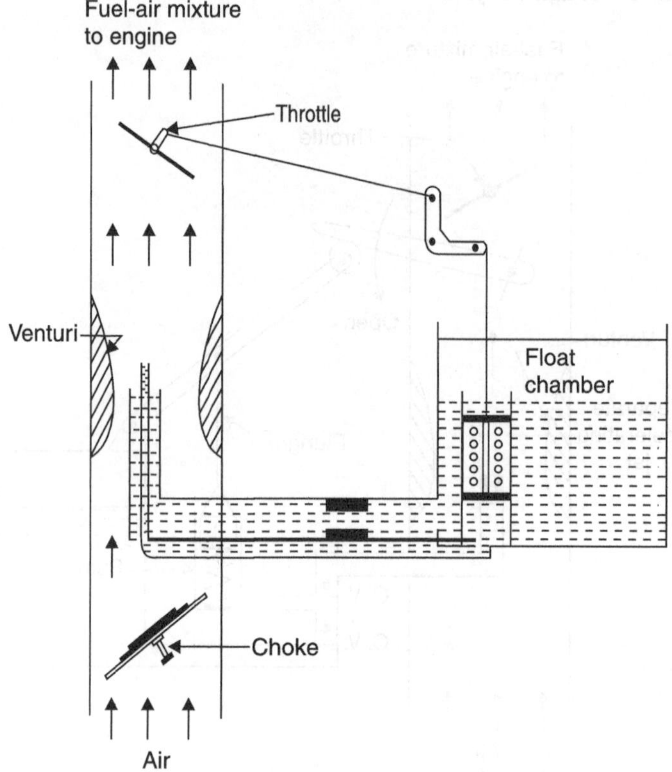

Fig. 3.19. Choke valve with spring-loaded by-pass.

- The provision of *auxiliary fuel jets* that are opened manually or automatically only as required, is an *alternative to the choke*.

3.10. CARBURETTORS

3.10.1. Essential Features of Good Commercial Carburettor for Automotive Engines

Carburettor is a *mixing device to supply the engine with air-fuel mixture. It atomizes the fuel and mixes it with air in varying proportions to meet the changing operating conditions of automotive engines.* It is required to provide the following *essential features* :

1. To meter and supply the proper quantity and proportion of air and fuel at correct strength under all conditions of load and speed of the engine of the car for

(*i*) starting it easily from cold.

(*ii*) providing a rich mixture for slow idling.

(*iii*) providing a rich mixture for acceleration,

(*iv*) providing a rich mixture for high speed, and

(*v*) providing a rich mixture for low speed when moving up-gradient.

2. To operate satisfactorily when cold, or when hot.
3. To operate satisfactorily both on level and hills.
4. To overcome air-cleaner restrictions.
5. To withstand vibrations and road jerks.

3.10.2. Types of Carburettors

Carburettors, basically, are of the following *two types* :

1. **Open choke type**

- Here, the main orifice known as the *choke tube* or *venturi* is of *fixed dimensions*, and *metering is affected by varying the pressure drop across it*. Almost all carburettors, except S.U. carburettor, belong to this category of carburettors.
- The *important examples* of this type of carburettor are :
  (*i*) Zenith carburettor
  (*ii*) Solex carburettor
  (*iii*) Carter carburettor
  (*iv*) Stromberg carburettor.

2. **Constant vacuum type**

- In this type of carburettor the *area of the air passage is varied automatically while the pressure drop is kept approximately constant.*
- *Example* : S.U. carburettor.

Basic forms of carburettors :

Refer Fig. 3.20. Carburettor may be of the following *three basic forms*.

(*i*) Updraught (*ii*) Downdraught
(*iii*) Horizontal.

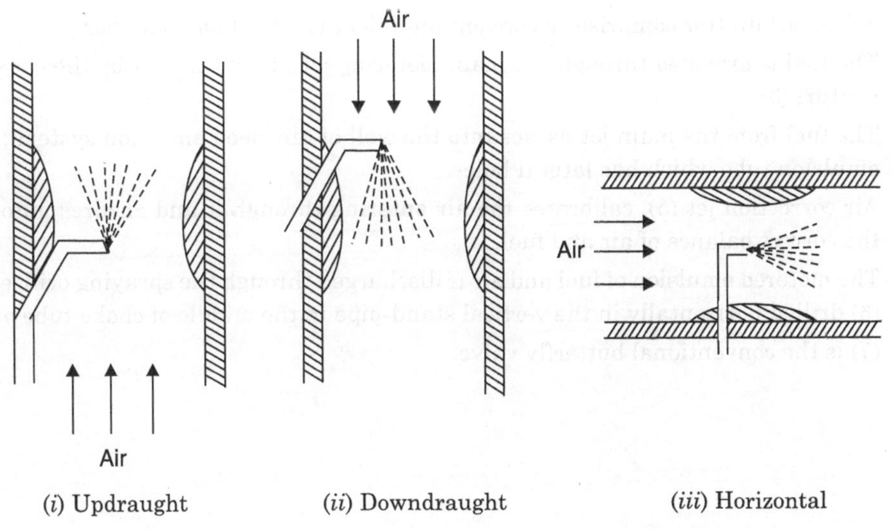

Fig. 3.20. Basic forms of carburettors.

- The *updraught variety* is now *obsolete* and is *only used where neither of the other types can be accommodated.*
- The *downdraught and horizontal types* of carburettors are *most widely used.*
  — The *advantage of downdraught variety is* that the *mixture is assisted by gravity in its passage into the engine induction tract,* and *at the same time the carburettor is usually reasonably accessible.*
  — The *horizontal type of carburettor has some advantage when under-bonnet space is limited.*

3.10.3. Description of Some Important Makes of Carburettors

Following carburettors will be described here :

1. Solex carburettors
2. Carter carburettors
3. S.U. Carburettor.

3.10.3.1. Solex Carburettor

This carburettor is made in various models and is used in Fiat, Standard and Willy's Jeep. It is famous for the following *characteristics* :

(*i*) *Easy starting ;*

(*ii*) *Good performance ;*

(*iii*) *Reliability.*

Fig. 3.21 shows the schematic arrangement of a solex carburettor. The unique feature of this carburettor is **Bi-starter** for *cold starting.*

The various components and the circuits for air and fuel for various ranges of operation are explained below :

1. **Normal running :**

- Solex carburettor comprises a conventional float (1) in a float chamber.
- The fuel is provided through the main metering jet (2) and the air by the choke tube or venturi (3).
- The fuel from the main jet passes into the well of air-bleed emulsion system ; (4) is the emulsion tube which has lateral holes.
- Air correction jet (5), calibrates the air entering through it and ensures automatically the correct balance of air and fuel.
- The metered emulsion of fuel and air is discharged through the spraying orifice or nozzle (6) drilled horizontally in the vertical stand-pipe in the middle of choke tube or venturi.
- (7) is the conventional butterfly valve.

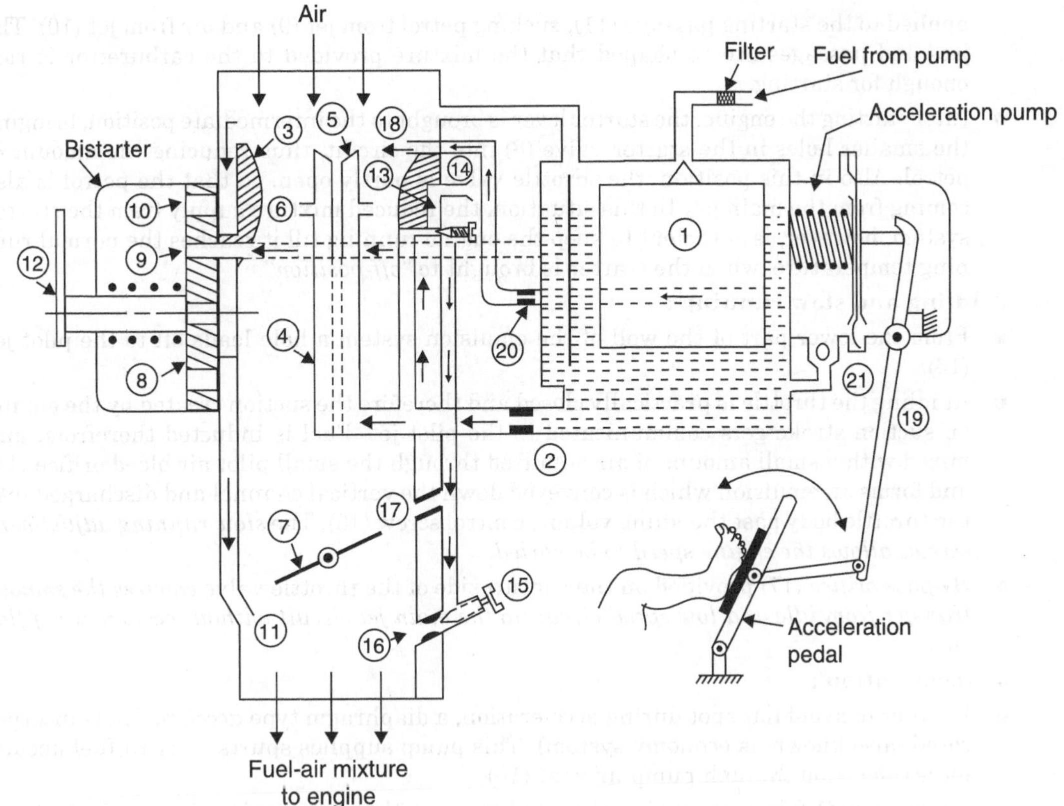

1. Conventional float	2. Main jet
3. Choke tube or venturi	4. Emulsion tube
5. Air correction jet	6. Spraying orifice or nozzles
7. Conventional butterfly valve	8. Flat disc with holes of different sizes
9. Starter petrol jet	10. Jet
11. Starting passage	12. Starter lever
13. Pilot jet	14. Small pilot air bleed orifice
15. Idling volume control screw	16. Idle port
17. By-pass orifice	18. Pump injector
19. Pump lever	20. Pump jet
21. Pump inlet valve.	

Fig. 3.21. Schematic arrangement of a solex carburettor.

2. **Cold starting and warming :**

The *provision of a bi-starter or a progressive starter is the unique feature of solex carburettors.*

- The starter valve is in the form of a flat disc (8) with holes of different sizes. These holes connect the starter petrol jet (9) and starter air jet sides to the passage which opens into a hole just below the throttle valve at (11). Either bigger or smaller holes come opposite the passage, depending upon the position of the starter lever (12). The starter lever is operated by flexible cable from the dash board control. Initially, for starting richer mixture is required and after the engine starts, the richness required decreases. In the start position when the starter control is pulled out fully, bigger holes are the connecting holes. The throttle valve being in closed position the whole of the engine suction is applied to the starting passage (11), sucking petrol from jet (9) and air from jet (10). The jets and passages are so shaped that the mixture provided to the carburettor is rich enough for starting.

- After starting the engine, the starter lever is brought to the intermediate position, bringing the smaller holes in the starter valve (8) into the circuit, thus reducing the amount of petrol. Also in this position, the throttle valve is partly open, so that the petrol is also coming from the main jet. In this situation, the reduced mixture supply from the starter system, however, is sufficient to keep the engine running till it reaches the normal running temperature, when the starter is brought to "*off-position*".

3. **Idling and slow running :**

- From the lower part of the well of the emulsion system a hole leads off to the pilot jet (13).
- At idling the throttle is practically closed and therefore the suction created by the engine on suction stroke gets communicated to the pilot jet. Fuel is inducted therefrom, and mixed with a small amount of air admitted through the small pilot air bleed orifice (14) and forms an emulsion which is conveyed down the vertical channel and discharged into the throttle body past the idling volume control screw (15). The *slow running adjustment screws allows the engine speed to be varied.*
- *By-pass orifice* (17) provided on the venturi side of the throttle valve *ensures the smooth transfer from idle and low speed circuit to the main jet circuit without occurrence of flat spot.*

4. **Acceleration :**

- In order to avoid flat spot during acceleration, a diaphragm type *acceleration* is incorporated (also known as economy system). This pump supplies spurts of extra fuel needed for acceleration through pump injector (18).
- Pump lever (19) is connected to the accelerator so that on pressing the pedal, the lever moves towards left, pressing the membrane towards left, thus forcing the petrol through pump jet (20) and injector (18). On making the pedal free, the lever moves the diaphragm back towards right creating vacuum towards left which opens the pump inlet valve (21) and thus admits the petrol from the chamber into pump.

3.10.3.2. Carter Carburettor

A *carter carburettor* is an American make carburettor and is used in *jeeps*. It is a *standard equipment on chevorolet and Pontiac series of cars.*

Fig. 3.22 shows the schematic arrangement of a downdraft type Carter carburettor. The brief description of the components and circuits is given below :

- The petrol (fuel) enters the conventional type float chamber (1).
- The air enters the carburettor from the top, a choke valve (2) in the passage remains open during normal working.
- This carburettor has a triple venturi diffusing type of choke, *i.e.*, it has *three venturies*, the smallest (3) lies above the level in the float chamber, and the remaining two venturies (6) and (5) are below the fuel level (in the float chamber), one below the other.
- At very low speeds, suction in primary venturi (3) is sufficient to draw the fuel. The nozzle (4) enters the primary venturi at an angle, and throws the fuel up against the air stream evenly, thereby providing finely divided atomised fuel. The mixture from venturi (3) passes centrally through the secondary venturi (5) where it is surrounded by a blanket of air stream and finally this leads to the third main venturi (6), where again the fresh air supply insulates the stream from the secondary venturi. The fuel-air mixture

enters the engine in well mixed atomised state. *The multiple venturi gives more homogeneous and better mixture at very low speeds resulting in steady and smooth operation at low speeds. This arrangement also ensures adequately formed mixture at high speeds.*

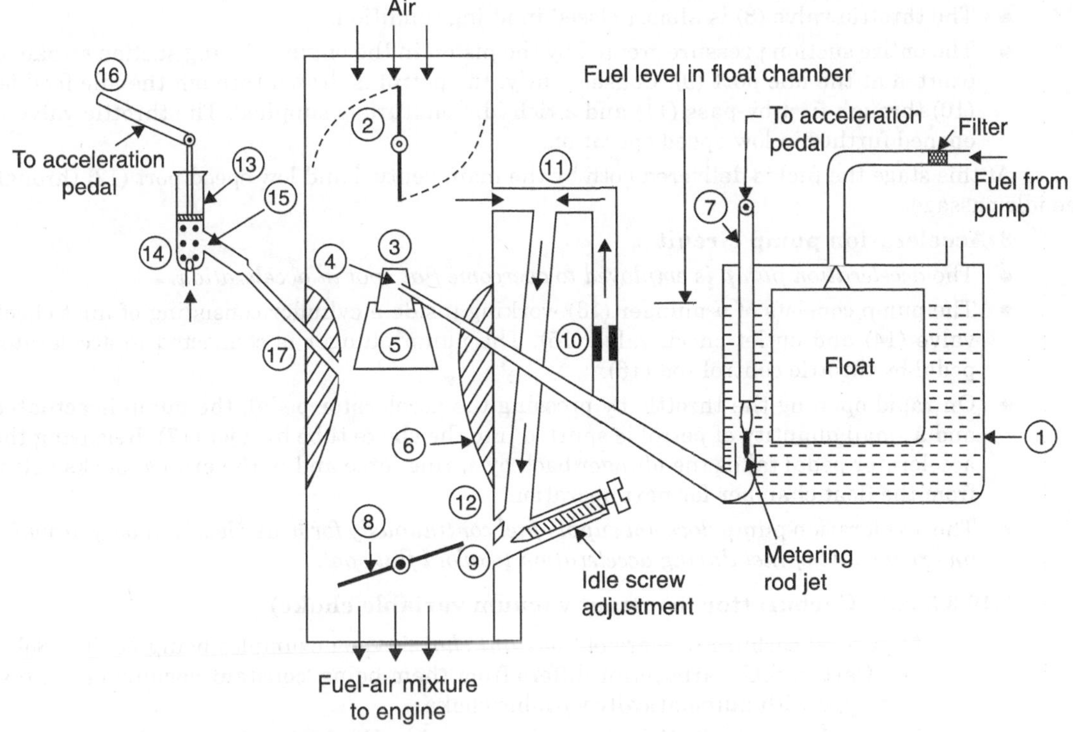

1. Float chamber
2. Choke valve
3. Primary venturi
4. Nozzle
5. Secondary venturi
6. Third (main) venturi
7. Metering rod
8. Throttle valve
9. Idle port
10. Idle feed jet
11. By pass
12. Low speed port
13. Plunger
14. Inlet check valve
15. Outlet check valve
16. Throttle control rod
17. Jet.

Fig. 3.22. Schematic arrangement of a downdraft Carter carburettor.

- In Carter carburettor mechanical metering method is used. In the fuel circuit there is a metering rod (7) (having two or more steps of diameter) which is actuated by a mechanism connected with the main throttle. *The amount of petrol drawn into the engine is governed by the area of opening between the metering rod jet and metering rod.*

1. **Starting circuit :**

- In order to start the engine, a *choke valve (2)* is incorporated in the air circuit. The choke valve is of butterfly type, one half of which is spring controlled. The valve is hinged at the centre.
  — When the engine is fully choked, the whole of the engine suction is applied at the main nozzle, which then delivers fuel. Since the airflow is quite small, very rich fuel-air mixture is supplied.

— When the engine starts/fires, the spring controlled half the choke valve is sucked open to provide correct quantity of air during the period of warm up.

2. **Idle and low speed circuit :**

- For *idling rich mixture in small quantity is required.*
- The throttle valve (8) is almost closed in idling condition.
- The entire suction pressure created by the piston in the engine, during suction stroke, is exerted at the idle port (9). Consequently, the petrol is drawn through the idle feed jet (10) through first by-pass (11) and a rich idle mixture is supplied. The throttle valve is opened further in low speed operation.

At this stage the fuel is delivered both by the main venturi and low speed port (12) through the idle passage.

3. **Acceleration pump circuit :**

- The *acceleration pump is employed to overcome flat spot in acceleration.*
- The pump consists of a plunger (13) working inside a cylinder consisting of inlet check valve (14) and outlet check valve (15). The pump plunger is connected to accelerator pedal by throttle control rod (16).
- On rapid opening the throttle by pressing the accelerator pedal, the pump is actuated and a small quantity of petrol is spurted into the choke tube by a jet (17). Releasing the accelerator pedal takes the plunger back by spring force and in the process sucks petrol from the float chamber for next operation.
- The acceleration pump *does not supply fuel continuously for heavy load but only provides an extra spurt of fuel during acceleration to avoid flat spot.*

3.10.3.3. S.U. Carburettor (constant vacuum variable choke)

— In general carburettors are of '*constant choke*' type ; examples being Zenith, Solex and Carter. S.U. carburettor differs from them being 'constant vacuum or depression' type with automatically variable choke.

— It is used in many British cars and was used in Hindustan Ambassador car.

Fig. 3.23 shows the S.U. carburettor schematically. The various components and circuits are described below :

1. **Normal operating conditions :**

- The full metering is accomplished by a **tapered needle** which is raised or lowered in a jet to alter the effective annular jet orifice, and hence fuel flow. The needle projects from underneath the flat face of the cylindrical air-valve, which alters the choke area as it is raised or lowered. The upper part of the air-valve is enlarged to form a piston which fits into the lower open end of the vacuum chamber. A spindle situated in the centre of the air-valve guides the assembly into the cylindrical vacuum chamber. To improve the accuracy and time response of the air-valve vertical movement with very small changes in engine demands, the friction between the air-valve spindle and guide is sometimes reduced by installing a ball-race between the two sliding surfaces.
- While the engine is running, the effect of the depression above the piston in the upper chamber, and the atmospheric pressure underneath, is to raise the air-valve and piston assembly against its own weight and the stiffness of the return spring. Since the doward load is almost constant, a constant depression is needed to keep the air-valve stationary in any raised position. *The amount the air-valve lifts depends on the flow rate of air which passes through the mixing chamber,* this being controlled by the engine speed and throttle opening position.

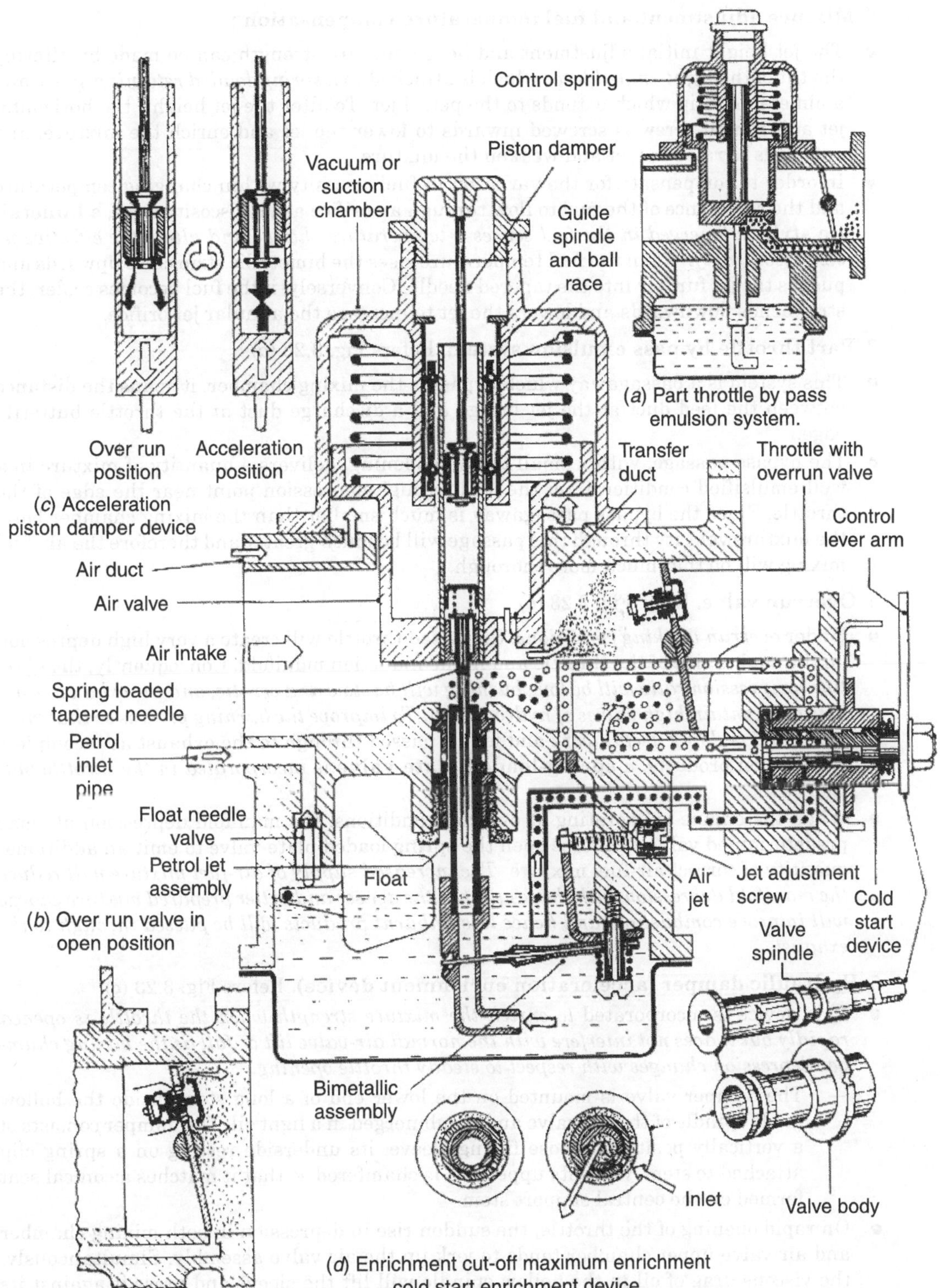

Fig. 3.23. S.U. carburettor (constant vacuum variable choke).

2. **Mixture adjustment and fuel temperature compensation :**

- The jet height initial adjustment and hence mixture strength can be made by altering the tilt of the *right angled lever* which is attached to a *spring-loaded retaining screw* and a bimetallic strip which extends to the petrol jet. To alter the jet height, the horizontal jet adjustment screw is screwed inwards to lower the jet and enrich the mixture, and outwards to raise the jet and weaken the mixture.
- In order to compensate for the variation in fuel viscosity within changing temperature and the reluctance of the fuel to flow through an orifice as its viscosity rises, a **bimetallic strip** *submerged in the fuel senses a temperature change and alters the effective jet size accordingly.* When the fuel temperature rises the bimetallic strip curls upwards and pushes the jet further into the tapered needle. Conversely, if the fuel becomes cooler, the strip bends downwards and lowers the jet to increase the annular jet orifice.

3. **Part throttle by pass emulsion system.** Refers Fig. 3.23 (*a*).

- This system is a passagaway which bypasses the mixing chamber, it spans the distance between the feed duct at the jet bridge and a discharge duct at the throttle butterfly edge.
- The bypass passage, with a small throttle opening, delivers a quantity of mixture in a well emulsified condition from the jet to a high depression point near the edge of the throttle. Since the bypass passagaway is much smaller than the mixing chamber bore, the mixture velocity through this passage will be much greater and therefore the air-fuel mixing will be that much more thorough.

4. **Overrun valve.** Refer Fig. 3.23 (*b*).

- Under *overrun working conditions,* the closed throttle will create a very high depression on the engine side of the throttle and in the induction manifold. Consequently, the *effective compression ratio will be low, burning will be slow and erratic, and the exhaust products will contain high valves of hydrocarbon. To improve the burning process so that more of the fuel is doing useful work* and less is passed through to the exhaust as incomplete combustion products, a **spring-loaded plate-valve** is *incorporated in the throttle butterfly disc.*
- When the engine is operating at overrun conditions, the manifold depression at some predetermined value will force open the spring loaded plate-valve to emit an additional quantity of correct air-fuel mixture. *The increased supply of air-fuel mixture will reduce the manifold depression with the result that the denser and better prepared mixture charge will improve combustion, and hence less unburnt products will be passed through to the exhaust.*

5. **Hydraulic damper (acceleration enrichment device).** Refers Fig. 3.23 (*c*).

- This device is incorporated *to enrich the mixture strength when the throttle is opened rapidly but it does not interfere with the normal air-valve lift or fall as the mixing chamber depression changes with respect to steady throttle opening.*
  — The damper valve is mounted on the lower end of a long stem inside the hollow guide spindle of the air-valve and is submerged in a light oil. The damper consists of a vertically positioned loose fitting sleeve, its underside resting on a spring clip attached to stem, while its upper end is chamfered so that it matches a conical seat formed on the central support stem.
- On rapid opening of the throttle, the sudden rise in depression in both mixing chamber and air valve upper chamber tends to jerk up the air valve assembly. Simultaneously, the viscous drag of oil in the hollow spindle will lift the sleeve and press it against its seat, and so the oil is thus temporarily trapped beneath the damper so that it prevents

any further upward movement of the air valve. For this brief period a temporary increase in the depression over the jet orifice is achieved, and more fuel will therefore be drawn to enrich the resultant mixture strength.

— When the change in engine speed steadies, the depression in the upper air-valve vaccum chamber will also stabilize and there will be a slight leakage of oil between the sleeve and its spindle bore. Consequently any oil pressure created underneath the sleeve damper will now be released enabling the sleeve to drop down onto spring clip-Oil will now move freely through the annular space made between the sleeve and its seat so that the *air-valve vertical movement can again react to small changes in demands of the engine.*

6. **Cold start device.** Refer Fig. 3.23 (*d*).

- A *cold start-device is in the form of a rotary-valve consisting of a cylindrical valve body, which has an annular groove in the middle region with a single radial hole drilled in its side.* Fitted inside the valve body is a spindle which has an axial hole bored half-way along from one end, while at the other end a control level is bolted. A double taper notched radial hole intersects the axial hole in the spindle. The whole assembly of the valve body and spindle is positioned in a larger hole made in the side of the floor chamber.
- In order to *cold start the engine* the *choke knob* situated on the instrument panel is *pulled out,* the interconnecting cable rotates the control lever and spindle to a position where the radial hole for both spindle and valve body are aligned. When the engine is cranked a high depression is created in the mixing chamber formed between the jet bridge and throttle valve, and this depression is conveyed to the axial hole in the control spindle where it then passes to the annular groove on the outside of the valve body. Here it divides and draws the fuel from the dip tube and atmospheric air from the float chamber by way of the air jet. The emulsified mixture is then drawn into the hollow spindle along the discharge passage duct and out into the mixing chamber.
- With warming up of the engine, the choke knob can be pushed back steadily, this rotates the control lever and spindle so that the *notched hole passageway becomes progressively smaller and thus restricts the quantity of air and petrol emulsion trying to enter the mixing chamber.*

3.10.3.4. Aircraft Carburettor

The *major difference* between an automobile carburettor and an aircraft carburettor is that whereas the *former operates at ground level conditions,* the latter ***operates at varying altitudes.*** With the increase in altitude the density of air decreases and A / F ratio which is proportional to air density/fuel density decreases, *i.e.,* the fuel-air mixture goes on becoming richer with increase in altitude. The mixture will be about 40% richer at an altitude of about 7000 metres (since at this altitude the air density is nearly one-half that at ground level and hence A / F ratio about 0.7 times the valve at ground level).

In view of the above, it is imperative to provide in aircraft carburettors an **altitude mixture correction device** to reduce the quantity of fuel progressively with altitude. For this purpose, the following methods are employed :

(*i*) Air bleeding ;
(*ii*) Back suction control ;
(*iii*) Incorporating a metering pin.

The aircraft carburettors entail the following other *special features :*

- For fuel level system a *special float chamber is required.*
- For controlling/eliminating the formation of ice in the choke tube and on the throttle valve due to *low temperature* an automatic *de-icing unit* is fitted or *hot engine oil is arranged to flow around the carburettor barrel and through the hollow throttle valve.*

3.11. PETROL INJECTION

3.11.1. Drawbacks of Modern Carburettors

The modern carburettors have the following *drawbacks* :

1. The mixture supplied to various cylinders of a multi-cylinder engine *varies in quality and quantity.* Also, due to fuel condensation in induction manifold, the *mixture proportion is affected.*

2. Due to presence of several wearing parts, the carburettors operate at a lower efficiency.

3. Reduced volumetric efficiency due to non-availability of a free flow passage for the mixture owing to the presence of choke tubes, throttle valves, jets, bends etc.

4. At low temperatures, freezing can occur (if special means to obviate this are not provided).

5. When the carburettor is tilted or during acrobatics in air-craft surging can occur (if means to avoid this are not provided).

6. In the absence of flame traps, backfiring may occur which may lead to ignition of fuel outside the carburettor.

3.11.2. Introduction to Fuel Injection

- The *function* of a fuel injection system is :

(i) *To monitor the engine's operating variables,*

(ii) *To transfer this information to a metering control, then*

(iii) *To discharge and atomise the fuel into the incoming air stream.*

- The position where the fuel is injected into the air charge considerably influences the performance of the engine.

3.11.3. Direct Injection. Refer to Fig. 3.24.

- In this type of layout the *fuel injectors are positioned in the cylinder-head so that fuel is directly discharged into each combustion chamber* (Fig. 3.24).
- With this arrangement it is essential that injection is timed to occur about 60° after T.D.C. on the induction stroke.
- Because of the shorter time period for fuel spray to mix with the incoming air charge, increased air turbulence is necessary. To compensate for the shorter permitted time for injection, atomising and mixing, the *injection pressure needs to be higher than for indirect injection.*
- *More overlap of exhaust and inlet valves can be utilized* compared with other carburetted or injected systems, so that incoming fresh air can assist in sweeping out any remaining exhaust gases from the combustion chambers.
- The injector nozzle and valve have to be designed to withstand the high operating pressures and temperatures of the combustion chamber, *this means that a more robust and costly injector unit is required.*
- *Generally, direct-injection air and fuel mixing is more thorough in large cylinders than in small ones because fuel droplet sizes do not scale down as the mixing space becomes smaller.*
- All condensation and wetting of the induction manifold and ports is *eliminated* but some spray may condense on the piston crown and cylinder walls.

3.11.4. Indirect Injection. Refer to Fig. 3.24.

- In this arrangement the fuel is injected into the air *stream prior to entering the combustion chamber.* Fuel spray may be delivered from a Single-Point Injection (S.P.I.) source, which is usually just upstream from the throttle (air intake side of the throttle), or it may be supplied from a Multi-Point Injection (M.P.I) source, where the *injectors are positioned in each induction manifold branch pipe just in front of the inlet port.* (Fig. 3.24).

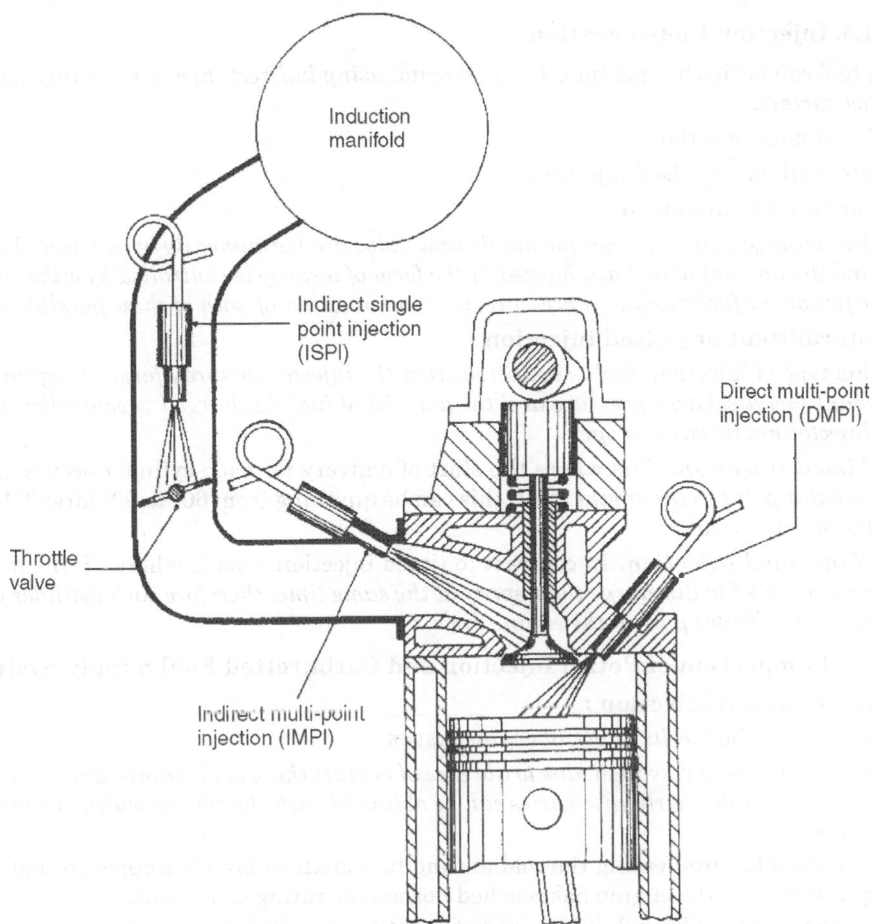

Fig. 3.24. Three-principal injector positions.

- Indirect injection can be discharged at relatively low pressure (2 to 6 bar) and need not be synchronized to the engine's induction cycle. Fuel can be discharged simultaneously to each induction pipe where it is mixed and stored until the inlet valve opens.
- Since indirect injection does not need to be timed, it *requires only low discharge pressures* and *the injectors are not exposed to combustion, the complexity of the operating mechanisms can be greatly reduced, which considerably lowers cost.*
- The single-point injection system has the same air and fuel mixing and distribution problems as a carburettor layout *but without venturi restriction* so that *higher engine volumetric efficiencies are obtained. High injection pressures,* compared with the carburettor discharge method of fuel delivery, *speed up and improve the atomization of the liquid spray.*
- The *multi-point injection* layout, in contrast to the single-point injection method *has no fuel distribution difficulties* since each injector discharges directly into its own induction port and the mixture then has only to move a short distance before it enters the cylinder. Since the induction manifold deals mainly with only induced air, the branch pipes can be enlarged and extended to maximize the ram effect of the incoming air charge.

A **major feature** *with petrol injection is that there is separate air and fuel metering and that fuel metering is precise under all engine operating condition's.*

3.11.5. Injection Considerations

The fuel can be discharged into the air stream, *using indirect injection arrangements,* by the following *two methods* :

1. Continuous injection.
2. Intermittent or pulsed injection.

1. **Continuous injection :**

In this arrangement, *the injector nozzle and valve are permanently open* while the engine is operating and the *amount of fuel discharged in the form of a spray is controlled by either varying the metering orifice or the fuel discharge pressure, or a combination of both of these possible variables.*

2. **Intermittent or pulsed injection :**

In this type of injection, *fuel is delivered from the injector in spray form at regular intervals with a constant fuel discharge pressure and the amount of fuel discharged is controlled by the time period the injector nozzle valve is open.*

- ***Timed injection.*** This where the start of delivery for each cylinder *occurs at the same angular point in the engine cycle,* this can be anything from 60° to 90° after T.D.C. on the induction stroke.
- ***Non-timed injection.*** In contrast to timed injection, this is where *all the injectors are programmed to discharge their spray at the same time, therefore each cylinder piston will be on a different part of the engine cycle.*

3.11.6. Comparison of Petrol Injection and Carburetted Fuel Supply Systems

Merits of petrol injection :

Following are the *merits of petrol engine system :*

1. In petrol injection system, *due to absence of venturi there is the minimum of air restriction so that higher engine volumetric efficiencies can be obtained with the corresponding improvement in power and torque.*

2. *The spots* for pre-heating the cold air and fuel mixture are *eliminated* so that *denser air enters the cylinder* when the engine has reached normal operating conditions.

3. As the manifold branch pipes are not greatly concerned with mixture preparation they *can be designed to utilize the inertia of the air charge to increase the engine's volumetric efficiency ;* (this does not apply for single point injection).

4. Because of *direct spray discharge* into each inlet port, *acceleration response is better.*

5. *Atomization of fuel droplets is generally improved* over normal speed and load driving conditons.

6. It is possible to *use greater inlet and exhaust valve overlap without poor idling, loss of fuel or increased exhaust pollution.*

7. The monitoring of engine operating parameters enables accurate matching of air and fuel requirements under normal speed and load conditions which *improves engine performance, fuel consumption and reduces exhaust pollution.*

8. Fuel injection equipment is *precise in metering injected fuel spray* into the intake ports over the complete engine speed, load and temperature operating range.

9. There is *precise fuel distribution* between engine cylinders even under full load conditions *with multi-point injection.*

10. *Multi-point injection does not require time for fuel transportation in the intake manifold and there is no manifold wall melting.*

11. With fuel injection, when cornering fast or due to heavy braking, fuel surge is eliminated.

12. The single point as well as multi-point injection systems are *particularly adaptable and suitable for supercharged engines.*

Demerits / Limitations of petrol injection :

Petrol injection system entails the following *demerits/limitations* :

1. Initial cost of equipment is high ; replacement parts are also expensive.
2. Increased care and attention required.
3. In order to diagnose fuel injection system faults and failures, special servicing equipment is necessary.
4. It is necessary to have considerably more mechanical and electrical knowledge to diagnose and rectify the faults of fuel equipment.
5. Injection equipment may be elaborately complicated, delicate to handle and impossible to service.
6. More electrical and mechanical components to go wrong.
7. Due to pumping and metering of the fuel there is increased mechanical and hydraulic noise.
8. Due to the fine working tolerances of the metering and discharging components, very careful filtration is needed.
9. To drive the fuel pressure pump or injection discharge devices, power (be it electrical or mechanical) is necessary.
10. More bulky and heavy (than that of a carburetted fuel supply system).

3.11.7. Electronic Fuel Injection

Fig. 3.25 shows the fuel injection system-L-Jetronic with air flow metering (developed by Robert Based Corp.) It consist of the following *units* :

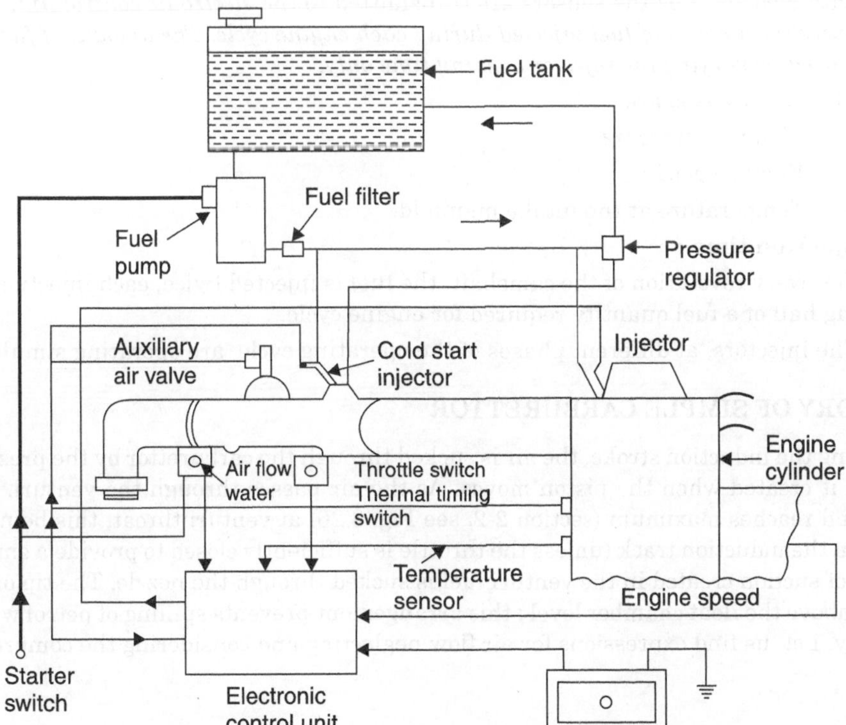

Fig. 3.25. Fuel injection system-L-Jetronic with air-flow metering.

1. Fuel delivery system
2. An induction system
3. Sensors and air flow control system
4. Electronic control unit.

1. **Fuel delivery system :**

- It consists of an electrically driven fuel pump which draws fuel from a *fuel tank*. The pump forces the oil through a *filter* into a line at the end of which is situated a *pressure regulator*, which in turn is connected to intake manifold.
- The pressure regulator keeps the pressure difference between the fuel pressure and the manifold pressure constant, so that the quantity of fuel injected is dependent on the injector open time only.

2. **Air induction system :**

- After passing the air filter, the incoming air flows through an air flow meter, which generates a voltage signal (depending on the quantity of air flow).
- Just behind the *throttle valve* is fitted a cold start *magnetic injection valve,* which injects additional fuel for cold start. This valve also supplies the extra fuel needed during warm-up period.
- An *auxiliary valve* (which by-passes the throttle valve) supplies the extra air required for idling (in addition to rich-air-fuel mixture). This extra air increases the engine speed after cold start to acceptable idling speed.
- To the throttle valve is attached a *throttle switch* equipped with a set of contacts which generate a sequence of voltage signals during the opening of throttle valve. The voltage *signals result in injection of additional fuel required for acceleration.*

3. **Electric control unit :**

- The *sensors* are incorporated to measure the operating data at different locations. The data measured by the sensors are transmitted to the *electronic control unit which computes the amount of fuel injected during each engine cycle. The amount of fuel injected is varied by varying the injector opening time only.*
- The *sensors* used are :
  - Manifold pressure ;
  - Engine speed ;
  - Temperature at the intake manifold.

4. **Injection time :**

- For every revolution of the camshaft, the fuel is injected twice, each injection contributing half of a fuel quantity required for engine cycle.
- The injectors, at different phases of the operating cycle, are operating simultaneously.

3.12. THEORY OF SIMPLE CARBURETTOR

During the induction stroke, the air is sucked through the carburettor by the pressure difference across it created when the piston moves. As the air passes through the venturi, its velocity increases and reaches maximum (section 2-2, see Fig. 3.26) at venturi throat, this being the minimum area in the induction track (unless the throttle is sufficiently closed to provide a smaller area). As a result of suction created in the venturi fuel is sucked through the nozzle. The tip of the nozzle is z metres above the float chamber level ; this arrangement prevents spilling of petrol when vehicle is stationary. Let us find expressions for air flow neglecting and considering the compressibility of air.

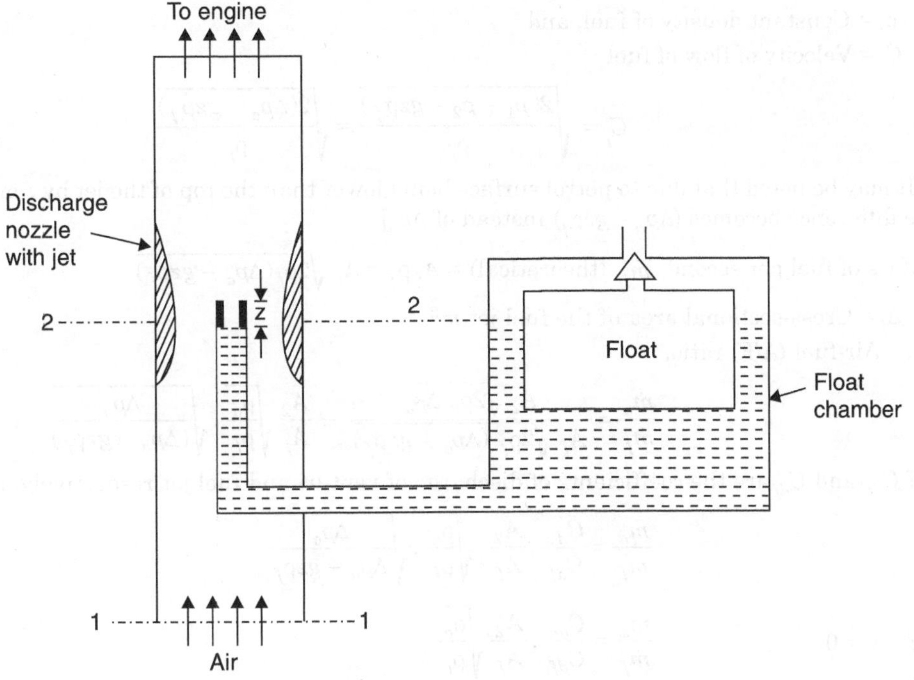

Fig. 3.26. Principle of a simple carburettor.

Case I. *Neglecting the compressibility of air*—**Approximate Analysis.**

Applying Bernoulli's equation at sections 1-1 and 2-2, the equation for air flow is given by

$$\frac{p_1}{\rho_a} + \frac{C_1^2}{2} = \frac{p_2}{\rho_a} + \frac{C_2^2}{2} \qquad ...(3.1)$$

where, ρ_a = Density of air kg/m³, and

p_1, p_2 = Pressure at sections 1-1 and 2-2 respectively.

$C_1 = C_2$ = Velocities at sections 1-1 and 2-2 respectively, m/s.

Assuming initial *velocity of air to be zero* ($C_1 = 0$), *density of air* (ρ_a) to be *constant,* since air is *assumed incompressible,* we have

$$\frac{p_1}{\rho_a} = \frac{p_2}{\rho_a} + \frac{C_2^2}{2} \qquad ...(3.2)$$

$$\therefore \qquad C_2 = \sqrt{\frac{2(p_1 - p_2)}{\rho_a}} = \sqrt{\frac{2\Delta p_a}{\rho_a}} \qquad ...(3.3)$$

where, $\Delta p_a = p_1 - p_2$

Mass of air per second $\dot{m}_a = C_2 A_2 \rho_a = A_2 \sqrt{2\rho_a \Delta p_a}$ ...(3.4)

where A_2 is area of venturi throat in m^2.

(The above equation gives *theoretical mass flow of air. The actual mass flow is obtained by multiplying the co-efficient of discharge of venturi*)

Similarly, for the flow of fuel, we have

$$\frac{p_1}{\rho_f} = \frac{p_2}{\rho_f} + \frac{C_f^2}{2} + gz \qquad ...(3.5)$$

where, ρ_f = Constant density of fuel, and
C_f = Velocity of flow of fuel.

$$\therefore \qquad C_f = \sqrt{\frac{2(p_1 - p_2 - gz\rho_f)}{\rho_f}} = \sqrt{\frac{2(\Delta p_a - gz\rho_f)}{\rho_f}} \qquad ...(3.6)$$

[It may be noted that due to petrol surface being lower than the top of the jet by z metres the pressure difference becomes $(\Delta p_a - gz\rho_f)$ instead of Δp_a]

Mass of fuel per second, $\dot{m}_f$ (theoretical) $= A_f \rho_f = A_f \sqrt{2\rho_f(\Delta p_a - gz\rho_f)}$ ...(3.7)

where, Δ_f = Cross-sectional area of the fuel jet m^2.

$\therefore$ Air-fuel (A/F) ratio,

$$\frac{\dot{m}_a}{\dot{m}_f} = \frac{A_2\sqrt{2\rho_a \Delta p_a}}{A_f\sqrt{2\rho_f(\Delta p_a - gz\rho_f)}} = \frac{A_2}{A_f}\sqrt{\frac{\rho_a}{\rho_f}} \cdot \sqrt{\frac{\Delta p_a}{(\Delta p_a - gz\rho_f)}} \qquad ...(3.8)$$

If C_{da} and C_{df} are the coefficients of discharge of venturi and fuel jet respectively, then

$$\frac{\dot{m}_a}{\dot{m}_f} = \frac{C_{da}}{C_{df}} \cdot \frac{A_2}{A_f} \cdot \sqrt{\frac{\rho_a}{\rho_f}} \cdot \sqrt{\frac{\Delta p_a}{\Delta p_a - gz\rho_f}} \qquad ...(3.9)$$

If $z = 0$,

$$\frac{\dot{m}_a}{\dot{m}_f} = \frac{C_{da}}{C_{df}} \cdot \frac{A_2}{A_f}\sqrt{\frac{\rho_a}{\rho_f}} \qquad ...(3.10)$$

Case II. *Taking into consideration the compressibility of air in account*—**Exact Analysis.**

When the compressibility of air is taken into account, the air flow will change but the fuel flow will remain unchanged. Applying Steady Flow Energy Equation (S.F.E.E.) at sections 1-1 and 2-2, we get,

$$h_1 + \frac{C_1^2}{2} + Q = h_2 + \frac{C_2^2}{2} + W$$

or

$$Q - W = (h_2 - h_1) + \frac{C_2^2 - C_1^2}{2}$$

where, h_1, h_2 = Enthalpies at sections 1-1 and 2-2 respectively.

Since $Q = 0$, $W = 0$ and $C_1 = 0$

$$\therefore \qquad C_2 = \sqrt{2(h_1 - h_2)}$$

Substituting $h_1 = c_pT_1$ and $h_2 = c_p T_2$, we get

$$C_2 = \sqrt{2c_pT_1\left(1 - \frac{T_2}{T_1}\right)} \qquad ...(3.11)$$

Since the flow process between the atmosphere and the venturi throat is isentropic,

$$\therefore \qquad \frac{T_2}{T_1} = \left(\frac{p_2}{p_1}\right)^{\frac{\gamma-1}{\gamma}} \qquad ...(3.12)$$

Substituting eqn. (3.6) in eqn. (3.5), we get

$$C_2 = \sqrt{2c_pT_1\left[1 - \left(\frac{p_2}{p_1}\right)^{(\gamma-1)/\gamma}\right]} \qquad ...(3.13)$$

Now, the mass of flow of air is constant from inlet to venturi throat, and is given by

$$\dot{m}_a = \frac{A_1 C_1}{v_1} = \frac{A_2 C_2}{v_2} \quad ...(3.14)$$

where v_1, v_2 = Specific volumes at sections 1-1 and 2-2 respectively.

Since $p_1 v_1^{\gamma} = p_2 v_2^{\gamma}$, $\quad \therefore \quad v_2 = v_1 \left(\frac{p_1}{p_2}\right)^{1/\gamma} = \frac{RT_1}{p_1}\left(\frac{p_1}{p_2}\right)^{1/\gamma}$ ...(3.15)

Substituting the values of C_2 and v_2 in eqn. 3.8, we get

$$\text{Theoretical mass flow of air, } \dot{m}_a = \frac{A_2 p_1}{RT_1}\left(\frac{p_2}{p_1}\right)^{1/\gamma}\sqrt{2c_p T_1\left[1-\left(\frac{p_2}{p_1}\right)^{(\gamma-1)/\gamma}\right]}$$

$$= \frac{A_2 p_1}{R\sqrt{T_1}}\sqrt{2c_p\left[\left(\frac{p_2}{p_1}\right)^{2/\gamma}-\left(\frac{p_2}{p_1}\right)^{(\gamma+1)/\gamma}\right]} \quad ...(3.16)$$

$$\therefore \quad \frac{\dot{m}_a}{\dot{m}_f} = \frac{C_{da}}{C_{df}} \times \frac{A_2}{A_f} \times \frac{p_1}{R\sqrt{T_1}} \times \frac{\sqrt{2c_p\left[\left(\frac{p_2}{p_1}\right)^{2/\gamma}-\left(\frac{p_2}{p_1}\right)^{(\gamma+1)/\gamma}\right]}}{\sqrt{2\rho_f\,(\Delta p_a - gz\rho_f)}} \quad ...(3.17)$$

Comments on Air-Fuel (A / F) ratio supplied by the carburettor :

1. From eqn (3.8), it may be observed that no fuel flow will take place when $\Delta p_a < gz\rho_f$. When $\Delta p_a > gz\rho_f$, the fuel flow will start and the mixture becomes progressively richer.

The minimum air velocity at throat to cause fuel flow, with given value of z, can be calculated as follows :

From eqn. 3.4, $\quad C_2 = \sqrt{\frac{2\Delta p_a}{\rho_a}} = \sqrt{\frac{2(gz\rho_f)}{\rho_a}}$ ...(3.18)

2. At higher air flows when $\Delta p_a >> gz\rho_f$ (*i.e.*, fraction $gz\rho_f/\Delta p_a$ becomes negligible), the air-fuel ratio approaches

$$\frac{\dot{m}_a}{\dot{m}_f} = \frac{A_2}{A_f}\cdot\frac{C_{da}}{C_{df}}\sqrt{\frac{\rho_a}{\rho_f}} \quad ...(3.19)$$

3. A reduction in the density of air reduces the value of A/F, *i.e.*, the mixture becomes richer. It happens at (*i*) high air flows when Δp_a is large, hence C_2 is large and ρ_2 becomes small, (*ii*) high altitudes where density of air is small.

Thus in a *simple carburettor the air-fuel mixture becomes progressively richer with increasing air flows and increasing altitudes* ; this is a *big drawback*. The other *drawbacks* are :

- It cannot supply rich mixture required during idling and low load operation.
- It cannot supply rich mixture for starting from cold.
- It cannot supply extra fuel needed during acceleration.

WORKED EXAMPLES

Example 3.1. *A four cylinder four-stroke engine having diameter and length of stroke as 100 mm and 120 mm respectively is running at 1800 r.p.m. Its carburettor venturi has a 28 mm throat. Assuming co-efficient of air flow 0.8, density of air 1.2 kg/m³ and volumetric efficiency of the engine as 75 per cent, determine the suction at the throat.*

Solution. *Given* : $D = 100$ mm $= 0.1$ m ; $L = 120$ mm $= 0.12$ m ; $N = 1800$ r.p.m. ;

Throat diameter, $d_2 = 28$ mm $= 0.028$ m ; $C_{da} = 0.8$; $\rho_a = 1.2$ kg/m³ ; $\eta_{vol.} = 75\%$.

Suction at the throat Δp_a :

Stroke volume $= \frac{\pi}{4} \times (0.1)^2 \times 0.12 \times 4 = 0.00377$ m³

Actual volume per strokes $= \eta_{vol.} \times 0.00377 = 0.75 \times 0.00377 = 0.00283$ m³

$\therefore$ Actual volume sucked per second

$$= 0.00283 \times \frac{1800}{2} \times \frac{1}{60} = 0.04245 \text{ m}^3/\text{s}$$

$$\dot{m}_a = 0.04245 \times 1.2 = 0.05094 \text{ kg/s}$$

As the initial temperature and pressure are not given, the problem is solved by approximate method *i.e.*, neglecting compressibility of the air.

$$\dot{m}_a = C_{da} \times A_2 \times \sqrt{2\rho_a \Delta p_a} \qquad \text{...[Eqn. (3.4)]}$$

$$0.05094 = 0.8 \times \frac{\pi}{4}(0.028)^2 \sqrt{2 \times 1.2 \times \Delta p_a}$$

$$= 7.63 \times 10^{-4} \sqrt{\Delta p_a}$$

$$\therefore \quad \Delta p_a = \left(\frac{0.05094}{7.63 \times 10^{-4}}\right)^2 = 4457 \text{ N/m}^2 = \mathbf{0.04457 \text{ bar.}} \quad \textbf{(Ans.)}$$

Example 3.2. *A spark ignition engine on test consumes 5 kg/h of petrol when running on an air-fuel ratio of 16 : 1. The engine uses a single-jet carburettor having a fuel orifice area of 2 sq mm and the tip of the jet is 5 mm above the level of petrol in the float chamber, when the engine is not running. Calculate the depression in the venturi throat to maintain the required fuel flow rate through the carburettor. Assume specific gravity of petrol as 0.75 and the coefficient of discharge of the fuel orifice as 0.8. What area of venturi throat will be required to maintain the desired flow rate ? Density of air is 1.20 kg/m³ and the coefficient of discharge for venturi throat is 0.8. Neglect compressibility of air.* **(Roorkee University)**

Solution. *Given* : $\dot{m}_f = \frac{5}{3600} = 0.001389$ kg/s ; A/F ratio = 16 : 1 ;

Fuel orifice area, $A_f = 2$ mm² $= 2 \times 10^{-6}$ m² ; $z = 5$ mm $= 0.005$ m ;

Sp. gr. of petrol = 0.75 ; $C_{df} = 0.8$, $\rho_a = 1.2$ kg/m³ ; $C_{da} = 0.8$.

Depression in venturi throat, Δp_a :

The actual fuel flow rate is given by,

$$\dot{m}_f = C_{df} \cdot A_f \sqrt{2\rho_f(\Delta p_a - gz\rho_f)} \qquad \text{...[Eqn. (3.7)]}$$

where Δp_a is in N/m².

or, $$0.001389 = 0.8 \times (2 \times 10^{-6}) \sqrt{2 \times (0.75 \times 1000)\,(\Delta p_a - 9.81 \times 0.005 \times 0.75 \times 1000)}$$

or, $$\frac{0.001389}{0.8 \times 2 \times 10^{-6}} = 38.73\sqrt{\Delta p_a - 36.79}$$

or, $$\Delta p_a - 36.79 = \left(\frac{0.001389}{0.8 \times 2 \times 10^{-6} \times 38.73}\right)^2$$

or, $$\mathbf{\Delta p_a = 539.2\ N/m^2.\quad (Ans.)}$$

Throat area, A_t :

Air flow rate, $$\dot{m}_a = \frac{5}{3600} \times 16 = 0.02222\ \text{kg/s}$$

Also, $$\dot{m}_a = C_{da} \times A_t\sqrt{2\rho_a \Delta p_a} \quad \text{...(Eqn. (3.4)}$$

(Here $A_2 = A_t$)

$$0.02222 = 0.8 \times A_t\sqrt{2 \times 1.2 \times 539.21}$$

$$\therefore \quad A_t = \frac{0.02222}{0.8\sqrt{2 \times 1.2 \times 539.21}} = 7.7209 \times 10^{-4}\ \text{m}^2$$

$$= \mathbf{7.7209\ cm^2.\quad (Ans.)}$$

Fig. 3.27

Example 3.3. *The following data relate to a petrol engine :*

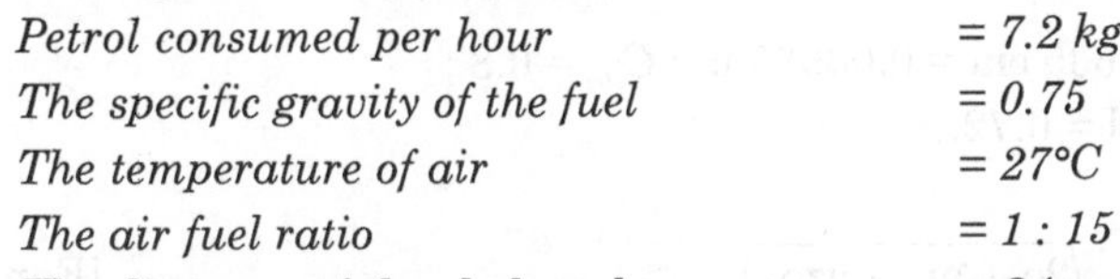

Petrol consumed per hour	*= 7.2 kg*
The specific gravity of the fuel	*= 0.75*
The temperature of air	*= 27°C*
The air fuel ratio	*= 1 : 15*
The diameter of the choke tube	*= 24 mm*

The height of top of the jet above the petrol level = 4.2 mm = 0.0042 m in the float chamber

The co-efficient of discharge for air	*= 0.8*
The co-efficient of discharge for fuel	*= 0.7*
Atmospheric pressure	*= 1.013 bar*

Calculate the diameter of the fuel jet of a simple carburettor.

Solution. *Given :* $\dot{m}_f = \frac{7.2}{3600}$ kg/s ; $\rho_f = 0.75 \times 1000 = 750$ kg/m³ ; $T_1 = 27 + 273 = 300$ K ;
A / F ratio = 1 : 15 ; d_2 = 24 mm = 0.024 m ; z = 4.2 mm = 0.0042 m ;
$C_{da} = 0.8$; $C_{df} = 0.7$; $p_1 = 1.013$ bar.

Diameter of the fuel jet, d_f :

We know that, $$\rho_a = \frac{p_1}{RT_1} = \frac{1.013 \times 10^5}{(0.287 \times 1000) \times 300} = 1.176\ \text{kg/m}^3$$

Air flow rate, $$\dot{m}_a = A_2\, C_{da}\sqrt{2\rho_a \Delta p_a}$$

$$\frac{15 \times 7.2}{3600} = \frac{\pi}{4} \times (0.024)^2 \times 0.8\sqrt{2 \times 1.176 \times \Delta p_a}$$

$$= 5.55 \times 10^{-4}\sqrt{\Delta p_a}$$

$$\therefore \quad \Delta p_a = \left(\frac{15 \times 7.2}{3600 \times 5.55 \times 10^{-4}}\right)^2 = 2922\ \text{N/m}^2$$

Fuel flow rate, $$\dot{m}_f = A_f\, C_{df}\sqrt{2\rho_f(\Delta p_a - gz\rho_f)} \quad \text{...[Eqn. 3.7]}$$

$$\frac{7.2}{3600} = \frac{\pi}{4}(d_f)^2 \times 0.7 \sqrt{2 \times 750\,(2922 - 9.81 \times 0.0042 \times 750)}$$

$$= 1144.89\,(d_f)^2$$

$$\therefore \quad \mathbf{d_f} = \left(\frac{7.2}{3600 \times 1144.89}\right)^{1/2} = 1.32 \times 10^{-3} \text{ m} \quad \text{or} \quad \mathbf{1.32 \text{ mm.}} \quad \textbf{(Ans.)}$$

Example 3.4. *A simple carburettor under a certain condition delivers 5.45 kg/h of petrol with an air-fuel ratio of 15. The fuel jet area is 2 mm² with a coefficient of discharge of 0.75. If the tip of the fuel jet is 0.635 cm above the level of petrol in the float chamber and the venturi throat coefficient of discharge is assumed to be 0.80, calculate :*

(i) The venturi depression in cm of H_2O necessary to cause air and fuel flow at the desired rate.

(ii) The venturi throat diameter.

(iii) The velocity of air across the venturi throat.

You may take density of air = 1.29 kg/m³ and specific gravity of period = 0.72.

(Madras University)

Solution. *Given :* $\dot{m}_f = \frac{5.45}{3600} = 0.001514$ kg/s ; A/F ratio = 15 ; $A_f = 2$ mm² $= 2 \times 10^{-6}$ m² ; $C_{df} = 0.75$; $z = 0.635$ cm $= 0.00635$ m ; $C_{da} = 0.8$; $\rho_a = 1.29$ kg/m³ ; Sp. gr. of petrol = 0.72.

(*i*) **Venturi depression, Δp_a :**

$$\dot{m}_f = C_{df} \cdot A_f \sqrt{2\rho_f(\Delta p_a - gz\rho_f)} \qquad \text{...[Eqn. (3.7)]}$$

where Δp_a is in N/m².

$$0.001514 = 0.75 \times 2 \times 10^{-6} \sqrt{2 \times (0.72 \times 1000)\,(\Delta p_a - 9.81 \times 0.00635 \times 0.72 \times 1000)}$$

$$\frac{0.001514}{0.75 \times 2 \times 10^{-6}} = 37.95\sqrt{(\Delta p_a - 44.85)}$$

or,
$$\Delta p_a - 44.85 = \left(\frac{0.001514}{0.75 \times 10^{-6} \times 37.95}\right)^2 = 707.37$$

$$\therefore \quad \Delta p_a = 752.22 \text{ N/m}^2 = \frac{752.22}{9810} \text{ m of water} = \mathbf{7.67 \text{ cm of } H_2O.} \quad \textbf{(Ans.)}$$

(*ii*) **Venturi throat diameter, D_t :**

$$\text{Air flow rate} = \frac{5.45}{3600} \times 15 = 0.02271 \text{ kg/s}$$

Also,
$$\dot{m}_a = C_{da} A_t \sqrt{2\rho_a \Delta p_a}$$

$$\therefore \quad 0.02271 = 0.8 \times A_t \sqrt{2 \times 1.29 \times 752.22}$$

or,
$$A_t = 6.444 \times 10^{-4} \text{ m}^2 = \frac{\pi}{4} D_t^2$$

$$\therefore \quad \Delta_t = \left[\frac{6.444 \times 10^{-4} \times 4}{\pi}\right]^{1/2} = 0.0286 \text{ m} = \mathbf{2.86 \text{ cm.}} \quad \textbf{(Ans.)}$$

(*iii*) **Velocity of air across the venturi throat C_t :**

$$C_t \ (or\ C_2) = \sqrt{\frac{2(gz\rho_f)}{\rho_a}} \qquad \text{...(Eqn. 3.18)}$$

$$= \sqrt{\frac{2 \times 9.81 \times 0.00635 \times (0.72 \times 1000)}{1.29}} = \textbf{8.34 m/s.} \quad \textbf{(Ans.)}$$

Example 3.5. *A carburettor ; tested in the laboratory has its float chamber vented to atmosphere. The main metering system is adjusted to give an air-fuel ratio of 15 : 1 at sea level conditions. The pressure at the venturi throat is 0.8 bar. The atmospheric pressure is 1 bar. The same carburettor is tested again when an air cleaner is fitted at the inlet to the carburettor. The pressure drop to air cleaner is found to be 30 mm of Hg when air flow at sea level condition is 240 kg/h. Assuming zero tip and constant coefficient of flow, calculate (i) the throat pressure when the air cleaner is fitted and (ii) air-fuel ratio when the air cleaner is fitted.* **(Bombay University)**

Solution. *Given* : A / F ratio = 15 : 1 at sea level conditions ;

$p_1 = 1$ bar ; $p_2 = 0.8$ bar ;

(*i*) **The throat pressure when the air cleaner is fitted :**

Quantity of air flowing is same in both the cases.

$$(\dot{m}_a)_{actual} = C_{da} A_t \sqrt{2\rho_a (p_1 - p_2)}$$

When there is *no air cleaner,*

$$\Delta p_a = p_1 - p_2 = 1 - 0.8 = 0.2 \text{ bar}$$

When the *air cleaner is fitted,* let p_t be the throat pressure, then

$$\Delta p_a' = \left[1 - \left(1000 \times 13.6 \times 9.81 \times \frac{30}{1000} \times 10^{-5}\right) - p_t\right] \text{bar}$$

$$= (0.96 - p_t) \text{ bar} \qquad (\because \ 1 \text{ bar} = 10^5 \text{ N/m}^2)$$

For the same air flow and constant coefficients,

$$\Delta p_a = \Delta p_a'$$

$$\therefore \quad 0.2 = 0.96 - p_t$$

or $$\mathbf{p_t = 0.76 \ bar.} \quad \textbf{(Ans.)}$$

(*ii*) **Air-fuel ratio when the air cleaner is fitted :**

Without air cleaner, $\Delta p_f = \Delta p_a = 0.2$ bar

With air cleaner fitted (with float-chamber still vented to atmosphere),

$$\Delta p_f = 1 - 0.76 = 0.24 \text{ bar}$$

As Δp_f has increased more fuel will flow making the mixture richer.

$$\text{New A / F ratio} = \text{A / F ratio when air cleaner is not fitted} \times \sqrt{\frac{\Delta p_f \text{ without air cleaner}}{\Delta p_t \text{ with air cleaner}}}$$

$$= 15 \times \sqrt{\frac{0.2}{0.24}} = \textbf{13.69.} \quad \textbf{(Ans.)}$$

Example 3.6. *A simple jet carburettor is required to supply 4.6 kg of air per minute. The pressure and temperature of air are 1.013 bar and 25°C respectively. Assuming flow to be isentropic and compressible and velocity coefficient as 0.8, calculate the throat diameter of the choke for air flow velocity of 80 m/s.*

Solution. *Given* : $\dot{m}_a = \frac{4.6}{60} = 0.0767$ kg/s ; $p_1 = 1.013$ bar, $T_1 = 25 + 273 = 298$ K ;

$C_1 = 0$; $C_2 = 80$ m/s ; $C_v = 0.8$.

Throat diameter d_2 :

Applying S.F.E.E. at sections 1-1 and 2-2, we have

$$h_1 + \frac{C_1^2}{2} + Q = h_2 + \frac{C_2^2}{2} + W$$

or, $$h_1 = h_2 + \frac{C_2^2}{2} \qquad \left(\because Q = 0 ; W = 0, \; C_1 = 0\right)$$

or, $$C_2 = \sqrt{2(h_1 - h_2)}$$

$$= \sqrt{2c_p(T_1 - T_2)}$$

$$= \sqrt{2c_p T_1\left(1 - \frac{T_2}{T_1}\right)}$$

$$= \sqrt{2c_p T_1\left[1 - \left(\frac{p_2}{p_1}\right)^{\frac{\gamma-1}{\gamma}}\right]}$$

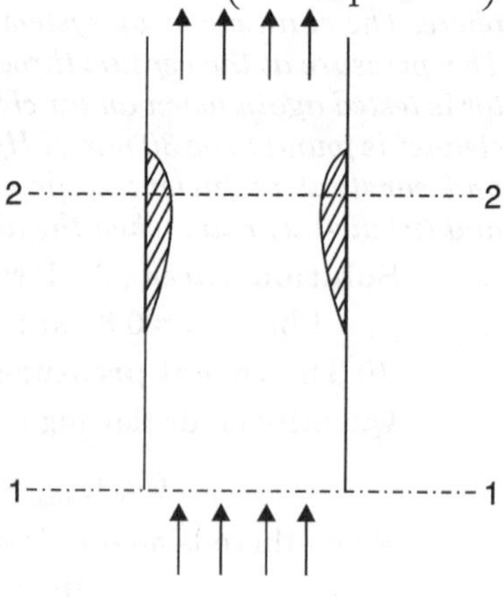

Fig. 3.28

$\therefore$ $$(C_2)_{\text{actual}} = C_v \sqrt{2c_p T_1\left[1 - \left(\frac{p_2}{p_1}\right)^{\frac{\gamma-1}{\gamma}}\right]}$$

$$c_p = R\left(\frac{\gamma}{\gamma - 1}\right) = 0.287 \times \frac{1.4}{4.1 - 1} = 1.005 \text{ kJ/kg K}$$

$$80 = 0.8\sqrt{2 \times 1.005 \times 1000 \times 298\left[1 - \left(\frac{p_2}{1.013}\right)^{\frac{1.4-1}{1.4}}\right]} = 619.15\sqrt{1 - \left(\frac{p_2}{1.013}\right)^{0.2857}}$$

or, $$1 - \left(\frac{p_2}{1.013}\right)^{0.2857} = \left(\frac{80}{619.15}\right)^2 = 0.01669$$

or, Throat pressure, $$p_2 = \left[(1 - 0.01669)^{\frac{1}{0.2857}}\right] \times 1.013 = 0.955 \text{ bar}$$

$$\rho_1 = \frac{p_1}{RT_1} = \frac{1.013 \times 10^5}{(0.287 \times 1000) \times 298} = 1.1844 \text{ kg/m}^3$$

Now, $pv^\gamma = \text{constant}$

or, $$\frac{p}{\rho^\gamma} = \text{constant}$$

or, $$\frac{p_1}{\rho_1^\gamma} = \frac{p_2}{\rho_2^\gamma}$$

or, $$\rho_2 = \rho_1\left(\frac{p_2}{p_1}\right)^{1/\gamma} = 1.1844\left(\frac{0.955}{1.013}\right)^{\frac{1}{1.4}} = 1.1356 \text{ kg/m}^3$$

We know that, $\dot{m}_a\ (= \rho AC) = \rho_2 A_2 C_2$

$$0.0767 = 1.1356 \times A_2 \times 80 \quad \text{(where } A_2 = \text{throat area)}$$

or, $$A_2 = 8.443 \times 10^{-4}\ \text{m}^2$$

or, $$A_2 = 8.443 \times 10^{-4} = \frac{\pi}{4} d_2^2$$

$\therefore$ *Throat diameter,* $$\mathbf{d_2} = \left(\frac{8.443 \times 10^{-4} \times 4}{\pi}\right)^{1/2} = 0.0328\ \text{m} \quad \text{or} \quad \mathbf{3.28\ cm.\ (Ans.)}$$

Example 3.7. *A simple jet carburettor is required to supply 6 kg of air per minute and 0.45 kg of fuel of density 740 kg/m³. The air is initially at 1.013 bar and 27°C.*

(i) Calculate the throat diameter of the choke for a flow velocity of 92 m/s. Velocity coefficient = 0.8.

(ii) If the pressure drop across the fuel metering orifice is 0.75 of that at the choke, calculate the orifice diameter assuming $C_d = 0.60$. **(Nagpur University)**

Solution. *Given :* $\dot{m}_a = \frac{6}{60} = 0.1$ kg/s ; $\dot{m}_f = \frac{0.45}{60} = 0.0075$ kg/s ; $\rho_f = 740$ kg/m³ ;

$p_1 = 1.013$ bar ; $T_1 = 27 + 273 = 300$ K ; $C_2 = 92$ m/s ; $C_{da} = 0.8$;

$C_{df} = 0.60$.

(i) **Throat diameter, D_2 :**

Velocity of air at venturi throat,

$$C_2 = C_{da}\sqrt{2c_pT_1\left[1 - \left(\frac{p_2}{p_1}\right)^{\frac{\gamma-1}{\gamma}}\right]}$$

$$92 = 0.8\sqrt{2 \times 1.005 \times 1000 \times 300\left[1 - \left(\frac{p_2}{p_1}\right)^{\frac{1.4-1}{1.4}}\right]}$$

or, $$1 - \left(\frac{p_2}{p_1}\right)^{\frac{0.4}{1.4}} = \left(\frac{92}{0.8}\right)^2 \times \frac{1}{2 \times 1.005 \times 1000 \times 300} = 0.021932$$

$\therefore$ $$\left(\frac{p_2}{p_1}\right)^{0.2857} = 0.97807\ , \quad \therefore \quad \frac{p_2}{p_1} = 0.925$$

or, $$p_2 = 1.013 \times 0.925 = 0.937\ \text{bar}$$

Now, $p_1 v_1^{\gamma} = p_2 v_2^{\gamma}$

$$v_2 = v_1\left(\frac{p_1}{p_2}\right)^{1/\gamma}$$

$$= \frac{RT_1}{p_1}\left(\frac{1}{0.925}\right)^{\frac{1}{1.4}}$$

$$= \frac{287 \times 300}{1.013 \times 10^5}\left(\frac{1}{0.925}\right)^{0.7143} = 0.898\ \text{m}^3/\text{kg}$$

Now, $\dot{m}_a = \dfrac{A_2 C_2}{v_2}$

$\therefore$ Throat area, $A_2 = \dfrac{\dot{m}_a \times v_2}{C_2} = \dfrac{0.1 \times 0.898}{92} = 9.76 \times 10^{-4}\ m^2 = 9.76\ cm^2$

But, $A_2 = \dfrac{\pi}{4} D_2^2 = 9.26$

$\therefore$ $\mathbf{D_2}$ (or D_t) $= \sqrt{\dfrac{9.76 \times 4}{\pi}}$ = **3.525 cm. (Ans.)**

(*ii*) **Orifice diameter, d_f :**

Pressure drop at venturi = 1.013 – 0.937 = 0.076 bar

Pressure drop at jet = 0.75 × 0.076 = 0.057 bar

Now, $\dot{m}_f = A_f C_{df} \sqrt{2\rho_f \Delta p}$

$$0.0075 = A_f \times 0.6\ \sqrt{2 \times 740 \times 0.057 \times 10^5} = 1742.68$$

$\therefore$ $A_f = 4.304 \times 10^{-6}\ m^2$ or $4.304\ mm^2$

But, $A_f = \dfrac{\pi}{4} d_f^2 = 4.304$

$\therefore$ $\mathbf{d_f} = \sqrt{\dfrac{4 \times 4.304}{\pi}}$ = **2.34 mm. (Ans.)**

Example 3.8. *The following data relate to a 4-stroke petrol engine of Hindustan Ambassador :*

Capacity of the petrol engine	*= 1489 c.c.*
Speed at which maximum power is developed	*= 4200 r.p.m.*
The volumetric efficiency (at the above speed)	*= 75 percent*
The air-fuel ratio	*= 13 : 1*
Theoretical air speed at choke (at peak power)	*= 85 m/s*
The co-efficient of discharge for venturi	*= 0.82*
The co-efficient of discharge of the main petrol jet	*= 0.65*
The specific gravity of petrol	*= 0.74*
Level of petrol surface below the choke	*= 6 mm*
Atmospheric pressure and temperature	*= 1.013 bar, 20°C respectively*

An allowance should be made for the emulsion tube, the diameter of which can be taken as 40 percent of the choke diameter.

Calculate the sizes of a suitable choke and main jet.

Solution. *Given* : V_s = 1489 c.c. = $1489 \times 10^{-6}\ m^3 = 0.001489\ m^3$; N = 4200 r.p.m ; $\eta_{vol.}$ = 75% ; A/F ratio = 13 : 1 ; $C_t\ (= C_2)$ = 85 m/s ; C_{da} = 0.82 ; C_{df} = 0.65 ; $\rho_f = 0.74 \times 1000 = 740\ kg/m^3$; $p_1\ (= p_a)$ = 1.013 bar ; $p_2\ (= p_t)$ = ? ; $T_1\ (= T_a)$ = 20 + 273 = 293 K ; z = 6 mm = 0.006 m ; $d = 0.4\ D$.

Volume of air induced = $\eta_{vol.} \times V_s$

$$= \frac{0.75 \times 0.001489 \times 4200}{2 \times 60} = 0.03909\ m^3/s$$

$\therefore$ Mass flow of air, $\dot{m}_a = \frac{p_1 v}{RT_1} = \frac{1.013 \times 10^5 \times 0.03909}{0.287 \times 10^3 \times 293} = 0.04709$ kg/s

For compressible flow, velocity at throat,

$$C_t = \sqrt{\left[2T_a c_p \left\{1 - \left(\frac{p_t}{p_a}\right)^{\frac{\gamma-1}{\gamma}}\right\}\right]} \quad \text{...[Eqn. (3.13)]}$$

$$85 = \sqrt{\left[2 \times 293 \times (1.005 \times 10^3) \left\{1 - \frac{p_t}{p_a}\right\}^{\frac{1.4-1}{1.4}}\right]}$$

$$85 = 767.4 \sqrt{\left\{1 - \left(\frac{p_t}{p_a}\right)^{0.2857}\right\}}$$

$$\therefore \quad \frac{p_t}{p_a} = \left[1 - \left(\frac{85}{767.4}\right)^2\right]^{\frac{1}{0.2857}} = 0.9576$$

or, $p_t = 1.013 \times 0.9576 = 0.97$ bar

Volume flow of air at choke,

$$v_t = 0.03909 \times \left(\frac{p_a}{p_t}\right)^{\frac{1}{\gamma}}$$

$$= 0.03909 \times \left(\frac{1.013}{0.97}\right)^{\frac{1}{1.4}} = 0.04032 \text{ m}^3\text{/s}$$

$$\therefore \quad A_t = \frac{v_t}{C_t \times C_{da}} = \frac{0.04032}{85 \times 0.82} = 0.0005785 \text{ m}^2 = 578.5 \text{ mm}^2$$

Now, $\frac{\pi}{4}(D^2 - d^2) = 578.5$

or, $\frac{\pi}{4}[D^2 - (0.4\,D)^2] = 578.5$

or, $\frac{\pi}{4} \times 0.84\, D^2 = 578.5$

or, Choke dia., **D = 29.61 mm. (Ans.)**

Mass flow of fuel, $\dot{m}_f = \frac{\dot{m}_a}{13} = \frac{0.04709}{13} = 0.003622$ kg/s

$$\dot{m}_f = C_{df} \, . \, A_j \sqrt{2\rho_f (\Delta p_a - gz\rho_f)} \quad \text{...[Eqn. (3.7)]}$$

$$0.003622 = 0.65 \times A_j \sqrt{2 \times 740 \,\{(1.013 - 0.97) \times 10^5 - 9.81 \times 0.006 \times 740\}}$$

$$= 1639.75\, A_j$$

$\therefore \quad A_j = 2.209 \times 10^{-6}\ \text{m}^2 \quad \text{or} \quad 2.209\ \text{mm}^2$

$$A_j = \frac{\pi}{4} D_{jet}^2 = 2.209$$

$\therefore$ $\mathbf{D_{jet} = 1.68\ mm.}$ **(Ans.)**

Example 3.9. *The following data refer to a simple carburettor :*

Throat diameter	*= 18 mm*
Diameter of fuel orifice	*= 1.2 mm*
Co-efficient of air flow	*= 0.82*
Co-efficient of fuel flow	*= 0.65*
Level of petrol surface below the throat	*= 6 mm*
Density of air	*= 1.2 kg/m³*
Density of fuel	*= 750 kg/m³.*

Calculate :

(*i*) *The A/F ratio for a pressure drop of 0.065 bar when the nozzle lip is neglected ;*

(*ii*) *The A/F ratio when the nozzle lip is taken into account ;*

(*iii*) *The minimum velocity of air or critical air velocity required to start the fuel flow when nozzle lip is provided.*

Solution. *Given* : $d_a = 18$ mm = 0.018 m ;

$d_f = 1.2$ mm = 0.0012 m ; $C_{da} = 0.82$; $C_{df} = 0.65$; $z = 6$ mm = 0.006 m ;

$\rho_a = 1.2$ kg/m³ ; $\rho_f = 750$ kg/m³

(*i*) **A / F ratio when the nozzle lip is neglected :**

Air flow, $\dot{m}_a = C_{da} A_a \sqrt{2\rho_a \Delta p_a}$ ...[Eqn. (3.4)]

Fuel flow, $\dot{m}_f = C_{df} A_f \sqrt{2\rho_f \Delta p_a}$

$$\therefore \quad \text{A / F ratio} = \frac{C_{da}}{C_{df}} \cdot \frac{A_a}{A_f} \sqrt{\frac{\rho_a}{\rho_f}}$$

$$= \frac{0.82}{0.65} \times \left(\frac{0.018}{0.0012}\right)^2 \sqrt{\frac{1.2}{750}} = \mathbf{11.35.} \quad \textbf{(Ans.)}$$

$$\left[\because \frac{A_a}{A_f} = \frac{\frac{\pi}{4} d_a^2}{\frac{\pi}{4} d_f^2} = \left(\frac{d_a}{d_f}\right)^2\right]$$

(*ii*) **A / F ratio when the nozzle lip is taken into account :**

The *air flow will remain same.* The *fuel flow* will become,

$$\dot{m}_f = C_{df} \times A_f \sqrt{2\rho_f (\Delta p_a - gz\rho_f)}$$ ...[Eqn. (3.7)]

$$\therefore \quad \text{A / F ratio} = \frac{C_{da}}{C_{df}} \cdot \frac{A_a}{A_f} \sqrt{\frac{\rho_a}{\rho_f}} \sqrt{\frac{\Delta p_a}{\Delta p_a - gz\rho_f}}$$

$$= \frac{0.82}{0.65} \times \left(\frac{0.018}{0.0012}\right)^2 \sqrt{\frac{1.2}{750}} \sqrt{\frac{0.065}{0.065 - (9.81 \times 0.006 \times 750/10^5)}}$$

$$= 11.35 \times \sqrt{\frac{0.065}{0.065 - 0.00044145}} = \mathbf{11.39.} \quad \textbf{(Ans.)}$$

(*iii*) **Minimum velocity of air, C_2 :**

The *flow of fuel when lip is provided will start only when the minimum velocity of air required to create requisite pressure difference for flow of fuel to overcome nozzle lip exists.*

∴ Pressure difference Δp_a must be *equal to* $gz\rho_f$. Assuming velocity at entrance of venturi, $C_1 = 0$,

$$\frac{p_1}{\rho_a} = \frac{p_2}{\rho_a} + \frac{C_2^2}{2}$$

or $$\frac{\Delta p_a}{\rho_a} = \frac{C_2^2}{2}$$

or $$\frac{C_2^2}{2} = \frac{gz\rho_f}{\rho_a}$$

or $$C_2 = \sqrt{\frac{2gz\rho_f}{\rho_a}} = \sqrt{\frac{2 \times 9.81 \times 0.006 \times 750}{1.2}} = \mathbf{8.58\ m/s.}\quad \textbf{(Ans.)}$$

Note. The **nozzle lip** is *the height of fuel nozzle in the throat above petrol surface in the carburettor. It is provided so that there is no spilling of fuel due to vibration or slight non-horizontal position of carburettor. This would avoid wastage of fuel and fire hazard.*

Example 3.10. *The following data refer to an eight-cylinder four-stroke petrol engine :*

Bore	= *110 mm*
Stroke	= *110 mm*
Composition of the fuel used	= *C = 84% ; H_2 = 16%*
Throat diameter of the choke tube	= *42 mm*
Volumetric efficiency at 300 r.p.m.	= *75% (referred to 0°C and 1.013 bar)*
The pressure depression	= *0.12 bar*
The temperature at the throat	= *15°C*
Characteristic gas constant : For air	= *287 J/kg K*
For fuel vapour	= *97 J/kg K*

If chemically correct air-fuel ratio is supplied for combustion, determine :

(*i*) *Fuel consumption in kg/h ;*

(*ii*) *The air velocity through the tube.*

Solution. *Given* : D = 110 mm = 0.11 m ; L = 110 mm = 0.11 m ; d_a = 42 mm = 0.042 m ; $\eta_{vol.}$ = 75%, N = 3000 r.p.m. ; R_a = 287 J / kg K ; R_v = 97 J/kg K ; Δp_a = 0.12 bar.

(*i*) **Fuel consumption in kg/h :**

The volume of mixture supplied at 0°C and 1.013 bar per minute

$$= \frac{\pi}{4} D^2 L \times 8 \times \frac{N}{2} \times \eta_{vol.}$$

$$= \frac{\pi}{4} \times (0.11)^2 \times 0.11 \times 8 \times \frac{3000}{2} \times 0.75 = 9.408\ \text{m}^3/\ \text{min.}$$

Also
$$\begin{array}{ccccc} C & + & O_2 & = & CO_2 \\ 12 & & 32 & & 44 \end{array}$$

and
$$\begin{array}{ccccc} 2H_2 & + & O_2 & = & 2H_2O \\ 4 & & 32 & & 36 \\ 1 & & 8 & & 9 \end{array}$$

Thus, air required for combustion of 1 kg of fuel

$$= \left(0.84 \times \frac{32}{12} + 0.16 \times 8\right) \times \frac{100}{23} = 15.3 \text{ kg}$$

Thus A / F ratio $= \frac{m_a}{m_f} = 15.3$

The volume of one kg of air at 0°C and 1.013 bar,

$$v_a = \frac{R_a T}{p} = \frac{287 \times 273}{1.013 \times 10^5} = 0.773 \text{ m}^3/\text{kg}$$

Similarly volume of 1 kg of fuel vapour at 0°C and 1.013 bar,

$$v_f = \frac{R_f T}{p} = \frac{97 \times 273}{1.013 \times 10^5} = 0.2614 \text{ m}^3/\text{kg}$$

Thus $m_a / \min \times v_a + m_f / \min \times v_f = 9.408$

$15.3\, m_f / \min \times v_a + m_f / \min \times v_f = 9.408$

Thus, $$m_f/\min = \frac{9.408}{15.3\, v_a + v_f} = \frac{9.408}{15.3 \times 0.773 + 0.2614} = 0.778 \text{ kg/min}$$

∴ Fuel consumption = 0.778 × 60 = **46.68 kg/h. (Ans.)**

(*ii*) **The air velocity through the tube, C_2 (= C_a)**

Density of air at the throat,

$$\rho_a = \frac{p_2}{R_a\, T_2}$$

$$= \frac{p_1 - \Delta p_a}{R_a\, T_2} \qquad (\because \; p_1 - p_2 = \Delta p_a)$$

$$= \frac{(1.013 - 0.12) \times 10^5}{287 \times (15 + 273)} = 1.08 \text{ kg/m}^3.$$

∴ Velocity at the throat in m/s,

$$\mathbf{C_a} = \frac{m_a}{A_a\, \rho_a} = \frac{15.3\, m_f}{\frac{\pi}{4} \times (0.42)^2 \times 1.08}$$

$$= \frac{15.3 \times (0.778 / 60)}{\frac{\pi}{4} \times (0.042)^2 \times 1.08} = \mathbf{132.59 \text{ m/s.}} \quad \textbf{(Ans.)}$$

Example 3.11. *Determine the air-fuel ratio supplied at 4500 m altitude by a carburettor which is adjusted to give an air-fuel ratio of 14 : 1 at sea level where air temperature is 25°C and pressure 1.013 bar.*

The temperature of air decreases with altitude as given by the expression,

$$t = t_s - 0.0064\, h$$

where h is the height in metres and t_s is sea level temperature in °C.

The pressure of air decreases with altitude as per relation :

$$h = 19300 \log_{10} \left(\frac{1.013}{p}\right)$$

where p is expressed in bar at altitude.

Solution.
$$t = t_s - 0.0064\,h$$
$$= 25 - 0.0064 \times 4500 = -3.8°\text{C}$$

Now,
$$h = 19300 \log_{10}\left(\frac{1.013}{p}\right)$$
$$4500 = 19300 \log_{10}\left(\frac{1.013}{p}\right)$$
$$\therefore \quad \log_{10}\left(\frac{1.013}{p}\right) = \frac{4500}{19300} = 0.2332$$

or,
$$\frac{1.013}{p} = 1.711$$

or,
$$p = \frac{1.013}{1.711} = 0.592 \text{ bar}$$

Now,
$$\frac{\text{A / F ratio at altitude}}{\text{A / F ratio at sea level}} = \sqrt{\frac{\rho_{\text{altitude}}}{\rho_{\text{sea level}}}}$$
$$\therefore \quad \textbf{A/F ratio at altitude} = 14\sqrt{\frac{p_{\text{alt.}}/RT_{\text{alt.}}}{p_{\text{sea}}/RT_{\text{sea}}}}$$
$$= 14\sqrt{\frac{p_{\text{alt.}} \times T_{\text{sea}}}{p_{\text{sea}} \times T_{\text{alt.}}}}$$
$$= 14\sqrt{\frac{0.592 \times (25 + 273)}{1.013 \times (-3.8 + 273)}}$$
$$= \mathbf{11.26.} \quad \textbf{(Ans.)}$$

II. C.I. Engines

3.13. INTRODUCTION TO FUEL INJECTION SYSTEMS FOR C.I. ENGINES

— In C.I. engines, the air is taken in during the suction stroke and compressed to a high pressure (28 to 70 bar) and high temperature (520° to 720°C) according to the compression ratio used (12 : 1 to 20 : 1). The high temperature of air at the end of stroke is sufficient to ignite the fuel.

— Fuel is injected into the cylinder at the end of the compression stroke ; the pressure of fuel injected lies between 100 to 200 bar. During the process of injection the fuel is broken into *very fine droplets*. The droplets vaporise taking the heat from the hot air and form a combustible mixture and start burning. As the burning starts, the vaporisation of fuel is accelerated as more heat is available. As the combustion progresses, the amount of oxygen available for burning reduces and therefore heat release is reduced.

— The period between the start of injection and start of ignition, called the *ignition delay*, is about 0.001 second for high speed engines and 0.002 second for low speed engines. The injection period covers about 25° of crank rotation. After the ignition the temperature and pressure rise rapidly. The *whole performance of engine is totally dependent on the delay period ; the lesser the delay period better is the engine performance.*

3.14. FUNCTIONAL REQUIREMENTS OF AN INJECTION SYSTEM

The *functional requirements of an injection system* are listed below :

1. Introduction of the fuel into the combustion chamber should take place within a precisely defined period of the cycle.
2. The metering of the amount of fuel injected per cycle should done very accurately.
3. The quantities of fuel metered should vary to meet the changing load and speed requirements.
4. The injection rate should be such that it results in the desired heat release pattern.
5. The injected fuel must be broken into very fine droplets.
6. The pattern of spray should be such as to ensure rapid mixing of fuel and air.
7. The beginning and end of the injection should be sharp.
8. The timing of injection, if desired, should change as per the requirements of load and speed.
9. The distribution of the metered fuel, in the case of multi-cylinder engines, should be uniform among various cylinders.
10. Besides above requirements, the weight and the size of the fuel injection system must be minimum. It should be cheaper to manufacture and least expensive to attend to ; adjust or repair.

For accomplishing these requirements the following **functional elements** are required in a fuel injection system :

1. *Pumping elements.* To move the fuel from the fuel tank to cylinder and piping etc.
2. *Metering elements.* To measure and supply the fuel at the rate demanded by the load and speed.
3. *Metering controls.* To adjust the rate of metering elements for changes in load and speed of the engine.
4. *Distributing elements.* To divide the metered fuel equally among the cylinders.
5. *Timing controls.* To adjust the start and the stop of injection.
6. *Mixing elements.* To atomise and distribute the fuel within the combustion chamber.

3.15. FUNCTIONS OF A FUEL INJECTION SYSTEM

The *main functions of a fuel injection system are :*

1. Filter the fuel.
2. Metre or measure the correct quantity of fuel to be injected.
3. Time the fuel injection.
4. Control the rate of fuel injection.
5. Automise or break up the fuel to fine particles.
6. Properly distribute the fuel in the combustion chamber.

- The injection systems are manufactured with *great accuracy,* especially the parts that actually meter and inject the fuel. Some of the tolerances between the moving parts are very small of the order of 1 micron. Such closely fitting parts require special attention during manufacture and hence the injection systems are costly.

3.16. FUEL INJECTION SYSTEMS

In compression ignition engines (diesel and semi-diesel) two methods of fuel injection are used. These are :

1. Air injection
2. Solid or airless injection.

3.16.1. Air Injection

In this method of fuel injection air is compressed in the compressor to a very high pressure (much higher than developed in the engine cylinder at the end of the compression stroke) and then injected through the fuel nozzle into the engine cylinder. *The rate of fuel admission can be controlled by varying the pressure of injection air.* Storage air bottles which are kept charged by an air compressor (driven by the engine) supply the high pressure air.

Advantages :

(*i*) It provides better automisation and distribution of fuel.

(*ii*) As the combustion is more complete, the b.m.e.p. is higher than with other types of injection systems.

(*iii*) Inferior fuels can be used.

Disadvantages :

This method is *not-used* now-a-days due to the following *reasons/disadvantages* :

(*i*) It requires a high pressure multi-stage compression. The large number of parts, the intercooler etc. make the system complicated and expensive.

(*ii*) A separate mechanical linkage is required to time the operation of fuel valve.

(*iii*) Due to the compression and the linkage the bulk of the engine increases. This also results in reduced B.P. due to power loss in operating the compression and linkage.

(*iv*) The fuel in the combustion chamber burns very near to injection nozzle which many times leads to overheating and burning of valve and its seat.

(*v*) The fuel valve sealing requires considerable skill.

(*vi*) In case of sticking of fuel valve the system becomes quite dangerous due to the presence of high pressure air.

3.16.2. Solid or Airless Injection

Injection of fuel directly into the combustion chamber without primary automisation is termed as **solid injection.** It is also termed as *mechanical injection.*

Main Components :

The *main components* of a fuel injection system are :

(*i*) *Fuel tank.*

(*ii*) *Fuel feed pump* to supply the fuel from the main fuel tank to the injection pump.

(*iii*) *Fuel filters* to prevent dust and abrasive particles from entering the pump and injectors.

(*iv*) *Injection pump* to meter and pressurise the fuel for injection.

(*v*) *Governor* to ensure that the amount of fuel is in accordance with variation in load.

(*vi*) *Fuel pipings and injectors* to take the fuel from the pump and distribute it in the combustion chamber by atomising it in fine droplets.

Main types of modern fuel injection systems :

1. Common-rail injection system.
2. Individual pump injection system.
3. Distributor system.

Atomisation of fuel oil has been secured by (*i*) *air blast* and (*ii*) *pressure spray*. Early diesel engines used air fuel injection at about 70 bar. This is sufficient not only to inject the oil, but also to atomise it for a rapid and thorough combustion. The expense of providing an air compressor and tank lead to the development of "*solid*" injection, using a liquid pressure of between 100 and 200 bar which is sufficiently high to atomise the oil it forces through spray nozzles. Great advances have been made in the field of solid injection of the fuel through research and progress in fuel pump, spray nozzles, and combustion chamber design.

1. **Common-rail injection system :**

Two types of common-rail injection systems are shown in Fig. 3.29 and 3.30 respectively.

- Refer Fig. 3.29. A single pump supplies high-pressure fuel to header, a relief valve holds pressure constant. The control wedge adjusts the lift of mechanical operated valve to set amount and time of injection.
- Refer Fig. 3.30. Controlled-pressure system has pump which maintains set head pressure. Pressure relief and timing valves regulate injection time and amount. Spring loaded spray valve acts merely as a check.

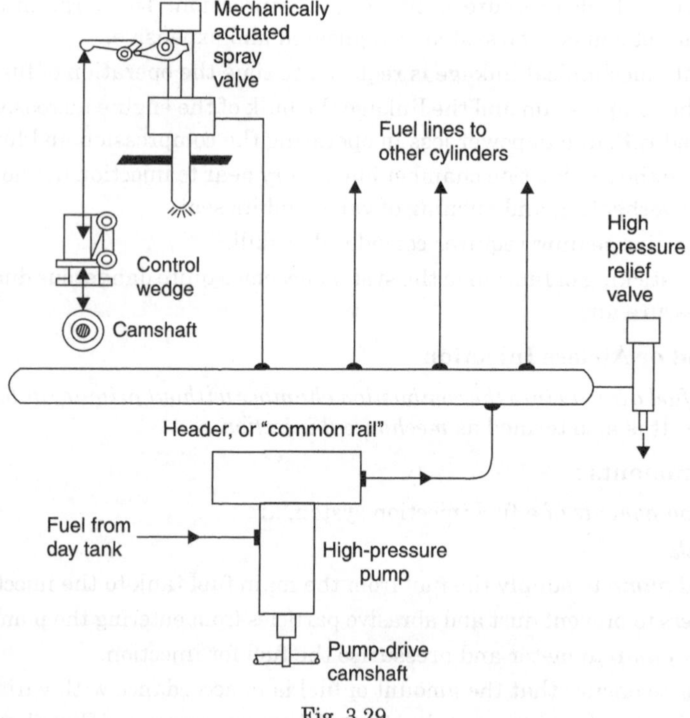

Fig. 3.29

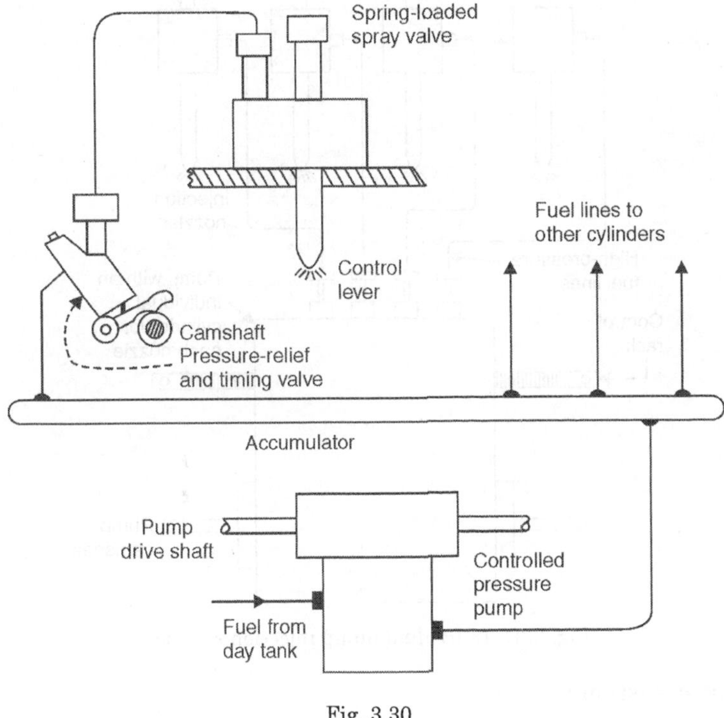

Fig. 3.30

Advantages :

(*i*) The system arrangement is simple and less maintenance cost.

(*ii*) Only one pump is sufficient for multi-cylinder engine.

(*iii*) It fulfills the requirements of either the constant load with variable speed or constant speed with variable load.

(*iv*) Variation in pump supply pressure will affect the cylinders uniformly.

Disadvantages :

(*i*) There is a tendency to develop leaks in the injection valve.

(*ii*) Very accurate design and workmanship are required.

2. Individual pump injection system :

- Refer Fig. 3.31. In this system an individual pump or pump cylinder connects directly to each fuel nozzle. Pump meters charge and control injection timing. Nozzles contain a delivery valve actuated by the fuel-oil pressure.
- The design of this type of pump must be very accurate and precise as the volume of fuel injected per cycle is 1/20,000 of the engine displacement at full load and 1/100,000 of the engine displacement during idling. The time allowed for injecting such a small quantity of fuel is very limited (about 1/450 second at 1500 r.p.m. of the engine providing injection through 20° crank angle). The pressure requirements vary from 100 to 300 bar.

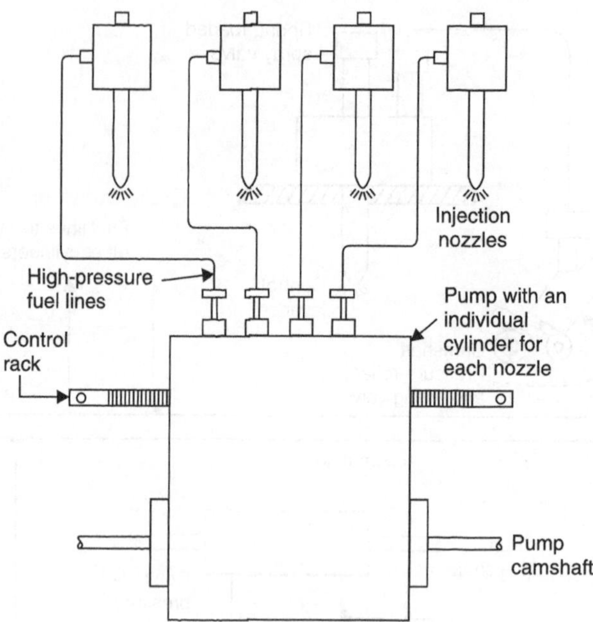

Fig. 3.31. Individual pump injection system.

3. Distributor system :

Refer Fig. 3.32. In this system, the fuel is metered at a central point ; a pump pressurises, meters the fuel and times the injection. From here, the fuel is distributed to cylinders in correct firing order by cam operated poppet valves which open to admit fuel to the nozzles.

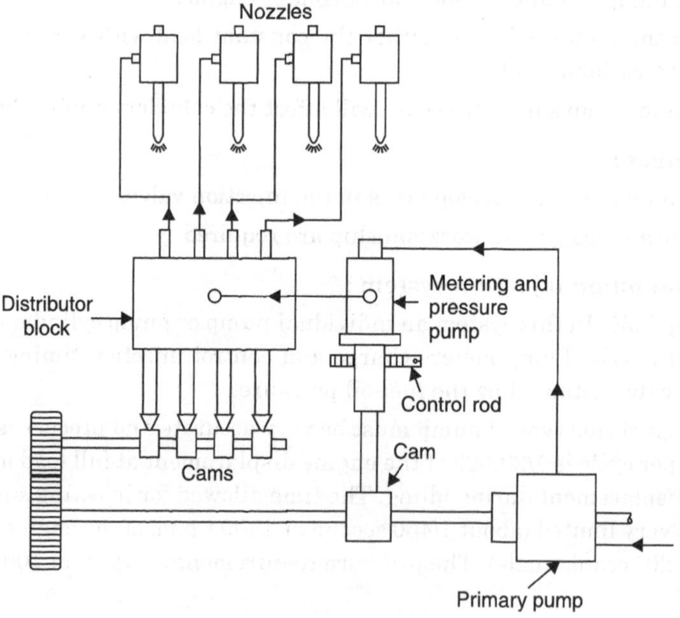

Fig. 3.32. Distributor system.

3.17. FUEL PUMP AND FUEL INJECTOR (ATOMISER)

3.17.1. Fuel Pump

A large number of igneous fuel pump designs have been developed by the manufacturers. Only one type fuel pump will be discussed here.

Bosch fuel injection pump : Refer Fig. 3.33.

- L is the **plunger** which is driven by a cam and tappet mechanism at the bottom (not shown), *B* is the **barrel** in which the plunger reciprocates. There is a rectangular vertical groove in the plunger which extends from top to another helical groove. *V* is the **delivery valve** which lifts off its seat under the liquid fuel pressure and the spring force(s). *The fuel pump is connected to fuel atomiser through the passage P. SP* and *Y* are the **spill** and **supply ports** respectively.

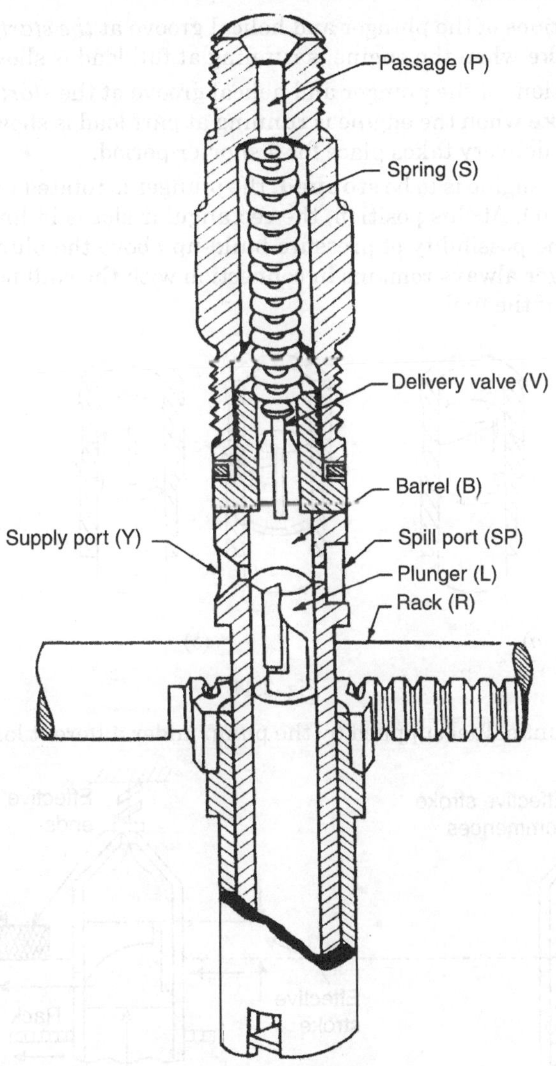

Fig. 3.33. Fuel pump.

- When the plunger is at its bottom stroke the ports *SP* and *Y* are uncovered (as shown in Fig. 3.33) oil from low pressure pump (not shown) after being filtered is forced into the barrel. When the plunger moves up due to cam and tappet mechanism, a stage reaches when both the ports *SP* and *Y* are closed and with the further upward movement of the plunger the fuel gets compressed. The high pressure thus developed lifts the delivery valve off its seats and fuel flows to atomiser through the passage *P*. With further rise of the plunger, at a certain moment, the port *SP* is connected to the fuel in the upper part of the plunger through the rectangular vertical groove by the helical groove, as a result of which a sudden drop in pressure occurs and the delivery valve falls back and occupies its seat against the spring force. The plunger is rotated by the rack *R* which is moved in or out by the governor. By changing the angular position of the helical groove (by rotating the plunger) of the plunger relative to the supply port, the length of stroke during which the oil is delivered can be varied and thereby quantity of fuel delivered to the engine is also varied accordingly.
    - — The positions of the plunger and helical groove at *the starting and end of the* **delivery stroke** when the engine is running at full load is shown in Fig. 3.34 (*a*).
    - — The positions of the plunger and helical groove at the *starting and end of the* **delivery stroke** when the engine is running at part load is shown in Fig. 3.34 (*b*). In this case, the delivery takes place for a shorter period.
    - — When the engine is to be **stopped,** the plunger is rotated to the position as shown in Fig. 3.34 (*c*). At this position, the rectangular slot is in line with the spill port and there is no possibility of pressure build-up above the plunger as the upper part of the plunger always remains in connection with the spill port. Therefore there is no delivery of the fuel.

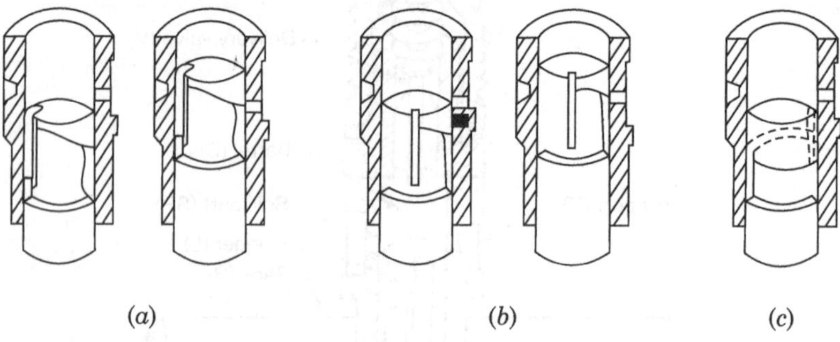

(*a*) (*b*) (*c*)

Fig. 3.34

- — The amount of fuel supplied by the pump under different loads is shown in Fig. 3.35.

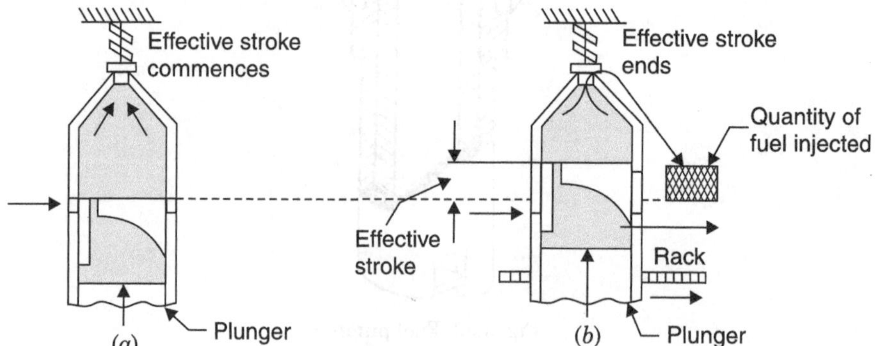

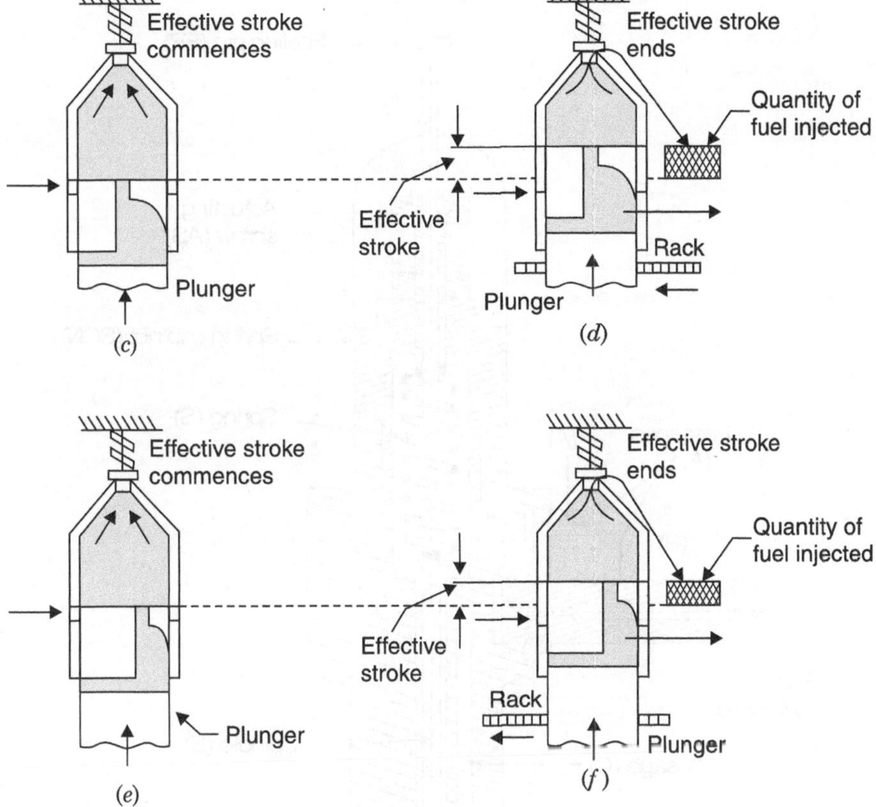

Fig. 3.35. Principle of helix bypass pump.

3.17.2. Fuel Atomiser or Injector : Refer Fig. 3.36.

- It consists of a nozzle valve (*NV*) fitted in the nozzle body (*NB*). The nozzle valve is held on its seat by a spring '*S*' which exerts pressure through the spindle *E*. '*AS*' is the adjusting screw by which the nozzle valve lift can be adjusted. Usually the nozzle valve is set to lift at 135 to 170 bar pressure. *FP* is the feeling pin which indicates whether valve is working properly or not.
- The fuel under pressure from the fuel pump enters the injector through the passages *B* and *C* and lifts the nozzle valve. The fuel travels down nozzle *N* and injected into the engine cylinder in the form of fine spray. Then the pressure of the oil falls, the nozzle valve occupies its seat under the spring force and fuel supply is cut-off. Any leakage of fuel accumulated above the valve is led to the fuel tank through the passage *A*. The leakage occurs when the nozzle valve is *worn out*.

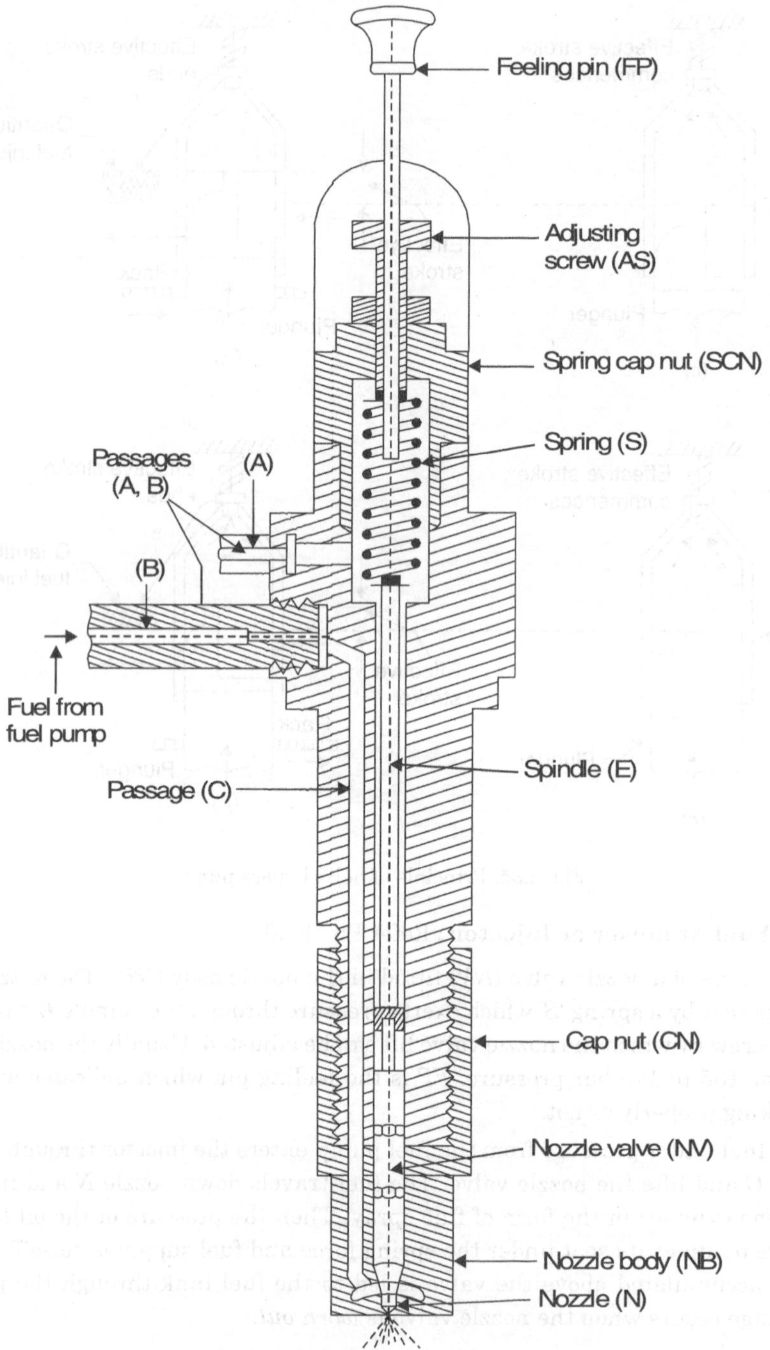

Fig. 3.36. Fuel atomiser or injector.

3.17.3. Faults, causes and remedies of injectors :

S. No.	Faults	Causes	Remedies
1.	*Pressure too high*	(*i*) Spring rate too high.	(*i*) Adjust if possible or replace spring with one having lower spring rate, or vary the number of shims.
		(*ii*) Needle valve seized in almost closed position due to dirt etc.	(*ii*) Soak in fuel oil. Then try to remove needle valve. If necessary replace with new needle valve and nozzle.
		(*iii*) Carbon blocking injector holes.	(*iii*) Clean with pricker or proke. Reverse flush using syringe and paraffin.
2.	*Pressure too low*	(*i*) Spring rate too low.	(*i*) Adjust if possible or replace spring with one having a higher spring rate.
		(*ii*) Spring broken	(*ii*) Replace with correct type.
		(*iii*) Needle valve seized in position.	(*iii*) Soak in fuel oil then try to remove the valve. If necessary replace with new needle nozzle.
3.	*Spray distorted*	(*i*) Damaged needle	(*i*) Replace.
		(*ii*) Nozzle or holes carboned altering shape of orifice.	(*ii*) Clean using pricker or proke. Reverse flush using syringe and paraffin.
4.	*Dribble from injector holes excessive*	Valve not seating correctly due to	
		(*i*) Dirt or carbon between needle seat and nozzle.	(*i*) Clean with paraffin ; do not wipe with fluffy cloth, but assemble wet from clean fuel. Hard carbon may be removed by using a solution of 60 grams caustic soda in 0.6 litres of water and add 15 grams of detergent. Boil for 1 to $1\frac{1}{2}$ hours.
		(*ii*) Scoring or pitting on needle seat or nozzle seat :	(*ii*) If possible lap needle and nozzle seat together.
		(*iii*) Needle corroded, or binding nozzle due probably to faulty spindle or misalignment of nozzle and nozzle holder.	(*iii*) Replace as necessary.
5.	*Excessive leakage off*	(*i*) Too much clearance between valve and hole in nozzle.	(*i*) Fit new nozzle and valve.
		(*ii*) Nozzle cap loose.	(*ii*) Tighten.
		(*iii*) Injector body and nozzle separated by dirt or burns caused by lack of care in handling.	(*iii*) Remove nozzle and examine pressure faces. Remove dirt or burns. Do not overtighten cap nut as a cure.

3.18. TYPES OF NOZZLES AND FUEL SPRAY PATTERNS

3.18.1. Main Requirements of an Injector Nozzle

The *main requirements of an injector nozzle are :*

1. To inject fuel at a sufficiently high pressure so that the fuel enters the cylinder with a *high velocity. Higher the velocity of the fuel smaller will be the droplet size. The momentum of smaller droplets is less, hence, penetration is also.*
2. Penetration should *not* be high so as to impinge on cylinder walls ; this may *result in poor starting.*
3. *Fuel supply and cut-off should be rapid ;* there should be *no dribbling.*

3.18.2. Classification and Description of Nozzles

The type of nozzle used is greatly dependent on the type of combustion chamber as open type or pre-combustion chamber. The nozzles are classified as per the type of orifice and its number used for injecting the fuel in the combustion chamber.

The nozzle are classified as :

1. Single hole nozzle.
2. Multi-hole nozzle
3. Circumferential nozzle
4. Pintle nozzle
5. Pintaux nozzle.

1. **Single hole nozzle.** Refer Fig. 3.37.

- This is the simplest type of nozzle and is *used in open combustion chambers.*
- It consists of a single hole bored centrally through the nozzle body and closed by the needle valve. The size of the hole is usually larger than 0.2 mm.
- Its spray cone angle varies from 5° to 15°. In some cases, a core is given a series of spiral grooves in order to impart a rotational motion to the fuel for bettering with air.

Fig. 3.37. Single hole nozzle.

Advantages :

Simple in construction and operation.

Disadvantages :

(*i*) *Very high injection pressure is required* because whole of the fuel passes through a single hole and, also, because the relative fuel velocity required is high.

(*ii*) This type of nozzle has a *tendency to dribble.*

(*iii*) As the spray angle is very narrow (usually about 15°), this does not facilitate good mixing unless higher air velocities are provided.

2. **Multi-hole nozzle.** Refer Fig. 3.38.

- This type of nozzle finds extensive use in automobile engines, particularly having open combustion chambers.

  It mixes the fuel with air properly even with slow air movement available with open combustion chambers.
- The number of holes varies from 4 to 18 ; the greater number provides better fuel distribution. The hole diameter lies between 0.25 to 0.35 mm and hole angle lies between 20° to 45°.

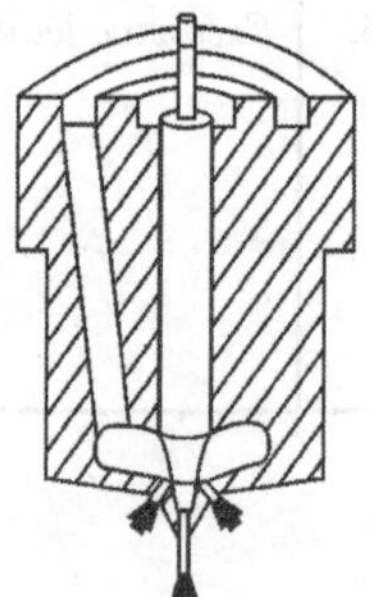

Fig. 3.38. Multi-hole nozzle.

- Usually the holes are drilled symmetrically but many times they are non-symmetrical to meet certain specific requirements of the combustion chamber.

Advantages :

(*i*) Gives *good automisation.*

(*ii*) *Distributes fuel properly* even with lower air motion available in open combustion chambers.

Disadvantages :

(*i*) Holes are small and *liable to clogging.*

(*ii*) *Dribbling* between injections.

(*iii*) *Very high injection pressures* (180 bar and above).

(*iv*) Close tolerance in manufacture (due to small holes) and hence *costly*).

3. **Circumferential nozzle.** Refer Fig. 3.39.

- Its spray characteristics are similar to a plate type opening.
- The injected fuel particles tend to be projected in the form of plane, with wide angle cone ; the purpose of which is to obtain as large an area of fuel spray as possible to come into contact with the air in the combustion chamber.

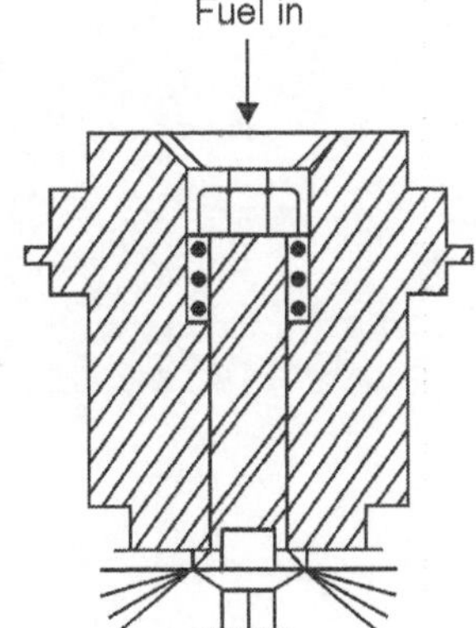

Fig. 3.39. Circumferential orifice.

4. **Pintle nozzle :**

Fig. 3.40 shows a Pintle nozzle.

- The stem of the nozzle valve is extended to form a pin or pintle which *protrudes through the mouth of the nozzle body*. It may be either cylindrical or conical in shape.
  - — The size and shape of the pintle can be varied according to requirement. The spray core angle is generally 60°.
- When the valve lifts, the pintle partially blocks the orifice and thus does not allow the pressure drop to be greater. As the lift of the valve increases the entire orifice is uncovered and full area for flow is available. Thus *dribbling is avoided.*
  - — *The spray obtained by the pintle nozzle is hollow conical spray.*

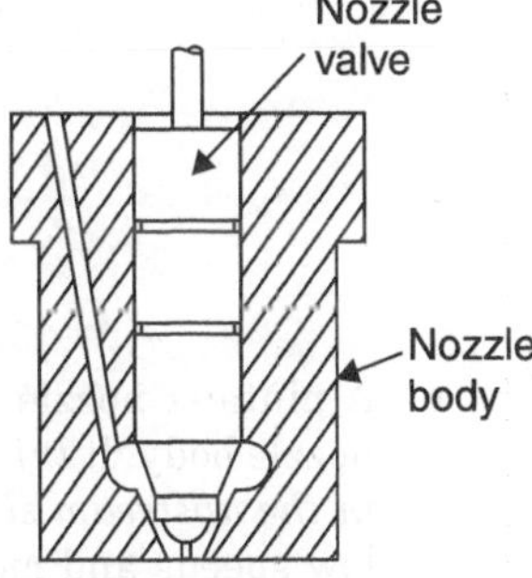

Fig. 3.40. Pintle nozzle.

Advantages :

(*i*) It is self cleaning type and prevents the carbon deposition on the nozzle hole.

(*ii*) It avoids weak injection and dribbling.

(*iii*) It results in good atomisation.

(*iv*) Its injection characteristics are more near the required one.

Disadvantage :

Distribution and penetration poor, hence not suitable for open combustion chambers.

5. **Pintaux nozzle :**

- In case the fuel is injected in a direction *upstream the direction of air*, the delay period is reduced due to increased heat transfer between air and fuel. This results in good *cold starting performance.* However, if whole of the fuel is injected in this manner the efficiency of combustion is greatly reduced by flow of products of combustion back into the injection path. Thus in order to improve cold starting performance without any detrimental effect on efficiency a *pintaux nozzle* as shown in Fig. 3.41 is used.

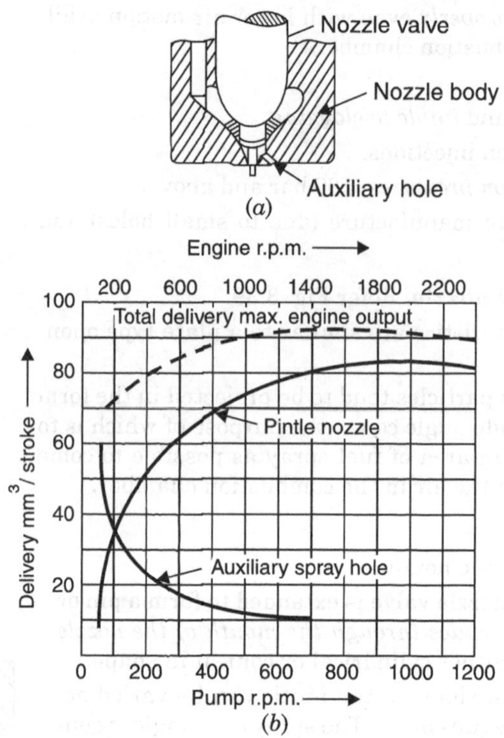

Fig. 3.41. Pintaux nozzle.

- A pintaux nozzle is a type of pintle nozzle which has an *auxiliary hole* drilled in the nozzle body. It *injects a small amount of fuel through this additional* hole (pilot injection) *in the upstream slightly before the main injection.* The needle valve does not lift fully at low speeds and most of the fuel is injected through the auxiliary hole, giving good cold starting performance.

Disadvantages / Drawbacks :

(*i*) The tendency of the auxiliary hole to choke.

(*ii*) The injections characteristics are even poorer than multi-hole nozzle.

Injection rate characteristics :

- *In order to avoid knocking in the engine it is always desirable to supply less quantity of fuel.*
- The characteristics of multi-hole, pintle and pintaux nozzles are shown respectively in Fig. 3.42 (*i*), (*ii*), (*iii*).
  — It is obvious from the figure the pintle nozzle gives desired characteristics as ($dm_f/d\theta$) which is smaller at the beginning compared with multi-hole nozzle. The characteristics of pintaux nozzle is totally different because of fuel supply through auxiliary and main orifices.

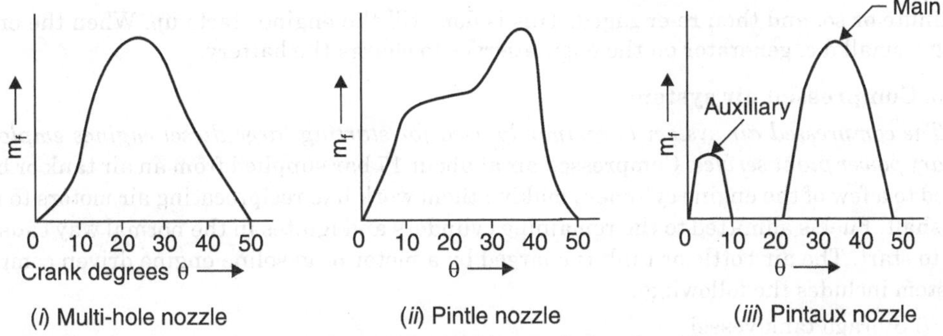

Fig. 3.42. Injection rate characteristics of nozzles.

Fuel spray pattern :

- For better and quick evaporation of fuel which is essential for better burning of fuel, the fuel pattern and relative direction of fuel particles with air are of significant importance.
- To give an idea, how the patterns help for better evaporation, two patterns are shown in Fig. 3.43.

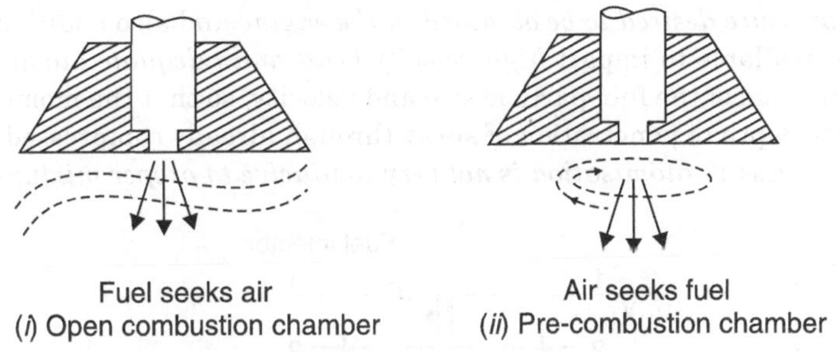

Fig. 3.43. Air and fuel movement for two types of combustion chambers.

3.19. ENGINE STARTING SYSTEMS

The following three are the commonly used starting systems in large and medium size engines :

1. Starting by an auxiliary engine
2. Use of electric motors or self starters
3. Compressed air system.

1. **Starting by an auxiliary engine (generally petrol driven) :**

In this system an auxiliary engine is mounted closed to the main engine and drives the latter through a clutch and gears. The clutch is first disengaged and the auxiliary engine started by hand or by a self starter motor. When it has warmed up and runs normally the drive gear is engaged through the clutch, and the main engine is cranked for starting. To avoid the danger of damage to drive gear it is desirable to have an over-running clutch or starter type drive.

2. **Use of electric motors or self starters :**

These are employed for small diesel and gasoline engines. A storage battery of 12 to 36 volts is used to supply power to an electric motor which is geared to the flywheel with arrangement for automatic disengagement after the engine has started. The motor draws a heavy current and is designed to be engaged continuously for about 30 seconds only, after which it is required to cool off, for a minute or so, and then re-engaged. This is done till the engine starts up. When the engine is running a small d.c. generator on the engine serves to charge the battery.

3. **Compressed air system :**

The compressed air system is commonly used for starting large diesel engines employed for stationary power plant service. Compressed air at about 17 bar supplied from an air tank or bottle is admitted to a few of the engine cylinders making them work like reciprocating air motors to run the engine shaft. Fuel is admitted to the remaining cylinders and ignites in the normal way causing the engine to start. The air bottle or tank is charged by a motor or gasoline engine driven compressor. The system includes the following :

(*i*) Storage tank/vessel
(*ii*) A safety valve
(*iii*) Interconnecting pipe work.

- *Small stationary engines of about 10 kW capacity are started by* ***hand cranking.***

3.20. FUEL INJECTION COMPUTATION IN C.I. ENGINES

- Refer Fig. 3.44. The fuel injector should develop a pressure which is *higher than the highest pressure desired to be obtained in the engine* and also additional pressure differential is available to impart *high velocity head and adequate atomisation.* There is an optimum value of the fuel particle size and velocity, so that the momentum is maximum and hence depth of penetration of spray through already compressed air in the cylinder is large. *Excessive atomisation is not very conducive to proper mixture formation.*

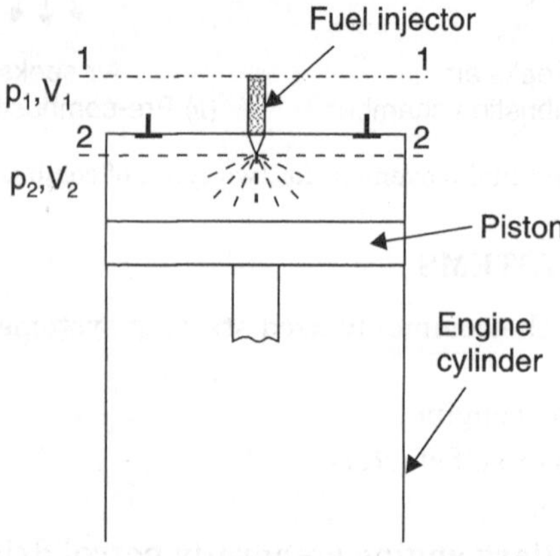

Fig. 3.44

Let p_1 = Injection pressure,
V_1 = Velocity at section 1-1,
p_2 = Pressure in the cylinder when injection of fuel takes place,
V_2 = Velocity at section 2-2,
ρ_f = Density of fuel, and
v_f = Specific volume of fuel (assumed in compressible).

Then, $$\frac{V_1^2}{2} + p_1 v_f = \frac{V_2^2}{2} + p_2 v_f \quad ...(3.20)$$

Neglecting V_1, being very small compared to V_2, we have

$$V_2 = \sqrt{2v_f\,(p_1 - p_2)} \quad ...(3.21)$$

$$= \sqrt{\frac{2\,(p_1 - p_2)}{\rho_f}} \quad ...(3.22)$$

or, if V_f = Actual fuel velocity of injection (the velocity of fuel for good atomisation is of the order of 400 m/s), and

C_f = Flow coefficient of orifice,

Then,
$$V_f = C_f\sqrt{\frac{2\,(p_1 - p_2)}{\rho_f}} \quad ...(3.23)$$

The volume of the fuel injected per second Q_f is given by :

Q_f = Area of all orifices × fuel jet velocity × time of injection × number of injections per second for one orifice

or,
$$Q_f = \left[\frac{\pi}{4}d_0^{\,2} \times n_0\right] \times V_f \times \left[\frac{\theta}{360} \times \frac{60}{N}\right] \times \frac{N_i}{60} \quad ...(3.24)$$

where, d_0 = Diameter of fuel orifice, m^2,

n_0 = Number of orifices,

V_f = Velocity of flow of fuel through orifice,

θ = Duration of the injection in degrees of crank angle,

N = r.p.m., and

N_i = Number of injection per min.

$$= \frac{\text{r.p.m.}}{2} \quad \text{... for 4-stroke cycle engine}$$

$$= \text{r.p.m.} \quad \text{for 2-stroke cycle engine}$$

The rate of fuel injection is usually expressed in mm³/degree crank angle/litre cylinder volume to avoid the effect of engine size.

- Since the injection pressures employed are very high (to the tune of 100 to 150 bar), the assumption of incompressibility of fuel may lead to errors. In order to account for compressibility, a factor called ***"Coefficient of Compressibility"*** is introduced.

Coefficient of compressibility,

$$C_c = \frac{[v_{f(\text{sump})} - v_{f_1}]}{v_{f(\text{sump})}\,(p_1 - p_{\text{sump}})} \quad ...(3.25)$$

For pressures expressed in bar, accepted valve for $C_c \simeq 80 \times 10^{-8}$ per bar

Work of fuel compression per kg,

$$W_c = \frac{1}{2}\,(p_1 - p_{\text{sump}}) \times [v_{f(\text{sump})} - v_{f_1}] \quad ...(3.26)$$

Work of delivering fuel,

$$W_d = (p_1 - p_2) \times \text{volume of fuel injected} \quad ...(3.27)$$

Then, fuel pump work per kg,

$$W_p = W_c + W_d \quad ...(3.28)$$

WORKED EXAMPLES

Example 3.12. *A six-cylinder, four-stroke diesel engine develops 125 kW at 3000 r.p.m. Its brake specific fuel consumption is 200 g/kWh. Calculate the quantity of fuel to be injected per cycle per cylinder. Specific gravity of the fuel may be taken as 0.85.*

Solution. *Given :* $n = 6$; B.P. = 125 kW ; $N = 3000$ r.p.m. ; b.s.f.c. = 2.00 g/kWh ;

Sp. gr. of fuel = 0.85.

Fuel consumption per hour = b.s.f.c. × B.P.

$$= \frac{200}{1000} \times 125 = 25 \text{ kg}$$

∴ Fuel consumption per cylinder $= \frac{25}{n} = \frac{25}{6} = 4.167$ kg/h

$$\text{Fuel consumption per cycle} = \frac{\text{Fuel consumption per cylinder per min.}}{\text{No. of cycles per min.}}$$

$$= \frac{(4.167/60)}{(3000/2)} = 4.63 \times 10^{-5} \text{ kg} = 0.0463 \text{ g}$$

∴ Volume of fuel injected per cycle

$$\therefore \quad = \frac{\text{Fuel consumption per cycle}}{\text{Specific gravity of fuel}}$$

$$= \frac{0.0463}{0.85} = \mathbf{0.05447 \text{ c.c.}} \quad \textbf{(Ans.)}$$

Example 3.13. *A 6-cylinder 4-stroke C.I. engine develops 220 kW at 1500 r.pm. with brake specific fuel consumption of 0.273 kg/kWh. Determine the size of the single hole injector nozzle if the injection pressure is 160 bar and the pressure in the combustion chamber is 40 bar. The period of injection is 30° of crank angle. Specific gravity of fuel = 0.85 and orifice discharge coefficient = 0.9.*

Solution. *Given :* $n = n_0 = 6$; $N = 1500$ r.p.m. ; B.P. = 220 kW, b.s.f.c. = 0.273 kg/kWh

$\theta = 30°$, Sp. gr. of oil = 0.85, $C_f = 0.9$, $\Delta p = p_1 - p_2 = 160 - 40 = 120$ bar.

Diameter of the nozzle orifice, d_0 :

We know that, *actual* fuel velocity of injection,

$$V_f = C_f \sqrt{\frac{2(p_1 - p_2)}{\rho_f}} = C_f \sqrt{\frac{2\Delta p}{\rho_f}} \quad \text{...(Eqn. 3.23)}$$

$$= 0.9 \times \sqrt{\frac{2 \times 120 \times 10^5}{(0.85 \times 1000)}} = 151.23 \text{ m/s}$$

Volume of fuel injected per second,

$$Q_f = \frac{0.273 \times 220}{(0.85 \times 1000) \times 3600} = 1.963 \times 10^{-5} \text{ m}^3/\text{s}$$

Also, volume of fuel injected per second,

$$Q_f = \left[\frac{\pi}{4} d_0^2 \times n_0\right] \times V_f \times \left[\frac{\theta}{360} \times \frac{60}{N}\right] \times \frac{N_i}{60} \quad \text{...(Eqn. (3.24)}$$

(where, N_i = No. of injection/min. $= \frac{1500}{2} = 750$)

$$1.963 \times 10^{-5} = \left[\frac{\pi}{4} d_0^2 \times 6\right] \times 151.23 \times \left[\frac{30}{360} \times \frac{60}{1500}\right] \times \frac{750}{60} = 29.694\ d_0^2$$

$$\therefore \quad \mathbf{d_0} = \left(\frac{1.963 \times 10^{-5}}{29.694}\right)^{1/2} = 8.13 \times 10^{-4}\,\text{m} \quad \text{or} \quad \mathbf{0.813\ mm.} \quad \textbf{(Ans.)}$$

Example 3.14. *Fuel injection in a single cylinder, 4-stroke cycle C.I. engine running at 650 r.p.m. takes place through a single orifice nozzle and occupies 28° of crank travel. The fuel consumption of the engine is 2.2 kg/hour and the fuel used has a specific gravity of 0.875. If injection pressure is 150 bar and the combustion chamber pressure is 32 bar estimate the volume of fuel injected per cycle and the diameter of the orifice. Take coefficient of discharge of orifice = 0.88.*

Solution. *Given* : $n = 1$, $N = 650$ r.p.m. ; $\theta = 28°$ of crank travel ;

Fuel consumption = 2.2 kg/h ; Sp. gr. = 0.875 ;

$$\Delta p = p_1 - p_2 = 150 - 32 = 118 \text{ bar} ; \; C_d = 0.88$$

Volume of fuel injected per cycle :

$$\text{Fuel to be injected per cycle} = \frac{\text{Fuel consumption per cylinder}}{\text{No. of cycles per min.}}$$

$$= \frac{(2.2/60)}{(650/2)} = 1.128 \times 10^{-4} \text{ kg}$$

$$\textit{Volume of fuel injected per cycle} = \frac{\text{Mass of fuel injected per cycle}}{\text{Density of fuel } (\rho_f)}$$

$$= \frac{1.128 \times 10^{-4}}{0.875 \times 1000} = 1.289 \times 10^{-7}$$

$$= \mathbf{0.1289\ cm^3.} \quad \textbf{(Ans.)}$$

Diameter of the orifice, d_0 :

$$\text{Time for fuel injection per cycle} = \left(\frac{\theta}{360} \times \frac{60}{N}\right) \text{sec.}$$

$$= \frac{28}{360} \times \frac{60}{650} = 0.00718\,\text{s}$$

Mass of fuel injected per second,

$$m_f = \frac{\text{Fuel injected per cycle}}{\text{Time for fuel injection}} = \frac{1.128 \times 10^{-4}}{0.00718} = 0.0157 \text{ kg/s}$$

Actual velocity of per cycle injection,

$$V_f = C_f \sqrt{\frac{2\Delta p}{\rho_f}} = 0.88 \sqrt{\frac{2 \times 118 \times 10^5}{(0.875 \times 1000)}} = 144.5 \text{ m/s}$$

Now, $$m_f = A_0 \times V_f \times \rho_f$$

or, $$0.0157 = \left(\frac{\pi}{4} d_0^2\right) \times 144.5 \times (0.875 \times 1000)$$

or, $$d_0 = \left[\frac{0.0157 \times 4}{\pi \times 144.5 \times (0.875 \times 1000)}\right]^{1/2} = 3.976 \times 10^{-4} \text{ m} \simeq \mathbf{0.4\ mm.} \quad \textbf{(Ans.)}$$

Example 3.15. *A four-stroke engine using 0.272 kg/kWh fuel of 32° API develops 15 kW per cylinder at 2000 r.p.m. The fuel injection pressure is 120 bar and the combustion chamber pressure is 30 bar. If the duration of injection is 30° of crank travel and velocity coefficient is 0.9 determine the diameter of the fuel orifice.*

$$\textit{Table Sp. gr.} = \frac{141.5}{131.5 + {}^\circ API}$$ **(Madras University)**

Solution. *Given* : s.f.c. = 0.272 kg/kWh, Power developed = 15 kW ;
$\Delta p = p_1 - p_2 = 120 - 30 = 90$ bar ; Duration of injection = 30° of crank angle ; $C_f = 0.9$

Diameter of the orifice, d_0 :

$$\text{Sp. gr.} = \frac{141.5}{131.5 + 32} = 0.8654$$

$$\text{Fuel consumption/cycle} = \frac{\text{s.f.c.} \times \text{kW}}{\text{cycle / hour}}$$

$$= \frac{0.272 \times 15}{\left(\frac{2000}{2}\right) \times 60} = 6.8 \times 10^{-5} \text{ kg}$$

$$\text{Duration of injection} = \frac{\theta}{360} \times \frac{60}{N} = \frac{30}{360} \times \frac{60}{2000} = 0.0025\text{s}$$

$$\therefore \quad m_f = \frac{6.8 \times 10^5}{0.0025} = 0.0272 \text{ kg/s}$$

Actual fuel velocity of injection,

$$V_f = C_f \sqrt{\frac{2\Delta p}{\rho_f}} = 0.9 \times \sqrt{\frac{2 \times 90 \times 10^5}{(0.8654 \times 1000)}} = 129.8 \text{ m/s}$$

Now,

$$m_f = A_f \times V_f \times \rho_f$$

$$= \frac{\pi}{4} d_0^2 \times V_f \times \rho_f$$

or

$$0.0272 = \frac{\pi}{4} d_0^2 \times 129.8 \times (0.8654 \times 1000)$$

$$\therefore \quad d_0 = \left[\frac{0.0272 \times 4}{\pi \times 129.8 \times (0.8654 \times 1000)}\right]^{1/2} = 5.55 \times 10^{-4} \text{ m}$$

$= 5.55 \times 10^{-4}$ m or **0.555 mm. (Ans.)**

Example 3.16. *A four-stroke cycle C.I. engine develops 11 kW per cylinder while running at 1800 r.p.m. and using fuel oil of 32° API. Fuel injection occupies 32° of crank travel and takes place through a fuel injection orifice 0.47 mm diameter with flwo coefficient of 0.9. Fuel is injected at a pressure of 118.2 bar into combustion chamber where the pressure is 31.38 bar.*

Estimate the quantity of fuel injected in kg/kWh. Specific gravity of fuel oil is given by :

$$\frac{141.5}{131.5 + {}^\circ API}.$$

Solution. *Given* : Power developed = 11 kW ; N = 1800 r.p.m., θ = 32° of crank travel ; $d_0 = 0.47$ mm ; $C_f = 0.9$; $\Delta p = p_1 - p_2 = 118.2 - 31.38 = 86.82$ bar

$$\text{Sp. gr.} = \frac{141.5}{131.5 + {}^\circ\text{API}} = \frac{141.5}{131.5 + 32} = 0.8654$$

Actual fuel velocity of injection,

$$V_f = C_f \sqrt{\frac{2\Delta p}{\rho_f}} = 0.9 \sqrt{\frac{2 \times 86.82 \times 10^5}{(0.8654 \times 1000)}} = 127.48 \text{ m/s}$$

Now,

$$m_f = A_0 \times V_f \times \rho_f$$

$$= \frac{\pi}{4} d_0^2 \times V_f \times \rho_f$$

$$= \frac{\pi}{4} \times \left(\frac{0.47}{1000}\right)^2 \times 127.48 \times (0.8654 \times 1000) = 0.01914 \text{ kg/s}$$

Time for fuel injection per cycle $= \frac{\theta}{360} \times \frac{60}{N} = \frac{32}{360} \times \frac{60}{1800} = 2.963 \times 10^{-3}\text{s}$

Mass of fuel injected per cycle $= 0.01914 \times 2.963 \times 10^{-3} = 5.671 \times 10^{-5}$ kg/cycle

Total number of cycles per hour $= \frac{1800}{2} \times 60 = 54000$

∴ **Fuel consumption in kg/kWh**

$$= 5.671 \times 10^{-5} \times 54000 \times \frac{1}{11} = \mathbf{0.278 \text{ kg/kWh.}} \quad \textbf{(Ans)}$$

Example 3.17. *An high-cylinder, four-stroke diesel engine has a power output of 386.4 kW at 800 r.p.m. The fuel consumption is 0.25 kg/kWh. The pressure in the cylinder at the beginning of injection is 32 bar and the maximum cylinder pressure is 55 bar. The injector is expected to be set at 207 bar and the maximum pressure at the injector is set to be about 595 bar. Calculate the orifice area required per injector if the injection takes place over 12° crank angle.*

Assume the following :

Specific gravity of fuel = 0.85 ; coefficient of discharge for the injector = 0.6 ; atmospheric pressure = 1.013 bar ; The effective pressure difference is the average pressure difference over the injection period. **(Madras University)**

Solution. *Given* : $n = 8$; Power output = 386.4 kW, $N = 800$ r.p.m.,

Fuel consumption = 0.25 kg / kWh, $\theta = 12°$ crank angle ; Sp. gr. = 0.85 ;

$C_f = 0.6$; $p_{atm.} = 1.013$ bar.

Orifice area reqd. per injection, A_0 :

$$\text{kW per cylinder} = \frac{386.4}{8} = 48.3$$

∴ Fuel consumption per cylinder = 48.3 × 0.25 = 12.075 kg / h or 0.2012 kg/ min.

Fuel to be injected per cycle $= \frac{0.2012}{(800/2)} = 5.03 \times 10^{-4}$ kg

Time for fuel injection per cycle $= \left(\frac{\theta}{360} \times \frac{60}{N}\right) = \frac{12}{360} \times \frac{60}{800} = 0.0025$ s

∴ Mass of fuel injected per second,

$$m_f = \frac{5.03 \times 10^{-4}}{0.0025} = 0.2012 \text{ kg/s}$$

Pressure difference at beginning $= 207 - 32 = 175$ bar

Pressure difference at end $= 595 - 55 = 540$ bar

Average pressure difference $= \frac{175 + 540}{2} = 357.5$ bar

Now,
$$m_f = A_0 \times V_f \times \rho_f$$
$$= A_0 \times \left[C_f \sqrt{\frac{2\Delta p}{\rho_f}} \right] \times \rho_f = A_0 \times C_f \sqrt{2\Delta p.\rho_f}$$

or,
$$0.2012 = A_0 \times 0.6 \sqrt{2 \times (357.5 \times 10^5) \times (0.85 \times 1000)}$$
$$= 147915.5\, A_0$$

$\therefore$ $\mathbf{A_0} = 1.36 \times 10^{-6}$ m^2 or **0.0136 cm^2. (Ans.)**

Example 3.18. *A six-cylinder, four-stroke oil engine operates on A/F ratio = 20. The diameter and stroke of the cylinder are 100 mm and 140 mm respectively. The volumetric efficiency is 80 percent. The condition of air at the beginning of compression are 1 bar, 27° C.*

(i) Determine the maximum amount of fuel that can be injected in each cylinder per second.

(ii) If the speed of the engine is 1500 r.p.m., injection pressure is 150 bar, air pressure during fuel injection is 40 bar and fuel injection is carried out for 20° crank angle, determine the diameter of the fuel orifice assuming only one orifice is used.

Take, $\rho_f = 860$ *kg/m*^{3} *;* $C_f = 0.67$. **(Roorkee University)**

Solution. *Given* : $n = n_0 = 6$; A / F ratio = 20 ; d = 100 mm = 0.1 m, l = 140 mm = 0.14 m ; $\eta_{vol.}$ = 80% ; p_a = 1 bar, T_a = 27 + 273 = 300 K, N = 1500 r.p.m. ; θ = 20° crank angle ; ρ_f = 960 kg/m^3 ; C_f = 0.67, Δ_p = 150 – 40 = 110 bar.

(*i*) **Amount of fuel injected into each cylinder per cycle :**

Volume of air supplied per cylinder per cycle

$$= \text{Stroke volume} \times \eta_{vol.}$$
$$= \frac{\pi}{4} d^2 \times l \times \eta_{vol.} = \frac{\pi}{4} \times (0.1)^2 \times 0.14 \times 0.8 \simeq 8.8 \times 10^{-4} m^3$$

Mass of this air at suction conditions,

$$m_a = \frac{p_a V_a}{RT_a} = \frac{1 \times 10^5\, (8.8 \times 10^{-4})}{287 \times 300} = 1.022 \times 10^{-3} \text{ kg/cycle}$$

$$\text{A / F ratio} = \frac{m_a}{m_f} = \frac{20}{1}$$

$$\therefore \quad m_f = \frac{m_a}{20} = \frac{1.022 \times 10^{-3}}{20} = 5.11 \times 10^{-5} \text{ kg/cycle}$$

Time taken for fuel injection per cycle

$$= \left(\frac{\theta}{360} \times \frac{60}{N} \right) = \frac{20}{360} \times \frac{60}{1500} = 0.00222 \text{s}$$

$\therefore$ *Amount/mass of fuel injected into each cylinder per second,*

$$\dot{m}_f = \frac{5.11 \times 10^{-5}}{0.00222} = \mathbf{0.023\ kg/s.} \quad \textbf{(Ans.)}$$

(*ii*) **Diameter of the fuel orifice, d$_0$:**

The mass of fuel injected into each cylinder per second,

$$\dot{m}_f = A_0 \times V_f \times \rho_f$$

or
$$0.023 = \frac{\pi}{4} {d_0}^2 \times C_f \sqrt{\frac{2\Delta p}{\rho_f}} \times \rho_f$$

$$= \frac{\pi}{4} d_0^2 \times C_f \sqrt{2\Delta p \times \rho_f}$$

$$= \frac{\pi}{4} d_0^2 \times 0.67 \sqrt{2 \times 110 \times 10^5 \times 960} = 72381.11\, d_0^2$$

or $d_0 = 5.637 \times 10^{-4}$ m or **0.5637 mm. (Ans.)**

Example 3.19. *In a diesel fuel injection pump, the volume of fuel in the pump barrel before commencement of the effective stroke is 7 c.c. The diameter of the fuel line from pump to injector is 3 mm and is 700 mm long. The fuel in the injection valve is 2 c.c.*

(i) To deliver 0.10 c.c. of fuel at a pressure of 150 bar, how much displacement the plunger undergoes ? Assume a pump inlet pressure of 1 bar ;

(ii) What is the effective stroke of the plunger if its diameter is 7 mm.

Assume coefficient of compressibility of oil as 78.8 × 10⁻⁶ per bar at atmospheric pressure.

Solution. *Given* : The volume of fuel in the pump barrel before commencement of the effective stroke = 7 c.c.

The diameter and length of the fuel line from pump to injector = 3 mm, 700 mm,

Volume of fuel in the injection valve = 2 c.c.

Volume of fuel to be delivered = 0.10 c.c.

The pressure at which fuel to be delivered, p_1 = 150 bar

Atmospheric pressure, p_2 = 1 bar

Coefficient of compressibility, $C_c = 78.8 \times 10^{-6}$ per bar at atmospheric pressure

Diameter of plunger, d_p = 7 mm

(*i*) **Displacement of plunger :**

Coefficient of compressibility of oil,

$$C_c = \frac{\text{Change in volume per unit volume}}{\text{Difference in pressure causing compression}} = \frac{(V_1 - V_2)}{V_1(p_1 - p_2)}$$

Total initial fuel volume,

V_1 = Volume of fuel in barrel + volume of fuel in the delivery line + volume of fuel in the injection valve

$$= 7 + \frac{\pi}{4}(0.3)^2 \times 70 + 2 = 13.95 \text{ c.c.}$$

No pressure is built up till the pump plunger closes the inlet port. Further advance of plunger will compress the fuel oil and raise the pressure to a required value. Once the delivery pressure is attained, further movement of plunger results in delivery of fuel oil at constant pressure.

Change in volume due to compression = $C_c(p_1 - p_2) \times V_1$

or $(V_1 - V_2) = 78.8 \times 10^{-6} \times (150 - 1) \times 13.95$

$= 0.16379$ c.c.

Total displacement of plunger

$= (V_1 - V_2) + 0.1 = 0.16379 + 0.1 =$ **0.26379 c.c. (Ans.)**

(*ii*) **Effective stroke of the plunger, l_p :**

$$\frac{\pi}{4} d_p^2 \times l_p = 0.26379 \quad \text{or} \quad \frac{\pi}{4} \times (0.7)^2 \times l_p = 0.26379$$

$$\therefore \quad l_p = \frac{0.26379 \times 4}{\pi \times (0.7)^2} = 0.6854 \text{ cm}$$ or **6.854 mm. (Ans.)**

Example 3.20. *At injection pressure of 145 bar, spray penetration of 22 cm in 16 milliseconds is obtained. Determine the time required for the spray to penetrate the same distance at an injection pressure of 235 bar. Assume the same orifice and the combustion chamber density. Combustion chamber pressure is 30 bar.*

Use the relation : $s = t\sqrt{\Delta p}$

where *s is the penetration in cm,*

t is the time in milliseconds, and

Δp *is the pressure difference between the injection pressure and combustion chamber pressure.*

Solution. *Given :* $p_1 = 145$ bar, $t_1 = 16$ milli-seconds, $p_2 = 230$ bar ;

$p_{cyl} = 30$ bar ; $s_1 = s_2 = 22$ cm

Here, $\Delta p_1 = p_1 - p_{cyl} = 145 - 30 = 115$ bar

$\Delta p_2 = p_2 - p_{cyl} = 235 - 30 = 205$ bar

From given relation $s = t\sqrt{\Delta p}$, we have

$$\frac{s_1}{s_2} = \frac{t_1}{t_2}\sqrt{\frac{\Delta p_1}{\Delta p_2}}$$

or

$$t_2 = \frac{s_2}{s_1} \times t_1\sqrt{\frac{\Delta p_1}{\Delta p_2}}$$

$$= 1 \times 16\sqrt{\frac{115}{205}} = \textbf{11.98 milli-seconds.} \quad \textbf{(Ans.)}$$

3.21. TROUBLESHOOTING OF A FUEL SYSTEM

Troubleshooting of a fuel system in a tabular form is given below :

S. No.	Trouble	Cause	Remedy
1.	*Engine will not start*	(*i*) Starting mixture too weak	(*i*) Check the size of the starter petrol jet against specification. Checks for air leaks and that the throttle has returned to idling position.
2.	*Engine fires but will not continue running*	(*i*) Volume of starting mixture insufficient	(*i*) Starter petrol jet and air jet smaller than specified. Check and fit correct sizes of starter petrol jet and air jet.
3.	*Engine will not idle*	(*i*) Air leaks	(*i*) Check joints between carburettor and manifold and engine. Check for worn throttle spindle.
4.	*Engine misfires at idling speed*	(*i*) Pilot jet smaller than specified.	(*i*) Fit pilot jet as specified and adjust volume control screw as necessary.
5.	*Engine "hunts" at idling speed*	(*i*) Pilot jet larger than specified.	(*i*) Fit pilot jet of specified size and adjust idling speed adjustment screw as necessary.

6.	*Engine top speed insufficient*	(*i*) Correction jet larger than specified (*ii*) Main jet smaller than specified (*iii*) Fuel supply restricted.	(*i*) Fit correction jet as specified (*ii*) Fit main jet as specified (*iii*) Check fuel pump pressure. Check for blockage in fuel supply pipe.
7.	*Engine knocks and overheats*	(*i*) Mixture too weak (Main jet smaller than specified)	(*i*) Fit main jet of specified size.
8.	*Heavy fuel comsumption*	(*i*) Air cleaner choked (*ii*) Bi-starter disc valve not closing completely (*iii*) Carburettor flooding internally (*iv*) Jet setting not to specifications	(*i*) Clean air cleaner thoroughly. (*ii*) Check for sufficient travel in the dashboard control wire. Check that spring is holding the disc valve flat against casting. (*iii*) Check that the float is not leaking. Shaking the float will show whether there is petrol inside. Needle valve may be worn or not moving due to the presence of gum ; wash thoroughly in methylated spirit. Check correct size of valve assembly and number and thickness of fibre washers used. Check lift pump pressure. If excessive it will force needle off the seating. Vent hole to atmosphere may be blocked ; clear it thoroughly. (*iv*) Check and alter where necessary.
9.	*Poor acceleration*	(*i*) Accelerator pump is not functioning. (*ii*) Pump jet smaller than specified.	(*i*) Check that valves are free, diaphragm spring is sound, passages (air and petrol) are clear. (*ii*) Fit pump jet of specified size.

3.22. TROUBLESHOOTING OF CARBURETTOR

Troubleshooting of carburettor, in a tabular form, is given below :

S. No.	Trouble	Cause	Remedy
1.	*Flooding of carburettor.*	(*i*) Excessive pressure of fuel by fuel pump. (*ii*) Float level set high. (*iii*) Punctured float ball. (*iv*) Restricted movement of float ball lever. (*v*) Worn out or broken gasket of float bowl. (*vi*) Loose, worn out or pitted seat of needle valve. (*vii*) Dirty valve seat. (*viii*) Loose main jet.	(*i*) Give thicker, packing between fuel pump body and block. (*ii*) Adjust float ball level correctly. (*iii*) Replace with a new one. (*iv*) Straighten lever if bent. See that the ball is moving free on pin. (*v*) Replace the gasket. (*vi*) Replace. (*vii*) Clean it and lap it with valve. (*viii*) Check up and tighten the same.

2.	*Low speed or idle speed circuit sending lean fuel.*	(*i*) Blocked metering hole. (*ii*) Blocked economizer hole. (*iii*) Blocked idle jet. (*iv*) Blocked idle passage in carburettor. (*v*) Torn out gasket of carburettor foundation. (*vi*) Any connection of intake manifold leaking. (*vii*) Wiper connecting tube leaking.	(*i*) Clean the passage. (*ii*) Clean the same. (*iii*) Clean the same. (*iv*) Clean. (*v*) Replace. (*vi*) Change the gasket. (*vii*) Change the tube or plug the hole.
3.	*Rich mixture on idle or slow speed.*	(*i*) Use of bigger size of jet for slow running. (*ii*) Slow speed jet hole as become bigger. (*iii*) Restricted passage of bypass or air bleed. (*iv*) Damaged idle port. (*v*) Worn out threads of idle screw. (*vi*) Wrong fitting of throttle valve.	(*i*) Use correct size jet. (*ii*) Replace the jet. (*iii*) Clean the carburettor. (*iv*) Get it repaired. (*v*) Replace the screw and in case of body threads get it repaired. (*vi*) Set it correctly.
4.	*Lean mixture in high speed.*	(*i*) Less fuel pump pressure. (*ii*) Obstruction in fuel line. (*iii*) Float ball level setting too low. (*iv*) Wrong size of metering rod. (*v*) Wrong size (small size) of jet of metering rod. (*vi*) Mis-adjustment of throttle opening metering rod. (*vii*) Obstruction in the passage of metering jet sleeve. (*viii*) Worn out throttle or its shaft.	(*i*) Leaky outlet valve of fuel pump and all such defects as mentioned in fuel pump chapter for pumping less fuel. (*ii*) Clean the pipe with air pressure. (*iii*) Set the float ball level correctly. (*iv*) Fix up the metering rod of correct size. (*v*) Use correct size of jet. (*vi*) Check up linkages : if worn out or bent replace these. (*vii*) Clean. (*viii*) Change throttle, shaft and bushes.
5.	*Rich fuel in high speed circuit*	(*i*) High fuel pump pressure. (*ii*) High setting of float ball level. (*iii*) Sagged metering rod spring or spring came out of its position. (*iv*) Worn out metering rod and its jet. (*v*) Bent metering rod. (*vi*) Sticky choke valve. (*vii*) Choked air cleaner. (*viii*) Obstruction in passage of air bleed hole to main jet. (*ix*) Crack or torn out float bowl gasket.	(*i*) Use thicker gasket between the block and body. (*ii*) Set float ball level correctly. (*iii*) Replace spring or fix it right. (*iv*) Replace the set. (*v*) Replace. (*vi*) Lubricate choke valve linkages. See that it operates freely. (*vii*) Clean the air cleaner. (*viii*) Clean the passage. (*ix*) Replace the gasket.

6.	*Accelerating pump sending lean mixture.*	(*i*) Weak plunger spring. (*ii*) Worn out plunger cup washer. (*iii*) Leaky feed pump valve. (*iv*) Obstruction in fuel feed line. (*v*) Blocked acceleration jet. (*vi*) Feed pump stroke defective.	(*i*) Replace the spring. (*ii*) Replace the cup washer. (*iii*) Replace the valve. (*iv*) Clean the pipe. (*v*) Clean the jet. (*vi*) Check up for worn out rocker or working of diaphragm.
7.	*Accelerating pump sending too rich mixture.*	(*i*) Bigger stroke of accelerator pump. (*ii*) Bigger hole of accelerator jet.	(*i*) Set the stroke correctly. (*ii*) Replace the jet. Use correct size jet.
8.	*Hard starting of engine under all conditions.*	(*i*) Insufficient fuel reaching carburettor.	(*i*) Disconnect spark-plug high tension lead, keep it little away. Crank the engine to see if there is spark. If spark is there, remove air cleaner. Operate throttle by hand, a steady stream of fuel should flow from discharge nozzle of carburettor. If it does not proceed as under, remove fuel line from pump to carburettor. Crank the engine. If no fuel is coming, feed pump is defective. If fuel is coming, carburettor is defective. Overhaul the carburettor.
9.	*Hard starting when engine is cold.*	(*i*) Choke stuck in open position. (*ii*) Binding choke operating mechanism. (*iii*) Less fuel reaching carburettor.	(*i*) Free the choke. (*ii*) Free the operating mechanism. (*iii*) Incorrect jet size. Use correct size jet. Low float ball level. Adjust. Less delivery by feed pump.
10.	*Hard starting when engine is hot.*	(*i*) Choke closes when engine is hot (automatic choke). (*ii*) Flooding of carburettor. (*iii*) Vapour lock.	(*i*) Inspect its operation. Binding choke linkage. (*ii*) Adjust float ball setting. Excessive pressure of feed pump. (*iii*) Keep the feed line away from exhaust manifold and exhaust pipe as well as wrap the pipe with asbestos.
11.	*Engine stalls.*	(*i*) Idle speed too slow. (*ii*) Idle fuel mixture too lean. (*iii*) Faulty choke operation. (*iv*) Float ball setting incorrect. (*v*) Dirt or water in fuel line or filter. (*vi*) Throttle linkage binding. (*vii*) Accelerator pump setting defective.	(*i*) Adjust as required. (*ii*) Adjust in conjunction with idle speed. (*iii*) Repair as required. (*iv*) Adjust as required. (*v*) Clean the fuel line. (*vi*) Free it. (*vii*) Adjust it properly.

3.23. COMPARATIVE DIESEL ENGINE FUEL SYSTEM DATA OF SOME INDIAN AUTOMOBILES

The comparative diesel engine fuel system date of some Indian automobiles, in tabular form, is given below :

S. No.	Item	Ashok Leyland Viking Passenger	Tata Diesel Vehicles 1210	Swaraj Mazda (T3500 series SL engine)
1.	*Fuel injection pump*	Hindustan Bosch in-line type HB-PE 6A 80B 320RS 2154	Hindustan Bosch in-line type HB-PES 6A 80B 410RS 2405	Diesel Kiki Co. Ltd. PE(S)-A (in-line type)
2.	*Injection pressure, kg/cm² (MPa)*	160 (15.7)	180 (17.6)	170 (16.7)
3.	*Nozzle holder (injector)*	Hindustan Bosch type HB-KBAL 83S 7/4	Hindustan Bosch type HB-KDAL 74S 3/19	Bracket type
4.	*Nozzle*	Hindustan Bosch type HB-DLL 140 S 215	Hindustan Bosch type HB-DLLA 150 S 187	Hole type
5.	*Feed pump*	HB-FP/KE 22 AD 223/2	HB-FP/KE 22 AD 280/2	KS (Single-acting type) Feeding pressure 3.1 kg/cm² (0.3 MPa)
6.	*Injection Timer*	—	Centrifugal HB-EP/SF 500–1400 B4R 1 (Max. advance angle 8°)	SCD M (Max. advance angle 7°)
7.	*Governor*	Centrifugal variable speed type HB-EP/ RSV 200–1200 A1A41R	Centrifugal variable speed type HB-EP/RSV 300–1400 A2B1062DL	RLD–K (all speed mechanical type with torque cam)
8.	*Idling speed r.p.m.*	400	600	620–700
9.	*Fuel filter*	Micro dual filter, paper element type	Micro dual filter, cloth insert as primary stage filter and paper element as final stage filter	Paper element type
10.	*Air cleaner*	Oil bath type with wire mesh element	Oil bath type with wire mesh element	Oil bath type with filter paper element
11.	*Fuel tank capacity, litres*		160	90

SHORT ANSWER QUESTIONS

Filter

Q. 1. Name the types of air cleaners used in a engine.

Ans. (*i*) Dry–fabric element air cleaners, (*ii*) Oil wetted wire woven element air cleaners, and (*iii*) Oil bath element air cleaners.

Q. 2. Which type of cleaner is more efficient ?

Ans. Paper element cleaner is more efficient.

Q. 3. Why should not the engine be cranked with the air cleaner off ?

Ans. If the engine backfires with the air cleaner off it would cause fire in the engine compartment.

Q. 4. What is a fuel filter ?

Ans. It is a part of the fuel pump, fitted to prevent dirt in fuel, from entering the pump of the carburettor.

Q. 5. What is a canister oil filter ?

Ans. It is that which uses a replaceable filter element inside a permanent metal canister. This has a reusable steel shell with a replaceable cartridge or element inside.

Q. 6. What is 'air lock' ?

Ans. Air trapped in the pipe line which prevents the flow of fuel is known as 'air lock'.

Carburettor

Q. 7. Name the vehicles in which gravity feed system is used.

Ans. Motorcycle, scooters, motor boats and other stationary engines.

Q. 8. Why is gravity feed system not suitable for cars and other heavy vehicles ?

Ans. Because it is not possible to place the fuel tank at a level higher than that of the carburettor.

Q. 9. What are the two main classes of carburettor ?

Ans. (*i*) Constant choke and (*ii*) Constant vacuum.

Q. 10. What is a constant choke carburettor ?

Ans. It is a type in which the orifice area is constant and the pressure difference or depression is varied.

Q. 11. What is a constant vacuum carburettor ?

Ans. It is a type in which area of the orifice is varied to meet the changing demands, the depression being kept constant.

Q. 12. Why do most engines use down draft carburettors ?

Ans. In the down draft carburettor, gravity assists the spraying of fuel and in filling the cylinder engine accessibility is improved.

Q. 13. What is a carburettor venturi ?

Ans. It is a miniature taper bore to increase the air velocity and create partial vacuum causing suction of petrol from the float chamber.

Q. 14. What is a carburettor jet ?

Ans. It is a very small hollow plug having a number of holes to spray a metered quantity of fuel.

Q. 15. How can fuel consumption be reduced ?

Ans. By using a jet of small size.

Q. 16. What is meant by a "submerged jet" and why is it used ?

Ans. A submerged jet is one where the orifice is below the fuel level. It gives a steady uniform discharge as it is not affected by air speed, or difference in float chamber fuel level.

Q. 17. How can a choked jet be cleaned ?

Ans. By blowing air through it in reverse direction of petrol flow.

Q. 18. What is a choke ?

Ans. It is a valve controlling the amount of air passing to the carburettor and is used when starting a cold engine.

Q. 19. What is another name for the choke ?

Ans. Strangler.

Q. 20. What is a choke tube ?

Ans. It is a small tube around the jet to increase the speed air for effective atomisation of fuel.

Q. 21. Why is the float chamber fitted in a carburettor ?

Ans. The float chamber is fitted in a carburettor to maintain and control the supply of fuel.

Q. 22. What are reasons for the 'engine missing' with reference to petrol flow ?

Ans. (*i*) Dust particle in petrol, (*ii*) Choked filter, (*iii*) Improperly adjusted ball in the float, and (*iv*) Presence of water in petrol.

Q. 23. What is meant by 'flooding' of carburettor ?

Ans. It means that the engine cylinders receive raw or liquid gasoline, or too rich a mixture.

Q. 24. State the cause and remedy for popping back in a carburettor.

Ans. Popping back in a carburettor is usually due to the mixture being too weak ; it can be cured by enriching the mixture.

Q. 25. What is an economiser ?

Ans. It is an air bleed fitted with a needle valve so that the amount of air entering the nozzle can be varied to suit running conditions.

Fuel injection

Q. 26. What are the usual methods of fuel injection ?

Ans. (*i*) Air injection, (*ii*) Mechanical injection, and (*iii*) Solid fuel injection.

Q. 27. What is the function of the fuel injection pump ?

Ans. The function of the fuel injection pump is to supply the engine with fuel in quantities exactly metered in proportion to the amount of power required and timed for smooth running with maximum economy.

Q. 28. What is meant by air injection fuel system ?

Ans. The use of compressed air which forces a definite amount of fuel, mixed with the air, into the cylinder is termed as air-injection.

Q. 29. What is an injector ?

Ans. It is a nozzle which ensures and actually supplies an atomized fuel into the engine cylinder.

Q. 30. What is a unit injector ?

Ans. A unit injector is a small high pressure pump and spray nozzle for an individual cylinder.

Q. 31. What is bleeding in injection system ?

Ans. Removal of air from the pipeline is termed as bleeding in injection system.

Q. 32. What are the usual types of diesel injectors ?

Ans. Single-hole, multi-hole, long-stem, pintle and pintaux types.

Q. 33. What is the speciality of a pintle nozzle ?

Ans. A pintle nozzle does not need the high energy jet, a characteristic of hole type nozzle. The orifice is an annulus formed by an extension of the valve needle-the pintle.

Q. 34. Where is the pintle nozzle used ?

Ans. It is used in the pre-chamber and swirl type of combustion system.

Q. 35. What is the speciality of a pintaux nozzle ?

Ans. A pintaux type was designed primarily to enable cold starting, without the use of a heater plug.

Q. 36. What causes a dripping injector ?

Ans. A carbon deposit or a sticking injector valve can cause the injector to become leaky.

Q. 37. How is fuel injection maintained evenly in all cylinders ?

Ans. By means of adjusting strews which control the time and flow of fuel into each cylinder.

Q. 38. Name two causes of misfiring.

Ans. (*i*) One or more nozzles not spraying correctly, and

(*ii*) Fuel not reaching the nozzles owing to choked filter, or an air lock in the delivery pipe.

Q. 39. What are the functions of a governor in an injection pump ?

Ans. The governor functions to avoid the engine getting over race on no load condition and also to supply enough fuel on load.

Exhaust system

Q. 40. Name four major parts that form as exhaust system in an automobile system.

Ans. (*i*) Exhaust manifold, (*ii*) Catalytic converters, (*iii*) Muffler and, (*iv*) tail pipe.

Q. 41. What is known as "Smog" in an automobile ?

Ans. Smog in an automobile is a mixture of particles of unburnt fuel and soot in the air.

Q. 42. What the word "Smog" refers to ?

Ans. The word "Smog" refers to "Smoke" and "Fog". Smog is a sort of fog with other substances mixed in.

Q. 43. What are the methods to clean the exhaust gas ?

Ans. By (*i*) controlling air fuel mixture, (*ii*) combustion, and (*iii*) blowing fresh air using a catalytic converter.

Q. 44. Why are mufflers necessary ?

Ans. Mufflers are attached to the exhaust manifold through a pipe line. This is to reduce the pressure of the exhaust gas and discharge it into the atmosphere with a minimum power loss, less noise, less vibration and transfer of heat.

HIGHLIGHTS

S.I. Engines

1. The process of preparing, in the S.I. engine, a combustible fuel-air mixture outside the engine cylinder is called *carburation*.
2. A *carburettor* is a device which atomises the fuel and mixes it with air.
3. The air-fuel ratio for maximum power is 13 : 1.
4. The following addition devices / systems are added to the simple carburettor :
   (*i*) Main metering system.
   (*ii*) Idling system.
   (*iii*) Power enrichment or economiser system.

(*iv*) Acceleration pump system.

(*v*) Choke.

5. Carburettors, basically, are of the following types :

(*i*) Open choke type

Examples : Zenith, Solex, Carter and Stromberg carburettors

(*ii*) Constant vacuum type

Example : S.U. carburettor.

C.I. Engines

1. Main types of modern fuel injection systems are :

(*i*) Common rail injection system.

(*ii*) Individual pump injection system.

(*iii*) Distributor system.

2. The main components of a fuel injection system are :

Fuel tank ; Fuel feed pump ; Fuel filters ; Injection pump ; Governor ; Fuel pipings and injectors.

3. The nozzle are classified as follows :

(*i*) Single hole nozzle (*ii*) Multi-hole nozzle

(*iii*) Circumferential nozzle (*iv*) Pintle nozzle

(*v*) Pintaux nozzle.

4. Volume of fuel injected per min.,

$$Q_f = \left[\frac{\pi}{4} d_0^2 \times n_0\right] \times V_f \times \left[\frac{\theta}{360} \times N\right] \times N_i$$

where, d_0 = Diameter of fuel orifice, m^2,

n_0 = Number of orifices,

V_f = Velocity of flow of fuel through orifice,

θ = Duration of the injection in degrees of crank angle,

N = r.p.m., and

N_i = Number of injection per min.

= r.p.m./2 = for four-stroke engine

= r.p.m. for two-stroke engine.

OBJECTIVE TYPE QUESTIONS

Fill in the Blanks or Say "Yes" or "No" :

S.I. Engines

1. Carburettor is used for ....
2. For maximum power of S.I. engines the fuel air mixture ratio should be ....
3. Typical pressure in the induction manifold at the start of induction stroke of S.I. engine, under idling conditions is ....
4. Relative fuel-air ratio (F_R) for maximum power in S.I. engine may be ....
5. Relative fuel-air ratio (F_R) for maximum thermal efficiency of S.I. engine may be ....
6. For best thermal efficiency of S.I. engine, the fuel-air mixture ratio should be ....
7. In S.I. engine the process of preparing a combustible fuel-air mixture outside the engine cylinder is called ....
8. The pipe that carries the prepared mixture to the engine cylinders is called the intake manifold.

9. The A / F ratio for maximum power is not the same as the A / F ratio for maximum economy.
10. The maximum power is obtained at about .... A / F ratio.
11. At full throttle, maximum efficiency occurs at an A / F ratio of about ....
12. In stationary engines the designed air-fuel ratio is that which gives the maximum economy.
13. The richening of mixture increases the probability of contact between fuel and air particles and thus improves ....
14. In the carburettor, complete automization and vaporization of the fuel is achieved.
15. The volatility of fuel significantly affects the starting and warm up characteristics of the engine.
16. The term 'acceleration' with regard to engines, is generally used to refer to an increase in engine speed resulting from opening the throttle.
17. The petrol engine is .... governed.
18. In the modern carburettors the mixture correction is done by air bleeding alone.
19. A common method of changing the air-fuel ratio in large carburettors is the back suction control.
20. Almost all carburettors, except S.U. carburettor, are of open choke type.
21. The important difference between an aircraft and automobile carburettor is that the former operates at varying altitudes whereas the latter operates mostly at ground level conditions.
22. Engines fitted with petrol injection system can be used in tilt position which will cause surge trouble in carburettors.
23. Injection systems generate less noise.
24. Weight and bulk of petrol injection system is more than that of a carburettor.
25. A petrol injection system has increased volumetric efficiency.

ANSWERS

1. S.I. engine	**2.** rich	**3.** 0.3 bar	**4.** 1.2	**5.** 0.8
6. lean	**7.** carburetion	**8.** Yes	**9.** Yes	**10.** 12.5 : 1
11. 17 : 1	**12.** Yes	**13.** combustion	**14.** No	**15.** Yes
16. Yes	**17.** quantity	**18.** Yes	**19.** Yes	**20.** Yes
21. Yes	**22.** Yes	**23.** No	**24.** Yes	**25.** Yes.

C.I. Engines

1. Fuel is injected into the cylinder at the end of ...... stroke.
2. The period between the start of injection and start of ignition is called ...... delay.
3. The beginning and end of the injection should be sharp.
4. The injected fuel may not be broken into very fine droplets.
5. Air injection system is largely used these days.
6. Inferior fuels can be used in air injection system.
7. Solid injection is also called ...... injection.
8. In ...... injection system only one pump is sufficient for multi-cylinder engine.
9. Very accurate design and workmanship are required in air injection system.
10. In ...... system, the fuel is metered at a central point ; a pump pressurises, meters the fuel and times the injection.
11. In ...... pump injection system an individual pump or pump cylinder connects directly to each fuel nozzle.
12. A multi-hole nozzle gives poor automisation.
13. A ...... nozzle is a type of pintle nozzle which has an auxiliary hole drilled in the nozzle body.

ANSWERS

1. compression	**2.** ignition	**3.** Yes	**4.** No	**5.** No
6. Yes	**7.** mechanical	**8.** common rail	**9.** Yes	**10.** distributor
11. individual	**12.** No	**13.** pintaux.		

THEORETICAL QUESTIONS

S.I. Engines

1. What do you mean by the term 'carburetion' ?
2. What is a carburettor ?
3. Draw a typical induction system of a petrol engine.
4. Enlist the factors which affect the process of carburetion.
5. Why is a choke used in a carburettor ?
6. Explain briefly the essential features of good commercial carburettor for automotive engines.
7. (*a*) Draw a curve representing variation of mixture requirements (fuel-air ratios) from no-load to full-load in a S.I. engine, mark the relative position of stoichiometric fuel-air ratio line and then explain why :
   (*i*) an idling engine requires a rich mixture ;
   (*ii*) a cruising engine requires an economy mixture ;
   (*iii*) maximum power demands a rich mixture ;
   (*b*) What will be the effects of prolonged running of such an engine on (*i*) the weakest mixture ; (*ii*) the richest mixture ?
8. With the help of a neat sketch explain the working principle of a simple carburettor.
9. A simple carburettor is inherently unsuitable to meet the varying mixture requirements of S.I. engine. What are the drawbacks of a simple carburettor ? How are they overcome by incorporating compensating devices ? Explain with the aid of suitable sketches wherever necessary.
10. Describe with suitable sketches the following systems of a modern carburettor :
    (*i*) Main metering system ;
    (*ii*) Idling system ;
    (*iii*) Economiser system ;
    (*iv*) Acceleration pump system.
11. With the help of a neat sketch describe the construction and working of a solex carburettor.
12. How the power and efficiency of the S.I. engine vary with (*a*) air-fuel ratio at full-load ; (*b*) part load ?
13. Sketch and explain the fuel consumption loop in the S.I. engine.
14. Why a rich mixture is required for idling ?
15. Why a rich mixture is required for maximum power ?
16. What do you understand by transient mixture requirements ?
17. Why multi-cylinder engines require richer mixture than single cylinder engines ?
18. What are the basic types of carburettors ?
19. Explain with neat sketches the following types of carburettors :
    (*i*) Carter carburettor
    (*ii*) Solex carburettor
    (*iii*) S.U. carburettor.
20. Device an expression for A / F ratio
    (*a*) neglecting compressibility ;
    (*b*) taking compressibility into account.

21. State the special requirements of an aircraft carburettor ?
22. What is petrol injection ?
23. State the advantages and limitations of petrol injection.
24. Explain briefly 'continued' and 'timed' injection systems.
25. With the help of a neat sketch, explain briefly any petrol injection system.
26. What is the difference between 'Direct injection' and 'Indirect injection' ?
27. Explain the difference between 'continuous injection' and ' Intermittent or pulsed injection'.
28. Give the comparison between petrol injection and carburetted fuel supply systems.

C.I. Engines

1. What are the functional requirements of an injection system ?
2. How are injection systems classified ? Describe them briefly. Why the air injection system is not used now-a-days ?
3. What are the essential requirements to be fulfilled by a fuel injection system for C.I. engines ? What is the most common injection system used in multi-cylinder diesel engines ?
4. What are the main functions of the nozzle ?
5. What are the most common types of nozzles used ?
6. What is the difference between air injection and solid injection ?
7. State the merits and demerits of multiple orifice nozzle and pintle type nozzle in C.I. engine injectors.
8. Bring out the differences in construction and working of printle and pintaux nozzles with the help of sketches and discuss their relative merits.
9. How are solid injection system classified ? Explain the working of common rail, individual pump and distributor systems with the help of neat sketches. Discuss their relative merits and demerits.

UNSOLVED EXAMPLES

S.I. Engines

1. A four-cylinder four-stroke engine having diameter and length of stroke as 100 mm and 120 mm respectively is running at 2000 r.p.m. Its carburettor venturi has a 30 mm throat. Assuming coefficient of air flow 0.8, density of air 1.2 kg/m^3 and volumetric efficiency of the engine as 70 percent, determine the suction at the throat. **[Ans.** 0.0339 bar]
2. A simple jet carburettor is required to supply 6 kg of air per minute and 0.45 kg of fuel of density 740 kg/m^3. The air is initially at 1.013 bar and 27°C. Calculate the throat diameter of the choke for a flow velocity of 91 m/s. Velocity coefficient = 0.8.

   If the pressure drop across the fuel metering orifice is 0.75 of that at the choke, calculate orifice diameter assuming $C_d = 0.6$. **[Ans.** 35.25 mm ; 2.34 mm]
3. A four-stroke petrol engine of Hindustan Ambassador has a capacity of 1489 c.c. It develops maximum power at 4200 r.p.m. The volumetric efficiency at this speed is 70 percent and the air/fuel ratio is 13 : 1. At peak power the theoretical air speed at choke is 90 m/s. The coefficient of discharge for venturi is 0.85 and that of the main petrol jet is 0.66. An allowance should be made for the emulsion tube, the diameter of which can be taken as 1/2.5 of the choke diameter. The petrol surface is 6 mm below the choke at this engine condition. The specific gravity of petrol is 0.74. Atmospheric pressure and temperature are 1.013 bar and 20°C respectively.

   Calculate the sizes of a suitable choke and main jet. **[Ans.** 27.35 mm ; 1.58 mm]
4. A single jet simple carburettor is to supply 6.11 kg/min. of air and 0.408 kg/min of petrol, density 768 kg/m^3. The air is initially at 1.027 bar and 15.5°C. Calculate the throat diameter of the venturi throat if the speed of air is 97.5 m/s, assuming a velocity coefficient of 0.84. Assume adiabatic expansion and γ for air as 1.4. If the drop across fuel metering orifice be 0.8 of the pressure at the throat ; calculate the orifice diameter assuming a coefficient as 0.66. **[Ans.** 2.05 mm]
5. An engine having a simple single jet carburettor consumes 6.5 kg of fuel/hour. The fuel density is 700 kg/m^3. The level of fuel in the float chamber is 3 mm below the top of the jet when the engine is not

running. Ambient conditions are 1.01325 bar and 17°C. The jet diameter is 1.25 mm and its discharge coefficient is 0.6. The discharge coefficient of air is 0.85. Air-fuel ratio is 15. Determine the critical air velocity and the throat diameter (effective). Express the pressure depression in cm of water. Neglect compressibility of air. **[Ans.** 4.945 m/s ; 19.9 mm ; 43.99 cm]

6. An eight-cylinder four-stroke petrol engine with bore and stroke of 100 mm each uses volatile fuel of composition C = 84%, H_2 = 16%. The throat diameter of choke tube is 40 mm. The volumetric efficiency at 3000 r.p.m. is 75 percent referred to 0°C and 1.01325 bar. The pressure depression is 0.116 bar and the temperature at throat is 16°C. If chemically correct A / F ratio is supplied for consumption, determine :

(*i*) Fuel consumption in kg / h ;

(*ii*) The air velocity through the tube.

Take characteristic gas-constant *R* for air and fuel as 287 J/kg K and 97 J/kg K, respectively.

[Ans. 35.1 kg/h ; 116 m/s]

C.I. Engines

1. A six-cylinder four-stroke diesel engine develops 89 kW at 2500 r.p.m. Its brake specific fuel consumption is 245 g/kWh. Calculate the quantity of fuel to be injected per cycle per cylinder. Specific gravity of the fuel may be taken as 0.84. **[Ans.** 0.0576 c.c.]

2. A six-cylinder, four-stroke oil engine develops 200 kW at 1200 r.p.m. and consumes 0.3 kg/kWh. Determine the diameter of a single orifice injector if the injection pressure is 200 bar and combustion chamber pressure is 40 bar. The injection is carried out for 30° rotation of a crank. Each nozzle on a cylinder is provided with single orifice.

Take ρ_f = 900 kg/m³ and C_f = 0.7. **[Ans.** 1.005 mm]

3. A four-stroke cycle C.I. engine develops 15 kW per cylinder at 2000 r.p.m. The specific fuel consumption is 0.275 kg/kWh fuel of 30° API. The injection pressure is 130 bar for 25° of crank angle. The pressure in the combustion chamber is 40 bar. The co-efficient of velocity = 0.875 and the specific gravity is given by : Sp. gr. = $\dfrac{141.5}{131.5 + {}^\circ API}$. Assuming fuel to be incompressible, determine the diameter of the fuel orifice.

[Ans. 0.62 mm]

4. An eight-cylinder, four-stroke diesel engine has a power output of 368 kW at 800 r.p.m. The fuel consumption is 0.238 kg/kWh. The pressure in the cylinder at the beginning of injection is 35 bar and the maximum cylinder pressure is 60 bar. The injector is expected to be set at 210 bar and maximum pressure at the injector is set at about 600 bar. Calculate the orifice area required per injector if the injection takes place over 12° crank angle.

Assume the following :

Specific gravity of the fuel = 0.85 ; Coefficient of discharge for the injector = 0.6 ; Atmospheric pressure = 1.013 bar ; The effective pressure difference is the average pressure difference over the injection period.

[Ans. 0.01234 cm²]

5. In a diesel fuel injection pump the volume of the fuel in the pump barrel before the commencement of the effective stroke is 7 c.c. The fuel line from the pump to injector is 3 mm in diameter and 700 mm long and the fuel in the injection valve is 3 c.c.

(*i*) Determine the pump displacement necessary to deliver 0.15 c.c. of fuel at a pressure of 200 bar. Assume a sump (atmospheric) pressure of 1 bar.

(*ii*) Calculate the effective stroke of the plunger if the diameter of the plunger is 8 mm.

Take the coefficient of compressibility of oil as 78.8×10^{-6} per bar at atmospheric pressure.

[Ans. (*i*) 0.384 c.c. ; (*ii*) 7.65 mm]

4

Engine Friction and Lubrication Systems

4.1. INTRODUCTION

- In an I.C. engine almost all machine parts have relative motion and rub against each other. To reduce this rubbing action lubrication is required, which increases the life of engine. The purpose of lubrication in I.C. engine is generally *two-fold* : (*i*) To reduce the rubbing action between different machine parts having relative motion with each other ; (*ii*) To remove the heat generated inside the cylinder.
- **Engine friction** *is defined as the difference between the indicated power (I.P.) (power developed inside the engine) and the brake power (B.P.) (power available at the crank-shaft) i.e.,*

  Frictional power, F.P. = I.P. – B.P. ...(4.1)
- It is impossible to totally remove all the friction loss but it can be reduced by using lubrication between the parts which have relative motion with each other. *Increase in friction is ultimately dissipated as heat to the cooling water and it further increases the pump and fan power requirements also.*
- The frictional resistance between two moving parts having relative motion is mostly dependent on the following *factors* :
  — Lubricating oil properties
  — Surface condition
  — Materials of the surfaces
  — Rate of relative motion
  — Nature of relative motion
  — Quantity of lubricating oil.

4.2. TOTAL ENGINE FRICTION

The difference between I.P. and B.P. is known as ***total engine friction loss.*** This includes the following losses :

1. Direct frictional losses
2. Pumping loss
3. Blowby losses
4. Valve throttling losses
5. Combustion chamber pump loss
6. Power loss to drive the auxiliaries.

1. **Direct frictional losses :**

It includes bearing losses (main bearing, camshaft bearing), piston and cylinder friction loss etc. In reciprocating I.C. engines the frictional losses are comparatively higher.

2. **Pumping loss :**

- In *four-stroke* cycle engines an ample amount of power is used during intake and exhaust processes (Fig. 4.1).

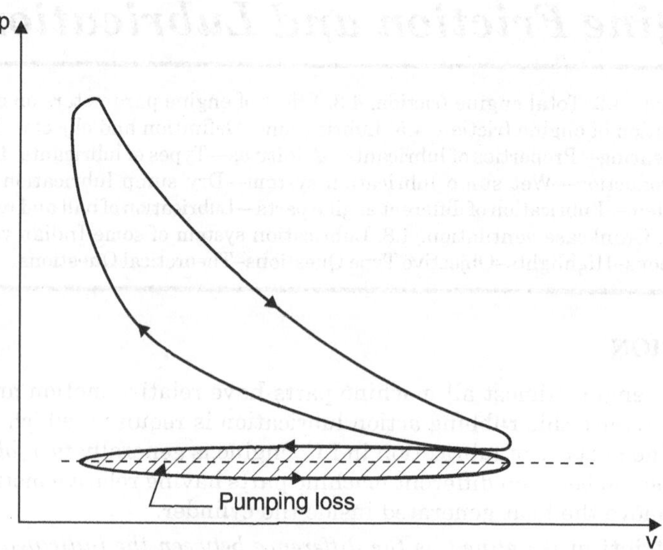

Fig. 4.1

- The pumping loss is *negligible* in *two-stroke* cycle engines since the incoming fresh mixture is used for scavenging the exhaust gases and charging the cylinder.

3. **Blowby losses :**

These losses are *caused due to the leakage of combustion products past the piston from the cylinder into the crankcase.*

— These losses *depend upon the inlet pressure and compression ratio.*

— These losses *increase directly with compression ratio but get reduced with an increase in engine speed.*

4. **Valve throttling losses :**

- The standard practice for sizing the *exhaust valve* is to make them a *certain percentage smaller than the inlet valves.* This usually results in an insufficiently sized exhaust valve and hence, results in exhaust pumping loss.
  - — If due attention is not given to the valve size, valve timing and valve flow coefficients there may be a substantial loss with the increase in engine speed.
- The *inlet throttling loss* occurs due to the *restrictions* imposed by air cleaner, carburettor venturi, throttle valve, intake manifold and intake valve. All these restrictions lead to pressure loss. Similarly some pressure loss is necessary to exhaust the combustion products.

5. **Combustion chamber pump loss :**

This type of loss is caused due *to the pumping work required to pump gases into and out of the pre-combustion chamber*. Its exact value depends upon the orifice size (connecting the fore-combustion chamber and the main chamber), and the speed. *Higher the speed greater is the loss and smaller the orifice size greater is the loss.*

6. **Power loss to drive the auxiliaries :**

In order to drive auxiliaries such as water pump, oil pump, fuel pump, cooling fan and generator some power is needed. This is also considered as loss since a part of engine power developed is used for these purposes.

4.3. EFFECT OF ENGINE PARAMETERS ON ENGINE FRICTION

Following *parameters exercise their influence on engine friction* as described below :

1. Stroke-to-bore ratio
2. Cylinder size and number of cylinders
3. Piston rings
4. Compression ratio
5. Engine speed
6. Engine load
7. Cooling water temperature
8. Oil viscosity.

1. **Stroke-to-bore ratio :**

- The effect of this parameter on engine friction and economy is not very significant.
- Lower stroke-to-bore ratio tends to decrease i.m.e.p. Its lower value reduces the friction losses as the surface area decreases with decreasing stroke-to-bore ratio with the same value of the stroke volume.

2. **Cylinder size and number of cylinders :**

- When a smaller number of large cylinders are used, the friction and economy improve. This is owing to the fact that the proportion between the working piston area and its friction producing area, *i.e.*, circumference is reduced.
- Fig. 4.2 shows the effect of number of cylinders on the variation of friction for the same piston displacement.

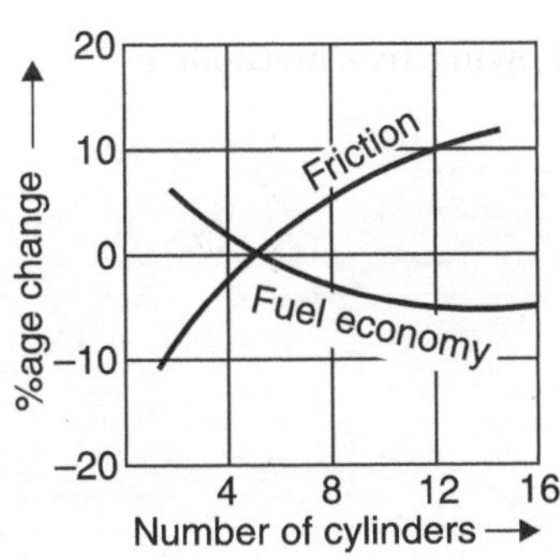

Fig. 4.2. Effect of number of cylinders on friction.

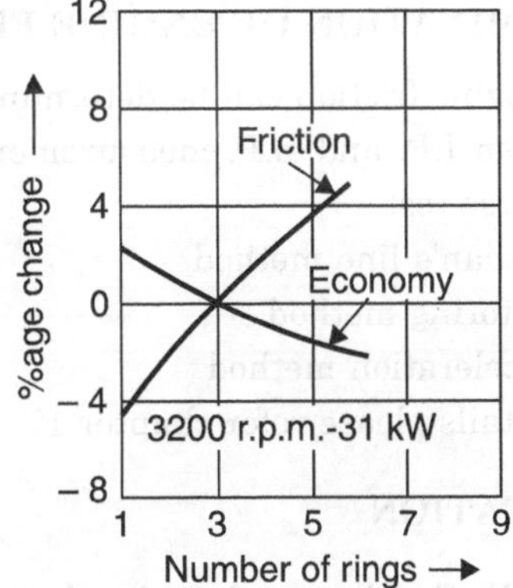

Fig. 4.3. Effect of number of rings on piston friction.

3. **Piston rings :**

The effect of number of piston rings on friction is not significant as the selection depends on the engine size, lightness required and material used for the rings. It is obvious from Fig. 4.3 that generally 3-rings provide best fuel economy.

4. **Compression ratio :**

The frictional mean effective pressure increases as the compression ratio is increased. But the *mechanical efficiency may even increase because of improvement in i.m.e.p.*

5. **Engine speed :**

The *mechanical friction increases with the increase in speed.* From Fig. 4.4. It is evident that there is nearly linear variation in f.m.e.p. with speed.

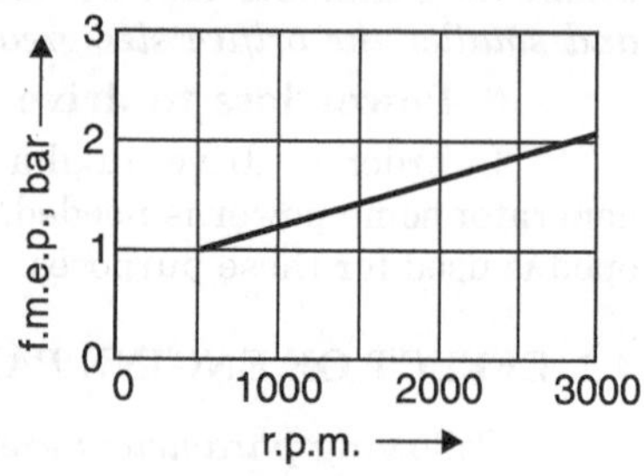

Fig. 4.4. Effect of speed on engine friction.

6. **Engine load :**

When the load on the engine increases, the i.m.e.p. also increases and friction loss also increases. However this increase in friction loss is compensated by decrease in viscosity of the lubricating oil due to higher temperature resulting from increased load.

7. **Cooling water temperature :**

- The *rise in cooling water temperature reduces the frictional loss as the viscosity of oil at higher temperature is lower which reduces the frictional loss.*

8. **Oil viscosity :**

Higher the viscosity of oil greater is the friction loss. As the oil temperature increases, the viscosity decreases and friction losses are reduced during a certain temperature range as shown in Fig. 4.5. If the temperature exceeds a certain value the local film is destroyed causing metal-to-metal contact.

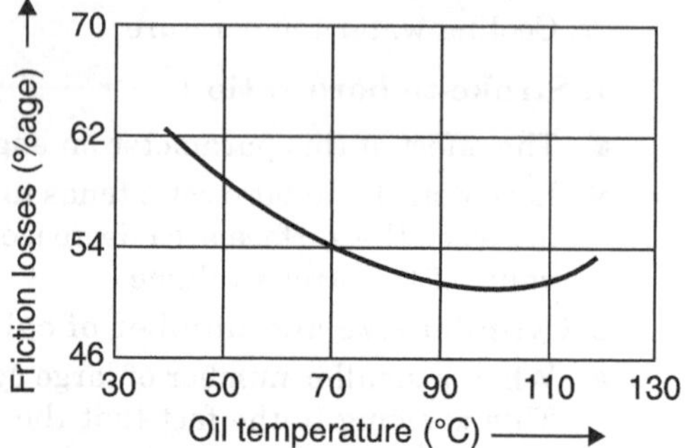

Fig. 4.5. Effect of oil temperature on friction.

4.4. DETERMINATION OF ENGINE FRICTION

The engine friction can be determined by the following five methods :

1. From I.P. and B.P. measurements
2. Morse test
3. Willian's line method
4. Motoring method
5. Deceleration method.

For details please refer chapter 17.

4.5. LUBRICATION

4.5.1. Definition and Objectives

Lubrication *is the admittance of oil between two surfaces having relative motion. The objectives of lubrication may be one or more of the following :*

1. To reduce friction between the parts having relative motion.
2. To reduce wear of the moving part.
3. To cool the surfaces by carrying away heat generated due to friction.

4. To seal a space adjoining the surfaces.
5. To absorb shocks between bearings and other parts and consequently reduce noise.
6. To remove dirt and grit that might have crept between the rubbing parts.

4.5.2. Behaviour of a Journal in its Bearing

Fig. 4.6 shows the behaviour of a journal rotating in a bearing, the clearance between the two being shown very much exaggerated. The clearance space is supposed to be completely filled with oil at all times, which is possible by supplying the oil as fast as it runs out. If the shaft does not rotate it will sink to the bottom of the clearance space due to the load W, and the journal and bearing will touch as shown in Fig. 4.6 (*a*). The way in which the rotating shaft will build up pressure in the oil sufficient to separate the surfaces is shown in Fig. 4.16 (*b*) and (*c*). As the shaft starts to turn it will climb the bearing wall as shown in Fig. 4.6 (*b*). But as the speed increases, the moving journal tends to put oil into the wedge shaped area between the shaft and the bearing. As a result the oil pressure on right side becomes more than on the left side, and the journal is, therefore, forced away from the bearing wall. An equilibrium position is finally reached as shown in Fig. 4.6 (*c*) with the surfaces separated by a film of oil whose minimum thickness is r_o. The magnitude of r_o and position of the line of centres in Fig. 4.6 (*c*) will depend *on the load, the fluid properties of oil, the size and speed of shaft, the clearance and length of the bearing.*

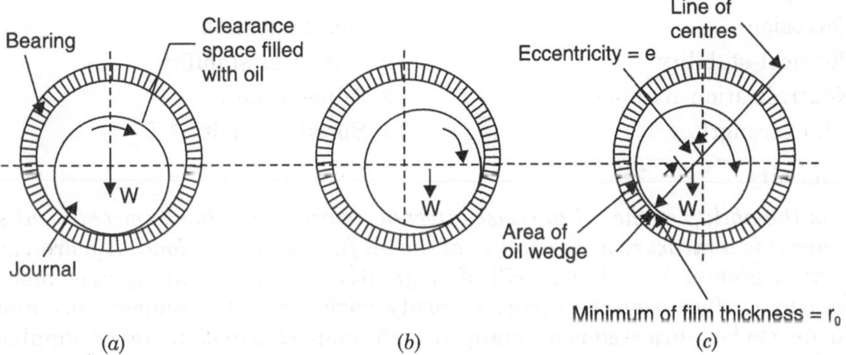

Fig. 4.6. Behaviour of a journal rotating in a bearing.

From above, the following points, about the items influencing bearing performance, can be concluded :

1. A slippery bearing material is desirable during starting and stopping, but once the oil film is established the bearing material is less important.
2. Higher is the speed of journal, more will be the oil pulled into the apex of the wedge of oil in the clearance space of Fig. 4.6 (*c*), and as a result, more supporting pressure will be developed.
3. An increase in supporting pressure will increase the oil film thickness r_o and decrease the eccentricity *e*.
4. If the eccentricity '*e*' is decreased, the supporting pressure must decrease because the sides of the wedge are more nearly parallel.
5. Oil film thickness r_o is also influenced by changing of clearance.

Film Lubrication. It is that type of lubrication in which *bearing surfaces are completely separated by a layer of film of lubricant and that the frictional resistance arises only due to relative movements of the lubricant layers.*

Boundary Lubrication :

- Under the hydro-dynamic condition the oil film supports the load. If the oil film becomes thin enough so as not to support the load without occasional metal to metal contact then journal friction developed is called *boundary friction* and the lubrication existing in this range is known as *boundary lubrication*. The word boundary friction is used because under this condition journal friction is neither completely dry and not completely fluid. In the boundary state, the kind of bearing material, the hardness and surface finish of the shaft, as well as type of lubricant all contribute to the amount of journal friction.
- When the load acting on the bearings is very high, the material itself deforms elastically against the pressure built up of the oil film. This type of lubrication, called **elasto-hydrodynamic** *lubrication*, occurs between *cams and followers, gear teeth, and rolling bearings when the contact pressures are extremely high.*

4.5.3. Properties of Lubricants

The chief qualities to be considered in selecting oil for lubrication are :

1. Viscosity
2. Flash point
3. Fire point
4. Cloud point
5. Pour point
6. Oiliness
7. Corrosion
8. Emulsification
9. Physical stability
10. Chemical stability
11. Neutralisation number
12. Adhesiveness
13. Film strength
14. Specific gravity.

1. **Viscosity**

- *It is the ability of the oil to resist internal deformation due to mechanical stresses and hence it is a measure of the ability of the oil film to carry a load.* A more viscous oil can carry a greater load, but it will offer greater friction to sliding movement of the one bearing surface over the other. Viscosity varies with the temperature and hence if a surface to be lubricated is normally at high temperature it should be supplied with oil of a higher viscosity.
- The viscosity is measured by *viscosimeter*. The important types of viscosimeters are :
  1. Saybolt universal viscosimeter
  2. Red wood viscosimeter
  3. Engler viscosimeter
  4. Barbey viscosimeter.

  The unit of viscosity is given as "*seconds saybolt*" or "*seconds redwood*" etc.
- The present-day method of expressing the rate at which the viscosity of an oil will change with temperature is by stating its *viscosity index* (V.I.). The oil is compared with *two reference oils having same viscosity at 99°C. One is paraffinic base oil (viscosity changes considerably with temperature), is arbitrarily assigned an index of* ***zero*** *and the other, a napthenic base oil (little change in viscosity with temperature) assigned an index of* ***100***.
  - — *A high viscosity index indicates relatively smaller changes in viscosity of the oil with the temperature.*
  - — The higher the viscosity index, the lower the rate at which its viscosity decreases with the increase in temperature. Although a high viscosity index is desirable in materials and much effort expended in improving the viscosity index of oils by

refining, it should be emphasized that viscosity-temperature characteristics are of little importance for oils that are to function at approximately constant temperature, such as *turbine oils.*

— *Viscosity index of oil is very important where extreme temperatures are encountered.*

- In order to improve the viscosity index of an oil certain compounds, called V.I. improvers, are added to it. These are long chain paraffinic compounds which enable to obtain an oil having easy starting characteristic of thin oils combined with good protection against high temperature.

2. **Flash point.** *It is defined as the lowest temperature at which the lubricating oil will flash when a small flame is passed across its surface.* The flash point of the oil should be *sufficiently high so as to avoid flashing of oil vapours at the temperatures occurring in common use.* High flash point oils are needed in air compressors.

3. **Fire point.** *It is the lowest temperature at which the oil burns continuously.* The fire point also *must be high in a lubricating oil, so that oil does not burn in service.*

4. **Cloud point.** When subject to low temperatures the oil changes from liquid state to a plastic or solid state. In some cases the oil starts solidifying which makes it to appear cloudy. The temperature at which this takes place is called the *cloud point.*

5. **Pour point.** *Pour point is the lowest temperature at which the lubricating oil will pour.* It is an indication of its ability to move at low temperatures. This property must be considered because of its effect on starting an engine in cold weather and on free circulation of oil through exterior feed pipes when pressure is not applied.

6. **Oiliness.** *This is the property which enables oil to spread over and adhere to the surface of the bearing.* It is most important in boundary lubrication.

7. **Corrosion.** A lubricant *should not corrode* the working parts and it must retain its properties even in the presence of foreign matter and additives.

8. **Emulsification.** A lubricating oil, when mixed with water is emulsified and loses its lubricating property. The *emulsification number is an index of the tendency of an oil to emulsify with water.*

9. **Physical stability.** A lubricating oil *must be stable physically at the lowest and highest temperatures between which the oil is to be used.* At the lowest temperature there should not be any separation of solids, and at the highest temperature it should not vapourise beyond a certain limit.

10. **Chemical stability.** A lubricating oil should also be stable chemically. There should not *be any tendency for oxide formation.*

11. **Neutralisation number.** An oil may contain certain impurities that are not removed during refining. The neutralisation number test is a simple procedure to determine acidity or alkalinity of an oil. It is the weight in milligrams of potassium hydroxide required to neutralise the acid content of one gram of oil.

12. **Adhesiveness.** *It is the property of lubricating oil due to which the oil particles stick with the metal surfaces.*

13. **Film strength.** *It is the property of a lubricating oil due to which the oil retains a thin film between the two surfaces even at high speed and load.* The film does not break and the two surfaces do not come in direct contact. Adhesiveness and film strength cause the lubricant to enter the metal pores and cling to the surfaces of the bearings and journals keeping them wet when the journals are at rest and presenting metal to metal contact until the film of lubricant is built up.

14. **Specific gravity.** It is a *measure of density of oil.* It is an indication regarding the grade of lubricant by comparing one lubricant with other. It is determined by a hydrometer which floats in the oil, and the gravity is read on the scale of the hydrometer at the surface of the oil.

4.5.3.1. Additives

Primarily, there are two types of oils : *Mineral* and *vegetable*. Both of them fulfill some basic requirements as lubricating oil. The vegetable oil is superior to mineral oil under some extreme conditions, but it is rarely used as it is too expensive oil. The *mineral oils are very commonly used for all lubricating purposes.*

— Simple mineral oils have most of the required characteristics as a good lubricant. However, varying operating conditions require some specific properties which it cannot meet, as high viscosity index and resistance to oxidation and corrosion. In order to achieve some required properties, different types of compounds, called *additives* are added.

1. **Detergents :**

- Control high-temperature deposits such as gums.
- If overbased, a detergent acts as effective acid neutraliser.

2. **Dispersants :**

- Control low-temperature deposits such as cold sludge and varnish deposits.

3. **Anti-wear additives :**

- These additives reduce wear and prevent scoring, galling and seizure.
- They provide an extra strength needed to ensure efficient lubrication under severe operating conditions.
- The additives used are : *Chlorine* and *phosphate compounds.*

4. **Rust inhibitors :**

- Reduce rusting by acid neutralisation of formation by protective film.

5. **Viscosity Index (V.I.) improver :**

- *It prevents undue thinning of the oil as the temperature increases.*
- A high viscosity index is always desired for lubricating oils as it can be used in severe winter and summer conditions. *High molecule polymers* are used for the purpose as they are soluble in the oil.
- Recently *multigrade oils* have been produced which consist of low viscosity oils within thicking polymers. Their special ability is to adopt to both summer and winter conditions.

6. **Pour point additives :**

- The base oil at extremely low temperatures change into buttery solids, due to formation of wax crystals. If left unchecked, the wax crystals cogulate into honeycomb like structure which first inhibit and then totally suppress the oil flow. *These additives reduce pour point of oil by interfering with wax cystallization.*
- *Polymerized phenols or esters are used upto 1 percent of the oil* for the purpose.

7. **Anti-foam agents :**

- Reduce oil foaming by causing collapse of bubbles due to air entrainment.
- *Silicon esters are used as antifoam agents.*

8. **Anti-oxidants :**

- Reduce oil oxidation to protect alloy bearings against corrosive attack.

9. **Oiliness improvement :**

- Some substances such as colloidal graphite and zinc oxide when added to the oil are valuable in maintaining the oil film.

4.5.3.2. Oil contamination and sludge formation

- The lubricating oil after a certain operating period gets contaminated to the extent that it becomes unsuitable for further use. The *contamination* of the oil takes place due to *oxidation, dilution, water, formation of carbon, lead compounds, metals, dust and dirt.* When these contaminants mix with the oil, sludge is formed in an engine.
- **Sludge** is a black, brown or grey deposit having the consistency of mud. *Its formation takes place as a result of operation at low engine temperatures during starting, warming up, and idling periods.*

4.5.4. Types of Lubricants

Most lubricants are *oils* or *greases.* However, in special circumstances other fluids like *water, air etc. and solids* such as *graphite* may perform the function of lubrication. Synthetic lubricants are also used occasionally.

1. **Oils :**

The different lubricating oils are : *Mineral, Fatty* and *Synthetic.*

- The **mineral lubricating oils** are obtained from the residual mass left during crude petroleum distillation. In this treatment gasoline, kerosene oil, and gas oil fractions are recovered from the distillate. Paraffinic crudes generally give a higher yield of lubricating oils than napthenic crudes. The residue is subject to vacuum distillation. In several cases steam is also introduced into the system. Bubble towers are used for fractionalisation of oil into two or three fractions of different viscosities.
- **Fatty oils** from animal and vegetable origin are sometimes used alone but are frequently mixed with mineral oils. The fatty oils exhibit poor keeping quality and undergo decomposition. Fatty oils however, exhibit more oiliness than mineral oils of the same viscosity. Non drying oils such as olive oil, castor oil, rapeseed oil, lard and fish oil are mainly used for lubricating purposes. They are generally compounded with mineral oil to the extent of about 2 to 10 percent in order to increase oiliness of mineral oils. *Olive oil* is employed for lubricating textile machinery, as it can be easily washed out from the cloth without leaving any stain. ***Rapeseed oil*** is used for making lubricants for *railway engines,* while lard oil is used for producing lubricants for *internal combustion engines.* ***Caster oil*** is usually compounded with heavy mineral oils for making *extreme pressure lubricants.*
- **Synthetic lubricants** are named synthetic because they are not obtained directly from petroleum. Various applications require oils which will have, for example, a lower point, or better viscosity-temperature characteristics, or a higher degree of resistance to oxidation or isothermal decomposition than petroleum derived lubricants do. The synthetic lubricants presently in use are *classified* as :

  (*i*) Dibasic acid esters ;
  (*ii*) Organo-phosphate esters ;
  (*iii*) Silicate esters ;
  (*iv*) Silicon polymers ;

(*v*) Polyglycolethers and related compounds ; and

(*vi*) Fluorinated and chlorinated hydrocarbon compounds.

Classification of lubricating oils

- The lubricating oils are *normally classified according to their viscosity.* The SAE (Society of Automotive Engineers) method of assigning number to different oils is used universally.
  - — SAE has assigned a number to an oil whose viscosity at given temperatures falls in certain range. There are two temperatures used as reference for assigning the number to oils – 18°C and 99°C.
  - — SAE, 5 W, 10 W, 20 W grades are defined in terms of viscosity at – *18°C* and are the oils *which render starting of engine in cold weather easy.* SAE, 20, 30, 40 and 50 grades are defined in terms of viscosity at *99°C* ; these oils *work satisfactorily in normal and hot climates.* These numbers are merely used for classification of oil according to viscosity and *do not indicate the quality of* oil since factors like stability, oiliness etc. are not considered (by these numbers).

Table 4.1. SAE classification of Lubricanting Oils

SAE Viscosity Number	Viscosity Units	Viscosity Range			
		At – 18°C		At 99°C	
		Min.	Max.	Min.	Max.
5 W	(*a*) Centipoise	—	1200	—	—
	(*b*) SUS	—	6000	—	—
10 W	(*a*) Centipoise	1200	2400	—	—
	(*b*) SUS	6000	12000	—	—
20 W	(*a*) Centipoise	2400	9600	—	—
	(*b*) SUS	12000	48000	—	—
20	(*a*) Centistokes	—	—	5.7	9.6
	(*b*) SUS	—	—	45	58
30	(*a*) Centistokes	—	—	9.6	12.9
	(*b*) SUS	—	—	58	70
40	(*a*) Centistokes	—	—	12.9	16.8
	(*b*) SUS	—	—	70	85
50	(*a*) Centistokes	—	—	16.8	22.7
	(*b*) SUS	—	—	85	110

- **Multigrade oils.** It is possible to develop an oil with more than one viscosity at different temperatures. Thus an oil may be in SAE-30 grade at 99°C and in the SAE-10 W grade at – 18°C, oils of this type are known as *multigrade oils* ; other possible grades are : 5 W/20, 20 W/20, 20 W/40.

Following are the "*advantages*" *of multigrade oils* :

(*i*) No necessity to change the oil as per the ambient temperature.

(*ii*) Owing to ease of cranking at low ambient temperature the life of battery is extended.

(*iii*) Long engine life ; required viscosity is maintained under different operating temperatures.

(*iv*) Easy starting and short warming up period.

(*v*) Reduced oil consumption.

(*vi*) Prolonged mileage between the decarbonisation.

(*vii*) Due to execellent thermal and oxidation stability of multigrade oils the engine parts are protected against rust, corrosion and wear.

2. Greases :

- ***Lubricating grease** is a solid to semi-solid dispersion of a thickening agent in liquid lubricant. Other ingredients imparting special properties may be included.*
- Greases are normally used under conditions of lubrication for which oil is not as suitable or convenient. *Greases preform better than oils under conditions requiring* :

  (*i*) High bearing loads and shock loads.

  (*ii*) Slow journal speed.

  (*iii*) Temperature extremes.

  (*iv*) Cleanliness or avoidance of splash or drip.

  (*v*) Minimum attention.

  (*vi*) A seal against external contaminants.

  (*vii*) Large bearing clearances.

The various *types of greases used for lubrication are* :

(*i*) Calcium soap greases ;

(*ii*) Sodium soap greases ;

(*iii*) Aluminium soap greases ;

(*iv*) Mixed soap greases ;

(*v*) Barium soap greases ;

(*vi*) Lithium soap greases ;

(*vii*) Pure petroleum greases.

4.6. LUBRICATION SYSTEMS

4.6.1. Introduction

The *main parts of an engine which need lubrication* are as given under :

(*i*) Main crankshaft bearings.

(*ii*) Big-end bearings.

(*iii*) Small end or gudgeon pin bearings.

(*iv*) Piston rings and cylinder walls.

(*v*) Timing gears.

(*vi*) Camshaft and camshaft bearings.

(*vii*) Valve mechanism.

(*viii*) Valve guides, valve tappets and rocker arms.

Various *lubrication systems used for I.C. engines may be classified as* :

1. Wet sump lubrication system.
2. Dry sump lubrication system.
3. Mist lubrication system.

4.6.2. Wet Sump Lubrication System

These systems employ a large capacity oil sump at the base of crankchamber, from which the oil is drawn by a low pressure oil pump and delivered to various parts. Oil there gradually returns back to the sump after serving the purpose.

(*a*) **Splash system :**

- This system (Fig. 4.7) is used on some *small four-stroke stationary engines*. In this case the caps on the big ends bearings of connecting rods are provided with scoops which, when the connecting rod is in the lowest position, just dip into oil troughs and thus direct the oil through holes in the caps to the big end bearings. Due to splash of oil it reaches the lower portion of the cylinder walls, crankshaft and other parts requiring lubrication. Surplus oil eventually flows back to the oil sump. Oil level in the troughs is maintained by means of a oil pump which takes oil from sump, through a filter.

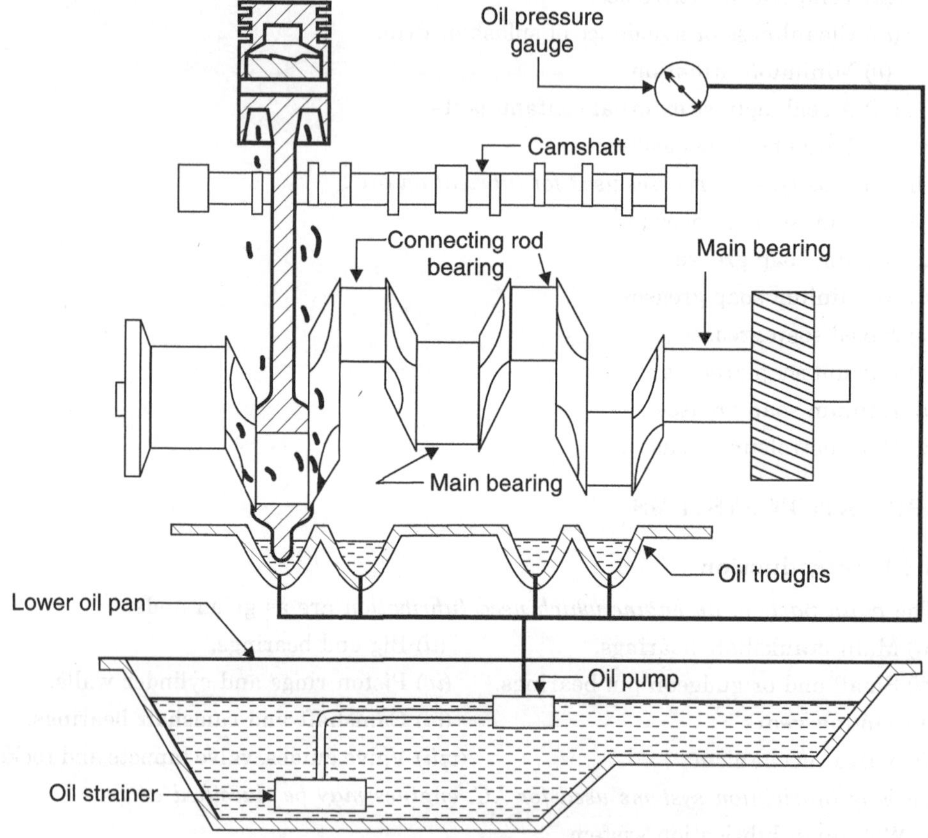

Fig. 4.7. Splash system.

- Splash system is suitable for *low and medium speed engines having moderate bearing load pressures*. For high performance engines, which normally operate at high bearing pressures and rubbing speeds this system does not serve the purpose.

(*b*) **Semi-pressure system :**

- This method is a combination of splash and pressure systems. It incorporates the advantages of both. In this case main supply of oil is located in the base of crank chamber. Oil is drawn from the lower portion of the sump through a filter and is delivered by means of a gear pump at pressure of about 1 bar to the main bearings. The big end bearings are lubricated by means of a spray through nozzles. Thus oil also lubricates the cams, crankshaft bearings, cylinder walls and timing gears. An oil pressure gauge is provided to indicate satisfactory oil supply.
- The system is *less costly to install as compared to pressure system. It enables higher bearing loads and engine speeds to be employed as compared to splash system.*

(*c*) **Full pressure system :**

- In this system, oil from oil sump is pumped under pressure to the various parts requiring lubrication. Refer Fig. 4.8. The oil is drawn from the sump through filter and pumped by means of a gear pump. Oil is delivered by the pressure pump at pressure ranging from 1.5 to 4 bar. The oil under pressure is supplied to main bearings of crankshaft and camshaft. Holes drilled through the main crankshafts bearing journals, communicate oil to the big end bearing and also small end bearings through hole drilled in connecting rods. A pressure gauge is provided to confirm the circulation of oil to the various parts. A pressure regulating valve is also provided on the delivery side of this pump to prevent excessive pressure.

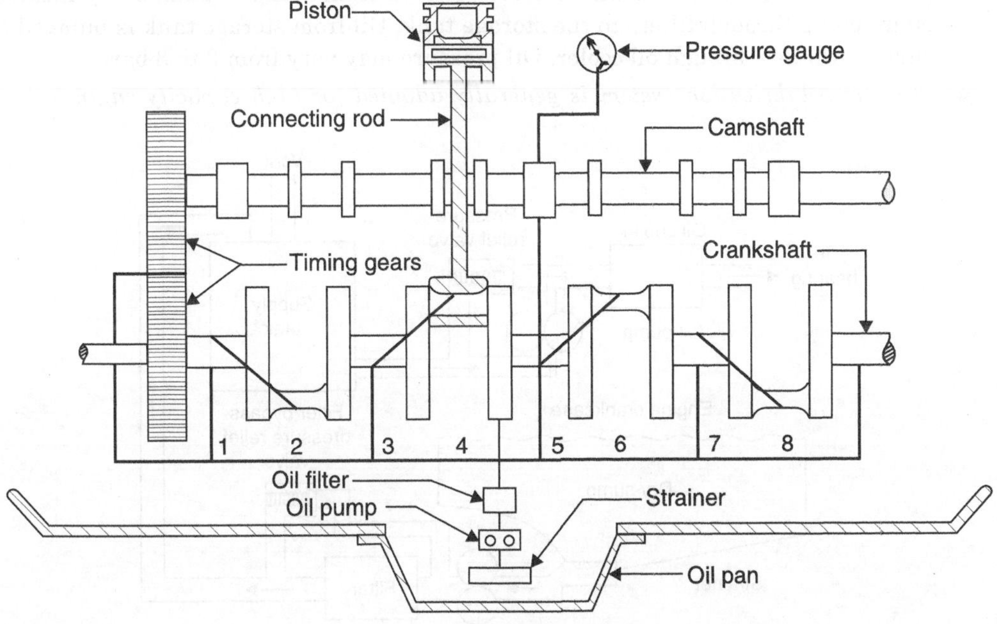

Fig. 4.8. Full pressure system.

- This system *finds favour from most of the engine manufacturers as it allows high bearing pressure and rubbing speeds.*
- The general arrangement of wet sump lubrication system is shown in Fig. 4.9. In this case oil is always contained in the sump which is drawn by the pump through a strainer.

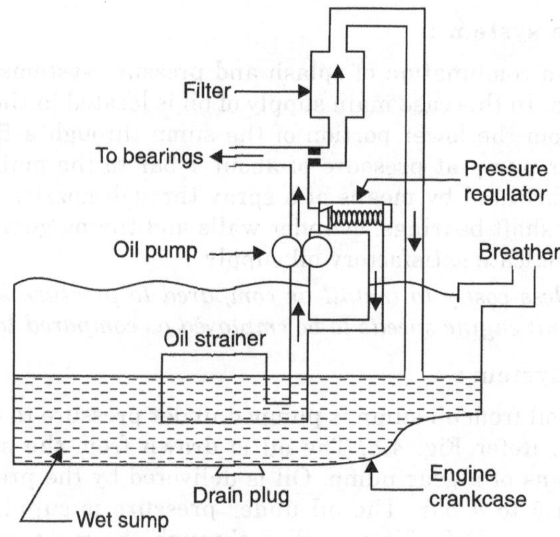

Fig. 4.9. Wet sump lubrication system.

4.6.3. Dry Sump Lubrication System

- Refer Fig. 4.10. In this system, the oil from the sump is carried to a separate storage tank outside the engine cylinder block. The oil from sump is pumped by means of a sump pump through filters to the storage tank. Oil from storage tank is pumped *to the engine cylinder* through oil cooler. Oil pressure may vary from 3 to 8 bar.
- *Dry sump lubrication system is generally adopted for high capacity engines.*

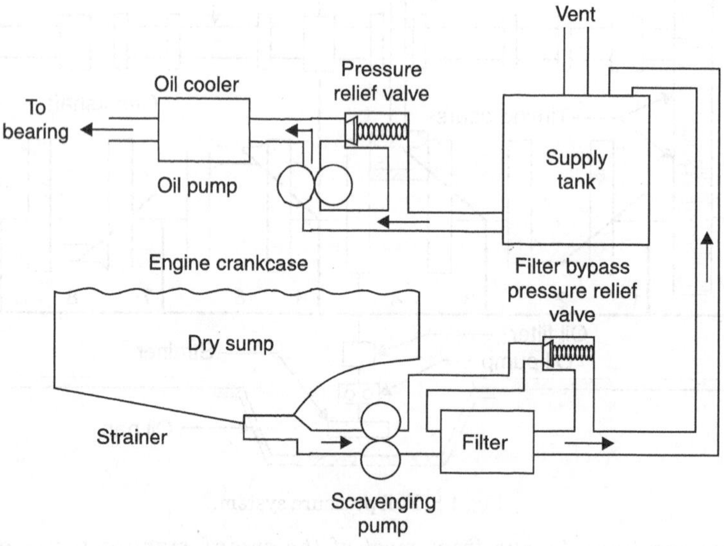

Fig. 4.10. Dry sump lubrication system.

4.6.4. Mist Lubrication System

- *This system is used for two stroke cycle engines.* Most of these engines are crankcharged, *i.e.*, they employ crankcase compression and thus, are not suitable for crankcase lubrication.
- These engines are lubricated by adding 2 to 3 per cent lubricating oil in the fuel tank. The oil and fuel mixture is induced through the carburettor. The gasoline is vaporised ; and the oil in the form of mist, goes via crankcase into the cylinder. The oil which impinges on the crankcase walls lubricates the main and connecting rod bearings, and rest of the oil which passes on the cylinder during charging and scavenging periods, lubricates the piston, piston rings and the cylinder.
- For good performance, F/A ratio used is also important. A F/A *ratio of 40 to 50 : 1 is optimum. Higher ratios increase the rate of wear and lower ratios result in spark plug fouling.*

Advantages :

1. System is simple.
2. Low cost (because no oil pump, filter etc. are required).

Disadvantages :

1. Some lubrication oil will burn and cause heavy exhaust emissions and deposits on piston crown, ring grooves and exhaust port and thus hamper the good performance of the engine.
2. Since the lubricating oil comes in contact with acidic vapours produced during the combustion process, it *rapidly loses its anti-corrosion properties* resulting in *corrosion damage of bearing.*
3. The oil and fuel must be thoroughly mixed for effective lubrication. This requires either separate mixing prior to use or use of some additive to give the oil good mixing characteristics.
4. Owing to higher exhaust temperature and less efficient scavening the crankcase oil is diluted. In addition some lubricating oil burns in combustion chamber. This results in *5 to 15 per cent higher lubricant consumption for two stroke engine of similar size.*
5. As there is no control over the lubricating oil, once introduced with fuel, most of the two stroke engines are *over-oiled most of the time.*

4.6.5. Lubrication of Different Engine Parts

1. **Lubrication of main bearings :**

- The *main bearings* are lubricated satisfactorily with the help of a ring (or chain) type feeder.
- The ring oiling system is shown in Fig. 4.11. It consists of a ring which is bigger in diameter than the shaft. There is one groove at the bearing cap and the ring is placed on the shaft. The lower part of the ring is dipped into an oil reservoir. Due to the rotation of the shaft, the ring rotates at a slower speed and carries the oil from the reservoir to the bearing. *This system works on the principle of adhesion.* In this system the lubricating oil can be fed only *when the shaft is rotating.* Instead of rings, sometimes *chains* are also used.

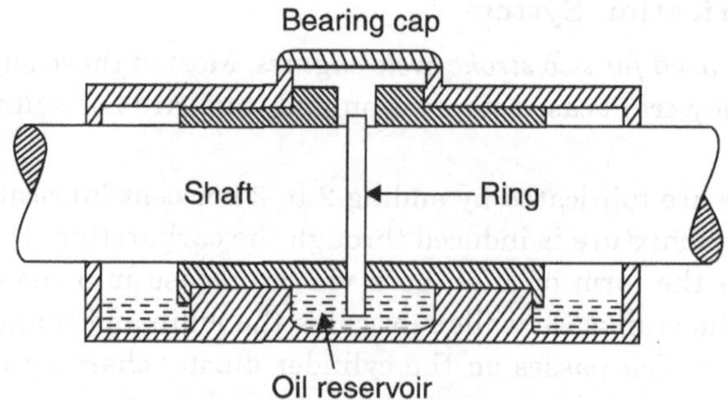

Fig. 4.11. Ring oiling system.

- This type of lubrication is *more useful for medium speed engines because at high speeds the oil will be thrown off due to centrifugal force and at low speeds, the amount of oil carried is inadequate.*

2. **Lubrication of cylinder and small end bearing of connecting rod :**

- The cylinder, small end bearing (gudgeon pin), valve gear pins, rocker shaft, crankpins etc. are lubricated by *drip system.*
- In drip system oil is fed to machine parts drop by drop, from an oil cup. Although it is not an efficient method yet it is often the most convenient way of lubricating the external parts of engines and machines. Valve gear pins, rocker shafts, main bearings of small engines, crank-pins, cross head pins, line shaft bearings, and many other machine and engine parts are lubricated in this fashion.

Fig. 4.12 shows a *drop feed oil cup.* It comprises of an oil reservoir or cup made of glass. At the top of the cup there is an opening called filling hole through which the lubrication oil is poured into the cup. The oil enters into the inner chamber through the openings and it falls drop by drop due to gravity by the nozzle. The passage of oil through the nozzle is controlled by a needle valve. A sight feed glass is provided to see the drops of oil.

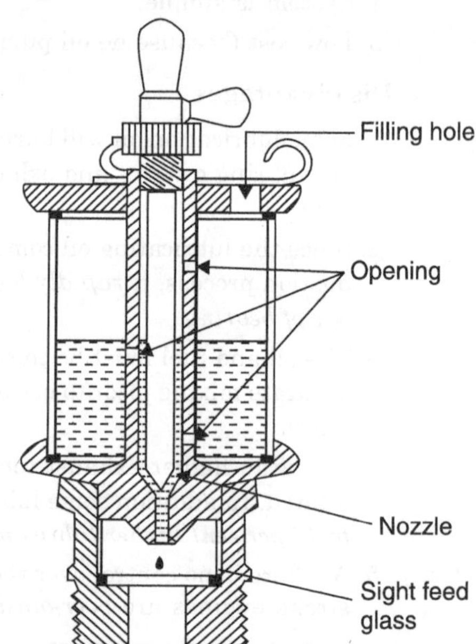

Fig. 4.12. Drop feed oil cup.

3. **Lubrication of crank and gudgeon pin :**

Fig. 4.13 shows *splash lubrication system.* The connecting rod is dipped into the oil of the crankcase. At the time of rotation the oil is splashed due to centrifugal force and it reaches the different parts requiring it. This type of lubrication is suitable for a crankshaft speed of 200 r.p.m. or more. The oil must be filtered or renewed periodically. Also the level of the oil should be maintained.

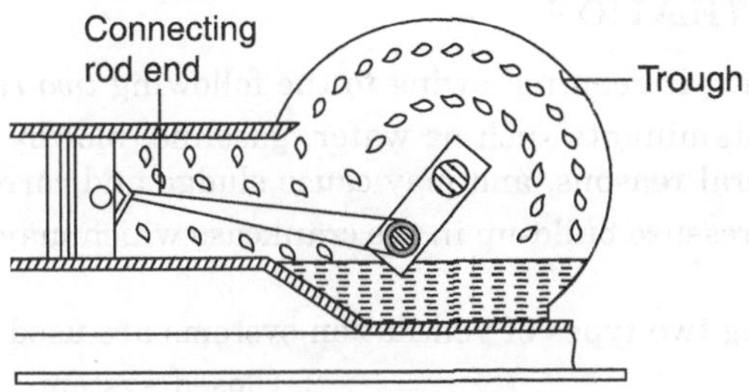

Fig. 4.13. Splash lubrication system.

Note. *Grease cup method* is used for lubricating rocker arms and reciprocating parts with a jerkey motion. It is also employed on parts that are not readily accessible and can be lubricated only at fairly long intervals.

4.6.6. Lubrication of Ball and Roller Bearings

The ball and roller bearings are lubricated for the following purposes :

1. To reduce friction and wear between the sliding parts of the bearing.
2. To prevent rusting or corrosion of bearing surfaces.
3. To protect the bearing surfaces from water, dirt etc.
4. To dissipate the heat.

For lubricating the ball and roller bearings, *generally oil or light grease is used.* Only pure mineral oil or a calcium-base grease should be used. If there is a possibility of moisture contact, then potassium or sodium base greases may be used. Another additional advantage of the grease is that it forms a seal to keep out dirt or any other foreign substance. The temperature should be kept below 90°C and in no case a bearing should operate above 150°C.

4.6.7. Oil Filters

All the lubricating oil (used for lubrication purpose) from the oil sump, must pass through an oil filter before it is supplied to the engine bearings. Bearings maintain very close tolerances and are likely to be damaged by any foreign abrasive material entering the lubrication line.

The filter arrangement may be of the following two types : (*i*) *By-pass type* ; (*ii*) *Full-flow type.*

1. **By-pass type filter arrangement :**

- In this arrangement only a small portion of the lubricating oil is passed through the filter and the remaining lubricating oil is directly supplied to the bearings by the oil pump at pre-set pressure, determined by the pressure regulating valve. Consequently a portion of the oil is continuously filtered.
- Since quantities of oil flowing through filter are *small*, a *very fine filter or a special filter impregnated with resin to avert disintegration due to moisture is used.* Such a fine paper/filter will remove all harmful contaminants.

2. **Full-flow type filter arrangement :**

- In this filter arrangement *whole of the oil is filtered* before it is supplied to various-bearings. Thus, the size of the filter is comparatively large.
- In this case, it is hardly possible to remove very fine particles because of high pressure required to pump oil through such filters.
- All the lubricating oil, in normal course, should be filtered approximately every half minute. A pressure relief valve is used to prevent excessive pressure build up after a cold start.

4.7. CRANKCASE VENTILATION

Crankcase ventilation is required owing to the following *two reasons* :

(*i*) The various contaminants such as water, gasoline, blowby gases etc. enter the crankcase due to several reasons, and may cause sludge and corrode metal parts.

(*ii*) To relieve any pressure build-up in the crankcase which may cause leakage of the crankshaft seal.

In practice, following two types of ventilation systems are used :

1. Open system 2. Closed system.

1. Open system :

- In this system, fresh air supply is inducted into the crankcase during the compression stroke (due to creation of small vacuum). The entering air picks up the contaminants (water vapour, gases and H_2SO_4 vapour) and discharge them to the atmosphere during expansion stroke.
- The main *disadvantage* of this system is that the *natural ventilation is quite inadequate during idling or running at low speeds.*

2. Closed system :

- In closed system the *fresh air supply is taken to the crankcase from the carburettor.*
- Air cleaner and the breather outlets are connected to the intake manifold through a PCV valve to ensure the burning of all the crankcase gases in combustion chamber.

4.8. LUBRICATION SYSTEM OF SOME INDIAN VEHICLES

S. No.	Make of the vehicle	Type of lubrication system	Oil pump	Oil filter	Oil pressure (kg/cm²)
1.	*Ambassador*	Wet sump (pressure type)	Eccentric	Full flow with paper element	3.5
2.	*Premier Padmini*	Wet sump (pressure type)	Gear type	(*i*) Full flow (*ii*) Bypass cartridge type	2.5
3.	*Maruti 800*	Wet sum (pressure type)	Internal Gear type	Full flow cartridge	3.0
4.	*Jeep Mahindra*	Wet sump	Gear type (pressure type)	Full flow	4.0
5.	*Maruti Gypsy*	Wet sump (pressure type)	Internal Gear type	Full flow cartridge type	4.5
6.	*Ashok Leyland Comet Passenger*	We sump (pressure type)	Gear type	Full flow	4.2
7.	*Tata Truck*	Wet sump	Gear type (pressure type)	Full flow	4.2
8.	*Santro*	Forced circulation system	Gear type driven by crankshaft	Full flow cartridge type with relief valve	1.5
9.	*Maruti Baleno*	Wet sump (pressure feeding type)	Trochoid type	Full flow cartridge	3.3
10.	*Maruti Wagon R.*	Wet sump (pressure feeding type)	Internal Gear type	Full flow cartridge	3.3

SHORT ANSWER QUESTIONS

Q.1. What are the various methods of oil lubrication to the engine parts ?

Ans. (*i*) Splash lubrication ; (*ii*) Pressure lubrication ; (*iii*) Pressure and splash lubrication.

Q. 2. How often should the filter be replaced ?

Ans. (*i*) Under clear operating conditions–after 6000 km ; (*ii*) Under dirty and dusty condition–after 4000 km.

Q. 3. What does S.A.E. indicate ?

Ans. S.A.E. indicates the number by which lubricating oils are graded. It means Society of Automotive Engineers.

Q. 4. Is S.A.E. oil less viscous than S.A.E. oil ?

Ans. No.

Q. 5. How are pistons usually lubricated ?

Ans. Piston lubrication is by oil, which is thrown from the ends of the big end bearings.

Q. 6. How is the connecting rod bearing lubricated ?

Ans. For connecting rod bearing lubrication, the lubricant is forced by a centrifugal force through holes drilled in the cap.

Q. 7. Does lead in fuel act as a lubricant ?

Ans. Yes.

Q. 8. Why should lubricating oil be changed ?

Ans. Lubricating oil should be changed otherwise it gets contaminated due to carbon, fuel, water and dirt.

Q. 9. How is oil supplied to the big end bearings in a forced feed system ?

Ans. In a forced feed system oil is supplied through a passage drilled in the crankshaft, leading from the main bearings.

Q. 10. What is meant by oil pan ?

Ans. Oil pan is the lower portion of the crankcase that holds the oil when it is not circulated.

HIGHLIGHTS

1. The difference between I.P. and B.P. is known as *total engine friction loss.*
2. *Lubrication* is the admittance of oil between two surfaces having relative motion.
3. *Film lubrication* is that type of lubrication in which bearing surfaces are completely separated by a layer of film of lubricant and that the frictional resistance arises only due to relative movement of the lubricant layers.
4. If the oil film becomes thin enough so as not to support the load without occasional metal-to-metal contact then journal friction developed is called *boundary friction* and the lubrication existing in this range is known as *boundary lubrication.*
5. *Viscosity* is the ability of the oil to resist internal deformation due to mechanical stresses and hence it is a measure of the ability of the oil film to carry a load.
6. *The mineral oils are very commonly used for all lubricating purposes.*
7. *Lubricating grease* is a solid to semi-solid dispersion of a thickening agent in liquid lubricant.
8. Lubrication systems are classified as follows :
   (*i*) Wet sump lubrication system
   (*ii*) Dry sump lubrication system
   (*iii*) Mist lubrication system used for *two stroke engines.*

OBJECTIVE TYPE QUESTIONS

Fill in the blanks or say "Yes" or "No" :

1. Engine friction is defined as the difference between I.P. and ....
2. The pumping loss in two-stroke cycle engines is quite significant.
3. .... losses are caused due to the leakage of combustion products past the piston from the cylinder into the crankcase.
4. Blow by losses decrease directly with compression ratio.
5. Combustion chamber pump loss is caused due to the pumping work required to pump gases into and out of the pre-combustion chamber.
6. Lower stroke-to-bore ratio tends to .... i.m.e.p.
7. When a small number of large cylinders are used the friction and economy improve.
8. The effect of number of piston rings on friction is not significant.
9. The frictional mean effective pressure increases as the compression ratio is increased.
10. The mechanical friction .... with the increase in speed.
11. The rise in cooling water temperature .... the frictional loss.
12. When the load on the engine increases, the i.m.e.p. also increases.
13. Higher the viscosity of oil lower is the friction loss.
14. .... is the admittance of oil between two surfaces having relative motion.
15. .... lubrication is that type of lubrication in which bearing surfaces are completely separated by a layer of film of lubricant and that the frictional resistance arises only due to relative movement of the lubricant layers.
16. When the load on the bearings is very high, the material itself deforms elastically against the pressure built up of the oil film. This type of lubrication is called .... lubrication.
17. .... is the ability of the oil to resist internal deformation due to mechanical stresses and hence it is a measure of the ability of the oil film to carry a load.
18. The viscosity is measured by ....
19. A more viscous oil can carry a greater load.
20. .... is the method of expressing the rate at which the viscosity of an oil will change with temperature.
21. A low viscosity index indicates relatively smaller changes in viscosity of the oil with the temperature.
22. .... point is the lowest temperature at which the lubricating oil will flash when a small flame is passed across its surface.
23. .... point is the lowest temperature at which oil burns continuously.
24. .... point is the lowest temperature at which the lubricating oil will pour.
25. .... is the property of an oil which enables it to spread over and adhere to the surface of the bearing.
26. The .... number is an index of the tendency of an oil to emulsify with water.
27. .... is the property of lubricating oil due to which the oil particles stick with the metal surfaces.
28. Specific gravity is a measure of density of oil.
29. The .... oils are very commonly used for all lubricating purposes.
30. Viscosity index improver prevents undue thinning of the oil as the temperature increases.
31. Silicon easters are used as anti foam agents.
32. .... is a black, brown or grey deposit having the consistency of mud.
33. The mineral lubricating oils are obtained from the residue mass left during crude petroleum distillation.
34. Splash system is used on some small four-stroke stationary engines.
35. Semi-pressure system is a combination of splash and pressure systems.
36. .... sump lubrication system is generally adopted for high capacity engines.
37. .... lubrication system is used for two-stroke cycle engines.
38. In .... crankcase ventilation system fresh air supply is taken to the crankcase from the carburettor.

39. In .... crankcase ventilation system fresh air supply is inducted into the crankcase during the compression stroke.
40. Lubricating grease is a solid to semi-solid dispersion of a thickening agent in liquid lubricant.

ANSWERS

1. B.P.	2. No	3. Blowby	4. No	5. Yes
6. decrease	7. Yes	8. Yes	9. Yes	10. increases
11. reduces	12. Yes	13. No	14. lubrication	15. Film
16. elasto-hydrodynamic		17. viscosity	18. viscosimeter	19. Yes
20. viscosity index	21. No	22. Flash	23. Fire	24. Pour
25. Oiliness	26. emulsification	27. Adhesiveness	28. Yes	29. mineral
30. Yes	31. Yes	32. Sludge	33. Yes	34. Yes
35. Yes	36. Dry	37. Mist	38. closed	39. open
40. Yes.				

THEORETICAL QUESTIONS

1. How is 'engine friction' defined ?
2. State the importance of engine friction.
3. Enumerate the factors on which the frictional resistance between two moving parts having relative motion is dependent.
4. Discuss the components into which the total engine friction can be divided.
5. Explain briefly the following :
   (*i*) Direct frictional losses
   (*ii*) Valve throttling losses
   (*iii*) Combustion chamber pump loss
   (*iv*) Blowby losses.
6. State the effect of the following engine parameters on engine friction :
   (*i*) Stroke-to-bore ratio
   (*ii*) Cylinder size and number of cylinders
   (*iii*) Piston rings
   (*iv*) Compression ratio
   (*v*) Engine speed
   (*vi*) Engine load
   (*vii*) Cooling water temperature
   (*viii*) Oil viscosity.
7. Enumerate the methods by which engine friction can be determined.
8. Define the term 'lubrication'.
9. What are the objects of lubrication ?
10. Discuss the behaviour of a journal in its bearing.
11. Explain briefly the following :
    (*i*) Film lubrication
    (*ii*) Elasto-hydrodynamic lubrication
    (*iii*) Boundary lubrication.
12. Enumerate and discuss the chief qualities to be considered in selecting oil for lubrication.
13. Explain briefly the following properties of a lubricant :
    (*i*) Viscosity
    (*ii*) Flash point
    (*iii*) Oiliness
    (*iv*) Emulsification
    (*v*) Neutralisation number
    (*vi*) Adhesiveness.
14. What are additives ?

15. What do you understand by 'oil contamination and sludge formation' ?
16. How are lubricating oil classified ?
17. What are "Multigrade oils" ? What are their advantages ?
18. What is a grease ?
19. What is the importance of lubrication in I.C. engines ?
20. Enumerate lubrication systems and explain wet sump lubrication system with the help of a neat sketch.
21. Where is dry sump lubrication system preferred and why ?
22. What is the role of lubricating oil filters of automobile engines ?
23. How the lubrication of two wheeler is done ?
24. Explain briefly "Mist lubrication system".
25. What are the various desired properties of a lubricant ? Explain how additives help to achieve the desired properties ?
26. Explain briefly "Elasto-hydrodynamic lubrication" and "Boundary lubrication".
27. What do you understand by full-flow type and bypass type oil filters ? When one is preferred over the other ?
28. What do you understand by "crankcase ventilation" ?
29. What is the difference between open crankcase and closed crankcase systems ?

5

Cooling Systems

5.1. NECESSITY OF ENGINE COOLING

In an I.C. engine, the temperature of the gases inside the engine cylinder may vary from 35°C or less to as high as 2750°C during the cycle. If an engine is allowed to run without external cooling, the cylinder walls, cylinder and pistons will tend to assume the average temperature of the gases to which they are exposed, which may be of the order of 1000 to 1500°C. Obviously at such high temperature ; the *metals will loose their characteristics and piston will expand considerably and sieze the liner. Of course theoretically thermal efficiency of the engine will improve without cooling but actually the engine will sieze to run.* If the cylinder wall temperature is allowed to rise above a certain limit, about 65°C, the lubricating oil will begin to evaporate rapidly and both cylinder and piston may be damaged. Also high temperature may cause excessive stress in some parts rendering them useless for further operation. In view of this, part of the heat generated inside the engine cylinder is allowed to be carried away by the cooling system. *Thus cooling system is provided on an engine for the following* **reasons :**

1. The even expansion of piston in the cylinder may result in seizure of the piston.
2. High temperatures reduce strength of piston and cylinder liner.
3. Overheated cylinder may lead to preignition of the charge, in case of spark ignition engines.
4. Physical and chemical changes may occur in lubricating oil which may cause sticking of piston rings and excessive wear of cylinder.
5. If the cylinder head temperature is high the volumetric efficiency and hence the power output of the engine is reduced.

Thus engine cooling is required to keep the temperature of the engine low in order to *avoid* :

(*i*) Loss of volumetric efficiency and hence power ;

(*ii*) Engine seizure ;

(*iii*) Danger of engine failure.

- *Almost 25 to 35 percent of total heat supplied in the fuel is removed by the* ***cooling medium.***
- Heat carried away by *lubricating oil and heat lost by radiation* amounts to *3 to 5 per cent of the total heat supplied.*

  It must be noted that *heat carried away by the coolant is a dead loss because not only no useful work can be obtained from it but a part of the engine power is also used to remove this heat.* Hence, it is of paramount importance that this loss is *kept minimum* by the designer.

Table 5.1 gives the heat balance of prime movers.

Table 5.1. Heat Balance of Prime Movers

S. No.	Prime mover	% of Fuel energy				Total
		To power	To coolant	To exhaust	To radiation	
1.	*4-stroke S.I. engine*	26	30	32	12	100
2.	*Diesel engines :*					
	(*i*) 2-stroke	30	21	37	12	100
	(*ii*) 4-stroke					
	(*a*) Naturally aspirated	31	26	30	13	100
	(*b*) Turbo-charged	35	22	29	14	100
3.	*Gas turbine :*					
	(*i*) Simple cycle	15	—	70	15	100
	(*ii*) Regenerative cycle	15	—	65	20	100

Demerits of overcooling. Overcooling of the engine is *harmful* because of the following *reasons* :

1. At very low temperature, starting of engine becomes difficult.
2. Due to overcooling, engine life is reduced due to corrosion.
3. If the engine is overcooled some of the heat which could be used to expand the gases will be *lost.*
4. The fuel will not vaporise properly and some of the gases produced by combustion will condense on the cylinder walls. This *leads to dilution of the oil in the pump and the addition of harmful corrosive acids.* Removal of the oil film from the cylinder wall by unvaporised fuel *leads to increased cylinder bore wear.*
5. Inadeque lubrication of the engine, due to oil not being warm enough to flow freely, results in *greater frictional losses.*

In general, *due to overcooling the* ***economy and life of the engine is reduced.***

Demerits of undercooling :

The following are the *demerits* of undercooling :

1. Undercooling can *cause engine seizure, or at least-shorten valve life and possible distortion of the cylinder block head or gasket.*
2. A hot-spot inside the combustion chamber may be sufficient to *cause pre-ignition, i.e.,* to ignite the fuel before the spark plug does, thus *causing loss of power and possible damage to the engine components.*
3. Water in cooling system may boil and evaporate, and *should the oil film burn away additional friction and wear will occur between cylinder and piston.*

5.2. AREAS OF HEAT FLOW IN ENGINES

- The transfer of heat takes place due to difference in temperature and from higher temperature to lower temperature. Thus, there is a heat transfer *to the gases* during intake stroke and the first part of the compression stroke, but during the combustion and expansion processes the heat transfer takes place *from the gases to the walls.*
- The *hot combustion gases* give part of their heat to the following components :

(*i*) Cylinder liner

(*ii*) Piston and piston rings

(*iii*) Cylinder head

(*iv*) Exhaust valves and exhaust ports.

— Most of this heat is carried away by the cooling system while some is lost by direct radiation from the engine surfaces.

— The heat going to the surrounding air, to the structure of the engine and to the lubricating oil is usually small (less than 10%), whereas most *of the heat rejected goes either to the cooling system or as heat in exhaust gases.*

5.3. GAS TEMPERATURE VARIATION

During different processes of the cycle there is an appreciable temperature difference of gases inside the engine cylinder.

— At the beginning of induction stroke the temperature is that of clearance gases.

— As the cool mixture is inducted in the engine cylinder the temperature falls rapidly.

— During compression process the temperature increases and attains its maximum value at the end of combustion process.

— During the expansion process the temperature decreases and then drops very rapidly during the release process.

— In *actual engines*, there is some temperature drop during the exhaust process.

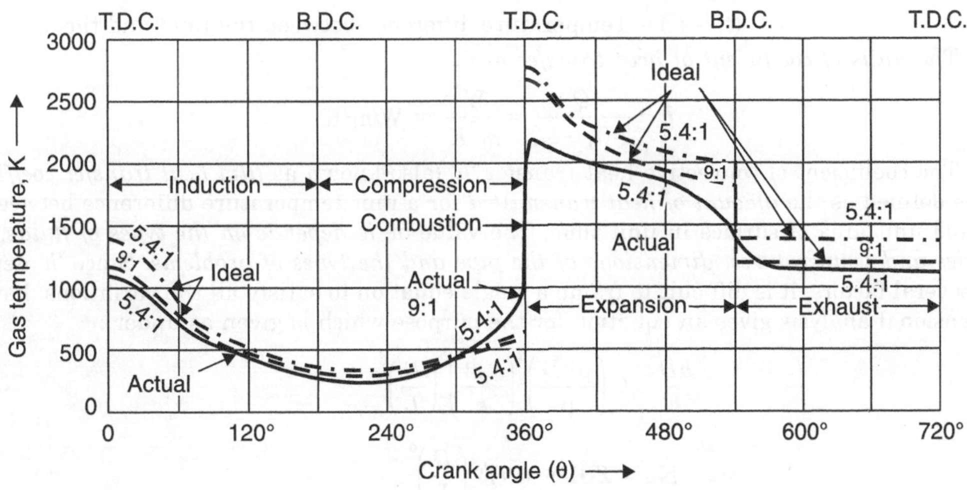

Fig. 5.1. Gas temperature variations for one cycle of 4-stroke engine.

Fig. 5.1 shows the gas temperature variations for one cycle (of a 4-stroke engine) and two compression ratios (*i.e.*, 5.4 : 1 and 9 : 1) for chemically correct mixture of C_8H_{18} and air, plus clearance gas.

5.4. HEAT TRANSFER, TEMPERATURE DISTRIBUTION AND TEMPERATURE PROFILES

5.4.1. Heat Transfer

Following are the processes by which heat transfer takes places :

Conduction. '*Conduction*' is the transfer of heat from one part of a substance to another part of the same substance, or one substance to another in physical contact with it, without appreciable displacement of molecules forming the substance.

Convection. '*Convection*' is the transfer of heat within a fluid by mixing of one portion of the fluid with another.

Fuel or natural convection. It occurs when the fluid circulates by virtue of the natural differences in densities of hot and cold fluids. The denser portions of the fluid move downward because of the greater force of gravity, as compared with the force on the less dense.

Forced convection. When the work is done to blow or pump the fluid, it is said to be forced convection.

Radiation. *Radiation* is the transfer of heat through space or matter by means other than conduction or convection. This phenomenon is not very significant with regard to reciprocation I.C. engines.

- *At least 95 per cent of the heat transfer between the working fluid and engine components and the engine components and cooling fluid is effected by* ***"Forced convection"***.

The transmission of heat per unit time from a surface by convection is given by :

$$Q = h\,A\,(t_1 - t_2)$$

where,
Q = Quantity of convective heat transferred,
h = Coefficient of convective heat transfer,
A = Area of surface, and
$(t_1 - t_2)$ = Temperature difference between the fluid and the surface.

The *units of coefficient of heat transfer* are :

$$h = \frac{Q}{A(t_1 - t_2)} = \frac{W}{m^2 K} = \text{W/m}^2\text{K}$$

The coefficient of convective heat transfer '*h*' (also known as *film heat transfer coefficient*) may be defined as the *amount of heat transmitted* for a unit temperature difference between the fluid and unit area of surface in unit time. The value of '*h*' *depends on the types of fluids, their velocities and temperatures, dimensions of the pipe and the types of problems.* Since '*h*' depends upon several factors, it is difficult to frame a single equation to satisfy all the variations, however a dimensional analysis gives an equation for the purpose which is given as under :

$$\frac{hD}{k} = C\left(\frac{\rho VD}{\mu}\right)^a \left(\frac{c_p \mu}{k}\right)^b \left(\frac{D}{L}\right)^c \quad \text{...(5.1)}$$

or

$$\text{Nu} = Z(\text{Re})^a\,(\text{Pr})^b \left(\frac{D}{L}\right)^c$$

where,
Nu = Nusselt number $\left(\frac{hD}{k}\right)$,

$$\text{Re} = \text{Reynold's number} \left(\frac{\rho VD}{\mu}\right),$$

$$\text{Pr} = \text{Prandtl number} \left(\frac{c_p\mu}{k}\right),$$

$\frac{D}{L}$ = Diameter to length ratio,

C = A constant to be determined experimentally,

c_p = Specific heat at constant pressure,

k = Thermal conductivity,

ρ = Density,

μ = Dynamic viscosity, and

V = Velocity.

The overall heat transfer coefficient :

While dealing with the problems of fluid to fluid heat transfer across a metal boundary, it is usual to adopt an overall heat transfer coefficient U which *gives the heat transmitted per unit area-unit time per degree temperature difference between the bulk fluids on each side of the metal.*

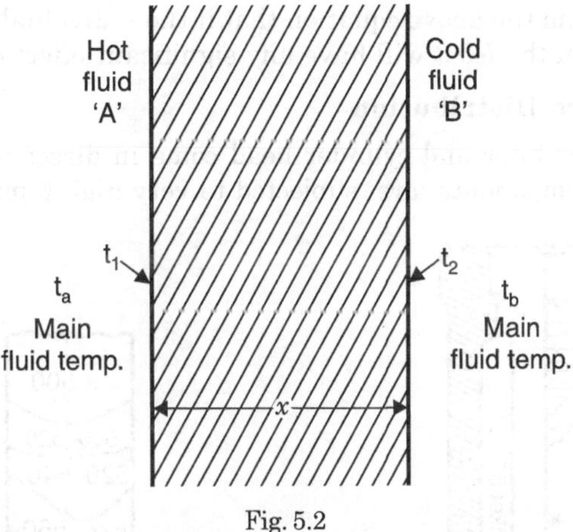

Fig. 5.2

Refer Fig. 5.2.

Let, h_a = Heat transfer coefficient from hot fluid to metal surface,

h_b = Heat transfer coefficient from metal surface to cold fluid, and

k = Thermal conductivity of metal wall.

The equations of heat flow through the fluids and the metal surface are as follows :

$$Q = h_a A\,(t_a - t_1) \quad ...(i)$$

$$Q = \frac{kA\,(t_1 - t_2)}{x} \quad ...(ii)$$

$$Q = h_b A\,(t_2 - t_b) \quad ...(iii)$$

By rearranging the equations (*i*), (*ii*) and (*iii*), we get

$$t_a - t_1 = \frac{Q}{h_a A} \quad ...(iv)$$

$$t_1 - t_2 = \frac{Qx}{kA} \quad ...(v)$$

$$t_2 - t_b = \frac{Q}{h_b \, . \, A} \quad ...(vi)$$

Adding equations (*iv*), (*v*) and (*vi*), we get

$$t_a - t_b = Q\left[\frac{1}{h_a \, . \, A} + \frac{x}{kA} + \frac{1}{h_b \, . \, A}\right]$$

$$Q = \frac{A(t_a - t_b)}{\frac{1}{h_a} + \frac{x}{k} + \frac{1}{h_b}} \quad ...(5.2)$$

If U is the overall coefficient of heat transfer then

$$Q = UA\,(t_a - t_b) = \frac{A\,(t_a - t_b)}{\frac{1}{h_a} + \frac{x}{k} + \frac{1}{h_b}}$$

or

$$U = \frac{1}{\frac{1}{h_a} + \frac{x}{k} + \frac{1}{h_b}} \quad ...(5.3)$$

It may be noticed from the above equation that if the individual coefficients differ greatly in magnitude only a change in the *least* will have any significant effect on the rate of heat transfer.

5.4.2. Temperature Distribution

Since piston cylinder liner and cylinder head come in direct contact with hot combustion gases, therefore, these components are subjected to very high temperatures. The temperature

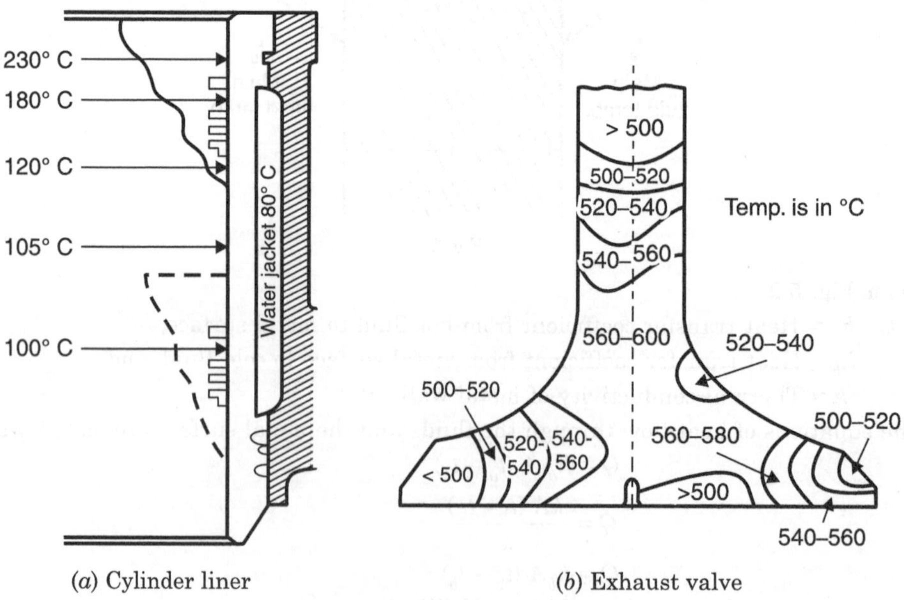

(*a*) Cylinder liner (*b*) Exhaust valve

Fig. 5.3. Temperature distribution.

distribution along a typical liner of a four-stroke diesel engine is shown in Fig. 5.3 (*a*). The temperature isotherms for a typical engine exhaust valve is shown in Fig. 5.3 (*b*). The very wide variations of temperature at different exhaust valve sites lead to development of large thermal stresses.

5.4.3. Temperature Profiles

- Fig. 5.4 shows the representative temperature profiles across the cylinder barrel wall of S.I. engine.

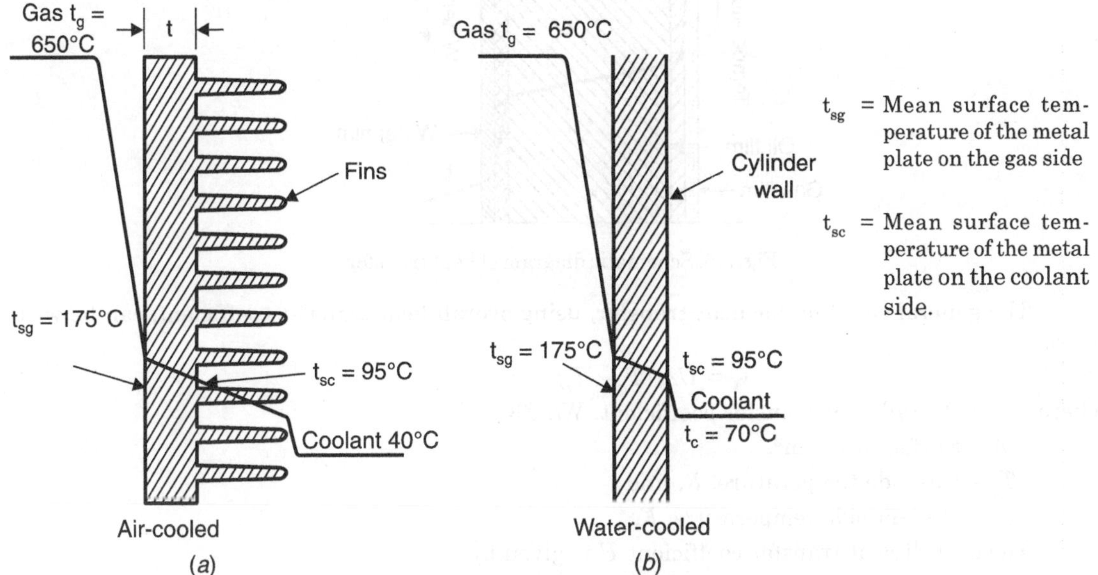

Fig. 5.4. Temperature profiles across cylinder barrel wall.

— It can be observed (from the Fig. 5.4) that there is a very large temperature fall in the 'boundary layer' of gas on the inner cylinder wall. This is due to *very low conductivity of the rather stagnant gas.* It is due to this stagnant layer of gas which acts as a protective layer that I.C. engine is feasible.

— In case of **air cooling** [Fig. 5.4 (*a*)] the boundary layer as on the gas side offers great resistance, but the effect of this is compensated largely by providing *more surface areas by way of cooling pins.*

— In case of **water cooling** [Fig. 5.4 (*b*)] there is also a boundary layer of the liquid coolant on the outside of the cylinder wall. Owing to high conductivity of water, the resistance of this boundary layer is small and as such comparatively small drop in temperature occurs.

For maintaining the inner surface temperature largely the same in both the cooling systems (air and water cooling), the *maximum rate of heat transfer from the gases* should be substantially the *same* with either type of cooling system.

- Heat transfer from hot gases to the coolant takes place by forced convection or by nucleate boiling when heat flux is high. Fig. 5.5 shows the cylinder (metal) wall surrounded by gas and coolant films, and also the variation of temperature from gas side to coolant side.

On the gas side are shown the gas film (stagnant gas layer) and oil film (lubricating oil layer). On the coolant side is the water film.

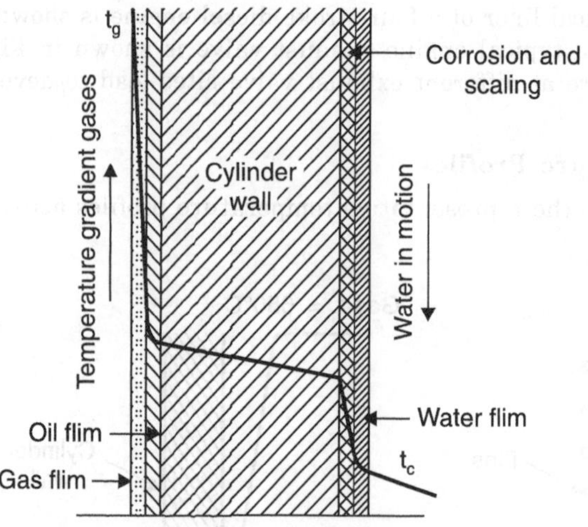

Fig. 5.5. Schematic diagram of heat transfer.

The general equation for heat transfer, using overall heat transfer coefficient, can be written as :

$$Q = UA\,(T_g - T_c)$$

where, U = Overall heat transfer coefficient, W/m²K,

A = Surface area, m²,

T_g = Gas side temperature, *K*, and

T_c = Coolant side temperature, *K*.

The overall heat transfer coefficient U is given by

$$U = \frac{1}{\dfrac{1}{h_g} + \dfrac{L}{k} + \dfrac{1}{h_c}} \qquad \text{...(5.4)}$$

where, h_g = Heat transfer coefficient on gas side, W/m²K,

h_c = Heat transfer coefficient on coolant side, kJ/m²K,

L = Thickness of the cylinder wall, *m*, and

k = Thermal conductivity in W/mK.

The maximum rate of heat transfer to external cooling medium is obtained when the following *requirements* are met with :

(*i*) The thickness of the gas film is minimum.

(*ii*) The surface film of oil is minimum.

(*iii*) Best conducting materials like aluminium and its alloys are used.

(*iv*) The wall of the cylinder is of no greater thickness than strength and wear consideration require.

(*v*) The external temperature is as low as possible relatively to the cycle mean temperature of the cylinder contents.

(*vi*) Surface deposits such as those of carbon which deposit on piston crown, corrosion and scale due to hard water etc. are reduced to minimum.

5.5. EFFECTS OF OPERATING VARIABLES ON ENGINE HEAT TRANSFER

1. **Compression ratio :**

- Fig. 5.6 shows that, in general, as the *compression ratio increases, there is marginal reduction in heat rejection.*

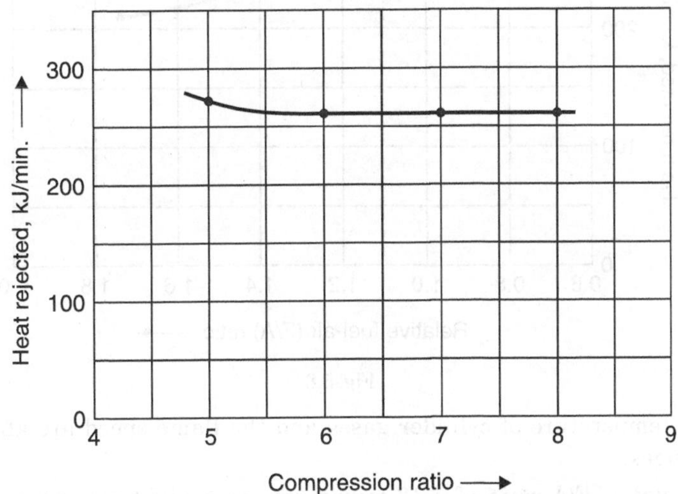

Fig. 5.6. Effect of compression ratio on heat rejected from the engine.

— As the compression ratio is increased the gas temperature near the T.D.C. increases slightly only. But due to greater expansion ratio, there is considerable decrease in gas temperature near the B.D.C. where large cylinder wall is exposed. Due to greater expansion the exhaust gas temperature is also reduced to a much lower value, and the heat rejected during blowdown is also less.

2. **Fuel-air ratio :**

- Air-fuel ratio, inlet temperature and exhaust pressure are the only variables which appear to influence gas temperature (t_g), it is illustrated in Fig. 5.7.

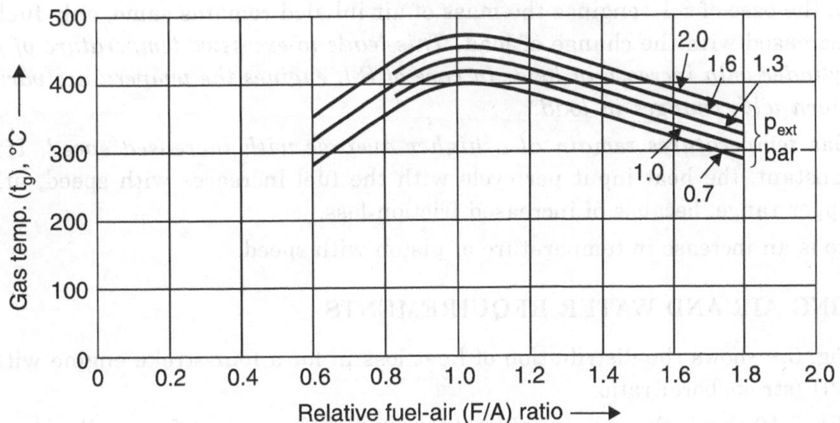

Fig. 5.7

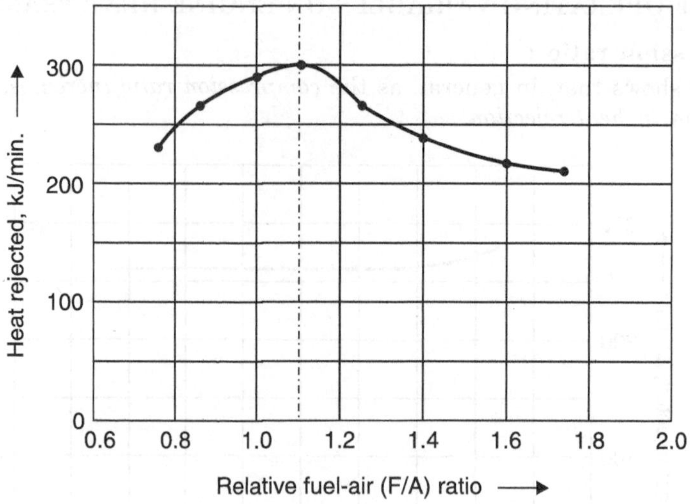

Fig. 5.8

— The temperature of cylinder gases and the flame speed are affected by F/A ratio changes.

— At *relative* F/A ratio of *1.12 maximum mean gas temperature occurs,* and *maximum heat rejection takes place.*

It is worthnoting that at leaner mixtures more heat is rejected and the engine has a tendency to overheat.

3. **Ignition timing :**

When the *spark advance is different from the optimum value the heat rejected to cooling system is increased.* It is due to the fact that any value of the spark advance other than the 'minimum spark advance' for best torque will reduce the power output in S.I. engines resulting in rejection of more quantity of heat.

4. **Load and speed :**

- In the case of S.I. engines the mass of air inhaled remains same, only fuel supplied is increased with the change of load. *This leads to excessive temperature of gases in the cylinder with increase in load. In case of S.I. engines the temperature variation is not much with changes in load.*
- Gas *temperatures remain at a higher average with increased speed.* If the load is constant, the heat input per cycle with the fuel increases with speed, atleast in the upper range, because of increased friction loss.

There is an increase in temperature of piston with speed.

5.6. COOLING AIR AND WATER REQUIREMENTS

- Fig. 5.9 shows the distribution of heat loss in for a four-stroke engine with respect to L/D (stroke/bore) ratio.
- Fig. 5.10 shows the amount of heat rejected to the coolant for small automotive diesel and petrol engines with cast-iron and aluminium cylinder heads (as dependent on speed).

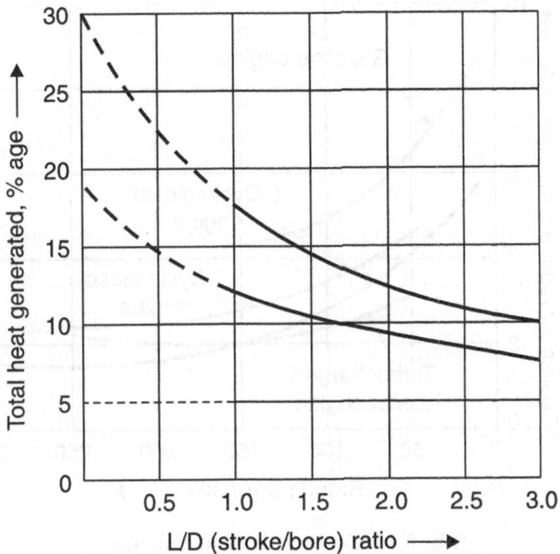

Fig. 5.9. Distribution of heat-loss in four-stroke engine.

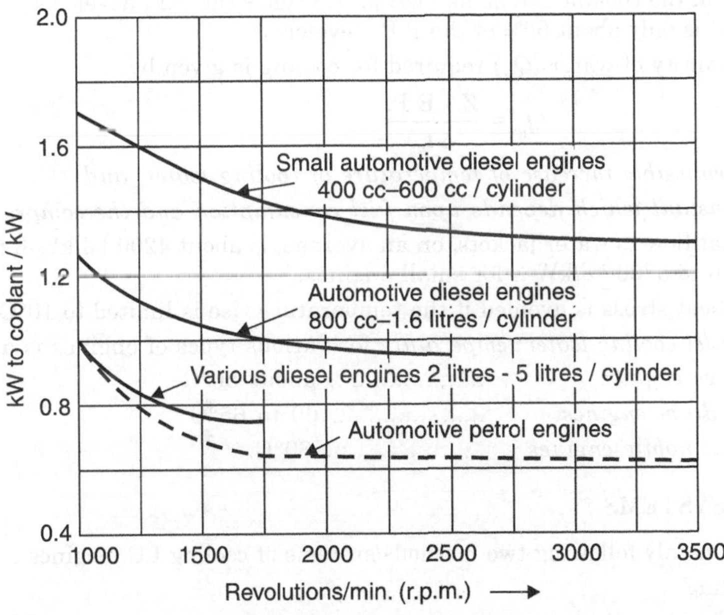

Fig. 5.10. Heat rejected by various I.C. engines.

Cooling air requirements :

- Fig. 5.11 shows the cooling air requirements for air-cooled radiator system of representative types of prime movers.
  - — It is obvious from the figure that the gasoline engine requires much more air than a diesel engine and that the turbo-charged diesel engine requires less cooling air than naturally aspirated diesel engines.

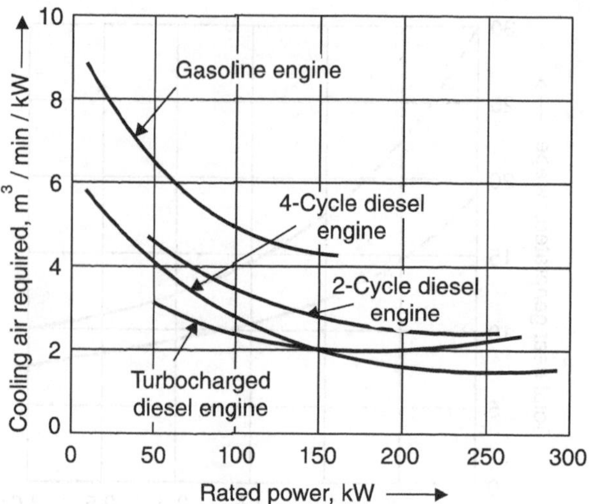

Fig. 5.11. Cooling air requirements.

Cooling water requirements. As is evident from Fig. 5.10, the heat rejected to the coolant greatly depends on the type of the engine. It can be seen that for a small high speed engine the heat rejected in the coolant can be as high as 1.3 times the B.P. developed, while for an open chamber engine it is only about 60% of the B.P. developed.

The quantity of water (Q_w) required for cooling is given by

$$Q_w = \frac{Z \times \text{B.P.}}{\Delta t_w} \qquad ...(5.5)$$

where, Δt_w = *Permissible increase of temperature of cooling water, and*

Z = *Constant which depends upon fuel consumption and the compression ratio.*

- The heat flow to water jackets, on an average, is about 4200 kJ/kW-h for large engines and 500 to 5700 kJ/kW-h for small engines.
- Large heat stress is avoided if the temperature rise is limited to 10°C–12°C.
- The *outlet cooling water temperature* for various types of engines is as follows :

  For large engines .... about 50°C

  For medium engines .... 60 to 65°C

  For automobile engines .... 80°C.

5.7. COOLING SYSTEMS

There are mainly following two methods/systems of cooling I.C. engines :

1. Air-cooling
2. Water/liquid cooling.

5.7.1. Air-Cooling System

- In this system, heat is carried away by the air flowing over and around the cylinder.
- Here fins are cast on the cylinder head and cylinder barrel which provide additional conductive and radiating surface (See Fig. 5.12). The fins are arranged at right angles to cylinder axis. The number and dimensions should be adequate to take care of the surplus heat dissipation. From all points of view, the truncated conical fin with rounded edges, as shown in Fig. 5.13, accomplish the purpose.

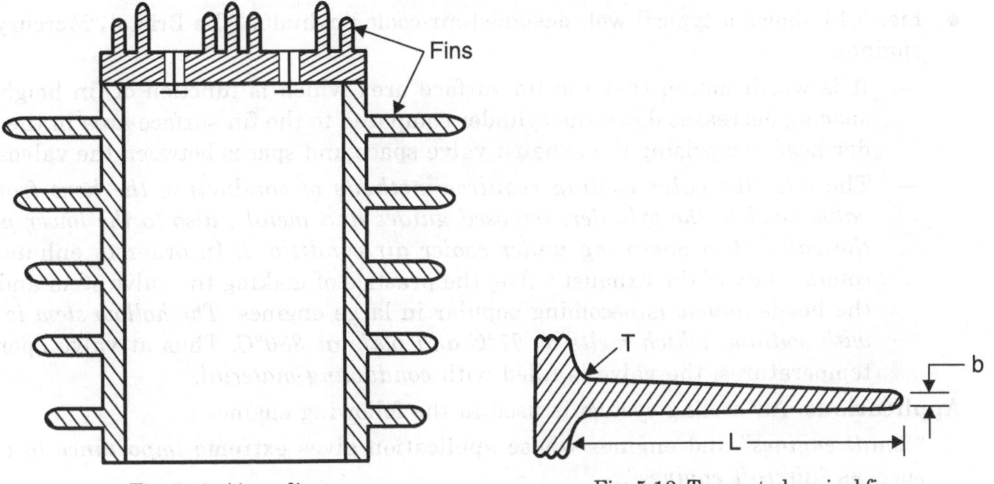

Fig. 5.12. Air cooling. Fig. 5.13. Truncated conical fin.

A fairly good relationship for fin proportions is as follows :

$$L = 0.35\ D\ ;\ p = 0.10\ D\ ;\ r = 0.018\ D\ ;\ b = 2 \text{ to } 2.5 \text{ mm.}$$

where L, D, p, r and b are fin length, cylinder diameter, fin pitch, rounded edge radius and width at the fin tip respectively.

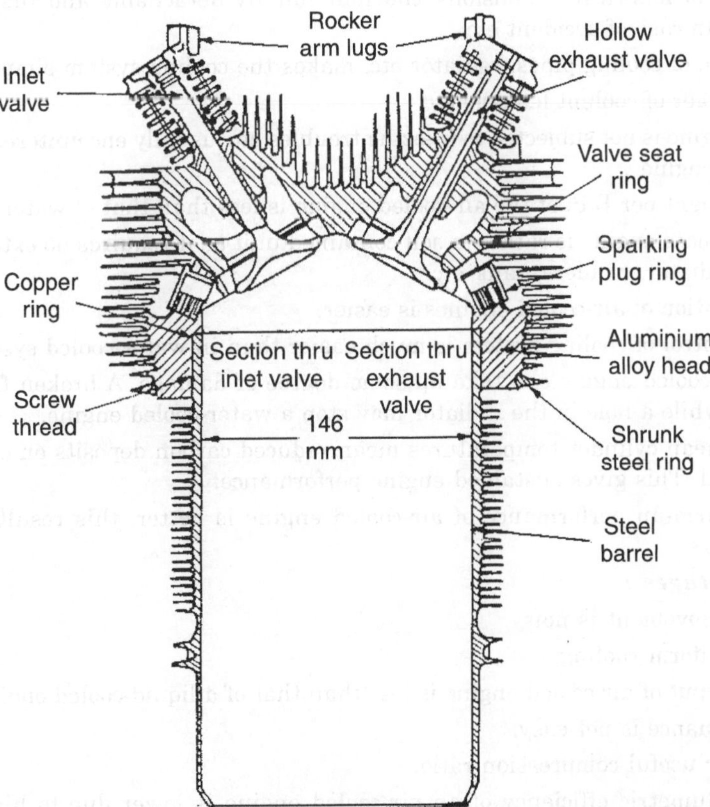

Fig. 5.14. Air-cooled cylinder (The Bristol "Mercury" S.I. engine).

- Fig. 5.14 shows a typical well designed air-cooled cylinder (the Bristol "Mercury" S.I. engine).
  - It is worth noting that the fin surface area which is function of fin height and spacing decreases down the cylinder compared to the fin surface area in the cylinder head comprising the exhaust valve space and spaces between the valves.
  - The *effective valve cooling requires methods of conducting the heat from the valve head to the cylinder, imposed guides and metal ; also to the lower part of the valve stem operating under cooler air conditions.* In order to enhance the conductivity of the exhaust valve, the practice of making the valve stem and even the heads *hollow* is becoming popular in large engines. *The hollow stem is filled with sodium, which melts at 97°C and boils at 880°C.* Thus at valve operating temperatures, the valve is filled with *conducting material.*

Applications. Air cooling system is used in the following engines :

1. "*Small engines*" and engines whose application gives extreme *importance to weight* such as "*aircraft engines*".
2. "*Industrial and agricultural engines*" where there can be a *strong objection to use of water as coolant.*

Advantages :

1. The design of the engine becomes simpler as no water jackets are required. The cylinder can be of identical dimensions and individually detachable and therefore cheaper to renew in case of accident etc.
2. Absence of cooling pipes, radiator etc. makes the cooling system simpler.
3. No danger of coolant leakage etc.
4. The engine is not subjected to freezing troubles etc. usually encountered in case of water-cooled engine.
5. The weight per B.P. of the air-cooled engine is less than that of water-cooled engine.
6. In this case engine is rather a self contained unit as it requires no external components *e.g.*, radiator, headers, tank etc.
7. Installation of air-cooled engines is easier.
8. The control of cooling system is much easier than in water-cooled system.
9. An air-cooled engine can take up some degree of damage. A broken fin does not affect much while a hole in the radiator may stop a water-cooled engine.
10. High mean cylinder temperatures mean reduced carbon deposits on combustion chamber wall. This gives sustained engine performance.
11. The warm-up performance of air-cooled engine is better, this results in low wear of cylinders.

Disadvantages :

1. Their movement is noisy.
2. Non-uniform cooling.
3. The output of air-cooled engine is less than that of a liquid-cooled engine.
4. Maintenance is not easy.
5. Smaller useful compression ratio.
6. The volumetric efficiency of an air-cooled engine is lower due to high cylinder head temperatures.

5.7.2. Water/Liquid Cooling System

In this method of cooling engines, the cylinder walls and heads are provided with jackets through which the cooling liquid can circulate. The heat is transferred from cylinder walls to the liquid by convection and conduction. The liquid becomes heated in its passage through the jackets and is itself cooled by means of an air-cooled radiator system. The heat from liquid in turn is transferred to air.

- The coolant to be employed in liquid-cooling system should have the following *characteristics* :

  (*i*) Low freezing temperature.

  (*ii*) A high boiling point.

  (*iii*) A large latent heat of vaporisation.

  (*iv*) Non-corrosive.

  (*v*) Easily and cheaply available.

Water possesses these properties, and is mostly employed in liquid-cooling system.

Role of anti-freeze solution in water-cooling system. When the engine is kept in unheated areas and it is not in operation, and the temperatures are below freezing, the water in the cooling system is liable to *get frozen.* The *solid mass of water due to expansion (10 percent increase in volume) may cause fracture of the cylinder block, water jackets and pipes. The radiator may also be damaged. Anti-freeze mixtures or solution are added to water in extreme winter seasons to lower its freezing temperatures below the danger point.* Commonly used *anti-freeze materials* are :

(*i*) Denatured alcohol (*ii*) Wood alcohol

(*iii*) Glycerine (*iv*) Kerosene

(*v*) Sugar solution (*vi*) Calcium or magnesium chloride

(*vii*) Ethylene glycol and propylene glycol.

In addition, *some oils are also used as anti-freeze solutions.*

Effect of warm-up period on S.I. engine fuel economy :

- When an engine is started from cold condition, it takes some time to reach its operating temperatures. Usual practice of warming up requires engine to run in idling condition for about 2 to 3 minutes. *This helps in an efficient and complete evaporation of fuel when it comes out of carburettor jet, and hence increases fuel economy.*
- Under cold conditions, a portion of fuel remains in the liquid state and get stuck to the cold walls of inlet manifold, cylinder and piston top, which may go in exhaust without or partial combustion. *Initial warm up period helps in complete evaporation of fuel, thereby improving fuel economy.*
- But *excessive warm up period is not desirable, as it decreases the efficiency.* Hence in *modern engines, heating of fresh charge during idling is provided. In water-cooled engines thermostat prevents water circulation during warm up period, which again reduces the warm up period.*

Methods used for circulating water around the engine cylinder :

Various methods are used for circulating water around the cylinder and cylinder head. These are :

1. Thermo-syphon cooling
2. Forced or pump cooling
3. Cooling with thermostatic regulator
4. Pressurised water cooling
5. Evaporative cooling.

1. **Thermo-syphon cooling**

- The *basic principle of thermo-syphon cooling* can be explained with reference to Fig. 5.15(*a*).

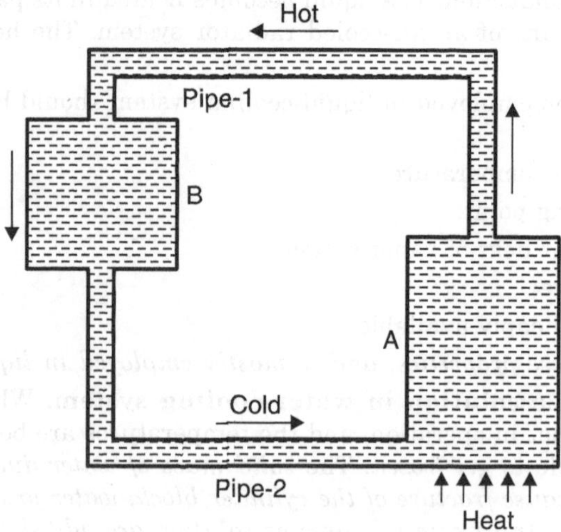

Fig. 5.15. (*a*) Principle of thermo-syphon.

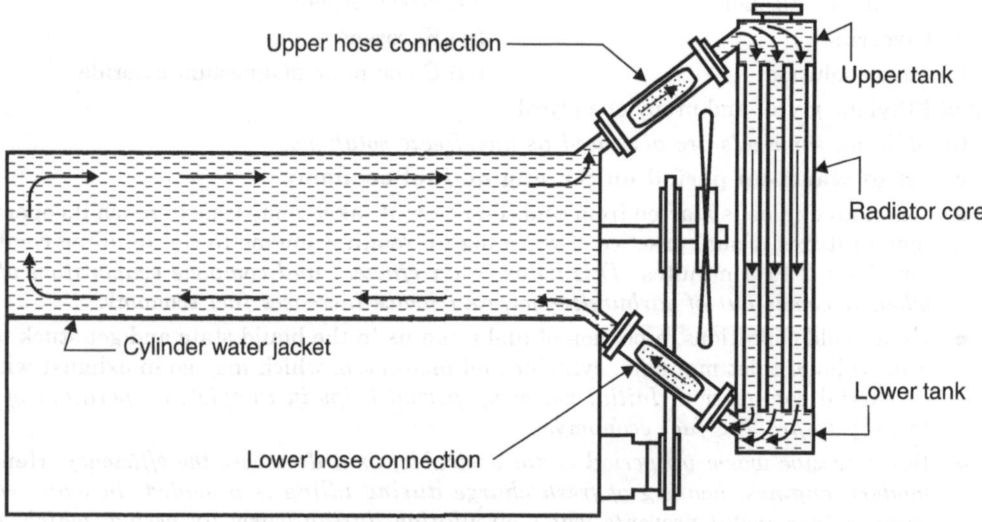

Fig. 5.15. (*b*) Thermo-syphon cooling system.

Heat is supplied to the fluid in the tank A. The hot fluid travels up, its place being taken up by comparatively cold fluid from the tank B through the pipe-2. The hot fluid flows through the pipe. Pipe-1 to the tank B where it gets cooled. Thus the fluid circulates through the system in the form of convection currents.

For engine application tank A represents *cylinder jackets* while tank B represents a *radiator*, and water acts as a *circulating fluid*. In order to ensure that coolest water is always made available to cylinder jackets, the water jackets are located *at a lower level than the radiator*.

- Fig. 5.15(*b*) shows the thermo-syphon cooling arrangement of an engine. The top of radiator is connected to the top of water jacket by a pipe, and the bottom of radiator to the bottom of the water jacket. Water travels down the radiator across which air is passed to cool it. The air flow can take place due to vehicle motion or a fan can be provided for the purpose.
- The *thermo-syphon cooling system cannot be used in high output engines due to its inability to meet the requirement of large flow rates of water.*

Limitations of thermo-syphon cooling system :

(*i*) The *cooling depends only on the temperature and is independent of the engine speed.*

(*ii*) The rate of circulation is slow and insufficient. The circulation of water starts only after the engine has *become hot enough to cause thermo-syphon action.*

(*iii*) This system requires that the radiator be above the engine for gravity flow of water to engine.

2. **Forced or pump cooling.** Refer to Fig. 5.16. In this system, a pump is used to cause positive circulation of water in the water jacket. Usually the pump is belt driven from the engine.

Merit. The main advantage of forced cooling is that *cooling is ensured under all conditions of operation.*

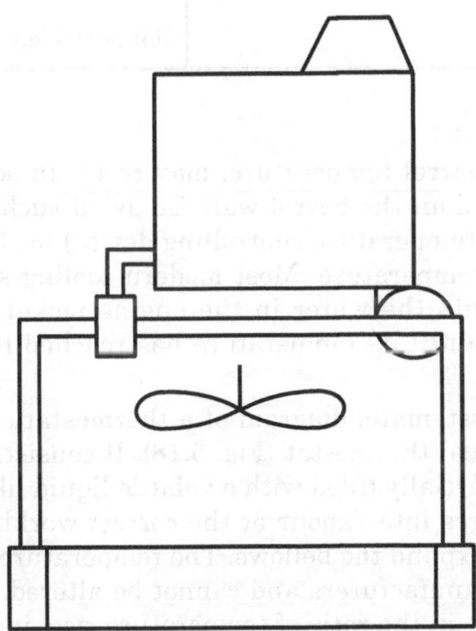

Fig. 5.16. Forced or pump cooling.

Demerits. (*i*) The cooling is not temperature dependent. Under certain circumstances, the engine may get overcooled.

(*ii*) The cooling requirement, while moving uphill, is increased because more fuel is burnt. The coolant circulation, however, is reduced which may lead to overheating of the engine.

(*iii*) The cooling ceases when the engine stops. This is undesirable since cooling must continue till the temperatures are reduced to normal values.

Comparison of thermo-syphon and forced cooling systems :

S. No.	Forced system	Thermo-syphon system
1.	Circulation of water by a centrifugal pump, belt driven from the engine.	Circulation of water by natural convection. No pump is used.
2.	Cooling is independent of temperature, but depends upon the engine speed.	Cooling depends only upon the temperature, and is independent of engine speed.
3.	Rate of circulation is fast.	Rate of cooling is slow and insufficient.
4.	The circulating water pump needs maintenance.	Simple, automatic, and no maintenance is required.
5.	Cooling is ensured under all conditions of operation.	The circulation of water starts only after the engine has become hot enough to cause thermosyphon action.
6.	Radiator position with respect to engine not restricted, can be placed anywhere.	The system requires that the radiator be placed above the engine for gravity flow of water to engine.
7.	It is costly.	It is cheaper.
8.	It is widely used.	It is not widely used.

3. **Thermostat Cooling :**

- Too lower cylinder barrel temperature, may result in severe corrosion damage due to condensation of acids on the barrel wall. To avoid such a situation it is customary to use a thermostat (a temperature controlling device) to stop flow of coolant below a preset cylinder barrel temperature. Most modern cooling system employ a thermo-static device which prevents the water in the engine jacket from circulating through the radiator for cooling until its temperature has reached to a value suitable for efficient engine operation.
- Fig. 5.17 shows a systematic diagram of a thermostatically controlled cooling system. Also shown is a typical thermostat (Fig. 5.18). It consists of bellows which are made of thin copper tubes, partially filled with a volatile liquid like ether or methyl alcohol. The volatile liquid changes into vapour at the correct working temperature, thus creating enough pressure to expand the bellows. The temperature at which the thermostat operates is set by the manufacturers and cannot be altered. The movement of the bellows opens the main valve in the ratio of temperature rise, increasing or restricting the flow of water from engine to the radiator. Hence when the normal temperature of the engine has been reached the valve opens and circulation of water commences. When the unit is closed the gas condenses and so the pressure falls. The bellows collapse and the thermostat seats on its seat and circulation around thermostat stops. When the thermostat valve is not open and the engine is running the water being pumped rises in pressure and causes the pressure relief valve to open. Thus the water completes its circulation through the by-pass as shown in Figs. 5.17 and 5.18. Now when the temperature of water around the engine-cylinder rises upto a certain limit, it causes the thermostat valve to open. The pressure of water being pumped falls and pressure relief valve closes. So the flow of cooling water in the normal circuit commences through the radiator. This accelerates the rise of temperature of the cylinder walls and water and more power is developed in a few moments of the starting of the engine.

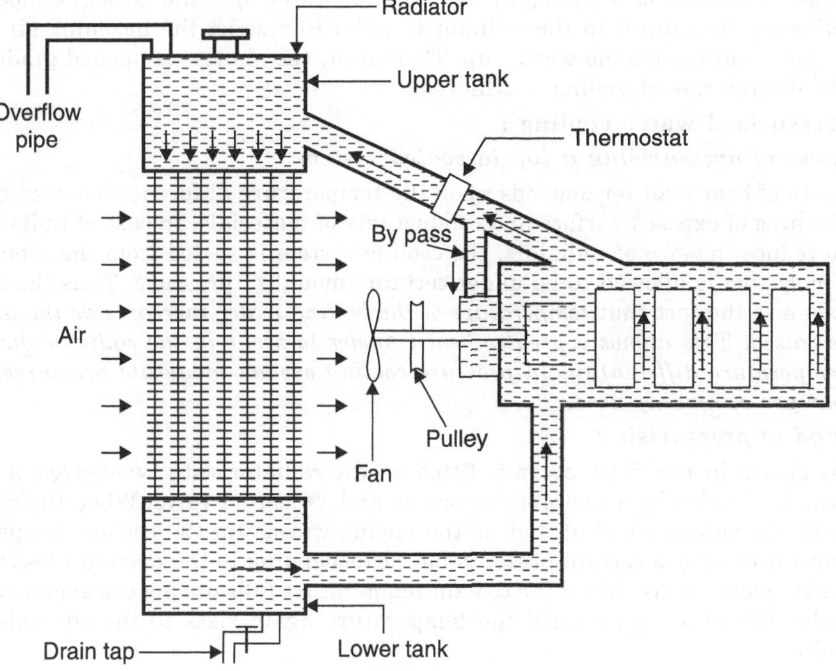

Fig. 5.17. Thermostatically controlled cooling system.

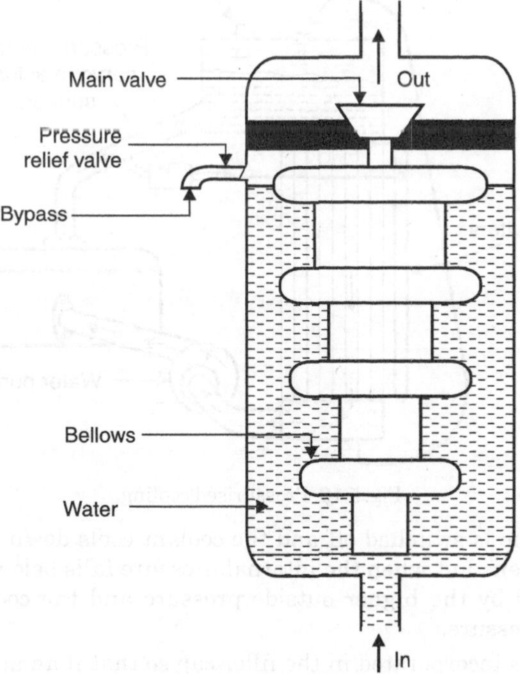

Fig. 5.18. Typical thermostat.

- Another method of warming up the radiator water upto the normal temperature is by utilising the shutter in the radiator in order to restrict the incoming air through the radiator till the engine warms up. Thereafter, the shutter is opened gradually so that the desired rate of cooling is achieved.

4. **Pressurised water cooling :**

Purpose of pressurising a liquid-cooled system

The rate of heat transfer depends upon the temperature difference between two medium, the area of exposed surface and conductivity of materials. In case of radiators, in order to reduce the size of radiators, the cooling system is sealed from the atmosphere, and the system is allowed to be under certain amount of pressure. Thus the advantage is taken of the fact that *temperature of the boiling water increases as the pressure on it is raised. This allows a greater heat transfer to occur in the radiator due to a larger temperature differential. In pressure cooling system moderate pressures, say upto 2 bar are commonly used.*

Method of pressurising

- As shown in Fig. 5.19, a cap is fitted on the radiator with two-valves, a safety valve which is loaded by a compression spring and a vacuum valve. When the coolant is cold both the valves are shut, but as the engine warms up the coolant temperature rises until it reaches a certain preset value corresponding to the desired pressure when the safety valve opens ; but if the coolant temperature falls during the engine operation the valve will close again until the temperature again rises to the equivalent pressure value.

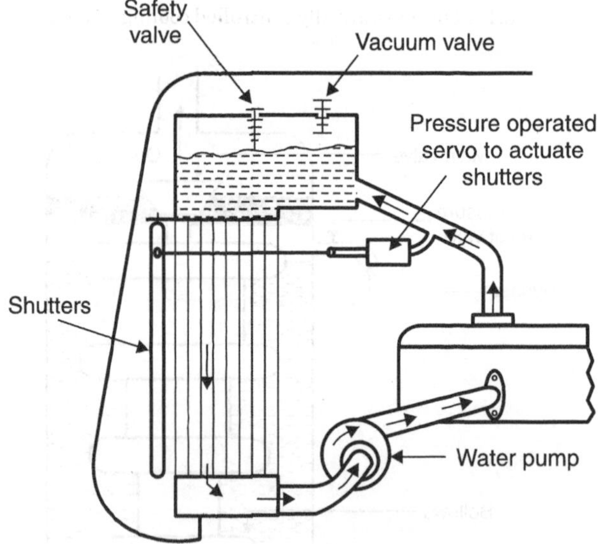

Fig. 5.19. Pressurised cooling.

- When the engine is switched off and the coolant cools down vacuum begins to form in the cooling system, but when the internal pressure falls below atmospheric the vacuum valve is opened by the higher outside pressure and the cooling system then attains atmospheric pressure.
- A safety valve is incorporated in the filler cap so that if an attempt is made to unscrew it while the system is under pressure, the first movement of the cap at once relieves the pressure and thus prevents the emission of steam or the blowing off the cap due to higher internal pressure.

Advantages of pressurised engine cooling over conventional thermo-syphon cooling system :

Following are the *advantages* of pressurised engine cooling over conventional thermo-syphon cooling system :

(*i*) Effective and positive cooling of all parts. Local overheating is avoided.

(*ii*) It can take overload easily because as the engine speed increases the water circulation also increases, and same effective cooling can be maintained at all the speeds.

(*iii*) In thermo-syphon system, the radiator should be kept well above the engine, to provide a height for natural circulation. There is no such requirement for pressurised forced pump system.

(*iv*) With pressurised system the coolant temperature is maintained higher. This reduces corrosion.

(*v*) Smaller coolant passages can be used. This reduces weight and bulk of the engine.

(*vi*) No loss of water by boiling and evaporation.

5. Evaporative cooling :

- In this system, also called steam or vapour cooling, the temperature of the cooling water is allowed to reach a temperature of 100°C. This method of cooling *utilises the high latent heat of vapourisation of water to obtain cooling with minimum of water.* Fig. 5.20 shows such a system. The cooling circuit is such that *coolant is always liquid but the steam formed is flashed off in the separate vessel.* The make up water so formed is sent back for cooling.
- This system is used for cooling of many types of *industrial engines.*

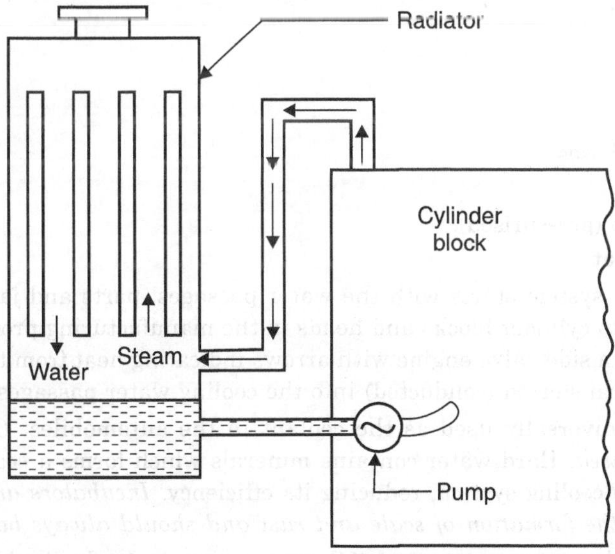

Fig. 5.20. Evaporating cooling.

Advantages and Disadvantages of liquid cooling :

Advantages :

1. Compact design of engine with appreciably smaller frontal area is possible.
2. The fuel consumption of high compression liquid-cooled engine is rather lower than for air-cooled one.

3. More even cooling of cylinder barrels and heads due to jacketing makes it easier to reduce the cylinder head and valve seating temperature.
4. In case of water-cooled engine installation is not necessary at the front of mobile vehicles, aircrafts etc. as the cooling system can be conveniently located wherever required. This is not possible in case of air-cooled engines.
5. The size of the engine does not involve serious problem as far as design of cooling system is concerned. In case of air-cooled engines particularly in high horse power range difficulty is encountered in circulation of required quantity of air for cooling purposes.
6. Volumetric efficiency of water-cooled engines is higher than that of air-cooled engines.

Disadvantages :

1. This is dependent system in which supply of water for circulation in the jacket is required.
2. Power absorbed by the pump for water circulation is considerably higher than that for cooling fans.
3. In the event of failure of cooling system serious damage may be caused to the engine.
4. Cost of system is considerably high.
5. System requires considerable attention for the maintenance of various parts of system.
6. The performance is weather sensitive.
7. The warm up performance is poor and has starting problems particularly in cold weather.

5.8. COMPONENTS OF WATER COOLING SYSTEM

The components of water-cooling system are enumerated briefly described below :

1. Water jacket
2. Water pump
3. Fan
4. Thermostat
5. Connecting hoses
6. Radiators
7. Radiator cap (pressurised).

1. **Water jacket :**

- The cooling system starts with the water passages, parts and jackets which are usually cast into cylinder blocks and heads in the manufacturing process. Fig. 5.21. shows a section of a side-valve engine with arrows indicating heat from the combustion chamber being transferred (conducted) into the cooling water passages.
- Water is universally used as the coolant in the automobiles. *Only clean soft water should be used.* Hard water contains minerals which forms a scale on the inside surfaces of the cooling system, reducing its efficiency. *Incubators are available to reduce or prevent the formation of scale and rust and should always be used as the coolant.*
- Above or below a temperature of 4°C, water expands. In the liquid state this expansion can be taken up by the movement of the water into a vacant space, including the additional volume created by the expansion of the cylinder block. When frozen, the solid mass expands and exerts a force against the internal surfaces of the cooling system. The result could be a fractured cylinder block or damaged radiator, although the latter is less likely, due to the mass of water involved, and engine cracks will be revealed when turns back to a liquid. Antifreeze *lowers the freezing point of the water and*

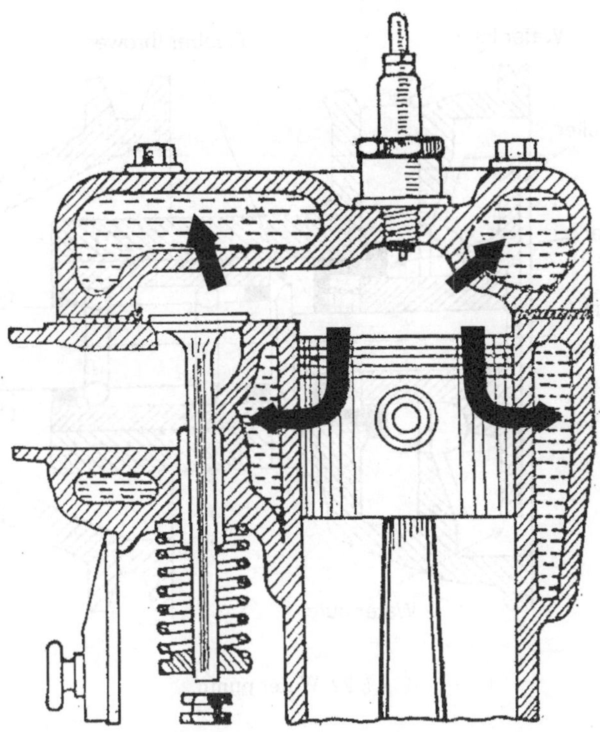

Fig. 5.21. Water jacket.

gives a margin of safety in cold condition. The proportion of antifreeze to water for various degree of protection is usually quoted by the manufacturers. An *ethylene-glycol based antifreeze combined with chemicals* to resist *corrosion* is quite popular.

2. **Water pump :**

- Water pumps entail the following *merits and functions :*

(*i*) Enables more efficient circulation and the radiator can be placed in a position which suits the body line of the vehicle, not necessarily at the fronts of the vehicle.

(*ii*) It is also not essential for the radiator header tank to be above the level of the engine cylinders, and the water passages can be reduced in size.

- Water pumps are usually of the *impeller* or *centrifugal type* and are driven from the crankshaft pulley, the degree of circulation achieved being in proportion to the engine *speed.*
- The impeller type has widely spaced vanes, so that should the impeller fail the vanes will offer little resistance to the natural thermo-syphon water flow.

  Centrifugal-type pumps are more powerful and provide positive water circulation, the water entering the pump at the centre of the rotor and being thrown outwards by the curved vanes, so creating a pressure difference between the enter and discharge parts of the pump.
- Fig. 5.22 shows a typical pump assembly steps between the belt and pulley and broken or damaged vanes can reduce the effectiveness of the pump and cause overheating of the engine. If leaks are apparent, the pump should be inspected for defective gaskets and seals, and at the same time the bearings should be checked for wear and the pump casing scraped to remove scale.

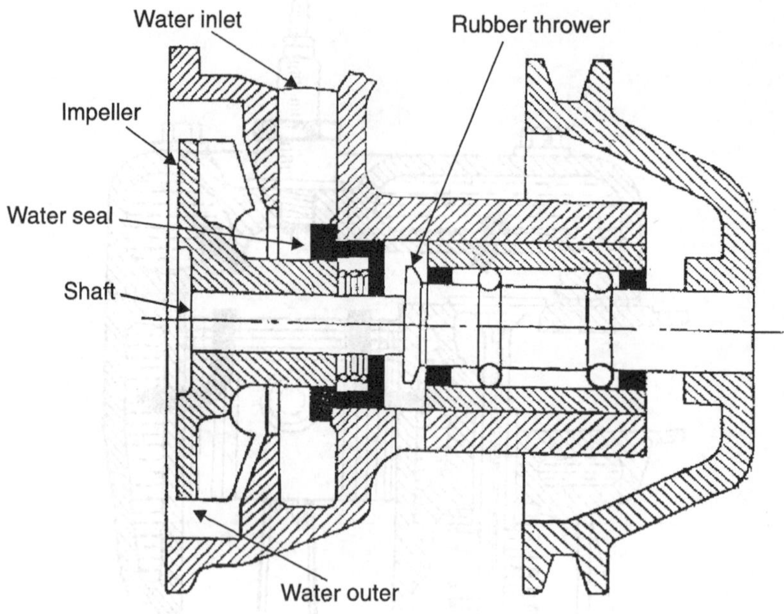

Fig. 5.22. Water pump.

3. **Fan :**

- The fan is generally fitted behind the radiator, usually being mounted on the water pump shaft and induces a flow of air through the radiator block. This function being especially important when vehicle is moving slowly, or when it is stationary and the engine is running. The fan is generally driven by a V-belt from the engine crankshaft and has a number of blades which are made from metal, nylon or some other plastic. As the number of blades increases, the fan size can be reduced.
- It is important that the fan belt is correctly adjusted, if it is too tight the dynamo and water pump bearing, may be damaged and excess power absorbed ; if it is too slack the efficiency of the cooling system and the dynamo output will be affected.
- On no account should the engine crankshaft be rotated by the fan blades as this will cause distortion to the blades and the assembly will run out of balance.

4. **Thermostat :**

- The thermostat is normally fitted either to the cylinder head water-outlet passage below the top hose, or in a special *hosing near the water pump.*
- The purpose of this device is primarily to restrict the circulation of water to cylinder head and block during the warm-up period. Then when a designed temperature is reached, the thermostat opens and the circulation of water starts in the radiator.
- The usual bronze bellow-type unit is shown in the Fig. 5.23. Expansion and contraction of vapour with the bellows gradually opens and closes the valve in proportion to the predetermined temperature range.

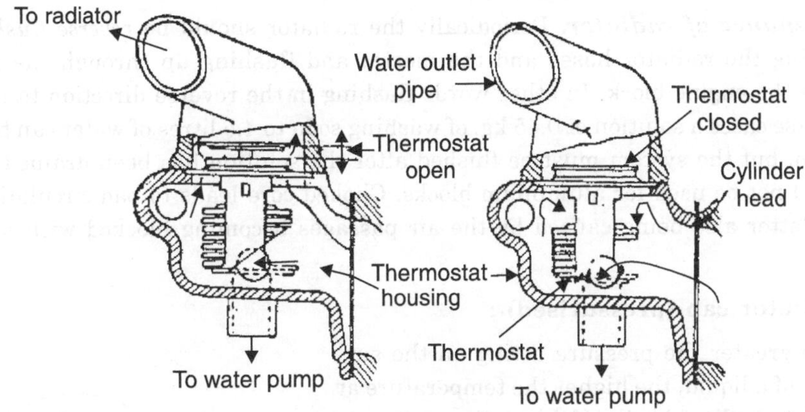

Fig. 5.23. Thermostat.

5. **Connecting hoses :**

- For convenience of connection and to isolate the radiator from engine movement and vibration, flexible hoses are used.
- They were originally of rubber, reinforced with canvas like large diameter garden hose-Neoprene and various plastics are now more suitable and are premoulded to the shapes and curves required for the installation.
- The ends of the hose are forced over the connecting spigots on the radiator and engine held in place by bore clamps or spring clips.

6. **Radiators :**

- The basic requirement of a radiator is to *provide a sufficiently large cooling area for transmission of heat from the coolant to the air*. The construction of the centre of the radiator or core varies, but in general the water passages terminate at a header tank at the bottom. In addition to an opening which enables the cooling system to be topped up the header tank allows for expansion and contraction of the coolant within the system.
- The principle types of radiator core are :
  (*i*) Film type
  (*ii*) Fin and tube
  (*iii*) Pack type.

Fig. 5.24 shows a radiator made of thin sheet brass and is typical of most cooling systems. As the water descends through hundreds of tubes or passage it is cooled by radiation to the air and returns to the engine via the bottom tank and the lower left connection. An overflow pipe from the filter neck can be clearly seen.

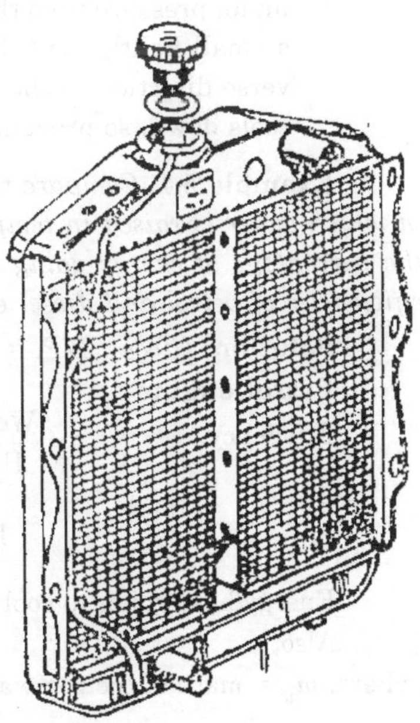

Fig. 5.24. Radiator.

Maintenance of radiator. Periodically the radiator should be *reverse flushed*. This involves removing the radiator hoses and thermostat and flushing up through the radiator and *down* through the engine block. In other words flushing in the reverse direction to normal water flow. For reverse cases a solution of 0.45 kg. of washing soda to 4.5 litres of water can be used in the cooling system, but the system must be flushed after the solution has been drained off, and this method should not be used for aluminium blocks. Choked core leads to bad circulation and overheating, the latter also being caused by the air passages becoming blocked with dirt and dead insects.

7. **Radiator cap (pressurised) :**

- The greater the pressure acting on the surface of a liquid, the higher the temperature at which the liquid boils. If the cooling system is sealed off, the expanding water will increase the pressure of air in the header tank ; any steam formed will assist in raising the pressure, and in consequence will require a higher temperature to cause the water to boil, which means that any *arduous driving, will not result in water loss due to evaporation.*
- Fig. 5.25 shows a radiator pressure cap. A safety device is incorporated in the cap to reduce the risk of accident when the cap is removed due to hot water a steam being ejected under pressure from the radiator. In addition, a small lightly-loaded valve operating in reverse direction to the main valve prevents a vacuum being created in the system as it cools down, so preventing collapse or damage to the radiator or the connecting hoses.

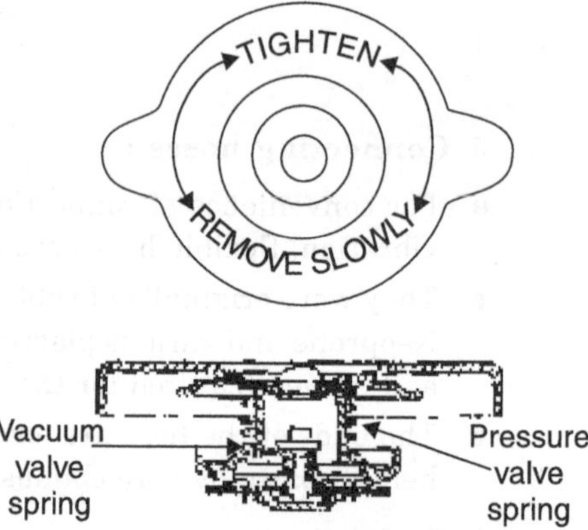

Fig. 5.25. Radiator pressure cap.

Example 5.1. *Compare the quantity of water required for 90 kW petrol and diesel engines in which water is raised in temperature by 27°C in passing through the jackets. In petrol engine the percentage of energy going to coolant is 32 per cent and in diesel engine 28 per cent. The efficiencies of petrol and diesel engines are 25 per cent and 30 per cent, respectively.*

Solution. *Given :* B.P. = 90 kW ; $\Delta t_w = 27°C$; $\eta_{petrol} = 25\%$, $\eta_{diesel} = 30\%$

Petrol engine :

Efficiency, $$\eta = \frac{\text{Work developed}}{\text{Heat supplied}}$$

$\therefore$ $$\text{Heat supplied} = \frac{\text{Work developed}}{\eta_{petrol}} = \frac{90}{0.25} = 360 \text{ kW or kJ/s}$$

Energy/heat going to cooling water = 360 × 0.32 = 115.2 kJ/s

Also, $$115.2 = m_w \times c_{pw} \times \Delta t_w$$

(where, m_w = mass of cooling water required ; c_{pw} = specific heat of water at constant pressure)

$\therefore$ $$m_w = \frac{115.2}{c_{pw} \times \Delta t_w} = \frac{115.2}{4.187 \times 27} = 1.019 \text{ kg/s or } \mathbf{3668 \text{ kg/h.}} \quad \textbf{(Ans.)}$$

Diesel engine :

Heat supplied $\frac{90}{0.3} = 300$ kW or kJ/s

Heat going to cooling water $= 300 \times 0.28 = 84$ kJ/s

Also, $84 = m_w \times 4.87 \times 27$

$\therefore \quad m_w = \frac{84}{4.187 \times 27} = 0.743$ kg/s or **2674.9 kg/h.** **(Ans.)**

5.9. SPECIFICATIONS FOR THE COOLING SYSTEM OF AN ENGINE

Radiator cap :

Location	Under hood
Relief valve pressure	0.5 kg/cm^2
Vacuum valve release	0.035—0.07 kg/cm^2
Thermostat :	
Type	Choke
Standard temperature :	
Starts to open	74°C
Fully open	87°C
Optional high temperature :	
Starts to open	82°C
Fully open	94°C
Water pump :	
Type	Centrifugal
Drive	V-belt
Bearing	Ball
Radiator :	
Type	Cellular tube and fin
Cooling system :	
Capacity	10
Flow rate	125 cm/s
Fan :	
Number of blades	4
Facing	76° and 104°
Diameter	38 cm
Ratio of fan to crankshaft	1.19 : 1
Bearing	Water pump
Drive belt :	
Angle of V	38°
Length	108 cm
Width	1.7 cm

5.10. TROUBLESHOOTING OF COOLING SYSTEM

S. No.	Trouble	Cause	Remedy
1.	*Overheating*	(*i*) Lack of coolant (*ii*) Non-operating thermostat (*iii*) Non-operating water pump (*iv*) Incorrect valve timing (*v*) Incorrect ignition timing (*vi*) Excessive piston blow-by (*vii*) Clogged radiator (*viii*) Broken fan belt (*ix*) Clogged core of radiator, prevented air passage (*x*) Excessive carbon formation (*xi*) Clogged muffler (*xii*) Bent exhaust pipe.	(*i*) Refill the radiator (*ii*) Replace the thermostat (*iii*) Overhaul the water pump. If it is not working, replace it (*iv*) Set the valve timing according to the specifications of the manufacturers (*v*) Correct the ignition timing (*vi*) Check pistons, rings and cylinder walls (*vii*) Clean the radiator by the reverse flushing method (*viii*) Replace the fan belt (*ix*) Clean the core with water and air pressure and remove the dirt and other materials (*x*) Remove the carbon deposits (*xi*) Replace the muffler (*xii*) Replace the exhaust pipe
2.	*Loss of cooling liquid*	(*i*) Very loose hose connections (*ii*) Damaged hose (*iii*) Leaking water pump (*iv*) Leaking radiator (*v*) Leaking cylinder head (*vi*) Crack in cylinder block or cylinder head	(*i*) Tighten the hose connections (*ii*) Replace the damaged hose (*iii*) Repair the water pump or replace it (*iv*) Remove the radiator and repair it (*v*) Replace the gasket for the cylinder head (*vi*) If the crack is small, it can be closed with the radiator sealer. If the crack is large, then it has to be taken to the workshop.

5.11. COOLING SYSTEM DATA OF SOME INDIAN VEHICLES

S. No.	Make of Vehicles	Component Part						
		Type of cooling system	Engine coolant temperature sensor	Type of radiator	Capacity in litres	Water pump	Thermostat	Anti-freeze mixture
1.	*Ambassador*	Water cooling	—	Tube and fin	8	Centrifugal	Metal bellows type	Ethylene glycol type
2.	*Premier Padmini*	Water cooling	—	Tube and fin	4.5	Centrifugal type	Metal bellows type	Natural glycerine or denatured alcohol
3.	*Maruti 800*	Water cooling	—	Tube and fin	3.6	Centrifugal type	Wax pallet type	Ethylene alcohol
4.	*Jeep Mahindra*	Water cooling	—	Cellular tube and fin	10	Centrifugal type	Choak type	Methyl alcohol
5.	*Maruti Gypsy*	Water cooling	—	Tube and fin	4	Centrifugal type	Wax pallet type	Ethylene glycol

6.	*Ashok Leyland*	Water cooling	—	Tube and fin	25	Centrifugal type	Bellow type	Ethylene glycol
7.	*Tata Truck*	Water cooling	—	Tube and fin	13	Centrifugal type	Bellow type	Ethylene glycol
8.	*Santro*	Water cooled pressurized, forced circulation with one electrical fan	Thermistor type	Pressurized corrugated fin type	6	Centrifugal type impeller	Wax pallet type with jiggle valve	Ethylene glycol
9.	*Wagon R*	Water-cooled	Thermistor type	Tube and fin type	3.5	Centrifugal type impeller	Wax pallet type	Ethylene glycol
10.	*Balleno*	Water-cooled	Thermistor type	Tube and fin type	4.2	Centrifugal type impeller	Wax pallet type	Ethylene glycol

SHORT ANSWER QUESTIONS

Q. 1. What is the distinct advantage of an air-cooled engine ?

Ans. Rapid warm up in the distinct feature in an air-cooled engine. Moreover there is no water jacket or radiator. The rapid warm up is used in avoiding sludge and crankcase dilution.

Q. 2. What are the advantages of water cooling system ?

Ans. (*i*) Less engine sound, (*ii*) Quick transfer of heat, and (*iii*) Easier method to control temperature.

Q. 3. What are the disadvantages of water cooling system ?

Ans. (*i*) Corrosion of metal part, (*ii*) Increase in weight of the vehicle, and (*iii*) Freezing difficulties in cold weather.

Q. 4. What is the purpose of a fan in the radiator ?

Ans. To assist the flow of air through the radiator, particularly when the vehicle is moving slowly.

Q. 5. Why is it desirable to flush out the radiator occasionally ?

Ans. If the radiator is not flushed out occasionally, particles or dirt may choke the radiator tubes.

Q. 6. What is the function of the thermostat valve in the cooling system ?

Ans. To reduce the cooling temperature of water by allowing some water to the radiator and bypassing the excess to the circulating pump.

Q. 7. How does the thermostat work ?

Ans. When the engine is too cold the thermostat closes the main valve, thus stopping the flow to the radiator. When the engine is too hot, it opens the main valve for normal circulation through radiator.

Q. 8. How is thermostat valve controlled ?

Ans. (*i*) By a heat sensitive spring, or (*ii*) A coil containing a liquid such as acetone.

Q. 9. What is a radiator ?

Ans. A very effective form of cooling surface in which heat is absorbed and transmitted by the circulating water.

Q. 10. What is the function of a radiator ?

Ans. To cool down the circulated water in the water jacket.

Q. 11. Is circulation in thermo-system cooling system through convection currents ?

Ans. Yes.

Q. 12. What is a pressure cap in the radiator ?

Ans. It is a seal in the radiator, that maintains a selected pressure in the cooling system, when the engine is hot.

Q. 13. What are the indications of an overheated engine ?

Ans. Pinking, or knocking, boiling, loss of power and eventual seizure.

Q. 14. Why should not an engine be allowed to run overcool ?

Ans. Overcooling results in loss of power, increased petrol consumption, and excessive cylinder wear.

Q. 15. What are loads acting on water pump bearings ?

Ans. (*i*) Radial load of the impeller, and (*ii*) Fan belt tension.

Q. 16. What precautions should be observed, when tightening the fan belt ?

Ans. (*i*) The fan blades do not foul, and (*ii*) The pulleys are in line.

Q. 17. Explain why anti-freeze solution is added in the liquid cooling system.

Ans. (*i*) To lower coolant freezing temperature, (*ii*) To raise the boiling point, and (*iii*) To provide water pump lubrication.

Q. 18. Why is bypass recirculation provided in the cooling system ?

Ans. It is mainly provided to prevent water entering the radiator for circulation during starting period and cold condition. With the thermostat valve closed, the bypass circulation builds up enough pressure and becomes heated up quickly.

HIGHLIGHTS

1. Almost 25% to 35% of total heat supplied in the fuel is removed by the cooling medium.
2. At least 95% of the total heat transfer between the working fluid and engine components and the engine components and cooling fluid is effected by "forced convection" process of heat transfer.
3. When the spark advance is different from the optimum value the heat rejected to cooling system is increased.
4. In *air-cooling* system, heat is carried away by the air flowing over and around the cylinder. Here fins are cast on the cylinder head and cylinder barrel which provide additional conductive and radiating surface.
5. In *water-cooling* system of cooling engines, the cylinder walls and heads are provided with jacket through which the cooling liquid can circulate.
6. The *methods* used for circulating water around the cylinder and cylinder head :
   (*i*) Thermo-syphon cooling
   (*ii*) Forced or pump cooling
   (*iii*) Cooling with thermostatic regulator
   (*iv*) Pressurised water cooling
   (*v*) Evaporative cooling.

OBJECTIVE TYPE QUESTIONS

Fill in the blanks or Say "Yes" or "No".

1. If the cylinder head temperature is high the volumetric efficiency and hence the power output of the engine is ....
2. Almost 25 to 35 percent of total heat supplied in the fuel is removed by the cooling medium.
3. Heat carried away by lubricating oil and heat lost by radiation amounts to 10 to 15 percent of the total heat supplied.
4. Heat carried away by the coolant is a dead loss.
5. At very low temperature, starting of engine becomes difficult.
6. Undercooling shortens valve life.
7. The overall heat transfer coefficient gives the heat transmitted per unit area-unit time per degree temperature difference between the bulk fluids on each side of the metal .
8. As the compression ratio is increased the gas temperature near the T.D.C. increases slightly only.
9. The temperature of cylinder gases and flame speed are not affected by F/A ratio changes.
10. At relative fuel-air ratio of 1.12 maximum mean gas temperature occurs, and maximum heat rejection takes place.
11. When the spark advance is different from the optimum value the heat rejected to cooling system is ....
12. In .... cooling system, heat is carried away by the air flowing over and around the cylinder.
13. The .... are arranged at right angles to cylinder axis.
14. Installation of air-cooled engines is easier.
15. Air-cooled engines are noisy.
16. The output of air-cooled engine is .... than that of water cooled engine.
17. .... mixtures or solutions are added to water in extreme winter seasons to lower its freezing temperatures below the danger point.
18. The ...... cooling system cannot be used in high output engines due to its inability to meet the requirement of large flow rates of water.
19. In thermo-syphon cooling system the cooling depends only on the .... and is independent of the engine speed.
20. The .... cooling system requires that the radiator be above the engine for gravity flow of water to engine.
21. In .... cooling there is no loss of water by boiling and evaporation.
22. .... cooling utilises the high latent heat of vaporisation of water to obtain cooling with minimum of water.
23. Evaporative cooling system is used for cooling of many types of industrial engines.
24. In a water-cooling system power absorbed by the pump for water circulation is less than that for cooling fans.
25. Volumetric efficiency of water-cooled engine is .... than that of air-cooled engines.

ANSWERS

1. reduced	2. Yes	3. No	4. Yes	5. Yes
6. No	7. Yes	8. Yes	9. No	10. Yes
11. increased	12. air	13. Fins	14. Yes	15. Yes
16. less	17. Antifreeze	18. thermo-syphon	19. temperature	20. thermo-syphon
21. pressurised	22. Evaporative	23. Yes	24. No	25. higher.

THEORETICAL QUESTIONS

1. Why is cooling necessary for I.C. engines ?
2. Why overheating and overcooling of I.C. engines is harmful ?
3. State the demerits of overcooling and undercooling.
4. Discuss briefly the areas of heat flow in engines.
5. Explain briefly gas temperature variations for a 4-stroke cycle engine.
6. Explain briefly with a neat sketches the representative temperature profiles across the cylinder barrel wall of S.I. engine.
7. Explain the effects of operating variables on engine heat transfer :
   (*i*) Compression ratio
   (*ii*) Fuel-air ratio
   (*iii*) Ignition timing.
8. Describe briefly 'cooling air' and 'cooling water' requirements for I.C. engines.
9. What is the film coefficient ? On what factors the film coefficient generally depends ?
10. What are the two main types of cooling systems ? Where these systems are used ?
11. How engines are air-cooled ? What is the purpose of the fins in an air-cooled system ? What is the size and spacing of fins ?
12. State the applications, advantages and disadvantages of air-cooling system.
13. Name the various types of liquid cooling systems.
14. How is the circulation accomplished in a thermo-syphon system ? What is the draw back of this system ?
15. Explain with a sketch the forced circulation system. State its merits and demerits.
16. What are the advantages and disadvantages of liquid-cooling system ?
17. Explain with a neat sketch 'thermostat cooling' method of cooling I.C. engines.
18. Describe with a neat sketch the construction and working of a thermostat.
19. Explain briefly the following methods of water coolings
   (*i*) Pressurised water cooling
   (*ii*) Evaporative cooling.
20. Describe the operation of the thermostat ? What is the main advantage of using a thermostat in the cooling system ?
21. What is the advantage of pressurised cooling ? How pressurising is accomplished ?
22. What is evaporative cooling ?
23. Compare the merits and demerits of air and water cooling systems.
24. What is the function of a radiator ?
25. What are the various type of radiators ? Explain any one in detail.
26. Explain the role anti-freeze solutions in water-cooling system.
27. Discuss the advantages of pressurised engine cooling over conventional thermo-syphon cooling.
28. What is the purpose of pressurising a liquid-cooled system ? How is it accomplished ?
29. What is the effect of warm-up period on S.I. engine fuel economy ? Discuss the provisions made in recent S.I. engines to reduce warm-up period.
30. Why is cooling necessary for I.C. engines ? What kind of cooling system is employed for mobile units like automobiles ? What are the effects of load and speed on the heat loss through cylinder walls ?

6

Chassis and Suspension

A. CHASSIS

6.1. INTRODUCTION TO CHASSIS

Chassis (also known as *carrying unit*) is a french term and was *initially used to denote the frame or main structure of a vehicle.*

The term **chassis** is now extensively used *to denote the complete vehicle except the body for the heavy vehicle having a separate body.* The chassis contains all the major units necessary to propel the vehicle, direct its motion, stop it, and allow it to run smoothly over uneven surfaces.

The chassis of an automobile *consists of the following components* suitably mounted (Fig. 6.1) :

(*i*) Frame (*ii*) Front axle
(*iii*) Steering system (*iv*) Rear-axle
(*v*) Suspension system (*vi*) Transmission
(*vii*) Brake system (*viii*) Engine
(*ix*) Electrical system.

The chassis is sub-divided into :

(*i*) Power plant (*ii*) Running gear.

The ***power plant*** includes the *engine assembly and power transmission assembly.*

The ***running gear*** includes the *frame, steering system, suspension system, brakes, wheels and tyres.*

— The *electrical system is part of both chassis and body.*

All the above mentioned components are mounted in either of the following two ways :

1. ***Conventional construction.*** In this case a *separate frame* is used.
2. ***Frameless or unitary construction.*** Here *no separate frame* is employed.

Out of these, the conventional construction is being used presently only for heavy vehicles whereas for cars the same has been replaced by the frameless type except of course by small manufactures, who still find it economical to use frame.

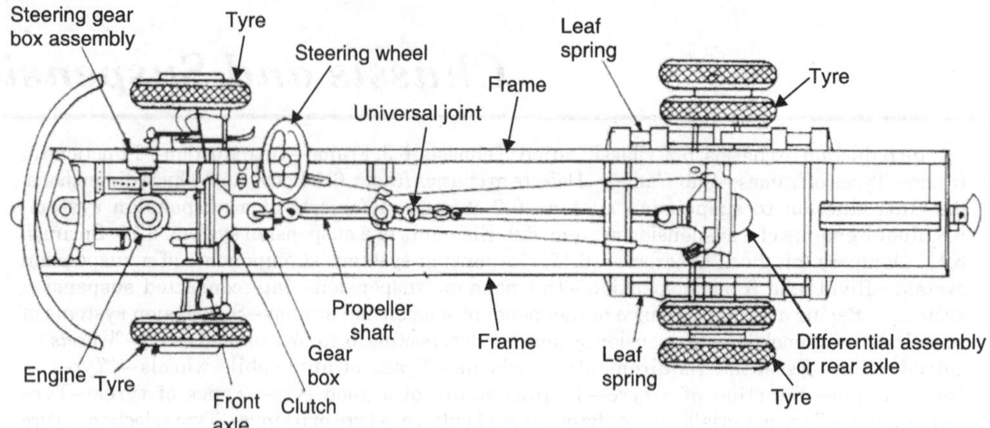

Fig. 6.1. Chassis (Tata truck).

6.2. CLASSIFICATION OF CHASSIS

The chassis may be *classified* as follows :

1. **According to its control :**

(*i*) ***Conventional chassis :***

- In this type of chassis, engine is fitted in front of the driver cabin or driver seat such as in *cars* and *previous model of Tata trucks*. Here, the driver sits behind the engine (*i.e.*, quite far off from the front axle) and as such he cannot see the road just in front of the front tyres. Owing to this reason slope is provided at the mudguard and bonnet to enable the driver to see close to the wheels as far as possible.
- In this arrangement portion of the chassis where engine is fitted cannot be utilised for carrying passengers and goods.

(*ii*) ***Semi-forward chassis :***

- This is such a chassis where half portion of the engine is in the driver cabin and remaining half is outside the cabin such as in *Standard*, *Bedford Pick-ups* and *Tata trucks*.
- In this arrangement a part of the chassis is utilised for carrying extra passengers.

(*iii*) ***Full-forward (or Bus) chassis :***

- In this type of chassis the complete engine is mounted inside the driver cabin.
- In this arrangement floor area is increased to the extent of accommodating 3 extra seats. Also, since the driver seat is just above the front wheels he can see full front road right from the front wheels.

2. **According to fitting of engine :**

(*i*) ***Engine at front :***

(***a***) Conventionally the engines are fitted at front and *drive is given to the wheels from the "rear"* (Fig. 6.2).

Advantages :

- Enough space is available for the luggage behind the rear seat.
- The weight of the vehicle is well balanced on all wheels.
- Adequate space is available for steering the front wheels.
- It is easier to inspect, repair and adjust the engine clutch and gear box.

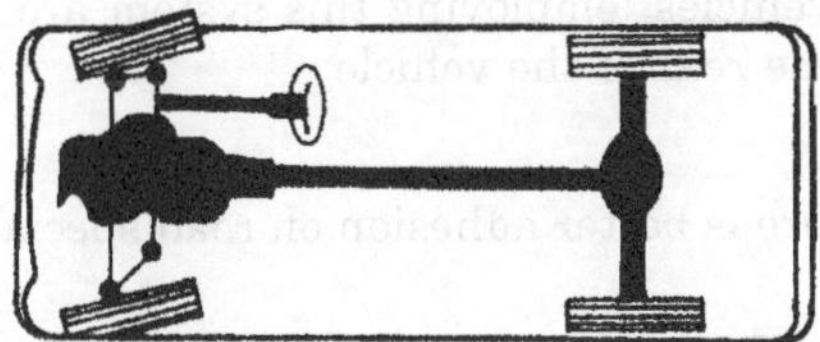

Fig. 6.2. Conventional drive.

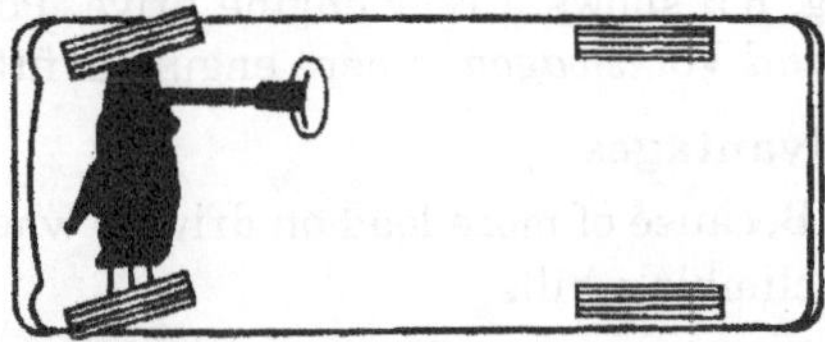

Fig. 6.3. Front engine drive.

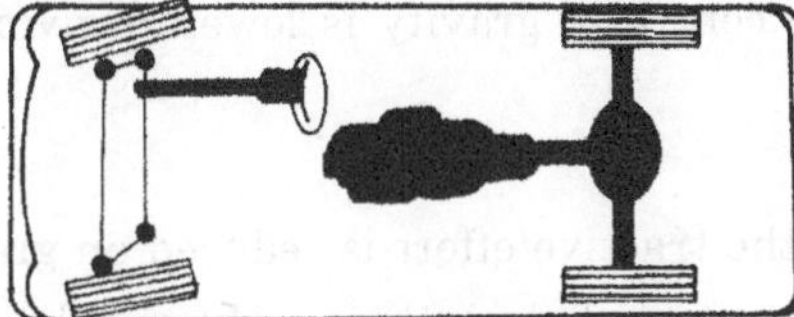

Fig. 6.4. Centre engine drive.

Fig. 6.5. Rear engine drive.

- Increased efficiency of cooling system (since radiator which is fitted in front gets natural draft of air due to forward movement of vehicle).

(*b*) In another arrangement, the *engine is fitted in front and drive is also given to the front wheels only* as in *Matador vehicles.*

Advantages :

- Low bed or floor is available since there is no propeller shaft and differential at the back.
- The vehicle has more adhesion on the road since more weight is placed on driving wheels. Due to good adhesion on road such vehicles have good road holding capacity even on the curves.
- The clutch, gear box and differential are usually made as one unit, thereby cost is reduced.
- At rear, luggage space is spacious.

Disadvantages :

- While accelerating or on a gradient, the weight of vehicle shifts to the rear wheels which is not desirable for better adhesion.
- Because of shifting load to rear, the tractive effort is reduced on gradient.
- On slippery gradients, under steer condition develops which is undesirable.

(*ii*) ***Engine fitted in front but crosswise :***

In this arrangement, engine is fitted in front not in conventional way but *crosswise* as in *Maruti, B.M.C. Mini* and *drive is given to front wheels only* (Fig. 6.3).

(*iii*) ***Engine fitted at the centre of the chassis :***

- In this case, the engine is fitted at the centre of the chassis *i.e.,* under the chassis (Fig. 6.4) as in *Royal Tiger World master buses* previously plied by Delhi Transport Corporation.
- This arrangement *provides full space of chassis floor for use.*

(*iv*) ***Engine fitted at the back :***

Fig. 6.5 shows a rear engine drive. Popular vehicles, employing this system are *Renault, Dolphin and Volkswagon,* where engine is fitted at the rear of the vehicle.

Advantages :

- Because of more load on driving wheels, there is better adhesion on road specially when climbing hills.
- In this system, during braking some of the weight of vehicle gets transferred to front wheels, it helps quick stopping when more load is placed at rear wheels.
- Flat floor is available since long propeller shafts are eliminated.
- With the elimination of propeller shaft the centre of gravity is lowered giving stable driving specially at high speed.

Disadvantages :

- Owing to shifting of weight on rear wheels the tractive effort is reduced on gradients.
- While climbing hills proper adhesion may be affected since the weight of vehicle moves to the rear, thereby reducing the weight on the front wheels.
- As a result of grouping of the engine with clutch, gear box and differential, the repair and adjustment become difficult due to congestion at the rear.
- In order to reduce wind resistance and give a better look at the road it becomes necessary to give more slope to the bonnet which reduces the space for luggage.

3. **According to the number of wheels fitted in the vehicles and the number of driving wheels :**

(*i*) ***4 × 2 drive chassis vehicle.*** Refer to Fig. 6.6 (Tyres in black colour are drive wheels)

- It consists of 4 wheels out of which 2 wheels are the driving wheels.

(*ii*) ***4 × 4 drive chassis vehicle.*** Refer to Fig. 6.7.

- It consists of 4 wheels and all of them are driving wheels.

(*iii*) ***6 × 2 drive chassis vehicle.*** Refer to Fig. 6.8.

- It consists 6 wheels out of which 2 wheels are the driving wheels.

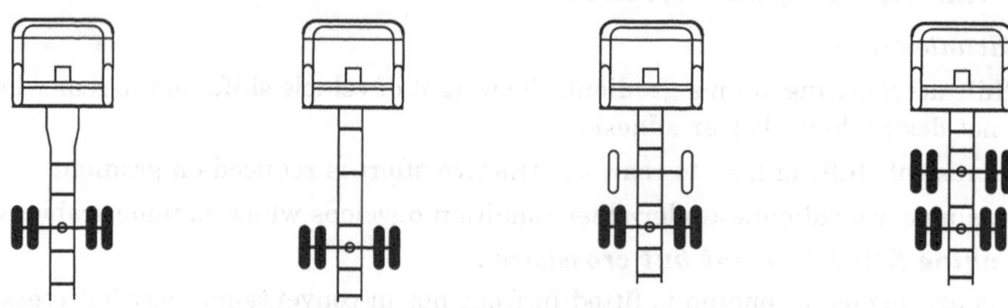

Fig. 6.6. 4 × 2 Drive. Fig. 6.7. 4 × 4 Drive. Fig. 6.8. 6 × 2 Drive. Fig. 6.9. 6 × 4 Drive.

(*iv*) ***6 × 4 drive chassis vehicle.*** Refer Fig. 6.9.

- It consists of 6 wheels out of which 4 wheels are the driving wheels.

It is to be noted that one side of the axle may consist of either one-wheel or two-wheels which will be considered as *one unit only*.

4. **According to wheel base size :** Refer to Fig. 6.10.

- In the *long wheel base chassis vehicle* ("*wheel base*" is *the distance between the centres of the front and rear wheels*), the distance is more thereby *providing more floor area of the chassis for passengers and goods.*
- In this type of chassis, the front overhang and particularly the rear overhang are also more.

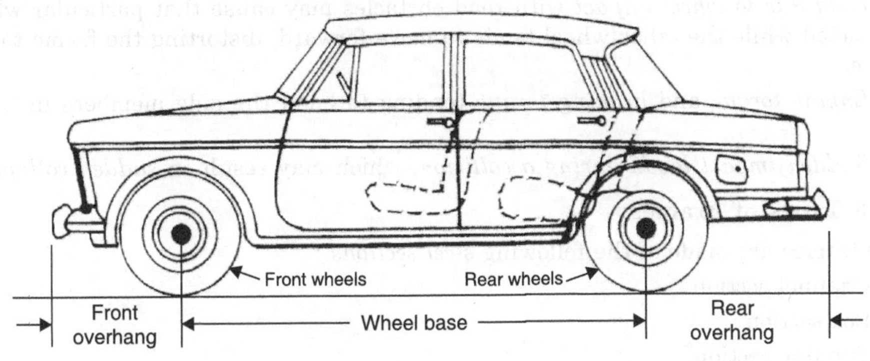

Fig. 6.10

6.3. FRAME

6.3.1. Introduction

- *The* ***frame*** or ***under body*** is *the main part of the chassis on which the remaining parts of the chassis are mounted.* It is a rigid structure that forms a skelton to hold all the major parts together.
  - The *engine* is mounted in the forward end of the frame and is connected to the clutch and transmission unit to form a complete power assembly.
  - At the rear end of the frame, the *rear axle* housing is attached through the rear spring.
  - The *wheel* and *tyre* assemblies support the frame.
  - The *steering system* has some parts bolted to the frame and some to the body.
  - The *petrol tank* is fastened to the rear of the frame.
- **Chassis frames** *are made of* ***"steel section"*** *so that they are strong enough to withstand the load and at the same time are also light in weight to reduce dead weight on the vehicle.*

The long sections which are right and left members of frame are called *long members.* There are joined together with cross members with the help of rivets or bolts or nuts. There are 5 to 6 cross members joining the long member to give the chassis frame a good strength. *In some of the models the long members are bent both at the front and rear to accommodate jumping of front and rear axels.*

6.3.2. Functions of Frame

The *main functions* of frame are :

1. To support the chassis components and the body.
2. To withstand static and dynamic loads with undue deflection or distortion.

The various loads dealt with by the frame are :

(*i*) *Weight* of the vehicle and the passengers, causing vertical bending of the side members.

(*ii*) *Vertical loads when the vehicle comes across a bump or hollow*, which results in longitudinal torsion due to one wheel lifted or lowered with other wheels at the usual road level.

(*iii*) *Loads due to road camber, side wind, cornering force* while taking a turn, which results in lateral bending of side members.

(*iv*) *Load due to wheel impact* with road obstacles may cause that particular wheel to remain obstructed while the other wheel tends to move forward, distorting the frame to *parallelogram shape*.

(*v*) *Engine torque* and *braking torque* tending to bend the side members in the vertical plane.

(*vi*) *Sudden impact loads during a collision*, which may result in *sudden collapse*.

6.3.3. Types of Frames

The frames are made of the following *steel sections* :

(*i*) Channel section
(*ii*) Box section
(*iii*) Tubular section.

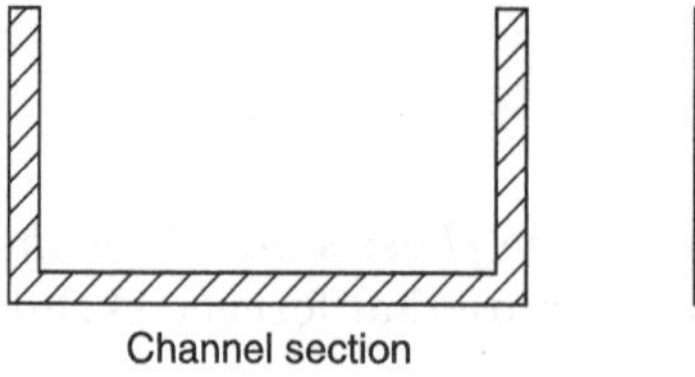

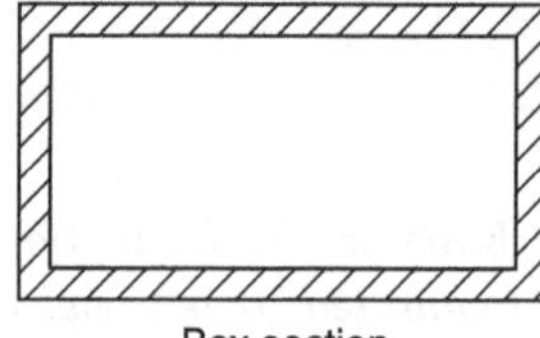

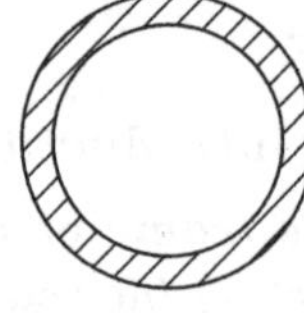

Fig. 6.11. Steel sections used in chassis frame.

- *Channel section* is used in *long members* of the frame.
- *Box section* is used in *short members* of the frame.
- *Tubular section* is used these days in *three-wheelers, scooters and matadors, pick-ups frames.*

— *The channel section is good is bending, tubular in torsion and box in bending and torsion.*

The frames should be strong enough to bear loads, sudden brakes and accidents.

Basically, there are following *three types of frames* :

1. Conventional frame.
2. Semi-integral frame.
3. Integral or unit frame.

1. **Conventional frame :**

- It is also known as *non-load carrying frame*. Here the loads on the vehicle are transferred to the suspension by the frame, which supports the engine, power train and the body.

- The body work is made of material like wood and completely isolated from the frame deflection with the help of rubber mountings. This type of construction is widely used in *trucks*.
- This type of frame is ***not much suited to resist torsion***. Use of a tubular or box-section backbone type of frame improves the torsional strength.

Fig. 6.12 (*a*) and (*b*) shows a typical frame used in cars, using longitudinal members of box-section.

— The complete frame is fabricated by the welding process.

— The frame tapers from the rear to the front to permit adequate movement of the steering wheel. The longitudinal members, by sweeping upwards at the rear end, allow for the vertical movement of the rear axle.

— The torsional rigidity of the frame is increased by providing tubular or box-section cross members.

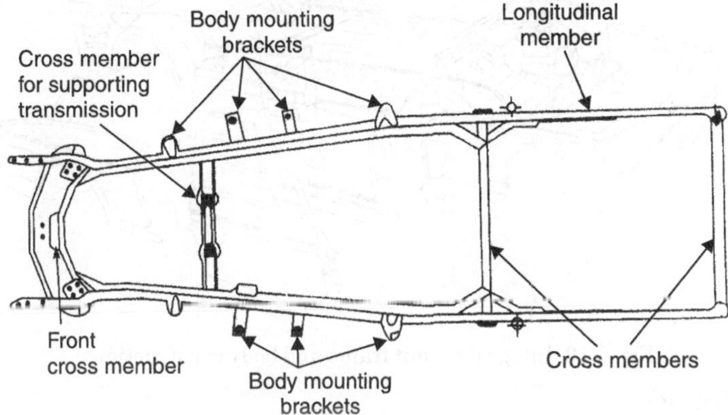

(*a*) Box-section type

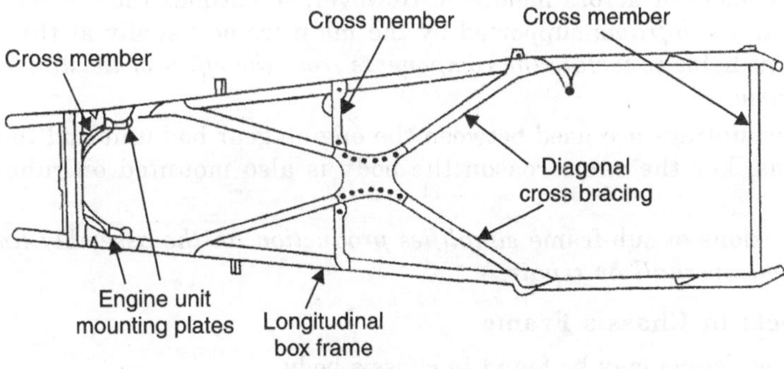

(*b*) X-member type

Fig. 6.12. Car frames.

2. **Semi-integral frame :**

- In this arrangement, the rubber body mounts are replaced by relatively stiff mounts so that a part of the frame load is transferred to the body structure also.
- This type of frame is quite popular for *small European cars and American cars*.
- This frame however, is heavy.

3. **Integral (or unitised) frame** : Refer to Fig. 6.13.

In this type of construction, there is no frame and all assembly units are attached to the body. The chassis, floor and body are assembled by welding from a large number of mild steel pressings.

In this arrangement, the chassis becomes a sub-frame and other sub-frames are used for the front and rear suspension units.

This design *reduces vehicle weight, lowers production costs and allows a lower floor.*

- *This is the modern form of construction for almost all cars and lighter commercial vehicles.*

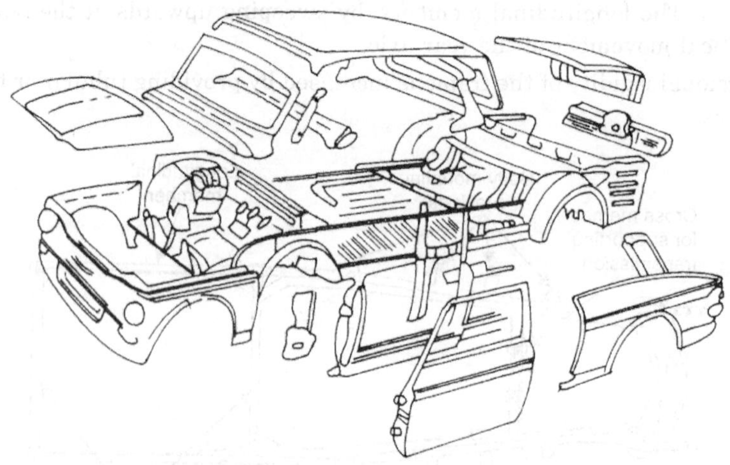

Fig. 6.13. Integral or unit frame and body construction.

6.3.4. Sub-frames

- Normally the various components of a vehicle are bolted directly either on the main frame members or across members. However, sometimes the engine and gear box are carried on a *sub-frame* supported by the main frame usually at three points. The arrangement helps to *isolate the components from the effect of twisting and flexing of the main frame.*
- Rubber mountings are used between the engine gear box unit and the frame to *reduce vibrations.* For the same reason the body is also mounted on rubber blocks on the frame.
- The provisions of sub-frame *simplifies production on the assembly line and facilitates subsequent overhall or repair.*

6.3.5. Defects in Chassis Frame

The following *defects* may be found in chassis body :

1. Cracks
2. Dislocated parts
3. Broken welds
4. Buckling.

- The chassis should be checked for the above mentioned defects and suitably rectified.
- Further, there will also be *corrosion and rusting of the chassis frame.* These can prove to be very dangerous. Therefore, the vehicle *should be guarded against corrosion and rusting.* To prevent rusting, the vehicle should be regularly painted. If this is not done, the chassis frame will be affected and soon *lose its strength.*

The following methods are used *to avoid rust formation* :

(*i*) Use of galvanised sheets.

(*ii*) Use of zinc enriched primer paint.

(*iii*) Use of plastic coating on panels.

(*iv*) Use of rubberised solution.

(*v*) Avoiding tight corners so that the moisture cannot get settled.

(*vi*) Providing plenty of drain holes to allow trapped water to run out.

- It is important that the chassis frame is *properly aligned*. After some repair is carried out on the frame, the alignment should be checked with the original measurements.

6.4. BODY

Body *is the super-structure of the vehicle. The chassis and body make the complete vehicle.*

In larger and heavier vehicles, the chassis and the body are each made as a separate unit and then bolted together.

The body is usually made from a large number of steel pressings which are welded together. The body is bolted to the chassis at *numerous points, rubber or felt strips being interposed to damp-down vibration and noise.*

The body of motor vehicle should fulfill the following *requirements* :

(*i*) Light in weight.

(*ii*) Cheap and easy in manufacturing.

(*iii*) Attractive in shape and colour.

(*iv*) Minimum number of components.

(*v*) Clear all-round vision through glass area.

(*vi*) Long fatigue life.

(*vii*) Minimum resistance to air.

(*viii*) Good access to the engine and suspension system.

(*ix*) Minimum vibrations during running of the vehicle.

(*x*) Adequate space for passengers and luggage.

All steel sections of the body are stamped out by dies separately and welded together. Decorative moulds and bands are formed around the window and door openings. The tendency of all steel body to drum or rumble is overcome by using insulating and sound deadening materials.

— The closed car body provided roof which may be used for luggage.

— The closed bus provides a large space for luggage.

— The open truck body does not have roof. It consists of surrounding side only. The rear side is only a half panel which may be opened down for loading and unloading. In some good trucks, the lower side panels may also be opened down for unloading.

- A car body is formed by a number of pressed steel panels welded together. For attaching the door, instrument panel, hood, truck lid, head lining etc. attachment brackets are welded to the body.

Different types of bodies for various vehicles are listed below. Refer to Fig. 6.14.

1. Two-door sedan
2. Four-door sedan
3. Convertible
4. Van
5. Truck Punjab body or Straight truck
6. Truck half-body

7. Truck platform type
8. Truck with articulated trailer
9. Tanker
10. Dumper truck
11. Jeep
12. Pick-up
13. Station wagon.

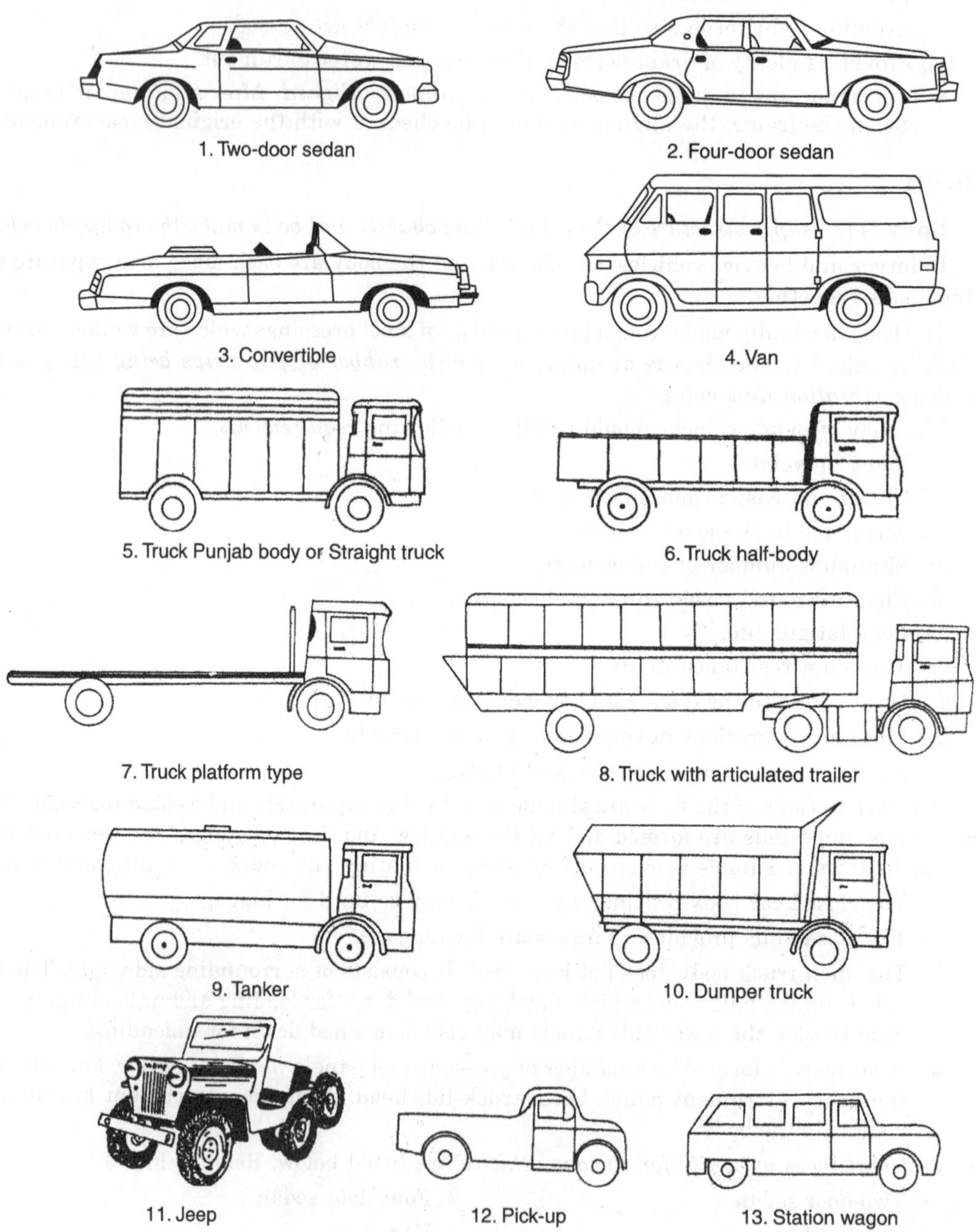

Fig. 6.14. Different types of bodies for various vehicles.

6.5. VEHICLE DIMENSIONS

Two important dimensions used to describe the size of an automobile of the framed or frameless type are :

1. Wheel base 2. Wheel track.

1. **Wheel base.** *This is the longitudinal distance between the centres of the front and rear axles* (Fig. 6.10).

2. **Wheel track.** *The distance between the centres of front wheels is known as "wheel track"*. It is shown as *"Front-wheel track"* at F in Fig. 6.15 and as *"Rear wheel track"* at F in Fig. 6.16.

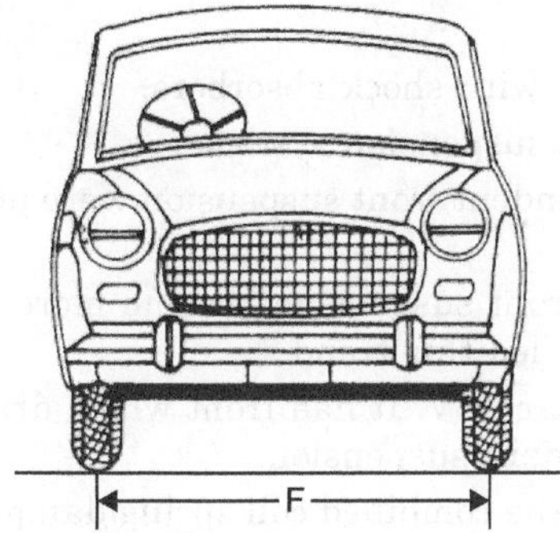
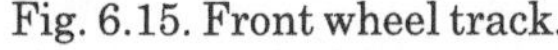

Fig. 6.15. Front wheel track.

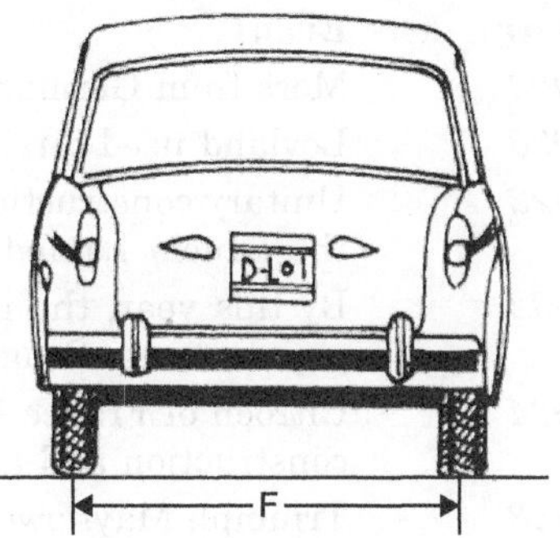

Fig. 6.16. Rear wheel track.

B. SUSPENSION

6.6. INTRODUCTION TO SUSPENSION SYSTEM

General Aspects :

The **suspension system** of *an automobile is one which separates the wheel/axle assembly from the body*. The *primary function* of the suspension system is to *isolate the vehicle structure from shocks and vibration due to irregularities of the road surface.*

Broadly speaking, suspension system consists of a *spring* and a *damper* (including also spring shackles, axles, wheels and stabilizer). The energy of road shock causes the spring to oscillate. These oscillations are restricted to a reasonable level by the damper, which is more commonly called a *shock absorber.*

— The suspension system of a motor vehicle is divided into the *rear-end* suspension and *front-end* suspension.

A good suspension system must have *springiness* and *damping.* Springiness is elastic resistance to a load. On application of sudden load the spring will compress/expand as the case may be without transmitting the same to the body. As the spring compression is complete it expands on rebound, and now damping becomes important since this will *absorb the work energy as heat energy and the continuous oscillations of the spring which normally would have taken place are absorbed.*

- The function of isolation of shocks and vibrations between the road and carriage is achieved by different elements at different sages as mentioned below :

(*i*) The first element/stage which takes the impact is the *tyre.* With pneumatic tyres, this is achieved by flexing and compression of pneumatic tyres at the contact point.

(*ii*) The second stage is between the axle/wheel system and the body. The elements incorporate *springs, dampers/shock absorbers, various linkages and tie bars. This part is called* **"suspension system".**

(iii) The last stage comprise the *seats* of the automobile which the passengers occupy. These are made of springs and foam/rubber cushions. They absorb all short amplitude high frequency vibrations which pass from the system to the passenger compartment.

— The *wheel and seats* usually overcome the high frequency and low amplitude vibrations.

— The *suspension system* tries to overcome the greater irregularities of the road which may impact all the wheels, both the wheels on one side, or one axle or only one wheel at a time, while transmitting minimum vibrations to the compartment of the passengers.

Brief History :

Year	***Event***
• *1903*	Mors from Germany fitted a car with shock absorbers.
• *1920*	Leyland used torsion bars in its suspension system.
• *1922*	Unitary construction and independent front suspension were pioneered on the Lancia Lambda.
• *1932*	By this year, the independent front suspension became more common in popular cars. Peugeot and Adler led this trend.
• *1934*	Citroen of France launched its car 7CV. It had front wheel drive, unitary construction and an all independent suspension.
• *1948*	Triumph Mayflower introduced the combined coil spring/damper unit and strut-type telescope damper.
• *1950*	Ford (Britain) adopted the McPherson strut independent front suspension on MK 1 consul.
• *1955*	Citroen used hydropneumatic suspension in its DS 19 Model.
• *1959, 1962*	BMC used independent rubber suspension in its Mini in 1959 and improved it in its 110 Model by using hydrostatic suspension in 1962.

6.7. FUNCTIONS/OBJECTS OF A SUSPENSION SYSTEM

Following are the *functions/objects of suspension system* :

1. To prevent the road shocks from being transmitted to the vehicle frame.
2. To preserve the stability of the vehicle in pitching or rolling, while in motion.
3. To safeguard the occupants from road shocks.
4. To maintain proper steering geometry.
5. To provide good road holding while driving, cornering and braking.

6.8. REQUIREMENTS OF A SUSPENSION SYSTEM

The *requirements of a suspension system* are :

1. Low initial cost.
2. Minimum weight.
3. Minimum wheel hop.
4. Minimum tyre wear.
5. Minimum deflection consistent with required stability.
6. Compatibility with vehicle components—tyre, frame, wheel base, steering linkage.
7. Low maintenance and operating costs.

6.9. ELEMENTS OF A SUSPENSION SYSTEM

Fig. 6.17 shows the schematic form of a suspension system. The **sprung weight** is *the weight of passenger carriage whereas the* **unsprung weight** *is the weight of the wheel-axle system.*

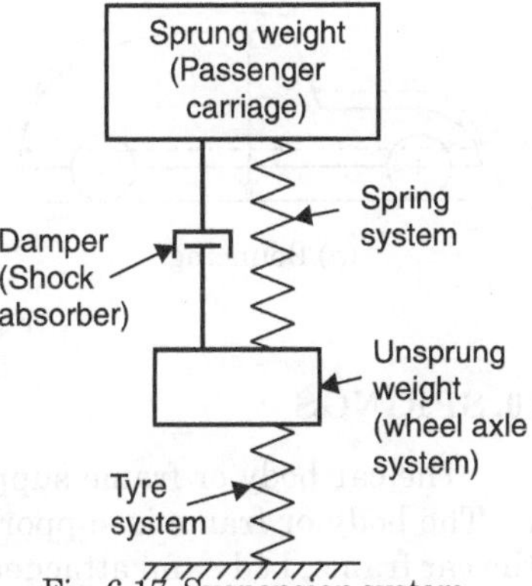

Fig. 6.17. Suspension system.

The important *elements of a suspension system* are :

1. Springs.
2. Dampers (or shock absorbers).

- The *springs* provide spring effect to a large extent ; the tyre, however, provides the spring effect to a smaller extent.
- The *dampers* (or shock absorbers) provide the damping effect to a large extent. However in case of leaf springs, the friction between the leaves in motion does generate some damping effect. Dampers perform the following *two functions* :

(*i*) They *reduce the tendency of the carriage unit to continue to "bounce"* [Fig. 6.18 (*a*)] *up and down* on its springs after the disturbance that caused the linear motion has ceased.

(*ii*) *They prevent excessive built up of amplitude of bounce* as a result of periodic excitation at a frequency identical to the natural frequency of vibration of the spring mass system.

Besides a simple bounce or vibration of the carriage unit as a whole, following two more types of vibration also exist :

(*i*) Rolling (*ii*) Pitching.

In **"rolling"**, [Fig. 6.18 (*b*)] *the carriage unit rolls about the longitudinal axis of the vehicle* while in **"pitching"**, *the carriage unit rolls about a transverse axis.*

— In rolling one side of the car goes down and the rear goes up and *vice versa.* The tendency to roll is checked by a *stabilizer.*

— *Pitching* [Fig. 6.18 (*c*)] is a more complex phenomenon and is affected by what is known as *"vibration coupling effect" i.e., interaction between front and rear suspension.* Since the picthing persists for a longer duration if the rear suspension has a lower natural frequency than the front suspension, therefore, the natural frequency of the rear suspension is normally made higher than that of the front.

— The pitching, in general, depends upon the following factors.

(*i*) The frequency of disturbances ;

(*ii*) Bumps over which the car rolls ;

(*iii*) Spacing of bumps ;

(*iv*) Speed of the vehicle ;

(*v*) Mass moment of inertia of the vehicle about the axis of pitch and its wheel base.

- *A combination of roll and pitch is called* ***diagonal pitch.***

  In order to control the above mentioned suspension movements, *antisway bars, stabilizers, pitch and roll control bars, mechanical levelling devices, hydroelastic systems etc. are employed.*

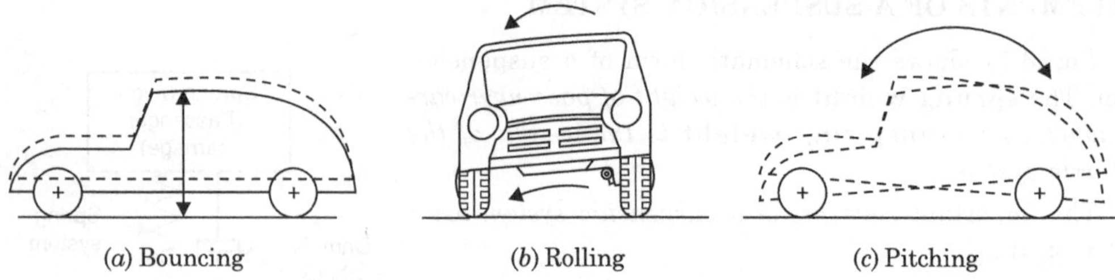

Fig. 6.18. Basic suspension movements.

6.10. SPRINGS

The car body or frame supports the weight of the engine, the power train, and the passengers. The body or frame is supported by the *springs*. There is a spring at each wheel. The weight of the car frame, body and attached parts applies an initial compression to the springs. The springs compress further as the car wheels hit bumps or expand as the wheels drop into the holes in the road. The springs cannot do the complete job of absorbing road shocks. The tyres absorb some of the irregularities in the road. The springs in the car seats also help to absorb shock. However, little shock from road bumps and holes is felt by the passengers.

- Springs are resilient members and as such act as reservoirs of energy. *They store the energy due to the sudden force which comes when vehicle encounters a bump or a ditch. This energy is released subsequently and with the action of dampers, the energy is converted into heat and bounce is avoided.*
- Springs used for suspension system should *absorb road shocks quickly and return to the original position slowly.* Now let us examine this, a soft spring will oscillate too much *i.e.*, it will go up and down many times making a vehicle move along with it, while a stiff spring will not oscillate too much and will give a rough ride. As such a compromise is made by *using a soft spring with a shock absorber to control its up and down movement as well as to absorb the road shocks.*

Types of Springs :

The various types of automotive springs are :

1. *Leaf (or laminated) springs :*

(*i*) Full elliptic–Two semi-elliptic springs connected to form the shape of ellipse.

(*ii*) Three quarter elliptic–One semi-elliptic spring connected over a quarter elliptic spring.

(*iii*) Semi-elliptic–Forming the shape of half ellipse.

(*iv*) Quarter elliptic–Half of semi-elliptic spring.

(*v*) Transverse–Semi-elliptic type spring has the saddle above forming a bow and is fitted parallel to the wheel axle.

2. *Coil springs.*
3. *Torsion bars.*
4. *Air and gas springs.*
5. *Rubber springs.*

The main factors governing the choice of springs used are :

(*i*) Total weight of suspension system.

(*ii*) Total cost of installation.

(*iii*) Relative capacity for storing energy.

(*iv*) Guide linkage required.

(*v*) Location.

(*vi*) Fatigue life.

1. **Leaf springs :**

- These springs are referred to as *laminated springs* since they use steel strips or laminations one over the other with reducing length (Fig. 6.19). They are also called *semi-elliptical springs* as they are bent in that form. However, these days they are *almost straight*.

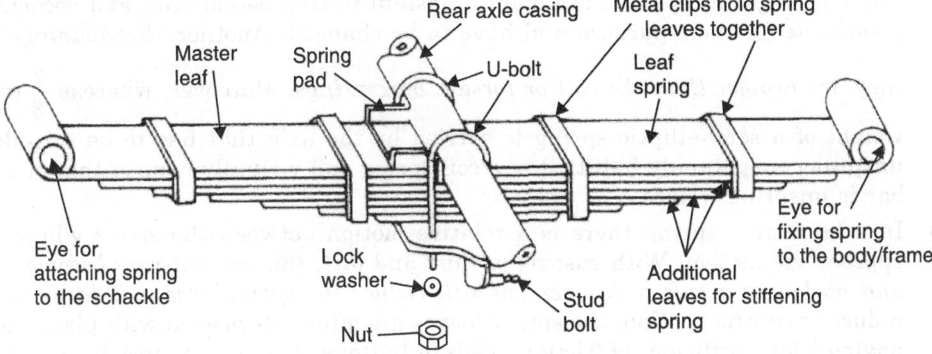

Fig. 6.19. Leaf spring with U-bolt and rear axle casing.

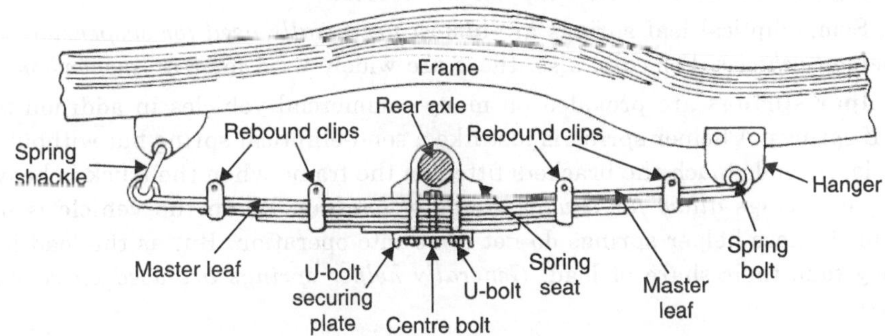

Fig. 6.20. Leaf spring attached to frame and axle.

- Leaf springs are made from flat strips of spring steel. Each strip is called a *leaf*. Several strips are placed one on the other. They are joined together by clamps (rebound clips) and a central bolt. The length of each leaf decreases so that the spring assembly acts as a flexible beam and is of uniform strength. The ends of the master leaf are formed into loops called spring eyes. One end of the spring is attached with frame through a spring bolt passing through the eye. The other end is secured through a **shackle** (a sort of link by means of which leaf spring is connected with the chassis frame) as shown in Fig. 6.20. The *shackle helps in accommodating the change in the length of the spring.*

When the wheel encounters a bump, the spring expands and increase in length takes place which is accommodated by the shackle. Similarly, the process of contraction is accommodated.

Bronze bushes are generally fitted into the spring eyes and through the bush, bolt passes. Sometimes rubber is used instead of bushes. The *rubber bushes are quiet in operation and require no lubrication.*

The leaf spring may be placed on the axle and kept in position by a U-bolt. The U-clamp in this case does not take any load but simply keeps the spring in position. In light duty cycle, the spring is placed below the axle and the load is transmitted through the U-bolts of the axle.

- The leaf springs in general have approximately *only a quarter of energy storage capacity for a given stress level of either coil spring or torsion bar.* Calculated on the basis of weight for a given energy storage capacity, the ratio is 3.9 : 1 in favour of coil or torsion bar springs.
- *The springiness of the leaf spring can be varied by varying the number of leaf laminations.* Increase in number would stiffen it while reduction will make it more flexible. These springs on repeated use due to constant flexing become soft and become *totally unsuitable* for their purpose and have to be changed. Another '*disadvantage*' is that they are *heavier than the coil or torsion bar springs.* Moreover, whereas $\frac{2}{3}$ rd of the weight of a semi-elliptic spring is carried by the axle that has to be considered as unsprung weight, only half that of a coil spring and virtually none of that of a torsion bar is unspring weight.
- In a laminated spring there is a relative motion between the strips where friction opposes the motion. With rusting or dust and dirt, this *motion may become excessive and no longer act as a damper but still reduce the springiness to a large extent.* To reduce or control friction, the spring leaves are often interleaved with plastic material having a low coefficient of friction ; pads or buttons of similar materials are interposed between the ends of adjacent leaves. In order to prevent the entry of dust and dirt the spring themselves are also wrapped and sealed.

Uses. Semi-elliptical leaf springs are *almost universally used for suspension in light and heavy commercial vehicles.* For *cars,* also, these are widely used for *rear suspension.*

- **Helper springs** are provided on many commercial vehicles in addition to the main leaf springs. A helper spring is just like a semi-elliptical spring but without eyes at the ends. Its ends touch the brackets fitted on the frame when the truck is heavily loaded. These springs *allow for a wide range of loading.* When the vehicle is only lightly loaded, these helper springs do not come into operation. But as the load is increased, they take their share of load. *Generally helper springs are used on rear suspension only.*

2. **Coil springs :**

A spring is shown in Fig. 6.21. It is made from a special spring steel wire. This spring is generally circular in cross-section and of suitable diameter to have the desired stiffness. The wire is wound in the shaft of coil. The spring is formed at high temperature, cooled and proper heat treatment is given to it in order to have the characteristics of elasticity.

When the wheel of the automobile experiences bump on the road side, the spring compresses to absorb the shock energy.

A ***helper coil spring*** is also sometimes used *to provide progressive stiffness against increasing load.*

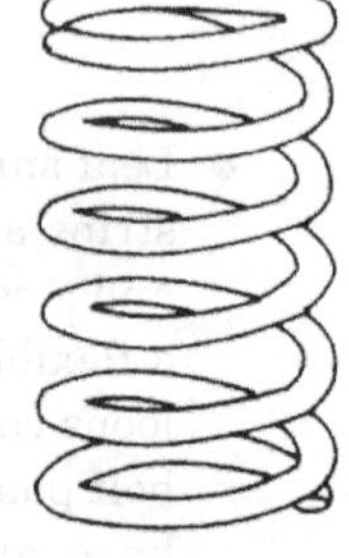

Fig. 6.21. Coil spring.

The *main advantages of coil springs* are as follows :

(*i*) The energy stored per unit volume is almost *double* in case of coil springs than the leaf springs.

(*ii*) They neither have noise problems, nor do they have static friction causing harshness of ride as in case of leaf springs.

(*iii*) Require less space for a given load as compared to a leaf spring.

(*iv*) Lighter for a given load as compared to leaf spring.

(*v*) Compact as a unit.

(*vi*) Since there is no friction damping effect as in leaf spring, they are more lively in action. As such good and powerful shock absorbers are required with coil springs.

The coil springs are **used** *mainly with independent suspension, though they have also been used in the conventional rigid axle suspension as they can be well accommodated in restricted spaces.*

- The *springs take the shear as well as bending stresses.* The coil springs, however, *cannot take torque reaction and side thrust,* for which alternative arrangements have to be provided.
- The life of coil springs is increased by *shot peening* their surfaces to induce compressive stresses in them and to reduce the effect of scratches in initiating fatigue cracks. Immediately after shot peening, such springs may be given an anticorrosion treatment again to increase their fatigue life.

3. **Torsion bars :**

Torsion bar is simply a *rod acting in torsion and taking shear stresses only.* These bars are made of *heat-treated alloy spring steel.* The *amount of energy stored per unit weight of material is nearly the same as for coil springs.*

The torsion bar performs the spring's action by its resistance to twisting. The bar is mounted transversally in some of the vehicles, whereas in other constructions, it is employed lengthwise along the frame.

Fig. 6.22 shows a frame and axle.

— The axle is supported by a lower plate and another plate fixed to the frame. The lower plate is connected to a torsion bar. The end of the torsion bar is fitted to the chassis frame.

— When the wheel moves up or down, the torsion bar gets slightly twisted. In this position it absorbs the vibrations. When the wheel is coming down, the torsion bar reaches its original position. It absorbs vibrations and shocks. By this arrangement it acts like a spring and maintains the stability of the vehicle.

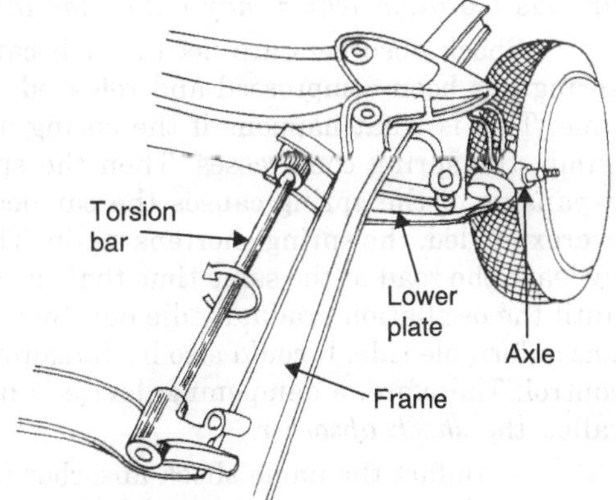

Fig. 6.22. Torsion bar.

- As in coil springs, shot peening and anticorrosion treatment is also given to the torsion bar to improve fatigue life.
- Torsion bars are *not very popular as suspension springs because their end fixings are more costly and provision has to be made for the adjustment of the ride height on the vehicle assembly line.* These are however used as *antiroll devices.*

4. **Air and gas springs :**

In these springs compressed air or gas is filled in the cylinder or bellows against which the wheel movement is transmitted through a diaphragm. As soon as the wheel passes over a road irregularity the compressed air pressure returns the system to its original position.

Advantage : Air springs give a high quality ride particularly with independent suspension.

Disadvantages :

(*i*) High cost.

(*ii*) Complexity of compressed air ancillary system.

(*iii*) Risk of breakdown.

(*iv*) Greater maintenance required.

(*v*) Freezing of moisture in air in cold weather.

(*vi*) These systems are generally too bulky and complex for cars.

- Air springs are fairly widely used in several countries in vehicles whose loaded and unloaded weights differ greatly *e.g., trucks, tractors, large draw bar trailers etc.*

5. **Rubber springs :**

- The use of rubber as a means of suspension has the following *advantages* : (*i*) Rubber is more reliable and as such a rubber suspension cannot suddenly fail like the metal springs ; (*ii*) The rubber has excellent vibration damping properties ; (*iii*) It can store greater energy per unit weight than steel, therefore, the rubber springing system can be made more compact ; (*iv*) The absence of squeaking which is always present in steel springs.
- Rubber springs have a *good fatigue strength in compression* (but worst in tension). They are *free of maintenance.*
- Rubber springs are **used** *more in commercial vehicles.*

6.11. DAMPERS (OR SHOCK ABSORBERS)

Dampers (or shock absorbers) *are used in the suspension system to check any continuous vibration which may follow the initial force on the system.*

Shock absorbers are necessary because springs do not "settle down" fast enough. After a spring has been compressed and released, it confines to *shorten and lengthen*, or *oscillate*, for a time. This is what happens if the spring at the wheel is not controlled. When the wheel hits a bump, the spring compresses. Then the spring expands after the wheel passes the bump. The expansion of the spring causes the car body and the frame to be thrown upward. But, having overexpanded, the spring shortens again. This action causes the wheel to move up and momentarily leave the road at the same time that the car body and frame drops down. The action is repeated until the oscillation gradually die out. Such spring action on car could produce a very bumpy and uncomfortable ride. It could also be dangerous, because a bouncing wheel makes the car difficult to control. Therefore, a dampening device is needed to control the spring oscillations. This device is called the *shock absorber.*

— In fact the name shock absorber is rather *misleading* since it is the spring and not the shock absorber that initially absorbs the shock. *The 'shock absorber' absorbs the energy of shock converted into vertical movement of the axle by providing damping and dissipating the same into heat.* Thus it *merely serves to control the amplitude and frequency of spring vibrations.* It cannot support weight and has *zero resistance.* Therefore, **'damper'** is a better term technically to describe the 'shock absorber'.

- In case of *"leaf spring suspension system", the friction between the leaves provides the damping effect.* But because of the uncertainty of lubrication conditions, the amount of friction also varies and hence the damping characteristics do not remain constant. Therefore, additional damping is provided by means of *dampers* or shock absorbers.
- Frequently, the shock absorber housing is linked to the frame cross member and shock absorber arm is connected to the spring, axle or suspension control arm.

Modern cars mostly have *hydraulic dampers* which are of the following two types :

1. Telescopic dampers
2. Rocking lever dampers.

In these dampers hydraulic fluid is used as damping agent ; the *principle of operation* is as follows :

"When a piston forces the fluid in a cylinder to pass through some hole, a high resistance to the movement of the piston is developed, which provides the damping effect."

An additional advantage of hydraulic damper is that the damping is *proportional to the square of speed.* So for small vibrations the damping is also small, while for larger ones the damping becomes automatically more.

1. Telescopic dampers :

A telescopic damper or shock absorber (direct acting) is shown in Fig. 6.23. The cylinder is filled with a fluid prescribed by the manufacturer. There are valve openings in the piston. The valve openings are also at the bottom of the cylinder tube. There is a reservoir tube full of the fluid.

— When the wheels meet bumps and pot holes the shock absorber lengthens and shortens. When this happens, a piston inside a cylinder of shock absorber moves up or down.

When the shock absorber is compressed, the fluid in the cylinder passes upwards through the restricted valves of the piston. At the same time the fluid passes down through a small valve of the cylinder tube. By this arrangement the piston is able to move against the resistance of the fluid. Thus the shock on the vehicle is absorbed by the fluid.

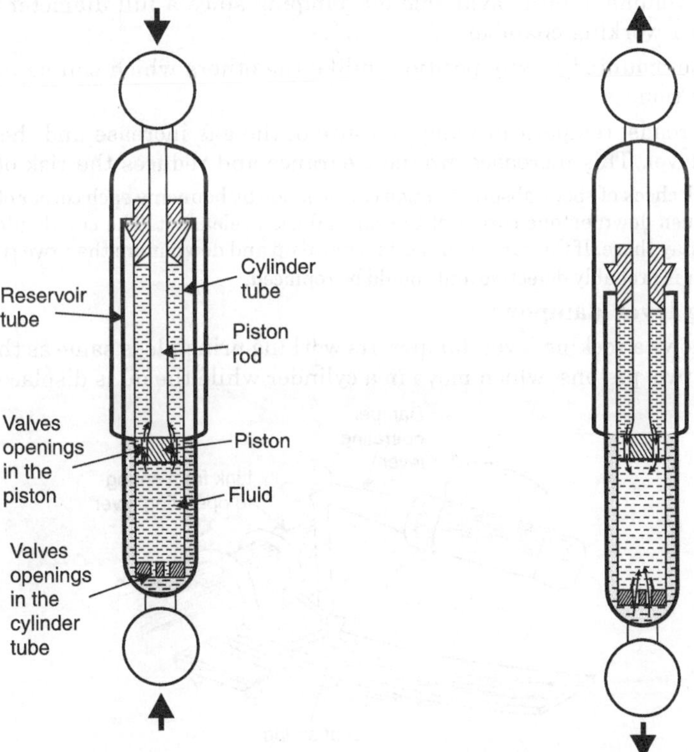

Fig. 6.23. Shock absorber under compression.

Fig. 6.24. Shock absorber under elongation.

— When the shock absorber is lengthening (Fig. 6.24), the liquid from the top portion of the piston is forced downward through the piston valves. At the same time the fluid from the reservoir tube enters through a valve at the bottom of the cylinder tube. By this arrangement the process of the shock absorber being lengthened is made very slow. This takes place after overcoming the resistance of the fluid.

Thus the shock absorber helps the springs of the axle from moving up and down at a very fast rate. It helps the oscillating springs of the car to settle down immediately.

Gas-filled shock absorber :

In this type of shock absorbers, *instead of only oil, a mixture of oil and gas is used for the damping effect.*

The main *advantages* of this type of shock absorber over the conventional type are as follows :

1. Because of the large piston, the specific pressures caused by damping are very much reduced, which has a positive effect on the valve components.
2. Provides longer life to tyres and other related components in the suspension (*e.g.*, springs bushes etc.)
3. It can tolerate heat to a greater degree ; this is an invaluable feature in tropical conditions.
4. The over-pressure in the working chamber produces an exact reaction of the valve components even at the lowest stroke and prevents the formation of bubbles (cavitation) at severe strain. This improves the rolling properties of the tyre and ensures a precise working of the shock absorber under any driving situation and road condition.
5. A large volume of oil is available for damping since a full diameter of the tube can be used as a working chamber.
6. It can be mounted in any position unlike the others which can be used only in a fixed 'top' position.
7. On bad roads, temperature and pressure of the gas increase and the body is lifted to a higher level. This increases ground clearance and reduces the risk of body damage.

Note. A quick check of shock absorber action can be made by bouncing each corner of the car. With the car on a level surface, push down on one corner of the car and then release it. The car should come back up to its original height and stay there. If the car continues to bounce up and down more than two times after you release it, the shock absorber is probably defective and should be replaced.

2. **Rocking lever damper :**

Fig. 6.25 shows a rocking lever damper. Its working principle is same as that of the telescopic damper but has two pistons which move in a cylinder while the oil is displaced through a valve.

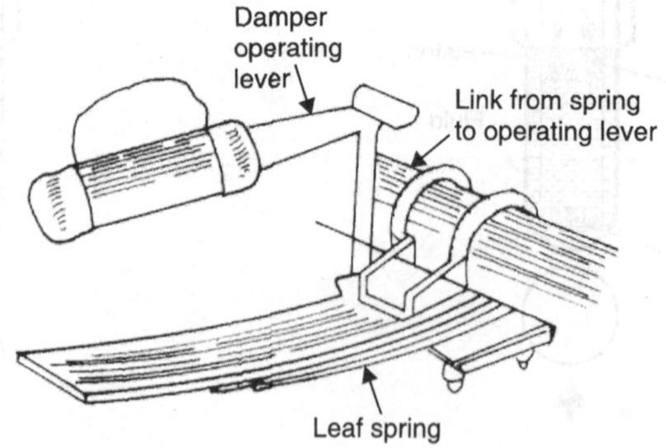

Fig. 6.25. Rocking lever damper/shock absorber.

The motion of the dual pistons takes place due to the motion of the wheels which is passed on to the pistons through the rocker levers.

Maintenance of damper/shock absorber :

When a shock absorber does not function properly it is better to remove it and replace it with a new one.

Sometimes an old *shock absorber can be properly serviced again, as per procedure given below* :

- Remove the shock absorber and dismantle its parts in a systematic manner.
- Clean these parts thoroughly.
- Fit new rubber gaskets.
- Then properly reassemble the parts.
- Pour in the fluid as recommended by the manufacturer for that particular shock absorber.

6.12. SUSPENSION SYSTEMS

6.12.1. Components of a Suspension System

A suspension system consists of the following *principle components* :

1. *Springs.* There neutralise the shocks from the road surface.

2. *Dampers (Shock absorbers).* They act to improve comfort by limiting the free oscillation of the springs.

3. *Stabilizer (Sway bar or antiroll bar).* It prevents lateral swaying of the car.

4. *A linkage system.* It acts to hold the above components in place and to control the longitudinal and lateral movements of the wheels.

Suspension may be *rigid axle suspension* or *independent suspension.*

6.12.2. Rigid Axle Front Suspension

A rigid axle suspension has the following *characteristics :*

1. It is durable enough for heavy-duty use.

2. The number of parts composing the suspension is small and the construction is simple ; therefore maintenance is simple.

3. While turning, there is a little tilting of the body.

4. Less tyre wear since there is little change in the alignment due to the up-and-down movement of the wheels.

5. Riding comfort is poor due to great unsprung weight.

6. Vibrations and oscillations occur rather easily since the movements of the left and right wheels mutually influence one another.

Fig. 6.26 shows a typical rigid axle front wheel suspension. This type of suspension was universally used before the introduction of independent front wheel suspension. It may use either two longitudinal leaf springs or transverse springs, usually in conjunction with shock absorbers. These assemblies are mounted similarly to rear leaf spring suspensions.

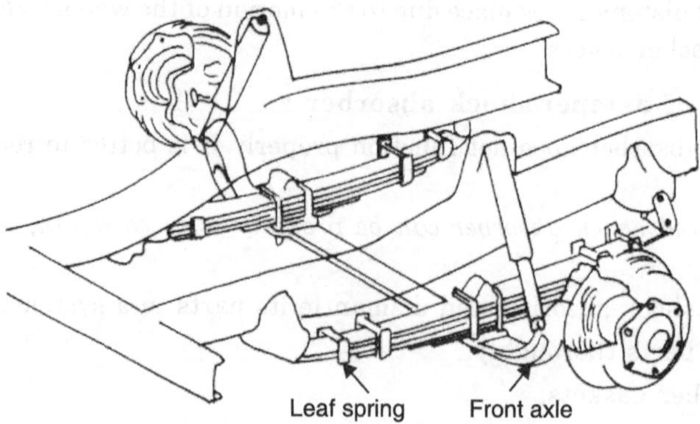

Fig. 6.26. Rigid axle front suspension.

6.12.3. Independent Suspension

"Independent suspension" *is a term used to describe any arrangement by which the wheels are connected to the carriage unit in a manner such that the rise and fall of one wheel has no effect on the others.* Almost all the passenger cars now use the independent front suspension, in which the coil spring arrangement is the most common.

When a vehicle with rigid axle suspension encounters road irregularities, the axle tilts and the wheels no longer remain vertical. This causes the whole of the vehicle to tilt to one side. Such a state of affair is *not desirable.* Besides causing *rough ride,* it causes *'wheel wobble'*. The road adhesion is also decreased. In order to avoid this the *wheels are sprung independent of each other, so that tilting of one does not affect the other.*

Advantages :

The independent suspension claims the following advantages over the rigid axle type suspension.

1. In independent system since the wheels more or less travel with their planes perpendicular to the road surface, the *gyroscopic effects are reduced to a minimum.*
2. The engine and chassis frame can be placed relatively lower which means engine position can be moved forward resulting in more space for passengers.
3. Provides a greater degree of vertical/springing movement.
4. Diminished wheel wobble and steering tramp.
5. Provides scope for use of springs of greater resilience giving much better springing action than most rigid axle vehicles.
6. Reduced unsprung weight and hence improved ride and better road holding while cornering and braking.
7. The frame and body do not tilt but remain horizontal and the wheels vertical when the wheel encounters a road bump.
8. Variations in caster angle are reduced.
9. It uses coil springs which can be placed closer to the wheel. This is definite advantage *vis-a-vis* leaf springs for a wheel to be steered.

Disadvantages :

Apart from the distinct advantages which the independent suspension possesses, it has the following *disadvantages :*

1. High initial cost.
2. Owing to larger number of bearings greater maintenance is required.

3. More rigid sub-frame or chassis frame required.

4. Forces due to unbalanced wheels are more pronounced and transmitted easily to the steering wheel.

5. In the event of body roll, the wheel chambers tilt outwards in case of wishbone type and inwards in case of MacPherson strut type, due to which cornering power is reduced.

6. Misalignment of steering geometry with the wear of components, thus requiring more frequent attention.

- In a car, the "*font axle*" is usually a *dead axle* (The dead axles do not contain differential and have no concern with the power transmission system of the automobile) although some cars do have front wheel drive. Therefore, ***the independent suspension systems have been almost universally adopted*** for the *front wheels.* The independent suspension system for the front wheels has to cope with the fact that they are to be *steered.*

"*Rear axle*" is usually a *live axle* (the rear axles are those axles which contain differential and through which rotary motion is transmitted to the wheels) with power being transmitted to the rear wheels. Thus independent suspension has not become popular for the rear wheels. Further rear wheels have to carry a lot of weight and while the weight on the front wheel remains more or less constant, in the case of rear wheels it makes a lot of difference when the car is running empty or when it is fully occupied. The suspension system has to cater to both these condition.

6.12.3.1. Front wheel (dead axle) independent suspension. The front suspension is more complicated than the rear suspension, *because the front wheels not only move up and down with respect to the car frame, but also swing at various angles to the car frame for steering.* In order to permit the front wheels to swing on one side or the other for steering, each wheel is supported on a spindle which is part of a steering knuckle. The steering knuckle is then supported through ball joints, by upper and lower control arms which are attached to the car frame.

Since the *front suspension* in a car has to bear a lot of forces particularly due to acceleration, braking and cornering, therefore, it must *adhere to the following conditions :*

(*i*) Not to allow the system to alter the tilt of the wheels to any serious degree.

(*ii*) Not to permit the various forces coming from road irregularities and cornering to deflect the car from its course of movement decided by the driver.

(*iii*) Not to allow the wheels to wobble, move any significant distance backwards or forwards or sideways.

The *following types of independent suspension systems are applicable to automobiles* :

1. Wishbone arm system.
2. Trailing link system.
3. Sliding pillar system.

1. **Wishbone arm system**

Fig. 6.27 shows the *double wish bone suspension system.* This is the *most popular type* of independent suspension system in which *coil springs are mostly used.* In European cars, torsion bars are quite popular in lieu of coil springs. In some automobiles, transverse leaf springs are used in this type of suspension. English car Humber contains transverse leaf spring in the front independent suspension system.

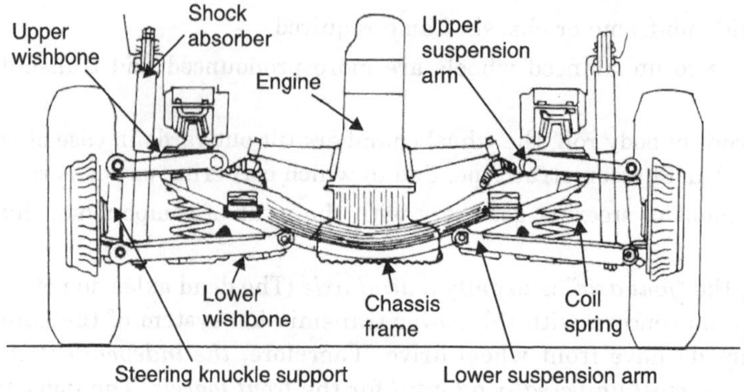

Fig. 6.27. Double wishbone suspension.

In this type of suspension, there are two suspension or control arms on each side of the vehicle. These arms are like the *two legs of chicken wishbone or letter V.* These wishbone arms are connected with chassis frame on the open end. One arm is *below* whereas other is *above* the frame. The closed ends of both upper and lower suspension arms are connected with the steering knuckle to support to which is attached steering knuckle by means of a king pin. A coil spring is placed between the frame and lower wishbone suspension arm. Mostly the open end of upper control arm is connected with the damper/shock absorber which is fitted at the frame. The upper and lower arms are connected in position, for the cradle.

When there is bump and the wheel tended to go up, the control arms move up and coil spring is compressed. Since the damper/shock absorber is fitted with the upper control arm, so it damps the vibrations set up in the coil spring due to road irregularities.

- The suspension using transverse leaf spring, consists of an upper wishbone, steering knuckle support and a transverse leaf spring. *The transverse spring serves two purposes of holding the lower end of steering knuckle support and of providing the spring action.* The basic construction is similar to regular wishbone design with the exception that no lower control arm is used.

— The wishbones may be *parallel and equal or unequal.* Early double wishbone suspension systems had both the link equal and parallel to each other ; they had the following *disadvantages* : (*i*) The wheels, on corners, lean outwards with the body resulting in undesirable steering effects ; (*ii*) Variation in track length resulted in adverse tyre wear. In order to avoid these undesirable effects in *modern automobile,* the two wishbones *were not only made unequal but also they were made non-parallel* (*usually the upper wishbone is made shorter so that the wheels do not remain upright but have a slight lean inwards*). In this arrangement the following *advantages* accrue : (*i*) The track length remains constant although a slight change in camber takes place which is better for tyre life ; (*ii*) Better cornering characteristics are obtained.

— In both the double wishbone suspension systems, when coil spring with telescopic damper is used they are usually installed *coaxially.* On some vehicles, the springs and dampers are *mounted separately,* so *that access could be gained more easily for servicing the damper.*

MacPherson strut assembly :

It is a single wishbone with a telescopic strut type system as shown in Fig. 6.28. The MacPherson system consists of a *telescopic strut, a single arm* and a *diagonal stay.* The strut is fixed to the body structure at the upper end through a flexible mounting and the lower part of the strut is connected at the bottom by a joint to the lower arm. The lower part of the strut also carries the stub axle, which in term carries the wheel. The steering motion is supplied to the lower part of the strut and it turns the whole strut. A coil and a hydraulic damper/shock absorber surround the upper part of the strut which takes care of the road irregularity shocks and vibrations.

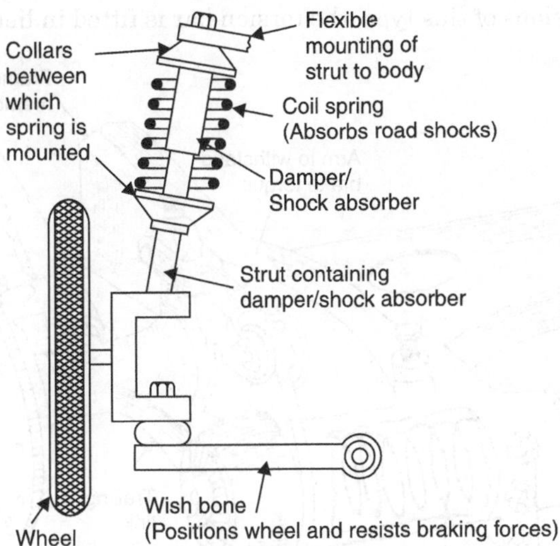

Fig. 6.28. MacPherson strut assembly.

Advantages of MacPherson system :

(*i*) Very easy maintenance.

(*ii*) Simple in mechanical construction.

(*iii*) Less variations in wheel camber.

(*iv*) Its light moving parts help the wheels to follow the road irregularities.

(*v*) Distinct advantages in case of transverse engines, since in that case there is no space or very little space for upper links to fit.

Disadvantages :

(*i*) During cornering and brake torque the radial loading comes on the piston due to the lateral forces.

(*ii*) In order to absorb the full suspension loads the body structure has to be really strong above the wheel arches where the struts are attached.

Maruti-800 suspension system :

Maruti-800 car incorporates the MacPherson type independent suspension system. It consists of *coil springs, front suspension struts, steering knuckles, a stabiliser bar and front suspension arms.* In this type of construction, shocks applied to wheels are distributed through steering knuckles from front suspension struts, coil springs, to front suspension arms and are absorbed.

- The spring rate of front coil spring and it's free length = 1.8 kg/mm and 336 mm respectively.
- The attenuation force of the suspension struts = 50 kg (extending side) and 24 kg (contracting side).
- The stroke of suspension struts = 135 mm.

2. **Trailing link system :** Refer to Fig. 6.29.

- These systems use parallelogram linkages lying beside the frame side members.
- Usually a horizontal coil spring is used in this type of suspension. During compression and rebound, the spring winds and unwinds like the balance spring in an ordinary watch.
- In some suspensions of this type, the torsion bar is fitted in lieu of horizontal coil spring.

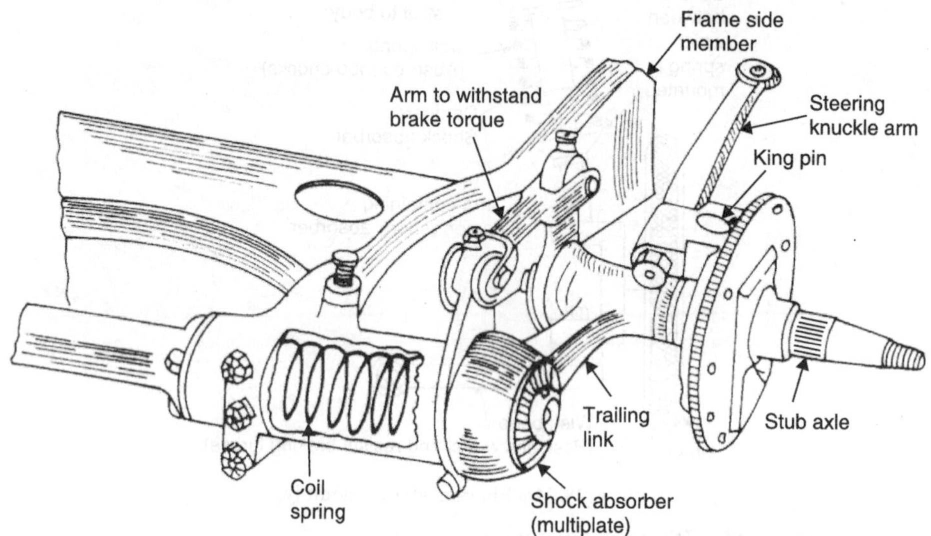

Fig. 6.29. Trailing link independent suspension.

3. **Sliding pillar type independent suspension.** Refer Fig. 6.30.

- This type of suspension was the first independent front end suspension ever used.
- In this system, the pillar or elongated king pin is attached to the wheel and slides up and down in the axle type beam affixed rigidly to the vehicle frame.

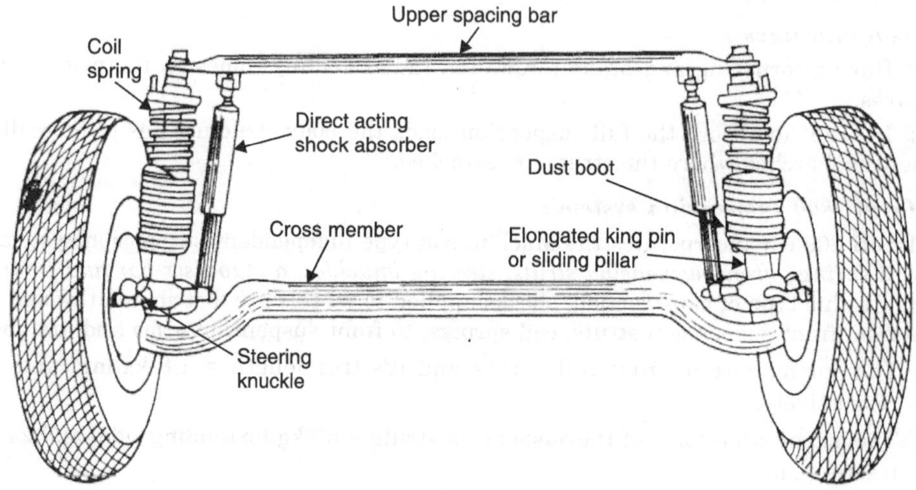

Fig. 6.30. Sliding pillar type of independent suspension.

Simplified diagrams of the independent front suspension :

Fig. 6.31 shows the simplified diagrams of the independent front suspensions using coil, torsion bar and leaf spring. Basically, the system is known as parallelogram type independent front suspension. It consists of upper and lower links connected by the stub axle carrier. In general, the lower link is larger than the upper and they may not be parallel. This arrangement maintains the track width as the wheels rise and fall and so minimize tyre wear caused by the wheel scrubbing sideways.

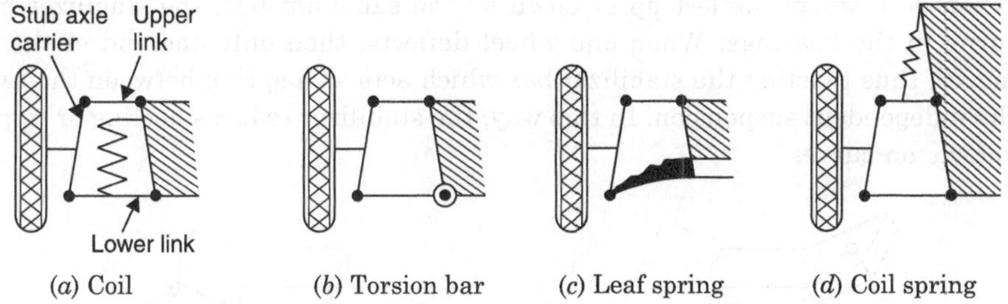

Fig. 6.31. Parallelogram independent front suspensions.

- "***Strut and link***" *type suspension system* (Fig. 6.32) is particularly for integral body construction, because the loading points are widely spaced. The normal top link is replaced by a flexible mounting and a telescopic damper acts as a the king pin. This system, *known as MacPherson system has little rolling action and absorbs shocks readily.*

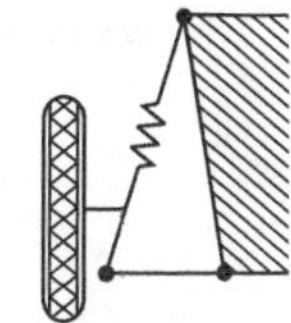

Fig. 6.32. Strut and link suspension system.

- *Trailing arm independent front suspension* (Fig. 6.33) maintains constant track and wheel altitude with a slight change in wheel base and caster angle. A coil spring is attached to the trailing arm which itself is attached to the shaft carrying the wheel hub. When the wheel moves up and down, it winds and unwinds the spring. In certain designs, a torsion bar has been used in place of the coil spring.

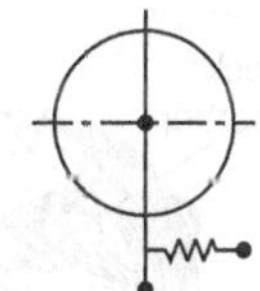

Fig. 6.33. Trailing arm suspension system.

- In ***sliding type*** *suspension system* (Fig. 6.34). The stub axle can move up and down as well as rotate in the frame members. Track, wheel altitude and wheel base remains unchanged throughout the rise and fall of the wheel.
- In ***vertical guide*** *suspension system*, the king pin is attached directly to the cross member of the frame. It can slide up and down, thus compressing and expanding the springs.

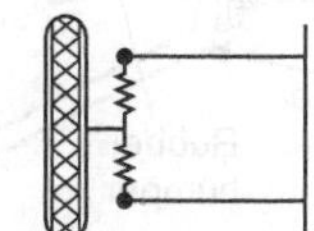

Fig. 6.34. Sliding type suspension system.

Torque rod :

- It is also known as *torque arm.* Its one end is rigidly fixed to the axle or axle housing, and the other end is attached to the frame by means of a pivoted mounting.
- It is used to *maintain correct alignment of the axle with the frame.* It also serves to remove all the stresses on the springs.

Stabilizer (or antiroll device) :

- A *stabilizer* or *a sway bar* (Fig. 6.35) is simply a bar of alloy steel with arms at each end connected to the lower wishbone of the independent suspension or axle. It is supported in bush bearings fixed to the frame, and is parallel to the cross member.
- It is necessarily used in all independent front end suspension. It *reduces the tendency of the vehicle to roll or tip on either side when taking a turn.* This tendency has been increased due to the use of softer springs and independent front end suspension.

— When both wheels deflect up or down by the same amount, the stabilizer bar simply turns in the bearings. When one wheel deflects, then only one end of the stabilizer moves, thus twisting the stabilizer bar which acts as a spring between the two sides of the independent suspension. In this way, the stabilizer reduces heeling or tipping of the vehicle on curves.

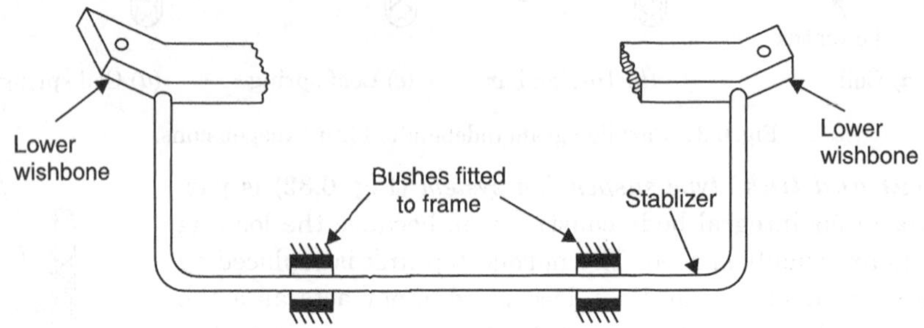

Fig. 6.35. Stabilizer.

Fig. 6.36 shows a stabilizer used in front suspension system.

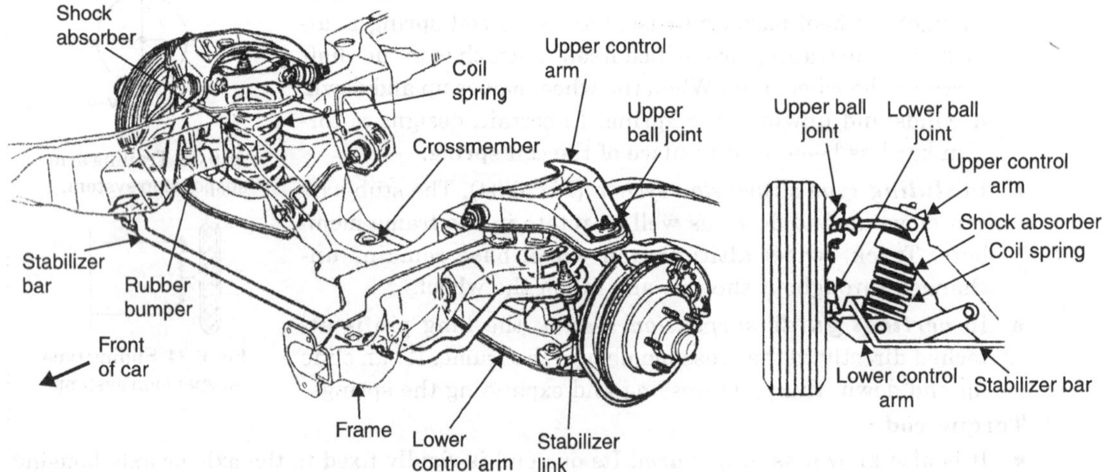

Fig. 6.36. Stabilizer used in front suspension system.

6.12.3.2. Rear wheel suspension system

Mostly rear axle is a live axle (except in front wheel drive cars). The rear axle even in case of four-wheel drive vehicles like jeeps mostly transmits the power. In case of heavy passenger vehicles or load carrying vehicles, the power is transmitted through the rear axle. A ***"live axle"*** *is one that either rotates or houses the shafts that rotate,* while a **"*dead axle*"** *simply carries at its ends the stub axles on which the wheels rotate.* A live axles performs the following *two functions* :

(*i*) It houses and supports the final drive, differential and shafts to the road wheel and reacts to the torques in both the input and output shafts.

(*ii*) It works as a beam that *carries* through the medium of springs and other suspension system, the weight of passenger compartment and its contents, and *transmits* these loads under dynamic conditions through the road wheels (rotating on its ends) to the ground.

The rear axle suspension system needs to be designed for overcoming the following forces :

(*i*) The weight of carriage unit (including contents).

(*ii*) Brake drag.

(*iii*) Torque reaction (both drive line and brakes).

(*iv*) Lateral forces.

(*v*) Driving thrust.

Both rigid suspension and independent rear wheel suspension have been designed in several ways and some are specifically known by the names of the car models in which they are used. Some of these are discussed below :

1. **Leaf-spring rear-suspension system (Rigid suspension) :**

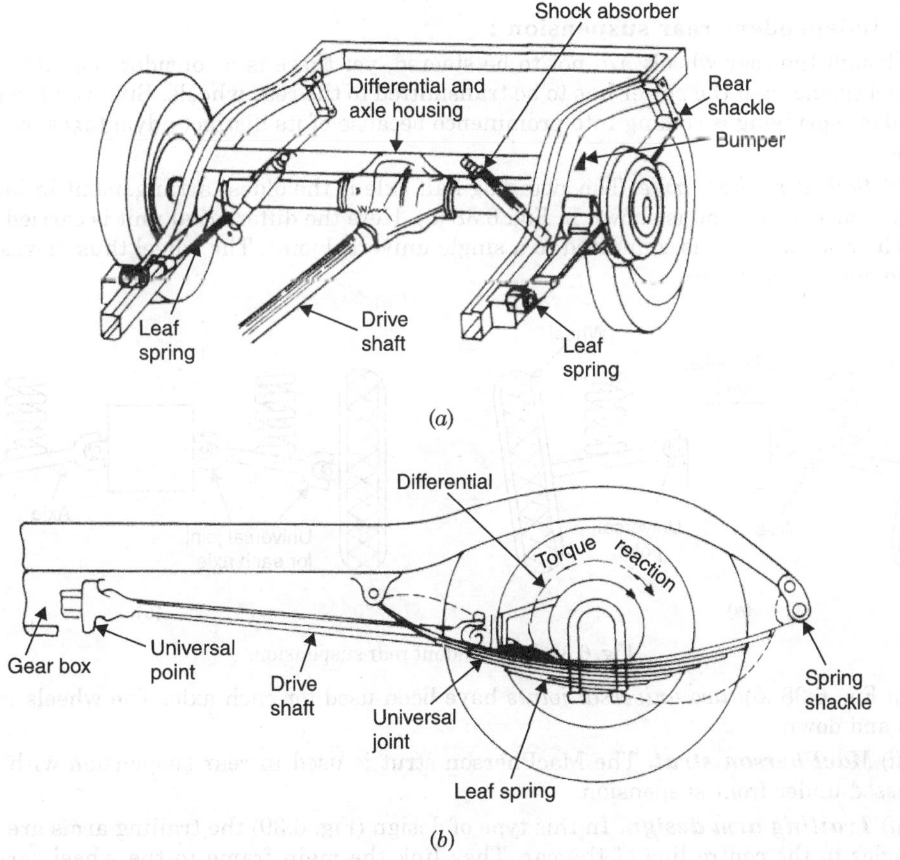

Fig. 6.37. Rear end rigid suspension (Hotchkiss drive).

Fig. 6.37 shows rear rigid suspension using leaf springs (Hotch kiss drive). It employs two leaf springs located as far as possible on the axle. These springs not only absorb shocks but also position the axle and the axle moves up and down with it. Near the leaf springs are attached the dampers (*i.e.*, shock absorbers), one each to the two sides of the rear axle. The axle is usually fitted exactly at the mid point of the spring. However, in some cases it is fixed a bit ahead of the mid point to give a downward tilt as the axle rise when riding over bumps. This reduces the amount by which the propeller shaft lifts on a bump and in turn minimises the height of the propeller shaft tunnel and the amount it intrudes into body of the car. The leaf springs are fitted to the body with a *rubber bush in the front* while the *rear end* of the spring is fitted to the body through a *shackle with rubber brushes.* By doing so the increase or decrease in length of the spring, as it flexes up and down, is accommodated.

A leaf spring has a demerit that it *tends to distort when the axle tries to turn during acceleration or braking.* The following modifications have been made to overcome some of the shortcomings associated with Hotchkiss drive.

(*i*) *Linking the axle with the main structure.* This arrangement limits the fore and aft distortion of the springs.

Radius rods trailing from mountings on the structure help to position the axle.

(*ii*) *Use of transverse rod.* A transverse rod (also known as *Panhard rod*) is pivoted on the body structure at one end and on the axle at the other which assists in holding the axle in the position.

2. **Independent rear suspension :**

Though the rear wheels are not to be steered, yet there is a considerable difficulty in the rear wheel springing if the power has to be transmitted to the rear wheels. But even the rear wheel independent springing is coming into prominence because of its distinct advantages over the rigid axle type.

(*i*) ***Swinging half axle.*** The swinging half axle is the oldest arrangement in independent rear wheel suspension and is shown in Fig. 6.38 (*a*). Here the differential unit is carried by the car frame. The rear half axle is connected to a single universal joint. The wheel thus moves vertically while moving over a bump.

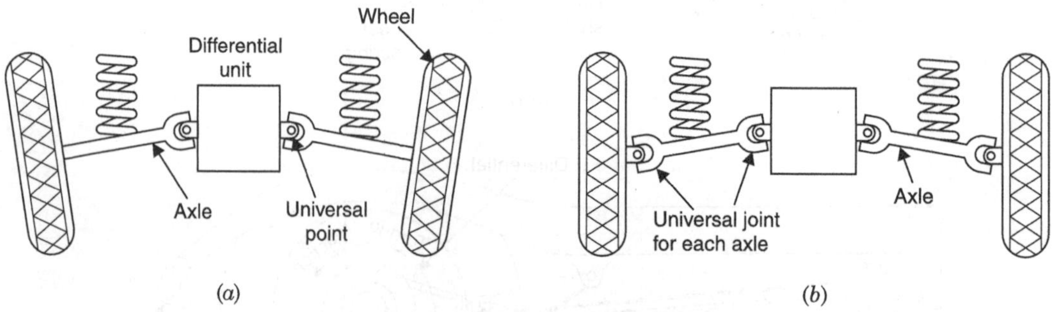

Fig. 6.38. Independent rear suspension.

In Fig. 6.38 (*b*), *two universal joints* have been used for each axle. The wheels move vertically up and down.

(*ii*) ***MacPherson strut.*** The MacPherson strut is used in rear suspension with half axles as discussed under front suspension.

(*iii*) ***Trailing arm design.*** In this type of design (Fig. 6.39) the trailing arms are pivoted at right angles to the centre line of the car. They link the main frame to the wheel carriers. This arrangement allows the wheels to move up and down.

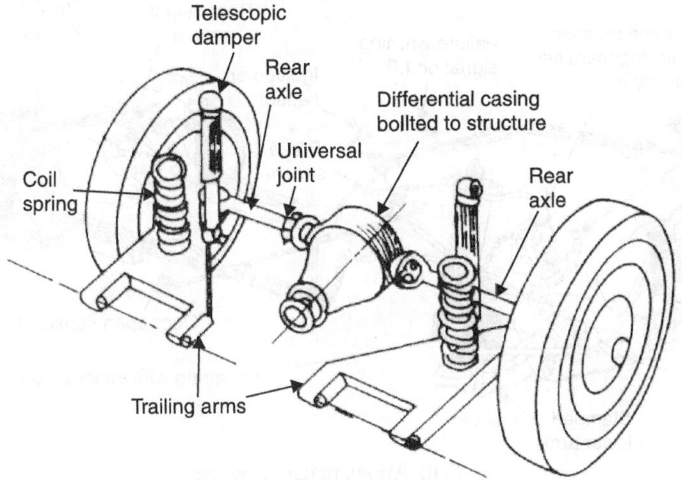

Fig. 6.39. Trailing arm.

6.12.4. Interconnected Suspension Systems

These are also called *'linked systems.'* In such systems, the front and rear suspension units or else the units on the two sides of the automobile are connected together.

The primary advantage of such a system is that tendency of the vehicle to bounce, pitch or roll is reduced and constant desired altitude of the wheel can be maintained.

The various types of three systems are :

1. Air suspension.
2. Hydroelastic suspension.
3. Hydragas suspension system.
4. The Daimler-Benz vehicle suspension.

1. Air suspension :

This is the simplest linked system. In air-suspension systems, the four steel springs are replaced by four rubber cylinders, or air springs (Fig. 6.40). Each rubber cylinder is filled with compressed air, which supports the car weight. When a wheel encounters a bump in the rod, the air is further compressed and absorbs the shock.

Fig. 6.40 shows an electronic air compression system which includes the following components :

(*i*) Electric air compressor,

(*ii*) A microcomputer control module [MCM),

(*iii*) Four air springs with built-in solenoid valves,

(*iv*) Three height sensors (two front and one rear), and

(*v*) The air-distribution system of lines and fillings.

The height sensors monitor the riding height or *vehicle trim height.* They signal the control module of any change. If the height is too high, the control module opens the solenoid valves in the springs with too much air. This allows some of the air to escape, lowering the car. If the height is too low, the control module turns on the air compressor. Then the control module opens the solenoid valves in the spring needing more air. Now more air flows into the spring until proper trim height is restored. The air-suspension system provides springing for all four wheels.

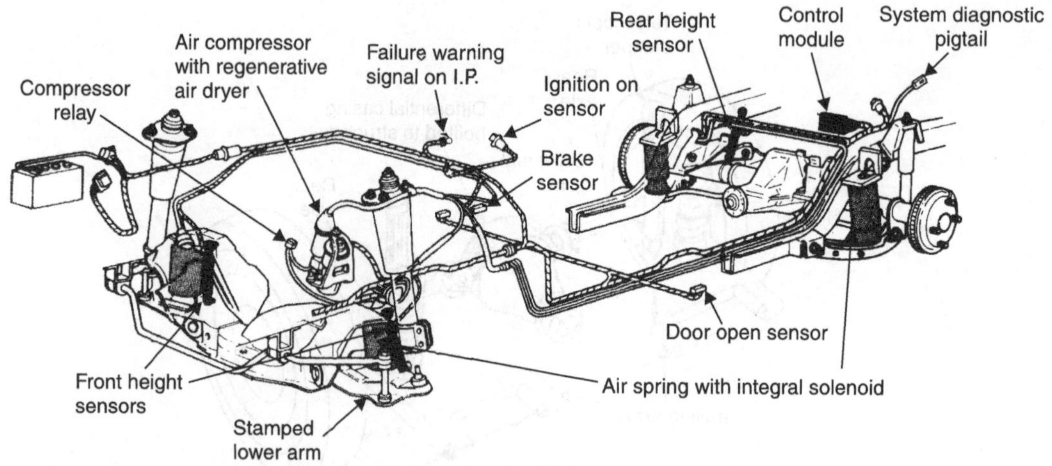

Fig. 6.40. Air suspension system.

Following are the ***advantages*** of *air suspension systems over the conventional metal springs* :

1. Reduced dynamic loading.

2. By virtue of the automatic control devices a variable space for wheel deflection is put to optimum use.

3. Owing to the improved standard of ride comfort and noise reduction attained with air springs the fatigue of the passengers and drive is reduced.

4. Due to the wheel altitude being constant, the changes in head lamp alignment due to carrying loads are avoided.

2. **Hydroelastic suspension :**

This system works on compression of fluid with the help of rubber washers or cups. There are four units fitted at four wheels with independent suspensions, all the units are hermetically sealed and connected to each other with the help of pipes. Such a system has been successfully used on BMW 1800 car wherein if one wheel rises over the stone, the fluid from this unit goes to the other units and raises the car thus maintains a perfect level.

Fig. 6.41 shows a cut-away view of a hydroelastic spring. Whenever a wheel encounters a bump, the tapered piston of the unit is pushed upwards. The spring is actuated by the fluid

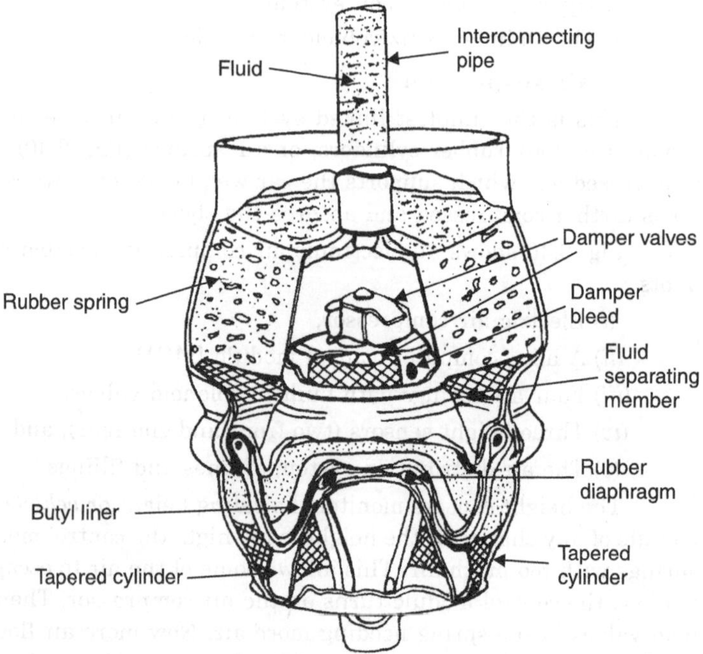

Fig. 6.41. Cut-away view of a hydroelastic spring.

which is beneath and in the centre of the rubber spring. This upward movement of the tapered piston also forces the fluid up through the interconnecting pipe to the other unit on the same side of the car. The other side of the car also gets lifted, thereby distributing the shock of the bump between the front and rear wheels of the car.

3. **Hydragas suspension system :**

This system uses *gas-filled spring units* (called *Hydragas springs*), one at each wheel. Each unit has a sealed chamber containing a quantity of nitrogen gas at high pressure. Below this chamber is a displacement chamber filled with water-based fluid (Fig. 6.42). When the wheel meets a bump, the fluid is pushed, compressing the gas. This action provides the springing effect.

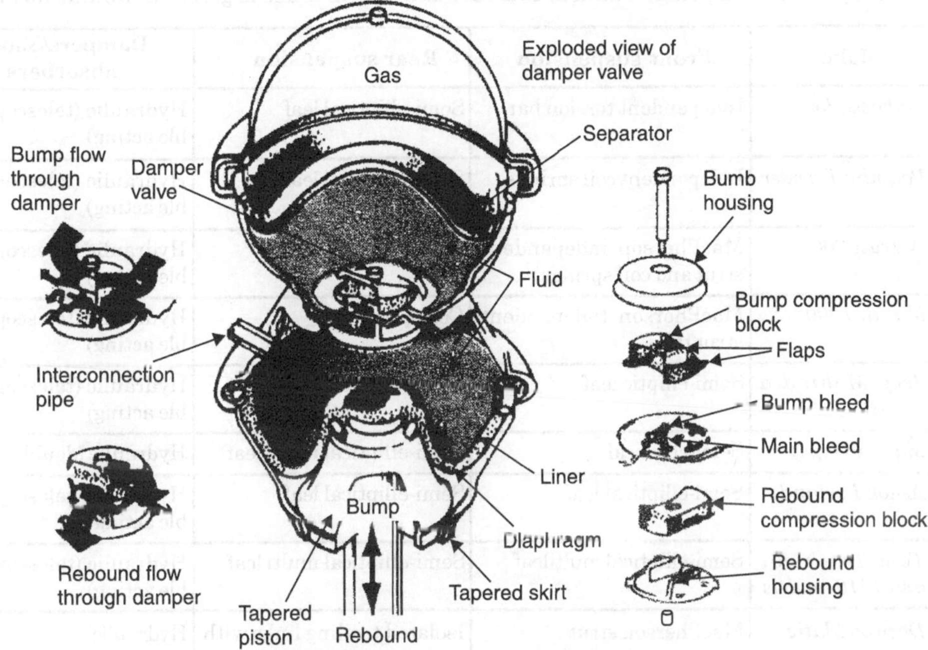

Fig. 6.42. Partly cut-away hydragas spring.

The two units on each side of the car are also interconnected front to back. Thus, when the left front wheel meets a bump, *for example*, a part of the fluid from the left front unit is forced through a pipe to the left rear unit. This action raises the left rear wheel also. Therefore the shock is distributed between the left front and left rear wheels and consequently the side is improved.

4. **The Daimler-Benz vehicle suspension :**

This suspension system (devised by Daimler-Benz A.G. of Germany) consists *primarily of one spring and incorporates resilient hydro-mechanical couplings to cause opposite wheels rise and fall together.*

Each wheel hub is attached to a piston and cylinder arrangement, by means of which vertical displacements of the wheel are transmitted hydraulically to a hydro-pneumatic spring containing an air chamber. Diagonally opposed wheels are served by pipe lines which pass through gear type hydraulic motor pumps, interconnected by torsion rods. The hydraulic pipes are themselves connected by a pipe that runs upto the spring.

6.12.5. Repair and Maintenance of Independent Suspension System

- In Independent suspension the upper and lower arms are fixed with the help of rubber bushes, which get worn out after some use. These bushes should be replaced on noticing first sign of wear with new bushes of good quality. There are two bronze bushes on swivel pin at upper and lower ends, they also get worn out quickly. In case they are not greased periodically, they require changing after dismantling upper and lower arms.
- During each service, apply grease to upper and lower links and also check up rubber bushes for deterioration or wear ; in case they are worn they should be replaced immediately.

6.12.6. Suspension Systems of Some Indian Automobiles

The comparative suspension data of some Indian automobiles is given in tabular form below :

S. No.	Make	Front suspension	Rear suspension	Dampers/Shock absorbers
1.	*Ambassador*	Independent torsion bar	Semi-elliptical leaf	Hydraulic (telescopic double acting)
2.	*Padmini Premier*	Independent coil springs	Semi-elliptical leaf	Hydraulic (telescopic double acting)
3.	*Maruti 800*	MacPherson independent strut and coil springs	Leaf spring	Hydraulic (telescopic double acting)
4.	*Maruti 1000*	MacPherson Independent strut type	Coil spring	Hydraulic (telescopic double acting)
5.	*Jeep Mahindra CJ-3B*	Semi-elliptic leaf	Semi-elliptical leaf	Hydraulic (telescopic double acting)
6.	*Maruti Gypsy*	Semi-elliptical	Semi-elliptical multi leaf	Hydraulic (double acting)
7.	*Ashok Leyland*	Semi-elliptical leaf	Semi-elliptical leaf	Hydraulic (telescopic double acting)
8.	*Tata Truck 407 and 1210 models*	Semi-elliptical multileaf	Semi-elliptical multi leaf	Hydraulic (telescopic double acting)
9.	*Daewoo Matiz*	MacPherson strut	Isolated trailing links with coil springs	Hydraulic
10.	*Hyundai Santro*	MacPherson strut with stabilizer bar	Torsion beam axle, three-link offset coil springs	Hydraulic
11.	*Telco Indica (Petrol)*	Independent, wishbone type with MacPherson strut	Independent semitrailing arm with coil springs	Hydraulic
12.	*Maruti Esteem*	MacPherson strut and coil spring	MacPherson strut and coil spring	Hydraulic (telescopic double acting)
13.	*Maruti Wagon R*	MacPherson strut with torsion type roll control device	Coil spring, gas-filled shock absorbers with three-link rigid axle and isolated trailing arms	Gas-filled
14.	*Maruti Alto*	MacPherson strut with torsion type anti-roll bar	Coil spring with double action telescopic shock absorbers	Hydraulic (telescopic double acting)
15.	*Maruti Baleno*	MacPherson strut and coil spring with an L shaped lower arm and anti-roll bar	MacPherson strut and coil spring with parallel link arrangement	Hydraulic (telescopic double acting)

6.12.7. Suspension System Troubleshooting

The troubleshooting in suspension system is given in tabular form below :

S. No.	Fault	Cause	Remedy
1.	*Front wheel low speed shimmy*	(*i*) Uneven tyre pressure (*ii*) Loose ball joints or king-pin (*iii*) Front springs soft (*iv*) Irregular tyre tread (*v*) Loose linkage (*vi*) Looseness in steering gear.	(*i*) Inflate to correct pressure (*ii*) Replace (*iii*) Replace (*iv*) Match treads (*v*) Readjust. Replace worn parts (*vi*) Replace worn parts.
2.	*Front wheel tramp*	(*i*) Wheels unbalanced (*ii*) Excessive wheel run out (*iii*) Shock absorber defective (*iv*) Also causes mentioned under item from wheel low speed shimmy.	(*i*) Rebalance (*ii*) Replace (*iii*) Replace
3.	*Rough ride*	(*i*) Excessive tyre pressure (*ii*) Shock absorber defective (*iii*) Excessive friction in suspension.	(*i*) Reduce to correct it (*ii*) Replace (*iii*) Lubricate parts.
4.	*Sway on turns*	(*i*) Stabiliser bar loose (*ii*) Sagging or weak springs (*iii*) Incorrect caster.	(*i*) Tighten as required (*ii*) Replace (*iii*) Re-adjust.
5.	*Breakage of spring*	(*i*) Overloading (*ii*) Loose centre bolt (*iii*) Loose U-bolt (*iv*) Defective shock absorber (*v*) Tight spring shackle.	(*i*) Avoid (*ii*) Tighten centre bolts (*iii*) Tighten U-bolt (*iv*) Replace (*v*) Loosen as necessary.
6.	*Sagging spring*	(*i*) Leaf broken (*ii*) Weak spring (*iii*) Short coil spring (*iv*) Defective shock absorber.	(*i*) Replace (*ii*) Replace (*iii*) Install shim or replace (*iv*) Replace.
7.	*Noises in the suspension system*	Commonly attributed to any loose, worn or unlubricated part such as U-bolts, rebound clips, shackles, shock absorber linkages, etc.	Examine the parts carefully. Locate the cause and correct as required.

6.13. WHEELS AND TYRES

The **"wheel assembly"** is generally thought to consist of *hub*, *disc* or *spokes*, *rim*, *tyre* and *tube*. Wheels are as important parts of a vehicle as the other parts. All the parts being perfectly in working order, the vehicle cannot move on the road without the wheels.

6.13.1. Wheels

6.13.1.1. Introduction

The wheels are legs of the vehicle which carry it to far-off distances. *They support the whole weight of the vehicle and convert rotary motion into longitudinal one.* A wheel also *resists side force, created by turning*. Resistance is also necessary so that wheel will absorb hard knocks and accidental damage.

6.13.1.2. Essential requirement of wheels

In case of automobiles the *essential requirements of wheels are as follows :*

1. Strong enough to take the weight of the vehicle. torque etc.
2. Flexible to absorb the road shocks.
3. Perfectly balanced statically as well as dynamically.
4. Lightest possible so that the unsprung weight is least.
5. Able to grip the road surface.
6. Mounted or removed easily.
7. Material of its construction should not deteriorate with weathering and age (To protect against corrosion, if required, suitable treatment must be given).
8. Not too expensive.

6.13.1.3. Types of automobile wheels

The automobiles wheels are of the following *three* types :

1. Pressed steel disc wheels.
2. Wire wheels.
3. Light alloy casting wheels.

1. Pressed steel disc wheels :

These wheels are most popular and most of the cars are fitted with this type of wheels. They have the following *merits/advantages* :

(*i*) Simple and robust in construction.
(*ii*) Easy to produce in large numbers at low cost.
(*iii*) Require negligible maintenance.
(*iv*) Ease in cleaning.

Fig. 6.43 shows a pressed steel disc wheel. If consists of a steel rim and a pressed steel disc. The rim is rolled section, sometimes riveted but usually welded to the flange of the disc. The disc performs the function of spokes. The disc is frequently dished to bring the point of ground contact under the large wheel bearing. The wheel assembly is bolted to the brake drum. The disc is often perforated with slots near the rim, which acts as a fan to blow air on the brakes. A hole in the rim serves to accommodate tube valve. A large chrome-plated or stainless steel hub cover can be sprung over the lugs in the disc. It improves the general appearance of wheel and acts as a dust and water excluder as well. This cover can be readily detached by a screw driver.

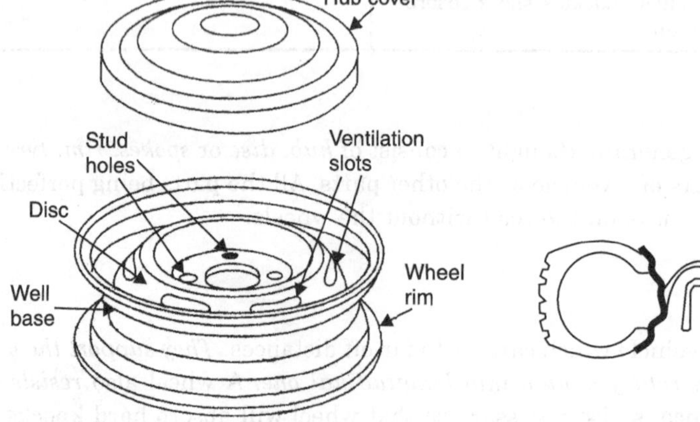

Fig. 6.43. Pressed steel disc wheel.

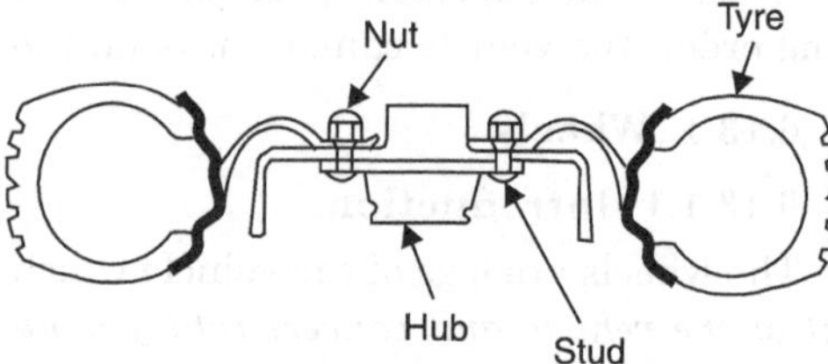

Fig. 6.44. Fitting of the disc wheel to the hub.

The wheel is fitted on the axle by bolting to a flange attached to the latter.

2. **Wire wheels :**

The wire wheel is the earliest type of wheel but presently its *use is limited to certain vintage sports and racing cars.* It is higher, heat dissipation is better and it can be fitted and removed very easily. However, tubeless tyres cannot be fitted over wire wheels which are also difficult to clean.

A wire wheel consists of a separate hub connected to the rim with a number of wires or spokes. The headed inner ends of the spokes fit in the hub holes and the threaded outer ends fit in the rim holes, where mushroom-headed tubular nuts are screwed through the rim holes to tighten the spokes. All the spokes must be of correct length and at correct tension to hold the rim, centrally around the hub. The spokes do not stick straight out as radii from the hub, but alternate spokes are screwed to slope forwards and backwards towards the rim (Fig. 6.45). This arrangement of spokes serves special purpose of the wheel.

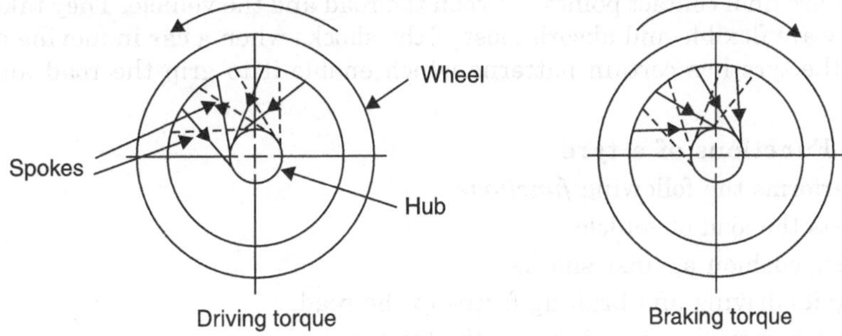

Fig. 6.45. Arrangement of spokes.

The forward-sloping spokes absorb braking torque and the rearward-sloping spokes convey driving torque.

The holes in the hub are arranged in inner and outer rows so that one set of spokes slope towards the rim from the outer row of the hub and the other set slopes outwards to the rim from the inner row of the hub. *These sideways inclinations of the spokes hold the wheel upright against cornering loads and side thrusts.* A rubber chafing band is fitted in the well of the rim to keep tube touching the spoke nuts.

The wire wheels allow free circulation of air around the brake drum.

3. **Light alloy casting wheels :**

The light alloy cast or forged wheel is the most recent type, whose use in ever increasing in both road-and sports cars. The use of light alloys (aluminium and magnesium alloys) makes it possible to *use wider rims, which allow low aspect ratio (i.e., wider tyres to be fitted), thus improving good adhesion, especially on corners.*

The advantages and disadvantages of such wheels may be summed up as follows :

Advantages :

(*i*) Light in weight.

(*ii*) Light alloys being good heat conductors dissipate heat produced by tyres and brakes more efficiently than steel.

(*iii*) Heavier sections can be used which improve the wheel stiffness and better stress distribution is obtained.

(*iv*) Rims with larger area can be used which results in the use of wider tyres with less diameter.

Disadvantages :

(*i*) Relatively costly.

(*ii*) More prone to corrosion.

- *Generally aluminium alloys are used for wheels of cars and commercial vehicles, whereas sports and racing cars usually have magnesium alloy wheels. Probably the higher cost is the only main disadvantage of light alloys.*

6.13.2. Tyres

6.13.2.1. Introduction

A **tyre** *is a cushion provided with an automobile wheel.* It consists of mainly the outer cover *i.e.,* the tyre proper and the tube inside. The tyre-tube assembly is mounted over the wheel rim. *The air inside the tube carries the entire load and provides the cushion.*

The tyres are final contact points between the road and the vehicle. They take all the load of the vehicle. They are flexible and absorb most of the shocks when a car in moving on rough roads. The surface of the tyre has certain patterns which enable it to grip the road and provide good traction.

6.13.2.2. Functions of a tyre

A tyre performs the following *functions* :

1. Supports the load of vehicle.
2. Provides cushion against shocks.
3. Transmits driving and braking forces to the road.
4. Provides cornering power for smooth steering.

6.13.2.3. Requirements of a good tyre

Some important *requirements of a good tyre* are :

1. To be strong enough to carry loads and resist damage.
2. To provide a comfortable ride to the motorists.
3. To provide good road grip for traction, cornering, accelerating and braking.
4. To provide quiet running.
5. To have a long life.
6. To respond accurately to steering without deflection by the ridges on the road.
7. To be flexible to cushion all shocks and impacts at least partly.
8. To be economical.
9. To meet its requirements in all weathers and on all surfaces without overheating.

Since none of the tyres can fulfill all the above mentioned requirements, therefore, a good tyre is at best is a compromise of some of the requirements/characteristics.

6.13.2.4. Types of tyres

The pneumatic tyres (the use of solid tyres on automobiles is now obsolete) are of the following two types :

1. Conventional tubed tyre.
2. Tubeless tyre.

1. Conventional tube tyre :

Fig. 6.46 shows a conventional tube tyre.

- The outer portion of the tyre which rolls on the road is made of synthetic rubber and is called *"tread"*.
- The tyre encloses a *tube* in which air is forced to a high pressure as a cushioning medium. A valve stem is attached to the tube for inflating or deflating the same.

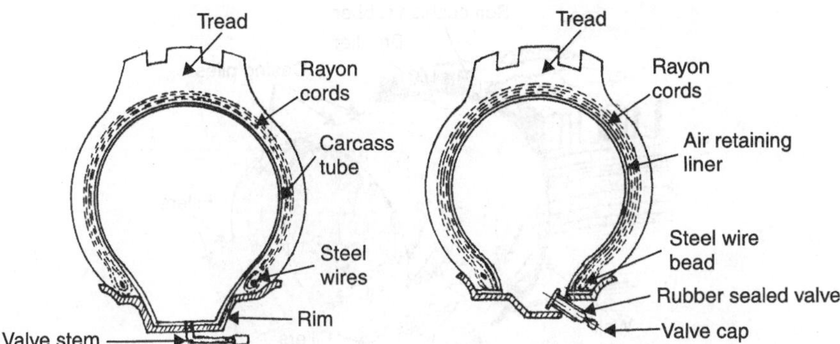

Fig. 6.46. Conventional tube tyre. Fig. 6.47. Tubeless tyre.

- In order to prevent the tyre from being thrown off the rim, the plies (formed from rayon cords) are attached to two rings of high tension steel wire. These rings are made to fit snugly against the wheel rim thereby anchoring the tyre to the rim. These rings are called *"beads"*.

2. **Tubeless tyre :**

- A tubeless tyre is shown in Fig. 6.47. This type of tyre *does not enclose a tube*, instead *the air under pressure is filled in the tyre itself for which purpose a non-return valve is fitted to the rim.*
- The inner construction of the type is almost same as that of the tubed tyre, except that it is lined with a special *air-retaining liner* made up of halogenated butyl rubber (*e.g.*, chlorobutyl or bromobutyl) for better air permeability together with heat and weather resistance.
- A tubeless tyre retains air for a longer period even after being punctured by nail, provided the nail remains in the tyre. Also, any hole in the tubeless tyre can be repaired simply by rubber plugging. Ordinary punctures can be repaired without removing the tyre from the wheel. It can be retreated in the same manner as the tube tyre.

The tubeless tyre has the following ***advantages*** *over a tubed tyre* :

(*i*) Easier to fit.
(*ii*) Slower leakage of air.
(*iii*) Lesser unsprung weight.
(*iv*) Better cooling.
(*v*) Improved safety.

6.13.2.5. Tyre construction

The construction of a tyre is shown in Fig. 6.48.

— The lower portion on both sides has a set of steel wires called ***bead wires.*** These wires house tyre snugly in rim and also do not allow it come out even in the event of tyre burst.

Bead wires are *surrounded by bead core and plies of canvass.* These days, instead of canvass, rayon or nylon is used to give more strength to the tyre.

— On the tyre side walls as on plies rubber is vulcanised which protects the plies from injuries due to pebbles or any such object.

— *The top portion is called* ***tread*** as it comes in contact with the road and is subject to wear. As such it is made thicker. Different designs are made in the tread for better grip and traction on different road conditions.

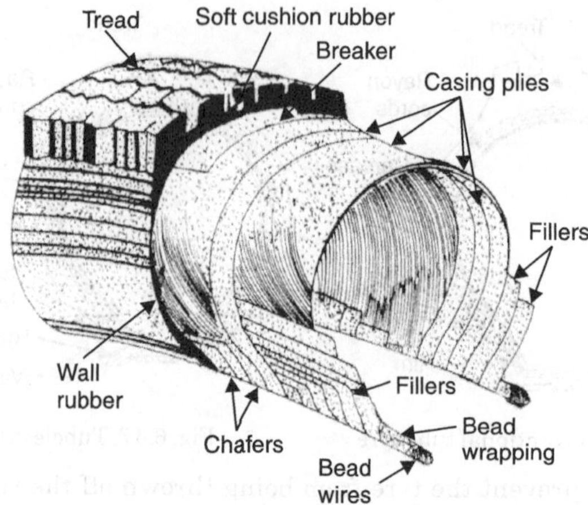

Fig. 6.48. Construction of a tyre.

— ***Breakers** are rubber-covered cords similar to plies.* The job of breaker *is to distribute road shock and prevent separation of tread from tyre.*

— ***Cushion*** is soft heat resistance rubber, which absorbs road shocks and also bind plies and breaker together.

— ***Cord plies*** give strength to resist internal pressure, to support load and road shocks.

Tyre construction is divided into the following three classes :

1. Cross ply construction.
2. Radial ply construction.
3. Belted-bias construction.

1. **Cross ply construction :**

In this type of tyre construction, the ply cords are woven at an angle (30° to 40°) to the tyre axis. There are two layers which run in opposite directions (Fig. 6.49). However, the cords are not worn like wrap and weft of ordinary cloth, because that would lead to rubbing of the two layers and thus produce heat which would damage material of the tyre.

- This type of tyres *have better wear and road holding characteristics.* But they must be fitted on the front wheels only. Fit radial tyres all around, or use cross ply on the front and radial ply tyres on the rear wheels otherwise covering characteristics will be seriously impaired.
- These tyres are *economical.* However they have *a tendency to squirm* as they go through the tyre foot print or contact patch.

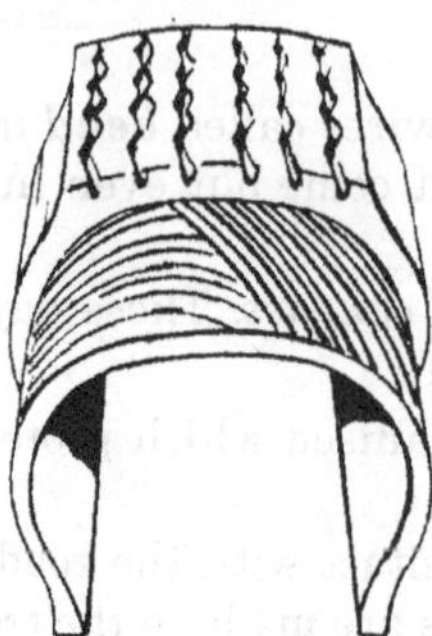
Fig. 6.49. Cross ply construction.

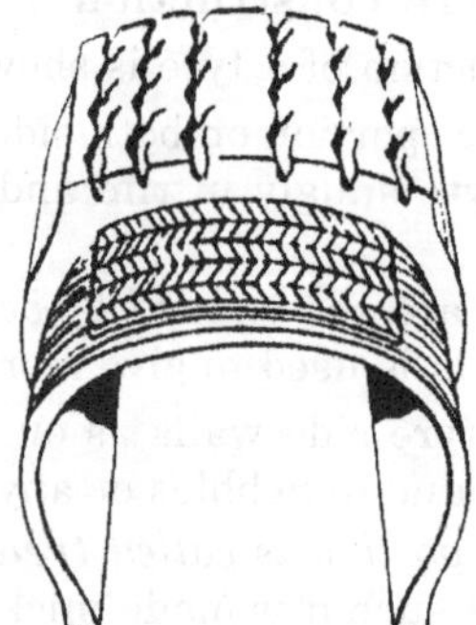
Fig. 6.50. Radial ply construction.

2. **Radial ply construction :**

These tyres have plies running from bead to bead across the crown at right angles to the rotation (Fig. 6.50). On the side walls the direction of these piles is radial and hence the name. However above the layers of these plies and below the tyre tread, there are belts of cord or breakers which run around the circumference. The angle between the cords varies from 18° to 22°. The number of layers depend upon the material used, the lateral stiffness needed and the load the tyre is required to carry.

Radial ply construction tyres are the most modern type and gradually replacing the other two.

Advantages and disadvantages :

Radial ply tyres possess the following *advantages over the bias-ply tyres* :

Advantages :

(*i*) They can absorb more bounce of rough roads and thus the ride is more comfortable at high speeds.

(*ii*) Improved acceleration and braking, operations (due to continuous flat contact–patch area with the road surface).

(*iii*) Lower rolling resistance and hysteresis loss, which ultimately means reduced fuel consumption.

(*iv*) Longer tread-life.

(*v*) For similar tread design the water removal efficiency and hence the braking efficiency on wet-roads is better in case of radial ply tyres.

(*vi*) Better steering characteristics.

(*vii*) Larger resistance to punctures, cuts and impacts in the tread area on account of the breaker belts.

(*viii*) Less tendency to distort and lift off the road from one side.

Disadvantages :

(*i*) Higher initial cost (about 20% more than cross-ply tyres).

(*ii*) At low speeds the ride is uncomfortable harsh.

(*iii*) Heavier steering at low speeds.

3. **Belted bias construction :**

This type of tyre construction is combination of the above two types. The basic construction is the bias-ply over which run a number of breaker belts. The belts improve the characteristics of the bias-ply tyre to a large extent. By keeping the tread shape these tyres show the following *advantages over cross ply* :

(*i*) Improved traction.

(*ii*) Run cooler.

(*iii*) Show greater road mileage.

However since they do not flex as easily as cross ply, the rider is harder because all road shocks are transmitted to the body. Therefore, for comfort, designed springs and suspension systems have to be employed to reduce road shock transfer to the passenger compartment.

6.13.2.6. Tyre materials

The various materials used for construction of tyres are as follows :

- *"Rubber" used in* ***tyres*** *is a blend of the natural and synthetic rubbers* to which various chemicals (carbon, sulphur etc.) are added *to obtain desired properties like wear resistance, less internal friction, reduced hysteresis etc.*

— *Carbon black* improves wear and abrasion resistance ;
— *Sulphur* acts as a vulcanising agent ;
— *Oil* improves road holding at the expense of tyre wear.

- *Bronze plated high tensile steel is used for* ***beads.*** Around and between the separate strands a special rubber insulation is squeezed. The wound bead is then covered with a strip of calendered fabric which provides additional strength and stability.
- The materials used for ***cords of the stabilizer belts are rayon, terylene, glass fibre,*** or *steel.*

— *Glass fibre* is very strong and extra classic but special techniques are required for spinning it into fibre cord and also achieving a satisfactory bond with rubber.
— *Steel* is very common in case of radial tyre belts but recently it has found use in the belted-bias tyres also.

6.13.2.7. Tyre shape

A tyre is usually indicated by a series number such 100, 83, 78, 70 and so on. This designation is derived from the *Aspect Ratio* which is numerically equal to the ratio of the section height to section width $\left(i.e., \dfrac{H}{W}\right)$ as shown in Fig. 6.51. For quite some time the conventional tyres has 100 percent aspect ratio (*i.e.*, $W = H$) and was designated the 100 series. With the improvement in the design of car, increasing speeds, and need for better road grip led to wider car tyres with *lower aspect ratios.* Gradually the aspects ratio dropped from 90 to 85 percent to 70 percent.

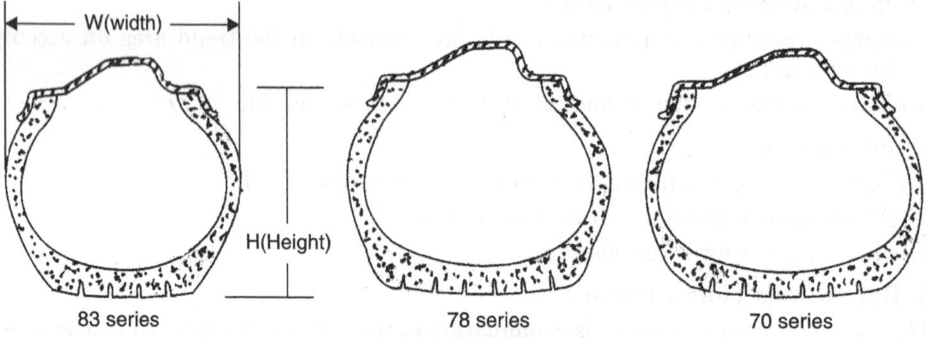

Fig. 6.51. Cross-section of tyre series.

— Passenger cars now-a-days have an aspect ratio of 70 percent in case of radial ply tyres. However ratios below 60 percent are difficult to accommodate in general automobiles and are used in racing cars (aspect ratio may as low as 35 to 45 percent).

The wider tyres claim the following *advantages* over the old symmetrical type :

(*i*) Vehicle handling is improved which gives a performance styling to the vehicle.
(*ii*) Better load carrying capacity.
(*iii*) Better cornering.
(*iv*) Better high speed performance.
(*v*) Longer life.

6.13.2.8. Tread pattern

For different road conditions there are different tread patterns.

- Fig. 6.52 (*a*) shows a tread pattern which is used on tyres that are employed on *hard, well surfaced roads.*

- Fig. 6.52 (*b*) shows a tread pattern which is only useful on *soft ground.* This type of pattern is not suitable for use on hard roads.

Fig. 6.52 (*c*) shows a tread pattern used *on both good roads as well as on unmade surfaces.*

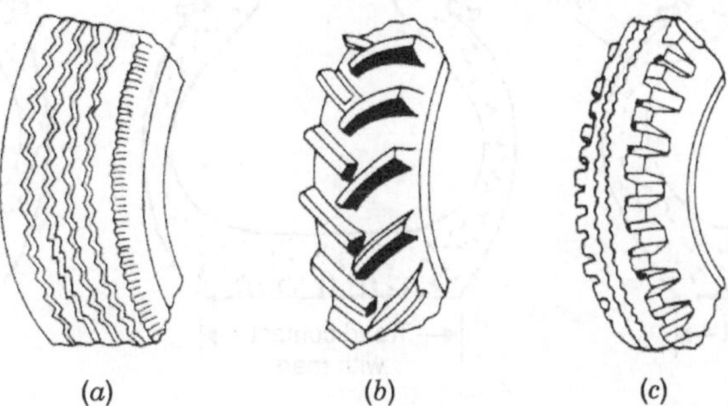

(*a*) (*b*) (*c*)

Fig. 6.52. Tread patterns.

6.13.2.9. Tyre markings

On every tyre certain size is written.

Example. 8.25 × 20 – 10 PR. It means :

- 8.25 (inches) is the width of tyre from shoulder to shoulder.
- 20 (inches) is the diameter of the bead circle, which fits on the rim.
- 10 PR is the ply rating, *i.e.*, in this tyre there are 10 plies.

— Scooter tyre consists of 1 to 4 plies, light truck tyres 6 to 10 plies and trucks tyres 10 to 22 plies. The number of plies makes the tyre hard to resist heavy loads, but the hard tyre does not absorb road shocks.

"*Metric designations*" are based on internationally recognized size nomenclature.

Example. P 205/75 R14.

- 'P' means that tyre is for a passenger car, its section width is 205 mm, aspect ratio 75 percent, it is of radial construction (R) and is meant for a rim of 14 inches diameter.

6.13.2.10. Tyre selection. The tyre selection depends upon the following *factors* :

1. The load acting per tyre.

2. The surface on which the vehicle is going to move. The selection is made on the basis of above specification.

- The selected tyre should be suitable for load capacity and road condition. It should also be prefer for the type of job for which the vehicle is intended.

6.13.2.11. Tyre inflation pressure

For every type of tyre, inflation pressures are specified by manufacturers. These recommended pressures should be strictly adhered to since they help to meet the requirements of :

(*i*) Good performance, (*ii*) Heat control,
(*iii*) Tread kilometre, (*iv*) Comfort, and
(*v*) Better vehicle control.

Incorrect pressure leads to :

— Overheating,
— Unsafe steering,
— Rapid tyre wear.

Fig. 6.53 shows stages of tyre inflation.

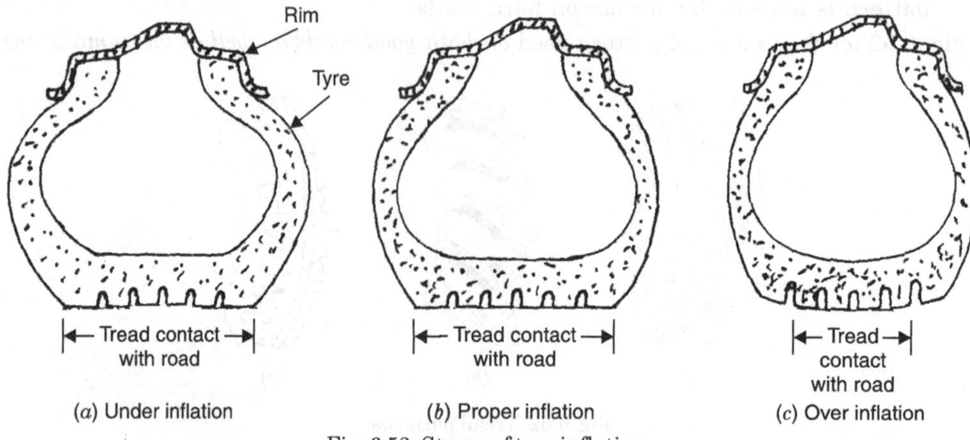

Fig. 6.53. Stages of tyre inflation.

- **Under inflation** *causes severe flexing of the tyre plies, side walls and other components, excessive heat and premature failure.* It causes the tyre to wear more on the edge than in the centre.
- **Over inflation** *overloads the components of the tyre and causes excessive tread wear and hard riding.* It causes *abnormal wear in the centre of the tyre.* Tyre *trends to puncture* when hitting a pointed object.

Both the above cases will *cause dangerous steering and control.*

High speed also causes the tyre wear and failure. When the vehicle is running at high speed, there will be high temperature produced by the distortion of tyre. Such distortion can be reduced by increasing the air pressure slightly above the normally recommended cold starting pressure.

6.13.2.12. Causes of tyre wear

Following are the *causes of tyre wear* :

(*i*) Incorrect inflation. (*ii*) Unequal tyres.
(*iii*) Bleeding of air in tyre. (*iv*) Incorrect rotation of tyres.
(*v*) Misalignment. (*vi*) Out of balance wheel.
(*vii*) Worn steering mechanism. (*viii*) Incorrect castor, camber, or toe in.
(*ix*) Excessive braking or violet acceleration.
(*x*) Wrong loading. (*xi*) Careless driving.
(*xii*) Toe-out incorrect on turn. (*xiii*) Defective brakes.
(*xiv*) Overloading. (*xv*) Worn kingpins.

6.13.2.13. Factors affecting tyre life

The following are the main factors which affect tyre performance and consequently their life :

1. Inflation.
2. Maintenance of vehicle.
3. Manner of driving.
4. Miscellaneous factors :

(*i*) Heat (*ii*) Position in which tyre in fitted
(*iii*) Road conditions (*iv*) Seasons.

6.13.2.14. Tyre maintenance

The systematic and correct tyre maintenance is one of the most important factors of safe vehicle operation. *Tyres must sustain the weight of a loaded vehicle, withstand more than ordinary rough service, provide maximum safety over all types of terrain, and furnish the medium on which the vehicle can be moved with ease.*

Besides other elements of tyre service, *inflation maintenance* is the most important and in many cases the most neglected. *The tyre pressure should be constantly maintained for safe operation.* An *'Under inflated'* tyre is dangerous as too much flexing causes breakage of the casing resulting in failure. *'Over inflation'* sometime may cause a blow-out.

The extent of tyre life and the ease and safety of vehicle control are influenced by the following *four factors* :

(*i*) Tyre pressure (*ii*) Tyre rotation

(*iii*) Wheel balance (*iv*) Wheel alignment.

Four of the most common tyre troubles are :

(*i*) Excessive wear around the *outer edges* resulting from under inflation.

(*ii*) Excessive wear in the *centre* of the tread resulting from over inflation.

(*iii*) Tyre tread worn on one side indicating wheels need *realigning.*

(*iv*) Cuplike depressions on one side of the tread indicating wheels need *balancing.*

In order to have a satisfactory four-wheel drive operation, a four-wheel drive vehicle *must be equipped with the same size tyres of equal circumference on all four wheels.* The tyres *must then be inflated to proper factory recommended pressures at all times.*

The practice of *cross-switching* the tyres every 10000 km evens out differences in wear and make a set of tyres last longer than they would without cross-switching.

Fig. 6.54 shows the actual *procedure for rotation of tyres of a car*

- Rear right wheel (1) is changed to the front left.
- Wheel (2) is changed to the rear left.
- Wheel (3) is changed to the front right.
- Wheel (4) is kept as a spare wheel.
- Spare wheel (5) is fitted to the rear right.

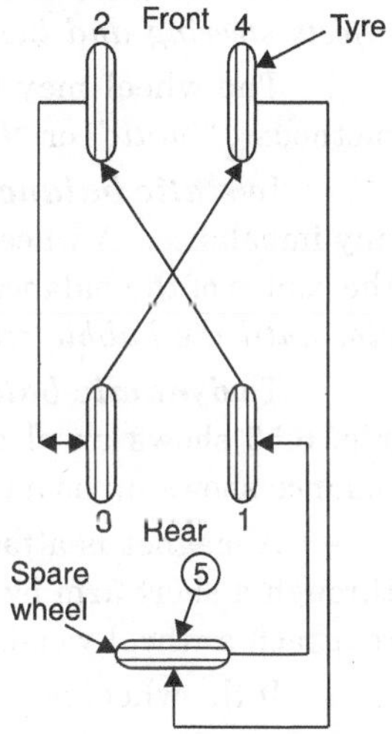

Fig. 6.54. Rotation of tyres.

6.13.2.15. How to enhance the tyre life ?

The life of a tyre can be enhanced/increased by following the basic maintenance rules given below :

1. Load the vehicle to the extent of its capacity fixed by the manufacturer.

2. During loading of the vehicle specially trucks, care should be taken to place the load in the centre of the body.

3. Tyres should neither be under inflated nor over inflated. They should be inflated to their proper capacity.

4. Air should never be bled from a hot tyre to set the pressure right.

5. Proper driving habits should be ensured.

6. In trucks and buses where dual wheels are fitted on the rear axle, the *height of both the tyres should be the same.* In case it is not possible then the difference in height should not be more

than about 12 mm failing which the bigger diameter will be carrying all the load. Due to being overloaded it will run hot and fail prematurely.

7. Reduce the mechanical faults to minimum possible, preferably nil.

Damages to tyres and their repair

The different types of damage to tyres and their repair are given below :

1. *The damage is not deep and confined to the tread rubber :*

Such damages do not need any repair. But care should be taken to see that the damage does not extend to the other parts of the tyre.

2. *The damage has extended to the outer casing cords :*

Vulcanised rubber should be used for repair. This prevents moisture and dirt from reaching the casing.

3. *The tyres are severely damaged :*

These tyres should be taken to specialists for necessary repair.

6.13.2.16. Wheel balance

The balance of a tyred wheel is essentially required *to avoid front wheel wobble which affects steering and increases tyre wear rates.*

The wheel may be checked for balance on or off the car. This is done by either of two methods : "*Static*" or "*Dynamic*".

In ***static balancing***, the wheel is taken off the car and put on a "*bubble*" balancer to detect any imbalance. A wheel that is out of balance is heavier in section. This will cause the bubble in the centre of the balancer to move off centre. To balance the wheel, *weights are added to the wheel rim until the bubble returns to centre.*

To ***dynamic-balance*** (or "*spin balance*") a wheel, the wheel is spun either on or off the car. Fig. 6.55 shows an electronic wheel balancer being used to balance a wheel on a car. Lack of balance shows up as a tendency for the wheel to move off centre or out of line as it spins.

A magnet is attached to the brake backing plate. Any movement of the magnet is sensed through a short arm by a vibration pickup. This causes the strobe light to flash indicating where to attach a wheel weight.

If the wheel is out of balance, one or more weights are installed on the wheel rim.

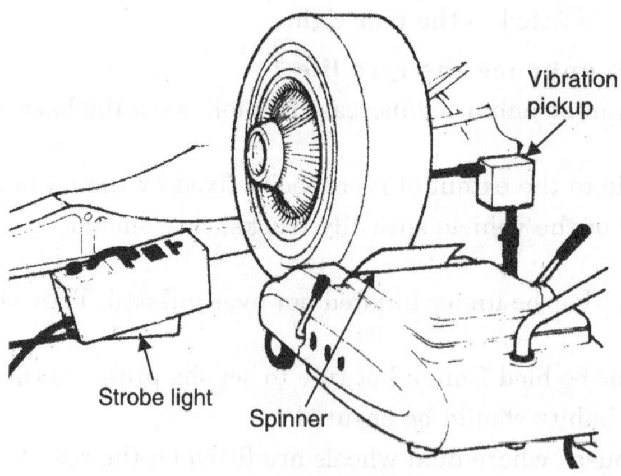

Fig. 6.55. Dynamic wheel balancer (Electronic type).

6.13.2.17. Comparative data of tyre in some Indian vehicles

The comparative data of tyre in some Indian vehicles is given below :

S. No.	Make of vehicle	Ply rating	Tyre size	Inflationary pressure	
				Front tyres (Psi)	Rear tyres (Psi)
1.	*Ambassador*	4	5.90×15^2	26	28
2.	*Premier*	4	5.20×14^2	24	26
3.	*Maruti 800*	4 or 6	4.50×12^2	22	22
4.	*Jeep CJ-3B*	4	6.00×16^2	26	28
5.	*Ashok Leyland*	12	9.00×20^2	85	85
6.	*Tata Truck 407*	14	7.50×16^2	55	100
7.	*Tata Truck 1210*	14	9.00×20^2	100	100
8.	*Telco Indica (Petrol)*	—	155/70 R13	22	22
9.	*Hyundai Santro*	—	155/70 R13	22	22
10.	*Daewoo Matiz*	—	145/70 R 13	22	22

16.13.2.18. Wheel and tyre troubleshooting

The *wheel and troubleshooting* is discussed in a tabular form below :

S. No.	Fault	Cause	Remedy
1.	*Wear on sides of tyre tread more than centre*	(*i*) *Under-inflation*	(*i*) Inflate to correct pressure.
2.	*Wear in centre of the tyre tread*	(*i*) Over-inflation	(*i*) Deflate to correct pressure.
3.	*'Filed' appearance of tyre tread*	(*i*) Misalignment of wheels	(*i*) Check and correct where necessary
4.	*One-sided wear of tyre tread*	(*i*) Incorrect camber angle (*ii*) Sagging axle (*iii*) Driving on steeply cambered roads	(*i*) Adjust (*ii*) Check and rectify (*iii*) Improve driving habits.
5.	*Uneven tyre tread wear*	(*i*) Under-inflation (*ii*) Wheel bent (*iii*) Wheel or tyre unbalanced	(*i*) Inflate to correct pressure (*ii*) Replace (*iii*) Check and correct as required.
6.	*Heel and toe wear of tyre tread*	(*i*) Eccentric brake drum (*ii*) Excessive braking and acceleration (*iii*) Overloading	(*i*) Grind or replace as required (*ii*) Avoid (*iii*) Avoid.
7.	*Cracking of casing*	(*i*) Under-inflation (*ii*) Over-inflation (*iii*) Overloading	(*i*) Inflate to correct pressure (*ii*) Deflate to correct pressure (*iii*) Load correctly.
8.	*Excessive loss of pressure*	(*i*) Porous tube (*ii*) Leaky valve (*iii*) Punctured tube (*iv*) Faulty patch on the tube.	(*i*) Replace (*ii*) Replace (*iii*) Repair (*iv*) Replace.

9.	*Worn spots on tyre tread*	(*i*) Improper brake adjustment (*ii*) Wheel bearing faulty (*iii*) Hub or axle assemblies faulty (*iv*) Loose steering linkage	(*i*) Re-adjustment (*ii*) Replace (*iii*) Check and correct as required. (*iv*) Check and correct as required.
10.	*Tyre tread or side wall scoring damage*	(*i*) Projecting bolts, spring shackles etc.	(*i*) Check and correct as required.
11.	*Kerbing damage*	(*i*) Tyre rubbing against kerb or other similar objects.	(*i*) Drive carefully
12.	*Tyre tread cracking*	(*i*) Overload (*ii*) Over-inflation (*iii*) Under-inflation.	(*i*) Avoid (*ii*) Deflate to correct pressure (*iii*) Inflate to correct pressure.
13.	*Tyre tread or ply separation*	(*i*) Overloading (*ii*) Over-inflation (*iii*) Under-inflation (*iv*) High speeds.	(*i*) Avoid (*ii*) Deflate to correct pressure (*iii*) Inflate to correct pressure (*iv*) Control speed.
14.	*Bead chafing and bead burst*	(*i*) Damaged wheel (*ii*) Overloading (*iii*) Under-inflation (*iv*) Insufficient twin tyre clearance.	(*i*) Replace (*ii*) Avoid (*iii*) Inflate to correct pressure (*iv*) Check that twin clearance is adequate.
15.	*Damaged beads*	(*i*) Careless use of tyre levers (*ii*) Faulty lock rings or flange.	(*i*) Handle properly (*ii*) Replace.
16.	*Punctures*	(*i*) Sharp objects penetrating carcass.	(*i*) Examine tyres regularly. Remove imbedded objects.
17.	*Twin tyre damage due to stones*	(*i*) Trapped stones	(*i*) Examine twin tyre frequently.

SHORT ANSWER QUESTIONS

CHASSIS

Q. 1. What do you mean by a chassis ?

Ans. The term *chassis* is used to denote the complete vehicle except the body for the heavy vehicle having a separate body.

Q. 2. What is a frame ?

Ans. The frame or underbody is the main part of the chassis, on which the remaining parts of the chassis are mounted.

Q. 3. What are the functions of a frame ?

Ans. The functions of a frame are :

(*i*) To support the chassis components and the body.

(*ii*) To withstand static and dynamic loads with undue deflection or distortion.

Q. 4. What are the types of frames ?

Ans. Types of frames are :

(*i*) Conventional frame (*ii*) Semi-integral frame

(*iii*) Integral or unit frame.

Q. 5. What are the types of sections used to make the frames ?

Ans. The frames are made of the following *sections* :

(*i*) Channel section (*ii*) Box section

(*iii*) Tubular section.

Q. 6. What is meant by 'dumb iron' in frame work ?

Ans. Lateral bending of the frame is taken by extra stiff to front and rear extension of the side members of the frame. These extensions are called as *dumb iron*.

Q. 7. What is an "X" frame ?

Ans. It is a frame consisting of large channels joined by a cross member at the middle.

Q. 8. What is the "backbone frame" ?

Ans. It is a frame in which heavy sections are run down at the centre.

Q. 9. Why is the frame narrowed at the front ?

Ans. The frame is narrowed at the front to have better steering lock, which gives *a smaller turning circle*.

Q. 10. Why is box section preferred in the design of chassis member to an open channel section ?

Ans. The box section has *greater torsional stiffness*.

SUSPENSION

Q. 11. What is a helper spring ?

Ans. A keeper spring is an additional spring to allow a wide range of loading and to provide progressive stiffness against increasing load. These are used *in rear suspension only*. They take a part of the load when it exceeds the maximum on the main spring.

Q. 12. What is the sprung weight ?

Ans. The sprung weight is the weight of *passenger carriage*.

Q. 13. What is unsprung weight ?

Ans. The unsprung weight is weight of the *wheel-axle system*.

Q. 14. What are the causes of rough ride ?

Ans. A *rough* (*or hard*) *ride* is due to excessive friction in spring suspension and/or improper operation of shock absorber.

Q. 15. What is bouncing in the suspension system ?

Ans. *Bouncing* is a vertical vibration of the complete body when the vehicle crosses an up and down surface of the road.

Q. 16. What is rolling in the suspension system ?

Ans. *Rolling* is the movement about a longitudinal axis, caused by a centrifugal force while turning a corner.

Q. 17. What is pitching in the suspension system ?

Ans. *Pitching* is a rocking action about a transverse axis through the vehicle, parallel to the ground. The front suspension moves out of phase with the rear, experiencing the rocking effect due to pitching.

Q. 18. What is the type of suspension used in heavy duty trucks ?

Ans. Longitudinal leaf spring rear suspension is used in heavy duty trucks.

Q. 19. What is the purpose of the tie rod in the rear end suspension ?

Ans. The tie rod is used to control fore-and-aft wheel movement and to position the spindle to provide toe setting in the rear end suspension.

Q. 20. What do you mean by "Independent suspension" ?

Ans. Independent suspension refers to the mounting of the wheel on a separate axle, so that road shocks affect only the particular wheel.

Q. 21. What is the difference between 'conventional suspension' and 'Independent suspension' systems ?

Ans. In *'Conventional system'* wheels are fitted on the beam type axles which are attached to the chassis through load spring. In *'Independent system'*, there is *no axle beam. Suspension for each wheel is of an independent unit and free from the effect of one another.*

Q. 22. Why is independent suspension normally applied to the front wheels and not to the rear wheels ?

Ans. If independent suspension is used in the rear wheels *additional universal joints* are necessary in order to transmit power to the wheels which move independently of each other.

Q. 23. What keeps the rise and fall of the front wheels in rest in an independent suspension system ?

Ans. Proportionate lengths of control arms keep the front wheels in rest.

Q. 24. What is referred to as control arm in the suspension system ?

Ans. *Control arm* is a horizontal rod that connects a wheel to the chassis. This allows the wheel to move up and down independently.

Q. 25. What is a 'wishbone' ?

Ans. 'Wishbone' is a triangular steel frame which connects the vehicle's body to each road wheel in the independent suspension system. It is like a leg bone of a chicken.

Q. 26. What is a half-shaft ?

Ans. Half-shaft is the type of drive axle that extends from the differential to the hub of an independently suspended system.

Q. 27. What is a 'roll axis' in suspension ?

Ans. Roll axis is a line which in a vertical plane contains the longitudinal axis of the car. When force acts on a car, the body may tilt or roll about this roll axis.

Q. 28. What is a panhard rod or track bar ?

Ans. A *panhard rod* is a bar or tube running from one side of the axle to the frame on the other side of the vehicle. It helps the leaf spring in keeping the axle centered under the body during turning a corner.

Q. 29. What is meant by 'shackle' in suspension ?

Ans. Shackle is a swinging connection for a road spring which permits it to vary in length as it deflects.

Q. 30. What is the purpose of a leaf spring in rear suspension ?

Ans. The leaf spring is provided to ensure good lateral stability and reduce midway movement which could contribute a well controlled ride.

Q. 31. What type of spring is normally used with solid axle beams ?

Ans. Leaf springs.

Q. 32. Why are leaf springs made of a series of steel strips ?

Ans. The leaf springs are made of a series of steel strips to overcome weakness at a centre of a single leaf spring.

Q. 33. What is semi-elliptic springs ?

Ans. The normal type of leaf springs, which is attached to the frame at each end and the axle about midway, is known as semi-elliptic string.

Q. 34. How are leaf springs lubricated ?

Ans. Leaf springs are lubricated by feeding lubricants from a centre grease nipple through grooved oil ways or by spraying oil on the leaves.

Q. 35. How is the laminated leaf spring self damping ?

Ans. The laminated leaf spring is highly cambered, pre-stressed and are longer with shackle end. Thus, the friction leaves provide damping.

Q. 36. What is a U-bolt ?

Ans. It is a U-shaped bolt which secures the leaf springs to the axle of the vehicle.

Q. 37. How is allowance provided for the deflection of the spring ?

Ans. Allowance is provided for deflection, by attaching one end to shackle as hinges on the chassis so that effect length of the spring is allowed to increases when loaded.

Q. 38. What type of shock absorber is mostly preferred in modern passenger cars ?

Ans. A hydraulic double-acting telescopic shock absorber is mostly preferred.

Q. 39. How is the shock absorber fitted in the vehicle ?

Ans. One end of the shock absorber is fitted to the frame and the other end is fitted to the axle or spring.

Q. 40. How do the upper and lower parts of the shock absorber function ?

Ans. The upper part controls the upward movement of the wheel support arm while the lower part controls the rebound movement.

Q. 41. How are the shock absorbers tested for replacement ?

Ans. Shock absorbers are checked by moving the front and rear of the vehicle up and down quickly. If the car does not return almost immediately is normal position, shock absorber is to be replaced.

Q. 42. What is meant by cruciform branching ?

Ans. Along the chassis, side members are joined midway by stiffening members in the form of a cross. This arrangement is known as cruciform branching.

Q. 43. What is a frame stabilizer ?

Ans. It is a device designed to resist any tendency on vehicles to dip on either sides, to resist centrifugal force which is the cause of slipping on curve.

WHEELS AND TYRES

Q. 44. To which wheels new tyres be fitted ?

Ans. The front wheels should be fitted with new tyres for the sake of safety.

Q. 45. Why is it important to paint the rims periodically ?

Ans. Rims must be painted periodically, otherwise rust will appear, and eventually the tyre bead may get destroyed.

Q. 46. What is a demountable rim ?

Ans. A demountable rim is one which can be removed easily from the wheel.

Q. 47. What is meant by "Wheel base" ?

Ans. Wheel base refers to the distance between the front and the rear wheel axes of the car, when in a straight ahead position.

Q. 48. Why are "mag wheels" so called ?

Ans. "Mag wheels" are called as such because they are *cast from magnesium.*

Q. 49. What is the cause for more wear on one side than the other, in the front wheels ?

Ans. Misalignment of the front wheels, causes an excessive amount of toe in, or excessive camber to cause more wear on one side.

Q. 50. What is tyre ?

Ans. Tyre refers to the assembly of casing and tread mounted on a car wheel to provide an air cushion and contact with the road.

Q. 51. How is tyre constructed ?

Ans. Tyre construction is a fabric casing with a rubber tread moulded on to it.

Q. 52. Why do tyres have tread pattern ?

Ans. Tyres have tread pattern to shed water when it rains and to increase the tread surface area, so the tyres cool better.

Q. 53. What are beads and why are these provided in a tyre ?

Ans. Steel wires fitted in circles are called *beads.* These keep the tyre fixed in the rim and also keep the tyre on the rim in case of burst or puncture.

Q. 54. State the factors which influence tyre tread contact with the road surface.

Ans. The condition of tyre tread, tyre inflation, wheel balance, weight on the wheels and shock absorber action.

Q. 55. What are the three major parts of a tubeless tyre ?

Ans. (*i*) Carcass, (*ii*) Tread, and (*iii*) Side valve.

Q. 56. State two advantages of a tubeless tyre.

Ans. Slow rate of loss of air, and (*ii*) Penetration of nails, etc., does not affect complete deflation.

Q. 57. What are plies in the tyre ?

Ans. Layers of chord shaped in the form of a tyre are called plies and fixed with wires.

Q. 58. What is ply rating of tyres ?

Ans. Ply rating is *an index of tyre strength.*

Q. 59. What is meant by section height of a tyre ?

Ans. Section height is the height of an inflated tyre from bottom of the bead to the top of the bead.

Q. 60. How is the 'aspect ratio' of a tyre defined ?

Ans. The ratio of section height to section width (H/W).

Q. 61. What is known as "Knife cut" in tyres ?

Ans. "Knife cuts" refer to the ribs of the tyre tread with slots to increase road grip.

Q. 62. What is meant by under-inflation ?

Ans. Under-inflation means that tyre is inflated to *lesser* than its normal pressure.

Q. 63. How often should the tyre pressure be checked ?

Ans. Tyre pressure should be checked weekly.

Q. 64. What are the reasons for uneven tyre wear ?

Ans. Misalignment, incorrect tracking, or worn stub axles and wheel bearings.

Q. 65. Is it necessary to run tyres at lower pressures in hot weather ?

Ans. No, as under-inflation alone may produce higher temperatures.

Q. 66. What are the bad effects of under-inflated tyres ?

Ans. Shoulders of the tyre get worn out, chord becomes loose and cracks are developed.

Q. 67. State three causes which reduce tyre life.

Ans. (*i*) High speeds, (*ii*) Running the tyres with underinflation, and (*iii*) High rates of cornering.

HIGHLIGHTS

1. The term *Chassis* is used to denote the complete vehicle except the body for the heavy vehicle having a separate body.
2. The *power plant* includes the engine assembly and power transmission assembly.
3. The *frame* or *underbody* is the main part of the chassis on which the remaining parts of the chassis are mounted.
4. The *channel section* is good in bending, *tubular* in torsion and *box* in bending and torsion.
5. *Body* is the superstructure of the vehicle.
6. The *suspension system* of an automobile is one which separates the wheel/axle assembly from the body.
7. The important elements of a suspension system are :
   (*i*) Springs (*ii*) Dampers (or shock absorbers).
8. The various tyres of automotive springs are :
   Leaf (or laminated) springs, Coil springs, Torsion bars, Air and gas springs and Rubber springs.
9. *Dampers* (or *shock absorbers*) are used in the suspension system to check any continuous vibration which may follow the initial force on the system.
10. Modern cars mostly have the following two types of hydraulic dampers :
    (*i*) Telescopic dampers (*ii*) Rocking lever dampers.
11. In a gas-filled shock absorber, instead of only oil, mixture of oil and gas is used for the damping effect.
12. The components of a suspension system are :
    Springs, Dampers (or shock absorbers), Stabilizer and a Linkage system.
13. Suspension may be *rigid axle suspension* or *independent suspension.*
14. "*Independent suspension*" is the term used to describe any arrangement by which the wheels are connected to the carriage unit in a manner such that the rise and fall of one wheel has no effect on the others.
15. The *MacPherson system* consists of a telescopic strut, a single arm and a diagonal stay.
16. A *stabilizer* or *sway bar* is simply a bar of alloy steel with arms at each end connected to the lower wishbone of the independent suspension or axle.
17. The primary advantage of an interconnected system is that tendency of the vehicle to bounce, pitch or roll is reduced and constant desired altitude of the wheel can be maintained.
18. In *air suspension systems*, the four steel springs are replaced by four rubber cylinders or air springs.
19. *Hydrostatic suspension* works on compression of fluid with the help of rubber washers or cups.
20. *Hydragas suspension system* uses gas-filled spring units (called Hydragas springs), one at each wheel.
21. *Wheels* are legs of the vehicle which carry it to far-off distances.
22. A *tyre* is a cushion provided with an automobile wheel.
23. *Aspect ratio* in reference to tyre shape is numerically equal to the ratio of the section height to section width.
24. The *balance of a tyred wheel* is essentially required to avoid front wheel wobble which affects steering and increases tyre wear rates.

OBJECTIVE TYPE QUESTIONS

Choose the Correct Answer :

1. The most effective section against bending is
   (*a*) rectangular bar (*b*) round bar
   (*c*) round hollow tube (*d*) square hollow section.

2. Weight of the vehicle produces in the side members of the frame
(*a*) vertical bending (*b*) horizontal bending (*c*) torsion (*d*) all of these.
3. Lateral bending of the frame side members may be caused on account of
(*a*) weight of passengers (*b*) side wind (*c*) engine torque (*d*) braking torque.
4. The number of points at which the engine-clutch-gear box unit is supported on the chassis frame is
(*a*) one (*b*) two (*c*) three (*d*) four.
5. The frame may get distorted to a parallelogram shape due to
(*a*) Weight of vehicle (*b*) Weight of passengers
(*c*) Cornering force (*d*) Wheel impact with road obstacle.
6. The purpose of tyre plies is to
(*a*) increase tread life (*b*) decrease noise level
(*c*) provide softer ride (*d*) increase traction.
7. Tread distortion is least on
(*a*) radial ply tyres (*b*) cross ply tyres
(*c*) crossply belted tyres (*d*) none of these.
8. An overinflated tyre will wear the tread most near the
(*a*) edges (*b*) corners (*c*) outside (*d*) centre.
9. Out of round brake drums cause the tyres to wear
(*a*) on the inside (*b*) on the outside
(*c*) in single spots (*d*) uniformly all round.
10. The type of wheel which cannot be used with a tubeless tyre is
(*a*) disc wheel (*b*) wire wheel
(*c*) light alloy wheel (*d*) composite wheel.
11. The type of wheels preferred in sports cars are
(*a*) disc wheel (*b*) wire wheel
(*c*) magnesium alloy wheel (*d*) aluminium alloy wheel.
12. The disadvantage of radial ply tyre compared to cross-ply tyre is
(*a*) uncomfortable ride at low speeds (*b*) uncomfortable ride at high speeds
(*c*) higher cornering power (*d*) lower rolling resistance.
13. An underinflated tyre will wear the tread most
(*a*) near centre (*b*) near the edges
(*c*) in the lateral direction (*d*) in the cross direction.
14. The purpose of tyre rotation on automobiles is to
(*a*) avoid ply separation (*b*) equalize wear
(*c*) get better ride (*d*) none of the above.
15. The permissible mixing of cross-ply and radial-ply automobile tyres allows
(*a*) cross-ply tyres on left wheels (*b*) cross-ply tyres on right wheels
(*c*) cross-ply tyres on front wheels (*d*) cross-ply tyres on rear wheels.
16. In case of a wire wheel the vehicle weight is supported by the wires in
(*a*) tension (*b*) compression (*c*) bending (*d*) shear.
17. The term 'ply rating' with reference to a tyre refers to the
(*a*) actual number of plies (*b*) recommended inflation pressure
(*c*) aspect ratio (*d*) rated strength.
18. 'Heel' and toe wear' in tyres is caused by
(*a*) excessive acceleration and braking (*b*) overinflation
(*c*) underinflation (*d*) excessive camber.

ANSWERS

1. (*d*)	2. (*a*)	3. (*b*)	4. (*c*)	5. (*d*)
6. (*d*)	7. (*a*)	8. (*d*)	9. (*c*)	10. (*b*)
11. (*c*)	12. (*a*)	13. (*b*)	14. (*b*)	15. (*c*)
16. (*a*)	17. (*d*)	18. (*a*).		

THEORETICAL QUESTIONS

Chassis

1. What is a chassis ?
2. List the various components of a chassis.
3. Give the classification of chassis.
4. What do you mean by 6 × 4 drive chassis vehicle ?
5. Define the following terms :
   (*i*) Wheel base (*ii*) Front overhang (*iii*) Rear overhang.
6. What is a frame ?
7. What are the functions of a frame ?
8. Explain briefly the steel sections used in chassis frame.
9. Describe briefly the following types of frames :
   (*i*) Conventional frame
   (*ii*) Semi-integral frame
   (*iii*) Integral or unit frame.
10. Explain briefly about the defects in chassis frame.
11. List the methods used to avoid rust formation in the body of vehicle.
12. What is 'body' of the motor vehicle ?
13. List the requirements which the body of a motor vehicle should fulfill.
14. Describe different types of motor vehicles.
15. Give classification of chassis with respect to engine fitting.
16. Advantages and disadvantages of fitting engine at rear.
17. Write short notes on the following :
   (*i*) Full-forward chassis (*ii*) Wheel track
   (*iii*) Chassis overhang (*iv*) Wheel base.
18. Name the materials used for frames :
   Ans. Steels used for pressed frame are mild sheet steel, carbon sheet steel and sheet nickel alloy steel. An aluminium alloy called '*Alpax*' has also been used as frame material.

Suspension

19. What is a "suspension system" ?
20. What are the functions/objects of a suspension system ?
21. List the requirements of a suspension system.
22. Explain briefly the elements of a suspension system.
23. Describe briefly the following three basic suspension movements of a car :
   (*i*) Bouncing (*ii*) Rolling (*iii*) Pitching.
24. What is spring ?
25. What are the various types of springs ?

26. Explain briefly the following springs :
    (*i*) Leaf springs (*ii*) Coil springs.
27. Describe briefly any *two* of the following :
    (*i*) Torsion bars (*ii*) Air and gas springs
    (*iii*) Rubber springs.
28. What is the main function of a damper/shock absorber ?
29. Name the two types of hydraulic dampers which the modern cars mostly have.
30. Give the description of a telescopic damper, with neat sketches.
31. What are the main advantages of a gas-filled shock absorber over the conventional type ?
32. Explain briefly the construction and working of a rocking lever damper.
33. Give briefly the procedure of maintenance of damper/shock absorber.
34. Enumerate the components of a suspension system and state their functions briefly.
35. List down the characteristics of 'Rigid axle front suspension."
36. What do you mean by the term "Independent suspension" ?
37. State the advantages and disadvantages of independent suspension over rigid axle type suspension.
38. Enumerate the types of independent suspension systems used in automobiles.
39. Explain briefly the following independent suspension systems used in automobiles :
    (*i*) Wishbone arm system. (*ii*) Trailing link system.
40. Describe the MacPherson strut assembly of independent suspension system.
41. What are the advantages of MacPherson system ?
42. Give the simplified diagrams of the independent front suspension.
43. What is the function of a stabilizer ?
44. What is the purpose of shock absorbers in the suspension system ?
45. Describe a leaf-spring rear suspension system.
46. Name the factors which govern the choice of springs.
47. List the advantages of coil springs over leaf springs.
48. Explain briefly the following designs of independent of rear suspension :
    (*i*) Swinging half axle (*ii*) MacPherson strut
    (*iii*) Trailing arm design.
49. What are interconnected suspension systems ?
50. Explain briefly any two of the following interconnected suspension systems :
    (*i*) Air suspension (*ii*) Hydroelastic suspension
    (*iii*) Hydragas suspension system.
51. State the advantages of air suspension systems over conventional metal springs.
52. Write a short note on maintenance of independent suspension system.

Wheels and Tyres

53. What are the functions of wheels in an automobile ?
54. State the essential requirements of wheels in case of automobiles.
55. Enumerate various types of automobiles' wheels.
56. Explain briefly pressed steel disc wheels.
57. List the merits/advantages of pressed steel disc wheels.
58. Give the description of a wire wheel.
59. Explain briefly light alloy casting wheels.
60. State the advantages and disadvantages of light alloy casting wheels.
61. What is a tyre ?
62. What are the functions of a tyre ?

63. State the requirements of a good tyre.

64. Explain briefly the following types of tyres :

(*i*) Conventional tube tyre (*ii*) Tubeless tyre.

65. What are the advantages of a tubeless tyre ?

66. Describe tyre construction in brief.

67. What are the results of under-inflation and over-inflation ?

68. List down the advantages of radial ply tyres over the bias-ply tyres.

69. Describe briefly 'tyre materials'.

70. Explain briefly 'tyre shape'.

71. What do you mean by 'Tyre marking' ? Explain briefly.

72. State the factors on which 'Tyre selection' depends.

73. What are the causes of tyre wear ?

74. List the factors which affect tyre life.

75. Explain the importance of tyre maintenance.

76. Enumerate the four most common tyre troubles.

77. What is tyre rotation and its purpose ?

78. How can the life of a tyre enhanced ?

79. Write a short note on 'Wheel balance'.

7

Transmission System

7.1. **Introduction to transmission system**—Definition—Requirements of transmission system—Types of transmission systems—Units of transmission system. 7.2. **Clutch**—Definition of a clutch—Functions of a clutch—Requirements of a clutch—Principle of operation of a clutch—Friction materials—Classification of clutches—Cone clutch—Single plate clutch—Multiplate clutch—Centrifugal switch—Semi-centrifugal switch—Free wheel or one-way clutch—Automatic clutches—Clutch actuating mechanism—Automatic transmission devices—Clutch adjustments—Troubleshooting of clutch—Comparative clutch data of some Indian automobiles. 7.3. **Gear box** (Transmissions)—Introduction to gear box (transmission)—Gear ratios—Types of gear boxes—Selective type gear boxes—Progressive type gear boxes—Maruti-800 gear box—Five-speed gear box—Gear shifting—Transfer case—Lubrication of gear box—Troubleshooting of gearbox—Gear boxes used in some Indian automobiles—Automatic transmission. 7.4. **Propeller shaft**—General aspects—Constructional features of propeller shaft—Troubleshooting of propeller shaft. 7.5. **Universal joints**—Introduction—Construction and working of universal joint—Types of universal joints. 7.6. **Final drive and differential**—Final drive—Differential. 7.7. **Rear axles**—Introduction—Types of rear axles—Axle shafts or half shafts—Causes of axle failures—Rear axle noises—Maintenance of rear axle—Comparative rear axle data of some Indian automobiles—Troubleshooting of rear axle—***Short Answer Questions***—Highlights—Objective Type Questions—Theoretical Questions.

7.1. INTRODUCTION TO TRANSMISSION SYSTEM

7.1.1. Definition

The mechanism that transmits the power developed by the engine of the automobile to the driving wheels is called the **transmission system** (or **Power train**). It is composed of clutch, the gear box, propeller shaft, universal joints, rear axle, wheel and tyres. The vehicles which have front wheel drive in addition include a second set of propeller shafts, universal joints, final drives and differentials for the front units.

7.1.2. Requirements of Transmission System

The transmission should fulfill the following *requirements* :

1. Provide means of connection and disconnection of engine with rest of power train without shock and smoothly.

2. Provide a varied leverage between the engine and the drive wheels.

3. Provide means to transfer power in the opposite direction.

4. Enable power transmission at varied angles and varied lengths.

5. Enable speed reduction between engine and the drive wheels in the ratio of about 5 : 1.

6. Enable diversion of power flow at right angles.

7. Provide means to drive the driving wheels at different speeds when required.

8. Bear the effects of torque reaction, driving thrust and braking effort effectively.

The above requirements are fulfilled by the following *main units of transmission system* :

1. Clutch
2. Gear box
3. Transfer case
4. Propeller shaft and universal joints.
5. Final drive
6. Differential
7. Torque tube
8. Road wheels.

7.1.3. Types of Transmission Systems

The transmission systems may categorized as follows :

1. *Electrical and electromagnetic transmission system :*

These have been used for motor vehicles but now-a-days they have become obsolete.

2. *Hydraulic transmission system :*

Hydraulic transmissions are now common in the shape of :

— Fluid coupling ;

— Torque converter ;

— *Automatic transmission* etc.

3. *Mechanical transmission system*

Mechanical transmissions can be divided into the following classes :

(*i*) Clutch, gearbox and line axle transmission.

(*ii*) Clutch, gearbox and dead axle transmission.

(*iii*) Clutch, gearbox, axleless transmission.

(*i*) Clutch, gearbox and line axle transmission :

This type of transmission is *employed by majority of the vehicles.* Power developed by the engine flows to the driving wheels through clutch, gearbox, propeller shaft, universal joints, final drive, differential, and half shafts.

This type of transmission can be split up into the following classes :

- Engine at front and drive to rear wheels on a live axle. ... *Two wheel drive*
- Engine at front and drive to all the four wheels mounted on front and rear live axles. ... *Four wheel drive*
- Engine at front and drive to front live axle and two rear live axles. ... *Six wheel drive*
- Engine at front or rear and power transmission to front or rear drive axle placed below the engine. ... *Power pack axles*

(*ii*) Clutch, gearbox and dead axle transmission :

In this type of transmission chain and sprockets are used to transmit power from transmission unit to dead axle over which the driving wheels are mounted. This arrangement is known as *chain drive* and was quite popular with early motor vehicles.

— There are certain compact units in which engine and transmission are enclosed to form a single one unit (as in "Vespa" scooter)

— Final reduction in the chain drive is obtained by putting bigger size sprockets on the dead axle.

(*iii*) Clutch, gear box and axleless transmission :

This type of transmission is used in the *vehicles employing independent suspension for the driving wheels.*

The differential housing is attached to the chassis frame. Drive from the differential to the individual stub axle of the driving wheels is given through carden shafts having universal joints at their ends. The power from the gear box flows to the differential through the propeller shaft and universal joints.

7.1.4. Units of Transmission System

The transmission system consists of the following *main units :*

1. **Clutch unit :**

It consists of :

- Clutch assembly ; or
- Fluid coupling.

2. **Transmission unit :**

It consists of :

- Gear box ;
- Transfer case ;
- Overdrive ;
- Free wheeling device ;
- Torque converter etc.

3. **Drive-line unit :**

It consists of :

- Propeller shaft ;
- Universal joints ;
- Sprockets and chains.

4. **Driving axle unit :**

It consists of :

- Final drive ;
- Differential ;
- Half shafts etc.

The various units of transmission system are discussed in the following articles.

7.2. CLUTCH

7.2.1. Definition of a Clutch

Definition. A ***clutch*** *is a mechanism which enables the rotary motion of one shaft to be transmitted at will to second shaft, whose axis is coincident with that of first.*

The clutch is *located between the engine and gear box.* When the clutch is *engaged,* the power flows from the engine to the rear wheels through the transmission system and the vehicle moves. When the clutch is *disengaged, the power is not transmitted to the rear wheels and the vehicle stops while the engine is still running.*

- The clutch is "***disengaged***" when (*i*) *starting the engine,* (*ii*) *shifting* the *gears,* (*iii*) *Stopping the vehicle* and (*iv*) *Idling the engine.*
- The clutch is "***engaged***" only when the vehicle is to move and is kept engaged when the vehicle is moving.

7.2.2. Functions of a Clutch

The *functions* performed by a clutch are as follows :

1. To permit engagement or disengagement of a gear when the vehicle is stationary and the engine is running.

2. To transmit the engine power to the road wheels smoothly without jolt/shock to the transmission system while setting the wheel in motion.

3. To permit the engaging of gears when the vehicle is in motion without damaging the gear wheels.

7.2.3. Requirements of a Clutch

The *requirements of a clutch* are as follows *:*

1. It should *engage gradually* to avoid sudden jerks.

2. It should be easily operated (*i.e.,* it should consume minimum physical effort at the time of engagement and disengagement).

3. It should be *dynamically balanced* (particularly required in case of high speed engine clutches).

4. It should be *free from slip* when engaged.

5. It should be as small as possible so that it will *occupy minimum space.*

6. The wearing surfaces should have *long life.*

7. It should be *easily accessible* and have *simple means of adjustment.*

8. It should have suitable mechanism to *damp vibrations* and to *eliminate noise* produced during the power transmission.

9. It should be able to *dissipate large amount of heat* which is generated during the clutch operation due to friction.

10. The driven members of the clutch should be made as light as possible so that they *will not continue to rotate for any length of time* after the clutch has been disengaged.

11. The clutch should have *free pedal play* in order to *reduce effective clamping load on the carbon thrust bearing and wear on it.*

7.2.4. Principle of Operation of a Clutch

The clutch principle is based on *friction.* When two friction surfaces are brought in contact with each other and pressed they are united due to friction between them. If now one is revolved, the other will also revolve. The friction between the two surfaces depends upon, (*i*) Area of the surface, (*ii*) Pressure applied upon them, and (*iii*) Coefficient of friction of the surface materials. One surface is considered as *driving* member and the other as *driven member.* The driving member is kept rotating. When the driven member is brought in contact with the driving member, it also starts rotating. When the driven member is separated from the driving member, it stops revolving. This is how, a clutch operates.

- The driving member of a clutch is the *flywheel* mounted on the crankshaft, the driven member is *pressure plate* mounted on the transmission shaft. Friction surfaces (clutch plates) are between the two members (driving and driven). On the engagement of the clutch, the engine is connected to the transmission (gear box) and the power flows from the engine to the rear wheels through the transmission system. When the clutch is disengaged by pressing a clutch pedal, the engine is disconnected from the transmission and consequently the power does not flow to the rear wheels while the engine is still running.

7.2.5. Friction Materials

There are three types of friction materials/linings, namely :

1. Woven type 2. Moulded or compression type 3. Mill board type.

1. The "*woven type*" is made by spinning threads from *asbestos fibres,* sometimes on brass wire, weaving this thread into a cloth and then impregnating it with a bonding material (such as vegetable grins, rubber, synthetic resins etc.).

2. The "*moulded* or *compression*" type of lining is composed of *asbestos fibres* in their natural state mixed with a *bonding material* and then moulded in dies under pressure and at elevated temperatures. Metallic wires are sometimes included but only to increase the wearing qualities and to eliminate scoring of the metal faces against which the lining rubs.

3. "*Mill board*" type friction materials mainly include asbestos sheets treated with different types of impregnants. They are cheap as well as quite satisfactory in operation.

"*Cotton*" is occasionally used instead of, or mixed with, asbestos and such fabrics can be made to give high co-efficients of friction (up to 0.6) but materials containing cotton cannot withstand a temperature exceeding about 150°C without being charred and ruined.

"*Cork*" is also used occasionally for clutches, almost always being arranged to run in oil. It has a coefficient of friction of about 0.3.

- The following materials are almost *universally used for clutch lining* :

(*i*) Leather ; (*ii*) Cork ;
(*iii*) Fabric ; (*iv*) Asbestos ;
(*v*) Reybestos and Ferodo.

Properties of a good clutch lining :

(*i*) High coefficient of friction.
(*ii*) Good wearing properties.
(*iii*) Cheap and easy to manufacture.
(*iv*) Good binder in it.
(*v*) High resistance to heat.

7.2.6. Classification of Clutches

The clutches can be *classified* as follows :

1. Positive clutches.
2. Gradual engagement clutches.

1. **Positive clutches.** In positive clutch the clutch is either "in" (Fig. 7.1(*i*)) so that the two shafts are rigidly connected and must revolve at the same speed ; or it is "out" (Fig. 7.1 (*ii*)), when the shafts are entirely disconnected. It is obvious, therefore that this type of clutch is *not suitable for use between the engine and the gear box.*

2. **Gradual engagement clutches.** In such clutches it is possible for one shaft to be revolving rapidly while the other is either stationary or revolving at a lower speed. As the engagement of the clutch proceeds the speed of the two shafts gradually become the same, and when the clutch is fully engaged the shafts revolve as one. Such clutches are called *friction clutches.*

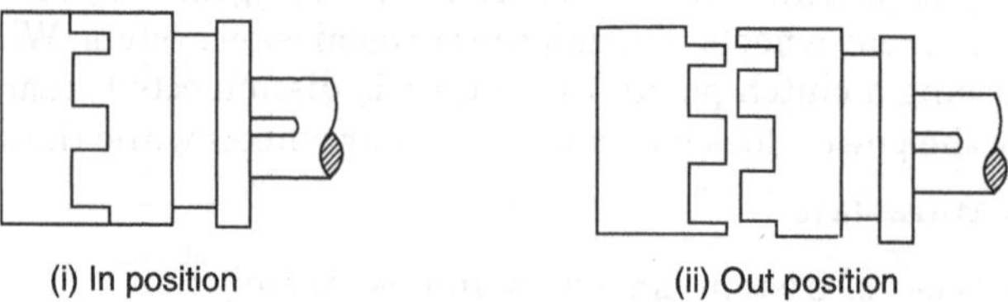

Fig. 7.1. Positive clutch.

(*a*) ***Friction clutches*** may be of the *three types :*

(*i*) Cone clutch.
(*ii*) Single plate clutch.
(*iii*) Multi-plate clutch.

— The single and multi-plate friction clutches are usually dry types. The "*dry type*" is used to distinguish these clutches from those that operate in a both of oil known as the "*wet plate*" type. In these clutches a cork-insert or phosphor bronze plates are fitted between steel plates, all the plates being immersed in oil. *Oil immersed clutches are, however, used in conjunction with, or as part of automatic transmission.*

(*b*) "*Other forms of clutches*" are also coming into use and generally form a part of the pre-selector, two-pedal or fully automatic transmission systems. These are :

- *Centrifugal and magnetic clutches ;*
- *Fluid flywheel (fluid hydraulic coupling) ;*
- *Hydraulic torque converter.*

7.2.7. Cone Clutch

Fig. 7.2 shows a cone clutch (in engaged position). It consists of a flywheel and a cone mounted on the driving and driven shafts respectively. The shape of the side of the flywheel facing the cone is such as to accommodate the cone readily when the clutch is engaged. The surfaces of contact are lined with the *friction lining* (Asbestos, leather etc.). The cone can be disengaged from flywheel by lever mechanism which operates in the groove of the cone.

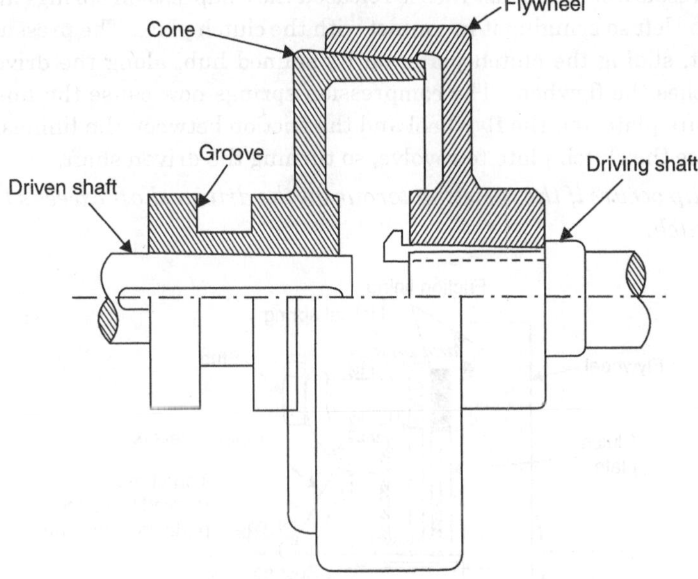

Fig. 7.2. Cone clutch.

The only *advantage* of this clutch is that the normal force acting on the friction surfaces is *greater than the axial force,* as compared to the single plate clutch is which the normal force acting on the friction surfaces is equal to the axial force.

The cone clutch is practically obsolete, due to the following *disadvantages.*

(*i*) If the cone angle is made smaller than 20°, the male cone tends to bind in the female cone and it is *difficult to disengage the clutch.*

(*ii*) A small amount of wear on the cone surfaces results in a *considerable amount of the axial movement* of the male cone for which it *will be difficult to allow.*

- It is only *used* in the synchromesh units of gear boxes, and in overdrives and some epicyclic gear boxes.

7.2.8. Single Plate Clutch

Fig. 7.3 shows a simplified schematic diagram of a single plate clutch which is commonly *used in most cars and small commercial vehicles.*

Construction. The *flywheel* is rigidly fixed to the engine shaft, the driven shaft to the gear box being supported at the engine end by a spigot bearing in the flywheel. A *clutch plate* is attached to a splined hub which is free to slide axially on the splines cut on the driven shaft. A *ring of friction lining* is attached to each side of the clutch plate. One end of each of a number of *helical compression springs* bear on the back of a pressure plate, the other ends of the springs pressing against a *cover, attached to the flywheel and rotating with it.* Three release levers on pivots mounted on the cover, bear on the bottoms of the nuts on the studs in the pressure plate and are actuated by the leftward movement of the thrust race which in turn is moved by the *clutch pedal mechanism.*

Working. The diagram shows the pressure plate pulled back by the release levers against the compression springs ; so that the friction linings on the clutch plate are free of flywheel and pressure plate. The flywheel then rotates without driving the clutch plate and hence the driven shaft.

When the pressure of the thrust race is released the compression springs are free to move the pressure plate to the left so bringing it in contact with the clutch plate. The pressure plate continues so move to the left, sliding the clutch plate, on its splined hub, along the driven shaft until the friction lining touches the flywheel. The compression springs now cause the linings to be gripped between the pressure plate and the flywheel and the friction between the linings and flywheel and pressure plate cause the clutch plate to revolve, so turning the driven shaft.

- *Clutch slip occurs if the resisting torque on the driven shaft exceeds the friction torque at the clutch.*

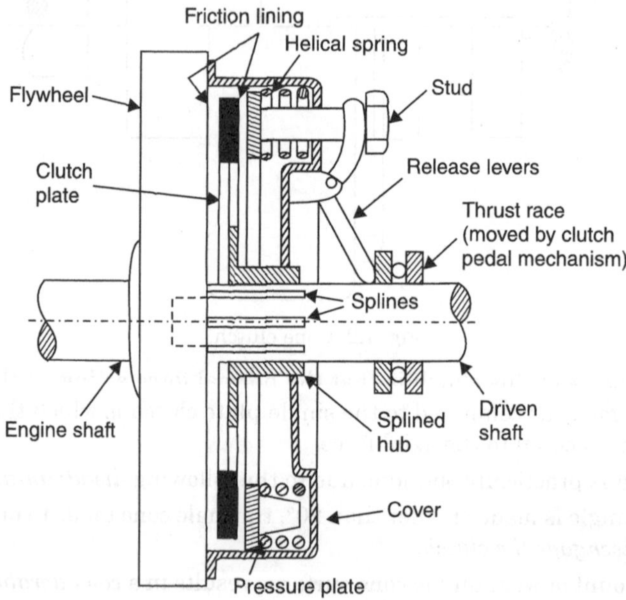

Fig. 7.3. Single plate clutch (Disengaged position).

Diaphragm spring type single plate clutch. This type of clutch is similar in construction to that of the single plate type of clutch described above except that here *diaphragm springs* (also called Belle ville springs) are used instead of the ordinary coil springs. In the free condition, the diaphragm spring is of conical form (Fig. 7.4) when assembled, it is constrained to an approximately flat condition because of which it exerts a load upon the pressure plate.

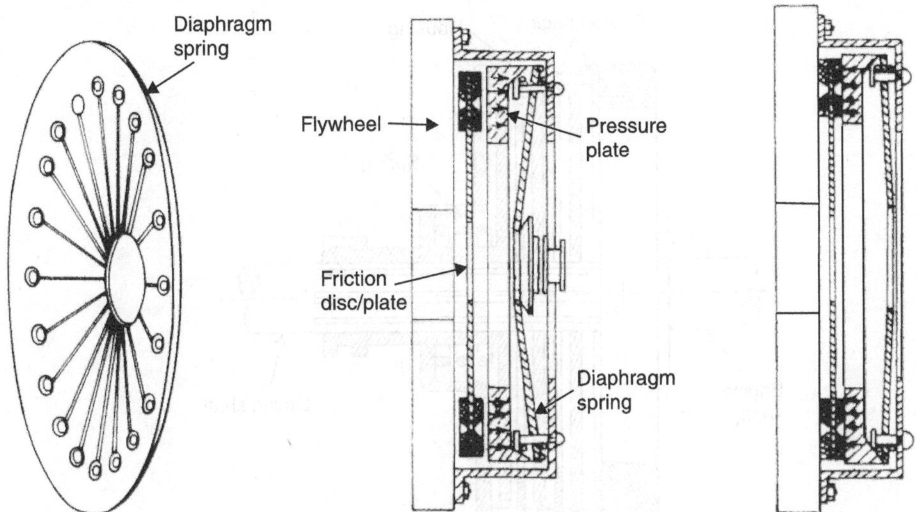

Fig. 7.4. Diaphragm spring used in diaphragm spring clutch.

Fig. 7.5. Diaphragm spring clutch (disengaged)

Fig. 7.6. Diaphragm spring clutch (engaged)

Diaphragm spring is shown in 'disengaged' and 'engaged' conditions in Fig. 7.5 and 7.6 respectively.

The diaphragm spring offers the following *advantages* :

(*i*) The operating load is practically uniform and constant on the driven plate.

(*ii*) It has a compact design, which results in smaller clutch housing.

(*iii*) Release levers are not required, since the diaphragm itself acts as a series of levers.

(*iv*) Squeaks, rattles and vibrations are mostly eliminated.

(*v*) It can withstand higher rotational speeds since the diaphragm is comparatively less affected by the centrifugal forces. On the other hand, coil springs have tendency to distort in the transverse direction at higher speeds.

7.2.9. Multiplate Clutch

When a great amount of torque is to be transmitted instead of single plate we can employ a *number of friction plates.* This will increase the number of mating friction surfaces, hence it is called "*multiplate clutch*".

These clutches are **used** in *heavy commercial vehicles, racing cars and motor cycles for transmitting high torque.*

Fig. 7.7 shows a multiplate clutch in which friction rings are splined on their outer circumferences to mate with corresponding splines on the bore of the housing and are free to slide on the splines. The friction material therefore rotates with the housing and engine shafts. Discs or plates are free to slide on the splines on the driven shaft and rotate with it. The disc on the right can be moved to the right against a powerful spring which, when the actuating force is removed, presses the disc into contact with friction rings. Torque is therefore transmitted between the engine shaft and the driven shaft.

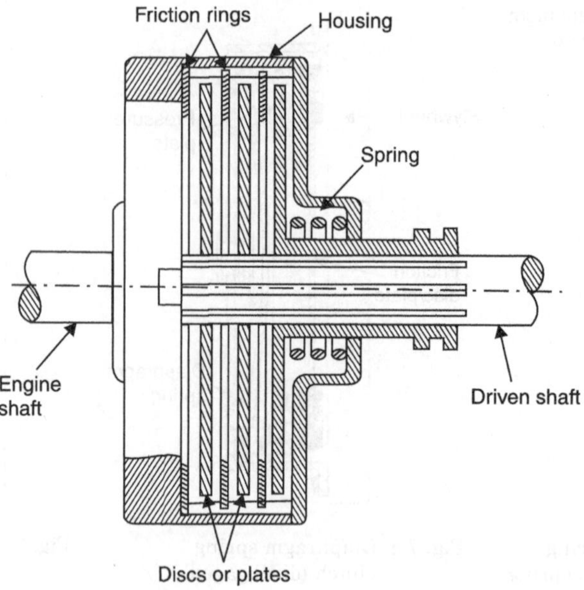

Fig. 7.7. Multiplate clutch.

In the diagram the left hand ring is attached to the flywheel so that only one side is effective. The total number of friction pair is then five. Each pair has the same axial thrust acting on it and this thrust is that exerted by the spring. The total torque is therefore five times that transmitted by one friction pair.

These clutches may be *dry* or *wet*.

Multiple wet clutch :

- It consists of a number of thin plates made of *steel* fitted to the engine shaft and those on gear shaft are made of *phosphor bronze*. These plates are immersed in a bath of *oil* and also have grooved surfaces for permitting the oil to flow through them. These grooves help to dissipate the heat generated during the engagement and release operations.
- These clutches are *generally used in conjunction with* or *as a part of the automatic transmission.*

Multiple dry clutch :

- In this type of clutch its different plates are lined with a frictional material similar to that used in case of a single plate clutch.

The clutches with metal plates are used in tractors or other light powered engine vehicles while cork inserted multiplate clutches are used in motorcycles. In order to avoid scorching or charging of cork-inserts due to frictional heat, it is necessary to have the clutch oil-cooled.

Description of parts of a plate clutch :

The principal parts of a plate clutch are :

1. Driving members.
2. Driven member (Clutch plate).
3. Operating members.

1. ***Driving members :***

One driving member consists of a cover which carries a *cast iron pressure plate* or *driving disc*, the *pressure springs and the releasing levers.* The entire assembly is bolted to the flywheel and rotates with it at all times. The flywheel acts as a second driving member ; the *flywheel and the pressure plate grip the driven member between them under the action of the pressure springs.*

- The clutch housing and the cover provided with openings dissipate the heat generated by the friction during the clutch operation.

2. ***Driven member :***

The driven member consists of a *disc or plate which is free to slide lengthwise on the splines of the clutch shaft but which drives the shaft through those splines.*

The clutch disc carries friction material on both bearing surfaces.

3. ***Operating members :***

The operating mechanism consists of the *foot pedal*, the *linkage*, the *release or throw-out bearing*, the *release lever* and springs necessary to ensure proper operation of the clutch.

Fig. 7.8 shows a typical single plate (coil spring) clutch in the engaged position. In the Fig. 7.9 are shown the major parts of a clutch (only three springs are shown for illustration).

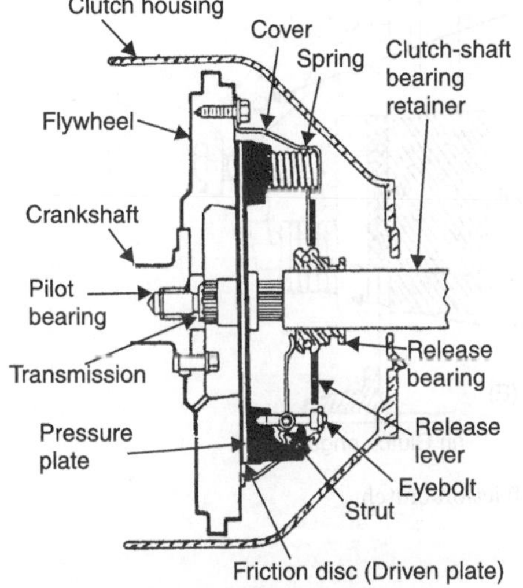

Fig. 7.8. Single plate (Coil-spring) clutch in engaged position.

Fig. 7.9. Major parts of the clutch.

- "*Pressure plate*" is accurately machined and presses the driven plate against the fly-wheel.
- "*Driven plate*" (or *friction disc*) is provided with annular facings and a spring-cushioned hub *i.e.*, springs and vibration damper rings.
- "*Release levers*" are provided with every lever mounted on an eye bolt locked on the cover by a nut and held in place by a retainment spring.
- "*Engagement springs*" force the pressure plate on driven plate.
- "*Sliding sleeve and thrust bearing*" controlled by a fork lever may slide until it presses with the thrust bearing on the release lever inner ends.
- "*Cover*" is of pressed sheet steel and is fixed to the flywheel by screws.

7.2.10. Centrifugal Clutch

In this type of clutch *centrifugal force*, instead of spring force, is used for *keeping it in engaged position*. Also, a centrifugal clutch does not require any clutch pedal for operating the clutch. *The clutch is operated automatically depending upon the engine speed.* The vehicle can be stopped in gear without stalling the engine. Similarly the vehicle can be started in any gear by pressing the accelerator pedal.

Fig. 7.10 shows a centrifugally, operated friction clutch. It consists of weights *A* pivoted at fixed pivot *B*. When the engine speed increases the weights fly off due to the centrifugal force, operating the bell crank levers, which press the pressure plate *C* through spring(s) *D* which ultimately presses the driven clutch plate *E* on the flywheel. This makes the clutch engaged.

At low engine speeds only the spring pressure acts, while at high engine speeds the force generated by the weights supplements the spring pressure. Thus smooth and acceptable gear changes are obtained under varying conditions of load and speed.

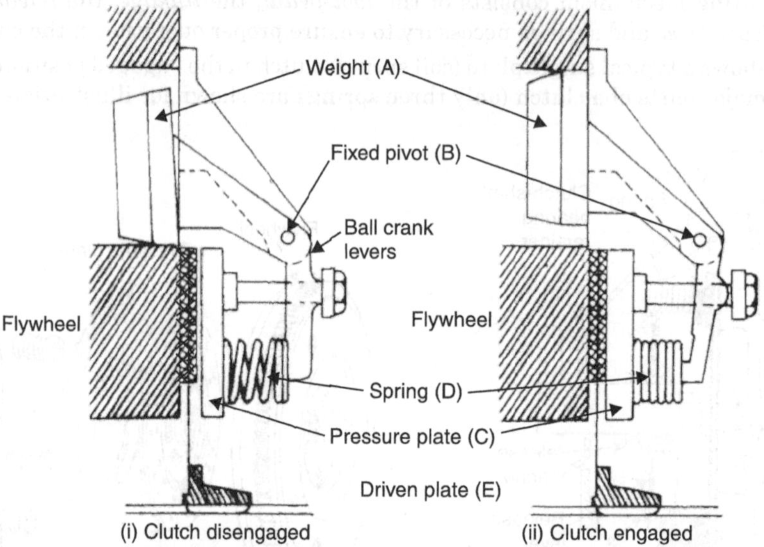

Fig. 7.10. Centrifugally operated friction clutch.

7.2.11. Semi-centrifugal Clutch

This type of clutch *uses centrifugal force as well as spring force for keeping it in engaged position. The springs are designed to transmit the torque at normal speeds, while the centrifugal force assists in torque transmission at higher speeds.* This type of construction permits the use of light coil pressure springs and the clutch can operate without too much pressure.

Fig. 7.11 shows a semi-centrifugal clutch. It consists of three hinged and weighted levers and three clutch springs alternately arranged at equal spaces on the pressure plate.

— At *normal speeds*, when the power transmission is low, the springs keep the clutch engaged, the weighted levers do not have any pressure on the pressure plate.

— At *high speed*, when the power transmission is high, the weights fly off and the levers also exert pressure on the plate, keeping the clutch firmly engaged. Thus, instead of having more stiff springs for keeping the clutch engaged firmly at high speeds, they are *less stiff*, so that the driver may not get any strain in operating the clutch.

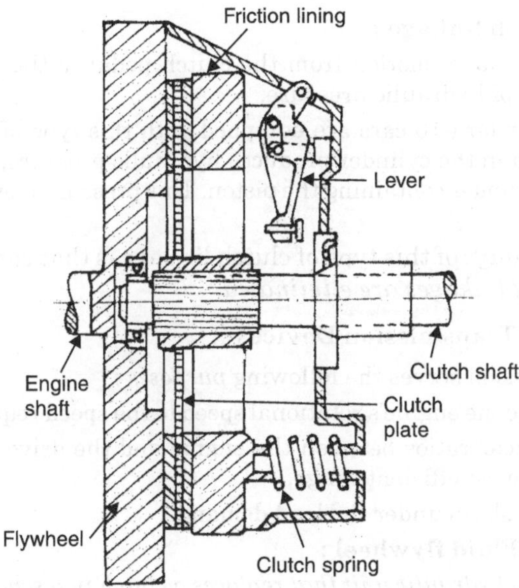

Fig. 7.11. Semi-centrifugal clutch.

— When the *speed decreases* the weights fall and the levers do not exert any pressure on the pressure plate. Only the spring pressure is exerted on the pressure plate which is adequate to keep the clutch engaged.

The centrifugal force on the pressure plate can be adjusted by an adjusting screw provided at the end of the lever.

7.2.12. Free Wheel or One-way Clutch

- It is device which *transmits power in one direction only.*
- It is fitted in transmission components just behind the gearbox so that transmission components beyond it can be overrun by components before it.

With this arrangement, the engine can idle without having to disengage the gears and also helps in gear changing. While going downhill on a long steep road also the engine can overrun the transmission components so that no power is transmitted from the engine.

7.2.13. Automatic Clutches

Recently several automatic clutch designs have become available. On certain Renault models and continental cars, Ferlec of French origin and the Smiths *electroshift design* have been used. It has a *conventional type of friction disc. When battery current is supplied, an electromagnet in the flywheel operates the pressure plate which engages the disc thus transmitting engine torque.*

7.2.14. Clutch Actuating Mechanism

It is through a clutch linkage that a driver is able to engage and disengage the clutch with a pedal. Following two types of clutch linkages are used :

1. Mechanical clutch linkage.
2. Hydraulic clutch linkage.

1. **Mechanical clutch linkage.** This type of clutch linkage has a system of levers that provides a force of about 180 N at the clutch pedal to release the clutch. The clutch linkage transmits the force applied to the clutch pedal to the throw out or release bearing and provides the necessary linkage to make the clutch operation possible with a reasonable amount of foot pressure.

2. **Hydraulic clutch linkage :**

- In this clutch linkage, motion from the clutch pedal to the pressure plate is obtained through the use of hydraulic pressure.

In our country, Standard-10 cars are equipped with this type of clutch operation. The foot pedal, operating the piston in the cylinder produces hydraulic pressure, which is then transmitted through a tube to a slave cylinder containing the piston. This pressure moves the slave piston thereby releasing the clutch.

- The *main advantage* of this type of clutch linkage is that *complicated linkages, vibration and wear of linkages are eliminated.*

7.2.15. Automatic Transmission Devices

Automatic transmission serves the following *purposes* :

(*i*) It helps to change the engine's rotational speed to the speed required at the driving wheels.

(*ii*) It changes the gear ratios between the engine and the drive wheels to keep the engine operating in its most efficient range.

(*iii*) It automatically shifts under the load demands.

1. **Fluid coupling/Fluid flywheel :**

"Fluid coupling" is a hydraulic unit that replaces a clutch in a semi or fully automatic system, and transmits the engine torque to a transmission system. Since the coupling is a major part of the engine flywheel assembly, it is also called a *'fluid flywheel'* or *'fluid drive' acting as an automatic clutch.*

It has *no mechanical connection or face to face contact*. The magnitudes of input and output torques are equal.

Construction. Refer to Fig. 7.12. It consists of the following two *rotating elements* :

(*i*) *Pump impeller* (*Driving unit*). It is attached to a driving shaft of the prime-mover.

(*ii*) *Turbine runner*. It is attached to a driven shaft.

Both the above units are enclosed in a single housing filled with a liquid, usually oil, because of its lubricating power, availability and stability. This oil serves to transmit torque from the pump impeller to the turbine runner. There is no direct contact between the driving and driven parts.

Working. As soon as the prime mover starts rotating, the pump impeller also starts rotating and throws the oil outward by centrifugal action. The oil then enters the turbine runner and exerts a force on the runner blades. The magnitude of the torque increases with an increase in the speed of the driving shaft and eventually when this torque *overcomes the inertia effects*, the turbine runner and the driven shaft begin to rotate. The oil from the runner then flows back into the pump impeller, thus a complete hydraulic (oil) circuit is established.

Expression for efficiency :

In a hydraulic coupling the power is transmitted *hydraulically* from the driving shaft to driven shaft and is free from engine vibrations. The efficiency of power transmission may be as high as 98 percent.

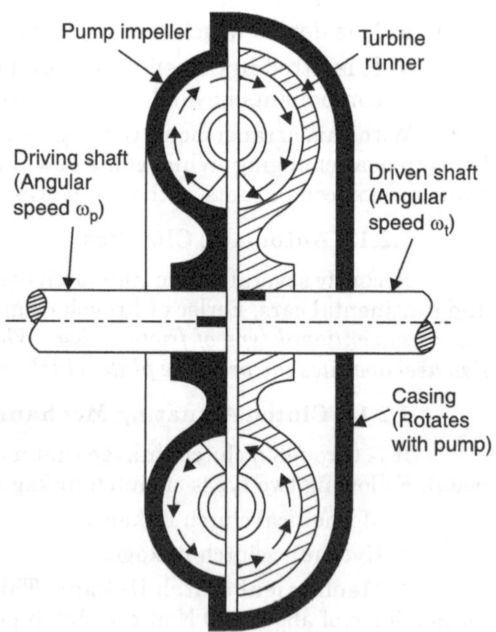

Fig. 7.12. Fluid coupling.

Let, T_p = Torque on the pump or driving shaft '1' (*i.e.*, input torque),

ω_p = Angular speed of the pump or the driving shaft '1',

T_t = Torque on the turbine or driven shaft '2' (*i.e.*, output torque), and

ω_t = Angular speed of the turbine shaft or driven shaft '2'.

$\therefore$ Power input, $P_i = T_p\omega_p$, ...(7.1)

Power output, $P_o = T_t\omega_t$ ...(7.2)

Since the tangential momentum due to the velocity of whirl suffers no change as the oil passes from impeller blades to the runner blades, therefore, the torque on the driven shaft (T_t) equals that on the driving shaft (T_p), *i.e.* $T_p = T_t$.

The efficiency (η) of the hydraulic coupling is defined as the *ratio of power output to power input, i.e.,*

$$\eta = \frac{\text{Power output } (P_0)}{\text{Power input } (P_i)} = \frac{T_t\,\omega_t}{T_p\,\omega_p} = \frac{\omega_t}{\omega_p} \quad ...(7.3)$$

$(\because \quad T_p = T_t)$

The ratio $\left(\frac{\omega_t}{\omega_p}\right)$ is known as *speed ratio.*

Slip of hydraulic or fluid coupling is defined as follows :

Slip, $$s = \frac{\omega_p - \omega_t}{\omega_p} = 1 - \frac{\omega_t}{\omega_p} = 1 - \eta \quad ...(7.4)$$

A typical *efficiency* (η) versus *speed ratio* $\left(\frac{\omega_t}{\omega_p}\right)$ curves for a fluid coupling is shown in Fig. 7.13. The efficiency is zero when the speed ratio is zero and it increases uniformly till the speed ratio is about 0.95, and then it rapidly reduces to zero.

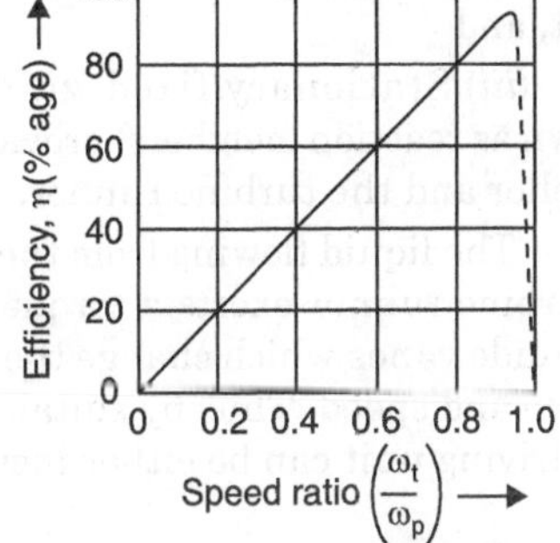

Fig. 7.13. Efficiency-speed ratio curve for hydraulic coupling.

Advantages. Following are the main advantages of a fluid coupling :

1. Power is transmitted smoothly from engine to gears.
2. Owing to absence of friction surface, less maintenance is required.
3. Clutch pedal is eliminated.
4. During braking or coming down a hill, the transmission shocks are absorbed by the fluid.
5. The torsional vibrations of the crankshaft are damped.

Disadvantage. The main disadvantage of fluid coupling is the presence of idling drag and that overloading of the fluid coupling not only slows down the turbine but also overloads the engine. Also slip is greater at lower speeds.

Uses :

A fluid coupling always slips by about 2 to 4 percent when transmitting full load. It means that the turbine is always running slightly slower than the impeller and as such complete disconnection of the drive is not possible. Thus the fluid coupling is *not suitable for ordinary gear box and is generally used with epicyclic gears to provide a semi or fully automatic gear box.*

- Although a fluid coupling has a low value of transmission efficiency when compared to mechanical coupling, yet it widely used in the following fields :
  (*i*) Automobiles, marine engines, ropeway cable drive units and such other applications *where driven shaft is required to run at a speed close to that of the driving shaft.*

(*ii*) These couplings are particularly useful where large initial loads are involved and smooth shock-free operations are required.

2. **Hydraulic torque converter :**

"Hydraulic torque converter" is a device used for transmitting increased or decreased power from one shaft to another. A variable torque is impressed on the driven member without the use of a gear train or clutch. The torque at the driven shaft may be increased about five times the torque available at the driving shaft with an efficiency of about 90 percent.

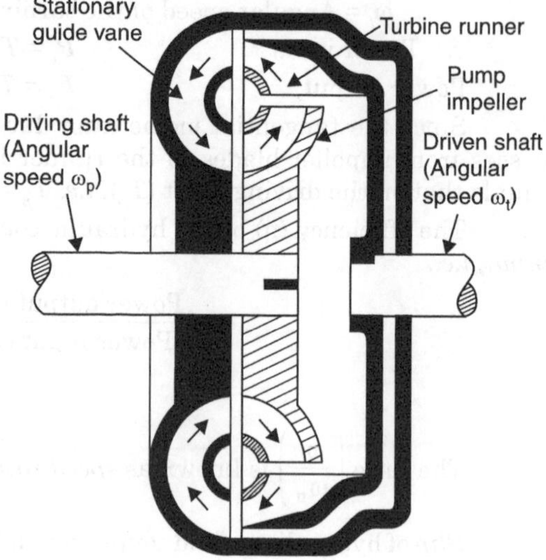

Fig. 7.14. Hydraulic torque converter.

Construction and working :

Refer to Fig. 7.14. The hydraulic torque converter consists of the following :

(*i*) Pump impeller coupled to the driving shaft,

(*ii*) Turbine runner coupled to the driven shaft, and

(*iii*) Stationary/fixed guide vanes (also known as reaction members) provided between the impeller and the turbine runner.

The liquid flowing from the pump impeller to turbine runner exerts a torque on the stationary guide vanes which change the direction of liquid, thereby making possible the transformation of torque and speed. Thus by suitably designing the stationary guide vanes the torque transmitted to the driving unit can be either increased or decreased. The torque relationship is given as :

$$T_t = T_p + T_v \qquad ...(7.5)$$

where, T_t = Torque transmitted to the *turbine shaft,*

T_p = Torque of *pump impeller*, and

T_v = Variation of torque caused by *fixed guide vanes.*

Power input, $P_i = T_p \, \omega_p$

Power output, $P_o = T_t \omega_t = (T_p + T_v)\, \omega_t$

$$\therefore \text{ Efficiency of torque converter, } \eta = \frac{\text{Power output } (P_o)}{\text{Power input } (P_i)} = \frac{(T_p + T_v)\,\omega_t}{T_p \omega_p}$$

$$= \frac{\omega_t}{\omega_p}\left(1 + \frac{T_o}{T_p}\right) \qquad ...(7.6)$$

Equation (7.4) indicates that when there are no guide vanes, torque converter reduces to flange coupling with $T_v = 0$ and then $\eta = \dfrac{\omega_t}{\omega_p} = 1 - s.$

The torque T_v depends upon the design of the stationary guide vanes ; in eqn. (7.5) it can be +ve or –ve. If the guide vanes are designed to receive a torque from the fluid which is in *opposite direction* to that exerted on the turbine (or driven) shaft, an *increased output* results. On the other hand if the shape of the guide vanes is such that they receive the torque in the *same* sense as that of driven shaft, a *torque reduction* results. Thus, since the torque in case of a hydraulic torque converter can be increased or decreased (in case of the hydraulic coupling $T_p = T_t$), therefore, it is comparable to an electric transformer. Usually the *torque converters are used for increasing the torque.*

When a large reduction in speed ($\omega_t << \omega_p$), and a large torque is required, the hydraulic torque converters are designed which utilize two or more sets of turbine runners, and fixed guide vanes located between the turbine runners.

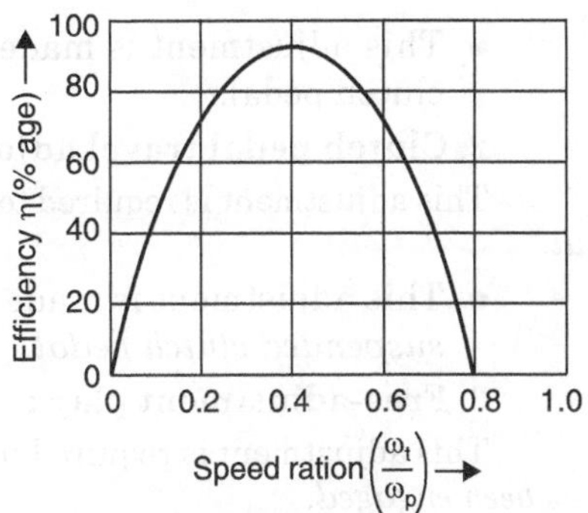

Fig. 7.15. Efficiency vs. speed ratio curve for the hydraulic torque converter.

Fig. 7.15, shows the efficiency versus speed ratio curve for a torque converter ; maximum value of efficiency occurs where speed ratio is approximately 0.5 ; at higher speed ratios the efficiency drops. From the efficiency curve, it is apparent that the efficiency of the torque converter is better at smaller speed than that of hydraulic coupling. The advantages of the hydraulic coupling and the torque converter can be obtained by *designing the system in such a way that at low speed ratios it acts as a converter and at high speed ratios as a coupling.*

Comparison between the "fluid coupling" and "hydraulic torque converter" :

The comparison between the fluid coupling and hydraulic torque converter is given below :

S. No.	Aspects	Fluid coupling	Hydraulic torque converter
1.	*Means of power transmission*	Through fluid.	Through fluid
2.	*Type of device*	Automatic.	Automatic
3.	*Main components*	Impeller and runner.	Pump, stator and turbine.
4.	*Type of unit*	It is simply a means to connect driving and driven members.	It is a torque multiplication unit.
5.	*Flow of oil*	Impeller and runner are locked up and movement of oil stops during engagement when centrifugal force is approximately the same on both members.	It never locks up and flow of oil never stops but continues.
6.	*Types of blades*	Merely fins.	The turbine blades are inclined having pitch.
7.	*Purpose served*	It serves the purpose of an automatic clutch.	It acts as an automatic clutch and serves the purpose of automatic gear box to increase torque.
8.	*Performance at highway speeds*	It is efficient at highway speeds.	It is not as efficient as fluid coupling at highway speeds but is slightly more efficient under load.
9.	*Assistance by a clutch*	It is not assisted by friction clutch.	It is usually used in conjunction with automatic clutch (mostly fluid flywheel) to eliminate the slight loss of efficiency at highway speeds.

7.2.16. Clutch Adjustment

The following four adjustments are made on most of the vehicles :

1. **Floor board clearance adjustment :**

This adjustment is required *to prevent the pedal arm from resting against the floor board when the clutch is engaged.*

- This adjustment is made with the *help of a screw* located near the lower end of the clutch pedal.

2. **Clutch pedal travel adjustment :**

This adjustment is required to *ensure complete clutch disengagement when the clutch is thrown out.*

- This adjustment is made by *raising or lowering the pedal assembly upper stop in the suspended clutch pedal.*

3. **Free-adjustment play :**

This adjustment is required to *keep a specified amount of free play in the pedal after the clutch has been engaged.*

- This adjustment is generally made by *changing the length of one rod located somewhere in the clutch linkage.* It should be made only after the correct floor board clearance or clutch pedal travel has been made.

4. **Clutch release lever adjustment :**

This adjustment is *made only after removing the clutch from the vehicle.* Also this adjustment should be made every time the clutch is removed from the vehicle.

- For making this adjustment to factory specification, a *clutch rebuilding machine equipped with a dial gauge* or *a gauge plate is used.*

7.2.17. Troubleshooting of Clutch

Troubleshooting of clutch is given below :

S. No.	Symptom	Cause	Remedy
1.	*Clutch slips* (*i.e.*, when the clutch is engaged, there is a lack of firm contact between the flywheel pressure plate and driven plate.	(*i*) Oil on the lining because of leaks in engine seal or gear box. (*ii*) Too small a clearance of clutch. (*iii*) Worn out facing. (*iv*) Weak clutch springs.	(*i*) Replace the lining and correct leaks. (*ii*) Adjust the clearance to the required back-lash of pedal. (*iii*) Install a new friction plate. (*iv*) Install a new set of springs.
2.	*Clutch fails to disengage*	(*i*) Excessive clutch clearance. (*ii*) Friction plate not moving freely on splined shaft. (*iii*) Friction plate.	(*i*) Properly adjust the clearance. (*ii*) Clean the splined shaft and hub and see that they move freely. (*iii*) Replace the facing on the friction plate.
3.	*Clutch grabs* (*i.e.*, the hub of the friction plate binds on the clutch shaft)	(*i*) Scored contact surfaces of flywheel or pressure plate warped. (*ii*) Warped friction plate. (*iii*) Release levers not evenly placed.	(*i*) Face the surfaces to make them smooth. (*ii*) Straighten or replace the friction plate. (*iii*) Adjust the levers properly.

7.2.18. Comparative Clutch Data of Some Indian Automobiles

Below is given the comparative clutch data of some Indian automobiles :

S. No.	Vehicle make	Type of clutch	Outside diameter (mm)
1.	*Ambassador*	Single dry plate	203.2
2.	*Fiat 1100*	Single dry plate	184.0
3.	*Premier Padmini*	Single dry plate	184.0
4.	*Fiat Uno*	Single dry plate	—
5.	*Maruti 800*	Dry single disc type	—
6.	*Jeep CJ-4, CJ-344 WD (Mahindra & Mahindra)*	Single dry plate	216
7.	*Tata Mobile – 201 pick and Tata 608*	Single dry plate, friction diaphragm type	228
8.	*Ashok Leyland Comet*	Single dry plate	330 mm
9.	*Dodge (passenger)* Forgo Model	Single dry plate	279.4
10.	*Swaraj Mazda*	Single dry plate (hydraulically operated)	260
11.	*Ford F 516 C Truck*	Single dry plate (hydraulically operated)	330
12.	*Royal Enfield, Bullet*	Multiplate	—
13.	*Yamaha RX-100*	Wet multiple disc type	—
14.	*Rajdoot 175 c.c.*	Wet multiple disc type	—
15.	*LML Vespa*	Wet multiplate	—
16.	*Luna*	Automatic centrifugal clutch	—

7.3. GEAR BOX (TRANSMISSION)

7.3.1. Introduction to Gear Box (Transmission)

Gearbox *is a speed and torque changing device between the engine and the driving wheels.*

It serves the following ***purposes*** in the transmission system of an automobile :

1. It exchanges engine power for greater torque and thus provides a mechanical advantage to drive the vehicle under different conditions.

2. It exchanges forward motion for reverse motion.

3. It provides a neutral position to disallow power flow to the rest of power train.

The automobile requires high torque when *climbing hills* and when *starting*, even though they are *performed at low speeds.* On the other hand, *when running at high speeds on level roads, high torque is not required because of momentum and it would be more preferable to have just the wheels alone turning at high speed.* However, the torque, which the engine can produce is limited to such as amount that the *engine by itself cannot develop the torque necessary for starting and for climbing hills.*

- The transmission (gear box) acts in accordance with the running conditions. *When driving power is required, it reduces the engine speed and transmits stronger torque to the driving wheels and when high running speed is desired, it transmits high speed low-torque to the wheels.* In addition, the transmission serves to *reverse the vehicle.* Since the engine can turn in one direction only, the transmission gears mesh in such a manner to allow running the vehicle in the reverse direction.

7.3.2. Gear Ratios

In automobiles various types of gears (Fig. 7.16) are used, which are basically similar. They all have teeth that mesh to transmit force and motion from one gear to another. The simplest gear is the spur gear, which is like a wheel with teeth. The gears used in transmissions are *helical gears*. In these gears, the teeth are set at angle to the gear centreline. The teeth have a wiping action which improves their contact and lubrication.

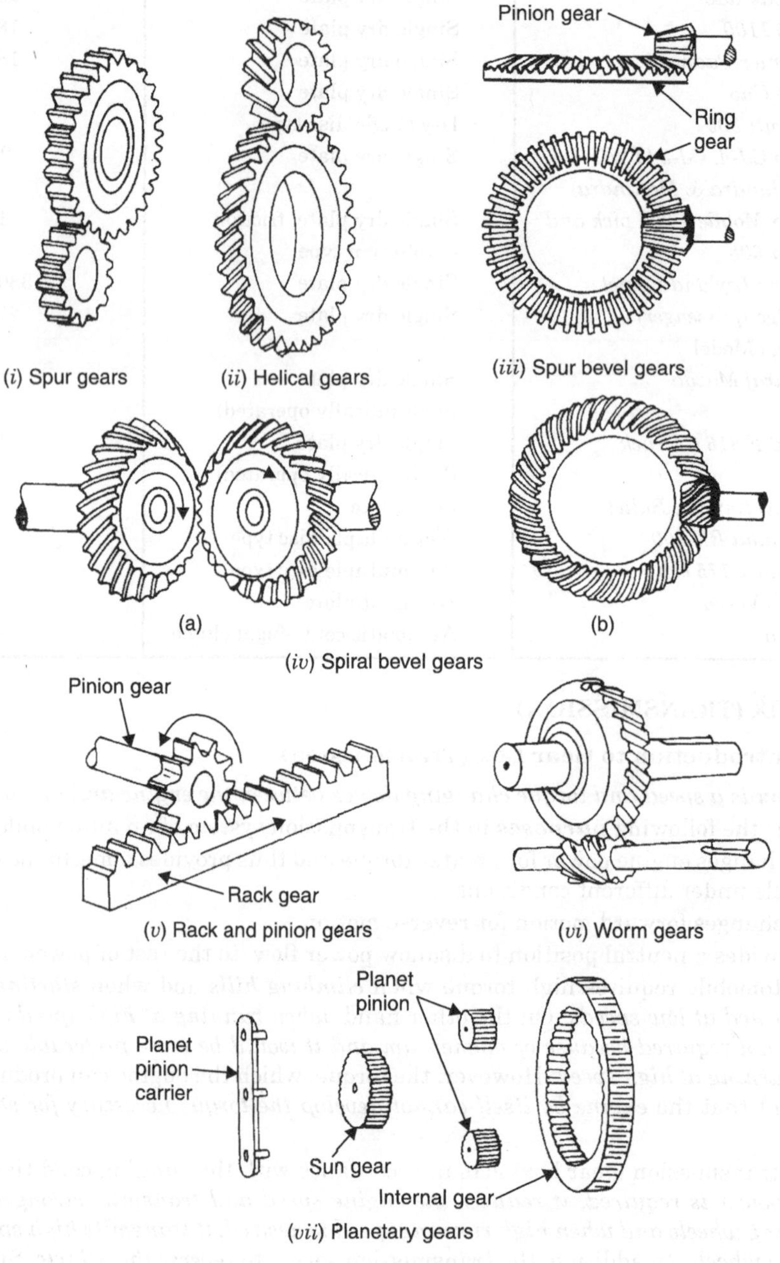

Fig. 7.16. Various types of gears.

When one meshing gear rotates, the teeth of that gear cause the teeth of the other gear to move so that the other gear also rotates. The relative speed of the two meshing gears (*gear ratio*) is determined by the number of teeth of the two gears.

The *gear ratio* is obtained by using the following formula :

$$\text{Gear ratio} = \frac{\text{No. of teeth on driven gear}}{\text{No. of teeth on driving gear}}$$

$$= \frac{\text{Speed of driving shaft}}{\text{Speed of driven shaft}}$$

$$= \frac{\text{Engine speed}}{\text{Propeller shaft speed}}.$$

Example. When two gears have the same number of teeth, they both rotate at the same speed. But if one gear has 15 teeth and other has 30 teeth (Fig. 7.17), the smaller gear will rotate twice for every revoluation of the larger gear. When the smaller gear is driving the larger gear, this is a two-to-one gear ratio (written 2 : 1). If the 15-tooth gear was driving with 45-tooth gear, the gear ratio would be 3 : 1.

The typical *gear box ratio* in a small car with a four-speed gear box is 3.5 : 1 in *first*, 2 : 1 in *second*, 1.4 : 1 in *third* and 1 : 1 in top. All these are multiplied by the axle ratio, which is taken as 4 : 1 to give the corresponding ratios between the engine speed and the road wheel speed.

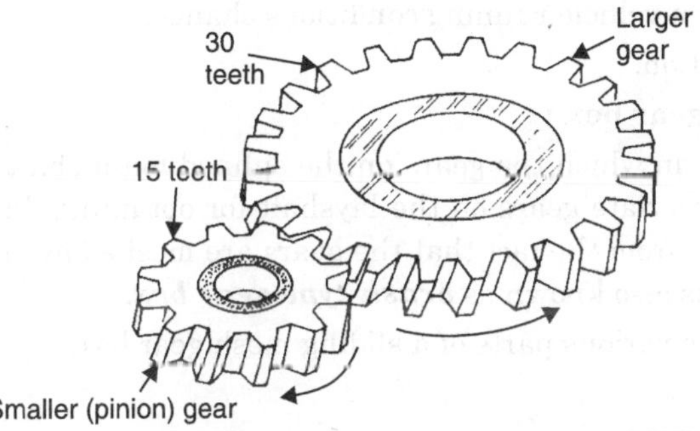

Fig. 7.17. Two meshing helical gears with different numbers of teeth.

The power available from the engine is directly related to the engine torque T and the gear box ratio G. Assuming no loss in transmission, torque available at the wheel, $T_w = T \times G$ and produces the driving force along the road. This driving force is known as *tractive effort* and the maximum amount which can be applied is limited by the coefficient of adhesion between the tyre and the road. The *tractive effort also varies with the vehicle speed as the engine torque varies with engine speed*. Usually the speed of the engine is constant. To vary the speed of the road wheels relative to the engine, the gear box provides a number of varying ratios-usually three to four gear ratios are sufficient for passenger cars. Gear wheels of different sizes are provided for engaging or disengaging them for transmission of motion and power.

In *gear combinations*, the gear ratios, *in order from the largest ratio*, are called, '*low*', '*second*', and '*third*'. The gear ratio at which the engine speed is the *same* as the propeller shaft speed is called '*top*'. When the *gear ratio is less than one*, that, is when the propeller shaft speed is *faster than* the engine speed is called "*overdrive*".

Torque and gear ratio. In Fig. 7.17 when the smaller gear is driving the larger gear, the *gear ratio* is 2 : 1. However, the *torque ratio* is 1 : 2. The larger gear turns at half the speed of the smaller gear. As a result, the larger gear will have twice the torque of the smaller gear. In gear systems, the *speed reduction means torque increase.*

7.3.3. Types of Gear Boxes

The following types of gear boxes are used in automobiles :

1. *Selective type* :

(*i*) Sliding mesh (*ii*) Constant mesh

(*iii*) Synchromesh.

2. Progressive type

3. Epicyclic or planetary type.

7.3.4. Selective Type Gear Boxes

It is that transmission in which any speed may be selected from the neutral position. In this type of transmission, neutral position has to be obtained before selecting any forward or reverse position.

Advantages of selective type gear boxes :

(*i*) Simple in construction. (*ii*) Relatively free from troubles.

(*iii*) Light and small. (*iv*) Low production costs.

Disadvantages :

(*i*) Gear ratios not being continuous but being in steps (3 to 5 steps), making it necessary to shift gears each time when vehicle running conditions change.

(*ii*) Noisy in operation.

1. **Sliding mesh gear box :**

It is that gear box in which the gears on the splined main shaft are moved right or left for meshing them with appropriate gears on the layshaft for obtaining different speeds. This type of gear box derives its name from the fact that the gears are meshed by sliding or crashing one on to the other. This gear box is also known as ***crash-type gear box.***

Fig. 7.18 shows the various parts of a sliding mesh gear box.

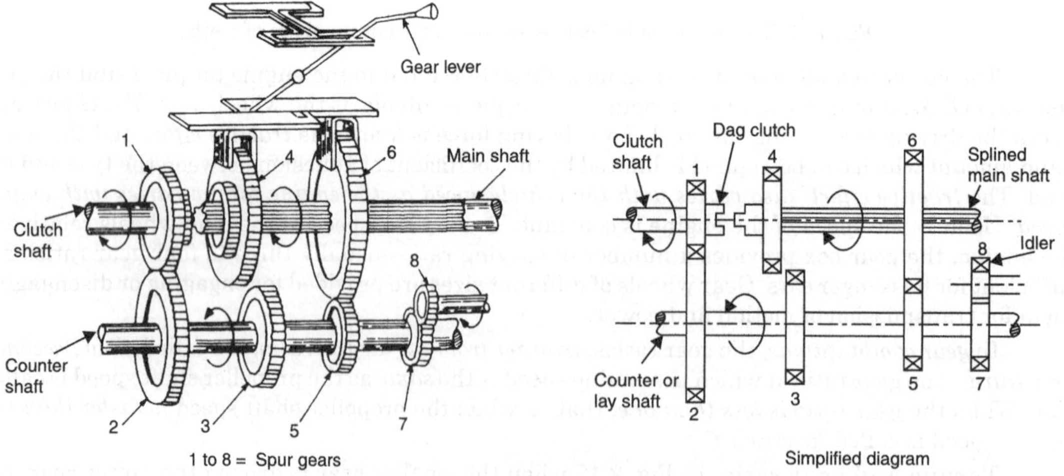

Fig. 7.18. Sliding mesh gear box – First gear position.

This gear box consists of a clutch shaft, counter or lay shaft and main shaft. The clutch shaft has *one gear*. The counter or lay shaft has *four gears*. All four gears form an *integral part* of the counter shaft. The main shaft has *two gears*. These two gears can slide horizontally along the splines of the main shaft. At the same time the gears rotate the main shaft. There is *no* such spline arrangement in the countershaft, as such the gears cannot move along it.

The various gear positions in a sliding gear box are described below.

First gear. Refer to Fig. 7.18.

When the gear shift lever is applied such that gear '6' meshes with gear '5', gear '4' does not mesh with gear '3'.

- When the clutch shaft is rotating, the drive is transmitted from gear '1' to gear '2'. Thus the counter shaft is also rotated. The direction of rotation of counter shaft is opposite to that of the clutch shaft.
- When the counter shaft is rotating, gear '5' rotates gear '6'. Therefore, gear '6' rotates the main shaft.

Thus the drive is transmitted along gears '1', '2', '5' and '6'. Gear '1' is *smaller* than gear '2', and gear '5' is smaller than gear '6'. Consequently, the *speed of the main shaft is reduced considerably.*

Second gear. Refer to Fig. 7.19.

Gear '4' is shown in mesh with gear '3'.

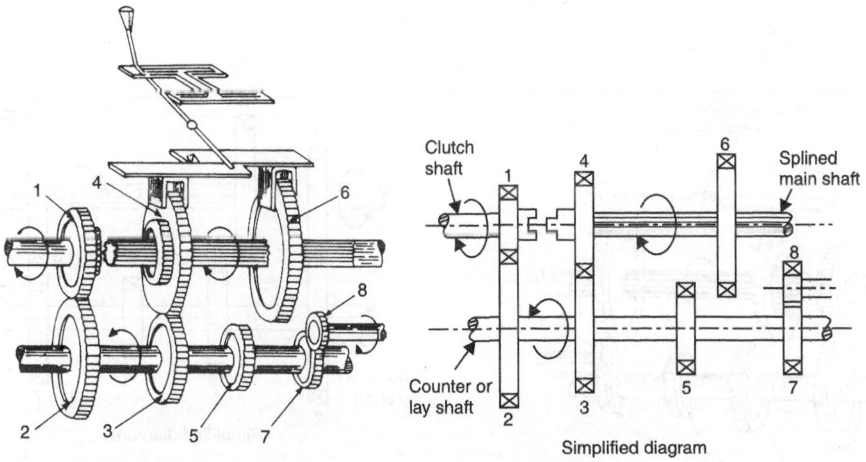

Fig. 7.19. Second gear position.

- When the clutch shaft is rotating, the transmission takes place between gears '1' and '2', and '3' and '4' ; the gear '4' rotates the main shaft.

The second gear is developed on the basis of the sizes of the gears.

Third gear. Refer to Fig. 7.20.

In this gear, the dog of gear '4' is directly meshed with gear '1'. The main shaft is in direct contact with the clutch shaft. Therefore, the main shaft acquires the same speed of the clutch shaft. This is the position for this gear.

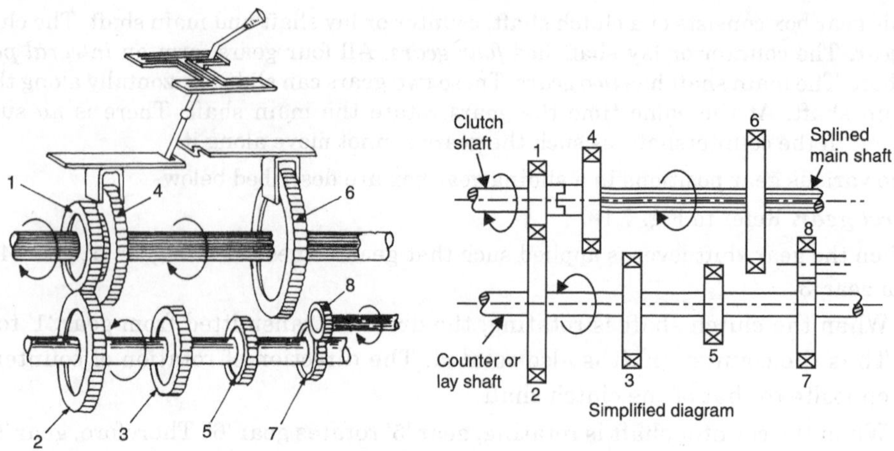

Fig. 7.20. Third gear position.

Reverse gear. Refer to Fig. 7.21.

In the reverse gear the transmission takes place form gears '1' to '2', '7' to '8' and then to '6'. The small intermediate gear '8' (idler) causes the reverse position of the gear box.

In this arrangement, the speed reduction is usually same as that in the first gear.

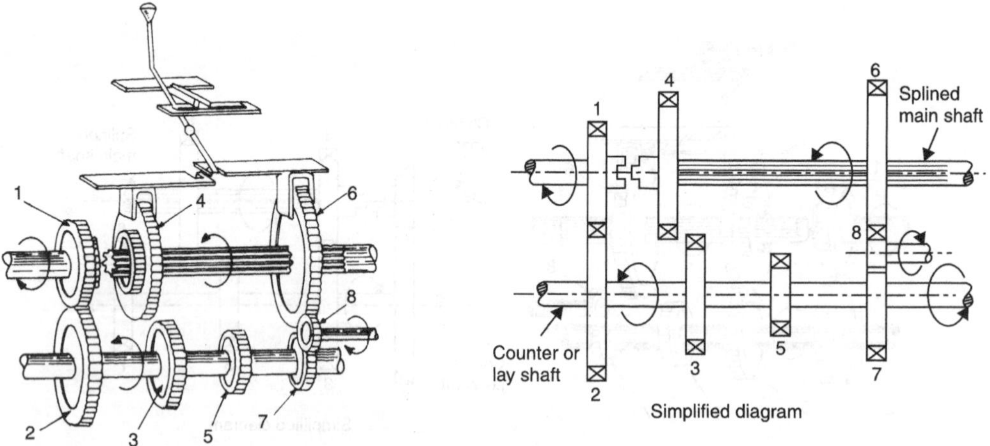

Fig. 7.21. Reverse gear position.

Neutral gear. Refer to Fig. 7.22.

In this arrangement the clutch shaft is in connection with the counter or lay shaft. As the layshaft is *not* connected to the main shaft, there is *no power transmission to the wheels.*

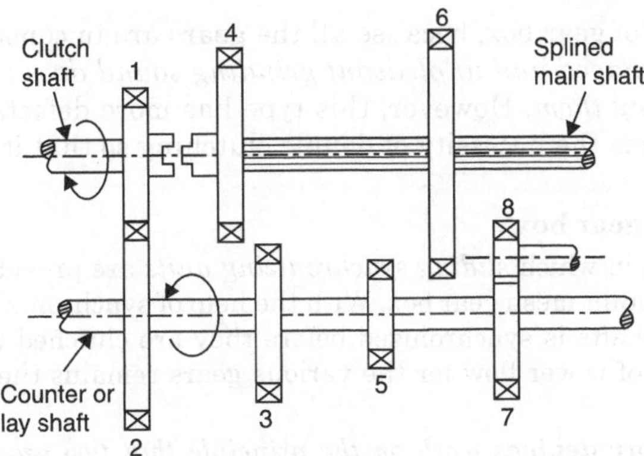

Fig. 7.22. Neutral gear position.

2. **Constant mesh gear box :**

Refer Fig. 7.23. It is that gearbox in which all the gears are in constant mesh with each other (hence the name constant mesh gear box) all the time and this gives a silent or *quiet operation*. Here, helical gears are used to make gear changing easier. The gears on the main shaft which is splined, are *free*. The gears on the counter or layshaft are, however, *fixed*. Two dog clutches are provided on the main shaft-one between the clutch gear and the second gear and the other between the low/first gear and reverse gear. Dog clutch can slide on the main shaft and rotates with it.

When the left-hand dog clutch is made to slide to the left by means of the gearshift lever, it meshes with the clutch gear and the top speed gear is obtained. When the dog clutch meshes with the second gear the second speed gear is obtained. Similarly by sliding the right-hand dog clutch to the left and right, the first speed gear and reverse gear are obtained respectively. However skillful handing is necessary on the part of the driver so that the speed of the locking dogs and respective pinion remain the same to effect a clash-free gear change.

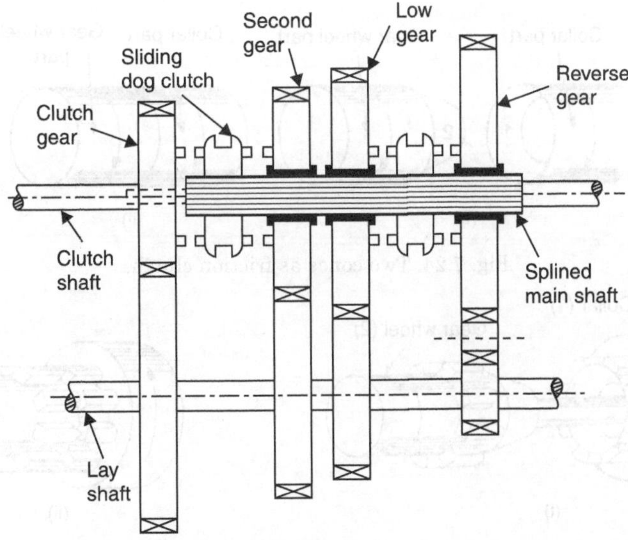

Fig. 7.23. Constant mesh gear box.

- In this of type of gear box, because all the gears are in constant mesh, they are *safe from being damaged and unpleasant grinding sound does not occur while engaging and disengaging them.* However, this type has more defects than the synchromesh type and there is the necessity of double clutching so that it is not used to any large extent.

3. **Synchromesh gear box :**

It is that gear box in which *sliding synchronizing units* are provided in place of sliding dog clutches as in case of constant mesh gear box. With the help of synchronizing unit, the speed of both the driving and driven shafts is synchronized before they are clutched together through train of gears. The arrangement of power flow for the various gears remains the same as in the constant mesh gear box.

Synchromesh gear devices *work on the principle that two gears to be engaged are first brought into frictional contact which equalises their speed after which they are engaged readily and smoothly.* The following types of such devices are mostly used in vehicles :

(*i*) Pin type ; (*ii*) Synchronizer ring type.

A synchronising system is used for *smooth meshing.* A synchromesh works like a *friction clutch.*

Refer to Fig. 7.24. Fig. 7.24 (*i*) shows two *conical* surfaces ; cone-1 is the part of the collar and cone-2 is part of the gear wheel. Cones 1, 2 are revolving at different speeds. While cone-2 is revolving, cone-1 gradually slides into it. Friction slows or speeds up the gear wheel. Finally both cones revolve at the same speed. They revolve one as shown in Fig. 7.24 (*ii*).

Refer to Fig. 7.25. In Fig. 7.25 (*i*) are shown collar (1) and gear wheel (2) as separate and revolving at different speeds. In Fig. 7.24 (*ii*), the internal cone of the collar comes in contact with the outer cone of the gear wheel. Here again, the friction slows or speeds up the gear wheel.

Refer to Fig. 7.26. In Fig. 7.26 (*i*) the collar and gear wheel rotate at the same speed. In Fig. 7.26 (*ii*) the *spring-loaded* outer ring of the collar is pushed forward. The dogs slide smoothly into mesh without clashing. The collar and gear wheel lock and revolve at the same speed. *This is the principle of* **synchromesh.**

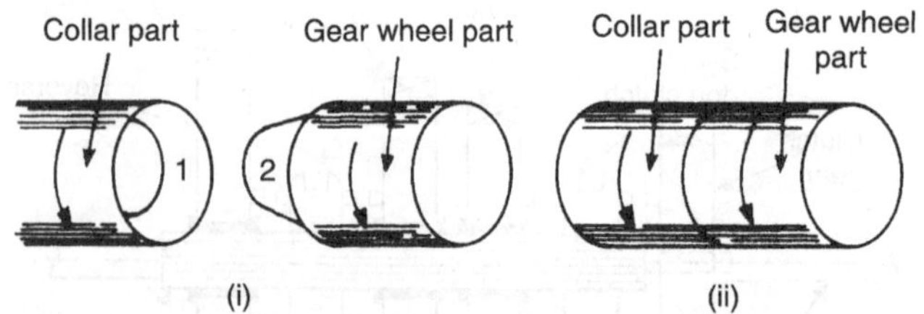

Fig. 7.24. Two cones as friction clutch.

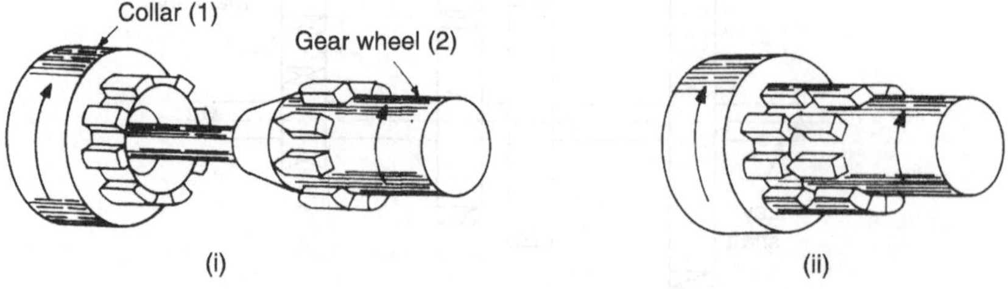

Fig. 7.25. Collar meshing with gear wheel.

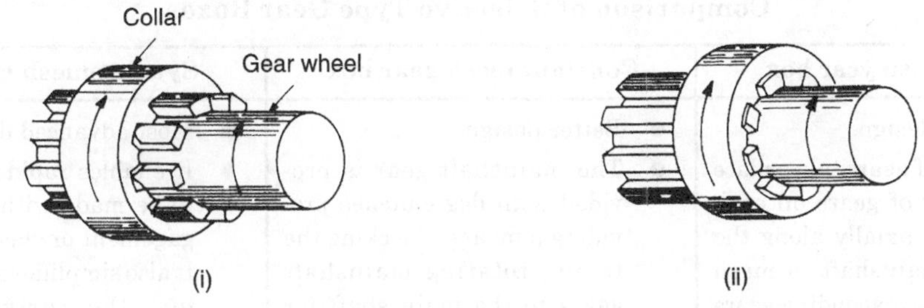

Fig. 7.26. Collar and gear wheel revolving together.

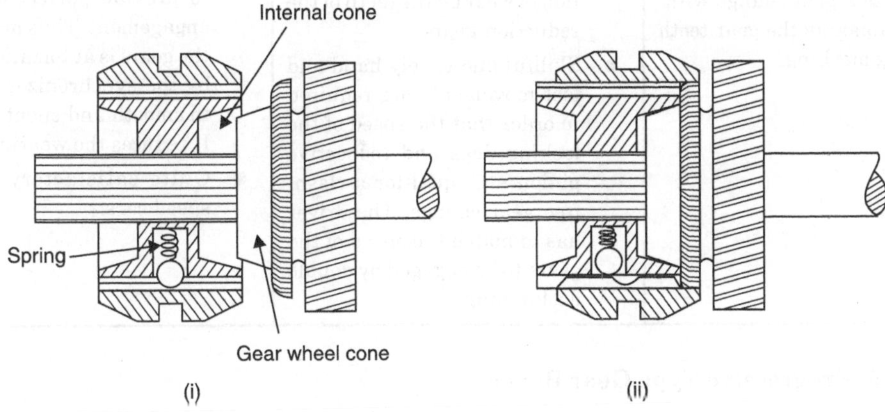

Fig. 7.27. Synchromesh unit.

Refer to Fig. 7.27. In Fig. 7.27 (*i*), the collar and gear wheel are shown close to each other. The cone of gear wheel and internal cone of the collar are in contact. The collar acquires the same speed as the gear wheel. In Fig. 7.27 (*ii*), a further movement of the collar makes the toothed outer ring slide into engagement.

- In most of the cars, the synchromesh devices are not fitted to all the gears. They are fitted only on the top gears. Reverse gear, and in some cases the first gear, do not have synchromesh device, since they are intended to be engaged when the vehicle is stationary.

Advantage. The synchromesh type of transmission has the big advantage of *allowing smooth and quick shifting of gears without danger of damaging the gears and without necessity for double clutching.*

The synchromesh gear box is *considered the most advanced and has been adopted in most cars.*

Comparison of Selective Type Gear Boxes

Sliding mesh year box	Constant mesh gear box	Synchromesh gear box
• Simplest design. • Meshing of gears takes place by sliding of gears on each other, *i.e.*, axially along the splined mainshaft to mesh with the corresponding gears on the layshaft. • Considerable skill is needed to make a gear change without damaging the gear teeth during meshing.	• Better design. • The mainshaft gear is provided with dog clutches providing a means of locking the freely, rotating mainshaft gears to the main shaft for making an easier gear change. • There is considerable reduction of wear on the teeth of the reduction gears. • Skillful and timely hand and foot movements are required in order that the speed of the locking dogs and respective pinion are equal for a clash-free gear change. The driver has to match the speed of the gears to be engaged by double declutching.	• Most advanced design. • It enables rapid gear changes to be made without noisy engagement or clashing of teeth. It also simplifies shifting action for the driver as the synchronizer itself adjust the members to the coupled so as to provide perfect dog teeth engagement. Thus matching of the gears is automatically done by the synchronizer resulting in smooth and silent coupling. It reduces the wearing of gear. • Quite satisfactory for high speed cars.

7.3.5. Progressive Type Gear Boxes

These gears boxes are usually used in motor cycles. In this type of transmission, the gears pass through the intervening speeds while shifting from one speed to another. There is a neutral position between two positions as shown below in the case of a four-speed gear box :

(4)	NEUTRAL	(3)	NEUTRAL	(2)	NEUTRAL	(1)

These gear boxes are a *combination of sliding and constant mesh gear boxes. The various gear speeds are obtained by sliding the dog clutch or gear to the required position.*

7.3.6. Epicyclic or Planetary Type Gear Box

This type of gear box/transmission *uses no sliding dogs or gears to engage but different gear speeds are obtained by merely tightening brake-bands on the gear drums.*

A planetary gear set (Fig. 7.28) consists of a *ring gear annular wheel, sun gear* and *planet gears with carrier.* In order to obtain different speeds any one of these units can be held from rotation by means of *brake bands.*

— The *ring gear* contains teeths on it inner circumference and is surrounded by a *brake band.* The brake band is operated by a gear stick or lever to grip the ring gear and hold its movement.

— The *sungear* is rotated by the *driving shaft from the engine* and thus moves along with the movement of the engine crankshaft.

— The *planet gears* are in constant mesh with both the sun gear and ring gear or annular wheel and are free to rotate on their axes carried by the carrier frame which in turn is connected to the *driven shaft*.

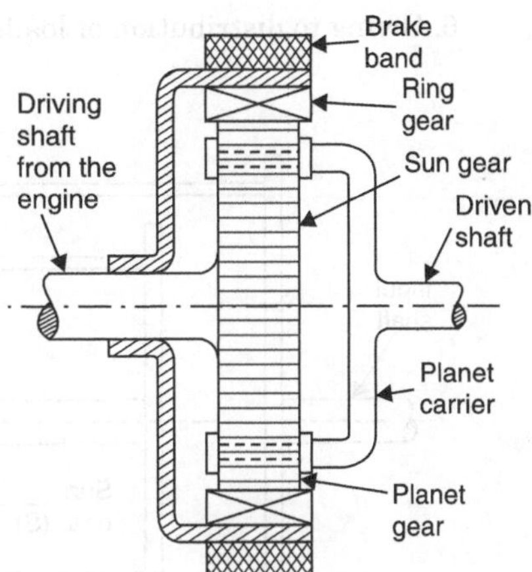

Fig. 7.28. Epicyclic gear set.

Working :

- When the ring gear is locked by the brake band, the rotating sun gear causes the planet gears to rotate. Since the ring gear cannot move, the planet gears are forced to climb over it. During this position, the *ring gear acts as track for the planet gears to move over*. The *driven shaft* which is connected to the planet gear carrier is thus *rotated*.
- When the ring gear is *released*, it is free to move in consequence to the rotation of planet gears which rotate around their axis. During this position, there is no movement of planet carries and hence the *driven shaft remains stationary*.

A planetary gear box contains a number of such units to obtain various speed reductions.

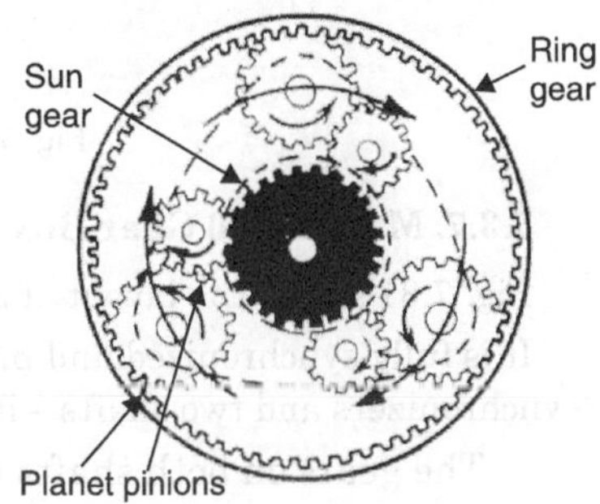

Fig. 7.29. Idler gear fitted in planetary system for reversing direction of rotation.

Fig. 7.29 shows idler gear fitted in planetary system for reversing direction of rotation.

Fig. 7.30 shows a typical two-speed epicyclic gear box. It is a two toward speed gear box.

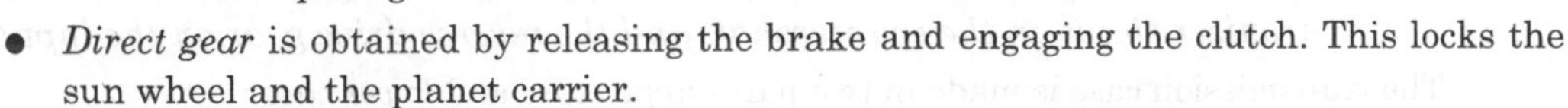

- *Direct gear* is obtained by releasing the brake and engaging the clutch. This locks the sun wheel and the planet carrier.
- *Lower gear* is obtained when brake is applied which locks sun gear S.

Advantages of epicyclic gear box :

Following are the *advantages* of epicyclic gear box :

1. It provides a *more compact unit operating about a common central axis*, because the planetary gear operate within a ring gear with its external surface of cylindrical form.

2. The planetary gears are in constant mesh and hence *dog clutches or sliding gears are not used.*

3. The gear and gear housings are comparatively *smaller in overall dimension.*

4. Instead of having the load on only one pair of gears, it is *distributed over several gear wheels.*

5. External contracting hand brackets or multiple clutches of relatively *small dimensions* are *used for changing the gears.*

6. Owing to distribution of loads, a generate area of *gear tooth contact can be used.*

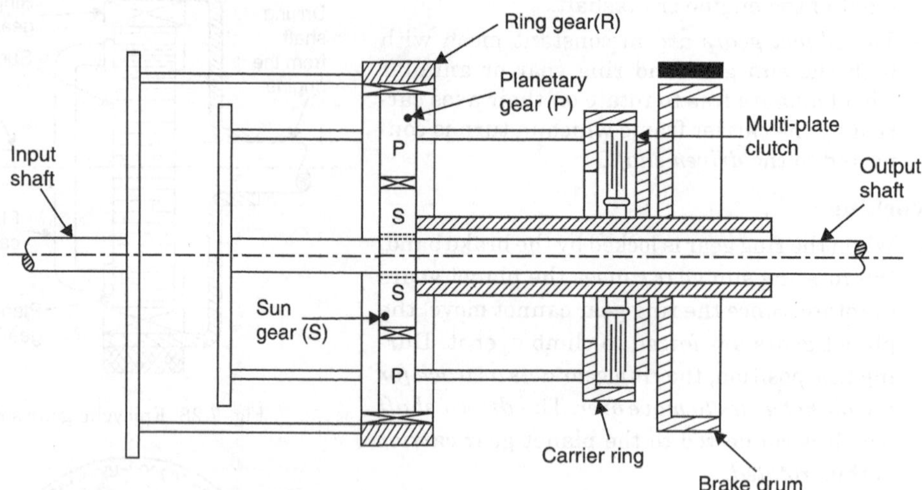

Fig. 7.30. Two-speed epicyclic gear box.

7.3.7. Maruti-800 Gear Box

Fig. 7.31 shows the layout of a Maruti-800 car gear box along with the differential.

It is fully synchronized and provides four forward speeds and one reverse speed by means of two synchronizers and two shafts—input shaft and countershaft.

— The gears on both shafts (input and counter) are always meshed.

— The *low speed synchronizer* on the countershaft is engaged either with the low driven gear or second driven gear, and the *high speed synchronizer* is engaged with the third driven gear or top driven gear.

— The *reverse idler gear* is of clash-meshing type and is *engaged with the low speed synchronizer sleeve on the countershaft and the reverse drive gear on the input shaft.*

The transmission case is made in two parts-*upper case* and *lower case* :

— The *upper case* houses the reverse shaft.

— The *lower case* has three fork shifting mechanisms, built into it.

In automobiles, transmission from engine to rear wheel flow through some forward and reverse gears, which are enumerated below :

1. **Low speed drive** —The drive is transmitted through the low drive gear on the input shaft and the low-driven gear on the countershaft to the gear of the final differential.

2. **Second speed drive** —The drive is transmitted through the second drive gear on the input shaft and second driven gear on the countershaft to the final gear of the differential.

3. **Third speed drive** —The drive is transmitted through the third drive gear on the input shaft and the third driven gear on the countershaft to the final gear of the differential.

4. **Top speed drive** —The drive is transmitted through the top drive gear on the input shaft and top driven gear on the countershaft to the final gear of the differential.

5. **Reverse drive** —The drive is transmitted through the reverse gear on the input shaft, reverse idle gear and low-speed synchronizer on the countershaft to the final gear of the differential.

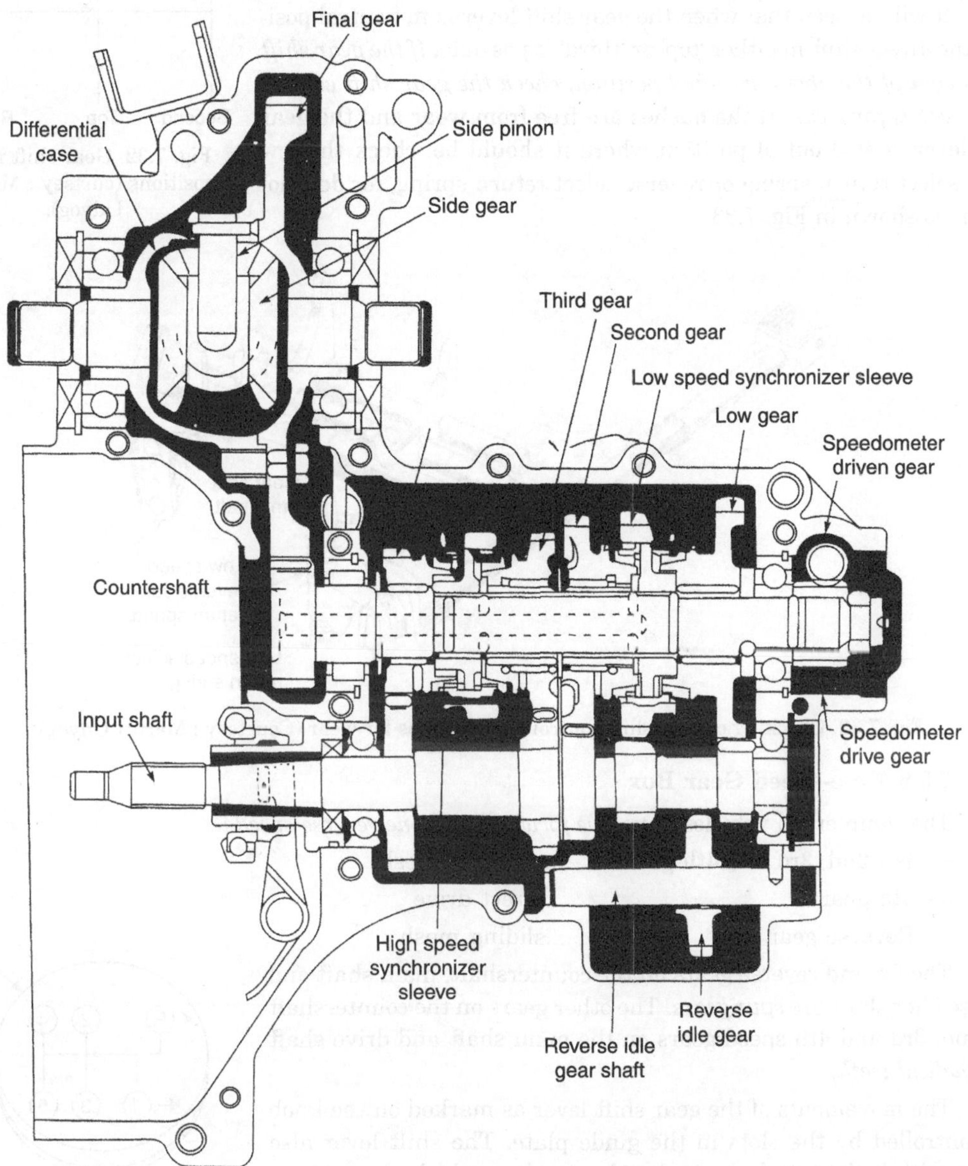

Fig. 7.31. Layout of Maruti-800 car gear box alongwith differential (Courtesy : Maruti Udyog Ltd.)

Fig. 7.32 shows the gear shift lever positions of the gear box used in maruti-800 car manufactured by M/S Maruti Udyog Ltd. in India.

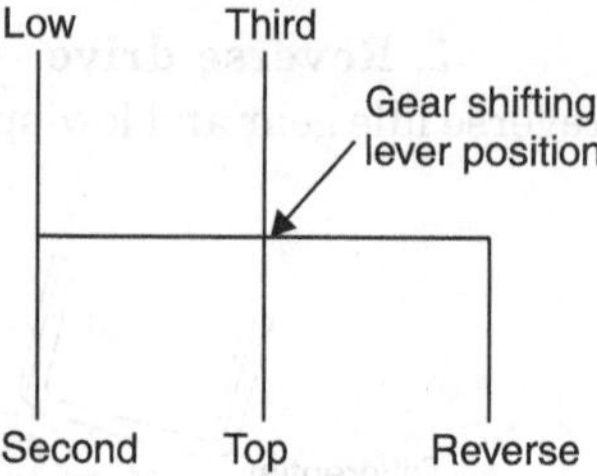

Fig. 7.32. Gear shift lever positions (curtsey : Maruti Udyog).

It will be seen that when the gear shift lever is in neutral position, the direct shift in either 'top' or 'third' is possible. *If the gear shift lever is out of the above specified position, check the gear shift control shaft bushes for wear.* If the bushes are free from wear and the gear shift lever is still out of position where it should be, check the low speed select return spring or reverse select return spring, for deterioration, as shown in Fig. 7.33.

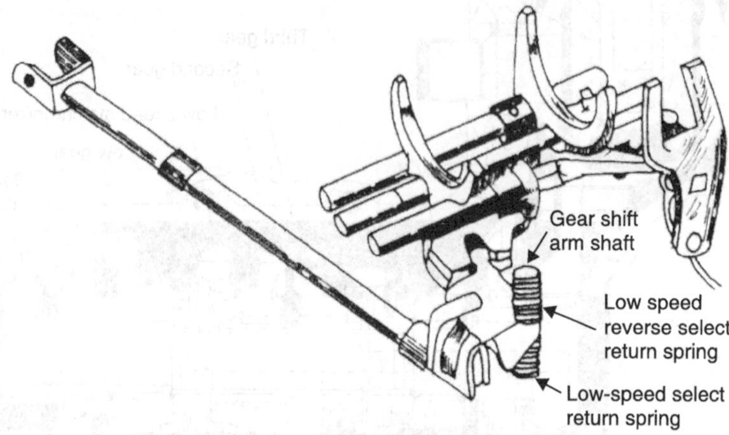

Fig. 7.33. Checking of gear shift control shaft bushes for wear (Courtesy : Maruti Udyog).

7.3.8. Five-speed Gear Box

The commercial vehicles have *five forward* and *one reverse speeds* :

— Ist, 2nd, 3rd and 4th gears ......constant mesh

— 5th gear ......direct drive

— Reverse gear ......sliding mesh

The Ist and reverse gears on the countershaft, main shaft and reverse idler shaft are *spur gears*. The other gears on the countershaft *i.e.*, 2nd, 3rd and 4th speed gears on the main shaft and drive shaft have *helical teeth*.

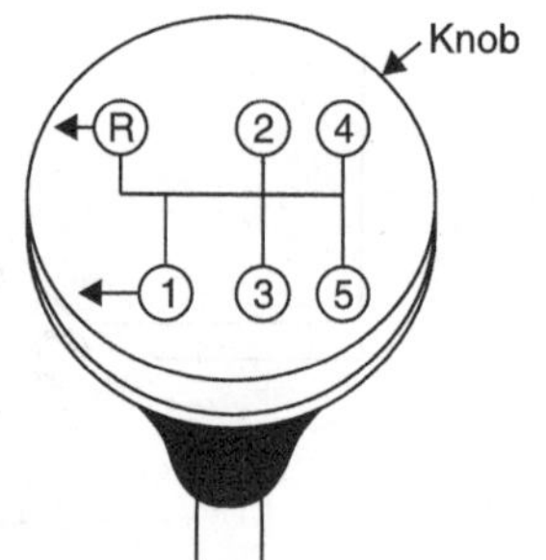

Fig. 7.34. Gear shift lever position (Courtesy : Telco).

The movements of the gear shift lever as marked on the knob are controlled by the slots in the guide plate. The shift lever also carries with it the interlocking plate having lugs which also slide in the slots of the guide plate. These lugs prevent simultaneous engagement of two gears at a time. The gear shift lever cannot be disengaged out of the reverse gear without a sidewise movement as indicated on the knob (Fig. 7.34).

- The gear box is fitted with a *magnetic drain plug*. The chips which collect around the magnet should be cleaned up while changing gear box oil.

- The transmission gear ratios, driving speeds and climbing ability of G-30 gears box are given below :

G-30 gears box	Ist gear	2nd gear	3rd gear	4th gear	5th gear	Reverse gear
Gear ratio	7.37 : 1	4.23 : 1	2.49 : 1	1.56 : 1	1 : 1	6.61 : 1
Speed in km/h	10.3	17.9	30.5	48.5	75.6	11.5
Climbing ability in % without trailer	23.5	12.9	7	4	2	—

7.3.9. Gear Shifting

Gear shifting is *the moving of the gears in transmission by means of the gear shift lever to change the gear mesh.*

In manually operated selective transmissions, the gear shift lever is located either on *steering column* or on the *floorboard* as shown in Fig. 7.35.

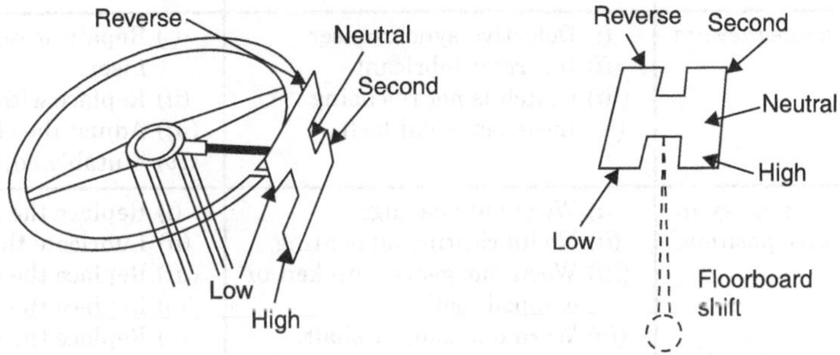

(*a*) Steering column gear shift lever (*b*) Floorboard gear shift lever

Fig. 7.35. Gear shift patterns.

To shift any gear, the clutch pedal is depressed to disconnect momentarily the engine from the transmission. Then the gear shift is moved first to demesh the previous gear and then to mesh the desired gear. The directions of the Ist, 2nd, 3rd, neutral and reverse gears are marked on the knob of gear shift lever.

7.3.10. Transfer Case

Transfer case is *a two-speed transmission located at the rear of the standard transmission which provides a low and direct gear, as in jeep.* It also *provides a means of connecting the power to the front axle.* The shifting mechanism is located on the transfer case.

On hard surface and level roads, disengage the front axle drive placing transfer case longer shift lever in forward position. The shorter shift lever controls the gear ratio—low and high. Low gear can only be engaged when the longer shift lever is in the engaged position for front drive.

- While operating vehicles in the 4-wheel drive more particularly in the low range, usually high noise is likely to be encountered is some vehicles. This noise may also be more pronounced at certain road speeds. This is *due to inherent resonance characteristics of the spur gear train used in transfer case. This noise is neither harmful in anyway nor will it affect the overall service life of the assembly.*

7.3.11. Lubrication of Gear Box

It is extremely important to ensure proper lubrication of the gear box. The transmission gears operate in a bath of lubricant. The gear box should always remain filled with lubricant. The lubricant prevents metal-to-metal contact (which would generate so much heat that the gear teeth would be burned away) and lubricates bearing of the gear box as well.

The oil level should be checked every week or every 1600 km. A reduction in level is very dangerous for the gears. Therefore, oil should be added whenever it falls below the required level. If the lubricant has become contaminated, the gear box should be drained, flushed and refilled with fresh lubricant.

7.3.12. Troubleshooting of Gear Box

Troubleshooting chart of gear box is given below :

	Symptom	Cause	Remedy
1.	*Oil leaks.*	(*i*) Too high lubricant level. (*ii*) Broken or missing gaskets. (*iii*) Damaged oil seals.	(*i*) Use just the right amount of oil. (*ii*) Replace the gaskets. (*iii*) Replace the oil seals.
2.	*Gears clash while they are shifted.*	(*i*) Defective synchroniser. (*ii*) Incorrect lubricant. (*iii*) Clutch is not releasing. (*iv*) Incorrect pedal lash.	(*i*) Repair or replace the synchroniser. (*ii*) Replace with correct lubricant. (*iii*) Adjust the clutch. (*iv*) Suitably adjust the pedal lash.
3.	*Transmission is noisy in the neutral speed position.*	(*i*) Worn out bearings. (*ii*) No lubricating oil bearing. (*iii*) Worn out gears ; broken or chipped teeth. (*iv*) Worn out counter shaft. (*v*) Bent or damaged counter shaft.	(*i*) Replace the bearings. (*ii*) Lubricate the bearings. (*iii*) Replace the damaged gears. (*iv*) Replace the worn out parts. (*v*) Replace the damaged parts.
4.	*Transmission sticks in gear.*	(*i*) Clutch is not releasing. (*ii*) Insufficient lubricant in gear box ; incorrect lubricant. (*iii*) Stuck synchronising unit.	(*i*) Adjust clutch pedal linkage from time to time. (*ii*) Add sufficient amount of lubricant ; replace with correct lubricant. (*iii*) Free the synchronising unit. Check for damage ; if some gears are damaged, replace them.
5.	*Transmission slips out of gear.*	(*i*) Worn out or defective synchronizer. (*ii*) Worn out bearings. (*iii*) Gear shift linkage is out of adjustment.	(*i*) Repair and, if necessary, replace the defective parts. (*ii*) Replace the bearings. (*iii*) Adjust the gear shift linkage.
6.	*Clutch spin, i.e.,* difficulty in changing the gear from neutral to first as a result of depressing the clutch pedal for a short period when the engine is idling.	(*i*) Pedal of clutch not travelling the full length. (*ii*) Wear on clutch plates or traces of oil on facings of clutch plates.	(*i*) Adjust the travel of the clutch pedal to give 2.5 cms free travel ; follow the specifications of the manufacturer. (*ii*) Remove the clutch outside, check and replace.

7.	*Worn or faulty gear selector mechanism.*	(*i*) The interlock pin of the third speed fork rod not moving freely. (*ii*) Bending of striking fork. (*iii*) Insufficient side clearance of striking fork pads.	(*i*) Replace the interlock pin. (*ii*) Remove the bend of the striking fork. (*iii*) Check for sufficient side clearance of the striking fork pads.
8.	*Development of noise on engagement of synchromesh gear.*	(*i*) Worn out synchronising ring and cone. (*ii*) Damaged synchronising ring or cone. (*iii*) Weak or broken springs in clutch hub.	(*i*) Replace the worn out parts. (*ii*) Replace the damaged parts. (*iii*) Replace the weak or broken springs.
9.	*Gear jumps out.*	(*i*) Worn out rollers and bearings in gear box. (*ii*) Worn out shafts of the selector mechanism. (*iii*) Incorrect setting of gear shift. (*iv*) Excessive wear of gear shift control.	(*i*) Replace the worn out parts. (*ii*) Replace the worn out parts. (*iii*) Set the gear shift correctly. (*iv*) Replace the gear shift control.

7.3.13. Gear Boxes used in Some Indian Automobiles

The comparative gear box data of some Indian automobiles is given below :

S. No.	Make of vehicle	Type of gear box	Gear ratio					
			I	II	III	IV	Top	Reverse
1.	*Ashok Leyland Comet Passenger*	Five speed and reverse	6.988 : 1	4.308 : 1	2.655 : 1	1.605 : 1	1 : 1	6.343 : 1
2.	*Tata mercedes Benz model 1210*	Five speed and reverse (*a*) Synchromesh type (*b*) Standard type	 8.82 : 1 7.37 : 1	 4.785 : 1 4.23 : 1	 2.736 : 1 2.49 : 1	 1.663 : 1 1.56 : 1	 1 : 1 1 : 1	 8.29 : 1 7.15 : 1
3.	*Dodge/Fargo*	Four speed and reverse synchromesh for second, third and top.	6.40 : 1	3.09 : 1	1.69 : 1	—	1 : 1	7.61 : 1
4.	*Hindustan Ambassador mark II, III*	For speed and reverse synchromesh for second, third and top.	3.807 : 1	2.253 : 1	1.506 : 1	—	1 : 1	3.807 : 1
5.	*Fiat 1100, Premier, President and Padmini*	Four speed and reverse synchromesh for second, third and top.	3.86 : 1	2.38 : 1	1.57 : 1	—	1 : 1	3.86 : 1
6.	*Jeep CJ-3B*	Three speed and reverse synchromesh for second and top.	2.798 : 1	1.551 : 1	—	—	1 : 1	3.798 : 1
7.	*Maruti 800*	Four forward all synchromesh and one reverse.	3.585 : 1	2.166 : 1	1.333 : 1	—	0.999 : 1	3.365 : 1

8.	*Maruti Gypsy*	Four forward all synchromesh and one reverse.	3.136 : 1	1.946 : 1	1.422 : 1	—	1.000 : 1	3.463 : 1
9.	*Fiat Uno*	Five speed all synchromesh with overdrive on five and one reverse.	3.909 : 1	2.238 : 1	1.280 : 1	1.029 : 1	0.838 : 1	—
10.	*Telco Indica (Petrol)*	Five speed all synchromesh with overdrive on five and one reverse.	3.420 : 1	1.950 : 1	1.360 : 1	0.950 : 1	—	—
11.	*Hyundai Santro*	Five speed all synchromesh overdrive on 4th and 5th and one reverse.	3.833 : 1	2.105 : 1	1.310 : 1	0.919 : 1	0.784 : 1	—
12.	*Daewoo Matiz*	Five speed all synchromesh and one reverse.	3.818 : 1	2.210 : 1	1.423 : 1	1.029 : 1	0.837 : 1	—
13.	*Maruti Zen*	Five speed all synchromesh and one reverse.	3.416 : 1	1.894 : 1	1.280 : 1	0.914 : 1	0.757 : 1	—

7.3.14. Automatic Transmission

Transmission in which various speeds are obtained automatically is known as **automatic transmission.** The driver merely selects the general car condition such as neutral, forward or reverse. The selection, timing and engagement of gear for the required gear speed are accomplished automatically when the accelerator is pressed or depressed.

Automatic transmission needs no gear change lever and clutch pedal since clutch and transmission is a combined unit and works automatically. (In *"semi-automatic transmission"* only the clutch is operated automatically). The transmission is operated hydraulically and is often known as *"hydramatic transmission"*. This transmission employs *fluid coupling, torque converter and epicyclic gear arrangement.* When all the different devices are combined into one unit, they perform their duties collaborately.

Automatic transmissions *operate basically by controlling the vehicle speed and engine load. Increasing vehicle speed needs changing gears upwards, whereas increasing engine load necessitate change from higher to lower gears. The engine load sensitive control unit is operated by the depression of the accelerator pedal.*

- In "**hydramatic gear box**", the planetary gear sets are placed *in series* so that power may be transmitted through them. A *centrifugal governor incorporated in the transmission selects the proper gear for each speed and throttle position.* The change from one gear to another is accomplished through *hydraulically operated pistons assisted by springs* which control brake bands on the planetary gear sets and clutches within the planetary unit. *The speed at which the various shifts occur, is governed by the throttle and centrifugal governor.*

Driver control is provided by a lever and sector segment which are usually located on the steering column, below the steering wheel. The control could be adjusted to any of the following *positions* :

Neutral – Drive – Low – Reverse

— In the *"drive position"*, the transmission shifts automatically for any one of the *forward speeds.*

— The *"low position"* is for *hill climbing* or muddy road or under heavy load, and only first and second gear speeds are provided.

— *"Reverse position"* is for *driving back.*

— In *"neutral position"*, there is no power transmission.

Advantages of automatic transmission :

Following are the *advantages of automatic transmission :*

1. Smoother running under all conditions due to hydraulic coupling and automatic gear change.
2. Noise free gear shifting.
3. No shocks or jerky driving.
4. Simplified driving control.
5. Less physical and mental fatigue to the driver.
6. Improved acceleration in hill climbing.
7. Less wear and tear due to planetary gearing.
8. No clutch pedal and gear lever and hence simplification of driver's compartment.
9. Reduced fuel consumption.
10. Longer life.

7.3.15. Overdrive

***Overdrive** is a device interposed between the transmission (gear box) and propeller shaft to permit the propeller shaft to turn faster than, or overdrive, the transmission main shaft.* It is so called because it provides a speed ratio *over* that of the high speed ratio.

- The overdrive permits the engine to operate only about 70 percent of the propeller shaft speed, when the vehicle is operating in the high speed ranges.
- The overdrive is suited to high powered cars employing three-speed gear boxes, since in order to produce flexible top gear performance a low gear final drive may be necessary, resulting in the engine running faster as high speeds than is desired.
- *Generally an overdrive is fitted to the top gear only*, but some sport cars have an over drive on 2nd, 3rd and top gear giving seven forward speeds.
- Overdrive is usually employed *to supplement conventional transmission*. It is *bolted to the rear of the transmission between the transmission and propeller shaft.*
- A *slightly higher rear-axle gear ratio* is employed with an overdrive than without one.

Advantages. The main *advantages* of an overdrive are :

1. Prolonged engine life since the engine turns slower for any given speed.
2. Fuel economy.
3. Reduced vibration and noises.

Construction and working of overdrive :

Fig. 7.36 shows the simplified diagram of an "*overdrive*".

Construction. It consists of an "*epicyclic gear train*" in which the sun gear is free to rotate on the input shaft, while the *carrier* can move on splines on the input shaft. A "*freewheel clutch*" is also fitted on the input shaft splines. The output shaft is connected to the ring-gear.

Working :

— When the sun gear is locked with the casing (*i.e.*, it becomes stationary), the speed of the output shaft is increased *i.e., overdrive is engaged*. However, when the sun gear is locked to the carrier to the ring gear, solid drive through the gear train is obtained.

— When the sun gear/wheel is kept free to rotate on the input shaft there is a direct drive through the freewheel clutch when the engine develops power. However, when the accelerator pedal is brought to zero position and the engine is simply idling, the output shaft will tend to override the input shaft. The rollers of freewheel clutch in this case no longer remain wedged and the car free wheels. Thus *for gear changing one simply has to lift his foot off the accelerator pedal ; the clutch need not to be operated.*

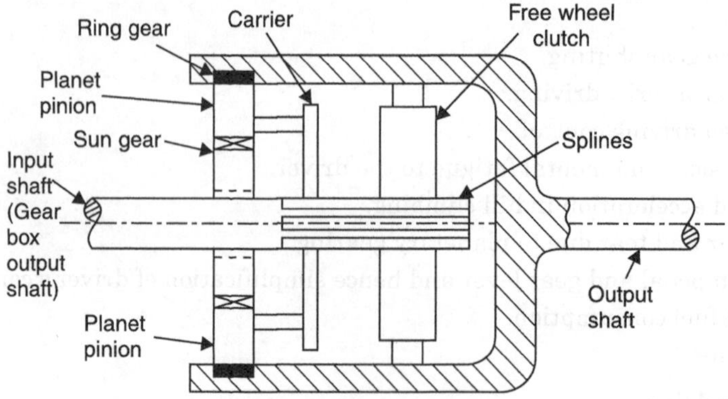

Fig. 7.36. An overdrive.

Fig. 7.37 shows a *cut-away view of the* ***overdrive assembly*** *when in position for overdrive operation.*

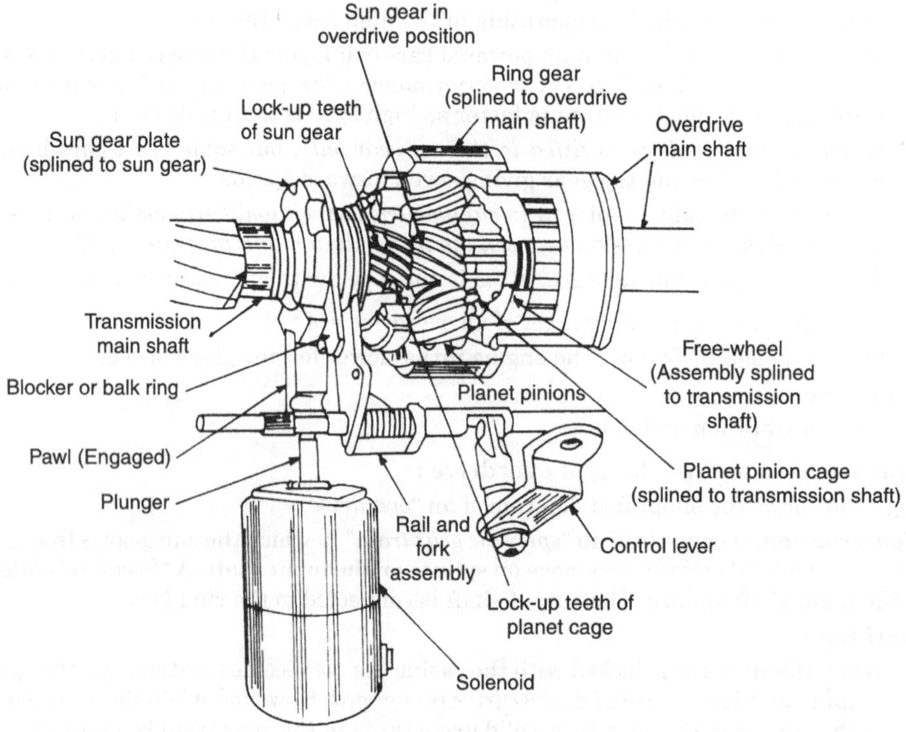

Fig. 7.37. Overdrive assembly.

Operational details :

- When the vehicle speed reaches 50 km/h, the electrical device operates and completes the operation to shift into overdrive. At the same time, the indicator lamp on the dashboard lights. This *speed at which the overdrive can be engaged* (in other words, the speed at which the indicator lamp lights) is called the ***"cut-in-speed"***.

- When the engine speed is dropping, the overdrive will not disengage at 50 km/h. There will be a difference of about 8 km/h, in other words, the overdrive will not disengage automatically until the speed drops to 42 km/h. This *speed at which the overdrive disengages automatically* is called the ***"cut-out speed"***.

There is a difference of 8 km/h between the cut-in and cut-out speeds and this difference is called "***hystresis***". Unless this difference is provided, a car running at 50 km/h will be subjected to complicated gear changes, alternating continuously between overdrive and normal top speed.

- When the control lever in front of the driver is pulled out, the lockout switch actuates. The overdrive electric circuit will be switched off and the overdrive will not be engaged even if the car speed rises above the cut-in-speed. The car will be running under normal direct drive.
- If while driving under overdrive, quick acceleration is desired or a hill has to be climbed, that is, if a large drive power is required, depressing the accelerator pedal all the way down will enable going out of overdrive automatically and allow the car to run under direct drive. This is called ***kick-down.***
- If the accelerator pedal is depressed all the way down while driving at high speed (about 110 km/h), the overdrive *will not disengage* because of the anti-kick-down device provided in the overdrive relay. The *purpose of this is to prevent the engine from over-running.*

Freewheel unit :

Freewheel unit, also known as *over running clutch, sprang clutch or one way clutch, is an essential part of every overdrive.* It transmits power in *one direction only*–from the transmission main shaft to the output shaft.

It enables the vehicle to freewheel or overrun while going doing a gradient or when the accelerator is released. It does not prevent the vehicle to be driven positively in forward direction. It is fitted at the back of the gear box and in case an overdrive is attached to the gear box, it is either incorporated in the overdrive or attached at its back. In every case, it is fitted between the transmission and the propeller shaft.

It works on the same principle as the freewheel in a bicycle.

Fig. 7.38 shows a free-wheeling device. It consists of a central *driving member* which is attached to the splined shaft forming rear end of the gearbox output shaft. An annular outer *driven member* is mounted over the driving member. The driven member is connected to the propeller shaft.

— In the *positive drive, the roller are forced to the right by the spring pad.* They become *wedged* between the driving and driven members as a result of spring pressure and driving effect of driving member.

— When the *driven member over-runs the driving member,* the rollers move to the *left* and the drive between them is disconnected.

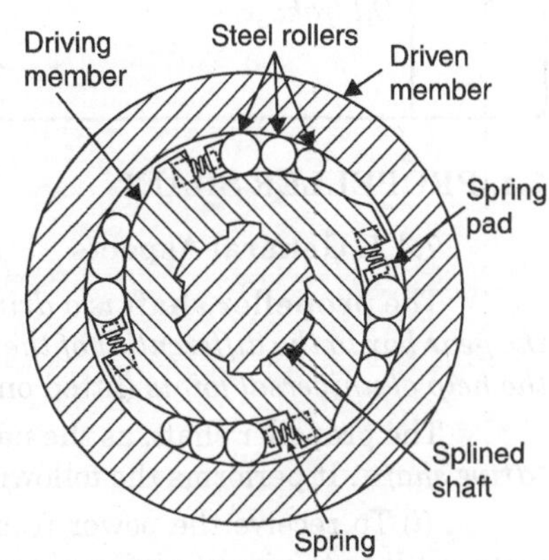

Fig. 7.38. Freewheel unit.

The freewheel unit is *locked through a dog-clutch type arrangement.*

- Freewheel is also used as *essential component of torque converter* in modern automatic transmission.

Overdrive troubles and their causes :

S. No.	Troubles	Causes
1.	*Not going into overdrive.*	(*i*) Defective wiring. (*ii*) Defective kick down switch. (*iii*) Defective solenoid. (*iv*) Defective governor. (*v*) Defective relay. (*vi*) Defective overturning clutch, jammed gears, excessive shaft end play. (*vii*) Linkage to control knob on dash out of adjustment.
2.	*Not coming out of overdrive.*	(*i*) Defective wiring. (*ii*) Defective kick-down switch. (*iii*) Defective solenoid. (*iv*) Jammed pawl. (*v*) Jammed sun gear.
3.	*Not shifting into reverse and jammed overdrive knob.*	(*i*) Defective solenoid. (*ii*) Defective relay. (*iii*) Defective kick down switch. (*iv*) Defective wiring. (*v*) Jammed sun-gear. (*vi*) Jammed pawl. (*vii*) Grounded governor. (*viii*) Grounded reverse lockout switch. (*ix*) Linkage to dash knob out of adjustment.
4.	*No power through overdrive.*	(*i*) Slipping over-running clutch. (*ii*) Broken planetary parts.
5.	*Noise in overdrive.*	(*i*) Worn gears. (*ii*) Worn main shaft bearing. (*iii*) Worn over-running clutch parts.
6.	*Oil leakage.*	(*i*) Defective or broken gasket or oil seals. (*ii*) Loose mounting. (*iii*) Excessive lubricant.

7.4. PROPELLER SHAFT

7.4.1. General Aspects

The **propeller shaft** *is a driving shaft that connects the transmission main or output shaft of the gear box to the differential of the rear axle. It transmits the power from gear box to rear axle with the help of universal joints* (fitted on both the ends of the shaft).

The propeller shaft, as the name suggests, *propels the vehicle.* It is also sometimes called the "*drive shaft*". It performs the following two *functions* :

(*i*) To receive the power from the gear box output shaft and without any change in speed transmit it to the input pinion of the differential for onward transmission to the rear axle and rear wheels.

(*ii*) To cope with the difference in line with the level of the gear box output shaft and the differential input pinion shaft.

The propeller shaft has to *operate at varied lengths and varied angles*. The engine of the automobile is somewhat rigidly attached to the frame by springs. As the vehicle moves on the road, there are jerks and bumps due to which the springs expand and contract. This changes the angle of drive between the propeller shaft and the transmission shaft. *The distance between the gear box and differential also changes due to the movement of the springs.*

The propeller shaft has to meet both of the changes *i.e.*, the *angular change* coming between the gear box and differential and the *change in length of the distance* between the gear box and differential. The propeller shaft is also not on the same line as the gear box output shaft but it runs to the rear axle at an angle because the level of the rear axle is lower than the gear box.

— To *adjust 'angular motion'*, ***universal joints*** *are provided.*

— To *adjust for the change in length of the propeller shaft, a* ***slip joint*** *is provided.*

- The propeller shaft has *to withstand the torsional stresses* of the transmitting torque, and yet *it must be light and well balanced so that vibrations and whip will not occur at high speeds*. For these reasons, it is made of a *strong steel* (*alloy*) *tube*. Solid propeller shafts are also used.
- The propeller shaft may be exposed to the atmosphere or protected by an outer tube.
- Some applications include bearings at or near the propeller shaft centre to support the shaft. In some applications, the propeller shaft is in two sections, supported by a centre bearing and coupled together by universal joint.

Fig. 7.39 shows a propeller shaft with *sliding joint* and *universal joints*.

Fig. 7.40 shows the fixing of propeller shaft to centre cross member. In the centre, cross member rubber rings are used on the centre bearing to take care of angular movement of shaft and also to absorb vibration.

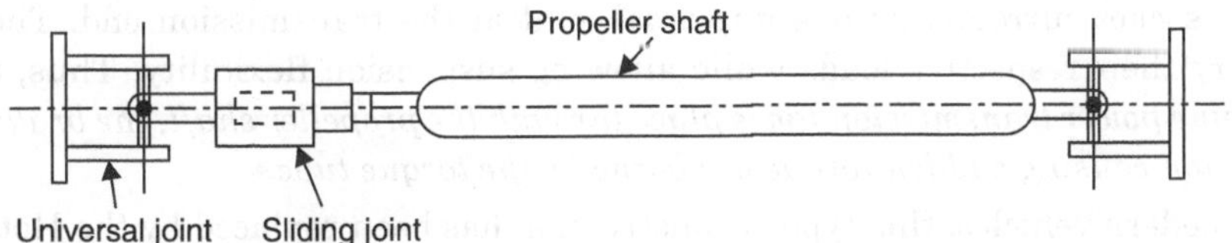

Fig. 7.39. Propeller shaft with sliding joint and universal joints.

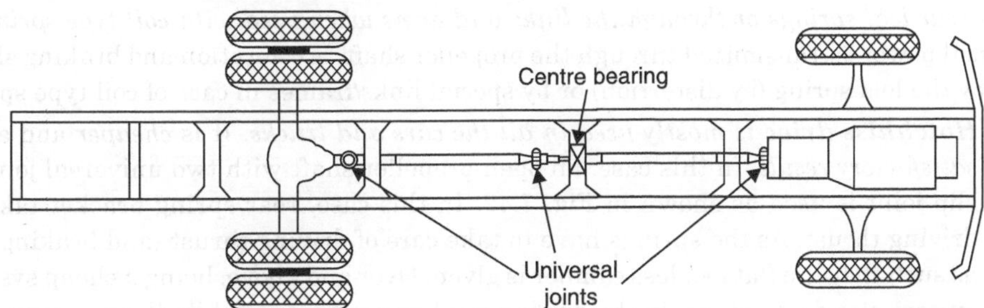

Fig. 7.40. Propeller shaft with centre bearing and universal joints.

7.4.2. Constructional Features of Propeller Shaft

The propeller shafts are subjected to driving and braking torques. When a front engine rear drive car starts from rest or is suddenly braked to slow down, the shock to the transmission is cushioned by the long propeller shaft which twists slightly and then untwists. These forces are taken care of by the following *two types of construction of propeller shaft* :

1. Torque tube type propeller shaft.
2. Hotchkiss type propeller shaft.

1. **Torque tube type propeller shaft :**

Fig. 7.41 shows a torque tube type propeller shaft.

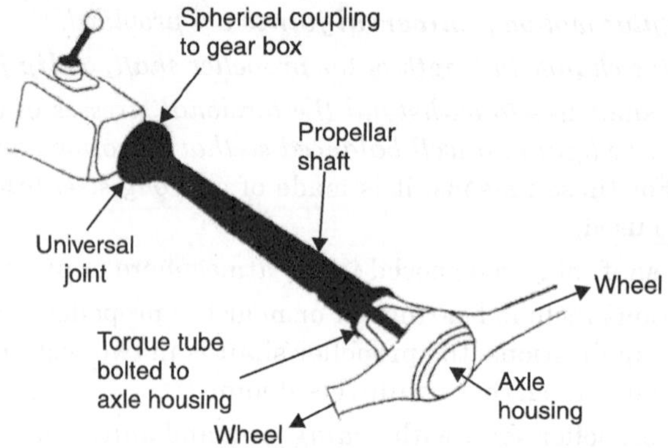

Fig. 7.41. Torque tube type propeller shaft.

- A torque tube is a large diameter tube fastened securely to the rear axle housing and *completely enclosing the propeller shaft.* The torque tube is fitted into a spherical ball and socket surrounding one universal joint at the transmission end. These two units carry their respective loads while allowing suspension flexibility. Thus, *while normal engine power transmission takes place through the propeller shaft, the braking and acceleration causing sudden torque are borne by the torque tube.*
- In modern vehicles, this type of construction has been replaced by the Hotchkiss drive.

2. **Hotchkiss type propeller shaft :**

In this type of propeller shaft, the driving and braking torques are *absorbed through the front half of the rear leaf springs or through the links and arms when used with coil type springs.* Thus while normal power is transmitted through the propeller shaft, acceleration and braking shocks are taken up by the leaf spring (by distortion) or by special links/frames in case of coil type springs.

- ***Hotchkiss drive*** *is mostly used in all the cars and trucks.* It is *cheaper* and also gives *satisfactory result.* In this case, an open propeller shaft with two universal joints and a slip joint is used as shown in Fig. 7.42. In this case, rear spring bracket take care of driving thrust. As the springs have to take care of driving thrust (and braking torques) as such they are flat *i.e.,* less camber is given. Over and above, being a cheap system, this system also does not strain the driving mechanism because of its flexible connection. In this system, the axle housing can rock slightly when sudden load is applied in letting the clutch in or when suddenly or quickly applying the brake.

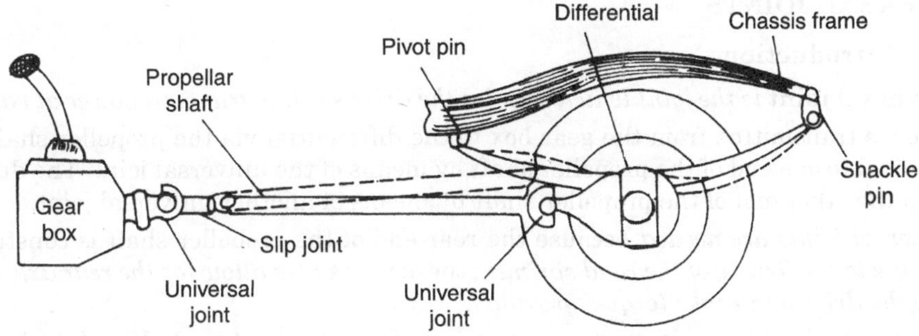

Fig. 7.42. Hotchkiss drive.

7.4.3. Troubleshooting of Propeller shaft

1. **Propeller shaft shakes :**

The shaking of propeller shaft is sometimes experienced at high speed with the result that it becomes difficult to drive at high speed as riding is uncomfortable.

To check up the cause of shaking, jack up the rear end of the vehicle, accelerate the engine in gear and inspect the propeller shaft rotation. Do not forget to lock the front wheels as safety measures.

The *reasons for shaky propeller shaft can be as under.*

(*i*) *Propeller shaft and splined yoke coupling are not properly connected.*

Remedy : Assemble the splined yoke so that yoke on rear end of the shaft and propeller shaft are in the same plane.

(*ii*) *Bent propeller shaft.*

— In case the shaft is badly bent, it should be replaced with new one.

(*iii*) *Worn out needle bearing of universal joint.*

— In case of slightly worn out needle bearing, replace the needle bearing which comes duly assembled in cap. In case, wear is more, replace the universal cross with new bearing.

2. **Noisy running of propeller shaft :**

While running the vehicle some noise is audible from propeller shaft ; the various *reasons* may be as under :

(*i*) *Slip joint splines worn out.*

— Measure the play with the help of dial gauge in direction of rotation ; in case play is more than 0.5 mm, replace the splined shaft and splined yoke.

(*ii*) *Universal joint being worn out.*

— Replace universal joint assembly.

(*iii*) *Loose flanged yoke.*

— Tighten it fully.

(*iv*) *Centre bearing loose or worn out.*

— If the bearing fitted in centre bracket is loose or its rubber cushion is worn out, it will amount to noisy running – replace worn out parts.

(*v*) *Centre bearing housing misaligned.*

— Align the centre bearing housing, tighten it holding down bolts.

7.5. UNIVERSAL JOINTS

7.5.1. Introduction

Universal joint *is the joint which enables the drive shaft to transmit power at varied angles.*

Power is transmitted from the gear box to the differential via the propeller shaft. The gear box is connected to one end of the propeller shaft by means of the universal joint. The differential is connected to the other end of the propeller shaft by means of another universal joint.

Universal joints are needed because the rear end of the propeller shaft is constantly rising and falling *due to the flexing of the road springs* ; these joints *also allow for the rear axle assembly to twist due to the drive and brake torque operation.*

The universal joint permits the torque transmission not only at angle, but also while this angle is changing.

7.5.2. Construction and Working of Universal Joint

Construction. Fig. 7.43 shows a simple universal joint. It consists of two Y-shaped *yokes*, one on the driving shaft and other on the driven shaft. The four arms of spider, known as *trunions*, are assembled into bearing in the ends of the two shaft yokes. The driving shaft and the driven shaft are at an angle to each other, the bearings in the yokes permit the yokes to swing around on the trunions with each revolution.

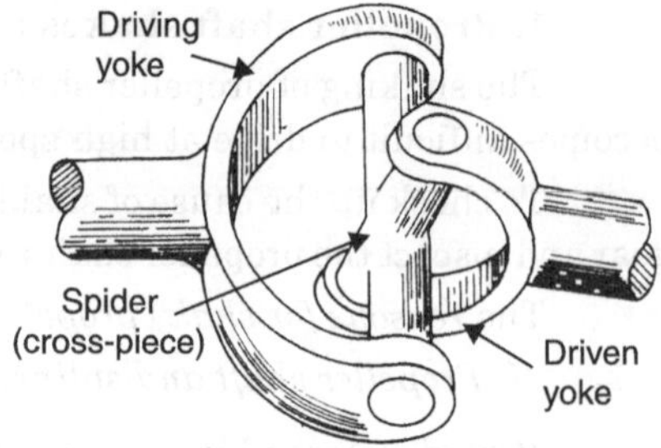

Fig. 7.43. Universal joint.

Working :

- When the driving shaft is rotating, the driven (propeller) shaft also rotates. At the same time the universal joint permits angular motion and propeller shaft can rotate at any angle. Thus the power is transmitted from the gear box to the propeller shaft at any particular angle.

A *slip joint* is attached to the driven yoke in order to increase or decrease the propeller shaft. There are splines on the outer side of the slip joint. These move along the internal splines of the propeller shaft. Due to this drive line is increased or decreased. Fig. 7.44 shows the position of the propeller shaft when the differential falls down. The length of the drive line is decreased.

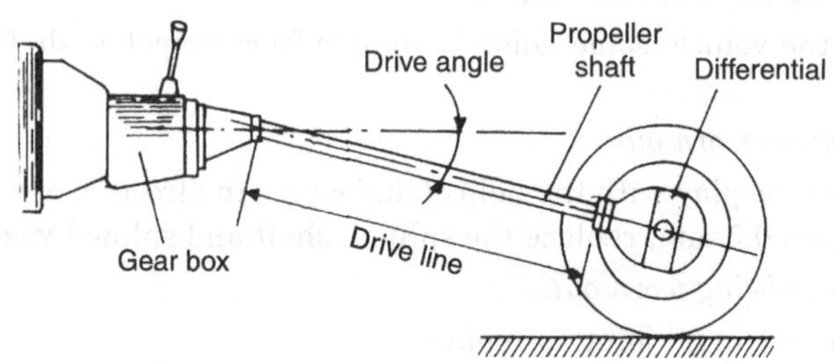

Fig. 7.44. Length of drive angle decreased.

- A simple universal joint does not transmit the motion uniformly when the shafts are operating at an angle, except in constant velocity type universal joint. Because the pivot pins do not revolve in the same plane, the driven shaft will increase to maximum and decrease to a minimum twice in each revolution. Although the degree of variation is a small, however, it may be minimized by the *use of two universal joints.* The two joints are arranged so that the non-uniform rotation of each joint tends to neutralize that of the other, as shown in Fig. 7.45.

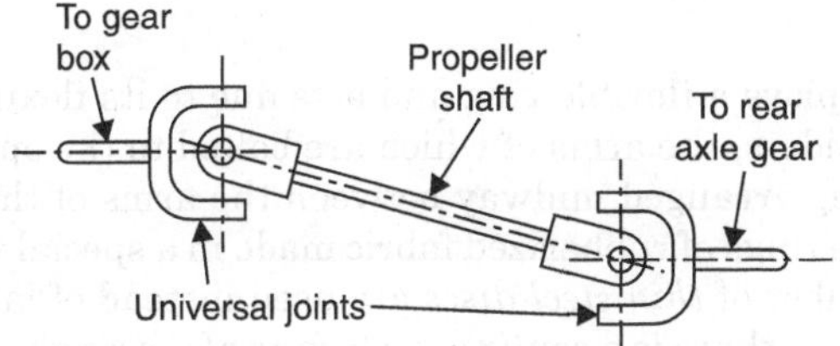

Fig. 7.45. Use of two universal joints.

7.5.3. Types of Universal Joints

Universal joints can be classified as under :

1. *Variable velocity joints :*

(*i*) Cross or spider type.

(*ii*) Ring type.

(*iii*) Ball and trunion type.

2. *Constant velocity joints :*

(*i*) Rzeppa

(*ii*) Bendix Weiss

(*iii*) Tracta.

1. Variable velocity joints :

- In these joints, the driving and driven shafts do not turn at the same speed through each part of a revolution although they turn at the same r.p.m. The driving and driven shafts should therefore be in a straight line so that they may turn at the same speed through each part of a revolution. But in an automobile, it is not feasible as the drive shaft is inclined. When there is an angle between the driven and driving shafts, the driven shaft turns lower than the driving shaft through half a revolution and faster than the driving shaft through the other half revolution. Thus the average speed of the driven shaft is equal to the driving shaft. *The speed variation in the driven shaft increases when the flex of angle of the universal joint is increased.* It is owing to this fact that *variable velocity joints are usually employed when the flex angle is small.*
- When two variable velocity universal joints are used in one drive line, the *yokes on the shafts connecting the universal joints should be in the same plane. It helps in balancing the speed variation.*

(*i*) ***Cross or spider type (Hooke's joint) :***

Refer to Figs. 7.42 and 7.43.

- It consists of a cross piece or spider and two yokes, therefore, it is known as cross type of spider and two-yoke type universal joint. There are four needle bearings, one for each trunion of the spider. The bearings are held in place by rings heat drop into under cuts in the yoke-bearing holes.
  — One commercial design of the cross type universal joint incorporates a *slip joint.* One yoke is integral with the hub that holds the female end of the slip joint. When the joint is used between the propeller shaft and rear axle gear shaft, the slip joint is omitted so that a direct connection is made between the two units.
- These types of joints are *mostly used with the drive shafts—"Hotchkiss drive'.*

(*ii*) ***Ring type :***

A ring type joint employs a flexible ring and acts due to its flexing. The shafts are provided with two or three armed spiders, the arms of which are bolted to the opposite faces of flexible ring, the arms of one spider being arranged midway between the arms of the other. The *flexible ring* is usually made of one or more rings of rubberized fabric made in a special way for providing necessary strength. Sometimes a number of *thin steel discs* are used instead of fabric rings. When the shafts are revolving about their axes, there is a *continuous flexing of the ring to enable drive through varied angles.*

Merits. (*i*) Can accommodate a considerable amount of axial movement of the shaft.

(*ii*) Helps in smoothing out torque fluctuations.

(*iii*) Needs no lubrication.

Demerit. Its main drawback is that the ring does not withstand for a long period.

(*iii*) ***Ball and trunnion type :***

- This joint combines both the universal and slip joint in one assembly. It allows axial movement within the joint, so no separate joint is necessary. The balls are mounted on a spider through needle bearings. Each ball fits into a partly cylindrical housing bore that runs in an axial direction. This provides a means to drive through the joint in a radial direction and allows slip in the axial direction.
- Later developments in this type are a *two ball and trunion joint and three ball and trunion joint.* In the latter, the use of three equally spaced balls causes the joint to transfer constant velocity motion to the output shaft. This type of joint is *ideal for the inside joint on front wheel drive vehicles because it eliminates the slip joint and transfers constant velocity motion.*

2. **Constant velocity joints :**

- In constant velocity joints, the *driven shaft turns at the same speed as the driving shaft, through each part of its revolution and at any degree of flex.*
- These joints are mostly used in *front drive axles* where the joint must transmit power through a large scale and at the same time not introduce a speed variation or vibration which would cause steering difficulty and excessive tyre wear.

(*i*) ***Rzeppa.*** It consists of spherical inner and outer ball races which have grooves cut parallel to the shaft. Steel balls are held in the grooves on the spherical recess. The torque is transmitted from one race to another by the balls. The circular pattern of balls results in both shafts to turn at the same velocity.

(*ii*) ***Bendix-Weiss :***

- It is a constant velocity joint and *also acts as a slip joint.* It consists of two spiders, each having three or four arms. Each spider arm is located between the adjacent arms of the second spider. The motion of one set of spider arms is transferred to the other spider arms by steel balls wedged in grooves that are cut at an angle so that the balls always operate in a plane that bisects the joint's operating angle. Thus it operates at a constant velocity.
- These joints are usually *used on front wheel drive vehicles*, where steering is another universal joint requirement.

(*iii*) ***Tracta :***

- In this joint there are four yokes ; two of them fastened to the shafts and the other two floating in the centre of the joint. The mating parts of the yokes are segments of a circle. The circular segments plus the floating action of the two yokes result in a constant velocity joint.
- It differs from the above two joints as it *separates off under tension or pull.*

7.6. FINAL DRIVE AND DIFFERENTIAL

7.6.1. Final Drive

The **final drive** *is the last stage in transferring power from the engine to the road wheels.* (It provides a *fixed reduction* between the drive shaft and the driving axles)

- It *brings down the speed* of the propeller shaft to that of the *wheels.*
- It *turns the drive* of the propeller shaft through a *right angle to drive the wheels.*

Fig. 7.46 shows a simplified diagram of the final drive system.

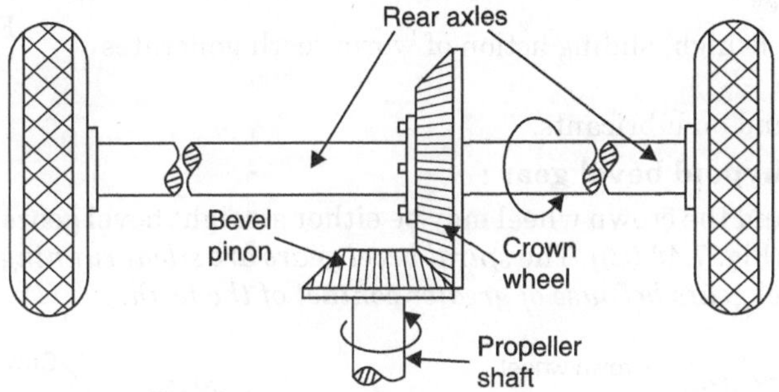

Fig. 7.46. Final drive system.

The propeller shaft has a small bevel pinion which is in mesh with the crown wheel. This crown wheel rotates the rear axles. The axles are in half portions.

— Since the pinion is bevel one, therefore, it rotates the crown wheel at right angles. The crown wheel in turn rotates the rear axles. Thus the rotation of the propeller shaft is such that it is converted at *right angles to the rear axle.*

— The crown wheel is larger than the bevel pinion, therefore, the speed of the crown wheel is *lower* than the speed of the bevel pinion.

Classes of final drive :

Final drive is of the following two main classes :

1. *Chain type and*
2. *Gear type.*

(*i*) Worm and wheel type

(*ii*) Bevel and hypoid bevel gear type.

1. **Chain type.** This type of final drive is obsolete in motor cars and trucks now-a-days. In this type, the drive wheel is connected with the gearbox by means of chains and sprockets. Motor cycles employ this type of drive.

2. **Gear type.** The gear type final drive consists of a *ring gear* and a *drive pinion*. The ring is riveted with the differential cage and the drive pinion is connected with the propeller shaft. Power from the propeller shaft flows towards the axle shafts through the drive pinion and ring gear.

(*i*) **Worm and wheel type.** Refer Fig. 7.47. The worm and worm wheel arrangement for final gear reduction is extremely strong. It gives high reductions without using large gears. Although the worm drives are expensive and heavier than bevel gears, their mechanical efficiency can be high provided great care is taken in their design and manufacture.

Advantages :

(*i*) High reduction ratio in single reduction, as high as 15 : 1.

(*ii*) High ground clearance—when worn is over slung.

(*iii*) Low drive line when the worm is under slung.

(*iv*) Ease of providing for a through drive to a second axle in tandem with the first.

(*v*) This arrangement runs silently.

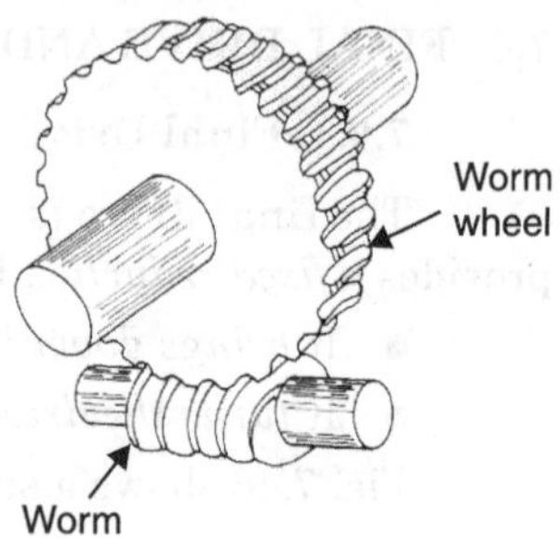

Fig. 7.47. Worm and worm wheel.

Disadvantages :

(*i*) High cost of manufacture.

(*ii*) Low efficiency.

(*iii*) If gear ratio is high, sliding action of worm teeth generates lot of heat.

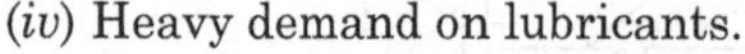

(*iv*) Heavy demand on lubricants.

(*ii*) **Bevel and hypoid bevel gear :**

- The pinion and the crown wheel may be either straight bevel gears (Fig. 7.48 (*i*)) or spiral bevel gears (Fig. 7.48 (*ii*)). The *spiral bevel gears are silent running and stronger than the straight bevel gears because of greater contact of the teeth.*

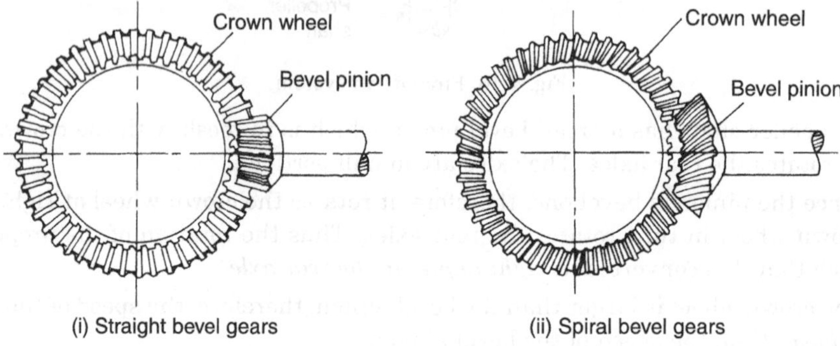

Fig. 7.48. Bevel gears.

- The spiral gears are used on some motor cars and commercial vehicles, although they are being replaced by the hypoid bevel gears.
  — In hypoid bevel gears (teeth are cut in a hyperbola curve) the pinion shaft is placed *below the axis of the crown wheel.* This permits a lower position of the propeller shaft, thus allowing a chassis height.
- The heavier type commercial vehicle having double reduction gearing often has a spiral bevel for the first reduction and double helical gearing for the final reduction. In double reduction, however, the size of the larger gear becomes too big.

The bevel and hypoid bevel final drives have the following *advantages :*

(*i*) Low sliding friction.

(*ii*) Low cost of manufacture.

(*iii*) Higher efficiency.

7.6.2. Differential

Differential *is the mechanism by means of which outer wheel runs faster than the inner wheel while taking a turn or moving over upheaval road.*

Differential is a part of the inner axle housing assembly, which includes the differential, rear axles, wheels and bearings. The differential consists of a *system of gears arranged in such a way that connects the propeller shaft with the rear axles.*

The **purpose** *of the differential is to provide the relative movement to the two rear wheels when the car is taking a turn. The torque transmitted to each wheel is, however, always equal.*

Differentials are *used* in rear drive axle of front-engine, rear-wheel drive vehicles. Differentials are also used in the trans axles on front-engine, front-wheel drive wheels. Also, four-wheel-drive vehicles have differential at both the front and rear wheels. In addition, some four-wheel-drive vehicles have a third differential in the transfer case.

Construction of differential :

Fig. 7.49 shows the various parts of a differential unit. The bevel pinion is fixed to the propeller shaft which rotates the crown wheel. The crown-wheel has another unit called the differential unit. It consists of two bevel gears (sun gears) and two bevel pinions (planet pinions). The bevel gears are in contact with the half shafts of the rear axle.

When the crown wheel is rotating, it rotates the differential unit. The bevel (sun) gears of the differential rotate the two half shafts.

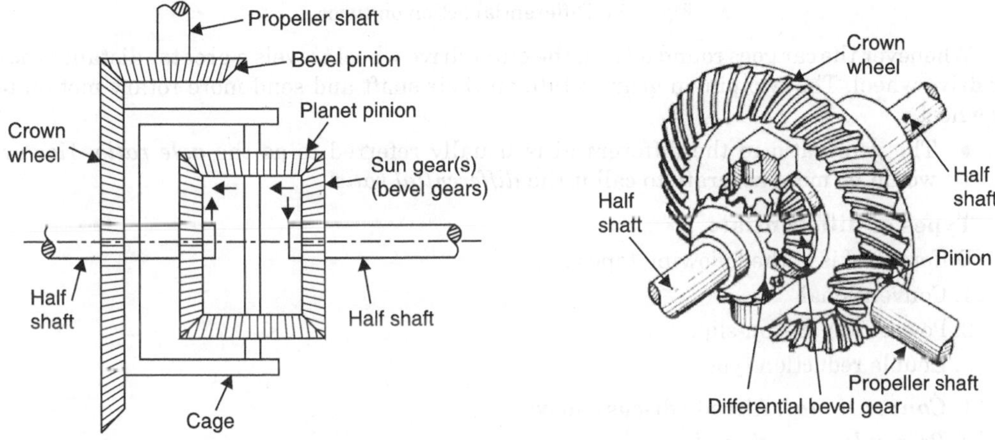

(*i*) Simple differential-simplified diagram (*ii*) Typical three-dimensional view

Fig. 7.49. Differential.

Operation of differential :

When the car is on a straight road, the ring gear, differential case, differential pinion gears, and two differential side gears all turn as a unit. The two differential pinion gears do not rotate on the pinion shaft. This is because they exert equal force on the two differential side gears. As a result, the side gears turn at the same speed as the ring gear, which causes both drive wheels to turn at the same speed also. However, when the car begins to round a curve, the *differential pinion gears rotate on the pinion shaft.* This *permits the outer wheel to turn faster than the inner wheel.*

Suppose one wheel turns slower than the other as the car rounds a curve. As the differential case rotates, the pinion gears must rotate on their shaft. This occurs because the pinion gears must walk around the slower-turning differential side gear. Therefore, the pinion gears carry additional rotary motion to the faster-turning outer wheel on the turn. The action in a typical turn is shown in Fig. 7.50. The differential speed is considered to be 100 percent. The rotating action of the pinion gears carries 90 percent of this speed to the slower-rotating inner wheel. It sends 110 percent of the speed to the faster-rotating outer wheel. This is how the differential allows one drive wheel to turn faster than the other.

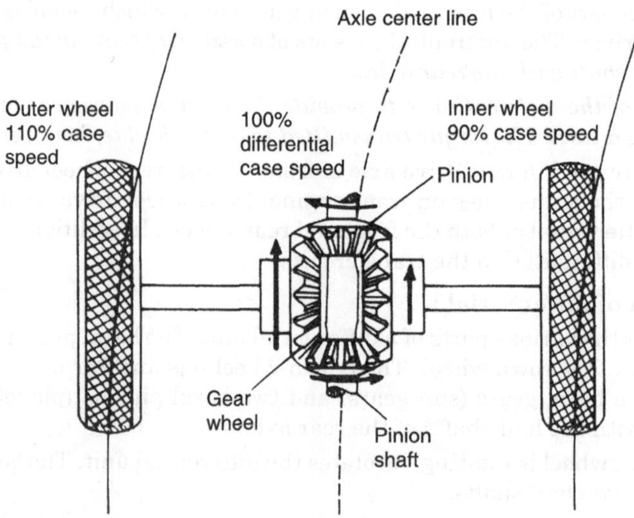

Fig. 7.50. Differential action on turns.

Whenever the car goes round a turn, the outer drive wheel travels a greater distance than the inner drive wheel. The two pinion gears rotate on their shaft and send more rotary motion to the outer wheel.

- The gear ratio in the differential is usually referred to as the *axle ratio.* However it would be more accurate to call it the *differential ratio.*

Types of differential :

Differential is of the following types :

1. Conventional.
2. Power-lock or Non-slip.
3. Double reduction type.

1. ***Conventional.*** Already discuss above.

2. ***Power-lock or Non-slip :***

The conventional differential delivers the same amount of torque to each wheel on the drive axle. If one wheel slips on a slippery road, mud or ice, the wheel becomes stationary, all the power being flowing to the slipping wheel. This results in no movement to the vehicle. In order to overcome this drawback, the vehicles are now *provided with a non-slip or self-locking differential.*

Non-slip differential is very much similar in construction to conventional type except that *it has two sets of clutch plates in addition.* Furthermore, the ends of pinion shafts lay loosely in notches, in two halves of the differential cage.

- *Straight head drive.* The rotating differential cage carries the sun gear shafts around with it. Since there is a considerable side thrust, the axle shafts in sun gears tend to slide up the sides of the notches in the two halves of the differential cage. As they slide up, they are forced outward and this force is transmitted to the clutch plates. Clutch plates thus lock the sun gears and the axle shafts to the differential cage. In this case, if one wheel encounters a muddy or slippery surface, *that causes it to lose traction temporarily avoiding it from slipping or spining* since it cannot turn faster than the other wheel as the half shaft is locked with differential cage by means of clutch plates.
- *Rounding a corner.* While taking a turn, the non-slip differential functions in the conventional way to tend the outer wheel to run faster than the inner wheel and thus to cover

more distance within the same time period. This action is permitted by the slipping of clutch plates.

3. ***Double reduction type :***

This type of differential provides *one additional gear reduction* between the engine and the driving wheels. Here the drive pinion meshes with a ring gear mounted on a straight shaft which contains a reduction gear drive set. The reduction-gear set drives a driven gear set attached to the differential cage. The driven gear set contains more number of gear teeth as compared to reduction-drive gear set. Gear reduction is thus obtained between the drive pinion and ring gear and also between the two-reduction gear sets.

- This is employed in *heavy duty vehicles* which *require higher gear ratio* between the engine and the driving wheels.

Troubleshooting of differential :

1. **Vehicle does not move when put in gear :**

Causes : (*i*) Broken propeller shaft.

(*ii*) Broken axle shaft.

(*iii*) Broken pinion flange key.

(*iv*) Broken teeth of crown wheel and pinion.

(*v*) Broken cross or star pinion.

2. **Noisy differential :**

Causes : (*i*) Less lubricating oil in the differential housing.

(*ii*) Low viscosity oil or use of poor quality oil.

(*iii*) Wrong adjustment of crown wheel and pinion teeth.

(*iv*) Worn out or broken teeth of crown wheel or pinion.

(*v*) Crown wheel misaligned on cage.

(*vi*) Broken or worn out thrust washers of star pinion or sun pinion.

7.7. REAR AXLES

7.7.1. Introduction

A ***rear axle*** *lies in between the differential and the driving wheels and transmits power from the differential to the driving wheels.* It consists of two halves connected by the differential, one part is known as *half shaft.* Inner end of the half shaft is connected to the sun gear of the differential, and the outer end to the driving wheel.

The rear axles perform the following *two* **functions :**

(*i*) Carry the weight of the vehicle.

(*ii*) To rotate and transmit the power from the engine to the wheels.

- In *rear wheel drive* vehicles, the rear wheels are the driving wheels, whereas in *front wheel drive* vehicles, the front wheels are the driving wheels.
- Almost all rear axles on modern passenger cars are *live axles,* that is they revolve with the wheels. *Dead axles* simply remain stationary, do not move the wheels.
- A "*housing*" completely encloses the rear axles and the differential, protecting them from water, dust and injury, in addition to mounting their inner bearings and providing a container for the lubricant.

Depending upon the arrangement of the bearings the rear axle or the axle casing must support :

(*i*) The total rear weight of the vehicle, causing both bending and shearing action.

(*ii*) Side thrust on the wheels when cornering, which imposes bending load and an end thrust or a pull.

(*iii*) Driving torque.

Fig. 7.51 shows a typical cut away differential and rear axle.

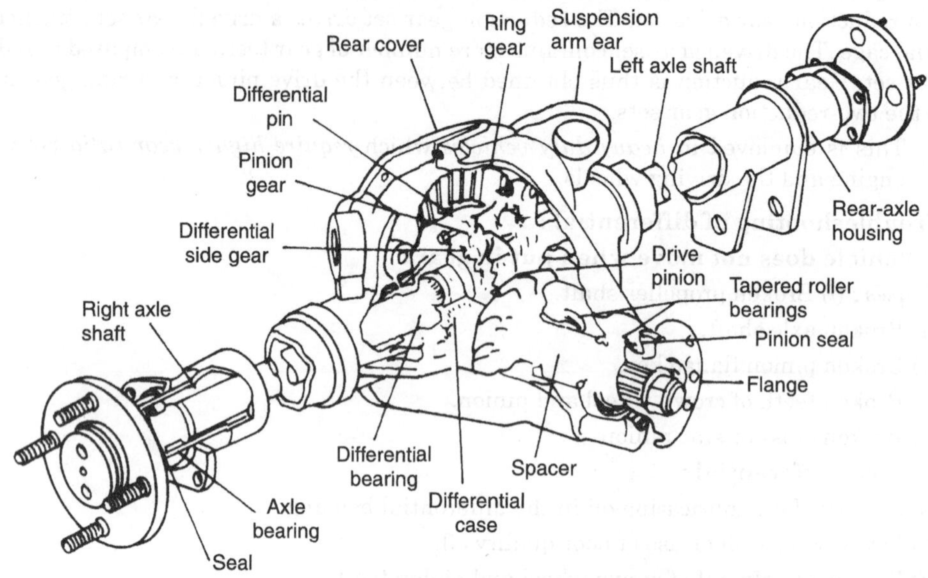

Fig. 7.51. A typical cut away differential and rear axle.

7.7.2. Types of Rear Axles

The rear drive axle differs from front drive axle in that it contains no joints at its outer ends. Its housing is either banjo type or split type.

- In the "*banjo axle*", the axle is a single unit and final drive assembly is carried in a separate casing which is bolted to the axle housing. The banjo construction is often used for axle cases of the *smaller and lighter vehicles.*
- In the "*split type axle*" arrangement a central housing contains the final drive and it is fitted with a tube on each side to carry the half axles and bearings.

Depending upon the methods of supporting the rear axles and mounting the rear wheels, the axles are of following *three* types :

1. Half floating rear axle.
2. Three-quarter floating rear axle.
3. Fully floating rear axle.

1. Half floating rear axle :

Fig. 7.52 shows the half floating rear axle.

The bearings are *inside the axle casing.* The axle of the wheel is at the centre of the axle casing. The wheels are fitted at the two ends of the axle by means of a key, bolt and nut.

The whole weigxhe vehicle is first transmitted to the suspension spring then to the axle case, rear axle, wheel and finally to the ground.

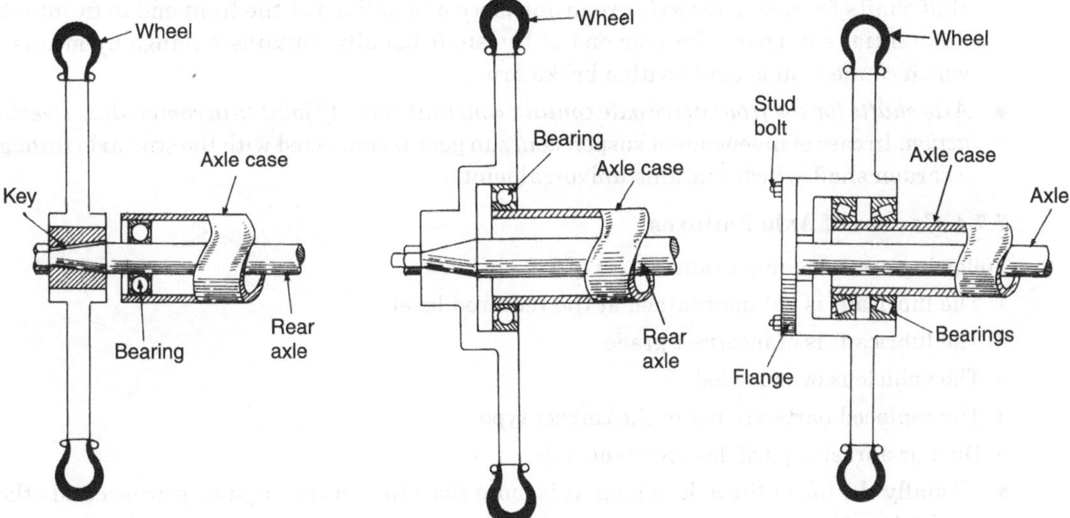

Fig. 7.52. Half floating rear axle.

Fig. 7.53. Three-quarter floating rear axle.

Fig. 7.54. Fully-floating axle with flange.

2. **Three-quarter floating rear axle :**

Fig. 7.53 shows a three-quarter floating rear axle.

The axle is fitted inside the axle casing. The *bearings are on the outer side between the wheel and the axle casing.* The wheel is fitted with the axle by means of a key, bolt and nut.

The weight of the vehicle is supported partly by the axle case and partly by the axle. The axle only takes care of the rotation and transmission of power. Power is transmitted from the engine to the wheel.

The main advantage of this type of axle over the half floating axle is that the *major part of the load is taken by the axle casing and not by the axle.*

3. **Fully-floating rear axle :**

Fig 7.54 shows the various parts of a fully-floating rear axle with flange.

The *bearings are between the wheel and axle case.* The axle is introduced inside the axle case. The axle end is fitted with the wheel by means of a flange, bolt and nut.

The *entire weight of the vehicle is supported by the wheel and axle case.* The axle is not supported by bearings at either end, and its position is maintained by the way that it is supported at both ends. Thus the axle is relieved of all strain caused by the weight of the vehicle on end thrust. It *transmits only the driving torque* ; for this reason it is called *full-floating.*

- This type of axle is *very strong* and is *used for heavy duty vehicles.* If an axle shaft breaks, there would be no danger of the wheel coming off and the *shaft can be removed and replaced easily.* But this is the *costliest* type of drive.

7.7.3. Axle Shafts or Half Shafts

- *These are the jack shafts through which power flows from the differential to the driving wheels.* There is a separate shaft for each half of the drive axle. A half shaft connects sun gear of differential with the brake drum over which wheel is mounted.

- Half shafts for rear drive axle are in one piece and splined at the front end to fit into the differential sun gear. The rear end of the shaft usually contains a flange by means of which connection is made with a brake drum.
- *Axle shafts for the front drive axle contain constant velocity joint to accommodate steering action.* In case of independent suspension, sun gear is connected with the stub axle through a carden shaft which contains universal joints.

7.7.4. Causes of Axle Failures

Following are some important *causes of axle failures :*

1. The lubricant is not maintained at the required level.
2. The lubricant is of incorrect grade.
3. The vehicle is over-loaded.
4. The replaced parts are not of the correct type.
5. Dust or abrasive particles enter the axle.

- Usually the life of the axle is long. It is more than that of the engine, gear box and other major components.

7.7.5. Rear Axle Noises

In the rear axle assembly, *noise* is an usual indication of trouble. Following may be the *causes* of noise :

1. Noise occurs when the *vehicle is pulling* :

Causes : (*i*) Too much clearance between the gears ;

(*ii*) End play in the pinion assembly.

2. Noise occurs *when vehicle is cornering.*

Cause. Faulty differential gears, either damaged teeth or excessive wear.

3. Noise occurs due to the *shortage of oil* or *incorrect viscosity of oil.*

- Sometimes the noise appears coming from the rear axle, but actually may come from the transmission, front wheel bearings or engine. The tyres also make a humming sound on certain types of road surface. Sometimes muffler noises may be considered as coming from the rear axle. Therefore, it is important to check these conditions before beginning to work on rear axle assembly.

7.7.6. Maintenance of Rear Axle

- In case of rear axle routine maintenance is limited to checking the oil level and topping up when necessary. Oil checks should be carried out every 5000 km.
- It is not possible to drain some modern rear axles, but ensure when topping up the oil level that the axle is not overfilled. Too much oil can result in a blown seal and oil contaminated rear brakes.
- In the rear axle, the only indications of problems are :
  — Abnormal noise ;
  — Fluid leaks.

These usually result from damaged oil seals, bearing failure and gear tooth wear in differentials.

7.7.7. Comparative Rear Axle Data of Some Indian Automobiles :

S. No.	Vehicle make	Type	Type of drive	Ratio
1.	*Hindustan Ambassador Mark-II*	Three-quarter floating	Hypoid	4.875 : 1
2.	*Fiat 1100*	Semi-floating	Hypoid	4.3 : 1
3.	*Jeep CJ Series*	Semi-floating	Hypoid	4.38 : 1
4.	*Ashok Leyland Comet Passenger*	Fully-floating	Spiral-bevel	6.216 : 1
5.	*Ashok Leyland Viking*	Fully-floating	Spiral-bevel	6.167 : 1
6.	*Dodge/Fargo Model 89M4*	Fully-floating	Hypoid	6.166 : 1
7.	*Tata 1210E Vehicles*	Fully-floating	Hypoid	6.857 : 1
8.	*Standard 20*	Semi-floating	Hypoid	4.625 : 1 (normal duty) 5.375 : 1 (heavy duty)
9.	*Swaraj Mazda*	Fully-floating		6.571 : 1
10.	*Eicher Mitsubishi 'Canter'*	Fully-floating	Hypoid	6.166 : 1
11.	*Ford FS 16C*	Fully-floating		6.5 : 1

7.7.8. Troubleshooting of Rear Axle

S. No.	Fault	Causes	Remedies
1.	*Axle noisy on acceleration.*	(*i*) Heavy heat contact on ring gear. (*ii*) Improper adjustment of pinion and ring gear. (*iii*) Rough pinion bearings. (*iv*) Loose pinion bearings.	(*i*) Correct it (*ii*) Re-adjust (*iii*) Replace (*iv*) Adjust.
2.	*Axle noisy on coast.*	(*i*) Excessive backlash in ring gear and pinion. (*ii*) End play in the pinion shaft. (*iii*) Heavy toe contact on ring gear. (*iv*) Rough bearings.	(*i*) Adjust (*ii*) Re-adjust (*iii*) Re-adjust (*iv*) Replace.
3.	*Axle noisy on both coast and acceleration.*	(*i*) Worn differential gears. (*ii*) Worn pinion and ring gears. (*iii*) Defective bearings. (*iv*) Excessive backlash between ring gear and pinion. (*v*) Pinion set too deep in ring gear. (*vi*) Pinion and ring gear too tight.	(*i*) Replace (*ii*) Replace (*iii*) Replace (*iv*) Adjust (*v*) Adjust (*vi*) Adjust.
4.	*Back lash.*	(i) Axle shaft splines worn. (*ii*) Axle shaft nut loose. (*iii*) Worn universal joints. (*iv*) Worn differential bearings. (*v*) Worn differential side gear thrust washers.	(*i*) Replace axle shaft (*ii*) Tighten as necessary (*iii*) Replace (*iv*) Replace (*v*) Replace.

SHORT ANSWER QUESTIONS

Q. 1. What is a transmission system ?

Ans. The mechanism that transmits the power developed by the engine of the automobile to the driving wheels is called the *transmission system*.

Q. 2. How are transmission systems categorized ?

Ans. Electrical and electromagnetic, hydraulic and mechanical transmission systems.

Q. 3. Name the units of transmission system.

Ans. Clutch unit, Transmission unit, Drive-line unit and Driving axle unit.

CLUTCH

Q. 4. What is a clutch ?

Ans. A *clutch* is a mechanism which enables the rotary motion of one shaft to be transmitted at will to second shaft, whose axis is coincident with that of first.

Q. 5. Where is clutch located ?

Ans. Between the engine and the gear box.

Q. 6. What are the features of a good quality clutch ?

Ans. A good quality clutch should (*i*) pick up load without any noise, (*ii*) have capacity to damp out vibrations, and (*iii*) need least pedal pressure to function.

Q. 7. What is a 'wet clutch' ?

Ans. A wet clutch is one which runs in oil in which the plates are lined with cork.

Q. 8. Why are clutch plates perforated ?

Ans. To dissipate the large amount of heat produced due to friction.

Q. 9. What is clutch drag ?

Ans. When the clutch is not disengaging fully and provides some difficulty in changing the gears then this defect is called as '*clutch drag.*'

Q. 10. What is the purpose of installing coil springs in a clutch plate ?

Ans. To provide a flexible torsional connection between the centre hub flange and the side plates. Moreover, with coil springs, desired load deflection pattern is also possible.

Q. 11. What will happen if all clutch springs in a multispring clutch are not of the same length and strength ?

Ans. Clutch chattering will take place.

Q. 12. What is "Belliville spring" in a clutch assembly ?

Ans. In a pressure plate, sometimes multiple coil springs are replaced by a single conical/flat spring which works like the bottom of an oil can. This is called *Belliville spring*.

Q. 13. Why is free play provided in clutch ?

Ans. Without free play, the release levers would be riding on the ends of the release levers, this would cause rapid wear of the bearing and levers.

Q. 14. Why is it considered a bad practice to rest the foot on the clutch pedal while driving ?

Ans. Resting of foot on the clutch pedal, causes a load to be placed on the thrust race, and it *increases the rate of wear*.

Q. 15. What is clutch slippage ?

Ans. Any increase in engine speed at the same vehicle speed is referred to as slippage of clutch.

Q. 16. What are the causes for clutch slippage ?

Ans. (*i*) Worn out friction disc, (*ii*) Broken return spring, (*iii*) Oil or grease on the plate, (*iv*) Incorrectly adjusted linkage and levers, and (*v*) Weak pressure plate.

Q. 17. What are main causes of clutch noise ?

Ans. (*i*) Worn throw out bearing or poor lubrication while disengaging, (*ii*) Loose friction disc hub on the clutch shaft and (*iv*) Worn out transmission bearing, shifting gears too fast in neutral.

Q. 18. What is a "Bell housing" ?

Ans. A large bell shaped metal casting mounted between the engine and gear box (transmission). It houses the clutch assembly and supports the gear box.

Q. 19. What is a "fluid coupling" ?

Ans. It is a hydraulic unit that replaces a clutch in a semi or fully automatic system and transmits the engine torque to a transmission system.

Q. 20. What is "hydraulic torque converter" ?

Ans. It is a device used for transmitting increased or decreased power from one shaft to another.

GEAR BOX (TRANSMISSION)

Q. 21. Why are helical gears preferred in transmission ?

Ans. (*i*) Their operation is silent and smooth, (*ii*) Can be made lighter to function at highest speeds without any vibration, and (*iii*) It is easier to distribute the load among two or more teeth simultaneously.

Q. 22. Why is sliding mesh gear box not preferred ?

Ans. (*i*) More force is required to engage the gears, and (*ii*) More noise is made while changing gears.

Q. 23. Why is constant mesh gear box so called ?

Ans. It is so called since all the gears to the countershaft and main shaft are in constant mesh with each other.

Q. 24. What is the function of a gear box ?

Ans. It allows the engine to run at different speeds relative to the road vehicles, so as to maintain its power and regulate the torque.

Q. 25. How is reverse gear obtained in the normal type of gear box ?

Ans. Reverse gear is obtained by providing an extra pinion meshing with an idler pinion. When engaging for reverse, the low gear pinion on the main shaft is moved into engagement with the idler pinion, thus reversing the direction of the mainshaft.

Q. 26. What is the principle of a "synchromesh" gear box ?

Ans. The basic principle of a synchromesh gear box is first to bring into frictional contact that member which is ultimately to be engaged positively and then as friction and speed are equalised, to make positive contact.

Q. 27. If a gear box is noisy what items should be checked ?

Ans. (*i*) Lubricant level and suitable viscosity ; (*ii*) Wear in the gears or bearings.

Q. 28. What are the advantages of the synchronism mechanism ?

Ans. (*i*) A smooth meshing is possible, (*ii*) Noise and jerk are eliminated, and (*iii*) Double declutching is not needed.

Q. 29. What is a 'Transfer case' ?

Ans. 'Transfer case' is a gear train coupled with the rear axle on one side and the front axle provided with differential and propeller shaft on the other side. This unit is meant to have *transmission to both the axles.*

Q. 30. What provisions are made in the gear box to ensure that only one set of gears are in mesh ?

Ans. A ball type locking of selector shaft is provided to ensure for meshing of one set of gears.

Q. 31. Why is oil of high viscosity used in the gear box ?

Ans. Engaging and disengaging of teeth during the running position cause a lot of heat generation. So thicker oil becomes thinner. So oil of higher viscosity is used.

Q. 32. Why are breather plugs fitted in some of gear boxes ?

Ans. Breather plugs allow air to get in or out depending on oil in the gear box being cold or hot respectively.

Q. 33. What is known as 'selective transmission' ?

Ans. It refers to the type of transmission in which it is possible to select any gear speed directly from the neutral position.

Q. 34. What is meant by sliding mesh type of gear ?

Ans. Gears that are changed by moving the pinions along the main shaft, to mesh with those on the layshaft are called sliding mesh type of gear.

Q. 35. What is a constant mesh gear box ?

Ans. A constant mesh gear box is a transmission in which all the gears of lay shaft and main shaft are in constant mesh with the respective gears.

Q. 36. What is a preselector gear box ?

Ans. A gear box fitted with a special mechanism which enables the driver to select the gear before hand is known as a preselector box.

Q. 37. Is it easier to drive over sandy road in top gear than in low gear ?

Ans. Yes.

Q. 38. How are different speeds obtained in a planetary gear box ?

Ans. In order to obtain different speeds any one of the ring gear, sun gear and planet with carrier, can be held from rotation by means of brake bands.

Q. 39. What are the advantages of auxiliary transmission when used along with the conventional one ?

Ans. (*i*) The number of speeds previously available is doubled,

(*ii*) The front axle and rear axle gear ratio are higher, and

(*iii*) The traction and hill climbing ability is increased.

Q. 40. What is the special use of power take off (auxiliary transmission) in a military truck ?

Ans. The special use of power take off is to drive a winch or dumb body hoist.

Q. 41. What is an overdrive ?

Ans. It is a drive interposed between the transmission and propeller shaft to permit the propeller shaft to run faster than, or overdrive, the transmission main shaft.

Q. 42. If no power is coming through the overdrive, what may be the causes ?

Ans. (*i*) Broken planetary parts ; (*ii*) Slipping over running clutch.

Q. 43. If overdrive creates noise, what may be the reasons ?

Ans. Worn gears, Worn main shaft bearings and worn overrunning clutch parts.

Q. 44. Where normally is the free wheel placed ?

Ans. Usually on the rear end of the gear box ; this makes the engine idle at controlled speed under overrun conditions, without the need for selecting neutral gear.

PROPELLER SHAFT, UNIVERSAL JOINT, FINAL DRIVE, DIFFERENTIAL AND REAR AXLES

Q. 45. What is the necessity of a propeller shaft ?

Ans. A propeller shaft is provided (*i*) to cope with and maintain the distance between the gear box and differential, and (*ii*) to adjust with the varied length and angle.

Q. 46. What is a Hotchkiss drive ?

Ans. It is an open propeller shaft, in which the torque reaction is taken by the springs.

Q. 47. What is a slip joint ?

Ans. A joint comprising an external splined end and internal splines of the joint integral with the universal joint hub is called as *slip joint.*

Q. 48. What is the purpose of a slip joint in the drive lines ?

Ans. A slip joint lengthens or shortens the drive shaft.

Q. 49. What is an universal joint ?

Ans. A joint which permits power to be transmitted from one shaft to another which are not in alignment, is known as an universal joint.

Q. 50. What are the various universal joints in use ?

Ans. (*i*) Cross type, (*ii*) Ball and trunion type, and (*iii*) Constant velocity type universal joints.

Q. 51. What is a constant velocity type universal joint ?

Ans. A joint which does not allow variation in angular velocity occurring in each revolution is known as constant velocity universal joint.

Q. 52. What is the advantage of the constant velocity universal joint ?

Ans. The elimination of transmitted speed fluctuations.

Q. 53. What is a spider ?

Ans. It is a cross-shaped member of a cross and yoke type universal joint.

Q. 54. What is a trunion ?

Ans. The four arms of the spider which connect two yokes of the universal joint are called trunions.

Q. 55. How many universal joints are there in the torque tube transmission and why ?

Ans. Only one universal joint is provided at the gear box end, as the torque tube is bolted to the axle casing and movement is confined to the front end.

Q. 56. What is meant by "double cardon" joint ?

Ans. This is a type of constant velocity joint with two cross and yoke V-joints mounted back to back with a linking yoke and a centring ball.

Q. 57. What is the final drive called as such ?

Ans. The final drive is the final stage in transmitting power from the engine to the road wheels. It provides final speed reduction between the propeller shaft and that of the wheels. So, it is called as final drive.

Q. 58. What are the different types of final drive ?

Ans. (*i*) Worm and wheel type, (*ii*) Spur bevel gear type, (*iii*) Spiral bevel gear type, and (*iv*) Hypoid type.

Q. 59. What is a "differential" ?

Ans. It is a mechanism housed in the drive axle, which drives the outer wheel faster than the inner one while taking a turn.

Q. 60. What will happen if differential is not used ?

Ans. If differential is not used, the vehicle cannot be controlled during a turn, the wheels will skid and tyres may wear-off quickly.

Q. 61. What is the main difference between a standard and limited slip differential ?

Ans. In case of a limited slip differential special friction material is provided. So it tends to keep both the rear axle shafts rotating at the same speed.

Q. 62. What is referred as an axle ?

Ans. It is a steel shaft that supports the weight of the vehicle and transmits torque from differential to the drive wheels.

Q. 63. What is dead axle ?

Ans. It is a solid axle mounted on the springs with a spindle at each end upon which the wheel turns.

Q. 64. What is live axle ?

Ans. Live axle is one which turns within a tubular housing.

Q. 65. What is the difference between dead and live axles ?

Ans. *"Dead axle"* only supports the wheel but *does not transmit power*, while a *"live axle"* apart from supporting the wheels *also transmits torque to the rear wheels.*

Q. 66. In which vehicle live axle is in front ?

Ans. Live axle is in front in military vehicles and jeeps.

Q. 67. What are the functions of the front axle ?

Ans. The front axle (*i*) bears the front weight of the vehicle, (*ii*) turns the front wheel easily, and (*iii*) provides cushioning effect through a spring.

Q. 68. What is a three-quarter floating type rear axle ?

Ans. It is that axle in which part load of the rear portion of the vehicle is bone by the axle tube and part by the axle itself. In this arrangement, one bearing of brake drum hub is fitted on the axle whereas the other on the axle tube.

Q. 69. What is a full-floating rear axle ?

Ans. It is that axle in which whole load of rear portion of the vehicle is borne by the axle housing or tube only ; the axle shaft acts as a driving shaft.

Q. 70. Which type of rear axle is in use for ?

(a) Heavy vehicle ; **(b) Light vehicle.**

Ans. (*a*) Full-floating ; (*b*) Semi-floating axle.

Q. 71. What is a swing axle ?

Ans. It is a semi-floating axle for independently suspended rear wheels. In this design only inner end of each axle is fitted with the universal joints.

Q. 72. What are Banjo type and split type axle housings ?

Ans. *Banjo type*—single piece housing in final drive ;

Split type—made in two pieces for easy maintenance.

Q. 73. How is the oil level in rear axle indicated ?

Ans. (*i*) A dip stick ; (*ii*) combined filler and level plug ;

(*iii*) A separate level plug is used to check the oil level.

Q. 74. What is Transaxle ?

Ans. It is a combination of transmission and rear axle as a single unit.

Q. 75. What are the advantages of the Transaxle ?

Ans. (*i*) Extremely rigid engine and drive components, (*ii*) Misalignment is avoided, (*iii*) Unit is compact, and (*iv*) Better riding qualities.

Q. 76. What is a double reduction axle ?

Ans. It refers to a drive axle provided with two sets of reduction gears for extra reduction of gear ratio.

HIGHLIGHTS

Clutch

1. The mechanism that transmits the power developed by the engine of the automobile to the driving wheels is called the *transmission system* (or Power train).
2. The transmission consists of the following main units :
   (*i*) Clutch unit (*ii*) Transmission unit
   (*iii*) Drive-line unit (*iv*) Driving axle unit.
3. A "*clutch*" is a mechanism which enables the rotary motion of one shaft to be transmitted at will to second shaft, whose axis is coincident with that of first.
4. Friction materials/linings of the following three types :
   (*i*) Woven type (*ii*) Moulded or compression type
   (*iii*) Mill board type.
5. The following materials are almost universally used for clutch lining :
   (*i*) Leather (*ii*) Cork (*iii*) Fabric
   (*iv*) Asbestos (*v*) Reybestos and Ferodo.
6. Clutch can be classified as follows :
   (*i*) Positive clutch
   (*ii*) Gradually engagement clutches
   (*a*) Friction clutches (*b*) Other forms of clutches.
7. Single plate clutch is commonly used in most cars and small commercial vehicles.
8. The principal parts of a plate clutch are :
   (*i*) Driving members (*ii*) Driven member (Clutch plate)
   (*iii*) Operating members.
9. Free wheel or one-way clutch transmits power in one direction only.
10. "*Fluid coupling*" is a hydraulic unit that replaces a clutch in a semi or fully automatic system, and transmits the engine torque to a transmission system.
11. "*Hydraulic torque converter*" is a device used for transmitting increased or decreased power from one shaft to another.

Gear box

12. *Gear box* is a speed and torque changing device between the engine and the driving wheels.
13. The gear ratio is obtained by using the following formula :

$$\text{Gear ratio} = \frac{\text{No. of teeth on driven gear}}{\text{No. of teeth on driving gear}}$$

$$= \frac{\text{Speed of the driving shaft}}{\text{Speed of the driven shaft}} = \frac{\text{Engine speed}}{\text{Propeller shaft speed}}$$

14. The following types of gear boxes are used in automobiles :
   1. Selective type :
   (*i*) Sliding mesh, (*ii*) Constant mesh, and (*iii*) Synchromesh.

2. Progressive type
3. Epicyclic or planetary type.

15. *Synchromesh gear devices* work on the *principle* that two gears to be engaged are first brought into frictional contact which equalises their speed after which they are engaged readily and smoothly.
16. *Epicyclic or planetary type gear box* uses no sliding dogs or gears to engage but different gear speeds are obtained by merely tightening brake-bands on gear drums.
17. *Gear shifting* is the moving of the gears in transmission by means of the gear shift lever to change the gear mesh.
18. *Transfer case* is a two-speed transmission located at the rear of the standard transmission which provides a low and direct gear (as in jeep).

Propeller shaft, Universal joint, Final drive and differential, Rear axles

19. The *propeller shaft* is a driving shaft that connects the transmission main or output shaft of the gear box to the differential of the rear axle.
20. *Universal joint* is the joint which enables the drive shaft to transmit power at varied angles. It permits the torque transmission not only at angle, but also while this angle is changing.
21. Types of universal joints :
    (*a*) Variable velocity joints : Cross or spider type ; Ring type ; Ball and trunion type.
    (*b*) Constant velocity joints : Tracta ; Bendix Weiss ; Rzeppa.
22. The *final drive* is the last stage in transferring power from the engine to the road wheels. It provides a fixed reduction between the drive shaft and the driving axles.
23. *Classes of final drive :*
    1. Chain type
    2. Gear type : Worm and wheel type ; Bevel and hypoid bevel gear type.
24. *Differential* is the mechanism by means of which outer wheel runs faster than the inner wheel while taking a turn or moving over upheaval road.
25. *Types of differential :*
    1. Conventional ; 2. Power lock or non-slip ; 3. Double reduction type.
26. A *rear axle* lies between the differential and the driving wheels and transmits power from the differential to the driving wheels.
27. *Types of rear axles :*
    1. Half-floating ; 2. Three-quarter floating ; 3. Fully floating rear axles.

OBJECTIVE TYPE QUESTIONS

Choose the Correct Answer :

Clutch

1. A clutch is usually designed to transmit maximum torque which is :
   (*a*) 90 percent of the maximum engine torque (*b*) equal to the maximum engine torque.
   (*c*) 150 percent of the maximum engine torque (*d*) none of the above.
2. The inertia of the rotating parts of the clutch should be
   (*a*) minimum (*b*) maximum (*c*) zero (*d*) none of these.
3. The coefficient of friction for the clutch facing is approximately
   (*a*) 0.1 (*b*) 0.4 (*c*) 0.8 (*d*) 1.2.
4. In a clutch with coil springs, the wear of the clutch facing will cause the clamping load to
   (*a*) increase (*b*) decrease (*c*) remain constant (*d*) become infinite.
5. Clutch facings are usually attached to the plate by
   (*a*) steel rivets (*b*) brass rivets (*c*) aluminium screws (*d*) steel screws.

6. Free pedal play in car clutches is about.
(*a*) 3 mm (*b*) 30 mm (*c*) 60 mm (*d*) 100 mm.
7. The maximum value of axial force at the clutch which a driver can apply while driving, without getting fatigued is approximately.
(*a*) 10 N (*b*) 100 N (*c*) 500 N (*d*) 5000 N.
8. The maximum intensity of pressure, which the clutch facing can withstand without being damaged is about
(*a*) 10 Pa (*b*) 10 kPa (*c*) 100 kPa (*d*) 1000 kPa.
9. The clutch is located between the transmission and the
(*a*) engine (*b*) rear axle (*c*) propeller shaft (*d*) differential.
10. The parts of the cover assembly that hold the pressure plate against the clutch plate are the
(*a*) release levers (*b*) thrust bearings (*c*) struts (*d*) springs.
11. Cushioning springs in clutch plate are meant to reduce
(*a*) torsional vibrations (*b*) vehicle speed (*c*) jerky starts (*d*) none of these.
12. The thrust bearing should come into contact with the release levers when the
(*a*) vehicle is stationary (*b*) vehicle is running very fast
(*c*) vehicle is driven very slow (*d*) clutch pedal is pressed.

Gear box (Transmission)

13. In a simple planetary gear set, the output member to increase torque is always the
(*a*) sun gear (*b*) ring gear (*c*) planet carrier (*d*) none of these.
14. In a simple epicyclic gear set, the output member to increase torque in reverse is always the
(*a*) ring gear (*b*) planet carrier (*c*) sun gear (*d*) none of these.
15. The central gear of an epicyclic gear set is called a
(*a*) ring gear (*b*) sun gear (*c*) planet gear (*d*) internal gear.
16. The component of the torque converter that allows multiplication of torque is the
(*a*) turbine (*b*) impeller (*c*) pump (*d*) stator.
17. The component of the torque converter that drives the oil is the
(*a*) turbine (*b*) impeller (*c*) freewheel (*d*) stator.
18. The maximum torque multiplication ratio in a torque converter is about
(*a*) 2.5 (*b*) 10 (*c*) 25 (*d*) 100.
19. The component of the torque converter that redirects the flow of oil to impeller is
(*a*) turbine (*b*) impeller (*c*) stator (*d*) freewheel.
20. In a torque converter maximum torque multiplication occurs at
(*a*) stop (*b*) low speed (*c*) medium speed (*d*) high speed.
21. The blades in a torque converter have a shape which is
(*a*) square (*b*) round (*c*) flat (*d*) curved.
22. The purpose of transmission in an automobile is
(*a*) to vary the speed of automobile (*b*) to vary the torque at the road wheels
(*c*) to vary the power of automobile (*d*) none of the above.
23. Increase of torque in a vehicle is obtained by
(*a*) decreasing speed (*b*) decreasing power
(*c*) decreasing petrol consumption (*d*) all of the above.
24. Two advantages of using helical gears rather than spur gears in a transmission are :
(*a*) strength and cost (*b*) strength and less end thrust
(*c*) noise level and strength (*d*) noise level and economy.

25. By using synchronizing device, the two involved adjacent gears have their speeds
(*a*) increased (*b*) reduced (*c*) equalized (*d*) unequalized.

26. The maximum torque multiplication ratio in a torque converter is about
(*a*) 2.5 (*b*) 8 (*c*) 20 (*d*) 80.

27. The component of the torque converter that redirects the flow of oil to impeller is
(*a*) turbine (*b*) impeller (*c*) stator (*d*) freewheel.

28. In a torque converter maximum torque multiplication occurs at
(*a*) stop (*b*) low speed (*c*) medium speed (*d*) high speed.

29. The blades in a torque converter have a shape which is
(*a*) square (*b*) round (*c*) flat (*d*) curved.

30. In a simple planetary gear set, the output member to increase torque is always the
(*a*) sun gear (*b*) ring gear (*c*) planet carrier (*d*) none of these.

31. In a simple epicyclic gear set, the output member to increase torque in reverse is always the
(*a*) ring gear (*b*) planet carrier (*c*) sun gear (*d*) none of the above.

32. The central gear of an epicyclic gear set is called a
(*a*) ring gear (*b*) sun gear (*c*) planet gear (*d*) internal gear.

33. The component of the torque converter that allows multiplication of torque is the
(*a*) turbine (*b*) impeller (*c*) pump (*d*) stator.

34. The component of the torque converter that drives the oil is the
(*a*) turbine (*b*) impeller (*c*) freewheel (*d*) stator.

35. When the rear wheels are jacked up and gears are in neutral, turning one rear wheel in a rear drive vehicle will cause the other wheel to
(*a*) turn backward (*b*) turn forward
(*c*) turn in either direction (*d*) remain stationary.

36. The type of rear axle on trucks is
(*a*) semi-floating (*b*) fully-floating
(*c*) three-quarter floating (*d*) none of above.

37. Salisbury type rear axle casing is also known as
(*a*) banjo type casing (*b*) unitized carrier casing
(*c*) separate carrier casing (*d*) none of these.

38. The function of a universal joint is to allow the propeller shaft to
(*a*) change length (*b*) bend sideways
(*c*) transfer torque at an angle (*d*) change inclination.

39. Central portion of a propeller shaft is made from a
(*a*) steel shaft (*b*) gun metal shaft (*c*) steel tube (*d*) cast iron rod.

40. A two-piece propeller shaft requires
(*a*) one universal joint (*b*) a centre support bearing
(*c*) the shaft to be solid (*d*) none of the above.

41. The inner end of the axle shaft is splined to the
(*a*) sun gear (*b*) planet pinion (*c*) crown wheel (*d*) differential cage.

42. The adjustment for backlash in a differential is provided between
(*a*) crown wheel and the sun gear (*b*) sun gear and the planet gear
(*c*) crown wheel and the drive pinion (*d*) crown wheel and the planet gear.

43. Hypoid gears require special lubricant because
(*a*) tooth is made of soft material (*b*) tooth is made of hard material
(*c*) such gears rotate faster (*d*) sliding action is there between the teeth.

44. The smallest gears inside the differential casing are
(*a*) pinion gears (*b*) sun gears (*c*) side gears (*d*) ring gears.

Answers

1. (*c*)	**2.** (*a*)	**3.** (*b*)	**4.** (*b*)	**5.** (*b*)	**6.** (*b*)	**7.** (*b*)
8. (*c*)	**9.** (*a*)	**10.** (*d*)	**11.** (*c*)	**12.** (*d*)	**13.** (*c*)	**14.** (*a*)
15. (*b*)	**16.** (*d*)	**17.** (*b*)	**18.** (*a*)	**19.** (*c*)	**20.** (*a*)	**21.** (*d*)
22. (*b*)	**23.** (*a*)	**24.** (*c*)	**25.** (*c*)	**26.** (*a*)	**27.** (*c*)	**28.** (*a*)
29. (*d*)	**30.** (*c*)	**31.** (*a*)	**32.** (*b*)	**33.** (*d*)	**34.** (*b*)	**35.** (*a*)
36. (*b*)	**37.** (*b*)	**38.** (*c*)	**39.** (*c*)	**40.** (*b*)	**41.** (*a*)	**42.** (*d*)
43. (*d*)	**44.** (*a*).					

THEORETICAL QUESTIONS

1. Define a 'transmission system'.
2. What are the requirements of a transmission system ?
3. How are transmission systems categorized ?
4. Enumerate units of transmission system.

Clutch

5. What is a clutch ?
6. What are the functions of a clutch ?
7. Explain briefly the principle of operation of a clutch.
8. Why is clutch placed between the flywheel and transmission ?
9. State the purpose of the following :
   (*i*) Driven plate (*ii*) Pressure plate (*iii*) Torsional springs
   (*iv*) Coil spring (*iv*) Clutch disc.
10. Describe briefly the various types of friction materials linings used in clutches.
11. What are the properties of a good clutch lining ?
12. How are clutches classified ?
13. Explain the working of a cone clutch.
14. Describe briefly with a neat sketch the construction and working of a single plate clutch.
15. Differentiate between : (*i*) Single and multiplate clutches, and (*ii*) Dry and wet clutch.
16. Give the working of a diaphragm spring type single plate clutch.
17. What are the advantages of a diaphragm spring ?
18. With the help of a neat sketch explain the construction and working of a multiplate clutch.
19. Describe briefly the parts of a plate clutch.
20. Explain the working of a centrifugal clutch with a suitable sketch.
21. What is the difference between the centrifugal and semi-centrifugal clutches.
22. What are two common types of clutch linkages ?
23. What is a fluid flywheel ? How does it work ?
24. Describe the working of a hydraulic torque converter.
25. What is a hydraulic clutch ? How does it differ from ordinary clutches ?
26. What is the difference between 'fluid coupling' and 'hydraulic torque converter' ?
27. Explain briefly the various causes of clutch troubles. How can those be remedied ?

Gear box (Transmission)

28. What is a gear box ?
29. What purposes are served by a gear box in the transmission system of an automobile ?
30. What is gear ratio ? How is it obtained ?

31. How are torque and gear ratio related ?
32. Name the types of the gear boxes used in automobiles.
33. Explain briefly any one of the following gear boxes :
    (*i*) Sliding mesh gear box (*ii*) Constant mesh gear box
    (*iii*) Synchromesh gear box.
34. What is the working principle of synchromesh gear devices ?
35. Give the comparison among sliding mesh, constant mesh and synchromesh gear boxes.
36. Explain briefly progressive type gear boxes.
37. What are the approximate gear ratios provided in a three forward speed gear box ?
38. What is a synchronizer ?
39. What is a transfer case ?
40. What should be the cause when the vehicle is running with excessive noise in gear box ?
41. What are the three active members of a planetary gear set ?
42. What are the advantages of epicyclic gear box ?
43. Name the components of an automatic transmission.
44. How does semi-automatic transmission differ from automatic transmission ?
45. What are the advantages of automatic transmission ?
46. Describe briefly the troubleshooting of a gear box.
47. What is an overdrive ? What are its advantages ?
48. How are constant mesh transmissions arranged for obtaining torque changes ? Discuss the advantages of a constant mesh box over the sliding mesh type.
49. Describe the working of a synchromesh gear box with the help of sketch(es).
50. Why is a gear box necessary in a motor car ? Draw neat sketch of a gear box that is normally used in a heavy duty commercial vehicle (bus or truck) and explain its action.
51. Sketch and describe the action of three-speed and reverse gear box used in any Indian car.
52. What are different types of selective type transmission ?
53. What are the advantages and disadvantages of freewheeling ?
54. What is free-wheeling device ? How does it work and what are its advantages ?
55. Explain briefly any two of the following ?
    (*i*) Transfer case (*ii*) Overdrive
    (*iii*) Automatic transmission (*iv*) Pre-selective gear box.
56. What is the difference between constant-mesh and synchromesh transmission ? Describe any one of them.
57. Describe the construction of a sliding mesh gear box. Show how the power flows in various speeds.

Propeller shaft, Universal joints, Final drive and differential, Rear axles

58. What is propeller shaft ?
59. What are the functions of a propeller shaft in the transmission system of a vehicle ?
60. How is the length of a propeller shaft varied automatically ?
61. What is the advantage of a two-piece propeller shaft ?
62. Explain briefly following two types of construction of propeller shaft :
    (*i*) Torque tube propeller shaft (*ii*) Hotchkiss type propeller shaft.
63. What is a universal joint ? Where is it used ?
64. What are the two main types of universal joints and how they differ from each other ?
65. Explain the different types of constant velocity universal joints.
66. Describe variable velocity universal joints. Why this type of joint is not suggested for front axle half shafts ?
67. How is drive from propeller shaft turned at right angles ?

68. What is the function of slip joint ?
69. What is purpose of the final drive ?
70. Why are differentials necessary ?
71. Describe briefly the various classes of final drive.
72. Explain the construction and working of a differential with a neat sketch.
73. Explain briefly any two types of the following differentials :
(*i*) Conventional differential
(*ii*) Power-lock or Non-slip differential
(*iii*) Double reduction type differential.
74. Illustrate the working of conventional differential.
75. What is a rear axle ? What are its functions ?
76. Differentiate between a live and a dead axle.
77. Describe different types of live rear axles and explain their advantages and disadvantages.
78. Explain briefly, with neat sketches, any two of the following rear axles :
(*i*) Half floating rear axle ; (*ii*) Three-quarter floating rear axle ;
(*iii*) Fully-floating rear axle.
79. What are axle shafts or half shafts ?
80. What are the causes of axle failures ?
81. Discuss briefly the causes of rear axle noises.

8

Steering and Front Axle

A. STEERING SYSTEM

8.1. PURPOSE OF A STEERING SYSTEM

The **steering system** *allows the driver to guide the car along the road and turn left or right as desired.*

The system includes the following :

(*i*) *The steering wheel* ...... which the driver controls.

(*ii*) *The steering gear* ...... which changes the rotary motion of the wheel into straight line motion, and

(*iii*) *The steering linkages* ...... which transmit the steering gear movement to the front wheels.

- The steering system configuration depends on vehicle design (the drive train and suspension system used, whether it is a passenger car or a commercial vehicle etc.). At present, the rack-and-pinion type and the recirculating-ball types are in use.
- Most steering systems were *manual* until a few year back. Then *power steering* became popular. It is now installed on almost all costly cars.

8.2. FUNCTIONS OF A STEERING SYSTEM

Following are the *functions* of a steering system :

1. The primary function of the steering system is *to achieve angular motion of the front wheels to negotiate a turn.*

2. To provide directional stability of the vehicle when going straight ahead.

3. To facilitate straight ahead recovery after completing a turn.

4. To minimise wear and tear of tyres.

5. To absorb a major part of the road shocks thereby preventing them to get transmitted to the hands of the driver.

6. To provide perfect rolling motion of the road wheels at all times.

8.3. REQUIREMENTS OF A GOOD STEERING SYSTEM

Following are the *requirements of a good steering system* :

1. Very accurate.
2. Easy to handle.
3. Provide directional stability.
4. Multiply the turning effort applied on the steering wheel by the driver.
5. Irreversible to a certain degree, so that the shocks of the road surface encountered by the wheels are not transmitted to driver's hands.

8.4. GENERAL ARRANGEMENT OF A STEERING SYSTEM

Fig. 8.1 shows the general arrangement of a steering system. The layout of steering system is shown in Fig. 8.2.

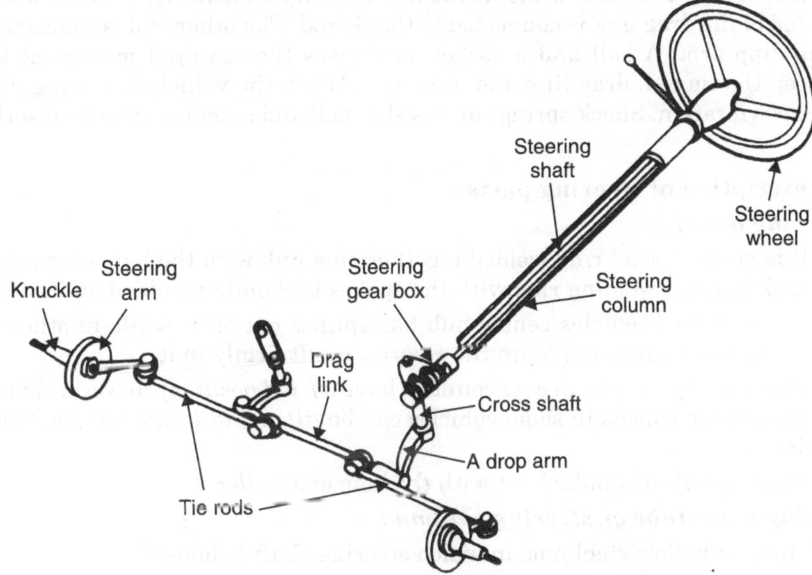

Fig. 8.1. General arrangement of a steering system.

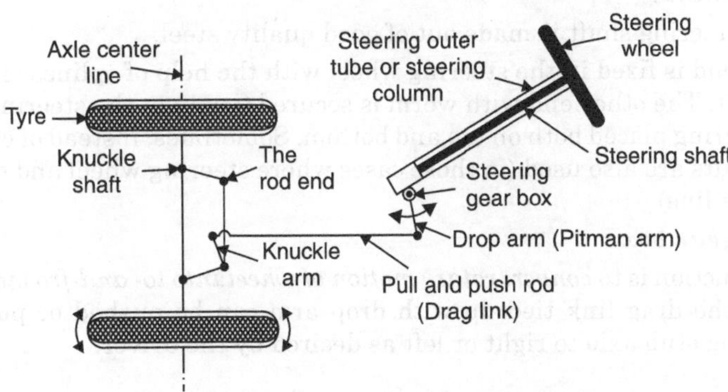

Fig. 8.2. Layout of steering system.

The *main parts of a steering system* are :

1. Steering wheel.
2. Steering column.
3. Steering shaft.
4. Steering gear box.
5. Steering drop arm (Pitman arm).
6. Pull and push rod (Drag link).
7. Knuckle arm.
8. Tie rod and tie rod end.

Working of steering mechanism :

- The *steering wheel* rotates the *steering column.* The *steering gear box* is fitted to the end of this column. Therefore, when the wheel is rotated, the *cross shaft* in the gear box oscillates. The cross shaft is connected to the *drop arm.* This arm is linked by means of a *drag link* to the *steering arms.* The steering arms on both wheels are connected by the *tie rods* to the drag link.
- When the steering wheel is operated, the knuckle moves to and fro, moving the wheels to the right or left. The ends of the tie rod and steering knuckle are connected to each other. One end of the drag link is connected to the tie rod. The other end is connected to the end of the drop arm. A ball and a socket joint gives the required movement to the joints between the tie rod, drag link and drop arm. When the vehicle is moving, the drop arm develops vibration. Shock springs are used in ball and socket system to absorb this vibration.

Brief description of steering parts :

1. ***Steering wheel :***
   - It is made of steel ring welded together on a hub with the help of two, three or four spokes. After welding ring with the spokes is ebonite moulded on it.
     — In certain vehicles centre hub has splines cut on it while in other cases a key groove is given to secure the steering shaft firmly in it.
   - The steering wheels, in over country, have a *fixed position.* However, in foreign countries, these wheels in some vehicles can be *tilted and located in position* to suit the driver.
   - Steering wheel is pulled out with the help of a *puller.*
2. ***Steering outer tube or steering column :***
   - This is a hollow steel pipe in which steering shaft is housed.
   - One end of the pipe is fixed on steering box, the other end is usually held with the help of bracket under the instrument panel.
3. ***Steering shaft :***
   - The steering shaft is made out of good quality steel.
   - One end is fixed in the steering wheel with the help of splines or key and kept tight by nut. The other end with worm is secured firmly in the steering box with the help of bearing placed both on top and bottom. Sometimes, instead of one shaft, two pieces of shafts are also used (in those cases where steering wheel and steering box are not in one line)
4. ***Steering gear box :***
   - Its function is to *convert rotary motion of wheel into to- and-fro motion of drop arm* so that the drag link tied up with drop arm can be pushed or pulled resulting into moving stub axle to right or left as desired by the driver.
5. ***Drop arm :***
   - It is forged out of good quality steel.

- Its one side is provided with splines which match the splines of sector shaft and held on sector shaft by nut. The other end has taper hole in which ball end is held tight with the help of nut.

8.5. STEERING GEARS

The steering gears perform the following *two functions* :

(*i*) Change the rotary motion of the steering wheel into straight line motion that will move the steering linkage.

(*ii*) Provide a gear reduction that will make the automobile easier to start.

There are many types and makes of steering gears in use for automobiles. However, only a few important ones out of them are enumerated and discussed (The working of other types is also similar) below :

1. Worm and worm wheel steering gear.
2. Worm and nut steering gear.
3. Worm and roller steering gear.
4. Recirculating ball type steering gear.
5. Rack and pinion steering gear.

1. **Worm and worm wheel steering gear :**

A simplified schematic diagram of a worm and worm wheel steering gear is shown in Fig. 8.3.

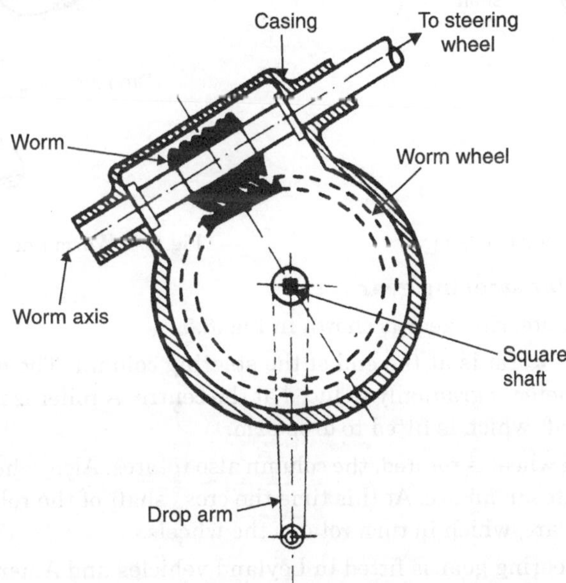

Fig. 8.3. Worm and worm wheel steering gear.

In this steering system there are square threads or worms on the steering rod end which engages in a worm wheel. The drop arm (pitman arm) is keyed to the same shaft as the worm wheel and works rigidly with it. Usually a square shaft is used for the worm wheel, so that as wearing of worm sector occurs, the wormwheel can be turned round to a new position. The arc of movement of the drop arm is usually from 60° to 90°. Many makers provide *only a sector of a wheel* for this purpose. This gives a smaller and lighter mechanism but has no provision for worm and wheel teeth wear.

- This system is commonly used in *tractors*.

2. **Worm and nut steering gear :**

Fig. 8.4 shows a worm and nut steering gear.

In this system when the steering wheel is rotated the steering column rotates. The worm is an integral part of the steering column. Therefore, when the steering column is rotated, the worm also rotates. Now the worm is in mesh with a nut arrangement. Thus when the worm rotates, the nut is able to move. This movement takes place along the axis of the column either up or down. This enables the cross shaft to rotate in an arc. This in turn helps the drop arm to also move in an arc. This arm transmits the steering to all wheels.

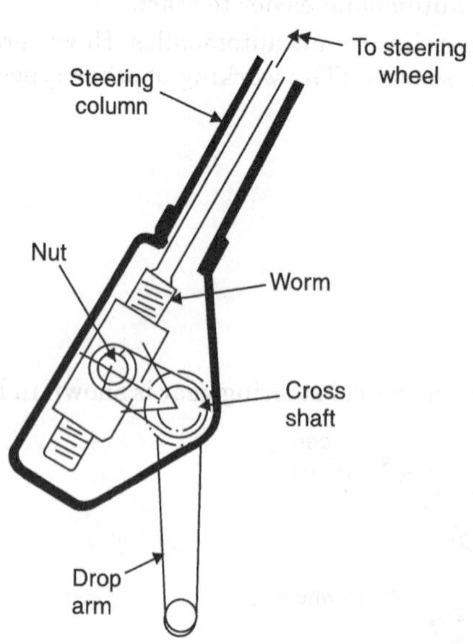

Fig. 8.4. Worm and nut steering gear.

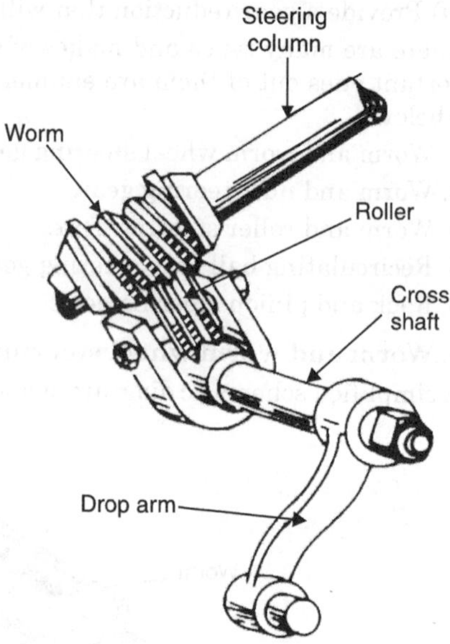

Fig. 8.5. Worm and roller steering gear.

3. **Worm and roller steering gear :**

A worm and roller steering gear is shown in Fig. 8.5.

In this system the worm is at the end of the steering column. The diameter of the worm is more at its ends. Its diameter is gradually reduced at the centre. A roller is in mesh with this worm. The roller has a cross shaft which is fitted to drop arm.

When the steering wheel is rotated, the column also rotates. Also, when the worm of the shaft is rotated, the roller rotates in an arc. At this time the cross shaft of the rollers moves. This makes the drop arm move in an arc, which in turn rotates the wheels.

- This type of steering gear is fitted in Leyland vehicles and American passenger cars.

4. **Recirculating ball type steering gear :**

Fig. 8.6 shows a recirculating ball type steering gear. It consists of a worm at the end of steering rod/shaft. A nut is mounted on the worm with two sets of balls in the grooves of the worm, in between the nut and the worm. The balls reduce the friction during the movement of the nut on the worm. The nut has a number of teeth on the outside, which mesh with the teeth on a worm wheel sector, on which is further mounted the drop arm, which steers the road wheels through the link rod and the steering arms.

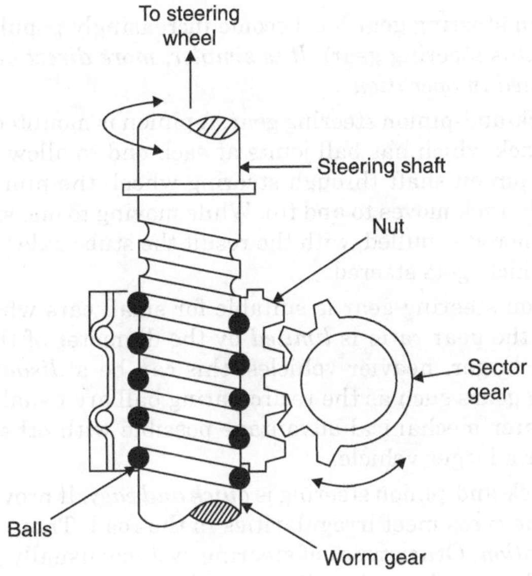

Fig. 8.6. Recirculating ball steering gear.

When the steering wheel is turned, the balls in the worm roll in the grooves and cause the nut to travel along the length of the worm. The balls which are in two sets, are recirculated through the guides. The movement of the nut causes the sector gear to turn at an angle and actuate the link rod through the drop arm, resulting in the desired steering of the wheels.

- In India, this type of steering gear is used in the Tata, Dodge/Fargo, Standard 20 vehicles.

5. **Rack-and-pinion steering gear :**

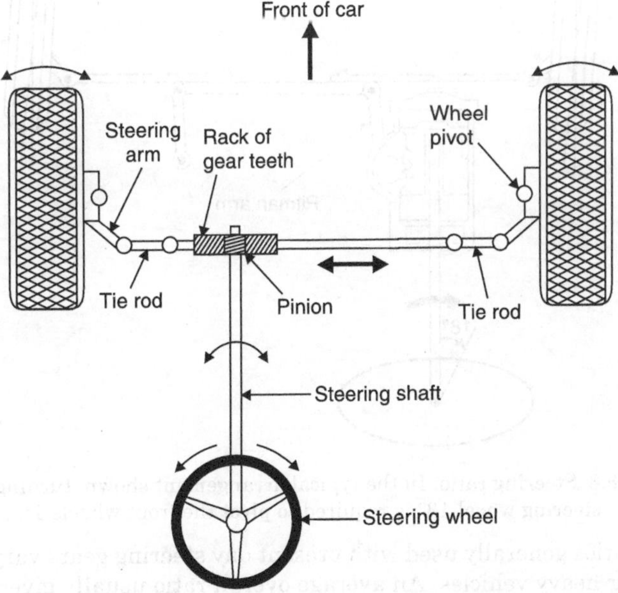

Fig. 8.7. Simplified "rack and pinion" steering system.

The rack-and-pinion steering gear has become increasingly popular for today's smaller cars (*Maruti-800* cars employ this steering gear). *It is simpler, more direct acting, and may be straight mechanical or power assisted in operation.*

Fig. 8.7 shows a rack-and-pinion steering gear. A pinion is mounted on the end of the steering shaft. It engages with a rack which has ball joints at each end to allow for the rise and fall of the wheels. On revolving the pinion shaft through steering wheel, the pinion revolves. As the pinion teeth mesh with rack teeth, rack moves to and fro. While moving to one side, it pushes the tie rod to that side while the other tie rod is pulled, with the result the stubs axle being tied up with these tie rods also move and the vehicle gets steered.

- A rack-and-pinion steering gear is suitable for small cars where the *steering forces are light*. However, the gear ratio is *limited* by the diameter of the steering wheel and the pinion gear. On larger, heavier vehicles, this can be a *disadvantage*. Therefore, other types of steering gears such as the recirculating ball are usually found on large cars and trucks. The greater mechanical advantage possible with other types reduces the effort required to steer a larger vehicle.
- In a small car, rack-and-pinion steering is *quick and easy*. It provides the maximum amount of road feel as the tyres meet irregularities in the road. There is *no damping out of road shocks and vibration*. Other types of steering systems usually *provide some damping action*. They can also be *adjusted to eliminate almost all play in the steering wheel.*

8.6. STEERING RATIO

The **steering ratio** *is the number of degrees that the steering wheel must be turned to pivot the front wheels 1 degree*. For example, in Fig. 8.8 the steering wheel must be turned 18 degrees to pivot the front wheels 1 degree. Therefore the steering ratio is 18 : 1.

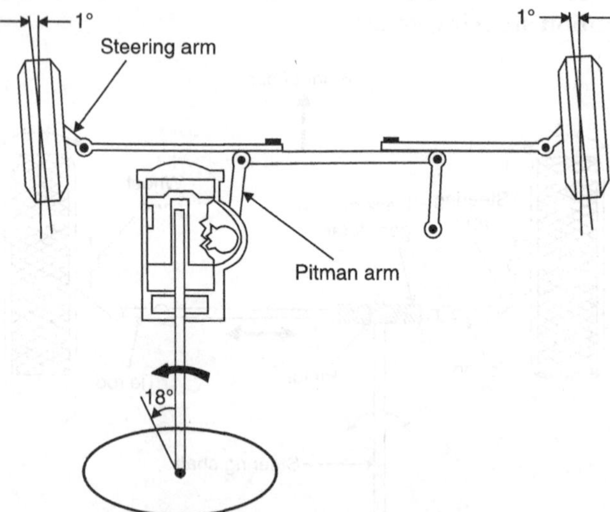

Fig. 8.8. Steering ratio. In the typical arrangement shown, turning the steering wheel 18° is required to pivot the front wheels 1°.

The steering ratios generally used with present day steering gears vary from about 12 : 1 for cars to about 35 : 1 for heavy vehicles. An average overall ratio usually gives about one and a half complete turns of the steering wheel each side of mid position to apply a full lock of 45 degrees each way on the wheels.

The higher the steering ratio, the easier it is to steer the car, all other things being equal. However, the higher the steering ratio, the more the steering wheel has to be turned to achieve steering.

- Actual steering ratios vary greatly, depending on the *type of vehicle and the type of operation*. Many cars with manual steering use steering ratios as high as 28 : 1 to minimise steering effort. Some lightweight sports cars have steering ratios as low as 0.1.
  - *High steering ratios* are often called *"slow steering"* because the steering wheel has to be turned several degrees to produce a small steering effect.
  - *Low steering ratios* called *'faster quick steering'*, require much less steering wheel movement to produce the desired steering effect.
- Steering ratio is determined by *two factors* :

  Steering-linkage ratio and *the gear ratio* in the steering gear. The steering-linkage ratio is determined by the relative length of the pitman arm and the steering arm. The steering arm is bolted to the wheel spindle at one end and connected to the steering linkage at the other end.
- The *vehicles equipped with 'power steering' have steering ratio 20% less than the vehicles with manual steering.*
- Many steering gears have a *variable gear ratio*, which is higher for the straight ahead range and lower for the outer range. It is obtained by *varying the pitch of the worm* or *cam so that it is flatter at the centre.*

8.7. REVERSIBILITY

The steering gear is said to be **reversible** *if deflection of the steered wheels due to road surface is transmitted through the steering linkage and steering gear box to the steering wheel.* If every small imperfection of the road surface causes the steering wheel to rotate, the driver would find much tiring and frustrating.

Some degree of irreversibility is desired to stop shocks sustained by the road wheels. Such a system is known as *semi-reversible.*

The steering system is said to be *irreversible* if the steered wheels do not transfer any deflection to the steering wheel. It would not tend to straighten out after negotiating a turn, and would not easily follow the course of rutted road without undue stress on the mechanism. Therefore, *semi-reversible steering gears are used in most of the passenger cars.*

8.8. STEERING GEOMETRY

When a car is moving along a curve, all its wheels should *roll truly without any lateral slip.* This can be achieved if the axes of all four wheels intersect at one point. This point will be the centre about which the vehicle will be turning at that instant. Fig. 8.9, shows the steering geometry of the four wheels of a vehicle. The rear wheels rotate along two circles. The centres of both these circles are at O. The front wheels 1 and 2 have different axes. They rotate along two other circles with the *same centre point. For correct functioning of any steering system, the centres of the wheels of the rear axles and of wheels 1, 2 should coincide.*

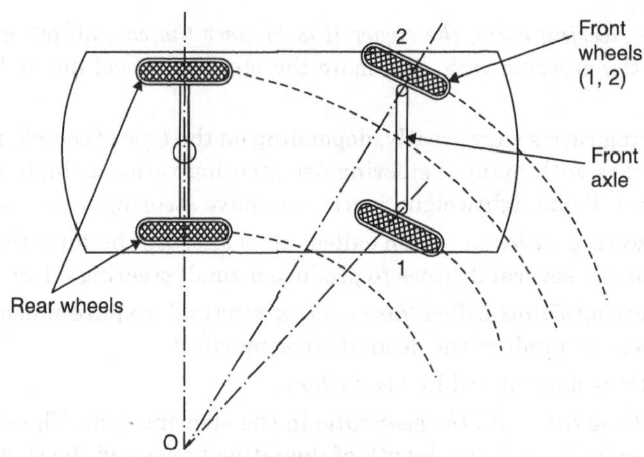

Fig. 8.9. Steering geometry.

8.9. WHEEL ALIGNMENT

Wheel alignment *refers to the positioning of the front wheels and steering mechanism that gives the vehicle directional stability, promotes ease of steering and reduces tire wear to a minimum.*

A wheel is said to have directional stability or control if it can :

— run straight down a road,

— enter and leave a turn easily, and

— resist road shocks.

The *front wheel alignment depends upon the following factors* :

I. ***Factors pertaining to steering geometry :***

1. Camber 2. King pin inclination 3. Caster 4. Toe-in 5. Toe-out.

II. ***Factors pertaining to front wheel condition :***

1. Balance of wheels 2. Inflation of tyres 3. Brake adjustments.

1. **Camber :**

Camber *is the tilting in or out of the front wheels from the vertical when viewed from the front of the vehicle.* If the top of the wheel tilts *out*, it has "*positive*" camber (Fig. 8.10). If the top of the wheel tilts *in*, it has "*negative*" camber. *The amount of tilt measured in degrees from the vertical, is called "camber angle".*

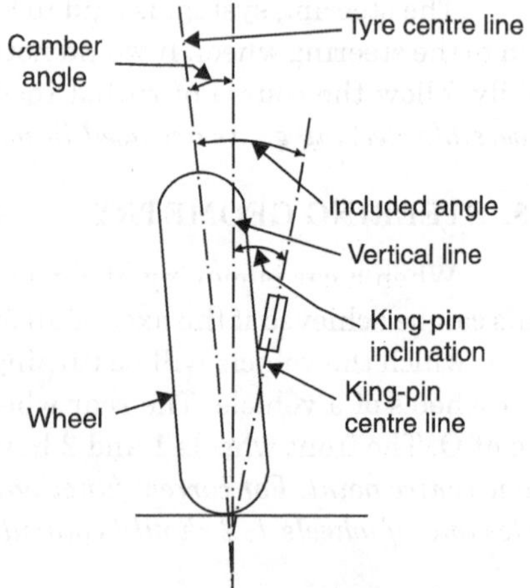

Fig. 8.10. Camber angle (positive) and king-pin inclination.

Any amount of camber, positive or negative, *tends to cause uneven or more tyre wear on one side than on the other side.* Camber should *not exceed two degrees.*

The front wheels are not usually mounted parallel to each, but are fitted outward slightly at the top and inward at the bottom to have positive camber. *The purpose of the camber is to prevent the top of the wheels from tilting inward too much because of excessive loads or play in king pins and wheel bearing.* When the vehicle is loaded and rolling along on the road, the load will just bring the wheels to a vertical position. The opposite is true in case of vehicles having negative chamber.

2. **King-pin inclination :**

The ***king-pin inclination*** *(or steering axle inclination) is the angle between the vertical line and centre of the kingpin or steering axle, when viewed from the front of the vehicle (Fig. 8.6).*

The king-pin inclination is absolutely necessary due to the following reasons :

(*i*) It helps the car to have steering stability.

(*ii*) It makes the operation of the steering quite easy particularly when the vehicle is stationary.

(*iii*) It helps in reducing the wear on the tyre.

Included angle :

The *combined camber and king-pin inclination is called the "**included angle**"*.

This angle is important because it determines the point of intersection of the wheel and king pin centre lines. This in turn determines whether the wheel will tend to toe-out or toe-in.

— If the point of intersection is *above the ground*, the wheel *tends to toe-in.*

— If it is *below the ground*, the wheel tends to *toe-out.*

— If it is at the ground, the wheel keeps its straight position without any tendency to toe-in or toe-out. In this position the steering is called *centre point steering.*

3. **Caster :**

The angle between the king-pin centre line (or steering axis) and the vertical, in the plane of the wheel is called ***Caster angle.*** If the king-pin centre line meets the ground at a point ahead of the vertical centre line, as is shown in Fig. 8.11 it is called *positive* caster while if it is behind the vertical, wheel centre line it is called *negative caster*. The caster angle in modern vehicles range from 2° to 8°.

The purpose of caster is to give a trailing effect to the front wheels. When the wheel trails the line of weight, that is, moves in the same direction as the vehicle, it is easy to steer a straight course.

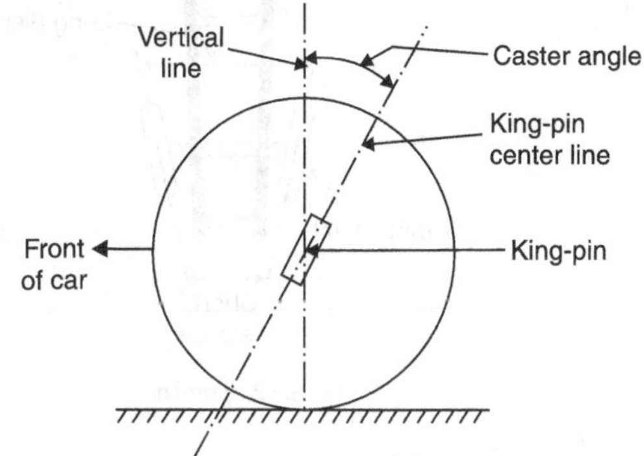

Fig. 8.11. Caster angle (Positive).

- The positive caster in wheels results in a natural tendency in the wheels to toe-in. The negative caster would have the opposite effect, *i.e.*, the wheels will tend to toe-out.
  — The *positive caster increases the effort required to steer and tries to keep the wheels straight ahead.* In the *heavy duty trucks negative caster is provided, this makes steering easier.*
- When both the wheels have the same positive caster, both will have equal tendencies to toe-in, which will be balanced by each other, because track rod is provided to maintain the distance between the wheels rigidly. However, when the caster at the two wheels is not equal, the tendency to toe-in at the wheel with larger caster will be more which will cause the vehicle to pull constantly towards the side of the wheel with lesser caster.
- Since the change of caster angle results in the change of the other angles of the steering geometry, *i.e.*, camber, king-pin inclination and toe-in or toe-out, it is very important that this angle is adjusted first of all, while doing the adjustment job.
- *About 3° of caster gives good results.*

4. **Toe-in**

The position of toe-in of a vehicle is shown in Fig. 8.12. The distance at the front end between the front wheels is *less* than the distance between them at the rear. The front wheels point inward. The actual amount of toe-in in only a few millimetres (usually 3 to 5 mm).

The toe-in serves the following *purposes :*

(*i*) It ensures parallel running of the front wheels.

(*ii*) It stabilises steering.

(*iii*) It prevents side slipping and excessive wear of the tyres.

(*iv*) It also serves to offset the small deflections in the wheel-support system which comes out when the car is moving forward.

Although the wheels are set to toe-in slightly when the car is standing still, they tend to roll parallel on the road when the car is moving forward. Some alignment specialists set the front wheels in "straight-away alignment" in preference to "toe-in adjustment".

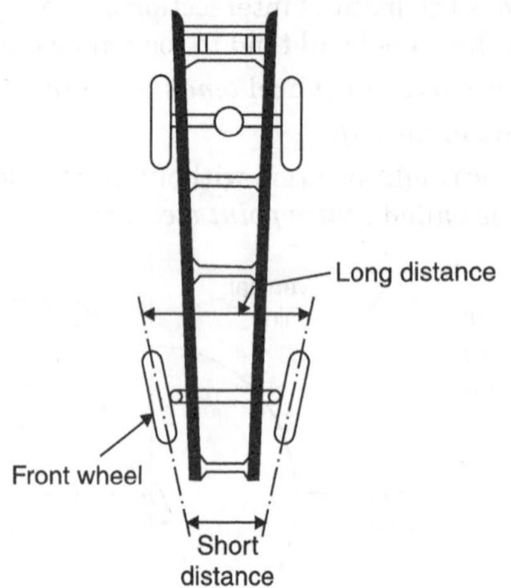

Fig. 8.12. Toe in.

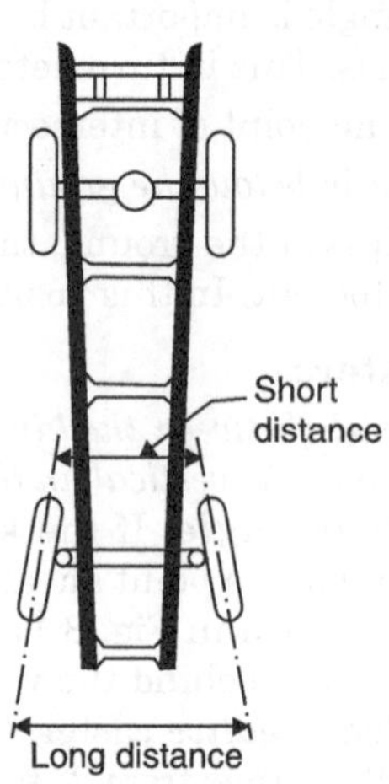

Fig. 8.13. Toe-out.

5. **Toe-out**

In Fig. 8.13 the distance between front wheels at front end is *greater* than the distance between them at the rear. When car is stationary, this position of the front wheel is called *toe-out*.

The *purpose of toe-out is to give correct turning alignment and to prevent excessive wear.*

Toe-out on turn **(Turning radius) :** Refer to Fig. 8.14.

"Toe-out" is the difference in angles between the two front wheels and the car frame during a turn. The steering is designed to turn the inside wheel through a larger angle than the outside wheel when making a turn. This condition causes the wheels to toe-out on turns, due to the difference in their turning angles. When the car is taking a turn, the outer wheels roll on a large radius than the inner wheel, and the circles on which the two front wheels must roll are *concentric*. Therefore, the inner wheel must make a large angle with a car frame than that the outer wheel makes. As shown in Fig. 8.14, when the front wheels are steered to make a turn, the inner wheel turns to an angle of 23° with the car frame, while the outer wheel turns only 20° with the car frame.

- The *toe-out is secured by providing the proper relationship between the knuckle arms, tie rods and pitman arm.*
- ***Turning radius*** is checked during wheel alignment. If it is incorrect it cannot be adjusted. A *vehicle with an incorrect turning radius will drag the wheels around a corner, causing fast wearing of tyres.* The turning radius of cars varies from 5 m to 7.5 m, whereas in case of buses or trucks it is as high as 13.85 m.

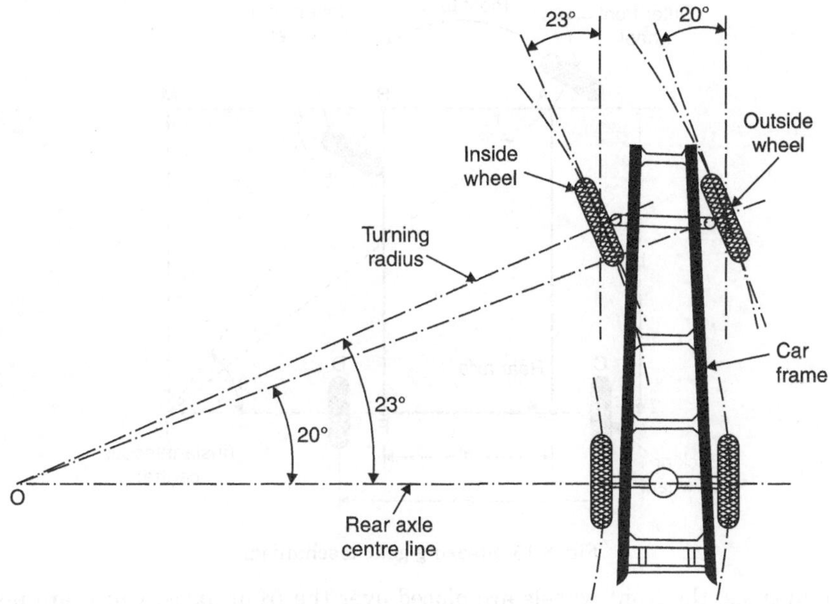

Fig 8.14. Toe-out on turn.

Checking of wheel alignment :

Before checking the wheel alignment, king-pin inclination, camber and caster angles and toe-in need to be checked.

The king-pin inclination and the camber angle are fixed in those vehicles which have axle beam. But the caster angle is adjustable by a caster plate. In those vehicles having independent suspension system, these angles are adjusted by a shim between the lower and upper arm brackets or by eccentric clamping bolts of the arm.

- The camber, caster king-pin inclination are checked by a particular type of gauge. This gauge consists of two parts : Level and angle gauge and turn table.
- For checking toe-in several gauges are available.
  - Optical gauge of Dunlop Tyre Company gives very accurate reading.
  - *Telescopic gauge* is commonly used in workshops. It consists of two pipes which can slide one inside the other. At the end of each pipe there is a rod which can be adjusted at any position on the pipe. Each rod has an adjustable pointer. To check the toe-in touch the pointers of the gauge with the front wheels and note the reading on the pipe. Similarly, touch the pointers of the gauge with the rear portion of the front wheels and note the reading. The *difference between the two readings is the toe-in.*

8.10. STEERING MECHANISMS

8.10.1. Fundamental Equation for Correct Steering

The **steering gear mechanism** is *used for changing the direction of two or more of the wheel axles with reference to the chassis, so as to move the automobile in any desired path.* Usually the two back or rear wheels have a common axis, which is fixed in direction with reference to the chassis and the steering is done by means of front wheels.

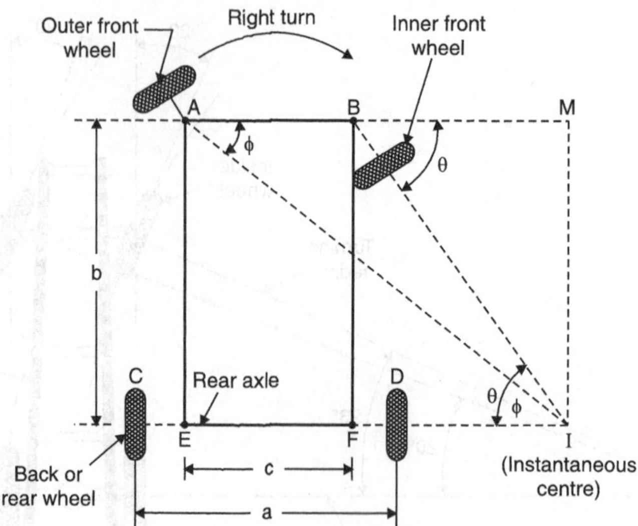

Fig. 8.15. Steering gear mechanism.

In automobiles, the front wheels are placed over the front axles, which are pivoted at the points A and B, as shown in Fig. 8.15. These points are fixed to the chassis. The back or rear wheels are placed over the back axle, at the two ends of the differential tube. When the vehicle takes a turn, the front wheels along with the respective axles turn about the respective pivoted points. The rear wheels remain straight and do not turn. Therefore, the steering is done by means of front wheels only.

In order to avoid skidding (*i.e.*, the slipping of the wheels tyres), the two front wheels must turn about the same instantaneous centre I which lies on the axis of the rear wheels. If the instantaneous centre of the two front wheels do not coincide with the instantaneous centre of the rear wheels, the skidding on the front or rear wheels will definitely take place, which will cause more wear and tear of the tyres.

Thus, the *condition for the correct steering is that all the four wheels must turn about the same instantaneous centre.* The axis of the inner wheel makes a larger turning angle θ than the angle ϕ subtended by the axis of outer wheel.

Let, a = Wheel track

b = Wheel base, and

c = Distance between the pivots A and B of the front axle.

Now, from ΔIBM, $$\cot \theta = \frac{BM}{IM}$$

and, from ΔIAM, $$\cot \phi = \frac{AM}{IM} = \frac{AB + BM}{IM} = \frac{AB}{IM} + \frac{BM}{IM} = \frac{c}{b} + \cot \theta \qquad (\because \ IM = b)$$

$$\therefore \qquad \cot \phi - \cot \theta = \frac{c}{b} \qquad ...(8.1)$$

This is the *fundamental equation for correct steering*. If this condition is satisfied, there will be no skidding of wheels, when the vehicle takes a turn.

8.10.2. Types of Steering Gear Mechanisms

The steering gear mechanisms are of the following *two* types :

1. Davis steering gear 2. Ackermann steering gear.

- "*Davis steering gear*" has *sliding pairs.* Since sliding pair has more friction (than the turning pair), therefore this type of gear will wear out earlier and become inaccurate after sometime.
- "*Ackermann steering gear*" has *turning pairs.* This type of gear is not mathematically accurate except in three positions contrary to the Davis steering gear which is mathematically correct in all position. However, the Ackermann steering fear is *preferred to the Davis steering gear.*

The whole mechanism of the Ackermann steering gear is on the *back* of the front wheels, whereas in Davis steering gear, it is in *front* of the wheels.

1. **Davis steering gear :**

Refer to Fig. 8.16. It consists of a cross link *CD* sliding parallel to another link *AB* and is connected to the stub axles of the two front wheels by means of two similar bell crank levers *LAC* and *MBD* pivoted at *A* and *B*, respectively. The cross link *CD* slides in the bearing and carries pairs at its ends *C* and *D.* The slide blocks are pivoted on these pins and move with the turning of bell crank levers as the steering wheel is operated.

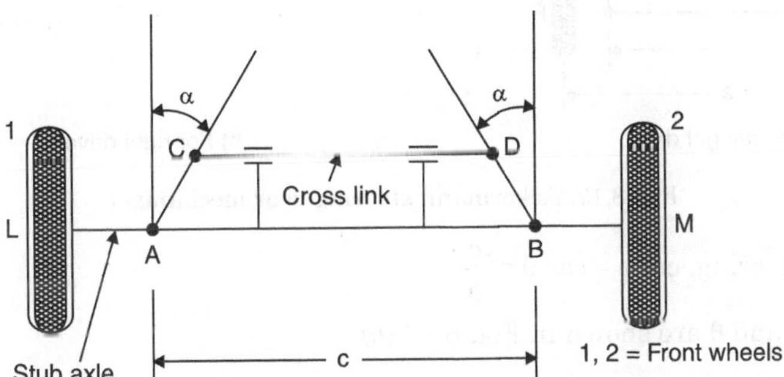

Fig. 8.16. Davis steering gear mechanism.

When the wheel is running straight, the gear is said to be in its mid position. The short arms *AC* and *BD* are inclined at an angle (90° + α) to their stub axles *AL* and *BM*, respectively.

The correct steering depends upon the suitable selection of cross-arm angle α, and is given by :

$$\tan \alpha = \frac{c}{2b}$$

where, c = Distance between the pivots of front axles, and

b = Wheel base.

The range of $\frac{c}{b}$ is 0.4 to 0.5, hence α lies between 11.3° and 14.1°.

Note. Davis steering gear is *theoretically correct*, but due to presence of more sliding members, the wear will be increased which produces slackness between the sliding surfaces, thus *eliminating the original accuracy.* Hence this gear is *not in common use.*

2. **Ackermann steering gear :**

Refer to Fig. 8.17. It consists of a cross link *CD* connected to the short axles *AL* and *BM* of the two front wheels through the short arms *AC* and *BD*, forming bell crank lowers *LAC* and *MBD*, respectively.

When the wheel is running straight [Fig. 8.17 (*a*)], the cross-link *CD* is parallel to *AB*, the short arms *AC* and *BD* both make angle α to the horizontal axis of chassis. In order to satisfy the fundamental equation for correct steering, the links *AC* and *BD* are suitably proportioned and angle α is the suitably selected.

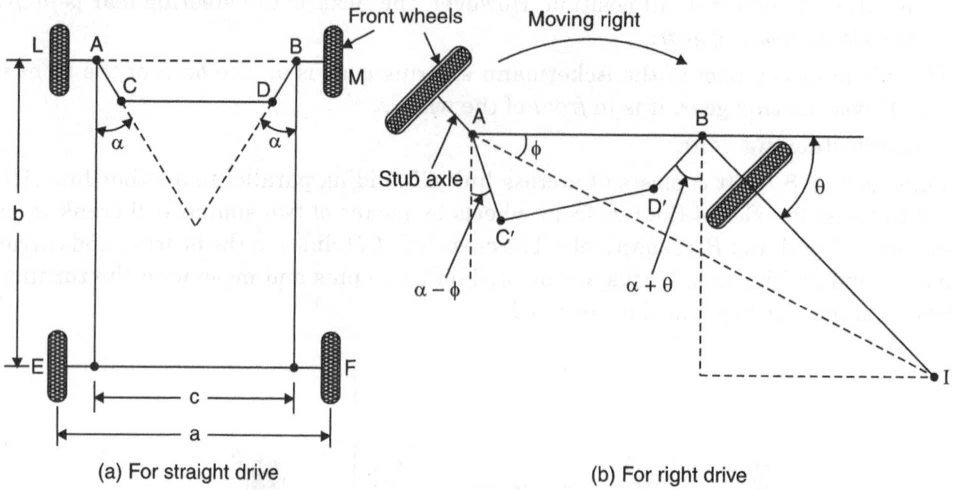

Fig. 8.17. Ackermann steering gear mechanism.

For correct steering, $\cot \phi - \cos \theta = \dfrac{c}{b}$

The angles ϕ and θ are shown in Fig. 8.17 (*b*).

The value of $\dfrac{c}{b}$ lies between 0.4 and 0.5, generally 0.455. The value of $\cot \phi - \cot \theta$ corresponds to the positions when steering is correct.

In fact there are *three* values of θ which give correct steering of the vehicle : *First,* while turning to 'right' ; *second,* while turning to left and *third* while it is running 'straight'.

8.11. UNDERSTEERING AND OVERSTEERING

When a vehicle takes a turn, a centrifugal force acts on it which produces a side thrust. In order to sustain that force the plane of the wheel must make some angle with the direction of motion of the vehicle. This is achieved by the distortion of the tyre which is flexible (Fig. 8.18). The *angle through which the wheel has to turn to sustain the side force is called* ***slip*** *or* ***creep angle*** and the force produced due to this (at right angles to the plane of the wheel) which counters the side thrust, is known as "***cornering force***".

The *value of the slip angle depends upon the following factors* :

(*i*) Amount of side force,

(*ii*) Flexibility of tyre,

(*iii*) Load carried by the wheel,

(*iv*) Camber angle, and

(*v*) Condition of the road surface, etc.

The magnitude of slip angle is small at low speeds and less sharp curves and increases at high speeds and on sharp turns, till on excess of speed and sharpness of curve the wheel skids sideways.

For the same slip angle, positive camber increases the cornering force while negative chamber decreases it. *The alteration in the cornering force, due to camber is usually called "**camber force**".*

The *ratio of side force sustained and slip angle* is *called the* ***cornering power***. The cornering power depends largely upon the *design of the tyre and its inflation pressure.*

When the slip angle of the front wheel is *greater* than the rear wheel, the wheel turns at a radius *larger* than that intended and the driver has to keep steering into the turn. This condition is known as ***understeer.***

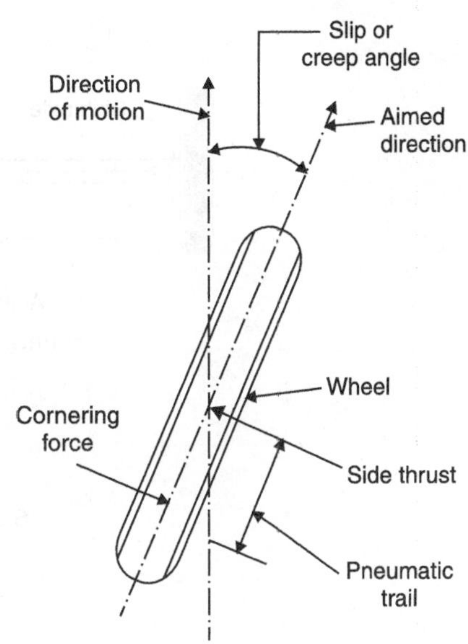

Fig. 8.18. Slip angle.

When the slip angle is greater at the rear wheel than the front, the vehicle tends to **oversteer,** that is to turn into the curve more than the driver intended.

When the front and rear slip angles are the same neutral steer occurs.

- Although both the understeer and oversteer are *unwanted*, yet *understeer is less undesirable* because the driver reacts naturally and positively by steering in the desired direction. In the case of oversteer, he is always conscious of losing the control and therefore greater care is needed. Owing to this reason, generally the vehicles are designed to provide understeer conditions at normal speeds. However, at higher speeds, the situation changes first to neutral steer and then to oversteer. The above explanation (of understeer and oversteer conditions) is equally valid even when the vehicle is going straight and is subjected to side force due to some cross-wind or camber.

8.12. STEERING LINKAGES

Steering linkage *is a connection of various links between the steering gear box and the front wheels.* It depends upon the type of the vehicle, whether it is a car which has independent front suspension or a commercial vehicle having generally a rigid axle type front suspension.

The steering linkage transfers the side-to-side or front-to-rear movement of the pitman arm into the left-to-right movement at the wheels.

Conventional steering linkage :

Fig. 8.19 shows a steering linkage for a conventional rigid axle suspension. It is commonly used in cars provided with rigid front axle.

— The *drop arm* (or Pitman arm) is rigidly connected to cross-shaft of the steering gear at its upper end, which its lower end is connected to the *link rod* through a ball joint. To the other end of the link rod is connected the link rod arm through a ball joint. Attached rigidly to the other end of the link rod arm is the *stub axle* on which the road wheel is mounted.

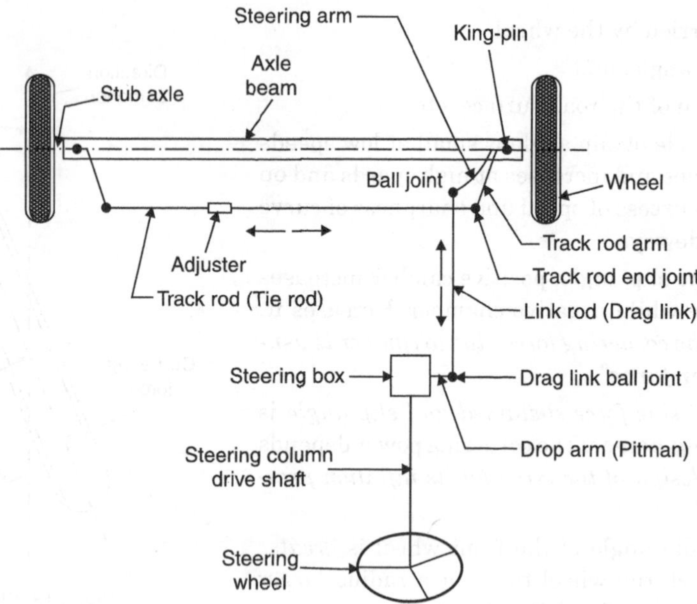

Fig. 8.19. Steering linkage for a conventional rigid axle suspension.

— Each stub axle has a forged track rod arm rigidly bolted to the wheel axis. The other ends of the track rod arms are connected to the track rod by means of ball joints. An adjuster is also provided in the track rod to change its length for adjusting wheel alignment.

Working. When the steering wheel is turned, the swinging action of the drop arm imparts a near linear movement to the link rod. This movement is transmitted through the link rod arm to the stub axle so as to turn the later about its pivot, which may be a king-pin or ball joints. The other wheel is steered through the track rod. Thus only one wheel is positively steered.

Steering linkages for independent front suspension :

Some steer linkages for independent front suspension are shown in Fig. 8.20. In these linkages, the ball joints are fitted between the linkage and the steering arms to allow for the independent movement of the wheels.

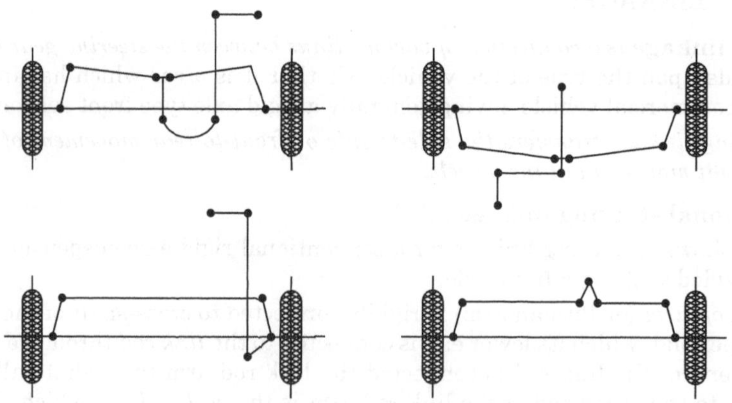

Fig. 8.20. Various types of steering linkages.

8.13. STEERING WHEEL AND COLUMN

- The **steering wheel** *consists of a hub with metal spokes and a rim fabricated* as *a single piece.* The diameter of the wheel rim should be large enough to convert the driver rim pull into maximum input torque but is limited by the comfort of driver. The rim is *elliptical* in cross-section with finger indentations on the undersuface and is so dimensioned that a good grip is obtained. In our country steering wheels have a fixed position, but in foreign countries, in some vehicles these wheels can be tilted and locked in any position to suit the driver. The size of the steering wheel should be such that in case of an accident the impact load is transmitted to the driver's shoulders and not to chest.
- The **steering column** on today's car is designed to do more than just house the steering shaft. As the steering is closest to the driver, a number controls such as horn, lights, turn signals, windshield wipers and washes, ignition key and variety of other devices are provided on it. For protecting the driver steering columns are designed to be a *collapsible steering column.* In such a steering, the shaft and the tube or column are made from two pieces that fit together like an extended telescope. *When a collision takes place, the column and shaft absorb energy and shaft collapses back into the column.*

The various types of steering columns employed in many foreign cars may be *summed up* as follows :

1. Energy absorbing steering column.
2. Tilt wheel steering column.
3. Tilt and telescopic steering column.
4. Steering column with anti-theft lock.

8.14. STEERING ARM

The steering arm (or drop arm) varies between 12.5 cm and 20 cm in case of motor *cars* and between 20 cm and 25 cm for *commercial vehicles* of the heavy type. However, the *exact length* of steering arm depends upon the following factors :

(*i*) Angular motion of steering knuckle,

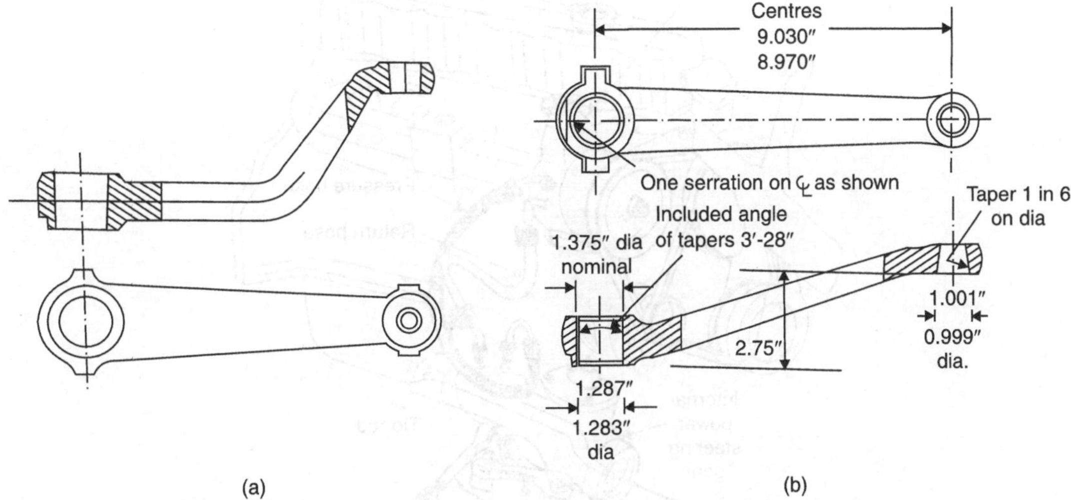

Fig. 8.21. Various types of drop arms (Courtesy : Ashok Leyland).

(*ii*) Angular motion of the steering arm, and

(*iii*) Length of knuckle arm to which the drag link gets connected.

Fig. 8.21 shows the two different types of drop arms fitted on commercial vehicles manufactured by M/s Ashok Leyland.

The arm is secured to the rocker shaft with the help of a tapered, serrated joint, on the tapered portion of the rocker shaft by means of nuts and washers.

8.15. DRAG LINK

- A **drag link** *connects the steering arm or drop arm with knuckle arm. It is of the same section as the tie rod and is made of tubular stock.*
- The section of a drag link and tie rod may be also different but it is advisable to have the same section because of two *reasons* : Firstly, having the same section will give *more safety factor* which is desirable in this part. Secondly, there are advantages from manufacturing point of view since the same tubular stock can be used for *maintaining similarity in the design of connectors.*

8.16. STEERING STOPS

- The *stops are provided to limit the angular deflections of the front wheels.* The *purpose of these stops is to avoid rubbing of tyres against the frame or against the fenders which would cause undue wear and tear of the tyres.*
- The steering stops can be provided at the following two different places :

(*i*) In the path of motion of the steering arm or drop arm.

(*ii*) In the path of motion of the steering knuckle.

8.17. POWER-STEERING

8.17.1. Fundamentals of Power-steering

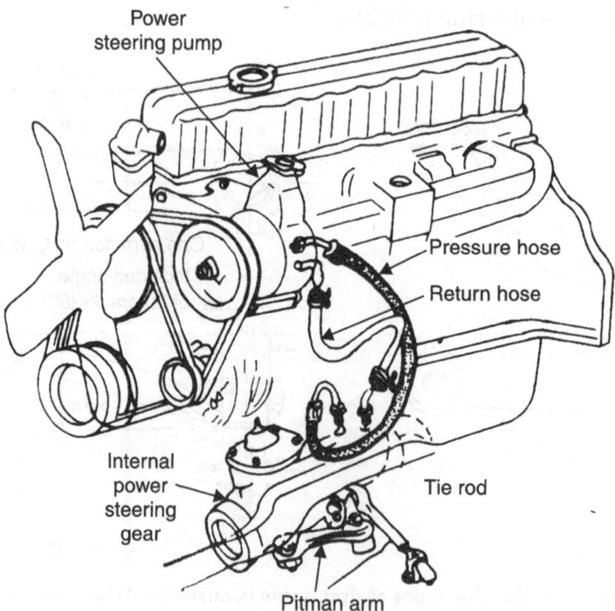

Fig. 8.22. A typical power steering installation.

- When a car is steered manually, the driver supplies all the steering force. Then, through the mechanical advantage of the steering gear and linkage, the front wheels are pointed to the right or left as desired by the driver. Large amount of torque is required to be applied by the driver for steering of medium and heavy vehicles. The *power steering system provides automatic hydraulic assistance to the turning effort applied to the manual steering.* The power system is designed to become operative when the effort at the wheel exceeds a predetermined value (say 10 N). The system is always so designed that in the event of the failure of the power system, the driver is able to steer the vehicle manually although with increased effort.
- In the **power-steering system,** *a continuously operating pump provides hydraulic pressure when needed* (Fig. 8.22). *At the steering wheel is turned, valves are operated to admit this hydraulic pressure to a cylinder that contains the power piston. Then the pressure causes the piston to move, and it provides most of the steering force.*

8.17.2. Types of Power-steering Systems

In automobiles, the following two general types of power-steering systems are found :

1. Integral power steering.
2. Linkage power steering.

1. **Integral power steering :**

In integral power steering (Fig. 8.23), the *power piston is built into the steering gear.*

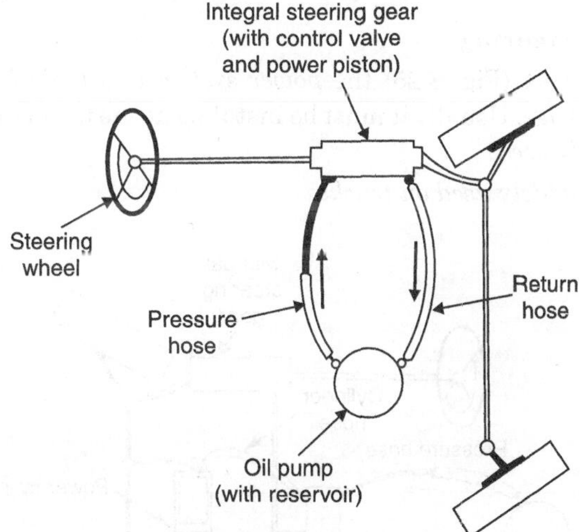

Fig. 8.23. Integral power steering (simplified diagram).

Fig. 8.24 shows a cut away rotary-valve integral power-steering gear.

It consists of a *spool valve, rotary valve,* and a *torsion bar.* The torsion bar is attached to the end of the worm shaft. When the steering wheel is turned, the torsion bar twists. This moves the spool valve slightly, opening passages which move fluid to the side of the piston where hydraulic assist is required. When there is no power-assist action, the spool valve is in the neutral or straight-ahead position.

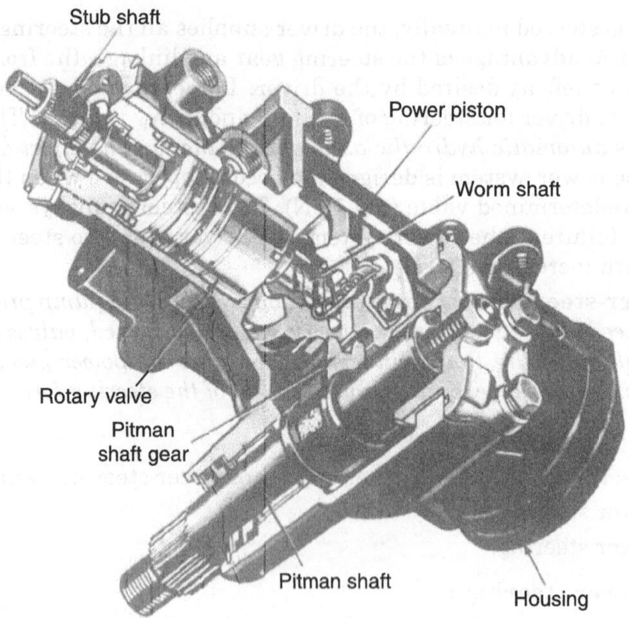

Fig. 8.24. Cut away rotary valve integral power steering gear.

2. **Linkage power steering :**

In this type of steering (Fig. 8.25) the *power system is attached between the frame of the vehicle and the steering linkage.* Usually it must be installed at the factory as the vehicle is built. The integral type is *more widely used.*

The *linkage type is widely used on trucks.*

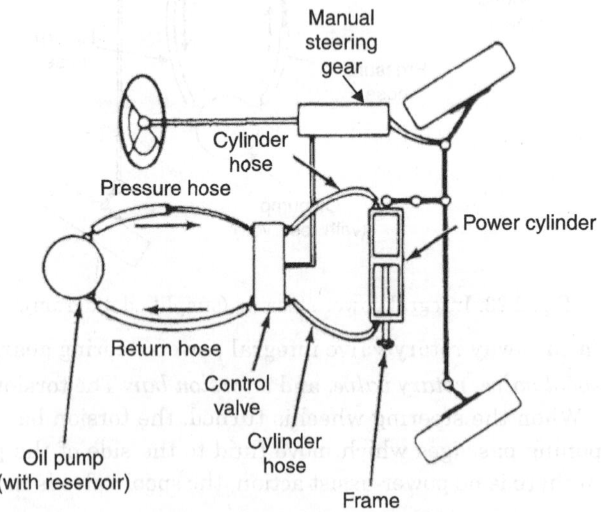

Fig. 8.25. Linkage power steering (simplified diagram)

Fig. 8.26 shows a typical linkage power steering (Ford Motor)

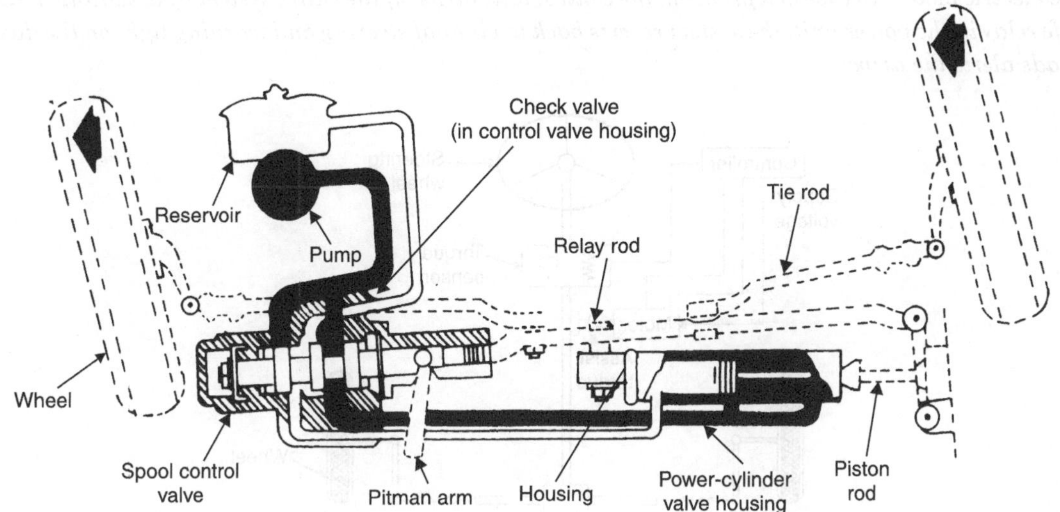

Fig. 8.26. Linkage power steering (Ford Motor Co.)

- When the vehicle is running "*straight ahead*" (*i.e.* in neutral position), a spool valve in the control valve assembly is held in the centre position by means of a centring spring. Oil flows to both sides of the piston in the power cylinder and around the valve bands and returns to the pump reservoir.
- While the vehicle is taking "*left turn*", and the force on the wheel exceeds about 20 N, the pitman arm moves the spool control valve with sufficient force to overcome the pressure of the centring spring so that the valve slides to the right hand side of the valve body. In this position it connects the left-hand side of power cylinder to return line leading to the pump reservoir and directs oil pressure to the right hand side of the power cylinder. This causes the cylinder housing to move to the right, which moves the relay rod to the right in order to turn the wheels to the left.
- In case of the wheel taking a "*right turn*", the conditions are reversed and the relay rod is forced to turn the wheels to the right.

Electronic power steering :

In this type of steering system (Fig. 8.27), steering censor consists of two sensors : "*Torque sensor*" and "*Rotation sensor*".

'*Torque sensor*' converts the steering torque input and its direction into voltage signals.

'*Rotation sensor*' converts the rotation speed and direction into voltage signals and is located on the input shaft of the steering gear box.

The inputs from the steering sensor and the vehicle speed sponsor are sent to a *microprocessor control unit* where these are compared with a preprogrammed force assist map. The control unit then sends out the appropriate command signal to the current controller which supplies the appropriate current to the electric motor which pushes the rack to the right or left depending on in which direction the current flows. The amount of power assist increases with the increase in current. There is a provision to protect the electric motor from being overloaded and also from the voltage surges from a faulty alternator or charging problem. The electronic steering control unit is capable of self-diagonising faults by monitoring the system's inputs and outputs and the driving current of the electric motor. In case of a *problem, the control unit turns off the entire system by actuating a fail-safe relay in the power unit, the system reverts back to manual steering and warning light on the dash boads alerts the driver.*

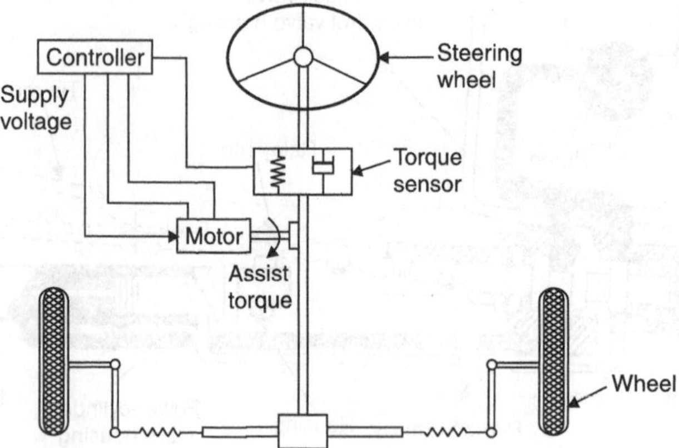

Fig. 8.27. Electronic power system.

Advantages :

Following are the *advantages of an electronic power steering over hydraulic power steering* :

1. Manual steering requires less force comparatively, since there is no fluid to be forced through the valves.
2. Energy is consumed only while steering.
3. Availability of steering assistance even when the engine is not running.
4. The problem of fluid leakage is eliminated.

- The following advance steering systems are under different stages of development :

1. Variable ratio rack and pinion steering system.
2. Speed sensitive steering.
3. Four wheel steering.

8.18. ADJUSTMENT OF STEERING GEOMETRY

The adjustment of steering geometry involves the following *steps* :

1. Levelling of springs.
2. Measurement of camber angle.
3. King-pin inclination and the combined angle.
4. Measurement and adjustment of caster angle.
5. Adjustment of camber angle.
6. Measurement and adjustment of toe-in.

8.19. STEERING TROUBLESHOOTING

Troubleshooting chart for steering mechanism is given below :

	Complaint	Causes	Remedy
1.	*Hard steering.*	(*i*) Low or uneven pressure. (*ii*) Incorrect lubricant. (*iii*) Insufficient lubricant. (*iv*) Broken or bent steering arms or knuckles, or suspension arms. (*v*) Too tight spherical ball joints.	(*i*) Inflate the tyres to correct pressure. (*ii*) Lubricate with recommended lubricants. (*iii*) Apply sufficient amount of lubricant in the steering box system. (*iv*) Replace the bent or broken parts. (*v*) Replace all tight ball joints.
2.	*Excessive play or looseness in the steering system.*	(*i*) Very loose steering linkage. (*ii*) Worn out steering linkage. (*iii*) Too loose front wheel bearings. (*iv*) Worn out front wheel bearings. (*v*) Loose steering gear flexible coupling. (*vi*) Worn out steering gear flexible coupling.	(*i*) Locate the points of excessive play and tighten the connection. (*ii*) Replace the worn out parts. (*iii*) Tighten the loose bearings. (*iv*) Replace the worn out parts. (*v*) Tighten the loose couplings. (*vi*) Replace the worn out coupling.
3.	*Erratic steering, especially when the brakes are applied.*	(*i*) Worn out brake lining. (*ii*) Brake lining choked with oil. (*iii*) Brake lining choked with brake fluid.	(*i*) Replace the brake lining. (*ii*) Locate and remove the cause for choke. (*iii*) Remove the cause for choke due to the brake fluid.
4.	*Car* wanders on road, i.e., substantial steering is required to keep it moving in a straight line.*	(*i*) Bent steering knuckle. (*ii*) Bent king-pin. (*iii*) Inoperative stabilizer.	(*i*) Replace the steering knuckle. (*ii*) Replace the king-pin. (*iii*) Repair the stabilizer.

* *Wander* is the tendency of a car to veer away from a straight path without driver control.

8.20. COMPARATIVE STEERING DATA OF SOME INDIAN AUTOMOBILES

The comparative steering data of some Indian automobiles is given in the chart below :

S. No.	Vehicle make	Type of steering	Camber	Caster	King pin inclination	Toe-in (mm)	Steering ratio	Turning circle (Metres)
1.	*Maruti (Suzuki) 800*	Rack & Pinion	1°21′	3°15′	12°50′	2 – 4	17.5 : 1	8.8
2.	*Hindustan Ambassador Mark II*	Rack & Pinion	$\frac{1°}{2}$ to 1°	$2\frac{1°}{2}$ to $3\frac{1°}{2}$	7° – 42′ to 8° – 42′	1.6 – 4.0	14 : 1	10.83 (Right) 10.74 (Left)
3.	*Fiat 1100*	Worm & Roller	(0° – 30′) ∓ 20′	(20° – 10′) + (10° – 30′)	7°	7 – 9 Unladen 1 – 3 Laden	16.4 : 1	

4.	*Jeep CJ Series*	Worm & Roller	$1\frac{1}{2}°$	3°	$7\frac{1}{2}°$	1.2 – 2.4	20 : 1	11.72
5.	*Ashok Leyland Comet Passenger*	Cam & Double Roller	$1\frac{1}{2}°$	1°	3°	Nil	24.7 : 1	20.7
6.	*Tata 1210 vehicles*	Recirculating Ball	1°	30° – (1° – 30′)	9° – 30′	1 – 3	34.2 : 1	
7.	*Dodge/Fargo Model 89 M4*	Recirculating Ball	$1\frac{1}{2}°$	$\frac{1}{2}°$	8°	0 – 3.165	24 : 1	
8.	*Standard 20*	Worm and Roller or Recirculating Ball	1° – 36′	Nil	5° – 58′	1.5		10.36
9.	*Swaraj Mazda*	Ball nut type with Tilt and Telescopic Steering Wheel	1° – 30′	2° – 30′	7°	0 – 6	24 – 30 : 1	

B. FRONT AXLE

8.21. INTRODUCTION TO FRONT AXLE

Front axle is *an axle over which the front wheels are mounted.* The front axle is fitted with wheels at its ends. A part of the weight of the vehicle is transmitted to the wheels through this axle.

In automobiles, the front wheels are fixed to the chassis in two ways, *i.e.,* either with the independent suspension system or with the help of axle beam with leaf spring.

The front axle performs the following **functions :**

1. It carries the weight of the front of the vehicle.
2. It carries the horizontal and vertical loads on bumpy roads.
3. It works as a cushion through its spring for a comfortable side.
4. In a four-wheel drive it also transmits power to the road vehicles.
5. When brakes are provided at the front wheels, it withstands bending stresses and torsional stresses.

8.22. CONSTRUCTION OF FRONT AXLE

The front axle must be robust in construction. It is usually a steel drop forging having 0.4% carbon steel or 1-3% nickel steel.

The front axle is a simple beam of steel which is changed depending on the stresses induced.

When the front wheel is *not braked,* the axle beam is usually a simple forging of I-section with suitable seats for attachment of the springs and with the ends suitably shaped to carry the stub axles (Fig. 8.28). I-section is adopted because it is *best adopted to withstand the bending action to which the beam is subjected.*

However, when the front wheels are *braked,* the horizontal bending action becomes considerable. In this case a *rectangular or circular section* may be more suitable than an I-section. Again, when the front wheels are braked, the portions of the axle beam between the stub axles and spring seats are subjected to a *twisting action*, the torque reaction of the braking torque applied to the brake drums. A *circular section is best adopted to withstand this twisting action.* Hence *when front wheels are braked, circular section may be used.*

In order to keep the low chassis heights, its centre portion is given a downward sweep (Fig. 8.28). The different *components of the front axles* are :

(*i*) Axle beam (*ii*) Stub axle

(*iii*) Swivel pin (*iv*) Track rod.

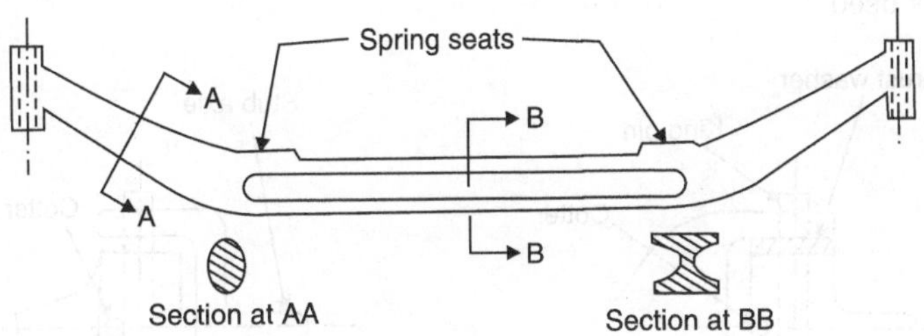

Fig. 8.28. Front axle.

A typical front axle with stub axle is shown in Fig. 8.29.

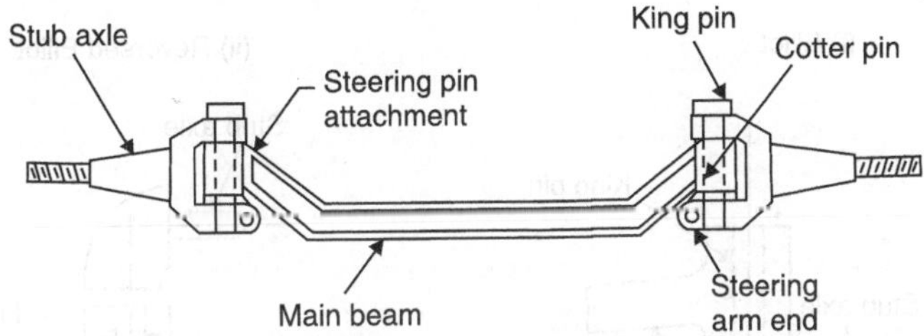

Fig. 8.29. A typical front axle.

8.23. TYPES OF FRONT AXLES

The front axles are mainly of the following two types :

1. Dead front axle. 2. Live front axle.

The front axles are usually *dead axles* since they do not rotate, in contrast to the *live axles* that they are used in the rear axle to transmit power to the rear wheels.

The *dead* front axle has sufficient rigidity and strength to transmit the weight of the vehicle front the springs to the front wheels.

- A *live* front axle, as compared to the dead axle, has *additional function of transmitting the driving power taken from a transfer gear box to the front wheels having a different swivelling mechanism*. The live front axles although resembling the gear boxes have some difference at the axle half shaft ends where the wheels are mounted.

8.24. STUB AXLES

The front wheels are mounted on the **stub axles**, *which are connected to the front axle by means of king pin.*

The stub axles are the forging of 3 percent nickel steel and alloy steels containing chromium and vanadium.

The stub axle turns on the pin which is a light drive fit in the axle beam eye located and locked by a taper pin or some similar arrangement.

Stub axle can have any of four shapes given in Fig. 8.30.

- The Elliot stub axle (Fig. 8.30 (*i*)) is attached to the front axle by placing it in the yoke end with a king pin and cotter to join the two together.
- In reversed Elliot type stub axle (Fig. 8.30 (*ii*)], the arrangement is reversed.
- In Lemoine type stub axle [Fig. 8.30 (*iii*) and (*iv*)], instead of yoke type hinge, a L-shaped spindle is used.

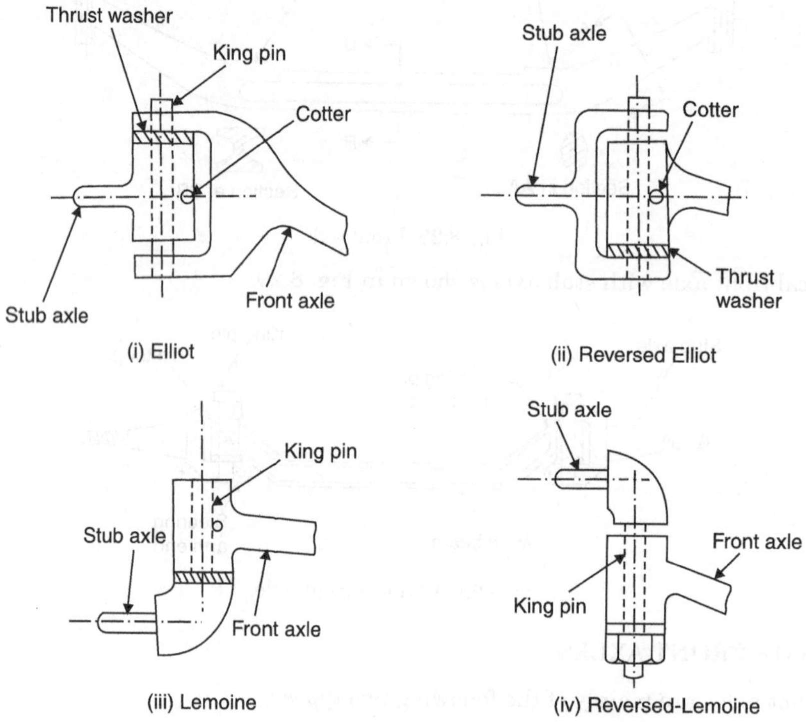

Fig. 8.30. Stub axles.

SHORT ANSWER QUESTIONS

Q. 1. What are the causes of hard steering ?

Ans. The causes of hard steering are : (*i*) Tight steering gear and linkage, (*ii*) Uneven tyre pressure, and (*iii*) Friction in steering gear and linkage.

Q. 2. Why do vehicles pull to one side ?

Ans. Due to (*i*) Incorrect pressure, (*ii*) Bent shock absorber, and (*iii*) Improper setting of steering geometry.

Q. 3. Why do vehicles wander on either side ?

Ans. Due to : (*i*) Unequal inflation of tyre, (*ii*) Loose king-pin, and (*iii*) Loose ball joint.

Q. 4. What are the causes of excessive steering play ?

Ans. (*i*) Wear or improper adjustment in the linkage ; (*ii*) Worn out ball joints, and (*iii*) Loose wheel bearing.

Q. 5. What is known as 'shimmy' in the steering system ?

Ans. *Shimmy refers to wear in the steering system or unbalanced road wheels causing a sideways wobbling of the front wheels.*

Q. 6. What is wheel wobbling ?

Ans. A rapid oscillation of steering wheel occurring at high speed is known as 'wheel wobbling'.

Q. 7. What are causes of wheel wobble ?

Ans. (*i*) Excessive play in the steering linkage, (*ii*) Wear or play in the wheel bearings and stub axle pins, and (*iii*) Incorrect tyre pressures.

Q. 8. What is camber ?

Ans. Camber is the tilting angle of the top of the front wheel from the vertical axis.

Q. 9. Why is camber angle provided ?

Ans. (*i*) To minimise the side thrust on the king pin, and (*ii*) To compensate for the shifting of the centre of the wheel plane outside the centre line of the king pin.

Q. 10. What is the effect of camber being more ?

Ans. More camber will cause more wear on the outer edge of the tyre.

Q. 11. How is camber setting adjusted ?

Ans. Camber setting is adjusted by rotating an eccentric cam on a bolt which holds the strut on to the steering knuckle.

Q. 12. What are the five important angles in steering ?

Ans. Caster, camber, toe-in, steering axle inclination, and toe-out.

Q. 13. What is caster ?

Ans. Caster is the forward or backward tilting of the steering axle to provide directional/steering ability.

Q. 14. What is the effect of excessive caster ?

Ans. When excessive caster is provided the vehicle will get lifted each time during running when the wheels are turned to right or left. This may cause hard steering and somo difficulty in straightoning the vehicle after turning.

Q. 15. What is positive and negative caster ?

Ans. Backward tilt is termed as positive caster, whereas the kingpin forward tilting at the top is called as negative caster.

Q. 16. What is zero caster ?

Ans. The king pin in a vertical position is known as zero caster.

Q. 17. What is a negative or reverse caster ?

Ans. Inclination of king pin towards front of the wheels is known as negative caster.

Q. 18. What is the condition necessary to achieve true rolling of the wheels ?

Ans. For true rolling, lines drawn though the centres of the front wheels should intersect at a point on the axis of the rear axle.

Q. 19. What is a steering knuckle ?

Ans. A cast spindle on which the wheel turns.

Q. 20. What is a drop (or pitman) arm ?

Ans. The arm connecting the steering box to the steering linkage, turning the wheels is known as pitman arm.

Q. 21. What is a 'tie rod' ?

Ans. A rod which joins the two front stub axles through an arm attached to it, is called as tie rod.

Q. 22. Why are king-pins inclined ?

Ans. King-pins are set inclined to decrease friction between tyre and road while turning.

Q. 23. What is positive and negative caster ?

Ans. Backward tilt is termed as positive caster, whereas the king-pin forward tilting at the top is called as negative caster.

Q. 24. What is a steering ratio ?

Ans. It is defined as the number of turns the steering wheel has to make for one turn of pitman arm. It is the ratio between the steering wheel angle and the corresponding angle in stub axle.

Q. 25. What is centre point turning ?

Ans. If the intersection point of the wheel centre line and the king pin centre is at the ground the wheel keeps its straight position, without any tendency of toe-in toe-out. This position of steering is called 'centre point steering'.

Q. 26. What do you mean by 'turning radius' ?

Ans. The radius of the circle on which the outer front wheel turns to its extreme outer position is called its turning radius.

Q. 27. What is a collapsible steering column ?

Ans. It is one that is designed to get shortened during a crash.

Q. 28. What is the purpose of steering linkage ?

Ans. Steering linkage transfers the different movement of the pitman arm into left-and-right movement at the wheels.

HIGHLIGHTS

1. The *steering* systems allows the driver to guide the car along the road and turn left or right as desired.
2. The function of *steering gear box* is to convert rotary motion of wheel into to-and-fro motion of drop arm so that the drag link tied up with drop arm can be pushed or pulled resulting into moving stub axle to right or left as desired by the driver.
3. The *steering ratio* is the number of degrees that the steering wheel must be turned to pivot the front wheels 1 degree.
4. *Wheel alignment* refers to the positioning of the front wheels and steering mechanism that gives the vehicle directional stability, promotes ease of steering and reduces tyre wear to minimum.
5. *Camber* is the tilting in or out of the front wheels from the vertical when viewed from the front of the vehicle.
6. The angle between the king-pin centre line (or steering axis) and the vertical, in the plane of the wheel is called *caster angle.*
7. *Toe-out* is the difference in angles between the two front wheels and the car frame during turns.
8. The condition for the correct steering is that all the four wheels must turn about the *same instantaneous centre.*
9. Davis steering gear has sliding pairs whereas Ackermann steering gear has turning pairs.
10. The angle through which the wheel has to turn to sustain side force is called *slip or creep angle.*
11. The ratio of side force sustained and slip angle is called the *cornering power.*
12. The *steering linkage* transfer the side-to-side, or front-to-rear movement of the pitman arm into the left-to-right movement at the wheels.
13. In the *power-steering system*, a continuously operating pump provides hydraulic pressure when needed.
14. *Wander* is the tendency of a car to veer away from a straight path without driver control.

OBJECTIVE TYPE QUESTIONS

Choose the correct answer :

1. Most popular manual steering gear for cars today is
   (*a*) rack and pinion type (*b*) worm and wheel type
   (*c*) cam and roller type (*d*) worm and nut type.
2. The steering ratio for manual steering of cars is approximately
   (*a*) 5 (*b*) 15 (*c*) 50 (*d*) 100.
3. The type of steering gear used in a Maruti 800 car is
   (*a*) rack and pinion (*b*) worm and roller (*c*) worm and wheel (*d*) none of these.
4. On cars having rack and pinion steering, the gear rack is attached to
   (*a*) cross shaft (*b*) pitman axis (*c*) cross shaft (*d*) tie rod.
5. The turning circle for a car is approximately
   (*a*) 1 metre (*b*) 2 metres (*c*) 10 metres (*d*) 30 metres.
6. Cornering force divided by slip angle is called
   (*a*) self-righting torque (*b*) cornering power
   (*c*) pneumatic trail (*d*) castor trail.
7. The angle formed by the line joining the stub axle-steering arm ball joints with the vertical, when this line slants forward at the top is called
   (*a*) positive camber (*b*) negative camber (*c*) positive caster (*d*) negative caster.
8. The included angle is the sum of the
   (*a*) camber and caster (*b*) caster and S.A.I. (*c*) camber and S.A.I. (*d*) camber and toe-in.
9. The side force sustained by the wheel during cornering divided by the slip angle is called
   (*a*) cornering force (*b*) camber force (*c*) cornering power (*d*) none of these.
10. Another name for the steering link rod is
    (*a*) track rod (*b*) tie rod (*c*) drag link (*d*) pitman.
11. One purpose of a recirculating ball type steering gear is to reduce the
    (*a*) operating friction (*b*) operating cost (*c*) toe-out during turns (*d*) number of parts.
12. In the recirculating ball type steering gear, the balls travel between the ball nut and the
    (*a*) gear rack (*b*) worm wheel (*c*) steering wheel shaft (*d*) worm shaft.

Answers

1. (*a*)	2. (*b*)	3. (*a*)	4. (*d*)	5. (*c*)	6. (*b*)	7. (*d*)
8. (*c*)	9. (*c*)	10. (*c*)	11. (*a*)	12. (*d*).		

THEORETICAL QUESTIONS

A. Steering

1. What is the purpose of a steering system ?
2. What are the functions of a steering system ?
3. State the requirements of a good steering system.
4. Give the layout of a steering system and label the various parts. Also explain the working of the steering system.
5. Give briefly the description of the following steering parts :
   (*i*) Steering wheel. (*ii*) Steering outer tube or steering column.
   (*iii*) Steering shaft. (*iv*) Steering gear box.
   (*v*) Drop arm.

6. State the functions of steering gears.
7. Explain briefly the following types of steering gears :
   (*i*) Worm and worm wheel steering gear. (*ii*) Worm and roller steering gear.
   (*iii*) Rack and pinion steering gear.
8. What is 'steering ratio' ?
9. What do you mean by 'slow steering' and 'fast steering' ?
10. What is 'reversibility' in a steering system ?
11. Give a description of steering geometry.
12. What is 'wheel alignment' ?
13. Name the factors which pertain to steering geometry.
14. Define and explain the following :
   (*i*) Camber angle (*ii*) Caster angle
   (*iii*) King-pin inclination, and (*iv*) Toe-in.
15. How does the king-pin inclination produce direction stability ?
16. Give the reason for the front wheels having toe-out on turns.
17. Explain the terms toe-in and toe-out.
18. What is 'turning radius' ? What is its significance ?
19. What is the purpose of a 'steering gear mechanism' ?
20. Explain briefly the following steering gear mechanisms :
   (*i*) Davis steering gear. (*ii*) Ackermann steering gear.
21. Explain briefly the following :
   (*i*) Slip or creep angle, (*ii*) Cornering force, and
   (*iii*) Cornering power.
22. What is Ackermann principle ? Explain it in detail.
23. What is a 'steering linkage' ?
24. (*a*) Explain briefly with a neat sketch the steering linkage for a conventional rigid axle suspension.
   (*b*) What is the 'understeering' and 'oversteering' ?
25. Explain briefly the steering linkages for independent front suspension.
26. Describe briefly steering wheel and column.
27. Name the various types of steering columns employed in many foreign cars.
28. What is a 'steering arm' ? On what factor does its exact length depend ?
29. Explain briefly the following :
   (*i*) Drop arm (*ii*) Drag link.
30. What is a 'steering stop' ?
31. From where is the power steering hydraulic pressure obtained ?
32. Explain briefly the following :
   (*i*) Integral power steering. (*ii*) Linkage power steering.
33. Describe briefly 'Electronic power steering'.
34. State the advantages of an electronic power steering over hydraulic power steering.
35. What are steering troubles, their causes and remedies ?

B. Front axle

36. What is the purpose of the front axle ?
37. State the functions which a front axle performs.
38. Give the construction of a front axle.
39. What is the difference between a 'dead front axle' and a 'live front axle' ?
40. Why are stub axles fitted in front-axles ? Sketch the different types of stub axles.

9

Braking System

9.1. INTRODUCTION TO BRAKING SYSTEM

This chapter deals with the construction and operation of the various types of braking systems used in automotive vehicles.

Brakes are one of the most important control components of the vehicle. They contribute very much in the running and control of the vehicle. On the efficiency of brakes depends the lives and comfort not only of driver and passengers but other persons moving on the road. Furthermore it is a fact that owing to recent improvements in the braking mechanism it has been possible to have increased speeds of the modern cars on the road.

The braking system used *most frequently operates hydraulically, by pressure applied through a liquid.* These are the foot-operated brakes that the driver normally uses to slow or stop the car. They are called the *"service brakes"*. In addition all cars have a *parking-brake system* which is mechanically operated by a *separate foot or hand lever.* On some trucks and buses, the braking system is operated by air pressure (*pneumatically*). This type of brake is an *air brake.* Many boat and camping trailers have *electric brakes.*

- *All the braking systems depend upon friction between moving parts and stationary parts for their stopping force.*
- ***Braking efficiency*** *is the ratio between the retarding force (or force of friction between the linings and the drum) and the weight of the vehicle.* It is expressed as percentage.

9.2. NECESSITY OF A BRAKING SYSTEM

In an automobile, if the pressure from accelerator pedal is removed, the vehicle tends to slow up because of wind resistance, drag of engine and road friction. These forces, of course, would stop the vehicle but in the present day traffic, this would be quite unpracticable and dangerous. *The braking system provides added friction to overcome motion and to slow up or to stop the vehicle. The momentum or kinetic energy developed by the vehicle when in motion is converted to heat energy by the friction of brake shoes and drums which is dissipated into the surrounding air.*

9.3. FUNCTIONS OF BRAKES

Brakes perform the following *functions* :

1. To stop the moving vehicle in the shortest possible time.

2. To help in controlling the speed of the vehicle and to reduce the speed at turnings and other crowded places.

3. To hold the vehicle in its stationary position, without the presence of the operator, after it has been brought to a stop.

— In a moving vehicle, the *friction* between brake drum and brake shoes (having lining riveted to it) slows down the rotation of wheel or stops the vehicle.

9.4. REQUIREMENTS OF A GOOD BRAKING SYSTEM

The *requirements of a good braking system* are as follows :

1. The brakes should stop the vehicle within a reasonable distance. The retardation shall be smooth and free from jerk or shudder.

2. The braking system should be very reliable to promote highest degree of safety on the road.

3. The braking system should not be affected by water, heat, road grit or dust etc.

4. Pedal effort applied by the driver should not be more so as not to strain the driver.

5. Brake should work equally good in all weathers.

6. The wear and tear of the material of the brake lining should be minimum for its longer life.

7. Due to the rubbing action of brake shoes alongwith lining against drum, large amount of heat is generated due to friction. The brake design system should be capable of dissipating this heat very quickly.

8. All the components and levers of the braking system should be strong enough to take the mechanical stresses and strains which are encountered during brake actuation.

9. No braking system can work at its best through worn out or incorrectly inflated tyres. Good, tyres are, therefore, a prime essential for efficient braking.

- The **capacity of a brake** depends upon the following *factors* :

(*i*) The unit pressure between the braking surfaces.

(*ii*) The coefficient of friction between the braking surfaces.

(*iii*) The peripheral velocity of the brake drum.

(*iv*) The projected area of the friction surfaces.

(*v*) The ability of the brake to dissipate heat equivalent to the energy being absorbed.

9.5. CLASSIFICATION OF BRAKES

Brakes may be *classified* as follows :

1. Mechanical brakes.

2. Hydraulic brakes.

3. Power brakes.

(*i*) Air brakes (*ii*) Air-hydraulic brakes

(*iii*) Vacuum brakes (*iv*) Electric brakes.

9.6. MECHANICAL BRAKES

9.6.1. Introduction

The *brakes which are operated mechanically by means of levers, linkages, pedals, cams, bell cranks, etc. are known as* **mechanical brakes.**

The external contracting brake which is usually hand brake in automobiles is mechanical brake. Automobiles contain *service brakes* operated mechanically.

Mechanical brakes were employed in olden days but now hydraulic and other types of braking system have taken its place.

9.6.2. Internal Expanding Mechanical Brake

Fig. 9.1 shows an *internal expanding brake (mechanical).*

Construction. It consists of two ***shoes*** S_1 and S_2. The outer surfaces of the shoes are lined with some ***friction material***, to increase the frictional coefficients and to prevent wearing away of the metal. Each shoe is pivoted at one end about a fixed ***fulcrum*** (O_1 and O_2) and made to contact a ***cam*** at the other end. When the cam rotates, the shoes are pushed outwards against the rim of the ***drum.*** The friction between the shoes and the drum produces the braking torque and consequently speed of the drum reduces. Brake shoe ***retracting spring*** which connects both the brake shoes at their loose end helps them in contracting after the brakes are released.

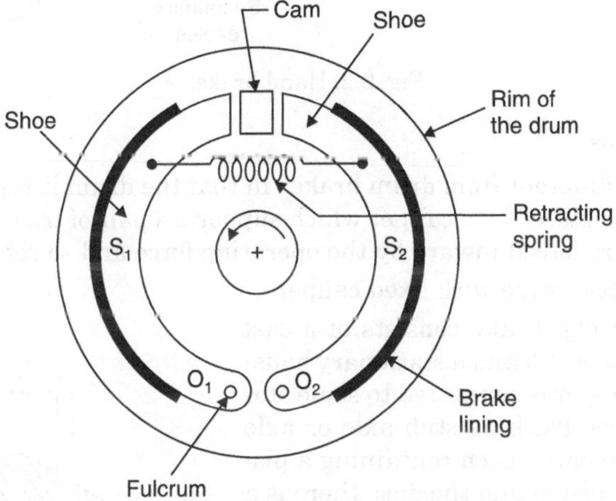

Fig. 9.1. Internal expanding brake.

Operation :

- When the brake pedal is *pressed down,* its motion is transmitted to the cam through various mechanical linkages. The motion of the cam tends to expand out the brake shoes. This inward motion of the brake shoes will try to slow down the motion of the rotating brake drum. Because the wheel is fixed to the brake drum, so automatically it will be held to move further.
- When brake pedal is *released,* the pedal will move up because of the tension of the return spring. A retracting spring draws the shoes away from the drum when the cam is moved to its initial position and hence the brake shoes are no longer in contact with the drum, which is now free to rotate.

9.6.3. Hand Brake (or Parking brake)

Hand brakes are usually the mechanical brakes. These brakes operate independently of the foot brakes. These are used for *parking on slopes and during emergency* and are also called "*secondary brakes*". Hand brake is generally located on the side of the driver's seat. On most of the vehicles hand brake applies only the rear brakes.

Fig 9.2 shows a schematic diagram of a typical hand brake.

In order to apply the brakes the ratched is released first by pressing the ratchet release handle, which causes the pawl to move up, disengaging the rotchet. Then the brake lever is pulled up, while further pulls the cable which in them operates the rear brakes mechanically through a linkage operating on the piston of the rear wheel cylinder, which is two halves. The ratchet release handle which had been pressed so far is released now, so that the pawl moves down the spring action and engages with the ratchet thus keeping the brakes applied.

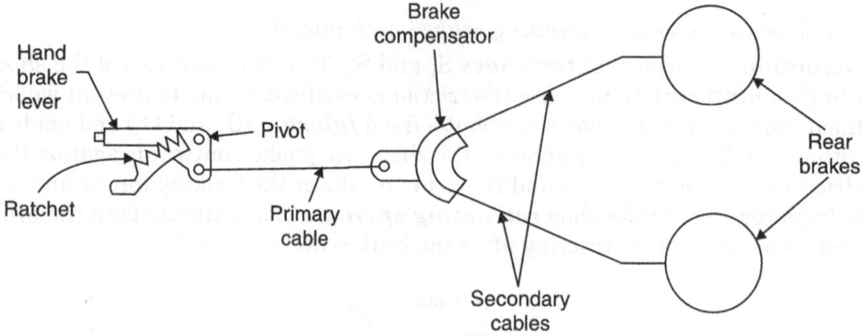

Fig. 9.2. Hand brake.

9.6.4. Disc Brakes

These brakes are different from drum brakes in that the drum is replaced by a circular plate and the brake shoes are replaced by a caliper which supports a pair of friction pads, one on each side of the disc. These pads are forced inward by the operating force and so retard the disc.

Fig. 9.3 shows a disc brake with fixed caliper.

Construction. A disc brake consists of a cast iron disc bolted to the wheel hub and a stationary housing called *caliper*. The caliper is connected to some stationary part of the vehicle (such as stub axle or axle casing) and is cast in two parts, each containing a piston. In between the each piston and the disc, there is a friction pad held in position by retaining pins, spring plates etc. Passages are drilled in the caliper for the fluid to enter or leave each housing. These passages are also connected to one another for bleeding. Each cylinder contains a rubber sealing ring between the cylinder and the piston.

Working. On the application of brakes, hydraulically actuated pistons move the friction pads into contact with the disc, applying equal and opposite forces on the later. When the brakes are released, the rubber sealing rings act as return springs and retract the pistons and the friction pads away from the disc.

Special types of disc brakes include the *swinging caliper type* and the *sliding caliper type*.

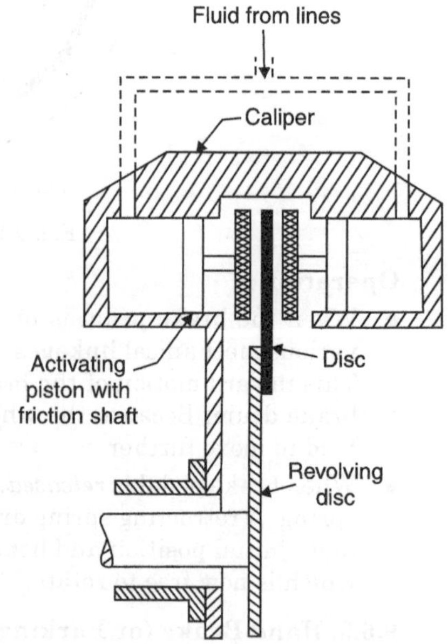

Fig. 9.3. Disc brake with fixed caliper.

The *advantages* and *disadvantages* of *disc brakes compared with drum brakes* are as follows :

Advantages :

(*i*) Lighter than drum brakes.

(*ii*) Better cooling (since the braking surface is exposed directly to air).

(*iii*) Offer better resistance to fade.

(*iv*) Uniform pressure distribution (since disc brakes have no self-servo effect).

(*v*) Brake pads can be easily replaced.

(*vi*) These brakes are self adjusting by design.

Disadvantages :

(*i*) Costlier than drum brakes.

(*ii*) For stopping the vehicle higher pedal pressure is required.

(*iii*) There is no servo action in these brakes.

(*iv*) It is difficult to install an adequate parking attachment.

- The *major drawback of a "mechanical brake system" is that it is very difficult to get simultaneous brake action on all the four wheels. Also lengths of various rods and cables vary and this causes unequal braking action.*

9.7. HYDRAULIC BRAKES

9.7.1. Introduction

Brakes which are operated by means of hydraulic pressure are known as **hydraulic brakes.**

In a *hydraulic system*, when the brakes are applied, the pressure in increased sufficiently in the system to produce equal and uniform backing action on all the four wheels. The hydraulic brakes function on the principle of *Pascal's law* which reads as follows :

"Pressure applied to a liquid is transmitted equally in all directions."

9.7.2. Hydraulic Braking System

The hydraulic braking system consists of *four wheel cylinders*, one at each of the four wheels of the vehicle as shown in Fig. 9.4. The system also consists of one *master cylinder*, which is connected

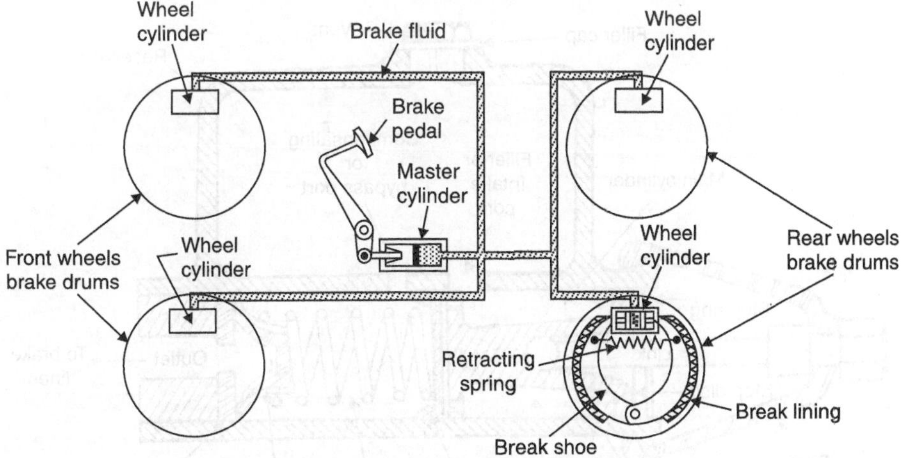

Fig. 9.4. Hydraulic brake system.

to the wheel cylinders by steel tubing. Each wheel cylinder contains *two pistons*, which will move out when the pressure will be applied through *brake fluid*. When the brakes are not in operation, the system is filled with brake fluid. Each wheel brake consists of a cylindrical brake drum which is mounted on the inner side of the wheel and revolves with it. There are *two brake shoes* mounted inside each of the brake drum but do not rotate with it.

When the brakes are to be applied, the driver presses down the brake pedal, the piston is forced into the master cylinder, thus increasing the pressure of the fluid in the master cylinder and in the entire hydraulic system. This pressure is conducted instantaneously to the wheel cylinders on each of the four brakes, where it forces the wheel cylinder pistons outwards. These pistons, in turn, force the brake shoes out against the brake drums. Thus the brakes are applied.

When the driver releases the brake pedal, the master cylinder piston returns to its original position due to the return spring pressure, and thus the fluid pressure in the entire system drops to its original low value, which allows retracting springs on wheel brakes to pull the brake shoes out of contact with the brake drums into their original positions. This causes the wheel cylinder pistons also to come back to their original inward position. Thus the brakes are released.

The hydraulic brake system contains two important components upon which the system is mostly dependent, these are :

1. Master cylinder
2. Wheel cylinder.

1. **Master cylinder :**

It is the *main cylinder* in the hydraulic brake system. It serves the following objects in the system.

(*i*) It builds up hydraulic pressure to operate the brakes.

(*ii*) It maintains a constant volume of fluid in the system owing to its reservoir.

(*iii*) It serves as a pump to bleed or force air out of the hydraulic system.

There are the following two types of master cylinders :

(*a*) *"Single master cylinder"* for all the front and rear wheel cylinders.

(*b*) *"Tandem master cylinder"* containing separate units for front and rear wheel cylinders.

(*a*) ***Single master cylinder :***

Construction :

The master cylinder (Fig. 9.5), named as the heart of the hydraulic brake systems, consists of two main chambers : (*i*) The *"fluid reservoir"* (which contains the fluid to supply

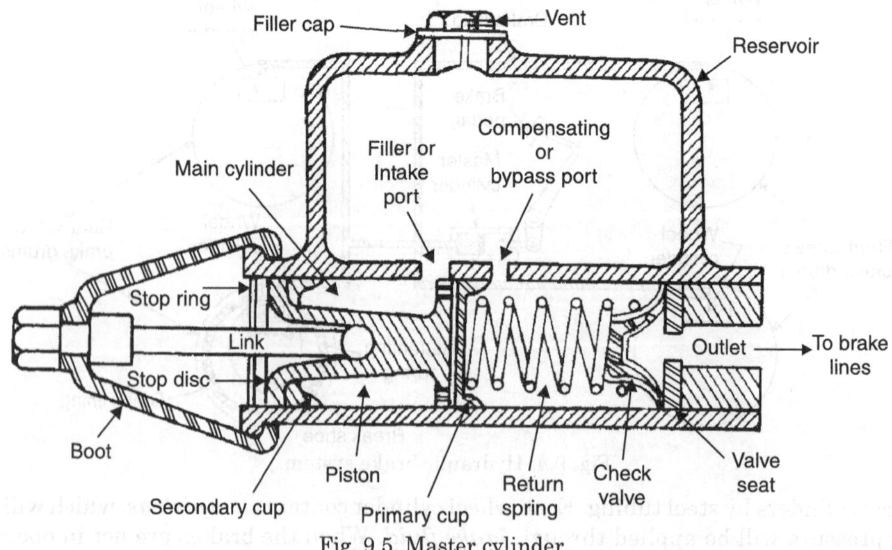

Fig. 9.5. Master cylinder.

to the brake system), and (*ii*) the "*compression chamber*" (in which piston operates). The reservoir supplies fluid to the brake system through two ports. The larger port is called the *filler or intake port* and is connected to the hollow portion of the piston (there are a number of holes in the piston on the primary or high pressure seal side) between the primary and secondary cups which act as piston seals. The smaller port is called the *compensating, bypass or relief port* which connects the reservoir *directly* with the cylinder and lines when the piston is in the *released position*. The reservoir is vented to the atmosphere so that atmospheric pressure causes the flow through the filler port. The vent is placed in the filler cap. The "*boot*" covers the push rod and the end of the cylinder to keep it free from foreign matter. Towards the brake lines side of the compression chamber, there is a fluid "*check valve*" with a rubber cup inside. It serves to retain the residual pressure in the brake lines even when the brakes are released.

Working :

- When the brake pedal is ***pressed*** piston of the master cylinder moves forward to force the liquid under pressure into the system. The relief port is sealed out of the system. The liquid pressure is conducted to the wheel cylinders, where it forces the wheel cylinder pistons outwards. These pistons force the brakes shoes out against the brake drums and the brakes are applied.
- As soon as the brake pedal is ***released,*** the *return spring* quickly forces the master cylinder piston back against the piston stop. Since the fluid in the lines returns rather slowly, a vacuum tends to form in the cylinder in front of the piston. This causes the primary cup to collapse/deflect to allow the liquid to flow from the reservoir through the filler port past the piston to fill the vacuum.
- When the pedal is in "*off position*", the liquid may flow from the reservoir through the relief port in the master cylinder, supply lines, and wheel cylinders to make up for any fluid that may be lost or to compensate for shrinkage cooling of the liquid. In this way, a complete column of liquid is always maintained between the master cylinder piston and wheel cylinder pistons.

In some makes of master cylinders a bleeder screw is also provided to bleed air out of the master cylinder.

(*b*) ***Tandem master cylinder :***

A simple master cylinder is mostly used in all small and medium type of vehicles. However, in some vehicles, the tandem master cylinder is used.

A "*tandem master cylinder*" is *the master cylinder with two separate cylinders and reservoirs in the same master cylinder assembly, one operating front brakes and the other cylinder operating rear brakes.* This master cylinder avoids the possibility of all the brakes of a vehicle being put out of order by a leak or fracture in the pipe line leading to one wheel cylinder.

It *ensures reliability with not much extra cost.* In this arrangement separate lines go to different sections of the brake system, say, the rear and the front brakes and it so arranged that if the front brake lines are damaged, the rear brakes will be still effective. Similarly if rear brake line is defective, at least front brakes will be applied.

Fig. 9.6 shows a single line diagram of a tandem master cylinder. Two pistons have been shown which are in line with each other. If line A fails, the piston A bottoms against the end of the cylinder while the piston B continues to develop pressure in the line B thus applying brakes to one set of wheels. If the line B fails piston B comes up against piston A, thus building up pressure in line A.

Fig. 9.7 shows the sectional view of a tandem master cylinder. As shown in diagram, the master cylinder contains two pistons, the rear being operated directly by the brake pedal. The space between the pistons is connected to the front brakes while the connection to rear brakes is made at the front end opposite to operating link.

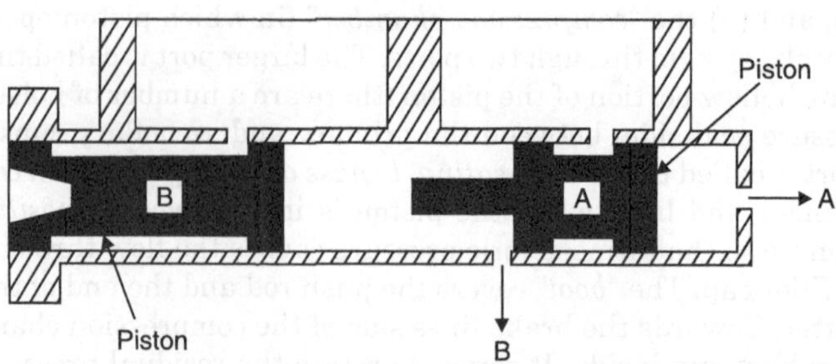

Fig. 9.6. Simple line diagram of tandem master cylinder.

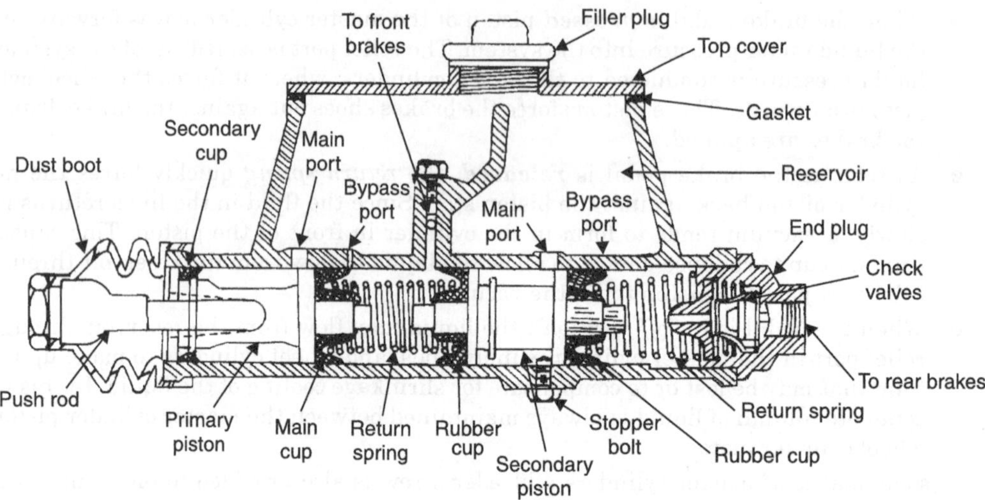

Fig. 9.7. Tandem master cylinder (Sectional view).

Depressing the brake pedal moves the rear piston inward developing pressure in the operating cylinder for front brakes. Since the front piston is free to move along the cylinder, so it also moves ahead developing an equal pressure in the operating cylinder for rear brakes. The return motion of the front piston is limited by the stopper screw.

2. **Wheel cylinder (or Slave cylinder) :**

Refer to Fig. 9.8. A wheel cylinder consists of a *cylinder*, *two pistons*, *two rubber cups* and a *spring*. The fluid presses against pistons. The pistons move outward in the cylinder. When the pistons

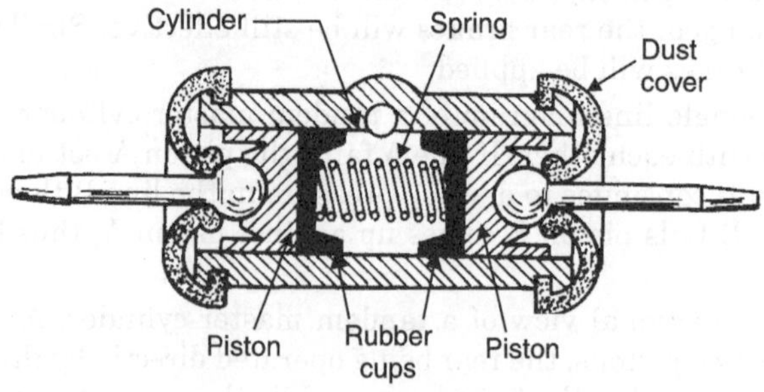

Fig. 9.8. Wheel cylinder.

come closer, the liquid is forced into the master cylinder. The spring between the two pistons holds the rubber cups in position.

- The copper-coated, tin-plated annealed steel tubing and flexible hoses are used to connect the master cylinder to the wheel cylinders. The hoses are used to connect the lines to the front wheel cylinder to permit the front wheel to be turned. Rear wheel cylinders are generally connected directly to a line fastened to the rear axle housing. The brake lines are attached directly or by means of brackets to the frame or axle housings.

9.7.3. Advantages and Disadvantages of Hydraulic Brakes

Advantages :

1. Equal braking effort to all the four wheels (since fluid exerts equal pressure every where in the circuit).

2. The system is simple in construction.

3. Less rate of wear (due to absence of joints compared to mechanical brakes).

4. The system is mostly self-lubricating.

5. Increased braking effort.

6. High mechanical advantage.

7. Flexibility in brake lines.

8. The hydraulic brakes can also provide differential braking action between the front and rear brakes by using the wheel cylinder of different size for the front and rear wheels.

Disadvantages :

1. Even slight leakage of air into the braking system makes it useless.

2. The brake shoes are liable to get ruined if the brake fluid leaks out.

3. This system is suitable only for applying brakes *intermittently.* For parking purpose separate mechanical linkage has to be employed.

9.7.4. Hydraulic Brake Fluid

The fluid used in the braking system is a special kind of fluid which has to be satisfactory under all conditions. Most fluids are based on *polyglycols* and *additives are added* to achieve the required properties. A 50 percent solution of caster oil in alcohol to which a neutraliser is added, meets the above mentioned requirements satisfactorily. The neutraliser is added to counter act the effect of any free acids which may be present in caster oil or alcohol.

The hydraulic brake fluid should possess the following ***characteristics :***

1. Should be non-compressible.

2. Must remain fluid at low temperature.

3. Should not rust corrode or rust metallic parts in the brake system.

4. Mix satisfactorily with other makes of hydraulic fluids.

5. Must be chemically stable.

6. Should not soften the rubber parts used in the hydraulic brake system.

7. Must act as a lubricant to the moving parts inside the system.

8. Must retain all its characteristics for a maximum long period.

9.7.5. Bleeding of Hydraulic Brakes

The *process of eliminating or removing air out of the braking system of an automobile is called* **"bleeding"**.

The hydraulic braking system must be bled to remove air from the system whenever :

(*i*) A pipeline or any part has been disconnected from the system to do any repair work or replace the parts.

(*ii*) There is a leak and the air has entered the system through that leak.

(*iii*) Whenever the fluid level in the master cylinder has become too low or the fluid reservoir in the master cylinder has been allowed to run dry (This may be the result of leak).

When the air traps into the system, it being *compressible, the effort of brake pedal goes waste in applying brakes.* Until and unless air from the system is removed, the brakes would not function properly. Hence, to remove air from the system it must be bled.

When bleeding the brakes, it is *advisable that the longest fluid line from the master cylinder is bled first, the next longest second, etc.* The master cylinder must be kept least $\frac{3}{4}$ full of hydraulic fluid during the bleeding operation.

Bleeding procedure :

1. After removing all dirt from around the master cylinder filler plug, remove the filler plug and fill the master cylinder to the lower edge of the filler neck.

2. Clean all bleeding connections at all four wheel cylinders. Attach bleeder hose and fixture to longer brake line wheel cylinder bleeder screw and place end of the bleeder tube in a glass jar containing some brake liquid.

3. Open the bleeder valve one-half to three-quarter to a turn. Depress the foot pedal, allowing it to return very slowly. Continue this pumping action to force the fluid through the line and out of the bleeder hose which carries with it any air in the system (Fig. 9.9). When bubbles cease to appear at the end of the bleeder hose, close the bleeder valve and remove the hose.

Fig. 9.9. Bleeding of brake.

4. When the bleeding operation at each wheel cylinder is completed, fill the master cylinder reservoir and replace the filler plug.

- Don't reuse the liquid which has been removed from the lines through the bleeding process because it contains air bubbles and dirt.
- Don't mix two brands of brake fluids.

9.7.6. Hill Holder

It is an *auxiliary braking device* used on hydraulic braking system. It *automatically prevents the car from sliding down on a grade* after it has been fully stopped by the use of service brake, and as long as the clutch is disengaged.

The hill holder essentially consists of a check in the fluid line just ahead the master cylinder which maintains the pressure as long as the clutch is held disengaged.

With this device in the hydraulic system, the driver can remove his foot from brake and use it on the accelerator as the clutch is engaged.

9.7.7. Self-energizing Brakes

Usually the hydraulic wheel brakes of drum type are provided with a *self-energizing or servo feature*, in which the *force of rotating drum is utilized to increase the brake pressure.*

The self-energizing brake shoe action is shown in Fig. 9.10 (The relative size of the arrows indicates the increase of brake force or pressure). When the vehicle is travelling forward, the drum is rotating in anticlockwise direction. When the brakes are applied, the primary shoe at the left, tends to move in the direction of drum's rotation, due to the friction of rotating drum. Because the primary shoe is linked to the secondary shoe at the bottom, the secondary shoe is forced against the anchor pin at the top. This action causes to force both the shoes into tighter contact with drum and the braking force is applied more uniformally. This combination of forces is the *self energizing action*. Since greater force is applied to the secondary shoe, the secondary shoe lining area is usually larger than the primary shoe lining area.

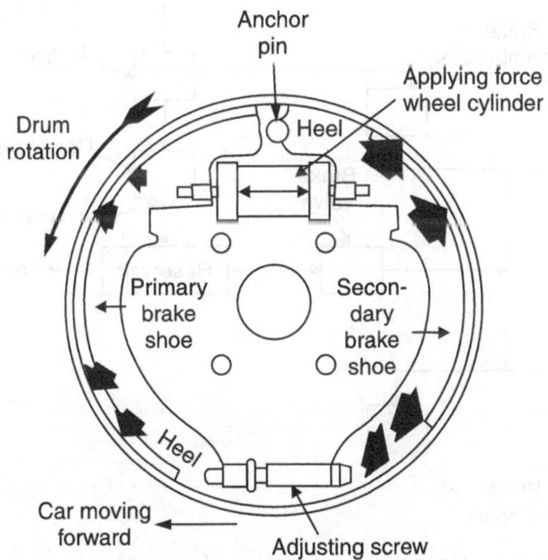

Fig. 9.10. Self-energizing brake shoe action.

9.8. POWER BRAKES

Brakes operated by power other than physical effort applied at brake pedal, are known as **power brakes**. The power may be exerted by air pressure, engine vacuum or electrical energy.

There are semi-power or power assisted brakes too, in which the *power assists or supplements the power in actuating the brakes. "Hydrovac brakes"* are semi-power brakes ; in these brakes the engine vacuum assists in the actuation of hydraulically operated brakes.

Main features

The main features of the power brake system are :

1. *"Time-lag"* or the time internal between the moment when pedal is pressed down and the moment when the brakes are put on *must be very small.*

2. The force applying the brakes should be *closely proportional* to the force exerted by the driver on the brake pedal.

- The brake system should be such that the driver may judge the intensity of the application of brakes fairly accurately.

These days the following systems are predominantly used for actuating the brakes :

1. Air brakes.
2. Air hydraulic brakes.
3. Vacuum brakes.
4. Electric brakes.

9.8.1. Air Brakes

9.8.1.1. Description of air brakes

In air brakes, the brake shoe operating cam is operated by means of air pressure which is developed by an air compressor driven by the engine. There are separate brake chambers for the separate brake shoe operating cams. The brake chambers are connected with the air reservoir by means of pipe line. A brake valve operated by the foot pedal, controls the pressure of air which affects the brake chambers. Fig. 9.11 shows the general arrangement (simple line diagram) of an air brake system. Fig. 9.12 shows the layout of an air brake system for bus or truck.

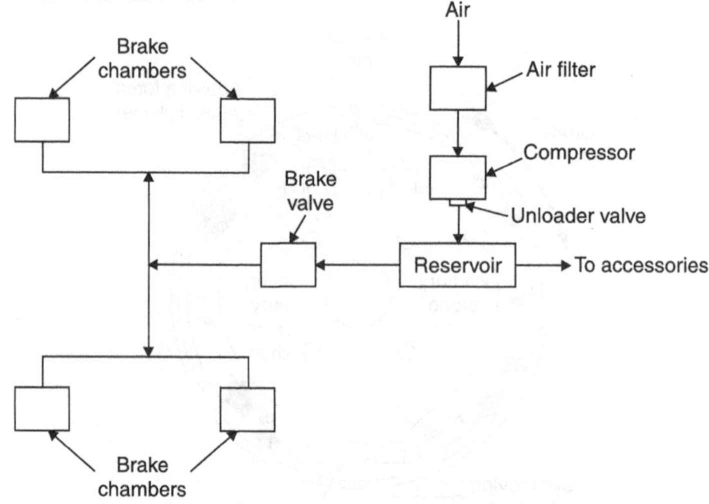

Fig. 9.11. General arrangement of an air brake system.

When the foot pedal is pressed down, air pressure acts on the diaphragm of brake chamber. The diaphragm is linked with the brake shoe operating camshaft. The diaphragm is pushed outward in the brake chamber causing movement of brake shoe operating cam. The brake shoe expands outwards and hold the moving brake drums as they come into its contact. The brakes are thus *applied.*

As soon as the pressure is released from the brake pedal, it comes back with the help of return spring. This results in closing of brake valve and release of pressure inside the brake chamber. The brake shoe operating cam moves in the reverse direction as a result of pressure release on the brake chamber. The brake shoe contracts inward with the help of retracting spring, releasing the brake drum of the binding effect. The brakes are thus *released.*

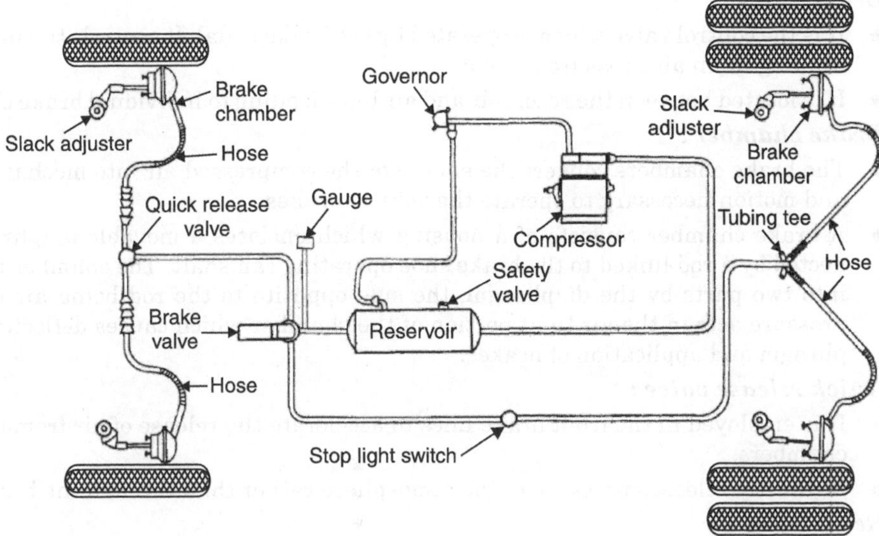

Fig. 9.12. Layout of an air brake system for bus or truck.

Main parts of an air brake :

Following are the main parts of an air brake :

(*i*) Air compressor.
(*ii*) Unloader valve.
(*iii*) Reservoir.
(*iv*) Brake valve.
(*v*) Brake chamber.
(*vi*) Quick release valve.
(*vii*) Relay valve.
(*viii*) Warning signal.

(*i*) ***Air compressor :***

- It builds up air pressure in reservoir.
- A piston type air compressor is commonly employed in the brake system. When piston moves downwards air is drawn into the cylinder through intake valve. When the piston moves upward, the intake valve is closed and outlet valve is opened by the air pressure and the air is forced out into the reservoir.

(*ii*) ***Unloader valve :***

- This valve is mounted in the air pressure system between the compressor and the reservoir to control the pressure of air in the reservoir. It relieves the compressor of its pumping load once the unloader cut-out pressure is obtained and seals the reservoir when the compressor has built up a pressure depending upon the setting of the adjusting screw which is normally 7.4 bar. The unloader then diverts the air delivered by the compressor to the atmosphere thus allowing the compressor to run right whilst the reservoir contains an adequate supply of air.
- It consists mainly of a *governor valve, an unloader plunger and a non-return valve.*

(*iii*) ***Reservoir :***

- *It stores compressed air at the specified pressure for brake application.*
- It is made of sheet steel and a *safety valve* is provided at the top of the reservoir to regulate the air pressure. A *drain plug* is also provided at the bottom for periodic draining of the reservoir, without which the lubricating oil from the compressor and moisture in the air would form emulsion which would damage the other brake units (*e.g.,* brake valve, brake chambers etc.).

(*iv*) ***Brake valve :***

- It is the control valve which is operated by the brake pedal. It controls the intensity of braking in an air pressure system.
- It is located between the reservoir and air lines leading to individual brake chambers.

(*v*) ***Brake chamber :***

- The brake chambers convert the energy of the compressed air into mechanical force and motion necessary to operate the vehicle brakes.
- A brake chamber consists of a housing which encloses a movable diaphragm connected by a rod linked to the brake shoe operating camshaft. The chamber is divided into two parts by the diaphragm, the side opposite to the rod being air tight. Air pressure acts in the air tight portion of the chamber which causes deflection of diaphragm and application of brakes.

(*vi*) ***Quick release valve :***

- It is employed in the front brake lines to accelerate the release of air from the brake chambers.
- It directly releases pressure to the atmosphere rather than through the brake valve.

(*vii*) ***Relay valve :***

- The relay valve speeds up the application and release of air from the brake chambers.
- It supplies air to the brake chambers directly from the reservoir for quick application of the brake. It also exhausts compressed air from the rear brake chambers directly to the atmosphere rather than through the brake valve.

(*viii*) ***Warning signal :***

It is a warning light or buzzer which warns low pressure in the circuit.

9.8.1.2. Advantages and disadvantages of air brakes

Advantages :

Air brakes entail the following *advantages* :

1. Much more powerful than the ordinary mechanical or hydraulic brakes (that is why these are exclusively used in *heavy vehicles*).
2. Simplified chassis design.
3. The compressed air from reservoir, apart from braking, can be used for tyre inflation, windscreen wipers, horns etc.

Disadvantages :

1. Involve relatively more parts.
2. The air compressor uses a certain amount of the engine power.

9.8.1.3. Fail safe brake system

In an unlikely event of sudden loss of air pressure due to failure of a unit in the brake system, a safety system has been developed for Viking the Cornet model, as a precautionary measure. It is called the *fail safe brake system* and has the following *essential features* :

(*i*) In case of failure in the air pressure system, the vehicle will automatically come to a stop due to the application of the spring brakes on the rear wheels.

(*ii*) It can be used for both hand brake and parking purpose and as such the mechanical hand brake assembly is eliminated.

(*iii*) It will not be possible for the driver to move the vehicle away unless sufficient air pressure (4 bar, minimum) is built up in the system.

This system consists of the following *major assemblies* :

(*i*) Hand control valve. (*ii*) Double check valve.

(*iii*) Spring brake chamber.

The circuit diagram of a fail safe brake system is shown in Fig. 9.13. The working of the system is as follows :

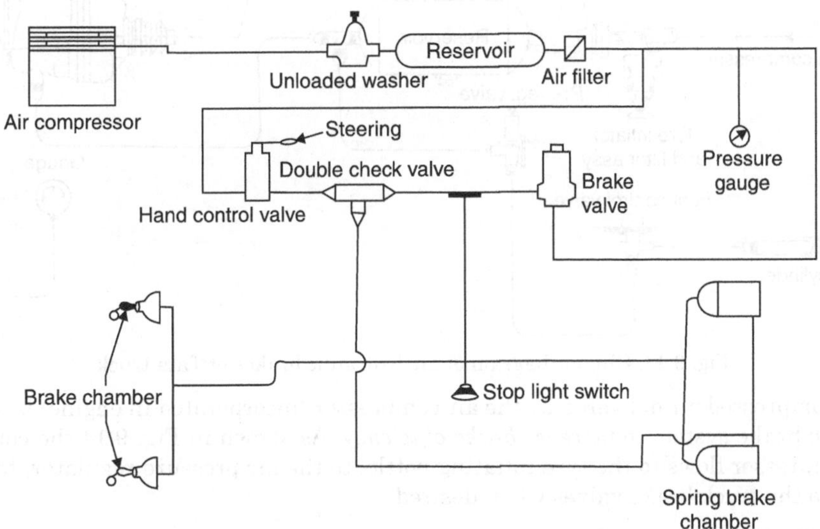

Fig. 9.13. Circuit diagram of a fail safe brake system.

- The movement of the vehicle takes place only when adequate pressure is built up in the air reservoir. When the air pressure in the system is inadequate, the spring brakes are applied. As soon as the air pressure starts building up with the engine compressor running this pressure is communicated to the spring brake chamber. When the air pressure exceeds the hold off pressure of the spring brake chamber, the springs are held off and the brakes are released. During the *service brake operation* the main spring in the spring brake chamber is held out of action by air pressure and does not affect the operation of the service brake.
- When the hand brake is applied by operating handle to "ON" position, the air supply to the spring brake chamber is cut-off and the air pressure in the chamber is released to the atmosphere through the hand brake valve.

As the air exhausts, the main spring forces the push rod out and applies the brake. When the hand brake is released, air supply is restored to the spring brake and the brake is released as the spring is held off by the air pressure.

- Under emergency condition when the system air pressure drops below the spring brake hold off pressure, the spring will exert force and apply the brakes. In case the brake valve fails, the hand control valve has to be operated to apply the spring brakes.

9.8.2. Air-hydraulic Brakes

In air-hydraulic brakes, the air pressure is converted into hydraulic pressure. In this system the air power cylinder is combined with the hydraulic master cylinder and the reservoir. With the help of this unit the conventional type hydraulic brakes are actuated by the air power.

Fig. 9.14 shows the circuit diagram of compressed air-hydraulic brakes of Tata truck braking system. The system consists of the following *parts* :

(i) Air compressor.
(ii) Tyre inflating bottle.
(iii) Air pressure regulator.
(iv) Air container.
(v) Truck brake valve.
(vi) Air pressure gauge.

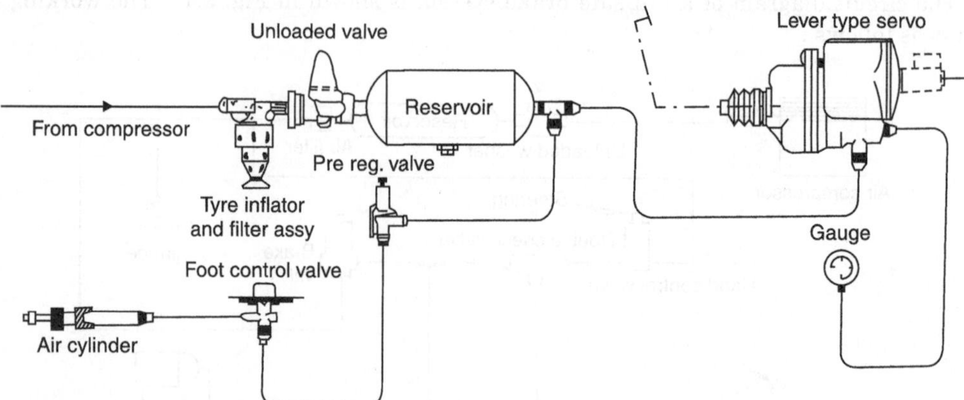

Fig. 9.14. Circuit diagram of air-hydraulic brakes of Tata truck.

The compressed air delivered by the air compressor (incorporated in engine) is used to assist the hydraulic brake system *to increase brake efficiency*. As shown in Fig. 9.14 the compressed air from the compressor flows to the tyre inflating bottle, to the air pressure regulator, to the air container, and to the truck brake valves when desired.

9.8.3. Engine Exhaust Brake

It is an *auxiliary brake* and should be used under the following *circumstances :*

(i) In traffic situations demanding slowing down over a long stretch of road.
(ii) While travelling on long downhill gradient.

The exhaust brake serves the following *purposes*.

(i) Improves the life and efficiency of the service brakes.
(ii) Increases the life of the service brake linings and drums.
(iii) Reduces overheating of service brakes.

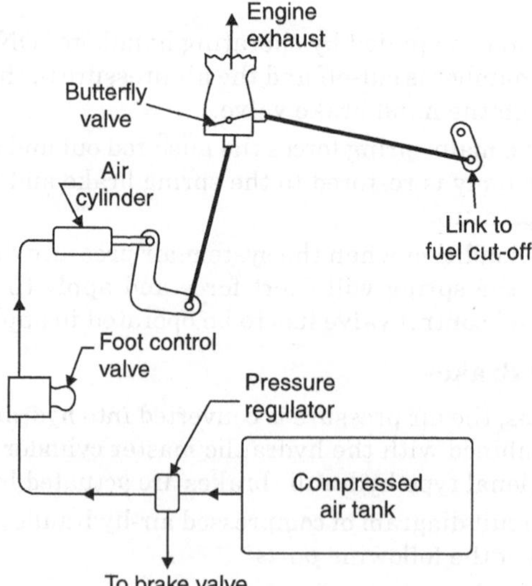

Fig. 9.15. Schematic diagram of engine exhaust brake.

The main components of the engine brake system (Fig. 9.15) are :

1. Pressure regulator.
2. Foot control valve.
3. Air cylinder.

This brake is operated by depressing a foot control valve located between the clutch and brake pedals and keeping it depressed as long it is required to apply the engine brake. By a linkage attached to exhaust brake air cylinder, the throttle valve at the exhaust manifold is closed ; at the same time, the accelerator lever comes to stop position and the accelerator pedal is locked thus cutting off the fuel. This brake is very effective below vehicle speed of 40 km/h. However, it must be remembered that it cannot stop the vehicle like a service brake.

As a *design improvement*, the exhaust operation has now been coupled with a service brake. Hence whenever brake pedal is depressed for applying service brakes, the engine exhaust brake also gets operated automatically. This improvement has been gradually introduced in Tata truck models 1210, 1510, 1612, 1613. The major *advantages* derived from the improved designs are :

(*i*) Optimum fuel economy.
(*ii*) Reduce maintenance time and cost.
(*iii*) Increased safety and better control of vehicle.
(*iv*) Improved life of brake liners, brake drums and tyres.

9.8.4. Vacuum Brakes

Any mechanism which adds to the driver's effort in applying the brakes is called a ***servo mechanism*** (It may be mentioned that servo is also used in place of power cylinders), although that effort remains a considerable part of the total braking effort required.

Initially mechanical servos were used, but after the introduction of vacuum operated servos these have becomes obsolete.

In the vacuum brakes the *suction from the engine inlet manifold is utilised for brake application.*

There are *two types* of vacuum brakes, both incorporating a piston or a diaphragm operating in a cylinder and provided with suitable linkage for brake application. A small vacuum reservoir is also there to provide enough vacuum for several brake applications even after the engine has stopped.

First type. In this system both sides of the piston are *exposed to atmosphere* when brakes are in the released position. For applying brakes, engine vacuum is applied on one side of the piston, subjecting the same to differential pressure and thus operating the linkage. This system is called "*atmospheric suspended*" system.

Second type. In this case both sides of the piston are *subjected to engine vacuum* in the brakes released position. To apply brakes, one side is exposed to atmosphere which provides the desired force on the piston. This system is called "*vacuum suspended*" system. This system is preferred over the first system since this is comparatively *more rapid in action.* These days *vacuum suspended brakes are predominantly used.*

In ***"vacuum suspended servo"*** the air is first exhausted from both sides of the piston in a large cylinder and during the application of brake, the air then admitted from the atmosphere to one side of the piston. The amount of air entering the cylinder is controlled by the driver. The piston of the cylinder moves under the pressure differential which in turn moves the hydraulic piston thus forcing the fluid into brake pipes into the wheel cylinders. The vacuum (about 0.35 bar) is obtained from the inlet manifold of the engine. In this system a vacuum reserve tank is fitted between servo and engine. The tank and servo are protected with the help of non-return valves in order to enable them to hold some vacuum in case of failure of other components. The *tank provides reserve of some vacuum which can be utilised for making number of stops in quick succession.*

- Till now engine vacuum was the only source of power for vacuum-assist brakes on cars. However, with the use of smaller engines and provision of emission controls, some vehicles do not have sufficient vacuum to operate the power brakes. In such cases, *separate vacuum pumps are* used.

9.8.5. Electric Brakes

In an **electric brake** *the current from the battery is utilized to energize an electromagnet within the brake drum. This actuates a cam to expand the brake shoes.* When current stops, the cam and brake shoes are returned to the release position by retractor springs. The *severity of braking is controlled by means of a rheostat,* which is operated by the driver through the foot pedal.

Hydraulic pressure has also been used to apply electric brakes. As pedal is pressed more, hydraulic pressure actuates the rheostat to increase current to the electromagnet.

Advantages :

Electric brakes claim the following *advantages* :

1. Simple in design and installation.
2. Simplified operating linkage (only one cable for each drum).
3. Less time-lag between the pressing of brake pedal and application of brakes at the wheels.
4. Much better control over the braking effort.
5. Absence of grab.
6. No adjustment required during its lining life.
7. For trailer brakes, this type offers much simplified connections ; simply one cable has to be carried to the trailer side.

- Electric brakes are *not very popular, as service brakes.* These are *commonly used on trailers.*

9.9. BRAKE EFFECTIVENESS

The following *factors contribute to the effectiveness of the brakes* :

1. Amount of pressure applied to the shoe brakes.
2. Area of brake linings.
3. Radius of brake drum.
4. Radius of car wheel.
5. Coefficient of friction of braking surfaces.
6. Coefficient of friction between tyre and road surface.

9.10. FACTORS CONTROLLING THE STOP OF AN AUTOMOBILE

The following *factors control the stop of an automobile* :

1. **Speed and load :**

- Lesser the speed and lesser the load, the less energy shall be absorbed to stop the vehicle as the brakes convert energy of motion into heat.
- The higher the speed and more the load, the more energy shall be absorbed for stopping a vehicle.

2. **Road surface.** The coefficient of friction between the road and tyre is affected by the nature of road. If the road is slippery, the vehicle shall stop at a greater distance than if the road is dry.

3. **Tyre tread.** A good tyre tread has good holding ability. It provides a greater coefficient of friction than a smooth tyre.

4. **Gradient :**

- When a vehicle is going up a gradient or hill, the gravitational force assists in stopping the movement of the vehicle.
- If the vehicle is going down a gradient or hill, the force of gravity tends to keep the vehicle moving and hence a greater braking force is required to overcome its movement.

5. **Number of wheels braking :**

- For getting maximum braking effort, all the wheels should be equipped with brakes.
- If any of the wheels is not equipped with brakes the time and distances of stopping the vehicle shall increase.

6. **Coefficient of friction of braking surfaces.** The ability to hold a moving wheel is controlled by the coefficient of friction between the brake lining and brake drum. If the linkages are glazed with oil or grease they will tend the brake drums to slip as the coefficient of friction is lowered down.

7. **Braking force of engine.** The engine is also used as a brake when going down a hill. The *braking effect of the engine is more in lower gear speed than in direct drive.*

8. **Pressure applied through energization.** More the self-energizing ability of the brake shoe, better the braking effect.

9. **Weight transfer.** On application of brakes, there is a natural transfer of weight from the rear wheels to the front wheels caused by the tendency of the mass to continue in its forward motion. The greater the deceleration, the greater is the load or weight transfer from rear to front wheels. During the stop, the weight holding the rear wheels to the road is lessened while the weight on the front wheel is increased by exactly the same amount. It is due to this fact that *brakes are designed so that applied effort is greater on front wheels than on the rear.*

9.11. ARRANGEMENT OF BRAKES IN DIFFERENT VEHICLES

In an automobile, there are two types of brakes *viz.* "*service brake*" and "*hand brake*". The arrangement of these brakes in different automobiles is usually as under :

1. Service or foot brake applicable to all front and rear wheels and hand brake applicable to rear wheels alone.

2. Service or foot brake on all wheels and hand brake on drum at transmission shaft.

9.12. BRAKE SHOES AND LININGS

- Brake shoes are made of metal to which a lining of friction material is attached. The lining has to be *tough*. During hard braking, the shoe may be pressed against the drum by force of 4500 N or higher. Since friction increases as the applied force increases, a strong frictional drag is produced on the brake drum. This produces the braking effect at the wheel.
  — A large quantity of heat is produced by the frictional contact between the brake shoes and the drum. Under heavy braking conditions, drum-brake temperatures may reach 260°C. Some heat flows through the brake linings to the shoes and braking plate. There the heat is carried away by the surrounding air. But most of the heat is absorbed by the brake drum. Some brake drums have cooling fins. They provide additional surface to get rid of the heat more quickly. To help provide drum cooling, some brake linings are grooved at the centre.
  — Excessive temperature may damage brakes and burn the brake linings. When the linings and drums are excessively hot, the *coefficient of friction changes.* Then less

effective braking action results. This is the reason that brakes ***"fade"*** *after continuous use, as driving down a mountain.*

- The ***brake linings*** are either of *"solid woven type"* or *"moulded type"*.
  - — The asbestos base non-metallic linings have an average coefficient of friction of 0.4 upto about 260°C. This maximum temperature resistance is about 350°C.

Zinc wire linings have better resistance to wear than the non-metallic type. Also zinc serves to conduct some heat away from the working surface.

  - — *"Moulded type linings"* are prepared directly from the *mix* which contains asbestos fibres, together with resin powders and fillers. These linings have *good wear resistance.* Their maximum temperature resistance is about 450°C. The average coefficient of friction is 0.4.
- The braking linings are attached (Fig. 9.16) with shoe brakes by synthetic resin adhesives or by riveting. The first method is preferable, due to the following **reasons** :

(*i*) Absence of riveting holes.
(*ii*) More contact surface.
(*iii*) Free from scoring action.
(*iv*) More effective wearing thickness.

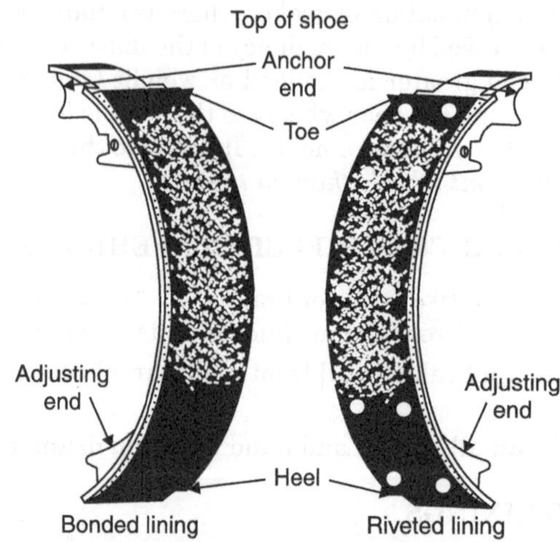

Fig. 9.16. Two methods of attaching the lining to the shoe.

9.13. BRAKE DRUM

The brake drum rotates with the road wheel and provides a contacting surface for the brake shoes to come into contact for braking action. The brake drums are usually made of *cast iron.* In some cases brake drums are made of *steel or aluminium having inner liner of cast iron.*

- In most cases, pins are provided at the outer surface of the brake drum for rapid dissipation of heat and thus help in its cooling.

9.14. BRAKE SHOE HOLDING DOWN ARRANGEMENTS

In internal expanding brakes the brake shoes are held in position by different ways in different designs of brakes. Some popular methods of holding down brake shoes are :

1. Single anchor arrangement.
2. Double anchor arrangement.
3. Self centring or adjusting arrangement.
4. Separate wheel cylinders for each brake shoe.
5. Articulating link arrangement.

1. **Single anchor arrangement :**

- In this case, both the primary and secondary brake shoes are anchored at their *bottom ends through a 'single anchor' or pivot attached to back plate.* The upper ends of the brake shoes rest over the wheel cylinder pistons or operating cam, as the case may be.
- The brake shoes are adjusted by snail cams at upper parts of the back plate and by a cam operated by the anchoring bolt at the lower end.

2. **Double anchor arrangement :**

In this arrangement, the individual brake shoes are anchored at their bottom ends through *separate anchor bolts.* The upper ends of brake shoes rest over the wheel cylinder pistons or operating cam. The adjustment of brake shoes is carried out as is done in single anchor arrangement.

— In some cases, adjustment of brake shoes at their upper ends is through notched collars provided at the wheel cylinder pistons. The collars are screwed in or out to adjust the brake shoes at their upper ends.

3. **Self centring or adjusting arrangement :**

In this case the brake shoes are not pivoted at their lower ends ; the lower ends simply rest over the brake adjuster. The brake shoes are held over the brake adjuster and wheel cylinder or brake expander by means of springs, one at the lower and the other at the upper end and holding down springs and clips. The brake adjuster may be of girling type or bendix or screw type.

— In certain cases, there is a peg or anchor placed between the uppermost ends of the brake shoes. In some cases the anchor is fixed whereas in some designs, it is adjustable. The wheel cylinder is located below that peg which assists in self-energization action and servo action of the brakes.

4. **Separate wheel cylinders for each brake shoe :**

- In this type of arrangement there are separate wheel cylinders for the individual brake shoes. One end of each brake shoe is anchored to one wheel cylinder which operates the opposite brake shoe. There are two retracting springs, one at the lower end and the other at the upper end.
- The adjustment of the brake shoes is through snail cams and anchor bolt cams.

5. **Articulating link arrangement :**

In this case, the brake shoes are connected to single anchor pin through articulating links, which connect the lower ends of the brake shoes together. The pivot pin for articulating link is anchored to the back plate. The upper ends of the brake shoes rest over the piston ends of the wheel cylinder. *The articulating links provide double lever action on the brake shoes which results in self energization.*

9.15. BRAKE TESTERS

The brake testers are of the following *two types* :

1. Static brake tester.
2. Dynamic brake tester.

1. **Static brake tester :**

One type of static tester has four tread plates and registering columns (Fig. 9.17). To make the tests, the car is driven onto the tread plates at a specified speed and the brakes are applied hard.

The stopping force at each wheel is registered on the four columns. If the *readings are too low, or are unequal, brake service is needed.*

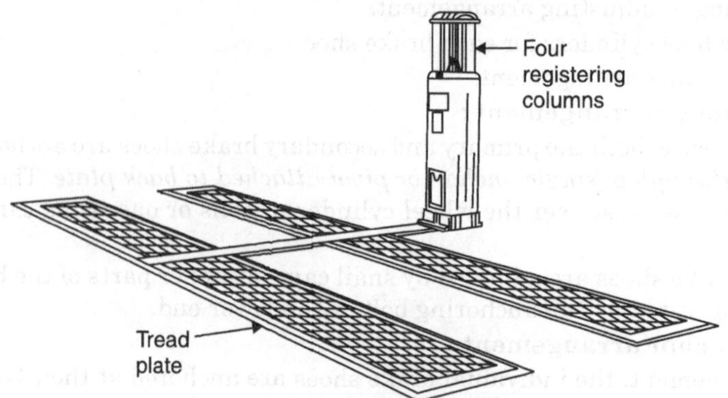

Fig. 9.17. Static brake tester.

2. **Dynamic brake tester :**

The dynamic brake tester has rollers in the floor. The two wheels for which brakes are to be tested are placed on the rollers (Fig. 9.18). If these wheels are drive wheels, the wheels are spun at the specified speed by the vehicle engine. For nondriving wheels, the rollers and wheels are spun by an electric motor. Then throttle is released or the electric motor is turned off and brakes are applied. The braking force at wheel registers on meters. *This shows if the brake performs normally or if they need service.*

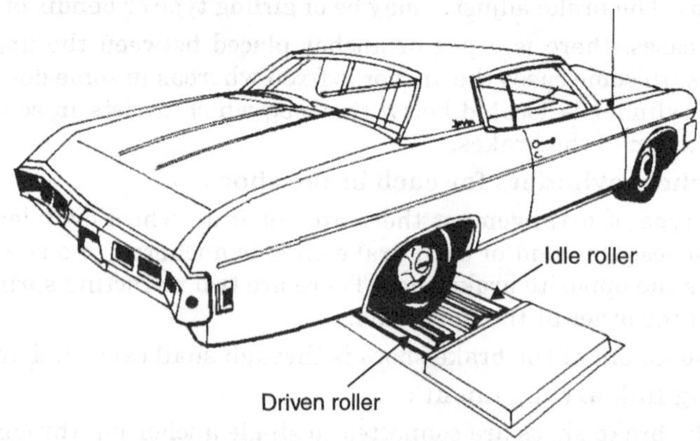

Fig. 9.18. Dynamic brake tester.

9.16. BRAKE SERVICE

If there is any complaint of faulty braking action it should be analysed to determine the cause.

- Sometimes, all that is necessary in the earlier drum brakes is a brake-shoe adjustment to compensate for lining wear. On the later drum brakes with *self adjusters, the brakes automatically adjust to compensate for lining wear. However, the self-adjusters may fail.*
- *Other brake services* include the following :

(*i*) Addition of brake fluid.

(*ii*) Bleeding the hydraulic system to remove air.

(*iii*) Repair or replacement of master cylinders, wheel cylinders, and calipers.
(*iv*) Replacement of brake linings.
(*v*) Refinishing of brake drums or disks.
(*vi*) Overhauling of power brake units.

9.17. TROUBLESHOOTING CHART OF HYDRAULIC BREAKS SYSTEM

Troubleshooting chart for hydraulic brake system is given below :

Faults	Causes	Remedies
1. *Brakes drag, i.e., they remain applied without pressing the pedal.*	(*i*) Weak or broken springs of brake shoes. (*ii*) Clogged by-pass port of master cylinder. (*iii*) Weak or broken pedal spring. (*iv*) Seized mechanism of hand brake. (*v*) Seized pistons in wheel cylinder.	(*i*) Replace them by new ones. (*ii*) Clean the port. (*iii*) Remove it and replace it by a new one. (*iv*) Clean the mechanism and lubricate it wherever necessary. (*v*) Clean the wheel cylinder ; replace the damaged pistons.
2. *Inefficient brake.*	(*i*) Greasy linings. (*ii*) Bad linings. (*iii*) Incorrect type of linings used.	(*i*) Remove the grease from the linings. (*ii*) Replace the lining. (*iii*) Replace it with correct type of linings.
3. *Pedals feel spongy when pressed.*	(*i*) Air in brake system. (*ii*) Worn out secondary cup of master cylinder (*iii*) Leakage in brake system.	(*i*) Remove the air by bleeding the system. (*ii*) Replace the cup. (*iii*) Rectify the leakage.
4. *Excessive pedal travel.*	(*i*) Leakage past main cup into master cylinder. (*ii*) Low oil level in master cylinder. (*iii*) Push rod of master cylinder out-of-adjustment. (*iv*) Brake shoes out-of-adjustment.	(*i*) Remove the main cup and clean it ; replace if necessary. (*ii*) Fill the master cylinder with sufficient oil. (*iii*) Adjust the push rod. (*iv*) Adjust the brake shoes ; if necessary, fit them with new linings.

9.18. TROUBLESHOOTING CHART OF AIR BRAKES

Troubleshooting chart for air brakes is given below :

Faults	Causes	Remedies
1. *Compressor does not unload.*	Unloader valve not working properly,	Dismantle the unloader valve. Clean it and check for fault.
2. *Pressure builds up very slowly.*	(*i*) Choked air filter.	(*i*) Check the air filter, clean and refit it. If it is a spoilt, replace it with a new one.

	(*ii*) Leakage in the pipeline between the compressor and the reservoir or between the reservoir and the trailer brake.	(*ii*) Find out and rectify the points of leakage.
	(*iii*) Defective unloader valve.	(*iii*) Check for escape of air ; if it is present, the unloader should be dismantled, cleaned and rectified properly.
	(*iv*) Faulty compressor.	(*iv*) Dismantle the compressor, overhaul it and check the valves and springs.
	(*v*) Leaking brake valve.	(*v*) Overhaul the brake valve.
3. *Brake application is very slow.*	(*i*) Stuck brake control valve.	(*i*) Dismantle the brake control valve and overhaul it ; see that the piston is moving freely.
	(*ii*) Leaking brake cylinder.	(*ii*) Overhaul the brake cylinder.
	(*iii*) Worn out brake linings.	(*iii*) Replace or adjust the brake linings.
4. *Brakes work inefficiently even though the reservoir pressure is sufficient.*	(*i*) Wrongly adjusted brake linings.	(*i*) Replace or adjust the brake linings.
	(*ii*) Worn out drums.	(*ii*) Check the worn out parts of the drum ; if necessary, true it.
	(*iii*) Incorrectly adjusted brakes.	(*iii*) Adjust the brakes correctly.
	(*iv*) Leaking brake lines.	(*iv*) Test for leakage ; rectify the leakage if it is found out.
5. *Brakes bind, i.e. they grip, but do not release the brake drum.*	(*i*) Brakes not adjusted properly.	(*i*) Adjust the brakes properly.
	(*ii*) Defective brake valve.	(*ii*) Overhaul the brake valve and then fit it.

9.19. TROUBLESHOOTING CHART OF BRAKE SHOES AND DRUMS

The troubleshooting chart for brake shoes and drums is given below :

Faults	Causes	Remedies
1. *Brakes fade, i.e., they fail to hold on.*	Brake drum getting too hot.	Allow the brake drum and brake linings to cool.
2. *Brake squeal.*	Rivet heads touching the brake drum.	Change the brake linings and use new rivets.
3. *Brake chatter.*	(*i*) Too much brake lining.	(*i*) Cut-off excess material.
	(*ii*) Poor brake adjustment.	(*ii*) Adjust brakes.
	(*iii*) Brake drum out-of-round.	(*iii*) Rebore or replace the drum.
4. *Brakes grab.*	(*i*) Damaged or broken brake lining.	(*i*) Replace the brake lining.
	(*ii*) Badly scored brake drum.	(*ii*) Rebore or replace the drum.
5. *Wheel lock due to brakes.*	(*i*) Broken brake shoe.	(*i*) Replace the shoe.
	(*ii*) Broken lining.	(*ii*) Replace the lining.
	(*iii*) Loose lining.	(*iii*) Tighten the lining.
6. *Brake drum wear.*	Brake pedal touching the foot board.	True up the brake drum ; replace it, if necessary.

9.20. COMPARATIVE BRAKE DATA OF SOME INDIAN VEHICLES

Comparative brake data of some Indian vehicles is given below :

Make of the vehicle	Type of brake	Drum of disc	Type of brake lining
1. *Ambassador*	Hydraulic	Drum	Ferodo
2. *Premier Padmini*	Hydraulic	Drum	Ferodo
3. *Maruti*	Hydraulic	Disc in front wheels and drum on rear wheels	Ferodo
4. *Gypsy*	Hydraulic	Disc in front wheels and drum on rear wheels	Ferodo
5. *Jeep Mahindra 6.*	(CJ—3B) Hydraulic	Drum	Ferodo
6. *Ashok Leyland Comet*	Compressed air Diaphragm operated.	Drum	Ferodo
7. *Telco Truck*	Air hydraulic	Drum	Ferodo
8. *Telco Indica Petrol*	Dual circuit diagonally spilt vacuum assisted with P and R valves	Disc in front and drum on rear	
9. *Hyundai Santro*	Servo assisted with dual circuit front and self adjusting rear brakes	Disc in front and drum on rear	

SHORT ANSWER QUESTIONS

Q. 1. What is the function of a braking system ?

Ans. To reduce the speed, stop the moving vehicle within a minimum possible time and a minimum possible distance and to hold the vehicle.

Q. 2. Why is an internally expanding brake preferred in automobile ?

Ans. (*i*) The brake effect is gradual, protection and care is possible and easy. (*ii*) It is convenient to fasten brake lining on brake shoes.

Q. 3. How many types of brakes are fitted in vehicle ?

Ans. (*i*) Service brake, and (*ii*) Auxiliary brake for parking.

Q. 4. What is service brake system ?

Ans. A control device which acts on the wheels and is operated by a foot pedal.

Q. 5. What is auxiliary brake system ?

Ans. A control device installed on the gear box main shaft and operated by a lever under hand control.

Q. 6. What is fading of brakes ?

Ans. Due to prolonged application of brakes the effectiveness decreases, this is called "*fading*".

Q. 7. What is the cause of fading of brakes ?

Ans. High temperature on brake linings (on account of reversible changes in friction properties).

Q. 8. What features should an efficient braking system possess ?

Ans. Smooth operation, stopping the vehicle within a reasonable distance, and retaining adjustment for long periods.

Q. 9. What is brake compensation ?

Ans. A process of ensuring equal braking effect at all the wheels inspite of unequal wear of lining.

Q. 10. What is meant by brake binding ?

Ans. When the brake drums become hot after a short run, brakes are said to be "*binding*".

Q. 11. What is the minor and major adjustments of brakes ?

Ans. "*Minor adjustment*" refers to the adjustment of brake shoes to *compensate lining wear without removing the wheel.* "*Major adjustment*" is *installation of new shoes or relining the old shoes.*

Q. 12. Define reaction time as referred to brake application.

Ans. Time needed to move the foot and apply the brake.

Q. 13. What is meant by weight transfer ?

Ans. Centre of gravity is above the wheel centres, so a sudden stop will transfer some weight from rear wheels and place it towards front wheels. This phenomenon is known as *weight transfer*.

Q. 14. What is self energization ?

Ans. Tendency of the rotating drum to drag the lining along with it.

Q. 15. What are the causes of poor braking ?

Ans. (*i*) Wet brake lining, and (*ii*) Burnt or worn out linings.

Q. 16. How does disc brake function ?

Ans. A disc brake consists of a rotating disc and two friction pads which are actuated by four hydraulic wheel pistons contained in two halves of an assembly called *caliper*.

Q. 17. What is major advantage of a disc brake ?

Ans. Its resistance to fading and its tendency to remain cooler even under repeated applications.

Q. 18. Why brake fading cannot occur in case of disc brakes ?

Ans. Because the friction surfaces are closer to the pad lining.

Q. 19. What is the advantage of using an aluminium brake drum ?

Ans. Aluminium conducts more heat than cast iron.

Q. 20. What is the function of a brake shoe ?

Ans. To control the speed of the vehicle or to lock the wheel, by rubbing against the wheel drum.

Q. 21. What is the cross-section of a brake shoe ?

Ans. T-section.

Q. 22. Why are hydraulic brakes preferred in vehicles ?

Ans. Because they are more efficient and need less pedal pressure, moreover, braking pressure is distributed to all the wheels.

Q. 23. How does the hydraulic brake function ?

Ans. When the foot pedal is pressed, the brake fluid in the master cylinder is forced through a pipeline to the individual wheel cylinders. The movement of the brake fluid in the wheel cylinder operates the expander mechanism to move the brake shoes against the drum.

Q. 24. What are the functions of a master cylinder in a hydraulic braking system ?

Ans. (*i*) To maintain a constant amount of brake fluid in the system as a reservoir ; (*ii*) To maintain a high uniform pressure in the system.

Q. 25. What is bleeding of brake ?

Ans. Removal of air bubbles present in the braking system is called *bleeding*.

Q. 26. Why are all hydraulic brakes naturally "self-equalising" ?

Ans. In hydraulic braking system all the brake shoes are forced to contact the drum by the pressure initially built up, well before increase of the pressure in the system. Thus all the wheel brakes are forced to act simultaneously. So hydraulic brakes are self-equalising.

Q. 27. How stop light switch placed in the hydraulic brake system work ?

Ans. A stop light is connected to the master cylinder, when the pressure in the system is increased on application of brakes the mechanism completes the circuits of connecting battery and stop light glows.

Q. 28. What causes vapour lock in a brake system ?

Ans. When the brake fluid becomes hot enough to vaporise, a gas is emitted so that a drop of fluid gets changed into a vapour form. These bubbles are the cause of vapour lock in a brake system.

Q. 29. What is the basic principle of the 'Air-braking system' ?

Ans. The compressed air is used to actuate the brake shoes, thus making the operation easy.

Q. 30. What is the function of an unloader valve ?

Ans. It serves to regulate the line pressure in air brake circuit.

Q. 31. What are the functions of brake lining ?

Ans. (*i*) To increase the coefficient of friction, (*ii*) To prevent wearing away of the metal.

Q. 32. If the vehicle pulls to one side when the brakes are applied, what does this indicate ?

Ans. The vehicle pulls to one side during brake application due to uneven operation of brakes or due to mal-adjustment or grease on the lining and drums.

Q. 33. What are 'power brakes' ?

Ans. Brakes powered by power other than physical effort applied at brake pedal are known as power brakes. The power may be exerted by air pressure, engine vacuum or electrical energy.

Q. 34. What is meant by servo brake ?

Ans. Servo brake is one in which the pedal effort is assisted by mechanical or vacuum effect.

Q. 35. What do you mean by 'Hydrovac' ?

Ans. *Hydrovac* is a combined hydraulic and vacuum power braking system, utilising the vacuum from the intake manifold as an operating source.

Q. 36. What is major cause of squealing of brake shoes ?

Ans. Squealing of brake shoes is due to vibration, in the drums or shoes, and due to lack of rigidity in their construction or mounting.

HIGHLIGHTS

1. All the braking systems depend upon friction between moving parts and stationary parts for their stopping force.
2. Braking efficiency is the ratio between the retarding force and the weight of the vehicle.
3. The brakes which are operated mechanically by means of levers, linkages, pedals, cams, bell cranks etc. are known as *mechanical brakes*.

4. Brakes which are operated by means of hydraulic pressure are known as *hydraulic brakes*.
5. The process of eliminating or removing air out of the braking system of an automobile is called *bleeding*.
6. *Hill holder* is an auxiliary braking device used on hydraulic braking system. It automatically prevents the car from sliding down on a grade after it has been fully stopped by the use of service brake and as long as the clutch is disengaged.
7. Brake operated by power other than physical effort applied at the brake pedal, are known as power brakes. The power may be exerted by air pressure, engine vacuum or electrical energy.
8. Any mechanism which adds to the driver's effort in applying the brakes is called a *servo mechanism*.
9. In *vacuum brakes* the suction from the engine inlet manifold is utilised for brake application.

OBJECTIVE TYPE QUESTIONS

Choose the Correct Answer :

1. The function of master cylinder is
(*a*) to increase pressure equally in all cylinders (*b*) to increase pressure unequally in all cylinders
(*c*) to decrease pressure unequally in all cylinders (*d*) to decrease power equally in all cylinders.
2. In trucks, the system of wheel braking employed is
(*a*) Mechanical (*b*) Hydraulic (*c*) Pneumatic (*d*) None of above.
3. In Jeep we use brakes operated
(*a*) Mechanically (*b*) Hydraulically (*c*) Pneumatically (*d*) None of these.
4. Brake lining consists mainly of
(*a*) asbestos (*b*) copper (*c*) cast iron (*d*) aluminium.
5. The maximum disc run out allowed on the vehicle is generally
(*a*) 1 mm (*b*) 0.5 mm (*c*) 0.1 mm (*d*) 0.01 mm.
6. The brake efficiency of a new vehicle is about
(*a*) 30 percent (*b*) 50 percent (*c*) 80 percent (*d*) 100 percent.
7. Brake fade is
(*a*) loss of pedal (*b*) loss of friction
(*c*) loss of hydraulic fluid (*d*) none of these.
8. Electric brakes are commonly used on
(*a*) two wheelers (*b*) cars (*c*) trucks (*d*) trailers.
9. On suspended vacuum brakes, there is vacuum on both sides of the piston during
(*a*) brake application (*b*) brake release
(*c*) part application of brakes (*d*) all of these.
10. During braking the push rod directly operates
(*a*) primary piston (*b*) secondary piston
(*c*) residual pressure valve (*d*) compensating port.
11. The brake bleeding process removes from system
(*a*) air (*b*) vacuum (*c*) excess fluid (*d*) excess pressure.
12. In Maruti car the brakes used are
(*a*) disc brakes only (*b*) drum brakes only
(*c*) disc on front drum on rear (*d*) drum or front and disc on rear.
13. Fading of brakes occurs
(*a*) at high speeds (*b*) at low speeds
(*c*) during continuous application (*d*) when brake lining is worn.
14. The brakes employed in cars are usually operated
(*a*) mechanically (*b*) hydraulically
(*c*) by means of engine vacuum (*d*) by compressed air.

15. Air brakes are mostly used in case of
(*a*) cars (*b*) jeeps (*c*) trucks (*d*) three-wheelers.

16. The hand brake usually operates on
(*a*) rear wheels (*b*) front wheels (*c*) right wheels (*d*) left wheels.

Answers

1. (*a*) 2. (*b*) 3. (*b*) 4. (*a*) 5. (*c*) 6. (*c*) 7. (*b*)
8. (*d*) 9. (*b*) 10. (*a*) 11. (*a*) 12. (*c*) 13. (*c*) 14. (*b*)
15. (*c*) 16. (*a*).

THEORETICAL QUESTIONS

1. Define the term 'braking efficiency'.
2. What is the necessity of a braking system ?
3. What are the functions of brakes ?
4. What are the requirements of a good braking system ?
5. How are brakes classified ?
6. Discuss the classification of brakes for vehicles. Describe shoe and drum type mechanical brakes with the help of simple sketches.
7. List the factors on which the capacity of a brake depends.
8. What are mechanical brakes ? Describe briefly an 'Internal expanding mechanical brake'.
9. Explain briefly a 'hand brake'.
10. With the help of a neat sketch explain the construction and working of a disc brake.
11. What are the advantages and disadvantages of disc brakes compared with drum brakes ?
12. What is the major drawback of a 'mechanical brake system' ?
13. Draw the sketch of a mechanical braking system and explain various parts and working of this braking system.
14. What are 'hydraulic brakes' ?
15. Explain with a neat sketch a 'hydraulic braking system'.
16. What are the essential differences between mechanical brakes and hydraulic braking system ?
17. Draw a simple wheel cylinder. Name its major parts and give their function.
18. (*a*) What are the features of an efficient braking system ?
(*b*) How do you adjust the brakes ?
19. Explain the functions of main parts of a master cylinder used in hydraulic brakes.
20. What is the function of master cylinder in hydraulic brakes ?
21. Why wheel cylinder necessary in hydraulic brakes ?
22. What are the advantages and disadvantages of hydraulic brakes ?
23. What do you mean by the self-energizing of the brakes ?
24. Explain the working of a master cylinder.
25. Describe the construction and working of a wheel cylinder used in a braking system of a car.
26. Describe briefly a 'tandem master cylinder' used in a hydraulic braking system.
27. What are desirable characteristics of a "hydraulic brake fluid" ?
28. What is the necessity of bleeding of brakes ? Explain bleeding process.
29. Explain briefly a 'hill holder'.
30. Describe briefly "Self-energizing brakes".
31. What are "power brakes" ?
32. What are the main features of the power brake system ?

33. Explain briefly the working of 'air brakes'.
34. Enumerate the main parts of an air brake.
35. What is a 'Fail safe brake system' ? Explain briefly with a circuit diagram.
36. Describe briefly Air-hydraulic brakes.
37. Explain briefly, with a neat sketch, the construction and working of an engine exhaust brake.
38. Give brief description of vacuum brakes.
39. How do the electric brakes operate ? What are their advantages ?
40. Enumerate the factors which contribute to the effectiveness of the brakes.
41. Explain briefly the factors which control the stop of an automobile.
42. Give the arrangement of brakes in different vehicles.
43. What is the function of brake shoes ?
44. Name the materials used for brake lining. What properties should they possess ?
45. Give the construction of brake linings. How are they fixed to the brake shoes.
46. Describe briefly brake shoe holding down arrangements.
47. Explain briefly the following brake tester :
    (*i*) Static brake tester ;
    (*ii*) Dynamic brake tester.
48. Write short note on 'Brake service'.
49. Draw a troubleshooting chart for brakes of an automobile giving symptom, causes and possible remedies.

10

Automotive Electrical System

10.1. INTRODUCTION TO ELECTRICAL SYSTEM

The major components of the automotive electrical system are shown in Fig. 10.1. The major systems include the following :

1. **Starting system :**
   - *Battery*
   - *Starting motor*
   - *Wiring*
   - *Switches.*
2. **Charging system :**
   - *Alternator and regulator, with wiring.*

3. **Ignition system :**
   - *Contact-point ignition system*
   - *Electronic ignition systems.*
4. **Lighting and Accessory system :**
   - *Lighting*
   - *Horns*
   - *Instrument-panel warning systems.*

Fig. 10.1 shows a typical automotive electrical system.

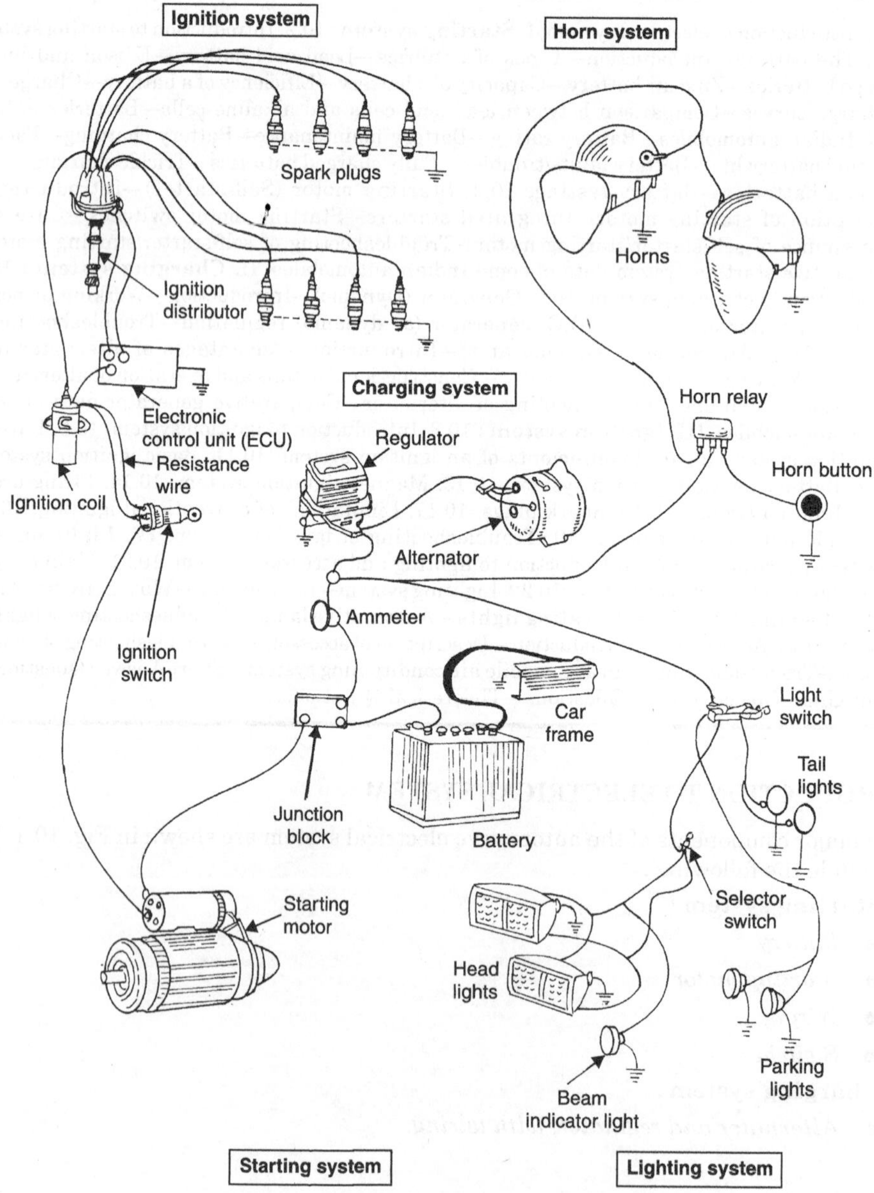

Fig. 10.1. Typical automotive electrical system.

I. STARTING SYSTEM

10.2. INTRODUCTION TO STARTING SYSTEM

- The I.C. engine is not capable of self starting. Some big engines used in trucks, tractors, and off-road and industrial equipment are *cranked by compressed air* or by a *small starting engine*. However automotive engines (both S.I. and C.I.) are cranked by a small but powerful electric motor. This motor is called '**cranking motor**', '**starter**' or '**starting motor**'.
  — When the driver turns the ignition switch to '*START*', the battery sends current to the starting motor. This causes a pinion gear in the starting motor to mesh with teeth on the ring gear around the engine wheel. The starting motor then rotates the engine crankshaft for starting.
- The *typical starting system includes the following* :
  (*i*) The battery,
  (*ii*) The starting motor and drive mechanism,
  (*iii*) The ignition switch,
  (*iv*) The starter relay or solenoid,
  (*v*) The neutral safety switch, and
  (*vi*) The wiring which connects the above components.

These components are described in the following articles.

10.3. THE BATTERY

10.3.1. Introduction

The "**battery**" is one of the most important components of electrical system and may be described as the *heart* of the electrical system of a vehicle.

It performs the following ***functions*** :

1. To supply current to operate the starting motor and ignition system when the *engine is being started.*

2. To act as *voltage stabilizer* by supplying currents for the lights, radio and other electrical accessories when the alternator is *not* handling the load.

The battery is an *electrochemical device*. This means it uses chemicals to produce electricity. The amount of electricity it can produce is *limited*. As the chemicals in the battery are "used up", the battery runs down, or is *discharged*. It can be *recharged* by supplying it with electric current from a battery charger, or from the vehicle alternator. The "used up" chemicals are then *returned to their original condition*, so the battery becomes recharged.

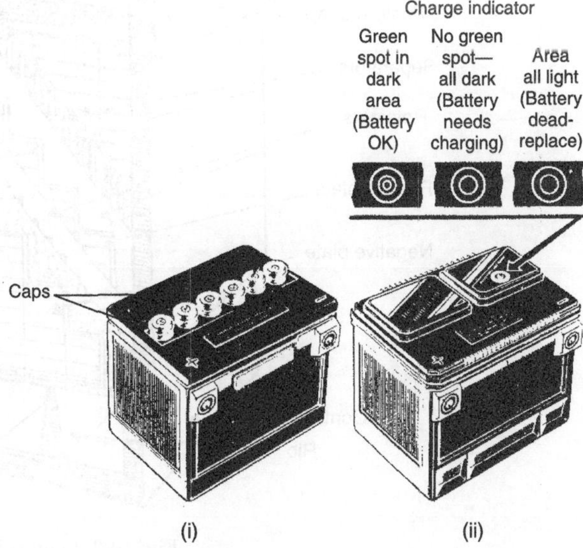

Fig. 10.2. Two types of automotive batteries.

- The term ***cell*** and ***battery*** are used interchangeably by incorrectly. A *battery* means a group of interconnected cells. Thus a *cell* is one unit of battery.

10.3.2. Types of Batteries

The batteries are of the following types :

1. Lead acid battery
2. Alkaline battery
   (*i*) Nickel-iron type (*ii*) Nickel-cadmium type
3. Zinc-air battery.

- Fig. 10.2 shows following two types of automotive batteries :

(*i*) This battery has caps which can be removed to check the battery state of charge and to *add water if needed.*

(*ii*) This type of battery with side terminals is a sealed maintenance-free type and *requires no water*. The charge indicator in the top shows the state of charge of the battery.

10.3.3. Lead Acid Battery

The lead acid battery is most widely used on all passenger cars. Due to its economical and electrical advantages, the lead acid battery is in general use for many purposes. The most common of which is for starting, ignition and lighting the modern automobile.

10.3.3.1. Components of a lead acid battery

Fig. 10.3 shows a lead acid battery.

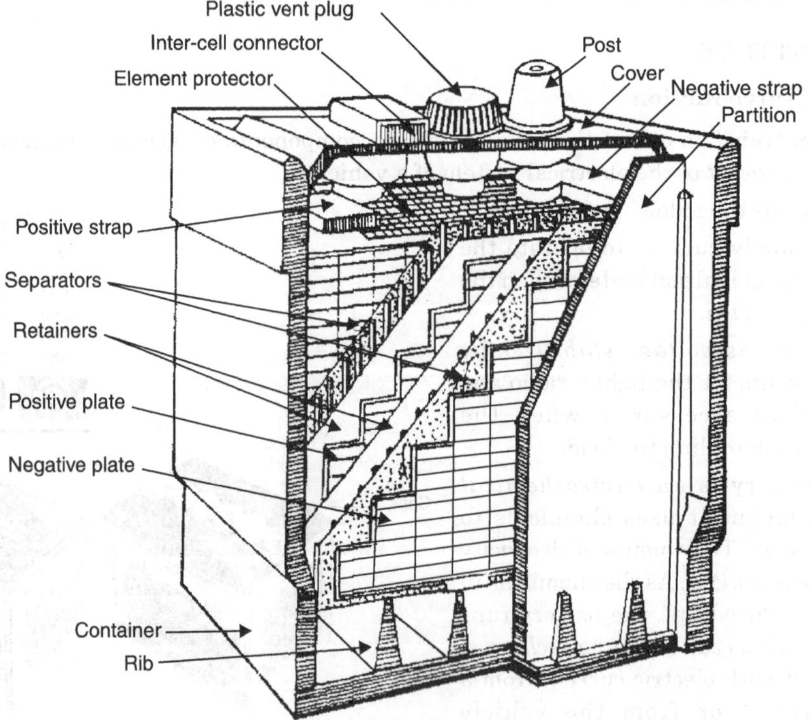

Fig. 10.3. Lead acid battery.

1. **Positive plate.** PbO_2 (*lead peroxide*), deposited on a grid frame of antimony lead alloy. (When the battery is in fully charged condition, the *positive plate is dark brown in colour*).

2. **Negative plate.** Pb (*porous spongy lead*), deposited on a grid frame (similar to the grid frame of positive plate). When the battery is in charged condition, the negative plate is *grey in colour*.

The number of negative plates in every battery is always *one more than the number of the positive plates so that action occurs on both sides of the positive plate.*

3. **Electrolyte.** Dilute sulphuric acid (Sp. gr. 1.28). It consists of 40% sulphuric acid and 60% distilled water. The level of electrolyte is about 10 mm above the top of plates.

4. **Separator.** The function of the separator is to keep the positive and negative plates *electrically apart.* The separators are usually made of specially treated wood, hard rubber, resin impregnated fibre alone or in combination with rubber or mats of glass fibres. Some batteries its have separators made of polyvinyl chloride or polyethylene saturated cellulose.

5. **Container.** The container is made of hard glass or hard rubber or other acid resistant materials. It must withstand extreme of heat and cold as well as mechanical shocks and must be resistant to the absorption of acid.

10.3.3.2. Chemistry of a lead acid battery

Negative plate (Pb). When the cell is producing current, lead atoms on the surface of the plate lose two electrons each, becoming Pb^{++} ions. These Pb^{++} ions do not dissolve into the liquid, but remain on the plate and attract SO_4^{--} ions from the sulphuric acid solution, thus forming an invisibly thin layer of $PbSO_4$ on the negative lead plate.

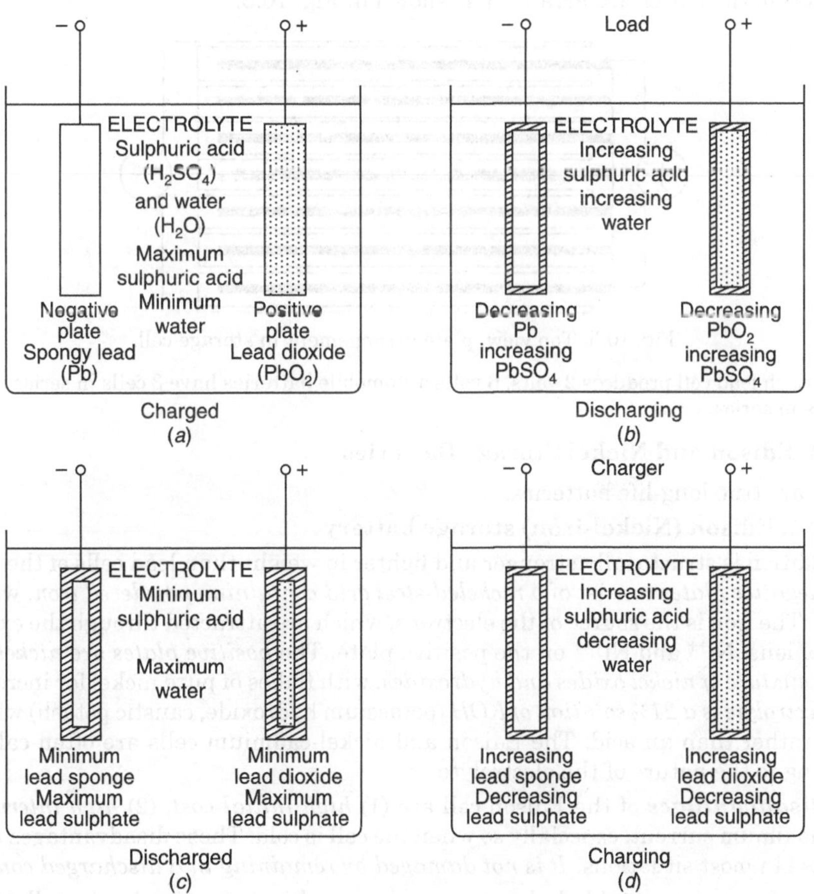

Fig. 10.4. Charging and discharging of a lead acid battery.

Positive plate (PbO_2). The positive plate consists of lead peroxide (PbO_2), in which each lead particle is lacking four electrons, which were given to the oxygen when the plate was formed. Each Pb^{++++} ion takes two electrons from the external circuit, becoming Pb^{++}.

The energy is obtained from the tendency of neutral lead atoms to give 2 electrons each to Pb^{++++} ions, both becoming Pb^{++} as a result of the transfer.

Incidentally, when Pb^{++++} ions of lead peroxide pick up the two electrons, they can no longer hold the oxygen, which goes into the solid solution and combines with hydrogen ions of the acid, forming water molecules. The lead Pb^{++} remains on the plate and picks up SO_4^{--} from the sulphuric acid solution, forming lead sulphate. These actions are shown in Fig. 10.4.

The charging and discharging of the cell can be represented by a single reversible equation given below :

$$\underset{\text{(Positive plate)}}{PbO_2} + 2H_2SO_4 + \underset{\text{(Negative plate)}}{Pb} \underset{\text{Charge}}{\overset{\text{Discharge}}{\rightleftharpoons}} \underset{\text{(Positive plate)}}{PbSO_4} + 2H_2O + \underset{\text{(Negative plate)}}{PbSO_4} + \text{electrical energy}$$

The chemical action can go on only where the plates are in contact with H_2SO_4 solution. In order to produce a large current, plates are made so that a lot of surface area is in contact with the solution. In a cell, the plates are arranged as shown in Fig. 10.5.

Fig. 10.5. Top view, plate arrangement in storage cell.

Note. Each lead cell produces 2 volts. 6 volts automobile batteries have 3 cells in series ; 12-volt batteries have 6 cells in series.

10.3.4. Edison and Nickel Storage Batteries

These are true long-life batteries.

10.3.4.1. Edison (Nickel-iron) storage battery

The Edition is structurally stronger and lighter in weight than lead cells of the same current rating. The *negative plates* consist of *a nickeled-steel grid containing powdered iron*, with some FeO and $Fe(OH)_2$. The iron is the source of the electrons, which are attracted through the external circuit toward nickel ions, Ni^{++} and Ni^{+++} on the positive plate. The *positive plates* are *nickel-plated tubes containing a mixture of nickel oxides and hydroxides*, with flakes of pure nickel for increased conductivity. The *electrolyte is a 21% solution of KOH* (potassium hydroxide, caustic potash) which is chemically a base rather than an acid. The Edison and nickel-cadmium cells are often called **alkaline cells,** referring to the nature of the electrolyte.

The ***disadvantages*** of the Edison cell are (1) *high initial cost,* (2) *high internal resistance* that limits maximum current, especially so when the cell is cold. These disadvantages are enough to prevent its use in most situations. *It is not damaged by remaining in a discharged condition.*

It is used in some portable lighting equipment and in a few marine installations, where it neither gets nor needs the attention that lead cells would.

The Edison battery is appropriate for running *electrical traction equipment,* such as mine locomotives and fork-lift trucks, *but not appropriate for starting gasoline and diesel engines, because its internal resistance limits the current too much.*

10.3.4.2. Nickel-cadmium battery

The nickel-cadmium battery followed a line of development that produced a battery not intended for frequent cycling, but rather a more general-purpose battery that enabled the user to draw *as many amperes as possible from a battery of given ampere-hour rating, without excessive falling-off of voltage. The starting of gasoline and diesel engines and the operation of signals, relays and controls are jobs for which the nickel-cadmium battery is suited.*

10.3.5. Zinc-Air Battery

This type of battery unlike the conventional battery in which the energy is stored and drawn as required from the active material of the electrodes, *draws its energy from the zinc within its cells and from oxygen of the atmosphere.*

It can be operated as a primary battery, that is to be thrown away after use, and as a secondary or rechargeable battery.

It consists of porous zinc oxide in the centre and a thin air-cathode on each side. Air cathodes are made from a hydrophobic material, which supports a conducting mesh and a layer of catalyst (to improve the rate of reduction of atmosphere oxygen). The hydrophobic material is *impervious to electrolyte* but *allows the air to pass through it* ; it is used as container also.

- *During discharge, cathode remains uncharged, while the anode is changed to zinc oxide.*
- The open-circuit voltage of this system is 1.4 volts.

Advantages. (*i*) Good discharge characteristics over a wide range of temperature.

(*ii*) Ancillary equipment required is minimum.

(*iii*) Raw materials used are readily available and cheap.

(*iv*) High energy densities.

10.3.6. Capacity of a Battery

The capacity of a battery is given in terms of *ampere-hours on discharge.* This is determined by the following factors :

(*i*) Final limiting voltage of the cells. (*ii*) Discharge rate.

(*iii*) Number, design and dimensions of plates.

(*iv*) Design of separators. (*v*) Quantity of electrolyte.

(*vi*) Density of electrolyte. (*vii*) Temperature etc.

10.3.7. Efficiency of a Battery

The **efficiency of a battery** *is defined* as, *"the ratio of the output of a cell or a battery to the input required to restore the initial state of charge under specified conditions of temperature, current rate and final voltage".*

1. *Ampere-hour efficiency* $= \dfrac{\text{Ampere-hours on discharge}}{\text{Ampere-hours on recharge}} \times 100\%$ (may be about 85% to 95%)

2. *Volt efficiency* $= \dfrac{\text{Average voltage during discharge}}{\text{Average voltage during recharge}}$

3. *Watt-hour efficiency* $= \dfrac{\text{Watt-hours output on discharge}}{\text{Watt-hours input on recharge}} \times 100\%$ (may be about 75%)

$$\text{or,}\quad \text{Energy efficiency} = \frac{\text{Discharge current} \times \text{average discharge V} \times \text{time}}{\text{Charging current} \times \text{average charging V} \times \text{time}} \times 100\%$$

10.3.8. Charge and Discharge Curves

- Fig. 10.6 shows a typical charge-voltage curve at constant charging rate.

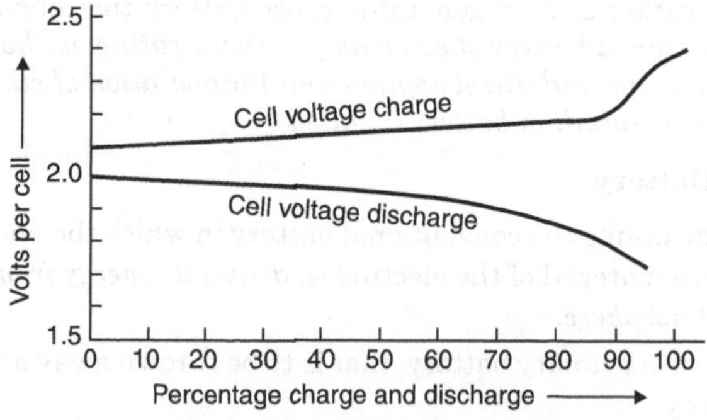

Fig. 10.6. A typical charge-voltage curve at constant charging rate.

- Fig. 10.7 shows typical discharge curves of a lead acid cell.

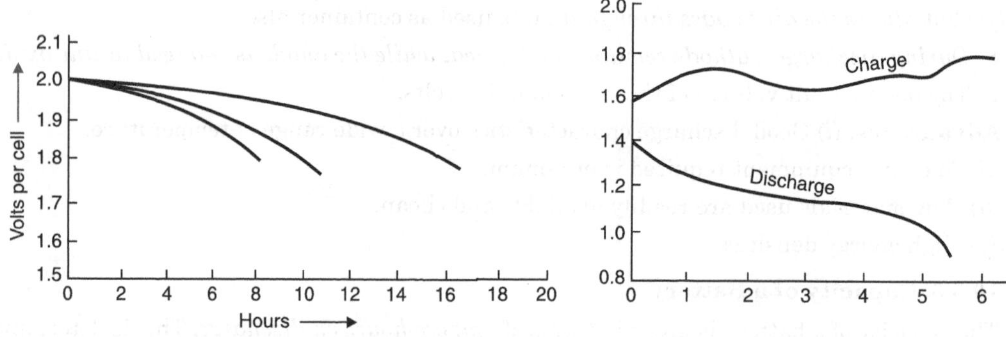

Fig. 10.7. Typical dischrage curves of a lead-acid cell.

Fig. 10.8. Charge and discharge curves for Nickel-iron (Edison) alkaline cell.

Fig 10.8. Charge and discharge curves for Nickel-iron (*Edison*) *alkaline cell.*

- Fig. 10.8 shows the charge and discharge curves for Nickel-iron alkaline cell.

10.3.9. Comparison between Lead-Acid Cells and Alkaline Cells

S. No.	Aspects	Lead-acid cell	Alkaline cell
1.	*Positive plate*	Lead peroxide (PbO_2). Dark chocolate brown in colour.	Perforated steel tubes into which is placed nickel hydroxide.
2.	*Negative plate*	Spongy lead (Pb). Dark grey in colour.	Steel grid into the pocket of which is placed powdered iron oxide.
3.	*Electrolyte*	Dilute solution of sulphuric acid (H_2SO_4).	Dilute solution of caustic potash (KOH) into which a small quantity of lithium hydroxide is added.
4.	*Average e.m.f. of the cell*	2.0 V	1.2 V.

5.	*Life*	1250 charges and discharges.	About 5 years.
6.	*Cost*	Cheaper than alkaline cell.	Costlier than the lead-acid cell.
7.	*Internal resistance*	Low	High
8.	Efficient (*a*) Ampere-hour (*b*) Watt-hour	 About 92% About 75%	 About 80% About 60%
9.	*Trickle charge*	Cells when not in use must be put on trickle charge.	No need of trickle charge.
10.	*Weight per kWh*	More weight	Lighter.
11.	*Discharged condition*	Should not be left in the discharged condition.	Can be left in the discharged condition.
12.	*Short circuits*	With short circuits the life of the cell decreases to much low values.	Short circuits do not reduce the life.
13.	*Advantages*	Used more in practice because of *higher ampere-hour capacities and voltages and higher efficiencies.*	(*i*) Mechanically more sturdy (*ii*) Do not evolve obnoxious fumes (*iii*) Less maintenance (*iv*) The plates do not buckle or swell.

PRIMARY AND SECONDARY CELLS

A. *Primary Cells

S. No.	Type	Voltage (V)	Remarks
1.	*Carbon-zinc*	1.5	• Low cost • Low current capacity • Used for flash lights and toys.
2.	Zinc-chloride	1.5	• Similar to carbon-zinc battery but higher current capacity.
3.	*Manganese-alkaline*	1.5	• Hydroxide electrolyte • High current capacity.
4.	*Silver oxide*	1.5	• Hydroxide electrolyte • Miniature button sizes.
5.	*Lithium*	2.95	• Long life • High cost • High energy density.
6.	*Mercury*	1.35	• Cathode is mercuric oxide.

B. Secondary Cells

S. No.	Type	Voltage (V)	Remarks
1.	*Lead-acid*	2.2	• Wet electrolyte • Lowest cost • Very low internal resistance • Very high current ratings.

* The simple voltage cell is a *primary cell.* A primary cell supplies current until the electrolyte is exhausted or the negative electrode is completely dissolved. *The shelf life of a primary cell is about one year.*

2.	*Nickel-iron (Edison cell)*	1.36	—Wet hydroxide electrolyte. —Industrial uses.
3.	*Nickel-cadmium*	1.25	—Most common rechargeable batteries.
4.	*Silver-zinc*	1.86	—Rechargeable dry cell. —High efficiency.
5.	*Silver-cadmium*	1.1	—Rechargeable dry cell. —High current rating. —High energy density.

The following points are worthnoting :

1. Matching a cell means making the lead resistance equal to the generator's internal resistance. The result is *maximum power delivered to the load* from the cell.

2. A *constant-voltage source has a very low internal resistance.* Output voltage is relatively constant with changing values of load because of small internal voltage drops.

3. A *constant-current source has a very high internal resistance.* This determines the constant value of current in the source circuit relatively independent of the load resistance.

10.3.10. Batteries Data of Some Indian Automobiles

S. No.	Make	Type	Capacity (amp. hr. at 20 hour rating)
1.	*Maruti 800*	NS-450	30
2.	*Hindustan Ambassador Mark II*	Exide 6XCV9R-12V	62
3.	*Fiat 1100*	Exide 6XCV9R-12V	62
4.	*Jeep Universal*	Exide 6XTA9R-12V	50
5.	*Dodge/Fargo Model 89M4 (Petrol type)*	Exide 6XCV9L-12V	62
6.	*Ashok Leyland Comet passenger*	Exide 6XCSI9L-24V	139
7.	*Standard 20*	11 plates, 12 volts	75
8.	*Tata Diesel vehicles*	12 volts battery	180

10.3.11. Battery Rating

Storage batteries are rated in many different ways, the most important of which are given below :

1. 20-hour rating :

20-hour rating in ampere-hours (an *ampere-hour* is the quantity of electrical energy furnished by a current of 1 ampere for 1 hour) has been prescribed by the S.A.E. (Society of Automobile Engineers) as an accepted standard for automobile batteries. This is the *amount of current that a battery can deliver for 20 hours without the cell voltage dropping below 1.75 volts with an electrolyte temperature of 26.7°C.*

Example. A battery that can deliver a current of 6 amperes for *20 hours is rated as having a 120 ampere-hour capacity* (20 × 6 = 120).

- *The capacity of a battery depends upon the amount of active material that can react with the electrolyte.* This depends upon the thickness and design of the plates. Hence the number of plates is not always an accurate index of the capacity of the battery.

2. **25-ampere rating :**

- It measures the battery performance at a moderate constant current output at 26.7°C to a final limiting voltage of 1.75 volts per cell.
- This figure indicates the ability of the battery to carry the electrical operating load (lights, ignition, etc.) when the generator is not operating.

3. **Cold-cranking rate :**

- *One* of the two cold-cranking rates is the number of amperes that a battery can deliver for 30 seconds when it is at – 17.8°C (0°F) without the cell voltage falling below 1.2 volts. A typical rating for a battery with a reserve capacity of 125 minutes would be 430 amperes. This figure *indicates the ability of the battery to crank the engine at low temperatures.*
- The *second* cold-cranking rate is measured at – 28.9°C (– 20°F). In this the final voltage is allowed to drop to 1 volt per cell. A typical rating for a battery with a reserve capacity of 125 minutes would be 320 amperes.

10.3.12. Battery Maintenance

Complete battery maintenance include the following :

1. Visually check the battery.
2. Check electrolyte level in cells on batteries with vent caps.
3. Add water in if the level is low (vent-cap batteries)
   — Do not add too much water. This can cause electrolyte to leak out and corrode the battery carrier and other metal nearby.
4. Clean off corrosion around battery terminals and from top of batteries with top terminals.
5. Check battery condition by testing its state of charge.
6. Recharge battery if it is low.

- If the battery is kept in *partially discharged condition* too long, the lead sulphate formed on the plates may change into a hard crystalline state, which reduces the active material since the resulting dense crystals of sulphates are no longer convertible into electrical energy. Heavy sulphation is difficult to get-rid off by recharging and it is, thoro important to maintain a battery in a full charged condition at all times.
- A completely discharged battery should not be left in the car or truck or any other vehicle on a *cold night*. This will lead to *freezing of electrolyte and battery damage* beyond *repair*.

10.3.13. Battery Charging

10.3.13.1. Necessity for recharging a battery

The battery, under normal operation, is kept fully charged by the car generator. The rate of charging from the generator to the battery depends upon the state of the battery, *i.e.*, high rate will be there for a discharged battery and low rate for a charged battery. Normally, there is no necessity for charging battery from external source, but under certain condition the need for recharging a battery arises :

(*i*) When the vehicle is parked for long periods with parking lights on ;

(*ii*) Use of accessories such as heater, fan and car radio when the vehicle is stationary ;

(*iii*) Frequent use of car starer that draws heavy current.

10.3.13.2. Methods of battery charging

The batteries can be charged by the following three methods :

1. Constant current charging.
2. Constant voltage or potential charging.
3. Quick charging.

(Methods 1–3: "*Slow charge*" methods)

- *Only direct current (D.C.) should be used for charging batteries.*

1. **Constant current charging :**

- In the *"slow charge"* method battery is supplied with a relatively small amount of current *i.e.*, at a relatively slow rate (1 ampere per positive plate or in amperes equal to 7% of its ampere-hour rating) for a *long period* depending on how much battery is discharged. In the *"Quick charge"* method the battery is supplied with a high current (as much as 100 amperes for 6-bolt battery and 50 amperes for 12-volt battery) for a short period of time (about 1 hour).
- In this method, the *current input* to the battery is *adjusted* to the valve as recommended by the battery manufacturer. The *charging is continued until the battery is gassing freely and there is no rise in gravity for two hours.*

Constant current chargers usually employ a *"rectifier" to convert A.C. to D.C.* and a rheostat to adjust current flow to the battery.

The following *procedure* should be followed to charge batteries *with constant chargers* :

(*i*) Remove filler plugs, check electrolyte level and add distilled water if the level is low.

(*ii*) Connect battery to the chargers. If *more than one battery is to be charged, connect the batteries in "series"* upto the capacity of charger.

(*iii*) Switch on the charger and *adjust the charging rate.*

(*iv*) Check gravity of batteries by means of *'hydrometer' after every two hours* until there is *no rise in specific gravity.*

(*v*) If a battery boils violently or overheats, remove it from the charging line and check up the defect.

Fig. 10.9 shows battery charging by constant current method.

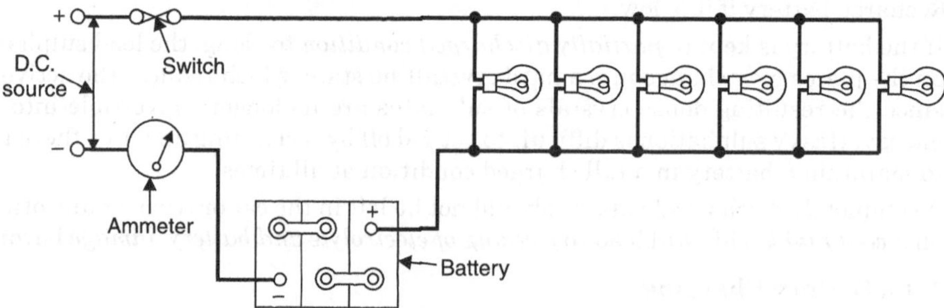

Fig. 10.9. Battery charging by constant current method.

Fig. 10.10 shows wiring diagram of a series hook-up for constant current battery charging. When charging on series (constant current) line, connect the batteries in series and *charge at the rate which is safe for the smallest battery.* Continued charging at high rate will result in high temperature, which should be avoided.

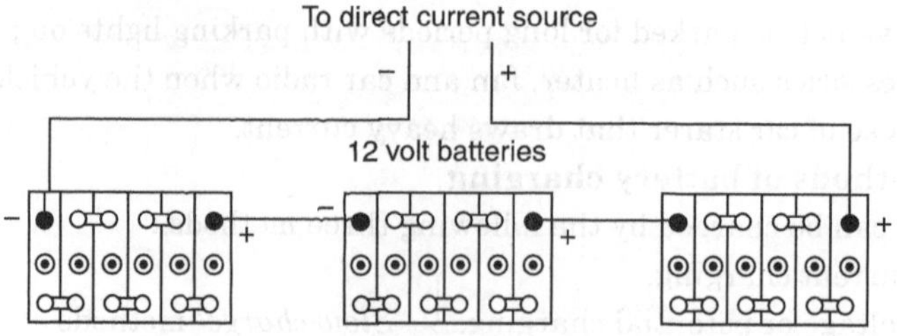

Fig. 10.10. Wiring diagram of a series hook-up for constant current battery charging.

2. **Constant voltage or potential charging :**

- In this method, the *charging voltage is kept constant* at a constant value. As the battery approaches a charged condition, it *resists to the charging current* which tapers off gradually. When the battery is fully charged, the input current has been reduced to few amperes.

In this method, the battery temperature remains within limits. If the temperature of battery electrolyte rises high, battery resistance shall fall down and the battery shall be damaged due to overcharging. In such a case the battery should be removed from the charging line to avoid damage.

- ***Constant voltage charger.*** It consists of a *motor-generator set* in which an A.C. motor drives a D.C. generator. The generator is usually rated at 7.5 volts for 6-volt batteries or 15-volts for 12-volt batteries.

The following *procedure* is followed to *charge batteries with constant voltage charger* :

(*i*) Remove filler plugs.

(*ii*) Check electrolyte and add distilled water if the level is low.

(*iii*) Connect battery to the charger. If more than one battery is to be charged, connect the batteries in "*parallel*".

(*iv*) Keep watch on battery boiling and temperature as there is possibility of overheating in this method. If the battery boils or overheats, remove it from the line.

(*v*) Stop charging when the gravity shows no further rise after an hour of charging.

This method is *not universally used.*

3. **Quick charging :**

In this method, the battery is supplied with a high current as much as 100 amperes for 6-volt battery and 50 amperes for 12-volt type battery for a short period of time. If takes about one hour to charge a battery by this method.

Quick chargers or boosters. They supply current to the battery at high rate and work on the theory that the battery will be brought upto a charged condition before excessive battery temperatures are reached. But as a rule the quick chargers cannot bring a battery up to *full* charge in a short time. The battery can be substantially charged or *boosted* but for bringing a battery to a fully charged condition, the charging cycle must be completed by charging at a low normal rate.

- "Quick charge" method *must not* be employed if any of the following conditions exists with battery :

(*i*) If the electrolyte is disoloured with a brownish segment, the quick charging method may produce an internal short and ruin the battery.

(*ii*) If the specific gravity readings are not uniform, the cell showing low readings may have an internal defect. During quick charging considerable heat developed may ruin the battery.

(*iii*) If the battery has been badly overcharged, then such a battery may quickly fail if placed on quick charge.

(*iv*) If the battery is badly sulphated then it will get overheated during quick charging. Such a battery requires charging *half* the normal 'Slow Charge' rate from 60 to 100 hours to reconvert the crystalline lead sulphate into active material.

(*v*) The cell voltages and colour of the electrolyte should be checked few minutes after the battery has been put on a quick charging method. If the voltage readings are not uniform within 0.2 volt or if the electrolyte has become discoloured with brownish sediment, the quick charging should be stopped immediately. Further charging of the battery may be continued by the 'slow charge' method.

Fig. 10.11 shows a three-wire battery charging circuit. There are two 7.5 V rectified supply leads. The 6-volt batteries are connected between the rectified positive 7.5 V lead and neutral lead. 12-volt batteries are connected between the positive and negative leads.

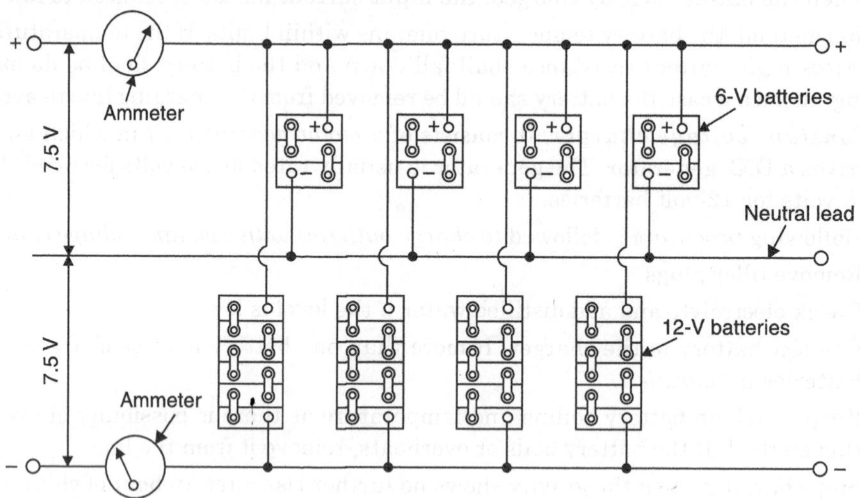

Fig. 10.11. A three-wire battery charging circuit—Constant voltage method.

Precautions to be observed during charging of a battery :

During charging of a battery following *precautions* should be observed :

- Battery overcharging must be avoided because it results in the damage to the battery. The positive plates of the battery are affected badly by overcharging.
- A battery charger must be connected in parallel with battery. The positive terminal of the charger must be connected to the positive terminal of the battery and the negative terminal of the charger should be connected to the negative terminal of the battery. In case a battery charger is connected in *reverse polarity*, the battery can be *ruined very quickly and may even explode due to the extreme heat developed in the electrolyte.*
  — Most modern batteries are equipped with a *protective device*. In case of charge *reverse polarity, a signal lamp will glow and emits light.* This device also prevents the flow of current until correct polarity is maintained.

10.3.14. Factors affecting Battery Life

The *life of a battery is affected by the following factors* :

1. Overcharging.
2. Undercharging.
3. Lack of water.
4. Loose hold-downs.
5. Excessive loads.
6. Freezing of electrolyte.

1. **Overcharging :**

If the battery frequently requires water, it is probably being overcharged. Too much overcurrent is being supplied to the battery. This is a damaging condition that overworks the active materials in the battery and shortens battery life. In addition, overcharging causes more rapid loss of water from the battery electrolyte. Unless this water is replaced frequently, the electrolyte level is likely to fall below the tops of the plates. This exposes the plates and the separators to the air and may ruin them.

Following is the summary of harmful effects of charging a battery *greatly in excess* of what is required :

(*i*) Overcharging decomposes water of electrolyte into hydrogen and oxygen gas. Gas bubbles tend to wash active material from the plates and carry moisture and acid from the cells as a fine mist.

Decomposition of water leaves acid more concentrated. The concentrated acid is harmful to separators and negative plate material at high temperatures over a prolonged period of time, charring the separators and making the negative material sandy and granular.

(*ii*) High internal heat is created, which accelerates corrosion of positive plate grids and damages separators. Also, containers may be softened and distorted and sealing compound displaced. Overcharging alone or in combination with the previous condition of undercharging may cause severe buckling and warping of positive plates with accompanying perforation of separators.

(*iii*) Overcharging may cause damage by corrosion to cradle, cables and other vital electrical and engine parts by forcing liquid from the cell if charge rates are excessive.

2. **Undercharging :**

- A battery operated with *insufficient charge* over a long period of time may develop a type of sulphate in plates which is dense, hard and crystalline and which cannot be electrochemically converted to normal active material again. Such lead sulphate being less dense than the active material from which it was formed will set up strains in the positive plates so that distortion or blowing of plates, called buckling may result. Severely buckled plates will pinch the separators at the plate corners or chafe the centre of the separators. This may *result in perforations of an unprotected separator and develop short circuit of the cell.*

Lead sulphate formed on the plates during discharging is relatively insoluble as long as specific gravity of electrolyte is kept above 1.125. Subsequent charging may convert these crystalline deposits to filamentous metallic lead which may short the positive and negative plates through the areas of the separator affected. These small shorts may cause a condition of low cell voltage when the battery is charging. *For this reason the battery should not be allowed to stand idle in a more than 75% discharged condition for very long periods of time.*

- A battery operated in an uncharged condition is *not only unable to deliver full power but is liable to freeze during severe winter weather.*

3. **Lack of water :**

Water (one of the four essential chemicals of a lead-acid storage battery) under normal conditions of operation is the only component of the battery which is lost as a result of charging. It should be replaced as soon as the liquid level falls to the top of the separators.

If water is not replaced, and the plates are soon exposed, the acid will reach a dangerously high concentration that may charge disintegrate the separators and may permanently sulphate and impair the performance of the plates. Plates cannot take full part in the battery action unless they are completely covered by the electrolyte.

- *Sulphuric acid should never be added to a cell unless it has been lost due to spillage.* Then it should be replaced only. No satisfactory substitute electrolyte has been found for the simple mixture of sulphuric acid in water ; no substitutes should be used.

4. **Loose hold-downs :**

If hold-downs are not properly adjusted, the battery may *bounce* around in the cradle. This may cause the bridges on which the elements rests to notch the bottoms of the separators and may cause the plates to notch the bridge tops, causing a severe disarrangement of the element. The bouncing of the battery may also crack or wear the container badly and cause sealing compound to open up and leak acid. Leaking acid corrodes terminals and cables and makes high resistance battery connections, thereby weakening the battery power and shortening its life.

Hold-downs on the other-end, can be *too tight,* distort or crack the container and loosen sealing compound, allowing loss of acid from the cells, and this may cause loss of battery capacity.

5. **Excessive loads :**

A *battery should never be used to propel the car by the use of the starting motor with clutch engaged except in a great emergency.* This may produce extremely high internal battery temperature and damage the starting motor.

6. **Freezing of electrolyte :**

Electrolyte of a battery in various stages of charge will start to freeze at temperatures indicated below :

Sp. gravity at 27°C	*Freezing temperature*
1.280	– 71°C
1.230	– 40°C
1.200	– 26.5°C
1.150	– 15°C
1.100	– 7°C

The given temperatures indicate the approximate points at which the first ice crystals begin to appear in the solution. The solution does not freeze solid until a lower temperature is reached, solid freezing of the electrolyte may crack the container and damage the positive plates.

- A 75% charged battery is in no danger of damage from freezing. Therefore, keep batteries better than 75% charged, specially in winter.

10.3.15. Battery Faults/Troubles

The following are the *common battery troubles* :

1. **Rapid loss of electrolyte :**

This can be caused by *overcharging, sulphation of plates* or the *battery being fitted in a position exposed* to *excessive engine heat.*

- Distilled water should always be used to topping up the correct level and proper gravity.

2. **Sulphate plates :**

This trouble may result due to the following *reasons* :

(*i*) Leaving the battery idle when discharged.

(*ii*) Using a battery for long periods in an undercharged state.

Symptoms of sulphation are a low specific gravity, high temperature and voltage on charging, reduced capacity and low output. Other causes may be *insufficient electrolyte in the battery.*

- If sulphation is not too severe, it may be removed by giving the battery a slow charge for a period of two or three times the normal, otherwise the only remedy is to replace the battery.

3. **Internal short circuit :**

This results from conducting material lodging between positive and negative plates or from faulty or damaged separations that allow the plates to touch, resulting in a constant internal discharge of current. It also causes sulphation of plates.

- If the battery is otherwise in good condition then clean the battery.

4. **Open circuit :**

This defect can be there due to broken cell bridge to which the plates are welded or by a loose cell connector. It *results in making the whole battery dead.*

Sometimes the fault is intermittent and it becomes difficult to locate it.

5. **Reversed polarity :**

This type of fault is not common. It is caused by one cell becoming discharged before the remainder, when the discharge current from other cells charge the faulty one in reverse direction. It *results in the overall battery voltage being reduced.*

- The *remedy* is to *give a long charge from an external source.* It will *revert the cell to its normal polarity.*

6. **Cracked container or jar :**

Cracks can develop in the container due to mechanical shock and vibrations when the car moves on a rough road. Cracks can also develop due to rough handling of the vehicle.

Two types of cracks may occur, one on the outer wall of the container and other on the interior walls. The electrolyte will leak through the outer wall. A crack in the interior wall of cell will cause the short ciruiting of the plates of the two adjoining cells.

- It is worthwhile to replace the cracked container than to do any repair.

7. **Deterioration of plates of the cells :**

The following *factors* can cause the deterioration of the plates of the cell and may damage them :

(*i*) Very high specific gravity of the electrolyte.

(*ii*) Very high rate of battery charging.

(*iii*) Freezing of electrolyte in the container of the battery.

(*iv*) The level of electrolyte in the container is not proper one.

8. **Mechanical faults :**

The mechanical faults may be as follows :

(*i*) *Loose terminal posts,* which require the fitting of a new cell top.

(*ii*) *Cracks in the cell tops or casing,* which can be repaired using a properitory battery sealing compound.

10.3.16. Fully Charged Battery

(*i*) The voltage of each cell should be from 2.5 to 2.7 V.

(*ii*) The specific gravity of the electrolyte should be 1.285.

(*iii*) All the cells should be freely gassing or bubbling.

If the conditions prevail then it may be said that the batteries are fully charged.

- The charger should be switched off before disconnecting any cell.

10.3.17. Trickle Charging

As the batteries left in storage for *long periods* are subjected to *self discharge,* continuous charging at *very low charging rate* (of about 5% of the normal charging current) has been employed for *keeping in proper condition the batteries for display purposes for vehicles doing little running.*

- The continuous trickle charging, however, should *not* be used and it is recommended that only during day time the battery should be connected to trickle charge while during night it should be disconnected.

10.3.18. Dry Charged Batteries

Dry charged batteries have charged positive and negative poles and do not have an electrolyte. They are kept in dry condition. These batteries may be stored without damage for *2 to 3 years.*

In the dry condition, the battery cannot supply any current. In order to get a supply of current, an acid of recommended density is added to the battery to a correct level. Only then does the battery supply current.

The recommended specific gravity of the battery acid (sulphuric acid) varies from 1.250 to 1.300. It varies depending upon who the manufacturers is.

The *procedure for servicing the battery is as follows* :

(*i*) Take sulphuric acid of specific gravity 1.835.

(*ii*) Add this acid to water to dilute it to the recommended specific gravity (***water should never be added to acid***, since it is highly *dangerous*).

When the acid is added to water, it will get hot. Now wait till the temperature of the mixture comes down to about 21°C.

(*iii*) Measure the specific gravity of the mixture with a hydrometer, it should be within 1.250 to 1.300.

(*iv*) Pour the liquid electrolyte in the dry battery.

The battery is ready for use.

10.3.19. Battery Testing

The following tests are conducted to *ascertain the conditions of a battery* :

1. Specific gravity test (Hydrometer test)
2. Open volt test.
3. High rate discharge test.
4. Cadmium test.

1. **Specific gravity test** (*Hydrometer test*) :

This test is carried out with the help of a *hydrometer*. By measuring specific gravity of the electrolyte with a hydrometer, an indication of the appropriate state of charge of the battery can be obtained.

While the chemical reactions are taking place in the battery during discharge, the electrolyte becomes dilute due to formation of water. The proportion of water goes on increasing as the discharging continuous. The relative amount of water and acid is determined by the specific gravity test. This is done by the use of a *hydrometer* contained in the syringe.

Fig. 10.12 shows a battery with a hydrometer. The electrolyte is drawn up into the syringe by the bulb and the hydrometer float sinks to a greater or lesser extent according to amount of sulphuric acid in the electrolyte. If the hydrometer reads 1.280, it indicates that the liquid is 1.280 times as heavy as water, and at the reading the battery is fully charged.

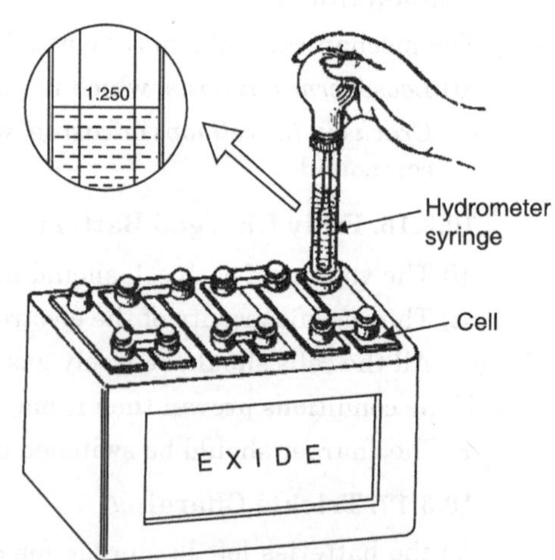

Fig. 10.12. Testing a battery's specific gravity with a hydrometer. The reading is taken on the graduated stem of the float.

The readings of specific gravity of the electrolyte are given below ; they change with temperature :

Climate below 32°C :

1.270 to 1.290 ... Cell is fully charged.

1.190 to 1.210 ... Cell is about half discharged.

1.110 to 1.130 ... Cell is fully discharged.

Climate frequently above 32°C :

1.210 to 1.230 ... Cell is fully charged.

1.130 to 1.150 ... Cell is about half discharged.

1.050 to 1.070 ... Cell is fully discharged.

- The scale of the hydrometer is read on the stem at the surface of the electrolyte when hydrometer is floating in it. The specific gravity of the electrolyte is affected by its temperature. Thus, temperature of the electrolyte is also recorded by a thermometer and then the specific gravity reading is connected accordingly.
- The specific gravity test should not be made while the battery is gassing. The test should also not be made immediately after the water has been added to adjust the electrolyte level, otherwise this would affect the reading.

2. **Open volt test :**

Where the hydrometer cannot be used to determine the state of the charge, the vehicles are now fitted with maintenance free (sealed) batteries. In such a case, the same can be determined from the *open volt test* of the battery.

The open circuit voltage of a fully charged battery cell is about 2.1 volts. This can be measured with the help of a voltmeter. It can be observed that a charge of 0.01 volt of open circuit voltage is equivalent to a charge of 0.010 in the specific gravity of the electrolyte. Thus, the gravity measurements can be indirectly made with a voltmeter also.

3. **High rate discharge test :**

The hydrometer test indicates only the chemical conditions of a battery and is not necessarily an indication of the actual battery condition. A high-rate discharge test with a voltmeter measures the actual capacity of a battery to convert the chemical energy into electrical energy. There are many designs of the high rate discharge test equipment, which are in use, this test using a cell tester is described below.

The high rate discharge test by a *cell tester* (Fig. 10.13) is made by using an individual cell tester consisting of a *low-reading voltmeter and heavy fixed resistance.* The individual cell tester indicates the voltage of each cell on the voltmeter.

In order to carry out the test, contact the voltmeter rods to the proper cell terminals *i.e.*, red to the positive and black to the negative using caution *not* to connect across more than one cell. For each cell reading, the point of the prongs will have to be pushed through the sealing compound to make contact with the burried link. The circuit is closed through the fixed resistance of the cell tester (for about 10 seconds). Under normal starting-motor load, the voltage of the cell is indicated on the voltmeter of the cell tester.

The individual-cell readings should not vary more than 0.15 volt between any two cells. If the variations are more than 0.15 volt between any two cells, then battery should be replaced.

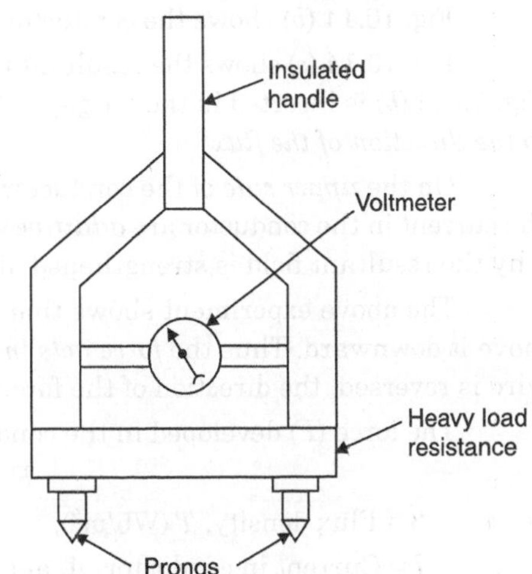

Fig. 10.13. High rate discharge tester.

If the cell tester is not available, high rate discharge test may also be made with the help of a *cranking motor when the ignition switch is off*. The voltage readings are taken with the help of an accurate voltmeter.

- *This test is a very severe test and should be made only on the fully charged or at least on half-fully charged batteries. Even then it should not be made continuously for more than 15 seconds, otherwise the battery plates will get permanently damaged.*

4. **Cadmium test :**

This test is carried out to *ascertain whether the battery plates are defective or not.*

This test is conducted with the help of a cadmium rod enclosed in a perporated ebonite tube. The rod is immersed in the electrolyte, and connected to the negative terminal of a voltmeter. Its positive terminal is connected alternately to the positive and negative terminals of the battery cell. When connected with *positive* terminals, the voltage reading *should not be less than 2.5 volts, a lower reading would indicate defective positive plates.* When connected with *negative* terminals, the voltage reading *should not be more than 0.2 volt, a higher reading would indicate defective negative plates.*

10.4. STARTING MOTOR (or SELF-STARTER)

10.4.1. Introduction

Self-starter (also known as *cranking motor* or *starting motor*-introduced in 1911 to replace laborious hand cranking method for starting of engines) *is an electrical machine (D.C. motor) which converts electrical energy into mechanical energy.* It can *crank the engine fast enough to give a quick start under all operating conditions.*

Principle working of D.C. motor :

The principle of motor action can be stated as follows :

"*Whenever a current carrying conductor is placed in a magnetic field, it experiences a force whose direction is given by Fleming's left hand rule*".

Fig. 10.14 illustrates this principle.

Fig. 10.14 (*a*) shows the field set up by the poles.

Fig. 10.14 (*b*) shows the conductor field due to flow of current in the conductor.

Fig. 10.14 (*c*) shows the resultant field produced when the current carrying conductor wire of Fig. 10.14 (*b*) is inserted in the air gap of Fig. 10.14 (*a*) with the *axis of the conductor at right angles to the direction of the flux.*

On the *upper side* of the conductor in Fig. 10.14 (*c*) the magnetizing forces of the field and of the current in the conductor are *additive* while on the *lower side* these are *subtractive*. This explains why the resultant field is strengthened above and weakened below the conductor (wire).

The above experiment shows that the wire in Fig. 10.14 (*c*) has a force on it which tends to move it downward. Thus the *force acts in the direction of the weaker field.* When the current in the wire is reversed, the direction of the force is also reversed, as in Fig. 10.14 (*d*)

The force (F) developed in the conductor is given by the relation,

$$F = BIl \text{ newtons}$$

where B = Flux density, T (Wb/m^2),

I = Current in conductor, A, and

l = Exposed length of conductor, m.

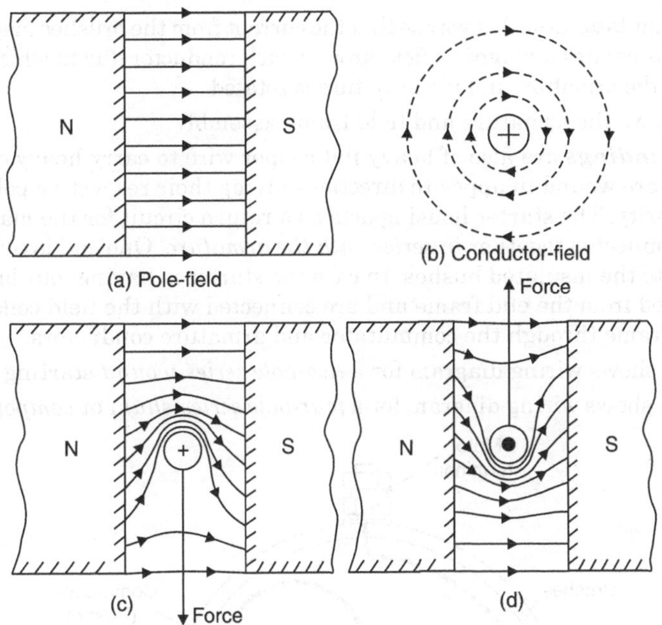

Fig. 10.14. The principle of motor action.

10.4.2. Description of Starting Motor

***Starting motor** is a low voltage D.C. motor incorporating a gear-drive mechanism which converts the electrical energy received from the storage battery, into mechanical energy.* When the circuit between the starting motor and the storage battery is completed through the starting switch, the starter *cranks the engine*. The starter drive gear is engaged with the *ring gear at engine flywheel*, manually and automatically. Once the engine starts, the starter is automatically disengaged or the starter armature is prevented from overspeeding until disengagement is affected.

The starting motor consists of the following two main *units* :

1. Motor unit
2. Drive unit

1. **Motor unit :**

Construction. The motor unit consists of the following *components :*

(*i*) Cylindrical housing (*ii*) Field coils (*iii*) Pole shoes
(*iv*) Armature (*v*) Brushes etc.

Steel pole shoes are securely attached to the inner surface of ***steel housing.*** The number of pole shoes varies from two to six but mostly there are four pole shoes. The pole shoes hold the ***field coils*** and become magnetised when current flows through the coils. The armature revolves between the pole shoes with a slight clearance.

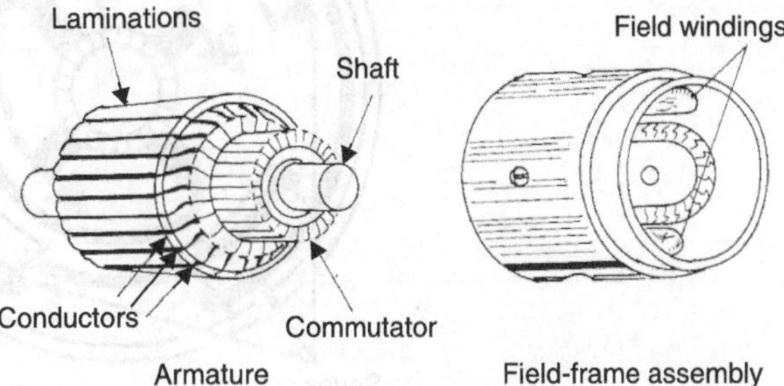

Fig. 10.15. The armature and field frame assembly.

The ***armature*** consists of a "*slotted iron core*", "*shaft*", "*commutator*" and "*armature windings*". The windings are made of heavy flat copper strip for carrying safely the heavy current which passes through them. The ***windings*** consists of a number of coils, each coil having single loop. The sides of the coils fit into the slots of the armature core and are insulated from the core. The coils are connected to the commutator in such a way so that the current from the brushes may flow through all the coils at one time. This creates a magnetic field around each conductor due to which a repulsive force is produced all around the armature and the armature is rotated.

Fig. 10.15 shows the armature and field frame assembly.

Field coil windings are also of heavy flat copper wire to carry heavy current. The coils on adjacent pole shoes are wound in apposite directions giving their respective pole shoes a north and south magnetic polarity. The starter housing acts as a return circuit for the magnetic lines of force. The field coils are connected together in *series with the armature.* One end is connected with starter terminal and other to the insulated bushes. In case the starter is having four brushes, two opposite brushes are insulated from the end frame and are connected with the field coils. The other two are earthed to the end frame through the commutator and armature conductors.

Fig. 10.16 (*i*) shows wiring diagram for a *two-pole series-wound* starting motor.

Fig. 10.16 (*ii*) shows wiring diagram for a *four-pole series-shunt* or *compound*, starting motor.

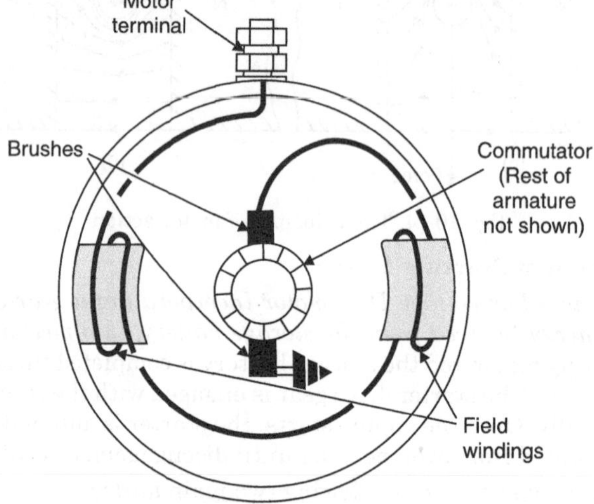

(*i*) Two-pole series wound starting motor.

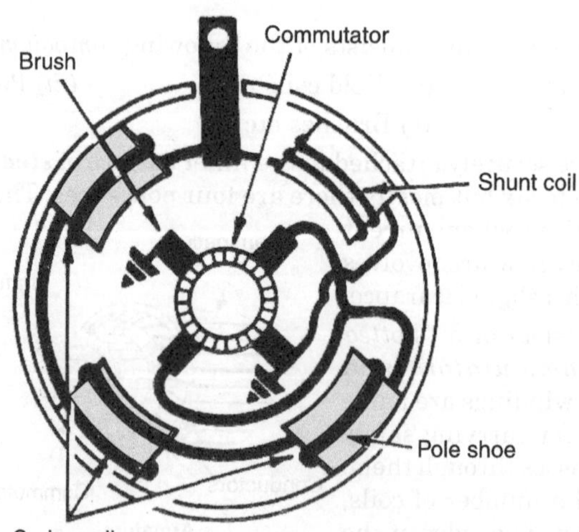

(*ii*) Four-pole series-shunt, or compound starting motor.

Fig. 10.16. Electric circuits in starting/cranking motor.

Starting motors are made in different sizes for various engines.

Working. When the starter switch is put on, current from battery flows to the starting motor and sets up a strong circular magnetic field around the armature coil. Due to the loop of the armature coil, the current flows in opposite directions in the two sides of the coil. The same current, inflowing through the field coils, makes pole shoes electromagnets due to which a strong magnetic is created between the faces of the shoes. *The reaction between the two magnetic fields tends the main magnetic fields across the pole shoes to become distorted or bent* as rubber is bent under tension (Fig. 10.14 (*c*), (*d*)). The bent magnetic lines of force always tend to *straighten out* and while so doing, they *exert a force* on the armature coil, *causing the armature coil to rotate between the pole shoes.* The greater the current flowing through the armature and field coils the greater the torque produced by the starting motor.

2. **Drive unit :**

The ***drive unit*** **or** ***drive mechanism*** of *the starting motor transmits the torque developed by the starting motor to the engine flywheel for cranking engine as starter armature revolves.*

A pinion is fitted on the armature shaft which gets meshed with the flywheel ring gear when starter is worked and thus rotates the same. Starter pinion is *not* permanently meshed with the ring gear but arranged in such a way that when starter switch is put on, due to running armature shaft, this pinion travels towards the ring gear, its teeth get meshed with the ring gear and thus rotates the same. When the engine starts and speeds up, the *speed of the ring gear being more than the speed of the pinion makes the pinion to move back due to threads made on armature shaft.* This system where pinion moves on the shaft of the self-starter/starting motor, meshes the ring gear and comes back when engine has started is known as ***"bendex drive"***.

The ratio between the drive gear (or *pinion*) on the starting motor and ring gear on the engine flywheel is about 1 : 15 as drive gear has few gear teeth as compared to number of gear teeth or flywheel ring gear. Thus a *gear* reduction is *provided which enables the torque developed* by *the starting motor to turn the engine over cranking speeds.* The starting motor drive mechanism permits quick disengagement of drive gear from the flywheel ring gear or the drive gear from the armature shaft in order to save the armature from ruin due to centrifugal force developed at engine speed.

The drive mechanism is of the following two types :

(*i*) Inertia drive ; (*ii*) Overrunning clutch.

(*i*) **Inertia drive :**

An inertia drive depends upon the inertia of the drive gear or pinion to produce meshing. *Inertia* is the property which all things have that resists any change in motion. When the drive gear is not rotating, it resists any force which tends to set it in motion.

"Bendix drive" *is known as inertia drive.* It is of the following two types :

(*a*) In-board drive

(*b*) Out-board drive.

(*a*) ***In-board drive :***

In this arrangement (Fig. 10.17) the Bendix pinion is fitted on the armature shaft in such a way that it is kept out from starter ring and on starting the pinion moves *towards armature.* The armature shaft has left hand (L.H.) threads and a heavy duty spring fitted at the end of armature shaft. The purpose of

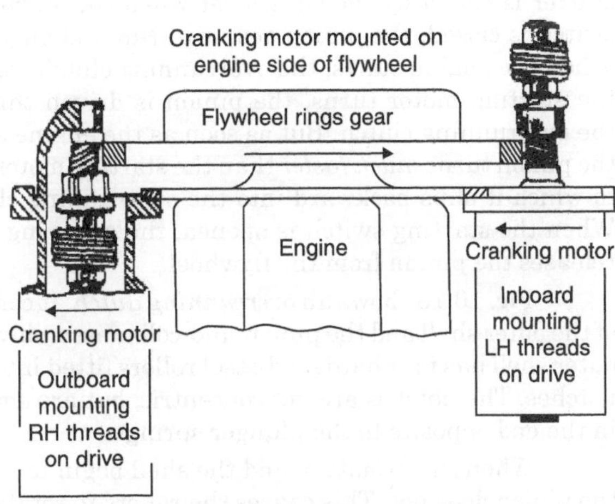

Fig. 10.17. In-board and out-board Bendix drive.

fitting is to get compressed slightly when the pinion has moved towards flywheel and its teeth not meshed. The compressing of spring gives the pinion slight pause and as the teeth are in alignments, the pinion moves inward for complete mesh of teeth, thus making a positive mesh of pinion and flywheel ring gear teeth.

(b) ***Out-board drive***

In this arrangement (Fig. 10.17) the pinion is fitted on the armature shaft and when drive is given, it *moves out* and meshes with ring gear. Such arrangement is *most commonly used in present day automobiles.*

2. Overrunning clutch :

Fig. 10.18 shows a *overrunning clutch drive.* The starter lever is linked to a starter pedal which extends into the driver's compartment and is operated by foot-pressure.

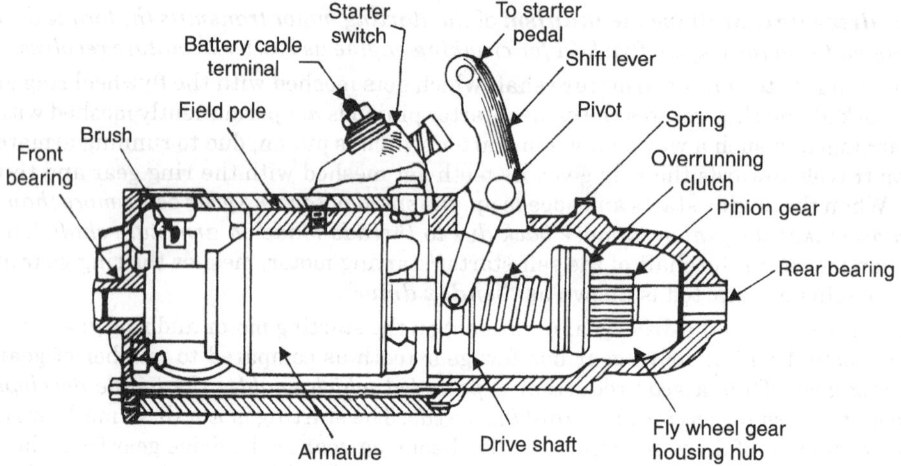

Fig. 10.18. Overrunning clutch drive.

When the starter pedal is pressed, the shift lever compresses the drive shaft and spring which ultimately pushes the overrunning clutch and pinion gear assembly toward the flywheel. The starter is *closed* by the shift lever when the starter pedal is *fully pressed.* As soon as the starter switch is closed, the pinion gear will run and engage with the flywheel, thus starting the engine. When the engine starts, the overrunning clutch comes into action. The unit is so designed that, as the starting motor turns, the pinion is driven through the overrunning clutch. But as soon as the engine starts the pinion turns *much faster* than the starting motor, due to which it slips backward into the overrunning clutch. When the starting switch is opened, the engaging lever releases the pinion from the flywheel.

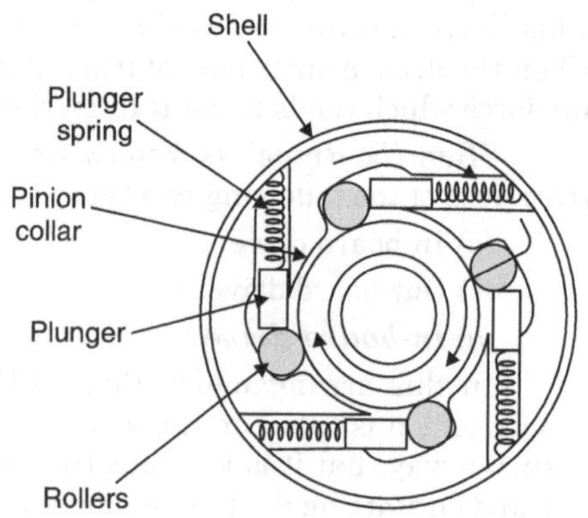

Fig. 10.19. Overrunning clutch.

Fig. 10.19 shows an *overrunning dutch.* It consists of the outer shell and the pinion-and-collar assembly. The outer shell has four hardened-steel rollers fitted into four notches. The notches are not concentric, but are smaller in the end opposite to the plunger springs.

When the armature and the shell begin to rotate, the pinion does not. This causes the rollers to rotate into the smaller sections of the notches, where they *jam tight. The pinion must now rotate with armature, cranking the engine*. After the engine starts, it spins the pinion *faster* than the armature is turning. The rollers are rotated into the larger sections of the notches, where they are *free*. This allows the pinion to spin independently of, or overrun, the remainder of the clutch. A spring on the shift lever pulls the pinion back out of mesh when the shift lever is released.

10.4.3. Starting Motor Switches

Starting motor switches are of the following types :

1. Manually operated switches.
2. Magnetic switch or relay.
3. Solenoid switch.

1. Manually operated switches :

These are simple switches which open and close the circuit and can be used with all types of starting motors.

- In overrunning clutch type starters, the shift lever operates this switch.
- In Bendix drive, the switch in a push type button.
- In some starting systems, the accelerator operates the starter switch when the engine is not running. When the engine is *running*, a *latch device operated by a 'vacuum control'* holds the starting motor control mechanism away from the accelerator.

Fig. 10.20 shows a *manually operated switch*. It is operated by hand. When the plunger is pressed, it makes the contact between the two terminals to connect the battery to the starting motor.

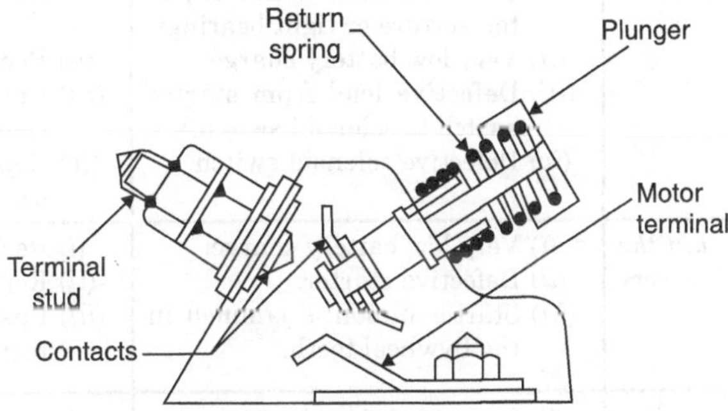

Fig. 10.20. Mannually operated switch.

2. Magnetic switch or relay :

In Bendix type self-starters, simple starting motor relays are used. These are controlled either by means of a remote control having push button type switch or vacuum switch. When the circuit is completed through the remote control switch, the coil in the relay (magnetic switch) becomes energised due to which the contract disc is drawn against the contact terminals and the battery current flows to the starting motor.

3. Solenoid switch :

This type of switch is used in self-starters employing over-running clutch. The *purpose of the solenoid is to shift the drive pinion magnetically into mesh with the flywheel and to complete the circuit between the battery and the self-starter.*

10.4.4. Integrated Starters

Integrated Starter Alternator Damper (ISAD) is a highly efficient and integrated system. Starting an engine with ISAD is noiseless with cranking speeds of upto 700 r.p.m.

In this system an *electric motor performs* the following *functions.*

(*i*) Serves as a starter motor for the engine.

(*ii*) Delivers electrical power to the batteries and the other electrical accessories. In this mode it can produce *upto 2 kW at idling speed.* It can *supply both A.C. and D.C. at different voltages.*

(*iii*) Transfers the drive from the engine to the transmission and also *acts as a vibration absorber unit.*

10.4.5. Care and Maintenance of Self-Starter/Starting Motor

- The starter button should not be pressed when the engine is running. At the same time the ignition starter key should not be turned to start position. If this is done, the *flywheel teeth get damaged.*
- For proper maintenance remove the metal band dust cover and examine the brushes every 15000 km run. Check whether brushes are worn out. Ensure that the brush works properly in its holder. If there is any difficulty, clean the brush with a petrol-moistened cloth.
- Check the commutator. If there is any dirt or grease, clean it with a soft cloth. If the surface is still not cleaned, use a cloth with a few drops of petrol.

10.4.6. Troubleshooting of Self-Starter/Starting Motor

S. No.	Troubles	Causes	Remedies
1.	*Starter will not crank the engine.*	(*i*) Engine is too stiff for operation. This may be due to piston seizure or tight bearings. (*ii*) Very low battery charge. (*iii*) Defective lead from starter switch to solenoid switch. (*iv*) Defective solenoid switch.	(*i*) Crank the engine by hand and check the defect. (*ii*) Recharge the batteries. (*iii*) Change the defective lead and check. (*iv*) Repair on replace the solenoid switch.
2.	*Starter works, but the engine is cranked very slowly.*	(*i*) Very low battery charge. (*ii*) Defective starter. (*iii*) Starter pinion is jammed in the flywheel teeth.	(*i*) Recharge the battery. (*ii*) Repair the starter. (*iii*) Push the car in reverse and rectify the defect.
3.	*Starter works, but does not crank the engine.*	(*i*) Starter pinion does not engage flywheel gear as there is dirt on the screwed sleeve.	(*i*) Clean the dirt.

10.4.7. Comparative Starting System Data of Some Indian Automobiles

S. No.	Make	Starting Motor	Starting Drive
1.	*Maruti 800*		Over-running clutch with solenoid shift
2.	*Hindustan Ambassador Mark II*	Lucas M35G	Bendix
3.	*Fiat 1100*	E76-0, 5/12	Overrunning clutch
4.	*Jeep Universal*	Prestolite MDU-7004	Bendix
5.	*Dodge/Fargo Model 89M4 (Petrol type)*	Prestolite MDU-5003	Bendix
6.	*Ashok Leyland Comet passenger*	Lucas GBS-524P190	Axial starter
7.	*Tata Vehicles*	Lucas-TVS Type GSP5-12B-25	Beacon type 512STR16 Axial starter

II. CHARGING SYSTEM

10.5. INTRODUCTION TO CHARGING SYSTEM

Necessity and function. An automobile battery supplies a sufficient amount of electricity to the automobile's electrical components such as starting motor, headlight and wipers. However, this battery is limited in capacity and is not capable of providing, on a continuous basis, all the power required by the automobile. Therefore, it is necessary for the battery to always be fully charged so that it supplies the necessary amount of electricity at the required time to each of the electrical components. Consequently, the *automobile requires a charging system to produce electricity and keep the battery charged.* Actually, *while the automobile's engine is in operation, the charging system produces electricity to both re-charge the battery and to supply the electrical components with the amount of electricity required.*

Thus the *charging system accomplishes the following two* **functions :**

1. To put back into the battery the current used to start the engine.
2. To handle the load of lights, ignition, radio and other electrical and electronic equipment while the engine is running.

Requirements of a charging system. The *main requirements of a charging system,* when the vehicle is running are :

1. To supply the current required for charging of battery under all conditions.
2. To supply the current demands of all the loads.
3. To supply constant voltage under all conditions.
4. To be reliable and quiet.
5. To have highest power to weight ratio.
6. To include a provision to indicate proper working.
7. To require least maintenance.

- The charging system *consists* of the following *components* : Refer to Fig. 10.21.

(*i*) **Generator**—To convert mechanical energy from the engine into electrical energy.
(*ii*) **Regulator**—To control the amount of electrical energy so produced.
(*iii*) **Relay**—To regulate the flow of the charging current from the generator to the battery equivalent to the state of charge of the battery.
(*iv*) **An armature or indicating lamp**—To indicate whether the system is operating or not.

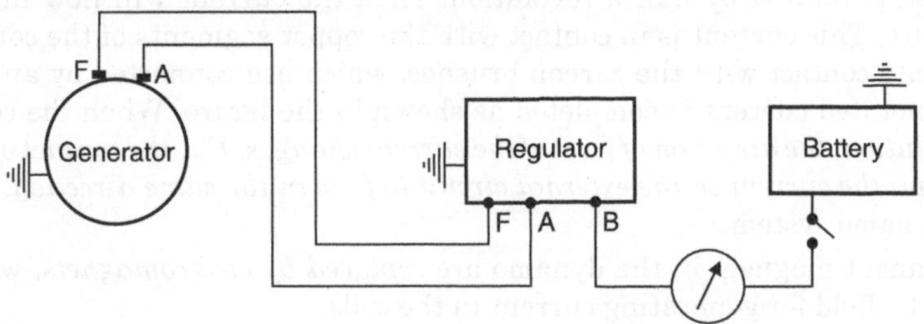

Fig. 10.21. Charging system.

10.6. GENERATOR (DYNAMO)

10.6.1. Introduction

A **generator** is *a machine which converts mechanical energy into electrical energy.* When rotated, it generates electricity by *electromagnetic induction.* In automobiles, it is driven by the engine through a belt pulley or sprocket and chain. It replaces in the battery the current used in starting the engine and also supplies current for operation of electrical devices such as ignition system, lights, radio etc.

- The automobile generator is a low voltage generator producing *direct current only* because the use of battery requires direct current.
- *Shunt wound type of "D.C. generator or dynamo"* is the one that has been used for automotive purposes till recently when it has given way to the **A.C. generator** or **Alternator** *whose output is rectified by in-built semiconductor rectifier diodes.*

In modern cars, the charging system consists of an alternator only.

10.6.2. Working Principle of D.C. Generator or Dynamo

The automobile generator or dynamo works on the following principle :

"When a conductor is moved in a magnetic field, a current is produced, the direction of which is determined by Fleming's right-hand rule".

The amount of current induced in a conductor depends upon the strength of magnetic field, the number of turns, cutting the field and speed with which they pass through the field.

Fig. 10.22 shows the principle of the dynamo.

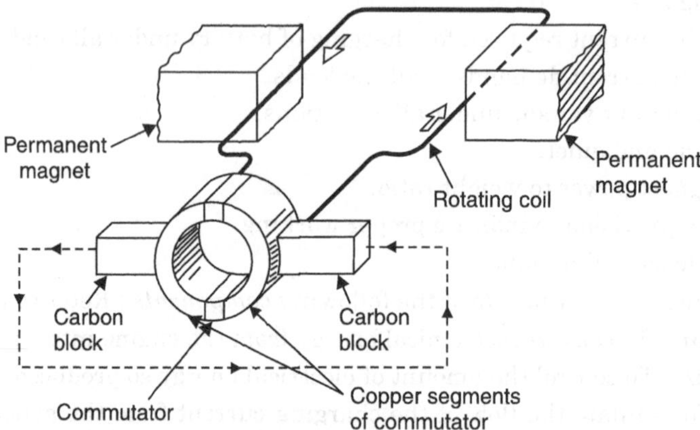

Fig. 10.22. Principle of a D.C. generator or dynamo.

Let the coil is rotated by half a revolution. Then the current will flow in the direction as shown in the figure. This current is in contact with the copper segments of the commutator. These segments come into contact with the carbon brushes, which are connected by an external circuit. Therefore the generated current is completed as shown in the figure. When the *coil is changed by another half revolution, the direction of flow of the current changes. But the contact of the commutator and brushes allows the current in the external circuit to flow in the same direction.* This principle is adopted in the dynamo system.

The permanent magnets in the dynamo are *replaced by electromagnets*, which provide the necessary magnetic field for generating current in the coils.

Parts of a dynamo :

The *main parts of a dynamo* are shown in Fig. 10.23 and are as under :

1. Armature
2. Field windings/coils
3. Frame
4. Brushes and brush gear.

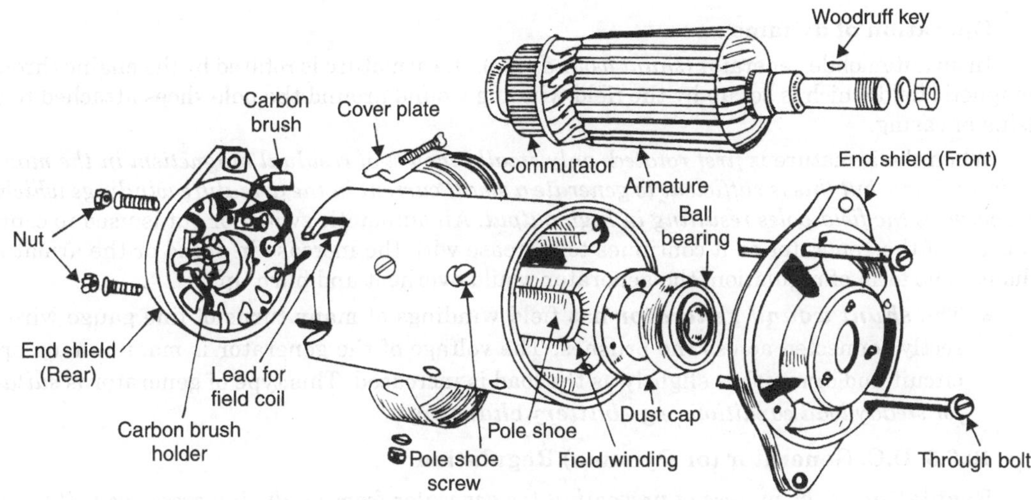

Fig. 10.23. Exploded view of a dynamo.

1. **Armature.** Armature assembly (central part of the dynamo) consists of the following :

(*i*) An armature shaft, (*ii*) Armature core,

(*iii*) Armature windings, and (*iv*) Commutator.

- The construction of armature is as follows :

A group of slotted cores are pressed on the shaft, and in the slots of this core, armature winding is wound with enamelled wire or cotton braided wire. For insulation, the winding with the core prespan insulating paper is placed in slots of core before winding. After having wound the armature, the slots fitted with windings are locked with strips of bamboo sticks to that winding does not come out of core during working. On the one end of armature is fitted a **commutator**, *which consists of taper brass strips clipped together on the armature shaft*. For insulating each strip called segment, *mica sheets* are placed in between each segment and the ends of wires of armature coil are soldered on these segments of commutator. After having made the commutator, the whole armature assembly is vanished to give it a good insulation.

2. **Field windings/coils.** Field coils are made out of enamelled wire wound in the shape of pole shoe and slightly bent to fit in the circular body shape. After having wound these coils, these are taped with cotton tape and varnished. These are placed over the pole shoe and pole shoe held in the body by pole shoe screw. The sleeves are placed on the ends of windings to avoid any risk of short-circuit. In dynamo two or four field windings are fixed.

3. **Frame.** The frame consists of magnet pole pieces which are fitted on its inner surface. The end covers support the armature bearings. These bearings are usually of ball race type. The brush gear is fitted on the commutator end of the dynamo cover. There is also a dust cover for the purpose of inspection.

4. **Brushes and brush gear.** Carbon brushes bear on the commutator. They collect current from the armature. These brushes should bear on the commutators very evenly. While bearing on the commutator, they should not develop friction. This would lead to wear and tear of the commutator.

The carbon brushes are kept in carbon holders, which operate and press on the commutator by means of spring tension. Two main brushes are provided at an angle of 180° for a two-pole dynamo. If there are four poles, four brushes are provided at angle of 90°.

Operation of dynamo :

In an automobile generator (*shunt wound type*), the armature is rotated by the engine through its magnetic field which is set up by the field winding wound around the pole shoes attached to the housing or casing.

When the armature is *first rotated, only small amount of residual magnetism in the magnet poles is available but this is sufficient to generate a small current in the armature windings which in turn energizes the field poles resulting in high output.* An automatic switch or out is used to control the output of the generator as it continues to increase with the increase in speed or the armature. Without some form of regulation, the generator would overheat and burn up.

- The ***shunt wound generator*** has field windings of many turns of fine gauge wire directly connected across the brushes. The voltage of the generator is maximum on open circuit and diminishes slightly as the load is increased. This type of generator is *suitable for steady load conditions e.g.,* ***battery charging.***

10.6.3. D.C. Generator (or Dynamo) Regulation

Regulation *is the process of preventing the generator from producing excessive voltage and current.*

With the increase in the dynamo speed the current supply to the shunt winding increases, which in turn increases the magnets flux. Consequently the output of the dynamo increases. The output goes on increasing with the increase in the dynamo speed. To avoid this type of damage, regulators are used.

Fig. 10.24 shows a resistor provided in the field winding, which regulates the current flowing through the field coil. This arrangement regulates the output of the dynamo within prescribed limit ; the current in the field coil does not go beyond a certain limit and the output of the dynamo is fully regulated.

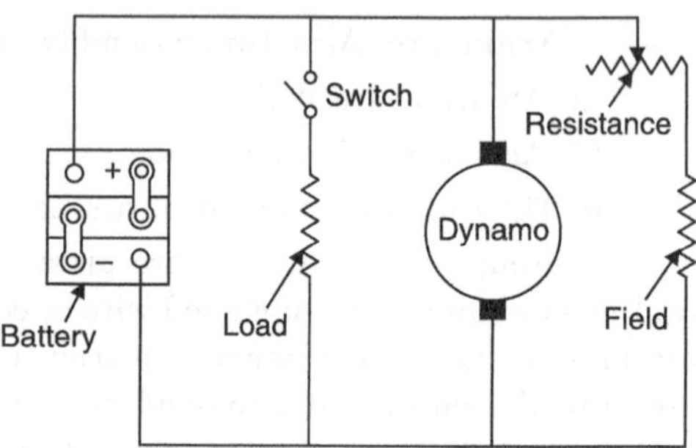

Fig. 10.24. Output regulation of dynamo (Shunt wound).

Third brush control :

In this system instead of two carbon brushes, an extra carbon brush usually called *third brush* is also placed as shown in Fig. 10.25. The position of third brush can be *adjusted by hand.* The third brush is also known as *adjusting brush* and is placed between negative and positive brushes.

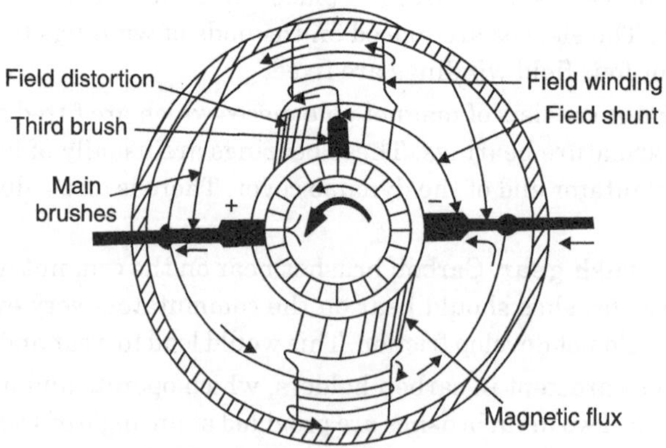

Fig. 10.25. Working of third brush generator.

In case we shift the position of third brush towards *positive brush*, it will *increase the charging rate* and if we shift it towards *negative brush* it will *decrease the charging rate* as the third brush is connected to the field winding and supplies it current according to it position.

This system was very popular in olden days, but these days it is not being used as every time charging rate has to be increased or decreased position of third brush has to be changed. Now its place has been *taken over by voltage and current regulator.*

Cut-out relay :

The cut-out relay (or circuit breaker) closes the circuit between the generator and the battery when the generator is producing the current. Also, it opens the circuit so that the battery cannot discharge back through the generator when the generator stops or slows down.

Fig. 10.26 shows the schematic arrangement of a cut-out relay. It is simply a magnetic switch. It consists of two windings, shunt winding and series winding, assembled round a soft iron and a flat steel armature mounted on a hinge above the core. The series winding consists of a few turns of fine wire. The armature is connected to the battery by contact points.

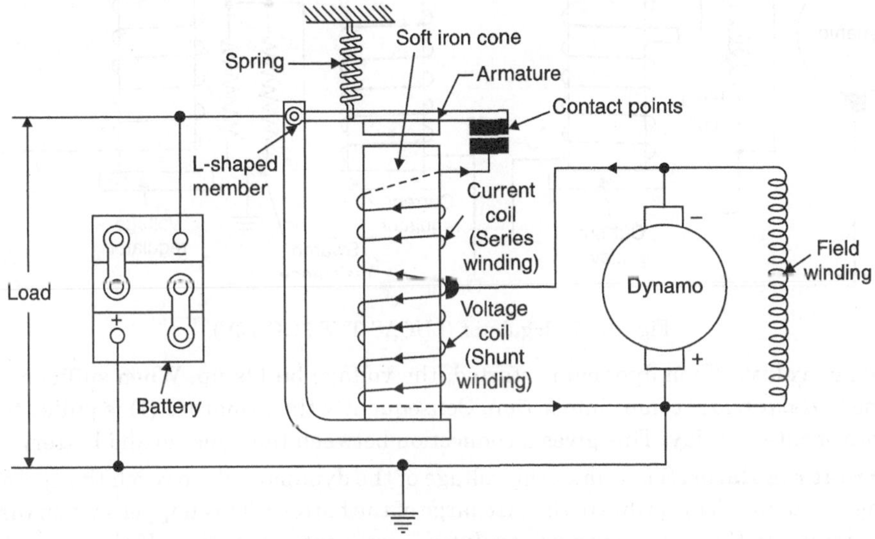

Fig. 10.26. Cut-out relay.

When the generator is *not operating*, a spring on the armature holds the contact points apart, this keeps the circuit between the generator and the battery open.

When the generator is *operating* the core winding creates magnetic field which attracts the armature, contact points are closed and generator is connected to the battery. Current flows from the generator to the battery, and passes through the series winding. But when the generator stops, the current begins to flow from battery to generator. This reverse direction of current flow in the series winding causes the magnetic field to reverse. Thus, the magnetic fields of the winding no longer aid and is so weakened that it cannot hold the armature down. The spring tension pulls the armature up, separating the contact points. This opens the circuit between the generator and the battery.

Regulators for D.C. generators :

Regulators for D.C. generators are made by different companies, the most famous of them are as follows :

1. Delco-Remy Regulator
2. Auto-lite regulator
3. Ford regulator.

The working principle of these regulators is the same. The ***voltage and current regulator*** consists of three major *components* :

1. Cut-out relay. 2. Current regulator. 3. Voltage regulator.

Fig. 10.27 shows the regulator.

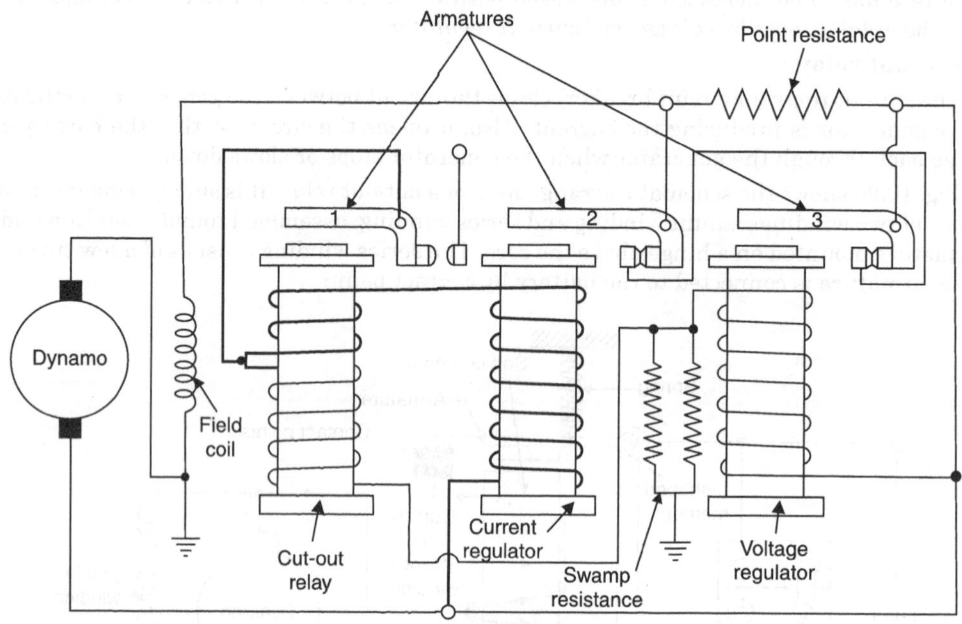

Fig. 10.27. Regulator (LUCAS-TVS 3GC-12V).

Cut-out relay. When dynamo is started, the voltage builds up. When sufficient voltage is reached, the "*cut-out relay*" comes into action. Consequently the armature (1) is pulled towards the soft iron core of cut-out relay. This gives a connection between the dynamo and battery.

Current regulator. Sometimes the voltage of the dynamo fails to reach the operating value of the voltage regulator. This is due to the discharge of the battery. This happens when there is *large electrical loading.* At this time "*current regulator*" comes into operation. If the current output is increasing, the current regulator pulls armature (2) closer. When the contact points of the current regulator open, point resistance is brought into the field circuit. Thus the current limit is maintained.

Voltage regulator. When the dynamo starts running at a high speed, the "*voltage regulator*" immediately comes into action. Armature (3) is pulled away from the contact points of the voltage regulator. The contact points are now opened. Point resistance is then introduced in the field coil, thereby bringing down the voltage.

By *combining the voltage and current regulators, the optimum specified values of the voltage and current outputs are maintained.*

Temperature compensation. The voltage regulators and also some current regulators are temperature compensated. They have higher settings when cold than when hot.

- Temperature compensation of the *voltage regulator* allows the increase of voltage with low temperature.

- The *current regulator* is compensated to permit a higher generator output when the current regulator is cold than when it is hot. *The generator output decreases as the temperature of the current regulator increases.*

10.6.4. Troubleshooting of Dynamo

Troubles	Causes
1. *Dynamo is not producing voltage.*	(*i*) Short or burn out field coil or armature winding. (*ii*) Loose or broken wire connection. (*iii*) Sticky carbon brushes in the holder. (*iv*) Short-circuit in field winding. (*v*) Short-circuit in armature winding. (*vi*) Loose wire connection on commutator. (*vii*) Defective switch. (*viii*) Blown out fuse of charging circuit. (*ix*) Broken or loose V-belt (*x*) Broken or sheared key of pulley on armature shaft.
2. *Dynamo is charging intermittently or charging less.*	(*i*) Loose earth connection. (*ii*) Loose connection at dynamo terminal. (*iii*) Dirty commutator. (*iv*) Presence of oil on carbon brushes. (*v*) Worn out carbon brushes. (*vi*) Sticking of carbon brush in holder. (*vii*) Weak carbon bruoh opring. (*viii*) Work out commutator. (*ix*) Loose battery connection. (*x*) Loss of electrolyte in battery. (*xi*) Weak battery.

10.7. ALTERNATOR (A.C. GENERATOR)

10.7.1. Introduction

Some cars (superior models) as well as heavier passenger vehicles require more electricity. These vehicles have electrically operated windows, sliding front seats, air-conditioners, electrical equipment for automatic transmissions, overdrives etc. In large cities the vehicles have to often move at very low speeds due to heavy traffic. Any direct current dynamo will not be able to charge the battery at such low speeds. Therefore A.C. generator or alternator is used.

An **alternator** or **A.C. generator** *is the generator which produces alternating current instead of the direct current as produced by a commutator type generator (or dynamo).* However, as the automotive electrical system requires only direct current, the alternating current produced in the alternator *has to be converted first into the direct current, which is easily done with the help of* ***diodes.***

- A.C. generator (alternator) differs from the D.C. generator or Dynamo (commutator type generator) in the method of voltage and current control and also in design. It uses *silicon "diodes"* to change the alternating current to direct current for the vehicle's electrical system. Also it *does not require any 'cut-out'* to prevent the battery from discharging through it, because the *diode rectifiers allow the current to flow in one direction only.*

10.7.2. Advantages of Alternator Over Dynamo

Following are the *advantages of an alternator over a dynamo (or d.c. generator)* :

1. Much smaller in size.
2. Lighter in weight (for the same current output)
3. At *idling speeds,* the alternator produces substantial current (about 20 A). A dynamo does not produce any current at these speeds.
4. There is no necessity for a cut-out unit.
5. Requires less maintenance and is more reliable.
6. The maximum permissible speed is generally about 30 percent higher than the dynamo (D.C. generator).

10.7.3. Working Principle of an Alternator

The D.C. generator rotates the conductors in a stationary magnet whereas *A.C. generator* (or alternator) *rotates a magnetic field so that stationary conductors may cut the moving lines of force.* This way the voltage is generated in the stationary coils. This has the *advantage that heavy currents do not have to pass through the brushes and collector rings.*

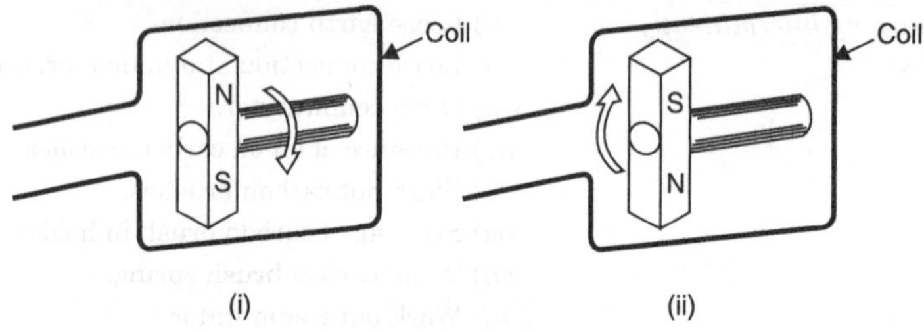

Fig. 10.28. Principle of production of alternating current.

Refer to Fig. 10.28 (i). A coil of wire is around an armature magnet. In this N-pole is at the top and S-pole at the bottom. When the armature magnet is rotated from top to bottom, by a half revolution, it produces current in the wire in one direction.

Refer to Fig. 10.28 (ii). The S-pole is at the top and N-pole at the bottom. When the magnet is rotated by another half revolution, the flow of current in the wire is *reversed.* Therefore, in one revolution of the armature magnet, the current flows in one direction and reverses in the opposite directions. This type of flow in both forward and reverse directions is called alternating current (A.C.).

10.7.4. Constructional Details and Operation of Alternator

The alternator is made of a *stator* (Fig. 10.29), *rotor* (Fig. 10.30), *slip ring* and *brush* assembly. Many modern alternators have regulator built into the housing as a complete unit.

The operation of an alternator is improved by placing the stators and rotor assembly inside an iron frame or housing which provides a conducting path for magnetic lines of force. The voltage can be greatly increased if the stator winding is formed into many coils or turns. A stator consisting of one winding of wire (regardless of number of coils of wire in that winding) is called a *single-phase stator.* Most alternators use three windings which is called a 3-phase stator and since the connections of the winding are like a Y, it is called a 3-phase *Y-connected stator.* The advantage of this stator is that the voltage is *more even between phases and that each phase reaches its highest voltage at a different time.*

An alternator consists of a *rotor assembly,* a *stator assembly* and a ***rectifier*** mounted in a *housing* (Fig. 10.31).

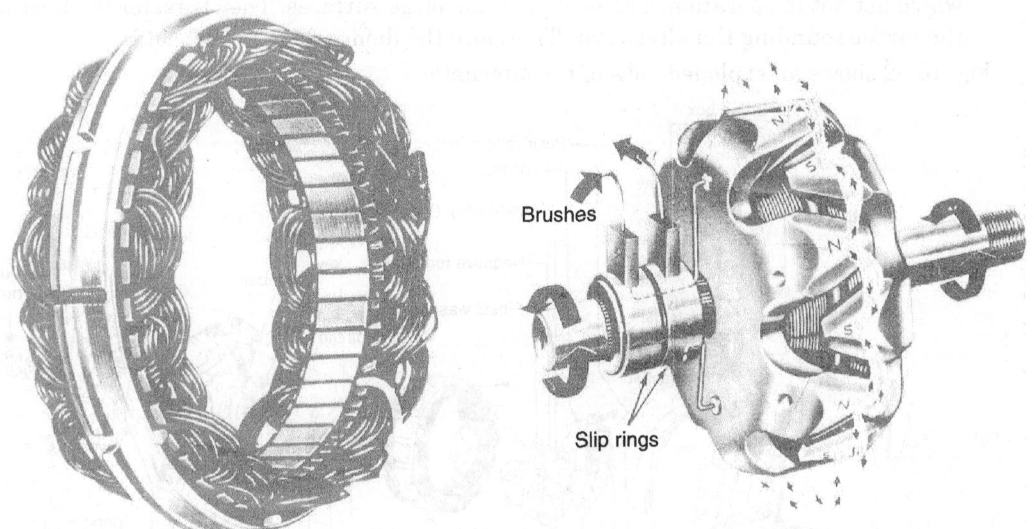

Fig. 10.29. Stator of an alternator. Fig. 10.30. Rotor of an alternator.

— The '*housing*' is usually made of two pieces of die cast aluminium (because it is non-magnetic, light weight and provides good heat dissipation). Bearings, supporting the rotor assembly are mounted in the front and rear housing.

— The '*rotor assembly*' consists of a rotor shaft, a winding around an iron core, two pole pieces and slip rings. The rotor shaft is pressed into the core. Two brushes are held against the slip rings by springs and are connected through a switch to the battery.

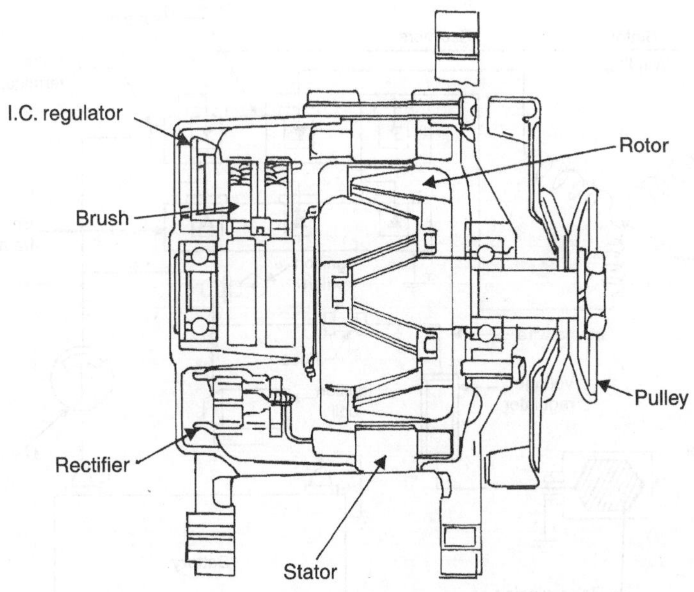

Fig. 10.31. Alternator.

— The '*rectifiers*' (*diodes*) are pressed in the slip end ring head or heat sinks (rectifier mounting plates) and are connected to the stator leads. The heat sink takes heat from the diodes, which get hot in operation. The heat sink has large surfaces. They transfer the heat into the air surrounding the alternator. Therefore the diodes do not overheat.

Fig. 10.32 shows an exploded view of the alternator.

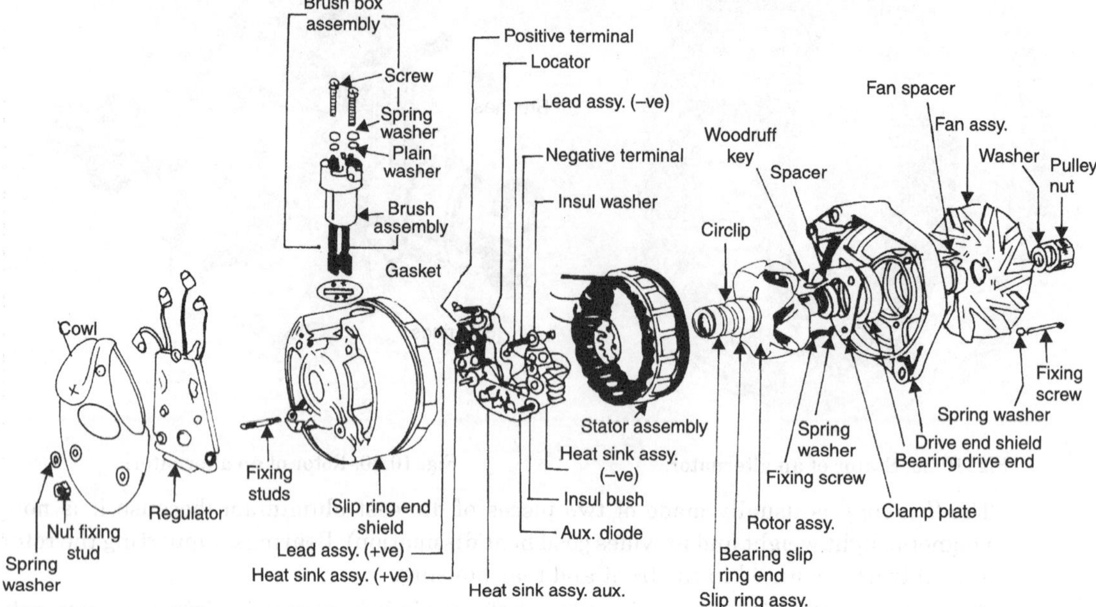

Fig. 10.32. Exploded view of the Alternator (Courtesy : Telco).

- Fig. 10.33 shows the charging circuit of alternator.

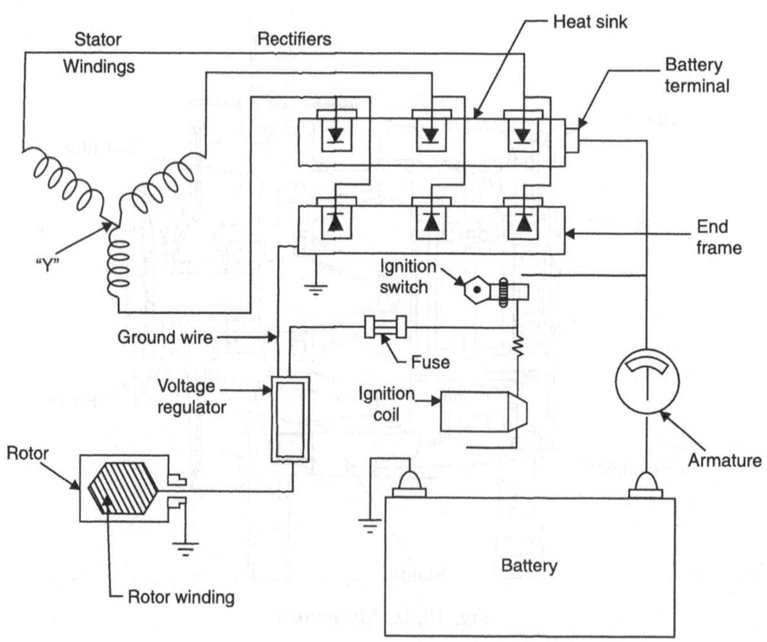

Fig. 10.33. Charging circuit of alternator.

When the ignition switch is turned on, the rotor receives current from the battery through the voltage regulator. This current produces the magnetic field of the rotor. As the rotor is turned by the pulley, current is induced in the stator winding. This alternating current is charged into direct current by *rectifiers*. The voltage depends upon the rotor speed. A voltage regulator is used to limit the generator voltage to a correct value.

10.7.5. Alternator Regulation

A variety of devices have been used to regulate alternators. When alternators were first introduced, many regulators included a field relay, an indicator-light relay, and a voltage regulator. Now alternator-regulator systems have been simplified. For example, some of the latest types have the regulator built into the alternator. Basically, the regulator limits the alternator field current as necessary to prevent excess alternator voltage. The stator remains permanently connected to the battery through the *diodes* ; they *prevent the battery from discharging back through the stator when the alternator is not operating* (so that cut-out relay is not needed). The field (rotor) is connected to the battery only when the alternator is operating. The connection is made through either the field relay or the ignition switch.

Several types of alternator regulators that are mounted outside the alternator have been used.

Transistorised regulator. Since the contact breaker points of the regulator open several times a second throughout its life, it causes arcing and pitting of the contact points. This reduces its working life and requires costly materials such as silver-tungsten combination for the contact points.

In order to meet with these contact point problems, *transistorized regulators* are used in two different methods. In one, the transistors are used to reduce the amount of currents *through the vibrating contacts* ; *i.e.*, it *acts as a switch actuated by a small current.* Such system are very economical.

In all transistor systems both the magnetic circuit and spring reference is replaced by a zaner diode, its breakdown voltage providing the reference.

A fully transistorized regulator has no moving parts and provides very close tolerances in voltage control regardless of the vibration, etc.

A semiconductor type regulator has been developed by Bosch and used in automobiles. It is known as *p-n*. junction regulator. The essential element of this regulator is the junction of the *p*-type and *n*-type materials. *The characteristic curve of transistorized regulator is similar to that of the current and voltage regulator but it has no current control chamber.*

10.7.6. Troubleshooting of Alternator

Troubles	Causes	Remedies
1. *Battery is not charged when engine is running.*	(*i*) Too loose alternator drive belt. (*ii*) Worn-out brushes. (*iii*) Worn-out slip rings. (*iv*) Open rectifiers.	(*i*) Adjust belt according to specifications. (*ii*) Install new brushes. (*iii*) Instal new rotor. (*iv*) Dismantle the alternator. Test and install new rectifiers, if necessary.
2. *The alternator continues to charge at high rate even after battery attains full charge.*	(*i*) Voltage regulator set too high. (*ii*) Stuck regulator contacts. (*iii*) Poor regulator—ground connection.	(*i*) Reset the regulator according to specifications. (*ii*) Install new voltage regulator. (*iii*) Correct ground connection.

3. *Alternator makes noise while running*	(*i*) Loose mounting. (*ii*) Worn out bearings. (*iii*) Interference between fan and starter leads.	(*i*) Tighten and adjust the mounting. (*ii*) Replace the bearing. (*iii*) Dismantle the alternator and correct the defect.

10.7.7. Comparative Generator Data of Some Indian Automobiles

Comparative generator data of some Indian automobiles is given below, in tabular form :

S. No.	Make	Generator type	Regulator type
1.	*Maruti 800*	Alternator	Vibrating voltage type
2.	*Hindustan Ambassador Mark II*	Lucas, Motolite C-40	Lucas Regulator vibration type
3.	*Fiat 1100*	Fiat Generator R90-180/12-2500	A/4-180/12 vibrator type
4.	*Jeep Universal*	Prestolite GJP-7432	Prestolite-VBO 4231 E-5A vibrator type
5.	*Dodge/Fargo Model 89M4 (Petrol type)*	Prestolite GJM-7001S	Prestolite-VRX 6301A vibrator type
6.	*Ashok Leyland Comet Passenger*	Lucas GGS 524/3	Lucas 3GC-24V vibrator type
7.	*Tata Diesel Vehicles*	Lucas-TVS Type C-40 LQ 22740B (20 Amps) (For SE/SK/SA/SD vehicles) or type GGS-12-1 (35 amps) (for LP/LP/LPO vehicles)	Lucas-TVS Type 3GC 37367 (20 amps) (For Se/SK/SA/SD vehicles) or Type 3GC-37379 (35 amps) (For IP/LPT/LPO vehicles)

III. IGNITION SYSTEM

10.8. INTRODUCTION TO IGNITION SYSTEM

- In S.I. engines the combustion process is initiated by a spark between the two electrodes of spark plug. This occurs just before the end of compression stroke. The ignition process must add necessary energy for *starting* and *sustaining* burning of the fuel till combustion takes place.
- Ignition is only *a pre-requisite of combustion. It does not influence the gross combustion process.* It is only a small scale phenomenon taking place within a specified small zone in the combustion chamber.
- Ignition only ensures initiation of combustion process and has no degree intensively or extensively.

Energy requirements for ignition :

— A spark energy below 10 millijoules is adequate to initiate combustion for A/F ratio 12-13 : 1 (Range of mixtures normally used) ; the duration of few micro-seconds is sufficient to start combustion.

— A spark can be struck between the gap in the two electrodes of the spark plug by sufficiently high voltage. There is a critical voltage called *breakdown voltage* below which no sparking would occur. In practice the pressure, temperature and density have a profound influence on the voltage required to cause the spark. Also, the striking voltage is increased due to the fouling factor of the electrodes owing to deposits and abrasion.

— For automotive engines, in normal practice, the spark energy to the tune of 40 millijoules and duration of about 0.5 millisecond is sufficient over entire range of operation.

10.9. PURPOSE OF IGNITION SYSTEM

The ignition system supplies high-voltage surges as high as 47,000 volts (in some electronic systems) to the spark plugs in the engine cylinders. The surges produce electric spark across the spark plugs. The heat from the spark ignites, or sets fire to, the compressed air-fuel mixture in the combustion chambers.

— When the engine is ***idling***, the spark appears at the plug gap *just as the piston nears T.D.C.* on the compression stroke.

— When the engine is ***operating at higher speeds***, or ***with part throttle***, the *spark is advanced*. It is moved ahead and occurs *earlier in the compression* stroke. This gives the compressed mixture *more time to burn and deliver its energy to the pistons.*

10.10. REQUIREMENTS OF AN IGNITION SYSTEM

For an ignition system to be acceptable it must be *moderately priced, reliable and its performance must be adequate to meet all the demands imposed on it by various operating conditions.*

An ignition system should *fulfil the following requirements* :

1. It should have an adequate reserve of secondary voltage and ignition energy over the entire operating speed range of the engine.

2. It should consume the minimum of power and convert it efficiently to a high-energy spark across the spark-plug electrode gap.

3. It should have a spark duration which is sufficient to establish burning of the air-fuel mixture under all operating conditions.

4. It should have the ability to produce an ignition spark when a shunt (short) is established over the spark-plug electrode insulator surface, due possibly to carbon, oil or lead deposits, liquid fuel or water condensation.

5. Good performance at high speeds.

6. Longer life of breaker points and spark plug.

7. Good starting when the breaker points open slowly at cranking speed.

8. Good reproducibility of secondary voltage rise and maximum rise.

9. Adjustment of spark advance with speed and load.

- The basic source of electrical energy is either *battery*, a *generator* or a *magneto.*
  - — The battery and generator normally provide 6 V or 12 V direct current, while the magneto provides an alternating current of higher voltage.
  - — The low voltage (6 V or 12 V) is boosted to a very high potential of about 10 kV to 20 kV, in order to overcome the spark gap resistance and to release enough energy to initiate self propagating flame from within the combustible mixture.

10.11. BASIC IGNITION SYSTEMS

The following *basic ignition systems are in use* :

1. **Battery ignition system**—Conventional, transistor assisted.

2. **Magneto ignition system**—Low tension, high tension.
3. **Electronic ignition system.**

The difference between Battery and Magneto ignition systems lies only in the source of electrical energy. Whereas 'Battery ignition system' uses a battery, 'Magneto ignition system' uses a magneto to supply low voltage, all other system components being similar.

10.12. BATTERY (OR COIL) IGNITION SYSTEM

It is a commonly used system because of *its combined cheapness, convenience of maintenance, attention and general suitability.*

Construction. This system consists of the following *components* :

1. Battery (6 or 12 volts)
2. Ignition switch
3. Induction coil
4. Circuit/contact breaker
5. Condenser
6. Distributor.

Refer to Fig. 10.34.

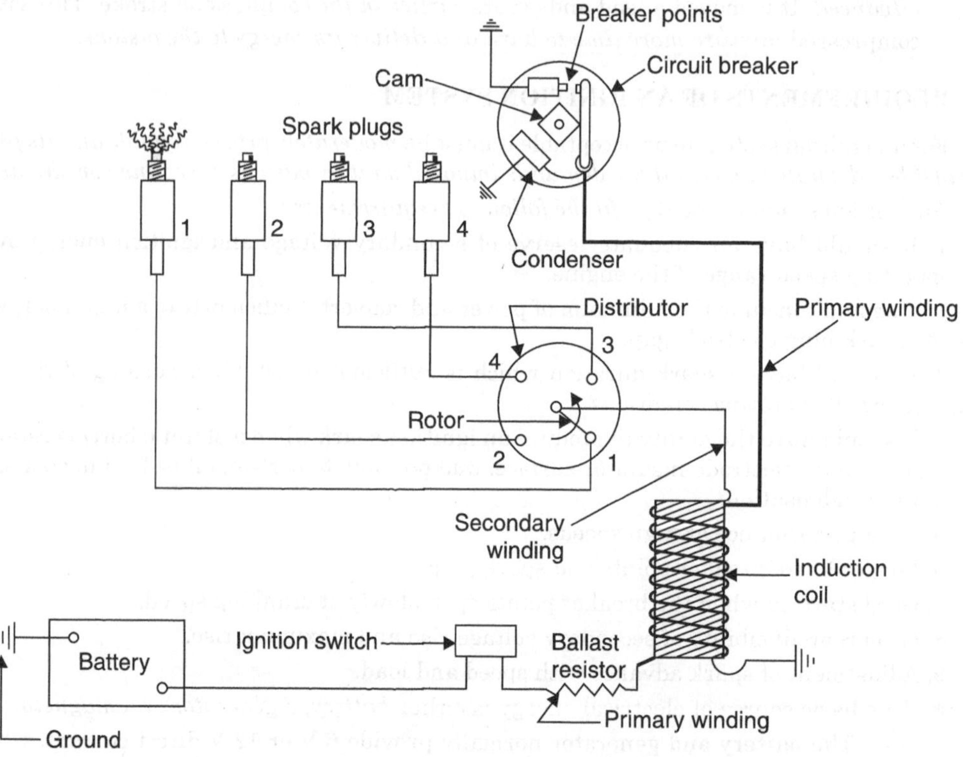

Fig. 10.34. Battery or coil ignition system.

— One terminal of the battery is ground to the frame of the engine, and other is connected through the ignition switch to one primary terminal of the ignition coil (consisting of a *comparatively few turns of thick wire* wound round an iron core). The other primary terminal is connected to one end of the contact points of the circuit breaker and through closed points to ground. The primary circuit of the ignition coil thus gets completed when contact points of the circuit breaker are together and switch is closed. The secondary terminal of the coil is connected to the central contact of the distributor and hence to distributor rotor. The secondary circuit consists of secondary winding (consisting of a *large number of turns of fine wire*) of the coil, distributor and four spark plugs. The contact breaker is driven by a cam whose speed is *half* the engine speed (for *four stroke engines*) and breaks the primary circuit one for each cylinder during one complete cycle of the engine.

— The breaker points are held on contact by a spring except when forced apart by lobs of the cam.

— A *ballast resistor* is provided in series with the primary winding to regulate primary current. For starting purposes this resistor is by passed so that more current can flow in the primary circuit.

Working :

— To start with, the ignition switch is made on and the engine is cranked *i.e.* turned by hand when the contacts touch, the current flows from battery through the switch, primary winding of the induction coil to circuit breaker points and the circuit is completed through the ground. A condenser connected across the terminals of the contact breaker points prevent the sparking at these points.

— The rotating cam breaks open the contacts immediately and breaking of this primary circuit brings about a change of magnetic field ; due to which a very high voltage to the tune of 8000 to 12000 V is produced across the secondary terminals. (The number of turns in the secondary winding may be 50 to 100 times than in primary winding). Due to high voltage the spark jumps across the gap in the spark plug and air fuel mixture is ignited in the cylinder.

- Fig. 10.35 shows the gradually building up of the primary current from the time the points close until they open.
- Fig. 10.36 shows a typical wave-form or pattern of the *normal ignition action*.

  — At point **L** the distributor opens and the magnetic field of the coil-primary winding *collapses* and consequently the secondary voltage, indicated by the firing line, rises to point **M**. The height of the firing line shows the voltage needed to jump the rotor gap and to ionize the gap between the spark plug electrodes. After the spark is initiated the gap becomes ionized resulting in decreased gap resistance and a smaller voltage is then required to maintain the arc across the gap. The lower voltage and the spark duration is represented by the height and length of the spark line NP.

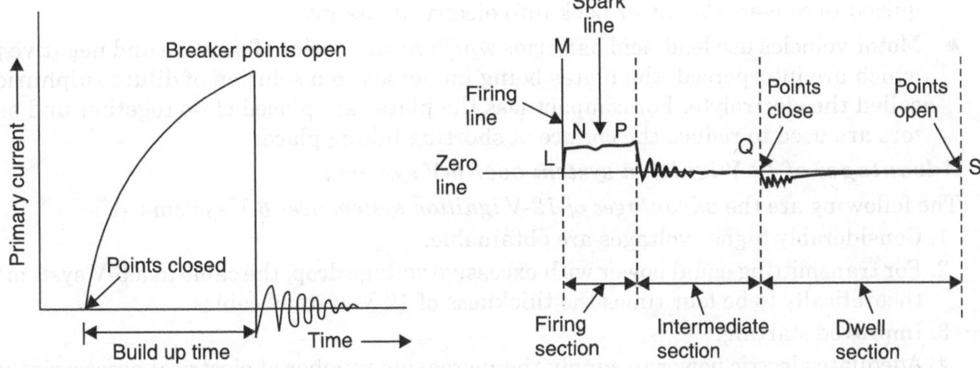

Fig. 10.35. Built up time for primary current. Fig. 10.36. Typical pattern of normal ignition action.

  — At point **P** the major portion of the energy of the coil is expended and consequently there is a drop in the secondary voltage which result in extinguishing of the spark.

— Due to spark extinction the circuit becomes open, the current flow is stopped, and, hence the magnetic field (produced in the secondary winding, during the firing period NP while the current was flowing in the secondary winding and across the spark gap to ground) collapses, thereby, inducing a current in the primary winding, which eventually *flows into the condenser and charge it.*

— When voltage in the condenser becomes higher than that in the primary winding, it discharges back in the primary winding. This results in collapsing of the magnetic field and rebuilding up of voltage in the secondary winding. This pulsing back and forth, weakened each time, continues till whole of the energy is dissipated (Refer to Fig. 10.36 intermediate section).

— At point *Q* the contact points close and remain so during *dwell period.* At the end of this period the points again open at S (there being no condenser action during the period, since it is shorted out across the closed points.)

Advantages :

1. It offers better sparks at low speeds, starting and for cranking purposes.
2. The initial cost of the system is low.
3. It is a reliable system and periodical maintenance required is negligible except for battery.
4. Items requiring attention can be easily located in more accessible position than those of magnetos.
5. The high speed engine drive is usually simpler than magneto drive.
6. Adjustment of spark timing has no deteremental effect over the complete ignition timing range.

Disadvantages :

1. With the increasing speed, sparking voltage drops.
2. Battery, the only unreliable component of the system needs regular attention. In case battery runs down, the engine cannot be started as induction coil fails to operate.
3. Because of battery, bulk of the system is high.

Description of Components of Battery Ignition System

1. **The battery :**

- The function of battery is to store electricity in the form of chemical energy, when required to convert the latter back into electrical energy.
- Motor vehicles use lead-acid batteries which have a series of positive and negative plates which are interpersed, the plates being immersed in a solution of dilute sulphuric acid, called the electrolyte. For compactness the plates are placed close together and separators are used to reduce the chance of shorting taking place.

Advantages of 12-V ignition system over 6-V system :

The following are the *advantages of 12-V ignition system over 6-V system* :

1. Considerably higher voltages are obtainable.
2. For transmitting equal power with excessive voltage drop, the cable in a 6-V system needs theoretically to be four times the thickness of 12-V system, cables.
3. Improved starting.
4. Adequate electric power to supply the increasing number of electrical accessories used.

2. **The induction/ignition coil :**

- To create an adequate spark across the gap of sparking plug high electrical pressure is needed. Electrical pressure is measured in volts and the 12 volts supplied by the battery is totally inadequate.

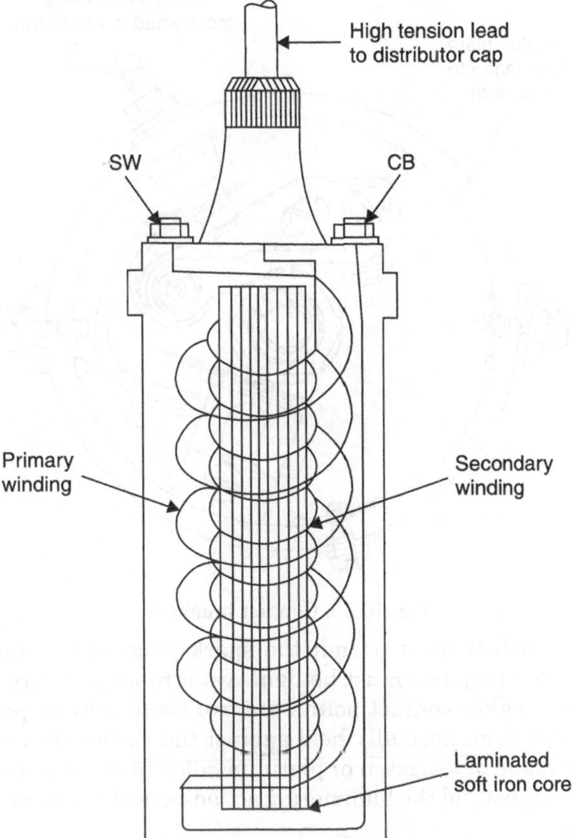

Fig. 10.37. The induction coil.

- *The function of ignition coil is to increase the voltage between 10,000 and 15,000 volts in some conditions*, although the voltage which occurs under normal running conditions is of the order of 4000–5000 volts.
- Two coils of insulated wire are wound on a laminated soft iron core. The *inner coil*, called the *secondary*, has more turns than the outer primary coil. There are about 20000 turns on the secondary and 400 turns on the primary.
- If a low voltage passing through the primary coil is switched off a higher voltage is induced in the secondary coil, the *increase being approximately in the same proportion to the number of turns of the two coils*. The core and windings are placed in an iron sheath. The entire assembly being housed in a sealed container (Fig. 10.37).
- A high tension lead from the centre of the coil carries the supply to the distributor. Two small terminals are situated either side of the high tension lead, one being connected to the contact breaker and marked CB and the other to the ignition switch identified by the letters SW.

3. **Contact breakers :**

- The distributor unit used on the modern motor vehicle contains the breaker contacts which make and break the primary current, and the distributor mechanism which supplies the secondary current to the plugs in their proper firing order. Fig. 10.38 shows how the distributor is connected into the typical battery ignition system.

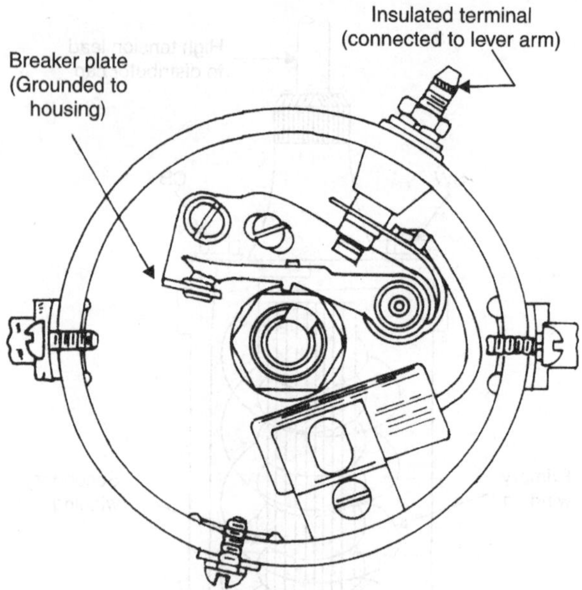

Fig. 10.38. Contact breakers.

- *Breaker contact points* must permit the spark plugs to fire accurately at high speed. Therefore the moving parts must be light, yet strong, and carefully built of high-grade materials. The breaker contact points are two small contact pieces one stationary and one on a movable arm, normally held against the stationary contact by spring tension. The points are made of tungsten or platinum alloy to resist burning and pitting, and are hard enough to withstand the hammering action caused by the rapid closing of the breaker at high speeds.

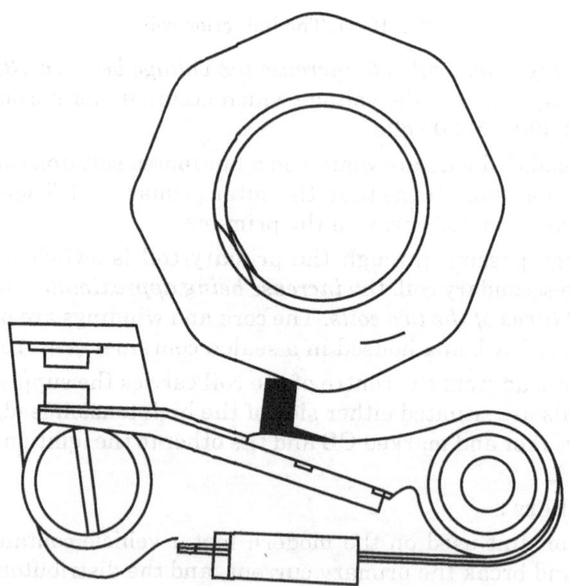

Fig. 10.39. Closed circuit type circuit breaker.

- A small cam, with a lobe for each time the breaker is required to open per revolution, revolves and pushes the movable breaker arm so that the contact points are separated. In this way, the current in the primary winding of the coil is interupted every time a spark is required in one of the cylinders.
- The usual circuit breaker used on modern vehicles is the closed-circuit type (Fig. 10.39). The contact points normally remain in the closed position, *being separated only when the breaker arm is lifted by a lobe of the breaker arm*. The closed circuit type is more adaptable to high speed engines because the points are in contact long enough to allow complete magnetisation of high tension coil. This results in especially good sparks at slow starting speeds, with less intense sparks at higher speeds when the time of the contact is shortened.

4. **The distributor :**

- In order for the coil to produce a high voltage from a low voltage supply the flow of electricity must be interrupted (switched off and on). The distributor contains the contact breaker and a cam, which is of rotary switch, that interrupts the supply. The *distributor cap* has a centre terminal which connects with the high tension terminal of the induction coil, and as many terminals equally spaced around it as there are spark plugs to fire. The cap is usually moulded of a highly resistant insulating material, such as bakelite which is moisture proof and possesses high insulating properties even under excessive heat. The terminals are of brass or metal alloy moulded in position, terminating on the underside either in the form of a button flush with the surface or in the form of a pin. The distributor head is usually held in place by two spring clips which snap on only when the head is in its proper position. Thus the head can be easily removed to inspect and adjust the rotor and breaker mechanism with no chance of replacing it incorrectly. The *rotor* or *distributor arm* is mounted on the upper end of the distributor shaft, on which the breaker cam is also located. The inside end of rotor makes contact with the centre terminal of the head, while the outside end in its rotation completes the circuit successively with the terminals leading to the spark plugs.
- The distributor may be considered as a revolving switch located in the secondary circuit, which connects the high-tension wire from coil to the proper spark plug at proper timo.
- The distributor rotor must be well insulated to prevent grounding of the high-tension current, consequently it is usually moulded from an insulating material similar to that used in the distributor cap.
- It is designed to fit over the end of the liner shaft in one position only, to prevent its installation in a position which would throw it out of time with the breaker.
- The distributor may be of gap type or contact type ; classified according to the method used in completing the circuit between the rotor and the distributor head terminals.

5. **The condenser :**

- When the primary circuit is broken the current tends to continue flowing and a spark jumps across the separated contact breaker points. This effect is characteristic of an inductive circuit and is undesirable for the following *reasons* :

  (*i*) *It causes arcing, burning of the contact breaker points.*

  (*ii*) *The effectiveness of the coil is reduced since a sudden interruption of the circuit is essential to produce a high voltage.*

- The condenser absorbs and stores this inductive flow of current and causes the current in the coil to die away rapidly which increases the voltage in the secondary coil.
- A condenser is constructed strips of sheets of tin foil insulated by thin sheets of paraffined paper or mica. The alternate layers of tin foil are connected in parallel, forming two

groups, each group provided with a terminal for external connection. Rolled or cylindrical condensers are made by winding alternate layers of the tin foil and the insulation into a tight roll (Fig. 10.40).

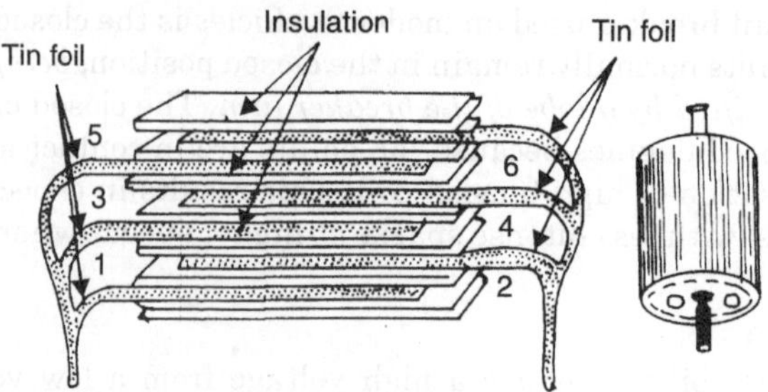

Fig. 10.40. The condenser.

Maintenance of ignition system :

Induction coil. (*i*) It is only necessary to keep the connections clean and tight.

(*ii*) Never leave the ignition switched on without the engine running. If the contact breaker points have come to rest in the closed position the coil will overheat and burn out.

Distributor. (*i*) Provide light grease or clean engine oil on the cam, but make sure it does not get into the contact faces of the contact breakers.

(*ii*) Wipe inside of distributor cap and the electrodes with a clean cloth.

(*iii*) Remove the rotor arm and inject light engine oil into the rotor arm spindle to lubricate the cam bearing.

(*iv*) Check that if the carbon brush is clean and moves freely in its holder.

(*v*) Burnt or blackened contact breaker points should be cleaned with a thin carborudum stone. Care must be taken that these surfaces are kept flat ; ensuring that the entire area of the points is in contact. Remove grease or metallic dust with a petrol moistened cloth.

(*vi*) Breaker gaps should be periodically checked, and if necessary reset.

(*vii*) If the engine does not fire disconnect the centre high tension lead and hold close to the cylinder head. Arrange the cam so that the points are closed, each time the points are separated, with the ignition switch on, a good spark should be obtained. This is only a rough check since the spark does not occur under the conditions preventing in the cylinder.

Condenser. The following symptoms may indicate a faulty condenser although they are usually checked by special equipment :

(*i*) Severe misfiring causing explosion in the silencer and carburettor.

(*ii*) Engine refuses to perform satisfactorily under load.

(*iii*) Arcing at contact breaker points causing blackened and pitted surfaces.

(*iv*) Weak spark.

Check connections and if fault still occurs replace the condenser.

High tension cables. Check that connections are clean and secure, and that the cable insulation is free from cracks.

Plugs. (*i*) Inspect and clean every 3000–4800 km and replace every 16000 km. When setting the gap adjust the side electrode and check with a wire gauge for preference.

(*ii*) It is not satisfactory to check the action of a plug by lying it on the cylinder head and rotating the engine. This is because it is not subjected to working conditions.

(*iii*) Plugs should be checked using a special machine which stimulates cylinder conditions and automatically gives the condition of the plugs.

10.13. MAGNETO IGNITION SYSTEM

The magneto ignition system is similar in principle to the battery system except that the magnetic field in the core of the primary and secondary windings is produced by a *rotating permanent magnet* (Fig. 10.41). As the magnet turns, the field is produced from a positive maximum to a negative maximum and back again. As this magnetic field falls from a positive maximum value, a voltage and current are induced in the primary winding. The primary current produces a magnetic field of its own which keeps the total magnetic field surrounding the primary and secondary windings approximately constant. When the permanent magnet has turned for enough so that its contribution to the total field is strongly negative, the breaker points are opened and the magnetic field about the secondary winding suddenly goes from a high positive value to a high negative value. This induces a high voltage in the secondary winding which is led to the proper spark plug by the distributor.

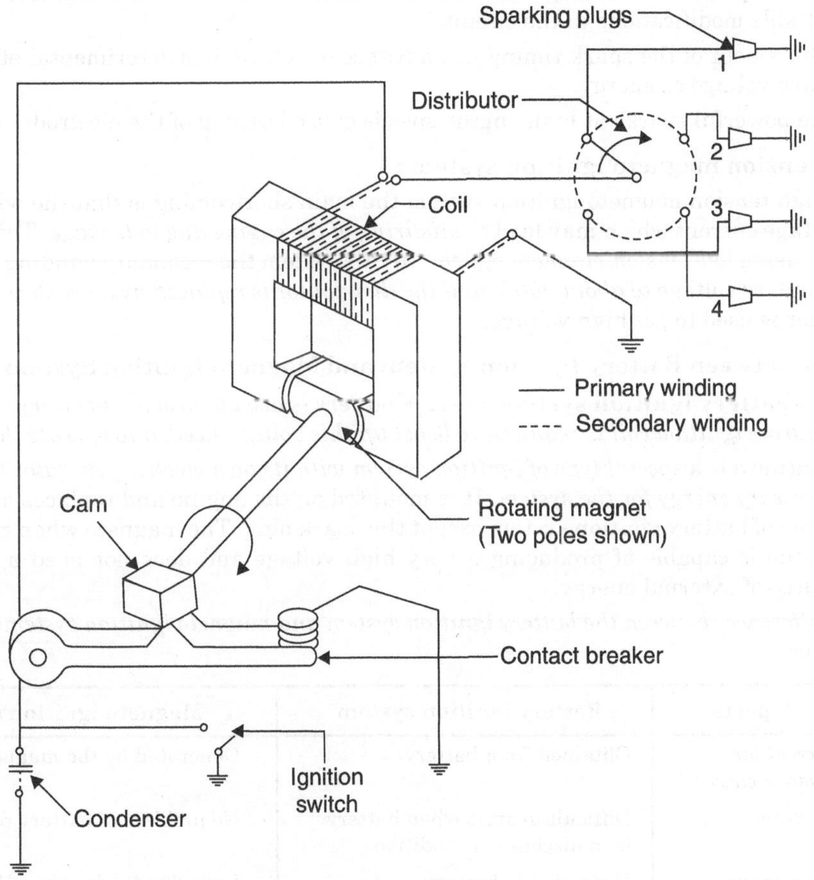

Fig. 10.41. Magneto ignition system.

- The magneto is an *efficient, reliable, self contained unit which is often preferred for aircraft engines because storage batteries are heavy and troublesome.* Special starting means are required, however, as the magneto will not furnish enough voltage for ignition at low speeds. Variation in ignition timing is more difficult with the magneto, since the breaker point must be opened when the rotating magnets are in the most favourable position. It is possible to change the engine crank angle at which the magneto points open without disturbing the relationship between point opening and magnet position by designing the attachment pad so that the entire magneto body may be rotated a few degrees about its own shaft. Obviously this method is *not as satisfactory as rotating a timer cam-plate.*

Advantages :

1. The system is more reliable as there is no battery or connecting cable.
2. The system is more suitable for medium and very high speed engines.
3. With use of cobalt steel and nickel-aluminium magnet metals very light and compact units can be made which require very little room.
4. With recent development this system has become fairly reliable.

Disadvantages :

1. At low speeds and during cranking the voltage is very low. This has been overcome by suitable modifications in the circuit.
2. Adjustment of the spark timing *i.e.,* advance or retard, has deterimental effect upon the spark voltage or energy.
3. The powerful sparks at high engine speeds cause burning of the electrodes.

Low tension magneto ignition system :

In a high tension magneto ignition system the main shortcoming is that the wirings carry a very high voltage current which may lead *to misfiring of the engine due to leakage.* This trouble can be avoided by using *low tension magneto system.* In this system the secondary winding is changed to limit the secondary voltage to *about 400 V and the distributor is replaced by a brush contact.* A step-up transformer is used to get high voltage.

Comparison Between Battery Ignition System and Magneto Ignition System :

- In a **battery ignition system** a *6-12 V battery is used to provide primary voltage, and a separate ignition coil is required to boost up this voltage needed to operate the spark plug.*
- **Magneto** is a *special type of ignition system with its own electric generator to provide the necessary energy for the system.* It is mounted on the engine and replaces all the components of battery ignition system except the spark plug. The magneto when rotated by the engine is capable of producing a very high voltage and does not need a battery as a source of external energy.

The *differences between the battery ignition system and magneto ignition systems* are given in the table below :

S. No.	Aspects	Battery ignition system	Magneto ignition system
1.	*Current for primary circuit*	Obtained from battery.	Generated by the magneto.
2.	*Starting*	Difficult to start when battery is in discharged condition.	No problem of battery discharge.
3.	*Maintenance problems*	More, due to battery.	Less, due to absence of battery.
4.	*Intensity of spark at low speed*	Good	Poor

5.	*Efficiency*	Efficiency of the system *decreases* with the reduction in spark intensity as the *engine speed increases.*	Efficiency of the system *improves* as the engine speed increases due to high intensity spark.
6.	*Cost*	Less	More
7.	*Space occupied*	More	Less
8.	*Applications*	In cars and light commercial vehicles.	In racing cars and two wheelers, aircrafts etc.

- Fig. 10.42 shows the plot between the breaker current vs. speed for battery and H.T. magneto ignition systems.

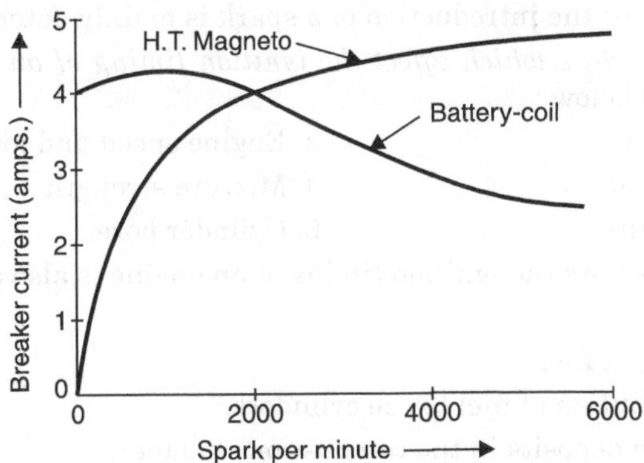

Fig. 10.42. Breaker current *vs* speed for battery coil and magneto ignition systems.

— It can be observed that at start, in case of magneto ignition system, the current generated is very low since the starting cranking speed is low. The current increases with increase in the engine speed. Owing to the poor starting characteristics and effect of the spark timing changes on voltage generated, *invariably battery ignition system is preferred to the magneto ignition system.*

— However, the *battery ignition system is heavier and requires more maintenance than magneto ignition system.*

10.14. FIRING ORDER

Firing order *is the order in which various cylinders of a multi-cylinder engine fire. The firing order is arranged to have power impulses equally spaced, and from the point of view of balancing.*

Firing orders for various engines are given below :

No. of cylinders	**Firing order**
Two	1, 2
Three	1, 3, 2
Four	1, 4, 3, 2 or 1, 3, 4, 2
Six	1, 5, 3, 6, 2, 4 or 1, 4, 2, 6, 3, 5
Eight	1, 6, 2, 5, 8, 3, 7, 4 or 1, 8, 7, 3, 6, 5, 4, 2.

Eight (Vee) : Either of the following alternatives.

(A) 1*L*, 1*R*, 4*L*, 4*R*, 2*R*, 3*L*, 3*R*, 2*L*

(B) 1*L*, 4*R*, 2*R*, 2*L*, 3*R*, 3*L*, 4*L*, 1*R*

(C) 1*L*, 3*L*, 2*R*, 4*R*, 3*R*, 2*L*, 4*L*, 1*R*

(D) 1*L*, 4*R*, 4*L*, 2*L*, 3*R*, 3*L*, 2*R*, 1*R*

(E) 1*L*, 3*L*, 3*R*, 2*L*, 2*R*, 1*R*, 4*L*, 4*R*

L and *R* indicate cylinder on left and right hand side respectively. The firing order for a four-stroke engine with its cylinder numbered consecutively from 1 to *n* will be 1, 3, 5, 7, ... to *n* for one revolution of the crankshaft and 2, 4, 6, 8, ... to (*n* – 1) for the next revolution.

10.15. IGNITION TIMING

Ignition timing is *the correct instant for the introduction of spark near the end of compression stroke in the cycle. The ignition timing is fixed to obtain maximum power from the engine.*

The correct instant for the introduction of a spark is mainly determined by the *ignition log.*

There are several *factors which affect the ignition timing of an engine.* Some of them are enumerated and discussed below :

1. Compression ratio
2. Engine speed and load
3. Quality of fuel used
4. Mixture strength
5. Engine temperature
6. Cylinder bore.

Besides the above factors the ignition timing of an engine is also affected by the following :

(*i*) Volatility of fuel.

(*ii*) Combustion rate of fuel.

(*iii*) Method of distribution of fuel to the cylinders.

(*iv*) Excessive carbon deposits in the combustion chamber.

(*v*) Design and shape of the combustion chamber.

(*vi*) Location of spark plugs in the combustion chamber.

(*vii*) Throttle opening.

1. **Compression ratio.** If the compression ratio is *low*, combustion will be slower (due to lesser temperature and pressure obtained during compression stroke), *more ignition advance* is required. And if the compression ratio is *high*, faster combustion will take place due to higher temperature and pressure obtained during the compression stroke which helps in complete combustion of the high pressure gases, *less ignition advance* is needed.

2. **Engine speed and load.** At *low* engine speeds and load, the combustion of fuel in the engine cylinder is slower and for complete combustion *more ignition advance* is required. But at *low engine speeds and full load*, the combustion is comparatively faster and *less ignition advance* is needed. And at *high engine speeds and full load, the combustion is slower and more ignition advance is required.*

3. **Quality of fuel used.** Depending upon the octane value of fuel used, for *low octane* value of fuel, the combustion rate is faster, thus *less ignition advance* is needed whereas for *high octane* value of fuel, combustion rate is slower, thus needing *more ignition advance.*

4. **Mixture strength.** In the case of *lean mixture* (air/fuel) the combustion rate is faster and the *less ignition advance* is required. And in the case of *rich mixture* the combustion rate will be slower thus requiring *more ignition advance.*

5. **Engine temperature.** The combustion in the engine cylinders depends upon the engine temperature. If the engine is hot, the combustion will take place at faster rate than when the engine is cold. *For faster combustion rate less ignition advance is required and vice versa.*

6. **Cylinder bore.** The *larger the cylinder bore* (diameter) slower the combustion rate and thus *more ignition advance* will be needed. The *smaller the cylinder diameter* (bore) faster the combustion rate and thus *less the ignition advance* will be required.

- When ignition occurs early in the compression stroke, the engine is said to be *advanced*. A *retarded ignition takes place when the piston is just near the compression stroke.*
  - — If the ignition is **advanced to much**, it will be complete before the end of the compression stroke. Under these conditions, the crankshaft and connecting rod will have to push the piston upward compressing the gases. In such a case the force might not be sufficient to overcome the pressure and the *engine would stop or stall.*
  - — If the ignition is **retarded too much**, the combustion of fuel will continue, during the power stroke of the piston (working stroke) and the *maximum pressure will not be developed and less work will be obtained from heat energy. Under these conditions comparatively more exhaust gases will be going out of the engine cylinder overheating the exhaust valve.*

Spark Advance Mechanisms :

It is of significance importance that the point in the cycle where the spark occurs must be regulated *automatically to ensure maximum power and economy at different loads and speeds.* Most of the engines are fitted with a mechanism which is integral with the distributor and automatically regulates the optimum spark advance to account for change of load and speed.

There are two general methods used in modern engines to advance and retard the ignition timing automatically in relation to engine speed and operating condition after initial timing is set manually, most manufacturers call these methods as "*Automatic advance mechanisms*".

1. Centrifugal spark advance mechanism.
2. Vacuum spark advance mechanism.

1. **Centrifugal spark advance mechanism :**

- The centrifugal spark advance is essential to compensate for the increase in speed of the engine.
- This mechanism consists of *two flyweights* (*advance weights*), *cam, spring,* and *base plate* (Fig. 10.43). The flyweights are carried by the distributor drive shaft through the base plate which is fixed to the drive shaft. The flyweights are pivoted on the base plate, and attached to the cam through springs. The cam is connected with the distributor drive shaft through the springs, flywheel and base plate.

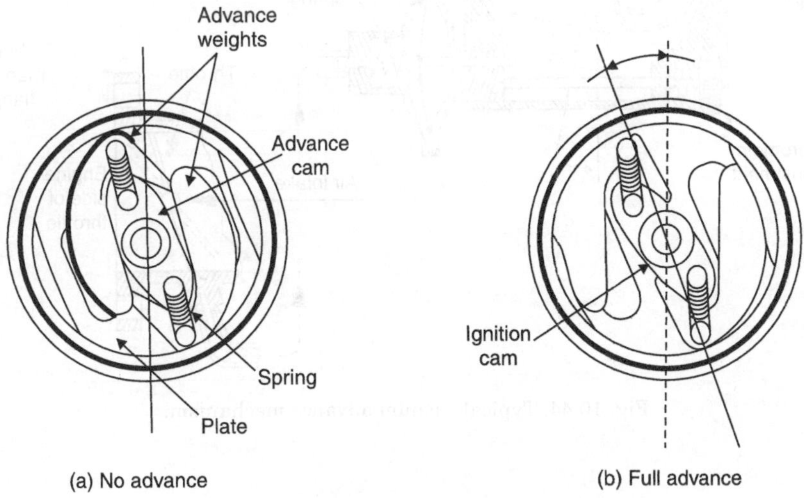

Fig. 10.43. Centrifugal spark advance mechanism.

— When the engine speed increases, the weights move out due to centrifugal force rotating the plate and cam in the anticlockwise direction. This movement affects the desired advance. Thus *timing of the spark varies from no advance at low speed to full advance at high speed.*

2. **Vacuum spark advance mechanism :**

During part throttle operation, a small amount of fuel-air mixture is sucked in and the compression pressure is low in the engine. With lower pressures, the mixture does not burn as quickly as it should. In order to obtain maximum efficiency under such conditions, the spark should be advanced more than that obtained by the centrifugal governor during low engine speed. The additional ignition advance is obtained by means of *vacuum advance mechanism.*

The 'vacuum advance mechanism' contains a spring loaded diaphragm which is connected with the distributor by means of a link and damp. The diaphragm is contained in a housing, the spring loaded side of which is air tight and is connected with the carburettor by means of pipeline. Connection at the carburettor is made at a point on the atmospheric side of fully closed throttle so that engine suction does not affect the diaphragm during idling speed when additional advance is not required.

When throttle is opened, suction of the engine affects the diaphragm through the suction line which causes the diaphragm to deflect. The movement of the diaphragm is transmitted to the distributor through the linkage. In some cases, the distributor is rotated, on its mounting while in other cases, the breaker plate is moved which results in ignition advance.

When distributor body is rotated *opposite to the rotation of rotor* ignition is *advanced*, while when the body is rotated in the *direction of rotation of motor*, ignition is *retarded.*

Fig. 10.44 shows a typical vacuum advance mechanism.

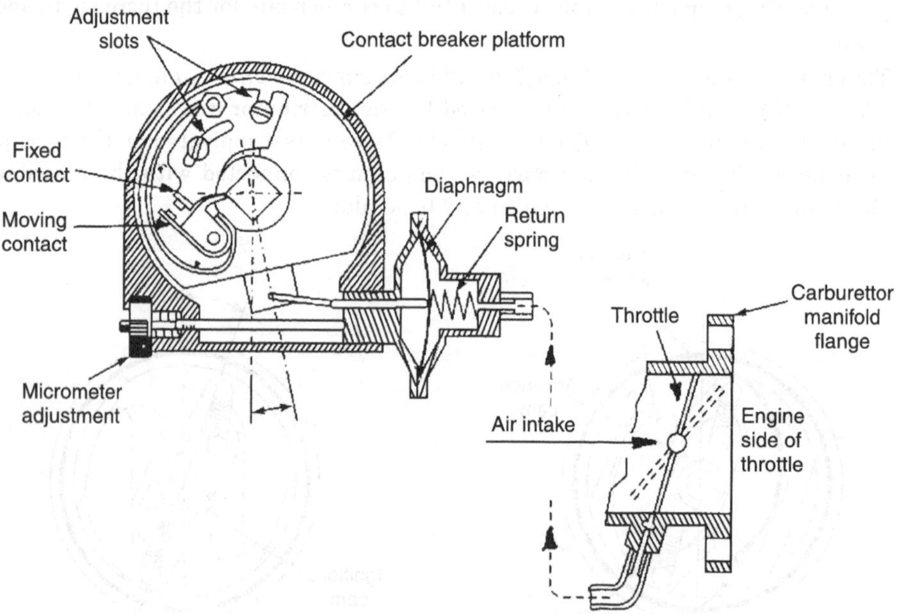

Fig. 10.44. Typical vacuum advance mechanism.

Combined centrifugal and vacuum advance mechanism :

It is seen that the '*centrifugal advance mechanism*' *takes much care of* "*speed only*" and *not* the load conditions, whereas the '*vacuum advance mechanism*' takes much care of "*load conditions*". Therefore, a combination of the two mechanisms is applied on the distributor. In this design, part of the ignition advance is due to the centrifugal force and part due to the vacuum produced in the intake manifold. The combination of the two mechanisms gives the practically perfect-spark timing for all driving conditions.

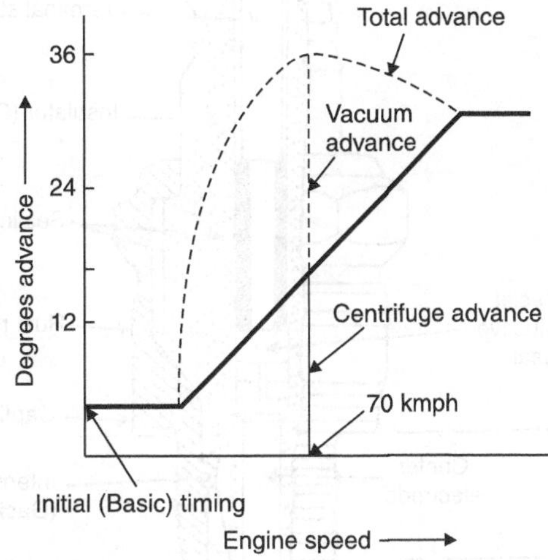

Fig. 10.45. Centrifugal and vacuum advance curves.

Fig. 10.45 shows the functioning of vacuum and centrifugal governor at various speeds.

10.16. SPARK PLUGS

- *The main* **function** *of a spark plug is to conduct the high potential from the ignition system into the combustion chamber. It provides the proper gap across which spark is produced by applying high voltage, to ignite the mixture.*
- A spark plug entails the following *requirements* :
  (*i*) It must withstand peak pressures up to atleast 55 bar.
  (*ii*) It must provide suitable insulation between two electrodes to prevent short circuiting.
  (*iii*) It must be capable of withstanding high temperatures to the tune of 2000°C to 2500°C over long periods of operation.
  (*iv*) It must offer maximum resistance to erosion burning away of the spark points irrespective of the nature of fuel used.
  (*v*) It must possess a high heat resistance so that the electrodes do not become sufficiently hot to cause the preignition of the charge within the engine cylinder.
  (*vi*) The insulating material must withstand satisfactorily the chemical reaction effects of the fuel and hot products of combustion.
  (*vii*) Gas tight joints between the insulator and metal parts are essential under all operating conditions.
  — The spark plug (Fig. 10.46) consists of a metal shell having two electrodes (oftenly made of nickel alloy with alloying elements of tungsten silicon or chromium) which are insulated from each other with an air gap. High tension current jumping from the supply electrode produces the necessary spark.

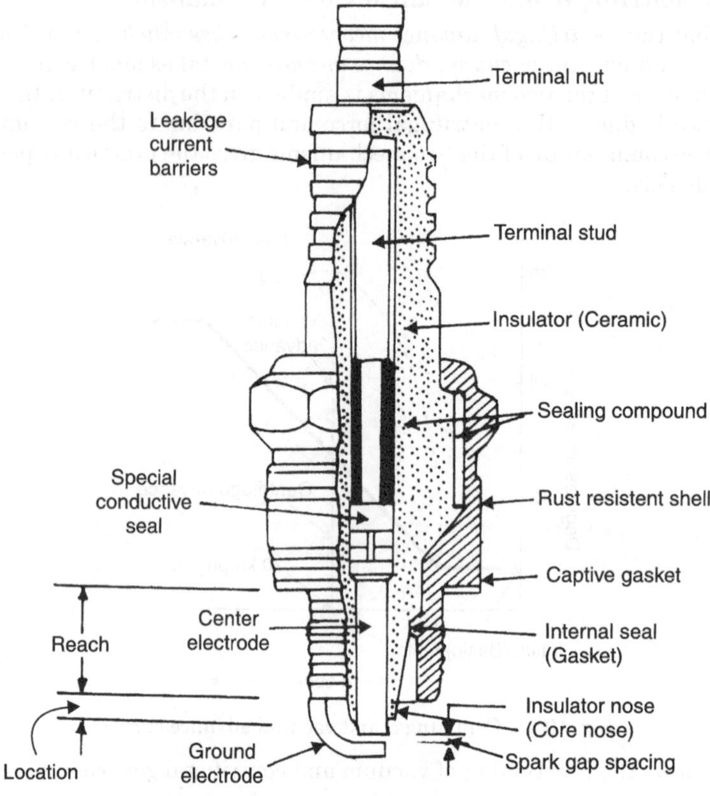

Fig. 10.46. Spark plug.

— *Plugs are sometimes identified by the heat range or the relative temperature obtained during operation. The correct type of plug with correct width of gap between the electrodes are important factors.*

— The spark plug gap can be easily checked by means of a *feeler gauge* and set as per manufacturer's specifications.

— It is most important that while adjusting the spark plug it is the outer earthed electrode *i.e.*, tip which is moved in or out gradually for proper setting of the gap. No bending force should be applied on the centre-electrode for adjusting the gap as this can cause crack and fracture of insulation and render the plug absolutely useless.

— *Porcelain* is commonly used as insulating material in spark plugs, as it is cheap and easy to manufacture. Mica can also be used as insulating material for spark plugs. Mica, however, cannot withstand high temperatures successfully.

Factors affecting establishment of arc across the air gap of the spark plug :

The problem of establishing an arc across the air gap of the spark plug is *affected by the following factors* :

(i) Air gap length. Greater the air gap, greater is the breakdown voltage.

(ii) Gap geometry. For pointed small electrodes less breakdown voltage is required.

(iii) Mixture density. High densities (*i.e.*, high throttle openings) require higher breakdown voltages. It follows the **Paschen's law** ; $V = K\rho h$, where K, ρ and *h* are a constant (for given substance and electrode shape), density of the material and gap distance, respectively.

(iv) Electrode temperature. High temperatures lead to low breakdown voltage.

(v) Insulator's leakage resistance. The carbon and metallic oxides from electrically-conductive coating on the insulator which, thus, shunt the secondary winding and reduce the maximum voltage that the secondary can impress across the spark plug.

(vi) Voltage increase rate at the gap. When the voltage is built up at a rapid rate by the ignition system, the effect of leakage gets minimized, and greater voltage is available.

(vii) Fuel-air ratio. The fuel-air ratio fixes the electrical properties of the mixture ; lean mixtures have higher breakdown voltage than those of slightly rich mixtures.

(viii) Electrode material.

(ix) The presence of ionised gases in the gap.

Difference between hot spark plug and cold spark plug. Refer Fig. 10.47

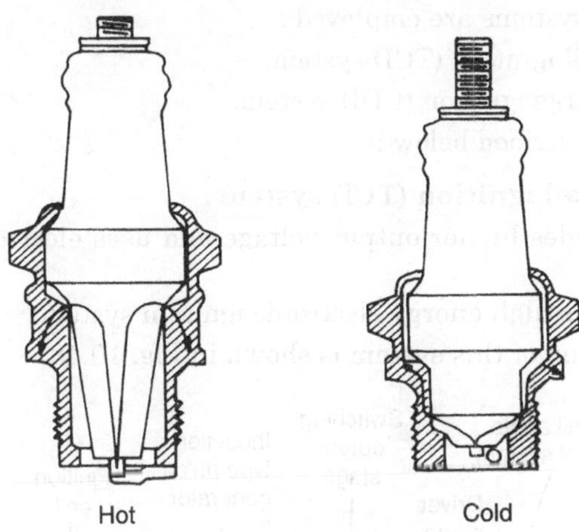

Fig. 10.47. Hot and cold spark plug.

The difference between hot and cold spark plugs is due to the *relative operating temperature range of the tip of the high tension electrodes*. The operating temperature is governed by the amount of heat transferred which depends upon the heat transfer path from the tip to the cylinder head and the amount of surface area exposed to the combustion gases. A **cold plug** *has a short heat transfer path and a small area exposed to the combustion gases, as compared to a* **hot plug**.

Spark plug failure :

Spark plug will fail in its function due to the following *reasons* :

1. Plug fouled by too rich mixture.
2. Plug fouled by engine oil entering combustion chamber.
3. Plug badly covered with carbon from poor ignition.
4. Plug gap incorrect.
5. Burned electrodes or broken lower insulator caused by overheating.
6. Red, brown or yellow oxide deposits on plug interior that short the insulator.
7. Accumulation of dirt or moisture on the outside of the insulator that short the plug by grounding the high voltage.
8. Cracked or broken insulator sealing.

10.17. LIMITATIONS OF CONVENTIONAL IGNITION

The conventional ignition system entails the following *limitations* :

1. Frequent servicing and replacement of contact points.

2. Weaker sparks (or sometimes misfiring).
3. Inefficient at low speeds.
4. Poor starting ability.
5. Reduced ability to fire fouled plugs (since most of the spark energy is lost as current flows through low-resistance fouling deposits).
6. Poor reproducibility of the secondary voltage rise, peak voltage and firing time (owing to inherent erratic operation of the mechanical points).

10.18. ELECTRONIC IGNITION SYSTEMS

The limitations / disadvantages of conventional ignition system has led to the development of new ignition system concepts using *solid state devices*. In modern automobiles the following two types of electronic ignition systems are employed :

1. Transistorised coil ignition (TCI) system.
2. Capacitive discharge ignition (CDI) system.

These systems are described below :

1. **Transistorised coil ignition (TCI) system :**

- TCI system provides higher output voltage and uses electronic triggering to keep the required timing.
- It is categorised as high energy electronic ignition system.
- The circuit diagram of this system is shown in Fig. 10.48.

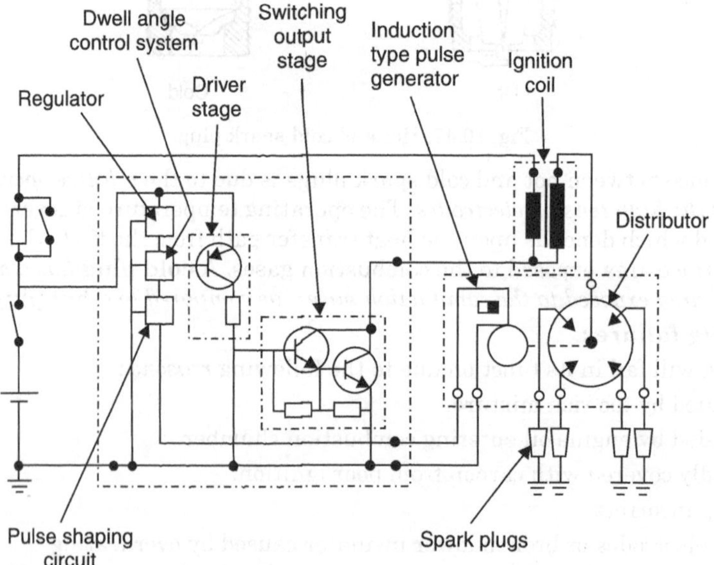

Fig. 10.48. Circuit diagram of TCI system.

— The function of *magnetic pulse generator* (*more oftenly used pulse generator*) *system is to detect the position of the distributor shaft and transmit an electrical pulse to an electronic control model.* This system replaces the contact breaker and the cam assembly of the conventional ignition system.

— The **module** switches off the flow of current inducing a high voltage in the secondary winding which is distributed to the spark plugs in the same fashion as in conventional breaker system.

— **Timing circuit** is contained in the **control model** that closes the primary circuit later so that the primary current build-up can occur for the next cycle.

Advantages :

The advantages of TCI system are :

1. Reduced wear of the components.
2. No problem faced in burning lean mixtures.
3. Reduced maintenance of the ignition system.
4. Increased reliability.
5. Increased spark plug life.

2. **Capacitive Discharge Ignition (CDI) system :**

Fig. 10.49 illustrates the capacitive discharge ignition system.

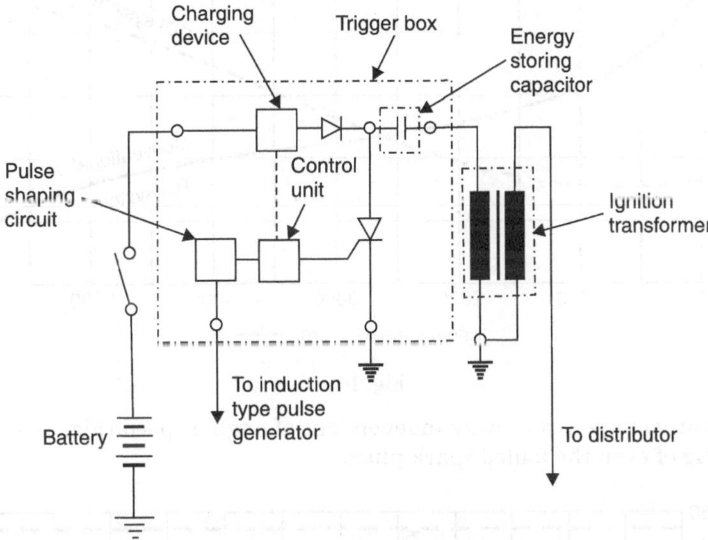

Fig. 10.49 Circuit diagram of CDI system.

- In this system a **capacitor** is used to store the energy, induction coil is *not* used. The magnitude of stored energy depends on the capacitance and the charging voltage of the capacitor.
- The primary voltage developed at the time of spark by the discharge of capacitor is boosted up by the **ignition transformer,** through the thyristor, to the high voltage required at the spark plug.
- The CDI **trigger box** comprises the *capacitor*, *thyristor power switch*, *charging device* to convert battery voltage to the charging voltage of 300 to 500 V, by means of pulses *via*, the voltage transformer, pulse shaping unit and control unit.

Advantages :

The *advantages of CDI system are* :

1. The capacitor in this system can store several thousands times more energy per unit of capacitance compared to that per unit inductance of the conventional TCI system. This permits a high output voltage even at high spark rates (speeds).

2. Due to the internal resistance being very small (about 50 Ω), faster voltage rise is obtainable and consequently the system is *relatively insensitive to side tracking.*

3. As in TCI system, the breaker points serve as a trigger only. This avoids frequent maintenance and increases life of the contact points.

4. As evident from Fig. 10.50, at low engine speeds in CDI systems low current is drawn from the battery and with increase in speed the current drawn also increases, unlike what happens in conventional or TCI system where high current is drawn at low speeds. This characteristic of CDI system *increases the low speed efficiency and allows easier cold starting.*

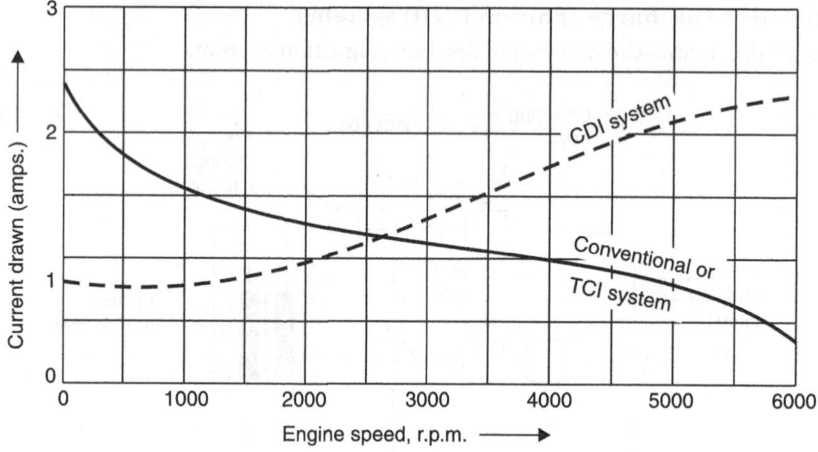

Fig. 10.50

5. The output voltage is relatively independent of engine speed (Fig. 10.51), this eliminates chances of misfiring of even the fouled spark plugs.

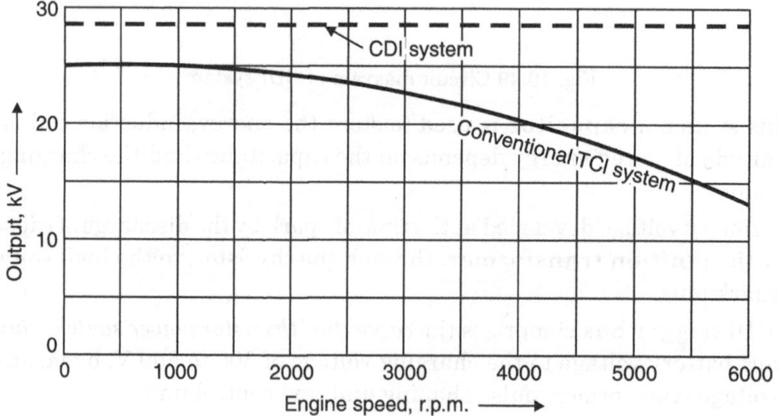

Fig. 10.51

Disadvantages :

The main *disadvantage* of CDI system is very rapid discharging of capacitor leading to strong spark but of *short duration* of about 0.1 ms. This may *cause ignition* problems when operating with rich mixtures.

Advantages of Breakless Electronic Ignition System (E.I.S.) :

Following are the *advantages of breakless electronic ignition system* :

1. Better starting with cold engine.
2. Longer life due to stable timing.
3. Less possibility of sparking at spark plug.
4. Longer life of spark plug, due to less electrode wear.
5. Periodic checks of ignition timing, are no longer required, once the basic timing has been set.
6. No change in ignition timing owing to absence of traditional mechanical devices such as C.B. circuit breaker points, cam, sliding block etc. which are exposed to wear and tear.
7. Produces a high tension peak (maximum output voltage) which is attained in a very short time and with a very limited discharge, resulting in a regular running, even with dirty/foul spark plug and wider plug gap.

Precautions to be taken with transistorised ignition system :

Following *precautions should be taken with transistorised ignition system* :

1. Keep the air gap of spark plug accurate as specified as too much gap will build up high voltage in the system and may puncture distributor cap.
2. Never test high voltage current from high tension lead keeping it at a far distance—keep it close to plug otherwise it may puncture the insulation of the H.T. lead.
3. Loose or bouncing points may be avoided.
4. Loose wire connection will cause ignition off and on at odd times.
5. Never use booster battery of high voltage.
6. Never keep the system's charging rate high.
7. Never run the vehicle with loose battery connection.
8. Never change the polarity of the system.
9. Avoid running the system with defective voltage regulator.

10.19. TROUBLESHOOTING OF IGNITION SYSTEM

The troubleshooting of ignition system and spark plug one discussed below (tabular form) :

A. Ignition system :

S.No.	Troubles	Causes	Remedies
1.	*Engine misfires on one cylinder.*	(*i*) Detached high tension cable. (*ii*) Broken wires in cable. (*iii*) Faulty spark plug. (*iv*) Large gaps in spark plugs.	(*i*) Attach high tension cable. (*ii*) Replace the wires. (*iii*) Replace the spark plug. (*iv*) Adjust the gap to the specified value.
2.	*Engine misfires irregularly on all cylinders.*	(*i*) Loose L.T. connection. (*ii*) Loose contact screw. (*iii*) Dirt on moisture on distributor or ignition coil.	(*i*) Tighten L.T. connection. (*ii*) Tighten contact screw. (*iii*) Clean the distributor or ignition coil and remove the dirt or moisture.

		(iv) Gaps of spark plug are more. (v) Leaking plugs. (vi) Worn out distributor brush and segments.	(iv) Adjust the gaps to the specified value. (v) Change the plugs. (v) Replace them.
3.	*Engine does not start. There is also no spark from the coil.*	(i) Discharged battery. (ii) Circuit is open in L.T. wiring. (iii) Dirty contacts. (iv) Loose contact screw. (v) Defective condenser. (vi) Defective ignition coil.	(i) Charge the batter. (ii) Close the circuit. (iii) Clean the contacts. (iv) Tighten it. (v) Replace it. (vi) Replace it.
4.	*Engine gets unduly hot.*	(i) Retarded ignition lever. (ii) Automatic governor not moving freely. (iii) Leaky plugs. (iv) Incorrect timing.	(i) Adjust ignition lever to the correct position. (ii) Clean the parts of the governor. (iii) Replace them. (iv) Adjust for correct timing.

B. Spark plug :

S. No.	Troubles	Causes	Remedies
1.	*Badly worn or burnt electrodes.*	(i) Incorrect gap. (ii) Incorrect distributor contact breaker points gap. (iii) Timing out. (iv) Plug has rendered its useful life. (v) Hot running conditions in engine.	(i) Reset gap correctly. (ii) Reset gap. (iii) Re-time. (iv) Replace. (v) Replace with cooler plug.
2.	*External insulator broken.*	(i) Careless handling.	(i) Install new plug.
3.	*Internal insulation cracked, broken at lower end.*	(i) Engine running hot. (ii) Carelessness in adjusting gap.	(ii) Install cooler plug. (iii) Adjust gap by bending the side electrode only.
4.	*Glassy or blistered internal insulator.*	(i) Engine running hot. (ii) Timing out.	(i) Install cooler plug. (ii) Re-time.
5.	*Dry black soot on internal insulator.*	(i) Engine running cold. (ii) Distributor contact breaker points gap incorrect.	(i) Install hotter plug. (ii) Reset gap.
6.	*Oil plug.*	(i) Timing out. (ii) Distributor contact breaker points gap incorrect.	(i) Re-time. (ii) Reset gap.

IV. LIGHTING AND ACCESSORY SYSTEM

10.20. INTRODUCTION TO LIGHTING AND ACCESSORY SYSTEM

The lighting and accessory system consists of the following :

Lighting system :

1. Head light
2. Side light
3. Rear light
4. Fog lamps
5. Number plate illumination lamp
6. Interior lights
7. Indicator flashers.

Accessories :

1. Horns
2. Wind screen wiper
3. Electric petrol pump
4. Petrol gauge
5. Radio sets
6. Cigar lighter
7. Heater
8. Wind screen defroster
9. Signalling devices.

10.21. MAIN CIRCUITS OF THE AUTOMOBILE ELECTRICAL SYSTEM

Fig. 10.52 shows a simplified lighting and accessory system. It is made up of many small circuits that *operate the lights and accessories like the headlight, spotlight, indicators, instrument panel lamps, gauges and horn etc.* The system also includes a lot of wiring that connect there devices to a power source and many control switches that operate them.

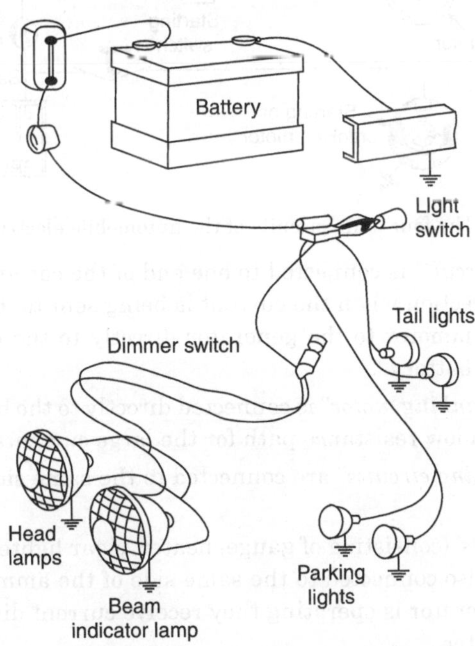

Fig. 10.52. Lighting and accessory system.

The electrical system of a car performs the following two *functions* :

(*i*) To supply electrical energy to start and operate the engine.

(*ii*) To provide the power to operate the lights, instruments and other electrical accessories.

The modern car has about 60 metres of wire joining its electrical components.

The electrical system is usually divided into the following *four main circuits* (Fig. 10.53).

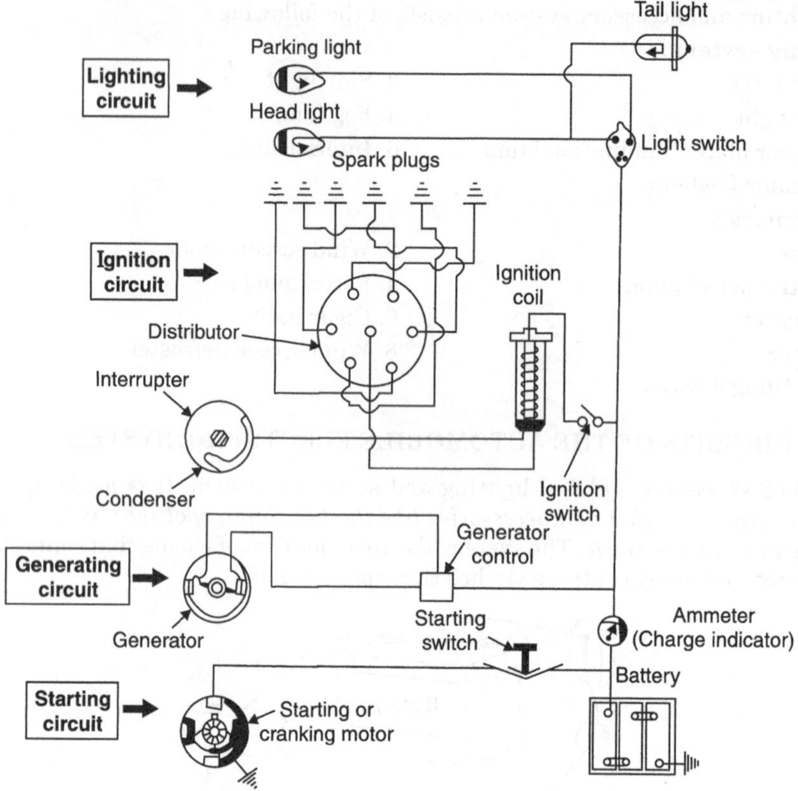

Fig. 10.53. The four main circuits of the automobile electrical system.

— The "*generating circuit*" is connected to one end of the car ammeter so that meter registers the charge direction when the current is being sent to battery. A wire connects the other end of the ammeter to the generator directly to the cable leading to the underground pole of the battery.

— The "*starting or cranking motor*" is connected directly to the battery through cables and a switch to provide a low resistance path for the large currents required by the motor.

— "*Ignition and lighting circuits*" are connected to the same side of the ammeter, as is the generator.

— The '*branch circuits*' (consisting of gauge, heater, cigar lighter, wind-shield wiper, signal devices, etc.) are also connected to the same side of the ammeter as is the generator, so that when the generator is operating they receive current directly from it without going through the ammeter.

When the generator is not running, these circuits draw current directly from the battery or through the ammeter which reads the discharge direction. When the demands of circuits exceed generator output, the extra current is supplied by battery directly or through the ammeter.

When the engine is being cranked for starting, the battery supplies current for the operation of the starting motor and ignition system. It also supplies current for light, radio, heater and several other accessory units when the generator is not operating fast enough to handle the electrical load. Thus the *battery circuit is the nerve centre of the whole system. The engine cannot be started without the starting motor, since only the battery circuit can supply the required large amount of current for the purpose.*

- Fig 10.54 shows simple wiring diagram of the electrical system of an automobile showing all the circuits. The actual diagram is, however, rather complicated.

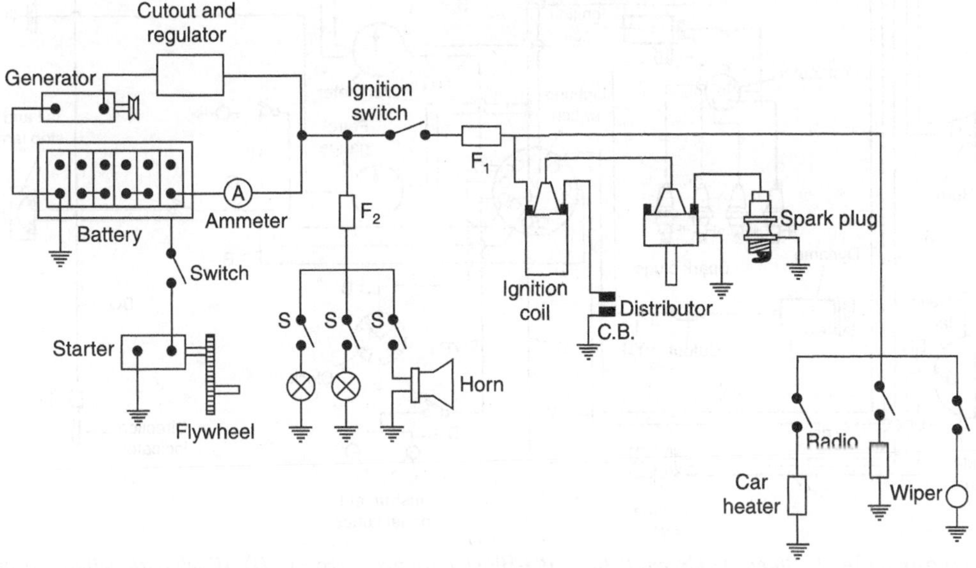

Fig. 10.54. Simple wiring diagram of the electrical system of an automobile.

- Fig. 10.55 shows a typical car wiring diagram.

Various circuits in the wiring diagram are given with different colour codings. The complete wiring diagram for the ignition circuit, charging circuit, starting circuit, lighting circuit and the circuit for accessories are indicated clearly. Whenever some repair has to be undertaken in the wiring circuits, the wiring diagram of the car should be followed, the fault can then be easily located and rectified.

Symbols used in electrical diagrams

Fig. 10.56 shows the symbols used in electrical diagrams.

Cable colours

The colours used by Lucas for wiring cables is given below :

Brown (battery circuit). The following are *coloured "brown"* :

— The lines from the battery to the ammeter and to control box ;

— Connection to the lighting and ignition switches from the control box terminal ;

— The line from the starter switch to the electric clock ;

— The line from the starter switch to the inspection sockets and battery auxiliary fuse ;

— The lines to the electric horns, cigarette lighter and interior lights.

White (ignition circuit). When the ignition is switched on, the electric petrol pump, solenoid shorter switch etc., all come into operation. All these wires are *"white"* in colour.

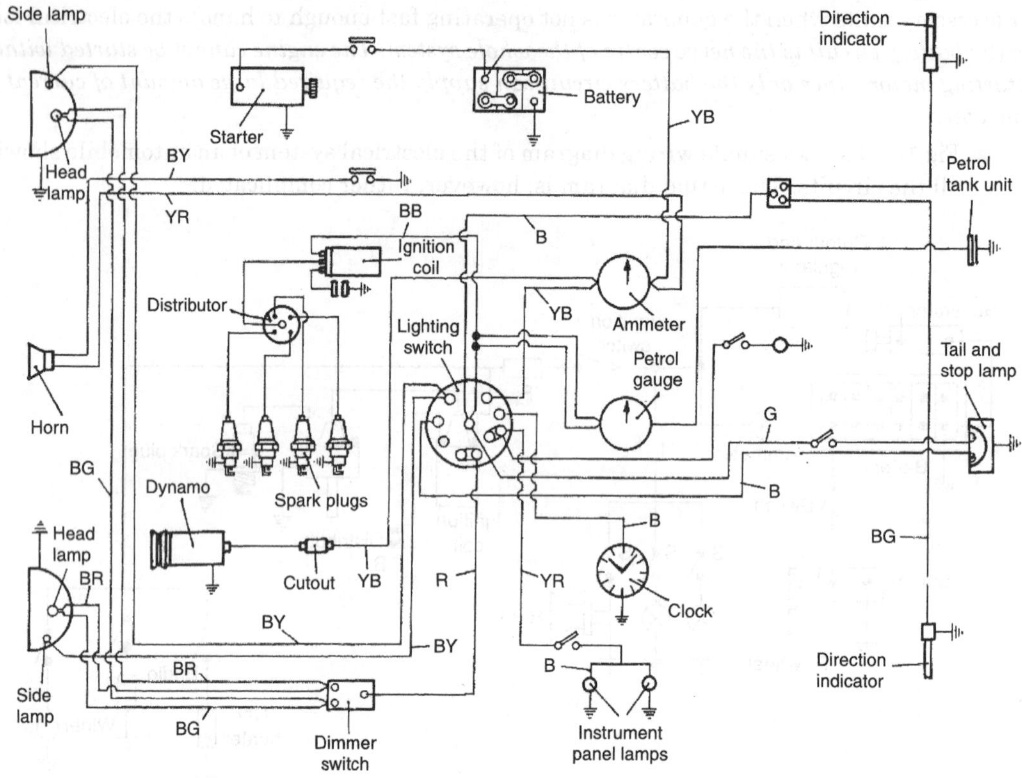

Cable colour code : *B-Black ; G-Green, R-Red ; BG-Black with green tracer ; BY-Black with yellow tracer ; BB-Black with blue cross tracer ; YR-Yellow with red tracer ; YB-Yellow with black tracer ; BR-Black with red tracer.*

Fig. 10.55. Typical car wiring diagram.

Yellow (dynamo circuit). The line from the terminal to the regular unit control box terminal and to ignition warning light is *"yellow"* in colour. This applies to the complete dynamo circuit.

Green (auxiliary circuits). Auxiliary circuits such as the stop lamp, fuel gauge, direction indicators, wind screen wipers, etc. and all *"green"*. These units start operating when the ignition switch is on.

Blue (head lamp circuit). The lines connecting the head lamps from the lighting switch are *"blue"* in colour.

Red (side and tail lamp circuits). The following are coloured *"red"* :

— The line from the panel light switch to the side head light and dip switch control ;

— The line connecting the side lights and rear lights ;

— Fog lamps, panel lights and other lamps.

Black (earth circuits). If a components is not earthed internally a cable is connected outside to metal chassis. This cable is given a *"black" colour*.

— The earthings of the fuel tank, rheostat for the fuel gauge, cigarette lighter, control box and ignition warning light are all *black* in colour.

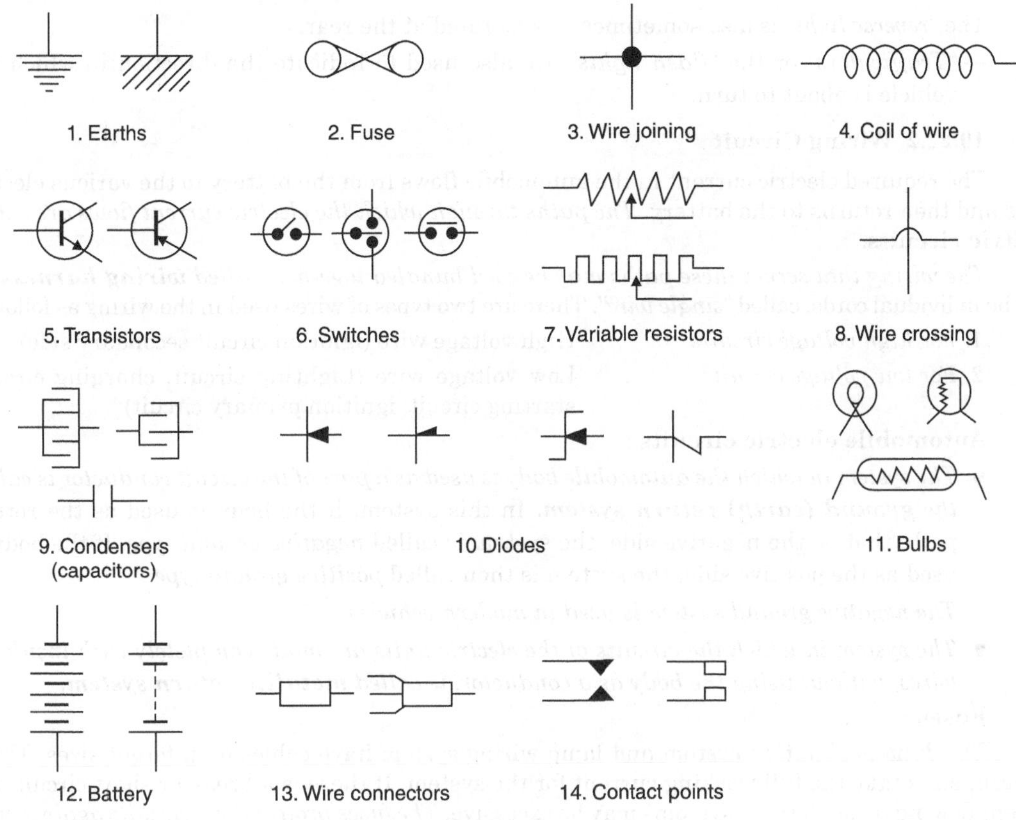

Fig. 10.56. Symbols used in electrical diagrams.

10.22. LIGHTING SYSTEM

10.22.1. Introduction

The lighting system of an automobile, as earlier stated, consists of *head lamps, side, tail, stop and reverse lamps.* In addition to this there are *instrument panel lights, direction indicator flash lights and the lights inside the body to light up passenger compartment.*

— The '*head lights*' enable the driver to drive at reasonable speeds at night with safety, but provision should also be made that the drivers of other vehicles coming from the opposite direction *do not experience a glare.*

— '*Side lights*' are the low intensity parking lamps.

— '*Tail lights*' are meant for illuminating back of the car so that the driver of the vehicle coming behind it will be able to see it. These are covered by red lenses so that they can be distinguished clearly by reflection of light from the automobile approaching from the rear, even when these are not on.

Side and tail lights help to see the vehicle and estimate its width. Further their colour serves to identify the front (white) or rear (red) of the vehicle.

— For licence plate there is usually a *separate light.*

— "*Stop lights*" are at rear and flash when brakes are applied.

The "*reverse light*" is also sometimes incorporated at the rear.

— "*Trafficators*" or the "*flash lights*" are also used to indicate the direction in which the vehicle is about to turn.

10.22.2. Wiring Circuit

The required electric current in the automobile flows from the battery to the various electric parts and then returns to the battery. *The paths through which the electric current flows are called* **electric circuits.**

The wiring that serves these paths may be cord bundled together, called ***wiring harness***, or may be individual cords, called "*single wire*". There are two types of wires used in the wiring as follows :

1. *For high voltage circuits* ... High voltage wire (ignition circuit secondary side)
2. *For low voltage circuits* ... Low voltage wire (Lighting circuit, charging circuit, starting circuit, ignition primary circuit)

Automobile electric circuits :

- *The system in which the automobile body is used as a part of the circuit conductor is called the* ***ground (earth) return system.*** In this system, if the body is used as the return path, that is, the negative side, the system is called *negative ground type.* If the body is used as the positive side, the system is then called *positive ground type.*

  The negative ground system is used in modern vehicles.

- *The system in which the circuits of the electric units are made completely with insulated wires, without using the body as a conductor, is called* ***metallic return system.***

Fuses :

The dynamo, ignition system and lamp wiring system have cables of different sizes. These different sizes take the full working current for the system. If there is a break or short circuit, the current flowing through these systems may be excessive. *The fuses protect these circuits against such excessive currents* (when current exceeding the allowable value flows through the circuit, the fuse heats up and melts, thereby cutting off the flow of current to the electric units).

The fuses are of *cartridge type.* The length of a fuse is very small. It is made of very fine lead alloy. It fuses at a lower current value than that of the circuit system. The fuse wire is soldered with the nickel-plated brass end caps. The whole unit is enclosed in a glass tube.

Fig. 10.57 shows a blown fuse. To replace the fuse, first switch off the current. Then remove the blown fuse from its spring clips and put a new fuse in its place.

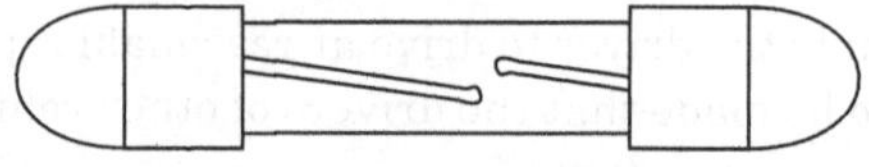

Fig. 10.57. Blown fuse.

— Auxiliaries such as *horn* are operated when the ignition switch is off. Such auxiliaries are protected by a type of fuse.

— Auxiliaries such as the ignition, gauges etc., etc. are operated only when the ignition switch is on and these auxiliaries are protected by another type of fuse.

— The starting motor circuit is *not* provided with a protective fuse.

The values of the fuses used in small cars is about 35 A.

Wires :

An electrical wire cable should have the following *properties* :

(*i*) Least electrical resistance.

(*ii*) Effective insulation throughout its life.

(*iii*) Flexibility so that it may be accommodated in the available spaces.

The electrical wire usually consists of a number of strands of copper or aluminium. A single conductor is never used because that would be comparatively less flexible and will easily undergo fatigue with continuous bending. On the other hand in a multi-strand wire even if some strands are damaged, the remaining strands would still share the electrical load.

Electrical wires are commonly *specified by the number and size of the strands.*

Example. 7/36 means seven strands of 36 gauge.

- For quick identification, insulation of various wires/cables in a circuit are assigned different colours. Various colour codes are in use for vehicle wiring.

10.22.3. Head Lights

The head light fitted to vehicle should be good enough to illuminate the road so that the driver can see the road clearly even in dark nights from a quite far distance. At the same time light beam should not be strong as to make other drivers coming from opposite side blind. At the same time there should be back light and stop light also emitting enough light so that the following vehicle can see the vehicle moving on the road. Apart from this there should be light inside the vehicle and in the dashboard so that instrument and gauges can be read.

When a vehicle is moving, the head lamp provides light on the road. The head lamp is in the form of a parabola (the best light beam from a lamp is obtained by the use of a parabolic reflector). There is a *focal point* at the rear of the parabolic reflector at which the light rays from the lamp are picked up by the polished surface of the reflector and directed in *parallel lines to give a beam with circular cross-section.* Any other position of the lamp will not give the light as confined beam but will scatter the light (see Fig. 10.58) where at *(i)* ***bulb is positioned at focal point (F) so it gives the corrected light rays.*** At *(ii)* bulb is placed behind the focal point so light rays are *scattered.* At *(iii)* bulb is placed ahead of focal point and the effect is that, rays are *intersecting* each other.

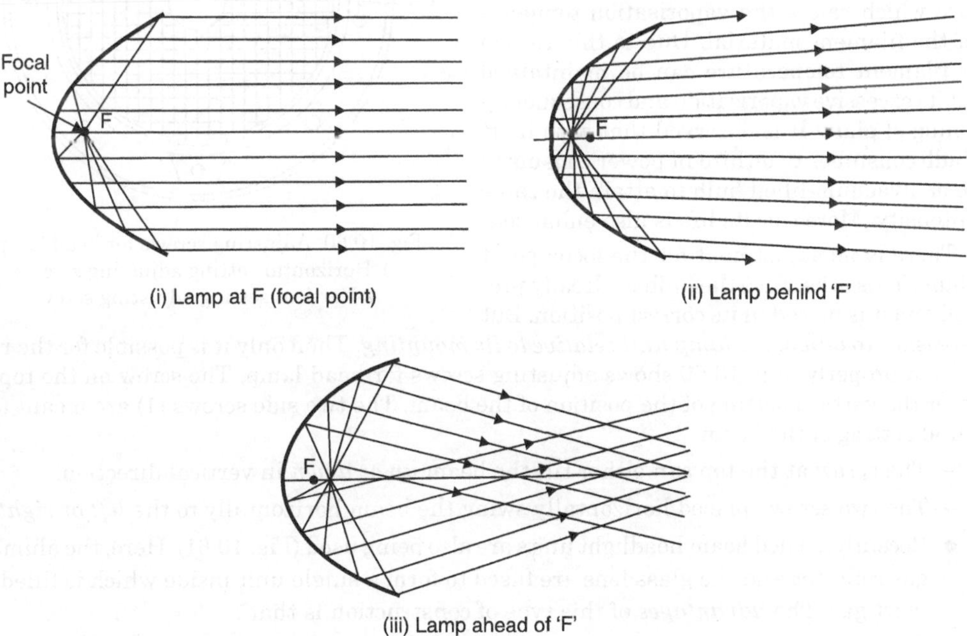

Fig. 10.58. Effect of lamp position in parabolic reflector.

Prismatic lens over the reflector is placed so that *(i)* light rays are properly distributed on the road for correct illumination, *(ii)* dirt cannot find its way to highly polished reflector, otherwise it will leave scoring mark resulting into dim light. The effect of the prismatic lens fitted to a parabolic reflector is shown in Fig. 10.59. The lens bends the parallel rays from the reflector so that light is evenly distributed on the road. The vertical flutes of the lens spread the light rays so that the beam is flattened with the edges thrown out towards the side of the high-way.

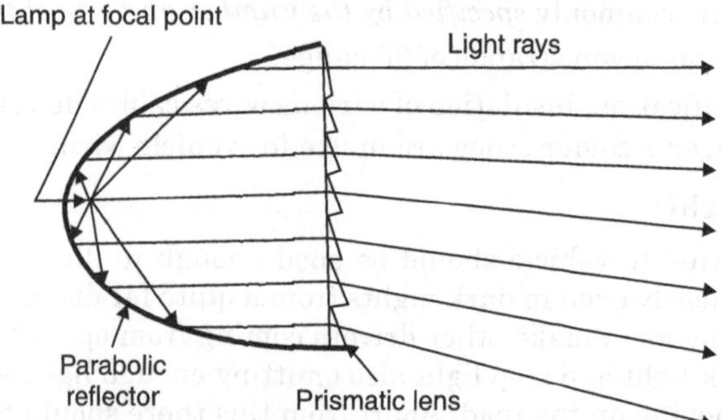

Fig. 10.59. Rays of light from headlight distributed by prismatic lens.

Pre-focus bulb head lamp. The headlights used in Indian vehicles consist of a combined aluminised metal reflector and front lens assembly. The *prefocussed bulb* is fitted in a sleeve attached to the rear of the reflector to ensure preset positioning of the filaments with respect to the focal point of the reflector. The bulb is fitted with some inert gas such as argon or nitrogen at some pressure which raises the vaporisation temperature of the filament material. Due to this reason higher filament temperature can be maintained without its excessive vaporisation and consequently blackening of glass. It is observed that with inert gas a bulb consumes two-third of power consumed in case of a vacuum-filled bulb to attain the same light intensity. Moreover its life is also enhanced.

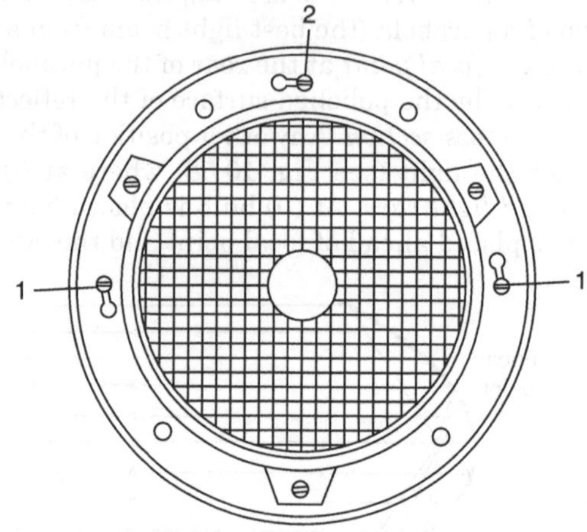

Fig. 10.60. Adjusting screws for head lamp : 1-Horizontal setting adjusting screws 2-Vertical setting adjusting screw.

There is no adjustment for the focus point of the bulb. This is because the bulb is already prefocussed and it is placed in its correct position. But it is necessary to *adjust the lamp unit relative to its mounting*. Then only it is possible for the rays to be reflected properly. Fig. 10.60 shows adjusting screws for head lamp. The screw on the top (2) is meant for the vertical setting of the position of the beam. The two side screws (1) are meant for the horizontal setting of the beam.

— The screw at the top can either tilt the beam *up or down* in vertical direction.

— The two screws placed horizontally swing the beam horizontally to the *left or right*.

- Recently, sealed beam headlight units are also being used (Fig. 10.61). Here, the aluminised gas reflector and the glass lens are fused to form a single unit inside which is filled some inert gas. The *advantages* of this type of construction is that :

(*i*) It can be manufactured to comparatively higher accuracy because the unit is not to be adjusted again in its lifetime.

(*ii*) Since in this case the area for the filament material vapour to deposit is large compared to a pre-focus type bulb, the glass lens is not blackened, which *increases* the unit life.

However, the *disadvantage* of this unit is that when any filament fails the *whole unit has to be replaced*.

Halogen-quartz lamps are also in use.

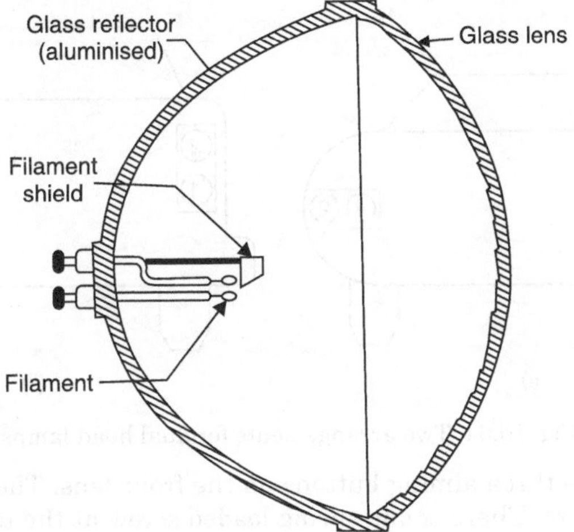

Fig. 10.61 Sealed beam headlight unit.

Head lamp double filament. In this arrangement, the *upper beam* filament is positioned *at the focal point* of the reflector (Fig. 10.62). It reflects the light beam at a long range distance.

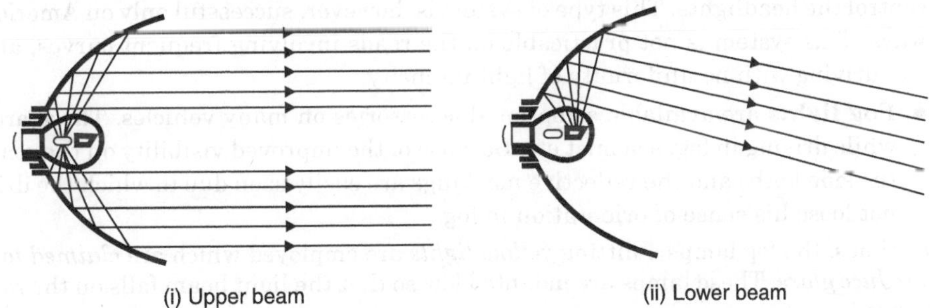

Fig. 10.62. Double filament lamp.

The filament for the *lower beam* is placed in *front of the focal point*. It is shielded at the bottom by a small hood. Therefore light beams are emitted only to the top. They are reflected to the bottom at an angle. This *prevents glare*.

By operating the light switch, the head lamp can produce either a *long range beam* (or main beam) or a *dipped beam* (Fig. 10.63).

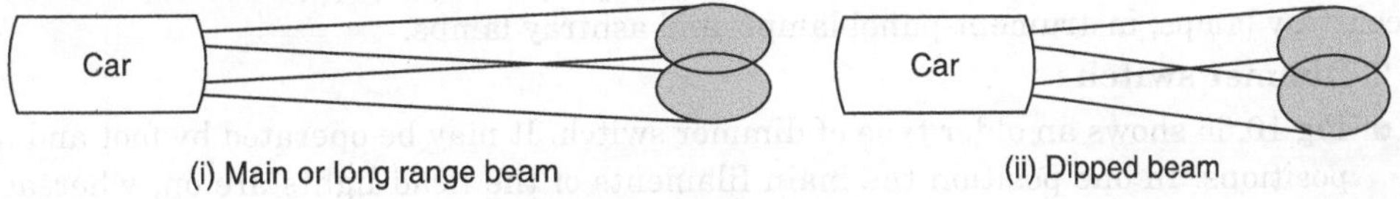

Fig. 10.63. Headlight beams.

- For correct headlight aiming, most American cars have four headlamps, two headlamps are arranged in each side (Fig. 10.64). The two headlamps can be arranged in a horizontal row or in a vertical row on each side. The head lamp may be either of one filament type or two filament type. The two-filament type is always placed outside or above the one filament type.

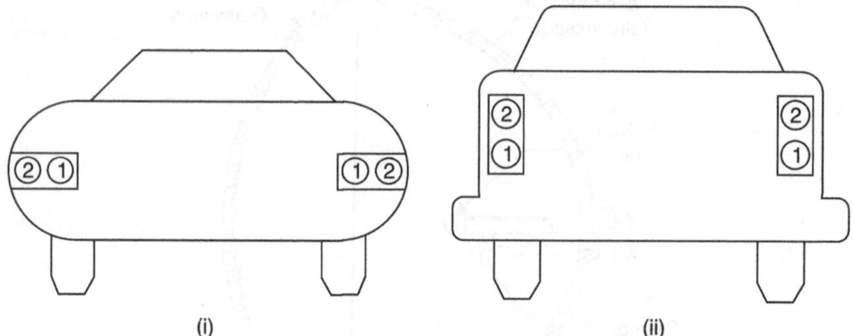

Fig. 10.64. Two arrangements for dual head lamps.

The headlights have three aiming buttons on the front lens. The adjustments are made by turning spring loaded screws. There is one spring loaded screw at the top for up and down adjustment and one at the side for left to right adjustment.

— Some American manufacturers have used photoelectric cells for automatic dipping of headlights. In the '*Autronic Eye*', the photoelectric cell senses the intensity of light from the oncoming vehicles and dips the headlights accordingly ; the driver does not have to control the headlights. This type of system is, however, successful only on American highways. This system is not practicable on the roads involving frequent curves, and in the city driving with no uniformity of light intensity.

- **Fog lights** are available as optional accessories on many vehicles. These are helpful while driving in fog, sea mist etc. Because of the improved visibility on the road ahead, the side kerbs and the reflective markings are easily seen due to which the driver does not loose his sense of orientation in fog.

Sometimes, the fog lamps emitting *yellow lights* are employed which are *claimed to improve vision and reduce glare.* These lamps are mounted low so that the light beam falls on the road about 10 metres ahead.

10.22.4. Lighting Switches

In an automobile, different types of switches are used in the lighting circuits, some important ones are enumerated and discussed below :

1. Light switch. 2. Dimmer switch. 3. Stop light switch.

1. **Light switch :**

This switch in most cars controls the head lamps, parking lamps, tail lamps, license-plate lamp, courtesy lamps, instrument-panel lamps, and ashtray lamps.

2. **Dimmer switch :**

- Fig 10.65 shows an older type of dimmer switch. It may be operated by foot and has two positions. In one position the main filaments of the head lights are on, whereas in the second position, the dipper light circuit is completed.
- In modern vehicles the switch is mounted on the steering column. By rotating the lever cap, the headlight, dimmer and the "off position" are obtained as desired.

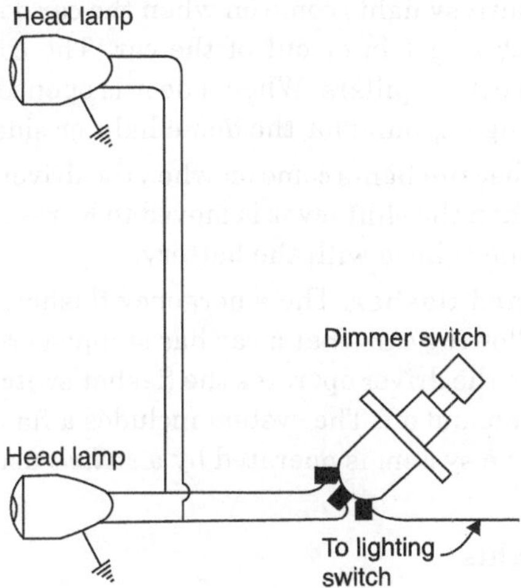

Fig. 10.65. Dimmer switch.

3. **Stop light switch :**

This switch is used to operate the stop lights at the rear of the vehicle. It gives an indication to the driver of other vehicle coming from the back so that he may slow down his vehicle's speed and stop if required.

The stop light switches may be either hydraulic or mechanical.

— Fig. 10.66 shows a mechanical stoplight switch. It is operated by the brake pedal. When the brake pedal is pressed by the leg, the brake pedal arm pushes a switch activating pin against the spring pressure which makes the contact points of the switch closed and the spotlight gets on.

— The hydraulic stop light brake system consists of a diaphragm that moves by the hydraulic pressure when the brakes are applied. This action closes a switch which connects the stop lights to the battery.

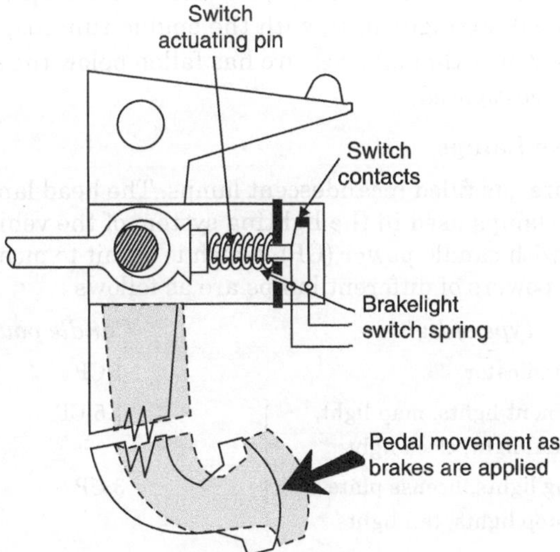

Fig. 10.66. Mechanical stop light switch.

- **Courtesy lights.** Courtesy lights come on when the car doors are opened so that passengers or driver can see to get in or out of the car. The courtesy lights are operated by switches in the door posts or pillars. When a door is opened, the switch closes to complete the circuit by providing a ground for the dome light or side lights.
- **Backup lights.** The backup lights come on when the driver shifts into reverse. The switch contacts are closed when the shift lever is moved to R (reverse). This connects the backup lights at the rear of the vehicle with the battery.
- **Emergency or hazard flasher.** The emergency flasher, or hazard-warning system, is designed to signal following cars that a car has stopped or stalled or has pulled up to the side of the road. When the driver operates the flasher switch, it causes all four directional signal lights to flash on and off. The system includes a flasher similar to the one used for directional signals. The system is operated by a *switch* usually located on the side of the steering column.

10.22.5. Indicating Lights

1. **Main beam warning light :**
   - This light glows when the main beams of the headlights are on.
   - This reminds the driver to dip the lights for the oncoming vehicles.
2. **Ignition warning light :**
   - The ignition warning light lights up when the ignition is switched on. It goes off when the engine speed is increased.
   - This lamp serves as a warning against leaving the ignition switched on when the engine is not running.
3. **Flashing indicator warning light.** This light which is also flashing, lights up when the flashing direction indictor is operative.
4. **Oil pressure warning light :**
   - This light lights up when the oil pressure in the lubricating system falls below the desired level. This lights up when the ignition is switched on.
   - When the engine is started and oil pressure is built up to the specified level, this warning lamp will extinguish. If, with the engine running, the light comes on, it is a warning to the driver that oil pressure has fallen below the safe limit which should be immediately investigated.

10.22.6. Automotive Lamps

Automotive lamps are gas filled incandescent lamps. The head lamps are of 40-50 watts in a medium priced car. All the lamps used in the lighting system of the vehicle operate on the battery voltage. They are marked with candle power (CP), which is a unit to measure the brightness of the source of light. The candle powers of different lamps are as follows :

Type of light	*Candle power (CP)*
Beam indicator	1 CP
Instrument lights, map light, radio dial light, clock light	1.5 CP
Parking lights, license plate light, stop lights, tail lights	3 CP
Head light	50000 to 75000 CP

10.22.7. Troubleshooting of Lighting System

In a lighting system, the following main *troubles* are experienced :

S. No.	Troubles	Causes	Remedies
1.	*No light.*	(*i*) Blown up fuse. (*ii*) Light switch not making proper contact. (*iii*) The battery may be disconnected or the circuit may be broken.	(*i*) After determining the basic cause, replace the fuse. (*ii*) Replace the switch, if found defective. (*iii*) Check the circuit and the defect, if any, remedied accordingly.
2.	*Dim light.*	(*i*) The battery may be discharged. (*ii*) The bulbs may be for a voltage higher than. The specified one or may be otherwise of a poor quality. (*iii*) Reflectors or lenses may be dirty. (*iv*) Bad contacts at terminals. (*v*) Bad contacts in the switches.	(*i*) Test the battery and recharge if necessary. (*ii*) In such a case the same should be replaced. (*iii*) May be cleaned at once. (*iv*) The contacts to be properly cleaned. (*v*) Replace the switch(es).
3.	*Lamps light, but gradually fade out.*	(*i*) May be due to discharged battery.	(*i*) Recharge the battery.
4.	*Lights flicker.*	(*i*) Loose connections.	(*i*) The connections may be inspected and tightened properly.
5.	*Frequent burning out of the bulbs.*	(*i*) Excessive battery voltage. (*ii*) The bulb may be for a voltage lower than specified one or may be otherwise of a poor quality.	(*i*) The battery voltage to be checked property and remedied, if required. (*ii*) The bulb may be replaced with one of proper voltage and good quality.

10.23. ACCESSORIES

10.23.1. Introduction

In an automobile, the accessories are divided into the following *two categories* :

1. **Essential accessories :**

(*i*) Fuel gauge
(*ii*) Horn
(*iii*) Windscreen wiper
(*iv*) Direction indicators
(*v*) Water temperature gauge
(*vi*) Speedometer and odometer.
(*vii*) Rear view mirrors etc.

2. **Desirable but not essential :**

(*i*) Spare wheel
(*ii*) Reserve fuel tank
(*iii*) Radio
(*iv*) Radiator blind
(*v*) Heater
(*vi*) Air-conditioner
(*vii*) Cassette player
(*viii*) Lift tail gate
(*ix*) Sliding roof
(*x*) Central locking system
(*xi*) Lockable tank cap
(*xii*) Roof carrier

(*xiii*) Winch (*xiv*) Adjustable head rest

(*xv*) Power windows.

10.23.2. Description of Accessories

Some important accessories are discussed below :

1. **Fuel gauge :**

A fuel gauge is fitted in all the motor vehicles to show the driver as to how much fuel is available in the tank. The fuel gauge circuit (Fig. 10.67) consists of *two units* connected with each other with wires ; these are :

(*i*) Gauge fitted on the dashboard ;

(*ii*) Measuring unit fitted in the tank.

In measuring unit there is a float ball attached to a rod which in turn is hinged to a lever. This lever moves over the resistance fitted in the cap. The float ball being empty ball always floats on the upper layer of fluid and travels up and down as the level of fuel goes up and down.

When the rod moves, the lever also moves on the resistance, the increase or decrease in resistance is measured on the gauge fitted on the dashboard which is calibrated to show the quantity of fuel available in the tank.

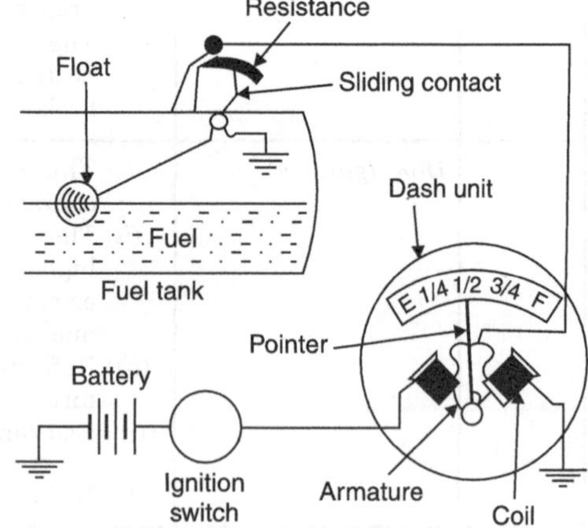

Fig. 10.67. Fuel gauge circuit.

Fig. 10.67 shows the working of the fuel gauge circuit.

2. **Horn :**

In automobiles, the following types of horns are used :

(*i*) Bulb horns ; (*ii*) Air pressure horns ;

(*iii*) Electric horns.

(*a*) Vibratory type (*b*) Wind tone type.

(*i*) ***Bulb horn :***

- It is a mechanical type of horn which is pumped by pressing the rubber bulb. This pumped air is made to pass through a need, which produces the sound. To increase the sound level a bugle is fixed after the need.
- Such type of horns are fitted in *buses and trucks.*

(*ii*) ***Air pressure horns :***

- In this type of horn, air is made to pass through reed which produces an alarming sound.
- *These horns are fitted to such vehicles where air tank is employed for air pressure brake system.* In this case, a flexible pipe is connected to air horn switch or valve which is operatable through a lever fitted under the steering wheel. Once the air pressure valve is pressed through this lever, it allows air to pass through the valve to horn reed, where sound is produced.

Since these horns produce an alarming sound, the other road users get confused. As such these horns have been banned, but nevertheless these horns can be seen usually fitted to vehicles playing on long routes as buses and trucks.

(*iii*) ***Electric horn :***

Electric horns are standard fittings on all the automobiles. These are classified in two major categories-*vibrating type* and *wind tone type.* As a matter of fact, both the horns work on *electromagnetic make and break system*, the only *difference being in the way in which sound is emitted out of diaphragm.* In vibratory type the sound is emitted *directly* from the diaphragm while in wind tone type, sound from the diaphragm is made to pass from a *wind pipe* having an opening as bugle.

Fig. 10.68 shows a simple type of high frequency horn.

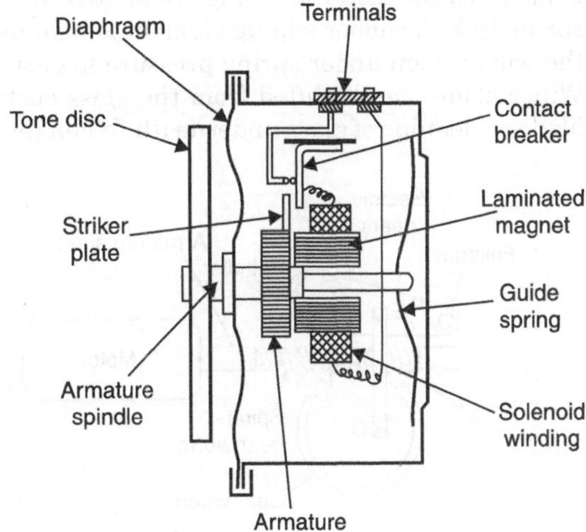

Fig. 10.68. Electric horn.

Construction. It consists of a laminated magnet having a wound core. The armature is attached to a central spindle which is supported at one end by a guide spring. The spindle is supported at the other end by a diaphragm. There is a second diaphragm called the tone disc. This disc is fitted only at its centre. In this horn system, there is a contact breaker, which is in series with the solenoid.

Working. When the horn switch is pressed the current flows through the contact breaker and solenoid. The circuit is completed through an earthed terminal. When this circuit is completed, the laminated core is magnetised. The magnet attracts the armature, which then moves towards the magnet. There is a protruding plate on the armature. This plate separates the contact points. Immediately after, the magnetic flux collapses and the armature then returns to its normal position due to the action of the spring. This cycle of operation takes place at a *high frequency.* Consequently vibrations develop in the armature and the diaphragm and the sound of the horn is produced. *Two notes* are produced in the horn : (*i*) A deep *note* is produced by the *low frequency of the diaphragm*, and (*ii*) A *higher frequency* note is developed by the *tone disc.*

Following factors will ensure an *efficient high frequency horn* :

(*i*) Correct setting of the breaker points.

(*ii*) Availability of proper voltage from the battery.

- In some systems, a '*relay' is also used to avoid carrying heavy current required by the horn through the steering column and rack.* The relay closes its contacts to connect the horn to the battery. This way, the voltage drop in the wiring from battery to horn is eliminated and higher voltage is available for operating the horn with better performance.

The horn relay serves a second purpose in some cars. It gives a *warning that the ignition key has been left in the ignition switch when the car door is opened.*

3. **Wind screen wiper :**

The wind screen wipers are used *to keep the windscreen clean during rain, snow etc. to ensure good visibility.* They are operated by a small electric motor (shunt wound type), mounted on the engine side of the bulk head.

The electric wiper machine consists of two major components, namely *electric motor* and *drive box.* The electric motor is a small motor having spiral gear/worm cut on its armature shaft. This gear/worm is in mesh with one big gear in drive unit. To deduce its speed to convert the rotary motion to angular motion, lever and sector gears are used as shown in Fig. 10.69 (*i*). On the spindle is mounted the motor arm (Fig. 10.69 (*ii*)). Wiper blade is attached to the wiper arm by means of a spring lock. A rubber wiping element is held in place in the wiper blade. Wiper arms pivot against the wind screen under spring pressure to ensure adequate wiper blade pressure against the glass. Wiper blades can be lifted from the glass surface against this spring pressure for replacement of blade or cleaning of glass underneath. When the wipers are switched off, they *reach the park position.*

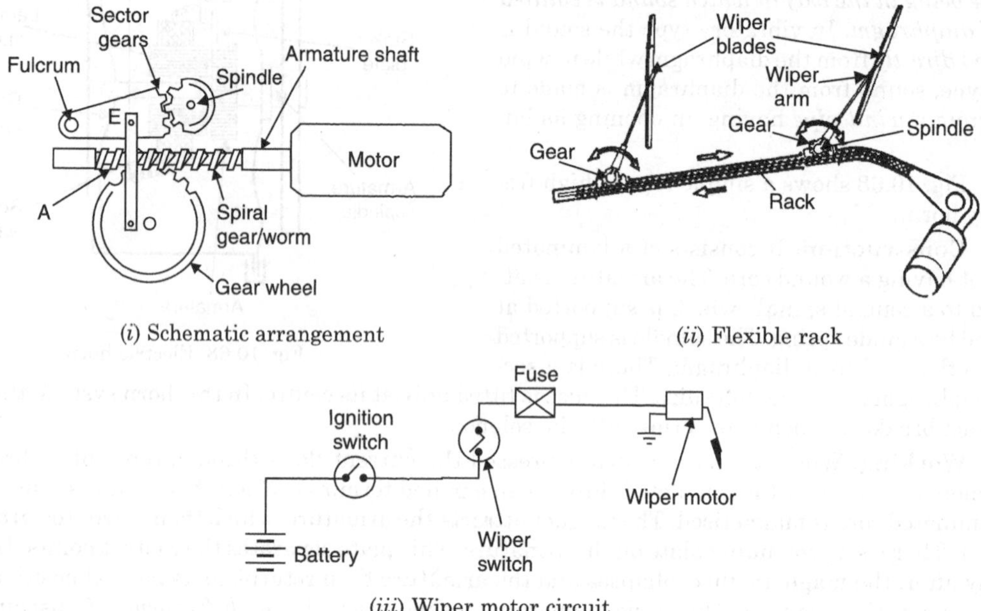

(*i*) Schematic arrangement (*ii*) Flexible rack

(*iii*) Wiper motor circuit

Fig. 10.69. Windscreen wiper.

The gears and the cable are fully lubricated by the manufacturers, therefore, there is no necessity for adjustment and maintenance of these wind screen wipers.

Fig. 10.69 (*iii*) shows the wiper motor circuit.

- Most cars have a windshield (or wind screen) washer as a part of the windshield-wiper system. When the driver presses a button, a liquid squirts on the windslive so that the blades can clean more effectively.

Many cars have a controlled-action wiper. One name for this system is *"multiplex pulse-wiper system"*. It includes a delay control which can be adjusted from zero delay to 12 seconds, delay between the blade movements. The wiper blades will move across and back, pause, and then repeat the wiper action.

4. **Direction indicators :**

In most cars directional indictors are used *to indicate the direction in which direction the driver tends to turn his car.* The lighting circuit of the directional signals is connected to that of the parking lights in the front and rear of the car, so that the attention of other drivers, coming from the front or rear, is carefully attracted while the car has to take a turn. It actually prevents the accidents. The light of the directional signal *flashes about 80 times per minute* to attract the other drivers. The directional signals are attached on each side of the car and flashing light is visible from both the front and rear sides.

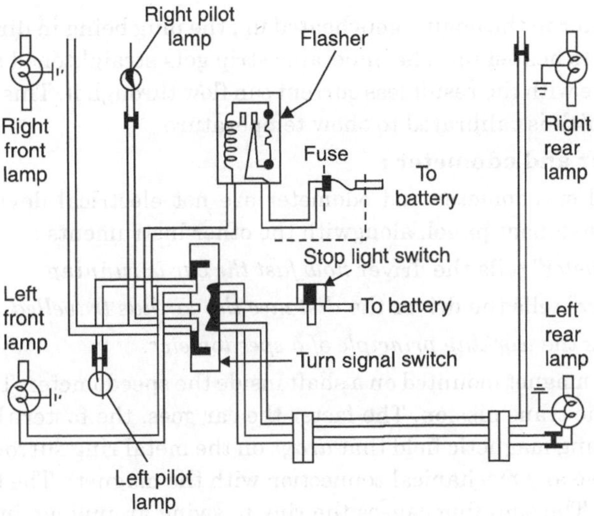

Fig. 10.70. Directional signal circuit.

Fig. 10.70 shows the directional signal circuit. The directional signal lights are operated either manually or automatically. The switch lever for operating the mechanism is located under the steering wheel opposite to the gear shaft lever, and is flipped down for left turn and up for right turn. The automatic turn off is accomplished by a mechanism that breaks the circuit when the wheel turns to straight-ahead position. The right and left pilot lamps on the instrument board in the form of arrows flash when the directional signals on the corresponding sides are in operation. The *"flasher unit", in series with the circuit, opens and closes the circuit automatically and periodically to produce the flashes.* Both the right and left turns can be signalled regardless of whether or not the brakes are applied.

5. Water temperature gauge :

The temperature water gauge is fitted in most of the automobiles to read the cooling system. The complete unit consists of a *"temperature plug"* and *a "gauge"*. The *plug is fitted in water gallery in the head or block while gauge is fitted on the dashboard.*

Fig. 10.71 shows the schematic arrangement of water temperature gauge.

— The plug has a fixed contact point duly earthed and movable point fixed on bimetal strip. On these bimetal strips heater coil is wound which in turn is connected to the terminal. The complete circuit of water temperature gauge is also shown, the resistance being shown fitted in between gauge and plug.

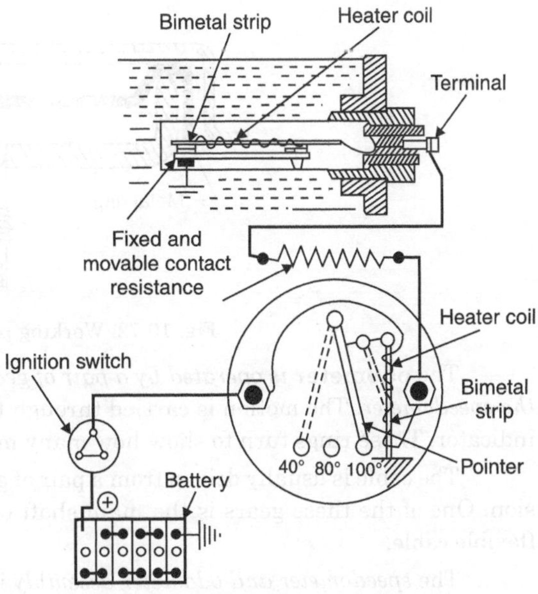

Fig. 10.71. Schematic arrangement of a water temperature gauge.

— When the water in the engine gets heated up, the plug being in direct contact with engine water also gets heated up. The bimetallic strip gets straightened and point attached to it open and close with the result less current can flow through it. This current is read through the gauge which is calibrated to show temperature.

6. **Speedometer and odometer :**

The mechanical speedometer and odometer are not electrical devices. However they are mounted on the car instrument panel, alongwith the other instruments :

— The "*speedometer*" tells the driver *how fast the car is running.*

— The "*odometer*" tells the driver the *distance the car has travelled.*

Fig. 10.72 shows the *working principle of a speedometer.*

There is a small magnet mounted on a shaft inside the speedometer. This magnet is driven by a flexible cable from the transmission. The faster the car goes, the faster the magnet *spring*. This action produces a rotating magnetic field that *drags* on the metal ring surrounding the magnet (the metal ring does not have any mechanical connection with the magnet). The faster the spinning, the more drag on the ring. The spinning causes the ring to swing around against the tension of a hair spring. This is turn, moves a pointer attached to the ring, which indicates car speed. The hair spring also serves to bring the pointer to zero and keep it there when the vehicle is brought to rest.

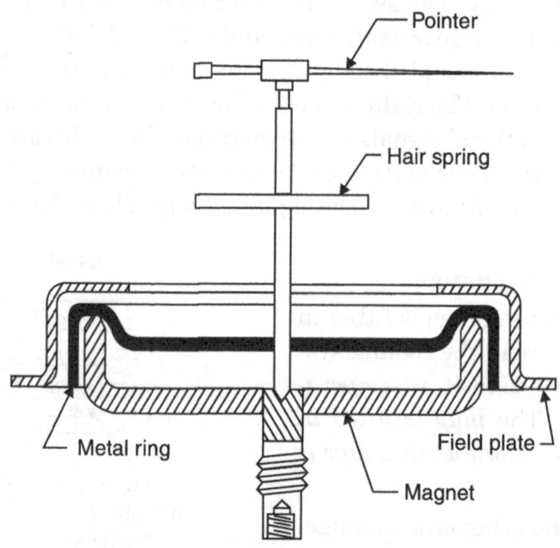

Fig. 10.72. Working principle of speedometer.

The **odometer** *is operated by a pair of gears from the same rotating flexible cable that drives the speedometer.* The motion is carried through the gears to the mileage/km rings on the odometer indicator. These rings turn to show how many miles/km the car has been driven.

The cable is usually driven from a pair of gears in the rear extension housing of the transmission. One of the these gears is the main shaft of the transmission. The other is on the end of the flexible cable.

The *speedometer and odometer assembly* is shown in Fig. 10.73.

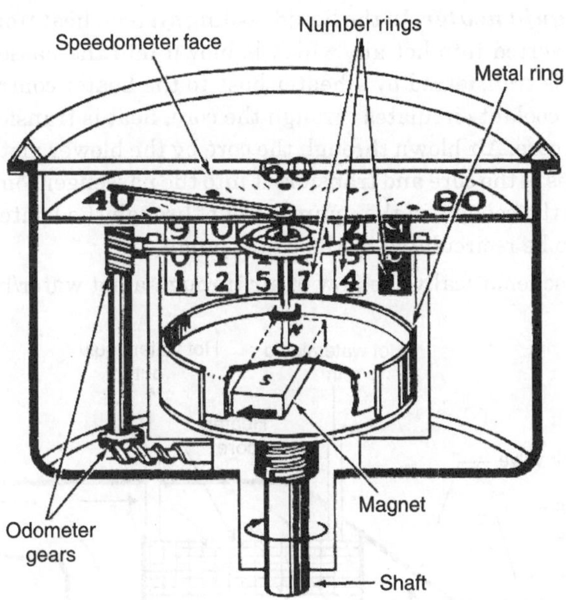

Fig. 10.73. Speedometer and odometer assembly.

7. **Ventilating system :**

A ***ventilating system*** is *provided on most vehicles to supply outside air to the passenger compartment through a number of vents.*

The air from outside enters the vehicle through openings in the front grill or through lowered opening near the base of the windscreen. The entry of this air is controlled by suitable valves or doors. The *air is circulated in the vehicle jointly due to the ram effect on account of the vehicle movement and due to a blower driven electrically.*

— The ducts for the distribution of air are invariably made of plastic.

— Doors or valves for air flow are controlled either manually by cables or by vacuum motors.

8. **Heating system :**

In most vehicles the *heating system is designed to work with the cooling system to maintain proper temperatures inside the vehicle.*

The main function of the heating system is to provide a comfortable passenger compartment temperature to keep vehicle windows clear of fog or frost.

Automotive heaters are discussed below :

(*i*) ***Hot air heater.*** This type of heater is usually used in case of *air-cooled engines.* It consists of a covering which encloses the exhaust manifold. Outside air is made to flow through the passage between the covering and exhaust manifold, either by means of ram effect or separate blower installed for the purpose. The air thus heated is circulated inside the car.

(*ii*) ***Petrol heater.*** A petrol heater consists of a combustion chamber and re-igniter. A mixture of air and petrol drawn from the carburettor burns inside the combustion chamber, from where the gases are taken to the re-igniter which helps them to burn continuously. Hot gases are made to flow over a heat exchanger through which air is circulated. The air thus heated is sent to warm the inside of the car.

The heater in put-off by cutting off the vacuum by means of dashboard control, which stops the supply of petrol and hence the combustion.

(iii) ***Hot water liquid heater.*** In the liquid-cooling system, heat from the coolant circulating inside the engine is converted into hot air, which is blown into the passenger compartment. Hot coolant from the engine is transferred by a heater hose to the heater control valve and then to the heater core inlet. As the coolant circulates through the core, heat is transferred from the coolant to the tubes and fins of the core. Air blown through the core by the blower motor and fan then picks up the heat from the surfaces of the core and transfers it into the passenger compartment of the vehicle. After giving up its heat, the coolant in then pumped out through the heater core outlet, where it is returned to the engine to be recirculated by the water pump.

Fig. 10.74 shows schematically the flow of air through a hot water/liquid type of heater.

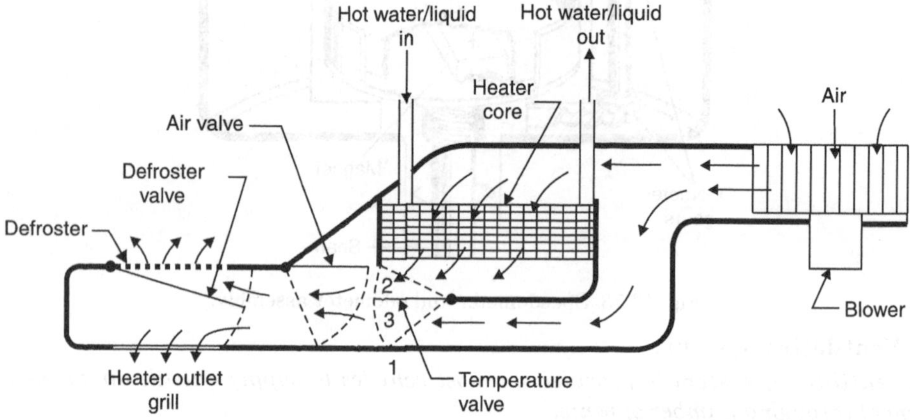

Fig. 10.74. Hot water/liquid heater.

The air is circulated with the help of a blower, the speed of which can be regulated by a control knob on the instrument panel. The position of the temperature valve then determines whether :

— the entire air will pass through the heater core and get heated : ... *'position 1'*;

— no air will pass through the heater, the outside air directly being circulated ... *'position 2'* ;

— a part of the air will pass through the heater, the rest going direct ...... 'position 3' ; the two streams then mix together.

The quantity of the heated air going further is regulated by means of air valve.

- The heater core is generally designed like a small radiator that functions exactly like the engine's radiator. Like the radiator, heater core tank, tubes and fins can become clogged over time with rust, scale and mineral deposits circulated by the coolant.
  - — Heater core failures are generally caused by *leakage* or *clogging*. Feel the heater inlet and outlet hoses while the engine is idling and warm with the heater temperature control on hot. If the hose downstream of the heater valve does not feel hot, the valve is not opening.

9. **Air-conditioning system :**

The main aim of ***"automotive air-conditioning"*** *is to control the temperature and humidity of the atmospheric air and circulate the same in the automobile so that a comfortable feeling is ob-*

tained by the passengers. The air-conditioning system is able to achieve this in conjunction with the heating and ventilating system.

In an automobile the air-conditioning system involves three processes of "*heating*", "*cooling*" and "*dehumidification*".

— In the "*heating system*", the heat needed to warm the automobile is derived from the engine coolant or circulating water after passing through a heating coil. The system either consists of the recirculatory heaters or fresh air intake system with provisions of recirculation.

— For producing a "*cooling effect*", an evaporator coil operated by a refrigeration system is placed inside the chamber of automobile. The air to be circulated inside the passenger compartment is cooled by the evaporator coil.

The dust particles are entrapped by the wet surfaces of the evaporator core and are drained of with condensed moisture. This provides clean and pure air for breathing.

Depending upon the speed of vehicle, frequency of stops, atmospheric and traffic conditions, the air-conditioning loads vary from instant to instant and place to place. Due to moving nature of the vehicle, the equipment used should be light, compact, quickly serviceable and unaffected by vibrations.

In comparison to domestic air-conditioning, very large amount of air-conditioning capacity is required for a car. This is due to the metal construction of the car, the flow of air around a moving car and relatively large glass areas in passenger compartment.

Typically, a car air-conditioning system may be between 1 to 4 tons.

A refrigerating machine driven by the car engine is used in the automobile air-conditioner to maintain the desired comfortable temperature level. For *heating purposes*, the warm water/liquid from the engine cooling system is used. The heat required to warm the automobile is generally provided by circulating warm water/liquid through a heating coil.

A **typical air-conditioning system** used in automobiles as shown in Fig. 10.75 consists of the following *components* :

(*i*) Compressor (*ii*) condenser
(*iii*) Receiver (*iv*) Dehrydrator
(*v*) Filter (*vi*) Sight glass
(*vii*) Blower (*viii*) Air distribution system
(*ix*) Certain type of capacity control.

The working of the system is as follows :

— The *refrigerant vapours from the evaporator is compressed to high pressure by the* "**compressor**" which is driven by means of a belt from the engine pulley, through an *electromagnetic clutch.* The clutch serves to engage and disengage the compressor as required. A variable-displacement A.C. compressor is sometimes used to match compressor capacity to varying cooling requirements.

— The high pressure refrigerant vapour from the compressor is then discharged to **condenser**, which is a simple heat exchanger. Air is caused to flow across the condenser by ram air while driving and by the action of engine fan or electric fan. *Cooling of the refrigerant vapour in the condenser converts it into liquid though still at high pressure.* This liquid is then collected in a "*receiver*" from where it is passed through a "*dehydrator*" to *extract any moisture.*

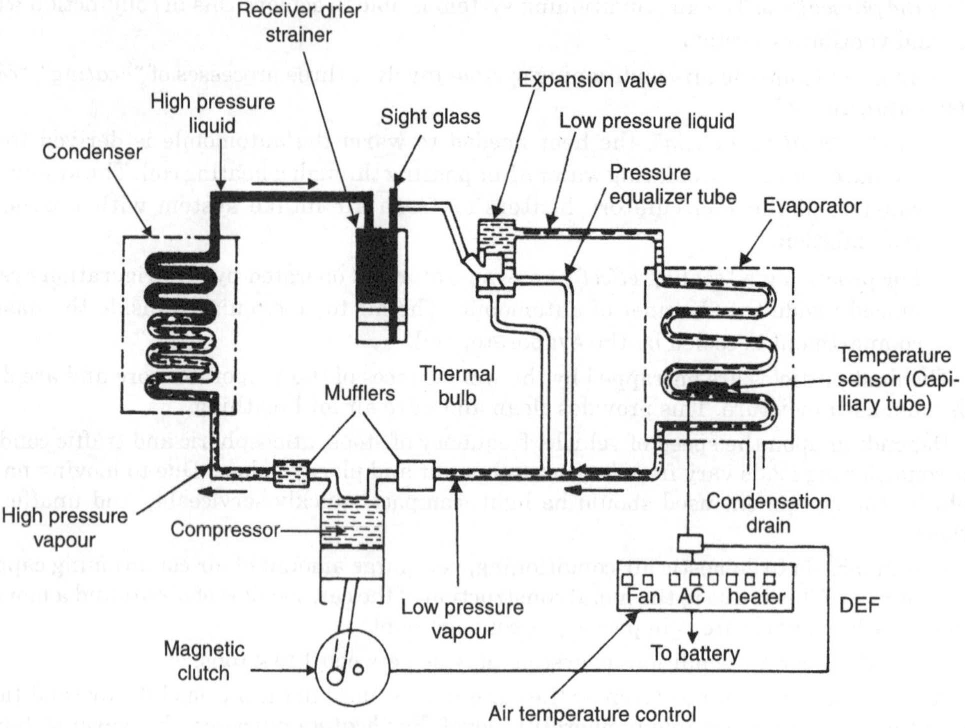

Fig. 10.75. Automobile air-conditioning system.

— Dry refrigerant liquid is then made to pass through **expansion valve** mounted at the inlet side of the evaporator. The expansion valve *allows the refrigerant liquid to expand to low pressure in the evaporator. The process of expansion to low pressure makes the refrigerant to evaporate and thereby cool the evaporator.* A sensing device (thermal sensor) called *'temperature tube'* signals the diaphragm in the expansion valve to vary its orifice size depending upon the refrigerant temperature at the evaporator outlet, thus achieving automatic temperature control.

Air from the blower is forced to flow over the evaporator coils which remove the latent heat of vaporisation, moisture and impurities from the air. Condensed moisture with the impurities from the evaporator drip into a tray from where the same are drained out.

- The various **controls** *in the air conditioning system are :*

  (*i*) The temperature control,

  (*ii*) Mixing of air control, and

  (*iii*) The selection of the cooling air to the desired outlets, *e.g.*, the defroster outlets, the floor outlets, the instrument panel outlets.

— These controls for mixing of air and selection of outlets are the same as for heating system.

— These controls may be manual, semi-automatic or fully automatic.

Fig. 10.76 shows the expansion valve type air-conditioning system.

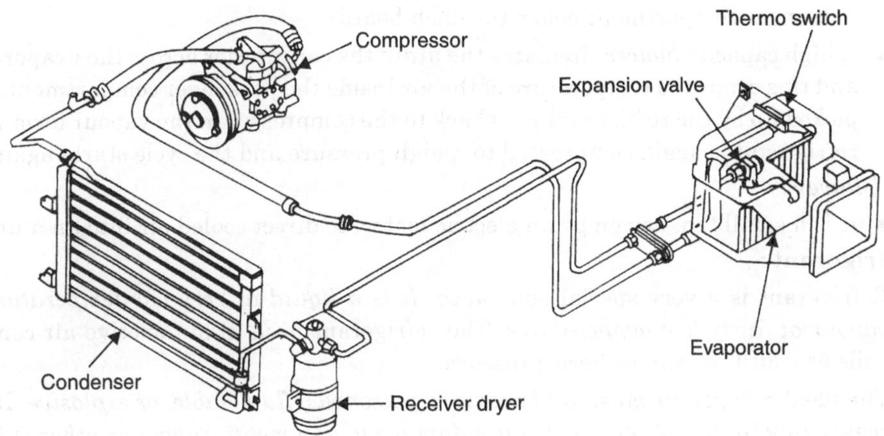

Fig. 10.76. Expansion valve type air-conditioning system.

Air-conditioning system components :

The *main components of an air-conditioning system* are discussed below :

(*i*) ***Compressor :***

- The compressor is the *heart* of air-conditioning system. It is a pump to circulate and increase the pressure of refrigerant in the system.
- It is fitted in front portion of the engine and is driven by a *V*-belt.
- It is operated through a *magnetic clutch* when the air-conditioner is switched on.

The compressors are usually of the following *three* types :

(*i*) Piston compressor. (*ii*) Swash plate compressor.

(*iii*) Rotary type of compressor.

(*ii*) ***Condenser :***

- The condenser is basically a heat exchanger. It consists of a series of tubes.
- It is mounted in front of the engine radiator.
- Air flowing over its fins takes away heat from the refrigerant and cools it down. The refrigerant *condenses* as it passes through the condenser.

(*iii*) ***Receiver/Drier :***

- This unit is placed between the condenser and the evaporator in the air-conditioning system.
- Its purpose is to *trap moisture* that enters the system during the original assembly or charging. Moisture is an enemy of air-conditioning systems because it can react with the refrigerant and corrode the inside of the system. Further it *assures the supply of pure liquid refrigerant* to the thermostatic expansion valve.

(*iv*) ***Expansion valve :***

- It is a thermostatic valve which is fitted at the inlet side of evaporator.
- It regulates refrigerant flow to the evaporator. Its regulating action is controlled by the temperature-sensing bulb placed on the suction line of evaporator.

(*v*) ***Evaporator :***

- The evaporator unit where the *cooling effect is obtained* is usually located inside the passenger compartment below the dash board.
- A high capacity *blower* circulates the air in the car interior across the evaporator coils, and this drops the temperature of the air inside the passenger compartment. The heat picked up by the refrigerant goes back to the compressor in the vapour form where the refrigerant is again compressed to a high pressure and the cycle starts again.

(*vi*) ***Blower :***

- It is a small fan, driven by an electric motor, to direct cooled air into an automobile.

Refrigerant :

- Refrigerant is a very special substance. *It is a liquid at very low temperatures and a vapour at fairly low temperatures.* The refrigerant used in automotive air-conditioners boils at – 30°C at atmospheric pressure.
- The ideal refrigerant *must not be toxic* or *poisonous, flammable, or explosive.* It must be reasonably inert and *must not cause damage to any metal, rubber* or *other substance it touches.* The refrigerant that has all these desirable qualities is special chemical. It is dichlorodifluoromethane (CCl_2F_2) and is commonly known as *Freon-12, Refrigerant-12* or simply Freon.

Note. Freon-12 being ozone-depleting is being replaced by alternative refrigerants.

10. **Mirrors :**

- In order to improve rear vision, mirrors are fitted both inside the car as well as outside it, because the inside mirror is bound to have some blind spots because of the car body pillars which hamper rear view.
- The mirrors used are generally of *convex type* to secure large field of vision.
- The internal mirror is placed above the windscreen at a suitable place while the external mirror is so located that the driver can adjust it from inside while sitting on his seat.

11. **Sun visors :**

- These are provided as a shield for driving in the sun.
- These can be moved towards the sides also if the sun rays are coming from that direction.

12. **Power windows :**

- There is an electric motor and a switch at every window besides a master switch at the dashboard which can control all motors simultaneously.
- The electric motor drives the window regulator mechanism forward or reverse, thus opening or closing the same.

13. **Central locking facility :**

- With this arrangement, all the doors and luggage compartment can be locked or unlocked simply by operating one key.
- A two-position electric solenoid actuator is installed in each door to electrically operate the door-locking mechanism. Each door can be locked or unlocked with one switch or key.
- In case of failure of electric locking system, the manual locking is still possible.

14. **Sliding roof :**
    - In order to enable the occupants to enjoy sun during good weather, a provision which is made available on many foreign cars as an optional extra, is the sliding roof which may be either raised open or slid open.
15. **Roof carrier :**
    - The additional load that cannot be accommodated in the luggage compartment can be carried at the car top, with the help of a roof carrier.
16. **Lockable tank cap :**
    - For checking/avoiding pilferage of the costly fuel, lockable cap at the fuel tank is quite effective.

10.23.3. Troubleshooting of Heating System

Troubleshooting of heating system is given, in tabular form, below :

S.No.	Troubles	Causes	Remedies
1.	*Little or no heat.*	(*i*) Air circulation insufficient. (*ii*) Coolant hose to heater blocked. (*iii*) Air in heater core. (*iv*) Heater core clogged. (*v*) Coolant valve or vacuum motor malfunctioning. (*vi*) Coolant level low in engine cooling system. (*vii*) Engine-cooling-system thermostat stuck open.	(*i*) Blower motor or switch at fault, air leaks from heater housing, temperature door or cable out of adjustment, loose carpet obstructing airflow. (*ii*) Unlink hose, replace defective hose. (*iii*) Bleed air out. (*iv*) Repair or replace core. (*v*) Repair or replace to permit coolant circulation. (*vi*) Add coolant, bleed air out of system, check system to locate and repair leaks. (*vii*) Replace thermostat.
2.	*Blower motor inoperative.*	(*i*) Blown fuse, poor electrical connections. (*ii*) Motor defective. (*iii*) Resistor open.	(*i*) Check for cause of blown fuse and correct, tighten connections. (*ii*) Replace. (*iii*) Replace.
3.	*Coolant leaks.*	Loose connections.	Check hoses, hose connections, heater core, water valve.
4.	*Too much heat.*	(*i*) Temperature-door cable out of adjustment. (*ii*) Engine-cooling system thermostat stuck closed.	(*i*) Readjust. (*ii*) Replace thermostat.
5.	*Insufficient defrosting.*	(*i*) Defrost-door control cable out of adjustment. (*ii*) Defrost outlets blocked. (*iii*) Any cause of little or no heat.	(*i*) Readjust cable. (*ii*) Remove obstructions. (*iii*) See item 1.

(Contd...)

6.	*Vent door does not operate.*	Defective vacuum motor, leaky vacuum connections, or a defective control assembly.	Repair/replace the components as required.
7.	*Controls had to operate.*	Loose or binding control cable or a sticky door.	Repair.
8.	*Odours from heater.*	Air leaks around blower case, coolant leaks around heater core.	Tighten bolts and see that seals and gaskets are in place ; remove and repair heater core.

10.23.4. Troubleshooting of an Automobile Air-conditioning System

Troubleshooting of an automobile air-conditioning system is given below :

S.No.	Troubles	Causes	Remedies
1.	*Insufficient cooling.*	(*i*) Low air flow. (*ii*) Defective heater temperature control valve. (*iii*) Heater controls or ventilator control not in the "off" position.	(*i*) Check blower option. Check for obstruction in air distribution system. Check for clogged evaporator. If iced, de-ice core and check adjustment and operation of suction throttle valve. (*ii*) Check operation of valve. Adjust or replace as necessary. (*iii*) Advise operator to correct operation of controls.
2.	*Compressor discharge pressure too high.*	(*i*) Engine overheated. (*ii*) Overcharge of refrigerant or air in system.	(*i*) Check for possible cause. (*ii*) Systems with excess discharge pressures should be slowly depressurized. • If discharge pressure *drops rapidly*, it indicates air (with possibility of moisture) in the system. When pressure drop levels but still indicates in excess of specifications, slowly bleed system until bubbles appear in the sight glass and stop. Add refrigerant until bubbles clear, then add one-half pound of refrigerant. Recheck operational pressures. If system pressures still remain above specifications, and the evaporator pressure is slightly above normal, then a *restriction exists in the high pressure side of the system.* • If discharge pressure *drop slowly*, it indicates, excessive refrigerant. If pressures drop to specifications and sight glass remains clear,

			stop depressurizing and recheck operational pressures. If pressures are satisfactory, depressurize until bubbles appear. Then add one-half pound refrigerant. Recheck operational pressures. • If discharge pressure *remains high after depressurizing the system*, continue depressurizing until bubbles appear in the sight glass. If evaporator pressures also remain high, there is a possibility of a *restriction* in the high pressure side of the refrigeration system, or the *suction throttle valve may require adjustment.*
		(*iii*) Restriction in condenser or receiver liquid indicator.	(*iii*) Remove parts, inspect, and clean or replace.
		(*iv*) Condenser air flow blocked.	(*iv*) Clean condenser and radiator core surfaces as well as the space between the condenser and radiator.
		(*v*) Evaporator pressure too high.	See 5. Evaporator pressure too high.
3.	*Compressor discharge pressure too low.*	(*i*) Insufficient refrigerant.	(*i*) Check for presence of bubbles or foam in liquid indicator. If bubbles or foam are noted (after five minutes of operation), check system for leaks. If no leaks are found, refrigerant should be added until the sight glass clears, then add an additional 1/2 lb.
		(*ii*) Low suction pressure.	(*ii*) See 4. Evaporator pressure too low.
		(*iii*) Defective compressor and/or broken compressor reed valves.	(*iii*) Repair compressor.
4.	*Evaporator pressure too low.*	(*i*) Valve capillary tube broken, inlet screen plugged, or valve otherwise failed.	(*i*) Replace valve or clean inlet screen of valve.
		(*ii*) Restriction in system tubes or hoses.	(*ii*) Replace kinked tube or restricted hose.
		(*iii*) Suction throttle valve adjusted improperly or defective.	(*iii*) Check operation of suction throttle valve. Repair if necessary.

5.	*Evaporator pressure too high.*	(*i*) Thermostatic expansion valve capillary-tube bulb not tight to evaporator outlet tube. (*ii*) Thermostatic expansion valve improperly adjusted or inoperative. (*iii*) Suction throttle valve adjusted improperly or defective. (*iv*) Vacuum modulator defective.	(*i*) Check for tightness. (*ii*) Replace valve. (*iii*) Check operation of suction throttle valve. Repair valve, if necessary. (*iv*) There should be no vacuum in the suction throttle valve when "cool" level on the instrument panel is on maximum "on" position. Replace vacuum modulator if defective. **Note :** If compressor suction line from suction throttle valve is extremely cold than the suction throttle valve inlet line from evaporator, this indicates that suction throttle valve outlet pressure is much lower than inlet pressure, and suction throttle valve may be defective.

SHORT ANSWER QUESTIONS

Battery

Q. 1. What is a cell ?

Ans. When two electrodes are connected through an external circuit and separated by a chemical electrolyte, it is called a *cell*.

Q. 2. What is a battery ?

Ans. It is an assembly of cells in series with the positive terminal of one connected to the negative of the other and so on.

Q. 3. Why is battery called storage battery ?

Ans. Because it stores electrical energy in the form of chemical energy.

Q. 4. What is a primary cell ?

Ans. It is a type of battery cell that cannot be recharged.

Q. 5. How many cells are there in a 12-V storage battery ?

Ans. Six cells.

Q. 6. What is 'capacity' of a battery ?

Ans. It is the product of the rate of discharge in ampere and time in hours required to discharge it completely. It is measured in ampere-hour.

Q. 7. What type of battery is used in automobiles ?

Ans. Lead acid battery.

Q. 8. What is an 'electrolyte' ?

Ans. It is a mixture of sulphuric acid and distilled water, used to complete the circuit inside the cell of battery.

Q. 9. Why ordinary type of water is not used in an electrolyte ?

Ans. Because the impurities in an ordinary water may short-circuit the cell of battery.

Q. 10. How is the electrolyte prepared for a battery ?

Ans. 40 percent sulphuric acid and 60% distilled water, mixed together.

Q. 11. What is meant by 'grid' in a battery ?

Ans. A lead framework or a structure that supports active materials like sponge lead or lead peroxide material in a battery.

Q. 12. How are corroded terminals cleaned up in a battery ?

Ans. With solution of baking soda and water.

Q. 13. What is the effect of neglecting to top-up the battery ?

Ans. The capacity of battery is reduced and the plates may be ruined.

Q. 14. What is the specific gravity for a fully charged battery ?

Ans. The specific gravity for a fully charged battery should lie between 1.23 and 1.28.

Q. 15. What is the specific gravity at a completely discharged position ?

Ans. About 1.19 to 1.21.

Q. 16. What instrument is used to measure the specific gravity of a liquid ?

Ans. Hydrometer.

Starting motor

Q. 17. What is the function of a starting motor ?

Ans. Starting motor is low voltage D.C. motor incorporating a gear-drive mechanism which converts the electrical energy received from the storage battery, into mechanical energy (cranks the engine.

Q. 18. Mention an important drive arrangement of the starting motor ?

Ans. Bendix drive.

Q. 19. How are starters linked to the engine flywheel ?

Ans. Through a small pinion, which acts as a gear reduction.

Q. 20. What is the approximate gear ratio between the cranking motor pinion and the engine flywheel ring gear ?

Ans. About 15 : 1.

Q. 21. For what purpose is the solenoid used in a starting circuit ?

Ans. (*i*) To connect the magnetic winding with the battery ;

(*ii*) To shift the starter pinion into agugement with the flywheel ring gear.

Q. 22. What is the function of the string in the Bendix drive ?

Ans. The spring is provided to relieve the sudden shock of start between the armature and the threaded sleeve, once the engine gets started.

Generator

Q. 23. What is the source of electric energy in an automobile ?

Ans. The dynamo/alternator.

Q. 24. What is the prime mover for the dynamo ?

Ans. The engine itself.

Q. 25. What type of dynamo is fitted to a motor vehicle ?

Ans. A two-pole two-brush shunt type dynamo, which has a rising voltage with increase of speed.

Q. 26. If the dynamo fails to charge, what items should be examined ?

Ans. (*i*) The dynamo field fuse, (*ii*) The connections to the dynamo, (*iii*) The commutator and brushes, and (*iv*) The drive to the dynamo if it is by belt.

Q. 27. What is a cut-out in an automobile ?

Ans. It is an automatic switch disconnecting the battery from the dynamo.

Q. 28. What three adjustments are available on a cut-out relay ?

Ans. Air gap, point opening and voltage required to close the contacts.

Q. 29. What will happen if a cut-out were not provided in an automobile ?

Ans. When the speed of the dynamo falls its voltage will fall below that of the battery, and it (battery) would discharge itself through the dynamo, if a cut-out were not fitted.

Q. 30. How is the cut-out coil tested ?

Ans. The cut-out coil is tested by connecting its each end to a six or twelve-volt battery and ammeter. When the circuit is completed the ammeter should show a reading if the coil is intact.

Q. 31. What is the purpose of a voltage regulator ?

Ans. It is used to prevent the circuit voltage from exceeding a predetermined value.

Q. 32. What is the purpose of a current regulator ?

Ans. It is used to protect the generator from overload.

Q. 33. Why is temperature compensation required on a voltage regulator ?

Ans. To avoid variation in voltage and current setting.

Q. 34. Why are cooling fans incorporated in the design of modern automobile alternators ?

Ans. To draw air through the case for cooling.

Q. 35. What is the function of a transistor in a Delco-Remy transistor regulator ?

Ans. To control the generator field current in the regulator.

Ignition System

Q. 36. How is heat for ignition obtained in the diesel engine cylinders ?

Ans. By compressing the air in the cylinders, until it has reached a temperature, high enough to ignite the injected tool.

Q. 37. What is the purpose of an ignition system in a petrol engine ?

Ans. It is intended to ignite the fuel-air mixture by electric spark plug in the cylinder at the end of compression stroke.

Q. 38. How are ignition systems classified ?

Ans. (*i*) Battery or coil ignition system ; (*ii*) Magneto-ignition system ; (*iii*) Electronic ignition system.

Q. 39. Name the three principal components of a coil ignition system.

Ans. (*i*) Battery (*ii*) Coil, and (*iii*) Distributor.

Q. 40. Why is a resistor used in the primary ignition circuit ?

Ans. It is used to connect the coil to the battery and control the load applied on battery.

Q. 41. What is "pre-ignition" ?

Ans. It refers to the ignition of the combustible mixture before occurrence of the spark.

Q. 42. What are the causes of pre-ignition ?

Ans. (*i*) Hot valves (*ii*) Hot spots that occur in the cylinder due to poor fitting of piston rings, and (*iii*) Incorrect insulation of the spark plug.

Q. 43. What do you understand by "detonation" ?

Ans. It is the phenomenon of producing a sound due to striking of waves, as a result of spontaneous combustion, on the cylinder walls.

Q. 44. What are the causes of detonation ?

Ans. (*i*) Poor quality fuel, (*ii*) Improper ignition timing, (*iii*) Too rich a mixture, and (*iv*) Excessive spark plug gap.

Q. 45. What is the purpose of an ignition coil ?

Ans. It is provided to step up the primary voltage (received from the battery and generator) of 12 V to approximately 20000 V.

Q. 46. What is the purpose of an ignition condenser ?

Ans. It prevents occurrence of an arc at the distributor contact points when the points first open.

Q. 47. When does a condenser get charged ?

Ans. A condenser gets charged when the contacts are opened.

Q. 48. When does the condenser get discharged ?

Ans. When the contact points are closed.

Q. 49. Why is condenser used in the primary circuit ?

Ans. It is used to store the field energy due to self induction current generated in the ignition coil primary winding during opening of contact breakers.

Q. 50. What is the essential difference between the magneto and the coil ignition system ?

Ans. The magneto is a self-contained source of electricity, whereas the coil system derives its initial current from the battery.

Q. 51. Is magneto-ignition system more reliable than coil ignition system ?

Ans. Yes.

Q. 52. What are the functions of the magneto in the scooter ?

Ans. The magneto helps for spark ignition and acts as a source for head lamp, tail lamp and horn.

Q. 53. What are the causes of weak current in a magneto ?

Ans. (*i*) Weak magnet, (*ii*) Weak condenser, and (*iii*) Incorrect contact breaker (C.B.) point.

Q. 54. What attention does the contact breaker require ?

Ans. Contact breaker points should be clean, meet squarely and open to correct gap.

Q. 55. What is the correct gap for contact breaker points ? How are the contact breaker points adjusted ?

Ans. The gap varies from 0.025 mm to 0.05 mm. To adjust the contact breaker points, the engine is cranked until the cam opens the points to their widest extent. By slackening the screw holding the contact plate, it can be moved until the gap is corrected.

Q. 56. What do you understand by firing order ?

Ans. The sequence at which the power stroke occurs in the cylinders, is called the firing order.

Q. 57. How many possible firing orders are there for a 4-cylinder, 4-cycle engine ?

Ans. Two : (*i*) 1 – 2 – 4 – 3 and (*ii*) 1 – 3 – 4 – 2.

Q. 58. How can the firing order of a unfamiliar engine be found ?

Ans. *By noting the order in which either the inlet or exhaust valves open.*

Q. 59. What is the voltage required in the spark plug, for igniting the combustible mixture ?

Ans. Between 18000 to 25000 volts.

Q. 60. What are the causes of the engine back-fire ?

Ans. (*i*) Late ignition timing, (*ii*) Spark plug of wrong heat range, and (*iii*) Excessive rich mixture.

Q. 61. What is meant by 'reach' of the spark plug ?

Ans. Reach refers to length of the threaded portion of the spark plug.

Q. 62. What are different foulings that occur in a spark plug ?

Ans. (*i*) Carbon fouling, (*ii*) High speed or load fouling, and (*iii*) Oil and carbon fouling.

Q. 63. What is a hot plug ?

Ans. A plug in which the heat path is of a longer distance to reach the cylinder head is called as hot plug.

Q. 64. What is meant by heat range of spark plug ?

Ans. The heat range of a spark plug is its stability to transfer heat from the firing end up through the insulator, gasket and shell to the cylinder head of water jacket.

Q. 65. What is a cold plug ?

Ans. A plug in which heat is required to travel a shorter distance from the insulator is called a cold plug.

Q. 66. What is the best way of cleaning the spark plug ?

Ans. By using sand blasting.

Q. 67. What are the reasons for no spark ?

Ans. (*i*) Pitted C.B. point (*ii*) Short-circuiting of coil or condenser circuit.

Lighting and Accessory System

Q. 68. Give a list of main lights in a modern vehicle.

Ans. Head light, stop light, parking light, backup light, direction signal light, tail light, blinker light and interior light.

Q. 69. What is the purpose of direction signal light ?

Ans. To indicate the direction in which the vehicle is to turn.

Q. 70. What is the use of tail lights ?

Ans. Tail lights illuminate back of the car in the night so that the other vehicles coming behind it are able to see it.

Q. 71. What is the use of stop lights ?

Ans. Stop lights are at the rear of the car and become "ON" when brakes are applied.

Q. 72. What is head light aiming ?

Ans. The head lights must be aimed correctly to the required direction to get proper light on the road and to prevent the vehicle from accident.

Q. 73. What are the major components included in automobile air-conditioning ?

Ans. Compressor, condenser, receiver-driver-thermostatic expansion valve and evaporator.

Q. 74. What refrigerant is used in an automobile air-conditioner ?

Ans. R-12.

Q. 75. What is a thermostatic expansion valve ?

Ans. It is a metering device which removes pressure from the liquid refrigerant so that it can expand and become refrigerant gas in the evaporator.

Q. 76. What is the function of evaporator ?

Ans. It absorbs heat from the passenger compartment.

Q. 77. What is purpose of the receiver-drier in automobile air-conditioning ?

Ans. It is a storage tank for liquid refrigerant. It also removes the foreign particles and moisture from the circulating refrigerant.

Q. 78. What is meant by "purging" the automobile air-conditioning ?

Ans. Purging is the process of discharging by releasing the refrigerant from the heigh and low sides of the system until no pressure exists.

HIGHLIGHTS

Battery

1. The major components of the electrical system are :
   (*i*) Starting system (*ii*) Charging system
   (*iii*) Ignition system (*iv*) Accessory system.
2. The battery is an electrochemical device.
3. The batteries are of the following types :
   (*i*) Lead acid battery (*ii*) Alkaline battery
   (*a*) Nickle-iron type (*b*) Nickle-cadmium type
   (*iii*) Zinc-air battery.
4. The *capacity* of a battery is given in terms of *ampere-hours* on discharge.
5. The *efficiency* of a battery is defined as, "The ratio of the output of a cell or a battery to the input required to restore the initial state of charge under specified conditions of temperature, current rate and final voltage."
6. The storage batteries are rated in the following ways :
   (*i*) 20-hour rating. (*ii*) 25-ampere rating.
   (*iii*) Cold-cranking rate.
7. The batteries can be *charged* by the following three methods :
   (*i*) Constant current charging. (*ii*) Constant voltage or potential charging
   (*iii*) Quick charging.
8. *Factors affecting battery life :*
   (*i*) Overcharging (*ii*) Undercharging
   (*iii*) Lack of water (*iv*) Loose hold-downs
   (*v*) Excessive loads (*vi*) Freezing of electrolyte.
9. *Battery faults/troubles :*
   (*i*) Rapid loss of electrolyte (*ii*) Sulphate plates
   (*iii*) Internal short-circuit (*iv*) Open circuit
   (*v*) Reversed polarity (*vi*) Cracked container or jar
   (*vii*) Deterioration of plates of the cells (*viii*) Mechanical faults.

10. The following tests are conducted to ascertain the conditions of a battery :
    (*i*) Specific gravity test (Hydrometer test) (*ii*) Open volt test
    (*iii*) High rate discharge test (*iv*) Cadmium test.

Starter

11. *Self-starter* (or starting motor) is an electrical machine (D.C. motor) which converts electrical energy into mechanical energy.
12. The starting motor consists of a motor unit and a drive unit.
13. The drive mechanism of the starting motor transmits the torque developed by the starting motor to the engine flywheel for cranking engine as starter armature revolves.
14. *Bendix drive* is known as inertia drive.
15. Starting motor switches are of the following types :
    (*i*) Manually operated switches, (*ii*) Magnetic switch or relay, and
    (*iii*) Solenoid switch.

Generator/Alternator

16. A *generator* is a machine which converts mechanical energy into electrical energy. The automobile generator is a low voltage generator (Dynamo) producing direct current only because the use of battery requires direct current.
    In modern cars, the charging system consists of an *alternator* only.
17. The *voltage and current regulator* consists of the following three major components.
    (*i*) Cut-out relay, (*ii*) Current regulator, and
    (*iii*) Voltage regulator.
18. The D.C. generator (or dynamo) rotates the conductors in a stationary magnet whereas A.C. generator (or alternator) rotates a magnetic field so that stationary conductors may cut the moving lines of force. This way the voltage is generated in the stationary coils.
19. Alternator differs from dynamo in the method of voltage and current control and also in design. It uses silicon *diodes* to change the alternating current to direct current for the vehicle's electrical system. Also it does not require 'cut-out' to prevent the battery from discharging through it *because the diode rectifiers allow the current to flow in one direction only.*

Ignition System

20. The *basic ignition systems* in use are :
    (*i*) Battery ignition system ; (*ii*) Magneto-ignition system ;
    (*iii*) Electronic ignition system.
21. The *main components* of a battery ignition system are :
    (*i*) Battery (6 or 12 V) (*ii*) Ignition switch
    (*iii*) Induction coil (*iv*) Circuit/contact breaker
    (*v*) Condenser (*vi*) Distributor.
22. The *battery ignition system* is commonly used because of its combined cheapness, convenience of maintenance, attention and general suitability.
23. The *magneto* is an efficient, reliable, self contained unit, which is often preferred for aircraft engines because storage batteries are heavy and troublesome.
24. *Firing order* is the order in which various cylinders of a multi-cylinder engine fire. The firing order is arranged to have power impulses equally spaced, and from the point of view of balancing.
25. *Ignition timing* is the correct instant for the introduction of spark near the end of compression stroke in the cycle. The ignition timing is fixed to obtain maximum power from the engine.
26. The main function of a *spark plug* is to conduct the high potential from the ignition system into the combustion chamber. It provides the proper gap across which spark is produced by applying high voltage to ignite the combustion chamber.
27. A *cold spark plug* has a short heat transfer path and a small area exposed to the combustion gases, as compared to a hot spark plug.

28. In modern automobiles, the following two types of electronic ignition system are employed :
(*i*) Transistorised Coil Ignition (TCI) system ;
(*ii*) Capacitor Discharge Ignition (CDI) system.

Lighting and Accessory System

29. The lighting and accessory system consists of the following :
Lighting system. Head light, sidelight, rear light, fog lamps, number plate illuminating lamp, interior lights and indicator flashers.
Accessories : Horns, wind screen wiper, electric petrol pump, petrol gauges, radio sets, cigar lighter, heater, wind screen defroster, signalling devices.

30. A typical air-conditioning system used in automobiles consists of the following components :
(*i*) Compressor, (*ii*) Condenser,
(*iii*) Receiver, (*iv*) Dehydrator,
(*v*) Filter, (*vi*) Sight glass,
(*vii*) Blower, (*viii*) Air distribution system, and
(*ix*) Certain type of capacity control.

OBJECTIVE TYPE QUESTIONS

Choose the Correct Answer :

Battery

1. The capacity of a lead-acid cell does *not* depend on its
(*a*) temperature (*b*) rate of charge
(*c*) rate of discharge (*d*) quantity of active material.
2. Cells are connected in series ion order to
(*a*) increase the voltage rating (*b*) increase the current rating
(*c*) increase the life of the cells (*d*) none of the above.
3. The capacity of a battery is expressed in terms of
(*a*) current rating (*b*) voltage rating
(*c*) ampere-hour rating (*d*) none of the above.
4. Electrolyte used in a lead acid cell is
(*a*) NaOH (*b*) only H_2SO_4
(*c*) only water (*d*) dilute H_2SO_4.
5. In a lead-acid cell, lead is called as
(*a*) positive active material (*b*) negative active material
(*c*) passive material (*d*) none of the above.
6. Overcharging a battery,
(*a*) will bring about chemical change in active materials
(*b*) will increase the capacity of the battery
(*c*) will raise the specific gravity of the electrolyte
(*d*) none of the above will occur.
7. Overcharging ......
(*a*) produces excessive gassing
(*b*) loosens the active material
(*c*) increases the temperature resulting in buckling of plates
(*d*) all above.

8. Undercharging ......
(*a*) reduces specific gravity of the electrolyte (*b*) increases specific gravity of the electrolyte
(*c*) produces excessive gassing (*d*) increases the temperature.

9. The specific gravity of electrolyte is measured by
(*a*) manometer (*b*) a mechanical gauge
(*c*) hydrometer (*d*) psychrometer.

10. The output voltage of a charger is
(*a*) less than the battery voltage (*b*) higher than the battery voltage
(*c*) the same as the battery voltage (*d*) none of these.

11. A discharged lead-acid battery has on its plates
(*a*) $PbSO_2$ (*b*) Pb
(*c*) Pb_3O_4 (*d*) $PbSO_4$.

12. A fully-charged 6 cell automotive battery should indicate
(*a*) 12 V (*b*) 12.6 V
(*c*) sq. gravity of 1.290 at 32°C (*d*) Both (*b*) and (*c*).

13. With the increase of battery temperature, the specific gravity of electrolyte
(*a*) increases (*b*) decreases
(*c*) stays the same (*d*) none of these.

14. The colour of positive plate of a lead-acid battery is
(*a*) brown (*b*) grey
(*c*) white (*d*) black.

15. The colour of negative plate of a lead-acid battery is
(*a*) brown (*b*) grey
(*c*) white (*d*) black.

16. A maintenance-free battery
(*a*) has lead-antimony plate grid (*b*) has lead-calcium plate grid
(*c*) does not contain acid (*d*) does not contain water.

17. The number of positive plates in a battery cell is
(*a*) one more than the negative plates (*b*) one less than the negative plates
(*c*) two less than the negative plates (*d*) two more than the negative plates.

18. The positive plate of a lead-acid battery has
(*a*) PbO_2 (*b*) Pb
(*c*) $PbSO_4$ (d) H_2SO_4.

19. The negative plate of a lead-acid battery has
(*a*) $PbSO_2$ (*b*) Pb
(*c*) $PbSO_4$ (*d*) H_2SO_4.

20. The capacity of a battery is determined by the number of plates per cell and
(*a*) number of cells (*b*) shape of plates
(*c*) size of plates (*d*) number of separators.

Starting Motor

21. Inertia type drives are commonly employed in
(*a*) cars (*b*) light transport vehicles
(*c*) heavy transport vehicles (*d*) cross-country vehicles.

22. Any short-circuit in the armature winding may be tested with a
(*a*) multimeter (*b*) growler
(*c*) voltmeter (*d*) ammeter.

23. Continuity in the field windings can be tested with a
(*a*) multimeter (*b*) growler
(*c*) voltmeter (*d*) none of the above.

24. Gear reduction between the starting motor and the flywheel is about
(*a*) 5 (*b*) 10
(*c*) 15 (*d*) 50.

25. In a starter motor the field windings are wound around
(*a*) armature (*b*) commutator
(*c*) brush (*d*) pole shoes.

26. Automotive starting motors are
(*a*) series wound (*b*) series-shunt wound
(*c*) shunt wound (*d*) both (*a*) and (*b*) above.

27. The stalling torques of starting motors for cars vary between
(*a*) 10 to 30 Nm (*b*) 30 to 60 Nm
(*c*) 60 to 100 Mm (*d*) 100 to 200 Nm.

Generator

28. The output of an alternator is controlled by
(*a*) voltage regulator (*b*) cutout relay
(*c*) current regulator (*d*) all of the above.

29. A thermistor is used in an alternator regulator to
(*a*) control maximum voltage (*b*) control maximum current
(*c*) compensate for temperature change (*d*) all of the above.

30. In a d.c. generator magnetic field is produced in the
(*a*) armature (*b*) commutator
(*c*) carbon brushes (*d*) stator.

31. The three units contained in a regulator for automobile d.c. generator are
(*a*) voltage regulator, current regulator and Zener diode
(*b*) voltage regulator, current regulator and temperature compensator
(*c*) voltage regulator, current regulator and cut-out relay
(*d*) any three of the above.

32. In an alternator the magnetic field is produced in the
(*a*) rotor (*b*) stator
(*c*) frame (*d*) regulator.

33. An alternator frame is made of
(*a*) cast iron (*b*) brass
(*c*) aluminium (*d*) copper.

34. The number of windings in the stator of an alternator is
(*a*) 1 (*b*) 2
(*c*) 3 (*d*) 4.

35. The number of diodes used in an alternator are
(*a*) 1 (*b*) 2
(*c*) 4 (*d*) 6.

36. The frequency of a vibrating contact type regulator for d.c. generator is
(*a*) 20 per second (*b*) 200 per second
(*c*) 20 per minute (*d*) 200 per minute.

37. The charging voltage at the start of the engine is ...... than when the engine is running on load.
(*a*) lower (*b*) higher
(*c*) same (*d*) any of the above.

Ignition system

38. A spark plug may be fouled by
(a) petrol (b) oil
(c) lead (d) all of the above.
39. Spark plug having white insulator tip indicates
(a) overadvanced ignition (b) retarded ignition
(c) leaded fuel (d) gas leak.
40. A spark plug with a black centre insulator indicates
(a) over advanced ignition (b) retarded ignition
(c) stuck-up choke valve in the carburettor (d) excessive dwell.
41. The primary winding of ignition coil consists of
(a) a few turns of thin wire (b) many turns of thin wire
(c) a few turns of thick wire (d) many turns of thick wire.
42. Contact breaker points are generally made of
(a) plastic (b) steel
(c) copper (d) tungsten.
43. The 'dwell' is
(a) the time for which the points remain closed (b) the distance between the cam lobes
(c) the angle at which the heat contacts the cam (d) none of the above.
44. The capacity of an automotive engine condenser is approximately
(a) 0.2 microfarads (b) 2 microfarads
(c) 20 microfarads (d) 2 millifarads.
45. The ratio of the distributor shaft and the crankshaft speeds is
(a) one (b) one-half
(c) one-fourth (d) two.
46. The centrifugal advance mechanism provides ignition advance proportional to
(a) engine load (b) engine speed
(c) both (a) and (b) (d) none of the above.
47. Excessive contact breaker gap results in
(a) advanced timing (b) increased dwell
(c) rapid burning of points (d) all of the above.
48. The spark occurs when the
(a) points close (b) points open
(c) ignition switch is on (d) none of the above.
49. Contact breaker points are opened by the cam and closed by the
(a) same cam (b) centrifugal force
(c) magnetic force (d) spring tension.
50. Vacuum advance mechanism operates at
(a) light engine load (b) heavy engine load
(c) high engine speed (d) any of the above.
51. Ignition timing can be adjusted by an
(a) accurate clock (b) stop watch
(c) stroboscopic light (d) vacuum gauge.
52. The commonly used material for insulator of spark plug is
(a) bakelite (b) asbestos
(c) alumina (d) copper.

53. Compared to the life of a spark plug of two stroke engine, the spark plug life of a four-stroke engine is approximately

(*a*) same (*b*) twice
(*c*) one-half (*d*) none of the above.

54. The three components of primary ignition circuit are

(*a*) the contact breaker, condenser and distributor cap
(*b*) the contact breaker, ignition coil and spark plugs
(*c*) the contact breaker, ignition switch and condenser
(*d*) the contact breaker, ignition switch and rotor.

55. The components of secondary ignition circuit include the secondary winding of ignition coil, distributor rotor, distributor cap and

(*a*) condenser (*b*) spark plugs
(*c*) ignition switch (*d*) distributor drive gear.

56. The dwell angle on a six-cylinder engine compared to a four-cylinder engine is

(*a*) more (*b*) less
(*c*) equal (*d*) sometimes less and sometimes more.

57. The contact breaker gap is set most accurately by means of a

(*a*) feeler gauge (*b*) steel rule
(*c*) dial gauge (*d*) dwell meter.

58. Contact breaker gap should be set

(*a*) before adjusting dwell angle (*b*) after adjusting dwell angle
(*c*) first after starting the engine (*d*) after setting spark plug gap.

Lighting and Accessory System

59. In a head lamp the metal shield is placed below tho

(*a*) dipper beam (*b*) main beam
(*c*) both the beams (*d*) none of these.

60. For aiming the headlights, the distance between the headlights and the screen should be

(*a*) 1.6 m (*b*) 4.6 m
(*c*) 7.6 m (*d*) 10.6 m.

61. In modern vehicles, the headlight dimming switch is mounted on the

(*a*) steering column (*b*) dash board
(*c*) flour board (*d*) side panel.

62. For identification, the colour of tail lights in a car is

(*a*) white (*b*) red
(*c*) green (*d*) yellow.

63. The two quantities used for specifying an electrical wire are

(*a*) the colour and length (*b*) the length and diameter
(*c*) the diameter and number of strands (*d*) the colour and diameter.

64. The type of reflector used for automobile head lamp is

(*a*) spherical (*b*) parabolic
(*c*) hyperbolic (*d*) none of these.

65. The most commonly used car heater is the

(*a*) hot air type (*b*) hot water type
(*c*) petrol type (*d*) diesel type.

66. The object of air conditioning a car is to control therein the

(*a*) temperature and pressure (*b*) pressure and humidity
(*c*) humidity and temperature (*d*) none of these.

67. Trafficators are lights used for

(*a*) heavy traffic (*b*) light traffic

(*c*) reversing the car in traffic (*d*) indicating the direction in which turning.

68. The frequency of light per minute flashing in a direction indicator is about

(*a*) 8 (*b*) 80

(*c*) 800 (*d*) none of these.

Answers

1. (*b*)	**2.** (*a*)	**3.** (*c*)	**4.** (*d*)	**5.** (*b*)	**6.** (*d*)	**7.** (*d*)
8. (*a*)	**9.** (*c*)	**10.** (*b*)	**11.** (*d*)	**12.** (*d*)	**13.** (*b*)	**14.** (*a*)
15. (*b*)	**16.** (*b*)	**17.** (*b*)	**18.** (*a*)	**19.** (*b*)	**20.** (*c*)	**21.** (*a*)
22. (*b*)	**23.** (*a*)	**24.** (*c*)	**25.** (*d*)	**26.** (*d*)	**27.** (*a*)	**28.** (*a*)
29. (*c*)	**30.** (*d*)	**31.** (*c*)	**32.** (*a*)	**33.** (*c*)	**34.** (*c*)	**35.** (*d*)
36. (*b*)	**37.** (*b*)	**38.** (*d*)	**39.** (*a*)	**40.** (*a*)	**41.** (*c*)	**42.** (*d*)
43. (*a*)	**44.** (*a*)	**45.** (*b*)	**46.** (*b*)	**47.** (*a*)	**48.** (*b*)	**49.** (*d*)
50. (*a*)	**51.** (*c*)	**52.** (*c*)	**53.** (*b*)	**54.** (*c*)	**55.** (*b*)	**56.** (*b*)
57. (*d*)	**58.** (*b*)	**59.** (*a*)	**60.** (*c*)	**61.** (*a*)	**62.** (*b*)	**63.** (*c*)
64. (*b*)	**65.** (*b*)	**66.** (*c*)	**67.** (*d*)	**68.** (*b*).		

THEORETICAL QUESTIONS

1. What are the major components of the automotive electrical system ?

2. What is a starting system ? What does it include ?

Battery

3. "Battery is the heart of electrical system in an automobile." Explain.

4. What are the different types of batteries ?

5. Briefly describe the construction of a lead-acid battery with the help of a neat sketch.

6. What is battery rating ? Discuss various ratings.

7. How does a battery get sulphated ? Can a sulphated battery be recharged ?

8. What are the various components of a 12-volt lead-acid battery. Indicate the chemical reactions taking place during

(*i*) Charging of the battery (*ii*) Discharging of the battery.

9. Explain the following terms :

(*i*) Battery voltage (*ii*) Battery capacity

(*iii*) Battery efficiency (*iv*) Batter rating.

10. What do you understand by recharging of a battery ?

11. What is a dry charged battery ?

12. What are slow rate and quick rate of charging ?

13. How will you check the state of a charged battery ?

14. What is the purpose of the following components ?

(*i*) Separator (*ii*) Grid.

15. How would you bring a dry charged battery into use ?

16. How would you check the specific gravity of an electrolyte ?

17. Describe how a battery should be maintained.

18. What are the battery troubles and their remedies ?

19. Explain briefly any two of the following tests conducted to ascertain the conditions of a battery :

(*i*) Specific gravity test (*ii*) Open volt test

(*iii*) High rate discharge test (*iv*) Cadmium test.

20. Enumerate the factors which affect battery life.
21. Explain briefly 'trickle charging'.
22. What is advantage of a dry charged battery ?
23. Why should only the distilled water be used in the battery ?
24. What electrolyte is used in an alkaline battery ?
25. What is the active material on positive plates in a nickle-cadimum battery ? What material is there on negative plates ?
26. Name the two types of alkaline batteries. Explain any one of them.
27. What materials are used for the cathode and anode of a zinc-air battery.?

Starting Motor

28. Discuss the construction and working of starting motor for automobiles.
29. What are different types of starting motor drives in current practice ?
30. Write short-note on integrated starters.
31. Explain briefly the following :
    (*i*) Bendix drive (*ii*) Over running clutch drive.
32. Explain briefly the following starting motor control switches :
    (*i*) Manually operated switch ; (*ii*) Solenoid switch.
33. Discuss the common troubles occurring in the starting system of an automobile engine. Suggest also suitable remedies.

Generator

34. Why is a generator required in the electrical system of a car ?
35. What is the working principle of a D.C. generator (dynamo) ?
36. Explain the working principle and construction of a dynamo.
37. Explain the working principle and briefly construction of an alternator.
38. In what respects, does a dynamo differ from an alternator ?
39. How do you control the generator output in the automobiles ? Explain the third brush control.
40. Explain briefly a cut-out relay as used in the battery generator circuit.
41. Discuss the construction and working of a combined voltage and current regulator. How is the temperature compensation achieved in such a regulator ?
42. Describe the common faults occurring in generator. Explain briefly how these fault may be detected and remedied.
43. State the purpose of the voltage regulator.
44. What is transistorised voltage regulator ?
45. What is the type of generator used in modern cars. Explain it briefly.

Ignition Systems

46. What do you mean by the term "Ignition" ? How is it related with "combustion" ?
47. What are the requirements of an ignition system for an I.C. engine ?
48. Enumerate the basic ignition systems and describe any of them.
49. Describe with the help of a neat sketch a battery ignition system.
50. State the functions of an ignition coil and a condenser in the battery ignition system of a multi-cylinder S.I. engine.
51. Name the various components of a battery ignition system and explain any three of them briefly.
52. State the advantages and disadvantages of a battery ignition system.
53. Describe a high tension magneto-ignition system. Also state its advantages and disadvantages.
54. What is the difference between 'ignition timing' and 'firing order' ?
55. What are the differences between battery and magneto-ignition systems ?

56. What do you understand by 'ignition timing' ? Enumerate the various factors which affect ignition timings.

57. What is the main difference between the battery and electronic systems ?

58. List various electronic ignition systems in use. Describe any one of them clearly stating its advantages over the conventional ignition system.

59. Differentiate between hot spark plug and cold spark plug.

60. Describe how the centrifugal advance system operates.

61. Explain briefly spark advance mechanism.

62. Explain briefly with a neat sketch the "Transistorised Coil Ignition" (TCI) system.

63. How does Capacitive Discharge Ignition' system differ from 'Transistorised Coil Ignition' system ?

Lighting and Accessary System

64. Draw a simplified wiring circuit for the lighting system of a car and discuss the same.

65. State the desirable properties in a wire for automobile electrical circuit.

66. Name different lights used on a modern car.

67. Describe briefly the following :
(*i*) Head lights (*ii*) Interior lights.

68. What do you understand by headlight aiming ?

69. With the help of a sketch, describe a stop light switch.

70. Write short note on "Automotive lamps".

71. What is sealed headlamp system ?

72. Why is headlight dimming necessary.

73. What is dual head light system.

74. What is an "Autronic eye" ?

75. Name common indicating lights in a car.

76. Name any four automobile essential accessories.

77. What is *'trafficator'* ?

78. Explain the working of the following accessories :
(*i*) Trafficators (*ii*) Flashing indicators.

79. Using single diagrams discuss the construction and working of the following accessories in an automobile :
(*i*) Windscreen wiper, (*ii*) Horn, and
(*iii*) Speedometer.

80. How many types of heaters are available for cars ? Explain the principle of each type and discuss the hot water type of heater in detail.

81. Draw the layout of the air-conditioning system for a car and explain its working.

11

Maintenance and Troubleshooting of Automobile

This chapter deals with the maintenance and troubleshooting of engine/automobile.

Since an automobile has to work under severe operational conditions, therefore, there is a great strain on its working parts. It is neither the age of an automobile nor the mileage which determines its usefulness.

Following are the two factors on which the life of an automobile depends :

(*i*) *How the vehicle has been driven ?*

(*ii*) *How the vehicle has been maintained* (*or looked after*) ?

The first factor can be taken care of by acquiring good driving habits. The second factor which relates to 'maintenance' of the automobile is very important one. *In order to preserve full working life of an automobile, its maintenance should be carried out religiously at the prescribed intervals and in the prescribed manner.*

11.1. MAINTENANCE

11.1.1. Introduction

Maintenance can be categorised into the following *two parts :*

1. Lubrication.
2. Cleaning, inspection and adjustment.

1. **Lubrication :**

The different parts of an automobile, depending on their state of operation need different types of lubricants :

— The parts (*e.g.,* engine bearings, cams and gears and transmission bearings and gears) moving at normal to high speeds need *lubricating oil.*

— The slow intermittently moving parts (*e.g.,* steering mechanism joints, pivot pins, spring shackles etc.) require a *heavier lubricant or grease.*

The lack of lubrication is most common cause of damage or failure. However, over-lubrication is also harmful. It is as important to use the *right amount* of lubricant as the *right lubricant* at the *right time* in the *right place.*

- *Excessive oil* in engine shall *lead to oiling-up and dark white smoke.*
- *Excessive oil* in the transmission shall *resist the movement of gears and shafts.*

- *Too much grease* on C.B. arm heel shall run on to the breaker points and *lead to fouling.*
- *Low engine oil* shall *lead to engine overheating in addition to damage to working parts.*
- *Lack of lubrication* at *steering linkage joints, pivot pins* etc. shall *result in hard steering.*
- *Poor lubrication* also *causes metallic noise in addition to increase in wear rate.*
- *Failure in oil supply* leads the engine to *seizure.*
- *Lack of lubrication at wheel hub bearings* leads the brake drums to *overheat and break away resulting in serious accident.*
- *Improper lubrication at speedometer* cable twists it to *break away.* Accelerator, clutch brake cables etc., stick into their outer casings and hold or restrict the movement of important controls.

Lubrication charts. The lubrication charts are supplied by the manufacturers in the operation as well as service manual as a guidance for the maintenance of the vehicle. These charts convey the following information :

(*i*) The points where lubrication is required.
(*ii*) The type of lubricant needed for a particular point.
(*iii*) The period (or interval) after which to lubricate.
(*iv*) The quantity and grade of lubricant required for different points.

- The lubricant's grade varies from condition to condition, region to region and season to season. The temperature affects the viscosity of oil.
  — During winter season when the temperature is low, thin oils are required and during summer season, thick ones are needed. Similarly the oil grade is different in cold regions than hot ones.
- Since the engine life much depends on the quality of engine oils, only high quality engine oils having excellent lubricating properties should be used.

Desirable properties of engine oils :

(*i*) They should have good detergent and dispersing effects.
(*ii*) They should have sufficient protection against corrosion, good resistance against deterioration, prevention of formation of foam and acid, etc.
(*iii*) The lubricant should provide a perfect bed (cushion) for the moving parts under different load conditions. Lubricant film should not break under these varied conditions.
  — If the lubricating film is subjected to *greater loads, speed and temperature,* the lubricant should be of *'higher viscosity'* and if it is subjected to *lower loads, speed and temperature,* the lubricant should be of *'lower viscosity.'*

The above mentioned requirements are fulfilled by heavy duty (HD) oils.

When topping or changing engine oil, the viscosity and quality should be kept in mind.

Automobile lubrication methods :

The following methods are employed in automobile lubrication :

(*i*) ***Pack method :***
- By this method lubricant is packed over the bearing surfaces.

(*ii*) ***Direct pouring method :***
- By this method, oil is directly filled into the machine housing.

(*iii*) ***Hand operated pump method :***
- By this method grease or oil is pumped out.

(*iv*) ***Simple grease gun method :***

- This method is used to feed grease through grease nipples.

(*v*) ***Power lubrication method :***

- By this method, the lubricant is fed under air pressure.

(*vi*) ***Automatic lubrication method :***

- This method is used to feed lubricant automatically to all the points.

2. **Cleaning, inspection and adjustment :**

The manufacturers, for this part of maintenance, provide *instruction manuals and charts* to carry out the job systematically. While doing the job, the maintenance instructions should be strictly followed.

- The following *"occasional inspections"* should be carried out :

(*i*) Check and if necessary, correct idle running adjustment. Inspect and clean all carburettor parts if required.

(*ii*) Check back-lash in steering gear ; adjust if necessary.

(*iii*) Check worm-screw roller bearings and adjust by removing shims between lower cover and box.

- The maintenance schedule differs from vehicle to vehicle, hence the particular instructions should be followed for the vehicle for which they are meant. Maintenance is done "***periodwise***" (*e.g.*, weekly, monthly, quarterly, half yearly, yearly etc.) or on the basis of "***mileage***".

  — Daily checks include the checking of :

Engine oil level ; water/liquid level in the radiator ; For belt tension ; Air cleaner ; Tyres' pressures ; Brake pedal toe-board clearance, Clutch pedal toe-board clearance ; Tightness of wheel nuts ; Lights and horns ; Tools and equipment.

- In case of private vehicles, where the daily distance to be covered by the vehicle is not fixed, periodwise maintenance is not proper and in such cases maintenance on mileage basis is quite in order.
- In case of vehicles which are used for short distance traffic or which have to work under unfavourable operating conditions (*e.g.*, threshers, crane crushers etc.) should not be maintained on mileage basis. In such cases oil changing should be done on the basis of *operational hours* or *fuel consumption should be used as a guide.*

11.1.2. Preventive Maintenance

Preventive maintenance is *the care to prevent troubles which often load to a major haul. It involves inspecting and replacing parts before they become problem.* There are good reasons for performing preventive maintenance. *Safety is greatly increased by changing worn parts.* ***Example :*** If the fan belt breaks, the water pump will not turn and the engine will overheat. Overheating damages the cylinders, pistons, heads, valves, head gaskets and possibly other parts as well. To avoid this serious damage of engine parts, the faulty or broken belt should be replaced.

- *Preventive maintenance is "cheaper".* It is regularly required for high performance of the engine/vehicle.

  — The parts changed in a regular tone-up keep horse power at the rated levels.

  — *Fuel mileage is increased dramatically through preventive maintenance.*

- Preventive maintenance is known by several names :

  — Many people simply call it a "*tune-up*" ;

— Most shops have a series of services done at specified kilometres. Manufacturers specify the service operations to be carried out every several thousand kilometres to keep warranties in effect as shown in table 11.1.

Table 11.1. Preventive Maintenance Schedule (Courtesy. Maruti Udyog Ltd.)

Interval	*Free services*																	
Maintenance service operation *km (× 1,000)*	*0.5-0.8*	*2.0-2.5*	*5.0-5.5*	*10*	*15*	*20*	*25*	*30*	*35*	*40*	*45*	*50*	*55*	*60*	*65*	*70*	*75*	*80*
ENGINE																		
1. Water pump drive belt (tension wear)	A	I	I	I	I	I	I	I	I	R	A	I	I	I	I	I	I	R
2. Engine coolant (level, leakage)	I	I	I	I	I	R	I	I	I	R	I	I	I	R	I	I	I	R
3. Engine oil (level leakage)	I	I	I	I	I	I	I	I	I	I	I	I	I	I	I	I	I	I
	API GRADE SC					Replace at an interval of 5,000 km.												
	API GRADE SF					Replace at an interval of 5,000 km.												
4. Cooling system hoses and connections (leakage, damage)	I	I	I	I	I	I	I	I	I	I	I	I	I	I	I	I	I	I
5. Engine oil filter (leakage)	I	I	I	R	I	R	I	R	I	R	I	R	I	R	I	R	I	R
6. Engine bolts (All cylinder head and manifold fixing)	—	—	T	—	T	—	T	—	T	—	T	—	T	—	T	—	T	—
7. Engine mounting (loose, damage)	—	—	I	—	T	—	T	—	T	—	T	—	T	—	T	—	T	—
8. Valve clearance	—	—	A	—	—	A	—	—	—	A	—	—	A	—	—	—	A	—
9. Camshaft belt (damage, wear)	—	—	I	—	—	I	—	—	—	I	—	—	—	I	—	—	—	I
10. Exhaust system, (noise, leakage or otherwise defective)	—	—	I	—	—	I	—	I	—	I	—	I	—	I	—	I	—	I
11. Positive crankcase ventilation system (Hoses, connections and valve)	—	—	I	—	—	I	—	—	—	I	—	—	—	I	—	—	—	I
IGNITION																		
1. Ignition wiring (damage, deterioration)	—	—	—	—	—	I	—	—	—	I	—	—	—	I	—	—	—	I
2. Distributor cap and rotor (wear deterioration)	—	—	—	—	—	I	—	—	—	I	—	—	—	I	—	—	—	I
3. Spark plugs (clean and adjust the gap)	—	—	I	R	—	R	—	R	—	R	—	R	—	R	—	R	—	R

4. Distributor breaker point	—	—	I	R	—	R	—	R	—	R	—	R	—	R	—	R	—	R
5. Ignition timing	—	A	A	A	—	A	—	A	—	A	—	A	—	A	—	A	—	A
6. Distributor advance	—	—	—	—	—	I	—	I	—	I	—	—	—	I	—	—	—	I
CLUTCH AND TRANSMISSION																		
1. Clutch pedal (play)	I	I	I	I	I	I	I	I	I	I	I	I	I	I	I	I	I	I
2. Clutch slipping (or excessive damage)	I	I	I	I	I	I	I	I	I	I	I	I	I	I	I	I	I	I
3. Transmission oil (level, leakage)	I	I	I	I	I	R	I	I	I	R	I	I	I	R	I	I	I	R

Note :
A- Adjust **T-Tighten to the specified torque**
R- Replace or change **I-Inspect and correct or replace if necessary**
L- Lubricate

Points relating to preventive maintenance :

For the trouble free working of the vehicle, the following points relating to preventive maintenance should be kept in view :

1. *Keep moisture out of fuel system :*

This can be accomplished as follows :

Keep the fuel tank full. Fill fuel tank up at the end of day's journey to drive out moisture laden air during winter or damp weather. Drain out about one litre fuel from the tank and fuel filter so that water trapped in there, is removed. The drained out fuel could be reused after filtering.

2. *Prevent freezing of cooling water/liquid :*

This is done as follows :

Drain out the cooling system completely. Wash the system with washing soda solution and flush with clean water. Fill the system with anti-freeze solution. Check up all joints and rectify immediately any leakage.

3. During winter, keeping in view the temperature, *use oils of lower viscosity* as prescribed by the manufacturer.

4. *Keep batteries well charged :*

— Keep the battery top and terminals clean to ensure minimum resistance and prevent corrosion.

— Don't bring any naked flame before the batteries lest the emitted gases catch fire.

5. Keep head lamp rings *tight*, lest water traps into the reflector and leads to rusting. Replace the defective gaskets.

6. Don't let water particles *stay under the floor mat* but keep clean lest rusting creep in.

7. Occasionally *remove the flints* which may have got embedded in the tyre treads.

8. *Spot paint external parts* from where paint has chipped off in order to avoid formation of rust.

9. Keep the *engine warm during extreme cold temperatures by blanketing the radiator with water-proof curtain rather than cutting out fan.*

10. In case of air brakes, *condensed water should be drained* out of compressed air tank everyday otherwise the braking efficiency shall be affected by the formation of ice in the pipelines and brake valves due to low atmospheric temperatures.

During extreme cold, anti-freeze material should be filled into the brake system.

11. *Mount anti-skid chains* on the rear wheels of a heavy vehicle to prevent skidding over icy, heavily snowed up or slippery roads.

12. After day's journey, *remove the batteries from the vehicle and store them in a heated* or *warm room when the atmospheric temperature is below 15°C* because *starting capacity of a warm battery is considerably higher.*

13. Dilute the diesel oil with superior kerosene in the following ratios for use in different temperature ranges.

(*i*) Outside temperature between 0°C and – 10°C : Diesel fuel = 80% to 90% ; Kerosene = 20% to 10%.

(*ii*) Outside temperature between – 10°C and – 18°C : Diesel fuel = 70% ; Kerosene = 30%.

11.1.3. Engine Maintenance

When the engine has no trouble, the automobile runs economically and without any difficulty. Hence every trouble should be removed properly.

- "*Repair and servicing*" should be done time-to-time for the engine maintenance, perfectly in working order—the procedures aimed at correct specific troubles.
- Another method of engine servicing is "*Engine re-building*". Some companies set up *special engine disassembly and re-building lines.* They bring in old, worn engines, disassemble them completely, repair or replace all worn parts and then completely rebuild the engine, using only the old parts that are still in good condition. By re-building the engine becomes as good as new.

The troubles and repairs of engine parts can be divided into the following *six groups* :

1. *Connecting rod assembly :*

(*i*) Connecting rod (*ii*) Connecting rod bearing

(*iii*) Piston pin and bearings (*iv*) Cylinder wall

(*v*) Piston and piston rings.

2. *Crankshaft assembly :*

(*i*) Crankshaft (*ii*) Main bearings

(*iii*) Flywheel (*iv*) Crankshaft gear or sprocket.

3. *Valve mechanism :*

(*i*) Valves (*ii*) Valve guides

(*iii*) Springs (*iv*) Seats and inserts

(*v*) Valve lifters and adjusting screws

(*vi*) Rocker arms and push rods for overhead valves

(*vii*) Hydraulic valve lifters (*viii*) Camshaft

(*ix*) Camshaft bearings (*x*) Camshaft gear or sprocket.

4. *Cylinder head and cylinder block*

5. *Oil pan*

6. *Various systems :*

(*i*) Ignition system (*ii*) Cooling system

(*iii*) Fuel system (*iv*) Lubricating system

(*v*) Starting and generator systems.

Fig. 11.1 shows the exploded view of main components of an engine.

11.1.3.1. Connecting rod assembly

1. **Connecting rod :**

- The first step is to *remove the cylinder head or the oil pan.*
  — In most cases, the connecting rod and piston assemblies are removed from the top of the engine. In few cases they are removed from the crankcase end.

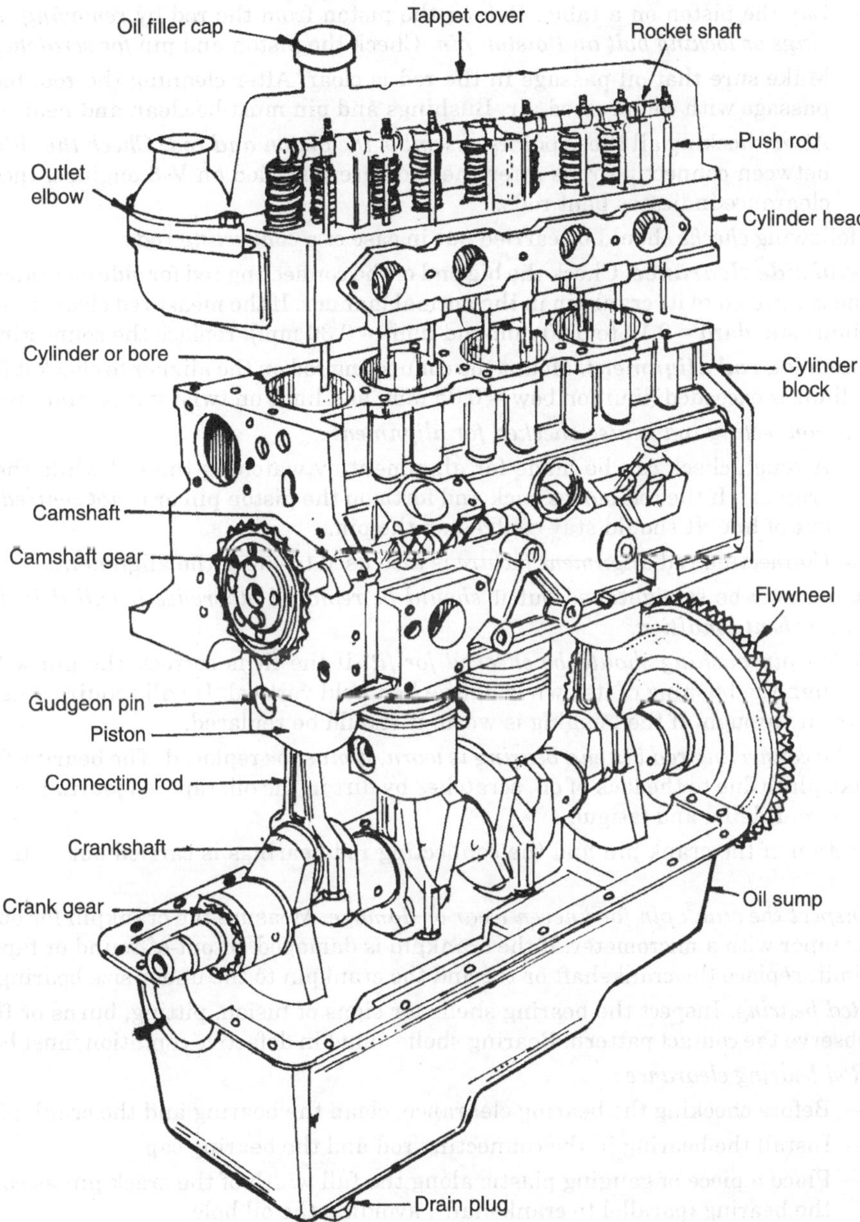

Fig. 11.1. Exploded view of main components of an engine.

- If the *piston is to be removed* from the cylinder head, examine the cylinder for wear. If the wear has taken place, there will be a *ridge* at the top of the cylinder that *marks the upper limit of the top ring travel.* This ridge is to be first removed by cutter so that piston may come out without damaging the piston rings. Crank the engine so that the piston is near the bottom. Remove the connecting rod nuts and cap. Again crank the engine so that the piston is near the top, take off the piston and rod from the cylinder.
  - Lay the piston on a table. Detach the piston from the rod by *removing the retainer rings or locking bolt and piston pin.* Check the piston and pin *for scratches and fits.*
  - Make sure that oil passage in the rod is clear. After cleaning the rod, blow out the passage with compressed air. Bushings and pin must be clean and neat.
  - After checking all these points *assemble the piston and rod. Check the side clearance* between connecting rods after the rods are installed on V-8 engines. Incorrect side clearance indicates bent rods.

The following *checks* should be carried out in case of a *connecting rod :*

Big end side clearance. Check the big end of the connecting rod for side clearance, with the rod fitted and connected to its crankpin in the normal manner. If the measured clearance is found to exceed its limit (standard = 0.10 to 0.20 mm and limit = 0.30 mm), replace the connecting rod.

Connecting rod alignment. Mount the connecting rod on the aligner to check it for bow and twist and if limit is exceeded (limit on bow = 0.05 mm, and limit on twist = 0.10 mm), replace it.

- The *connecting rod is also checked for alignment :*
  - A rough check can be made for alignment by watching the rod while the engine is cranked. If the rod moves back and forth on the piston pin or is *not centred*, the rod is out of line. It should stay centred on the pin.
  - Connecting rod *alignment fixture* is also used to check the alignment.

Bent rod may be straightened but it *should be replaced otherwise it will drift back to the permanent set or bent condition.*

- *Piston pin bushing should be checked for fit.* If the fit is correct, the pin will not drop through the bushing of its own weight when held vertical. It will require a light push to force it through. If the bushing is worn, it should be replaced.
- If the *connecting rod big end bearing is worn*, it *must be* replaced. The bearing failure may take place due to the lack of oil, scratches by dirt in the oil, tapered journal, poor seating, in bore, ridging and fatigue.

Inspection of the crank pin and the connecting rod bearings is carried out in the following ways :

(*i*) *Inspect the crank pin for uneven wear or damage.* Measure the crankpin for out-of-round or taper with a micrometer. If the crankpin is damaged, or out-of-round or taper is out of limit, replace the crankshaft or regrind the crankpin to the under size bearing.

(*ii*) *Red bearing.* Inspect the bearing shells for signs of fusion, pitting, burns or flaking and observe the contact pattern. Bearing shells found in defective condition must be replaced.

(*iii*) *Rod bearing clearance :*

- Before checking the bearing clearance, clean the bearing and the crank pin.
- Install the bearing in the connecting rod and the bearing cap.
- Place a piece of gauging plastic along the full width of the crack pin as contacted by the bearing (parallel to crankshaft), avoiding the oil hole.
- Install the rod-bearing cap on the connecting rod. When installing the cap, be sure to point the arrow mark on the cap to the crankshaft pulley side. After applying engine oil to the rod bolts, tighten the cap nuts to the specified torque. Do not turn the crankshaft with the gauging plastic installed.

— Remove the cap using a scale on the gauging plastic envelop, and measure the gauging plastic width at the widest point (clearance). If the clearance exceeds its limit (standard = 0.020 to 0.040 mm and limit = 0.080 mm), use a new standard size bearing and remeasure the clearance.

If the clearance cannot be brought to within its limit even by using new standard size bearing, regrind the crank pin to undersize and use a 0.25 mm undersize bearing.

It is made very clear that *getting the connecting rod checked is a must at the time of engine overhaul as slight bend in the connecting rod will make the piston rub against the wall of the cylinder and slight twist will put under pressure on the big end shells or small end bearing resulting in its premature failure.*

Fig. 11.2 shows checking parallelism of connecting rod.

Fig. 11.3 shows checking squareness of connecting rod with the piston assembled.

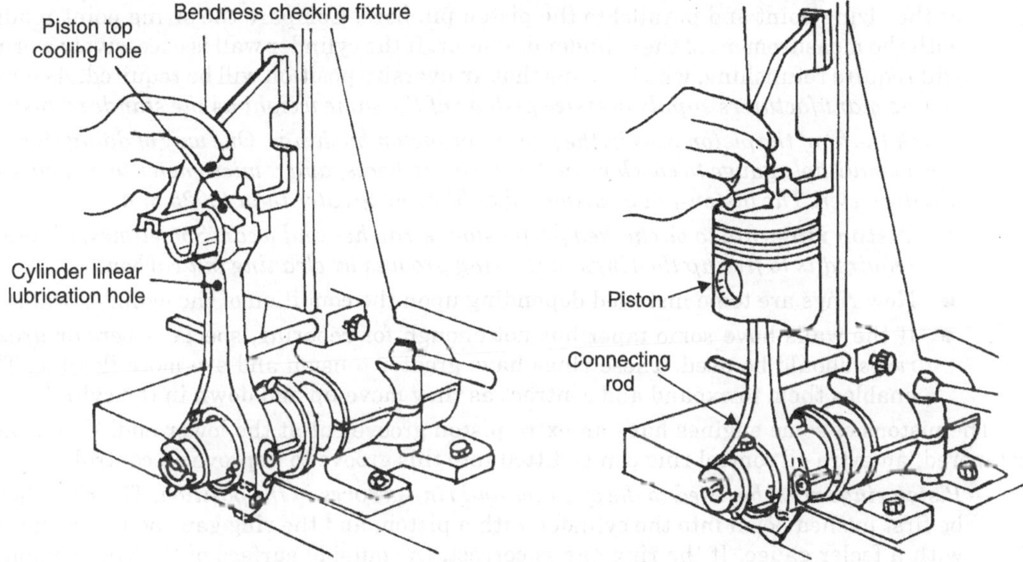

Fig. 11.2. Checking parallelism of connecting rod.

Fig. 11.3. Checking squareness of connecting rod with piston assembled.

Lubrication of connecting rod small end bush. The following *three methods* are used to lubricate the small end bush :

(*i*) By splash of oil thrown by the connecting rod big end from the oil sump to the liner and gudgeon pin bush.

(*ii*) By drilling a hole in the head of connecting rod in such a position that when this hole is registered with the oil hole in crankshaft, oil is injected to the gudgeon pin and cylinder liner.

(*iii*) By providing a positive supply line to the gudgeon pin bush. This is accomplished by drilling a small hole through the shank joining small end and big end head.

2. **Piston and rings :**

After the piston-and-rod assemblies are removed from the engine, the pistons and rods should be separated. Then remove the rings from the pistons by a special *ring tool.*

— *Remove carbon and varnish carefully from* ***piston*** *surfaces.* Do not use a caustic cleaning solution or wire brush. These could damage the piston-shirt finish. Use the cleaning method provided in your shop to clean the pistons.

— *Clean ring grooves* with a clean-out tool. The end of a broken piston ring filed to a sharp edge can also be used. Oil ring slots, or holes, must be clean so that oil can drain back through them. Use a drill of the proper size. Do not remove metal when cleaning the slots or holes.

— Examine the pistons carefully for wear, scuffs, scored skirts, worn ring grooves, and cracks. Look for cracks at the ring lands, skirts, pin bosses, and heads. Any defects require replacement of the piston, with these *exceptions* : (*i*) Worn ring grooves can sometimes be repaired by cutting the grooves larger and using ring-groove spacers ; (*ii*) Piston-skirt wear or collapse (reduction in skirt diameter) can sometimes be corrected by knurling the piston skirt.

— *Check the piston diameter* by taking measurements perpendicular to the piston pin bore at the sizing point and parallel to the piston pin bore. Compare the sizing point reading with the measurement of the cylinder diameter. If the cylinder wall is excessively worn, it will require refinishing, which means that an oversize position will be required. *As a rule, engine manufacturers supply oversize pistons of the same weight as the standard pistons.*

— *Check the fit of the piston pins in the pistons or piston bushings. One way of doing this is to use a small-hole gauge to check he piston-bushing bores, and a micrometer to measure the pin diameter. The piston-pin clearance should be no greater than 0.025 mm.*

- **Piston rings** are *to be checked for tension, scratches and wear. Sometimes, all that is required is to free up the rings in the ring grooves by cleaning out carbon.*
- New rings are to be installed depending upon the condition of the cylinder walls.
- If the walls have some taper but not enough for reboring, special *severe* or *drastic rings* should be used. These rings have greater tension and are more flexible. This enables them to expand and contract as they move up and down in the cylinder.

The pistons of some engines have an extra piston groove cut at the lower end. When walls wear tapered, an extra oil control ring can be fitted into this groove to improve oil control.

— *Piston rings must be fitted to the cylinder and ring grooves in the position.* The ring should be first pushed down into the cylinder with a piston, and the ring gap should be checked with a feeler gauge. If the ring gap is correct, the outside surface of the ring should be inserted into the proper ring groove in the piston and the ring rolled around in the groove to make sure that the ring has a free fit around the entire piston circumference. After the rings are installed in the grooves, *fit should be tested.* This test is made by *inserting a feeler gauge* between the rings and the side of the groove.

— In order to install the piston in the cylinder, compress, the rings in their grooves so that they will enter cylinder. It is done by a *piston* ring compressor. The compressor clamps around the rings, compressing them into the groove so that the piston can be pushed into the cylinder. Care should be taken to instal the piston facing in the right direction. *Many pistons have a notch or other marking that must face to the front of the engine.*

- **Piston resizing.** Resizing of collapsed or worn pistons is not recommended by automotive manufactures. *Excessive resizing can weaken the piston.*

— One piston-reconditioning procedure for older pistons is *"knurling"*. The piston skirt is run between a supporting wheel and a knurling tool. This displaces metal which expands the diameter of the piston skirt. The indentations form little pockets that can hold lubricating oil to reduce scuffing.

Fig. 11.4 shows using a micrometer to measure piston diameter.

Fig. 11.5 shows checking the fit of the piston ring in the ring groove.

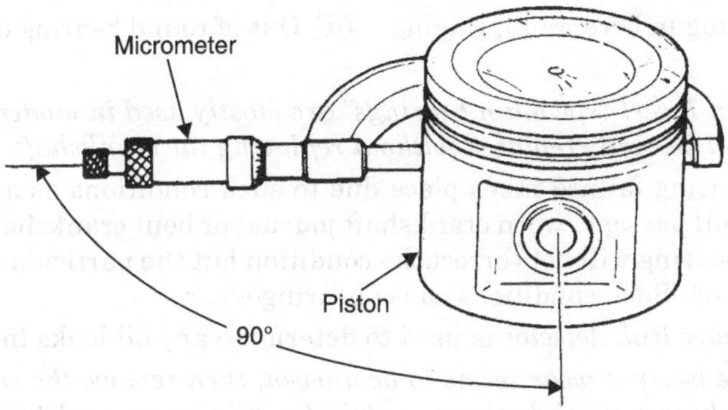

Fig. 11.4. Using a micrometer to measure piston diameter.

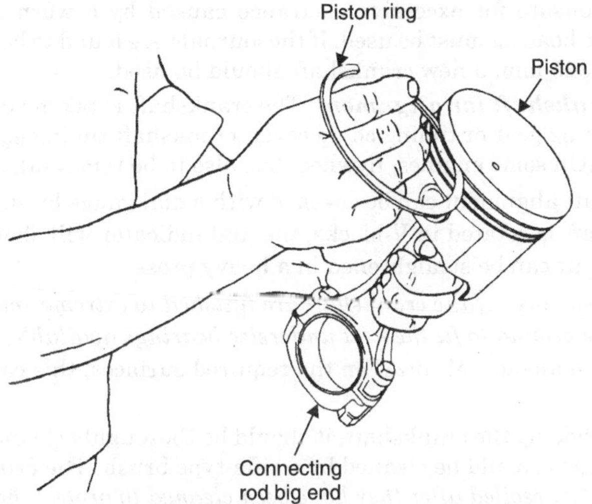

Fig. 11.5. Checking the fit of a piston ring in the ring groove.

11.1.3.2. Crankshaft assembly

1. **Crankshaft and main bearings :**

- Crankshaft is the backbone of the engine and is made of high carbon steel by forging process. It converts up down motion of piston to rotary motion. It can have a number of *throws* depending upon the number of pistons, normally one throw for each piston. These throws are arranged at specified angle keeping in mind the firing order for smooth working.
  - — The *main parts of a crankshaft* are :
    - (*i*) Journals (or main pins) (*ii*) Throws (or big end pins)
    - (*iii*) Webs, and (*iv*) Weights.
  - — The crankshafts usually come in two *types* :
    - (*i*) Built in crankshaft ; (*ii*) Single piece crankshaft.
- Wearing of crankshaft main bearings and journals occurs due to the following *reasons* :
  - (*i*) Lack of oil (*ii*) Scratches by dirt in the oil

(*iii*) Poor seating in bore, radii, ridging (*iv*) Out of round bearing or journals

(*v*) Fatigue.

— *"Precision Insert-type main bearings" are mostly used in modern automotive engines, which can be easily replaced without replacing the crankshaft.*

— If the bearing failure takes place due to such conditions as a warped engine block, stopped-oil passage, worn crankshaft journal or bent crankshaft, then simply replacing the bearing will not correct the condition but the particular difficulty will have to be removed. Such conditions cause bearing wear.

— *Oil pressure leak detector* is used to determine any oil leaks in the main bearings.

— *When the bearing wear seems to be uneven, then remove the crankshaft from the engine block and check both separately for alignment and for clogged air passages.* Semifinished main bearings require finishing and fitting after installation, so that the crankshaft must be taken out.

— To compensate for excessive clearance caused by a worn crankshaft journal, an undersize bearing must be used. If the journals are found to be worn more than 0.0125 mm to 0.325 mm, a new crankshaft should be used.

Inspecting crankshaft for alignment. The crankshaft is taken out by removing the parts such as the oil pan, timing gear or timing chain cover, crankshaft timing gear or sprocket, interfering oil lines, oil pump (On some engines, flywheel has also to be removed), and bearing caps.

— The crankshaft alignment can be checked with a dial gauge by supporting it on V-blocks. The crankshaft is rotated in V-blocks, the dial indicator will show any misalignment. A bent crankshaft can be straightened in a heavy press.

— *The bearing surfaces of the crankshaft are finished to extreme smoothness. Journals and crankpins are ground to fit the next undersize bearings available.* The griding wheel may leave a certain amount of "fuzz" on the required surfaces, this can be removed with fine crocus cloth.

- While servicing the crankshaft, it should be thoroughly cleaned in a suitable solvent. Oil passages should be cleaned by a rifle-type brush. The *bearing surfaces should be immediately reoiled after they have been cleaned to protect them from rusting.*

Fig. 11.6 shows the checking alignment of crankshaft, main bearings and connecting rod journals using dial indicator.

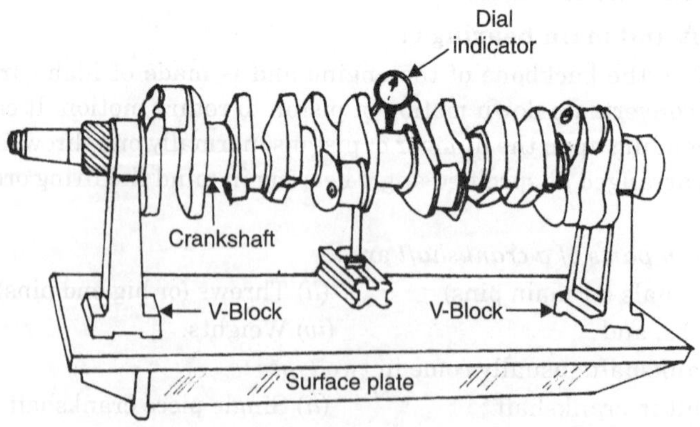

Fig. 11.6. Checking alignment of crankshaft, main bearings, connecting rod journals using dial indicator.

2. **Flywheel :**

- It must be ensured that the flywheel is tightly fitted on the crankshaft. If it looses, it will cause a heavy single knock, which can be detected by racing the engine slightly, then turn off the ignition switch. Just when the engine is about to stop, turn on the ignition switch, and at this time the flywheel knock will be heard. In order to *tighten the flywheel* on the crankshaft the clutch has to be removed. Before removing the flywheel, make two punch marks side by side, one on the flywheel and other on the crankshaft. While tightening it these marks should be *lined up.*

Fig. 11.7 shows the crankshaft, flywheel and the arrangement of crank throws (for 8-cylinder in-line engine).

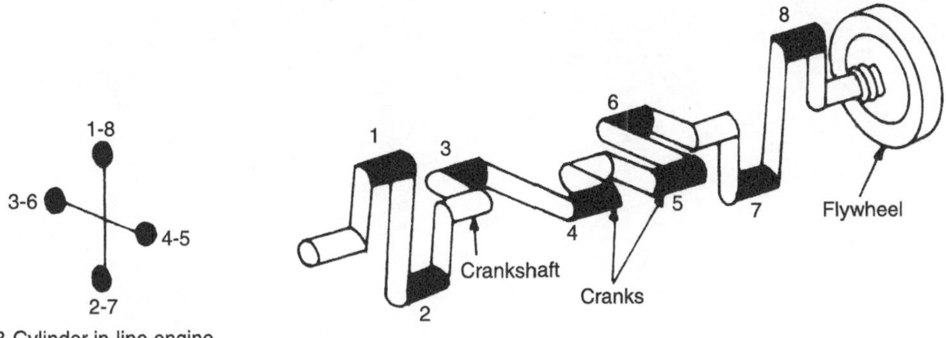

Fig. 11.7. Crankshaft, flywheel and the arrangement of crank throws (for 8-cylinder in-line engine).

11.1.3.3. Valve mechanism

Fig 11.8 shows the cylinder head from V-8 engine with related parts.

1. **Requirements for valve operation :**

The *requirements for valve operation* are as follows :

(*i*) It must be properly timed.

(*ii*) It must operate without lag.

(*iii*) The valve must seat tightly and properly on its seat.

(*iv*) Valve-tapped clearance and valve stem and guide clearance must be correct.

2. **Valve troubles :**

The various troubles are enumerated and discussed below :

(*i*) Valve sticking

(*ii*) Valve burning

(*iii*) Valve breakage

(*iv*) Valve face wear

(*v*) Valve deposits.

(*i*) ***Valve sticking :***

- Valve sticking is *caused by gum or carbon deposits on the valve stem.*
  - Excessive valve-steam clearance (worn valve guides) causes carbon deposits on the valves and stems. In every intake stroke, the oil will be pulled through this clearance into the combustion chamber, where it will burn. This causes excessive oil consumption, engine deposits, pre-ignition, clogged piston rings and fouled spark plugs.
  - Another cause of valve sticking is *warped stems.* These could result from overheating, an eccentric seat, or a cocked spring or retainer (which tends to bend the stem).

Fig. 11.8. Cylinder head for a V-8 engine with related parts.

— Valve sticking also sometimes takes place due to *insufficient oil.*

— Sometimes valves stick when cold, but work free when the engine warms up.

- When valves and piston rings are badly clogged with deposits, an engine overhaul is usually required. However, there are certain compounds that can be put into the fuel or oil to help free rings and valves.

(*ii*) ***Valve burning :***

Any condition that overheats the air-fuel mixture, or prevents normal valve seating, may lead to *valve burning.*

The normal valve cooling may be *prevented* due to the following :

— Poor seating ;

— A worn valve guide ;

— Clogged passages which prevent coolant circulation around the valve seat, leading to seat distribution, poor seating and overheated valves ;

— Engine overloading or overheating ;

— A lean air-fuel mixture ;

— Sometimes, valve burning occurs due to *guttering*. This is caused by carbon deposits on the valve face and seat. Parts of the deposits break off to form a path through which exhaust gas can pass when the valve is closed. This burns channel in the valve face.

(*iii*) ***Valve breakage :***

- Any condition that causes valve to overheat may cause it to break. Heavy pounding (as from excessive tappet clearance or from detonation) may cause valves to break. Excessive tappet clearance permits heavy impact seating.
- If the seat is eccentric to stem or if the valve spring or retainer is cocked, then the valve will be subjected to side movement every time it seats. This will cause the valve to fatigue and break. A scratch on the stem may serve as a starting point for a crack and a break in the stem.

(*iv*) ***Valve face wear :***

Valve face wear can take place due to excessive *tappet clearance* or *dirt on the valve face or seat.*

Excessive tappet clearance causes heavy impact seating. This wears the valve and may cause valve breakage.

— *Dirt* may cause valve-face wear if the engine operates in dusty conditions, or if the air cleaner is not functioning properly. The dust enters the engine with the intake air. Some dust settles on the valve seat.

Dust also causes bearing, cylinder wall, and piston and ring wear.

(*v*) ***Valve deposits :***

A fuel having excessive amounts of gum may deposit some of the gum on the intake valve ; this happens as the air-fuel mixture passes the valve on the way to engine cylinder.

The various *causes of carbon deposits on the valve* may be summed up as follows :

(*i*) Improper combustion due to rich mixture ;

(*ii*) Defective ignition system ;

(*iii*) Loss of compression in the engine.

(*iv*) Cold engine.

3. **Valve service :**

- The *valve service includes the following* :

(*i*) Adjusting valve-tappet clearances.

(*ii*) Cleaning carbon from all parts.

(*iii*) Refinishing valves and valve seats.

(*iv*) Installing new valve-seat inserts.

(*v*) Cleaning, knurling, reaming, or replacing valve guides.

(*vi*) Replacing the camshaft and crankshaft bearings.

(*vii*) Replacing the camshaft gear, or camshaft sprocket and chain or belt.

A *complete valve job*, or *valve grind*, includes the following :

(*i*) Refinishing the valve faces and valve seats ; checking the run out of the valve head and valve seat ;

(*ii*) Checking the clearance between the valve stem and valve guide ;

(*iii*) Checking valve-spring tension ;

(*iv*) Checking the valve-spring installed height ;

(*v*) Adjusting the valves (if applicable) ;

(*vi*) Adjusting ignition timing ;

(*vii*) Adjusting idle speed.

- The '*valve-tappet clearance*' is measured by a *feeler gauge. A two step "Go" and "Not Go" feeler gauge of specified thickness can be used. If the "Go" step fits the clearance, the adjustment is correct. If it is nôt correct turn the adjusting screw in or out as necessary to correct it. If it is not correct, turn the adjusting screw in or out as necessary to correct it.*
  - *On some engines, the measurement is made with engine cold and not running. The engine is turned over until the valve lifter is on the low point of the cam, and clearance is then checked. On other, engine is warmed up on idling.*
- *Valves and valve seats are ground to correct size and shape so that the valve may seat properly on the seat.*
  - For effective valve seating and sealing, the valve face must be concentric with the valve stem, and the valve guide must be concentric with the valve face.
  - Also the valve face angle must match the valve seat angle.
  - The *value-seat insert* is replaced if it is badly worn, or has been ground down previously so that there is no sufficient metal for another grind.
- *Check the valve guides for wear :*
  - Clean, replace or ream for large guides as necessary. A wire brush or adjustable-blade cleaner can be used to clean the guide.
  - If the guide is worn, it should be replaced.
- The *timing gears, or sprocket and chain are marked to establish the proper positions and correct valve timing.* Some engines have another marking system for checking valve timing. This marking is on the flywheel or vibration damper, near the ignition timing markings.

4. **Camshaft :**

Inspection of camshafts for alignment. The camshaft is checked for alignment (Fig. 11.9) by rotating it in *V*-blocks, and using a dial indicator to check for eccentricity.

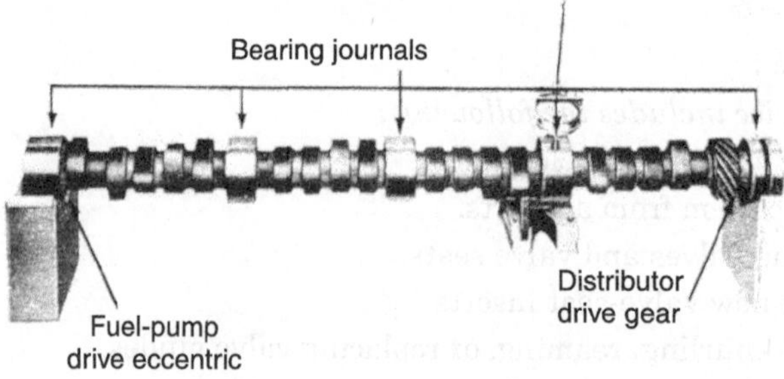

Fig. 11.9. Checking alignment of a camshaft with a dial indicator.

— The journal diameters should be checked with a micrometer and the bearings with a telescope gauge.

— A bent crankshaft may be straightened in hydraulic press.

— If the *bearings* are worm, they should be replaced.

5. **Hydraulic valve lifters :**

- *Non-hydraulic valve lifters* give little trouble and can be easily removed from the engine for replacement.
- *Hydraulic valve lifters* very seldom require attention. These are very simple in design and readjustments are not necessary.

On some engines, a *"leak down"* test is used to determine the condition of the hydraulic valve lifters.

— One way to make this is to insert a thickness gauge between the rocker arm and valve stem. Then, note the time it takes the valve lifter to leak enough oil to seat the valve. As the valve seats, the thickness gauge becomes loose. This indicates the end of the test.

— A more accurate leak-down test is made with the lifter out of the engine and installed in a leak-down tester. With this tester, the time required for a standard weight on the end of a lever to force the lifter plunger to bottom is measured. If the time is too short, the lifter is defective.

If the hydraulic valve lifter is defective, it is probably cheaper to replace it than to disassemble and service.

— To service the lifter, disassemble it and clean all parts in solvent. If any part is defective, the complete lifter should be replaced. On reassembly, fill the lifter with clean, light engine oil.

- When servicing and handling hydraulic valve lifters it is essential to keep everything clean, otherwise, *it takes only one tiny particle of dirt cause a lifter to malfunction.*

11.1.3.4. Cylinder head cylinder block

1. **Cylinder head :**

- The trouble that occurs in cylinder head is *warping*. If the warpage is slight, the head can be machined. A thicker cylinder gasket is used to compensate for the metal removed in machining the head. The best remedy is to put a new head.

  — When the head is to be replaced, it should be tightened down with a torque wrench to the manufacturers specifications. The torque wrench measures the twist or turning force used to tighten a nut or bolt. If it is not tightened properly, it will cause distortion of the cylinder block, which in turn will not allow the valves to seat squarely, eventually warping them with loss of compression. The distortion will also cause excessive wear of piston, cylinder wall and rings.

2. **Cylinder block :**

Cylinder wear. The following *factors tend to cause cylinder-wall wear :*

(*i*) The piston and ring movement.

(*ii*) The high temperatures and pressures of combustion.

(*iii*) The washing action of gasoline entering the cylinder.

— At the start of the power stroke, pressures are the greatest. The compression rings are forced with the greatest pressure against the cylinder wall. At the same time, the temperatures are highest while the oil film is least effective in protecting the cylinder walls. Therefore, the most wear ("taper") takes place at the top of the cylinder. As the piston moves down on the power stroke, the combustion pressure and temperature decrease. Less wear take place. The result is that the *cylinder wears irregularly.*

— The cylinder also *tends to wear slightly oval shaped.* This is due to the *side thrust of the piston* as it moves down in the cylinder on the power stroke. The side thrust results from the swing, from vertical, of the connecting rod. However, cylinders that are oval when cold will probably be less oval when hot.

— At times the air-fuel mixture is not perfectly blended. Small droplets of gasoline, still unvaporised, enter the cylinder and strike the cylinder wall, washing away the oil film. Thus, this area wear more rapidly.

Cleaning and inspecting cylinder block :

- During an engine overhaul, the cylinder block should be *thoroughly cleaned.* After removing the expansion plugs, oil-gallery plugs and any switches or other removable parts, the cylinder block should be cleaned in a hot tank or other suitable method.
- *Inspect* the cylinder bores for cracks, grooves, scratches or discolouration.
  - — Inspect the crankshaft bearing bores for signs of a spun bearing.
  - — Inspect for cracks across the top of the block between cylinders, between boltholes, on the outside of the block, and in the main-bearing bore webs.

Some limited types of block damage are repairable.

- If everything looks okay, measure the cylinder bores with a micrometer. If the bores are not excessively worn, and can be honed or rebored within specified limits, then clean and service the block. However, cylinder-block service is often performed by an automotive machine shop.

Cylinder service :

— The cylinders, upto certain limits, may wear tapered or out of round and not require refinishing.

— Some replacement piston-ring set can help control compression leakage and oil pumping in tapered and out-of-round cylinders. But when *wear becomes excessive, loss of compression, high oil consumption, poor performance, and heavy carbon accumulations in the cylinders will result. The cylinders must be refinished by honing or boring to restore proper engine operation.*

When the cylinders are honed or bored to a larger size, new pistons must be installed at the same time.

- *Wipe "walls"* and examine them for scores and spotty wear. Scores or spots will require honing or boring of the cylinder.
  - — Next, measure the cylinders for wear, taper, and out-of-round. This can be done with an inside *micrometer, a telescope gauge and an outside micrometer, or a cylinder-bore gauge.* The cylinder-bore gauge is moved up and down in the cylinder and rotated at various points to detect wear.
  - — A quick way to estimate *cylinder taper* is to push a compression ring down to the lower limit of ring travel. Measure the ring gap with a thickness gauge. Then pull the ring up to the upper limit of ring travel. Remeasure the ring gap. *The top gap minus the bottom gap divided by three gives the approximate cylinder taper.*

11.1.3.5. Oil pan

Oil pan is to be removed to clean it, to replace crankcase oil and gasket. After removing the oil pan, clean it thoroughly with gasoline to get all the heavy slug out so that it will not clog the oil system. Also, clean the oil screen and oil pump before replacing the pan. Scrap carefully the gasket material from the pan and block gasket surfaces. Apply new gas cement and lay the gasket in the place. Put the pan in the place and tighten the bolts or nuts to the proper tension.

11.1.3.6. Misassembly

During the assembly of engine, the following *precautions* should be taken :

1. In some of the engines, connecting rods are marked front, ensure that these are *fitted correctly*.

2. In some of the engines shims are placed between the cap and housing. Ensure that equal amount of shims is placed on both sides.

3. Connecting rod head and cap are mostly numbered to match each other, while assembling engine ensure that, the correct cap goes to the connecting rod.

4. While fitting bearing in cap, ensure that the locating lug fits snugly in the cap.

5. In main bearings, the upper half bearing is drilled to connect oil supply. While fitting ensure correct assembly otherwise, bearing will fail due to oil starvation.

6. While assembling ensure that oil hole drilled in the bearing matches with the hole in the block failing which the bearing will starve of oil and fail prematurely.

11.1.3.7. Maintenance of car engine-Useful hints

In order to maintain the engine in good condition, the following should be checked regularly at stipulated intervals :

1. Cylinder pressure.
2. Crankcase ventilation.
3. Air cleaner.
4. Carburettor.
5. The fuel pump.
6. Pollution by evaporation from the petrol tank and leaky fuel lines.
7. The ignition system.
8. Exhaust system.
9. The running temperature of the engine.

1. **Cylinder pressure :**

The cylinder pressure or compression should be maintained at prescribed level otherwise engine cannot run effectively and deliver the desired power. The pollutant level in the exhaust gases can never be kept at prescribed levels.

— Compression in the cylinders can easily be checked with a *compression gauge.*

2. **Crankcase ventilation :**

- All manufacturers provide for engine crankcase ventilation as well as absorption of the oil fumes by the air cleaner to be drawn into the engine alongwith air-fuel mixture.
- The pipes, passages and filler cap provided with breather passages or holes should be kept clean.

3. **Air cleaner :**

- The "*air cleaner*" *cleans the air going to the engine, silences the noise while the air is drawn into the engine and is also a flame arrester when engine back fires through the carburettor.*
- *A choked dry paper air cleaner or an oil both air cleaner with insufficient oil or unclean one, will result in the engine running on a rich mixture resulting in higher pollutants levels in the exhaust gases.* So it is necessary to have an air cleaner and maintain it in a good condition.
- Change the dry filter element ; and clean and refill oil bath every 1500 km.

4. **Carburettor :**

- The carburettor plays a significant role in the composition and levels of pollutants in the exhaust gases. It helps in vaporising the petrol as a fine mist, supply the correct strength of mixture at different conditions such as starting, idling, cruising and accelerating to enable the engine develop maximum possible power/speed with a possible economy on fuel.

— The choke is to be used only to give a rich mixture for starting in cold weather. Once the engine runs smoothly, the choke lever should be pushed back to original position and it should also be ensured that choke valve is fully open, otherwise the engine will continue to run on a rich mixture with the level of pollutants high in the exhaust gases.

- Carburettors should be adjusted correctly to allow engine to idle smoothly at 500 to 550 r.p.m. The engine *will not idle smoothly* due to the following *reasons* :

  (*i*) Advanced or retarded timing ;
  (*ii*) Incorrect gap at C.B. points or spark plugs ;
  (*iii*) Incorrect adjustment of volume control screw (air screw) in carburettor ;
  (*iv*) Checked exhaust silencer ;
  (*v*) Flooding of the carburettor due to faulty flood level or valve or higher discharge pressure of fuel pump due to sagged diaphragm besides poor and uneven compression in cylinders ;
  (*vi*) Improper valve clearance ;
  (*vii*) Rich mixture choked cleaner ;
  (*viii*) Burnt exhaust or inlet valves.

5. **The fuel pump :**

- The fuel pump should be serviced at prescribed levels say once in 3 months.
- If the fuel pump discharges pressure in excess (more than 2.2 to 2.8 psi in all the cars) then it might-either flood the carburettor and increase the fuel consumption even without flooding and result in a rich mixture being fed to the engine which will increase the level of pollutants in the exhaust gases.

6. **Pollution by evaporation from the petrol tank and leaky fuel lines :**

- Make sure that the tank is *never filled to the brim.* Always keep the level of fuel *above half the capacity of the tank to reduce evaporation loss and the pollution as well.*

7. **The ignition system :**

- The ignition system liberally contributes to the upkeep of the engine and its efficiency.
- The C.B. points should be clean and free from pitting or uneven surface.
- The spark plug used should be of the proper heat range recommended by the manufacturer and must be kept clean at every 5000 km.
- Care should be taken to ensure that the low or high tension circuit in the ignition system does not leak or short circuit.

8. **Exhaust system :**

- The exhaust system consists of the following :

  (*i*) Exhaust manifold, (*ii*) Exhaust pipe,
  (*iii*) Silencer, and (*iv*) Tail pipe.

— A choked silencer will create back pressure on the engine and exhaust valves. Consequently the engine will have to work extra to deliver the power it normally expects to deliver. This will *result in more fuel burnt resulting in more pollutants in exhaust gases.* This could be *detected when engine vibrates at cruising speed.*

9. **The running temperature of the engine :**

- The running temperature of the engine in all makes of cars should be between 75°C to 90°C.

The following *factors* other than the power unit, may cause more fuel to be used than necessary :

(*i*) Slipping clutch
(*ii*) Dragging brakes
(*iii*) Worn tyres, under or over inflated tyres
(*iv*) Tight wheel bearings
(*v*) Improper wheel alignment
(*vi*) Speed in excess of 50-60 kmph
(*vii*) Overloading.

11.1.3.8. Engine tuning

***Engine tuning** is a procedure to bring an engine in good mechanical condition to operate at peak efficiency and power.* Every unit of the engine is made to function at its best.

- Engine tuning is more frequently required in modern high compression engines which are more sensitive to variations from specifications than the earlier slow speed, low output engines.
- Often during the engine tuning, several conditions will be found which must be corrected to restore the engine to like new condition.

Tuning procedure :

Below is given a typical tuning or tune up procedure (some of the checks are not related to the engine but should be done for the safety purpose) :

1. After making the spark plugs loose, start the engine to blow out carbon and dirt, shut off engine and remove plugs.
2. Test engine compression. If the compression ratio is not upto specifications, perform engine services that will eliminate the trouble. If the compression is in order re-install the spark plugs.
3. Remove distributor cap. Clean it, and visually check it for carbon tracks, chips and corroded terminals. Replace it, if it is not in good condition.
4. Clean and inspect rotor. Replace it if it is not in good condition.
5. Inspect the high tension leads. If they have cracks, or frayed insulation or wires are damaged, replace them.
6. Check distributor centrifugal advance.
7. Test the vacuum advance.
8. Check distributor contact points and clean them. Read just the point opening.
9. Re-install distributor cap and replace wiring.
10. Check battery state of charge, water and hold-down clamps.
11. Check battery cables for damage, corrosion and loose connections and make necessary corrections.
12. If the battery has been overcharged or undercharged, check the alternator and regulator.
13. Check drive belts and tighten or replace them as required.
14. Check the condition of the manifold heat control valve, making sure that it is free to operate.

15. Check the intake manifold bolts for tightness to proper specifications. Even a slight leak will reduce engine performance.
16. Check fuel lines for tight connections and kinks, beads or leaks.
17. Check the cooling system for leaks, weak or collapsed hoses, correct coolant level and anti-freeze protection.
18. Check and adjust the accelerator linkage, if necessary.
19. Check crankcase ventilation system.
20. Check intake manifold and air injection system.
21. Remove carburettor, air cleaner, and check choke valve to make sure choke is working normally. Clean or replace air filter element, if necessary.
22. Check and adjust contact point dwell and ignition timing.
23. Adjust idle speed and mixture to specifications.
24. Check the doorjamb sticker to see if lubrication is required.
25. Check the working lights and horn. Check headlight adjustment.
26. Check steering system for looseness and ease of action.
27. Check suspension system and shock absorbers for looseness, excessive play and wear.
28. Check front wheels and ball joints for excessive wear or loose bearings.
29. Other tests that can be done include :
    (*i*) Cylinder balance test to find a weak cylinder ;
    (*ii*) Cranking motor operation ;
    (*iii*) Conditions of ignition coil and condenser ;
    (*iv*) Tightness of mounting bolts ;
    (*v*) Oil level in engine ;
    (*vi*) Air pressure in tyres ;
    (*vii*) Condition of tyres ;
    (*viii*) Efficiency of the brakes.

- **Engine vibrations.** The engine vibrations may be due to the following *possible causes* :

(*i*) Too stiff engine supports.
(*ii*) Misaligned engine-transmission-propeller shaft.
(*iii*) Crankshaft out of balance.
(*iv*) Clutch out of balance.
(*v*) Defective carburetion.
(*vi*) Defective ignition.
(*vii*) Maladjusted valves.
(*viii*) Connecting rods and pistons of unequal weight.

- **Crankshaft thumping.** "*Crankshaft thumping*" *means metallic, heavy and dull knocks which increase in frequency by the increase of rotational speed.*

Most commonly, thumping is caused by *excessive clearance* at one or more main bearings. An *excessive axial play* of crankshaft gives a sharper and higher noise at irregular intervals.

Following are the *causes of crankshaft thumping :*

(*i*) Excessive main bearing clearance.

(*ii*) Excessive crankshaft axial play.
(*iii*) Crankshaft out of balance.
(*iv*) Out-of-round or ovalized bearings.
(*v*) Loose flywheel.
(*vi*) Low oil pressure.
(*vii*) Insufficient lubrication.

- **Piston noise.** The piston noise, usually the *"thumping,"* is due to the *oscillation of piston in barrel.*
  - — It is heard as a dull and low sound.
  - — It is more evident at low r.p.m. rates under load.
  - — If the noise is slight when the engine is cold and disappear as the engine warms up, pistons do not require replacement.
- **Piston pin noise.** The piston pin noise is the *sharp and metallic knock easily heard when engine idles.*

The various *causes of piston noise are :*

(*i*) Excessive clearance between pin and piston boss bore.
(*ii*) Excessive clearance between pin and bush.
(*iii*) Bush loose in connecting rod small end.

- **Connecting rod noise :**
  - — The noise due to connecting rod is usually more intense than that originated by main bearings.
  - — The noise is more evident with engine idling and transmission in neutral and becomes high as the engine speed increases.

The following are the possible *causes of connecting rod noise* :

(*i*) Misaligned connecting rod. (*ii*) Excessive big end bearing clearance.
(*iii*) Ovalized or tapered crankpins. (*iv*) Low oil pressure.
(*v*) Insufficient lubrication.

- **Valve and tappet noise.** The valve and tappet give '*clicking*' noise. Its frequency is *lower* than that of any other noise of engine. If only one or two valves are noisy, the sound will be intermittent, otherwise it may be continuous.

The following are the *possible causes* of valve and tappet noise :

(*i*) Broken valve springs.
(*ii*) Excessive tappet clearance.
(*iii*) Excessive push rod-to-guide clearance.
(*iv*) Excessive wear of one or more cams of camshaft.
(*v*) Worn thread of set screws.
(*vi*) Excessive valve stem-to-guide bore clearance.

- **Oil low pressure.** Normal oil pressure in a new or overhauled engine must be 2.5 to 3 bar. *No pressure at all indicates a defective pump or oil leakage.*

The *possible causes* of *insufficient oil pressure* are :

(*i*) Defective relief valve. (*ii*) Defective oil pressure gauge.

(*iii*) Worn or damaged oil pump gears. (*iv*) Spoiled oil.

(*v*) Unsuitable grade of oil.

(*vi*) Excessive clearance between crankpin journals and bearings.

11.1.3.9. Auto inspection and testing

I. Inspection :

The *main object of inspection is to check up and verify* :

(*i*) Assemblies made up from different components such as engines, gear boxes, axles, steering boxes etc.

(*ii*) Mechanical build up of chassis or frame *i.e.*, the putting together of different assemblies.

(*iii*) Body build up, *e.g.*, painting, glazing, weather proofing and finish.

(*iv*) Electrical equipment.

(*v*) Running test, mechanical performance, steering and brakes.

(*vi*) Complaints, rectification and repairs under guarantee.

(*vii*) Diagnosis of mechanical and electrical faults as well as testing of the vehicle after repairs.

(*viii*) Inspection and reporting on accident damage after repair and repainting.

Inspection cards. The inspection is carried out systematically and recorded on a card.

The various types of inspection sheets/cards used are as follows.

Chassis Inspection Card

Particulars	**Rect. by**	**Passed**
Steering assembly		
Steering-wheel nut		
Steering ball joint		
Steering clamp bolts		
Brake system		
Complete check-top		
Complete check-bottom		
Incharge	Incharge Inspection department section **Complete Inspection**	

Vehicle make and year ...... **Capacity or horse power ......** **Regd. number**

Engine and Accessories :

Compression ......

Tappets : clearance ; noise ......

Valve gear : noise ; cover gaskets ......

Timing gears or Chain : noise ; backlash ......

Cylinder head : nuts ; gaskets ......

Manifolds : securing nuts : gaskets ; air or gas leaks ......

Driving belts : fan ; dynamo ; accessories ......

Water pump : leakage ; effectiveness ......

Radiator : water level ; leakage ; filler cap ; condition of core ; shutters ; relief valve ; overflow pipe ......

Oil leaks : crankcase ; pipes and unions ; timing covers, main bearings ; drain plugs ......

Oil filters : sump ; external ......

Engine mountings ......

Carburettor and Fuel pump :

Jets ......

Float chamber level ......

Chock : thermostat setting ......

Easy starting device : interconnecting linkage ; thermostat : dashboard control ......

Fuel pump : effectiveness ; delivery pressure ......

Filters : carburettor ; petrol pump ; petrol tank ; air cleaner ......

Ignition system :

Contact breaker points : gap ; condition ......

Rotor : condition ......

Distributor cap : cleanliness ; cracks ; tracking ......

Timing control : manual ; centrifugal ; suction operated ......

Coil and Wiring : efficiency ; condition of high and low tension wiring ; support ......

Magneto : efficiency ; impulse starter ; earthing wire ......

Sparking plugs : gap at points ; condition of ports and insulators ; washers ......

Electrical system :

Dynamo and Alternator : charging rate ......

Dynamo brushes and Commutator : burning ; chatter and sparking ......

Dynamo or Alternator drive : condition of belt ; slack in chain or gear ......

Voltage regulator ......

Cut out and Fuses ......

Ammeter : accuracy ......

Battery : acid level ; specific gravity ; voltage of each cell ; connections ; corrosion ......

Light :

Headlamps : focus and alignment ; condition of reflectors ......

Side lamps ......

Tail lamps ......

Stop lamps ......

Other lamps : ignition warning ; interior ; trafficators ......

Switch and Fuses ......

Electrical accessories :

Windscreen wipers : efficiency ; noise condition of blades ......

Trafficators ......

Starter : efficiency ; condition of pinion and flywheel teeth ......

Horns ......

Interior heater ......

Defroster ......

Radio ......

Steering :

Lost motion ......

Stiffness ......

Front axle alignment : toe in ; caster and camber angles ......

King pins and Bushes ......

Front hub bearings ......

Steering Linkage and joint ......

Record sheet :

...... Speedometer reading ...... Date of Inspection ......

Suspension :

Springs : condition of leaves ; broken leaves ; clips and shackles ......

Shock absorbers : adjustment ; fluid level ; security of attachments ......

Torque reaction roads or cables : adjustment ; security of attachment ......

Brakes :

Pedal travel ......

Condition of rods or cables ......

Hydraulic system : fluid level ; leakage from pipe lines ; choked pipes ; air in system ......

Handbrake : adjustment ; condition of ratchet ......

Servo mechanism : vacuum or air cylinder ; hydraulic booster pump ; valve mechanism accumulator ; pipe lines ......

Transmission :

Clutch : adjustment ......

Backlash : in universal joints ; in back axle ; in gearbox ......

Rear axle : leakage ......

Wheels :

Rims : buckling ; rust ......

Spokes : buckling ; looseness ......

Wheel nuts : tightness ; condition of studs ; splines on knock off wheels ......

Tyres :

Pressures ......

Tread wear ......

Condition of sidewalls ......

Valves and caps ......

Body work :

General appearance ......

Windows : winding controls ; glass ; channels ......

Doors : alignment ; buffers ; locks ; check straps ......

Sunshine roof : operation ; water leaks ......

Seats : condition of springs ; adjustments ......

Rear blind ......

Bumpers : condition ; attachment ......

Tool kit ......

Fire extinguisher ......

Road Test

Engine :

Ease of starting ......

Idling ......

Maximum speed ......

Oil consumption ......

Exhaust : noise ; restriction ;

Blue smoke ......

General noise ......

Oil pressure when hot ......

Acceleration ......

Fuel consumption ......

Engine temperature ......

Fumes ......

Clutch :

Slip ......

Spinning ......

Fierceness ......

Unusual noise ......

Gearbox :

Noisy gears ......

Pre-selector operation ......

Gears jumping out of mesh ......

Synchromesh operation ......

Vibration on gear lever ......

Propeller shaft :

Vibration ......

Noise ......

Back axle :

Noise on drive ......

Noise on over-run ......

Noise on over-run ......

Instruments :

Functioning ......

Accuracy ......

Suspension :

Comfort ......

Roll on corners ......

Brakes :

Power ......

Smoothness ......

Binding ......

Balance ......

Judder or squeal ......

Chassis :

Whip ...... Rattles ......

Steering :

Positiveness ...... Caster action ......

Wander ...... Road shocks at steering wheel.

Final Inspection Card

Vehicle type ... Chassis... Body No ...... Engine No ...... Date ......

S. No.		Particulars	Remarks	Rect. by	Passed by
1.	(*a*)	Check carpets.			
	(*b*)	Check front seats for adjustment.			
2.	(*a*)	Check doors and windows for operation.			
	(*b*)	Check boot lid and bonnet.			
3.		Check car for damaged paint work.			
4.		Check all touching up.			
5.		Check all glass and inside body for clean finish.			
6.		Check C.P. parts.			
7.		Check electrical equipment as per specifications.			
8.		Check history cards for clearance.			
9.		Check car plate.			
10.		Check for cleanliness of interior, without any surplus parts.			

Chief Inspector

Following inspection aspects should carefully dealt with :

- *Repair inspection.*
- *Accident inspection.*
- *Diagnosing engine faults and their removal.*
- *Excessive consumption.*

II. Testing :

"***Testing***" *provides the final proof of the satisfactory performance of the automobile and its engine.*

— The performance of an automobile is checked by carrying out both the *laboratory testing or test-bed and road testing of the vehicle.*

— The *laboratory test* consists not only the testing of the rated power output but also the tests are carried out after making final adjustments to check fuel consumption, acceleration, freedom from vibration and leakages as well as all other items required for making the engine acceptable for running.

A *laboratory test bed* consists of the following :

(*i*) A dynamometer (for measuring the engine output).

(*ii*) A carbon shaft or a flexible drive for connection of dynamometer to engine.

(*iii*) An adjustable engine stand or cradle.

(*iv*) Running-in equipment usually as separate motor.

(*v*) A control station (for providing necessary control levers and instruments for the engine as well as test data record or writing space).

(*vi*) Fuel supply equipment including a device for measuring the fuel consumption of the engine.

(*vii*) Equipment for exhaust gas disposal, water piping to and from the engine and dynamometer, engine lifting blocks accessories.

For the testing of automobiles and engines, the following *testing instruments* and *components* are used :

(*i*) Tachometer
(*ii*) Cylinder compression and leakage testers
(*iii*) Engine vacuum gauge
(*iv*) Exhaust gas analyser
(*v*) Full mileage and fuel pump testers
(*vi*) Flexiscope
(*vii*) Ignition timing electrical checks
(*viii*) Chassis dynamometer etc.

Road testing of an automobile :

The road testing of a new vehicle is carried out in the actual running conditions and environments. Road testing consists of the following two types of tests, which are *complementary* to each other :

1. Production test
2. Development test.

1. ***Production test :***

The main purpose of this test is to maintain a good standard of the automobile and take care to ensure that the finished products leave the floor in good standard conditions.

- In this test, a test route adjacent to the factory is mapped. The distances run vary considerably to the type of the vehicle under test. The road testing which is in general an inspection procedure should be as simple as possible. A report sheet in the form of a questionnaire is provided for each component from radiator to rear axle.
  — There are in general *two test runs* : First run performed by rough testers and final run performed by experienced and technical capable person after proper rectification.

2. ***Development test :***

These tests which are carried out by the research and development department are necessitated to put the ideas of the designer into practice for producing a vehicle to provide a trouble free service in different environments.

— These tests include a wide range of exhaustive trials carried out over a prolonged period. They are generally performed on the prototype of certain vehicles.

The following types of tests are generally performed :

(*i*) *Performance tests* :
— Preliminary tests
— Air resistance test
— Maximum speed test
— Acceleration test, etc.

(*ii*) *Fuel consumption tests* :
— Constant speed consumption test
— Overall consumption test
— Mixture strength test
— Cold start test, etc.

(*iii*) *Cooling system tests* :
— Radiator
— Cooling design
— Correct speed for fan and water pump
— Thermostatic control
— Fan design and ventilation tests.

(*iv*) *Brake tests* :
— Brake performance test
— Fade test
— Recovery test
— Racing conditions tests
— Brake down temperature recording
— Drum ventilation
— Linear wear tests.

(v) *Stearing and suspension tests* :
- — Roll and altitude angles
- — Turning effort
- — Shock absorber settings
- — Tests on rough surfaces
- — Stresses in the structure.

(vi) *Noise problem test.*

(vii) *Water proofing.*

(viii) *Dust sealing.*

(ix) *Final tests.* Accessory testing (lighting equipment, radio, heater, screen wipers etc.)

- **Road testing** *is carried out to find the following factors* :

(i) Tyre pressures
(ii) Conditions of tyres
(iii) Wheel lock
(iv) Caster, right and left
(v) Chamber, right and left
(vi) Toe in
(vii) Worn joints in steering linkage
(viii) Brake adjustment
(ix) Brake drag
(x) Spring clips
(xi) Shackles in good order
(xii) Steering gear mounting height
(xiii) Steering gear in mid-position when wheels are straight ahead
(xiv) Out of round brake drums
(xv) Wheel balance run out and bearing adjustment correct
(xvi) Axle location correct, front or rear
(xvii) Distorted rear axle housing, right or left
(xviii) Bent frame.

Engine test :

The engine testing can be carried out by fitting the engine to the vehicle and trying it out on the road. *Since the engine testing on road is quite difficult and expensive, therefore, for convenience and cheapness engines are tested in laboratory by mounting them in a frame or on a test bed with a controlled load provided to them.* In order to check their performance and behaviour, the following factors are checked and recorded (using different devices such as indicators, dynamometer etc.) :

(i) Engine speed
(ii) Load
(iii) Fuel consumption
(iv) Temperature of engine oil and working parts
(v) Heat given off by the cooling system
(vi) Exhaust temperature etc.

Standard operating conditions for an automotive engine (as per ISI) :

(i) *Mean barometric pressure* :	736 mm mercury corresponding to an altitude of 300 metres above mean sea level.
(ii) *Intake air temperature* :	35°C
(iii) *Humidity* :	Water vapour pressure of 27.4 mm of mercury, corresponding to a relative humidity of 65 percent at 35°C.
(iv) *Intake air depression and exhaust back pressure* :	Equal to obtaining with intake and exhaust systems normally fitted to engine.
(v) *Auxiliaries* :	The engine shall be driving all continuously running auxiliaries which are in service, mechanically connected to it and which are necessary for engine operation, such as coolant and lubricating oil pumps, scavenging blowers and in case of air-cooled engine only, the cooling fans.

- The *various tests to be carried out for performance of constant speed internal combustion engine for general purposes, are* :

(i) *Fuel consumption test.*

(ii) *12 hour rating test* consisting of a period of 11 hours of continuous run at rated fuel load at the rated speed to be followed immediately by one hour run at 10 percent overload. Observations are to be made every half an hour.

(iii) *Governing test.* While the test is in progress, readings are taken from various instruments and recorded in log sheet as recommended for the purpose. The results derived from the log sheet are plotted as performance curve.

Recommended Engine log sheet for I.S. specification for performance test of constant speed internal combustion engines for general purposes :

Engine manufacturer ......
Type ......
Model ...... Serial of ......
Rated r.p.m. ......
Bore ...... mm Stroke ...... mm
Place of test ...... Date ......
Testers ......
Observers ......

Sheet No. ...... of ......
Altitude ......
No. of cylinders ......
Nominal compression ratio ......
Fuel specification ......
(Including calorific value and specific gravity)
Cubic capacity ...... litre.
Oil specification ......
Dynameter : Type ......
Arm length ......
Constant ......

		Air tem. °C										*Fuel consumption*				*Cooling water*				
(1) Time	(2) Barometer mm of Hg	(3) Web bulb	(4) Dry bulb	(5) Engine intake	(6) Relative humidity	(7) Exhauste pressure, kg/cm^2	(8) Total correction factor	(9) Observed r.p.m.	(10) Brake load in kg	(11) Calculated b.h.p.	(12) Corrected b.h.p.	(13) Fuel consumed	(14) Fuel consumption time	(15) Fuel consumption rate g/hr	(16) Sp. fuel consumption g/b.h.p.	(17) Temp. inlet, °C	(18) Temp. outlet, °C	(19) Oil temp., °C	(20) Oil pressure, kg/cm^2	(21) Exhaust gas temp., °C

11.1.3.10. Maintenance of automobile body

In early days automobiles had wooden bodies with wooden framework and laminated material of wood like plywood, mica etc. With the increase in the demand of automobiles, new stamping techniques for sheet metal were developed with the help of different dies and heavy presses, so as to get the right shape using jigs and fixtures. *Different body panels and stampings are assembled together by spot welding in an all-steel-body shell.* It is important to take good care of vehicles to protect them from corrosion. Repair damage to the paint or protective coating as soon as possible.

Causes of corrosion and its prevention :

- Some common "*causes of corrosion*" are :
  (*i*) Accumulation of road salt, dust, moisture or chemicals in hard to reach areas of the vehicle underframe or frame.
  (*ii*) Chipping, scratches and any damage to treated or painted metal surfaces resulting from minor accidents or abrasion by stones and gravel.

The corrosion is accelerated by the following *environmental conditions* :

— High humidity particularly when the temperature range in just above the freezing point.

— Moisture in certain areas of a vehicle for an extended period of time.

— High temperatures cause an accelerated rate of corrosion to parts of the vehicle which are not well ventilated to permit quick drying.

— Road salt, dust chemicals, sea air or industrial pollution.

- The following methods are used *to prevent corrosion* :
  (*i*) Wash the vehicle frequently to preserve the finish and to help avoid corrosion. If the vehicle is driven on salted roads, it should be washed once a month in winter.
  (*ii*) Remove foreign material deposits such as salts, chemicals, road oil or tar, tree sap, bird droppings and industrial fall out which may damage the finish of the vehicle if left on painted surfaces. Remove these types of deposits as quickly as possible. Use a cleaner which is not harmful to the painted surface.
  (*iii*) Repair any damage to finish carefully and in time to the painted surfaces.
  (*iv*) Keep the passenger and luggage compartments clean ; moisture, dirt or mud can accumulate under the floor mats and cause corrosion.

Body care and maintenance :

- *Body care and maintenance* involve the following :

(*i*) Protecting the car	(*ii*) Paint body
(*iii*) Under carriage	(*iv*) Interior
(*v*) Windows	(*vi*) Engine compartment
(*vii*) Cleaning plastic parts	(*viii*) Vehicle store up recommendations.

- ***Vehicle care*** should be carried out regularly with the following *points* in mind :
  (*i*) *Store the vehicle in a dry, well ventilated area.*
  (*ii*) *Protect the vehicle with a cover* if a garage is not available.
  (*iii*) *Ensure safety :*
  - Do not wash and wax the vehicle with the engine running.
  - While cleaning the underside of the body and tender, avoid injury to your hand from sharp edged parts. Use hand gloves.
  - After washing, check up brakes before driving for safety.

(iv) *Clean the vehicle* carefully and properly.

- When cleaning the outside or inside of the vehicle, do not use volatile cleaning solvents such as acetone, lacquer thinners, enamel reducers, nail polish removers, laundry soaps, bleaching or reducing agents etc.
- Never use carbon tetrachloride, gasoline, benzene or naptha for any cleaning.
- Use proper cleaning techniques and cleaners while cleaning the interior for which vacuum cleaners may be used for dirt.
- Glass surfaces should be cleaned on a regular basis. Use of a glass cleaner will remove tobacco smoke and dust films. Never use abrasive cleaners on any vehicle glass to avoid scratches.
- The exterior paint finish providing beauty should be taken care of. Frequent washings with luke warm or cold water will help. Do not use a strong soap or chemical detergent. Periodic polishing and waxing will help to maintain the shine.

(v) *Washing the vehicle.* Depending upon the driving conditions, the vehicle should be washed at variable intervals. For washing a *car* it is lifted on a car lift and washed by a car washer. The car washer consists of a reciprocating pump run by a motor. The pump delivers water with pressure through a hose pipe and nozzle.

For *washing a car proceed as follows :*

- Lift the car on a car lift.
- Wash the bottom of car including wheels.
- Wash all body panels.
- Complete the washing with a sponge, rubbing gently at first to prevent scratching the finish with the entrained dirt and then more vigorously, using plenty of running water all the time.
- Dry car thoroughly with a clean chamois. No trace of water must remain on the finish.
- Rub a polishing compound of good commercial grade with a cloth.
- Grease, oil and tar stains on the body finish may be removed with some gasoline followed by immediate wiping with dry cloth.

(vi) *Foreign material deposits* (*e.g.,* calcium chloride and other salts, road oil and tar, tree and bird droppings, chemicals from industrial chimneys, etc., may damage the vehicle finish if left on the unpainted surface. Prompt washing may not completely remove all of these deposits. Other cleaners may be needed which may be safe for use on painted surfaces.

(vii) *Damage to finishing.* Any stone chips, fracture or deep scratches in the finish should be repaired promptly.

— Large areas of finish damage can be corrected by a body service dealer and point shop.

(viii) Underbody parts such as fuel lines frame, floor fans and the exhaust system should be *saved from corrosion.*

— Clean any areas where mud and debris can collect.

- **Painting work** involves the following *steps* :

(i) *Preparation of surface of steel body* ("*scrapping*").
(ii) *Application of red-oxider primer.* (iii) *Surface application.*
(iv) *Colour application.* (v) *Finish and polish.*

— Manual painting with the help of brushes does not give uniform thickness of paint and a smooth finish. As a result the paint comes off and body becomes spotted. Therefore, the *guns for spray painting using compressed air* have been devised for this process.

11.1.3.11. Automotive Vehicle Maintenance-General Hints

The general hints on the maintenance of automotive vehicle are summarised below :

1. Follow maintenance instructions given in the operation and maintenance manual supplied with the vehicle.
2. Always use lubricants of correct grade and quality at the correct interval as specified by the manufacturer.
3. Replace the engine oil as under :
   — Drain out the engine oil while warm ;
   — Fill in flushing oil upto "Min" (minimum) mark on dip stick and run the engine at about 1000 r.p.m. for 10 minutes ;
   — Drain out flushing oil ;
   — Refill with engine oil of specified grade.
4. Use purest available water in the radiator.
5. Be careful in using descaling compounds based on inorganic salts (*e.g.,* borates, carbonates, phosphates etc.), lest aluminium cylinder heads are affected by them.
6. In order to avoid crack of cylinder block, do not top up radiator with cold water after switching off engine when water level is very low and the engine is rather hot.
7. The radiator cap should be removed in two steps to check up water level in case engine is hot. Initially turn the cap to allow the discharge of pressure from the cooling system and then, only when the pressure is released.
8. For replacing water in the cooling system, drain water through the drain cocks located under the radiator and on cylinder block. Flush out radiator thoroughly if drained water is very dirty and full of sediments.

   Flush out the radiator as follows :
   — Drain out water from the cooling system.
   — Fill in water containing a 4 percent solution of sodium bicarbonate and run the engine at about 1000 r.p.m. for ten minutes.
   — Switch off engine and let the water stand in the radiator for half an hour.
   — Start the engine again and drain out water while the engine is running.
   — After draining water, switch off engine and let the engine cool down.
   — Keeping drain cocks open, push in water hose from external water supply into the radiator neck and circulate running water in the radiator and engine water jackets.
   — Close the cocks and fill radiator with clean water.
   — Run the engine for a while and drain the radiator again.
   — Close the cocks and fill radiator to normal level.
9. Use good quality antifreeze mixture containing corrosion initiator when engine temperature reaches 0°C.
10. Never use sand paper for refacing C.B. points but use oil stone.
11. While adjusting C.B. points and lubricating C.B. arm, be careful lest the points are upset resulting in arcing, rapid wear and ignition irregularities.
12. Keep the clutch pedal properly adjusted to avoid clutch slip which may result in overheating, unnecessary wear of facing and release bearing.
13. In case of excessive free play in brake pedal and braking unbalance, get the brake system checked and adjusted.
14. Use non-mineral brake fluid which should not corrode or affect rubber parts.

15. In order to avoid excessive tyre wear and steering irregularities, get the wheel alignment checked. In a properly adjusted steering system, the vehicle will continue to run straight ahead when hands are taken off the steering wheel and steering wheel shall return to straight ahead position after turning.
16. Keep tyres inflated to specific pressures.
17. Don't spill water on battery top but keep it clean and dry.
18. Add distilled water in battery when cold.
19. During garaging the vehicle for long periods recharge the battery once a month.
20. Check the wear and contact condition of dynamo and self-starter brushes and replace them if required.
21. Don't temper with cut out regulator adjustments.
22. Substitute burnt bulbs with bulbs of same type and capacity (wattage).
23. *Wash and clean the vehicle as under* :
    — Wash and clean the lower part of the vehicle first, including wheels, with water jet, using a sponge.
    — Wash the body shell avoiding excessive pressure water spray. Use sponge in washing. Rub it gently to avoid stretching.
    — Dry the vehicle with a chamois.
    — Remove grease, oil and tar spots from the painted parts by wetting with petrol and wiping immediately with a dry cloth.
    — For cleaning chrome plated parts, apply some natural vaseline and then rub with a soft and clean rag as well as chamois.
    — Clean windows and windshield with a linen cloth or a very soft chamois. Use windshield washer solvent or water containing some alcohol for very dirty glass panes.
    — Clean rubber mats with soap and water.
    — Clean leather seats or other parts with a wet and soapy sponge or cloth. Rinse carefully with clean water and dry well with chamois.
24. Avoid parking of vehicle in sun especially in summer.
25. Avoid spilling of petrol, alcohol, sodium compounds, windshield solvent or brake fluid on body finish.
26. Periodically clean water drain-holes at the bottom of doors to avoid rust formation.
27. Paint the under-chassis at least once a year to protect it from corrosion.
28. Never wash or polish the vehicle in sun especially in summer.
29. Park the vehicle under cover during nights, rainy days and winter months to avoid excessive condensation of water on vehicle body which would affect its finish.
30. Wipe out all drops of water penetrated inside the body as a result of moisture condensation otherwise water drops shall lead to rusting.
31. Touch the chipped off or scratched point to avoid corrosion.
32. Don't scuff painted surfaces with any rough object which may create scratches.
33. Dust off the body before washing or wiping with moist cloth.
34. Polish the vehicle with good quality wax polish after washing.

11.2. TROUBLESHOOTING

For keeping the vehicle rolling, the trouble must be shot (removed) immediately. The trouble can be discovered and removed if one is familiar with the construction and working of the automobile/vehicle.

An **automobile** is composed of the following *components/parts* :

(*i*) Frame and body
(*ii*) Springs
(*iii*) Shock absorbers
(*iv*) Axles and wheels
(*v*) Engine, clutch gear box and propeller shaft
(*vi*) Steering and linkages
(*vii*) Battery, horn, fuel tank and radiator
(*viii*) Accelerator, brake and clutch pedals
(*ix*) Switches, controls and gauges etc.

A **petrol vehicle** contains the following *systems* :

(*i*) Fuel system
(*ii*) Ignition system
(*iii*) Lubrication system
(*iv*) Cooling system
(*v*) Transmission system
(*vi*) Suspension system
(*vii*) Steering system
(*viii*) Brake system
(*ix*) Lighting and horn system
(*x*) Starting and charging system.

The manufacturers supply fault-finding charts for different makes/models of the above systems. These charts *act as a guide in troubleshooting.*

In order to give an idea of usual troubles in automobiles, their probable causes and remedies, a general chart of *petrol vehicles* is given below; however, *actual diagnosis and removal of defect needs thorough knowledge, practical experience and scientific approach to the problem.*

Troubleshooting of Automobile

S. No.	Troubles	Causes	Remedies
1.	***Engine does not start.***	(*i*) No fuel in tank. (*ii*) Choked fuel supply. (*iii*) Loose or leaking fuel line connection. (*iv*) Sticking needle valve in carburettor. (*v*) Flooded carburettor.	(*i*) Fill fuel. (*ii*) Clean and clear the obstruction in fuel line. (*iii*) Tighten loose connection or remove leakage. (*iv*) Release. (*v*) • Release sticking float. • Repair or replace punctured float. • Repair or replace defective needle valve. • Check and adjust float level. • Open choke valve if close. • Remove liquid petrol from induction manifold.
		(*vi*) Ignition off. (*vii*) Spark plugs dirty or bridged by soot or oil. (*viii*) Damaged spark plugs. (*ix*) Loose or dirty battery terminals. (*x*) Discharged battery. (*xi*) Loose or leaking H.T. lead from ignition coil. (*xii*) Defective or missing rotor. (*xiii*) Dirty, burned C.B. points. (*xiv*) C.B. points not opening. (*xv*) Defective ignition coil or condenser.	(*vi*) Switch on ignition/replace defective switch. (*vii*) Clean, check and adjust gap. (*viii*) Replace. (*ix*) Clean and tighten battery terminals. (*x*) Recharge. (*xi*) Tighten or replace. (*xii*) Replace/fix on rotor. (*xiii*) Clean or replace. (*xiv*) • Adjust gap. • Repair or replace defective C.B. arm. (*xv*) Replace.

		(*xvi*) Loose or leaking connection in ignition system. (*xvii*) Upset ignition-timing. (*xviii*) Distributor cap cracked. (*xix*) Distributor drive gear pin sheared. (*xx*) Poor compression.	(*xvi*) Remove leakage or tighten loose connection. (*xvii*) Reset. (*xviii*) Replace. (*xix*) Replace pin. (*xx*) • Tighten loose spark-plugs, cylinder head. Replace defective gaskets. • Check and adjust tappet clearance. • Replace worn out or broken piston rings. • Repair or replace sticking or leaking valves. • Replace worn out pistons.
2.	***Engine starts but stops immediately.***	(*i*) Faulty fuel pump. (*ii*) Restricted or leaking fuel line. (*iii*) Choked air-vent in fuel tank cap. (*iv*) Choked fuel filter. (*v*) Carburettor overflow. (*vi*) Sticking C.B. points. (*vii*) Loose connection in ignition circuits. (*viii*) Choked silencer.	(*i*) Check up and repair. (*ii*) • Clean/clear restriction. • Tighten loose connection. • Repair or replace defective pipe line. (*iii*) Clear the air-vent. (*iv*) Clean or replace filtering element. (*v*) Check as in trouble No. 1. (*vi*) Repair or replace. (*vii*) Tighten. (*viii*) Repair or replace
3.	***Starter does not turn engine.***	(*i*) Discharged battery. (*ii*) Loose connections in the starting system. (*iii*) Defective push button or solenoid switch. (*iv*) Defective self-starter. (*v*) Sticking starter pinion. (*vi*) Defective leads. (*vii*) Frozen water pump. (*viii*) Frozen engine oil. (*ix*) Seized engine. (*x*) Engaged gear.	(*i*) Recharge. (*ii*) Check up and tighten. (*iii*) Repair or replace. (*iv*) Repair. (*v*) Release or repair. (*vi*) Repair or replace. (*vii*) • Warm the engine and use anti-freeze mixture in the cooling system. • Repair defective water pump. (*viii*) • Warm the engine and replace engine oil. • Use correct grade of oil. (*ix*) Disassemble, find out fault and carry out necessary repairs and replacements. (*x*) Disengage.

4.	***Engine misfiring.***	(*i*) Faulty carburettor. (*ii*) Clogged air cleaner. (*iii*) Restricted fuel supply. (*iv*) Choked silencer. (*v*) Misadjusted valves. (*vi*) Sticking or leaking valves. (*vii*) Leaking cylinder head, induction manifold or carburettor flange. (*viii*) Defective spark plugs. (*ix*) Spark plugs of incorrect heat-value. (*x*) Dirty or pitted C.B. points. (*xi*) Defective ignition coil, condenser, distributer cap, rotor or H.T. leads. (*xii*) Loose connection or defective wiring. (*xiii*) Incorrect timing. (*xiv*) Defective automatic advance and retard mechanism. (*xv*) Water in fuel. (*xvi*) Unbalanced compression. (*xvii*) Tight engine. (*xviii*) Carbon deposit in engine combustion-chamber. (*xix*) Vacuum pipe connections loose at induction manifold.	(*i*) Clean and adjust. (*ii*) Clean or replace defective element. (*iii*) Clean and clear restriction. (*iv*) Clean. (*v*) Adjust tappet clearance. (*vi*) Repair. (*vii*) • Tighten loose bolts or nuts. • Replace defective gaskets. • Clean and adjust gap. (*viii*) Replace defective spark plugs. (*ix*) Use plugs of prescribed heat-value. (*x*) Clean/reface and adjust gap. (*xi*) Replace the defective parts. (*xii*) Repair or replace. (*xiii*) Adjust. (*xiv*) Repair or replace. (*xv*) Clean the fuel system and fill in fresh petrol. (*xvi*) Check up the cause and repair accordingly. (*xvii*) • Check up the cause of partial seizure and repair accordingly. • Check up cause for lack of lubrication and make necessary corrections. (*xviii*) Decarbonise. (*xix*) Tighten.
5.	***Engine continues to run after switching off.***	(*i*) Spark plugs of incorrect heat-value. (*ii*) Carbon deposit in combustion chamber. (*iii*) Defective ignition switch. (*iv*) Badly burnt or over-heated exhaust valves. (*v*) Engine overheating.	(*i*) Use spark-plugs of correct grade. (*ii*) Decarbonise. (*iii*) Replace. (*iv*) Replace (*v*) Check up cause of overheating and make necessary repairs.
6.	***Engine overheating.***	(*i*) Retarded timing. (*ii*) Distributor governor weights stuck in retarded position. (*iii*) Too lean fuel mixture. (*iv*) Less water in radiator. (*v*) Leakage of water in the cooling system.	(*i*) Check up and adjust. (*ii*) Check up, clean, release and lubricate sticking fly-weights. (*iii*) Clean and adjust carburettor. (*iv*) Add to level. (*v*) Repair the leakage.

		(*vi*) Choked radiator or water jackets.	(*vi*) Flush out and clean.
		(*vii*) Loose fan belt.	(*vii*) Adjust.
		(*viii*) Defective thermostat valve or water pump.	(*viii*) Repair or replace.
		(*ix*) Lack of lubrication.	(*ix*) ● Top up engine oil. ● Use correct grade and quality of oil. ● Clean or replace choked filter element. ● Check up serviceability of oil pump.
		(*x*) Tight engine.	(*x*) If new or overhauled engine, run carefully using correct grade of oil. If otherwise tight, check up mechanical failure and carry out necessary repairs.
		(*xi*) Weak compression.	(*xi*) ● Tighten loose cylinder head, spark plugs and manifold. ● Replace defective gaskets for above. ● Adjust tappet clearance. ● Repair or replace defective valves. ● Replace defective pistons and rings.
		(*xii*) Choked silencer.	(*xii*) Clean.
		(*xiii*) Clutch slipping.	(*xiii*) Adjust. If still slipping, dismantle and make out necessary repairs and replacements.
		(*xiv*) Brakes binding.	(*xiv*) ● Adjust. ● Lubricate the link and cable joints. ● Replace defective return/retracting springs. ● Replace defective rubber cups in master or wheel cylinders. ● Clean air vent and compensating port in master cylinder. ● Replace choked brakes hose pipes.
		(*xv*) Heat control valve sticking.	(*xv*) Repair or replace.
		(*xvi*) Excessive carbon deposit in engine.	(*xvi*) Decarbonise.
		(*xvii*) Incorrect use of gears.	(*xvii*) Avoid driving for too long period in low gears unless road conditions demand.
		(*xviii*) Inferior grade of petrol used, leading to detonation.	(*xviii*) Use proper grade of petrol.
		(*xix*) Exhaust valves burned.	(*xix*) Replace.
		(*xx*) Exhaust valves not opening to full extent.	(*xx*) Adjust tappet clearance.
		(*xxi*) Radiator partly blanked off.	(*xxi*) Remove curtain from the radiator core.

7.	***Engine producing abnormal noises.***	(*i*) Worn out gudgeon pins and small end bearings. (*ii*) Worn out piston and rings. (*iii*) Piston rings broken in grooves. (*iv*) Gudgeon pin circlips broken. (*v*) Worn out big-end bearings. (*vi*) Worn out crank-shaft main bearings. (*vii*) Excessive carbon deposit in engine. (*viii*) Ignition too far advanced. (*ix*) Inferior grade of fuel. (*x*) Wrong grade of spark-plugs. (*xi*) Loose flywheel or crank pulley. (*xii*) Worn out tappets and valve guides. (*xiii*) Worn rockers (O.H.V. engine) and shaft or bent push rods. (*xiv*) Excessive tappet clearance. (*xv*) Worn out timing gears or chain and sprockets. (*xvi*) Worn out distributor or oil pump drive gears. (*xvii*) Loose timing chain. (*xviii*) Dry generator or fan bearing. (*xix*) Defective water pump bearing. (*xx*) Lack of lubrication at distributor shaft bearing. (*xxi*) Hard or glazed dynamo brushes. (*xxii*) Fan blades striking radiator or hose. (*xxiii*) Slipping fan belt. (*xxiv*) Cranked or loose engine mountings.	(*i*) Replace. (*ii*) Replace. (*iii*) Make necessary repairs and replacements. (*iv*) Repairs and replacements. (*v*) Replace. (*vi*) Replace. (*vii*) Decarbonise. (*viii*) Adjust. (*ix*) Use proper grade of fuel. (*x*) Use spark plugs of correct heat value. (*xi*) Tighten. (*xii*) Replace. (*xiii*) Replace. (*xiv*) Adjust. (*xv*) Replace. (*xvi*) Replace. (*xvii*) ● Adjust. ● Replace defective chain tensioner. (*xviii*) Lubricate. (*xix*) Replace. (*xx*) Lubricate. (*xxi*) Lap and replace. (*xxii*) Make necessary adjustments. (*xxiii*) ● Adjust. ● Replace torn out fan belt. ● Align buckling crank pulley. (*xxiv*) Replace or tighten.
8.	***Excessive fuel consumption***	(*i*) Improper adjustments of carburettor. (*ii*) Dirty air cleaner. (*iii*) Leakage in the fuel line.	(*i*) Make necessary adjustments properly. (*ii*) Clean. (*iii*) ● Check up and remove leakage at different connections. ● Replace leaking pipes. ● Repair or remove leakage at fuel tank, filter, pump and carburettor.

		(*iv*) Sticking controls.	(*iv*) • Release sticking choke and throttle in carburettor. • Lubricate the link joints and shafts. • Make necessary adjustments.
		(*v*) Excessive idling.	(*v*) Do not allow the engine to idle unnecessary for long periods when vehicle is stationary.
		(*vi*) Engine overheating.	(*vi*) Avoid overheating.
		(*vii*) Dragging brakes.	(*vii*) Adjust or make necessary repairs and replacements.
		(*viii*) Underinflated tyres.	(*viii*) Inflate to proper pressure.
		(*ix*) Engine not tuned properly.	(*ix*) Tune up properly.
		(*x*) Vehicle overloaded.	(*x*) Avoid overloading.
		(*xi*) Unnecessary use of low gear.	(*xi*) Use gears intelligently.
		(*xii*) Operating with partially closed choke.	(*xii*) Check and adjust choke.
		(*xiii*) Oval or worn out jets in the carburettor.	(*xiii*) Replace.
		(*xiv*) Fierce acceleration and sudden braking.	(*xiv*) Drive carefully.
		(*xv*) Tight wheel bearings.	(*xv*) Adjust.
		(*xvi*) Slipping clutch.	(*xvi*) Adjust or repair.
		(*xvii*) Incorrect wheel alignment.	(*xvii*) Correct the alignment.
		(*xviii*) Worn out piston and rings.	(*xviii*) Replace.
		(*xix*) Incorrect spark plugs.	(*xix*) Use plugs of proper range.
9.	***Low or high oil pressure and excessive oil consumption in the engine.***	(*i*) Low oil pressure.	(*i*) • Top up engine oil if level low. • Ensure that proper grade and quality of oil is used. • Check up oil pressure gauge or indicator and replace if defective. • Remove any restriction in pipe leading to oil gauge. • Clean and adjust oil relief valve. • Replace defective ball seat or spiring in relief valve. • Adjust or replace loose or worn big-end and main bearings. • Clean oil strainer or suction side of oil pump. • Replace or repair worn out oil pump. • Remove leakage on suction side of oil pump. • Repair or replace leaking oil pipes.
		(*ii*) High oil pressure.	(*ii*) • Replace thick oil in winter. • Check up oil pressure gauge or indicator and replace if defective.

			• Check up adjustment of oil relief valve. • Remove restriction or replace choked oil pipe. • Check up engine bearing failure and as a result blocking of oil passages.
		(*iii*) Excessive oil consumption.	(*iii*) • Replace worn pistons and rings. • Rebore worn or scored cylinders. • Replace broken piston rings. • Scatter away if ring gaps in line. • Replace worn valve guides. • Replace defective oil seals on valve stems in O.H.V. engines. • Tighten bolts or replace valve chamber or oil sump gaskets if oil leaks through them. • Replace front and rear main bearing oil seals if oil leaks through them. • Replace damaged gasket of oil-sump drain plug. • Remove oil leakage at mechanical fuel pump flange. • Repair or replace defective positive crankcase ventilation (PVC) valve. • Remove any other external oil leakage. • Drain out excess oil. Maintain oil at the proper level.
10.	***Engine does not run slow.***	(*i*) Throttle valve open. (*ii*) Choke valve partly closed. (*iii*) Idle passage in carburettor restricted. (*iv*) Incorrect idle adjustment. (*v*) Worn out pilot slow running jet. (*vi*) Damaged or worn idle adjustment screw. (*vii*) Air cleaner dirty. (*viii*) Leakage at carburettor flange or induction manifold. (*ix*) Restriction in air-bleed holes in carburettor. (*x*) Damaged idle passage. (*xi*) Incorrect float level adjustment.	(*i*) Adjust. (*ii*) • Push down choke. • Adjust if still closed. (*iii*) clear off restriction. (*iv*) Adjust correctly. (*v*) Replace. (*vi*) Replace. (*vii*) Clean. (*viii*) • Tighten bolts and nuts • Replace damaged gaskets. (*ix*) Clean and blow out. (*x*) Repair. (*xi*) Adjust correctly.

11.	***Engine does not pick up speed.***	(*i*) Restriction in throttle valve operation. (*ii*) Metering rod in carburettor set too low. (*iii*) Restriction in fuel supply. (*iv*) Float level too low. (*v*) Carburettor overflowing. (*vi*) Restriction in main or high speed circuit of carburettor. (*vii*) Water in fuel. (*viii*) Retarded ignition. (*ix*) Wrong size main-jet installed. (*x*) Poor compression.	(*i*) Lubricate and remove any restriction. (*ii*) Adjust properly. (*iii*) Check up and remove. (*iv*) Adjust. (*v*) ● Adjust float level. ● Replace worn out needle valve. ● Replace punctured float. (*vi*) Clean and blow out restriction. (*vii*) ● Drain out fuel. ● Clean the fuel system. ● Fill new petrol. (*viii*) Adjust timing. (*ix*) Replace by correct size. (*x*) ● Remove leakage at cylinder head. ● Replace worn out pistons and rings. ● Replace leaking valves.
12.	***Engine popping and spilling (Explosions in carburettor and silencer).***	(*i*) Incorrect carburettor adjustment. (*ii*) Leaking valves (*iii*) Incorrect tappet clearance. (*iv*) Incorrect ignition timing. (*v*) Incorrect or defective spark plugs. (*vi*) Excessive carbon deposit in engine and silencer. (*vii*) Inferior grade of fuel. (*viii*) Lean mixture. (*ix*) Weak valve springs. (*x*) Defective ignition coil, condenser or C.B. points. (*xi*) Leakage at exhaust joint. (*xii*) Heat control valve defective or seized. (*xiii*) Exhaust valve head too thin, leading to overheating.	(*i*) Adjust properly. (*ii*) Repair or replace. (*iii*) Adjust. (*iv*) Adjust. (*v*) Replace with plugs of correct heat value. (*vi*) Decarbonise. (*vii*) Use proper grade fuel. (*viii*) Adjust correct mixture ratio. (*ix*) Replace. (*x*) Replace. (*xi*) Remove leakage. (*xii*) ● Replace. ● Install spring properly and free up valve. (*xiii*) Replace.
13.	***Clutch slipping.***	(*i*) No free play in pedal. (*ii*) Weak thrust springs. (*iii*) Worn clutch plate. (*iv*) Worn splines on primary shaft (clutch shaft). (*v*) Oil, grease or other such material on clutch facings. (*vi*) Pressure plate binding in cover.	(*i*) Adjust pedal free travel (about 1"). (*ii*) Replace. (*iii*) Replace. (*iv*) Replace shaft. (*v*) ● Clean clutch plate. ● Replace oil soaked clutch plate. (*vi*) Free up and lubricate.

		(*vii*) Excessively worn pressure-plate and flywheel. (*viii*) Sticking or misadjusted release fingers. (*ix*) Seized clutch linkage or release-mechanism.	(*vii*) Reface or replace. (*viii*) Free up, lubricate and adjust. (*ix*) Free up and lubricate.
14.	***Clutch dragging or spinning.***	(*i*) Oil or grease on clutch plate-linings. (*ii*) Clutch shaft out of line with engine. (*iii*) Warped or damaged pressure plate. (*iv*) Clutch plate hub binding on splines. (*v*) Clutch plate distorted. (*vi*) Cushion springs damaged.	(*i*) • Clean. • Replace oil soaked plate. (*ii*) • Replace worn out pilot bearing in crankshaft tail. • Replace worn out primary shaft bearing. • Tighten loose holding down bolts of gearbox. (*iii*) Replace. (*iv*) Replace. (*v*) Replace. (*vi*) Replace clutch plate.
15.	***Clutch judder.***	(*i*) Linings not making even contact. (*ii*) Pressure plate not parallel with flywheel. (*iii*) Buckled clutch plate. (*iv*) Bent clutch shaft.	(*i*) • Replace clutch plate. • Reface flywheel and pressure plate. (*ii*) • Tighten loose bolts holding pressure plate with cover. • Replace distorted pressure plate cover. (*iii*) Replace. (*iv*) Replace.
16.	***Clutch rattling.***	(*i*) Worn release bearing. (*ii*) Worn release bearing carrier. (*iii*) Worn release fork. (*iv*) Disconnected pedal return spring. (*v*) Release fork loose on ball stud.	(*i*) Replace. (*ii*) Replace. (*iii*) Replace. (*iv*) Connect. (*v*) Replace ball stud.
17.	***Gear slipping out of mesh.***	(*i*) Damaged ball, plunger or spring locking selector shaft or fork. (*ii*) Worn grooves of selector shaft. (*iii*) Worn shifter fork or gears in case of partial engagement. (*iv*) Damaged main-shaft pilot bearing. (*v*) Bent shifter fork.	(*i*) • Replace damaged ball. • Replace weak or broken spring. (*ii*) Repair or replace defective shaft. (*iii*) Replace. (*iv*) Replace. (*v*) Replace.
18.	***Gear changing difficult.***	(*i*) Defective synchronizing unit. (*ii*) Worn gearshift interlock guide plate. (*iii*) Worn reverse gear latch in case of reverse gear.	(*i*) Repair or replace. (*ii*) Replace. (*iii*) Replace.

		(*iv*) Gear shift lever lower end worn. (*v*) Damaged spring or positioning pins at the ball joint (fulcrum) of shift lever. (*vi*) Clutch dragging. (*vii*) Faulty selector mechanism. (*viii*) Gearbox cover misaligned. (*ix*) Misaligned linkage in case of remote control gear lever. (*x*) Binding in linkage of steering column gear lever.	(*iv*) Replace. (*v*) Replace. (*vi*) Remove trouble. (*vii*) • Lock the loose selector fork. • Lock the loose selector rods. (*viii*) Lift cover, fix in position and tighten bolts. (*ix*) Adjust and align properly. (*x*) Free up and lubricate.
19.	***Noisy gearbox operation.***	(*i*) Worn bearings. (*ii*) Worn gears. (*iii*) Foreign matter in gearbox. (*iv*) Damaged ball races. (*v*) Worn thrust washers. (*vi*) Lack of lubrication. (*vii*) Gearbox loose and out of alignment. (*viii*) Gears loose on shafts. (*ix*) Clutch shaft out of alignment. (*x*) Damaged rollers/needle bearing at the front end of main shaft.	(*i*) Replace. (*ii*) Replace. (*iii*) Clean the gearbox. (*iv*) Replace. (*v*) Replace. (*vi*) • Top up oil. • Use oil of specified grade. (*vii*) Tighten holding down bolts of gearbox and clutch housing. (*viii*) Replace gears and shafts. (*ix*) • Replace worn bearings. • Replace defective shaft. (*x*) Replace.
20.	***Propeller shaft gives noise and vibrates at speed.***	(*i*) Bent propeller shaft. (*ii*) Universal joints misaligned. (*iii*) Worn hanger bearing or cushion. (*iv*) Worn splines in slip joint. (*v*) Unbalanced propeller shaft. (*vi*) Loose 'U'-joint bolts.	(*i*) Replace. (*ii*) • Check up fitting and correct. • Replace worn spider and needle bearings. (*iii*) Replace. (*iv*) Replace shaft. (*v*) Replace. (*vi*) Tighten.
21.	***Noisy drive axle.***	(*i*) Worn gear and pinions. (*ii*) Improper adjustment of final drive. (*iii*) Lack of lubrication. (*iv*) Worn bearings. (*v*) Foreign matter in axle casing. (*vi*) Worn splines of axle shafts. (*vii*) Loose axle shafts. (*viii*) Loose or worn wheel bearings. (*ix*) Worn thrust washers on differential pinions. (*x*) Differential cage bolts loose.	(*i*) Replace. (*ii*) Adjust back-lash. (*iii*) • Replace defective oil seal. • Top up oil. (*iv*) Replace. (*v*) Remove and clean. (*vi*) Replace. (*vii*) Tighten and hold properly. (*viii*) Adjust or replace. (*ix*) Replace. (*x*) Tighten and lock properly.

22.	***Hard steering.***	(*i*) Front axle shifted away from seat. (*ii*) Lack of lubrication. (*iii*) Improper front-end alignment. (*iv*) Bent frame. (*v*) Steering gear tight. (*vi*) Tyres underinflated. (*vii*) Worn swivel pins and bushes. (*viii*) Pull and push rod joints tight. (*ix*) Bent or twisted steering arm. (*x*) Distorted steering column.	(*i*) Put on seat, tighten and lock loose "U" bolt nuts. (*ii*) • Replace leaking oil seal of steering gear box. • Top up oil in steering gear box • Lubricate steering knuckles and steering linkage. (*iii*) Adjust properly. (*iv*) Straighten. (*v*) Adjust back-lash. (*vi*) Inflate to specified pressure. (*vii*) Replace. (*viii*) Adjust. (*ix*) Replace. (*x*) • Align steering gearbox with mountings. • Tighten loose bolts.
23.	***Steering skimmy or wheel wobble.***	(*i*) Knuckle bearings loose or worn. (*ii*) Loose tie rod. (*iii*) Front axle shifted. (*iv*) Insufficient toe-in. (*v*) Excessive or insufficient caster. (*vi*) Improper load distribution. (*vii*) Unevenly inflated tyres. (*viii*) Excessive back-lash in steering gear. (*ix*) Unbalanced wheels. (*x*) Wheel rims buckled. (*xi*) Worn steering linkage connections. (*xii*) Uneven wear of tyres.	(*i*) Adjust or replace. (*ii*) Tighten or replace worn ends. (*iii*) Re-locate and tighten "U"-bolt nuts. (*iv*) Adjust. (*v*) Adjust. (*vi*) Equalize load. (*vii*) Inflate as specified. (*viii*) Adjust. (*ix*) Balance and fit correct weights. (*x*) Align. (*xi*) Adjust or replace. (*xii*) • Rotate tyres. • Adjust camber.
24.	***Vehicle wanders.***	(*i*) Low or unequal tyre pressure at front and rear. (*ii*) Uneven tyre pressure. (*iii*) Rear tyre pressures lower than front. (*iv*) Vehicle overloaded at rear. (*v*) Binding at some point in steering linkage. (*vi*) Insufficient caster angle. (*vii*) Front and rear axles shifted away. (*viii*) Tight steering gear. (*ix*) Front wheel bearings loose.	(*i*) Inflate as specified. (*ii*) —Do— (*iii*) —Do— (*iv*) Equalise load. (*v*) Check up and adjust/lubricate. (*vi*) Check up and adjust. (*vii*) Relocate and tighten U-bolts. (*viii*) • Adjust backlash. • Top up oil in steering gear-box. (*ix*) Adjust.

		(*x*) Twisted axle. (*xi*) Loose or worn spring shackles. (*xii*) Worn kingpins and bushes. (*xiii*) Loose steering gear-box (*xiv*) Loose wheel nuts. (*xv*) Pitman or drop arm loose.	(*x*) Replace. (*xi*) Adjust or replace worn parts. (*xii*) Replace. (*xiii*) Tight mounting-bolts. (*xiv*) Tighten. (*xv*) Tighten.
25.	***Poor (inefficient) brakes.***	(*i*) Excessive clearance between brake shoes and drum. (*ii*) Air in hydraulic brake system. (*iii*) New linings not bedded in. (*iv*) Oil or grease on brake linings. (*v*) Badly scored brake drums. (*vi*) Worn linings. (*vii*) Low fluid level in master cylinder. (*viii*) Rubber cups in masterand wheel cylinders damaged. (*ix*) Worn pistons in wheel-cylinders and master-cylinder. (*x*) Leakage in the break lines. (*xi*) Binding cables and brake expanders. (*xii*) Shoes incorrectly centred in brake drums. (*xiii*) Brakes not balanced. (*xiv*) Excessive brake pedal free play. (*xv*) Excessive play in wheel bearings.	(*i*) Adjust. (*ii*) Bleed out air. (*iii*) Reface. (*iv*) ● Clean. ● Replace oil soaked linings. (*v*) Skim or replace. (*vi*) Replace. (*vii*) Fill in fluid. (*viii*) Replace damaged cups. (*ix*) Replace worn pistons. (*x*) Remove leakage. (*xi*) Free up and lubricate. (*xii*) Centre properly. (*xiii*) Adjust. (*xiv*) Adjust. (*xv*) Adjust.
26.	***Brakes dragging.***	(*i*) Improper adjustment of brake linkage. (*ii*) Weak pedal return spring. (*iii*) Weak return spring in master cylinder. (*iv*) Mineral oil in the system. (*v*) Weak or broken brake shoe return spring in case of one wheel dragging. (*vi*) Tight brake shoe anchoring pins. (*vii*) Improper adjustment of brake shoes. (*viii*) Improper adjustment of wheel bearings. (*ix*) Swollen or distorted piston cups. (*x*) Restricted or damaged brake hoses or brake pipes. (*xi*) Oil or grease soaked linings.	(*i*) Adjust properly. (*ii*) Replace. (*iii*) Replace. (*iv*) Flush out system and use correct grade of fluid. (*v*) Replace return spring. (*vi*) ● Clean and lubricate. ● Replace defective parts. (*vii*) Adjust properly. (*viii*) Adjust properly. (*ix*) Replace rubber cups. (*x*) Replace. (*xi*) Replace.

		(*xii*) Choked air vent in master-cylinder filler cap. (*xiii*) No free play in brake pedal.	(*xii*) Clean. (*xiii*) Adjust.
27.	***Brake judder or squeal.*** ("*Judder*" is caused by low-frequency vibrations and "*squeal*" is caused by high frequency vibrations)	(*i*) Grease or oil on linings. (*ii*) Leading edge of lining picking up. (*iii*) Improper fitting of brake linings. (*iv*) Brake shoes out of line. (*v*) Rivets on shoes touching the brake drums. (*vi*) Worn shoe anchor-pins. (*vii*) Loose back plate. (*viii*) Weak brake drums due to excessive grinding.	(*i*) Clean or replace oil soaked linings. (*ii*) Chamfer the edge. (*iii*) Fit-in properly. If damaged, replace linings. (*iv*) Align. (*v*) Replace linings. (*vi*) Replace. (*vii*) Tighten. (*viii*) Replace.
28.	***Brakes spongy.***	(*i*) Air in hydraulic brake system. (*ii*) Improper brake fluid. (*iii*) Misadjusted brake shoes.	(*i*) Bleed out air. (*ii*) Flush out brake system and use proper grade of fluid. (*iii*) Adjust properly.
29.	***Noisy brakes.***	(*i*) Bent or distorted back plate. (*ii*) Bent or distorted brake shoes. (*iii*) Dirty linings. (*iv*) Loose lining rivets. (*v*) Improper linings. (*vi*) Worn drums.	(*i*) Repair or replace. (*ii*) Replace. (*iii*) Clean. (*iv*) Replace. (*v*) Replace with genuine ones. (*vi*) Skim or replace.
30.	***Vehicle pulls to one side.***	(*i*) Unequal adjustment of brake shoes. (*ii*) Oil or grease soaked linings of one wheel or one side. (*iii*) Loose back plate on one side. (*iv*) Wheel bearings improperly adjusted. (*v*) Improper tyre inflation. (*vi*) Improper linings. (*vii*) Loose leaf spring clips or U-bolts. (*viii*) Dust and grit in drums. (*ix*) Oval or scored drums. (*x*) Broken spring centre-bolt.	(*i*) Adjust evenly. (*ii*) Replace. (*iii*) Tighten. (*iv*) Adjust properly. (*v*) Inflate as specified. (*vi*) Replace with genuine ones. (*vii*) Avoid shifting of axles by tightening loose clips or U-bolts. (*viii*) Clean. (*ix*) Skim or replace. (*x*) Replace.
31.	***Vehicle sags to one side.***	(*i*) Weak road springs. (*ii*) Broken spring leaves. (*iii*) Incorrect adjustment of coil springs or torsion bars. (*iv*) Defective suspension units. (*v*) Defective shock absorbers.	(*i*) Replace. (*ii*) Replace. (*iii*) Adjust properly. (*iv*) Repair or replace. (*v*) Replace.

32.	***Harsh suspension or excessive road shocks transmitted to vehicle***	(*i*) Spring leaves corroded causing excessive friction. (*ii*) Wear depression in spring leaves. (*iii*) Wrong fitting of extra-load leaves. (*iv*) Seized shackle pins. (*v*) Defective shock-absorbers. (*vi*) Suspension arm pivots seized. (*vii*) Torsion bars incorrectly adjusted. (*viii*) Coil springs stiff.	(*i*) ● Clean and lubricate. ● Replace. (*ii*) Replace springs. (*iii*) Fit properly. (*iv*) ● Free up, clean and lubricate. ● Replace damaged parts. (*v*) Replace. (*iv*) ● Free up, clean and lubricate. ● Replace defective pins and bushes (*vii*) Adjust properly. (*viii*) Replace.
33.	***Excessive tyre wear.***	(*i*) Tyres improperly inflated. (*ii*) Buckled wheel. (*iii*) Incorrect toe-in. (*iv*) Uneven caster. (*v*) Fierce braking and acceleration. (*vi*) Loose wheel nuts. (*vii*) Sheared centre bolts of springs. (*viii*) Loose U-bolts of springs and axle. (*ix*) Oil soaked tyres. (*x*) Steering knuckle bearings worn. (*xi*) Loose steering connections. (*xii*) Loose or worn spring shackles. (*xiii*) Bent or twisted front axle. (*xiv*) Unbalanced wheels. (*xv*) Dragging brakes. (*xvi*) Worn or loose wheel bearings. (*xvii*) Bent frame. (*xviii*) Vehicle overloaded. (*xix*) Worn kingpins and bushes. (*xx*) Bent steering knuckles.	(*i*) Inflate as specified. (*ii*) Repair or replace. (*iii*) Adjust. (*iv*) Adjust. (*v*) Cultivate good driving habits. (*vi*) Tighten. (*vii*) Replace. (*viii*) Tighten. (*ix*) Replace. (*x*) Replace. (*xi*) Adjust, tighten or replace. (*xii*) Tighten or replace. (*xiii*) Replace. (*xiv*) Balance properly. (*xv*) Repair. (*xvi*) Replace or adjust. (*xvii*) Straighten or repair. (*xviii*) Keep load under permissible limits. (*xix*) Replace. (*xx*) Replace.
34.	***Discharged battery.***	(*i*) Ignition switch left on. (*ii*) Loose or dirty battery terminals. (*iii*) Low electrolyte level. (*iv*) Excessive use of self-starter. (*v*) Generator not charging. (*vi*) Leakage in wiring.	(*i*) Remove switch key for safety. (*ii*) Clean and tighten terminals. (*iii*) Fill in distilled water to correct level. (*iv*) Check up cause of difficult engine starting. (*v*) Check up generator and voltage regulator. (*vi*) Check up and repair.

		(*vii*) Cut out points stuck.	(*vii*) ● Check up and clean. ● Replace defective cut out regulator.
		(*viii*) Damaged battery.	(*viii*) Replace.
		(*ix*) Misadjusted or defective voltage regulator.	(*ix*) Adjust or replace.
		(*x*) Wrong connections of regulator.	(*x*) Set right.
35.	***Ammeter shows no charge.***	(*i*) Break in charging circuit.	(*i*) ● Check up any loose connection or short circuit and repair. ● Replace leaking or damaged leads.
		(*ii*) Generator producing weak or no current.	(*ii*) ● Clean dirty or greasy commutator. ● Replace worn out brushes or weak springs. ● Release sticking brushes in holder. ● Resolder broken connections at commutator. ● True up oval or worn commutator and undercut mica. ● Replace damaged commutator. ● Repair break in field coil circuit. ● Replace burnt or damaged field coils. ● Clean out slots in commutator bars. ● Replace damaged armature.
		(*iii*) Cut out points burnt.	(*iii*) Replace cut-out.
		(*iv*) Cut out points remain open.	(*iv*) ● Replace damaged cut-out. ● Clean dirty commutator.
		(*v*) Defective ammeter.	(*v*) Replace.
		(*vi*) Loose generator drive belt.	(*vi*) Tighten.
36.	***Ammeter shows overcharge.***	(*i*) High setting of voltage regulator.	(*i*) Check up and adjust.
		(*ii*) Generator field-windings shorted or grounded.	(*ii*) Repair.
		(*iii*) Short or open circuit in regulator.	(*iii*) Repair or replace regulator.
		(*iv*) Defective ammeter.	(*iv*) Replace.
37.	***Lighting system troubles.***	(*i*) Lamps give insufficient illuminations.	(*i*) ● Charge, repair or replace run-down or defective battery. ● Check up alignment of lamps and focusing of bulbs and correct if required. ● Clean or replace dirty or damaged reflectors.

			• Replace discoloured bulbs. • Check up earthing of lamps and clean dirty contact surfaces between lamp and mounting ; and mounting and chassis/body.
		(*ii*) Lamps gradually fade out after switching on.	(*ii*) Charge, repair or replace run-down or defective battery.
		(*iii*) Lights flicker.	(*iii*) Check up loose connection and tighten.
		(*iv*) Brilliance varies with vehicle speed.	(*iv*) • Charge, repair or replace run-down or defective battery. • Tighten loose bulbs or connections. • Repair or replace defective cables.
		(*v*) No lights.	(*v*) • Replace blown out fuses. • Tighten loos connection at supply point of lighting switch. • Replace defective switch.
38.	***Horn does not sound or sounds improperly.***	(*i*) Discharged battery. (*ii*) Loose connection. (*iii*) Break in circuit. (*iv*) Short circuit. (*v*) Defective switch. (*vi*) Blown off fuse. (*vii*) Poor earthing. (*viii*) Damaged C.B. points, resistor or field coil in horn. (*ix*) Incorrect tone adjustment. (*x*) Dust or water on horn diaphragm. (*xi*) Cracked diaphragm. (*xii*) Weak electromagnet.	(*i*) Recharge. (*ii*) Tighten. (*iii*) Repair. (*iv*) Trace out and repair. (*v*) Repair or replace. (*vi*) Replace. (*vii*) Clean the base and clamp horn properly. (*viii*) Replace damaged parts. (*ix*) Adjust properly. (*x*) Clean. (*xi*) Replace. (*xii*) Replace.

SHORT ANSWER QUESTIONS

Q. 1. What is meant by overhauling ?

Ans. Overhauling is a general process consisting of cleaning, inspection, diagnosis, re-adjustment, minor replacement and tuning up of engine.

Q. 2. When does an engine require overhaul ?

Ans. Engine requires overhaul when there is (*i*) loss of power, (*ii*) heavy oil consumption, and (*iii*) increased mechanical noise.

Q. 3. List out the steps to be adopted for complete overhauling of the propeller shaft.

Ans. (*i*) Removing the shaft, (*ii*) Dismantling, (*iii*) Inspection and service, (*iii*) Reassembling, and (*v*) Refitting.

Q. 4. What is the difference between minor overhauling and major overhauling ?

Ans. Minor overhauling consists of the removal of half shafts bearings and oil seals followed by inspection, adjustment and replacement. Major overhauling includes the removal, inspection, repair and reassembly of differential gears.

Q. 5. Name three reasons for preventive maintenance.

Ans. Reliability, service life and resale.

Q. 6. List the factors which cause higher oil consumption.

Ans. (*i*) Worn out piston rings, (*ii*) Leakage though gaskets, and (*iii*) Worn out valve guides.

Q. 7. If an engine races, what may be the causes ?

Ans. (*i*) The governor not working properly, (*ii*) The lost motion in the linkage between the governor and the fuel supply, (*iii*) The valves not working properly, and (*iv*) The injection pump not working properly.

Q. 8. What are the main causes of overheating ?

Ans. (*i*) Worn out fan belt or slip, (*ii*) Thermostat stuck ; (*iii*) External leak in cylinder gasket, (*iii*) Loose spark plug end, and (*iv*) Air leak in shroud seals.

Q. 9. What are the causes of excessive vibration in an engine ?

Ans. (*i*) Unbalance in firing arrangement, (*ii*) Unbalance in mechanical construction and (*iii*) Operation at the vibratory critical speed.

Q. 10. What is the main cause of chassis vibration ?

Ans. Unbalance in tyres and in propeller shaft.

Q. 11. What is the purpose of engine tune up ?

Ans. Engine tune-up is mainly to restore the engine's best level of performance, to maintain good fuel economy and minimum exhaust emission.

Q. 12. What is meant by engine tune-up ?

Ans. Engine tune-up refers to the procedure for inspecting, testing and adjusting an engine and replacing any worn out parts, to restore the engine to its best performance.

Q. 13. Name some tune-up instruments.

Ans. Vacuum gauge, compression gauge, distributor point gauge, oscilloscope, electric meters for testing the charging system, and chassis dynamometer.

HIGHLIGHTS

1. In order to preserve fuel working life of an automobile, its *maintenance* should be carried out religiously at the prescribed intervals and in the prescribed manner.
2. *Preventive maintenance* is the care to prevent troubles which often lead to a major overhaul. Preventive maintenance is cheaper.
3. The troubles and repairs of engine parts can be divided into the following six groups :
   (*i*) Connecting rod assembly
   (*ii*) Crankshaft assembly
   (*iii*) Valve mechanism
   (*iv*) Cylinder and cylinder block
   (*v*) Oil pan
   (*vi*) Various systems : Ignition system, cooling system, fuel system, lubricating system and starting and generator system.
4. *"Testing"* provides the final proof of the satisfactory performance of the automobile and its engine.
5. Road testing of an automobile consists of the following two types of tests, which are complementary to each other :
   (*i*) Production test ; (*ii*) Development test.

6. For keeping the vehicle rolling, the trouble must be shot (removed) immediately. The manufacturers supply fault-finding charts for different makes/models of various vehicles. These charts act as a guide in troubleshooting.

OBJECTIVE TYPE QUESTIONS

Fill in the blanks or Say 'yes' or 'No' :

1. Excessive oil in engine leads to oiling-up and dark white smoke.
2. Failure in oil supply leads the engine to ....
3. Too much grease on C.B. arm heel runs on to the breaker points and lead to fouling.
4. ...... maintenance is the care to prevent troubles which often lead to a major haul.
5. Preventive maintenance is costly.
6. When the engine has no trouble, the automobile runs economically and without any difficulty.
7. Resizing of collapsed or worn pistons is liberally recommended by the manufacturers.
8. Excessive resizing can ...... the piston.
9. Crankshaft is the backbone of the engine.
10. Precision Insert-type main bearings are rarely used in modern automotive engines.
11. Excessive valve-stem clearance (worn valve guides) causes carbon deposits on the valves and stems.
12. One of the causes of valve striking is warped ....
13. Any condition that causes valve to overheat may cause it to break.
14. Excessive tappet clearance causes heavy ...... seating.
15. The valve-tappet clearance is measured by a ...... gauge.
16. Non-hydraulic valve lifters give little trouble and can be easily removed from the engine for replacement.
17. Hydraulic valve lifters require a lot of attention.
18. If the hydraulic valve lifter is defective, it is probably cheaper to replace it than to disassemble and service.
19. When the cylinders are honed or bored to a larger size, new pistons must be installed at the same time.
20. For keeping the vehicle rolling, the trouble must be shot (removed) immediately.

Answers

1. Yes	**2.** seizure	**3.** Yes	**4.** Preventive	**5.** No
6. Yes	**7.** No	**8.** weaken	**9.** Yes	**10.** No
11. Yes	**12.** stems	**13.** Yes	**14.** impact	**15.** feeler
16. Yes	**17.** No	**18.** Yes	**19.** Yes	**20.** Yes.

THEORETICAL QUESTIONS

1. What are the factors on which the life of an automobile depends ?
2. Describe briefly the following two parts into which 'maintenance' can be categorised :
   (*i*) Lubrication
   (*ii*) Cleaning, inspection and adjustment.
3. Enumerate the automobile lubrication methods.
4. What do you mean by "Preventive maintenance" ? Explain briefly.
5. Is preventive maintenance cheaper ? Explain.
6. List the points relating to preventive maintenance which should be kept in mind for the trouble free working of the vehicle.

7. Explain briefly the troubles and repairs of engine parts relating to the following :
   (*i*) Connecting rod assembly
   (*ii*) Crankshaft assembly
   (*iii*) Cylinder head and cylinder block.
8. List some useful limits regarding maintenance of car engine.
9. What is engine 'tuning' ? Explain briefly the tuning procedure.
10. Write short note on Auto inspection and testing.
11. Explain briefly the road testing of an automobile.
12. Explain briefly the following tests relating to automobile testing :
   (*i*) Production test ; (*ii*) Development test.
13. Write short note on "Maintenance of automobile body".
14. State the causes of corrosion and its prevention, relating to automobile body.
15. List the points which should be kept in mind relating to "vehicle care".
16. Give a summary of general hints on the "maintenance of automotive vehicle".
17. Explain briefly "troubleshooting of automobile".

12

Miscellany

12.1. GAS POWER CYCLES

12.1.1. Definition of a Cycle

A **cycle** is defined as a *repeated series of operations occurring in a certain order*. It may be repeated by repeating the processes in the same order. The cycle may be of imaginary perfect engine or actual engine. The former is called **ideal cycle** and the latter **actual cycle.** In ideal cycle all accidental heat losses are prevented and the working substance is assumed to behave like a perfect working substance.

12.1.2. Air Standard Efficiency

To compare the effects of different cycles, it is of paramount importance that the effect of the calorific value of the fuel is altogether eliminated and this can be achieved by considering air (which is assumed to behave as a perfect gas) as the working substance in the engine cylinder. *The efficiency of engine using air as the working medium is known as an* "**Air standard efficiency**". This efficiency is oftenly called **ideal efficiency.**

The actual efficiency of a cycle is always less than the air-standard efficiency of that cycle under ideal conditions. This is taken into account by introducing a new term "**Relative efficiency**" which is defined as :

$$\eta_{relative} = \frac{\text{Actual thermal efficiency}}{\text{Air standard efficiency}} \qquad ...(12.1)$$

The analysis of all air standard cycles is based upon the following *assumptions* :

Assumptions :

1. The gas in the engine cylinder is a perfect gas *i.e.*, it obeys the gas laws and has constant specific heats.
2. The physical constants of the gas in the cylinder are the same as those of air at moderate temperatures *i.e.*, the molecular weight of cylinder gas is 29.

$$c_p = 1.005 \text{ kJ/kg-K}, \; c_v = 0.718 \text{ kJ/kg-K}.$$

3. The compression and expansion processes are adiabatic and they take place without internal friction, *i.e.*, these processes are isentropic.
4. No chemical reaction takes place in the cylinder. Heat is supplied or rejected by bringing a hot body or a cold body in contact with cylinder at appropriate points during the process.
5. The cycle is considered closed with the same 'air' always remaining in the cylinder to repeat the cycle.

12.1.3. The Carnot Cycle

This cycle has the *highest possible efficiency* and consists of *four* simple operations namely,

(*a*) Isothermal expansion (*b*) Adiabatic expansion

(*c*) Isothermal compression (*d*) Adiabatic compression.

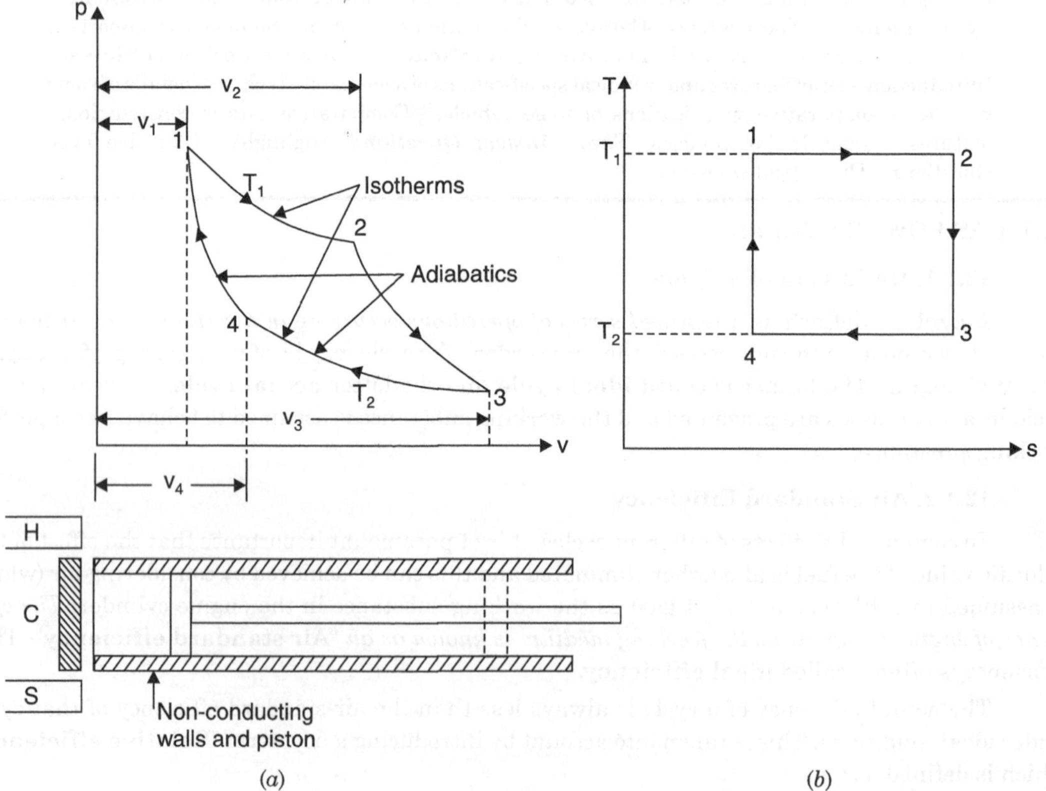

(*a*) (*b*)

Fig. 12.1

The condition of the Carnot cycle may be imagined to occur in the following way :

One kg of a air is enclosed in the cylinder which (except at the end) is made of perfect non-conducting material. A source of heat '*H*' is supposed to provide unlimited quantity of heat, non-conducting cover '*C*' and a sump '*S*' which is of infinite capacity so that its temperature remains unchanged irrespective of the fact how much heat is supplied to it. The temperature of source *H* is T_1 and the same is of the working substance. The working substance while rejecting heat to sump '*S*' has the temperature T_2 *i.e.*, the same as that of sump *S*.

Following are the *four stages* of the Carnot cycle. Refer to Fig. 12.1 (*a*).

Stage (1). Line 1-2 [Fig. 12.1 (*a*)] represents the isothermal expansion which takes place at temperature T_1 when source of heat *H* is applied to the end of cylinder. Heat supplied in this case is given by $RT_1 \log_e r$ and where *r* is the ratio of expansion.

Stage (2). Line 2-3 represents the application of non-conducting cover to the end of the cylinder. This is followed by the adiabatic expansion and the temperature falls from T_1 to T_2.

Stage (3). Line 3-4 represents the isothermal compression which takes place when sump '*S*' is applied to the end of cylinder. Heat is rejected during this operation whose value is given by $RT_2 \log_e r$ where *r* is the ratio of compression.

Stage (4). Line 4-1 represents repeated application of non-conducting cover and adiabatic compression due to which temperature increases from T_2 to T_1.

It may be noted that ratio of expansion during isothermal 1-2 and ratio of compression during isothermal 3-4 must be equal to get a closed cycle.

Fig. 12.1 (*b*) represents the Carnot cycle on *T-s* co-ordinates.

Now according to law of conservation of energy,

Heat supplied = Work done + heat rejected

Work done = Heat supplied – heat rejected

$$= RT_1 \,.\, \log_e r - RT_2 \log_e r$$

Efficiency of cycle

$$= \frac{\text{Work done}}{\text{Heat supplied}} = \frac{R \log_e r\,(T_1 - T_2)}{RT_1 \,.\, \log_e r}$$

$$= \frac{T_1 - T_2}{T_1} \quad \text{...(12.2)}$$

From this equation, it is quite obvious that if temperature T_2 decreases efficiency increases and it becomes 100% if T_2 becomes absolute zero which, of course is impossible to attain. Further more *it is not possible to produce an engine that should work on Carnot's cycle as it would necessitate the piston to travel very slowly during first portion of the forward stroke (isothermal expansion) and to travel more quickly during the remainder of the stroke (adiabatic expansion) which however is not practicable.*

Example 12.1. *A Carnot engine working between 400°C and 40°C produces 130 kJ of work. Determine :*

(*i*) *The engine thermal efficiency.*

(*ii*) *The entropy changes during heat rejection process.*

(*iii*) *The heat added.*

Solution. Temperature, $T_1 = T_2 = 400 + 273 = 673$ K

Temperature, $T_3 = T_4 = 40 + 273 = 313$ K

Work produced, $W = 130$ kJ.

(*i*) **Engine thermal efficiency,**

$$\eta_{th.} = \frac{673 - 313}{673} = \mathbf{0.535 \text{ or } 53.5\%.} \quad \textbf{(Ans.)}$$

(*ii*) **Heat added :**

$$\eta_{th.} = \frac{\text{Work done}}{\text{Heat added}}$$

i.e.,

$$0.535 = \frac{130}{\text{Heat added}}$$

$\therefore$ Heat added $= \dfrac{130}{0.535} = \mathbf{243 \text{ kJ.}} \quad \textbf{(Ans.)}$

(*iii*) **Entropy change during the heat rejection process, $(s_3 - s_4)$:**

Heat rejected = Heat added – work done

= 243 – 130 = 113 kJ

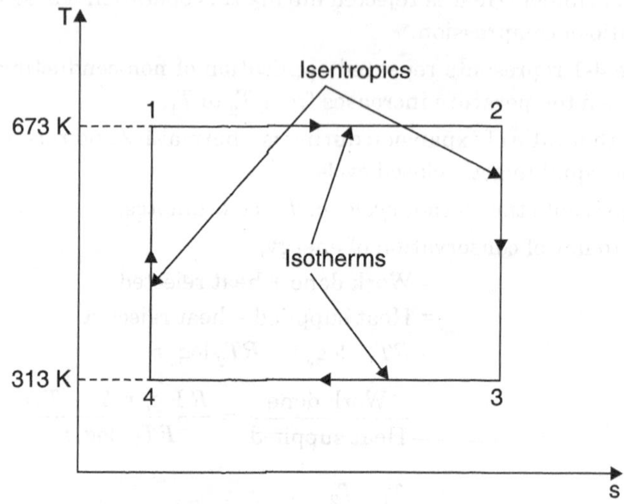

Fig. 12.2

Heat rejected $= T_3 (s_3 - s_4) = 113$

$\therefore$ $(s_3 - s_4) = \dfrac{113}{T_3} = \dfrac{113}{313} = \mathbf{0.361 \text{ kJ/K.}} \quad \textbf{(Ans.)}$

Example 12.2. *0.5 kg of air (ideal gas) executes a Carnot power cycle having a thermal efficiency of 50 percent. The heat transfer to the air during the isothermal expansion is 40 kJ. At the beginning of the isothermal expansion the pressure is 7 bar and the volume is 0.12 m*^{3}*. Determine :*

(*i*) *The maximum and minimum temperatures for the cycle in K.*

(*ii*) *The volume at the end of isothermal expansion in m*^{3}*.*

(*iii*) *The heat transfer for each of the four processes in kJ.*

For air $c_v = 0.721$ *kJ/kg K, and* $c_p = 1.008$ *kJ/kg K.*

Solution. Refer Fig. 12.3.

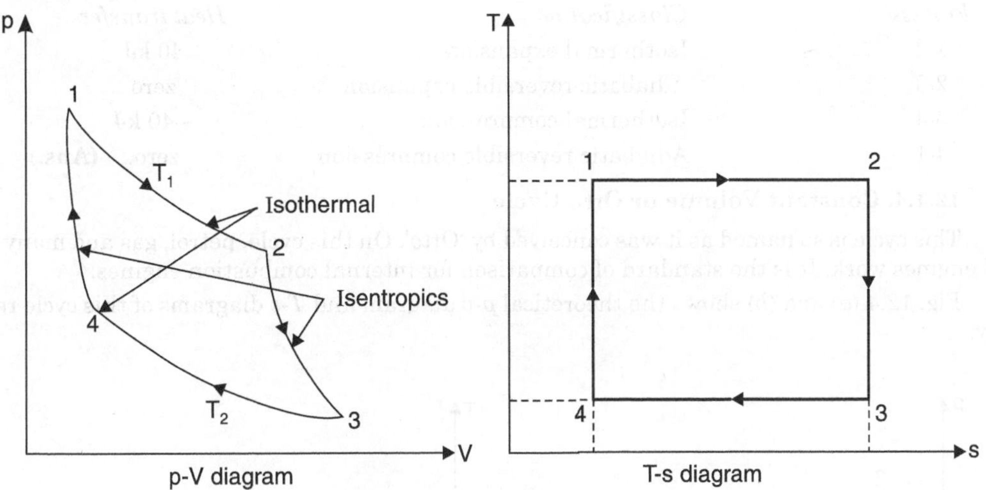

Fig. 12.3. Carnot cycle.

Given : $m = 0.5$ kg ; $\eta_{th} = 50\%$; Heat transferred during isothermal expansion = 40 kJ ;

$p_1 = 7$ bar, $V_1 = 0.12$ m^3 ; $c_v = 0.721$ kJ/kg K ; $c_p = 1.008$ kJ/kg K.

(*i*) **The maximum and minimum temperatures, $\mathbf{T_1, T_2}$:**

$$p_1V_1 = mRT_1$$

$$7 \times 10^5 \times 0.12 = 0.5 \times 287 \times T_1$$

$\therefore$ Maximum temperature, $\mathbf{T_1} = \dfrac{7 \times 10^5 \times 0.12}{0.5 \times 287} = \mathbf{585.4\ K.}$ **(Ans.)**

$$\eta_{cycle} = \frac{T_1 - T_2}{T_1}, \quad \text{or,} \quad 0.5 = \frac{585.4 - T_2}{585.4}$$

$\therefore$ Minimum temperature, $\mathbf{T_2} = 585.4 - 0.5 \times 585.4 = \mathbf{292.7\ K.}$ **(Ans.)**

(*ii*) **The volume at the end of isothermal expansion, $\mathbf{V_2}$:**

Heat transferred during isothermal expansion

$$= p_1V_1 \ln (r) = mRT_1 \ln \left(\frac{V_2}{V_1}\right) = 40 \times 10^3 \qquad ...(Given)$$

or, $$0.5 \times 287 \times 585.4 \ln \left(\frac{V_2}{0.12}\right) = 40 \times 10^3$$

or, $$\ln \left(\frac{V_2}{0.12}\right) = \frac{40 \times 10^3}{0.5 \times 287 \times 585.4} = 0.476$$

or, $$\mathbf{V_2} = 0.12 \times (e)^{0.476} = \mathbf{0.193\ m^3.} \quad \textbf{(Ans.)}$$

***(iii)* The heat transfer for each of the four processes :**

Process	*Classification*	*Heat transfer*
1-2	Isothermal expansion	40 kJ
2-3	Adiabatic reversible expansion	zero
3-4	Isothermal compression	– 40 kJ
4-1	Adiabatic reversible compression	zero. **(Ans.)**

12.1.4. Constant Volume or Otto Cycle

This cycle is so named as it was conceived by 'Otto'. On this cycle, petrol, gas and many types of oil engines work. It is the standard of comparison for internal combustion engines.

Fig. 12.4 (*a*) and (*b*) shows the theoretical *p-v* diagram and *T-s* diagrams of this cycle respectively.

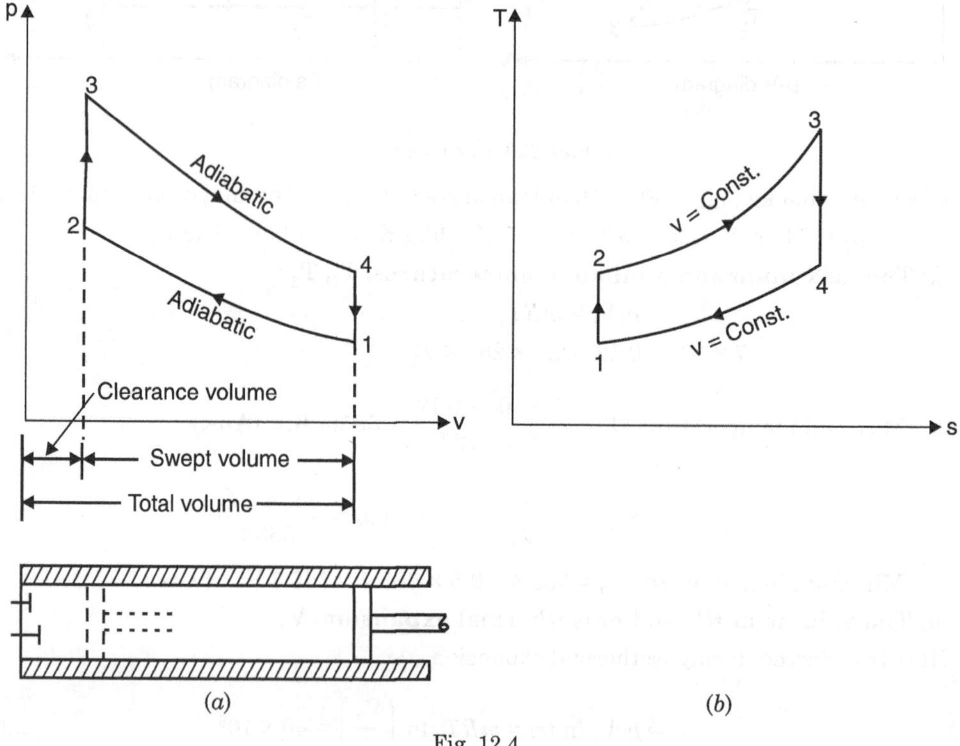

(*a*) (*b*)

Fig. 12.4

The point 1 represents that cylinder is full of air with volume v_1, pressure p_1 and absolute temperature T_1.

Line 1-2 represents the adiabatic compression of air due to which p_1, v_1 and T_1 change to p_2, v_2 and T_2 respectively.

Line 2-3 shows the supply of heat to the air at constant volume so that p_2 and T_2 change to p_3 and T_3 (v_3 being the same as v_2).

Line 3-4 represents the adiabatic expansion of the air. During expansion p_3, v_3 and T_3 change to a final value of p_4, V_4 or V_1 and T_4 respectively.

Line 4-1 shows the rejection of heat by air at constant volume till original state (point 1) reaches.

Consider 1 kg of air (working substance).

Heat supplied at constant volume = $c_v(T_3 - T_2)$.

Heat rejected at constant volume = $c_v(T_4 - T_1)$.

But, work done = Heat supplied – heat rejected

$$= c_v(T_3 - T_2) - c_v(T_4 - T_1)$$

$$\therefore \quad \text{Efficiency} = \frac{\text{Work done}}{\text{Heat supplied}} = \frac{c_v(T_3 - T_2) - c_v(T_4 - T_1)}{c_v(T_3 - T_2)}$$

$$= 1 - \frac{T_4 - T_1}{T_3 - T_2} \quad ...(i)$$

Let compression ratio, $r_c\ (= r) = \frac{v_1}{v_2}$

and, expansion ratio, $r_e\ (= r) = \frac{v_4}{v_3}$

(These two ratios are *same in this cycle*)

As, $\frac{T_2}{T_1} = \left(\frac{v_1}{v_2}\right)^{\gamma - 1}$

Then, $T_2 = T_1 \cdot (r)^{\gamma - 1}$

Similarly, $\frac{T_3}{T_4} = \left(\frac{v_4}{v_3}\right)^{\gamma - 1}$

or $T_3 = T_4 \cdot (r)^{\gamma - 1}$

Inserting the values of T_2 and T_3 in eqn. (*i*), we get

$$\eta_{otto} = 1 - \frac{T_4 - T_1}{T_4 \cdot (r)^{\gamma - 1} - T_1 \cdot (r)^{\gamma - 1}} = 1 - \frac{T_4 - T_1}{r^{\gamma - 1}(T_4 - T_1)}$$

$$= 1 - \frac{1}{(r)^{\gamma - 1}} \quad ...(12.3)$$

This expression is known as the air standard efficiency of the Otto cycle.

It is clear from the above expression that efficiency increases with the increase in the value of r, which means we can have maximum efficiency by increasing r to a considerable extent, but *due to practical difficulties its value is limited to about 7.*

The *net work done per kg* in the Otto cycle can also be expressed in terms of pv. If p is expressed in bar *i.e.*, 10^5 N/m², then work done

$$W = \left(\frac{p_3v_3 - p_4v_4}{\gamma - 1} - \frac{p_2v_2 - p_1v_1}{\gamma - 1}\right) \times 10^2 \text{ kJ} \quad ...(12.4)$$

Also, $\frac{p_3}{p_4} = r^\gamma = \frac{p_2}{p_1}$

$\therefore \quad \frac{p_3}{p_2} = \frac{p_4}{p_1} = r_p$

where r_p stands for *pressure ratio.*

and $v_1 = rv_2 = v_4 = rv_3 \quad \left[\because \frac{v_1}{v_2} = \frac{v_4}{v_3} = r\right]$

$$\therefore \quad W = \frac{1}{\gamma - 1}\left[p_4v_4\left(\frac{p_3v_3}{p_4v_4} - 1\right) - p_1v_1\left(\frac{p_2v_2}{p_1v_1} - 1\right)\right]$$

$$= \frac{1}{\gamma - 1}\left[p_4v_4\left(\frac{p_3}{p_4r} - 1\right) - p_1v_1\left(\frac{p_2}{p_1r} - 1\right)\right]$$

$$= \frac{v_1}{\gamma - 1}\left[p_4\,(r^{\gamma-1} - 1) - p_1\,(r^{\gamma-1} - 1)\right]$$

$$= \frac{v_1}{\gamma - 1}\left[(r^{\gamma-1} - 1)(p_4 - p_1)\right]$$

$$= \frac{p_1v_1}{\gamma - 1}\left[(r^{\gamma-1} - 1)(r_p - 1)\right] \qquad \text{...[12.4 (a)]}$$

Mean effective pressure ($\mathbf{p_m}$) is given by :

$$p_m = \left[\left(\frac{p_3v_3 - p_4v_4}{\gamma - 1} - \frac{p_2v_2 - p_1v_1}{\gamma - 1}\right) \div (v_1 - v_2)\right] \text{bar} \qquad \text{...(12.5)}$$

Also,

$$p_m = \frac{\left[\dfrac{p_1v_1}{\gamma - 1}(r^{\gamma-1} - 1)(r_p - 1)\right]}{(v_1 - v_2)}$$

$$= \frac{\dfrac{p_1v_1}{\gamma - 1}[(r^{\gamma-1} - 1)(r_p - 1)]}{(v_1 - v_2)} = \frac{\dfrac{p_1v_1}{\gamma - 1}[(r^{\gamma-1} - 1)(r_p - 1)]}{v_1 - \dfrac{v_1}{r}}$$

$$= \frac{\dfrac{p_1v_1}{\gamma - 1}[(r^{\gamma-1} - 1)(r_p - 1)]}{v_1\left(\dfrac{r-1}{r}\right)}$$

i.e.,

$$\mathbf{p_m} = \frac{p_1r[(r^{\gamma-1} - 1)(r_p - 1)]}{(\gamma - 1)(r - 1)} \qquad \text{...(12.6)}$$

Example 12.3. *The efficiency of an Otto cycle is 60% and $\gamma = 1.5$. What is the compression ratio ?*

Solution. Efficiency of Otto cycle, $\eta = 60\%$

Ratio of specific heats, $\gamma = 1.5$

Compression ratio, $r = ?$

Efficiency of Otto cycle is given by

$$\eta_{Otto} = 1 - \frac{1}{(r)^{\gamma-1}}$$

$$0.6 = 1 - \frac{1}{(r)^{1.5-1}}$$

or

$$\frac{1}{(r)^{0.5}} = 0.4, \quad \text{or,} \quad (r)^{0.5} = \frac{1}{0.4} = 2.5, \quad \text{or,} \quad r = 6.25$$

Hence, *compression ratio* $= \mathbf{6.25}$. **(Ans.)**

Example 12.4. *The minimum pressure and temperature in an Otto cycle are 100 kPa and 27°C. The amount of heat added to the air per cycle is 1500 kJ/kg.*

(i) Determine the pressures and temperatures at all points of the air standard Otto cycle.

(ii) Also calculate the specific work and thermal efficiency of the cycle for a compression ratio of 8 : 1.

Take for air : $c_v = 0.72$ *kJ/kg K, and* $\gamma = 1.4$.

Solution. Refer Fig. 12.5.

Given : $p_1 = 100$ kPa $= 10^5$ N/m² or 1 bar ; $T_1 = 27 + 273 = 300$ K ;

Heat added = 1500 kJ/kg ; $r = 8 : 1$; $c_v = 0.72$ kJ/kg ; $\gamma = 1.4$.

Consider 1 kg of air. Refer to Fig. 12.5.

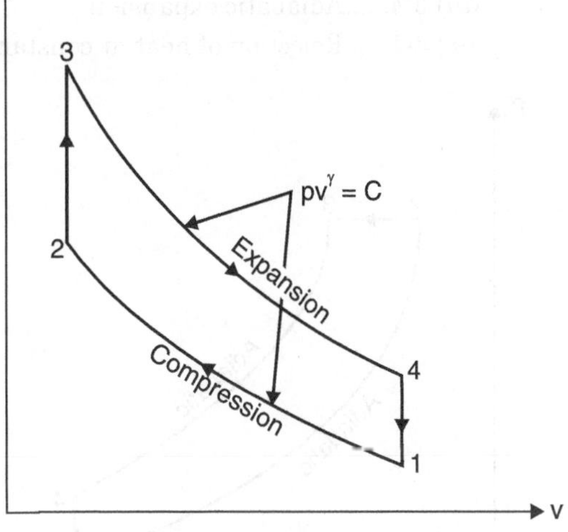

Fig. 12.5

(i) **Pressures and temperatures at all points :**

Adiabatic compression process 1-2 :

$$\frac{T_2}{T_1} = \left(\frac{v_1}{v_2}\right)^{\gamma - 1} = (r)^{\gamma - 1} = (8)^{1.4 - 1} = 2.297$$

$\therefore$ $\mathbf{T_2} = 300 \times 2.297 = \mathbf{689.1\ K.}$ **(Ans.)**

Also, $p_1 v_1^{\gamma} = p_2 v_2^{\gamma}$

or,
$$\frac{p_2}{p_1} = \left(\frac{v_1}{v_2}\right)^{\gamma} = (8)^{1.4} = 18.379$$

$\therefore$ $\mathbf{p_2} = 1 \times 18.379 = \mathbf{18.379\ bar.}$ **(Ans.)**

Constant volume process 2-3 :

Heat added during the process,

$$c_v (T_3 - T_2) = 1500$$

or,
$$0.72 (T_3 - 689.1) = 1500, \quad \text{or,} \quad \mathbf{T_3} = \frac{1500}{0.72} + 689.1 = \mathbf{2772.4\ K.} \ \textbf{(Ans.)}$$

Also,
$$\frac{p_2}{T_2} = \frac{p_3}{T_3}, \quad \text{or,} \quad \mathbf{p_3} = \frac{p_2 T_3}{T_2} = \frac{18.379 \times 2772.4}{689.1} = \mathbf{73.94\ bar.} \ \textbf{(Ans.)}$$

Adiabatic expansion process 3-4 :

$$\frac{T_3}{T_4} = \left(\frac{v_4}{v_3}\right)^{\gamma - 1} = (r)^{\gamma - 1} = (8)^{1.4 - 1} = 2.297$$

$$\therefore \quad \mathbf{T_4} = \frac{T_3}{2.297} = \frac{2772.4}{2.297} = \mathbf{1206.9\ K.} \ \textbf{(Ans.)}$$

Also,
$$p_3 v_3^{\gamma} = p_4 v_4^{\gamma}, \quad \text{or,} \quad \mathbf{p_4} = p_3 \times \left(\frac{v_3}{v_4}\right)^{\gamma} = 73.94 \times \left(\frac{1}{8}\right)^{1.4} = \mathbf{4.023\ bar.} \ \textbf{(Ans.)}$$

(ii) **Specific work and thermal efficiency :**

Specific work = Heat added – heat rejected

$$= c_v (T_3 - T_2) - c_v (T_4 - T_1) = c_v [(T_3 - T_2) - (T_4 - T_1)]$$

$$= 0.72 [(2772.4 - 689.1) - (1206.9 - 300)] = \mathbf{847\ kJ/kg.} \ \textbf{(Ans.)}$$

Thermal efficiency, $\eta_{th} = 1 - \frac{1}{(r)^{\gamma-1}} = 1 - \frac{1}{(8)^{1.4-1}} = 0.5647$ or **56.47%. (Ans.)**

12.1.5. Constant Pressure or Diesel Cycle

This cycle was introduced by Dr. R. Diesel in 1897. It differs from Otto cycle, in that heat is supplied at constant pressure instead of at constant volume. Fig. 12.6 (*a* and *b*) shows the *p-v* and *T-s* diagrams of this cycle respectively.

This cycle comprises of the following *operations* :

(*i*) 1-2......Adiabatic compression.

(*ii*) 2-3......Addition of heat at constant pressure.

(*iii*) 3-4......Adiabatic expansion.

(*iv*) 4-1......Rejection of heat at constant volume.

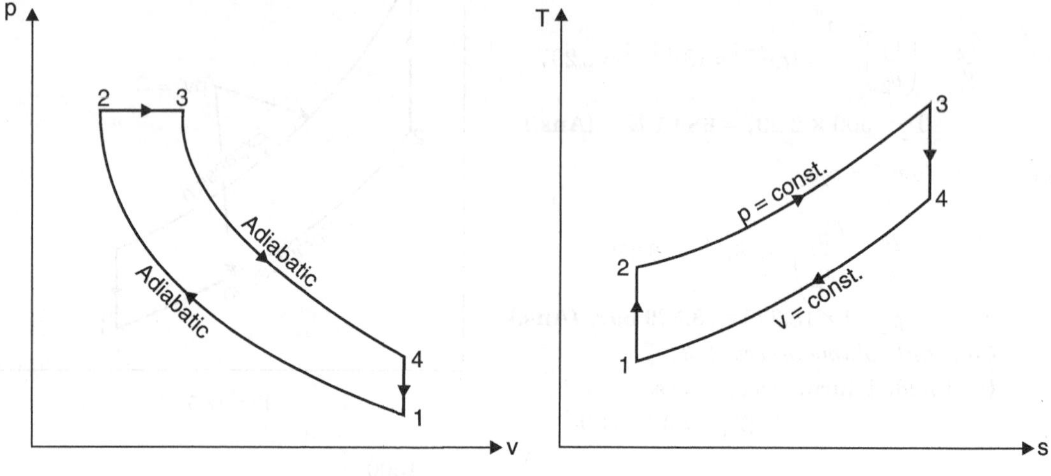

Fig. 12.6

Point 1 represents that the cylinder is full of air. Let p_1, v_1 and T_1 be the corresponding pressure, volume and absolute temperature. The piston then compresses the air adiabatically (*i.e.,* pv^γ = constant) till the values become p_2, v_2 and T_2, respectively (at the end of the stroke) at point 2. Heat is then added from a hot body at a constant pressure. During this addition of heat let volume increases from v_2 to v_3 and temperature T_2 to T_3, corresponding to point 3. This point (3) is called the *point of cut off*. The air then expands adiabatically to the conditions p_4, v_4 and T_4 respectively corresponding to point 4. Finally, the air rejects the heat to the cold body at constant volume till the point 1 where it returns to its original state.

Consider 1 kg of air.

Heat supplied at constant pressure $= c_p(T_3 - T_2)$

Heat rejected at constant volume $= c_v(T_4 - T_1)$

Work done $=$ Heat supplied – heat rejected

$$= c_p(T_3 - T_2) - c_v(T_4 - T_1)$$

$$\therefore \qquad \eta_{diesel} = \frac{\text{Work done}}{\text{Heat supplied}}$$

$$= \frac{c_p(T_3 - T_2) - c_v(T_4 - T_1)}{c_p(T_3 - T_2)}$$

$$= 1 - \frac{(T_4 - T_1)}{\gamma(T_3 - T_2)} \qquad \text{...}(i) \left[\because \frac{c_p}{c_v} = \gamma\right]$$

Let compression ratio, $r = \frac{v_1}{v_2}$

and, cut-off ratio, $\rho = \frac{v_3}{v_2}$ *i.e.,* $\frac{\text{Volume at cut-off}}{\text{Clearance volume}}$

Now, during *adiabatic compression 1-2,*

$$\frac{T_2}{T_1} = \left(\frac{v_1}{v_2}\right)^{\gamma - 1} = (r)^{\gamma - 1}, \text{ or, } T_2 = T_1 \, . \, (r)^{\gamma - 1}$$

During *constant pressure process 2-3,*

$$\frac{T_3}{T_2} = \frac{v_3}{v_2} = \rho, \text{ or, } T_3 = \rho \, . \, T_2 = \rho \, . \, T_1 \, . \, (r)^{\gamma - 1}$$

During *adiabatic expansion 3-4,*

$$\frac{T_3}{T_4} = \left(\frac{v_4}{v_3}\right)^{\gamma - 1}$$

$$= \left(\frac{r}{\rho}\right)^{\gamma - 1} \qquad \left(\because \frac{v_4}{v_3} = \frac{v_1}{v_3} = \frac{v_1}{v_2} \times \frac{v_2}{v_3} = \frac{r}{\rho}\right)$$

$$\therefore \quad T_4 = \frac{T_3}{\left(\frac{r}{\rho}\right)^{\gamma - 1}} = \frac{\rho \, . \, T_1 (r)^{\gamma - 1}}{\left(\frac{r}{\rho}\right)^{\gamma - 1}} = T_1 \, . \, \rho^{\gamma}$$

By inserting values of T_2, T_3 and T_4 in eqn. (*i*), we get

$$\eta_{\text{diesel}} = 1 - \frac{(T_1 \, . \, \rho^{\gamma} - T_1)}{\rho[(\rho \, . \, T_1 \, . \, (r)^{\gamma - 1} - T_1 \, . \, (r)^{\gamma - 1})]} = 1 - \frac{(\rho^{\gamma} - 1)}{\gamma(r)^{\gamma - 1}(\rho - 1)}$$

or,

$$\eta_{\text{diesel}} = 1 - \frac{1}{\gamma(r)^{\gamma - 1}} \left[\frac{\rho^{\gamma} - 1}{\rho - 1}\right] \qquad \text{...(12.7)}$$

It may be observed that eqn. (12.7) for efficiency of diesel cycle is different from that of the Otto cycle only in bracketed factor. This factor is always greater than unity, because $\rho > 1$. Hence *for a given compression ratio, the Otto cycle is more efficient.*

The *net work* for diesel cycle can be expressed in terms of *pv* as follows :

$$W = p_2(v_3 - v_2) + \frac{p_3 v_3 - p_4 v_4}{\gamma - 1} - \frac{p_2 v_2 - p_1 v_1}{\gamma - 1}$$

$$= p_2 \, (\rho v_2 - v_2) + \frac{p_3 \rho v_2 - p_4 r v_2}{\gamma - 1} - \frac{p_2 v_2 - p_1 r v_2}{\gamma - 1}$$

$$\left[\begin{array}{l} \because \frac{v_3}{v_2} = \rho \quad \therefore \; v_3 = \rho v_2 \text{ and } \frac{v_1}{v_2} = r \quad \therefore \; v_1 = r v_2 \\ \text{But } v_4 = v_1 \quad \therefore \quad v_4 = r v_2 \end{array}\right]$$

$$= p_2v_2(\rho - 1) + \frac{p_3\rho v_2 - p_4 r v_2}{\gamma - 1} - \frac{p_2v_2 - p_1 r v_2}{\gamma - 1}$$

$$= \frac{v_2[p_2(\rho - 1)(\gamma - 1) + p_3\rho - p_4 r - (p_2 - p_1 r)]}{\gamma - 1}$$

$$= \frac{v_2\left[p_2(\rho - 1)(\gamma - 1) + p_3\left(\rho - \frac{p_4 r}{p_3}\right) - p_2\left(1 - \frac{p_1 r}{p_2}\right)\right]}{\gamma - 1}$$

$$= \frac{p_2v_2[(\rho - 1)(\gamma - 1) + \rho - \rho^\gamma . r^{1-\gamma} - (1 - r^{1-\gamma})}{\gamma - 1}$$

$$\left[\because \frac{p_4}{p_3} = \left(\frac{v_3}{v_4}\right)^\gamma = \left(\frac{\rho}{r}\right)^\gamma = \rho^\gamma r^{-\gamma}\right]$$

$$= \frac{p_1v_1 r^{\gamma - 1}[(\rho - 1)(\gamma - 1) + \rho - \rho^\gamma r^{1-\gamma} - (1 - r^{1-\gamma})}{\gamma - 1}$$

$$\left[\because \frac{p_2}{p_1} = \left(\frac{v_1}{v_2}\right)^\gamma \text{ or } p_2 = p_1 . r^\gamma \text{ and } \frac{v_1}{v_2} = r \text{ or } v_2 = v_1 r^{-1}\right]$$

$$= \frac{p_1v_1 r^{\gamma - 1}[\gamma(\rho - 1) - r^{1-\gamma}(\rho^\gamma - 1)]}{(\gamma - 1)} \quad ...(12.8)$$

Mean effective pressure $\mathbf{p_m}$ is given by :

$$p_m = \frac{p_1v_1 r^{\gamma - 1}[\gamma(\rho - 1) - r^{1-\gamma}(\rho^\gamma - 1)]}{(\gamma - 1)v_1\left(\frac{r - 1}{r}\right)}$$

or,

$$\mathbf{p_m} = \frac{p_1 r^\gamma[\gamma(\rho - 1) - r^{1-\gamma}(\rho^\gamma - 1)]}{(\gamma - 1)(r - 1)} . \quad ...(12.9)$$

Example 12.5. *A diesel engine has a compression ratio of 15 and heat addition at constant pressure takes place at 6% of stroke. Find the air standard efficiency of the engine.*

Take γ for air as 1.4.

Solution. Refer to Fig. 12.7.

Compression ratio, $r \left(= \frac{V_1}{V_2}\right) = 15$

γ for air = 1.4

Air standard efficiency of diesel cycle is given by

$$\eta_{\text{diesel}} = 1 - \frac{1}{\gamma(r)^{\gamma - 1}}\left[\frac{\rho^\gamma - 1}{\rho - 1}\right] \quad ...(i)$$

where, ρ = Cut-off ratio = $\frac{V_3}{V_2}$

But, $V_3 - V_2 = \frac{6}{100} V_s$ (V_s = stroke volume)

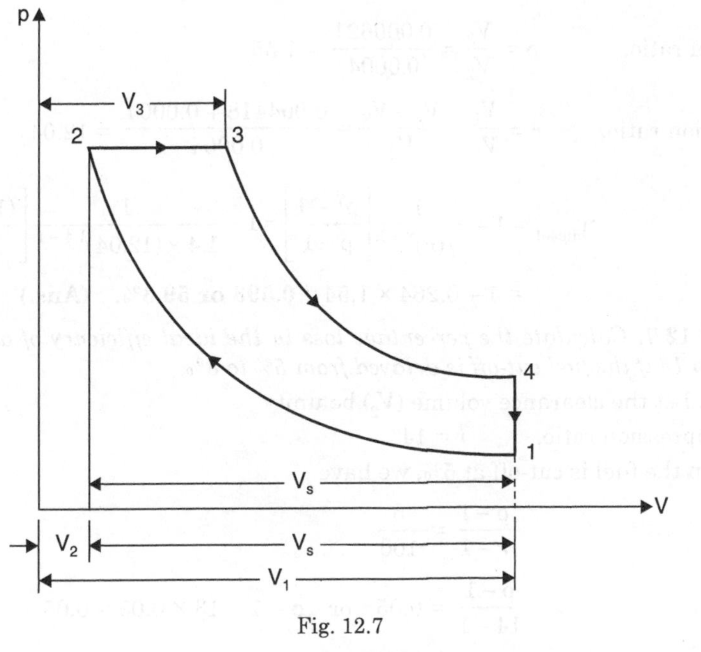

Fig. 12.7

$$= 0.06\ (V_1 - V_2) = 0.06\ (15\ V_2 - V_2) = 0.84\ V_2$$

or, $$V_3 = 1.84\ V_2$$

$\therefore$ $$\rho = \frac{V_3}{V_2} = \frac{1.84\ V_2}{V_2} = 1.84$$

Putting the value in eqn. (*i*), we get

$$\eta_{diesel} = 1 - \frac{1}{1.4\ (15)^{1.4-1}} \left[\frac{(1.84)^{1.4} - 1}{1.84 - 1} \right]$$

$$= 1 - 0.2417 \times 1.605 = \mathbf{0.612\ or\ 61.2\%.} \quad \textbf{(Ans.)}$$

Example 12.6. *The stroke and cylinder diameter of a compression ignition engine are 250 mm and 150 mm respectively. If the clearance volume is 0.0004 m³ and fuel injection takes place at constant pressure for 5 per cent of the stroke determine the efficiency of the engine. Assume the engine working on the diesel cycle.*

Solution. Refer to Fig. 12.7.

Length of stroke, $L = 250\ \text{mm} = 0.25\ \text{m}$

Diameter of cylinder, $D = 150\ \text{mm} = 0.15\ \text{m}$

Clearance volume, $V_2 = 0.0004\ \text{m}^3$

Swept volume, $V_s = \frac{\pi}{4} D^2 L = \frac{\pi}{4} \times 0.15^2 \times 0.25 = 0.004418\ \text{m}^3$

Total cylinder volume $= \text{Swept volume} + \text{clearance volume}$

$$= 0.004418 + 0.0004 = 0.004818\ \text{m}^3$$

Volume at point of cut-off, $$V_3 = V_2 + \frac{5}{100} V_s$$

$$= 0.0004 + \frac{5}{100} \times 0.004418 = 0.000621\ \text{m}^3$$

∴ Cut-off ratio, $\rho = \frac{V_3}{V_2} = \frac{0.000621}{0.0004} = 1.55$

Compression ratio, $r = \frac{V_1}{V_2} = \frac{V_s + V_2}{V_2} = \frac{0.004418 + 0.0004}{0.0004} = 12.04$

Hence, $\eta_{diesel} = 1 - \frac{1}{\gamma(r)^{\gamma-1}}\left[\frac{\rho^\gamma - 1}{\rho - 1}\right] = 1 - \frac{1}{1.4 \times (12.04)^{1.4-1}}\left[\frac{(1.55)^{1.4} - 1}{1.55 - 1}\right]$

$= 1 - 0.264 \times 1.54 = \mathbf{0.593}$ **or 59.3%. (Ans.)**

Example 12.7. *Calculate the percentage loss in the ideal efficiency of a diesel engine with compression ratio 14 if the fuel cut-off is delayed from 5% to 8%.*

Solution. Let the clearance volume (V_2) be unity.

Then, compression ratio, $r = 14$

Now, when the fuel is cut-off at 5%, we have

$$\frac{\rho - 1}{r - 1} = \frac{5}{100}$$

or, $\frac{\rho - 1}{14 - 1} = 0.05$ or $\rho - 1 = 13 \times 0.05 = 0.65$

∴ $\rho = 1.65$

$$\eta_{diesel} = 1 - \frac{1}{\gamma(r)^{\gamma-1}}\left[\frac{\rho^\gamma - 1}{\rho - 1}\right] = 1 - \frac{1}{1.4(14)^{1.4-1}}\left[\frac{(1.65)^{1.4} - 1}{1.65 - 1}\right]$$

$= 1 - 0.248 \times 1.563 = 0.612$ or 61.2%

When the fuel is cut-off at 8%, we have

$$\frac{\rho - 1}{r - 1} = \frac{8}{100}, \text{ or, } \frac{\rho - 1}{14 - 1} = \frac{8}{100} = 0.08$$

∴ $\rho = 1 + 1.04 = 2.04$

$$\eta_{diesel} = 1 - \frac{1}{\gamma(r)^{\gamma-1}}\left[\frac{\rho^\gamma - 1}{\rho - 1}\right] = 1 - \frac{1}{1.4(14)^{1.4-1}}\left[\frac{(2.04)^{1.4} - 1}{2.04 - 1}\right]$$

$= 1 - 0.248 \times 1.647 = \mathbf{0.591}$ or **59.1%. (Ans.)**

Hence percentage loss in efficiency due to delay in fuel cut-off

$= 61.2 - 59.1 = \mathbf{2.1\%}$. **(Ans.)**

Example 12.8. *An engine with 200 mm cylinder diameter and 300 mm stroke works on theoretical diesel cycle. The initial pressure and temperature of air used are 1 bar and 27°C. The cut-off is 8% of the stroke. Determine :*

(i) Pressures and temperatures at all salient points.

(ii) Theoretical air standard efficiency.

(iii) Mean effective pressure.

(iv) Power of the engine if the working cycles per minute are 380.

Assume that compression ratio is 15 and working fluid is air.

Consider all conditions to be ideal.

Solution. Refer to Fig. 12.8.

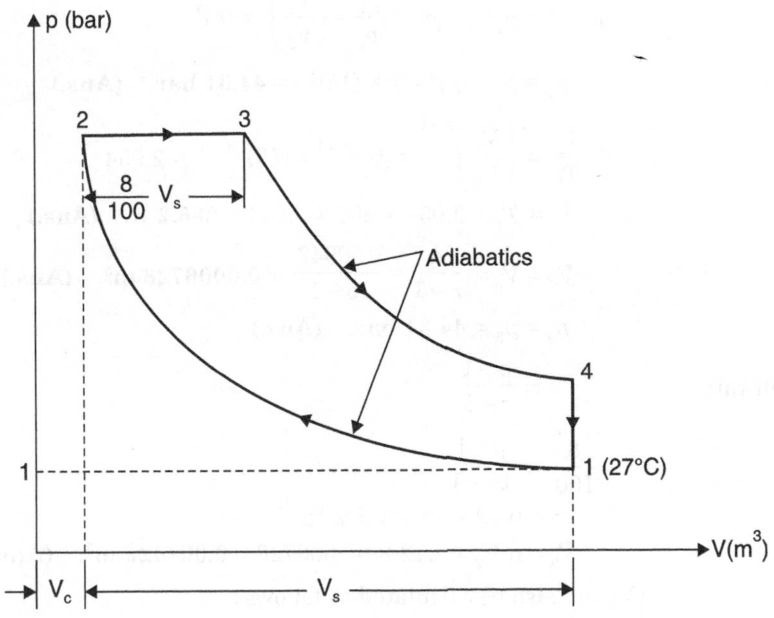

Fig. 12.8

Cylinder diameter, $D = 200$ mm or 0.2 m

Stroke length, $L = 300$ mm or 0.3 m

Initial pressure, $p_1 = 1.0$ bar

Initial temperature, $T_1 = 27 + 273 = 300$ K

Cut-off $= \frac{8}{100} V_s = 0.08\, V_s$

(*i*) **Pressures and temperatures at salient points :**

Now, stroke volume,
$$V_s = \frac{\pi}{4} D^2 L = \frac{\pi}{4} \times 0.2^2 \times 0.3 = 0.00942 \text{ m}^3$$

$$V_1 = V_s + V_c = V_s + \frac{V_s}{r-1} \qquad \left[\because V_c = \frac{V_s}{r-1}\right]$$

$$= V_s\left(1 + \frac{1}{r-1}\right) = \frac{r}{r-1} \times V_s$$

i.e.,
$$V_1 = \frac{15}{15-1} \times V_s = \frac{15}{14} \times 0.00942 = \mathbf{0.0101\ m^3}. \quad \textbf{(Ans.)}$$

Mass of the air in the cylinder can be calculated by using the gas equation,

$$p_1V_1 = mRT_1$$

$$m = \frac{p_1V_1}{RT_1} = \frac{1 \times 10^5 \times 0.0101}{287 \times 300} = 0.0117 \text{ kg/cycle}$$

For the *adiabatic (or isentropic) process 1-2*

$$p_1V_1^{\gamma} = p_2V_2^{\gamma} \quad \text{or} \quad \frac{p_2}{p_1} = \left(\frac{V_1}{V_2}\right)^{\gamma} = (r)^{\gamma}$$

$\therefore$ $p_2 = p_1 \cdot (r)^{\gamma} = 1 \times (15)^{1.4} = \mathbf{44.31\ bar.}$ **(Ans.)**

Also, $\dfrac{T_2}{T_1} = \left(\dfrac{V_1}{V_2}\right)^{\gamma - 1} = (r)^{\gamma - 1} = (15)^{1.4 - 1} = 2.954$

$\therefore$ $T_2 = T_1 \times 2.954 = 300 \times 2.954 = \mathbf{886.2\ K.}$ **(Ans.)**

$$V_2 = V_c = \frac{V_s}{r - 1} = \frac{0.00942}{15 - 1} = \mathbf{0.0006728\ m^3.}\ \textbf{(Ans.)}$$

$p_2 = p_3 = \mathbf{44.31\ bar.}$ **(Ans.)**

% Cut-off ratio $= \dfrac{\rho - 1}{r - 1}$

$$\frac{8}{100} = \frac{\rho - 1}{15 - 1}$$

i.e., $\rho = 0.08 \times 14 + 1 = 2.12$

$\therefore$ $V_3 = \rho\ V_2 = 2.12 \times 0.0006728 = \mathbf{0.001426\ m^3.}$ **(Ans.)**

$$\left[\begin{array}{l} V_3 \text{ can also be calculated as follows:} \\ V_3 = 0.08V_s + V_c = 0.08 \times 0.00942 + 0.0006728 = 0.001426\ \text{m}^3 \end{array}\right]$$

For the *constant pressure process 2-3,*

$$\frac{V_3}{T_3} = \frac{V_2}{T_2}$$

$\therefore$ $T_3 = T_2 \times \dfrac{V_3}{V_2} = 886.2 \times \dfrac{0.001426}{0.0006728} = \mathbf{1878.3\ K.}$ **(Ans.)**

For the *isentropic process 3-4,*

$$p_3V_3^{\gamma} = p_4V_4^{\gamma}$$

$$p_4 = p_3 \times \left(\frac{V_3}{V_4}\right)^{\gamma} = p_3 \times \frac{1}{(7.07)^{1.4}}$$

$$= \frac{44.31}{(7.07)^{1.4}} = \mathbf{2.866\ bar.}\ \textbf{(Ans.)}$$

$$\left[\because \frac{V_4}{V_3} = \frac{V_4}{V_2} \times \frac{V_2}{V_3} = \frac{V_1}{V_2} \times \frac{V_2}{V_3} = \frac{r}{\rho} = \frac{15}{2.12} = 7.07 \quad (\because V_4 = V_1)\right]$$

Also, $\dfrac{T_4}{T_3} = \left(\dfrac{V_3}{V_4}\right)^{\gamma - 1} = \left(\dfrac{1}{7.07}\right)^{1.4 - 1} = 0.457$

$\therefore$ $T_4 = T_3 \times 0.457 = 1878.3 \times 0.457 = \mathbf{858.38\ K.}$ **(Ans.)**

$V_4 = V_1 = \mathbf{0.0101\ m^3.}$ **(Ans.)**

(*ii*) **Theoretical air standard efficiency :**

$$\eta_{\text{diesel}} = 1 - \frac{1}{\gamma(r)^{\gamma - 1}}\left[\frac{\rho^{\gamma} - 1}{\rho - 1}\right] = 1 - \frac{1}{1.4\,(15)^{1.4 - 1}}\left[\frac{(2.12)^{1.4} - 1}{2.12 - 1}\right]$$

$= 1 - 0.2418 \times 1.663 = \mathbf{0.598\ or\ 59.8\%.}$ **(Ans.)**

(*iii*) **Mean effective pressure, p_m :**

Mean effective pressure of diesel cycle is given by

$$p_m = \frac{p_1(r)^{\gamma}[\gamma(\rho - 1) - r^{1-\gamma}(\rho^{\gamma} - 1)]}{(\gamma - 1)(r - 1)}$$

$$= \frac{1 \times (15)^{1.4}[1.4(2.12 - 1) - (15)^{1-1.4}(2.12^{1.4} - 1)]}{(1.4 - 1)(15 - 1)}$$

$$= \frac{44.31[1.568 - 0.338 \times 1.863]}{0.4 \times 14} = 7.424 \text{ bar.} \quad \textbf{(Ans.)}$$

(*iv*) **Power of the engine, P :**

Work done per cycle $= p_m V_s = \frac{7.424 \times 10^5 \times 0.00942}{10^3} = 6.99$ kJ/cycle

Work done per second = Work done per cycle × no. of cycles per second

$= 6.99 \times 380/60 = 44.27$ kJ/s $= 44.27$ kW

Hence *power of the engine* = **44.27 kW. (Ans.)**

12.1.6. Dual Combustion Cycle

This cycle (also called the *limited pressure cycle or mixed cycle*) is a combination of Otto and Diesel cycles, in a way, that heat is added partly at constant volume and partly at constant pressure; *the advantage of which is that more time is available to fuel (which is injected into the engine cylinder before the end of compression stroke) for combustion. Because of lagging characteristics of fuel this cycle is invariably used for diesel and hot spot ignition engines.*

The dual combustion cycle (Fig. 12.9) consists of the following *operations* :

(*i*) 1-2—Adiabatic compression

(*ii*) 2-3—Addition of heat at constant volume

(*iii*) 3-4—Addition of heat at constant pressure

(*iv*) 4-5—Adiabatic expansion

(*v*) 5-1—Rejection of heat at constant volume.

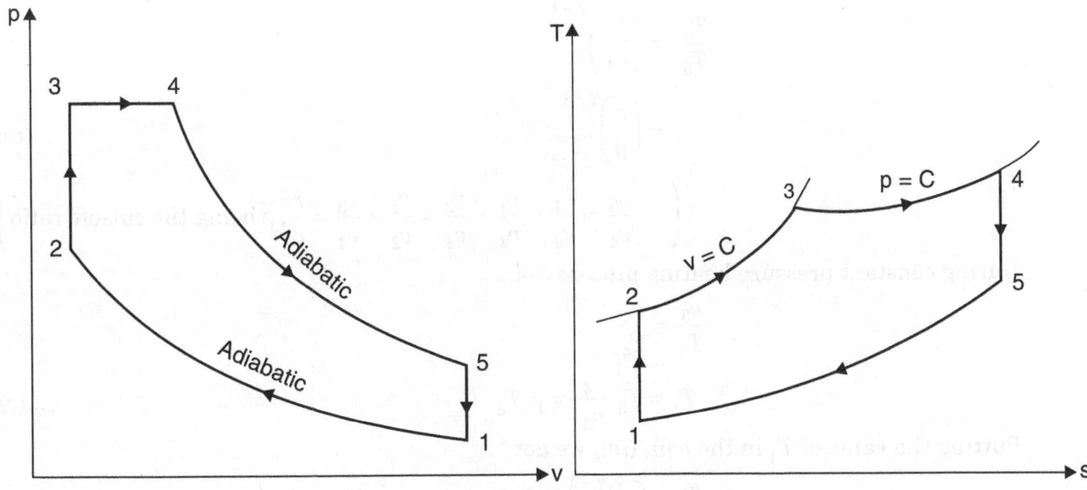

Fig. 12.9

Consider 1 kg of air.

Total heat supplied = Heat supplied during the operation 2-3 + Heat supplied during the operation 3-4

$$= c_v(T_3 - T_2) + c_p(T_4 - T_3)$$

Heat rejected during operation 5-1 $= c_v(T_5 - T_1)$

Work done = Heat supplied – Heat rejected

$$= [c_v(T_3 - T_2) + c_p(T_4 - T_3)] - c_v(T_5 - T_1)$$

$$\eta_{dual} = \frac{\text{Work done}}{\text{Heat supplied}} = \frac{[c_v (T_3 - T_2) + c_p (T_4 - T_3)] - c_v (T_5 - T_1)}{[c_v (T_3 - T_2) + c_p (T_4 - T_3)]}$$

$$= 1 - \frac{c_v (T_5 - T_1)}{c_v (T_3 - T_2) + c_p (T_4 - T_3)}$$

$$= 1 - \frac{c_v (T_5 - T_1)}{(T_3 - T_2) + \gamma (T_4 - T_3)} \quad ...(i) \quad \left(\because \ \gamma = \frac{c_p}{c_v}\right)$$

Compression ratio, $r = \dfrac{v_1}{v_2}$

During *adiabatic compression process 1-2,*

$$\frac{T_2}{T_1} = \left(\frac{v_1}{v_2}\right)^{\gamma - 1} = (r)^{\gamma - 1} \quad ...(ii)$$

During *constant volume heating process 2-3,*

$$\frac{p_3}{T_3} = \frac{p_2}{T_2}$$

or, $\dfrac{T_3}{T_2} = \dfrac{p_3}{p_2} = \beta$, where β is known as **pressure or explosion ratio.**

or, $$T_2 = \frac{T_3}{\beta} \quad ...(iii)$$

During adiabatic expansion process 4-5,

$$\frac{T_4}{T_5} = \left(\frac{v_5}{v_4}\right)^{\gamma - 1}$$

$$= \left(\frac{r}{\rho}\right)^{\gamma - 1} \quad ...(iv)$$

$$\left(\because \ \frac{v_5}{v_4} = \frac{v_1}{v_4} = \frac{v_1}{v_2} \times \frac{v_2}{v_4} = \frac{v_1}{v_2} \times \frac{v_3}{v_4} = \frac{r}{\rho}, \ \rho \text{ being the cut-off ratio}\right)$$

During constant pressure heating process 3-4,

$$\frac{v_3}{T_3} = \frac{v_4}{T_4}$$

$$T_4 = T_3 \frac{v_4}{v_3} = \rho \, T_3 \quad ...(v)$$

Putting the value of T_4 in the eqn. (*iv*), we get

$$\frac{\rho T_3}{T_5} = \left(\frac{r}{\rho}\right)^{\gamma - 1}, \quad \text{or,} \quad T_5 = \rho \, . \, T_3 \, . \left(\frac{\rho}{r}\right)^{\gamma - 1}$$

Putting the value of T_2 in eqn. (*ii*), we get

$$\frac{\frac{T_3}{\beta}}{T_1} = (r)^{\gamma - 1}$$

$$T_1 = \frac{T_3}{\beta} \cdot \frac{1}{(r)^{\gamma - 1}}$$

Now inserting the values of T_1, T_2, T_4 and T_5 in eqn. (*i*), we get

$$\eta_{dual} = 1 - \frac{\left[\rho . T_3 \left(\frac{\rho}{r}\right)^{\gamma - 1} - \frac{T_3}{\beta} \cdot \frac{1}{(r)^{\gamma - 1}}\right]}{\left[\left(T_3 - \frac{T_3}{\beta}\right) + \gamma(\rho T_3 - T_3)\right]} = 1 - \frac{\frac{1}{(r)^{\gamma - 1}}\left(\rho^{\gamma} - \frac{1}{\beta}\right)}{\left(1 - \frac{1}{\beta}\right) + \gamma(\rho - 1)}$$

i.e.,

$$\eta_{dual} = 1 - \frac{1}{(r)^{\gamma - 1}} \cdot \frac{(\beta . \rho^{\gamma} - 1)}{[(\beta - 1) + \beta\gamma(\rho - 1)]} \qquad ...(12.10)$$

Work done is given by,

$$W = p_3(v_4 - v_3) + \frac{p_4 v_4 - p_5 v_5}{\gamma - 1} - \frac{p_2 v_2 - p_1 v_1}{\gamma - 1}$$

$$= p_3 v_3(\rho - 1) + \frac{(p_4 \rho v_3 - p_5 r v_3) - (p_2 v_3 - p_1 r v_3)}{\gamma - 1}$$

$$= \frac{p_3 v_3(\rho - 1)(\gamma - 1) + p_4 v_3\left(\rho - \frac{p_5}{p_4} r\right) - p_2 v_3\left(1 - \frac{p_1}{p_2} r\right)}{\gamma - 1}$$

Also, $\quad \frac{p_5}{p_4} = \left(\frac{v_4}{v_5}\right)^{\gamma} = \left(\frac{\rho}{r}\right)^{\gamma}$, and, $\frac{p_2}{p_1} = \left(\frac{v_1}{v_2}\right)^{\gamma} = r^{\gamma}$

Also, $\quad p_3 = p_4,\ v_2 = v_3,\ v_5 = v_1$

$$\therefore \quad W = \frac{v_3[p_3(\rho - 1)(\gamma - 1) + p_3(\rho - \rho^{\gamma} r^{1-\gamma}) - p_2(1 - r^{1-\gamma})]}{(\gamma - 1)}$$

$$= \frac{p_2 v_2[\beta(\rho - 1)(\gamma - 1) + \beta(\rho - \rho^{\gamma} r^{1-\gamma}) - (1 - r^{1-\gamma})]}{(\gamma - 1)}$$

$$= \frac{p_1 (r)^{\gamma} v_1 / r[\beta\gamma(\rho - 1) + (\beta - 1) - r^{1-\gamma}(\beta\rho^{\gamma} - 1)]}{\gamma - 1}$$

$$= \frac{p_1 v_1 r^{\gamma - 1}[\beta\gamma(\rho - 1) + (\beta - 1) - r^{\gamma - 1}(\beta\rho^{\gamma} - 1)]}{\gamma - 1} \qquad ...(12.11)$$

Mean effective pressure (p_m) is given by,

$$p_m = \frac{W}{v_1 - v_2} = \frac{W}{v_1\left(\frac{r-1}{r}\right)}$$

$$= \frac{p_1 v_1 [r^{1-\gamma} \beta\gamma(\rho - 1) + (\beta - 1) - r^{1-\gamma}(\beta\rho^{\gamma} - 1)]}{(\gamma - 1)\, v_1 \left(\frac{r-1}{r}\right)}$$

$$\mathbf{p_m} = \frac{p_1 (r)^{\gamma}[\beta(\rho - 1) + (\beta - 1) - r^{1-\gamma}(\beta\rho^{\gamma} - 1)]}{(\gamma - 1)(r - 1)} \quad ...(12.12)$$

Example 12.9. *The swept volume of a diesel engine working on dual cycle is 0.0053 m³ and clearance volume is 0.00035 m³. The maximum pressure is 65 bar. Fuel injection ends at 5 per cent of the stroke. The temperature and pressure at the start of the compression are 80°C and 0.9 bar. Determine the air standard efficiency of the cycle. Take γ for air = 1.4.*

Solution. Refer to Fig. 12.10.

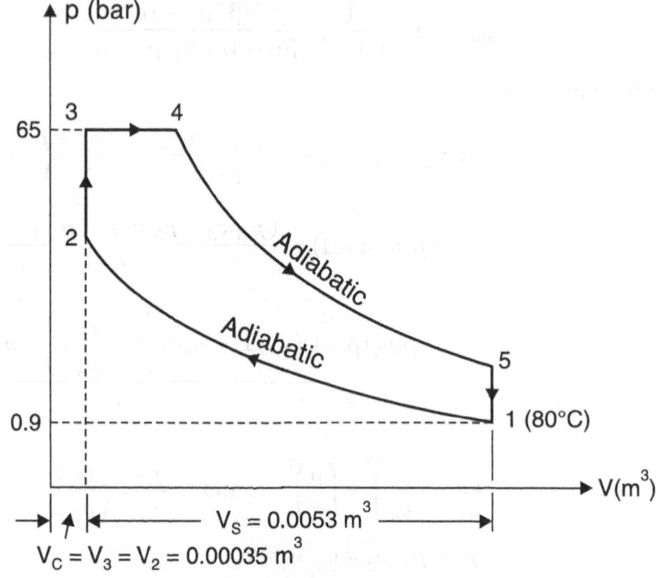

Fig. 12.10

Swept volume, $V_s = 0.0053$ m³

Clearance volume, $V_c = V_3 = V_2 = 0.00035$ m³

Maximum pressure, $p_3 = p_4 = 65$ bar

Initial temperature, $T_1 = 80 + 273 = 353$ K

Initial pressure, $p_1 = 0.9$ bar

η_{dual} :

The efficiency of a dual combustion cycle is given by

$$\eta_{dual} = 1 - \frac{1}{(r)^{\gamma - 1}}\left[\frac{\beta . \rho^{\gamma} - 1}{(\beta - 1) + \beta\gamma(\rho - 1)}\right]$$

Compression ratio, $r = \frac{V_1}{V_2} = \frac{V_s + V_c}{V_c} = \frac{0.0053 + 0.00035}{0.00035} = 16.14$

[$\because$ $V_2 = V_c$ = Clearance volume]

Cut-off ratio, $\rho = \frac{V_4}{V_3} = \frac{\frac{5}{100}V_s + V_3}{V_3} = \frac{0.05V_s + V_c}{V_c}$ $(\because\ V_2 = V_3 = V_c)$

$$= \frac{0.05 \times 0.0053 + 0.00035}{0.00035} = 1.757 \text{ say } 1.76$$

Also during the *compression process 1-2,*

$$p_1 V_1^{\gamma} = p_2 V_2^{\gamma}$$

or $$\frac{p_2}{p_1} = \left(\frac{V_1}{V_2}\right)^{\gamma} = (16.15)^{1.4} = 49.14$$

or $$p_2 = p_1 \times 49.14 = 0.9 \times 49.14 = 44.22 \text{ bar}$$

Pressure or explosion ratio,

$$\beta = \frac{p_3}{p_2} = \frac{65}{44.22} = 1.47$$

Putting the value of r, ρ and β in eqn. (*i*), we get

$$\eta_{dual} = 1 - \frac{1}{(10.14)^{1.4-1}}\left[\frac{1.47 \times (1.76)^{1.4} - 1}{(1.47 - 1) + 1.47 \times 1.4\,(1.76 - 1)}\right]$$

$$= 7 - 0.328\left[\frac{3.243 - 1}{0.47 + 1.564}\right] = \mathbf{0.6383 \text{ or } 63.83\%.} \quad \textbf{(Ans.)}$$

12.1.7. Comparison of Otto, Diesel and Dual Combustion Cycles

Following are the important variable factors which are used as a basis for comparison of the cycles :

- Compression ratio
- Maximum pressure
- Heat supplied
- Heat rejected
- Net work.

Some of the above mentioned variables are fixed when the performance of otto, diesel and dual combustion cycles is to be compared.

12.1.7.1. Efficiency versus compression ratio

Fig. 12.11 shows the comparison for the air standard efficiencies of the otto, diesel and dual combustion cycles at various compression ratios and with given cut-off ratio for the diesel and dual combustion cycles. It is evident from the Fig. 12.11 that the air standard efficiencies increase with the increase in the compression ratio. *For a given compression ratio otto cycle is the most efficient while the diesel cycle is the least efficient.* ($\eta_{otto} > \eta_{dual} > \eta_{diesel}$).

Note. The maximum compression ratio for the petrol engine is limited by detonation. In their respective ratio ranges, the diesel cycle is more efficient than the otto cycle.

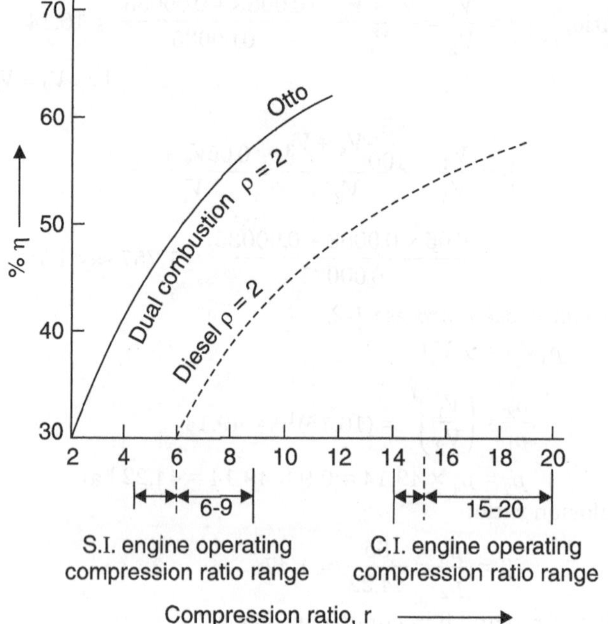

Fig. 12.11. Comparison of efficiency at various compression ratio.

12.1.7.2. For the same compression ratio and the same heat input

A comparison of the cycles (otto, diesel and dual) on the *p-v* and *T-s* diagrams for the *same compression ratio and heat supplied* is shown in the Fig. 12.12.

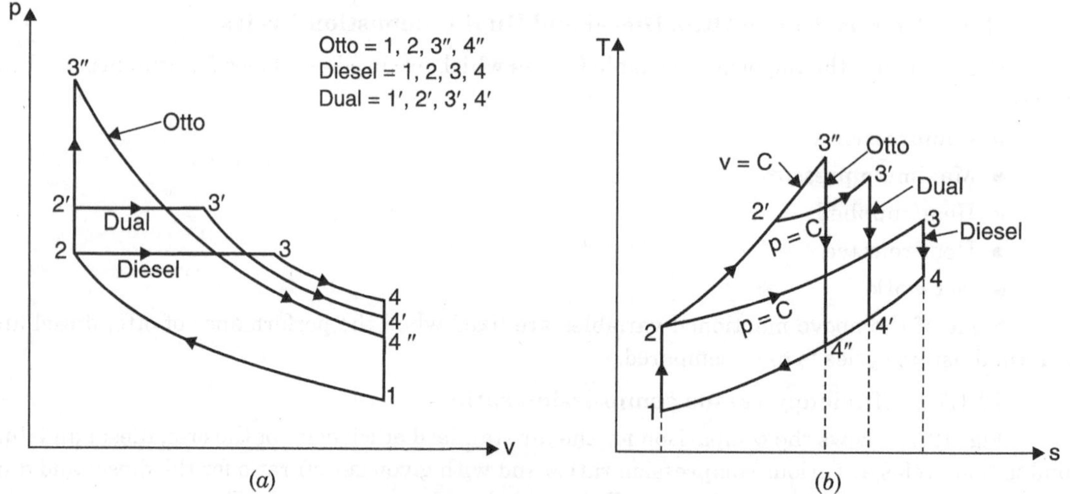

Fig. 12.12. (*a*) *p-v* diagram, (*b*) *T-s* diagram.

We know that,
$$\eta = 1 - \frac{\textbf{Heat rejected}}{\textbf{Heat supplied}} \quad ...(12.13)$$

Since all the cycles reject their heat at the same specific volume process line from state 4 to 1, the quantity of *heat rejected from each cycle is represented by the appropriate area under the line 4 to*

1 on the T-s diagram. As is evident from the eqn. (12.13) the cycle which has the least heat rejected will have the highest efficiency. Thus, otto cycle is the most efficient and diesel cycle is the least efficient of the three cycles.

i.e., $$\eta_{otto} > \eta_{dual} > \eta_{diesel}.$$

12.1.7.3. For constant maximum pressure and heat supplied

Fig. 12.13 shows the otto and diesel cycles on *p-v* and *T-s* diagrams for constant maximum pressure and heat input respectively.

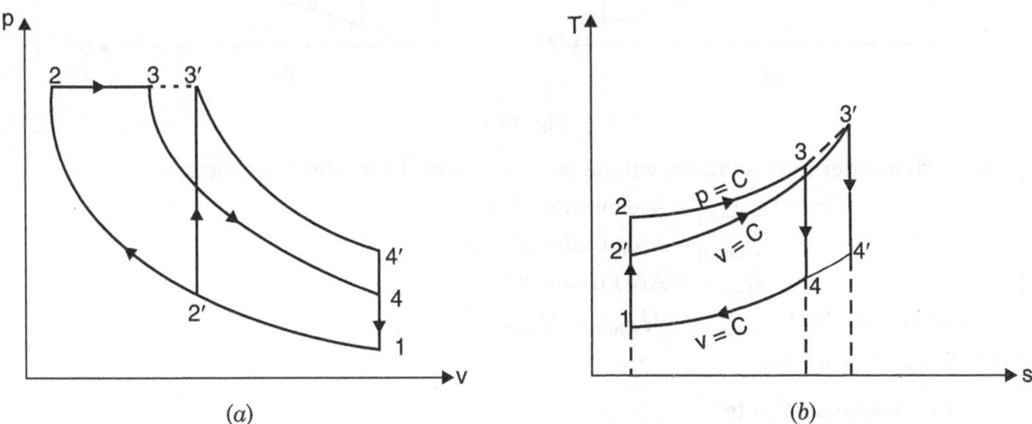

Fig. 12.13. (*a*) *p-v* diagram, (*b*) *T-s* diagram.

— For the maximum pressure the points 3 and 3′ must lie on a constant pressure line.

— On *T-s* diagram the heat rejected from the diesel cycle is represented by the area under the line 4 to 1 and this area is less than the otto cycle area under the curve 4′ to 1 ; hence the *diesel cycle is more efficient than the otto cycle for the condition of maximum pressure and heat supplied.*

Example 12.10. *With the help of p-v and T-s diagrams compare the cold air standard otto, diesel and dual combustion cycles for same maximum pressure and maximum temperature.*

Solution. Refer to Fig. 12.14 (*a, b*).

The air-standard otto, dual and diesel cycles are drawn on common *p-v* and *T-s* diagrams for the same maximum pressure and maximum temperature, for the purpose of comparison.

otto 1-2-3-4-1, dual 1-2′-3′-3-4-1, diesel 1-2″-3-4-1 (Fig. 12.14 (*a*)).

Slope of constant volume lines on *T-s* diagram is higher than that of constant pressure lines. [Fig. 12.14 (*b*)].

Here the otto cycle must be limited to a low compression ratio (*r*) to fulfill the condition that point 3 (same maximum pressure and temperature) is to be a common state for all the three cycles.

The construction of cycles on *T-s* diagram proves that for the given conditions the heat rejected is same for all the three cycles (area under process line 4-1). Since, by definition,

$$\eta = 1 - \frac{\text{Heat rejected, } Q_R}{\text{Heat supplied, } Q_S} = 1 - \frac{\text{Const.}}{Q_s}$$

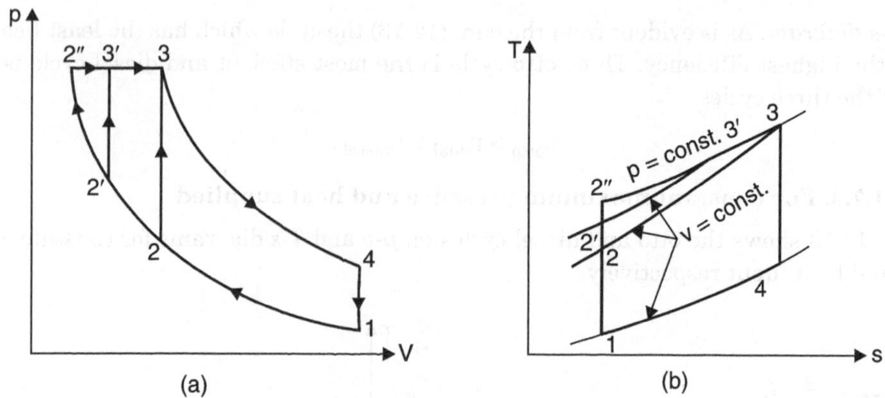

Fig. 12.14

the cycle, with greater heat addition will be more efficient. From the *T-s* diagram,

$$Q_{s(\text{diesel})} = \text{Area under } 2''\text{-}3$$

$$Q_{s(\text{dual})} = \text{Area under } 2'\text{-}3'\text{-}3$$

$$Q_{s(\text{otto})} = \text{Area under } 2\text{-}3.$$

It can be seen that, $Q_{s(\text{diesel})} > Q_{s(\text{dual})} > Q_{s(\text{otto})}$

and thus, $\eta_{\text{diesel}} > \eta_{\text{dual}} > \eta_{\text{otto}}$.

12.1.8. Atkinson Cycle

This cycle consists of *two adiabatics*, a *constant volume* and a *constant pressure process.*

p-V diagram of this cycle is shown in Fig. 12.15. It consists of the following four operations :

(*i*) 1-2—Heat rejection at constant pressure.

(*ii*) 2-3—Adiabatic compression.

(*iii*) 3-4—Addition of heat at constant volume.

(*iv*) 4-1—Adiabatic expansion.

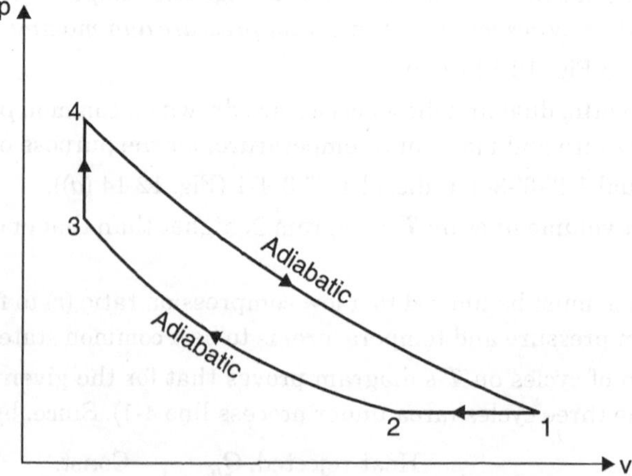

Fig. 12.15. Atkinson cycle.

Considering *1 kg of air.*

Compression ratio $= \dfrac{v_2}{v_3} = \alpha$

Expansion ratio $= \dfrac{v_1}{v_4} = r$

Heat supplied at constant volume $= c_v(T_4 - T_3)$

Heat rejected $= c_v(T_1 - T_2)$

Work done = Heat supplied – heat rejected

$$= c_v(T_4 - T_3) - c_v(T_1 - T_2)$$

$$\eta = \frac{\text{Work done}}{\text{Heat supplied}} = \frac{c_v\,(T_4 - T_3) - c_p(T_1 - T_2)}{c_v\,(T_4 - T_3)}$$

$$= 1 - \gamma \,.\, \frac{(T_1 - T_2)}{(T_4 - T_3)} \quad ...(i)$$

During *adiabatic compression 2-3,*

$$\frac{T_3}{T_2} = \left(\frac{v_2}{v_3}\right)^{\gamma - 1} = (\alpha)^{\gamma - 1}$$

or

$$T_3 = T_2\,(\alpha)^{\gamma - 1} \quad ...(ii)$$

During *constant pressure operation 1-2,*

$$\frac{v_1}{T_1} = \frac{v_2}{T_2}$$

or

$$\frac{T_2}{T_1} = \frac{v_2}{v_1} = \frac{\alpha}{r} \quad ...(iii) \quad \left(\frac{v_2}{v_1} = \frac{v_2}{v_3} \times \frac{v_3}{v_1} = \frac{v_2}{v_3} \times \frac{v_4}{v_1} = \frac{\alpha}{r}\right)$$

During *adiabatic expansion 4-1,*

$$\frac{T_4}{T_1} = \left(\frac{v_1}{v_4}\right)^{\gamma - 1} = (r)^{\gamma - 1}$$

$$T_1 = \frac{T_4}{(r)^{\gamma - 1}} \quad ...(iv)$$

Putting the value of T_1 in eqn. (*iii*), we get

$$T_2 = \frac{T_4}{(r)^{\gamma - 1}} \cdot \frac{\alpha}{r}$$

$$= \frac{\alpha T_4}{r^{\gamma}} \quad ...(v)$$

Substituting the value of T_2 in eqn. (*ii*), we get

$$T_3 = \frac{\alpha T_4}{r^{\gamma}}\,(\alpha)^{\gamma - 1} = \left(\frac{\alpha}{r}\right)^{\gamma} \cdot T_4$$

Finally putting the values of T_1, T_2 and T_3 in eqn. (*i*), we get

$$\eta = 1 - \gamma \left(\frac{\frac{T_4}{r^{\gamma-1}} - \frac{\alpha T_4}{(r)^\gamma}}{T_4 - \left(\frac{\alpha}{r}\right)^\gamma . T_4} \right) = 1 - \gamma \left(\frac{r - \alpha}{r^\gamma - \alpha^\gamma} \right)$$

Hence, air standard efficiency $= 1 - \gamma . \left(\frac{r - \alpha}{r^\gamma - \alpha^\gamma} \right)$ ...(12.14)

12.1.9. Ericsson Cycle

It is so named as it was invented by Ericsson. Fig. 12.16 shows *p-V* diagram of this cycle.

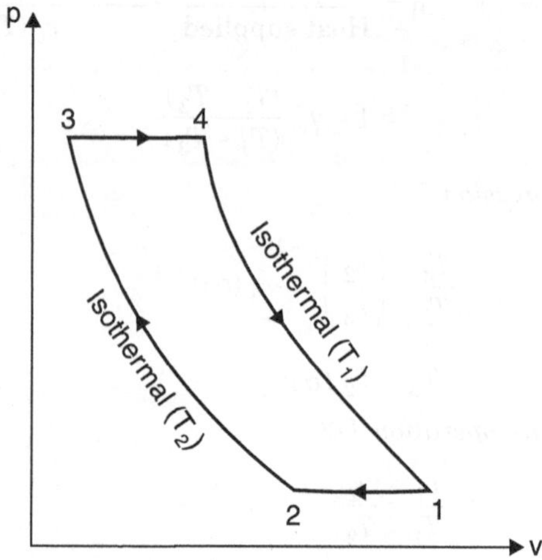

Fig. 12.16. Ericsson cycle.

It comprises of the following operations :

(*i*) 1-2—Rejection of heat at constant pressure

(*ii*) 2-3—Isothermal compression

(*iii*) 3-4—Addition of heat at constant pressure

(*iv*) 4-1—Isothermal expansion.

Considering *1 kg of air*.

Volume ratio, $r = \frac{v_2}{v_3} = \frac{v_1}{v_4}$

Heat supplied to air from an external source

= Heat supplied during the isothermal expansion 4-1

$= RT_1 \log_e r$

Heat rejected by air to an external source

$= RT_2 . \log_e r$

Work done = Heat supplied – heat rejected

$= RT_1 . \log_e r - RT_2 . \log_e r = R \log_e r \, (T_1 - T_2)$

$$\eta = \frac{\text{Work done}}{\text{Heat supplied}} = \frac{R \log_e r (T_1 - T_2)}{RT_1 . \log_e r}$$

$$= \frac{T_1 - T_2}{T_1} \quad ...(12.15)$$

which is the same as Carnot cycle.

12.2. AUTOMOTIVE GAS TURBINES

12.2.1. Introduction

A **turbine** *is a machine in which a rotary motion is obtained by the gradual change of momentum of a fluid.*

Probably a wind-mill was the first turbine to produce useful work, wherein there is no precompression and no combustion. The characteristic features of a "*gas turbine*" as we think of the name today include a *compression process* and a *heat addition* (or *combustion*) *process.*

The turbine represents perhaps the most satisfactory way of producing very large quantities of power in a self-contained and compact unit. For smaller gas turbine units, the inefficiencies in compression and expansion processes become greater and to improve the thermal efficiency it necessary to use a *heat exchanger.*

Gas turbines are now being used to propel *automobile vehicles, boats, aeroplanes* as well as other *stationary power plants.* These engines initially designed for aircraft, stationary purposes, are also being used as *automobile power plants.*

A gas turbine in simple terms is a heat engine utilising the expansion from the combustion of fuel and air in a combustion chamber. The *energy is directly transformed either into thrust or shaft power to push automobile wheel for turning.*

12.2.2. Historical Background

- George Brayton in 1873 invented the basic cycle of the modern gas turbine, known as "*Brayton cycle*". The first Brayton engine was a reciprocating two cylinder engine with compression of fuel, air being carried out in a small cylinder and the expansion of products of the combustion occurring in a large second cylinder. Today the term *Brayton cycle engine* refers to gas turbine engine (GTE) only.

For a long time there was no development of gas turbine due to lack of materials to withstand high temperatures and also lack of aerodynamic and thermodynamic knowledge for producing efficient turbines and compressors.

- *Chrysler* conducted surveys before World War II to analyse the suitability of the gas turbine for *vehicular applications.*
- *Turunen* assessed the automotive gas turbine in 1949. *Superior torque output characteristics, its ability to burn a wide variety of fuels, absence of cooling, vibration and noise free operation were some of the main advantages* which lead to greater interest in this type of power plant.
- '*Rover*' in 1950 demonstrated the *world's first gas turbine powered passenger car.*
- "*General Motors*" announced its first automotive turbine in *January 1954* and "*Chrysler corporation*" in *March, 1954.*
- In 1955, "*General Motors*" installed *first regenerative gas turbine engine,* the GT-304, in Firebird II. At the same time, "*Chrysler Corporation*" and "Leyland" tested *gas turbine powered heavy duty trucks.*
  - — Chrysler corporation tried a set of 50 turbine cars between 1955-56.
  - — "*Ford*" analysed a three shaft gas turbine arrangement.

Williams Research corporation. Volkswagon and many others are also working towards development of a gas turbine passenger car.

12.2.3. Merits and Limitations of Gas Turbines

Merits over I.C. engines :

1. The mechanical efficiency of a gas turbine (95%) is quite high as compared with I.C. engine (85%) since the I.C. engine has a large number of sliding parts.
2. A gas turbine does not require a flywheel as the torque on the shaft is continuous and uniform. Whereas a flywheel is a must in case of an I.C. engine.
3. The weight of gas turbine per H.P. developed is less than that of an I.C. engine.
4. The gas turbine can be driven at a very high speeds (40000 r.p.m.) whereas this is not possible with I.C. engines.
5. The work developed by a gas turbine per kg of air is more as compared to an I.C. engine. This is due to the fact that gases can be expanded upto atmospheric pressure in case of a gas turbine whereas in an I.C. engine expansion upto atmospheric pressure is not possible.
6. The components of the gas turbine can be made lighter since the pressures used in it are very low, say 5 bar compared with I.C. engine, say 60 bar.
7. In the gas turbine the ignition and lubrication systems are much simpler as compared with I.C. engines.
8. Cheaper fuels such as paraffin type, residue oils or powdered coal can be used whereas special grade fuels are employed in petrol engine to check knocking or pinking.
9. The exhaust from gas turbine is less polluting comparatively since excess air is used for combustion.
10. Because of low specific weight the gas turbines are particularly suitable for use in aircrafts.

Demerits of gas turbines :

1. The thermal efficiency of a simple turbine cycle is low (15 to 20%) as compared with I.C. engines (25 to 30%).
2. With wide operating speeds the fuel control is comparatively difficult.
3. Due to higher operating speeds of the turbine, it is imperative to have a speed reduction device.
4. It is difficult to start a gas turbine as compared to an I.C. engine.
5. The gas turbine blades need a special cooling system.
6. One of the main demerits of a gas turbine is its *very poor thermal efficiency at part loads*, as the quantity of air remains same irrespective of load, and output is reduced by reducing the quantity of fuel supplied.
7. Owing to the use of nickel-chromium alloy, the manufacture of the blades is difficult and costly.
8. For the same output the gas turbine produces five times exhaust gases than I.C. engine.
9. Because of prevalence of high temperature (1000 K for blades and 2500 K for combustion chamber) and centrifugal force the life of the combustion chamber and blades is short/small.

- *Due to very high speed the gas turbine causes serious problems in obtaining an effective transmission speed suitable for road speeds.*

Limitations of gas turbines :

The gas turbines have the following *limitations* :

(*i*) *They are not self starting.*

(*ii*) *Low efficiencies at part loads.*

(*iii*) *Non-reversibility.*

(*iv*) *Higher rotor speeds.*

(*v*) *Overall efficiency of the plant low.*

12.2.4. Ideal Brayton Cycle (Gas turbine cycle)

Brayton cycle is a constant pressure cycle for a perfect gas. It is also called ***Joule cycle.*** The heat transfers are achieved in reversible constant pressure heat exchangers. An ideal gas turbine plant would perform the processes that make up a Brayton cycle. The cycle is shown in the Fig. 12.17 (*a*) and it is represented on *p-v* and *T-s* diagrams as shown in Fig. 12.17 (*b, c*).

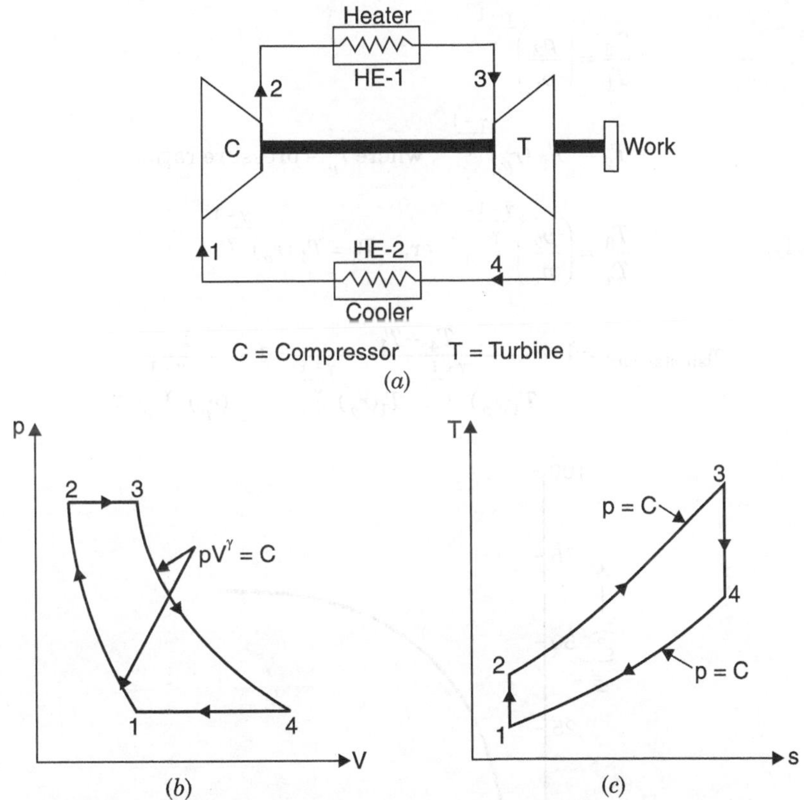

Fig. 12.17. Brayton cycle : (*a*) Basic components of a gas turbine power plant (*b*) *p-V* diagram (*c*) *T-s* diagram.

The various *operations* are as follows :

Operation 1-2. The air is compressed isentropically from the lower pressure p_1 to the upper pressure p_2, the temperature rising from T_1 to T_2. No heat flow occurs.

Operation 2-3. Heat flows into the system increasing the volume from V_2 to V_3 and temperature from T_2 to T_3 whilst the pressure remains constant at p_2. Heat received = $mc_p (T_3 - T_2)$.

Operation 3-4. The air is expanded isentropically from p_2 to p_1, the temperature falling from T_3 to T_4. No heat flow occurs.

Operation 4-1. Heat is rejected from the system as the volume decreases from V_4 to V_1 and the temperature from T_4 to T_1 whilst the pressure remains constant at p_1. Heat rejected = mc_p $(T_4 - T_1)$.

$$\eta_{\text{air-standard}} = \frac{\text{Work done}}{\text{Heat received}}$$

$$= \frac{\text{Heat received/cycle} - \text{Heat rejected/cycle}}{\text{Heat received/cycle}}$$

$$= \frac{mc_p\,(T_3 - T_2) - mc_p\,(T_4 - T_1)}{mc_p\,(T_3 - T_2)} = 1 - \frac{T_4 - T_1}{T_3 - T_2}$$

Now, from *isentropic compression 1,-2,*

$$\frac{T_2}{T_1} = \left(\frac{p_2}{p_1}\right)^{\frac{\gamma-1}{\gamma}}$$

$$T_2 = T_1\,(r_p)^{\frac{\gamma-1}{\gamma}}, \text{ where } r_p = \text{pressure ratio.}$$

Similarly, $$\frac{T_3}{T_4} = \left(\frac{p_2}{p_1}\right)^{\frac{\gamma-1}{\gamma}}, \quad \text{or,} \quad T_3 = T_4\,(r_p)^{\frac{\gamma-1}{\gamma}}$$

$$\therefore \quad \eta_{\text{air-standard}} = 1 - \frac{T_4 - T_1}{T_4(r_p)^{\frac{\gamma-1}{\gamma}} - T_1(r_p)^{\frac{\gamma-1}{\gamma}}} = 1 - \frac{1}{(r_p)^{\frac{\gamma-1}{\gamma}}} \quad \text{...(12.16)}$$

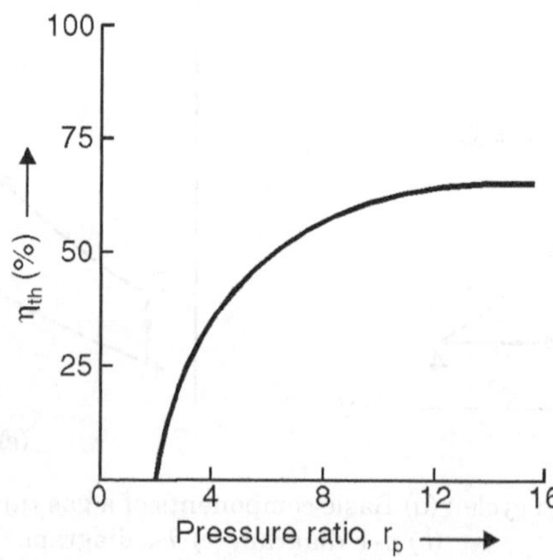

Fig. 12.18. Effect of pressure ratio on the efficiency of Brayton cycle.

The eqn. (12.16) shows that the *efficiency of the ideal joule cycle increases with the pressure ratio. The absolute limit of upper pressure is determined by the limiting temperature of the material of the turbine at the point at which this temperature is reached by the compression process alone, no further heating of the gas in the combustion chamber would be permissible and the work of expansion would ideally just balance the work of compression so that no excess work would be available for external use.*

Pressure ratio for maximum work :

Now we shall prove that the *pressure ratio for maximum work is a function of the limiting temperature ratio.*

Work output during the cycle

$$= \text{Heat received/cycle} - \text{heat rejected/cycle}$$

$$= mc_p (T_3 - T_2) - mc_p (T_4 - T_1)$$

$$= mc_p (T_3 - T_4) - mc_p (T_2 - T_1)$$

$$= mc_p \, T_3 \left(1 - \frac{T_4}{T_3}\right) - T_1 \left(\frac{T_2}{T_1} - 1\right)$$

In case of a given turbine the minimum temperature T_1 and the maximum temperature T_3 are prescribed, T_1 being the temperature of the atmosphere and T_3 the maximum temperature which the metals of turbine would withstand. Consider the specific heat at constant pressure c_p to be constant. Then,

Since, $$\frac{T_3}{T_4} = (r_p)^{\frac{\gamma-1}{\gamma}} = \frac{T_2}{T_1}$$

Using the constant $$'z' = \frac{\gamma - 1}{\gamma},$$

we have, work output/cycle

$$W = K \left[T_3 \left(1 - \frac{1}{r_p{}^z}\right) - T_1 (r_p{}^z - 1) \right]$$

Differentiating with respect to r_p

$$\frac{dW}{dr_p} = K \left[T_3 \times \frac{z}{r_p(z+1)} - T_1 \, z r_p{}^{(z-1)} \right] = 0 \text{ for a maximum}$$

$$\therefore \quad \frac{zT_3}{r_p{}^{(z+1)}} = T_1 z (r_p)^{(z-1)}$$

$$\therefore \quad r_p{}^{2z} = \frac{T_3}{T_1}$$

$$\therefore \quad r_p = (T_3/T_1)^{1/2z} \quad i.e., \quad r_p = (T_3/T_1)^{\frac{\gamma}{2(\gamma-1)}} \qquad \ldots(12.17)$$

Thus, the *pressure ratio for maximum work is a function of the limiting temperature ratio.*

Work ratio:

Work ratio is defined as the *ratio of net work output to the work done by the turbine.*

$$\therefore \quad \text{Work ratio} = \frac{W_T - W_C}{W_T}$$

$$\left[\begin{array}{ll} \text{where,} & W_T = \text{Work obtained from this turbine,} \\ \text{and,} & W_C = \text{Work supplied to the compressor} \end{array} \right]$$

$$= \frac{mc_p(T_3 - T_4) - mc_p(T_2 - T_1)}{mc_p\,(T_3 - T_4)} = 1 - \frac{T_2 - T_1}{T_3 - T_4}$$

$$= 1 - \frac{T_1}{T_3}\left[\frac{(r_p)^{\frac{\gamma-1}{\gamma}} - 1}{1 - \frac{1}{(r_p)^{\frac{\gamma-1}{\gamma}}}}\right] = 1 - \frac{T_1}{T_3}(r_p)^{\frac{\gamma-1}{\gamma}} \qquad \text{...(12.18)}$$

12.2.5. Classification of Gas Turbines

The gas turbines are mainly divided into *two groups* :

1. **Constant pressure combustion gas turbine**

(*a*) Open cycle constant pressure gas turbine.

(*b*) Closed cycle constant pressure gas turbine.

2. **Constant volume combustion gas turbine.**

In almost *all the fields open cycle gas turbine plants are used. Closed cycle plants were introduced at one stage because of their ability to burn cheap fuel.* In between their progress remained slow because of availability of cheap oil and natural gas. Because of rising oil prices, now again, the attention is being paid to closed cycle plants.

12.2.6. Constant Pressure Combustion Gas Turbines

12.2.6.1. Open cycle gas turbine

Refer to Fig. 12.19. The fundamental gas turbine unit is one operating on the open cycle in which a **rotary compressor** and a **turbine** are mounted on a common shaft. Air is drawn into the compressor and after compression passes to a combustion chamber. Energy is supplied in the combustion chamber by spraying fuel into the air stream, and the resulting hot gases expand through

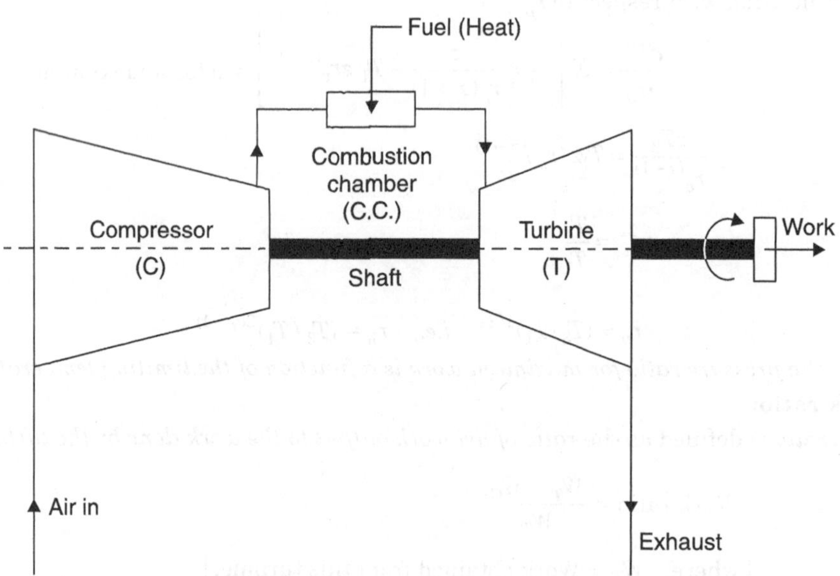

Fig. 12.19. Open cycle gas turbine.

the turbine to the atmosphere. In order to achieve net work output from the unit, the turbine must develop more gross work output than is required to drive the compressor and to overcome mechanical losses in the drive. The products of combustion coming out from the turbine are exhausted to the atmosphere as they cannot be used any more. The working fluids (air and fuel) must be replaced continuously as they are exhausted into the atmosphere.

Energy cycle for a simple-cycle gas turbine :

Fig. 12.20 shows an energy-flow diagram for a simple-cycle gas turbine, the description of which is as follows :

— The air brings in minute amount of energy (measured above 0°C).

— Compressor adds considerable amount of energy.

— Fuel carries major input to cycle.

— Sum of fuel and compressed-air energy leaves combustor to enter turbine.

— In turbine smallest part of entering energy goes to useful output, largest part leaves in exhaust.

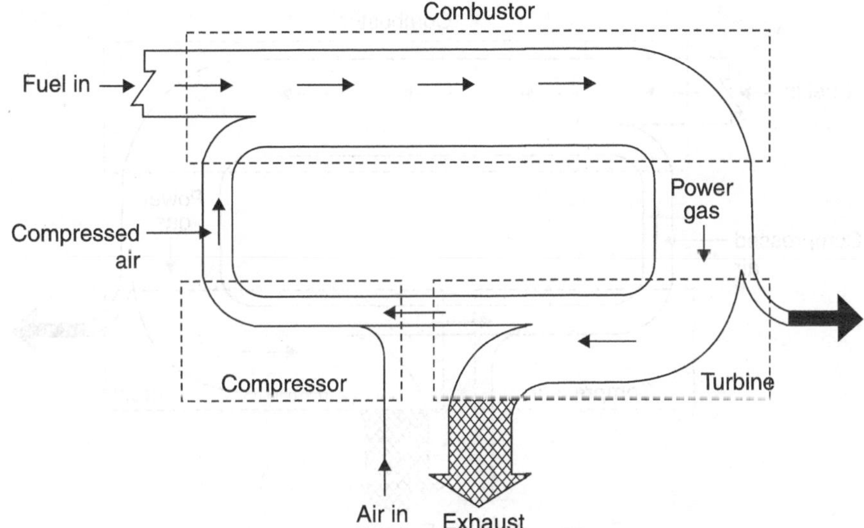

Fig. 12.20. Energy flow diagram for gas-turbine unit.

Shaft energy to drive compressor is about twice as much as the useful shaft output.

Actually the shaft energy keeps circulating in the cycle as long as the turbine runs. The important comparison is *the size of the output with the fuel input*. For the simple-cycle gas turbine the output may run about 20% of the fuel input for certain pressure and temperature conditions at turbine inlet. This means 80% of the fuel energy is wasted. While the 20% thermal efficiency is not too bad, it can be improved by including *additional heat recovery apparatus.*

12.2.6.2. Methods for improvement of thermal efficiency of open cycle gas turbine plant

The following methods are employed to increase the specific output and thermal efficiency of the plant :

1. Intercooling 2. Reheating 3. Regeneration.

1. Intercooling. A compressor in a gas turbine cycle utilises the major percentage of power developed by the gas turbine. The work required by the compressor can be *reduced* by compressing

the air in two stages and incorporating an *intercooler* between the two as shown in Fig. 12.21. The actual processes take place as follows :

1 – 2′ ... L.P. (Low pressure) compression
2′ – 3 ... Intercooling
3 – 4′ ... H.P. (High pressure) compression
4′ – 5 ... C.C. (Combustion chamber)–heating
5 – 6′ ... T (Turbine)–expansion.

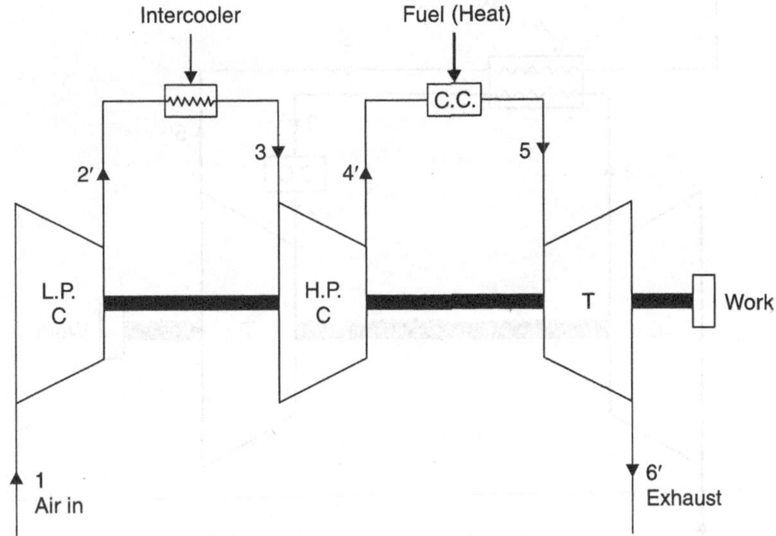

Fig. 12.21. Turbine plant with intercooler.

2. Reheating. The output of a gas turbine can be amply improved by expanding the gases in two stages with a reheater between the two as shown in Fig. 12.22. *The H.P. turbine drives the compressor and the L.P. turbine provides the useful power output.*

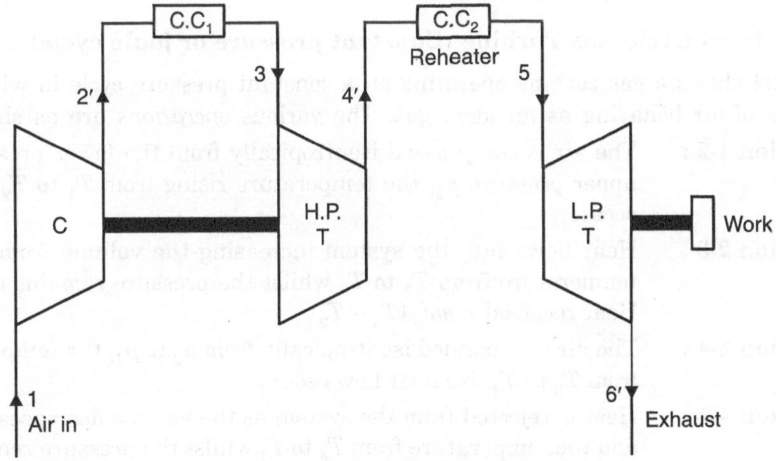

Fig. 12.22. Gas turbine with reheater.

3. Regeneration. The exhaust gases from a gas turbine carry a large quantity of heat with them since their temperature is far above the ambient temperature. They can be used to heat the air coming from the compressor thereby reducing the mass of fuel supplied in the combustion chamber. Fig. 12.23 shows a gas turbine plant with a regenerator.

Effectiveness, $\varepsilon = \dfrac{\text{Increase in enthalpy per kg of air}}{\text{Available increase in enthalpy per kg of air}}$

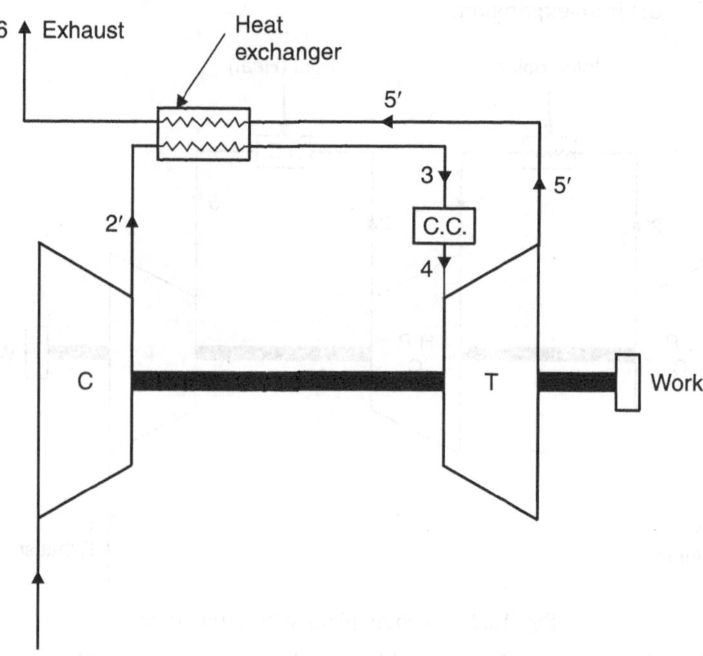

Fig. 12.23. Gas turbine with regenerator.

A heat exchanger is usually used in large gas turbine units for marine propulsion or industrial power.

12.2.7. Closed Cycle Gas Turbine (Constant pressure or joule cycle)

Fig. 12.24 shows a gas turbine operating on a constant pressure cycle in which the closed system consists of air behaving as an ideal gas. The various *operations* are as shown :

Operation 1-2 : The air is compressed isentropically from the lower pressure p_1 to the upper pressure p_2, the temperature rising from T_1 to T_2. No heat flow occurs.

Operation 2-3 : Heat flows into the system increasing the volume from V_2 to V_3 and temperature from T_2 to T_3 whilst the pressure remains constant at p_2. Heat received $= mc_p (T_3 - T_2)$.

Operation 3-4 : The air is expanded isentropically from p_2 to p_1, the temperature falling from T_3 to T_4. No heat flow occurs.

Operation 4-1 : Heat is rejected from the system as the volume decreases from V_4 to V_1 and the temperature from T_4 to T_1 whilst the pressure remains constant at p_1. Heat rejected $= mc_p (T_4 - T_1)$.

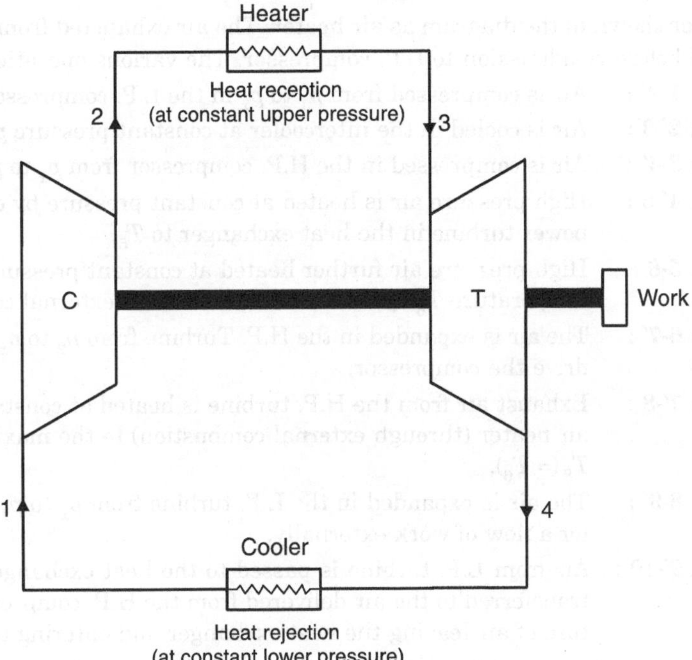

Fig. 12.24. Closed cycle gas turbine.

Fig. 12.25 shows the *p-V* diagram.

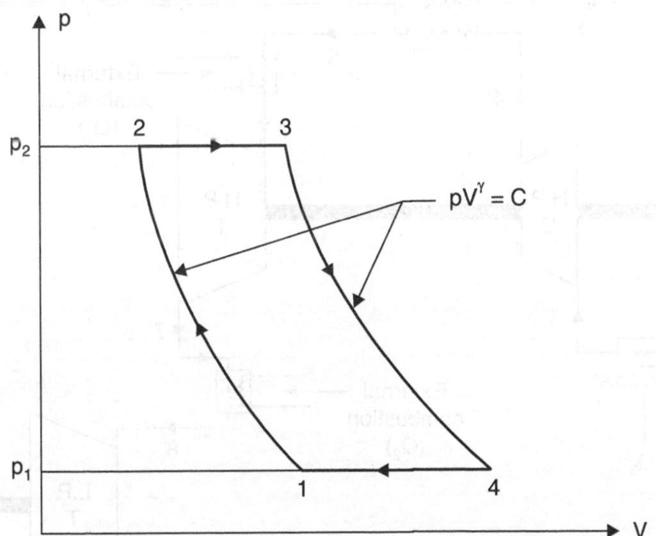

Fig. 12.25. *p-V* diagram.

- Fig. 12.26 shows an arrangement of closed cycle stationary gas turbine plant in which air is continuously circulated. This ensures that the air is not polluted by the addition of combustion waste product, since the heating of air is carried out in the form of heat exchanger shown in the diagram as air heater. The air exhausted from the power turbine is cooled before re-admission to L.P. compressor. The various operations are as follows :

Operation 1-2′ : Air is compressed from p_1 to p_x in the L.P. compressor.

Operation 2′-3 : Air is cooled in the intercooler at constant pressure p_x.

Operation 3-4′ : Air is compressed in the H.P. compressor from p_x to p_2.

Operation 4′-5 : High pressure air is heated at constant pressure by exhaust gases from power turbine in the heat exchanger to T_5.

Operation 5-6 : High pressure air further heated at constant pressure to the maximum temperature T_6 by an air heater (through external combustion).

Operation 6-7′ : The air is expanded in the H.P. Turbine from p_2 to p_x producing work to drive the compressor.

Operation 7′-8 : Exhaust air from the H.P. turbine is heated at constant pressure in the air heater (through external combustion) to the maximum temperature $T_8 (= T_6)$.

Operation 8-9′ : The air is expanded in the L.P. turbine from p_x to p_1, producing energy for a flow of work externally.

Operation 9′-10 : Air from L.P. turbine is passed to the heat exchanger where energy is transferred to the air delivered from the H.P. compressor. The temperature of air leaving the heat exchanger and entering the cooler is T_{10}.

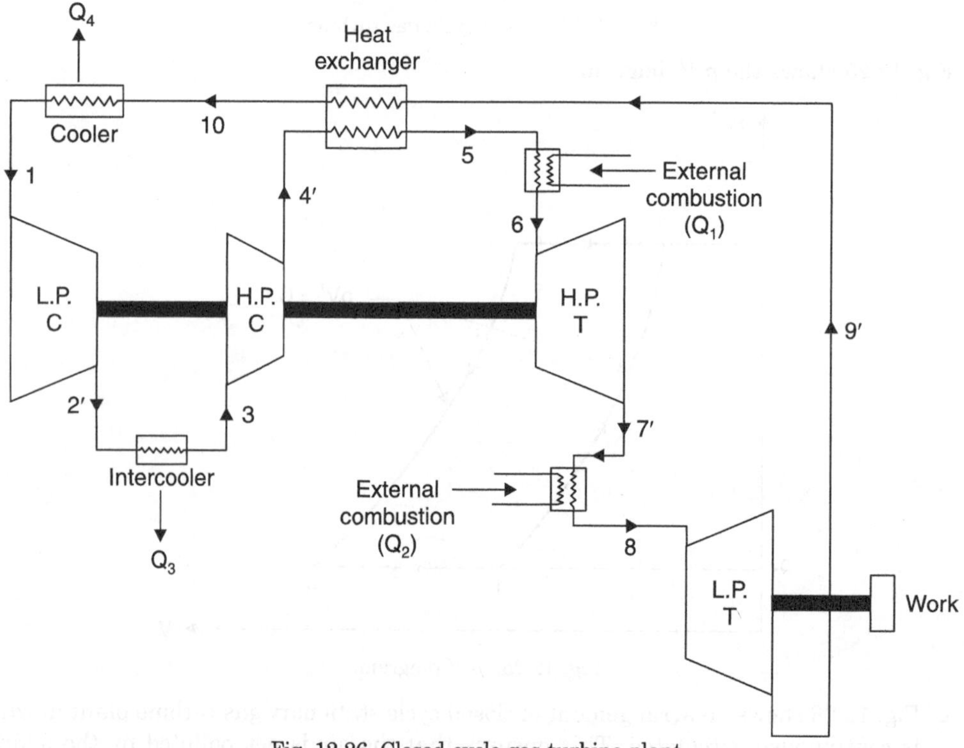

Fig. 12.26. Closed cycle gas turbine plant.

In a closed cycle plant, in practice, the control of power output is achieved by varying the mass flow by the use of a reservoir in the circuit. The *reservoir maintains the design pressure and temperature and therefore achieves an approximately constant level of efficiency for varying* loads. In this cycle, since it is closed, *gases other than air with favourable properties can be used ;* furthermore it is possible to burn solid fuels in the combustion heaters. *The major factor responsible for inefficiency in this cycle is the large irreversible temperature drop which occurs in the air heaters between the furnace and circulating gas.*

Note. 1. In a closed cycle gas turbines, although air has been extensively used, the use of 'helium' which though of a lower density, has been inviting the attention of manufacturers for its use, for large output gas turbine units. *The specific heat of helium at constant pressure is about 'five times' that of air, therefore for each kg mass flow the heat drop and hence energy dealt with in helium machines is nearly five times of those in case of air.* The *surface area of the heat exchanger for helium can be kept as low as 1/3* of that *required for gas turbine plant using air as working medium.* For the same temperature ratio and for the plants of the same output the *cross-sectional area required for helium is much less than that for air.* It may therefore be concluded that the *size of helium unit is considerably small comparatively.*

2. Some gas turbine plants work on a combination of two cycles, the open cycle and the closed cycle. Such a combination is called the *semi-closed cycle. Here a part of the working fluid is confined within the plant and another part flows from and to atmosphere.*

Merits and demerits of closed cycle turbine over open cycle gas turbine :

Merits of closed cycle :

1. Higher thermal efficiency.
2. Reduced size.
3. No contamination.
4. Improved heat transmission.
5. Improved part load efficiency.
6. Lesser fluid friction.
7. No loss of working medium.
8. Greater output.
9. Inexpensive fuel.

Demerits of closed cycle :

1. Complexity.
2. Large amount of cooling water is required. This limits its use to stationary installation or marine use where water in abundance.
3. Dependent system.
4. The weight of the system per H.P. developed is high comparatively, therefore not economical for moving vehicles.
5. Requires the use of a very large air heater.

12.2.8. Constant Volume Combustion Turbines

Refer to Fig. 12.27. In a constant volume combustion turbine, the compressed air from an air compressor *C* is admitted into the combustion chamber *D* through the valve *A*. When the valve *A* is closed, the fuel is admitted into the chamber by means of a fuel pump *P*. Then the mixture is ignited by means of a spark plug *S*. The combustion takes place at constant volume with increase of pressure. The valve *B* opens and the hot gases flow to the turbine *T*, and finally, they are discharged, into atmosphere. The energy of the hot gases is thereby converted into mechanical energy. For continuous running of the turbine these operations are repeated.

The main demerit associated with this type of turbine is that the *pressure difference and velocities of hot gases are not constant ; so the turbine speed fluctuates.*

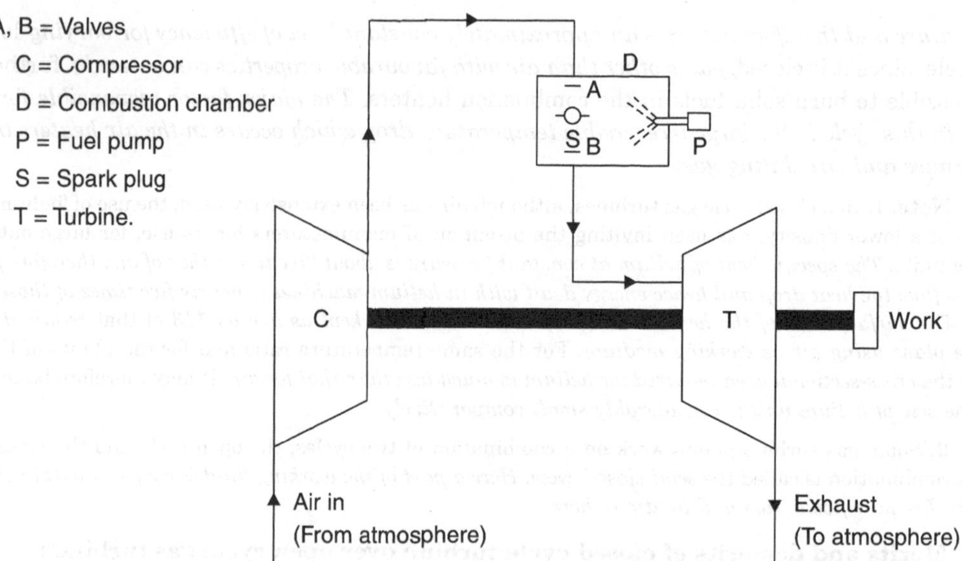

Fig. 12.27. Constant volume combustion gas turbine.

12.2.9. Uses of Gas Turbines

Gas turbines find wide applications in the following *fields* :

1. Supercharging.
2. Turbo-jet and turbo-propeller engines.
3. Marine field.
4. Railway.
5. Road transport.
6. Electric power generation.
7. Industry.

12.2.10. Gas Turbine Fuels

The various fuels used in gas turbines are enumerated and discussed below :

1. Gaseous fuels.
2. Liquid fuels.
3. Solid fuels.

1. **Gaseous fuels.** *Natural gas is the ideal fuel for gas turbines*, but this is not available everywhere.

Blast furnace and *producer gases* may also be used for gas turbine power plants.

2. **Liquid fuels.** Liquid fuels of petroleum origin such as distillate oils or residual oils are most commonly used for gas turbine plant. The essential qualities of these fuels include *proper volatility, viscosity and calorific value*. At the same time it *should be* free *from any contents of moisture* and *suspended impurities that would log the small passages of the nozzles and damage valves and plungers of the fuel pumps.*

Minerals like *sodium, vanadium* and *calcium* prove *very harmful* for the *turbine blading* as these build deposits or corrode the blades. The sodium in ash should be less than 30% of the vanadium content as otherwise the ratio tends to be critical. The actual sodium content may be between 5 ppm to 10 ppm (part per million). If the vanadium is over 2 ppm, the magnesium in ash tends to become critical. *It is necessary that the magnesium in ash is at least three times the quantity of vanadium.* The content of calcium and lead should not be over 10 ppm and 5 ppm respectively.

Sodium is removed from residual oils by mixing with 5% of water and then double centrifuging when sodium leaves with water. Magnesium is added to the washed oil in the form of epsom salts, before the oil is sent into the combustor. This checks the corrosive action of vanadium. Residual oils burn with less ease than distillate oils and the latter are often used to start the unit from cold, after which the residual oils are fed in the combustor. In cold conditions residual oils need to be preheated.

3. **Solid fuels.** The use of solid fuels such as coal in pulverised form in gas turbines presents several difficulties most of which have been only partially overcome yet. The pulverising plant for coal in gas turbines applications is much lighter and smaller than its counterpart in steam generators. *Introduction of fuel in the combustion chamber of a gas turbine is required to be done against a high pressure whereas the pressure in the furnace of a steam plant is atmospheric.* Furthermore, the *degree of completeness of combustion in gas turbine applications has to be very high as otherwise soot and dust in gas would deposit on the turbine blading.*

Some practical applications of solid fuel burning in turbine combustors have been commercially made available in recent years. In one such design finely crushed coal is used instead of pulverised fuel. This fuel is carried in steam of air tangentially into one end of a cylindrical furnace while gas comes out at the centre of opposite end. As the fuel particles roll around the circumference of the furnace they are burnt and a high temperature of about 1650°C is maintained which causes the mineral matter of fuel to be converted into a liquid slag. The slag covers the walls of the furnace and runs out through a top hole in the bottom. The result is that fly ash is reduced to a very small content in the gases. In *another design* a regenerator is used to transfer the heat to air, the combustion chamber being located on the outlet of the turbine, and the combustion is carried out in the turbine exhaust stream. The advantage is that only clean air is handled by the turbine.

12.2.11. Gas Turbine for Automotive Vehicles

A gas turbine essentially consists of the following *two sections* :

1. Gasifer section
2. Power section.

The *gasifer section* burns fuel in a burner and delivers the resulting gas to the power section, where it spins the power turbine. The power turbine then turns the vehicle wheels through a series of gears.

- The **gasifier** consists of a compressor which has a rotor with a series of blades around its outer edge.
  - — As the rotor rotates, air between the blades is carried around and thrown out by centrifugal force into the burner. Consequently, air pressure rises in the turbine.
  - — The fuel is injected in the burner where it burns and the pressure is further raised.
  - — The high pressure and high temperature gas then passes through the *gassifier nozzle diaphragm.* A series of stationary blades direct this gas against a series of curved blades on the outer edge of the *'gasifier turbine rotor'* causing it to rotate at a high speed. The gasifier turbine rotor and the compressor rotor are mounted on the same shaft, thus the compressor rotor also rotates at high speed. This action continues to supply the burner with a sufficient amount of compressed air. The action continues as long as fuel is supplied to the burner.
- The high pressure, high temperature gas, after leaving the gasifier section, enter the *'power turbine'*. Here, the gas strikes another series of stationary curved blades which directs it against a series of curved blades on the outer edges of the *'power turbine rotor'* causing it to rotate at high speed (about 50000 r.p.m.). This high speed is reduced by a series of transmission gears before the power is supplied to the wheels of the vehicle.

Major components of automotive gas turbine :

The major components of automotive gas turbine are briefly discussed below :

1. ***Compressor :***

- For automotive gas turbines, either the positive displacement or rotary compressors are used.
- Out of rotary compressors, the axial flow compressor is not preferred due to its low pressure ratio (1.2 : 1). A multistage compressor would be needed to achieve pressure ratio of 4 : 1 desired for automobile applications. This pressure ratio can also be achieved by a centrifugal compressor.

2. ***Turbine :***

- For automotive applications, the turbines may be axial flow and radial flow types because they have *good air handling characteristics.*
  - — A radial flow turbine has an efficiency *higher* than that of the axial flow turbine over a wide operating range.
  - — The radial flow turbines are *much more speed sensitive* than the axial flow turbines.

3. ***Regenerator :***

- In automotive turbines, regenerators are used for *heat exchange.*
  - — A rotary type periodic flow regenerator is commonly used. In a *disc generator*, the compressor air flows axially through a sector of slowly rotating disc, the hot gases flowing in the opposite direction through rest of the disc.
  - — In the *drum regenerator* the compressor air flows radially inward through the matrix and the hot gases radially outward through the matrix on the other side of a bulk head through which the drum slowly rotates.

4. ***Combustion chamber :***

"*Can type combustion chamber*" is most commonly used. It consists of an inner chamber which admits the primary air through swirl vanes near the fuel jets of the injector nozzle. The combustion chamber can be divided into the following *three zones—'primary'*, *'intermediate'* and *'secondary'*.

— The main combustion occurs in the *primary zone* ;

— In the *intermediate zone,* the combustion reactions continue to release energy ;

— In the *secondary zone*, an extra quantity of air is supplied to lower down the temperature of combustion products so that the turbine inlet temperature is within the permitted zone.

5. ***Transmission :***

Since the turbine speed is much more than the conventional automotive engines, a transmission system is employed to get the speed reduction. Following transmission designs are considered for the gas turbine engines :

(*i*) Multiple gearbox with controlled slip clutch ; (*ii*) Continuously variable transmission ; (*iii*) Variable ratio belt drive ; (*iv*) Electric ; (*v*) Hydraulic torque converter ; (*vi*) Hydromechanical drive.

- A *turbine powered vehicle* developed by "*Chrysler Corporation*" has the following brief details :
  - — Single stage centrifuge compressor with a pressure ratio of 4 : 1, compressor air flow of 1 kg/sec. ;
  - — A single can type combustion chamber ;

— Six generation gas turbines ;

— The first and second stage turbines are of axial type and the second stage turbine has variable geometry stator ;

— Two rotating discs type regenerator runs at 22 r.p.m., regeneration effectiveness is 87 per cent ;

— The efficiencies of compressor, first stage and second stage are 77, 88 and 71 per cent respectively.

— The maximum turbine speed is about 45700 r.p.m., while maximum output speed is 4700 r.p.m.

— The rated power of the engine is 150 bhp at 5000 r.p.m. output shaft speed.

12.3. AUTOMOBILE BODY AND SAFETY CONSIDERATIONS

12.3.1. Requirements of an Automobile Body

The body of an automobile, especially of car which is of framless type, must satisfy the following requirements (In case of the heavier vehicles of the conventional frame type construction, the body and frame should jointly meet these requirements) :

1. **Strength :**

- The body must be *strong* enough to withstand various types of forces such as :
  — Weight of the car ;
  — Passengers and the luggage ;
  — Inertia ;
  — Braking and side forces.
- It should able to cope with impact loads of reasonable magnitude.

2. **Stiffness :**

- The body of the car may be considered as a beam supported on wheels at each end. It should be stiff enough to prevent excessive sagging in the middle.

3. **Torsional stiffness :**

- The body should be rigid enough to withstand twisting on bad roads.

4. **Space :**

- The space in the body should be adequate to accommodate the passengers as well as the luggage conveniently.

5. **Air drag :**

- The air resistance during running depends on the shape of the body and *increases directly as the square of the vehicle speed.*
- The body shape should be such that the air drag is minimum.

6. **Resistance to corrosion :**

- The shape of the body should be designed in such a fashion that no moisture is accumulated which would produce *rust.*
- The materials used should also be such that corrosion does not take place.

7. **Protection against weather :**

- The body should be designed in such a way that complete protection is available to the occupants and luggage from bad weather.

8. **Protection in accident :**

- The design of the automobile body should be such that the driver is at maximum ease so that there are least chances of the accident taking place due to fatigue.

- However, if the accident does take place, the automobile body must protect the occupants.

12.3.2. Safety Considerations

The protection of occupant(s) in a vehicle is of prime importance. *Seat belts and other features work together to protect the occupants of the car during a crash.* When working properly, this can reduce the chances of serious injury or death in an accident. The seats and head restraint play a major role in the safety of passengers. The person driving the car should also know the use of hazard light and controls, and driver's air bag facility.

The safety of an automobile can be ensured by the following *two main considerations* :

1. The design of the automobile should be such that *chances of accident happening are minimum.*

2. If the accident takes place, the *injury to the occupants should be minimum.*

1. **Preventive design :**

The automobile's design should aim at driver comfort and safety of its various systems. It should take care of the *following points* :

(*i*) There should be largest possible glass area with minimum blind spots caused by body pillars. This will increase driver's efficiency especially for parking and when driving in heavy traffic.

The accessories such as windscreen wiper, windscreen washer, defroster nozzles, defroster heater etc. should be employed so that visibility is not decreased by the bad weather.

(*ii*) The design of the instrument panel should be such that the driver should be able to read all meters and see all indicators without any distraction.

(*iii*) The location of the controls should be within the reach of the driver, for reducing his fatigue to the minimum.

(*iv*) The seat of the driver should be comfortable and adjustable according to his requirements.

(*v*) The vehicle bodywork should be so designed that noise is minimum.

— The suspension for engine, gear box and front axle should be insulated against noise from engine and running noises.

— The exhaust system should have good noise attenuation characteristics.

(*vi*) The suspension and steering systems should so designed as to get better road-holding and improved characteristics while cornering.

(*vii*) The brakes must be so designed that the vehicle does not swerve, skid or veer out.

— Anti-skid system (ABS) is now standard on many cars. It prevents locking the wheel even when the brakes are applied hard and the vehicle remains steerable and controllable.

2. **Designing for minimum injury in case of accident :**

(*i*) The body design should be such as to have a rigid box for the occupants with weaker ends which crumple on collision thereby absorbing the energy of impact and bringing the vehicle to stop gradually.

— In order to test the designed body good manufacturers resort to crash tests to supplement the knowledge gained from accident statistics. The cars are subjected to head-on-crashes, road-end crashes, side sweeps etc. against massive concrete blocks and using lifetime dummies packed with electronic sensor devices.

— For providing protection against side sweeps, the doors may be strengthened by steel bars.

(*ii*) The fuel tank should be suitably located to reduce the risk of spilled fuel and consequent fire in case of rear end collision.

(*iii*) The doors should be fitted with suitable safety locks, so that in case of accident they stay property closed until the car comes to rest.

(*iv*) The cars' interiors should be suitably upholstered with adequate padding, to protect the occupants from injuries due to impact.

(*v*) The steering column and wheel should be so designed that in case of a collision, they should give way and thus avoid causing injury to the driver.

(*vi*) The windscreen should be made of laminated safety glass which is tough enough to take up impact forces without shattering and still remain in one piece.

(*vii*) The front seats should be so firmly fixed to the floor that will not become loose even when subjected to very high forces. The front seats should be provided with adjustable head restraints.

(*viii*) The various switches and controls etc. should be so shaped that they are not protruding excessively as to cause injury to the occupants at the front.

(*ix*) In order to reduce the risk of injury appreciably, the seat belts should be worn both by the driver as well as the passengers.

— For added safety, seat belt may be supplemented by **air bag system**. The *air bag* consists of an instantly inflatable cushion housed in a special compartment in the centre of the steering wheel on the driver's side and in the dash board on the passenger side. If there is a collision of such magnitude that it exceeds the setting of the special sensor, the mechanism is activated, the cushion inflates instantly and acts as a soft protecting barrier between the front passenger side seat, the passengers, and the structures in front of them that could cause injury.

Gadgets fitted in cars for safe operation :

The following gadgets are fitted for safe operation.

1. *Driver's air bag.*

2. *Antilock braking system (ABS).* The electronic ABS control unit analyses incoming signals from the wear-free wheel-speed sensors to determine the speeds of both the individual wheels and the vehicle as a whole. The ABS hydraulic modulator responds to in incipient wheel-lock by regulating the hydraulic pressure to the affected wheel's brakes. *By preventing the wheels from locking, ABS maintains stable tracking and steering control, even under braking.*

3. *Pollution control and environment protection.*

4. *Ventilation.*

5. *Turn signals and hazard warning indicators.*

6. *Other devices :*

(*i*) Power window switch.

(*ii*) Inside rear view mirror.

(*iii*) Outside rear view mirror.

(*iv*) Tiltable steering wheel.

(*v*) Lambda sensor (It ensures that the air and fuel are constantly mixed in correct proportion).

(*vi*) Three-way catalytic converter.

(*vii*) First aid kit.

(*viii*) Central locking and child safety locks.

Do's and Dont's for safe driving :

1. **Starting :**

- Adjust the driver's seat for proper driving posture.

- Adjust the rear view mirror so as to obtain the best possible rear view.
- Before moving off, look forward and back to confirm safety.
- Don't start quickly for it is dangerous and wastes fuel.

2. **General driving :**
   - Be sure to stop before the spot light and stop signal. While moving into the intersection without any traffic lights or signs, drive slowly to confirm safety.
   - Always follow other vehicles at a safe distance in order to prevent a rear end collision, in case of sudden stopping by the car ahead.
   - Turn 'ON' the turn signal at least 30 metres before making a turn or changing lane so as not to be hit from behind.
   - Before entering a corner, decelerate to a safe speed. Do not apply brakes during cornering avoiding a cornering skid.
   - When over-taking other cars, watch out for the coming car and carefully confirm safety.
   - Don't attempt zigzag driving which will hinder your control over the car and cause an accident.
3. **Braking :**
   - Use the parking brake when parking your car and shift the gearshift lever into the first gear or reverse gear position for sake of safety.
   - Don't use hand braking unless unavoidable to avoid skidding. It is specially dangerous with worn tyres for they allow a larger skid.
   - Use foot brakes. Warn the car behind you. Gradually apply the brakes or bring the car to a halt.
   - When driving downhill, do not try to apply brakes but use the engine brake effectively. Overuse of foot brake may result in total brake failure.
   - Avoid neckless high speed driving and try to drive at a safe speed suitable for the road conditions while maintaining a constant speed.
     - — Never indulge in sharp handling during high speed driving to avoid loosing control.
     - — When overtaking or changing the lane at high speed keep an ample car-to-car distance.
4. **Long distance driving :**
   - Before starting on a trip be sure to *perform safety checks.*
   - Take rest at certain intervals to prevent an accident which may occur if you are tired or sleeping.
5. **Night time driving :**
   - Drive more slowly at night than in daytime since visual range is restricted at night.
   - Don't overtake other cars at night. Darkness effects your sense of speed and judgement.
   - Don't use the high headlight beam unless the use is inevitable to avoid dazzle to others.
   - Always keep the window glass clean. Don't operate the windshield wiper when the glass is dry.
6. **Margin for safety :**
   - It is important to allow yourself a margin for safety during driving so that you can cope with the erroneous or unexpected driving of other drivers.

Some more points for safe driving :

(*i*) For personal safety and avoiding accidents take care of air pressure in your vehicle's tyres including wheel alignment. *Underinflated tyres are prone to failures due to high temperature.*

(*ii*) Tyres should not be of different sizes. Excessively worn out tyres can cause trouble.

(*iii*) The reverse gear should be engaged only when the car is stationary.

(*iv*) Do not wear your seat belt overhard or breakable objects in pockets or on your clothing. If an accident occurs, objects *e.g.* glasses, pen etc. under the seat belt can cause injuries.

(*v*) Never remove the cap from the coolant reservoir when the engine is hot else hot water with steam may come on you.

(*vi*) During normal operation the battery generates gas which is explosive in nature. A spark or open flame can cause the battery to explode, causing very serious injuries.

(*vii*) Make sure that the bonnet is fully closed and latched before driving to avoid its fly-up unexpectedly during driving, obstructing your view and resulting in an accident.

(*viii*) Avoid jacking up the vehicle on sloping surface. Use the jack properly. Passengers should not remain in a jacked vehicle particularly if the engine is running.

(*ix*) Make sure that the highly corrosive brake fluid does not drop on to the paint work. If it does, wash it off immediately with water. Brake fluid is poisonous. In the event of accidental contact, wash the affected part with water and mild soap rinse. Swallowing the fluid is highly dangerous.

(*x*) Heavy loads which are not secured/anchored could seriously injure passengers in the event of an accident.

(*xi*) Remove the key from the ignition switch only when parked.

(*xii*) Don't let the engine run in an unventilated area. It consumes oxygen and emits CO_2 and other gases which are harmful for the organism.

12.4. CAR DRIVING

12.4.1. Car Operation

Car operation is briefly discussed below :

When the driver turns on the starter switch, it releases electric current from a battery and sets into action an electrical starting motor which runs a flywheel connected to the engine crankshaft. Immediately the driver steps on a pedal which controls the rate of flow of air-fuel mixture into the inlet manifold. During the outward stroke of the piston this mixture is sucked into the cylinder. During the inward stroke this mixture is compressed and an electrical spark by a plug sets fire to it. The resulting explosions act high pressure on the piston which again moves outwards turning the crankshaft. Lastly during the final stroke the burnt gases leave the cylinder. As the process is repeated over and over again, the crankshaft is turned continuously and so the engine starts running.

— *The motion of the crankshaft is transmitted to the rear wheel through the clutch, gearbox and differential.*

— Through the mechanism of *clutch*, the crankshaft is connected to a set of gears. By means of gearshift lever, the gears of the transmission can be shifted to suit driving conditions for changing the speed.

— The motion from the gearbox to the *differential* is transmitted through a propeller shaft. The differential is arranged in the rear axle to permit one rear wheel to turn faster than the other while turning a corner. The rear wheels are directly connected to the gear axle and get motion as the rear axle turns and as such the vehicle is set into motion.

— The motor vehicle, rolling on the road, is directed by a *steering gear* which controls the direction in which the front wheels are pointed.

— The *braking system* is provided to stop the vehicle or slow it down as required by the driver.

— The uneveness of the road surfaces and jerks are partially controlled by *front and rear springs*, by *shock absorbers*, by *padding and springs in the car seats*, by *chassis cross members* which decrease sideways, and by the *distribution of chassis and body weight between front and rear wheels.*

12.4.2. Basic Requirements to Drive

In order to learn to drive the following are required :

(*i*) A car not too old.

(*ii*) 'L' plates in front and rear.

(*iii*) A driving range or open field.

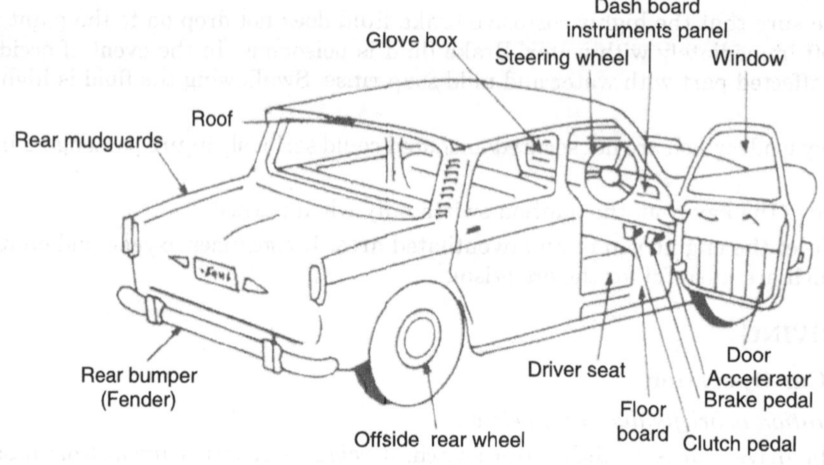

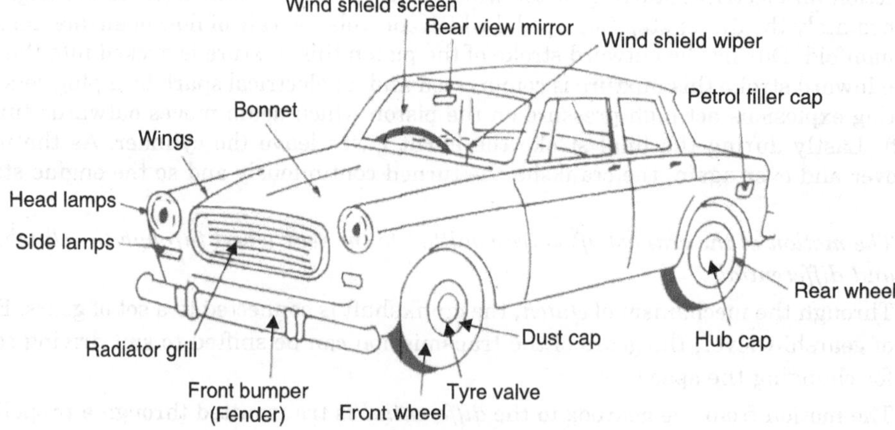

Fig. 12.28. Points of a car.

(*iv*) Learner's license

(*v*) Licensed driver by the learner's side.

— Additional 'L' signs on the rear and off sides of the car will add to security while practising traffic driving.

- *The learner should get familiar with the names of different parts of a car and terms normally used by motorists, filling station attendants, mechanics and others.*
- The major **driving controls** are :

(*i*) *Foot controls.* Accelerator pedal ; Clutch pedal ; Foot brake pedal.

(*ii*) *Hand controls.* Hand brake lever ; Steering wheel ; Gear shift level.

Fig. 12.28 shows points of a car.

12.4.3. Preliminary Inspection

Before going to driver seat the following *must be checke*d :

(*i*) Coolant in the radiator.
(*ii*) Engine oil level.
(*iii*) Battery levels.
(*iv*) Brake fluid level.
(*v*) Tyre pressures.
(*vi*) Fuel in the tank.

12.4.4. Seating, Starting, Moving and Stopping

You must be seated, resting firmly against the back of the seat, naturally and comfortably. Having seated yourself properly, *tick off the following* :

Starting :

(*i*) All doors are secured.

(*ii*) Check mirror for adjustment.

(*iii*) Seat belt is secure.

(*iv*) Hand brake is on.

(*v*) Gear is in neutral.

(*vi*) Choke if necessary.

(*vii*) Switch on the check ammeter. It should register discharge.

(*viii*) Hold the steering wheel at 9 o'clock and 3 o'clock, lightly but firmly.

(*ix*) Depress the clutch pedal fully.

(*x*) Press and release starter as engine comes alive, delay will damage starter and flywheel.

(*xi*) Raise the engine speed about 1000 r.p.m. and not more.

(*xii*) Let the clutch out.

(*xiii*) Cancel choke when engine runs evenly.

(*xiv*) Check ammeter.

(*xv*) Check temperature gauge.

(*xvi*) Check oil pressure gauge.

(*xvii*) Check air pressure gauge if you have air operated brakes.

(*xviii*) Check fuel gauge.

(*xix*) After pressing the gas pedal a wee bit, release it gradually to bring engine down to idling speed. Watch engine pick over smoothly.

Moving :

(*xx*) Depress clutch pedal fully.

(*xxi*) Move lever into first gear smoothly and return hand to steering gear.

(*xxii*) Let the clutch pedal out slowly till you feel car is anxious to move forward. Hold the clutch there.

(*xxiii*) Give the 'Intend to move out' signal.

(*xxiv*) Check mirrors for traffic coming from behind you.

(*xxv*) Look round over your right shoulder.

(*xxvi*) Both hands on the steering wheel. Look ahead. Left foot off the clutch pedal and on the floor board. Let the car move steadily forward for a few metres while you get the 'feel' of the gas pedal.

Stopping :

(*xxvii*) Check mirrors.

(*xxviii*) Signal intention to stop.

(*xxix*) Righ foot off gas pedal, gradually. Avoid jerky movements.

(*xxx*) Push brake pedal down, gradually increasing pressure.

(*xxxi*) Just a little before the car stops, push the clutch pedal down fully. If not done properly, engine will stop running.

(*xxxii*) Ease pressure on brake pedal just before the car stops. The car should roll free for the last few centimetres and stop.

(*xxxiii*) Apply hand brake.

(*xxxiv*) Shift gear lever into neutral.

(*xxxv*) Take both feet off pedals.

(*xxxvi*) Switch off and relax.

- Now repeat starting, moving forward and stopping. Concentrate on different 'feel' of the controls.
- Then, practice starting, moving in reverse and stopping. While reversing, look around over your left shoulder through the rear window.

12.4.5. Police Signals

Refer to Fig. 12.29.

1. To stop a vehicle approaching from behind.
2. To stop a vehicle coming from front.
3. To stop vehicles approaching simultaneously from front and behind.
4. To stop traffic approaching from left and wanting to turn right.
5. To stop traffic approaching from the right to allow traffic from the left to turn right.
6. To allow traffic coming from the right and turning right by stopping traffic approaching from the left.
7. Warning signal closing all traffic.
8. Beckoning on vehicles approaching from left.
9. Beckoning on vehicles approaching from right.
10. Beckoning on vehicles from front.

Fig. 12.29. Police signals.

12.4.6. Traffic Signs

The Schedule of the Motor Vehicle Act 1988.

Note. (1) The figure 20 is given merely as an example. The actual figures will be as prescribed in each case where this sign is issued.

2. The general design of the post is given for guidance.

3. Where the speed limit is, or is to be, imposed only on a certain class or classes of motor vehicles the class or classes will be specified on the 'definition plate'. While disc and special speed limit together with the class or classes of vehicle to which it applies will be specific on the "definition plate".

4. The paints to be used on the traffic signs should be of reflecting kind.

Note. Sign No. 5 as here set forth may be amplified by instructions inscribed upon a definition plate placed below it as in the general arrangement set forth in sign No. 1 of this part. Upon the definition plate may be set forth the time during which parking is prohibited. In like manner an arrow head inscribed on the definition plate will indicate that parking prohibited on that part of the street or road lying to the side of the sign to which the arrow head points.

TRAFFIC SIGNS
Part A–Mandatory Signs

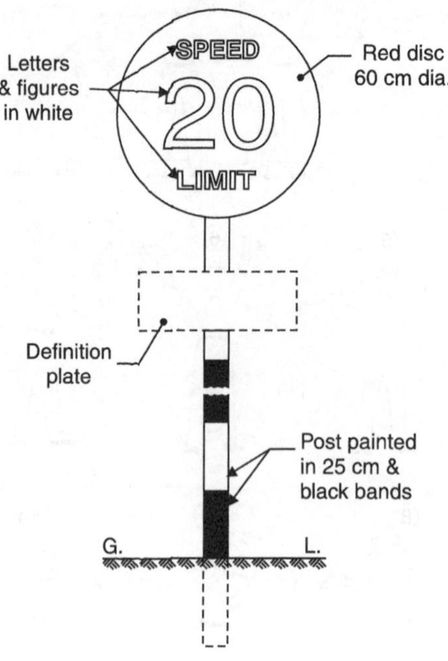

No. 1 Speed limit

Fig. A

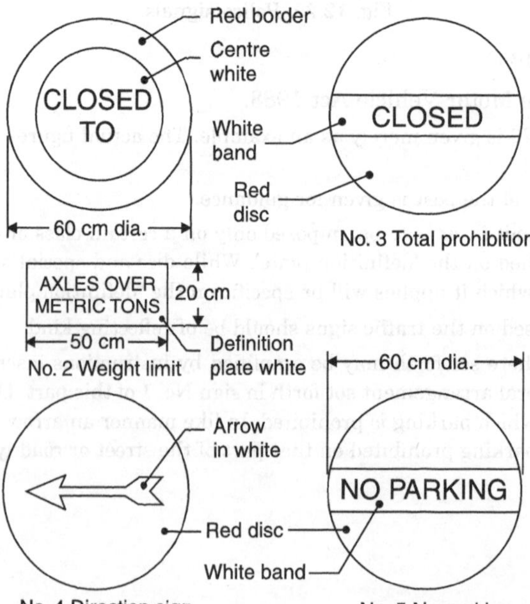

Fig. B

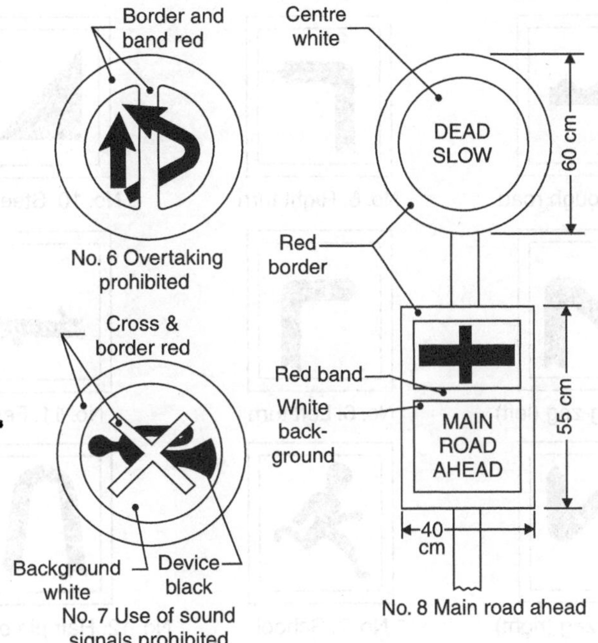

Fig. C

Part B-Cautionary Signs

The signs of this part shall be used in conjunction with a red triangular plate, the centre of which shall be either hollow or painted white, in the manner indicated in the general design reproduced below.

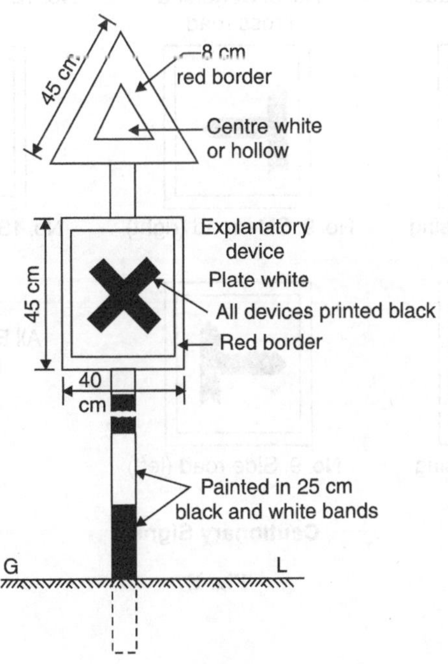

Fig. D

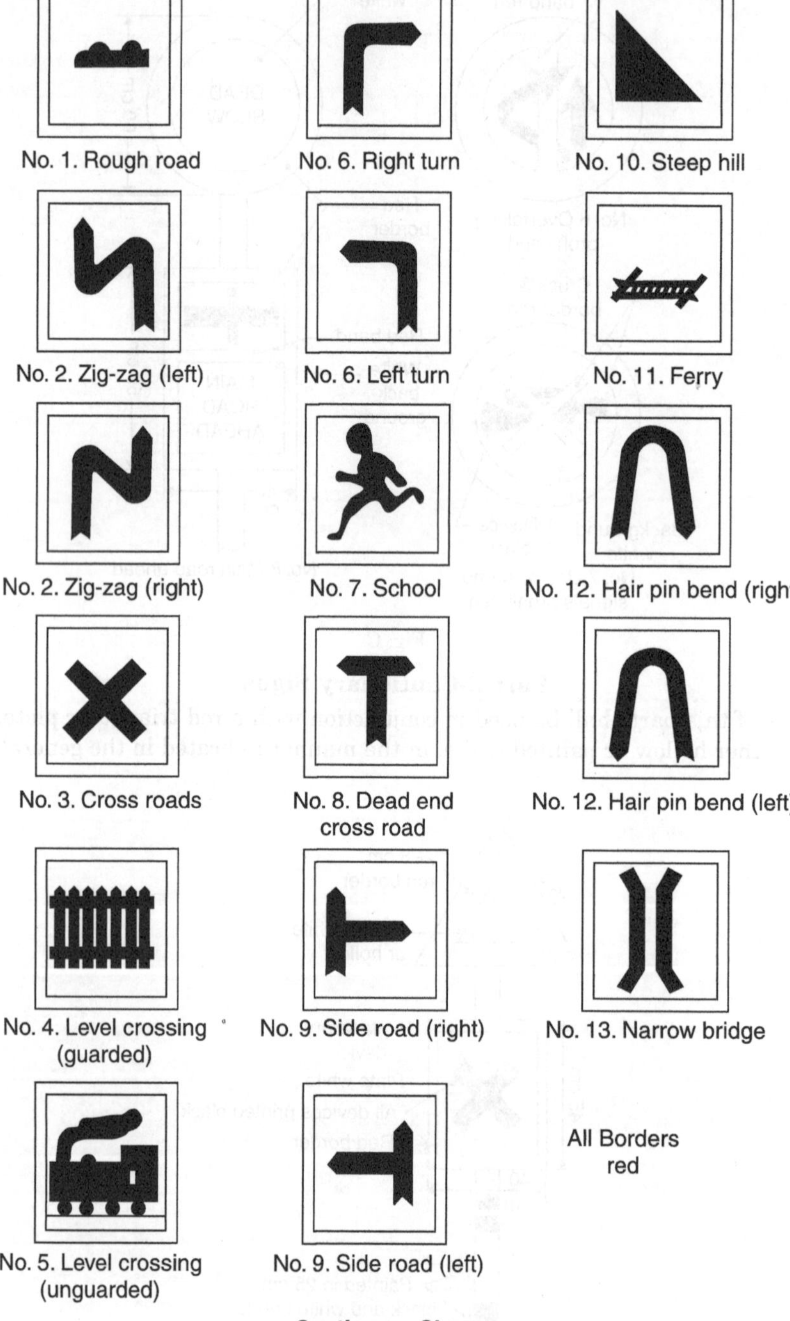

Cautionary Signs

Fig. E

Part C-Informatory Signs

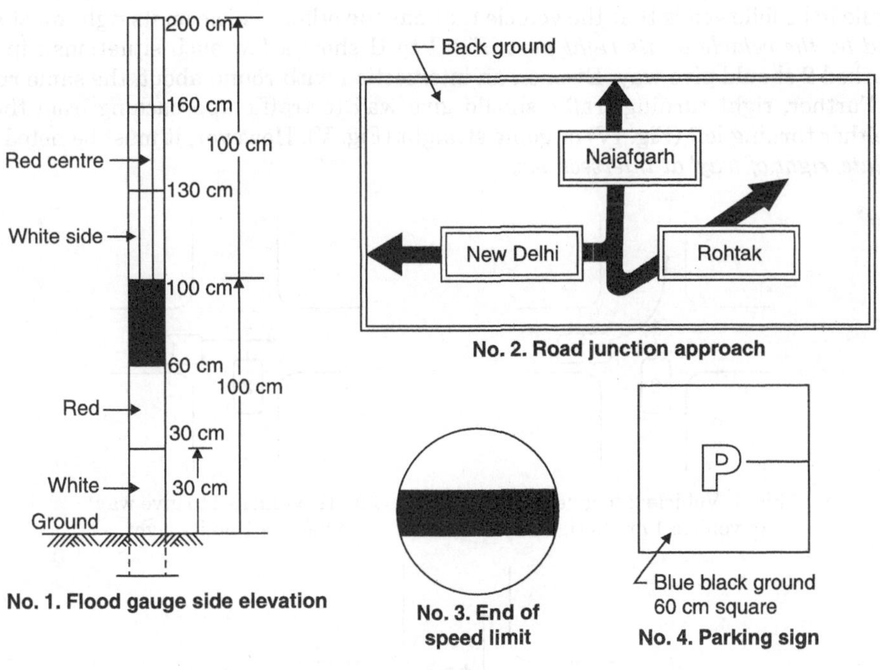

Fig. F

12.4.7. Driving Practice of Four Wheeler

A. Basic road manoenvers :

(*i*) Before you take out a vehicle, it must be ensured that it is in a good condition. *Check up lights, brakes, steering, tyres, tyre pressure, windscreen, windscreen wipers and washers. Also check up the lubricating oil level, fuel position, level of battery electrolyte, coolant level and the level of brake fluid.*

(*ii*) Before moving off, *adjust the position of rear view mirror* so as to have a good view of the rear. Also look over your shoulder as a final check, for the road traffic, particularly for small vehicles like bicycles, motor cycles, scooters etc. Signal a 'right turn' to show you are moving off.

(*iii*) The basic rule is to *keep to the left.* Do not drive along the middle of the road. When the road is divided into lanes, you may drive in any lane to the left of the centre. But it is a safe practice to keep in the left lane unless you are overtaking.

(*iv*) A signal of intention must be given before turning to the right or left, changing lanes, pulling out from the kerb or even generally diverging to the right or left. Any signal should be given at least 30 metres ahead as far as possible and *make sure the turn indicators are turned off after completing each manoeuvre.*

(*v*) Always drive such so that *you can stop without hard braking within the distance you can see.* Going down a long steep hill, *use a lower gear to ease strain on the brakes.* It is a good guide to use the same gear as you would use for climbing the same hill.

(*vi*) Leave enough space between you and the vehicle in front so that you can stop safely if it suddenly stops or slows down. As a rough guide, under good conditions, *leave a gap of at least two seconds.* As the vehicle in front of you passes a fixed point, say an electric pole, start to count out loud. Say, "one thousand and one, one thousand and two." If your vehicle reaches the fixed point before you finish saying," one thousand and two", you are following too close ; drop further back. To be safe, it is better to leave a longer gap at night.

(*vii*) One must approach the intersection cautiously particularly when there are no traffic lights. When two vehicles are approaching at an intersection and would collide if they continue as they are, the basic rule to be followed is that the vehicle that has the other vehicle on its right must *slow down or stop and let the vehicle on its right pass*. Fig. I to II show a few such situations ; in each case vehicle marked 2 should give way. Even on an intersection with round about, the same rule applies (Fig. III). Further, right turning traffic should give way to traffic approaching from the opposite direction either turning left (Fig. IV) or going straight (Fig. V). However, it must be noted that *there is no absolute 'right of way' at intersections.*

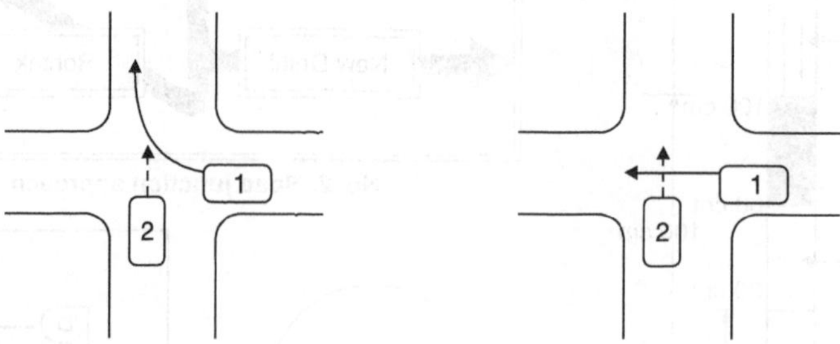

Fig. I. Vehicle 2 to give way to vehicle 1 on its right.

Fig. II. Vehicle 2 to give way to vehicle 1 on its right.

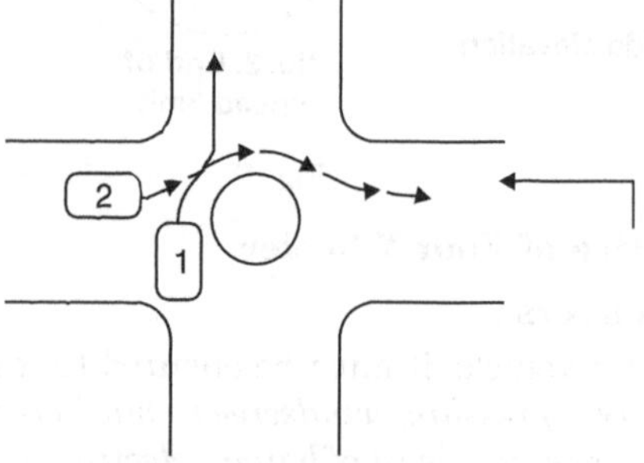

Fig. III. On intersection with round about, vehicle 2 should give way to vehicle.

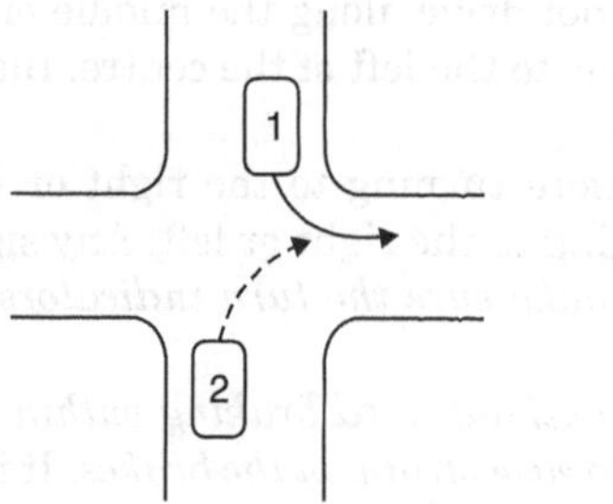

Fig. IV. Vehicle 2 to give way to vehicle 1.

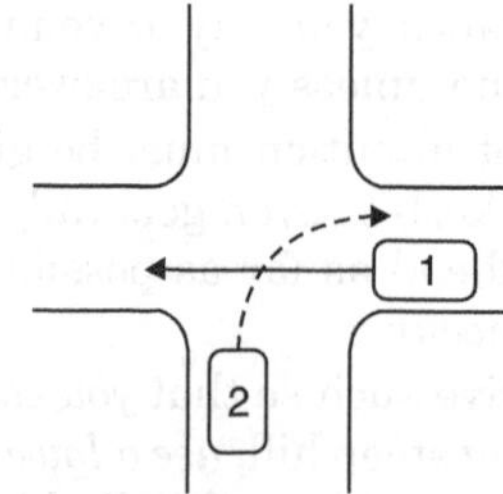

Fig. V. Vehicle 2 to give way to vehicle 1.

(*viii*) *Do not overtake unless you are sure of doing so safely.* Vehicles should normally be overtaken on the right. Before moving to overtake, check mirror and change gear if necessary. You must then signal your intention by blowing horn or flashing your headlights. Do not cut in after overtaking, but make sure that you are well clear. Signal your intention and move gradually back into the left.

When being overtaken, keep left and allow the overtaking vehicle room to pass and move back in. Do not move to the right or speed up till the overtaking vehicle is clear.

(*ix*) Before reversing, make sure that there is no one behind you. Look out particularly for children. If you are not certain, get someone else to help you or get out and check for yourself.

Never reverse from a minor road into a major road.

B. Driving techniques for special situations :

1. ***Driving in traffic :***

It needs what is called, defensive driving. The basis of defensive driving is to see and understand all that is happening around and to drive so that possible dangerous situations do not develop into actual dangers. Defensive or safe driving is usually smooth driving avoiding harsh and sudden use of accelerator, brakes and steering. A golden rule to follow in such a situation is to *keep allowance for the likely errors of other drivers.*

2. ***Driving at night :***

Driving is more difficult at night when road users are had to see, *e.g.*, pedestrians, cyclists, motor cyclists, etc., may become almost invisible. For night driving, the first thing to do is to *make sure your windscreen and head and tail light lenses are clean. Put your head lights on even when you may not need them to see, they help others to see you.* High beam may be used on any road, but you must dip your headlights,

- within 200 m of oncoming vehicle.
- when the lights of oncoming vehicle are dipped
- when you are behind within 200 m of another vehicle.

— You must adjust the vehicle speed so that you are able to stop within the distance you can see. If you are dazzled by the lights of the oncoming traffic, slow down until your eyes recover.

While parking at night, keep the parking lights on unless your vehicle is clearly visible under street light. Better still, park off the road if possible.

— One easily gets tired at night. Particularly after a hearty meal one feels more tired and sleepy. In such a situation, let *some one else drive for a while* or pull of the road and rest until you feel O.K.

3. ***Driving in slippery conditions :***

— Rain makes the road slippery and also makes it harder to see. To drive safely *increase the following distance from usual two seconds to four seconds behind the vehicle in front of you.*

— To improve visibility, *turn on the dipper beam of headlights.* First sprinkling of rain are particularly dangerous ; so are oil or grime on the road or dew or fog. Your vehicle will take longer to stop.

— To prevent skidding, drive slowly and brake easily and gently.

— Use wipers to keep the windscreen clear.

C. Emergency situations :

1. ***Skidding :***

— When one or more tyres lose grip on the road, skidding occurs. Skidding may occur on account of road being slippery due to rain, sand, gravel, ice, etc., or by sudden turns, acceleration or braking.

— If the rear wheels skid, the car will turn sideways. Turn the steering wheel to get the vehicle pointing in the right direction, but do not overdo it. Excessive steering may cause a skid in the opposite direction. Please remember to keep your feet completely off the accelerator and brake pedals till you have regained the control.

— In case front wheels skid, the vehicle will tend to go straight instead of responding to steering wheel. If caused by excessive braking, ease pressure on the brake pedal and centre the steering. If caused by excessive speed, take your foot off the accelerator pedal and centre the steering. Apply brake gently.

2. ***Tyre blowout :***

— In this situation, the car will pull to one side. Use the brakes as little as possible and definitely avoid heavy braking, which will make the things worse.

— Take your foot off the accelerator pedal, hold the steering wheel firmly and wait to stop by the side of the road.

3. ***Failure of brakes :***

— In case of brake failure, pump the pedal, change to lower gear and apply the parking brake gradually. If you have to go off the road to avoid collision, remember, earth banks and ditches are safer to hit than poles or trees.

— Even after driving through deep water, effectiveness of brakes is reduced appreciably due to some water staying inside. Therefore, *after driving out of water drive slowly with one foot gently on the brake pedal till the brakes are again dry.*

4. ***On the collision path :***

— It may so happen that the driver of some oncoming vehicle may fall asleep or loose control otherwise and come straight in your direction. In such a situation, do not panic and never drive towards the wrong side even if it may appear to be safe because the driver of the oncoming vehicle may change its course any time suddenly. Instead flash your lights and sound your horn.

— Look for room to the left, apply brakes gently and keep to the left. If required, drive off the road.

12.5. FUEL SAVING

Research, in India and abroad, show that you can save your 10 percent of fuel consumption in *scooters*, *motorcycles, mopeds* and *autorickshaws* by the following *methods* :

(*i*) Drive slow and steady.

(*ii*) Keep engine healthy.

(*iii*) Use the brakes sparingly.

(*iv*) Let go off your clutch.

(*v*) Clean air filter regularly.

(*vi*) Watch tyre pressure.

(*vii*) Stop fuel leaks.

(*viii*) Stop the engine if stoppage is for more than 2 minutes.

(*ix*) Use the light lubrication.

(*x*) Switch off light in the day time.

1. **Drive slow and steady :**

- Sudden starts and 'stop-and-go' driving waste fuel.
- Research shows that fast and rash driving can increase fuel economy by 50 percent.

Following speeds are recommended for fuel economy :

Scooters and auto-rickshaws	... 40 km/h
Motorcycles	... 50 km/h
Mopeds	... 20 to 30 km/h

2. **Keep engine healthy :**

- Test on a large number of vehicles shows that you can save over 6 percent by *regular tuning.*
- If engine of the vehicle emits smoke or has poor pulling power, have it checked immediately at a reputed garage.

— *Incorrect spark plug gap and faulty ignition timing reduce power output and increase fuel consumption.*

— Change spark plug and contact breaker points as recommended by the manufacturer.

— *Do not delay engine overhauls*. The cost of petrol that would be wasted may be more than the cost of an overhaul.

3. **Use the brakes sparingly :**

- A good driver anticipates stops. When you slam on the brakes, you waste a lot of useful energy. Anticipate corners—you will be able to take them without braking.
- Don't overtighten brakes or drive with your foot on the brake. These habits can cost you 5 percent more in fuel consumption.

4. **Let go off your clutch :**

- Unnecessary use of the clutch leads to loss of power.
- *Replace liners immediately when the clutch is worn out.*

5. **Clean air filter regularly :**

- Dust causes rapid wear of engine components and increases fuel consumption.

— Choked exhaust parts, and silencer make your engine struggle to breathe, causing poor consumption.

- *Dechoke the silencer and exhaust parts regularly,* as recommended by the manufacturer.

6. **Watch tyre pressure :**

- *Under-inflated tyres increase rolling resistance and petrol consumption.*

— Tests show that under-inflation of tyre by 25 percent cause a 5 percent increase in fuel consumption and a 25 percent decrease in tyre life.

- Watch for slow leaks.

7. **Stop fuel leaks :**

- Check your fuel tanks, carburettor and fuel lines daily for signs of leaking fuel. A leak of just drop for second costs you 500 litres of precious petrol a year.

— Shut off the fuel cock when you turn your motor off.

— Keep the fuel tank cap closed tightly.

— Park the vehicle in shade to prevent evaporation of fuel.

8. **Stop the engine if the stoppage is for more than 2 minutes :**

- Idling can cost money and fuel. Stop your engine at long waits (say Railway crossing, etc.)

— Always keep your engine ready to start.

9. **Use the light lubrication :**

- Mix the recommended grade of oil with petrol, in the ratio recommended by the manufacturer.

— Most engines need less than 5 percent oil in their fuel, but it must be thoroughly mixed.

10. **Switch off light in the day time :**

- Driving with the lights on in day light consumes more fuel.

Adjustments needed to reduce the consumption of fuel in cars :

(*i*) **Air filer.** A badly clogged air filter or restricted passage, strangulates the engine by restricting air supply. Less supply of air causes incomplete combustion. The engine not only loses power but begins to gulp down the field in uncanny amounts. A badly clogged air filter can cause as *much as 15 percent drop in fuel* economy of your car besides other consequential damages.

— If the vehicle operates in dusty conditions with a high rate of humidity, the air filter should be cleaned more often than specified.

(*ii*) **Thermostat :**

- *Thermostat is the key to quick engine warm up* (when the engine is cold, the fuel mixture tends to recondense and cling to the cylinder and manifold walls). Improper vaporisation leads to poor air-fuel mixture which refuses to burn, eventually resulting in starting problems.
- A car achieves the best fuel efficiency if an engine is operating between 84°C to 90°C.

(*iii*) **Cool air.** Cool air *improves volumetric efficiency.*

(*iv*) **Ignition system-engine time up :**

- Two thirds of today's mechanical shortcomings reported for poor performance and economy of petrol engine, can be attributed to ignition fault.
- *Ignition factors* that cause the greatest loss of performance and fuel economy are *'retarded spark timing'* and *'misfiring spark'* in cylinder. The cause may be as simple as a *bad terminal connection or a fouled plug.*

(*iii*) **Clutch slippage :**

- The clutch is a device by means of which power of the engine and driving mechanism may be connected and disconnected at the will of the driver. Its prolonged service, and through abuse and misuses (most common abuse is the driver constantly riding on the clutch pedal). The clutch disc is likely to require replacement, though not very regularly.
- The more common complaint on clutch is "*slippage*". That means *it fails to deliver the full power produced by the engine* to the rear axle through transmission. In the early stage of the clutch slippage, the defect is difficult to detect, unless noticed by an expert mechanic, on an experienced driver.

(*iv*) **Choked exhaust :**

- Engine inhales through the air cleaner via the carburettor and exhales through the muffler and tail pipe. Obviously, any sort of congestion in this system can certainly affect efficiency. Heavy carbon deposit, alongwith rust can clog the system as the car ages.

Badly dented muffler and exhaust pipe under the car can also restrict the passage and affect the smooth flow of gases. In technical jargon, this is commonly known as "*back pressure*". The higher the speed, the more the back pressure produced, impeding engine efficiency.

- *A badly congested system of a car robs as much as 3 to 5 H.P., at high speed.* An unhealthy engine is a sure sign of poor economy.

(*v*) **Under-inflated tyres :**

Out of all the causes of fuel wastage tested earlier under-inflated tyres is probably the greatest culprit. Softer the tyre, the more it flexes and more H.P. is absorbed out, at the cost of fuel.

— If *the tyres are under-inflated by 25 percent of the recommended tyre pressure, the fuel efficiency could suffer by about 5 to 10 percent and the tyre life will be reduced by 25 percent.*

- **The radial tyres roll better :**

— The radial type of tyres is different from the usual bias ply tyres. The major difference is in the construction of its body's structure, and due to the built-in characteristics of radial tyres, the *rolling resistance is considerably reduced* and *consequently better mileage is obtained than the normal bias ply tyres.*

— This type of tyre has a *better covering ability due to perfect road grip*, hence the car's stability is far superior at high speed. It *lasts longer than bias ply tyres.*

(*vi*) **Tight wheels :**

- *The wheels that do not roll freely waste fuel.* The engine has to do hard work to counteract the *drag* caused by tight wheel bearings, bent axle, improper adjusted brakes, lack of lubrication improper grade of lubricants etc.
- The slight drag imposed by one or a combination of the conditions mentioned above, can amply affect fuel economy.

Economy through driving habits :

- The researchers of motor industry have established that thoughtless driving can increase fuel consumption by as much as 50 percent (this fact has been proved by the economy rally drivers).
- According to experts, the *mini cars* only are more economical in certain types of operation. In many applications, *medium size family cars* are more economical and are certainly more safer than minis, especially on Indian roads. Besides, *compact cars at moderate speed have double the life of the little car's engine, which has to sweat out to keep pace with elder brother.*

12.6. GARAGE TOOLS AND EQUIPMENT

12.6.1. Types of Garages

The various types of garages are :

1. Small garages.
2. Medium garages.
3. Large garages.
4. Service station.

1. **Small garages :**

These garages deal with :

- *Replacement and adjustment of most of the major components for a particular range of vehicles ;*
- *Stock of large range of tools, parts and accessories.*

2. **Medium garages :**

- They provide all types of *routine services and repair work including tune-up and brake-testing*.
- They also have *parking facilities* and *suitable reception areas* for customers.

3. **Large garages :**

- These garages have *well trained and qualified staff* to carry out inspection, diagnosing, servicing and urgent repairs without advance booking. Body work repairs are of *high standard*.
- Adequate and *comfortable waiting and parking facilities* are available.
- In advanced countries, new automotive service centres called *"diagnostic centres"* or *"diagnostic lanes"* have been established with *latest automotive testing equipment*.

— A thorough analysis is made of each system of automobiles and the test report, indicating condition of the various components and how soon they will need repair, is given to the customer.

4. **Service station :**

- The service station, in general, includes facilities in addition to the care of motor vehicles like mechanical service and minor service *e.g.*, providing petrol and lubricating oils.
- Some stations provide comprehensive repair service also for vehicles.

Location of garages/service stations :

While *selecting the site of a garage* the following *factors* should be taken into consideration.

(*i*) Easily approachable.

(*ii*) Nearness to market and highway.

(*iii*) Traffic flow.

(*iv*) Adequate car population.

(*v*) Availability of facilities for drainage and sanitation.

(*vi*) Availability of qualified and experienced technical people near the surroundings.

- Sites near blind corners and busy cross-roads and also where it is difficult to turn into the area are not suitable.

12.6.2. Garage Tools

Garage tools—Part-I

A good mechanic tool kit contains the following tools : Refer to Fig. 12.30.

Refer to Fig. 12.30 (*i*). *Minimum number of tools for a mechanic tool kit :*

1. One ball peen type hammer, weighing 0.5 kg.
2. One pair of pliers.
3. One screw driver (250-260 mm in length).
4. One flat cold chisel (203.2 mm × 19 mm).
5. One set of double-ended set spanners (4.76 mm to 19 mm).
6. One set of box spanners (4.76 mm to 19 mm) with tommy bar.
7. One adjustable spanner (25 cm).

Refer to Fig. 12.30 (*ii*). *Additional tools for a mechanic tool kit ;*

1. 24 cm pressure grips (foot-print type).
2. One set of distributor contact spanners.
3. One adjustable hacksaw frame.
4. One small socket set (4.76 mm × 12.7 mm) with a ratchet type handle.
5. One 15 cm punch.
6. One centre punch.
7. Two box spanners (one for 14 mm spark plug and other for 18 mm spark plug)
8. One set of feeler gauges.

Refer to Fig. 12.30 (*iii*). *Further additional tools for a good mechanic tool kit ;*

1. One steel rule.
2. One pressure outside caliper
3. One pressure inside caliper.
4. One pressure hermaphrodite caliper.
5. One set thread pitch gauge.
6. One set screw driver (electrician type).
7. One cross-cut cold chisel.
8. One set ring spanner.
9. One vernier gauge.
10. One outside micrometer.
11. One set of scrapers.

- *Miscellaneous tools required by a mechanic from time-to-time :*

1. Files
2. Drills
3. Hacksaws
4. Punches (Hollow)
5. Reamers
6. Rivet sets.

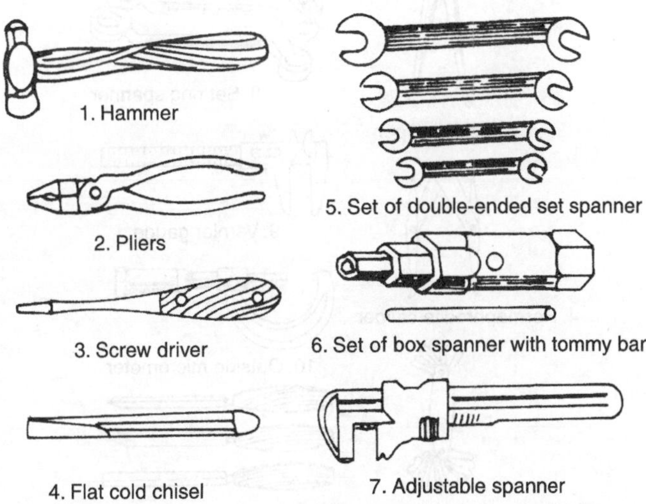

(*i*) Minimum number of tools for a mechanic tool kit.

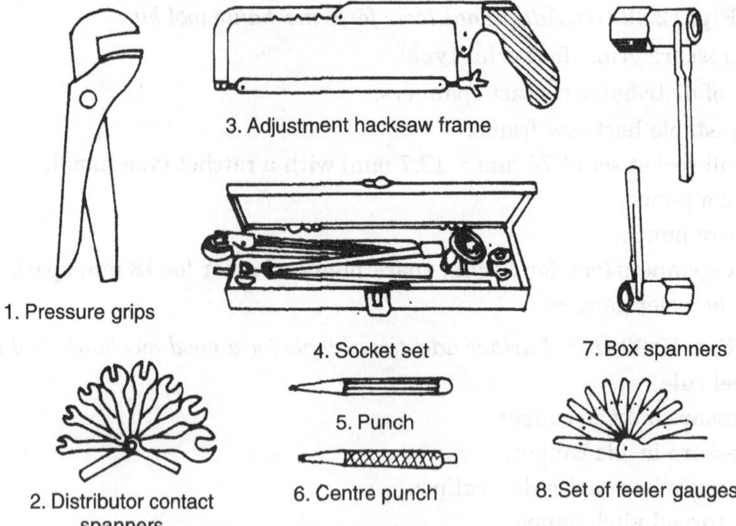

(*ii*) Additional tools for a mechanic tool kit.

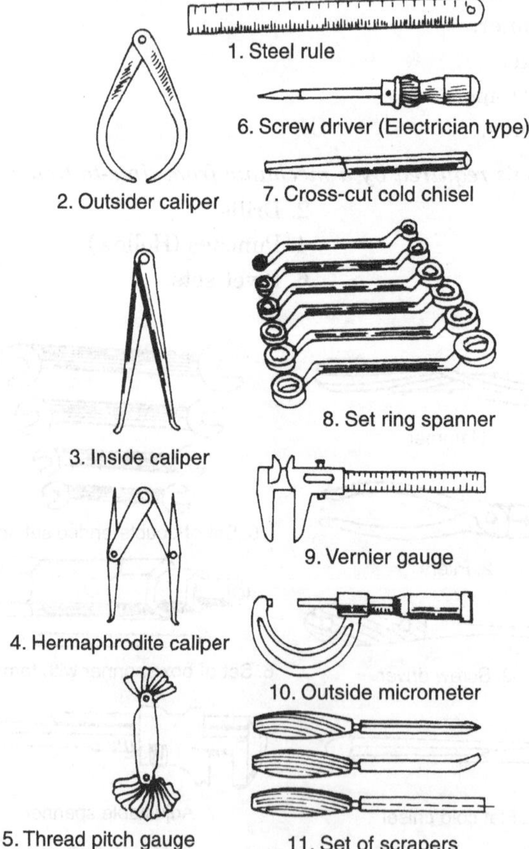

(*iii*) Further additional tools for a good mechanic tool kit.

Fig. 12.30. Garage tools.

- **Universal extractor tool :**

A universal extractor tool is used to remove pulleys and gears for dismantling the engines, gearboxes and back axles.

Fig 12.31 shows the various parts of a universal extractor tool.

The lower ends of the projecting lips grip the parts to be extracted. When the central screw is turned, it presses the central shaft which is kept in touch with the screw with the aid of suitable adapters.

This tool is employed to remove :

— Crankshaft gear,

— Fan and dynamo pulley,

— Propeller shaft flanges, and

— Bevel pinion flanges.

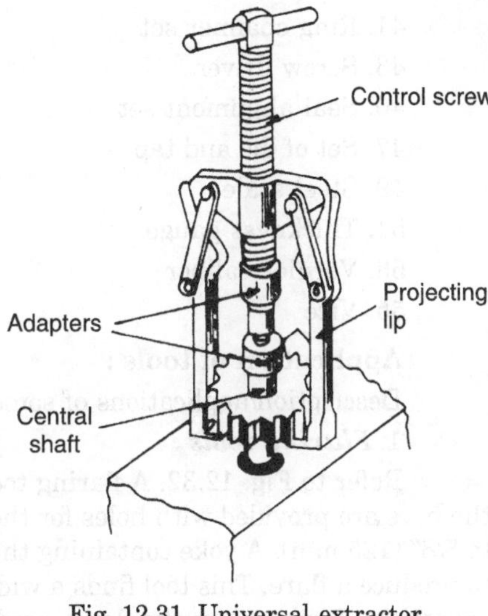

Fig. 12.31. Universal extractor.

- *Following tools are also required at times :*

1. Screwing tackle
2. Snips
3. Surface plate
4. V-blocks and surface gauge with dial indicator
5. Torsion wrench
6. Wheel brace
7. Valve lifters.

Garage tools—Part-II :

A list of refrigeration and air-conditioning tools commonly used is given below :

1. Adjustable wrench
2. Allen key set
3. Ampere meter
4. Blow lamp
5. Box spanner
6. Brushes
7. Centre punch
8. Charging line
9. Chisel set
10. Double-ended spanner
11. Drilling machines
12. Drills
13. Feeler gauge
14. Files
15. Flaring tools
16. Flexible charging line
17. Gas charging stand
18. Gas cylinder
19. Gauge manifold
20. Gauges—compound gauges, pressure gauge
21. Hacksaw Frame
22. Halide leak detector
23. Hammers
24. Hand shut-off valve
25. Hollow punch
26. Knife
27. Magnifying glass
28. Mallet
29. Measuring tape
30. Multimeter
31. Oil gun
32. Phase tester
33. Pinch-off tool
34. Pipe benders
35. Pipe wrenches
36. Pliers
37. Pulley puller
38. Ratchet wrench
39. Refacing tool
40. Revolution counter

41. Ring spanner set
42. Scissors
43. Screw driver
44. Screw wrench
45. Seal alignment set
46. Seal set puller
47. Set of die and tap
48. Soldering iron
49. Steel scale
50. Swaging tool
51. Thickness gauge
52. Tube cutter
53. Vernier caliper
54. Voltmeter
55. Vice
56. Thermometer.

Application of tools :

Description/applications of some important tools is given below :

1. ***Flaring tools :***

Refer to Fig. 12.32. A flaring tool consists of two bars held together by a wing nut and bolt ; the bars are provided with holes for the various sizes of tubing (usually varying from 3/16″ (38 mm) to 5/8″ (125 mm). A yoke containing the forming die is slipped on the bars and the handle is rotated to produce a flare. This tool finds a wide use because of its simplicity, ease of performing the flaring operation ; furthermore the flares produced by it are uniformly excellent in producing a tight seal.

Prior to application of pressure with the forming tool, place a drop of oil on the end of the tube. Make sure that only refrigeration oil is used (any other type oil, if used, may enter the tube and give trouble by reacting with the refrigerant at a later date).

Fig. 12.33 shows the proper use of the flaring tool.

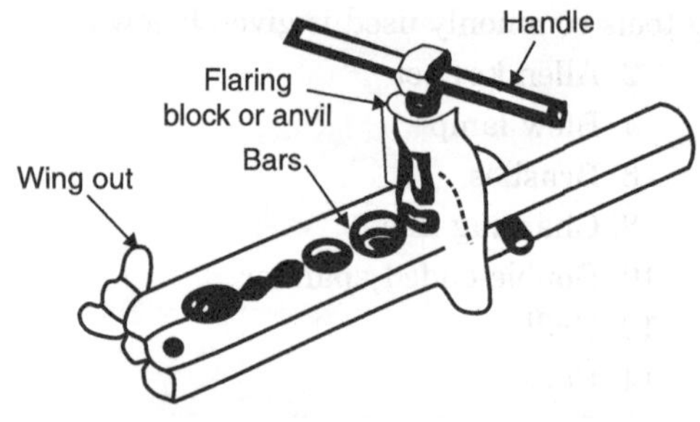

Fig. 12.32. Flaring tool.

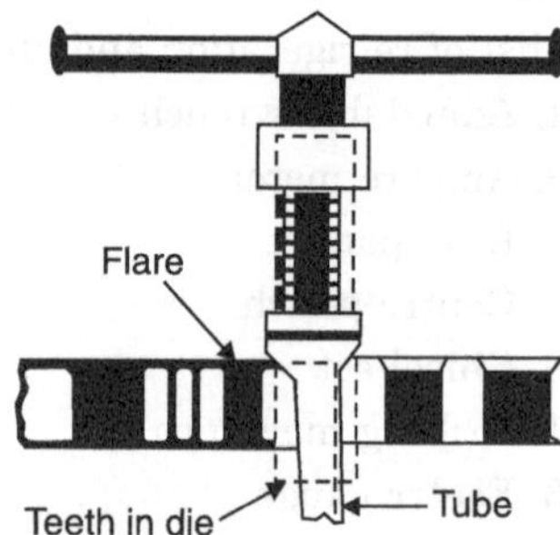

Fig. 12.33. Use of the flaring tool.

2. ***Swaging tool :***

Refer to Fig. 12.34. The process of shaping copper tubing so that two pieces may be joined without the use of a fitting is called *swaging* and the tool used for this process is known as *swaging tool*. The swaging tool can be of punch type or screw type. To carry out the swaging operation, hold the tube in the flaring block tool and set bar in vice, the tubing to be extended above the bar a distance of its diameter plus 3 mm and clamp. Now a specially designed punch is hammered into the tube to a depth equal to its original diameter. Now remove the punch and insert other piece of tubing in this swaged portion. Apply heat to the joint and add solder to make a proper joint.

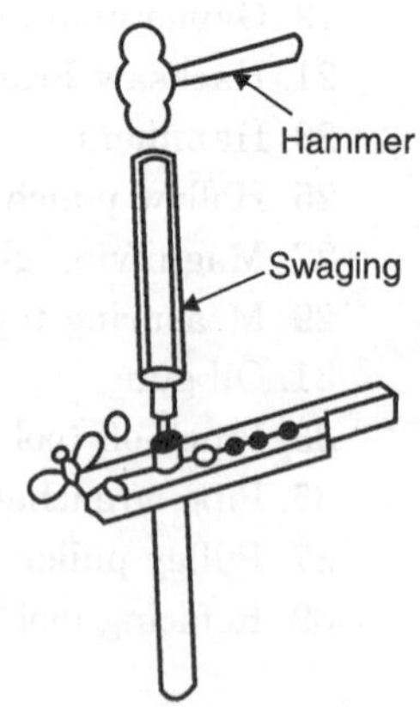

Fig. 12.34. Swaging tool.

3. ***Tube cutter :***

The copper tubing to be used in refrigeration and air-conditioning system may be cut in two ways : (*i*) By a *tube cutter* or (*ii*) By a *hacksaw.*

Normally the smaller machines utilise soft copper tubing, this tubing can be easily installed because it can be bent around obstructions and elimination of many elbows is also possible. While cutting soft copper tubing an easy and quick method is to apply a tube cutter. The tube to be cut is held in between the 'Roller' and the 'cutter wheel' and high pressure is put on the lead screw. Revolve the cutter slowly around the tubing so that the sharp cutting wheel feeds gradually into the tubing making a clean right angle cut. Fig. 12.35 illustrates the use of a tube cutter to cut tubing.

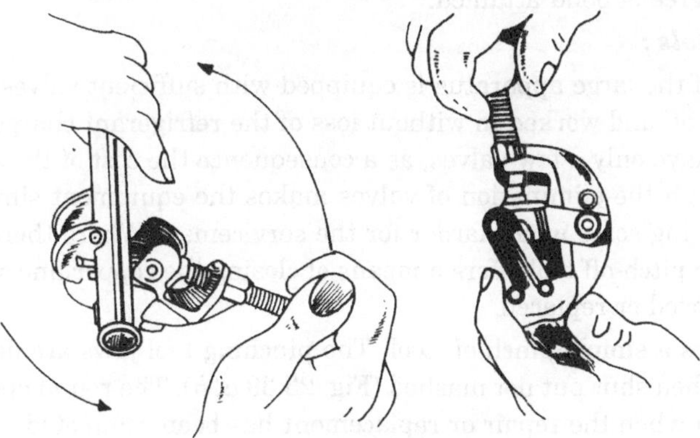

Fig. 12.35. Tube cutter in use.

Hard copper tubing can be cut by a hacksaw when it is held in the flaring block. Fig. 12.36 shows the method of cutting with hacksaw in a flaring block.

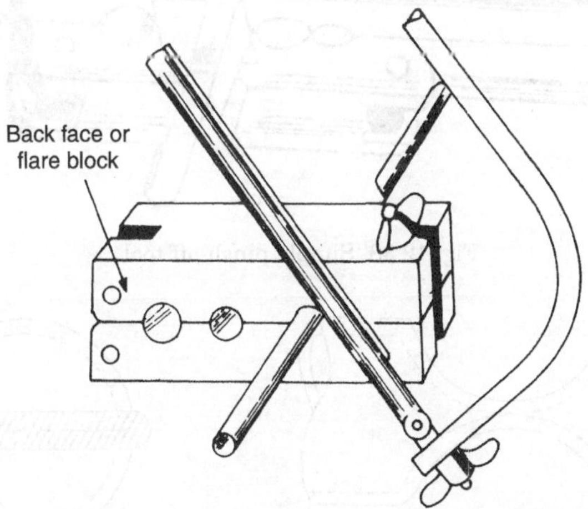

Fig. 12.36. Method of cutting tubing with hacksaw in a flaring block.

4. ***Tube bender :***

For carrying out bending operation several special tools are available. While using *spring type bender* it should be borne in mind that the tubing should be bent somewhat farther than it is required and then bend up to the proper angle. Thus the spring is loosened and can be easily removed. Fig. 12.37 shows spring type bender.

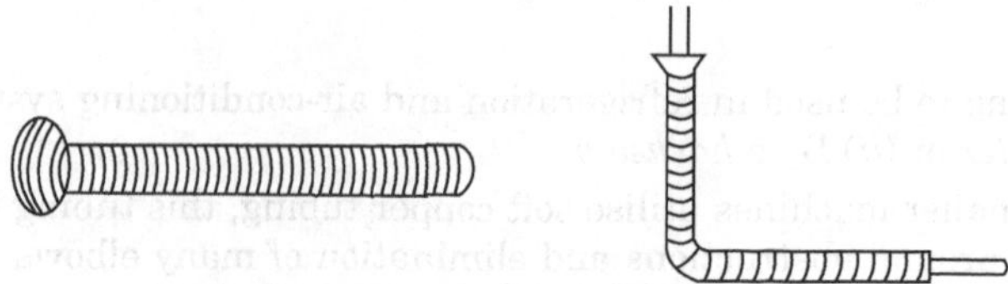

Fig. 12.37. Spring type tube bender.

In order to form bends neatly and accurately with buckling the tubing, a lever type cutter is used. These tools form bends up to 180° in one continuous operation. The forming wheel is calibrated to show the degree of bend attained.

5. ***Pinch-off tools :***

Usually most of the large apparatus is equipped with sufficient valves to enable any portion or section to be closed off and worked in without loss of the refrigerant charge. Some of the smaller and simpler systems have only a few valves, as a consequence the cost of the apparatus can be kept to a minimum. Although the elimination of valves makes the equipment simpler and lower priced yet it makes the servicing some what harder for the serviceman. Where there are only a few or no valves in a system, the pitch-off tool offers a means of closing the copper line so that the defective or worn part can be removed or replaced.

Fig. 12.38 shows a simple pinch-off tool. The pinching tool jaws are designed in such a way that the tubing is pinched shut put not mashed (Fig. 23.39 *a*, *b*). The round opening is used to round up the pinched section when the repair or replacement has been completed.

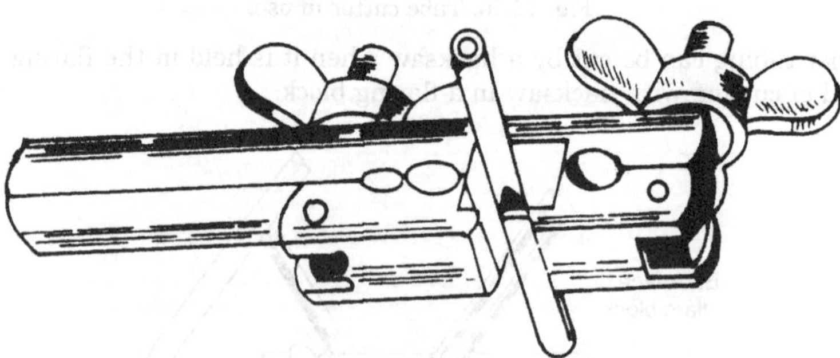

Fig. 12.38. Simple pinch-off tool.

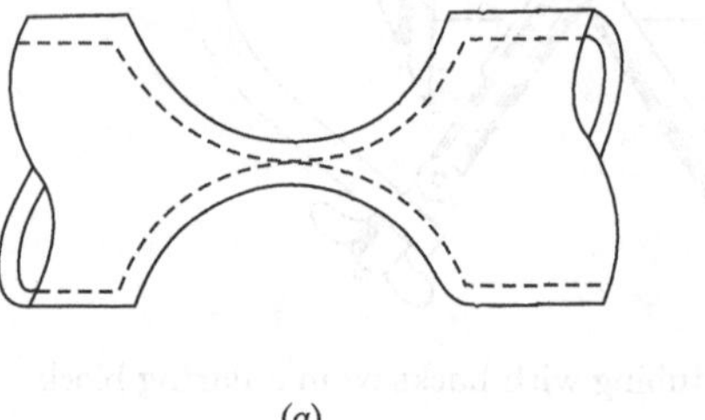

(*a*)

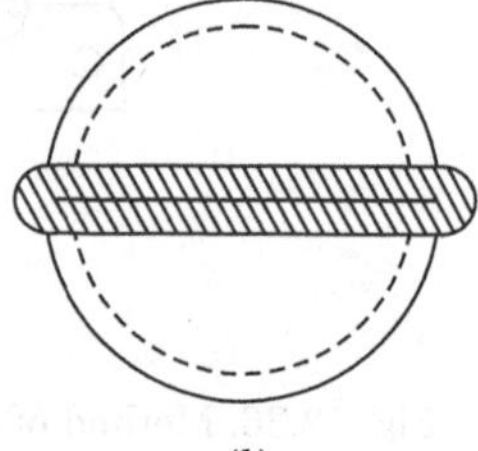

(*b*)

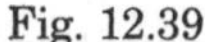

Fig. 12.39

6. ***Refacing tool :***

Fig. 12.40 shows a refacing tool. It is used for refacing seats of different types of connectors.

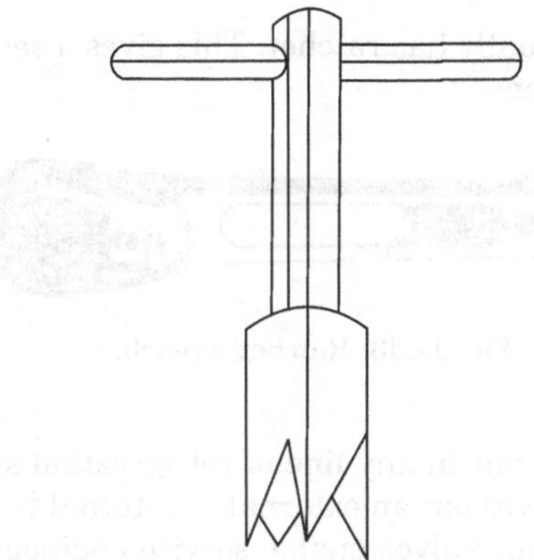

Fig. 12.40. Refacing tool.

7. ***Valve keys :***

Some refrigeration apparatus is equipped with valves which are without handles. The stems are provided with a square end, so that a handle can be applied or a key fitted to the valve stem as shown in Fig. 12.41.

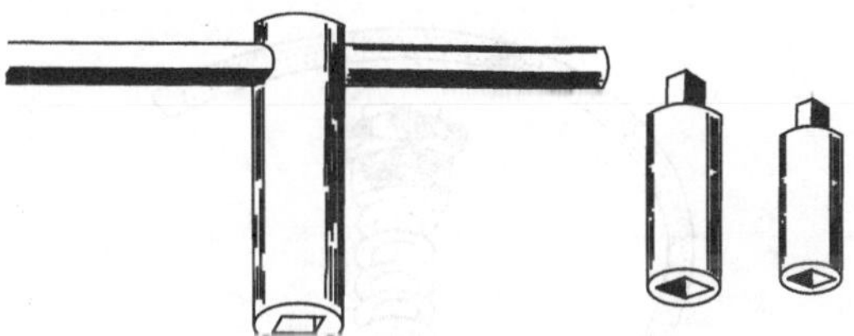

Fig. 12.41. Valve keys.

8. ***Packing keys :***

Some valves are packed with such a material that serves to prevent the refrigerant from leaking out from the stem. A packing nut can be tightened by means of a wrench. When tightening external packing nuts, use a 12-point wrench if one is available, or make use of the proper open-end wrench. The packing nut is generally the internal type and is provided with a slot across its face for tightening.

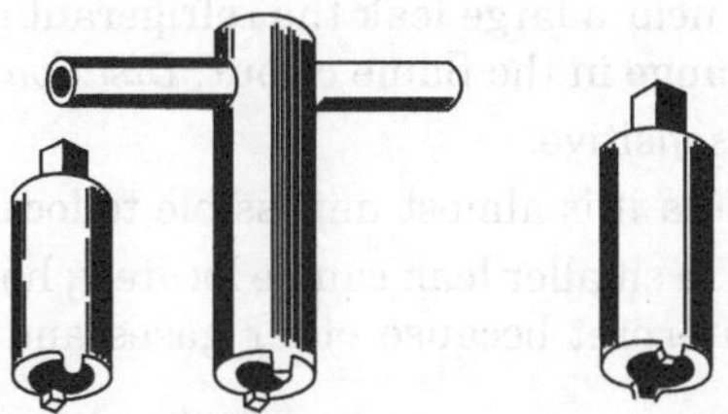

Fig. 12.42. Packing keys.

The stem of the valve projects through this nut and makes it necessary to use a special gland or packing nut key, such as shown in Fig. 12.42. This key is hollow in the centre, permitting the valve stem to extend through the tool, and is provided with two teeth that engage the slot in the packing nut.

9. ***Ratchet wrench :***

Refer Fig. 12.43. This tool usually has ratchet. This gives a service engineer rapid control for slight opening or closing of valve stem.

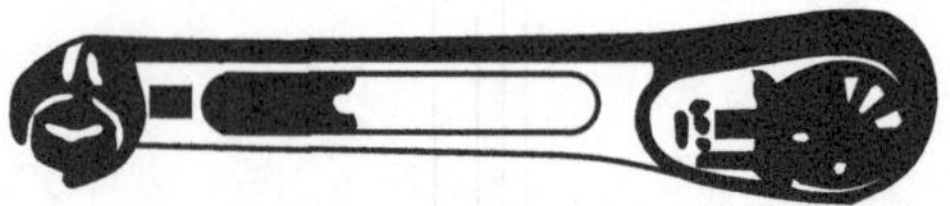

Fig. 12.43. Ratchet wrench.

10. ***Hand shut-off valve :***

A hand shut-off valve can be put in any line of refrigeration system. These valves can withstand frequent opening and closing without an external or internal leak. These can be installed with the help of flare nut or can be brazed. Valves enable service engineer to determine operation pressure, to charge a system with a refrigerant, to test in for leakage, or to vacuumize a system depending upon its location.

11. ***Halide leak detector :***

Refer Fig. 12.44. The halide leak detector may be used to detect leaks in Freon refrigerant system. It is named after the group of hydrocarbons known as the halogens. It consists of a combustible gas supply with a small burner on top. A tube from the bottom of this burner forms the probe.

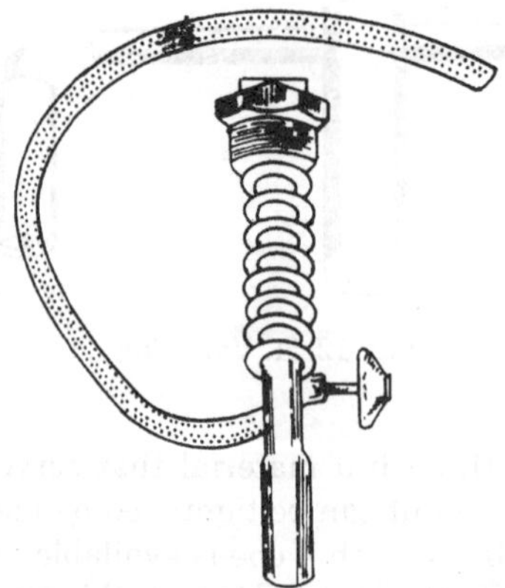

Fig. 12.44. Halide leak detector.

When this probe is moved near a large leak the refrigerant going into the tube combines with the flame to cause perceptible change in the flame colour. *Disadvantages* of using this detector are :

(*i*) Slow and relatively insensitive.

(*ii*) In a brightly lighted area it is almost impossible to locate smaller leaks.

(*iii*) If the area is darkened, a smaller leak can be located ; however, the change in colour of the flame is difficult to interpret because other gases and dust cause the flame to change colour.

12. ***Electronic type leak detector :***

As compared to halide leak detector, an electronic type leak detector is *more sensitive, faster responding* and is capable of detecting a leak even though the surrounding air may be contaminated.

It consists of an internal pump which draws air and leaking halogen tracer gases through a probe and hose as the probe is passed over leaking joints, gaskets, welds etc. The air gas mixture is drawn between two electrodes in the control unit—the *ion-emitter* and *collector*. The presence of halogen gas in the inter-electrode space causes an increase in interelectrode conductance that is proportional to the amount of halogen gas present, which results in greater positive ion current. In the electronic detector, this emission current is electronically amplified and the amplifier output causes the leak indicating lamp to flash. In order to adjust the sensitivity to detect leaks a knob on the control unit is provided to do so. This detector is specially designed for use in atmosphere containing a relatively high concentration of halogen gas.

To detect leaks by this type of detector, following procedure is followed : (*i*) Plug the test detector into an AC outlet (220/110V) ; (*ii*) Turn the sensitivity knob to the right and allow a 1-minute warm up ; (*iii*) Check the operation by turning the sensitivity knob quickly form one position to the another ; This should light the probe lamp, (*iv*) Probe for leaks, starting from maximum sensitivity ; (*v*) If the probe lamp-lights twice for each leak ; reduce the sensitivity ; (*vi*) recheck the leaks for confirmation.

13. ***Screw drivers :***

In refrigeration service work screw drivers are employed to loosen and tighten the screws. A screw drive primarily consists of a handle, ferrule and the blade. The length of a screw driver is measured from the blade tip to the handle. Handles are not measured.

Several types of screw drivers are used in refrigeration and air conditioning field out of which following two are most important : 1. Flat type and 2. Phillips type. The flat is a standard screw driver (Fig. 12.45) being very similar to chisel. The tip of the screw driver is applied on the job and head is moved to remove or insert the screw where as the *phillips screw driver* (Fig. 12.46) is made with a specially shaped tip to deal with cross slot screws.

Fig. 12.45. Flat type screw driver.

Fig. 12.46. Phillips screw driver.

14. ***Socket wrenches :***

Where the nut or bolt head has enough room around it, the six or twelve point wrench is best used. These sockets are generally made of chromium-vanadium steel and are turned by handles that have a 1/4, 3/8 or 1/2 inch square drive.

15. ***Box wrenches :***

Where it is not possible to use a socket wrench due to closed space around the nut or bolt head a box wrench is successfully employed. Most of the box wrenches are double ended (Fig. 12.47) ; both the ends may be of the same or different sizes.

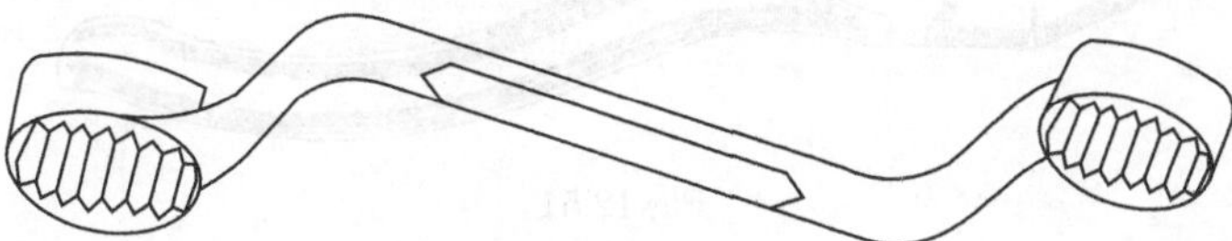

Fig. 12.47. Box wrench.

16. ***Adjustable wrenches :***

Refer to Fig. 12.48. These wrenches are used when odd sizes of nuts and bolts heads are to be handled. As these wrenches are not so strong (as other solid types), so their use should be restricted when forces used are of large magnitudes. In order to ensure that they are not cracked or bent out of shape, the jaw and adjustment screw must be checked up periodically.

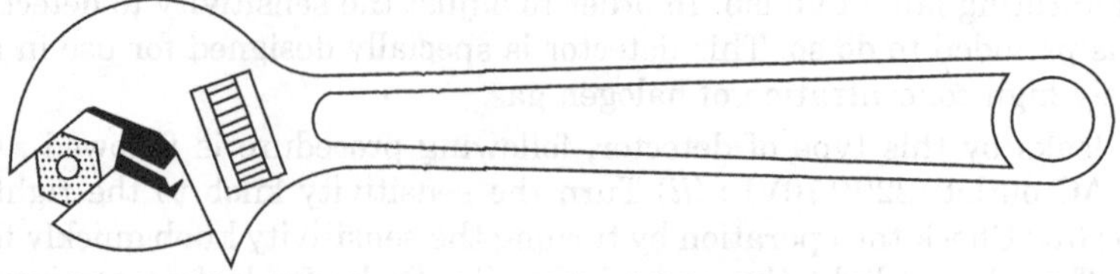

Fig. 12.48. Adjustable wrench.

17. ***Open ended wrench/spanners :***

These wrenches are the non-adjustable type with openings in each end. This type of wrench can slide on the nut or bolt head from the side. These wrenches are available in sets, each set containing about ten wrenches with openings of different sizes. These are used in close spaces on unions and other places where the socket wrench and box wrench cannot be used. Fig. 12.49 shows a typical double ended wrench/spanner.

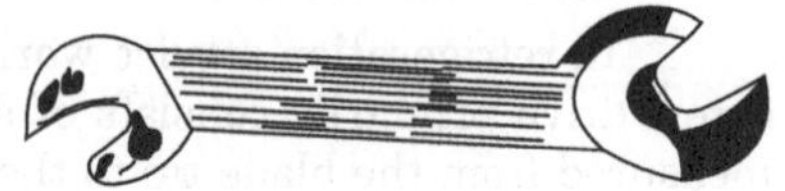

Fig. 12.49. Double ended wrench/spanner.

18. ***Pliers :***

Pliers of different designs are available, each for a particular purpose. For example *(i)* *'Cutting pliers'* are used for skinning a cable insulation and cutting wires when working on the wiring at a refrigeration equipment ; *(ii)* *'Round nose pliers'* are used to shape wire into loops and to bend sheet metal edges ; *(iii)* *'Slip nose plier'* and *'duck bill pliers'* are used in hard to reach plates.

Fig. 12.50, 12.51 and 12.52 show some of the commonly used pliers.

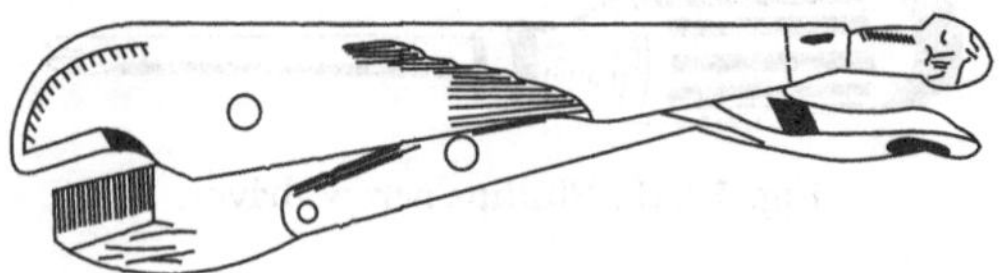

Fig. 12.50

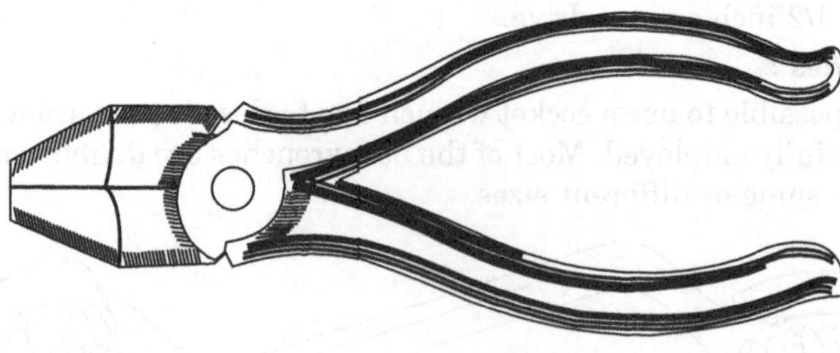

Fig 12.51

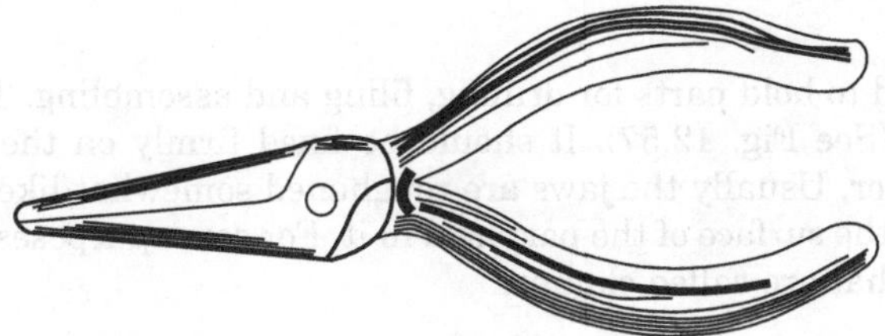

Fig. 12.52

19. ***Punches :***

The following four types of punches are used for carrying out different jobs.

1. Centre punch
2. Drift punch
3. Pin punch
4. Prick punch.

The *centre punch* (See Fig. 12.53) is made of steel and the point is hardened. The angle can vary from 60° to 90°. One of the main uses of a centre punch is to mark the centres of holes for drilling. For this purpose a much deeper mark is required than is necessary for simply marking lines. Such punches may also be used to make alignment marks on part before dismantling.

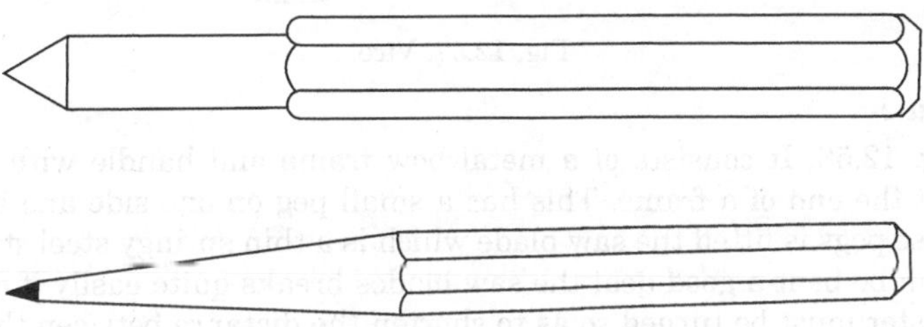

Fig. 12.53. Centre punch.

A *drift punch* (See Fig. 12.54) is employed to line up holes in the mating surfaces and drive out the keys.

Fig. 12.54. Drift punch.

A *pin punch* (See Fig. 12.55) is used for driving retainer pins in or out.

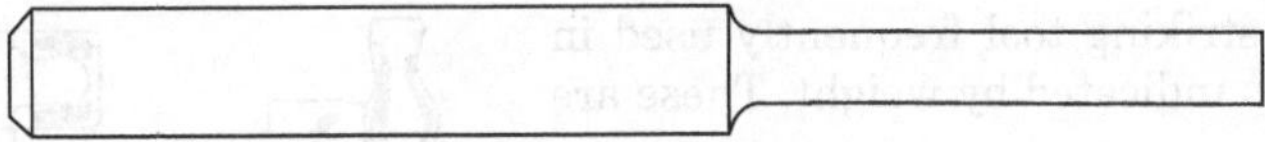

Fig. 12.55. Pin punch.

A *prick punch* (See Fig. 12.56) has a long, sharp point and is used exclusively for layout work.

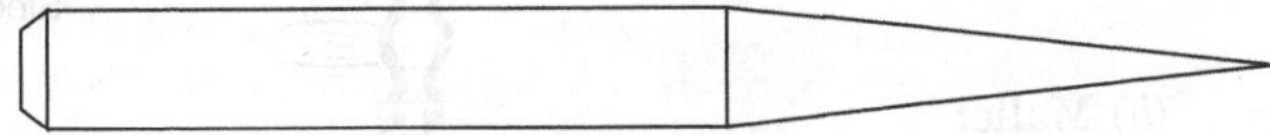

Fig. 12.56. Prick punch.

20. ***Vice :***

A vice is employed to hold parts for drilling, filing and assembling. The most useful type is a vice with parallel jaws (See Fig. 12.57). It should be fixed firmly on the bench at the front and preferably near one corner. Usually the jaws are roughened somewhat like a file. This gives a good grip but it also damages the surface of the part held in it. For some purposes, therefore it is desirable to cover the jaws with what are called clamps.

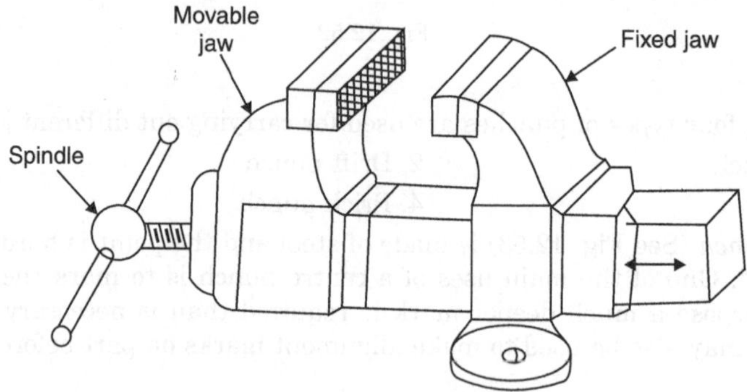

Fig. 12.57. Vice.

21. ***Hacksaw :***

Refer to Fig. 12.58. It consists of a metal bow frame and handle with some sort of screw adjuster, usually at the end of a frame. This has a small peg on one side and a similar peg at the handle end. On these pegs is fitted the saw blade which is a thin springy steel strip with teeth along side. Although it can be bent a good deal the saw blades breaks quite easily if mishandled. To fit a new blade the adjuster must be turned so as to shorten the distance between the two pegs.

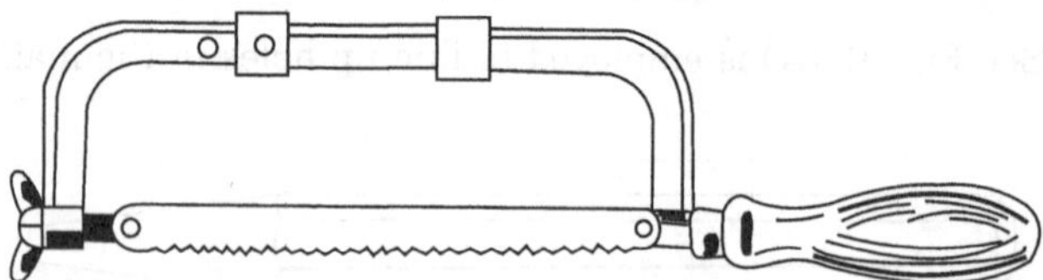

Fig. 12.58. Hacksaw.

The hacksaws are used to cut rods, strips, bars and pipes into desired lengths.

22. ***Hammer :***

Hammer is a striking tool frequently used in bench work. Its size is indicated by weight. These are classified as follows :

1. *Hard* :

(*a*) Ball peen (*b*) Cross peen

(*c*) Straight peen (*d*) Sledge.

2. *Soft* :

(*a*) Raw hide (*b*) Mallet

(*c*) Brass (*d*) Plastic

(*e*) Hard rubber.

Fig. 12.59 shows some of the commonly used hammers.

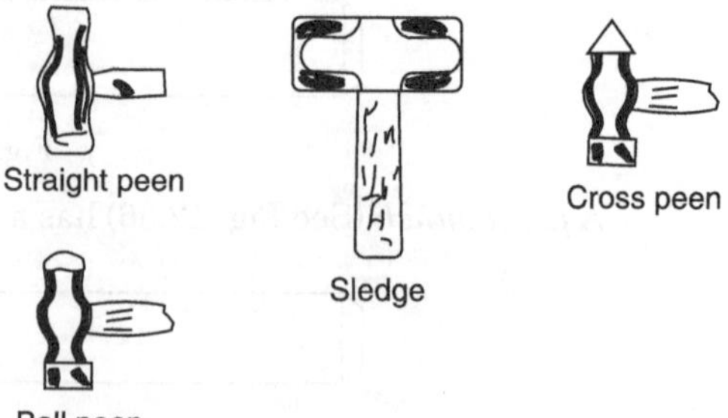

Fig. 12.59. Commonly used hammers.

23. ***Files :***

Files are used to abrade or smooth other substances such as metal, wood etc. These are classified on the basis of size, cut of teeth and shapes as follows : (*i*) *Flat file* (used for general work), (*ii*) *Round file* (used to enlarge the holes or circular openings and to file concave surface, (*iii*) *Square file* (used to file key ways, slots etc.), (*iv*) *Three square file* (used to file internal angles and clean out corners).

Fig. 12.60 shows some files.

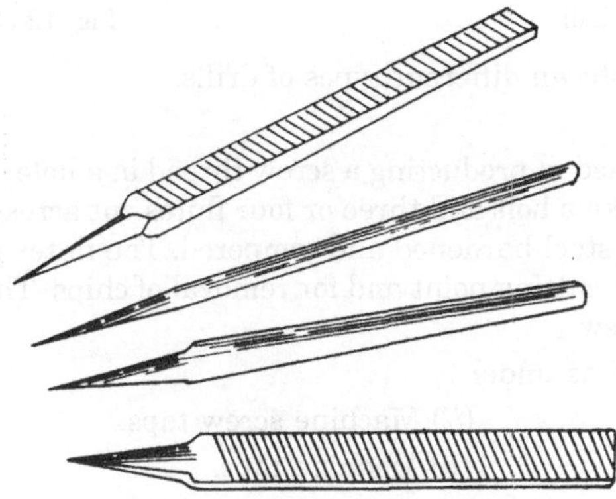

Fig. 12.60. Some files.

24. ***Chisels :***

Chisels are used for cutting or clipping metal. These are made of good grade tool steel (carbon steel) with hardened cutting edge. They are classified as given below :

(*i*) Flat chisel (*ii*) Cross-cut chisel (*iii*) Round nose chisel
(*iv*) Diamond point chisel (*v*) Side chisel.

The above mentioned chisels are shown in Fig. 12.61.

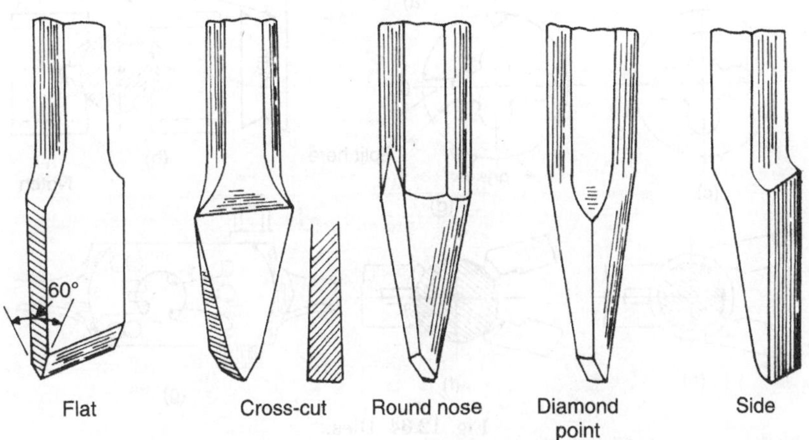

Fig. 12.61. Chisels.

25. ***Drills :***

Drill is a tool used for the production of round holes in metal or other solid material. Fig. 12.62 shows the leading characteristics of a twist drill.

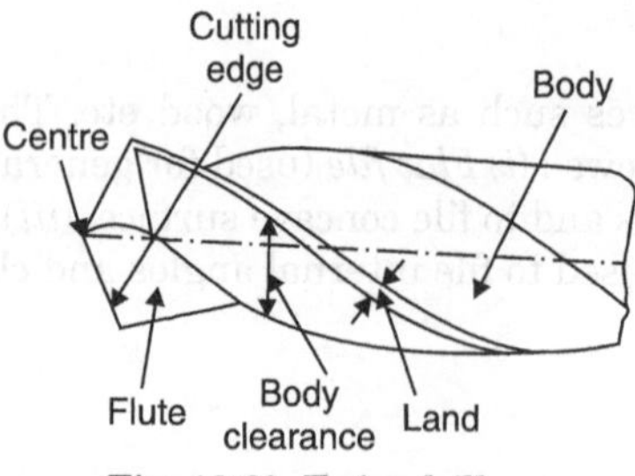

Fig. 12.62. Twist drill.

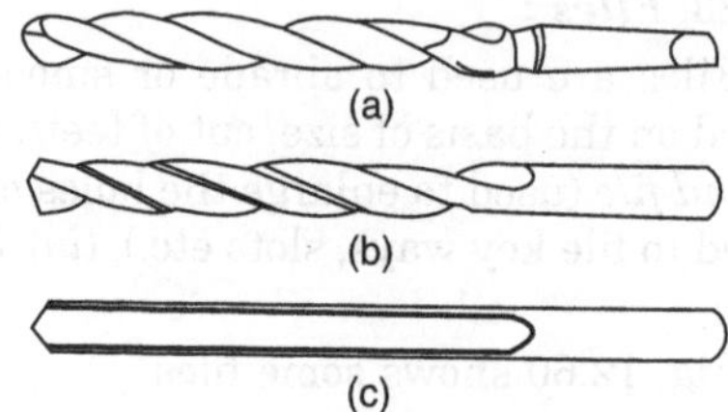

Fig. 12.63. Types of drills.

In the Fig. 12.63 are shown different types of drills.

26. ***Taps :***

A screw tap is a tool used in producting a screw thread in a hole, *i.e.*, an internal thread. A tap is a tool which has threads like a bolt and three or four flutes cut across the threads. Taps are made of carbon steel or high speed steel hardened and tempered. The flutes provide cutting edges as well as passage for oil to reach the cutting point and for removal of chips. The square shank is for the tap handle which rotates the screw.

The taps are classified as under :

(*i*) Hand taps (*ii*) Machine screw taps (*iii*) Tapered taps
(*iv*) Pulley taps (*v*) Machine taps.

27. ***Dies :***

A screw die is a tool used when threading cylindrical pieces beyond the capacity of a die plate. Dies are more efficient than die plates or die nuts and are made in many different designs both for hand and machine use.

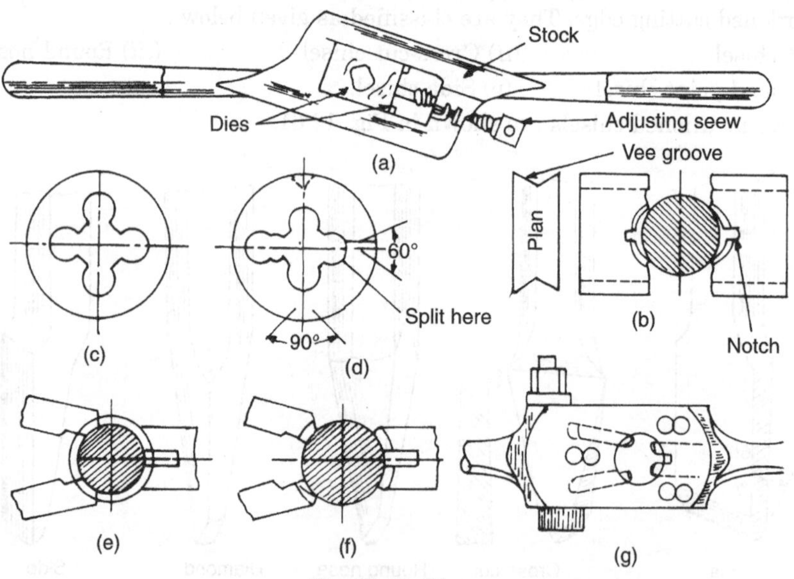

Fig. 12.64. Dies.

Refer to Fig. 12.64 :

(*a*), (*b*) : Stock and dies—the holder, heavier adjustable.

(*c*), (*d*) : Non-adjustable and adjustable circular dies.

(*e*), (*f*), (*g*) : Inserted chasers.

Dies will cut only in one direction of rotation, so it is no good, after running down the thread once with stocks and dies, to tighten up the dies still further in the hope that more will be removed while winding the stocks back.

28. ***Dividers :***

These are two-legged instruments of compass-like form, sometimes called *points*. They are used for measuring distances between points, for scribing arcs or circles on metals, for transferring lengths from scales, etc. Dividers may be ordinary firm joint or spring type. For smaller and more accurate jobs spring dividers are preferred. In this pattern the legs have a spring joint and the necessary adjustment is obtained by a knurled nut and screw (Refer Fig. 12.65).

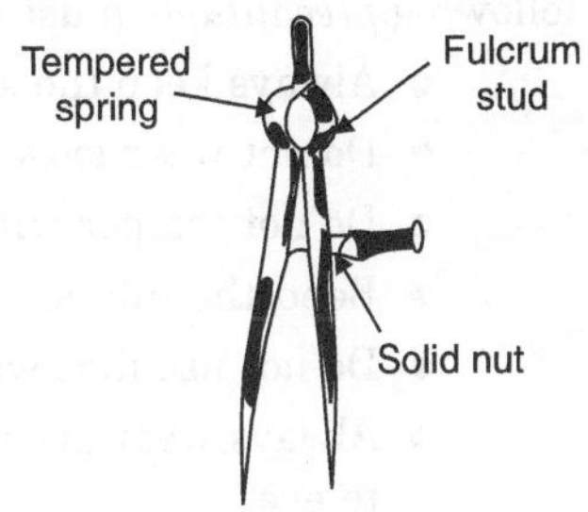

Fig. 12.65. Divider.

29. ***Caliper :***

Caliper is an instrument used for measuring distance between or over surfaces, or for comparing dimensions of workpieces with such standards as plug gauges, graduated rules, etc.

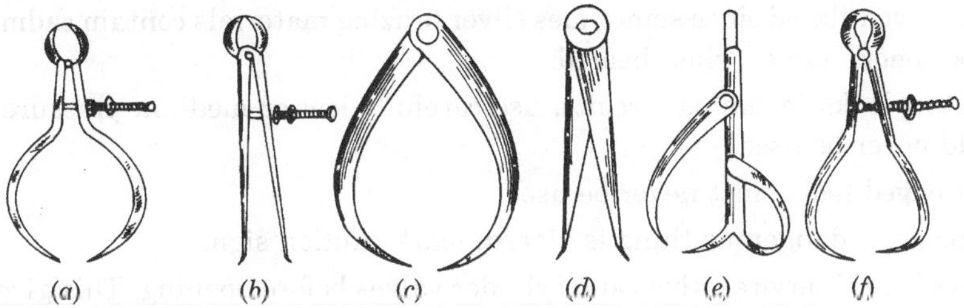

Fig. 12.66. Calipers.

Fig. 12.66 shows the various types of calipers in use :

(*a*) and (*b*) : external and internal calipers with adjusting screws.

(*c*) and (*d*) : firm-joint external and internal calipers, adjusted by tapping a leg or the joint.

(*e*) : Blacksmith's caliper.

(*f*) : Wide-jaw spring caliper used for measurement of major diameters of threaded work.

30. ***Feeler gauge :***

Feeler gauge is a leaf-gauge, rather like a multi-bladed pen knife, the various leaves or blades all having different thickness. It is used to estimate, by sense of touch, the clearance or tolerance between two components by inserting deferent blades until the one which will just go between the two surfaces.

A typical feeler gauge, as shown in Fig. 12.67, has twenty-two leaves, varying in thickness by thousands of an inch, running from 0.004 to 0.025 in, the thickness of each leaf in thousand of an inch being designed by the number engraved upon it. The leaves may be used singly or in combination with others, and any thickness within their limits can easily be obtained. Feeler gauges are also made in metre sizes form 5/100 to 1 mm thickness.

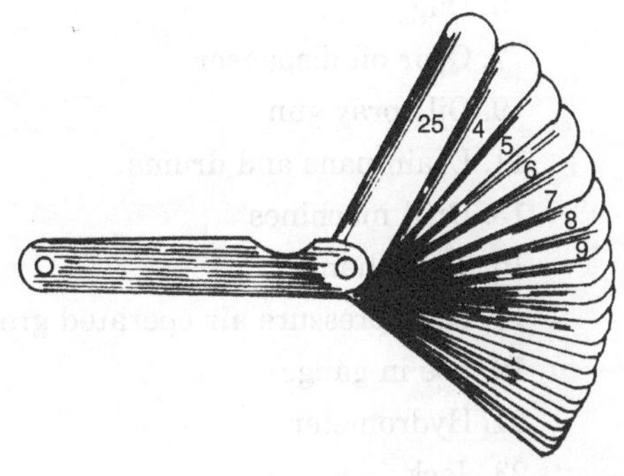

Fig. 12.67. Feeler gauge.

Safety precautions :

While working with tools and equipment in a refrigeration and air-conditioning shop the following *precautions* must be observed :

- Always keep the shop neat and tidy.
- Do not wear loose dress.
- Do not temper with the safety devices provided on the machines.
- Bend the tubing as large as possible.
- Do not use files without handles (as the tang may injure the hand).
- Always wear goggles while drilling, otherwise the flying chips may cause serious injury to eye.
- 'Mushroom' heads should be ground from chisels and punches as these tools are likely to fly when struck with hammer thus causing serious injury to the operator.
- Do not use emery cloth to clean tubing prior to soldering as it may leave an oily deposit.
- During the use of silver brazing materials it should be ensured that the work place is properly ventilated since sometimes silver brazing materials contain cadmium which give out poisonous fumes when heated.
- To detect leaks in tubing circuits, use carefully low or medium pressure CO_2 or N_2, O_2 should never be used.
- Blunt edged tools must never be used.
- Whereever a dangerous thing is placed, mark caution sign.
- Always 'crack' service valves and cylinder valves before opening. This gives quick control to the flow gases if there is any danger.
- Keep everything connected with refrigerating mechanism thoroughly dry.
- Cylinders containing refrigerant should never be filled more that 85% of their capacity as they are likely to burst due to hydrostatic pressure if overfilled.

12.6.3. Garage Service Station Equipment

1. Air compressor
2. Spark plug tester machine
3. Tyre inflator
4. Hydraulic lift
5. Dug
6. Engine oil dispenser
7. Gear oil dispenser
8. Rear axle oil dispenser
9. Oil spray gun
10. Car washer with gun
11. Drain pans and drums
12. Hand tools
13. Drill machines
14. Grinder
15. Tyre changer
16. Tube volcanizing machine
17. High pressure air operated grease gun
18. Wheel braces, etc.
19. Toe in gauge
20. Tune-up equipment
21. Hydrometer
22. Battery charger
23. Jack
24. Wooden horses.

- **Special Tools/Equipment :**

1. Air Compressor (1)
2. Battery charger (1)

3. Cell tester (1)
4. Hydrometer (1)
5. Brake tester (1)
6. Brake drum lathe (1)
7. Brake shoe riveter (1)
8. Degreasing plant (1)
9. Distributor tester (1)
10. Electric spot welding machine (1)
11. Electric welding plant (1)
12. Gas welding plant
13. Caster camber gauge
14. Toe-in toe-out gauge
15. Hydraulic trolley jack
16. Lathe machine
17. Drill machine
18. Grinder
19. Spark plug tester
20. Compression gauge
21. Vacuum tester
22. Petrol consumption tester
23. Timing light
24. Wheel balancer
25. Tap and die set (English and mm size)
26. Smoke tester
27. Growler
28. Voltage and current setting gauge
29. Spray paint gun
30. Buffers and grinders
31. Tyre changer.

- **Special Tools and Equipment for Engine Repair :**

1. Crankshaft grinding machine
2. Camshaft grinding machine
3. Piston cam grinder
4. Connecting rod alignment jig
5. Valve seat refacer
6. Valve grinder
7. Cylinder boring machine
8. Honing kit with drill machine
9. Hydrualic tester for block and cylinder head leaks
10. Lathe machine
11. Cylinder block line boring machine
12. Hydraulic press
13. Portable hoist
14. Crack detecting machine
15. Torque wrench
16. Hand tool kit
17. Micrometer set
18. Inside micrometre
19. Dial gauge
20. Cylinder bore gauge
21. Inside and outside callipers
22. Connecting rod small end boring machine
23. Chain pulley block
24. Drill machine
25. Vices
26. Reamer set
27. Stud extracter
28. Stud puller
29. Cylinder head refacer
30. Crank shaft balancing machine.

- **Special Tools and Equipment for Fuel Injection Room :**

1. Injection pump calibrating machine
2. Injector tester
3. Nozzle grinding and lapping machine
4. Hand tools
5. Stand for dismantling pump
6. Special fixture for mounting different pumps on test bench
7. Cleaning tank
8. Tap and die set
9. Grinder
10. Drill machine

11. Set of laps and lapping compound, special tools for dismantling and assembling pumps
12. Injection pipe flaring machine (nipple forming tools)
13. Bench vice
14. Bench grinder with buffer
15. Special tools for overhauling injection pumps
16. Nozzle viewer (microscope)
17. Arbour press for mounting and dismounting bearing.

12.6.4. Summary of Important Services and Equipment

Some important services and the tools equipment required for servicing are :

1. **Brake service :**

(*i*) Brake drum turning and lining equipment.
(*ii*) Brake relining and riveting machines.
(*iii*) Special tools, housing sets, plug gauges etc.

2. **Wheel alignment service :**

(*i*) Special equipment for wheel alignment and wheel balancing.

3. **Battery service :**

(*i*) Battery testers and analysers.
(*ii*) Battery charging unit including hydrometer, voltmeter, high rate discharge tester etc.

4. **Engine tune-up including carburettor, fuel pump service :**

(*i*) Special tools for adjusting carburettor fuel pump setting, tube cutting and flaring tools etc.
(*ii*) Ignition and electrical testing instruments for analysing the conditions of the engines :

- Neon timing light for checking ignition timing ;
- Spark plug testing tools (cleaners, wrenches, etc.) ;
- Tools for adjusting valves.

5. **Engine conditioning :**

Engine conditioning needs engine overhaul equipment comprising :

- Engine stands
- Cylinder boring
- Grinding and re-sleeving plant
- Connecting rod alignment jig.
- Piston and piston ring service tools
- Various cylinder gauges, micrometers, etc.
- Valve conditioning *i.e.*, value seat grinding and reseating equipment
- Main bearing and big-end bearing boring equipment.

6. **Welding and cutting service :**

- Oxyacetylene and metallic arc welding equipment.

7. **Radiator cleaning service :**

(*i*) Flushing equipment
(*ii*) Cleaning outfit
(*iii*) Test tank
(*iv*) Soldering etc.

8. **Body repair equipment :**

(*i*) Special tools and jigs for checking car dimensions

(*ii*) Straightening jacks (*iii*) Frame straightening

(*iv*) Sanders (Pneumatic or electric) (*v*) Paint spraying equipment

(*vi*) Sheet metal cutting, folding and rolling equipment.

Other requirements :

- Portable crane or hoists for lifting vehicles or engines.
- Shot-blasting cabinet for cleaning parts which have cavities in difficult positions.
- Fire fighting and first-aid equipment.

Fig. 12.68 to 12.74 show respectively the following equipment :

Air compressor, Hydraulic jack, Scissor jack, lifting tackle, cylinder boring bar, Brake relining machine and Moto balance 5000.

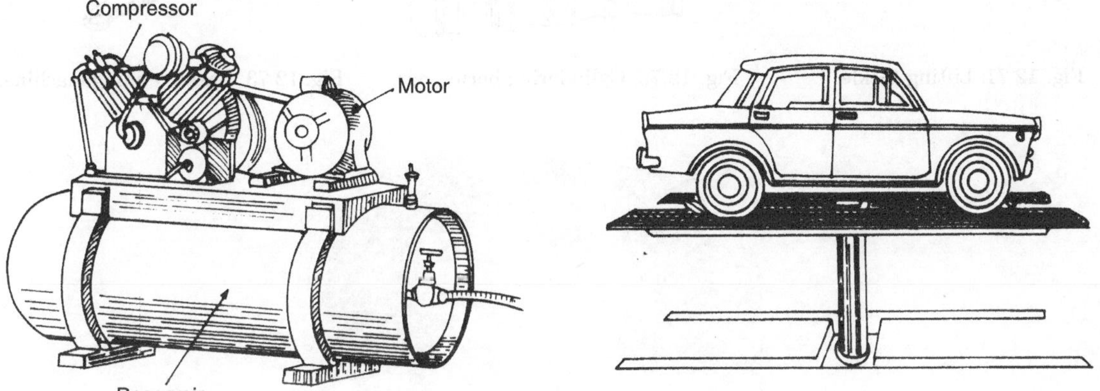

Fig. 12.68. Air compressor.

Fig. 12.69. Hydraulic lift.

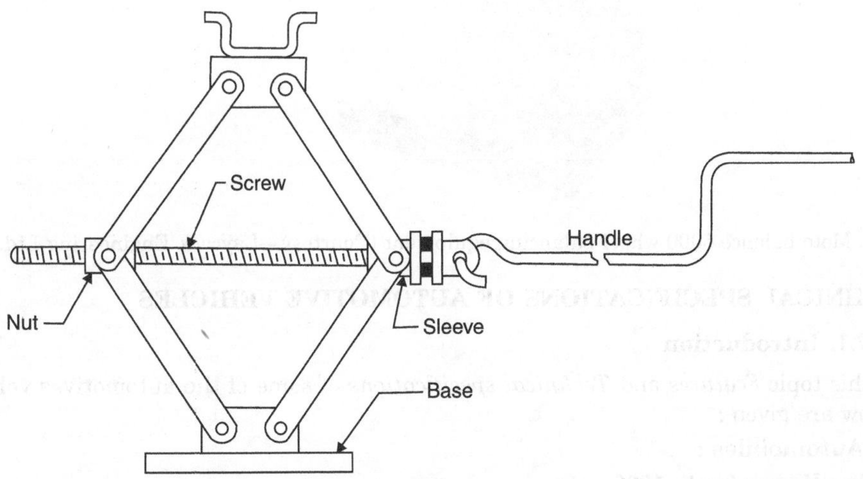

Fig. 12.70. Scissor jack.

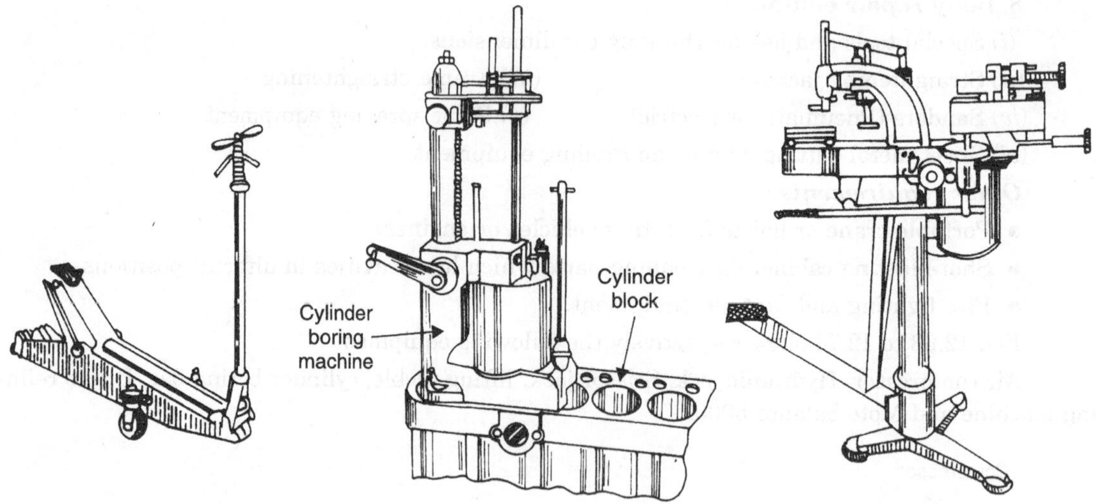

Fig. 12.71. Lifting tackle. Fig. 12.72. Cylindering boring bar. Fig. 12.73. Brake relining machine.

Fig. 12.74. Moto balance 5000 wheel balancing equipment (Courtesy—Laycack Engineering Ltd. England).

12.7. TECHNICAL SPECIFICATIONS OF AUTOMOTIVE VEHICLES

12.7.1. Introduction

In this topic *Features* and *Technical specifications* of some of the automotives vehicles mentioned below are given :

I. Indian Automobiles :

1. ***Small cars (upto 1000 cc)*** :

- Maruti-800
- Maruti Zen
- Alto LX
- Dolphin
- Fiat Uno
- Hyundai Santro
- Daewoo Matiz.

2. ***Medium cars (1000—2000 cc) :***

- Ambassador 2000 DSL
- Contessa Classic 2.0 DLX
- Esteem
- Wagon R
- Versa
- Baleno
- Alto VX
- Altura
- Mitsubishi Lancer (Petrol and Diesel)
- Premium Padmini
- Fiat Palio (Petrol and Diesel)
- Hyundai Getz
- Hyundai Accent
- Sonata
- Daewoo Cielo
- Daewoo Nexia
- Tata India (Petrol and Diesel)
- Indigo (Petrol and Diesel)
- Safari (Petrol and Diesel)
- Sierra
- Honda City
- Accord
- Opal Astra (Petrol and Diesel)
- Opal Corsa
- Forester
- Ford Ikon
- Mondeo
- Escort (Diesel)
- Skoda Octavia
- Daimler Chryeler C-200 K.

3. ***Large cars (more than 2000 cc) :***

- Opal Vectra
- Maruti Vitara
- Merceeds Benz E 220
- E 250 diesel, S-320 L
- Daimler Chrysler C-200 CDI.

4. ***Utility vehicles :***

- Mahindra MM-540 DP
- Armada
- Classic
- Commander NC 640 DP
- CT 4-A (Petrol) Jeep
- Scorpio (Petrol and Diesel)
- Maruti Gypsy King and Gypsy E
- Tata Sumo
- Ford Endeavour
- Toyota Qualis.

5. ***Transport vehicles :***

- Tata 207 DI pick up truck
- Tata LPT-1210 D truck
- Ashok Leyland Comet
- Panther and Hybrid vehicle
- Allwyn Nissan Cabstar
- Eicher Mitsubishi Canter
- Swaraj Mazda LTV
- Bajaj Tempo Excel 4 pick up truck
- Volvo FM 74 × 2 tractor.

6. ***Electric/Hybrid vehicles :***

- Maini Reva

II. Foreign Automobiles :

1. ***Small cars (upto 1000 cc) :***

- Revault type RZ 106

2. ***Medium Cars (1000—2000 cc) :***

- Chrysler Average Delux and Alpine S
- Volkswagen Golf GTI 16V
- Caddy and Polo (Petrol and Diesel)
- Mercedez Benz 200 and 200 D
- Skoda Octavia (Petrol and Diesel)
- Audi A4-1.87
- Alfa Remeo 2000 Spider Veloce
- 2.0 T. SPARK 'USA 83'

3. ***Large cars (more than 2000 cc) :***

- Rolls Royce Camargne
- Morgan Plus 8 sports
- Daimler Double-six
- Citraen CX 2400 Safari
- Audi 100 SE
- A 4-2.6 Quattro

- Porsche 911 Carrera
- Hyundai XG Saloon
- Volvo 740 (Petrol and Diesel)
- Ferrari F 2002

12.7.2. Brief Features and Technical Specifications of Some Cars

1. **Maruti-"800" :**

- **Efficient engine :**

— The three-pod power plant is efficient and returns a good mileage, at the same time delivering enough pop to help the car stay with the pack.

— Breathing through the conventional carburettor system, the engine continues to exhibit good reliability.

- **Low maintenance :**

— The engine is refined, silent and happy with regular oil changes and servicing.

— It has MacPherson strut suspension at the front and leaf/coil at the rear. The car's suspension is reliable, even on tough and uneven Indian roads.

— Most of all, spares are cheap and built for quality.

- **Cheap to buy and run :**

— Though the competion has arrived, there appears no direct threat to the Maruti-800 as none can match the low cost structure of this car.

— Automobile technicians have grown familiar with the car and it can be repaired almost anywhere, with easily available and reasonably priced spares.

- **Build quality :**

— For its unparalleled price, the car offers good build quality when compared with old generation cars priced approximately at the same level.

— Most of its components make it a modern car, offering ease of driving, reliability and various other features.

This car was responsible for triggering the automobile revolution in the country. It brought back the pleasure of driving, while at the same time returning excellent fuel averages.

Technical Specifications :

	Petrol
Engine :	Three-cylinder, in-line
Displacement :	796 cc
Maximum power :	39.592 bhp @ 5500 rpm
Maximum torque :	5.7 kgm @ 2500 rpm
Transmission:	Four-speed manual
Length :	3300 mm
Width :	1410 mm
Wheelbase :	2175 mm
Weight :	620 kg

2. **Maruti-"ZEN" :**

- **Efficient Engine :**

— The peppy, all-aluminium 993cc unit of Zen vehicle makes it a very satisfying car to drive.

— Lightweight aluminium transforms this delightful hatchback, capable of returning a fuel average in the high twenties per litre or even more.

— The powerful and efficient 1500 cc diesel unit is renowned Peugeot design and returns excellent fuel consumption figures.

- **Cheap to run :**

— The car's price tag may seem a bit exaggerated but given its ability to return excellent fuel averages, be it petrol or diesel, it is cheap to run.

— The engine and other mechanicals are reliable with regular service.

- **Good build quality :**

— The Zen is a new generation design and a result of modern engineering, which makes cars very reliable.

— The body is a monocoque construction, uses thinner gauge metal but exhibits good resistance to corrosion, thereby ensuring long life.

- **Ease of service :**

— On the road from 1994, the Zen is a familiar car to deal with. Spares are available at reasonable cost and almost anywhere.

— A MPFI system model is instrumental in elevating its appeal even more while equipping it to tackle emission norms.

- **Resale value :**

— The Zen commands a good resale value, due to its reliability, efficiency and light controls.

Technical Specifications :

	Petrol	Diesel
Engine :	Four-cylinder, in-line	Four-cylinder, in-line
Displacement :	993 cc	1527 cc
Maximum power :	50 bhp @ 6500 rpm	57 bhp @ 5000 rpm
Maximum torque :	7.2 kgm @ 4500 rpm	9.69 kgm @ 2250 rpm
Transmission :	Five-speed manual	Five-speed manual
Length :	3495 mm	3495 mm
Width :	1495 mm	1495 mm
Wheelbase :	2335 mm	2335 mm
Weight :	730 kg	830 kg

3. **Maruti-"Esteem" :**

- **Efficient engine :**

— The engine is punchy, catapulating the car ahead with the lightest touch of foot on pedal.

— The all-aluminium unit of 1200 cc displacement with its hollow camshaft is very light and extremely responsive.

- Its free-rewing character makes it a very refined unit giving the car phenomenal acceleration.

- **Cheap to run :**

— The Esteem comes with local content, but has managed to keep up the quality.

— Inspite of the competition, it has been able to offer good value, while at the same time returning a decent fuel consumption.

- **Value for money :**

— With rising competition, the Esteem has broken up at variant levels-LX, VX etc.—adding more equipment and value.

— Audio system, central locking and power windows are standard. Also, the introduction of new colours as well as the update carried out have made the car face younger.

- **Ease of service :**

— With a comprehensive service network located all over the country, Maruti products are given a decisive advantage and the Esteem is no exception.

— Spares are available at most places in the country other than at the dealers and are reasonably priced, making it easier to maintain this car than any other in the mid-segment.

- **Best-seller in the class :**

— This car, with its tempting offerings, is considerably cheaper than the competition.

— It is a familiar sight on Indian roads and has earned itself a name for reliability, all of which have helped it remain the best-selling car in the mid-segment category.

Technical Specifications

	Petrol
Engine :	Four-cylinder, in-line
Displacement :	1298 cc
Maximum power :	65 bhp @ 6000 rpm
Maximum torque :	10.1 kgm @ 4000 rpm
Transmission :	Five-speed manual
Length :	4080 mm
Width :	1575 mm
Wheelbase :	2365 mm
Weight :	815 kg

4. **Fiat Auto-"SIENA" :**

- **Chic European looks :**

— The Siena is a car adorned with the Italian styling that makes it distinctive. The front droops low while the rear has been treated with an interesting high 'deck' feel.

— The greenhouse is typically European and gives an aggressive feel.

— The slightly bulbous sides grant shoulder room for a car of this size.

- **Peppy petrol engine :**

— The 1242 cc engine seems to nullify the effects of car's weight ; power comes in seamless fashion, throwing it ahead with firm dexterity.

— This unit, though smaller in capacity than the other cars in its class, sees the vehicle in good stead because of its hard working characteristics.

— Power is available at low r.p.m., making Siena a more enjoyable car to drive in the city.

- **Comfortable interior :**

— The interior of the Siena is well provided with seats as well as doors draped in attractive fabric.

— The design of the seats makes them very comfortable to travel in, and there is plenty of legroom for the front and rear passengers.

- **Cavernous boot :**

— A high-decked styling of the rear has made room for a cavernous boot.

— A good loading height combined with efficient space management, the boot offers a whopping 500 litres of storage area.

- **Handling and suspension :**

— The highlight of the Siena is its strong suspension and taut handling.

— In addition to the line chassis, the suspension with anti-roll bars at both ends has been designed to give it an edge over its rivals in the handling department.

— Which ever way you throw the car around, it will not deter from its course and continue to return a positive feedback.

Technical Specifications :

	Petrol	**Diesel**
Engine :	Four-cylinder, in-line	Four-cylinder, in-line
Displacement :	1242 cc	1698 cc
Maximum power :	72 bhp @ 6000 rpm	64 bhp @ 4500 rpm
Maximum torque :	10.2 kgm @ 3000 rpm	12.2 kgm @ 2500 rpm
Transmission :	Five-speed manual	Five-speed manual
Length :	4100 mm	4100 mm
Width :	1614 mm	1614 mm
Wheelbase :	2360 mm	2360 mm
Weight :	1035 kg	1100 kg

5. **Fiat Auto-"UNO" :**

- **Compact looks :**

— The Uno boasts of the angular looks of the eighties, with a 'compact on the outside but roomy inside' image.

— The long front end overhang helps generate admirable space for the engine as well as in the cabin.

- **Wide interior :**

— The interior offers plenty of useful space when compared to other cars in this class.

— The upright sitting position and the steering wheel located high-up create enough above average, interior space.

— The tailgate with a 'punched out' shape, helps create spacious luggage space that is the largest amongst cars of its category.

- **Efficient engines :**

— The 1000 cc petrol unit delivers a respectable 45 B.H.P. while the 1700 cc diesel bestows exceptional performance, making it the most powerful diesel car in its class.

— Besides performance, both the units return excellent fuel charges.

- **Comfort :**

— The upright sitting position makes sitting comfortable. Four adults can travel for hours without feeling uncomfortable while the car can accommodate five without much complaint.

— The big car 'size' on the inside offers good head, shoulder and legroom.

— Overall, the Uno is a comfortable car to travel in.

- **Value for money :**

— Uno offers much more than many of its competitors car, especially the diesel version.

— In addition to the attention to detail, the interior is roomy enough to impart a 'big car feel' while making available good loading area.

Technical Specifications :

	Petrol	**Diesel**
Engine :	Four-cylinder, in-line	Four-cylinder, in-line
Displacement :	999 cc	1700 cc
Maximum power :	45 bhp @ 5600 rpm	54 bhp @ 4600 rpm
Maximum torque :	7.1 kgm @ 3200 rpm	10.1 kgm @ 3000 rpm
Transmission :	Five-speed manual	Five-speed manual
Length :	3688 mm	3688 mm
Width :	1562 mm	1562 mm
Wheelbase :	2362 mm	2360 mm
Weight :	775 kg	925 kg

6. **Ford India-"IKON" :**

- **Magnificent engines :**

— Bad fuel quality has no effect on the magnificent engines.

— The 1597 cc petrol and 1753 cc diesel units deliver good power and torque to give the car commendable performance in its class. Both the units are refined, develop torque at low r.p.m. to aid in-town drive ability and enough power to propel them into three-digit territory on the speedometer.

- **Expensive interior :**

— The attractive interior endows the feel of a bigger car and the fabric used for the seats and other areas radiates quality.

— Other than being attractive, the interior design is functional, with ample legroom for passengers and slightly laid back seating arrangements.

- **Balanced handling :**

— Behind the wheel, Ikon with its European origin, feels widely bigger than its actually is.

— The three-box design is derived from the Ford Fiesta and gives it an edge in handling over other cars in its category.

- **Solid body :**

— With a solid build, the Ikon presents itself in a mould where driveability and balance between comfort and handling takes front seat.

— Style, reliability and efficiency, when coupled with price, should see the Ikon in good stead.

- **Value for money :**

— Abiding by its policy, Ford's homework will not only help the Ikon invade the Indian car market but also refresh it.

Technical Specifications :

	Petrol	**Diesel**
Engine :	Four-cylinder, in-line	Four-cylinder, in-line
Displacement :	1597 cc	1753 cc
Maximum Power :	92 bhp @ 5600 rpm	58 bhp @ 4800 rpm
Maximum Torque :	13.2 kgm @ 3400 rpm	11.02 kgm @ 2200 rpm
Transmission :	Five-speed manual	Five-speed manual
Length :	4140 mm	4410 mm
Width :	1634 mm	1634 mm
Wheelbase :	2486 mm	2486 mm
Weight :	971 kg	1077 kg

7. **General Motors India-"OPEL ASTRA" :**

- **Solid build :**

— The solid, German engineering behind the Astra is evident from its build quality, which is superb.

- **Decent ergonomics :**

— Behind the wheel, the Astra's functionality and excellent ergonomics are apparent.

— All the controls feel good and are delightful to operate, while the seats are very comfortable and allow for hours of fatigue-free morning.

- **Exceptional safety :**

— Safety on this car has been given priority, and the solid build of the Astra contributes towards the same.

— The front end is designed to accommodate crumple zones, impact absorbing longitudinal members and a rigid transverse member. In addition, the B pillars, sills, and the central tunnel are made tough and head restraints are available even on rear seats.

- **Attention to detail :**

— Small but significant touches such as air-conditioning vent in the glove box and the use of high quality materials add all the possible trappings of a luxury car to the Astra-only the Mercedes Benz has more luxury features.

- **Driveability :**

— The Astra is powered by a 1600 cc engine which is tuned more for bottom end torque than outright power.

— The car's power-to-weight ratio presents it with a brisk performance that of a long distance cruiser.

Technical Specifications :

	Petrol	**Diesel**
Engine :	Four-cylinder, in-line	Four-cylinder, in-line
Displacement :	1597 cc	1700 cc
Maximum power :	75 bhp @ 5600 rpm	68 bhp @ 4500 rpm
Maximum torque :	12.2 kgm @ 3200 rpm	13.4 kgm @ 2400 rpm
Transmission :	Five-speed manual	Five-speed manual
Length :	4239 mm	4239 mm
Width :	1696 mm	1696 mm
Wheelbase :	2517 mm	2517 mm
Weight :	960 kg	1140 kg

8. **Hind Motors-"MITSUBISHI LANCER" :**

- **Noble looks :**

— A product of the Japanese school of design, the Lancer radiates noble yet mildly aggressive looks with flared front wings, fine sidelines, low front bumper and admirable rear light clusters.

- **Refined performer :**

— The performance of the Lancer is the highlight factor of this vehicle, with a seamless surge of 87 B.H.P. unfolding itself from way down the r.p.m. band, with reasonably proficient torque.

— The MPFI system, low-friction roller bearings and three valves per cylinder, places the car amongst the most refined in its class.

- **Taut handling :**

— The suspension, with sub-frame mounted lower arms at the front and passive track rods at the rear, results in great straight-line stability.

— A truly rough-edged, easy-to-tame beast with elegant handling characteristics, it offers a rewarding driving experience.

- **Interior and ergonomics :**

— The Lancer packs upmarket interior feel with a combination of fabric and plastic.

— While driving, the essential controls fall to hand easily while details like the useful floor-mounted oddments box, pockets, central arm rest on rear seat, add value.

- **Fine build quality :**

— The build quality of Lancer demonstrates the reliability of the car.

— The fit-and-finish of materials and quality of fabric used in among the best available.

— The paint job is of high calibre.

In short, everything about the car reflects the company's philosophy-the care truly means a reliable set of wheels.

Technical Specifications :

	Petrol	Diesel
Engine :	Four-cylinder, in-line	Four-cylinder, in-line
Displacement :	1468 cc	1998 cc
Maximum power :	87 bhp @ 5500 rpm	68 bhp @ 3500 rpm
Maximum torque :	13.4 kgm @ 3300 rpm	12.5 kgm @ 3000 rpm
Transmission :	Five-speed manual	Five-speed manual
Length :	4290 mm	4290 mm
Width :	1690 mm	1690 mm
Wheelbase :	2500 mm	2500 mm
Weight :	1010 kg	1095 kg

9. **Hind Motors-"AMBASSADOR" :**

- **Archetypal looks :**

— Originally the Morris Oxford, the Ambassador continues to display its solid feel of a car from the fifties.

— It has a bulky image with bulbous wings, bulgy bonnet and a curvy boot.

— Once used to driving it, other car owners respect it and gingerly move aside, making it the most preferred car for government officials, executives and classic car buffs alike.

- **Solid body :**

— The bulky build of this car consumes a lot of metal with a heavy construction that is solid and suggestive of an armoured carrier.

— The body uses a thick metal gauge that helps it traverse bad roads with ease.

- **Comfortable inside :**

— The bulbous styling contributes to the overall feeling of comfort in the cabin.

— The seats are well padded and feel homelike.

— Coupled with ample legroom, and easy sitting position, the interior continues to offer a high level of satisfaction.

- **Supply suspension :**

— The overall suspension set-up of the car is supple.

— At the front, it is equipped with torsion bar while at the rear leaf springs govern undulations on the road. The rear is fitted with gas shock absorbers that add to the degree of suppleness, while the 15-inch wheels continue to guard against loss of ground clearance.

- **Exemplary service :**

— Having criss-crossed the Indian subcontinent several thousand times, this car is just about the friendliest one for any automobile mechanic. The same is true with the spare parts that are available at almost any spare retailer, that at near rock-bottoms prices.

Technical Specifications :

	Petrol	Diesel
Engine :	Four-cylinder, in-line	Four-cylinder, in-line
Displacement :	1817 cc	1489 cc
Maximum power :	75 bhp @ 5000 rpm	40 bhp @ 4000 rpm
Maximum torque :	13.7 kgm @ 3300 rpm	8.6 kgm @ 2500 rpm
Transmission :	Five-speed manual	Four-speed manual
Length :	4312 mm	4312 mm
Width :	1676 mm	1676 mm
Wheelbase :	2464 mm	2464 mm
Weight :	1200 kg	1104 kg

10. **Hind Motors-"CONTESSA" :**

- **Classic looks :**

— With a long and low-slung body, the Contessa radiates the classic looks of the late and 60s and early 70s.

— For loyal fans of this design era characterised by large and muscular body panels it looks European and feels strong and solid.

- **Safest car :**

— The sixties design may have exempted the car from stringent safety standards but nevertheless this car offers excellent safety.

— There is a good deal of metal on the front and rear overhang, that is occupied by large crumple zone (probably the largest amongst all India cars) making it one of the safest *deshi* cars.

- **Comfortable interiors :**

— This car offers superior levels of comfort as there is ample legroom at the front and rear.

— The seats are well padded, contoured and offer good lumbar support, all this adds upto comfortable long distance travel. The central armrest on the rear seat is a luxury feature.

- **Punchy engine :**

— With 74 bhp, the 1800 cc engine demonstrates the ability to sprint all day long with speedometer reading three digits.

— Coupled to five-speed gearbox, power comes senselessly with a firm feel, making the car a marvellous long distance traveller.

— The optional two-litre diesel unit offers good fuel economy.

- **Easy service :**

— Spares are available with most retailers and are priced reasonably when compared with new-generation cars in the market.

— Most drivetrain and engine parts exhibit quality and make the car reliable.

— The engine is simple, breathes through a carburettor, while the electronic breakerless ignition keeps service intervals for apart.

Technical Specifications :

	Petrol	Diesel
Engine :	Four-cylinder, in-line	Four-cylinder, in-line
Displacement :	1817 cc	1995 cc
Maximum power :	75 bhp @ 5000 rpm	52 bhp @ 4500 rpm
Maximum torque :	13.7 kgm @ 3000 rpm	11.0 kgm @ 2500 rpm
Transmission :	Five-speed manual	Five-speed manual
Length :	4591 mm	4591 mm
Width :	1699 mm	1699 mm
Wheelbase :	2667 mm	2667 mm
Weight :	1138 kg	1270 kg

11. **Hyundai MotorS-"SANTRO" :**

- **Efficient engine :**

— Santro's four-cylinder 999 cc engine delivers a healthy 55 bhp while at the same time exhibiting good fuel efficiency and a raspy note.

— The power band cuts in at low r.p.m. and aids in good pick-up.

— The Multi-Point Fuel Injection system (MPFI) assures efficient running and also meets tight emission norms.

- **Easy ingress and egress :**

— The 'tall boy' design offers extremely good shoulder and head room. This feature also makes ingress and egress easy for men and women alike.

— The slim-high loading features and erect seating arrangement places the Santro near the MPV category, indicating an unusual design.

- **Tight turning circle :**

— Typical of a small MPV, the rear wheels of the Santro are placed way back, almost at the respective ends while at the front, the wheels are designed to turn generously from lock-to-lock offering a tight turning circle. The light steering with power assist option, helps even further.

- **Decent ergonomics :**

— The controls are well placed and fall to hand without the need to stretch while rushing on the speed way.

— The instrument pods are clear, easy to read and the switches for lights, wiper, air-conditioning, etc. operate with positive clicks.

- **Value for money :**

— Available in three models, standard, Deluxe and Deluxe with power steering, the Santro offers a deal for buyers with varied budget.

— Also, features like power windows and power steering provide the buyer with a comprehensive package.

Technical Specifications :

	Petrol
Engine :	Four-cylinder, in-line
Displacement :	999 cc
Maximum power :	55 bhp 5500 rmp
Maximum torque :	8.1 kgm @ 3100 rpm
Transmission :	Four-speed manual
Length :	3495 mm
Width :	1495 mm
Wheelbase :	2380 mm
Weight :	796 kg

12. **Hyundai Motors-"ACCENT" :**

- **Good looks :**

— The Accent packs a magnificent style yet looks compact.

— The front bumper and smiling grille followed by sweeping contours gives the Accent its unusual style and confirms a good entry level model.

— From the side, the Accent has a distinctive waste-line and at the rear a slightly high deck envelops the tail lamps, giving the car a fresh appearance.

- **Impressive performance :**

— The petrol engine of the Accent offers peppy performance with the engine churning out a decent 94 bhp.

— The power band cuts in at low r.p.m. give it the ability to be first to "take off" from signals.

— The SOHC engine stimulates three valves per cylinder for better efficiency.

— Coupled to the engine is the MPFI system, which makes the car reliable and puts it in a better position to handle emission norms.

- **Slick ride and handling :**

— The car exhibits good road holding while handling is troublefree.

— The independently sprung front and rear are sufficiently firm and help the Accent handle predictably through corners, maintaining a tight line.

- **Roomy interior :**

— The Accent cashes in on the advantage offered by the three-box design, as it makes available a good amount of interior space with enough legroom at both front and rear.

— The seats at the front and rear offer a good level of comfort with significant back and thigh support.

— Elbow room is generous too, which makes the Accent a relaxing car to travel in.

- **Air-conditioning :**

— The Accent offers progressive air-conditioning, with smartly laid out ducts to sufficiently chill the cabin. The duct located on the dashboard directs the air to the roof and allows it to bounce rearward to effectively cool rear-seat passengers.

Technical Specifications :

	Petrol	Diesel
Engine :	Four-cylinder, in-line	Four-cylinder, in-line
Displacement :	1495 cc	1527 cc
Maximum power :	94 bhp @ 5500 rpm	57 bhp @ 5000 rpm
Maximum torque :	12.2 kgm @ 3500 rpm	9.5 kgm @ 2250 rpm
Transmission :	Five-speed manual	Five-speed manual
Length :	4250 mm	4250 mm
Width :	1670 mm	1670 mm
Wheelbase :	2440 mm	2440 mm
Weight :	1006 kg	1006 kg

13. **Honda SIEL-"CITY" :**

- **Well-bread engine :**

— The city boasts the most refined power plant in the mid-segment category.

— The engines of both the 1300 cc and 1500 cc variants are extremely responsive, punchy, superbly balanced and silky smooth. Not only do they deliver 100 bhp and 90 bhp respectively, but give exceptional fuel mileage as well. In short, the engines are ages ahead of the competition-a Honda hallmark.

- **Build quality :**

— The body has been built with a check on weight to preserve a good power-to-weight ratio characteristic but without any sacrifice in quality.

— The build quality is among the best, with reinforcements added at vital points.

— In addition, the top notch paint job, tight shut lines and high finish levels indicate high build quality.

- **Handling :**

— The front and rear are independently sprung with detailed attention to ride and handling. Stress has been placed on handling and the car excels in this, with the low-slung body, finely tuned front suspension and precise steering aiding in tight and reassuring handling.

— The car manages to create a big grin on the face of the driver.

- **Refinement :**

— Considerable attention has been paid to make the car refined.

— The engine and gearbox show a high degree of refinement characterised by silky smooth operation. The asphalt-layered cabin floor, insulation inside the pillars, doors etc., serve to heighten the sense of refinement.

- **Air-conditioning :**

— The City comes with an excellent air-conditioning system. The power characteristics of both engines help in this, as not much strain is felt on either performance or efficiency with the air-conditioning on.

— The units are effectively designed to work in the tropics.

Technical Specifications :

	Petrol	Diesel
Engine :	Four-cylinder, in-line	Four-cylinder, in-line
Displacement :	1343 cc	1493 cc
Maximum power :	90 bhp @ 6400 rpm	100 bhp @ 6400 rpm
Maximum torque :	11.3 kgm @ 4700 rpm	13.0 kgm @ 4700 rpm
Transmission :	Five-speed manual	Five-speed manual
Length :	4225 mm	4225 mm
Width :	1690 mm	1690 mm
Wheelbase :	2500 mm	2500 mm
Weight :	980 kg	985 kg

14. **MERCEDES-BENZ-E-CLASS :**

- **Contemporary quality :**

— The build quality of this car has no parallel in India.

— The paint job, the perfect shut lines and panel gaps shout quality, while every nut and bolt says so in more detail. This is part of the company philosophy-pay meticulous attention to detail.

- **Majestic looks :**

— With the oval-eyed E-Class replacing the old solid-looking car, there has been a new freshness in the looks.

— The E-Class is cheekier, sexier, while the low-slung 'longish' body structure add weights to its looks.

— The head lamp treatment, mild spoiler at the rear and the rounded greenhouse complete the list.

- **Performance and refinement :**

— The performance of the E-Class stands unparalleled.

— The two petrol variants and one diesel variant come with engine and drivetrain that speak of refinement like no other.

— The specification as well as equipment list is impressive, while the car easily whisks the occupants away at unimaginable speeds that are plainly understated by the near pin-drop silence in the cabin.

- **Steady handling :**

— For a car of this size, the E-Class handles with amazing verve and as the speed rises the car seems to gain agility.

— The suspension has remarkable composure and refuses to surrender, giving an almost consistent ride even whilst encountering the worst of silently-dampened on-road bumps.

— Features like ABS add safely.

- **Status symbol :**

— With the price tag of this car including so many zeroes, it ferries around status for its owner, who will most likely be cuddled in the cosy and luxurious rear seat.

— Compared to this, any other Indian car looks inadequate, and imperfect.

Technical Specifications :

	Petrol E-200/230	**Diesel E-250**
Engine :	Four-cylinder, in-line	Five-cylinder, in-line
Displacement :	1998 cc/2295 c	2497 cc
Maximum power :	135 bhp @ 5500 rpm/ 150 bhp @ 5400 rpm	113 bhp @ 5000 rpm
Maximum torque :	19.4 kgm @ 4000 rpm/22.4 kgm @ 3700 rpm	17.3 kgm @ 3200-4600 rpm
Transmission :	Five-speed manual	Five-speed manual
Length :	4795 mm	4795 mm
Width :	1799 mm	1799 mm
Wheelbase :	1475 mm	1540 mm
Weight :	2833 kg	1540 kg

15. **Telco-INDICA :**

- **Solid Italian looks :**

— Topping the scales at 980 kg this car is solidly built.

— The thicker gauge metal employed makes this the safest and most sure-footed car in this class.

— The unmistakably Italian looks make it stand out from the crowd with its bold head lamps and tail lamp clusters built into the rear pillar.

- **Powerful engines :**

— Both the petrol and the diesel engines share the same block and hence the same displacement.

— The petrol engine is the more powerful of the two and imparts a lively performance. The diesel stands second in terms of performance but offers a good balance between top-end power and low-end torque.

- **Predictable handling :**

— The high weight helps the Indica to be more sure footed.

— The struts at front and coil at the rear give it a good handling capabilities with a balance between ride and comfort.

— A good deal of suspension travel and taut feel lets the car be thrown around any corner.

- **Almost 100 percent indigenous :**

— Whereas most cars started on foreign componentry, this car started on the premise of being totally Indian.

— The build quality of the India is on par with most new generation cars, making the Indian manufacturers, from components to the automobile itself proud of their job.

- **Cheap to buy :**

— Being indigenous and backed by Tata empire, both the variants have redefined the small car category.

— By introducing well-finished and built products at rock bottom prices, Tata has made buying a car easier.

Technical Specifications :

	Petrol	**Diesel**
Engine :	Four-cylinder, in-line	Five-cylinder, in-line
Displacement :	1405 cc	1405 cc
Maximum power :	60 bhp @ 4500 rpm	53.5 bhp @ 5000 rpm
Maximum torque :	10.7 kgm @ 2500 rpm	8.7 kgm @ 2500 rpm
Transmission :	Four-speed manual	4/5 speed manual
Length :	3660 mm	3660 mm
Width :	1625 mm	1625 mm
Wheelbase :	2400 mm	2400 mm
Weight :	980 kg	980/1005 kg

ENGINE SPECIFICATIONS FOR SOME OTHER CARS VEHICLES :

1. Wagon R :

Manufacturer :	Maruti Udyog Ltd. India
Engine type :	4-cylinder, in line, ferrous engine F10D
Bore :	68.5 mm
Stroke :	72.0 mm
Capacity :	1061 cc
Maximum power :	46.2 kW (62 bhp) at 6000 rpm
Maximum torque :	80.4 Nm (8.2 kgm) at 3500 rpm
Power to weight ratio :	73.8 bhp/tonne
Torque to weight ratio :	9.7 kgm/tonne
Specific output :	58.6 bhp/litre
Fuel system :	Multi-point fuel injection
Fuel tank capacity :	35 litres
Ignition system :	Electronic, transistorised.

2. "Versa" Car :

Manufacturer :	Maruti Udyog Ltd., India
Engine type :	4-cylinder, in line, longitudinal, all aluminium
Bore :	74.0 mm
Stroke :	75.5 mm
Capacity :	1298 cc
Compression ratio :	(9.0 ± 0.2) : 1
Maximum power :	61.2 kW (82 bhp) at 6000 rpm
Maximum torque :	102 Nm (10.4 kgm) at 3000 rpm

Power to weight ratio :	84 bhp/tonne
Torque to weight ratio :	10.7 kgm/tonne
Valve gear :	4 valves per cylinder, SOHC
Fuel system :	Multi-point Fuel Injection
Fuel tank capacity/range :	40 litres/500 km
Ignition system :	Electronic.

3. "Baleno" Car :

Manufacturer :	Maruti Udyog Ltd., India
Engine type :	4-cylinder, in line, all aluminium
Bore :	75.0 mm
Stroke :	90.0 mm
Capacity :	1590 cc
Compression ratio :	(9.2 ± 0.2)
Maximum power :	70 kW (94 bhp) at 6000 rpm
Maximum torque :	132 Nm at 3000 rpm
Valve gear :	16 valves, SOHC
Fuel system :	Multi-point Fuel Injection
Fuel tank capacity :	51 litres
Ignition system :	Electronic.

4. Getz 1.3 :

Manufacturer :	Hyundai Motors India Ltd.
Engine type :	4-cylinder, in line, petrol
Bore :	71.5 mm
Stroke :	83.5 mm
Displacement :	1341 cc
Compression ratio :	9.5
Maximum power :	60.4 kW (81 bhp) at 5800 rpm
Maximum torque :	115.8 Nm (11.8 kgm) at 3200 rpm
Power to weight ratio :	73.97 bhp/tonne
Torque to weight ratio :	10.78 kgm/tonne
Specific output :	60.4 bhp/litre.

5. "Sonata" Car :

Manufacturer :	Hyundai Motors India Ltd.
Engine type :	4-cylinder, in line, petrol
Bore :	85 mm
Stroke :	88 mm
Displacement :	1997 cc

Compression ratio :	10.0
Maximum power :	98.5 kW (132 bhp) at 6000 rpm
Maximum torque :	164 Nm (16.7 kgm) at 4500 rpm
Power to weight ratio :	92 bhp/tonne
Torque to weight ratio :	11.64 kgm/tonne
Specific output :	66 php/litre.

6. "Matiz" Car :

Manufacturer :	Daewoo Motors India Limited
Engine type :	3 cylinder, in line, petrol engine, transverse mounted
Bore :	68.5 mm
Stroke :	72 mm
Capacity :	796 cc
Compression ratio :	8.5
Valve gear :	Two valves per cylinder, SOHC
Engine control :	8 bit electronic control unit with MPFI
Maximum power :	39 kW (52 bhp) at 6000 rpm
Maximum torque :	71.5 Nm (7.3 kgm) at 3500 rpm
Fuel tank capacity :	35 litres.

7. "Cielo" Car :

Manufacturer :	Daewoo Motors India Limited
Engine type :	4 cylinder in line, SOHC, with MPFI
Displacement :	1498 cc
Maximum power :	80 PS at 5400 rpm
Maximum torque :	127.5 Nm (13 kgm) at 3200 rpm
Fuel tank capacity :	50 litres.

8. "Nexia" Car :

Manufacturer :	Daewoo Motors India Ltd.
Engine type :	4-cylinder, in line, petrol engine, cast iron engine block and cast aluminium cylinder head
Bore :	N.A.
Stroke :	N.A.
Displacement :	1998 cc
Compression ratio :	N.A.
Maximum power :	68.6 kW (92 bhp) at 5600 rpm
Maximum torque :	129.5 Nm (13.2 kgm) at 3400 rpm
Power to weight ratio :	84.66 bhp/tonne
Torque to weight ratio :	12.1 kgm/tonne

Specific output :	61.4 bhp/litre
Valve gear :	4-valves per cylinder
Ignition and fuel system :	Electronic control with MPFI.

9. "Safari" (Petrol) Car :

Manufacturer :	Tata Motors Ltd., India
Engine type :	4-cylinder, in line, petrol, longitudinal, rear wheel drive, iron block and alloy head
Bore :	66.0 mm
Stroke :	90.0 mm
Displacement :	2092 cc
Compression ratio :	8.5
Maximum power :	kW (135 bhp) at 5630 rpm
Maximum torque :	Nm (19.8 kgm) at 3750 rpm
Power to weight ratio :	70.68 bhp/tonne
Torque to weight ratio :	10.36 kgm/tonne
Specific output :	64.5 bhp/litre
Valve gear :	4 valves per cylinder, DOHC
Fuel supply :	Sequential MPFI
Fuel tank capacity/range :	65 litres/490 km.

10. "Safari" (Diesel) Car :

Manufacturer :	Tata Motors, India
Type :	TATA 483 Turbo diesel, Intercooled, 4-cylinder, in line
Bore :	83 mm
Stroke :	90 mm
Displacement :	1948 cc
Compression ratio :	21
Maximum power :	90 PS at 4300 rpm (DIN 70020)
Maximum torque :	186.4 Nm (19 kgm) at 2500 rpm (DIN 70020)
Air cleaner :	Dry (paper) type
Oil filter :	Spin-on full flow paper type
Fuel filter :	Two stage, fine filteration
Capacity of cooling system :	10 litres
Oil sump capacity :	7.7 litres
Weight of complete engine :	220 kg (dry)
Fuel tank capacity :	65 litres
Battery :	12 V, N 70 Z, negative ground
Alternator max. output :	135 amperes.

11. "Sumo/Sierra/Estate" :

Manufacturer :	Tata Motors, India
Model :	TATA 483 DL
No. of cylinders :	4, in-line
Capacity :	1948 cc
Maximum power :	68 PS at 4500 rpm
Maximum torque :	117.7 Nm (12 mkg) at 2500 rpm
Fuel tank capacity :	50 litres
Fuel supply :	Fuel injection pump—Rotary type with electric stop solenoid., Governor—Mechanical, Fuel filter—two stage, fine filtration
Battery :	12 V, N50 Z
Alternator :	65 Amps (–ve earth).

12. Tata 207 DI pick up vehicle :

Manufacturer :	Tata Motors, India
Engine type :	Tata 4997SP27, 4-cylinder in-line, direct injection diesel engine
Bore :	97 mm
Stroke :	100 mm
Displacement :	2956 cc
Compression ratio :	17.0
Maximum power :	45 kW (60.3 bhp) at 2800 rpm
Maximum torque :	180 Nm (18.3 kgm) at 1500-2000 rpm
Fuel injection pump :	MICO-in-line type
Governor :	Centrifugal type, variable speed
Fuel tank capacity :	60 litres.

13. Tata LPT-1210D truck :

Manufacturer :	Tata Engineering and Locomotive Co. Ltd. India.
Engine model and type :	692DI diesel engine, 6-cylinder, in line
Bore :	92 mm
Stroke :	120 mm
Capacity :	4788 cc
Compression ratio :	17.1 : 1
Brake power :	Taxable H.P. 31.5, Maximum (DIN) B.P. 83.5 kW at 2800 r.p.m.
Torque :	Maximum 294 Nm at 1800-2000 r.p.m.
Compression pressure :	1.960 MPa at 150-200 r.p.m.
Cooling system :	Water cooling, capacity 20 litres. Cooling water temperature 75°C to 95°C.

Fuel supply :	Fuel injection pump—Mico Bosch. Fuel injection begins 23° before T.D.C. Opening pressure of injection nozzles 19.6 + 1.0 MPa (New nozzles) Min 17.6 MPa (used nozzles)
Air cleaner :	Oil bath
No. of main bearings :	7.

14. Accord 2.4 VTi-L :

Manufacturer :	Honda Siel Cars Ltd., India
Engine type :	4-cylinder, in line, petrol, transverse, alloy block and head
Bore :	87.0 mm
Stroke :	99.0 mm
Displacement :	2354 cc
Compression ratio :	9.3 ± 0.1
Maximum power :	105 kW (141 bhp) at 5700 rpm
Maximum torque :	191 Nm (19.5 kgm) at 4800 rpm
Power to weight ratio :	99.3 bhp/tonne
Torque to weight ratio :	13.73 kgm/tonne
Specific output :	59.9 php/litre
Valve gear :	4 valves per cylinder, DOHC
Ignition and fuel :	Honda PGM-FI
Fuel tank capacity/range :	65 litres/630 km (Highway).

15. Cheverolet Forester 2.0 AWD :

Manufacturer :	General Motors, India
Engine type :	4-cylinder, horizontally opposed, petrol longitudinal, aluminium alloy block and head
Bore :	92 mm
Stroke :	75 mm
Displacement :	1994 cc
Compression ratio :	10 ± 0.2
Maximum power :	89.5 kW (120 bhp) at 5000 rpm
Maximum torque :	178 Nm (18.15 kgm) at 3600 rpm
Power to weight ratio :	87.27 bhp/tonne
Torque to weight ratio :	13.2 kgm/tonne
Specific output :	60.18 bhp/litre
Valve gear :	4 valves per cylinder, SOHC
Ignition and fuel system :	Electronic MPFI.

16. Ford Escort Diesel Car :

Manufacturer :	Ford India Ltd.
Model :	1.8 DNA
Type :	In-line 4-cylinder transverse mounted
Bore :	82.5 mm
Stroke :	82.0 mm
Displacement :	1753 cc
Compression ratio :	21.5
Valve gear :	Two valves per cylinder, SOHC
Fuel delivery :	Distributor type diesel fuel injection
Maximum power :	44.8 kW (60 bhp) at 4800 rpm
Maximum torque :	110 Nm (11.2 kgm) at 2500 rpm
Fuel tank capacity :	55 litres.

17. Skoda Superb (Petrol) Car :

Manufacturer :	Skoda India Ltd.
Engine type :	6-cylinder, V-type, alloy block and head, longitudinal
Bore :	N.A.
Stroke :	N.A.
Displacement :	2771 cc
Compression ratio :	N.A.
Maximum power :	144 kW (193 bhp) at 6000 rpm
Maximum torque :	279.6 Nm (28.5 kgm) at 3200 rpm
Power to weight ratio :	128.5 bhp/tonne
Torque to weight ratio :	18.97 kgm/tonne
Specific output :	69.64 bhp/litre
Valve gear :	5 per cylinder, DOHC
Fuel tank capacity :	62 litres.

18. Skoda Superb (Diesel) Car :

Manufacturer :	Skoda India Ltd.
Engine type :	6-cylinder, V-type, alloy block and head, longitudinal
Bore :	N.A.
Stroke :	N.A.
Displacement :	2496 cc
Compression ratio :	N.A.
Maximum power :	115.5 kW (154.9 bhp) at 4000 rpm
Maximum torque :	310 Nm (31.6 kgm) at 1400 rpm

Power to weight ratio :	98.97 bhp/tonne
Torque to weight ratio :	20.2 kgm/tonne
Specific output :	62.05 bhp/litre
Valve gear :	2 per cylinder, DOHC
Fuel tank capacity :	62 litres.

19. Octavia 1.9 TDI Car :

Manufacturer :	Skoda India Ltd.
Engine type :	4-cylinder, in line, turbodiesel, alloy block iron head
Bore :	79.5 mm
Stroke :	95.5 mm
Capacity :	1896 cc
Compression ratio :	19.5 : 1
Maximum power :	67 kW (90 bhp) at 4000 rpm
Maximum torque :	210 Nm (21.4 kgm) at 1900 rpm
Power to weight ratio :	67.66 bhp/tonne
Torque to weight ratio :	16.1 kgm/tonne
Specific output :	47.46 bhp/litre
Valve gear :	2 valves per cylinder, SOHC
Fuel system :	Direct Injection
Fuel tank capacity/range :	55 litres/770 km.

20. Qualis (Petrol) MPV :

Manufacturer :	Toyota Kirloskar Motor Ltd., India
Engine type :	1 RT-E, 4-cylinder in-line, petrol, longitudinal, iron block and alloy head
Bore :	86.0 mm
Stroke :	86.0 mm
Displacement :	1998 cc
Compression ratio :	9.0
Maximum power :	69 kW (94 PS) at 4800 rpm
Maximum torque :	150 Nm (15.3 kgm) at 2800 rpm
Power to weight ratio :	62.66 bhp/tonne
Torque to weight ratio :	10.2 kgm/tonne
Specific output :	47.05 bhp/litre
Valve gear :	2 valves per cylinder, SOHC
Fuel supply system :	Electronic Multi-point Fuel Injection
Fuel tank capacity :	53 litres
Emission performance :	Bharat II (Euro II of India).

21. Qualis (Diesel) MPV :

Manufacturer :	Toyota Kirloskar Motor Ltd., India
Engine type :	4-cylinder in-line, diesel
Bore :	92.0 mm
Stroke :	92.0 mm
Displacement :	2446 cc
Compression ratio :	22.2
Maximum power :	54 kW (75 PS) at 4200 rpm
Maximum torque :	151 Nm (15.4 kgm) at 2400 rpm
Fuel supply system :	Distribution type fuel injection
Fuel tank capacity :	53 litres.

22. Mahindra CJ 4-A Jeep (Petrol) :

Manufacturer :	Mahindra and Mahindra Limited., India
Engine model and type :	F4-134, 4-cylinder, in-line petrol
Bore :	79.37 mm
Stroke :	111.12 mm
Capacity :	2200 cc
Compression ratio :	7.4 : 1
Brake power :	R.A.C. rating 15.63, Maximum B.P 53.7 kW at 4000 r.p.m.
Torque :	Maximum 154 Nm at 2000 r.p.m.
Cooling system :	Fin and tube radiator, capacity 10 litres.

23. Mahindra MM-540 DP (Diesel)/Armada/Classic/commander 740 DP/NC 640 DP Jeep :

Manufacturer :	Mahindra and Mahindra Limited, India
Engine model and type :	Diesel XPD-4.90, 4 cylinder, in line.
Bore :	90 mm
Stroke :	83 mm
Capacity :	2112 cc
Compression ratio :	22.4 : 1
Brake power :	R.A.C. rating 20.0, Maximum B.P. 46.2 at 4500 r.p.m. (DIN)
Torque :	Maximum 121 Nm at 2000 r.p.m. (DIN)
Cooling :	Centrifugal water pump type
Fuel injection system :	Rotary pump
Weight of engine :	184 kg.

24. "Scorpio" Petrol S.U.V. :

Manufacturer :	Mahindra and Mahindra Limited, India
Engine type :	4 cylinder in-line, petrol, iron block and alloy head
Bore :	82.7 mm
Stroke :	93.0 mm
Displacement :	1998 cc
Maximum power :	86.5 kW (116 bhp) at 5500 rpm
Maximum torque :	183.4 Nm (18.7 kgm) at 3800 rpm
Power to weight ratio :	67.0 bhp/tonne
Torque to weight ratio :	10.8 kgm/tonne
Specific output :	58.06 bhp/litre
Valve gear :	4 valves per cylinder, DOHC
Fuel system :	Sequential MPFI

25. "Scorpio" Turbo 2.6 S.U.V. :

Manufacturer :	Mahindra and Mahindra Ltd.,, India
Engine type :	4-cylinder in-line, turbodiesel, iron block and alloy head
Bore :	94.0 mm
Stroke :	94.0 mm
Displacement :	2609 cc
Compression ratio :	18.5
Maximum power :	81.3 kW (109 bhp) at 3800 rpm
Maximum torque :	255 Nm (26 kgm) at 1800 rpm
Power to Weight ratio :	57.06 bhp/tonne
Torque to weight ratio :	13.6 kgm/tonne
Specific output :	41.8 bhp/litre
Valve gear :	2 valves per cylinder, SOHC
Fuel system :	Distributor type direct diesel injection

26. Rolls-Royce Camargue Car :

Manufacturer :	Rolls Royce Motors Ltd., England
Type :	8 cylinder petrol, 90°V.
Bore :	104.1 mm
Stroke :	99.1 mm
Cubic capacity :	6750 cc
Compression ratio :	8.1 :1
Valve arrangement :	Overhead
Tappet :	Self-adjusting hydraulic
Firing order :	$A_1, B_1, A_4, B_4, B_2, A_3, B_3, A_2$

Cylinder block :	Aluminium-silicon alloy, wet cast iron cylinder liners
Cylinder head :	Aluminium alloy
Crankshaft bearings :	5
Cooling system :	50% water/antifreeze, capacity 16.0 litres
Fuel supply :	Sump capacity-8.0 litres, full flow oil filter.
Lubrication :	Fuel tank capacity-107 litres. Solex 4 A1 compound 4-choke carburettor, 2 SU electric fuel pump.
Air cleaner :	AC Delco, paper element type.

12.7.3. Technical Details of Some Vehicles

Manufacturer	*Fuel*	*Cyl.*	*Engine capacity (cc)*	*Steer-ing*	*Max. power b.h.p., r.p.m.*	*Max. torque*	*Wheels size*	*Wheels driven*	*Fuel cap*
A. MULTI-UTILITY VEHICLE									
Mahindra and Mahindra									
Voyager	DL	4	2498	B&N	72.5@4000	15.5@2500	5J × 14	2	55 Ltrs.
Voyager AC	DL	4	2498	B&N	72.5@4000	15.5@2500	5J × 14	2	55 Ltrs.
Mercedes-Benz									
MB 140 D 2.9	DL	5	2874	PWR	95	185@3000	5.5J × 16	2	75 Ltrs.
MB 100 D 2.9	DL	5	2874	PWR	95	185@3000	5.5J × 16	2	75 Ltrs.
B. CARS									
1. Daewoo Motors									
Cielo Executive	PL	4	1498	PWR	80@5400	13.0@3200	05J × 13	2	50 Ltrs.
Matiz Standard	PL	3	796	R&P	52@6000	7.3@3500	4.5J × 13	2	35 Ltrs.
Matiz Delux	PL	3	796	R&P	52@6000	7.3@3500	4.5J × 13	2	35 Ltrs.
Matiz Executive	PL	3	796	R&P	52@6000	7.3@3500	4.5J × 13	2	35 Ltrs.
Matiz Premium	PL	3	796	PWR	52@6000	7.3@3500	4.5J × 13	2	35 Ltrs.
Nexia	PL	4	1498	PWR	92@5600	13.5@3400	5J × 14	2	50 Ltrs.
2. Fiat India									
Siena 75 ELX P.S.	PL	4	1242	PWR	72@6000	10.2@3000	5J × 13	2	47 Ltrs.
Siena 75 ELX P.S. MET	PL	4	1242	PWR	72@6000	10.2@3000	5J × 13	2	47 Ltrs.
Siena 75 SP-ELX	PL	4	1242	PWR	72@6000	10.2@3000	5J × 13	2	47 Ltrs.
Siena 75 SP-ELX MET	PL	4	1242	PWR	72@6000	10.2@3000	5J × 13	2	47 Ltrs.
Siena 60 EL	DL	4	1697	PWR	64@4500	12.2@2500	5J × 13	2	47 Ltrs.
Siena 60 EL Metallic	DL	4	1697	PWR	64@4500	12.2@2500	5J × 13	2	47 Ltrs.
Siena 60 ELX	DL	4	1697	PWR	64@4500	12.2@2500	5J × 13	2	47 Ltrs.

Siena 60 ELX Metallic	DL	4	1697	PWR	64@4500	12.2@2500	5J × 13	2	47 Ltrs.
Siena 60 SP ELX	DL	4	1697	PWR	64@4500	12.2@2500	5J × 13	2	47 Ltrs.
Siena 60 SP ELX Metallic	DL	4	1697	PWR	64@4500	12.2@2500	5J × 13	2	47 Ltrs.
3. Ford India									
Escrot 1.6 (Alpha)	PL	4	1598	PWR	84@5500	12.64@3000	5J × 13	2	55 Ltrs.
Escort 1.8 (Orion) DLX	DL	4	1753	PWR	60@4800	11.2@2500	5J × 13	2	55 Ltrs.
Sport E	PL	4	1598	PWR	84@5500	12.64@3000	5J × 13	2	55 Ltrs.
Ikon 1.6 CLxi	PL	4	1597	PWR	91@5500	13.3@2500		2	45 Ltrs.
Ikon 1.6 CLxi	PL	4	1597	R&P	91@5500	13.3@2500		2	45 Ltrs.
Ikon 1.6 CLxi PS	PL	4	1597	R&P	91@5500	13.3@2500		2	45 Ltrs.
Ikon 1.8 Zxi	DL	4	1753	PWR	53@4800	11.1@2200		2	45 Ltrs.
4. Hindustan Motors									
Contessa Classic 1.8 GLX	PL	4	1817	PWR	75@5000	13.8@3000	5J × 13	2	65 Ltrs.
Contessa Classic 2.0 DSL	DL	4	1995	R&P	52@4500	11@2500	5J × 13	2	65 Ltrs.
AMB Nova DSL BE + DLX + AC + TG + RDL	DL	4	1489	R&P	36@4000	8.5@2000	4J × 15	2	
AMB DSL 2000 BE + AC + TG + RDL	DL	4	1995	R&P	55@4500	11@2500	4J × 15	2	42 Ltrs.
AMB DSL 2000 BE + AC + TG + RDL	DL	4	1995	R&P	55@4500	11@2500	4J × 15	2	42 Ltrs.
AMB ISZ BE + RDL + TG	PL	4	1817	R&P	75@5000	13.8@3000	4J × 15	2	42 Ltrs.
AMB ISZ BE + RDL + TG + AC	PL	4	1817	R&P	75@4000	13.8@3000	4J × 15	2	42 Ltrs.
AMB ISZ BU + TG + RDL	PL	4	1817	R&P	75@4000	13.8@3000	4J × 15	2	42 Ltrs.
AMB ISZ BU + AC + TG + RDL	PL	4	1817	R&P	75@4000	13.8@3000	4J × 15	2	42 Ltrs.
Lancer Reg	PL	4	1468	PWR	87@5500	13.4@3300	5J × 13	2	50 Ltrs.
Lancer Spl	PL	4	1468	PWR	87@5500	13.4@3300	5J × 13	2	50 Ltrs.
Lancer Premium	PL	4	1468	PWR	87@5500	13.4@3300	5J × 13	2	50 Ltrs.
Lancer Reg	DL	4	1998	PWR	68@4500	12.5@3000	5J × 13	2	50 Ltrs.
Lancer Spl	DL	4	1998	PWR	68@4500	12.5@3000	5J × 13	2	50 Ltrs.
Lancer Premium	DL	4	1998	PWR	68@4500	12.5@3000	5J × 13	2	50 Ltrs.

5. Honda Motors									
City 1.3 LXI	PL	4	1343	R&P	90&6500	11.3@4700	5J × 13	2	45 Ltrs.
City 1.3 EXI	PL	4	1343	PWR	90&6500	11.3@4700	5J × 13	2	45 Ltrs.
City 1.5 EXI	PL	4	1493	PWR	100@6500	13.1@4600	5J × 13	2	45 Ltrs.
City 1.5 EXI (A/T)	PL	4	1493	PWR	100@6500	13.1@4600	5J × 13	2	45 Ltrs.
6. Hyundai Motors									
				Euro I					
Santro Std	PL	4	999	R&P	55@5500	8.4@2500	4.0J × 13	2	35 Ltrs.
Santro Std Metallic	PL	4	999	R&P	55@5500	8.4@2500	4.0J × 13	2	35 Ltrs.
Santro GLS1	PL	4	999	R&P	55@5500	8.4@2500	4.0J × 13	2	35 Ltrs.
Santro GLS1 Met	PL	4	999	R&P	55@5500	8.4@2500	4.0J × 13	2	35 Ltrs.
Santro GLS2	PL	4	999	PWR	55@5500	8.4@2500	4.0J × 13	2	35 Ltrs.
Santro GLS2 Met	PL	4	999	PWR	55@5500	8.4@2500	4.0J × 13	2	35 Ltrs.
				Euro II					
Santro Std	PL	4	999	R&P	55@5500	8.4@2500	4.0J × 13	2	35 Ltrs.
Santro Std Metallic	PL	4	999	R&P	55@5500	8.4@2500	4.0J × 13	2	35 Ltrs.
Santro GLS1	PL	4	999	R&P	55@5500	8.4@2500	4.0J × 13	2	35 Ltrs.
Santro GLS1 Met	PL	4	999	R&P	55@5500	8.4@2500	4.0J × 13	2	35 Ltrs.
Santro GLS2	PL	4	999	PWR	55@5500	8.4@2500	4.0J × 13	2	35 Ltrs.
Santro GLS2 Met	PL	4	999	PWR	55@5500	8.4@2500	4.0J × 13	2	35 Ltrs.
Accent GLE 1	PL	4	1495	R&P	94@5500	12.2@3500	4.5J × 13	2	45 Ltrs.
Accent GLE 2	PL	4	1495	R&P	94@5500	12.2@3500	4.5J × 13	2	45 Ltrs.
Accent GLS	PL	4	1495	PWR	94@5500	12.2@3500	4.5J × 13	2	45 Ltrs.
7. Fiat India Auto									
Fiat Uno DLX	PL	4	999	R&P	45@5500	7.2@3500	4.5J × 13	2	35 Ltrs.
Fiat Uno DLX Met	PL	4	999	R&P	45@5500	7.2@3500	4.5J × 13	2	35 Ltrs.
Fiat Uno DLX A/C	PL	4	999	R&P	45@5500	7.2@3500	4.5J × 13	2	35 Ltrs.
Fiat Uno DLX A/C Met	PL	4	999	R&P	45@5500	7.2@3500	4.5J × 13	2	35 Ltrs.
Fiat Uno Jubilee A/C	PL	4	999	PWR	45@5500	7.2@3500	4.5J × 13	2	35 Ltrs.

Fiat Uno Jubilee A/C Met	PL	4	999	PWR	45@5500	7.2@3500	4.5J × 13	2	35 Ltrs.
Fiat Uno DLX	DL	4	1700	R&P	57@4600	10@3000	4.5J × 13	2	42 Ltrs.
Fiat Uno DLX Met	DL	4	1700	R&P	57@4600	10@3000	4.5J × 13	2	42 Ltrs.
Fiat Uno DLX A/C	DL	4	1700	R&P	57@4600	10@3000	4.5J × 13	2	42 Ltrs.
Fiat Uno DLX A/C Met	DL	4	1700	R&P	57@4600	10@3000	4.5J × 13	2	42 Ltrs.
Fiat Uno P.S.	DL	4	1700	PWR	57@4600	10@3000	4.5J × 13	2	42 Ltrs.
Fiat Uno P.S Met	DL	4	1700	PWR	57@4600	10@3000	4.5J × 13	2	42 Ltrs.
Fiat Uno Jubilee A/C	DL	4	1700	PWR	57@4600	10@3000	4.5J × 13	2	42 Ltrs.
Fiat Uno Jubilee A/C Met	DL	4	1700	PWR	57@4600	10@3000	4.5J × 13	2	42 Ltrs.
8. Maruti Udyog									
Esteem LX	PL	4	1298	R&P	65@6000	10.1@4000	4J × 13	2	40 Ltrs.
Esteem VX	PL	4	1298	PWR	65@6000	10.1@4000	4J × 13	2	40 Ltrs.
Esteem AX A/T	PL	4	1298	PWR	65@6000	10.1@4000	4J × 13	2	40 Ltrs.
1000 Std	PL	4	970	R&P	46@5500	7.2@3500	4J × 13	2	40 Ltrs.
1000 AC	PL	4	970	R&P	46@5500	7.2@3500	4J × 13	2	40 Ltrs.
800 Std	PL	3	796	R&P	39.5@5500	5.7@2500	4J × 12	2	30 Ltrs.
800 Std Metallic	PL	3	796	R&P	39.5@5500	5.7@2500	4J × 13	2	30 Ltrs.
800 EX	PL	3	796	R&P	39.5@5500	5.7@2500	4J × 13	2	30 Ltrs.
800 EX Met	PL	3	796	R&P	39.5@5500	5.7@2500	4J × 13	2	30 Ltrs.
800 AC/TG	PL	3	796	R&P	39.5@5500	5.7@2500	4J × 13	2	30 Ltrs.
800 AC/TG	PL	3	796	R&P	39.5@5500	5.7@2500	4J × 13	2	30 Ltrs.
Zen Classic	PL	4	993	R&P	50@5000	7.2@4500	4J × 13	2	35 Ltrs.
Zen Classic Metallic	PL	4	993	R&P	50@5000	7.2@4500	4J × 13	2	35 Ltrs.
Zen LX	PL	4	993	R&P	50@5000	7.2@4500	4J × 13	2	35 Ltrs.
Zen LX Metallic	PL	4	993	R&P	50@5000	7.2@4500	4J × 13	2	35 Ltrs.
Zen VX	PL	4	993	R&P	50@5000	7.2@4500	4J × 13	2	35 Ltrs.
Zen VX Metallic	PL	4	993	R&P	50@5000	7.2@4500	4J × 13	2	35 Ltrs.
Zen A/T	PL	4	993	R&P	50@5000	7.2@4500	4J × 13	2	35 Ltrs.

Zen A/T Metallic	PL	4	993	R&P	50@5000	7.2@4500	4J × 13	2	35 Ltrs.
Zen DL	DL	4	1527	R&P	57@5000	9.7@2250	4J × 13	2	35 Ltrs.
Zen DL Metallic	DL	4	1527	R&P	57@5000	9.7@2250	4J × 13	2	35 Ltrs.
Zen-Vxi	PL	4	993	PWR (Elec)	50@5000	7.2@4500	4J × 13	2	35 Ltrs.
Zen-Vxi (MPE)	PL	4	993	PWR (Elec)	50@5000	7.2@4500	4J × 13	2	35 Ltrs.
Zen-Vxi (MPFI) Met	PL	4	993	PWR (Elec)	50@5000	7.2@4500	4J × 13	2	35 Ltrs.
9. Euro I									
800 Std	PL	3	796	R&P	39.5@5500	5.7@2500	4J × 12	2	30 Ltrs.
800 Std Metallic	PL	3	796	R&P	39.5@5500	5.7@2500	4J × 12	2	30 Ltrs.
800 EX	PL	3	796	R&P	39.5@5500	5.7@2500	4J × 12	2	30 Ltrs.
800 EX Met	PL	3	796	R&P	39.5@5500	5.7@2500	4J × 12	2	30 Ltrs.
800 AC/TG	PL	3	796	R&P	39.5@5500	5.7@2500	4J × 12	2	30 Ltrs.
800 AC/TG Metallic	PL	3	796	R&P	39.5@5500	5.7@2500	4J × 12	2	30 Ltrs.
Zen-Vxi	PL	4	993	PWR (Elec)	50@5000	7.2@4500	4J × 13	2	35 Ltrs.
Zen-Vxi Met	PL	4	993	PWR (Elec)	50@5000	7.2@4500	4J × 13	2	35 Ltrs.
Zen LX	PL	4	993	R&P	50@5000	7.2@4500	4J × 13	2	35 Ltrs.
Zen LX Metallic	PL	4	993	PWR	50@5000	7.2@4500	4J × 13	2	35 Ltrs.
Zen Classic	PL	4	993	R&P	50@5000	7.2@4500	4J × 13	2	35 Ltrs.
Zen Classic Metallic	PL	4	993	R&P	50@5000	7.2@4500	4J × 13	2	35 Ltrs.
10. Mercedes-Benz									
E-200	PL	4	1998	PWR	95@5500	180@4500	7J × 15	2	65 Ltrs.
E-230	PL	4	2295	PWR	107@5400	213@4500	7J × 15	2	65 Ltrs.
E-250 DL	DL	5	2497	PWR	83@5000	170@4600	7J × 15	2	65 Ltrs.
11. Opel									
Astra Metro Solid	PL	4	1598	PWR	76@5400	121@2800	5.5J × 14	2	52 Ltrs.
Astra Metro Met	PL	4	1598	PWR	76@5400	121@2800	5.5J × 14	2	52 Ltrs.
Astra MY99 Solid	PL	4	1598	PWR	76@5400	121@2800	5.5J × 14	2	52 Ltrs.

Astra MY99 Met	PL	4	1598	PWR	76@5400	121@2800	5.5J × 14	2	52 Ltrs.
Astra MY99 A/T Solid	PL	4	1598	PWR	76@5400	121@2800	5.5J × 14	2	52 Ltrs.
Astra MY99 A/T Met	PL	4	1595	PWR	76@5400	121@2800	5.5J × 14	2	52 Ltrs.
Astra Club (Met)	PL	4	1595	PWR	76@5400	121@2800	5.5J × 14	2	52 Ltrs.
Astra Club A/T (Met)	PL	4	1595	PWR	76@5400	121@2800	5.5J × 14	2	52 Ltrs.
Opel 100 (Met)	PL	4	1595	PWR	76@5400	121@2800	5.5J × 14	2	52 Ltrs.
Astra Metro Solid	DL	4	1700	PWR	65@4500	132@2400	5.5J × 14	2	52 Ltrs.
Astra Metro Met	DL	4	1700	PWR	65@4500	132@2400	5.5J × 14	2	52 Ltrs.
Astra MY99 Solid	DL	4	1700	PWR	65@4500	132@2400	5.5J × 14	2	52 Ltrs.
Astra BY99 Met	DL	4	1700	PWR	65@4500	132@2400	5.5J × 14	2	52 Ltrs.
Astra Club (Met)	DL	4	1700	PWR	65@4500	132@2400	5.5J × 14	2	52 Ltrs.
Opel 100 (Met)	DL	4	1700	PWR	65@4500	132@2400	5.5J × 14	2	52 Ltrs.
12. Telco									
Indica Petrol Std	PL	4	1405	R&P	53.5	8.7	4.5J × 13	2	37 Ltrs.
Indica Diesel Std	DL	4	1405	R&P	53.5	8.7	4.5J × 13	2	37 Ltrs.
Indica A/C (DLE)	DL	4	1405	R&P	53.5	8.7	4.5J × 13	2	37 Ltrs.
Indica DL A/C (DLE) Metallic	DL	4	1405	R&P	53.5	8.7	4.5J × 13	2	37 Ltrs.
Indica DL DLX	DL	4	1405	PWR	53.5	8.7	4.5J × 13	2	37 Ltrs.

12.7.4. Comparative Data of Constructional Features of Engines of Some Indian Vehicles

S. No.	Item	Maruti 800	Jeep Gypsy	Hindustan Ambassa-dor	Premier Padmini	Jeep (Mahindra) Universal	Santro Epsilon Engine	Maruti Esteem	Maruti Baleno	Maruti Zen
1.	*No of cylinders and arrangements*	3 in-line	4 in-line	4 in-line	4 in-line	4 in-line	4 in-line	4 in-line	4 in-line	4 in-line
2.	*Firing order*	1,3,2	1,3,4,2	1,3,4,2	1,3,4,2	1,3,4,2	1,3,4,2	1,3,4,2	1,3,4,2	1,3,4,2
3.	*Bore*	68.5 mm	65.5 mm	73.02 mm	68 mm	79.3 mm	66 mm	74 mm	75 mm	72 mm
4.	*Stroke*	72 mm	72 mm	88.9 mm	75 mm	111.12 mm	73 mm	75.5 mm	90 mm	61 mm
5.	*Cubic capacity*	796 cc	970 cc	1489 cc	1089 cc	2199 cc	999 cc	1298 cc	1590	993 cc
6.	*Compression ratio*	8.7 : 1	8.8 : 1	7.2 : 1	7.3 : 1	7.4 : 1	8.9 : 1	9 : 1	9 : 1	8.8 : 1
7.	*Cylinder liner*	No separate cylinder liner	No cylinder liner	No cylinder liner, originally	Originally, no separate cylinder	No cylinder liner	No cylinder liner	No cylinder liner	No cylinder liner	No cylinder liner
8.	*Piston*	Cast aluminium alloy cum concave skirt	Aluminium alloy con-cave crown split skirt	Aluminium alloy oval tapered split skirt	Aluminium alloy ground T-slot	Aluminium Alloy cam grounded	Cast aluminium alloy	Cast aluminium alloy	Cast aluminium alloy	
9.	*Compression rings*	2	2	3	1	2	2	2	2	2
10.	*Oil Control rings*	1	1	1	2	1	1	1	1	1

11.	*Gudgeon pin connection*	Fully floating	Fully floating	Clamped in connecting rod	Fully floating	Clamped in connecting rod	Semi floating	Fully floating	Fully Floating	Fully floating
12.	*No. of crankshaft main bearings*	4	5	3	3	3	5	5	5	5
13.	*No. of camshaft bearings*	4	5	3	3	3	5	5	5	5
14.	*Camshaft drive*	Grooved rubber belt drive	Grooved rubber belt drive	Chain	Chain	Single helical gear	Cogged belt drive	Grooved belt drive	Grooved belt drive	Grooved belt drive
15.	*Valve type*	Poppet	Poppet	Poppet	Poppet	Poppet	Poppet	Poppet	Poppet	Poppet
16.	*Valve mechanism*	Overhead	Overhead	Overhead	Overhead	Overhead inlet and side exhaust	Single overhead cam (SOHC)	Single overhead cam (SOHC)	Single overhead cam (SOHC)	Single overhead cam (SOHC)
17.	*Valve seat*	Originally no separate valve seat insert is fitted.	Originally no separate valve seat insert is fitted.	Originally no separate valve seat insert is fitted.	Cadmium steel insert.	Eatonite E.M.S. 58 insert for exhaust is fitted.	No separate valve seat	No separate valve seat	No separate valve seat	No separate valve seat
18.	*Valve seat angle*	45°	45°	45°	45°	45°				

SHORT ANSWER QUESTIONS

Q. 1. What is a cycle ?

Ans. A cycle is a repeated series of operations occurring in a certain order.

Q. 2. What is "air standard efficiency" ?

Ans. It is the efficiency of an engine using air as the working medium.

Q. 3. What is relative efficiency ?

Ans. It is the ratio of actual thermal efficiency to air standard efficiency.

Q. 4. How are gas turbines mainly classified ?

Ans. (*i*) Constant pressure or continuous combustion type and (*ii*) Constant volume or explosion type.

Q. 5. What are the two types of gas turbine cycles ?

Ans. (*i*) Open cycle, and (*ii*) closed cycle.

Q. 6. Give the two advantages of gas turbine.

Ans. (*i*) Smooth in operation and continuous in performance, and (*ii*) It is free from vibrations (since it does not have reciprocating parts).

Q. 7. Give two disadvantages of gas turbine.

Ans. (*i*) It consumes 70 percent more fuel than a petrol engine for the same power, (*ii*) Due to very high speed, the gas turbine causes serious problems in obtaining an effective transmission speed suitable for road speeds.

Q. 8. Give two advantages of closed cycle gas turbine.

Ans. (*i*) The combustion products do not mix with the working fluid, because the heating is external. Hence any fluid of high calorific value may be used, and (*ii*) There is no corrosion and accumulation of deposits of carbon or tar on the blades and nozzles of the turbine. There is no need of internal cleaning.

Q. 9. What are the major components of automotive gas turbine ?

Ans. (*i*) Compressor, (*ii*) Turbine, (*iii*) Regenerator, (*iv*) Combustion chamber, (*v*) Transmission, and (*vi*) Fuel and start-up controls.

HIGHLIGHTS

1. A cycle is defined as a repeated series of operations occurring in a certain order.
2. The efficiency of an engine using air as the working medium is known as an **'Air standard efficiency'**.
3. Relative efficiency, $\eta_{relative} = \dfrac{\text{Actual thermal efficiency}}{\text{Air standard efficiency}}$.
4. **Carnot cycle** efficiency, $\eta_{Carnot} = \dfrac{T_1 - T_2}{T_1}$.
5. **Otto cycle** efficiency, $\eta_{Otto} = 1 - \dfrac{1}{(r)^{\gamma-1}}$.

   Mean effective pressure, $p_{m(Otto)} = \dfrac{p_1 r[(r^{\gamma-1} - 1)(r_p - 1)]}{(\gamma - 1)(r - 1)}$
6. **Diesel cycle** efficiency, $\eta_{Diesel} = 1 - \dfrac{1}{\gamma(r)^{\gamma-1}}\left[\dfrac{\rho^{\gamma} - 1}{\rho - 1}\right]$

Mean effective pressure, $p_{m\,(\text{Diesel})} = \dfrac{p_1 r^{\gamma}\,[\gamma(\rho - 1) - r^{1-\gamma}(\rho^{\gamma} - 1)]}{(\gamma - 1)(r - 1)}$

7. **Dual cycle** efficiency, $\eta_{\text{dual}} = 1 - \dfrac{1}{(r)^{\gamma-1}}\left[\dfrac{(\beta.\rho^{\gamma} - 1)}{(\beta - 1) + \beta\gamma(\rho - 1)}\right]$

Mean effective pressure, $p_{m\,(\text{dual})} = \dfrac{p_1 r^{\gamma}\,[\beta(\rho - 1) + (\beta - 1) - r^{1-\gamma}(\beta\rho^{\gamma} - 1)}{(\gamma - 1)(r - 1)}.$

8. Atkinson cycle efficiency, $\eta_{\text{Atkinson}} = 1 - \gamma \cdot \dfrac{(r - \alpha)}{r^{\gamma} - \alpha^{\gamma}}$

where, α = compression ratio, and
r = expansion ratio.

9. The gas turbines are mainly divided into two groups :

(*a*) *Constant pressure combustion gas turbines :*

(*i*) Open cycle constant pressure gas turbine.

(*ii*) Closed cycle constant pressure gas turbine.

(*b*) *Constant volume combustion gas turbine.*

10. The following methods are employed to increase the specific output and thermal efficiency of the plant :

(*i*) Intercooling ; (*ii*) Reheating ; (*iii*) Regeneration.

11. The main demerit associated with constant volume combustion turbine is that the pressure difference and velocities of hot gases are not constant ; so the turbine speed fluctuates.

12. Various types of garages are :

(*i*) Small garages, (*ii*) Medium garages, (*iii*) Large garages, and (*iv*) Service station.

OBJECTIVE TYPE QUESTIONS

Choose the Correct Answer :

1. The air standard Otto cycle comprises

(*a*) two constant pressure processes and two constant volume processes

(*b*) two constant pressure and two constant entropy processes

(*c*) two constant volume processes and two constant entropy processes

(*d*) none of the above.

2. The air standard efficiency of Otto cycle is given by

(*a*) $\eta = 1 + \dfrac{1}{(r)^{\gamma+1}}$ (*b*) $\eta = 1 - \dfrac{1}{(r)^{\gamma-1}}$ (*c*) $\eta = 1 - \dfrac{1}{(r)^{\gamma+1}}$ (*d*) $\eta = 2 - \dfrac{1}{(r)^{\gamma-1}}$.

3. The thermal efficiency of theoretical Otto cycle

(*a*) increases with increase in compression ratio (*b*) increases with increase in isentropic index γ

(*c*) does not depend upon the pressure ratio (*d*) follows all the above.

4. The work output of theoretical Otto cycle

(*a*) increases with increase in compression ratio (*b*) increases with increase in pressure ratio

(*c*) increases with increase in adiabatic index γ (*d*) follows all the above.

5. For same compression ratio

(*a*) thermal efficiency of Otto cycle is greater than that of Diesel cycle

(*b*) thermal efficiency of Otto cycle is less than that of Diesel cycle

(*c*) thermal efficiency of Otto cycle is same as that for Diesel cycle

(*d*) thermal efficiency of Otto cycle cannot be predicted.

6. In air standard Diesel cycle, at fixed compression ratio and fixed value of adiabatic index (γ)
(*a*) thermal efficiency increases with increase in heat addition cut off ratio
(*b*) thermal efficiency decreases with increase in heat addition cut off ratio
(*c*) thermal efficiency remains same with increase in heat addition cut off ratio
(*d*) none of the above.

7. Thermal efficiency of a gas turbine plant as compared to diesel engine plant is
(*a*) higher (*b*) lower
(*c*) same (*d*) may be higher or lower.

8. Mechanical efficiency of a gas turbine as compared to internal combustion reciprocating engine is
(*a*) higher (*b*) lower (*c*) same (*d*) unpredictable.

9. For a gas turbine the pressure ratio may be in the range
(*a*) 2 to 3 (*b*) 3 to 5 (*c*) 16 to 18 (*d*) 18 to 22.

10. The air standard efficiency of closed gas turbine cycle is given by (r_p = pressure ratio for the compressor and turbine)
(*a*) $\eta = 1 - \frac{1}{(r_p)^{\gamma-1}}$ (*b*) $\eta = 1 - (r_p)^{\gamma-1}$ (*c*) $\eta = 1 - \left(\frac{1}{r_p}\right)^{\frac{\gamma-1}{\gamma}}$ (*d*) $\eta = 1 - \frac{1}{(r_p)^{\frac{\gamma-1}{\gamma}}}$.

11. The work ratio of closed cycle gas turbine plant depends upon
(*a*) pressure ratio of the cycle and specific heat ratio
(*b*) temperature ratio of the cycle and specific heat ratio
(*c*) pressure ratio, temperature ratio and specific heat ratio
(*d*) only on pressure ratio.

12. Thermal efficiency of closed cycle gas turbine plant increases by
(*a*) reheating (*b*) intercooling (*c*) regenerator (*d*) all of the above.

13. With the increase in pressure ratio thermal efficiency of a simple gas turbine plant with fixed turbine inlet temperature
(*a*) decreases (*b*) increases
(*c*) first increases and then decreases (*d*) first decreases and then increases.

14. The thermal efficiency of a gas turbine cycle with ideal regenerative heat exchanger is
(*a*) equal to work ratio (*b*) is less than work ratio
(*c*) is more than work ratio (*d*) unpredictable.

15. In a two stage gas turbine plant reheating after first stage
(*a*) decreases thermal efficiency (*b*) increases thermal efficiency
(*c*) does not affect thermal efficiency (*d*) none of the above.

16. In a two stage gas turbine plant, reheating after first stage
(*a*) increases work ratio (*b*) decreases work ratio
(*c*) does not affect work ratio (*d*) none of the above.

17. In a two stage gas turbine plant, with intercooling and reheating
(*a*) both work ratio and thermal efficiency improve
(*b*) work ratio improves but thermal efficiency decreases
(*c*) thermal efficiency improves but work ratio decreases
(*d*) both work ratio and thermal efficiency decrease.

18. For a jet propulsion unit, ideally the compressor work and turbine work are
(*a*) equal (*b*) unequal
(*c*) not related to each other (*d*) unpredictable.

19. Greater the difference between jet velocity and aeroplane velocity

(*a*) greater the propulsive efficiency (*b*) less the propulsive efficiency

(*c*) unaffected is the propulsive efficiency (*d*) none of the above.

ANSWERS

1. (*b*)	**2.** (*b*)	**3.** (*d*)	**4.** (*d*)	**5.** (*a*)	**6.** (*b*)	**7.** (*b*)
8. (*a*)	**9.** (*b*)	**10.** (*d*)	**11.** (*a*)	**12.** (*d*)	**13.** (*b*)	**14.** (*b*)
15. (*b*)	**16.** (*d*)	**17.** (*a*)	**18.** (*a*)	**19.** (*b*).		

THEORETICAL QUESTIONS

1. What is a cycle ? What is the difference between an ideal and actual cycle ?

2. What is an air-standard efficiency ?

3. What is relative efficiency ?

4. Derive expressions of efficiency in the following cases :

(*i*) Carnot cycle (*ii*) Diesel cycle (*iii*) Dual combustion cycle.

5. Explain "Air standard analysis" which has been adopted for I.C. engine cycles. State the assumptions made for air standard cycles.

6. Derive an expression for 'Atkinson cycle'.

7. Explain briefly Brayton cycle. Derive expression for optimum pressure ratio.

8. Describe with neat sketches the working of a simple constant pressure open cycle gas turbine.

9. Discuss briefly the methods employed for improvement of thermal efficiency of open cycle gas turbine plant.

10. Describe with neat diagram a closed cycle gas turbine. State also its merits and demerits.

11. What are the major fields of application of gas turbines ?

12. State the limitations of gas turbines.

13. How are gas turbines classified ?

14. State merits of gas turbines over I.C. engines and steam turbines.

15. Explain briefly an open cycle gas turbine. Also draw an energy flow diagram for gas-turbine unit.

16. Explain briefly the following methods for improvement of thermal efficiency of open cycle gas turbine plant :

(*i*) Intercooling ; (*ii*) Reheating ; (*iii*) Regeneration.

17. Describe briefly a closed cycle gas turbine.

18. State the merits and demerits of closed cycle gas turbine over open cycle gas turbine.

19. Explain briefly with a neat sketch a constant volume combustion turbine.

20. Write short notes on gas turbine fuels.

21. Briefly describe the major components of automotive gas turbine.

22. What are the requirements of an automobile body ? Explain them briefly.

23. Explain briefly the main considerations for safety of an automobile.

24. Name the gadgets which are fitted in cars for its safe operation.

25. List down some important Do's and Dont's for safe driving.

26. Write short note on car driving.

27. Before going to driver's seat, what must be checked as preliminary inspection of the vehicle ?

28. Explain briefly "police signals".

29. Write a short note on "Traffic signs".

30. Discuss briefly "Driving practice of four wheeler".

31. Explain briefly the methods by which fuel consumption can be reduced in scooters, motor cycles, mopeds and autorickshaws.

32. Discuss briefly the adjustments needed to reduce the consumption of fuel in cars.

33. What factors should be taken into consideration while selecting a site for a garage ?

33. List the number of tools required for the kit of a mechanic.

34. Name the types of equipment which are used in an automobile garage.

35. Name the tools/equipment required for following :

(*i*) Brake service.

(*ii*) Wheel alignment service.

(*iii*) Battery service.

(*iv*) Engine tune-up including carburettor, fuel pump service.

(*v*) Engine conditioning.

(*vi*) Welding and cutting service.

36. Give technical specifications of any four commonly used cars in India.

13

"Universities' Questions (Latest) With Answers/Solutions"

SECTION-A: "SHORT ANSWERS QUESTIONS"

Q. 1. *What do you mean by automobile engineering?*

Ans. ***Automobile engineering*** is a branch of engineering incorporating elements of mechanical, electrical, electronics and safety engineering as applied to the design, manufacture and operation of buses, cars, trucks, motorcycles, etc.

Q. 2. *State the advantages and disadvantages of frameless construction.*

Ans. The ***frameless construction*** claims the following *advantages* over the conventional framed construction:

Advantages :

(*i*) Lower manufacturing cost.

(*ii*) During collision the body absorbs shock due to inpact.

Disadvantages :

(*i*) Cost of repair increases during accident.

(*ii*) Reduction of strength and durability.

Q. 3. *What are the functions of an air cleaner?*

Ans. The *functions* of an ***air cleaner*** are:

(*i*) To clean the air which goes into cylinder of an engine.

(*ii*) It acts as a silencer for carburettor systems.

(*iii*) It also acts as flame arrester.

Q. 4. *What is the function of a thermostat?*

Ans. A ***thermostat*** is a device which regulates the temperature of a system so that the system's temperature is maintained near a desired set point temperature. It does this by switching heating or cooling devices on, or off, or regulating the flow of a heat transfer fluid as needed, to maintain the correct temperature.

A thermostat may be a control unit for a heating or cooling system or a component part of a heater or an air-conditioner. Thermostats can be constructed in several ways and may use a variety of sensors to measure the temperature. The output of the sensor then controls the heating or cooling apparatus.

Q. 5. *What do you mean by a 'clutch'?*

Ans. A ***clutch*** is a mechanism which enables the rotary motion of one shaft to be transmitted at will to second shaft, whose axis is coincident with that of first.

The clutch is *located between the engine and gear box.* When the clutch is *engaged*, the power flows from the engine to the rear wheels through the transmission system and the vehicle moves. When the clutch is *disengaged*, the power is *not* transmitted to the rear wheels and the vehicle stops while the engine is still running.

Q. 6. *Define camber of steering system.*

Ans. ***Camber*** is the inward or outward tilt of a wheel from the vertical. A wheel which tilts out at the top has *positive* camber and if 'in', then it has *negative* camber.

Q. 7. *What is the difference between the functions of a spring and a shock absorber?*

Ans. ***Springs*** are resilent members and as such act as reservoirs of energy. They store the energy due to the sudden force which comes when vehicle encounters a bump or a ditch. This energy is released subsequently, and with the action of *dampers (or shock absorbers) the energy is converted into heat and bounce is avoided.*

"Dampers or shock absorbers" are used in suspension system to *check any continuous vibration* which may follow the initial force on the system.

Q. 8. *Define 'servo brakes'.*

Ans. Mechanism which assists the driver's effort for applying brakes is called "servo mechanism". ***Servo brakes*** are generally used when vehicle load is above 3 tonnes.

Q. 9. *What is the function of 'injection pump'?*

Ans. The *function* of ***injection pump*** is to supply correct quantity of fuel to the injector under high pressure at the correct instant and in the correct sequence.

Q. 10. *What are the purposes of 'Automatic transmission'?*

Ans. ***Automatic transmission*** serves the following *purposes :*

(*i*) It helps to change the engine's rotational speed to the speed required at the driving wheels.

(*ii*) It changes the gear ratios between the engine and the drive wheels to keep the engine operating in its most efficient range.

(*iii*) It automatically shifts under the load demands. This system contains hydraulic pistons accomplished with spring control brakes bonds on the planetary gear sets and clutches within the planetary gear base arrangement.

Neutral → Drive → Low → Reverse

In this type of transmission, the *'neutral'* shows that engine is disconnected with transmission system. *'Drive'* shows when vehicle moves straight on road, *'low'* shows when vehicle moves on hilly roads or muddy area and *'reverse'* position shows when vehicle requires reversed motion.

Q. 11. *Define 'carburetion' and 'carburettor'.*

Ans. The process of preparing, in S.I. engine, a combustible fuel-air mixture outside the engine cylinder is called ***carburetion***. This complicated process is achieved in the induction system.

A ***carburettor*** is a device which atomises the fuel and mixes it with air. It is the most important part of the induction system.

Q. 12. *What are the basic requirements for lubricants?*

Ans. The basic *requirements* for good lubricants are :

(*i*) *Viscosity*: The lubricant should ensure that hydraulic lubrication action takes place.

(*ii*) *Chemical stability*: It should be chemically stable under different temperature conditions.

(*iii*) *Resistance against corrosion*: It should be good corrosion resistance agent.

(*iv*) *Pour point*: It should be *less* so that lowest temperature occurred in the engine.

(*v*) *Physical stability*: It should be stable under different conditions of temperature.

(*vi*) *Flash point*: It should be *high* to avoid flashing of oil vapours.

Q. 13. *What do you mean by 'battery rating' testing?*

Ans. *Battery rating*: Following are the most important ways in which storage batteries are rated :

(*i*) "*20-hour rating (in ampere-hours)*": This is the amount of current that a battery can deliver for 20 hours without the cell voltage dropping below 1.75 volts with an electrolyte temperature of 26.7°C.

(*ii*) "*25-ampere rating*": It measures the battery performance at a moderate constant current output at 26.7°C to a final limiting voltage of 1.75 volts per cell.

(*iii*) "*Cold-cracking rate*" :

— One of the two cold-cracking rates is the number of amperes that a battery can deliver for 30 seconds when it is at –17.8°C (0°F) without the cell voltage falling below 1.2 volts.

— The second cold-cracking rate is measured at –28.9°C (–20°F). In this the final voltage is allowed to drop to 1 volt per cell.

Q. 14. *Explain 'capacity of a battery'.*

Ans. The ***capacity of a battery*** is given in terms of *ampere-hours on discharge*. This is determined by the following *factors* :

(*i*) Final limiting voltage of the cells;

(*ii*) Discharge rate;

(*iii*) Number, design and dimensions of plates;

(*iv*) Design of separators;

(*v*) Quantity of electrolyte;

(*vi*) Design of electrolyte;

(*vii*) Temperature, etc.

Q. 15. *What do you understand by 'Efficiency of a battery'?*

Ans. The ***efficiency of a battery*** is defined as the *ratio of the output of a cell or a battery to the input required to restore the initial state of charge under specified conditions of temperature, current rate and final voltage.*

Q. 16. *List the factors which affect battery life.*

Ans. The life of a battery is affected by the following *factors* :

(*i*) Over charging;
(*ii*) Under charging;
(*iii*) Lack of water;
(*iv*) Loose hold-downs;
(*v*) Excessive loads;
(*vi*) Freezing of electrolyte.

Q. 17. *What is engine tuning?*

Ans. *Engine tuning* is the procedure to bring an engine in good mechanical condition to operate at peak efficiency and power. Every unit of the engine is made to function at its best.

- Engine tuning is more frequently required in modern high compression engines which are more sensitive to variations from specifications than the earlier slow speed, low output engine.

- Often during the engine tuning, engine conditions would be found which must be corrected to restore the engine to like new condition.

Q. 18. *What is the use of a 'pressure cap'?*

Ans. ***Pressure cap*** is used as the cover to radiator filter neck. The main advantage of using pressure cap is that the engine can operate at higher temperatures without boiling the coolant. Loss of coolant due to evaporation is also prevented. Over-heating of the engine is prevented at higher altitude with the help of pressure cap.

Q. 19. *What are primary and secondary filters?*

Ans. ***Primary filters*:** Primary filters are standard on most engines. They are also called *"full flow"* because 100% of the engine oil passes through them in normal operation. If blockage occurs in the filter, bypass valve opens. This allows oil to go around the filter and back to the engine. In this situation, lubrication with unfiltered oil is better than none at all.

***Secondary filters*:** The secondary filters take a small portion of the normal oil flow, usually less than 10% and quite often only 1% and subject it to additional cleaning. These filters are better known as *"bypass"* filters, but they act separately from the primary filter and have nothing to do with its bypass valves.

Q. 20. *What are the special features of modern carburettors?*

Ans. The various *features* of modern carburettors with their functions are :

(*i*) ***Throttle return check*:** It is connected to the throttle linkage. It helps in slowing down the closing movement of the throttle valve when the engine is running at very high speeds and thus gradually reduces the speed to idle.

(*ii*) ***Anti-diesling solenoid*:** On turning off the ignition, anti-dieseling solenoid is de-energised, which allows the throttle valve to close completely. Thus, mixture remains too rich to after fire as it prevents the air from getting into the engine.

(*iii*) ***Automatic mixture control*:** In this a computer controlled solenoid-operated dithering valve automatically adjusts the air-fuel ratio in the carburettor mixture.

Q. 21. *What is the function of an injector?*

Ans. The main function of an ***injector*** is to inject the fuel in the cylinder in the atomised form in proper quantity. It is also known as *atomiser*.

Q. 22. *Name the materials used for clutch lining.*

Ans. The following materials are almost universally used for clutch lining:

(*i*) Leather; (*ii*) Cork;

(*iii*) Fabric; (*iv*) Asbestos;

(*v*) Reybestos and ferodo.

Q. 23. *Why frame is narrow at the front?*

Ans. Frame is narrow at the front, so that turning radius should be small or minimum.

Q. 24. *What is 'toe-in' and 'toe-out'?*

Ans. ***Toe-in*** is the amount by which front wheels are more closer together at the front than the rear wheels in stationary state. In case of ***Toe-out*** the wheels may be closer at the rear than at the front. The difference of the distances between the front wheels at the front and at the rear is called ***toe-out***.

Q. 25. *What are the functions of a condenser in an ignition system?*

Ans. The *functions* of a **condenser** in an ignition system are :

(*i*) Its main function is to minimise arcing and pitting of contact points of breaker.

(*ii*) It intensifies the spark.

(*iii*) It stores energy during circuit breaking and supplies energy during making of circuit.

Q. 26. *What is the necessity for providing 'cooling system on an engine'?*

Ans. The ***cooling system*** is provided on an engine for the following *reasons* :

(*i*) The even expansion of piston in the cylinder may result in seizure of the piston.

(*ii*) High temperatures reduce strength of piston and cylinder liner.

(*iii*) Overheated cylinder may lead to pre-ignition of the charge in case of S.I. engines.

(*iv*) Physical and chemical changes may cause sticking of piston rings and excessive wear of cylinder.

(*v*) If the cylinder head temperature is high the volumetric efficiency and hence the power output of the engine is reduced.

Thus engine cooling is required to keep the temperature of the engine low in order to *avoid* : (*a*) Loss of volumetric efficiency and hence power; (*b*) Engine seizure; (*c*) Danger of engine failure.

Q. 27. *How does a radial ply tyre contributes to saving of fuel?*

Ans. A ***radial ply tyre*** provides better fuel economy because it has less rolling resistance and hence less engine power is required to roll the tyre.

Q. 28. *What is the function of 'universal joint (Hooker's joint)'?*

Ans. The function of ***universal joint*** is to adjust 'angular motion'.

Q. 29. *What is the function of a 'slip joint'?*

Ans. A ***slip joint*** is provided to adjust for the change in length of the propeller shaft.

Q. 30. *What is the function of an 'air brake valve'?*

Ans. An ***air brake value*** is the control valve which is operated by the brake pedal. It *controls the intensity of braking* in an air pressure system.

It is located between the reservoir and air lines leading to individual brake chambers.

Q. 31. *Why are all hydraulic brakes naturally "self-equalising"?*

Ans. In hydraulic braking system all the brake shoes are forced to contact the drum by the pressure initially built up, well before increase of the pressure in the system. Thus all the wheel brakes are forced to act simultaneously. So hydraulic brakes are 'self-equalising'.

Q. 32. *What is the 'cause of squealing of brake shoes'?*

Ans. "Squealing of brake shoes" is due to vibration in the drums or shoes, and due to lack of rigidity in their construction or mounting.

Q. 33. *What do you mean by 'servo brake'?*

Ans. ***Servo brake*** is one in which the pedal effort is assisted by mechanical or vacuum effect.

Q. 34. *What is the function of the spring in the Bendix drive?*

Ans. The spring is provided to relieve the sudden shock of start between the armature and the threaded sleeve, once the engine get started.

Q. 35. *What is meant by heat range of spark plug?*

Ans. The "heat range of a spark plug" is its *stability to transfer heat* from the firing end up through the insulator, gasket and shell to the cylinder head of water jacket.

Q. 36. *If an engine races, what may be the causes?*

Ans. An engine races due to the following *causes*:

(*i*) The governor not working properly.

(*ii*) The lost motion in the linkage between the governor and the fuel supply.

(*iii*) The valves not working properly.

(*iv*) The injection pump not working properly.

Q. 37. *What is an 'automobile'?*

Ans. An ***automobile*** is a *self-propelled vehicle* which is used for the transportation of passengers and goods on the ground.

Q. 38. *What is the purpose of supercharging?*

Ans. The purpose of "supercharging" is to increase the power which may be required at intervals and give more power where space does not permit larger engines.

Q. 39. *What is vapour lock?*

Ans. A ***vapour lock*** is the condition in the fuel system when fuel vaporizes, *causing* the presence of bubbles in the gasoline vapour *disturbing fuel delivery to the carburettor.*

Q. 40. *Name various methods/systems of engine cooling.*

Ans. Following are two methods/systems of cooling I.C. engines.

1. *Air cooling.*
2. *Water/liquid cooling* :

(*i*) Thermo-syphon cooling; (*ii*) Forced or pump cooling;

(*iii*) Thermostat cooling; (*iv*) Pressurised water cooling;

(*v*) Evaporative cooling.

Q. 41. *What is the function of a thermostat?*

Ans. A ***thermostat*** is a device for regulating the temperature of a system so that the system's temperature is maintained near a desired set point temperature. It does this by switching heating or cooling devices on or off, or regulating the flow of a heat transfer fluid as needed, to maintain the correct temperature.

A thermostat may be a control unit for a heating or cooling system or a component part of a heater or air-conditioner.

Q. 42. *What is the function of king-pin?*

Ans. The function of ***king-pin*** is to help the straightway ahead recovery of vehicle, which helps to provide directional stability also. When vehicle turns, king-pin causes body to move-up according to wheel's.

Q. 43. *What is ply rating of tyre?*

Ans. ***Ply rating*** is an *index of tyre strength.*

Q. 44. *What are beads and why are these provided in a tyre?*

Ans. *Steel wires fitted in circles are called* ***beads***. These keep the tyre fixed on the rim and also keep the tyre on the run in case of burst or puncture.

Q. 45. *What is 'wheel wobbling'?*

Ans. A rapid oscillation of steering wheel occurring at high speed is known as ***wheel wobbling***.

Q. 46. *What is 'zero caster'?*

Ans. The king-pin in a vertical position is known as ***zero caster***.

Q. 47. *What is a 'steering ratio'?*

Ans. ***Steering ratio*** is defined as the *number of turns the steering wheel has to make for one turn of pitman arm*. It is the ratio between the steering wheel angle and the corresponding angle in stub axle.

Q. 48. *Name the units of transmission system.*

Ans. The units of transmission system are :

(*i*) Clutch unit; (*ii*) Transmission unit;

(*iii*) Drive-line; (*iv*) Driving axle unit.

Q. 49. *What is a 'Transfer case'?*

Ans. ***Transfer case*** is a gear train coupled with the rear axle on one side and the front axle provided with differential and propeller shaft on the other side. This unit is meant to have *transmission to both the axles.*

Q. 50. *Where is the free wheel normally placed?*

Ans. The "free wheel" is placed usually on the rear end of the gear box; this makes the engine idle at controlled speed under overrun conditions, without the need for selecting neutral gear.

Q. 51. *What is a 'double cardon joint'?*

Ans. A ***double cardon*** joint is a type of constant velocity joint with two cross and yoke V-joints mounted back to back with a linking yoke and a centering ball.

Q. 52. *What is a full-floating rear axle?*

Ans. A *'full-floating axle'* is that axle in which whole load of rear portion of the vehicle is borne by the axle housing or tube only; the axle acts as a driving shaft.

Q. 53. *What is 'pitching' in the suspension system?*

Ans. ***Pitching*** is a rocking action about a transverse axis through the vehicle, parallel to the ground. The front suspension moves out of phase with the rear, experiencing the rocking effect due to pitching.

Q. 54. *What do you mean by power plant of automobile?*

Ans. An *"automobile power plant"* includes an engine, a transmission, a propeller shaft, etc.

Q. 55. *What is carburetion? Which factors influence the process of carburetion?*

Ans. The process of preparating in the S.I. engine a combustible fuel-air mixture outside the engine cylinder is called ***carburetion***. This complicated process is achieved in the induction system.

The various factors which influence the process of carburetion are :

(*i*) The engine speed.

(*ii*) The vaporisation characteristic of fuel.

(*iii*) The temperature of the incoming air.

(*iv*) The design of the carburettor.

Q. 56. *What is meant by a 'submerged jet' and why is it used?*

Ans. A ***submerged jet*** is one where the orifice is below the fuel level. It gives a steady uniform discharge as it is not affected by air speed, or difference in float chamber fuel level.

Q. 57. *Why are mufflers necessary?*

Ans. 'Mufflers' are attached to the exhaust manifold through a pipeline. This is to reduce the pressure of the exhaust gas and discharge it into the atmosphere with a minimum power loss, less noise, less vibration and transfer of heat.

Q. 58. *What is the function of Torsion bar' in suspension system?*

Ans. ***Torsion bar*** is used to *take shear stress* during suspension. It is simply a rod acting in torsion, made up of heat treated alloy spring steel.

Q. 59. *What is the function of fluid flywheel in automatic transmission?*

Ans. *Fluid flywheel* in automatic transmission is *used to transmit the same torque as given to it by the engine shaft.*

Q. 60. *What do you mean by 'oversteer' and 'understeer'?*

Ans. *Oversteer.* In this case slip angles of the front wheels are *less* than those of rear wheels, which causes radius to be *decreased* while turning. Here, vehicle is turned more sharply than rotation given to steering wheel.

Understeer. When the slip angles of front wheels are *more* than those of rear wheels, radius of turn *increases*; this position is called 'understeer'.

Q. 61. *What do you mean by bleeding of brakes?*

Ans. *Bleeding* means to drive out air from braking system. Because if air enters into braking system there is no actuation of brakes.

Q. 62. *What is non-slip differential?*

Ans. A ***non-slip differential*** is a type of differential gear arrangement that allows for some difference in angular velocity of the output shafts, but imposes a mechanical bound on the disparity. In an automobile such differentials are sometimes used in place of a standard differential, where they convey certain dynamic advantages at the expense of greater complexity.

Non-slip (or self-locking) differential is very much similar in construction to conventional type except that it *has two sets of clutch plates in addition.* Further more, the ends of pinion shafts lay *loosely* in notches, in two halves of the differential cage.

Q. 63. *What are helper springs in suspension system?*

Ans. *Helper springs* are provided on many commercial vehicles in addition to the main leaf springs.

A '*helper spring*' is just like a semi-elliptical spring but *without eyes* at the ends. Its ends touch the buckets fitted on the frame when the truck is heavily loaded. These springs *allow for wide range of loading*. When the vehicle is only lightly loaded, these helper springs *do not* come into operation. But as the load is increased, they take their share of load.

Helper springs are *generally used on 'rear suspension only'*.

Q. 64. *What are the different mechanisms of brake actuation?*

Ans. Following are the different mechanism of brake actuation.

(*i*) Mechanical brakes; (*ii*) Hydraulic brakes;

(*iii*) Vacuum brakes; (*iv*) Air brakes.

Q. 65. *What are oil additives?*

Ans. "Oil additives" are the oils which are added to lubricating oil to provide some extra properties other than the natural properties.

SECTION-B: "CONVENTIONAL QUESTIONS WITH ANSWERS/SOLUTIONS"

Q. 1. *Explain briefly the various methods to effect fuel economy.*

Ans. Following are *top ten economy methods / tips* to effect fuel economy :

1. **Clean the air filter :**

— A clean air filter is the *key* to good fuel economy. A dirty air filter *restricts* the flow of air into the engine, which harms performance and economy. For every litre of petrol used air filter cleans 10,000 litres of air.

— *Replacing a clogged air filter can improve car's economy by as much as 10 per cent.*

2. **Keep engine tuned :**

— A *fouled spark plug* or dirty fuel injector can *reduce fuel efficiency by 30 per cent* and rob car of performance.

— Ensure that spark plugs are checked at the manufacturer's recommended interval. Worn or incorrectly gapped spark plugs will increase the chances of *irregular firing*, resulting in uneven acceleration and a rough ride.

— Mileage can be improved by the use of recommended grade of motor oil.

3. **Maintain the right tyre pressure :** An under-inflated tyre increases rolling resistance, which cuts a vehicle's fuel efficiency. Buy a reliable type gauge, check your tyres and keep them properly inflated.

4. **Clear the clutter:** The more the weight that your car has to tow, the more fuel it needs to do more work. Remove heavy objects from your boot that you don't need immediately.

5. **Reduce air-conditioning :** Your car's air-conditioning drains your engine's power. When the air-conditioning is on, the engine works harder, increasing fuel consumption by about 20 per cent. The air-conditioner reduces fuel economy most during acceleration. Wait to turn on your air conditioner until you have reached cruising speed.

6. **Avoid wider tyres :** New wheels and tyres may look cool and they can certainly improve handling. But if they are wider than stock tyres, chances are they will create move rolling resistance and decrease fuel economy.

7. **Use the right octane :** Most new cars will run well on any grade of fuel, but many claim of older cars better economy from higher-octane fuels.

8. **Fuel up during the coolest hours :** Buy fuel during the early morning hours or evening when the air temperature is coolest. Fuel is densest during the cooler hours. Fuel pumps measure volumes of fuel and not the density. You will get more for your money during the cooler hours.

9. **Slow down :** The greatest enemy is aggressive acceleration and braking. *Minimise braking.* Driving aggressively and at high speed uses more fuel and causes more pollution. Aggressive driving style can increase fuel consumption by upto to 10 per cent.

Q. 2. *(a) What do you understand be the term 'engine performance'?*

(b) Name and show the various parts of an automobile with the help of a neat diagram.

Ans. (*a*) Relationship between power output, fuel or fluid consumption at *ambient conditions* in which an engine operates is called an **engine performance**.

(*b*) Every automobile consists of the following parts :

1. *Machine portion :*

(*i*) The "chassis and transmission".

(*ii*) The "engine".

(*iii*) The "electrical equipment".

2. *Carriage portion i.e., body.*

Fig. 1 shows the various parts of an automobile.

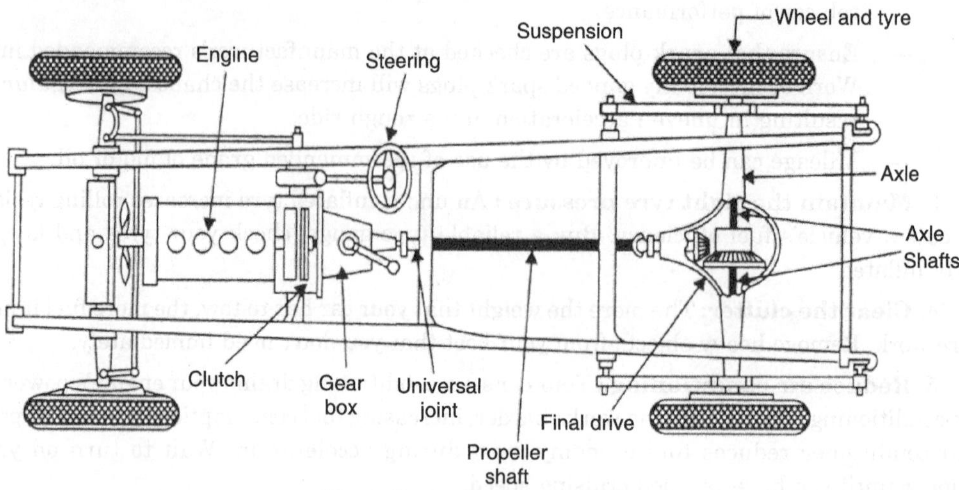

Fig. 1. Parts of an automobile.

Q. 3. *What is a 'Three-way catalytic converter'? Discuss.*

Ans. A ***catalytic converter*** is *a device used to convert toxic exhaust emission from an I.C. engine into non-toxic substances.*

Three-way catalytic converter: If an engine is operated at all times with an air fuel ratio close to stoichiometric, then both NO reduction and CO and HC oxidation can be done in a single catalyst bed. The catalyst effectively brings the exhaust gas composition to a near equilibrium state at these exhaust conditions, *i.e.*, a composition of CO_2, H_2O and N_2. Enough reducing gas will be present to reduce NO, and enough O_2 to oxidize the CO and hydrocarbons (HC). Such a converter is called ***three-way catalytic converter***, *since it removes all the three pollutants*. There is a narrow band of air-fuel ratios near stoichiometric in which high conversion efficiencies for all three pollutants are available. *Commercial three-way catalysts contain platinum, rhodium with some A_2O_3, NiO and CeO_2.* Alumina is the preferred support material.

Q. 4. *Explain the various types of modifications done in simple carburettor.*

Ans. For meeting the demand of the engine under all conditions of operation, the following additional devices/systems are *added* to the carburettor:

1. Main metering system.
2. Idling system.
3. Power enrichment or economiser system.
4. Acceleration pump system.
5. Choke.

1. **Main metering system :** The main metering system of a carburettor should be so designed as to supply a *nearly constant fuel-air ratio over a wide range of operation.* This F/A ratio is approximately equal to 0.064 (A/F ratio $\simeq$ 15.6) for best economy at full throttle. In order to correct the tendency of the simple carburettor to give progressively richer mixtures with load speed, the following *automatic compensating devices* are incorporated in the main metering system :

(*i*) Compensating jet device.

(*ii*) Emulsion tube or air bleeding device.

(*iii*) Back suction control or pressure reduction method.

(*iv*) Auxiliary valve carburettor.

(*v*) Auxiliary port carburettor.

2. **Idling system :** At idling and low load an engine requires a rich mixture having about air-fuel ratio 12 : 1. The main metering system not only fails to supply the enrich mixture at low air flows but also cannot supply any fuel during idling operation. It is due to the reason that a *separate idling jet must be incorporated in the basic carburettor.*

3. **Power enrichment or economiser system :** At the maximum power range of operation from 75% to 100% load, a device should be available to allow richer mixture (F/A about 0.08) to be supplied despite the compensating leanness. Meter rod *economiser* is such a device. It simply *provides a large orifice opening to the main jet* when the throttle is opened beyond specified limit. The rod may be tapered or stepped.

4. **Accelerating pump system :** Acceleration is a transient phenomenon. In order to accelerate the engine rapidly, a very rich mixture is required which a simple carburettor may not be able to supply. Rapid opening of throttle will be immediately followed by an increased air flow, but the inertia of liquid fuel (gasoline) will give momentarily lean mixture. Thus acceleration mixture required may not be met with in practice. To overcome this difficult solution an *accelerating pump* is incorporated.

5. **Choke :** Starting of a vehicle which is kept stationary for a long period (may be overnight) during cool winter seasons is often more difficult. At low cranking speeds and intake temperatures a very rich mixture is required to initiate combustion. Sometimes as high as *five to ten times* more fuel (than usual mixtures) is required. The main reason is that very large fraction of fuel may remain liquid suspended in air even in the cylinder, and only the vapour fraction can provide a combustible mixture with air. The most popular method of providing such mixture is by the use of *choke*. A choke is simply butterfly valve located between the extrance to the carburettor and the venturi throat.

Q. 5. *(a) What are the drawbacks of modern carburettors?*

(b) What are the merits of 'petrol injection system'?

Ans. (*a*) Drawbacks of modern carburettors:

The modern carburettors have the following *drawbacks* :

1. The mixture supplied to various cylinders of a multi-cylinder engine *varies in quality and quantity.* Also, due to fuel condensation in induction manifold, the *mixture proportion is affected.*

2. Due to presence of several wearing parts, the carburettors operate at a lower efficiency.

3. Reduced volumetric efficiency due to non-availability of a free flow passage for the mixture owing to the presence of choke tubes, throttle valves, jets, bends, etc.

4. At low temperatures, freezing can occur (if special means to obviate this are not provided).

5. When the carburettor is tilted or during acrobatics in aircraft surging can occur (if means to avoid this are not provided).

6. In the absence of flame traps, backfiring may occur which may lead to ignition of fuel outside the carburettor.

(*b*) Merits of petrol injection :

Following are the *merits of petrol injection system :*

1. In petrol injection system, *due to absence of venturi there is the minimum of air restriction so that higher engine volumetric efficiencies can be obtained with the corresponding improvement in power and torque.*

2. *The spots* for pre-heating the cold air and fuel mixture are *eliminated* so that *denser air enters the cylinder* when the engine has reached normal operating conditions.

3. As the manifold branch pipes are not greatly concerned with mixture preparation they *can be designed to utilize the inertia of the air charge to increase the engine's volumetric efficiency* (this does not apply for single point injection).

4. Because of *direct spray discharge* into each inlet port, *acceleration response is better.*

5. *Atomisation of fuel droplets is generally improved* over normal speed and load driving conditions.

6. It is possible to *use greater inlet and exhaust valve overlap without poor idling, loss of fuel or increased exhaust pollution.*

7. The monitoring of engine operating parameters enables accurate matching of air and fuel requirements under normal speed and load conditions which *improves engine performance, fuel consumption and reduces exhaust pollution.*

8. Fuel injection equipment is *precise in metering injected fuel spray* into the intake ports over the complete engine speed, load and temperature operating range.

9. There is *precise fuel distribution* between engine cylinders even under full-load conditions *with multi-point injection.*

10. *Multi-point injection does not require time for fuel transportation in the intake manifold and there is no manifold wall melting.*

Q. 6. *Discuss in detail the working of 'Electronic fuel injection system'.*

Ans. Electronic fuel injection system : Fig. 2 shows the fuel injection system-L-Jetronic with air flow metering (developed by Robert Based Corp.) It consist of the following *units* :

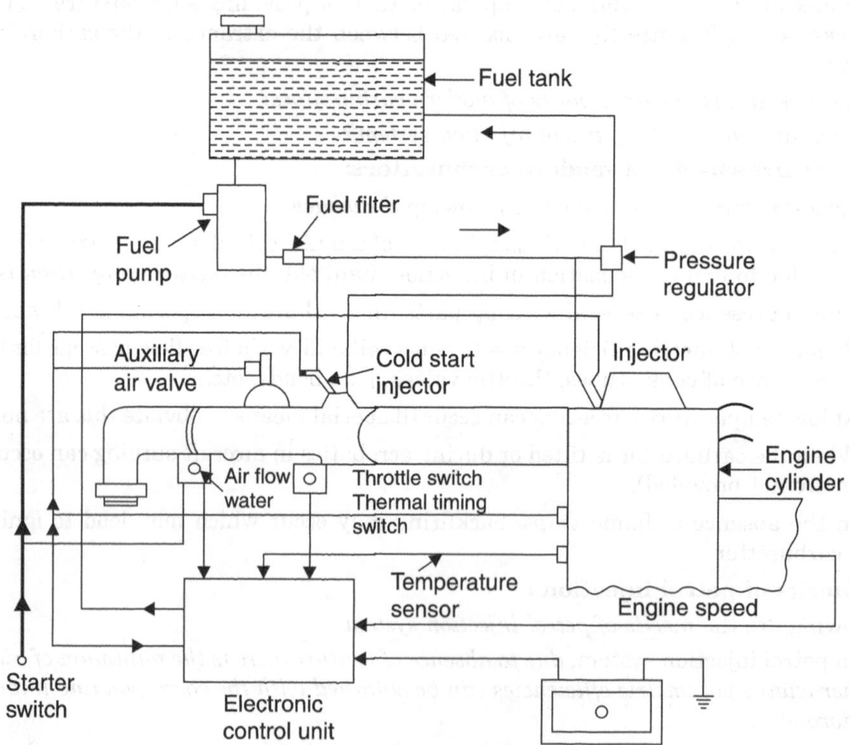

Fig. 2. Fuel injection system-L-Jetronic with air-flow metering.

1. Fuel delivery system; 2. An induction system;
3. Sensors and air flow control system; 4. Electronic control unit.

1. **Fuel delivery system :**

- It consists of an electrically driven fuel pump which draws fuel from a *fuel tank.* The pump forces the oil through a *filter* into a line at the end of which is situated a *pressure regulator*, which in turn is connected to intake manifold.
- The pressure regulator keeps the pressure difference between the fuel pressure and the manifold pressure constant, so that the quantity of fuel injected is dependent on the injector open time only.

2. **Air induction system :**

- After passing the air filter, the incoming air flows through an air flow meter, which generates a voltage signal (depending on the quantity of air flow).
- Just behind the *throttle valve* is fitted a cold start *magnetic injection valve,* which injects additional fuel for cold start. This valve also supplies the extra fuel needed during warm-up period.
- An *auxiliary valve* (which by-passes the throttle valve) supplies the extra air required for idling (in addition to rich-air-fuel mixture). This extra air increases the engine speed after cold start to acceptable idling speed.
- To the throttle valve is attached a *throttle switch* equipped with a set of contacts which generate a sequence of voltage signals during the opening of throttle valve. The voltage *signals result in injection of additional fuel required for acceleration.*

3. **Electric control unit :**

- The *sensors* are incorporated to measure the operating data at different locations. The data measured by the sensors are transmitted to the *electronic control unit which computes the amount of fuel injected during each engine cycle. The amount of fuel injected is varied by varying the injector opening time only.*
- The *sensors* used are :
  - Manifold pressure ;
  - Engine speed ;
  - Temperature at the intake manifold.

4. **Injection time :**

- For every revolution of the camshaft, the fuel is injected twice, each injection contributing half of a fuel quantity required for engine cycle.
- The injectors, at different phases of the operating cycle, are operating simultaneously.

Q. 7. *A 4-cylinder 4-stroke engine having diameter and length of stroke as 80 mm and 100 mm respectively is running at 1500 r.p.m. Its carburettor venturi has a 25 mm throat. Assuming coefficient of air flow 0.8, density of air 1.2 kg/m³ and volumetric efficiency of the engine as 78%, determine suction at the throat.*

Ans. *Given* : D = 80 mm or 0.08 m ; L = 100 mm or 0.1 m ; N = 1500 r.p.m ; Throat diameter, d_t = 25 mm = 0.025 m ; Coefficient of air flow, C_{da} = 0.8 ; Air density, ρ_a = 1.2 kg/m^3 ; $\eta_{vol.}$ = 78%.

Suction at the throat, Δp_a :

$$\text{Suction volume} = \frac{\pi}{4} \times D^2 \times L \times 4 = \frac{\pi}{4} \times (0.08)^2 \times 0.1 \times 4 = 0.00201 \text{ m}^3$$

$$\text{Actual volume per stroke} = \eta_{vol.} \times 0.00201 = 0.78 \times 0.00201 = 0.00157 \text{ m}^3$$

Actual volume sucked per second

$$= 0.00157 \times \frac{1500}{2} \times \frac{1}{60} = 0.01962 \text{ m}^3/\text{s}$$

Mass of air sucked, $m_a = 0.01962 \times 1.2 = 0.02354$ kg/s

As the initial temperature and pressure are not given, the problem is solved by approximate method *i.e.,* neglecting compressibility of the air.

$$M_a = C_{da} \times A_t \times \sqrt{2\rho_a \, \Delta p_a}$$

$$0.02354 = 0.8 \times \left(\frac{\pi}{4} \times 0.025^2\right) \times \sqrt{2 \times 1.2 \times \Delta p_a}$$

$$= 6.084 \times 10^{-4} \sqrt{\Delta p_a}$$

$$\therefore \quad \Delta p_a = 1497 \text{ N/m}^2 \text{ or } \mathbf{0.01497 \text{ bar}} \quad \textbf{(Ans.)}$$

Q. 8. *Name the various types of modern fuel injection system? Explain with a neat sketch the working of individual pump injection system.*

Ans. The various *types of modern fuel injection system* are :

1. Common-rail injection system.
2. Individual pump injection system.
3. Distributor system.

Individual pump injection system :

- Refer to Fig. 3. In this system an individual pump or pump cylinder connects directly to each fuel nozzle. Pump meters charge and controls injection timing. Nozzles contain a delivery valve actuated by the fuel-oil pressure.

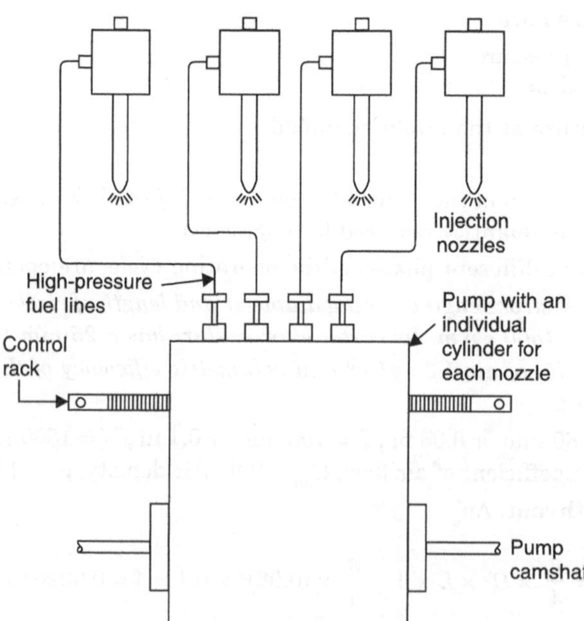

Fig. 3. Individual pump injection system.

- The design of this type of pump must be very accurate and precise as the volume of fuel injected per cycle is 1/20,000 of the engine displacement at full load and 1/100,000 of the engine displacement during idling. The time allowed for injecting such a small quantity of fuel is very limited (about 1/450 second at 1500 r.p.m. of the engine providing injection through 20° crank angle). The pressure requirements vary from 100 to 300 bar.

Q. 9. *With the helps of a neat sketch explain A.C. mechanical fuel pump.*

Ans. A.C. mechanical fuel pump : Refer to Fig. 4. These diaphragm pumps have now become standardized fitting on all vehicles.

The pump is usually bolted to the crankcase of the engine and operated directly by an ecentric on the crankshaft, or by push rod. As the crankshaft rotates, the eccentric lifts the rocker-arm which pulls the connecting rod, together with the diaphragm, downward against the pressure of the return spring, thus creating a vacuum in the pump chamber. Fuel from the tank is then sucked through the inlet connection, into the sediment chamber, through the gauge filter and inlet valve into the pump chamber. On the return stroke the pressure of the return spring pushes the diaphragm upward forcing fuel from the pump chamber through the outlet or delivery valve and outlet connection to the carburettor.

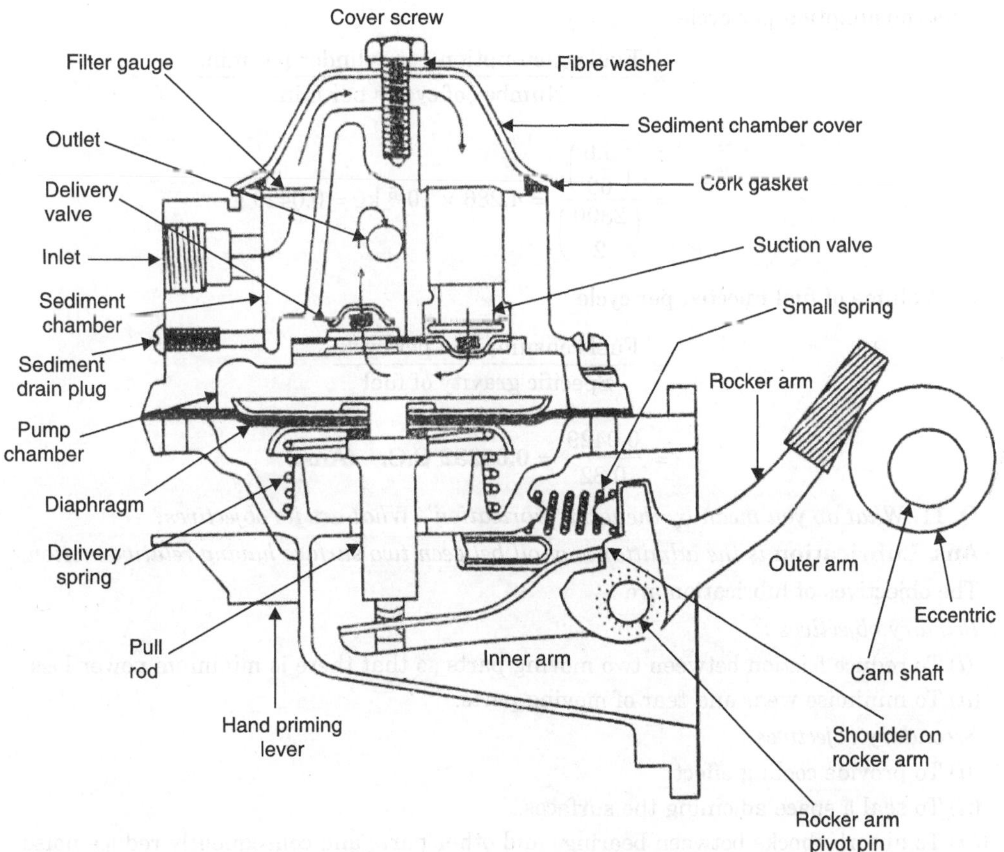

Fig. 4. A.C. mechanical fuel pump.

When the correct level to the carburettor float chamber is reached, the needle valve will close, thus creating a back pressure in the pump chamber. This pressure will hold the diaphragm downward against the return spring and it will remain in this position until the carburettor requires more fuel and the float chamber needle valve opens.

The small spring at the rocker arm shoulder is intended to keep the rocker arm in constant contact with the eccentric, to eliminate noise.

Q. 10. *A six-cylinder, four-stroke diesel engine develops 120 kW at 2800 r.p.m. Its brake specific fuel consumption is 180 g/kWh. Calculate the volume of fuel to be injected per cycle per cylinder. Specific gravity of the fuel may be taken as 0.82.*

Ans. *Given* : Number of cylinders, $n = 6$; Brake power, B.P. = 120 kW; Speed, $N = 2800$ r.p.m; Brake specific fuel consumption, *b.s.f.c.* = 180 g/kWh; Sp.gr. of fuel = 0.82.

Fuel consumption per hour = *b.s.f.c* × B.P.

$$= \frac{180}{1000} \times 120 = 21.6 \text{ kg}$$

$$\text{Fuel consumption per cylinder} = \frac{21.6}{n} = \frac{21.6}{6} = 3.6 \text{ kg/h}$$

Fuel consumption per cycle

$$= \frac{\text{Fuel consumption per cylinder per min.}}{\text{Number of cycles per min.}}$$

$$= \frac{\left(\frac{3.6}{60}\right)}{\left(\frac{2800}{2}\right)} = 4.286 \times 10^{-5} \text{ kg} = 0.0429 \text{ g}$$

∴ Volume of fuel injected per cycle

$$= \frac{\text{Fuel consumption per cycle}}{\text{Specific gravity of fuel}}$$

$$= \frac{0.0429}{0.82} = \mathbf{0.05232 \text{ C.C.}} \quad \textbf{(Ans.)}$$

Q. 11. *What do you mean by the term 'lubrication'? What are its objectives?*

Ans. Lubrication *is the admittance of oil between two surface having relative motion.*

The objectives of lubrication are :

Primary objectives :

(*i*) To reduce friction between two moving parts so that there is minimum power loss.

(*ii*) To minimise wear and tear of moving parts.

Secondary objectives :

(*i*) To provide cooling effect.

(*ii*) To seal a space adjoining the surfaces.

(*iii*) To absorb shocks between bearings and other parts and consequently reduce noise.

(*iv*) To remove dirt and grid that have crept between the rubbing parts.

Q. 12. *Explain, with the help of a sketch, the full pressure system of lubrication used in automobiles.*

Ans. Full pressure system of lubrication : Refer to Fig. 5.

In this system, oil from oil sump is pumped under pressure to the various parts requiring lubrication. The oil is drawn from the sump through filter and pumped by means of a gear pump. Oil is delivered by the pressure pump at pressure ranging from 1.5 to 4 bar. The oil under pressure is supplied to main bearings of crankshaft and camshaft. Holes drilled through the main crankshafts bearing journals, communicate oil to the big end bearing and also small end bearings through hole drilled in connecting rods. A pressure gauge is provided to confirm the circulation of oil to the various parts. A pressure regulating valve is also provided on the delivery side of this pump to prevent excessive pressure.

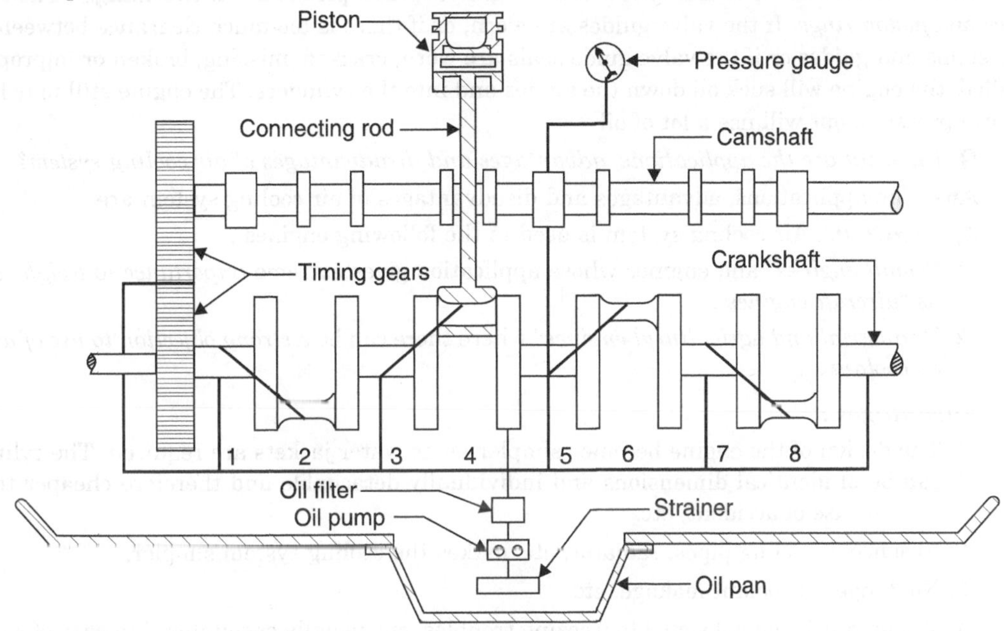

Fig. 5. Full pressure system.

- This system *finds favour from most of the engine manufacturers as it allows high bearing pressure and rubbing speeds.*

Q. 13. *What is 'crankcase ventilation' and how it is achieved in automobile engine?*

Ans. Crankcase Ventilation :

Crankcase ventilation is required owing to the following *two reasons* :

(*i*) The various contaminants such as water, gasoline, blowby gases, etc. enter the crankcase due to several reasons, and may cause sludge and corrode metal parts.

(*ii*) To relieve any pressure build-up in the crankcase which may cause leakage of the crankshaft seal.

In practice, following two types of ventilation systems are used :

1. Open system; 2. Closed system.

1. **Open system :**

- In this system, fresh air supply is inducted into the crankcase during the compression stroke (due to creation of small vacuum). The entering air picks up the contaminants (water vapour, gases and H_2SO_4 vapour) and discharges them to the atmosphere during expansion stroke.

- The main *disadvantage* of this system is that the *natural ventilation is quite inadequate during idling or running at low speeds.*

2. **Closed system :**

- In closed system the *fresh air supply is taken to the crankcase from the carburettor.*
- Air cleaner and the breather outlets are connected to the intake manifold through a PCV valve to ensure the burning of all the crankcase gases in combustion chamber.

Q. 14. *Discuss the factors on which the consumption of lubircating oil in an engine depends.*

Ans. Lubricating oil consumption in an engine depends primarily on two things : The *valve guides* and *piston rings.* If the valve guides are worn, or if there is too much clearance between the valve stems and guides or if the valve guide seals are worn, cracked, missing, broken or improperly installed, the engine will suck oil down the guides and into the cylinders. The engine still may have good compression, but will use a lot of oil.

Q. 15. *What are the applications, advantages and disadvantages of air cooling system?*

Ans. The applications, advantages and disadvantages of air cooling system are :

Applications. Air cooling system is used in the following engines :

1. "*Small engines*" and engines whose application gives extreme *importance to weight* such as "*aircraft engines*".
2. "*Industrial and agricultural engines*" where there can be a *strong objection to use of water as coolant.*

Advantages :

1. The design of the engine becomes simpler as no water jackets are required. The cylinder can be of identical dimensions and individually detachable and therefore cheaper to renew in case of accident, etc.
2. Absence of cooling pipes, radiator, etc. makes the cooling system simpler.
3. No danger of coolant leakage, etc.
4. The engine is not subjected to freezing troubles, etc. usually encountered in case of water-cooled engine.
5. The weight per B.P. of the air-cooled engine is less than that of water-cooled engine.
6. In this case engine is rather a self contained unit as it requires no external components *e.g.*, radiator, headers, tank, etc.
7. Installation of air-cooled engines is easier.
8. The control of cooling system is much easier than in water-cooled system.
9. An air-cooled engine can take up some degree of damage. A broken fin does not affect much while a hole in the radiator may stop a water-cooled engine.
10. High mean cylinder temperatures mean reduced carbon deposits on combustion chamber wall. This gives sustained engine performance.
11. The warm-up performance of air-cooled engine is better, this results in low wear of cylinders.

Disadvantages :

1. Their movement is noisy.
2. Non-uniform cooling.

3. The output of air-cooled engine is less than that of a liquid-cooled engine.
4. Maintenance is not easy.
5. Smaller useful compression ratio.
6. The volumetric efficiency of an air-cooled engine is lower due to high cylinder head temperatures.

Q. 16. *Explain with a heat sketch a "thermostatically controlled cooling system".*

Ans. Thermostatically controlled cooling system :

Too lower cylinder barrel temperature, may result in severe corrosion damage due to condensation of acids on the barrel wall. To avoid such a situation it is customary to use a thermostat (a temperature controlling device) to stop flow of coolant below a pre-set cylinder barrel temperature. Most modern cooling system employ a thermo-static device which prevents the water in the engine jacket from circulating through the radiator for cooling until its temperature has reached to a value suitable for efficient engine operation.

- Fig. 6 shows a systematic diagram of a thermostatically controlled cooling system. Also shown is a typical thermostat (Fig. 7). It consists of bellows which are made of thin copper tubes, partially filled with a volatile liquid like ether or methyl alcohol. The volatile liquid changes into vapour at the correct working temperature, thus creating enough pressure to expand the bellows. The temperature at which the thermostat operates is set by the manufacturers and cannot be altered. The movement of the bellows opens the main valve in the ratio of temperature rise, increasing or restricting the flow of water from engine to the radiator. Hence when the normal temperature of the engine has been reached the valve opens and circulation of water commences. When the unit is closed the gas condenses and so the pressure falls. The bellows collapse and the thermostat seats on its seat and circulation around thermostat stops. When the thermostat valve is not

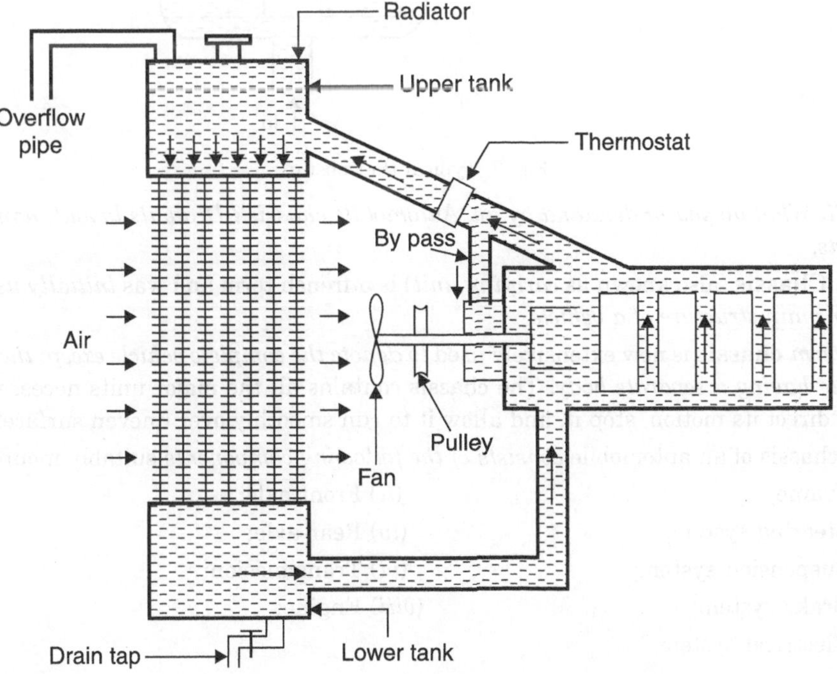

Fig. 6. Thermostatically controlled cooling system.

open and the engine is running the water being pumped rises in pressure and causes the pressure relief valve to open. Thus the water completes its circulation through the by-pass as shown in Figs. 6 and 7. Now when the temperature of water around the engine-cylinder rises upto a certain limit, it causes the thermostat valve to open. The pressure of water being pumped falls and pressure relief valve closes. So the flow of cooling water in the normal circuit commences through the radiator. This accelerates the rise of temperature of the cylinder walls and water and more power is developed in a few moments of the starting of the engine.

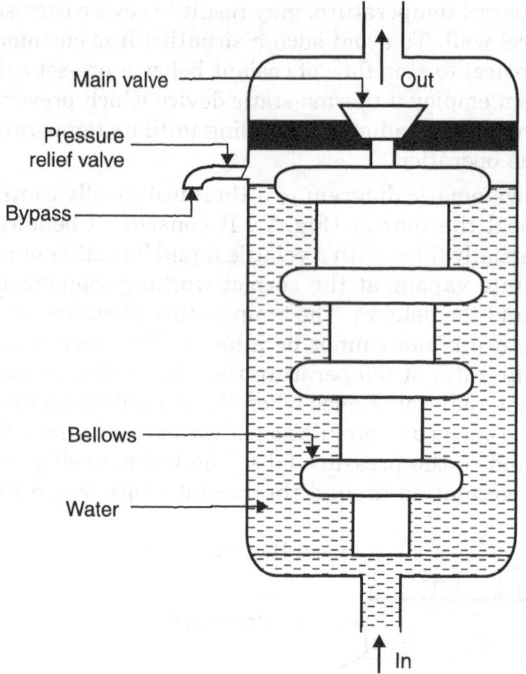

Fig. 7. Typical thermostat.

Q. 17. *What do you understand by an 'Automobile chassis'? Draw its layout, name and show various parts.*

Ans. Chassis (also known as *carrying unit*) is a french term and was *initially used to denote the frame or main structure of a vehicle.*

The term **chassis** is now extensively used *to denote the complete vehicle except the body for the heavy vehicle having a separate body.* The chassis contains all the major units necessary to propel the vehicle, direct its motion, stop it, and allow it to run smoothly over uneven surfaces.

The chassis of an automobile *consists of the following components* suitably mounted (Fig. 8) :

(*i*) Frame; (*ii*) Front axle;

(*iii*) Steering system; (*iv*) Rear-axle;

(*v*) Suspension system; (*vi*) Transmission;

(*vii*) Brake system; (*viii*) Engine;

(*ix*) Electrical system.

The chassis is sub-divided into :

(*i*) Power plant; (*ii*) Running gear.

The ***power plant*** includes the *engine assembly and power transmission assembly.*

The ***running gear*** includes the *frame, steering system, suspension system, brakes, wheels and tyres.*

— The *electrical system is part of both chassis and body.*

All the above mentioned components are mounted in either of the following two ways :

1. ***Conventional construction.*** In this case a *separate frame* is used.

2. ***Frameless or unitary construction.*** Here *no separate frame* is employed.

Out of these, the conventional construction is being used presently only for heavy vehicles whereas for cars the same has been replaced by the frameless type except of course by small manufactures, who still find it economical to use frame.

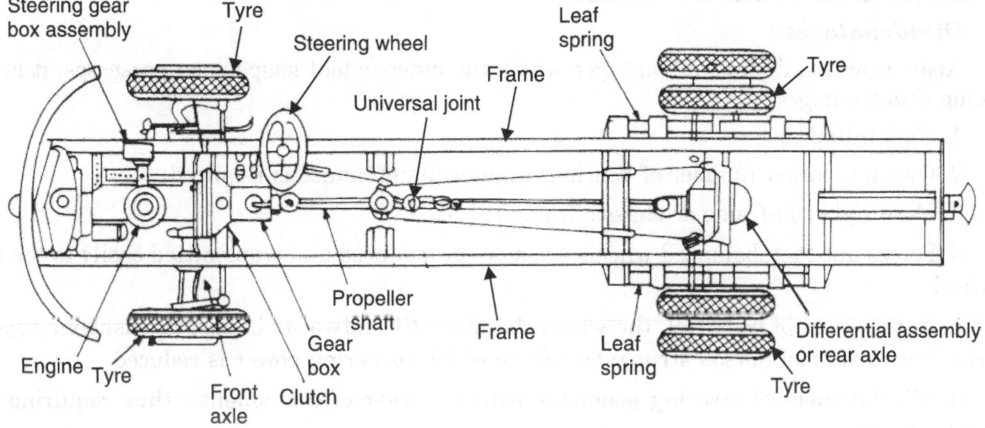

Fig. 8. Chassis (Tata truck).

Q. 18. *What do you understand by independent suspension system in a vehicle? What are its advantages and disadvantages?*

Ans. Independent suspension *is a term used to describe any arrangement by which the wheels are connected to the carriage unit in a manner such that the rise and fall of one wheel has no effect on the others.* Almost all the passenger cars now use the independent front suspension, in which the coil spring arrangement is the most common.

When a vehicle with rigid axle suspension encounters road irregularities, the axle tilts and the wheels no longer remain vertical. This causes the whole of the vehicle to tilt to one side. Such a state of affair is *not desirable.* Besides causing *rough ride*, it causes '*wheel wobble*'. The road adhesion is also decreased. In order to avoid this the *wheels are sprung independent of each other, so that tilting of one does not affect the other.*

Advantages :

The independent suspension claims the following *advantages* over the rigid axle type suspension.

1. In independent system since the wheels more or less travel with their planes perpendicular to the road surface, the *gyroscopic effects are reduced to a minimum.*

2. The engine and chassis frame can be placed relatively lower which means engine position can be moved forward resulting in more space for passengers.

3. Provides a greater degree of vertical/springing movement.

4. Diminished wheel wobble and steering tramp.

5. Provides scope for use of springs of greater resilience giving much better springing action than most rigid axle vehicles.

6. Reduced unsprung weight and hence improved ride and better road holding while cornering and braking.

7. The frame and body do not tilt but remain horizontal and the wheels vertical when the wheel encounters a road bump.

8. Variations in caster angle are reduced.

9. It uses coil springs which can be placed closer to the wheel. This is definite advantage *vis-a-vis* leaf springs for a wheel to be steered.

Disadvantages :

Apart from the distinct advantages which the independent suspension possesses, it has the following *disadvantages :*

1. High initial cost.

2. Owing to larger number of bearings greater maintenance is required.

3. More rigid sub-frame or chassis frame required.

4. Forces due to unbalanced wheels are more pronounced and transmitted easily to the steering wheel.

5. In the event of body roll, the wheel chambers tilt outwards in case of wishbone type and inwards in case of MacPherson strut type, due to which cornering power is reduced.

6. Misalignment of steering geometry with the wear of components, thus requiring more frequent attention.

- In a car, the *"font axle"* is usually a *dead axle* (The dead axles do not contain differential and have no concern with the power transmission system of the automobile) although some cars do have front wheel drive. Therefore, *the independent suspension systems have been almost universally adopted* for the *front wheels*. The independent suspension system for the front wheels has to cope with the fact that they are to be *steered.*

"Rear axle" is usually a *live axle* (the rear axles are those axles which contain differential and through which rotary motion is transmitted to the wheels) with power being transmitted to the rear wheels. Thus independent suspension has not become popular for the rear wheels. Further rear wheels have to carry a lot of weight and while the weight on the front wheel remains more or less constant, in the case of rear wheels it makes a lot of difference when the car is running empty or when it is fully occupied. The suspension system has to cater to both these conditions.

Q. 19. *With the help of neat diagram discuss the construction of MacPherson strut type suspension system for front wheels.*

Ans. MacPherson strut assembly :

It is a single wishbone with a telescopic strut type system as shown in Fig. 9. The MacPherson system consists of a *telescopic strut, a single arm* and a *diagonal stay*. The strut is fixed to the body structure at the upper end through a flexible mounting and the lower part of the strut is connected at the bottom by a joint to the lower arm. The lower part of the strut also carries the stub axle, which in term carries the wheel. The steering motion is supplied to the lower part of the strut and it turns the whole strut. A coil and a hydraulic damper/shock absorber surround the upper part of the strut which takes care of the road irregularity shocks and vibrations.

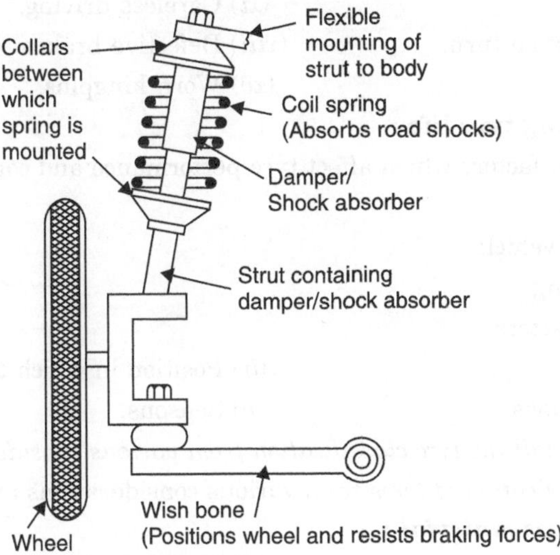

Fig. 9. MacPherson strut assembly.

Advantages of MacPherson system :

(*i*) Very easy maintenance.

(*ii*) Simple in mechanical construction.

(*iii*) Less variations in wheel camber.

(*iv*) Its light moving parts help the wheels to follow the road irregularities.

(*v*) Distinct advantages in case of transverse engines, since in that case there is no space or very little space for upper links to fit.

Disadvantages :

(*i*) During cornering and brake torque the radial loading comes on the piston due to the lateral forces.

(*ii*) In order to absorb the full suspension loads the body structure has to be really strong above the wheel arches where the struts are attached.

Maruti-800 suspension system :

Maruti-800 car incorporates the MacPherson type independent suspension system. It consists of *coil springs, front suspension struts, steering knuckles, a stabiliser bar and front suspension arms*. In this type of construction, shocks applied to wheels are distributed through steering knuckles from front suspension struts, coil springs, to front suspension arms and are absorbed.

Q. 20. *(a) What are the causes of tyre wear?*

(b) List the main factors which affect tyre life.

Ans. (*a*) Causes of tyre wear :

Following are the *causes of tyre wear :*

(*i*) Incorrect inflation. (*ii*) Unequal tyres.

(*iii*) Bleeding of air in tyre. (*iv*) Incorrect rotation of tyres.

(*v*) Misalignment. (*vi*) Out of balance wheel.
(*vii*) Worn steering mechanism. (*viii*) Incorrect castor, camber, or toe in.
(*ix*) Excessive braking or violet acceleration.
(*x*) Wrong loading. (*xi*) Careless driving.
(*xii*) Toe-out incorrect on turn. (*xiii*) Defective brakes.
(*xiv*) Overloading (*xv*) Worn kingpins.

(*b*) Factors affecting tyre life :

The following are the main factors which affect tyre performance and consequently their life.

1. Inflation.
2. Maintenance of vehicle.
3. Manner of driving.
4. Miscellaneous factors :
   (*i*) Heat (*ii*) Position in which tyre in fitted
   (*iii*) Road conditions (*iv*) Seasons.

Q. 21. *Discuss in detail the tyre classification from various considerations.*

Ans. Detailed *classification of tyres* from various considerations are :

1. ***Based on carcass material :***
(*i*) Nylon. (*ii*) Steel belted.
(*iii*) All steel. (*iv*) Polyster belted.
(*v*) Rayone.

2. ***Based on construction :***
(*i*) Radial } — Tube type.
(*ii*) Bias } — Tubeless.

3. ***Based on application :***
(*i*) Truck/bus. (*ii*) Light truck.
(*iii*) Passenger car. (*iv*) 2/3 wheelers.
(*v*) Agriculture

4. ***Based on pattern design :***
(*i*) Lug. (*ii*) Rib.
(*iii*) Semilug.

Q. 22. *Explain the working of single plate clutch with the help of suitable diagram.*

Ans. Single plate clutch :

Fig. 10 shows a simplified schematic diagram of a single plate clutch which is commonly *used in most cars and small commercial vehicles.*

Construction. The *flywheel* is rigidly fixed to the engine shaft, the driven shaft to the gear box being supported at the engine end by a spigot bearing in the flywheel. A *clutch plate* is attached to a splined hub which is free to slide axially on the splines cut on the driven shaft. A *ring of friction lining* is attached to each side of the clutch plate. One end of each of a number of *helical compression springs* bear on the back of a pressure plate, the other ends of the springs pressing against a *cover, attached to the flywheel and rotating with it.* Three release levers on pivots mounted on the cover, bear on the bottoms of the nuts on the studs in the pressure plate and are actuated by the leftward movement of the thrust race which in turn is moved by the *clutch pedal mechanism.*

Working. The diagram shows the pressure plate pulled back by the release levers against the compression springs, so that the friction linings on the clutch plate are free of flywheel and pressure plate. The flywheel then rotates without driving the clutch plate and hence the driven shaft.

When the pressure of the thrust race is released the compression springs are free to move the pressure plate to the left so bringing it in contact with the clutch plate. The pressure plate continues to move to the left, sliding the clutch plate, on its splined hub, along the driven shaft until the friction lining touches the flywheel. The compression springs now cause the linings to be gripped between the pressure plate and the flywheel and the friction between the linings and flywheel and pressure plate causes the clutch plate to revolve, so turning the driven shaft.

- *Clutch slip occurs if the resisting torque on the driven shaft exceeds the friction torque at the clutch.*

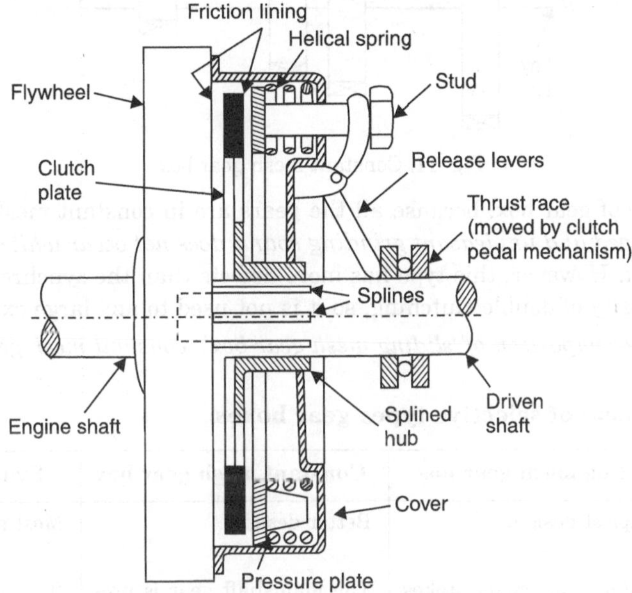

Fig. 10. Single plate clutch (Disengaged position).

Q. 23. *Explain with the help of neat-sketch 'Constant mesh gear box'.*

Ans. Constant mesh gear box :

Refer to Fig. 11. It is that gearbox in which all the gears are in constant mesh with each other (hence the name constant mesh gear box) all the time and this gives a silent or *quiet operation*. Here, helical gears are used to make gear changing easier. The gears on the main shaft which is splined, are *free*. The gears on the counter or layshaft are, however, *fixed*. Two dog clutches are provided on the main shaft-one between the clutch gear and the second gear and the other between the low/first gear and reverse gear. Dog clutch can slide on the main shaft and rotates with it.

When the left-hand dog clutch is made to slide to the left by means of the gearshift lever, it meshes with the clutch gear and the top speed gear is obtained. When the dog clutch meshes with the second gear the second speed gear is obtained. Similarly by sliding the right-hand dog clutch to the left and right, the first speed gear and reverse gear are obtained respectively. However skillful handing is necessary on the part of the driver so that the speed of the locking dogs and respective pinion remain the same to effect a clash-free gear change.

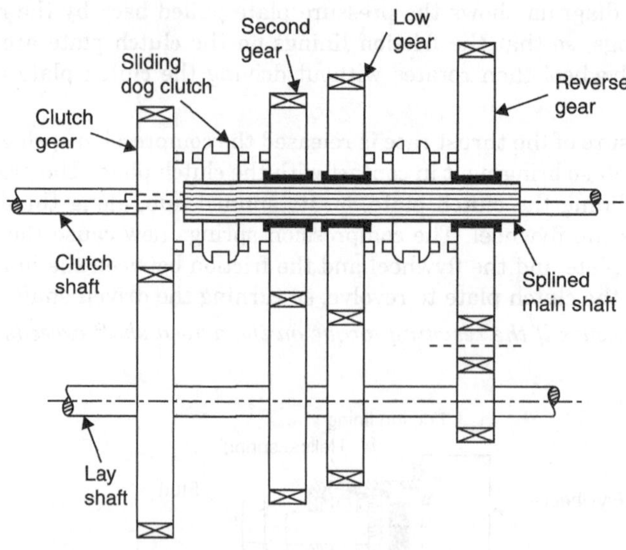

Fig. 11. Constant mesh gear box.

- In this type of gear box, because all the gears are in constant mesh, they are *safe from being damaged and unpleasant grinding sound does not occur while engaging and disengaging them.* However, this type has more defects than the synchromesh type and there is the necessity of double clutching, so it is not used to any large extent.

Q. 24. *Give the comparison of 'sliding mesh gear box', 'constant mesh gear box' and 'synchromesh gear box'.*

Ans. Comparison of selective types gear boxes:

Aspects	Sliding mesh gear box	Constant mesh gear box	Synchromesh gear box
1. *Type of design*	Simplest design.	Better design.	Most advanced design.
2. *Meshing of gears*	Meshing of gears takes place by sliding of gears on each other, *i.e.*, axially along the splined mainshaft to mesh with the corresponding gears on the layshaft.	The mainshaft gear is provided with dog clutches providing a means of locking the freely, rotating mainshaft gears to the main shaft for making an easier gear change.	It enables rapid gear changes to be made without noisy engagement or clashing of teeth. It also simplifies shifting action for the driver as the synchronizer itself adjusts the members to be coupled so as to provide perfect dog teeth engagement. Thus matching of the gears is automatically done by the synchronizer resulting in smooth and silent coupling. It reduces the wearing of gear.

3. *Skill needed*	Considerable skill is needed to make a gear change without damaging the gear teeth during meshing.	— There is considerable reduction of wear on the teeth of the reduction gears. — Skillful and timely hand and foot movements are required in order that the speeds of the locking dogs and respective pinion are equal for a clash-free gear change. The driver has to match the speed of the gears to be engaged by double declutching.	Quite satisfactory for high speed cars.

Q. 25. *Explain the principle of automatic transmission.*

Ans. Automatic transmission : *Transmission in which various speeds are obtained automatically is known as* ***automatic transmission.*** The driver merely selects the general car condition such as neutral, forward or reverse. The selection, timing and engagement of gear for the required gear speed are accomplished automatically when the accelerator is pressed or depressed.

Automatic transmission needs no gear change lever and clutch pedal since clutch and transmission is a combined unit and works automatically. (In "*semi-automatic transmission*" only the clutch is operated automatically). The transmission is operated hydraulically and is often known as "*hydramatic transmission*". This transmission employs *fluid coupling, torque converter and epicyclic gear arrangement.* When all the different devices are combined into one unit, they perform their duties collaborately.

Automatic transmissions *operate basically by controlling the vehicle speed and engine load. Increasing vehicle speed needs changing gears upwards, whereas increasing engine load necessitate change from higher to lower gears. The engine load sensitive control unit is operated by the depression of the accelerator pedal.*

- In ***hydramatic gear box***, the planetary gear sets are placed *in series* so that power may be transmitted through them. A *centrifugal governor incorporated in the transmission selects the proper gear for each speed and throttle position.* The change from one gear to another is accomplished through *hydraulically operated pistons assisted by springs* which control brake bands on the planetary gear sets and clutches within the planetary unit. *The speed at which the various shifts occur, is governed by the throttle and centrifugal governor.*

Driver control is provided by a lever and sector segment which are usually located on the steering column, below the steering wheel. The control could be adjusted to any of the following *positions* :

Neutral – Drive – Low – Reverse

— In the "*drive position*", the transmission shifts automatically for any one of the *forward speeds.*

— The "*low position*" is for *hill climbing* or muddy road or under heavy load, and only first and second gear speeds are provided.

— *"Reverse position"* is for *driving back*.

— In *"neutral position"*, there is no power transmission.

Q. 26. *Differentiate between torque tube and Hotchkiss type propeller shafts.*

Ans. The propeller shafts are subjected to driving and braking torques. When a front engine rear drive car starts from rest or is suddenly braked to slow down, the shock to the transmission is cushioned by the long propeller shaft which twists slightly and then untwists. These forces are taken care of by the following *two types of construction of propeller shaft* :

1. Torque tube type propeller shaft.
2. Hotchkiss type propeller shaft.

1. **Torque tube type propeller shaft :**

Fig. 12 shows a torque tube type propeller shaft.

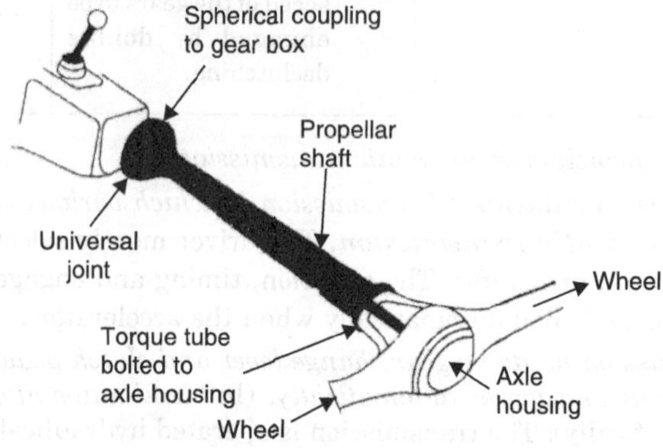

Fig. 12. Torque tube type propeller shaft.

- A torque tube is a large diameter tube fastened securely to the rear axle housing and *completely enclosing the propeller shaft*. The torque tube is fitted into a spherical ball and socket surrounding one universal joint at the transmission end. These two units carry their respective loads while allowing suspension flexibility. Thus, *while normal engine power transmission takes place through the propeller shaft, the braking and acceleration causing sudden torque are borne by the torque tube.*
- In modern vehicles, this type of construction has been replaced by the Hotchkiss drive.

2. **Hotchkiss type propeller shaft :**

In this type of propeller shaft, the driving and braking torques are *absorbed through the front half of the rear leaf springs or through the links and arms when used with coil type springs*. Thus while normal power is transmitted through the propeller shaft, acceleration and braking shocks are taken up by the leaf spring (by distortion) or by special links/frames in case of coil type springs.

- ***Hotchkiss drive** is mostly used in all the cars and trucks.* It is *cheaper* and also gives *satisfactory result.* In this case, an open propeller shaft with two universal joints and a slip joint is used as shown in Fig. 13. In this case, rear spring bracket take care of driving thrust. As the springs have to take care of driving thrust (and braking torques) as such they are flat *i.e.,* less camber is given. Over and above, being a cheap system, this system also does not strain the driving mechanism because of its flexible connection. In this system, the axle housing can rock slightly when sudden load is applied in letting the clutch in or when suddenly or quickly applying the brake.

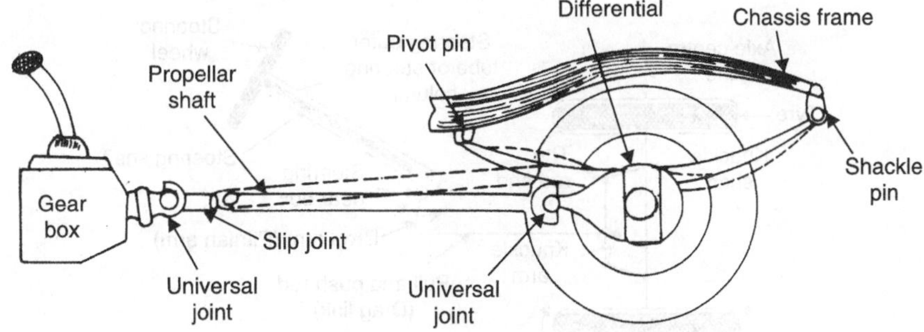

Fig. 13. Hotchkiss drive.

Q. 27. *What is final drive? Discuss in brief functions of various components of final drive in automobile transmission.*

Ans. Final drive : The ***final drive*** *is the last stage in transferring power from the engine to the road wheels.* It provides a *fixed reduction* between the drive shaft and the driving axles.

- It *brings down the speed* of the propeller shaft to that of the *wheels.*
- It *turns the drive* of the propeller shaft through a *right angle to drive the wheels.*

Fig. 14 shows a simplified diagram of the final drive system.

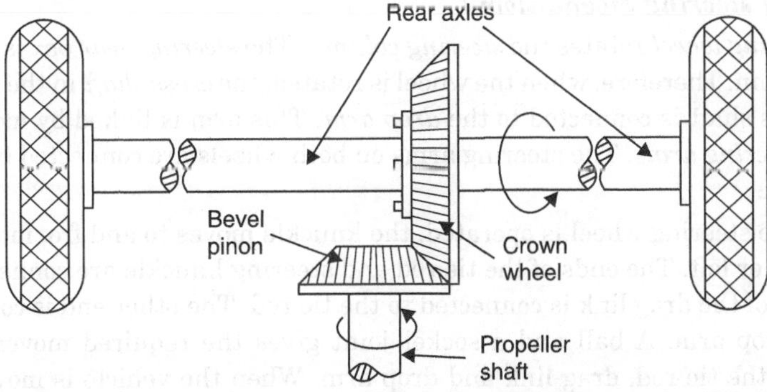

Fig. 14. Final drive system.

The propeller shaft has a small bevel pinion which is in mesh with the crown wheel. This crown wheel rotates the rear axles. The axles are in half portions.

— Since the pinion is bevel one, therefore, it rotates the crown wheel at right angles. The crown wheel in turn rotates the rear axles. Thus the rotation of the propeller shaft is such that it is converted at *right angles to the rear axle.*

— The crown wheel is larger than the bevel pinion, therefore, the speed of the crown wheel is *lower* than the speed of the bevel pinion.

Q. 28. *Draw a neat diagram of 'layout of steering system' and list the main parts of a steering system. Also discuss the working of steering mechanism.*

Ans. Layout of steering system :

Fig. 15 shows the layout of steering system.

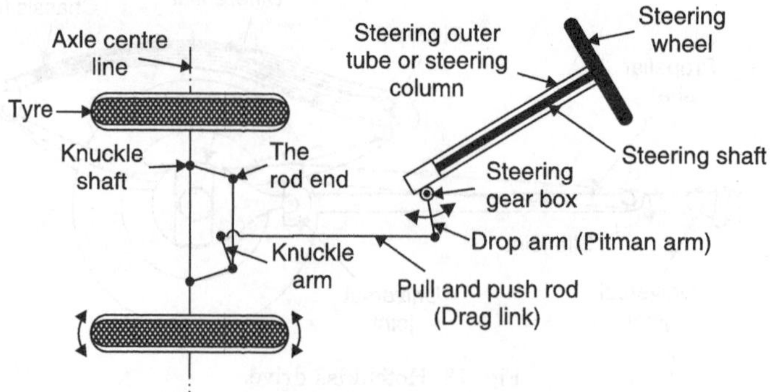

Fig. 15. Layout of steering system.

Main parts :

The *main parts of a steering system* are :

1. Steering wheel.
2. Steering column.
3. Steering shaft.
4. Steering gear box.
5. Steering drop arm (Pitman arm).
6. Pull and push rod (Drag link).
7. Knuckle arm.
8. Tie rod and tie rod end.

Working of steering mechanism :

- The *steering wheel* rotates the *steering column.* The *steering gear box* is fitted to the end of this column. Therefore, when the wheel is rotated, the *cross shaft* in the gear box oscillates. The cross shaft is connected to the *drop arm.* This arm is linked by means of a *drag link* to the *steering arms.* The steering arms on both wheels are connected by the *tie rod* to the drag link.
- When the steering wheel is operated, the knuckle moves to and fro, moving the wheels to the right or left. The ends of the tie rod and steering knuckle are connected to each other. One end of the drag link is connected to the tie rod. The other end is connected to the end of the drop arm. A ball and a socket joint gives the required movement to the joints between the tie rod, drag link and drop arm. When the vehicle is moving, the drop arm develops vibration. Shock springs are used in ball and socket system to absorb this vibration.

Q. 29. *Explain the construction and working of a differential used in automobile.*

Ans. Differential *is the mechanism by means of which outer wheel runs faster than the inner wheel while taking a turn or moving over upheaval road.*

Construction of differential :

Fig. 16 shows the various parts of a differential unit. The bevel pinion is fixed to the propeller shaft which rotates the crown wheel. The crown-wheel has another unit called the differential unit. It consists of two bevel gears (sun gears) and two bevel pinions (planet pinions). The bevel gears are in contact with the half shafts of the rear axle.

When the crown wheel is rotating, it rotates the differential unit. The bevel (sun) gears of the differential rotate the two half shafts.

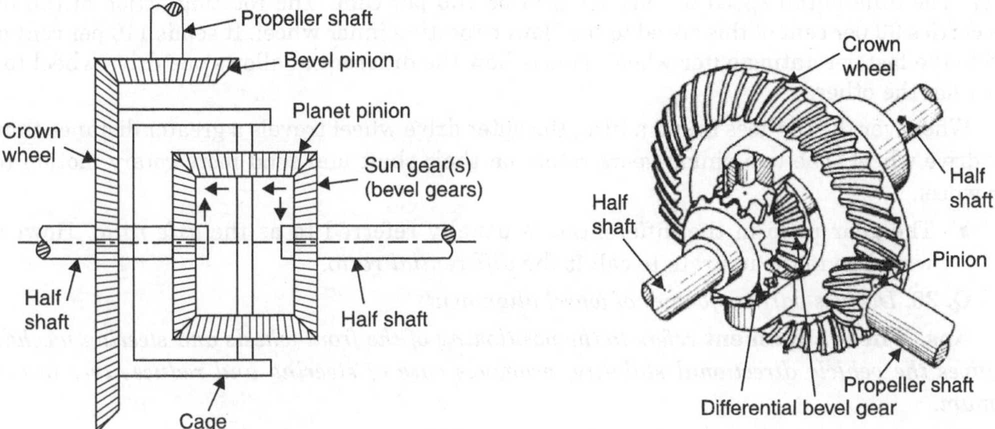

(*i*) Simple differential-simplified diagram (*ii*) Typical three-dimensional view

Fig. 16. Differential.

Operation of differential : When the car is on a straight road, the ring gear, differential case, differential pinion gears, and two differential side gears all turn as a unit. The two differential pinion gears do not rotate on the pinion shaft. This is because they exert equal force on the two differential side gears. As a result, the side gears turn at the same speed as the ring gear, which causes both drive wheels to turn at the same speed also. However, when the car begins to round a curve, the *differential pinion gears rotate on the pinion shaft*. This *permits the outer wheel to turn faster than the inner wheel.*

Suppose one wheel turns slower than the other as the car rounds a curve. As the differential case rotates, the pinion gears must rotate on their shaft. This occurs because the pinion gears must walk around the slower-turning differential side gear. Therefore, the pinion gears carry additional rotary motion to the faster-turning outer wheel on the turn. The action in a typical turn is shown in

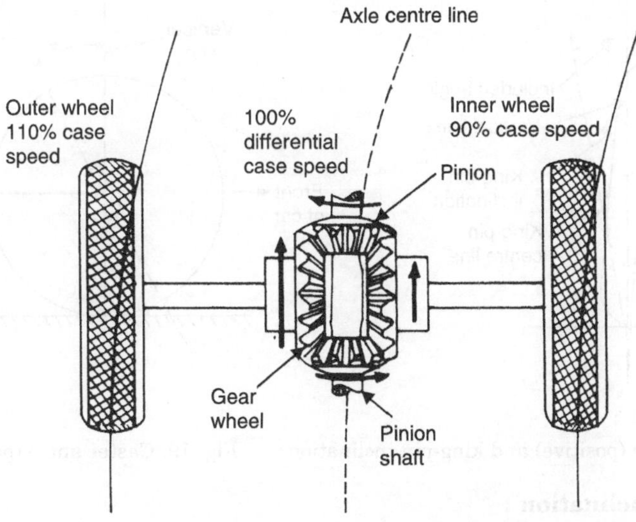

Fig. 17. Differential action on turns.

Fig. 17. The differential speed is considered to be 100 per cent. The rotating action of the pinion gears carries 90 per cent of this speed to the slower-rotating inner wheel. It sends 110 per cent of the speed to the faster-rotating outer wheel. This is how the differential allows one drive wheel to turn faster than the other.

Whenever the car goes round a turn, the outer drive wheel travels a greater distance than the inner drive wheel. The two pinion gears rotate on their shaft and send more rotary motion to the outer wheel.

- The gear ratio in the differential is usually referred to as the *axle ratio*. However it would be more accurate to call it the *differential ratio*.

Q. 30. *Discuss various factors of wheel alignment.*

Ans. Wheel alignment *refers to the positioning of the front wheels and steering mechanism that gives the vehicle directional stability, promotes ease of steering and reduces tyre wear to a minimum.*

The front wheel alignment depends upon the following *factors* :

I. ***Factors pertaining to steering geometry :***

1. Camber 2. King pin inclination 3. Caster 4. Toe-in 5. Toe-out.

II. ***Factors pertaining to front wheel condition :***

1. Balance of wheels 2. Inflation of tyres 3. Brake adjustments.

1. **Camber :**

Camber *is the tilting in or out of the front wheels from the vertical when viewed from the front of the vehicle.* If the top of the wheel tilts *out*, it has "*positive*" camber (Fig. 18). If the top of the wheel tilts *in*, it has "*negative*" camber. *The amount of tilt measured in degrees from the vertical, is called "camber angle".*

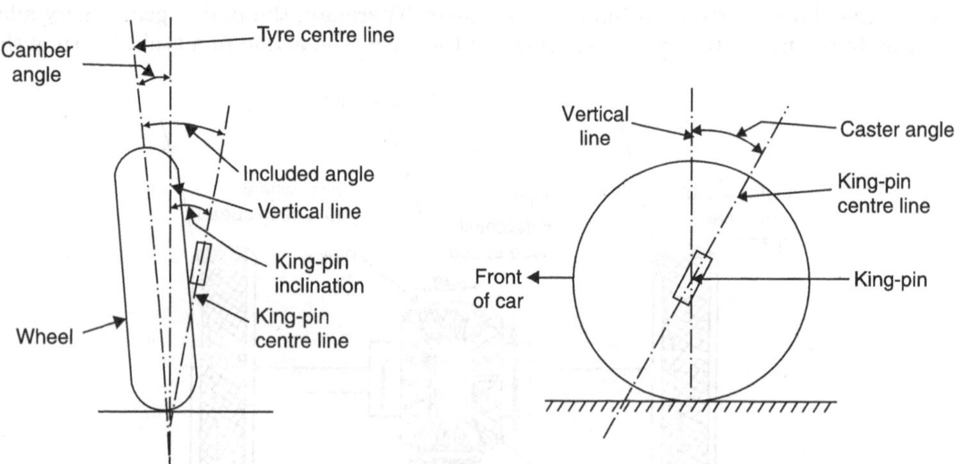

Fig. 18. Camber angle (positive) and king-pin inclination. Fig. 19. Caster angle (positive).

2. **King-pin inclination :**

The ***king-pin inclination*** (*or steering axle inclination*) *is the angle between the vertical line and centre of the kingpin or steering axle, when viewed from the front of the vehicle.*

The king-pin inclination is absolutely necessary due to the following reasons :

(*i*) It helps the car to have steering stability.

(*ii*) It makes the operation of the steering quite easy particularly when the vehicle is stationary.

(*iii*) It helps in reducing the wear on the tyre.

- The combined camber and king-pin inclination is called the *"included angle"*.

3. **Caster :**

The angle between the king-pin centre line (or steering axis) and the vertical, in the plane of the wheel, is called ***Caster angle.*** If the king-pin centre line meets the ground at a point ahead of the vertical centre line, as is shown in Fig. 19 it is called *positive* caster while if it is behind the vertical, wheel centre line it is called *negative* caster. The caster angle in modern vehicles range from 2° to 8°.

The purpose of caster is to give a trailing effect to the front wheels. When the wheel trails the line of weight, that is, moves in the same direction as the vehicle, it is easy to steer a straight course.

4. **Toe-in**

The position of toe-in of a vehicle is shown in Fig. 20. The distance at the front end between the front wheels is *less* than the distance between them at the rear. The front wheels point inward. The actual amount of toe-in in only a few millimetres (usually 3 to 5 mm).

The toe-in serves the following *purposes :*

(*i*) It ensures parallel running of the front wheels.

(*ii*) It stabilises steering.

(*iii*) It prevents side slipping and excessive wear of the tyres.

(*iv*) It also serves to offset the small deflection in the wheel-support system which comes out when the car is moving forward.

Although the wheels are set to toe-in slightly when the car is standing still, they tend to roll parallel on the road when the car is moving forward. Some alignment specialists set the front wheels in "straight-away alignment" in preference to "toe in adjustment".

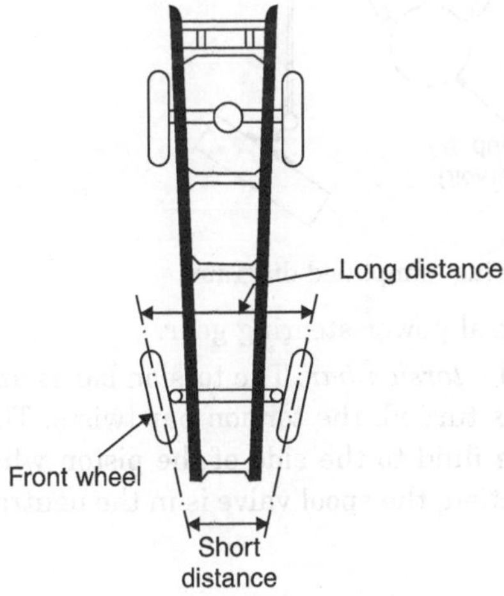

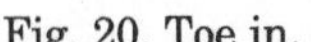

Fig. 20. Toe in.

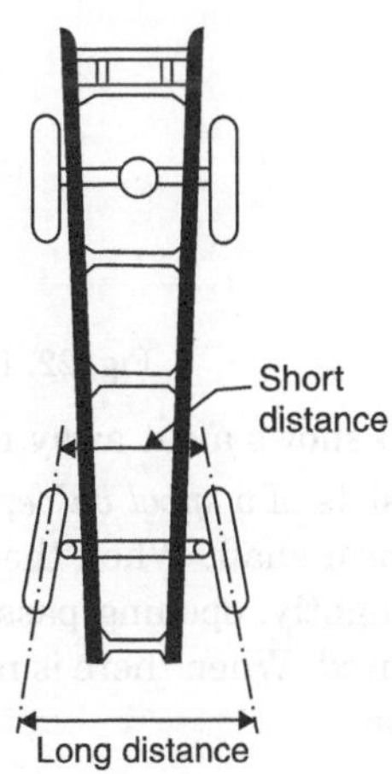

Fig. 21. Toe-out.

5. **Toe-out**

In Fig. 21 the distance between front wheels at front end is *greater* than the distance between them at the rear. When car is stationary, this position of the front wheel is called *toe-out*.

The *purpose of toe-out is to give correct turning alignment and to prevent excessive wear.*

Q. 31. *Explain the construction and principle of working of any one type of power-steering systems used in automobile.*

Ans. Power steering system :

In the ***power-steering system,*** *a continuously operating pump provides hydraulic pressure when needed. At the steering wheel is turned, valves are operated to admit this hydraulic pressure to a cylinder that contains the power piston. Then the pressure causes the piston to move, and it provides most of the steering force.*

Types of power-steering systems :

In automobiles, the following *two* general types of power-steering systems are found :

1. Integral power steering.
2. Linkage power steering.

Integral power steering :

In integral power steering (Fig. 22), the *power piston is built into the steering gear.*

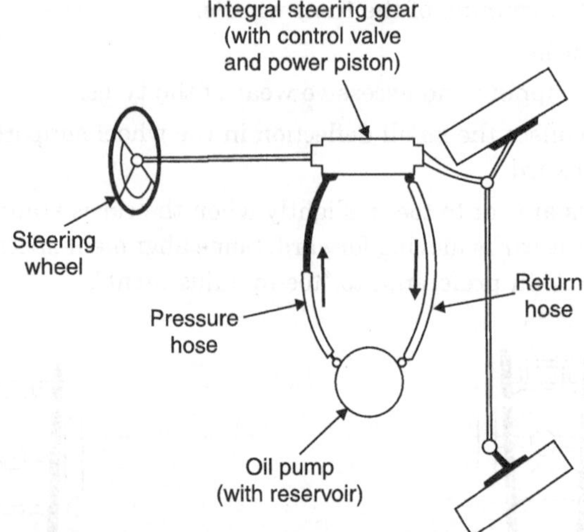

Fig. 22. Integral power steering (simplified diagram).

Fig. 23 shows a cut away rotary-valve integral power-steering gear.

It consists of a *spool valve, rotary valve,* and a *torsion bar*. The torsion bar is attached to the end of the worm shaft. When the steering wheel is turned, the torsion bar twists. This moves the spool valve slightly, opening passages which move fluid to the side of the piston where hydraulic assist is required. When there is no power-assist action, the spool valve is in the neutral or straight-ahead position.

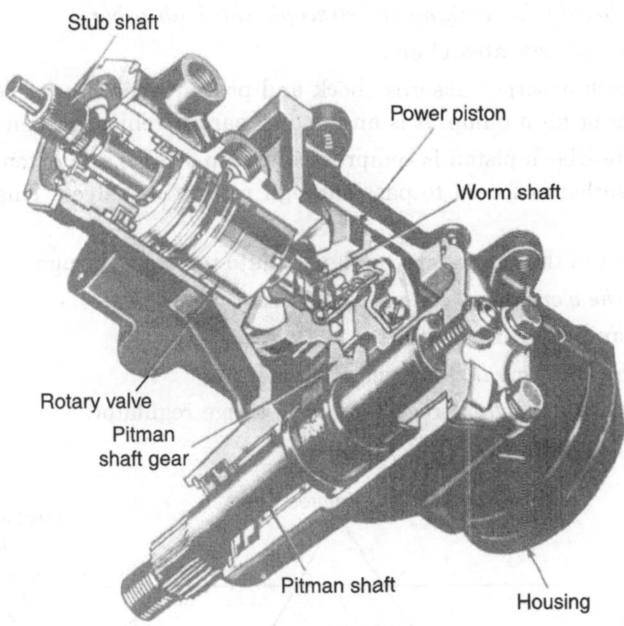

Fig. 23. Cut away rotary valve integral power steering gear.

Q. 32. *Discuss the requirements of a good braking system.*

Ans. Requirements of a good braking system :

The *requirements of a good braking system* are :

1. The brakes should stop the vehicle within a reasonable distance. The retardation shall be smooth and free from jerk or shudder.

2. The braking system should be very reliable to promote highest degree of safety on the road.

3. The braking system should not be affected by water, heat, road grit or dust etc.

4. Pedal effort applied by the driver should not be more so as not to strain the driver.

5. Brake should work equally good in all weathers.

6. The wear and tear of the material of the brake lining should be minimum for its longer life.

7. Due to the rubbing action of brake shoes along with lining against drum, large amount of heat is generated due to friction. The brake design system should be capable of dissipating this heat very quickly.

8. All the components and levers of the braking system should be strong enough to take the mechanical stresses and strains which are encountered during brake actuation.

9. No braking system can work at its best through worn out or incorrectly inflated tyres. Good, tyres are, therefore, a prime essential for efficient braking.

- The **capacity of a brake** depends upon the following *factors* :

(*i*) The unit pressure between the braking surfaces.

(*ii*) The coefficient of friction between the braking surfaces.

(*iii*) The peripheral velocity of the brake drum.

(*iv*) The projected area of the friction surfaces.

(*v*) The ability of the brake to dissipate heat equivalent to the energy being absorbed.

Q. 33. *Explain briefly the working of Telescopic shock absorber.*

Ans. Telescopic shock absorber :

A telescopic shock absorber absorbs shock and prevents continuous bounce every time you accelarate, stop, corner or hit a bump. It is an integral part of vehicle suspension.

As the shock absorber's piston is compressed (bump stroke) and extended (rebound stroke), fluid in the shock absorber is forced to pass through restrictive valves. This generates hydraulic resistance.

Now a days most of the shocks are configured into telescopic design.

Q. 34. *Discuss the working of 'voltage and current regulator'.*

Ans. Voltage and current regulator :

It consists of the following *three* major components:

1. Cut-out relay. 2. Current regulator. 3. Voltage regulator.

Fig. 24 shows the regulator.

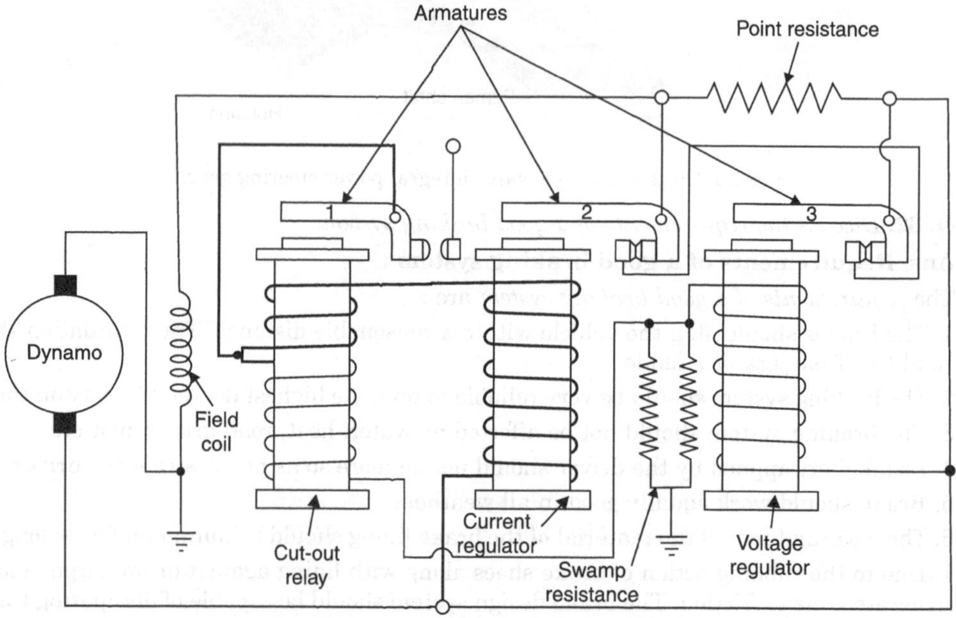

Fig. 24. Regulator (LUCAS-TVS 3GC-12V).

Cut-out relay. When dynamo is started, the voltage builds up. When sufficient voltage is reached, the *"cut-out relay"* comes into action. Consequently the armature (1) is pulled towards the soft iron core of cut-out relay. This gives a connection between the dynamo and battery.

Current regulator. Sometimes the voltage of the dynamo fails to reach the operating value of the voltage regulator. This is due to the discharge of the battery. This happens when there is *large electrical loading*. At this time *"current regulator"* comes into operation. If the current output is increasing, the current regulator pulls armature (2) closer. When the contact points of the current regulator open, point resistance is brought into the field circuit. Thus the current limit is maintained.

Voltage regulator. When the dynamo starts running at a high speed, the *"voltage regulator"* immediately comes into action. Armature (3) is pulled away from the contact points of the voltage regulator. The contact points are now opened. Point resistance is then introduced in the field coil, thereby bringing down the voltage.

By *combining the voltage and current regulators, the optimum specified values of the voltage and current outputs are maintained.*

Temperature compensation. The voltage regulators and also some current regulators are temperature compensated. They have higher settings when cold than when hot.

- Temperature compensation of the *voltage regulator* allows the increase of voltage with low temperature.
- The current regulator is compensated to permit a higher generator output when the current regulator is cold than when it is hot. The *generator output decreases as the temperature of the current regulator increases.*

Q. 35. *Explain with the help of a neat diagram the conventional battery ignition system.*

Ans. Battery or coil ignition system :

It is a commonly used system because of *its combined cheapness, convenience of maintenance, attention and general suitability*.

Construction. This system consists of the following *components* :

1. Battery (6 or 12 volts)
2. Ignition switch
3. Induction coil
4. Circuit/contact breaker
5. Condenser
6. Distributor.

Refer to Fig. 25.

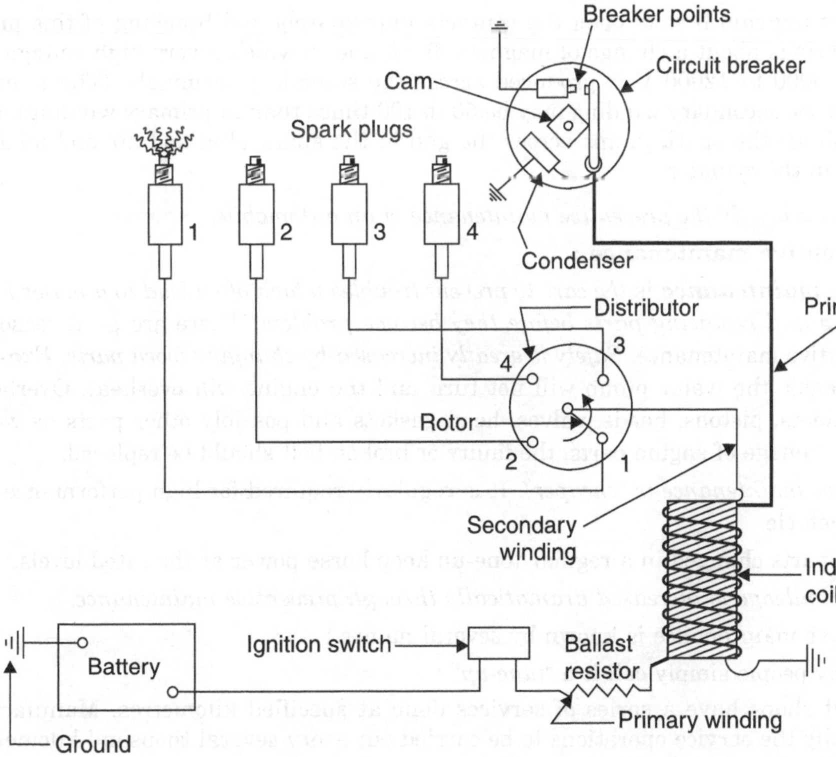

Fig. 25. Battery or coil ignition system.

— One terminal of the battery is ground to the frame of the engine, and other is connected through the ignition switch to one primary terminal of the ignition coil (consisting of a *comparatively few turns of thick wire* wound round an iron core). The other primary terminal is connected to one end of the contact points of the circuit breaker and through closed points to ground. The primary circuit of the ignition coil thus gets completed when contact points of the circuit breaker are together and switch is closed. The secondary terminal of the coil is connected to the central contact of the distributor and hence to distributor rotor. The secondary circuit consists of secondary winding (consisting of a *large number of turns of fine wire*) of the coil, distributor and four spark plugs. The contact breaker is driven by a cam whose speed is *half* the engine speed (for *four stroke engines*) and breaks the primary circuit one for each cylinder during one complete cycle of the engine.

— The breaker points are held on contact by a spring except when forced apart by lobs of the cam.

— A *ballast resistor* is provided in series with the primary winding to regulate primary current. For starting purposes this resistor is by passed so that more current can flow in the primary circuit.

Working :

— To start with, the ignition switch is made on and the engine is cranked *i.e.* turned by hand when the contacts touch, the current flows from battery through the switch, primary winding of the induction coil to circuit breaker points and the circuit is completed through the ground. A condenser connected across the terminals of the contact breaker points prevent the sparking at these points.

— The rotating cam breaks open the contacts immediately and breaking of this primary circuit brings about a change of magnetic field, due to which a very high voltage to the tune of 8000 to 12000 V is produced across the secondary terminals. (The number of turns in the secondary winding may be 50 to 100 times than in primary winding). Due to high voltage the spark jumps across the gap in the spark plug and air fuel mixture is ignited in the cylinder.

Q. 36. *Discuss briefly the preventive maintenance of an automobile.*

Ans. Preventive maintenance :

Preventive maintenance is *the care to prevent troubles which often lead to a major haul. It involves inspecting and replacing parts before they become problem.* There are good reasons for performing preventive maintenance. *Safety is greatly increased by changing worn parts.* ***Example :*** If the fan belt breaks, the water pump will not turn and the engine will overheat. Overheating damages the cylinders, pistons, heads, valves, head gaskets and possibly other parts as well. To avoid this serious damage of engine parts, the faulty or broken belt should be replaced.

- *Preventive maintenance is "cheaper".* It is regularly required for high performance of the engine/vehicle.
  - The parts changed in a regular tone-up keep horse power at the rated levels.
  - *Fuel mileage is increased dramatically through preventive maintenance.*
- Preventive maintenance is known by several names :
  - Many people simply call it a "*tune-up*" ;
  - Most shops have a series of services done at specified kilometres. Manufacturers specify the service operations to be carried out every several thousand kilometres to keep warranties in effect.

ADDITIONAL OBJECTIVE TYPE QUESTIONS' BANK

A. Choose the Correct Answer.

B. Match List I with List II.

C. Competitive Examinations Questions (With Solutions-Comments).

D. Fill in the Blanks.

ADDITIONAL OBJECTIVE TYPE QUESTIONS

(Selected from Competitive Examinations Question Papers)

A. Choose the Correct Answer :

1. The specific gravity of diesel is
   (*a*) 1 (*b*) 0.7 (*c*) 0.85
   (*d*) 0.5 (*e*) 1.25.
2. The most popular firing order in the four-cylinder in line I.C. engine is
   (*a*) 1—2—3—4 (*b*) 1—3—2—4 (*c*) 1—3—4—2 (*d*) 1—2—4—3.
3. The ratio of indicated thermal efficiency to the corresponding air standard cycle efficiency is called
   (*a*) efficiency ratio (*b*) relative efficiency
   (*c*) overall efficiency (*d*) mechanical efficiency.
4. If the compression ratio of an engine working on Otto cycle is increased from 5 to 7, the percentage increase in efficiency will be
   (*a*) 2% (*b*) 4% (*c*) 8% (*d*) 13.70%.
5. The pressure at the end of compression in a C.I. engine is of the order of
   (*a*) 10 bar (*b*) 16 bar (*c*) 25 bar (*d*) 35 bar.
6. The thermal efficiency of a diesel cycle having fixed compression ratio with increase of cut-off ratio will
   (*a*) increase (*b*) decrease (*c*) remain unaffected (*d*) none of the above.
7. The fuel in diesel engine is normally injected at pressure of
   (*a*) 5—10 bar (*b*) 20—25 bar (*c*) 60—80 bar (*d*) 90—130 bar.
8. In a petrol engine the spark plug gap is in the order of
   (*a*) 0.10 mm (*b*) 0.6 mm (*c*) 0.1 mm (*d*) 0.15 mm.
9. For maximum power generation, the air-fuel ratio for a petrol engine is of the order of
   (*a*) 9 : 1 (*b*) 12 : 1 (*c*) 16 : 1
   (*d*) 18 : 1 (*e*) 20 : 1.
10. Piston rings are usually made of
    (*a*) cast-iron (*b*) aluminium
    (*c*) phosphor bronze (*d*) carbon steel.
11. The self ignition temperature of diesel compared to petrol
    (*a*) is higher (*b*) is lower
    (*c*) is same (*d*) depends on quality of fuel.
12. The compression ratio of motor car is in the order of
    (*a*) 5 (*b*) 7 (*c*) 10
    (*d*) 16 (*e*) 19.
13. The rotation of camshaft with respect to rotation of crankshaft for 4-stroke IC engine is
    (*a*) 1/2 (*b*) 1 (*c*) 2 (*d*) 1/4.
14. Fuel consumption with increase in back pressure will
    (*a*) increase (*b*) decrease
    (*c*) remain unaffected (*d*) none of the above.

15. The specific gravity of petrol is

(*a*) 1 (*b*) 0.82 (*c*) 0.75
(*d*) 0.50 (*e*) 0.24.

16. The most popular firing order in the six-cylinder in-line IC engine is

(*a*) 1 — 2 — 3 — 4 — 5 — 6 (*b*) 1 — 3 — 5 — 4 — 6 — 2
(*c*) 1 — 3 — 6 — 5 — 4 — 2 (*d*) 1 — 5 — 3 — 6 — 2 — 4.

17. For same compression ratio and heat input, the cycle which has maximum efficiency may be

(*a*) Diesel cycle (*b*) Dual cycle (*c*) Otto cycle (*d*) None of the above.

18. For the same maximum pressure and heat supplied, the efficiency is maximum for

(*a*) Otto cycle (*b*) Diesel cycle (*c*) Dual cycle (*d*) None of the above.

19. If the working substance in case of air-standard cycle is changed from air to argon for the same compression ratio and heat input at constant volume, the efficiency will

(*a*) decrease (*b*) increase (*c*) remain constant (*d*) none of the above.

20. The pressure at the end of the compression in case of motor car (S.I. engine) is of the order of

(*a*) 7 bar (*b*) 10 bar (*c*) 15.5 bar (*d*) 20 bar.

21. The thermal efficiency of Otto cycle, having same heat input and working substance will

(*a*) increase (*b*) decrease (*c*) remain constant
(*d*) none of the above with increase of compression ratio.

22. In a petrol engine the high voltage for spark is in the order of

(*a*) 1000 V (*b*) 2000 V (*c*) 11 kV (*d*) 22 kV.

23. The material for centre electrode in spark plug is

(*a*) carbon (*b*) platinum (*c*) platinum-tungsten alloy
(*d*) nickel alloy (*e*) none of the above.

24. For economy (minimum fuel consumption), the air-fuel ratio for petrol engine is of the order of

(*a*) 9 : 1 (*b*) 12 : 1 (*c*) 16 : 1 (*d*) 20 : 1.

25. Material for piston in case of petrol engine is

(*a*) cast-iron (*b*) aluminium
(*c*) phosphorus-bronze (*d*) cast steel.

26. The ratings of C.I. engine fuel is given by

(*a*) octane number (*b*) performance number
(*c*) cetane number (*d*) none of the above.

27. The high-vapour pressure fuel of gas turbine is

(*a*) JP-3 (*b*) JP-4 (*c*) JP-5 (*d*) none of the above.

28. The compression ratio of diesel pump engine is in the order of

(*a*) 5 (*b*) 10 (*c*) 16 (*d*) 18.

29. The mechanical efficiency (η_m) of an IC engine is equal to

(*a*) IHP/BHP (*b*) BHP/IHP (*c*) BHP/FHP (*d*) FHP/BHP.

30. The ratio of the indicated thermal efficiency to the corresponding ideal air-standard efficiency is called

(*a*) brake thermal efficiency (*b*) indicated thermal efficiency
(*c*) volumetric efficiency (*d*) relative efficiency.

31. If the specific fuel consumption per BHP hour is approximately 0.2 kg, the engine is
(*a*) diesel engine (*b*) petrol engine
(*c*) steam engine (*d*) none of the above.

32. By reducing compression ratio, the knocking tendency in compression ignition engine will
(*a*) increase (*b*) decrease (*c*) not take place (*d*) none of the above.

33. The fuel which detonates easily, is
(*a*) n-heptane (*b*) iso-octane (*c*) benzene (*d*) alcohol.

34. The increase in the cut-off ratio of a Diesel cycle with fixed compression ratio would
(*a*) decrease m.e.p. (*b*) increase m.e.p.
(*c*) keep same m.e.p. (*d*) none of the above.

35. In a typical medium speed four-stroke cycle diesel engine intake valve
(*a*) opens at TDC and closes at BDC
(*b*) opens at 20°C before TDC and closes at 35°C after BDC
(*c*) opens at 10ºC after TDC and closes at 20ºC before BDC
(*d*) none of the above.

36. Addition of normal heptane, (C_7H_{16})
(*a*) resists auto-ignition (*b*) accelerates auto-ignition
(*c*) does not affect auto-ignition (*d*) none of the above.

37. Piston rings are usually made of
(*a*) aluminium (*b*) cast iron (*c*) carbon steel (*d*) phosphor bronze.

38. The thermal efficiency of two-stroke petrol engine with crankcase scavenging as compared to four-stroke petrol engine with same compression ratio will be
(*a*) higher (*b*) same (*c*) lower (*d*) unpredictable.

39. 'Knock' in the C.I. engine is characterised by
(*a*) sudden auto-ignition of the mixture at the very beginning of the combustion process
(*b*) sudden auto-ignition of the mixture near the end of the combustion period
(*c*) knock does not occur in C.I. engines
(*d*) none of these.

40. The highest flame speed will be obtained with an air-fuel ratio
(*a*) somewhat richer than chemically correct (*b*) stoichiometric
(*c*) very rich (*d*) lean.

41. Which one of the following is *correct* ?
(*a*) Late intake valve closing increases the compression ratio
(*b*) Late intake valve closing lowers the compression ratio
(*c*) Late intake valve closing does not affect the compression ratio
(*d*) None of the above statements is correct.

42. In four-stroke engines, the camshaft connected to the crankshaft by gears or chain rotates at
(*a*) half the crankshaft speed (*b*) three-fourth the crankshaft speed
(*c*) double the crankshaft speed (*d*) equal to the crankshaft speed.

43. Which one of the following statements is *correct* for S.I. engines ?
(*a*) The flame front progresses relatively faster at the beginning and near the conclusion of combustion, and slower during the intermediate portion
(*b*) The flame front progresses relatively slowly at the beginning and near the conclusion of combustion, and faster during the intermediate portion

(*c*) The flame front is steady throughout
(*d*) None of the above.

44. The mixture requirement of a S.I. engine under normal running on road is
(*a*) a stoichiometric mixture (*b*) a rich mixture
(*c*) a lean mixture (*d*) none of the above.

45. Compensating devices are provided in carburettors
(*a*) to change the quantity of mixture depending upon load
(*b*) to provide always an economy mixture
(*c*) to modify the mixture strength depending upon requirements under various operating conditions
(*d*) to supply extra fuel during acceleration only.

46. Auto-ignition in a S.I. engine means
(*a*) automatic ignition of the charge at the end of compression
(*b*) ignition induced by the passage of a spark
(*c*) ignition of the charge before the passage of the flame front
(*d*) ignition induced to supplement the process of normal combustion.

47. The best fuels for C.I. engines are
(*a*) straight chain paraffins (*b*) aromatics
(*c*) branched chain paraffins (*d*) naphthanes.

48. The compression ratio used in a high-speed C.I. engine is of the order of
(*a*) 4 (*b*) 8 (*c*) 12 (*d*) 20.

49. The pressure in the combustion chamber in a diesel engine at the end of compression is of the order of
(*a*) 5 bar (*b*) 10 bar (*c*) 15 bar (*d*) 22 bar.

50. In computing engine performance, the heating value of fuel used is
(*a*) the higher heating value (*b*) the lower heating value
(*c*) the average of lower and higher heating values
(*d*) none of the above.

51. Normally a lubricant is selected for an engine on the basis of
(*a*) SAE viscosity rating number (*b*) Saybolt seconds
(*c*) Redwood seconds (*d*) Viscosity in stokes.

52. The best possible location for the spark plug is
(*a*) near the inlet valve (*b*) near the exhaust valve
(*c*) at the centre of the cylinder head (*d*) any place of the combustion chamber.

53. If one of the two spark plugs provided at two ends of the cylinder fails
(*a*) the engine will stop (*b*) knocking will increase
(*c*) knocking will decrease (*d*) there will be no change.

54. The quantity of heat lost to the cooling water in an I.C. engine is about
(*a*) 10% (*b*) 30% (*c*) 50% (*d*) 70%.

55. Hydrocarbon emission is maximum in
(*a*) 4-stroke cycle petrol engine (*b*) 4-stroke cycle diesel engine
(*c*) 2-stroke cycle petrol engine (*d*) 2-stroke cycle diesel engine.

56. The thermal efficiency of high speed diesel engine is the order of
(*a*) 20% (*b*) 35% (*c*) 50% (*d*) 70%.

57. The thermal efficiency of an air standard diesel cycle having fixed compression ratio, with increase in cut-off ratio will
(*a*) increase (*b*) decrease (*c*) independent (*d*) none of the above.

58. The pressure at the end of compression in the case of diesel engine is in the order of
(*a*) 12 bar (*b*) 20 bar (*c*) 28 bar (*d*) 35 bar.

59. Supercharging of IC engines is essential for
(*a*) marine engine (*b*) aircraft engine
(*c*) stationary engine (*d*) none of the above.

60. The most effective method of determining FHP for multicylinder engines is
(*a*) Morse test (*b*) Willan's line method
(*c*) mechanical indicator (*d*) electronic indicator.

61. If the clearance volume of IC engines is increased, the compression ratio will
(*a*) increase (*b*) decrease (*c*) remain constant (*d*) any of these.

62. The camshaft of a four-stroke IC engines should be running at 1200 rpm will run at
(*a*) 1200 rpm (*b*) 600 rpm (*c*) 2400 rpm (*d*) none of the above.

63. In a C.I. engine higher combustion chamber wall temperature will
(*a*) reduce knocking tendency (*b*) increase knocking tendency
(*c*) reduce exhaust temperature (*d*) have no influence on combustion process.

64. Anti-knock property of C.I. engine fuel can be improved by adding
(*a*) tetra-ethyl lead (*b*) trimethyl pentane
(*c*) amylnitrate (*d*) hexadecane.

65. For the same output and compression ratio, four-stroke engines as compared to two-stroke engines have
(*a*) higher fuel consumption (*b*) higher exhaust temperature
(*c*) lower thermal efficiency (*d*) higher thermal efficiency.

66. If the pressures at the beginning and end of compression in an air-standard Otto cycle (r = 1.4) are 1 bar and 8.1 bar respectively, the efficiency will be
(*a*) 45% (*b*) 50% (*c*) 40% (*d*) 57%.

67. For combustion of 1 kg of carbon, the quantity of air required by mass is
(*a*) 12 kg (*b*) 2.67 kg (*c*) 15 kg (*d*) 11.6 kg.

68. Crankcase dilution means
(*a*) dilution of fuel in the crankcase
(*b*) dilution of mixture while passing through the crankcase in two-stroke engines
(*c*) dilution of lubricating oil in the crankcase
(*d*) addition of water in the crankcase.

69. Orsat apparatus is used for determining
(*a*) the calorific value of fuel
(*b*) volumetric analysis of the dry products of combustion
(*c*) volumetric analysis of the wet products of combustion
(*d*) gravimetric analysis of the products of combustion.

70. IC engine pistons are usually made of
(*a*) cast steel (*b*) forged steel (*c*) phosphor bronze (*d*) aluminium alloy.

71. Inlet valve of a four-stroke IC engine remains open for about
(*a*) 180° (*b*) 150° (*c*) 260° (*d*) 230°.

72. The purpose of thermostate in water cooling system is to
(*a*) avoid steam formation
(*b*) close the water passage when the engine is cold
(*c*) prevent abnormal rise in the engine temperature
(*d*) maintain the operating temperature of water.

73. For same maximum pressure and temperature
(*a*) Diesel cycle is more efficient than Otto cycle
(*b*) Otto cycle is more efficient than Diesel cycle
(*c*) Both Otto cycle and Diesel cycle are equally efficient
(*d*) None of the above.

74. The cetane rating of diesel fuel is of the order of
(*a*) 25 (*b*) 45 (*c*) 70 (*d*) 90.

75. Many vegetable oils can be used as a fuel in
(*a*) petrol engines (*b*) diesel engines
(*c*) both petrol engines and diesel engines
(*d*) neither in petrol engines nor in diesel engines.

76. In carburettors, the top of the fuel jet with reference to the petrol level in the float chamber
(*a*) is kept at same level (*b*) is kept at slightly higher level
(*c*) is kept at slightly lower lever (*d*) may be anywhere
(*e*) varies from situation to situation.

77. Economiser system is provided in the carburettors
(*a*) to achieve economy in fuel consumption
(*b*) to allow richer mixture for maximum power range
(*c*) to facilitate easy starting (*d*) to accelerate the engine rapidly.

78. On which factor out of the following volumetric efficiency do not depend ?
(*a*) Speed of engine (*b*) Compression ratio
(*c*) Clearance volume (*d*) Cylinder dimensions.

79. A stratified charge engine suffers from following disadvantage :
(*a*) It cannot tolerate a wide quality of fuels
(*b*) It does not have low exhaust emission levels
(*c*) It cannot be manufactured by the existing technology
(*d*) It results in reduced power for a given engine size.

80. The volumetric efficiency of a well designed IC engine may be
(*a*) 30 to 40% (*b*) 40 to 60%
(*c*) 60 to 70% (*d*) 75 to 90%.

81. The knocking tendency in C.I. engines for a given fuel will be
(*a*) enhanced by decreasing compression ratio
(*b*) enhanced by increasing compression ratio
(*c*) unaffected by change in compression ratio
(*d*) None of the above.

82. The knocking in diesel engines may be prevented by
(*a*) reducing the delay period (*b*) raising the compression ratio
(*c*) increasing the inlet pressure of air (*d*) increasing the injection pressure
(*e*) all of the above.

83. The knocking tendency in S.I. engines may be decreased by
(*a*) controlling the air-fuel ratio (*b*) controlling the ignition timing
(*c*) controlling the exhaust temperature (*d*) reducing the compression ratio
(*e*) All of the above.

84. In order to prevent knocking in S.I. engines, the charge away from spark plug should have
(*a*) low density (*b*) low temperature
(*c*) long ignition delay (*d*) rich mixture
(*e*) all of the above.

85. The secondary voltage in a spark ignition engine is of the order of
(*a*) 200 to 500 volts (*b*) 1000 to 2000 volts
(*c*) 5000 to 8000 volts (*d*) 15000 to 20000 volts.

86. Engine friction increases very rapidly with increase in
(*a*) stroke/bore ratio (*b*) compression ratio
(*c*) engine speed (*d*) engine load.

87. Oxides of nitrogen emission is maximum in
(*a* two-stroke air scavenged compression ignition engines
(*b*) four-stroke turbo-charged compression ignition engines
(*c*) four-stroke normally aspirated medium speed C.I. engines
(*d*) four-stroke normally aspirated high speed C.I. engines.

88. Blue white smoke is caused in diesel engine exhaust by
(*a*) incomplete diesel combustion
(*b*) liquid droplets of lubricating oil or fuel oil while starting from cold
(*c*) very lean air-fuel mixture (*d*) very high speed operation.

89. Morse test is used in multicylinder S.I. engine to determine
(*a*) thermal efficiency (*b*) mechanical efficiency
(*c*) volumetric efficiency (*d*) relative efficiency.

90. The thermal efficiency of a good IC engine at rated load is in the range of
(*a*) 10 to 20% (*b*) 30 to 35% (*c*) 60 to 70% (*d*) 80 to 90%.

91. Iso-octane content in a fuel for spark ignition engine
(*a*) accelerates auto-ignition (*b*) retards auto-ignition
(*c*) does not affect auto-ignition (*d*) none of the above.

92. The work output of theoretical Otto cycle
(*a*) increases with increase in compression ratio
(*b*) increases with increase in pressure ratio
(*c*) increases with increase in adiabatic index
(*d*) follows all the above.

93. The brake specific fuel consumption is expressed as the fuel consumed
(*a*) per unit time (*b*) per hour per unit brake horse power
(*c*) per km distance travelled (*d*) per hour per unit indicated horse power.

94. When an engine is idling, it requires
(*a*) no fuel in air (*b*) lean fuel-air mixture
(*c*) rich fuel-air mixture (*d*) stoichiometric mixture.

95. Cetane number is the measure of
(*a*) viscosity of fuel (*b*) auto-ignition temperature
(*c*) ignition quality (*d*) calorific value of fuel.

96. Morse test in a multi-cylinder engine is used to determine
(*a*) BHP (*b*) FHP
(*c*) FHP and IHP (*d*) volumetric efficiency.

97. The pressure at the end of compression in the case of diesel engine is of the order of the
(*a*) 58 N/cm^2 (*b*) 118 N/cm^2 (*c*) 195 N/cm^2 (*d*) 340 N/cm^2.

98. If the compression ratio of an engine working on Otto cycle is increased from 5 to 7, the %age increase in air standard efficiency will be
(*a*) 2% (*b*) 4% (*c*) 8% (*d*) 14%.

99. Which is more viscous lubricating oil ?
(*a*) SAE 30 (*b*) SAE 40 (*c*) SAE 50 (*d*) SAE 70.

100. Mean effective pressure of an IC engine can be accurately measured by
(*a*) mechanical indicator (*b*) electrical indicator
(*c*) electronic indicator (*d*) none of the above.

101. The most popular device of measuring engine output is
(*a*) electrical dynamometer (*b*) hydraulic dynamometer
(*c*) fan brake dynamometer (*c*) pony brake dynamometer.

102. In a cyclic heat engine operating between source and sink temperatures of 600° C and 20° C respectively, the least rate of heat rejection per kW output of the engine is
(*a*) 0.460 kW (*b*) 0.505 kW (*c*) 0.588 kW (*d*) 0.650 kW.

103. Otto cycle efficiency is higher than diesel cycle efficiency for the same compression ratio and heat input because, in Otto cycle
(*a*) combustion is at constant volume (*b*) expansion and compression are isentropic
(*c*) maximum temperature is higher (*d*) heat rejection is lower.

104. In case of diesel cycle, increasing the cut-off ratio will increase
(*a*) efficiency (*b*) the maximum pressure
(*c*) mean effective pressure (*d*) engine weight.

105. The mass of air required for complete combustion of unit mass of fuel can always be calculated from the formula, where C, H, O, and S are in percentage
(*a*) 0.1152 C + 0.3456 H (*b*) 0.1152 C + 0.3456 (H – 0.125 O)
(*c*) 0.1152 C + 0.3456 (H + 0.125 O) + 0.0432 S
(*d*) 0.1152 C + 0.3456 (H – 0.125 O) + 0.0432 S.

106. Orsat flue gas analyses gives
(*a*) mass analysis of wet products of combustion
(*b*) volumetric analysis of dry products of combustion
(*c*) mass analysis of dry products of combustion
(*d*) volumetric analysis of wet products of combustion.

107. Reference fuels for knock rating of S.I. engine fuels would include
(*a*) iso-octane and alpha-methyl naphthalene
(*b*) normal octane and aniline
(*c*) iso-octane and *n*-heptane
(*d*) iso-octane and *n*-hexane.

108. Which one of the following events would reduce the volumetric efficiency of a vertical C.I. engine ?
(*a*) inlet valve closing after BDC (*b*) inlet valve closing before BDC
(*c*) inlet valve opening before TDC (*d*) exhaust valve closing after TDC.

109. Generally, in Bosch type fuel injection pumps, the quantity of fuel is increased or decreased with change in load, due to change in
(*a*) timing of start of fuel injection (*b*) timing of end of fuel injection
(*c*) injection pressure of fuel (*d*) velocity of flow of fuel.

110. By higher octane number of S.I. engine fuels, it is meant that the fuel has
(*a*) higher heating value (*b*) higher flash point
(*c*) longer ignition delay (*d*) lower volatility.

111. Which one of the following quantities is assumed constant for an IC engine while estimating its friction power by extrapolation through Willan's line ?
(*a*) Mechanical efficiency (*b*) Indicated thermal efficiency
(*c*) Brake thermal efficiency (*d*) Volumetric efficiency.

112. In battery ignition system, the voltage across the spark gap is about
(*a*) 12 V (*b*) 120 V (*c*) 1200 V (*d*) 12000 V.

113. Energy taken away by the cooling system of an IC engine is about
(*a*) 10% (*b*) 20% (*c*) 30% (*d*) 50%.

114. Carbon monoxide emission from S.I. engine are higher when the engine is
(*a*) idling (*b*) cruising (*c*) accelerating (*d*) decelerating.

115. The camshaft of four-stroke IC engine running at 2000 r.p.m. will run at
(*a*) 1000 r.p.m. (*b*) 2000 r.p.m. (*c*) 4000 r.p.m. (*d*) none.

116. The most effective method of determining mechanical efficiency of an IC engine is
(*a*) Morse test (*b*) electronic indicator
(*c*) extending the graph of fuel consumption versus bhp on the negative side of x-axis
(*d*) motoring.

117. Out of the following variables, which one has predominant effect on detonation in an S.I. engine ?
(*a*) Compression ratio (*b*) A / F ratio
(*c*) Spark timing (*d*) Speed.

118. Stirling engine can be called as
(*a*) internal combustion engine
(*b*) external combustion engine
(*c*) combination of internal and external combustion engine
(*d*) none.

119. Which is the most effective alternative fuel for IC engine in rural area ?
(*a*) CNG (compressed natural gas) (*b*) Bio-gas
(*c*) Alcohol (*d*) Hydrogen.

120. The function of a carburettor of an S.I. engine is to control
(*a*) A / F ratio (*b*) amount of mixture
(*c*) A / F ratio and amount of mixture (*d*) compression ratio.

121. The indicated specific fuel consumption is expressed as the fuel consumed
(*a*) per unit time (*b*) per hour per unit brake horse power
(*c*) per km distance travelled (*d*) per hour per indicated horse power.

122. The specific gravity of petrol is about
(*a*) 0.65 (*b*) 0.75 (*c*) 0.85 (*d*) 0.95.

123. In a carburettor idling system is used
(*a*) to compensate for dilution of charge due to residual gases
(*b*) for cold starting

(*c*) for meeting maximum power requirements
(*d*) for rapid opening of throttle.

124. By higher octane number of S.I. engine fuel, it is meant that the fuel has
(*a*) high heating value (*b*) lower volatility
(*c*) higher flash point (*d*) longer ignition delay.

125. The size of inlet valve of an engine in comparison to exhaust valve
(*a*) is more (*b*) is less
(*c*) is same (*d*) varies from design to design.

126. Indicated horse power of a multicylinder S.I. engine can be determined by the use of
(*a*) Prony brake test (*b*) motoring test
(*c*) Morse test (*d*) Willan's line method.

127. Consider the following measures :
1. Increasing the compression ratio
2. Increasing the intake temperature
3. Increasing stroke-to-bore ratio of the cylinder
4. Increasing the engine speed.
The measures necessary to reduce the tendency to knock in C.I. engines would include
(*a*) 1, 2 and 3 (*b*) 1, 2 and 4
(*c*) 2, 3 and 4 (*d*) 1, 3 and 4.

128. An engine indicator diagram is used to determine
(*a*) speed (*b*) BHP
(*c*) temperature (*d*) mean effective pressure.

129. The brake specific fuel consumption of a petrol engine at fuel load is nearly
(*a*) 0.15 kg/BHP hr (*b*) 0.20 kg/BHP hr
(*c*) 0.25 kg/BHP hr (*d*) 0.30 kg/BHP hr.

130. Carbon deposits in I.C. engine cylinder results in an increase of
(*a*) clearance volume (*b*) volumetric efficiency
(*c*) combustion duration (*d*) effective compression ratio.

131. A 100 cc engine has the following parameter as 100 cc
(*a*) fuel tank capacity (*b*) clearance volume
(*c*) swept volume (*d*) cylinder volume.

132. Air-fuel ratio for idling of an S.I. engine is approximately
(*a*) 5 : 1 (*b*) 10 : 1 (*c*) 15 : 1 (*d*) 20 : 1.

133. Compression ratio (CR) in petrol engine is kept less than in diesel engine because
(*a*) we want to make petro engine lighter
(*b*) higher or equivalent CR in petrol engines is not possible due to pre-ignition
(*c*) less CR gives better performance
(*d*) it is just customary to have less CR in petrol engines.

134. Cylinder of petrol engine gets rich mixture when it is
(*a*) idling (*b*) operating at cruising speed
(*c*) operating at maximum speed (*d*) none of these.

135. For minimum fuel consumption per bhp hour an automobile engine (C.I.) should be run
(*a*) at its maximum speed (*b*) at the speed where bhp is maximum
(*c*) at the speed where torque is maximum
(*d*) at a particular speed which is usually less than the speed of maximum bhp.

136. Tetramethyl lead is a better additive than tetraethyl lead because
(*a*) TML has a lower boiling point (*b*) TEL has a lower boiling point
(*c*) TML has better mixing property (*d*) TEL has a better mixing property.

137. As the engine speed increases it is desirable to
(*a*) advance the ignition timing (*b*) retard the ignition timing.

138. At full throttle operation it is necessary to
(*a*) advance the spark (*b*) retard the spark.

139. The function of a distributor in an S.I. engine is to
(*a*) produce the high voltage for sparking
(*b*) distribute the fuel to the appropriate cylinder
(*c*) allow the exhaust gases to escape from the appropriate cylinder
(*d*) provide the correct firing order in the engine.

140. The pump used for circulating lubricating oil in the engine is
(*a*) of centrifugal type (*b*) of plunger type
(*c*) a gear pump (*d*) any of these.

141. If instead of 4-stroke, we use 2-stroke for the completion of an I.C. engine cycle, there would be a loss of efficiency
(*a*) more in S.I. (*b*) more in C.I. (*c*) equal (*d*) any of these.

142. For the same size and weight, a 2-stroke cycle engine would deliver power as compared to that of a 4-stroke
(*a*) about twice (*b*) about 1.7 times (*c*) about 1.9 times (*d*) nearly equal.

143. The purpose of venturi in the carburettor is to work as
(*a*) pump (*b*) compressor (*c*) ejector (*d*) none of these.

144. Vapour lock is caused due to
(*a*) locking carburettor jets due to high vapour pressure
(*b*) excess fuel supply to engine due to faster vaporisation
(*c*) complete or partial stopage of fuel supply due to the vaporisation of fuel in supply system
(*d*) supply of liquid fuel particles to engine.

145. The octane number of compressed natural gas (CNG) is approximately
(*a*) 97 (*b*) 120 (*c*) 87 (*d*) 77.

146. The inlet valve closes after BDC for a low speed engine at
(*a*) 10° (*b*) 30° (*c*) 55° (*d*) 40°.

147. The increase of volumetric efficiency of a C.I. engine will increase
(*a*) B.P. (*b*) brake thermal efficiency
(*c*) bmep (*d*) CO.

148. Morse test can be easily applied to determine I.P. of
(*a*) single cylinder C.I. engine (*b*) multi-cylinder S.I. engine
(*c*) single cylinder S.I. engine (*d*) multi-cylinder C.I. engine.

149. Diesel engines as compared to petrol engines require
(*a*) bigger flywheel (*b*) smaller flywheel
(*c*) same size of flywheel (*d*) no flywheel.

150. The tendency of a petrol engine to knock increases by
(*a*) reducing the spark advance (*b*) scavenging
(*c*) increasing cetane number of fuel (*d*) supercharging
(*e*) both (*c*) and (*d*).

151. An engine indicator is used to determine the following
(*a*) imep (*b*) bmep (*c*) ihp (*d*) bhp.

152. The lubrication system is used to
(*a*) lubricate the components (*b*) cool the components
(*c*) decrease F.P. (*d*) all of these.

153. A S.I. engine can be run with maximum compression ratio about 14 if petrol is substituted with
(*a*) Compressed Natural Gas (CNG) (*b*) Liquified Natural Gas (LNG)
(*c*) LPG (*d*) methanol.
(*e*) both (*a*), (*b*).

154. The best method of measuring, F.P. of a single cylinder S.I. engine is
(*a*) Motoring (*b*) Morse test
(*c*) Willan's line method (*d*) indicator diagram.

155. The maximum brake thermal efficiency of a C.I. engine is about
(*a*) 20% (*b*) 40%
(*c*) 60% (*d*) 50%.

156. Wankel rotary engine is not generally used because of it has
(*a*) high speed (*b*) gas leakage problem
(*c*) difficult construction (*d*) high power.

157. Stirling engine is an engine having
(*a*) internal combustion (*b*) external combustion
(*c*) internal and external combustion (*d*) any of these.

158. The catalytic converter is used to control
(*a*) B.P. (*b*) I.P.
(*c*) oxides of nitrogen (*d*) hydrocarbon
(*e*) both (*c*) and (*d*).

159. Gas engines operate on
(*a*) Otto cycle (*b*) Diesel cycle (*c*) Dual cycle (*d*) Carnot cycle.

160. Overhead valve engine is also known as
(*a*) T-head (*b*) L-head (*c*) F-head (*d*) I-head.

161. Carburettor is designed, for operation during cruising range, at air/fuel ratio of
(*a*) 16 : 1 (*b*) 1 : 16 (*c*) 1 : 10 (*d*) 5 : 1.

162. Knocking tendency of a C.I. engine decreases with
(*a*) increase in speed (*b*) decrease in speed
(*c*) decrease in compression ratio (*d*) decrease in jacket water temperature.

163. A pre-combustion chamber gives
(*a*) low mechanical efficiency (*b*) clean exhaust
(*c*) high brake thermal efficiency (*d*) high volumetric efficiency.

164. Charge stratification permits
(*a*) high compression ratios (*b*) use of low jacket water temperatures
(*c*) use of high octane fuel (*d*) use of rich mixtures.

165. Percentage heat rejection to jacket water of an I.C. engine, at full-load, is around
(*a*) 10% (*b*) 15% (*c*) 30% (*d*) 80%.

166. Air cooling of engines is preferred because
(*a*) it is more efficient (*b*) it is compact
(*c*) cooling rate can be controlled (*d*) none of the above.

167. Scavenging efficiency of uniflow scavenging system is around
(*a*) 30% (*b*) 50% (*c*) 10% (*d*) 75%.

168. Supercharging of C.I. engines leads to
(*a*) lower s.f.c. (*b*) higher s.f.c.
(*c*) more exhaust pollution (*d*) rough engine run.

169. Morse test is meant for
(*a*) estimating s.f.c. (*b*) estimating volumetric efficiency
(*c*) estimating IP (*d*) estimating exhaust losses.

170. An engine-indicator gives a plot of
(*a*) T-V diagram (*b*) p-T diagram (*c*) p-V diagram (*d*) p-S diagram.

171. Wankel engine has
(*a*) high specific weight (*b*) low frictional losses
(*c*) low s.f.c. (*d*) high specific power.

172. The camshaft of a four-stroke diesel engine running at 1000 rpm will run at
(*a*) 1000 rpm (*b*) 500 rpm (*c*) 2000 rpm (*d*) none of the above.

173. Scavenging is usually done to increase
(*a*) fuel consumption (*b*) speed
(*c*) power output (*d*) none of these.

174. The function of a carburettor in a S.I. engine is to control
(*a*) air-fuel ratio (*b*) mixture of air and fuel
(*c*) speed (*d*) pressure drop between venturi and nozzle tip.

175. The amount of diesel in a C.I. engine is controlled by
(*a*) rack and pinion arrangement (*b*) throttle
(*c*) governor (*d*) nozzle.

176. The supercharging in a S.I. engine has a tendency to
(*a*) increase knocking (*b*) decrease knocking
(*c*) decrease volumetric efficiency (*d*) none of the above.

177. The lower heating value of gasoline is in the order of
(*a*) 40 000 kJ/kg (*b*) 44 000 kJ/kg
(*c*) 50 000 kJ/kg (*d*) 30 000 kJ/kg.

178. The function of a decompression lever is to
(*a*) start a diesel engine (*b*) start a petrol engine
(*c*) start both diesel and petrol engines (*d*) none of the above.

179. The Stirling engine may be called as an
(*a*) internal combustion engine (*b*) external combustion engine
(*c*) combination of internal and external combustion engines
(*d*) none of the above.

180. The most effective method of controlling S.I. engine exhaust emission by
(*a*) recirculating exhaust (*b*) using catalytic converter
(*c*) using some additives in the fuel (*d*) none of the above.

181. In a two-stroke S.I. engine lubrication is done by
(*a*) splash system (*b*) full pressure system
(*c*) mist lubrication system (*d*) dry sump lubrication system.

182. Gas chromatograph is used to measure
(*a*) oxygen content in the flue gas (*b*) carbon-dioxide content in flue gas
(*c*) amount of various elements in an alloy
(*d*) concentration of individual gases in a mixture of gases.

183. The Ringlemann chart is associated with the measurement of
(*a*) smoke density (*b*) concentration of carbon monoxide
(*c*) concentration of sulphur dioxide (*d*) viscosity of lubricating oil.

184. Engine misfiring is likely to result from
(*a*) spark plug gap too small (*b*) spark plug gap too wide
(*c*) vapour lock in the fuel line (*d*) incorrect fuel air mixture.

185. By advancing the spark timing in a S.I. engine, the possibility of knock will
(*a*) increase (*b*) decrease
(*c*) not change (*d*) be eliminated.

186. The inlet valve of a four-stroke cycle I.C. engine remains open for nearly
(*a*) 235° (*b*) 180° (*c*) 200° (*d*) 275°.

187. The accumulation of carbon in a cylinder results in increase of
(*a*) clearance volume (*b*) ignition delay
(*c*) effective compression ratio (*d*) volumetric efficiency.

188. In a cycle, the spark lasts roughly for
(*a*) 1 sec (*b*) 0.1 sec
(*c*) 0.01 sec (*d*) 0.001 sec.

189. The firing order in a six-cylinder S.I. engine is
(*a*) 1–5–3–6–2–4 (*b*) 1–3–6–5–2–4
(*c*) 1–6–2–5–4–3 (*d*) 1–5–2–6–3–4.

190. For low load operation, most economical engine is
(*a*) a S.I. engine (*b*) a C.I. engine
(*c*) a two-stroke engine (*d*) both S.I. and C.I. engines are equally good.

191. Thermal efficiency of a S.I. engine operating on lean mixture is
(*a*) higher
(*b*) lower
(*c*) independent of mixture ratio
(*d*) higher or lower depending on engine rating.

192. The concentration of oxides of nitrogen in the exhaust of a S.I. engine will be maximum when
(*a*) the fuel air mixture is 10% lean (*b*) the fuel air mixture is stoichiometric
(*c*) the fuel air mixture is 10% rich (*d*) the fuel air mixture is 20% rich.

193. The process of increasing the density of air before it enters the engine cylinder is known as
(*a*) scavenging (*b*) supercharging (*c*) knocking (*d*) pre-heating.

194. The ratio of specific heats for a real gas
(*a*) decreases with increase in temperature
(*b*) increases with increase in temperature
(*c*) is independent of change in temperature
(*d*) increases with temperature to a limit and then decreases with further increase in temperature.

195. In C.I. engines, ignition accelerators are added to
(*a*) increase combustion knock (*b*) decrease combustion temperature
(*c*) decrease cetane number (*d*) decrease chemical and physical delay.

196. HUCR is the highest compression ratio at which the
(*a*) engine can be run (*b*) engine gives maximum power output
(*c*) engine is most efficient
(*d*) fuel can be used in a test engine without knocking.

197. In a four-stroke cycle engine, the four operations namely suction, compression, expansion and exhaust are completed in the number of revolutions of crankshaft equal to
(*a*) four (*b*) three (*c*) two (*d*) one.

198. In a two-stroke cycle engine, the operations namely suction, compression, expansion and exhaust are completed in the number of revolutions of crankshaft equal to
(*a*) four (*b*) three (*c*) two (*d*) one.

199. In a four-stroke cycle S.I. engine the camshaft runs
(*a*) at the same speed as crankshaft (*b*) at half the speed of crankshaft
(*c*) at twice the speed of crankshaft
(*d*) at any speed irrespective of crankshaft speed.

200. The following is an S.I. engine
(*a*) diesel engine (*b*) petrol engine
(*c*) gas engine (*d*) none of the above.

201. The following is C.I. engine
(*a*) diesel engine (*b*) petrol engine
(*c*) gas engine (*d*) none of the above.

202. In a four-stroke cycle petrol engine, during suction stroke
(*a*) only air is sucked in (*b*) only petrol is sucked in
(*c*) mixture of petrol and air is sucked in (*d*) none of the above.

203. In a four-stroke cycle diesel engine, during suction stroke
(*a*) only air is sucked in (*b*) only fuel is sucked in
(*c*) mixture of fuel and air is sucked in (*d*) none of the above.

204. The two stroke cycle engine has
(*a*) one suction valve and one exhaust valve operated by one cam
(*b*) one suction valve and one exhaust valve operated by two cams
(*c*) only ports covered and uncovered by piston to effect charging and exhausting
(*d*) none of the above.

205. For same output, same speed and same compression ratio the thermal efficiency of a two-stroke cycle petrol engine as compared to that for four-stoke cycle petrol engine is
(*a*) more (*b*) less
(*c*) same as long as compression ratio is same
(*d*) same as long as output is same.

206. The ratio of brake power to indicated power of an I.C. engine is called
(*a*) mechanical efficiency (*b*) thermal efficiency
(*c*) volumetric efficiency (*d*) relative efficiency.

207. The specific fuel consumption of a diesel engine as compared to that for petrol engines is
(*a*) lower (*b*) higher
(*c*) same for same output (*d*) none of the above.

208. The thermal efficiency of petrol engine as compared to diesel engine is
(*a*) lower (*b*) higher
(*c*) same for same output (*d*) same for same speed.

209. Compression ratio of petrol engines is in the range of
(*a*) 2 to 3 (*b*) 7 to 10
(*c*) 16 to 20 (*d*) none of the above.

210. Compression ratio of diesel engines may have a range
(*a*) 8 to 10 (*b*) 10 to 15
(*c*) 16 to 20 (*d*) none of the above.

211. The thermal efficiency of good I.C. engine at the rated load is in the range of
(*a*) 80 to 90% (*b*) 60 to 70%
(*c*) 30 to 35% (*d*) 10 to 20%.

212. In case of S.I. engine, to have best thermal efficiency the fuel air mixture ratio should be
(*a*) lean (*b*) rich
(*c*) may be lean or rich (*d*) chemically correct.

213. The fuel air ratio, for maximum power of S.I. engines, should be
(*a*) lean (*b*) rich
(*c*) may be lean or rich (*d*) chemically correct.

214. In case of petrol engine, at starting
(*a*) rich fuel-air ratio is needed (*b*) weak fuel-air ratio is needed
(*c*) chemically correct fuel-air ratio is needed
(*d*) any fuel-air ratio will do.

215. Carburettor is used for
(*a*) S.I. engines (*b*) Gas engines
(*c*) C.I. engines (*d*) none of the above.

216. Fuel injector is used in
(*a*) S.I. engines (*b*) Gas engines
(*c*) C.I. engines (*d*) None of the above.

217. Very high speed engines are generally
(*a*) Gas engines (*b*) S.I. engines
(*c*) C.I. engines (*d*) Steam engines.

218. In S.I. engine, to develop high voltage for spark plug
(*a*) battery is installed (*b*) distributor is installed
(*c*) carburettor is installed (*d*) ignition coil is installed.

219. In S.I. engine, to obtain required firing order
(*a*) battery is installed (*b*) distributor is installed
(*c*) carburettor is installed (*d*) ignition coil is installed.

220. For petrol engines, the method of governing employed is
(*a*) quantity governing (*b*) quality governing
(*c*) hit and miss governing (*d*) none of the above.

221. For diesel engines, the method of governing employed is
(*a*) quantity governing (*b*) quality governing
(*c*) hit and miss governing (*d*) none of the above.

222. Voltage developed to strike spark in the spark plug is in the range :
(*a*) 6 to 12 volts (*b*) 1000 to 2000 volts
(*c*) 20000 to 25000 volts (*d*) none of the above.

223. In a 4-cylinder petrol engine the standard firing order is
(*a*) 1–2–3–4 (*b*) 1–4–3–2
(*c*) 1–3–2–4 (*d*) 1–3–4–2.

224. The torque developed by the engine is maximum
(*a*) at minimum speed of engine (*b*) at maximum speed of engine
(*c*) at maximum volumetric efficiency speed of engine
(*d*) at maximum power speed of engine.

225. Iso-octane content in a fuel for S.I. engines
(*a*) retards auto-ignition (*b*) accelerates auto-ignition
(*c*) does not affect auto-ignition (*d*) none of the above.

226. Normal heptane content in a fuel for S.I. engines
(*a*) retards auto-ignition (*b*) accelerates auto-ignition
(*c*) does not affect auto-ignition (*d*) none of the above.

227. The knocking in S.I. engines increase with
(*a*) increase in inlet air temperature (*b*) increase in compression ratio
(*c*) increase in cooling water temperature (*d*) all of the above.

228. The knocking in S.I. engines gets reduced
(*a*) by increasing the compression ratio (*b*) by retarding the spark advance
(*c*) by increasing inlet air temperature (*d*) by increasing the cooling water temperature.

229. Increasing the compression ratio in S.I. engines
(*a*) increases the tendency for knocking (*b*) decreases tendency for knocking
(*c*) does not affect knocking (*d*) none of the above.

230. The knocking tendency in petrol engines will increase when
(*a*) speed is decreased (*b*) speed is increased
(*c*) fuel-air ratio is made rich (*d*) fuel-air ratio made lean.

231. The ignition quality of fuels for S.I. engines is determined by
(*a*) cetane number rating (*b*) octane number rating
(*c*) calorific value rating (*d*) volatility of the fuel.

232. Petrol commercially available in India for Indian passenger cars has octane number in the range
(*a*) 40 to 50 (*b*) 60 to 70 (*c*) 80 to 85 (*d*) 95 to 100.

233. Cetane number of the fuel used commercially for diesel engine in India is in the range
(*a*) 80 to 90 (*b*) 60 to 80 (*c*) 60 to 70 (*d*) 40 to 45.

234. The knocking tendency in C.I. engines increases with
(*a*) decrease of compression ratio (*b*) increase of compression ratio
(*c*) increasing the temperature of inlet air (*d*) increasing cooling water temperature.

235. Desirable characteristic of combustion chamber for S.I. engines to avoid knock is
(*a*) small bore (*b*) short ratio of flame path to bore
(*c*) absence of hot surfaces in the end region of gas
(*d*) all of the above.

236. Accumulation of carbon deposits on the cylinder head of an I.C. engine leads to increase in ...
(*a*) piston displacement (*b*) clearance volume
(*c*) compression ratio (*d*) swept volume.

237. Which of the following motor cycles has more than one cylinder ?
(*a*) Bullet (*b*) Yezdi
(*c*) Rajdoot (*d*) Yamah.

238. A diesel engine is generally more efficient than a petrol engine because of
(*a*) proper air fuel mixing and combustion (*b*) high calorific value of diesel fuel
(*c*) knock-free operation (*d*) high compression ratio.

239. Vapour lock refers to which of the following ?
(*a*) Excess supply of fuel to engine (*b*) Blocking of carburettor jets
(*c*) Complete or partial stoppage of fuel supply due to vapourisation of fuel in the supply line
(*d*) Supply of air-fuel mixture containing liquid particles.

240. White deposits on the face of a spark plug indicates that
(*a*) the engine is excessively advanced (*b*) mixture is too rich
(*c*) gap between the electrodes is too large (*d*) the rating of the spark plug is too high.

241. The injection presence in diesel engines is of the order of
(*a*) 30-40 bar (*b*) 100-150 bar (*c*) 170-220 bar (*d*) 400-600 bar.

242. The ignition temperature of diesel fuel is about
(*a*) 200°C (*b*) 400°C (*c*) 550°C (*d*) 700°C.

243. In a petrol engine the delay period is of the order of
(*a*) 0.001 s (*b*) 0.002 s (*c*) 0.015 s (*d*) 0.06 s.

244. ......is not the effect of detonation.
(*a*) Loud and pulsating noise (*b*) High local stresses
(*c*) High operating temperature (*d*) Loss in efficiency and power output.

245. The ignition quality of a petrol engine fuel is expressed as
(*a*) octane number (*b*) cetane number
(*c*) API gravity (*d*) SAE rating.

246. The use of tetraethyl lead in gasoline is being gradually discontinued since its presence
(*a*) decreases the engine speed (*b*) blocks the catalytic converter
(*c*) makes the fuel costly (*d*) gives bad odour.

247. Which of the following fuels has a cetane number of 100 ?
(*a*) Normal heptane (*b*) Ethyl fluid
(*c*) Cetane (*d*) α-methyl napthalene.

248. During idling a petrol engine requires
(*a*) lean mixture (*b*) rich mixture
(*c*) variable mixture (*d*) chemically correct mixture.

249. Stoichiometric ratio is
(*a*) actual ratio of air to fuel for maximum efficiency
(*b*) chemically correct air-fuel ratio by weight
(*c*) chemically correct air-fuel ratio by volume
(*d*) none of the above.

250. The capacity of most of the mopads in India is
(*a*) 50 CC (*b*) 150 CC (*c*) 200 CC (*d*) 250 CC.

251. Due to which of the following reasons diesel engines are preferred for road transport ?
(*a*) Complete combustion of charge (*b*) Low operating cost
(*c*) Low specific fuel consumption over a large range of load
(*d*) Easy starting.

252. ......is used for the insulating body of a spark plug
(*a*) Dolomite (*b*) Alumina (*c*) Glass (*d*) Silica.

253. The spark advance is usually specified in terms of
(*a*) degrees of crank rotation (*b*) time in seconds
(*c*) engine speed in rev./sec. (*d*) none of the above.

254. In an I.C. engine if intake air temperature increases, its efficiency will
(*a*) decrease (*b*) increase
(*c*) remain same (*d*) cannot be predicted.

255. Which of the following statements is *incorrect* ?
(*a*) Petrol engines work on Otto cycle
(*b*) For same power output petrol engines occupy more space than diesel engines
(*c*) In a four-stroke engine a power stroke is obtained in four strokes
(*d*) Thermal efficiency of four-stroke engine is more due to positive scavenging.

256. In a diesel engine if one of the cylinders receives more fuel than the others then which of the following will happen for that cylinder ?
(*a*) Exhaust temperature will be high (*b*) Exhaust will be smoky
(*c*) Piston rings would stick into piston grooves
(*d*) Engine will start overheating (*e*) All of the above.

257. The carbon accumulation in an engine cylinder results in
(*a*) increase of effective compression ratio (*b*) increase of volumetric efficiency
(*c*) increase of clearance volume (*d*) increase of ignition time.

258. The compression ratio in diesel engine is ... in comparison to expansion ratio
(*a*) less (*b*) more (*c*) same (*d*) variable.

259. In an automobile the magneto is basically
(*a*) d.c. generator (*b*) a.c. generator (*c*) transformer (*d*) capacitor.

260. Scavenging is usually done to increase which of the following ?
(*a*) Power output (*b*) Fuel consumption
(*c*) Thermal efficiency (*d*) Speed.

261. For a petrol engine for vehicles the air fuel ratio for maximum power generation is of the order of
(*a*) 8 : 1 (*b*) 12 : 1 (*c*) 18 : 1 (*d*) 20 : 1.

262. In loop scavenging the top of the piston is
(*a*) convex shaped (*b*) depressed (*c*) slanted (*d*) contoured.

263. Which of the following statements is *correct* regarding normal heptane ?
(*a*) It retards auto-ignition (*b*) It accelerates auto-ignition
(*c*) It helps to resist auto-ignition (*d*) It does not affect auto-ignition.

264. Due to which of the following reasons the piston rings are plated with chromium, cadmium or phosphate ?
(*a*) To prevent clogging (*b*) To improve heat transfer
(*c*) To reduce wear and eliminate scuffing (*d*) To improve surface finish.

265. The specific gravity of diesel oil is
(*a*) 0.6 (*b*) 0.75 (*c*) 0.85 (*d*) 1.2.

266. Detonation can be controlled by
(*a*) reducing the r.p.m. (*b*) retarding the spark timing
(*c*) varying compression ratio (*d*) any of the above.

267. Due to which of the following, the tendency of a diesel engine to knock increases ?
(*a*) Increase in engine speed (*b*) Increase in compression ratio
(*c*) Increase in octane value of fuel (*d*) Increase in engine power.

268. The part load efficiency of a carburettor is
(*a*) constant (*b*) maximum (*c*) optimum (*d*) poor.

269. ... can work on very lean mixture.
(*a*) C.I. engine (*b*) S.I. engine
(*c*) Two stroke engine (*d*) Four stroke engine.

270. Thermal efficiency of I.C. engine on weak mixture is
(*a*) lower (*b*) higher (*c*) unaffected (*d*) unpredictable.

271. Cetane number is the measure of
(*a*) viscosity of fuel (*b*) ignition quality
(*c*) calorific value of fuel (*d*) auto-ignition temperature.

272. In a S.I. engine an ignition coil performs which of the following functions ?
(*a*) Regulates battery voltage (*b*) Avoids sparking
(*c*) Controls spark (*d*) Supplies high voltage to the spark plug.

273. ... does not assist in getting higher output from diesel engine.
(*a*) High fuel air ratio (*b*) High compression ratio
(*c*) High excess air (*d*) Fine atomisation of fuel.

274. Which of the following factors does not promote detonation in S.I. engines ?
(*a*) High self ignition temperature of fuel
(*b*) Increase in inlet pressure and temperature of charge
(*c*) Higher compression ratio (*d*) Advanced spark timing.

275. Which of the following statements is correct ? The phenomenon of pre-ignition
(*a*) always occurs in petrol engines (*b*) always occurs in diesel engines
(*c*) never occurs in diesel engines (*d*) increases the power output of an engine.

276. The octane rating of the commercially available petrol in India is
(*a*) 15-35 (*b*) 45-55 (*c*) 60-70 (*d*) 85-90.

277. .... lubrication technique is used for lubrication of the cylinder of a scooter engine.
(*a*) Petrol (*b*) Splash (*c*) Gravity feed (*d*) Forced feed.

278. In 4-stroke engines the camshaft rotates at ... the crankshaft speed
(*a*) half (*b*) three-fourth (*c*) equal (*d*) double.

279. Which of the following is the distinctive features of an I.C. engine ?
(*a*) Easy and instantaneous starting (*b*) High overall efficiency
(*c*) Low weight to power ratio
(*d*) Combustion and conversion of heat energy into mechanical work occur inside a cylinder
(*e*) All of the above.

280. The minimum number of rings in a piston are
(*a*) two (*b*) three (*c*) four (*d*) six.

281. ... process is not associated with Diesel cycle.
(*a*) Constant pressure (*b*) Constant volume
(*c*) Adiabatic (*d*) Isothermal.

282. Highest useful compression ratio is the compression ratio at which
(*a*) the engine consumes minimum fuel for a particular power output
(*b*) the engine gives maximum power output

(*c*) the engine maintains operating pressures and temperatures within prescribed limits
(*d*) the engine can operate without detonation.

283. A 2-stroke cycle engine as compared to 4-stroke cycle engine
(*a*) has lower fuel consumption (*b*) can be easily started
(*c*) is smaller in size for the same output (*d*) has lesser shocks and vibrations.

284. For which of the following engines a prony brake is used to measure brake power ?
(*a*) single cylinder engine (*b*) low speed engine
(*c*) low power engine (*d*) variable speed engine.

285. What does scavenging air mean ?
(*a*) Burnt air containing combustion products
(*b*) Air sent under compression
(*c*) Forced air for cooling the engine cylinder
(*d*) Air used for forcing the burnt gases out of cylinder during the exhaust period.

286. ..... is the basic requirement of a good combustion chamber.
(*a*) Low volumetric efficiency
(*b*) High compression ratio
(*c*) Low compression ratio
(*d*) High power output and high thermal efficiency.

287. ....is the method of governing used in petrol engine.
(*a*) Quality governing (*b*) Hit and miss governing
(*c*) Quantity governing (*d*) Partial governing.

288. ..... is the method of governing used in Diesel engine.
(*a*) Quality governing (*b*) Hit and miss governing
(*c*) Quantity governing (*d*) Any of the above.

289. Hunting occurs due to which of the following ?
(*a*) Faulty governor (*b*) Poor-control by the governor
(*c*) Over-control by the governor (*d*) Bad engine design.

290. Maximum torque is generated by an engine when
(*a*) it runs at lowest speed (*b*) it develops maximum power
(*c*) it consumes maximum fuel (*d*) it runs at maximum speed.

291. With an increase of the number of cylinders in a multicylinder engine the power to weight ratio
(*a*) decreases (*b*) increases
(*c*) remains unaffected (*d*) none of the above.

292. What will happen if petrol is used in diesel engine ?
(*a*) Black smoke will be produced (*b*) Low power will be produced
(*c*) Higher knocking will occur (*d*) Efficiency will be low.

293. What will happen if diesel is fed by mistake in the oil tank of a petrol engine ?
(*a*) The engine will not run (*b*) The engine will knock
(*c*) The engine will detonate (*d*) The engine will give lot of smoke.

294. With which of the following tendency of detonation is S.I. engines increases ?
(*a*) Increase of compression ratio (*b*) Decrease of compression ratio
(*c*) Increase of engine speed (*d*) Decrease of engine speed.

295. Performance number are ......
(*a*) indicative of the fuels having anti-knock qualities superior to iso-octane
(*b*) indicative of the fuels having anti-knock qualities superior to cetane

(c) indices of efficiency of petrol engines
(d) indices of efficiency of diesel engines.

296. The bi-fuel engine uses which of the following ?
(a) Gas fuel during start up and liquid fuel as the basic fuel
(b) Liquid fuel during start up and gas as the basic fuel
(c) Two fuels used in two combustion chambers
(d) None of the above.

297. Lean air-fuel mixture is required for
(a) idling (b) acceleration (c) starting (d) cruising.

298. ....is not a part of petrol engine.
(a) Air filter (b) Induction coil
(c) Value mechanism (d) Fuel injector.

299. Regarding 2-stroke engines which of the following statements is *correct* ?
(a) There is only one valve for inlet and exhaust
(b) Charge enters the engine cylinder through ports only
(c) A diesel engine cannot operate on 2-stroke cycle
(d) Compression ratio is always lower than that of 4-stroke cycle engine.

300. Which of the following statements is *incorrect* ?
(a) Choke is kept open when cranking a cold engine for starting
(b) A carburettor prepares a homogeneous air-fuel mixture by atomising and vapourising the fuel
(c) Throttle valve controls the supply of air-fuel mixture
(d) Vaccum at the throat of venturi sucks the fuel through the fuel jet.

301. .... of heat supplied in the form of fuel in a 4-stroke engine is carried away by exhaust gases.
(a) 3-7 percent (b) 8-12 percent (c) 20-35 percent (d) 45-55 percent.

302. Petrol engines are adjusted to give minimum brake specific fuel consumption at
(a) no load (b) 20-30 percent of full load
(c) about 70 percent of full load (d) near full load.

303. Regarding contact breaker which of the following statements is *incorrect* ?
(a) Spark takes place when the points are open
(b) Excessive gas results in rapid burning of points
(c) Contact points are generally made of tungsten
(d) Points are opened by the cam and closed by the spring tension.

304. ...... regulates the pressure strokes in the fuel injection pump of a diesel engine.
(a) Pump shaft (b) Control rack
(c) Lift of plunger (d) Needle valve.

305. Which of the following could be the probable reason of power loss in a diesel engine ?
(a) Low injection pressure (b) Restricted exhaust
(c) Ineffective cooling (d) Clogging of aircleaner.

306. .... is not a part of magneto-ignition system.
(a) Condenser (b) Induction coil
(c) Battery (d) Circuit breaker.

307. Due to which of the following reasons the petrol engines are usually not supercharged ?
(*a*) Drop in volumetric efficiency (*b*) Increased knocking
(*c*) Increased specific fuel consumption (*d*) Power loss.

308. .... identifies the anti-knock quality of diesel fuel.
(*a*) Octane number (*b*) Cetane number
(*c*) either of the above (*d*) none of the above.

309. Volumetric efficiency of a well designed engine may be in the range
(*a*) below 20 percent (*b*) 30-40 percent
(*c*) 50-70 percent (*d*) 75-90 percent.

310. The advancing of spark timing in a S.I. engine will
(*a*) reduce knocking tendency (*b*) increase knocking tendency
(*c*) not have any effect (*d*) none of the above.

311. Free piston engines find application in
(*a*) gas turbines (*b*) mining installations
(*c*) compressed air supply (*d*) supercharging of diesel engines.

312. In a C.I. engine higher combustion chamber wall temperature will
(*a*) reduce exhaust temperature (*b*) reduce knocking tendency
(*c*) increase knocking tendency (*d*) have no effect.

313. ... acts as ignition accelerator for C.I. engines fuel.
(*a*) Hydrogen peroxide (*b*) Acetone peroxide
(*c*) *n* heptane (*d*) none of the above.

314. Why are fuel ignition accelerators added in C.I. engine ?
(*a*) To reduce combustion chamber temperature
(*b*) To reduce combustion knock (*c*) To accelerate combustion knock
(*d*) To increase delay period.

315. In which of the following engines crankcase explosion occurs ?
(*a*) S.I. engines (*b*) 4-stroke S.I. engines
(*c*) 2-stroke C.I. engines (*d*) 4-stroke C.I. engines.

316. Highest useful compression ratio is the compression ratio at which
(*a*) an engine operates smoothly (*b*) detonation first becomes audible
(*c*) an engine can be safely operated
(*d*) an engine gives maximum thermal efficiency.

317. Crankshafts are generally
(*a*) die cast (*b*) sand cast
(*c*) forged (*d*) turned from bar stock.

318. .... has maximum resistance to detonation.
(*a*) Alcohol (*b*) Benzene (*c*) Toulene (*d*) Iso-octane.

319. Why are oil rings slotted ?
(*a*) To minimise friction (*b*) To remove oil from cylinder
(*c*) To reduce the bulk
(*d*) To provide an escape for the oil that the slot edges cut from the cylinder wall.

320. In isochronous governors the speed drop is
(*a*) zero (*b*) 5 percent (*c*) 30 percent (*d*) 50 percent.

321. The top ring nearest to the piston crown is known as
(*a*) compression ring (*b*) oil ring
(*c*) scrapper ring (*d*) groove ring.

322. A diesel engine as compared to petrol engine (both running at full load) is
(*a*) less efficient (*b*) more efficient
(*c*) equally efficient (*d*) none of the above.

323. The level of oil in engine cylinder should be checked when the engine is
(*a*) running (*a*) not running (*c*) during starting (*d*) during cranking.

324. Endurance for I.C. engines is conducted for
(*a*) 200 hours (*b*) 300 hours (*c*) 400 hours (*d*) 500 hours.

325. Movement of air inside engine cylinder does not help in
(*a*) reducing noise (*b*) mixing of fuel with air
(*c*) distribution of fuel (*d*) reduction of after burning.

326. What is swirl in C.I. engines ?
(*a*) Circular motion imparted to suction air
(*b*) Radial motion imparted to fuel-air mixture
(*c*) Directional movement of fuel spray
(*d*) Circular motion imparted to gases after combustion.

327. In a C.I. engine squish is created
(*a*) towards the end of compression stroke (*b*) at the end of suction stroke
(*c*) at the beginning of suction stroke (*d*) during combustion.

328. An increase in the mean effective pressure of a diesel engine with fixed compression ratio can be obtained with increase in
(*a*) cut-off ratio (*b*) engine speed
(*c*) back pressure (*d*) charge density.

329. By which of the following a 2 stroke engine is usually identified ?
(*a*) Absence of valves (*b*) Size of flywheel
(*c*) Location of fuel tank (*d*) Weight of engine.

330. The knocking tendency in S.I. engines can be decreased by
(*a*) adding benzole (*b*) decreasing compression ratio
(*c*) controlling ignition timing (*d*) adding dopes (like tetraethyl lead etc.).

331. Due to which of the following injection lag in diesel engines is caused ?
(*a*) Leakage past the fuel-oil plunger (*b*) Compressibility of fuel
(*c*) Expansion of fuel-oil discharge lines under high pressure
(*d*) All of the above.

332. For reducing wear and eliminate scuffing, the piston rings are
(*a*) lubricated (*b*) made of cast iron
(*c*) provided with stepped groove (*d*) plated with chromium or cadmium.

333. Due to which of the following reasons a diesel engine gives a smoky exhaust ?
(*a*) Water in the fuel (*b*) Fuel injection is late
(*c*) Fuel is not distributed equally to all the cylinders
(*d*) Exhaust valve receives too much lube oil
(*e*) All of the above.

334. Compared to a diesel engine the compression ratio in petrol engine is kept low because
(*a*) engine design becomes simpler (*b*) it provides fuel economy
(*c*) petrol engine is a light engine
(*d*) higher compression ratio (in petrol engine) would lead to pre-ignition of fuel.

335. Compared to petrol engines, diesel engines require
(*a*) smaller flywheel (*b*) bigger flywheel
(*c*) same size flywheel (*d*) no flywheel.

336. Free acids in diesel oil for diesel engine leas to which of the following ?
(*a*) Excessive fuel consumption (*b*) Deposition on engine parts
(*c*) Damaging of both the storage tank and the engine
(*d*) Excessive engine wear.

337. Which type of cleaner in case of diesel engines is most effective ?
(*a*) Wet type (*b*) Dry type (*c*) Oil bath type (*d*) Whirl type.

338. What happens when cooling water temperature in petrol engine is increased ?
(*a*) The knocking tendency increases (*b*) The knocking tendency decreases
(*c*) The knocking tendency remains unaffected
(*d*) Unpredictable.

339. In a petrol engine .... gas gets exhausted out without burning and without transformation.
(*a*) CO (*b*) CO_2 (*c*) Nitrogen (*d*) O_2.

340. ...... smooths out power impulses from an I.C. engine.
(*a*) Flywheel (*b*) Gear box (*c*) Governor (*d*) Crankshaft.

341. What is the advantage of reversing the flow of air in an air cleaner ?
(*a*) The velocity of air is increased (*b*) The velocity of air is reduced
(*c*) The air flow is increased
(*d*) A large percentage of foreign matter is thrown out.

342. In scooters fins are provided over engine cylinder for
(*a*) good appearance (*b*) higher efficiency
(*c*) higher strength of cylinder (*d*) better cooling.

343. Which of the following systems of lubrication is used for motor cycles and scooters ?
(*a*) Wet sump method (*b*) Splash lubrication
(*c*) Forced lubrication system
(*d*) Mixing about 5 percent lub. oil with petrol.

344. How can the ignition timing of a multicylinder petrol engine be adjusted ?
(*a*) By adjusting ignition coil position (*b*) By rotating the crank
(*c*) By rotating the distributor (*d*) By adjusting the spark plug gap.

345. In I.C. engines leakage past the piston rings and valves ...... with increase in speed.
(*a*) decreases (*b*) increases
(*c*) remains same (*d*) none of the above.

346. Why are the exhaust pipes of engines covered with insulating material ?
(*a*) To conserve heat (*b*) To keep the exhaust pipes warm
(*c*) To reduce heat transfer to the engine room
(*d*) To increase the engine efficiency.

347. A fuel will detonate less if
(*a*) it has constant self ignition temperature
(*b*) it has lower self ignition temperature
(*c*) it has higher self ignition temperature
(*d*) none of the above.

348. Why are the pistons usually given a coating (*e.g.*, tin coating) ?
(*a*) To increase lubrication effect (*b*) To conduct heat efficiently
(*c*) To reduce weight (*d*) To reduce possibility of scoring.

349. For preventing knock in S.I. engines, the charge away from spark plug should have
(*a*) rich mixture (*b*) low temperature
(*c*) long ignition delay (*d*) low density
(*e*) all of the above.

350. Due to which of the following violent sound pulsations within the cylinder of an I.C. engine are caused ?
(*a*) Pre-ignition (*b*) Heavy supercharging
(*c*) Detonation (*d*) Heavy turbulence.

351. By which of the following air is supplied in a naturally aspirated diesel engine ?
(*a*) A centrifugal blower (*b*) A super charger
(*c*) Forced chamber (*d*) A vacuum chamber.

352. ....is the highest and most volatile liquid fuel.
(*a*) Kerosene (*b*) Gasoline (*c*) Diesel (*d*) Fuel oil.

353. By installing a supercharger on a 4-stroke cycle diesel engine upto....percent increase in power can result in.
(*a*) 20 (*b*) 40 (*c*) 70 (*d*) 100.

354. In.... engines supercharging is essential.
(*a*) marine (*b*) petrol (*c*) aircraft (*d*) diesel.

355. By which of the following methods diesel smoke can be reduced ?
(*a*) Adherence to proper fuel specification (*b*) Using additives in the fuel
(*c*) Avoidance of overloading (*d*) Reducing maximum flow of fuel
(*e*) All of the above.

356. The exhaust valve of an engine is ...in size in comparison to inlet valve.
(*a*) smaller (*b*) bigger
(*c*) same (*d*) varies from design to design.

357. On which of the following cycles are most high speed engines experated ?
(*a*) Two stroke cycle (*b*) Carnot cycle
(*c*) Diesel cycle (*d*) Dual cycle.

358. Aluminium alloy is usually used to make engines pistons because it
(*a*) wears less (*b*) is lighter
(*c*) is stronger (*d*) absorbs shocks.

359. With which of the following engines is the term scavenging associated ?
(*a*) Aero engines (*b*) Diesel engines
(*c*) High efficiency engines (*d*) None of the above.

360. The rating of a diesel engine will...with increase in air-inlet temperature
(*a*) increase parabolically (*b*) increase linearly
(*c*) decrease parabolically (*d*) decrease linearly.

361. Which of the following statements is *correct* ? For the same compression ratio
(*a*) Diesel cycle is more efficient than Otto cycle
(*b*) Otto cycle is more efficient than diesel cycle
(*c*) Both diesel and otto cycles are equally efficient
(*d*) Compression ratio has no relation with efficiency.

362. Due to which of the following compression loss in I.C. engines occurs ?
(*a*) Clogged air-inlet slots (*b*) Use of thick head gasket
(*c*) Leaking piston rings (*d*) All of the above.

363. An engine indicator is used to determine
(*a*) temperature (*b*) m.e.p. and I.P.
(*c*) speed (*d*) volume of cylinder.

364. The camshaft of a 4-stroke I.C. engine running at 2000 r.p.m. will run at
(*a*) 2000 r.p.m. (*b*) 1500 r.p.m. (*c*) 1000 r.p.m. (*d*) 500 r.p.m.

365. In a cycle the spark lasts for
(*a*) 0.001 s (*b*) 0.01 s (*c*) 0.1 s (*d*) 1 s.

366. By which of the following is the air pressure produced in the crankcase method of scaveng-ing ?
(*a*) Natural aspiration (*b*) Movement of engine piston
(*c*) Supercharger (*d*) None of the above.

AUTOMOBILE ENGINEERING

367. ...... is not the material for automobile pistons.
(*a*) Cast iron (*b*) Cast steel
(*c*) Steel forgings (*d*) Aluminium alloys.

368. By which of the following methods are automobile connecting rods mass produced ?
(*a*) Die casting (*b*) Forging
(*c*) Cold heading (*d*) Fine sand casting.

369. Most cars have ...... engine.
(*a*) free piston (*b*) rotary Wankel
(*c*) two stroke cycle (*d*) four stroke cycle.

370. Which of the following types of car batteries are generally used in India ?
(*a*) Lead-acid battery (*b*) Dry battery
(*c*) Nickel-cadmium battery (*d*) Nickel-iron battery.

371. An automobile engine is usually mounted at
(*a*) two points (*b*) three points
(*c*) four points (*d*) five points.

372. Which of the following statements regarding four-wheel drive is *correct* ?
(*a*) All the four wheels are powered (*b*) All the four wheels can be steered
(*c*) Vehicle has four wheels (*d*) None of the above.

373. In the differential unit of a passenger car the gear ratio is of the order of
(*a*) 3 : 1 (*b*) 6 : 1 (*c*) 8 : 1 (*d*) 10 : 1.

374. ...... batteries are generally used in automobiles.
(*a*) 6 V (*b*) 12 V (*c*) 24 V (*d*) 48 V.

375. Which of the following acids is used in automobile battery ?
(*a*) Hydrochloric acid (*b*) Nitric acid
(*c*) Sulphuric acid (*d*) Any of the above.

376. Due to which of the following reasons automobile engines are usually designed as multicylinder engines ?
(*a*) High efficiency (*b*) Lower fuel consumption
(*c*) Better balance, uniform torque output (*d*) None of the above.

377. Due to which of the following reasons engines are usually rubber mounted ?
(*a*) To prevent road shocks from reaching the engine
(*b*) To reduce the transmission of vibration between the engine and body

(*c*) To prevent flow of electricity between the engine and body
(*d*) To prevent the heat from passing between the engine and body.

378. ....gear boxes are used in four wheel drive.
(*a*) Two (*b*) Three (*c*) Four (*d*) Six.

379. Which automobile car engine has three cylinders ?
(*a*) Standard (*b*) Ambassador (*c*) Maruti-800 (*d*) Premier Padmini.

380. The drive from the gear box to the near axle is taken by
(*a*) clutch (*b*) universal joint
(*c*) propeller shaft (*d*) differential gear.

381. Where is the Hook's joint used in an automobile car ?
(*a*) Between gear box and propeller shaft (*b*) Between flywheel and clutch
(*c*) Between differential gear and wheels (*d*) Between clutch and gear box.

382. ..... is the part of the vehicle which holds the passengers and the cargo to be transported.
(*a*) Hull (*b*) Cabin (*c*) Chasis (*d*) Aft.

383. In V-8 engine there are ...... exhaust manifolds.
(*a*) two (*b*) three (*c*) four (*d*) six.

384. ...... is usually used to drive engine dynamo.
(*a*) Gear drive (*b*) Chain drive
(*c*) V-belt drive (*d*) Flat belt drive.

385. .....is generally provided with four wheel drive.
(*a*) Metador (*b*) Padmini car
(*c*) Ambassador car (*d*) Jeep.

386. .... is commonly used antifreeze solution in automobiles.
(*a*) Freon-12 (*b*) Liquid ammonia
(*c*) Glycol (*d*) Carbon disulphide.

387. The dynamo in an automobile
(*a*) converts mechanical energy into electrical energy
(*b*) continually recharge the battery
(*c*) acts as a reservoir of electrical energy
(*d*) supplies electric power.

388. What is wheel base of a vehicle ?
(*a*) It is width of tyres (*b*) It is the distance between front tyres
(*c*) It is the distance between front and rear axles
(*d*) It is the extreme length of the vehicle.

389. Which of the following is the diesel engined vehicle ?
(*a*) Maruti (*b*) Premier Padmini
(*c*) Standard Gazel (*d*) None of the above.

390. The number of cylinders in a Ambassador car is
(*a*) 3 (*b*) 4 (*c*) 6 (*d*) 8.

391. ......is not a part of the hydraulic braking system.
(*a*) Wheel cylinder (*b*) Master cylinder
(*c*) Steering mechanism (*d*) Brake pedal.

392. What is the efficiency of hydraulic braking system ?
(*a*) 20 to 30 percent (*b*) 40 to 50 percent
(*c*) 65 to 75 percent (*d*) about 90 percent.

393. What is the efficiency of mechanical brakes ?
(*a*) 25-35 percent (*b*) 60-70 percent
(*c*) 90-95 percent (*d*) 100 percent.

394. When does an automobile type wear rapidly ?
(*a*) If it is incorrectly inflated (*b*) If it is misaligned
(*c*) If it is overloaded
(*d*) If it is more frequently subjected to braking
(*e*) any of the above.

395. On which of the following brake lining is mounted ?
(*a*) Master cylinder (*b*) Wheel cylinder
(*c*) Brake shoe (*d*) Brake drum.

396. ...tractor has an air cooled engine.
(*a*) HMT (*b*) Eicher (*c*) Hindustan (*d*) Ford.

397. Chetak slippage while clutch is engaged is particularly noticeable
(*a*) during braking (*b*) at low speed
(*c*) during idling (*d*) during acceleration.

398. Radiator tubes are generally made of
(*a*) cast iron (*b*) steel (*c*) plastics (*d*) brass.

399. Automobile gears are generally made of
(*a*) cast iron (*b*) stainless steel (*c*) alloy steel (*d*) mild steel.

400. ...is used to check the state of charge of a battery.
(*a*) Battery charger (*b*) Hygrometer
(*c*) Hydrometer (*d*) Battery eliminator.

401. The level of electrolyte in automobile's battery should be
(*a*) exactly at the level of the plates (*b*) 5-10 mm below the top of plates
(*c*) 15 mm below the top of plates (*d*) 10-15 mm above top of plates.

402. Which could be the probable cause for hard steering in a vehicle ?
(*a*) Excessive castor (*b*) Bent wheel spindle
(*c*) Low tyre pressure (*d*) Tie rod ends tight
(*e*) Any of the above.

403. With which of the following the term master cylinder is associated ?
(*a*) Starting mechanism (*b*) Cooling cylinder
(*c*) Steering mechanism (*d*) Braking starting.

404. Odometer is an instrument used for
(*a*) smoke analysis (*b*) B.P. measurement
(*c*) distance measurement (*d*) any of the above.

405. In automobile engines a thermostat is provided for
(*a*) regulating the temperature of suction air
(*b*) regulating the temperature of lubricating oil
(*c*) controlling the temperature of the cooling system
(*d*) regulating the temperature of exhaust gases.

406. What will happen if there is no king pin offset in a vehicle ?
(*a*) Braking effort will be high (*b*) Starting steering effort will be zero
(*c*) Starting steering effort will be high (*d*) Wobbling of wheels will increase.

407. Dynamo in automobile is a
(*a*) shunt generator (*b*) series generator
(*c*) either of the above (*d*) none of the above.

408. By which of the following methods the problem of exhaust can be handled ?
(*a*) Treatment of exhaust gases (*b*) Modification of engine design
(*c*) Fuel modification (*d*) All of the above.

409. Automobile horns are ...... type.
(*a*) sinusoidal and digital (*b*) primary and secondary
(*c*) wind cone and discone (*d*) any of the above.

410. On which of the following vehicles power assisted brakes are often provided ?
(*a*) Slow speed vehicles (*b*) Heavy duty vehicles
(*c*) Motor cycles (*d*) Air cooled engine vehicles.

411. What is the tilt of the car wheels from the vertical called ?
(*a*) Slip angle (*b*) Camber, wheel rake
(*c*) Castor (*d*) None of the above.

412. What is the function of differential in automobiles ?
(*a*) It allows rear wheel movement
(*b*) It permits two rear wheels to have flexibility of relative speed, whenever it is required
(*c*) It permits two rear wheels to run independently
(*d*) It reduces speed of propeller shaft to suit the requirement of wheel axes.

413. A temperaturo indicator provided for automobiles indicates temperature of
(*a*) engine piston (*b*) engine cylinder walls
(*c*) air surrounding radiator (*d*) jacket cooling water.

414. ...is a front wheel drive.
(*a*) Standerd gazel (*b*) Maruti car
(*c*) Premier Padmini (*d*) Ambassador car.

415. ......are used to hold the pressure plate against the clutch plate ?
(*a*) Springs (*b*) Levers (*c*) Struts (*d*) Thrust bearings.

416. The majority of low and medium powered engines have ...... cylinders.
(*a*) two (*b*) three (*c*) four (*d*) six.

417. The clutch is located between the engine and
(*a*) universal joint (*b*) differential gear
(*c*) gear box (*d*) rear axle.

418. The driving wheels of a vehicle are carried by
(*a*) crank and slider (*b*) crankshaft
(*c*) axles (*d*) propeller shaft.

419. What is the maximum pressure intensity which the clutch facing can withstand without being damaged ?
(*a*) 20 kPa (*b*) 50 kPa (*c*) 150 kPa (*d*) 500 kPa.

420. Which type of engine should be used where the overhead clearance is small ?
(*a*) Radial engine (*b*) Vertical engine
(*c*) V-type engine (*d*) Horizontal engine.

421. Regarding a torque convertor which of the following statements is *incorrect* ?
(*a*) Maximum torque multiplication occurs at low speed
(*b*) The oil is driven by the impeller unit

(*c*) The blades have a curved shape
(*d*) The stator unit redirects the flow of oil to the impeller.

422. Engine and braking torque cause
(*a*) lateral bending of side members (*b*) longitudinal torsion
(*c*) distortion of frame to parallelogram shape
(*d*) bending of side members in vertical plane.

423. In a tractor the springs provided for the rear wheels are
(*a*) helical springs (*b*) leaf spring
(*c*) either (*a*) or (*b*) (*d*) no springs are provided.

424. Which type of springs is widely used for suspension in light and heavy commercial vehicles ?
(*a*) Semi-elliptic leaf spring (*b*) Tapered leaf spring
(*c*) Coil spring (*d*) None of the above.

425. The word castor is associated with which of the following ?
(*a*) Transmission system (*b*) Braking system
(*c*) Front axle alignment (*d*) None of the above.

426. What is the requirement of a good steering system ?
(*a*) It should provide directional stability (*b*) It must be accurate and easy to handle
(*c*) It should require minimum efforts to steer
(*d*) All of the above.

427. A suspension system is employed to
(*a*) safeguard occupants from road shocks
(*b*) keep the vehicle stable in pitching or rolling, while in motion
(*c*) check the road shocks being transmitted to the vehicle components
(*d*) all of the above.

428. The front wheel drive as compared to rear wheel drive
(*a*) requires longer propeller shaft (*b*) gives better riding performance
(*c*) has a greater skidding tendency
(*d*) provides increased tractive effort when going up steep gradient.

429. On which of the following vehicles two speed reverse gear arrangement is generally provided ?
(*a*) Jeep and military vehicles (*b*) Tractors
(*c*) Cars (*d*) None of the above.

430. ...... gears are not used in the final drive gearing system.
(*a*) Spur (*b*) Straight bevel
(*c*) Hypoid (*d*) Spiral.

431. A shackle with a leaf spring
(*a*) prevents squeaking sound (*b*) allows the spring length to change
(*c*) provides good traction (*d*) allows pivoting of spring end.

432. What is the disadvantage of the radial ply tyre as compared to cross ply tyre ?
(*a*) Uneven braking (*b*) Higher cornering power
(*c*) Lower rolling resistance (*d*) Uncomfortable ride at low speeds.

433. For cars which is the most popular manual steering gear ?
(*a*) Worm and nut type (*b*) Worm and wheel type
(*c*) Cam and roller type (*d*) Rack and pinion type.

434. A brake lining is usually made of
(*a*) fabric (*b*) leather (*c*) cork (*d*) asbestos.

435. The brake bleeding system serves to free the system from

(*a*) excess pressure (*b*) excess fluid (*c*) air (*d*) none of the above.

436. Circumferential grooves are provided on automobile tyres to

(*a*) reduce danger of skidding (*b*) increase load carrying capacity

(*c*) prevent good traction (*d*) all of the above.

(B) Match List I with List II and select the correct answer using the codes given below the lists.

437.

List I	List II
A. Farm equipment	1. Maruti
B. Public conveyance	2. Tata
C. Passenger car	3. HMT
D. Good transportation	4. Hero

Codes :

	A	B	C	D
(*a*)	2	4	3	1
(*b*)	3	4	1	2
(*c*)	1	2	3	4
(*d*)	1	3	4	2

438.

List I	List II
A. Combustion process at constant volume occurs in	1. Auto-ignition
B. Combustion process at constant pressure occurs in	2. Combustion knock
C. ....in the I.C. engines is produced by the spontaneous combustion or auto-ignition of an appreciable portion of the charge	3. S.I. or otto cycle
D. ......temperature of an air fuel mixture is the lowest temperature at which chemical reaction proceeds at a rate sufficient to result eventually in inflammation	4. C.I. or diesel cycle

Codes :

	A	B	C	D
(*a*)	3	4	2	1
(*b*)	1	2	4	3
(*c*)	1	3	2	4
(*d*)	4	2	3	1

439.

List I	List II
A. The principal source of exhaust CO is...combustion	1. Nozzle
B. ...distribute the air or the air and fuel to the various cylinders of multicylinder engines	2. Engine power
C. Supercharging permits more fuel to be burned and is practical means to greater......	3. Intake manifolds

D. For diesel engines the fuel injection system usually consists of a pump fuel line, and...... — 4. Rich mixture

Codes :

	A	B	C	D
(*a*)	4	3	2	1
(*b*)	2	1	4	3
(*c*)	2	3	1	4
(*d*)	3	4	1	2

440.

List I	List II
A. Bicycle	1. Kirloslar
B. Compressor	2. Atlas
C. Scooter	3. Escorts
D. Tractor	4. Bajaj

Codes :

	A	B	C	D
(*a*)	1	2	4	3
(*b*)	2	3	1	4
(*c*)	2	1	4	3
(*d*)	3	4	2	1

441.

List I	List II
A. Spark plugs are usually located near ......so that flame progresses towards the cooler part of combustion chamber.	1. Vapour lock
B. ....refers to complete or partial stoppage of fuel supply due to vapourisation of fuel in the supply line.	2. Low injection pressure
C. Under idling conditions the throttle valve of a petrol engine is more or less closed. That provides......	3. Exhaust valve
D. The probable reason of power loss in a diesel engine is......	4. Rich mixture

Codes :

	A	B	C	D
(*a*)	1	2	4	3
(*b*)	3	1	4	2
(*c*)	2	3	1	4
(*d*)	2	4	3	1

442.

List I	List II
A. ......regulates the pressure stroke in the fuel injection pump of a diesel engine.	1. minimum
B. A normal diesel engine will need minimum changes if made to run on.....	2. carburettor

C. Maximum power air fuel ratios require......spark advance — 3. kerosene

D. The...atomises and mixes the fuel with the air flowing to the engine. — 4. control rack

Codes :

	A	B	C	D
(*a*)	1	2	3	4
(*b*)	2	3	4	1
(*c*)	4	3	1	2
(*d*)	4	3	2	1

C. COMPETITIVE EXAMINATIONS QUESTIONS (With Solutions-Comments)

443. Consider the following statements comparing I.C. engines and gas turbines

1. Gas turbines are simple, compact and light in weight.
2. Complete expansion of working substance is possible in I.C. engines and not in gas turbines.
3. There is flexibility in the design of different components of gas turbines as different processes take place in different components.
4. Even low grade fuels can be burnt in gas turbines.

Of these statements

(*a*) 1, 2 and 3 are correct (*b*) 1, 3 and 4 are correct

(*c*) 2, 3 and 4 are correct (*d*) 1, 2 and 4 are correct.

444. Which of the following factors increase detonation in the S.I. engine ?

1. Increased spark advance.
2. Increased speed.
3. Increased air-fuel ratio beyond stoichiometric strength.
4. Increased compression ratio.

Select the correct answer using the codes given below :

Codes :

(*a*) 1 and 3 (*b*) 2 and 4

(*c*) 1, 2 and 4 (*d*) 1 and 4.

445. Consider the following statements :

I. The performance of an S.I. engine can be improved by increasing the compression ratio.

II. Fuels of higher octane number can be employed at higher compression ratio.

Of these statements

(*a*) both I and II are true (*b*) both I and II are false

(*c*) I is true but II is false (*d*) I is false but II is true.

446. Besides mean effective pressure, the data needed for determining the indicated power of an engine would include

(*a*) piston diameter, length of stroke and calorific value of fuel

(*b*) piston diameter, specific fuel consumption and calorific value of fuel

(*c*) piston diameter, length of stroke and speed of rotation

(*d*) specific fuel consumption, speed or rotation and torque.

447. Which one of the following curves is a proper representation of pressure differential (y-axis) vs velocity of air (x-axis) at the throat of a carburettor ?

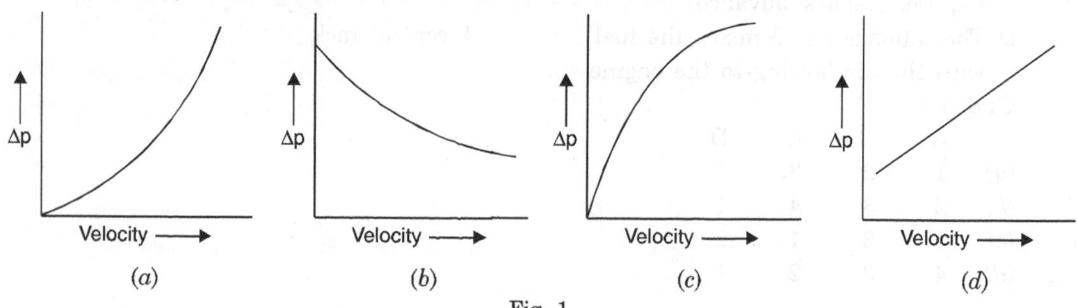

Fig. 1

448. For a typical automobile C.I. engine, for conditions of increasing engine speed match List I with List II and select the correct answer using the codes given below the lists :

List I *(Performance parameter)*	List II *(Tendency qualitatively)*
A. Power output	1. Increasing and then increasing
B. Torque	2. Decreasing and then increasing
C. Brake specific fuel consumption	3. Increasing throughout the range
	4. Decreasing throughout the range

Codes :

	A	B	C
(*a*)	1	2	3
(*b*)	1	4	3
(*c*)	2	3	4
(*d*)	3	1	2

449. Match List I with List II and select the correct answer using the codes given below the lists :

List I	List II
A. Pre-combustion chamber	1. Compression swirl
B. Turbulent chamber	2. Masked inlet valve
C. Open combustion chamber	3. Spark ignition
D. F-head combustion chamber	4. Combustion induced swirl
	5. M-chamber

Codes :

	A	B	C	D
(*a*)	4	5	3	2
(*b*)	1	3	5	2
(*c*)	2	3	1	5
(*d*)	4	1	2	3

450. Match List I with List II and select the correct answer using the codes given below the lists :

List I *(S.I. Engine problem)*	List II *(Characteristic of fuel responsible for the problem)*
A. Cold starting	1. Front end volatility
B. Carburettor icing	2. Mild-range volatility
C. Crankcase dilution	3. Tail end volatility.

Codes :

	A	B	C
(*a*)	1	2	3
(*b*)	1	3	2
(*c*)	2	3	1
(*d*)	3	1	2

***451.** **Match List I and List II and select the correct answer using the codes given below the lists :**

List I	**List II**
(Elements of a complete carburettor)	*(Rich-mixture requirement)*
A. Idling system	1. To compensate for dilution of charge
B. Economiser	2. For cold starting
C. Acceleration pump	3. For meeting maximum power range of operation
D. Choke	4. For meeting rapid opening of throttle

Codes :

	A	B	C	D
(*a*)	1	2	3	4
(*b*)	1	3	4	2
(*c*)	2	3	4	1
(*d*)	4	1	2	3

452. Consider the following statements regarding *n*-cetane :

1. It is a standard fuel used for knock rating of diesel engines
2. Its chemical name is *n*-hexadecane
3. It is a saturated hydrocarbon of paraffin series
4. It has long carbon chain structure.

Of the above statements :

(*a*) 1, 3 and 4 are correct (*b*) 1, 2 and 3 are correct

(*c*) 1, 2 and 4 are correct (*d*) 2, 3 and 4 are correct.

453. List I gives the different terms related to combustion while List II gives the outcome of the events that follow. Match List I with List II and select the correct answer using the codes given below the lists :

List I	**List II**
A. Association	1. Pseudo shock
B. Dissociation	2. Knock
C. Flame front	3. Endothermic
D. Abnormal combustion	4. Exothermic

Codes :

	A	B	C	D
(*a*)	3	4	1	2
(*b*)	4	3	1	2
(*c*)	3	4	2	1
(*d*)	4	3	2	1

***454.** Which one of the following engines will have heavier flywheel than the remaining ones ?

(*a*) 40 H.P. four-stroke petrol engine running at 1500 rpm

(*b*) 40 H.P. two-stroke petrol engine running at 1500 rpm

(*c*) 40 H.P. two-stroke diesel engine running at 750 rpm

(*d*) 40 H.P. four-stroke diesel engine running at 750 rpm.

455. Consider the following statements :

Knock in the S.I. engine can be reduced by

1. supercharging
2. retarding the spark
3. using a fuel of long straight chain structure
4. increasing the engine speed

Of these statements :

(*a*) 1 and 2 are correct (*b*) 2 and 3 are correct

(*c*) 1, 3 and 4 are correct (*d*) 2 and 4 are correct.

456. Consider the following statements :

The injector nozzle of a CI engine is required to inject fuel at a sufficiently high pressure in order to

1. be able to inject fuel in a chamber of high pressure at the end of the compression stroke
2. inject fuel at high velocity to facilitate atomisation
3. ensure that penetration is not high

Of the above statements :

(*a*) 1 and 2 are correct (*b*) 1 and 3 are correct

(*c*) 2 and 3 are correct (*d*) 1, 2 and 3 are correct.

457. Match List I with List II and select the correct answer using the codes given below the lists :

List I *(S.I. engine operating mode)*	**List II** *(Desired air-fuel ratio)*
A. Idling	1. 13.0
B. Cold starting	2. 4.0
C. Cruising	3. 16.0
D. Full throttle	4. 9.0

Codes :

	A	B	C	D
(*a*)	4	3	2	1
(*b*)	2	4	1	3
(*c*)	4	2	1	3
(*d*)	2	4	3	1.

458. Compensating jet in a carburettor supplies almost constant amount of petrol at all speeds because

(*a*) the jet area is automatically varied depending on the suction

(*b*) the flow from the main jet is diverted to the compensating jet with increase in speed

(*c*) the diameter of the jet is constant and the discharge coefficient is invariant

(*d*) the flow is produced due to the static head in the float chamber.

459. In the context of performance evaluation of I.C. Engine, match List I with List II and select the correct answer using the codes given below lists :

List I *(Parameter)*	**List II** *(Equipment for measurement)*
A. Brake power (B.P.)	1. Bomb calorimeter
B. Engine speed	2. Electrical tachometer
C. Calorific value of fuel	3. Hydraulic dynamometer
D. Exhaust emissions	4. Flame ionisation detector

Codes :

	A	B	C	D
(*a*)	3	1	2	4
(*b*)	4	2	1	3
(*c*)	3	2	1	4
(*d*)	2	3	4	1.

460. Consider the following statements :

In open cycle turbo-jet engines used in military aircraft, reheating the exhaust gas from the turbine by burning, more fuel is used to increase

1. thrust
2. the efficiency of engine
3. the range of aircraft

Of these statements :

(*a*) 1 and 3 are correct (*b*) 1 and 2 are correct
(*c*) 2 and 3 are correct (*d*) 1, 2 and 3 are correct.

461. In a turbojet engine, subsequent to heat addition to compressed air, to get the power output, the working substance is expanded in

(*a*) turbine blades, which is essentially an isentropic process
(*b*) turbine blades, which is essentially an isothermal process
(*c*) exit nozzle, which is essentially an isentropic process
(*d*) exit nozzle, which is a constant volume process.

462. Consider the following statements relating to rocket engines :

1. The combustion chamber in a rocket engine is directly analogous to the reservoir of a supersonic wind tunnel
2. Stagnation conditions exist at the combustion chamber
3. The exit velocities of exhaust gases are much higher than those in jet engines
4. Efficiency of rocket engines is higher than that of jet engines

Of these statements :

(*a*) 1, 3 and 4 are correct (*b*) 2, 3 and 4 are correct
(*c*) 1, 2 and 3 are correct (*d*) 1, 2 and 4 are correct.

463. Only rocket engines can be propelled to 'SPACE' because

(*a*) they can generate very high thrust (*b*) they have high propulsion efficiency
(*c*) these engines can work on several fuels (*d*) they are not air-breathing engines.

464. Items given in List I and List II pertain to gas analysis. Match List I with List II and select the correct answer using the codes given below the lists :

List I	**List II**
A. CO_2	1. Alkaline pyrogallol
B. Orsat apparatus	2. KOH solution

C. CO	3. Wet analysis
D. O_2	4. Ammonical cuprous chloride
	5. Dry analysis

Codes :

	A	B	C	D
(*a*)	2	3	1	4
(*b*)	1	3	2	4
(*c*)	1	5	4	2
(*d*)	2	5	4	1.

465. Which of the following factors are responsible for the formation of NO_x in spark ignition engine combustion ?

1. Incomplete combustion
2. High temperature
3. Availability of oxygen

Select the correct answer using the codes given below :

(*a*) 2 and 3 (*b*) 1 and 3 (*c*) 1 and 2 (*d*) 1, 2 and 3.

466. Consider the following statements :

1. Gas cooled thermal reactors use CO_2 or helium as coolant and require no separate moderator.
2. Fast reactors use heavy water as moderator and coolant.
3. Liquid metal fast breeder reactors use molten sodium as coolant.

Of these statements :

(*a*) 1 and 3 are correct (*b*) 2 and 4 are correct
(*c*) 3 and 4 are correct (*d*) 1 and 2 are correct.

467. Match List I with List II and select the correct answer using the codes given below the lists :

List I	List II
A. Plutonium-239	1. Fissile material
B. Thorium-233	2. Fissionable material
C. Cadmium	3. Moderator
D. Graphite	4. Poison

Codes :

	A	B	C	D
(*a*)	1	2	3	4
(*b*)	2	1	3	4
(*c*)	1	2	4	3
(*d*)	2	1	4	3.

468. If methane undergoes combustion with the stoichiometric quantity of air, the air-fuel ratio on molar basis would be

(*a*) 15.22 : 1 (*b*) 12.30 : 1 (*c*) 14.56 : 1 (*d*) 9.52 : 1.

469. The presence of nitrogen in the products of combustion ensure that

(*a*) complete combustion of fuel takes place
(*b*) incomplete combustion of fuel occurs
(*c*) dry products of combustion are analysed
(*d*) air is used for the combustion.

470. For maximum specific output of a constant volume cycle (otto cycle)
(*a*) the working fluid should be air
(*b*) the speed should be high
(*c*) suction temperature should be high
(*d*) temperature of the working fluid at the end of compression and expansion should be equal.

471. A two-stroke engine has a speed of 750 rpm. A four-stroke engine having an identical cylinder size runs at 1500 rpm. The theoretical output of the two-stroke engine will
(*a*) be twice that of the four-stroke engine
(*b*) be half that of the four-stroke engine
(*c*) be the same as that of the four-stroke engine
(*d*) depend upon whether it is C.I. or S.I. engine.

472. For same power output and same compression, as compared to two-stroke engines, four-stroke S.I. engines have
(*a*) higher fuel consumption
(*b*) lower thermal efficiency
(*c*) higher exhaust temperatures
(*d*) higher thermal efficiency.

473. In a S.I. engine, which one of the following is the correct order of the fuels with increasing detonation tendency ?
(*a*) Paraffins, Olefins, Naphthenes, Paraffins, Olefins
(*b*) Aromatics, Naphthenes, Paraffins, Olefins
(*c*) Naphthenes, Olefins, Aromatics, Paraffins
(*d*) Aromatics, Naphthenes, Olefins, Paraffins.

474. Consider the following statements :
Detonation in the S.I. engine can be suppressed by
1. retarding the spark timing
2. increasing the engine speed
3. using 10% rich mixture
Of these statements :
(*a*) 1 and 3 are correct
(*b*) 2 and 3 are correct
(*c*) 1, 2 and 3 are correct
(*d*) 1 and 2 are correct.

475. Which one of the following figures correctly represents the variation of thermal efficiency (y-axis) with mixture strength (x-axis) ?

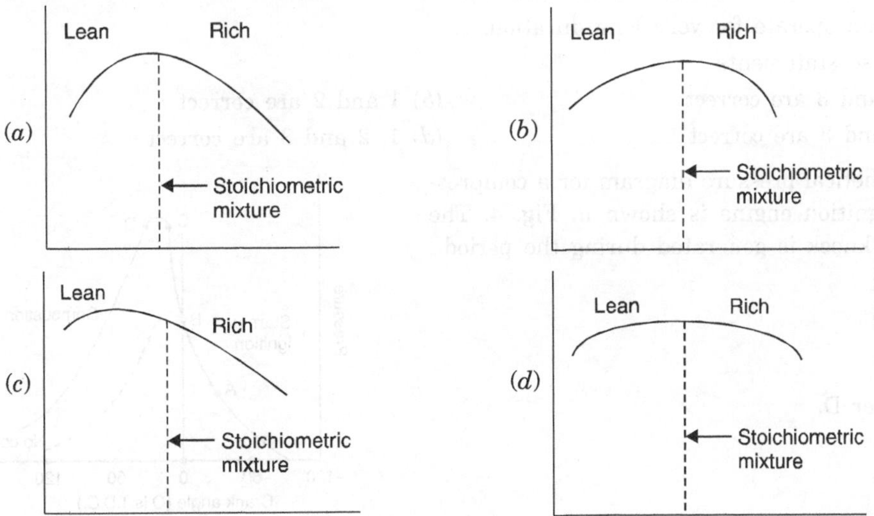

Fig. 2

476. Match List I with the performance curves and select the correct answer using the codes given below the List :

List I
(Performance parameter of an I.C. engine)

A. Indicated
B. Volumetric efficiency
C. Brake power
D. Specific fuel consumption

List II
(Performance curves)

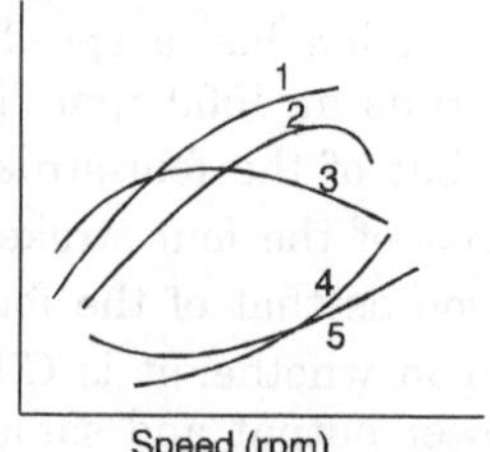

Fig. 3

Codes :

	A	B	C	D
(*a*)	1	3	2	5
(*b*)	1	3	2	4
(*c*)	1	2	3	5
(*d*)	2	1	4	3.

477. Consider the following statements :

1. Volumetric efficiency of diesel engines is higher than that of S.I. engines.
2. When a S.I. engine is throttled, its mechanical efficiency decreases.
3. Specific fuel consumption increases as the power capacity of the engine increases.
4. Inspite of higher compression ratios, the exhaust temperature in diesel engines is much lower than that in S.I. engines.

On these statements :

(*a*) 1, 2, 3 and 4 are correct
(*b*) 1, 2 and 3 are correct
(*c*) 3 and 4 are correct
(*d*) 1, 2 and 4 are correct.

478. Consider the following statements about a rocket engine :

1. It is very simple in construction and operation.
2. It can attain very high vehicle velocity.
3. It can operate for very long duration.

Of these statements :

(*a*) 1 and 3 are correct
(*b*) 1 and 2 are correct
(*c*) 2 and 3 are correct
(*d*) 1, 2 and 3 are correct.

479. Hypothetical pressure diagram for a compression ignition engine is shown in Fig. 4. The diesel knock is generated during the period

(*a*) AB
(*b*) BC
(*c*) CD
(*d*) after D.

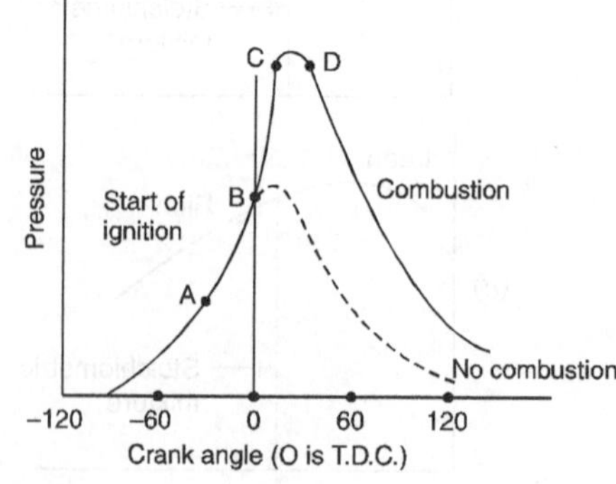

Fig. 4

ANSWERS

I. Choose the Correct Answer

1. (*c*) **2.** (*c*) **3.** (*b*) **4.** (*d*) **5.** (*d*) **6.** (*b*)
7. (*d*) **8.** (*b*) **9.** (*b*) **10.** (*a*) **11.** (*a*) **12.** (*b*)
13. (*a*) **14.** (*b*) **15.** (*c*) **16.** (*d*) **17.** (*c*) **18.** (*b*)
19. (*b*) **20.** (*b*) **21.** (*a*) **22.** (*d*) **23.** (*c*) **24.** (*d*)
25. (*a*) **26.** (*c*) **27.** (*b*) **28.** (*d*) **29.** (*b*) **30.** (*d*)
31. (*b*) **32.** (*a*) **33.** (*a*) **34.** (*b*) **35.** (*b*) **36.** (*b*)
37. (*b*) **38.** (*c*) **39.** (*a*) **40.** (*a*) **41.** (*b*) **42.** (*a*)
43. (*b*) **44.** (*c*) **45.** (*c*) **46.** (*c*) **47.** (*d*) **48.** (*d*)
49. (*d*) **50.** (*b*) **51.** (*a*) **52.** (*b*) **53.** (*b*) **54.** (*b*)
55. (*c*) **56.** (*b*) **57.** (*b*) **58.** (*d*) **59.** (*b*) **60.** (*a*)
61. (*b*) **62.** (*b*) **63.** (*a*) **64.** (*c*) **65.** (*b*) **66.** (*a*)
67. (*d*) **68.** (*c*) **69.** (*b*) **70.** (*d*) **71.** (*d*) **72.** (*d*)
73. (*a*) **74.** (*b*) **75.** (*b*) **76.** (*b*) **77.** (*b*) **78.** (*d*)
79. (*d*) **80.** (*d*) **81.** (*a*) **82.** (*e*) **83.** (*e*) **84.** (*e*)
85. (*d*) **86.** (*c*) **87.** (*b*) **88.** (*b*) **89.** (*b*) **90.** (*b*)
91. (*b*) **92.** (*d*) **93.** (*b*) **94.** (*c*) **95.** (*c*) **96.** (*c*)
97. (*d*) **98.** (*d*) **99.** (*d*) **100.** (*c*) **101.** (*b*) **102.** (*b*)
103. (*c*) **104.** (*c*) **105.** (*d*) **106.** (*b*) **107.** (*c*) **108.** (*c*)
109. (*b*) **110.** (*c*) **111.** (*b*) **112.** (*d*) **113.** (*c*) **114.** (*a*)
115. (*a*) **116.** (*d*) **117.** (*d*) **118.** (*b*) **119.** (*b*) **120.** (*c*)
121. (*d*) **122.** (*b*) **123.** (*a*) **124.** (*d*) **125.** (*d*) **126.** (*c*)
127. (*a*) **128.** (*d*) **129.** (*b*) **130.** (*d*) **131.** (*c*) **132.** (*b*)
133. (*b*) **134.** (*a*) **135.** (*d*) **136.** (*a*) **137.** (*a*) **138.** (*a*)
139. (*d*) **140.** (*c*) **141.** (*a*) **142.** (*b*) **143.** (*a*) **144.** (*c*)
145. (*b*) **146.** (*a*) **147.** (*a*) **148.** (*b*) **149.** (*a*) **150.** (*e*)
151. (*a*) **152.** (*d*) **153.** (*e*) **154.** (*a*) **155.** (*b*) **156.** (*b*)
157. (*b*) **158.** (*e*) **159.** (*a*) **160.** (*d*) **161.** (*a*) **162.** (*b*)
163. (*d*) **164.** (*a*) **165.** (*c*) **166.** (*b*) **167.** (*d*) **168.** (*a*)
169. (*c*) **170.** (*c*) **171.** (*b*) **172.** (*b*) **173.** (*c*) **174.** (*b*)
175. (*a*) **176.** (*a*) **177.** (*b*) **178.** (*a*) **179.** (*b*) **180.** (*b*)
181. (*c*) **182.** (*d*) **183.** (*a*) **184.** (*b*) **185.** (*a*) **186.** (*a*)
187. (*c*) **188.** (*d*) **189.** (*a*) **190.** (*b*) **191.** (*a*) **192.** (*b*)
193. (*b*) **194.** (*a*) **195.** (*d*) **196.** (*d*) **197.** (*c*) **198.** (*d*)
199. (*b*) **200.** (*b*) **201.** (*a*) **202.** (*c*) **203.** (*a*) **204.** (*c*)
205. (*b*) **206.** (*a*) **207.** (*a*) **208.** (*a*) **209.** (*b*) **210.** (*c*)
211. (*c*) **212.** (*a*) **213.** (*b*) **214.** (*a*) **215.** (*a*) **216.** (*c*)
217. (*b*) **218.** (*b*) **219.** (*b*) **220.** (*a*) **221.** (*b*) **222.** (*c*)
223. (*d*) **224.** (*c*) **225.** (*a*) **226.** (*b*) **227.** (*d*) **228.** (*b*)
229. (*a*) **230.** (*a*) **231.** (*b*) **232.** (*c*) **233.** (*d*) **234.** (*a*)
235. (*d*) **236.** (*c*) **237.** (*d*) **238.** (*d*) **239.** (*c*) **240.** (*a*)
241. (*b*) **242.** (*b*) **243.** (*b*) **244.** (*d*) **245.** (*a*) **246.** (*b*)
247. (*c*) **248.** (*b*) **249.** (*b*) **250.** (*a*) **251.** (*c*) **252.** (*b*)
253. (*a*) **254.** (*a*) **255.** (*b*) **256.** (*e*) **257.** (*a*) **258.** (*b*)
259. (*a*) **260.** (*a*) **261.** (*b*) **262.** (*d*) **263.** (*b*) **264.** (*c*)
265. (*c*) **266.** (*b*) **267.** (*b*) **268.** (*d*) **269.** (*a*) **270.** (*b*)

271. (*b*)	**272.** (*d*)	**273.** (*c*)	**274.** (*a*)	**275.** (*c*)	**276.** (*d*)
277. (*a*)	**278.** (*a*)	**279.** (*e*)	**280.** (*a*)	**281.** (*d*)	**282.** (*d*)
283. (*c*)	**284.** (*c*)	**285.** (*d*)	**286.** (*d*)	**287.** (*c*)	**288.** (*a*)
289. (*c*)	**290.** (*a*)	**291.** (*a*)	**292.** (*c*)	**293.** (*a*)	**294.** (*a*)
295. (*a*)	**296.** (*b*)	**297.** (*d*)	**298.** (*d*)	**299.** (*b*)	**300.** (*a*)
301. (*c*)	**302.** (*d*)	**303.** (*b*)	**304.** (*b*)	**305.** (*a*)	**306.** (*c*)
307. (*b*)	**308.** (*b*)	**309.** (*d*)	**310.** (*b*)	**311.** (*a*)	**312.** (*b*)
313. (*b*)	**314.** (*b*)	**315.** (*a*)	**316.** (*b*)	**317.** (*c*)	**318.** (*d*)
319. (*d*)	**320.** (*a*)	**321.** (*a*)	**322.** (*b*)	**323.** (*b*)	**324.** (*d*)
325. (*a*)	**326.** (*a*)	**327.** (*a*)	**328.** (*a*)	**329.** (*a*)	**330.** (*d*)
331. (*d*)	**332.** (*d*)	**333.** (*e*)	**334.** (*d*)	**335.** (*b*)	**336.** (*c*)
337. (*c*)	**338.** (*a*)	**339.** (*c*)	**340.** (*a*)	**341.** (*d*)	**342.** (*d*)
343. (*d*)	**344.** (*c*)	**345.** (*a*)	**346.** (*c*)	**347.** (*c*)	**348.** (*d*)
349. (*e*)	**350.** (*c*)	**351.** (*d*)	**352.** (*b*)	**353.** (*d*)	**354.** (*c*)
355. (*e*)	**356.** (*b*)	**357.** (*d*)	**358.** (*b*)	**359.** (*c*)	**360.** (*d*)
361. (*b*)	**362.** (*d*)	**363.** (*b*)	**364.** (*c*)	**365.** (*a*)	**366.** (*b*)
367. (*c*)	**368.** (*b*)	**369.** (*d*)	**370.** (*a*)	**371.** (*b*)	**372.** (*a*)
373. (*a*)	**374.** (*b*)	**375.** (*c*)	**376.** (*c*)	**377.** (*b*)	**378.** (*a*)
379. (*c*)	**380.** (*c*)	**381.** (*a*)	**381.** (*a*)	**382.** (*a*)	**383.** (*a*)
384. (*c*)	**385.** (*d*)	**386.** (*c*)	**387.** (*b*)	**388.** (*c*)	**389.** (*d*)
390. (*b*)	**391.** (*c*)	**392.** (*d*)	**393.** (*b*)	**394.** (*e*)	**395.** (*c*)
396. (*b*)	**397.** (*d*)	**398.** (*d*)	**399.** (*c*)	**400.** (*c*)	**401.** (*d*)
402. (*e*)	**403.** (*d*)	**404.** (*c*)	**405.** (*c*)	**406.** (*c*)	**407.** (*a*)
408. (*d*)	**409.** (*c*)	**410.** (*b*)	**411.** (*b*)	**412.** (*b*)	**413.** (*d*)
414. (*b*)	**415.** (*a*)	**416.** (*c*)	**417.** (*c*)	**418.** (*c*)	**419.** (*b*)
420. (*d*)	**421.** (*a*)	**422.** (*d*)	**423.** (*d*)	**424.** (*a*)	**425.** (*c*)
426. (*d*)	**427.** (*d*)	**428.** (*b*)	**429.** (*c*)	**430.** (*a*)	**431.** (*b*)
432. (*d*)	**433.** (*d*)	**434.** (*d*)	**435.** (*c*)	**436.** (*a*).	

B. Match the List I and II

437. (*b*)	**438.** (*a*)	**439.** (*a*)	**440.** (*c*)	**441.** (*b*)	**442.** (*c*).

C. Competitive Examination Questions

443. (*a*)	**444.** (*d*)	**445.** (*d*)	**446.** (*c*)	**447.** (*c*)	**448.** (*d*)
449. (*a*)	**450.** (*a*)	**451.** (*b*)	**452.** (*b*)	**453.** (*b*)	**454.** (*a*)
455. (*d*)	**456.** (*d*)	**457.** (*a*)	**458.** (*d*)	**459.** (*c*)	**460.** (*a*)
461. (*a*)	**462.** (*c*)	**463.** (*d*)	**464.** (*d*)	**465.** (*a*)	**466.** (*c*)
467. (*c*)	**468.** (*a*)	**469.** (*d*)	**470.** (*d*)	**471.** (*c*)	**472.** (*d*)
473. (*d*)	**474.** (*c*)	**475.** (*c*)	**476.** (*b*)	**477.** (*a*)	**478.** (*b*)
479. (*b*).					

SOLUTIONS-COMMENTS

443. 1, 2 and 3 statements are correct, Statement 4 is incorrect because low grade fuel cannot be burnt in gas turbines. Thus choice (*a*) is correct.

444. In S.I. Engines, detonation increases due to increase in spark advance and increase in compression ratio. Increase in speed and increase in air-fuel ratio beyond stoichiometric strength do not effect detonation much.

445. Statement I is false since the performance of an S.I. engine *cannot* be improved by increasing the compression. Statement II is true ; since high octane number tends to suppress detonation, therefore, to some extent fuels of higher octane number will prove useful at higher compression ratio. Thus (*d*) is the correct choice.

446. The formula indicated of power (I.P.) involves p_mLAN ; *i.e.*, I.P. depends upon mean effective pressure (p_m), length of stroke (L), piston diameter $\left(\text{Area A} = \frac{\pi}{4} D^2\right)$, and speed rotation (*n*). Thus (*c*) is the correct choice.

447. $\Delta p \propto v^2$, this relationship is shown in curve (*c*).

451. **Idling System** compensates dilution of charge ; economiser is used for meeting maximum power range of operation ; **acceleration pump** for meeting rapid opening of throttle, and choke for cold starting, thus (*b*) is the correct choice.

454. Because four-stroke engines require heavier flywheels as power stroke comes only once every four strokes and also petrol engine is running at the highest r.p.m.

D. Fill in the Blanks

1. Detonation in S.I. engines is caused by the .... of the charge to burn, while knock in C.I. engines is caused by the .... of the charge to burn.
2. Of all the three-phases of combustion process in a C.I. engine, the .... is the most important.
3. While volatility of the fuel is a determining factor in S.I. engines, the .... of the fuel is the determining factor in C.I. engines.
4. Octane number of fuel means the percentage of .... in a mixture of .... and ....
5. .... and .... are reference fuels for measuring octane number of S.I. engine fuels.
6. .... and .... are reference fuels for measuring cetane number of C.I. engine fuels.
7. .... is done for increasing the .... efficiency of a diesel engine.
8. The quantity of fuel in a .... engine is controlled by the rotation of fuel pump plunger by .... and .... arrangement.
9. The function of a carburettor is to control .... ratio and .... of mixture.
10. Crankcase dilutation is caused if the S.I. engine fuels are ...... volatile and vapour lock characteristics are caused if the S.I. engine fuels are .... volatile.
11. The Stirling engines are .... combustion engines and would be popular in......... sector.
12. Wankel rotary engines are of very .... speed but have some .... problem.
13. Engine exhaust emissions can be measured accurately by an .... exhaust gas ....
14. Chemically correct air-fuel ratio is called .... ratio and the ratio of actual mass of air to the theoretical mass of air in a diesel engine is called .... efficiency.
15. Ignition delay of fuel .... as the carbon-hydrogen ratio in the molecules increases.
16. The general formulae for paraffins is ....
17. I.C. engine .... is designed to remove about 30 per cent of the heat produced in the .... chamber.
18. The most commonly used firing order for a six-cylinder four-stroke engine is ....
19. The vibration of the .... induced by a variable torque is called .... vibration.
20. The chemically correct air-fuel ratio is called ....
21. Iso-octane is arbitrarily rated .... octane number.
22. The efficiency of a 4-stroke engine is .... than that of a 2-stroke engine.
23. By supercharging the .... and .... of a Diesel engine can be increased.

24. The procedure of increasing the inducted mass of air in an engine is called ....
25. The cetane number of diesel is ....
26. The unit of the measurement of exhaust emission such as the oxides of nitrogen is ....
27. The most popular device of controlling engine exhaust emissions is ....
28. Alcohol is a .... energy.
29. Supercharging .... the power output of an I.C. engine for a given .... of the engine.
30. In S.I. engine detonation occurs near the .... of combustion, whereas in C.I. engine detonation occurs near the .... of combustion.
31. During operation of an engine the piston is mainly subjected to .... force and .... force.
32. Willan's line method of determining frictional power of I.C. engines is applicable only to .... engines.
33. In dual-fuel engine the liquid fuel injected towards the end of compression is called .... fuel and acts as a .... of ignition.
34. A diesel engine with pre-combustion chamber produces less NO_x than a direct injection engine due to .... temperature.
35. The amount of fuel in the diesel engine is controlled by .... and .... arrangement of the fuel pump.
36. In an I.C. engine, the temperature of the cooling system is kept constant with the help of instrument called ....
37. In the 2-stroke S.I. engine, the lubrication is dine with the help of .... system.
38. The octane number of CNG (Compressed Natural Gas) is ....
39. Normal heptane is arbitrarily rated as .... octane number.
40. It is desirable to use .... free gasoline for the better life of noble metal catalytic converter.
41. The main exhaust emissions of an S.I. engine are ....
42. The most harmful pollution of a C.I. engine is ....
43. The most popular device for measuring engine output is ....
44. Standard firing order for a 4-cylinder petrol engine is ....
45. Petrol engines employ .... governing.
46. The 4-stroke cycle is completed in .... revolutions of the crankshaft.
47. Advancing spark timing in S.I. engines .... the tendency to knock.
48. Knocking tendency in C.I. engines increases with .... of compression ratio.
49. Cetane number of diesel commercially available in India is ....
50. Early or late injection of fuel in C.I. engine would result in .... of power.
51. Optimum spark advance is the spark timing at which the engine develops maximum ....
52. For low hydrocarbon emissions from an engine, the surface to volume ratio should be ....
53. In an S.I. engine, the specific fuel consumption is minimum at .... mixture.
54. A compensating jet device is used in a carburettor for ....
55. For S.I. engine knock occurs at the .... of combustion and for C.I. engines knock occurs at the .... of combustion.
56. In Morse test the .... is kept constant while any cylinder is cut-off.
57. Brake specific fuel consumption and brake thermal efficiency are related to each other as ....
58. The maximum fuel economy occurs at an A/F ratio of about .... (F/A = ...).
59. .... engines produce more environmental pollution.

60. The antiknock quality of a fuel when used in a S.I. engine appears to be poorest in the normal .... and best in......... .
61. The two basic ignition systems are .... ignition system and .... ignition system.
62. The two basic methods of circulation of the coolant are .... circulation and .... circulation.
63. The cetane number of diesel and alcohol are about .... and .... respectively.
64. For lubrication the wet sump system is employed in relatively .... engines whereas the dry pump system is employed in .... stationary engines.
65. Octane number of a fuel .... with increase in SIT.
66. Specific power of a 2-stroke engine is about .... times that of 4-stroke engine.
67. A turbo-charger is driven by the .... energy of the engine.
68. Tappet clearance is provided to take care of .... of valve steam.
69. Use of leaner mixtures .... the CO in exhaust of S.I. engines.
70. Exhaust smoke level .... drastically at full-load of C.I. engine.
71. The reference fuels for defining octane number are .... and ....
72. The catalytic converter is to control three main autoexhaust emissions .... , ... , and ....
73. The Wankel engine is called .... engine.
74. The diesel engine is manually started by using a ....
75. .... is done for increasing the .... efficiency of a C.I. engine.
76. Ignition delay increases as the inlet air temperature ....
77. In a S.I. engine, the flame speed is maximum at .... mixture.
78. Spark is generally provided at .... TDC.
79. Best fuel economy in a S.I. engine is obtained at .... mixture.
80. Octane number of normal heptane is ....

ANSWERS

D. Fill in the Blanks

1. auto-ignition, delay
2. second phase
3. evaporation
4. iso-octane, iso-octane and n-heptane
5. iso-octane and n-heptane
6. cetane and α-methyl-napthalene
7. supercharging, volumetric
8. diesel, rack and pinion
9. air-fuel, quantity
10. less, more
11. external, automobile
12. high, leakage
13. infrared, analyser
14. stoichiometric, volumetric
15. decreases
16. C_nH_{2n+2}
17. cooling system, combustion
18. 1-5-3-6-2-4
19. crankshaft, torsional
20. stoichiometric
21. 100
22. greater
23. efficiency, power output
24. supercharging
25. between 40 and 60
26. ppm (parts per million)
27. catalytic converter
28. renewable
29. increases, size
30. end, beginning
31. inertia ; gas pressure
32. C.I.
33. pilot, source
34. lower peak
35. rack, pinion
36. thermostat
37. mist
38. above 100
39. zero
40. lead
41. CO, NO_x, HC
42. NO_x
43. hydraulic dynamometer
44. 1-3-4-2
45. throttle
46. two
47. increases
48. decrease
49. 45

50. loss	**51.** power	**52.** low
53. lean	**54.** mixture, compensation	**55.** end, beginning
56. speed	**57.** s.f.c. $\propto \dfrac{1}{\eta_{th}}$	**58.** 17 : 1, 0.06
59. Diesel	**60.** heptane, iso-octane	**61.** battery, magneto
62. natural, forced	**63.** 40, 5	**64.** high speed, slow speed
65. increases	**66.** two	**67.** exhaust
68. thermal expansion	**69.** reduces	**70.** increases
71. iso-octane, *n*-heptane	**72.** NO_2, CO, HC	**73.** rotary I.C.
74. decompression lever	**75.** Supercharging, volumetric	**76.** decreases
77. about 10 percent rich	**78.** 20° before	**79.** about 10 percent lean
80. zero.		

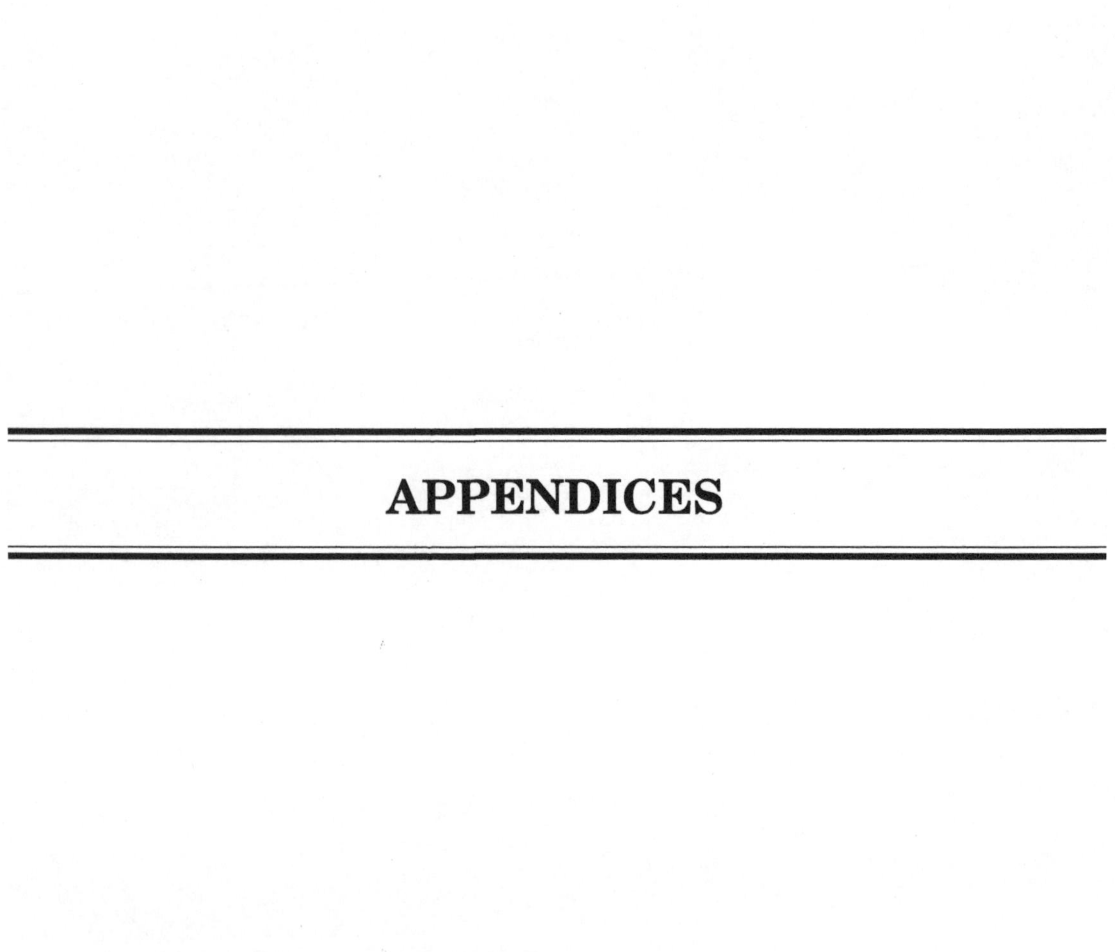

APPENDICES

Appendices

APPENDIX-I
GAS TABLES FOR AIR (SI UNITS)

$T(K)$	$h(kJ/kg)$	p_r	$u(kJ/kg)$	v_r	$\phi(kJ/kg\ K)$
200	199.97	0.3363	142.56	1707	1.29559
210	209.97	0.3987	149.69	1512	1.34444
220	219.97	0.4690	156.82	1346	1.39105
230	230.02	0.5477	164.00	1205	1.43557
240	240.02	0.6355	171.13	1084	1.47824
250	250.05	0.7329	178.28	979	1.51917
260	260.09	0.8405	185.45	887.8	1.55848
270	270.11	0.9590	192.60	808.0	1.59634
280	280.13	1.0889	199.75	738.0	1.63279
285	285.14	1.1584	203.33	706.1	1.65055
290	290.16	1.2311	206.91	676.1	1.66802
295	295.17	1.3068	210.49	647.9	1.68515
300	300.19	1.3860	214.07	621.2	1.70203
305	305.22	1.4686	217.67	596.0	1.71865
310	310.24	1.5546	221.25	572.3	1.73498
315	315.27	1.6442	224.85	549.8	1.75106
320	320.29	1.7375	228.42	528.5	1.76690
325	325.31	1.8345	232.02	508.4	1.78249
330	330.34	1.9352	235.61	489.4	1.79783
340	340.42	2.149	242.82	454.1	1.82790
350	350.49	2.379	250.02	422.2	1.85708
360	360.58	2.626	257.24	393.4	1.88543
370	370.67	2.892	264.46	367.2	1.91313
380	380.77	3.176	271.69	343.4	1.94001
390	290.88	3.481	278.93	321.5	1.96633
400	400.98	3.806	286.16	301.6	1.99194
410	411.12	4.153	293.43	283.3	2.01699
420	421.26	4.522	300.69	266.6	2.04142
430	431.43	4.915	307.99	251.1	2.06533
440	441.61	5.332	315.20	236.8	2.08870
450	451.80	5.775	322.62	223.6	2.11161
460	462.02	6.245	329.97	211.4	2.13407
470	472.24	6.742	337.32	200.1	2.15604
480	482.49	7.268	344.70	189.5	2.17760
490	492.74	7.824	352.08	179.7	2.19876
500	503.02	8.411	359.48	170.6	2.21952

T(K)	h(kJ/kg)	p_r	u(kJ/kg)	v_r	φ(kJ/kg K)
510	513.32	9.031	366.92	162.1	2.23993
520	523.63	9.684	374.36	154.1	2.25997
530	533.98	10.37	381.84	146.7	2.27967
540	544.35	11.10	389.34	139.7	2.29906
550	554.74	11.86	396.86	133.1	2.31809
560	565.17	12.66	404.42	127.0	2.33685
570	575.59	13.50	411.97	121.2	2.35531
580	586.04	14.38	419.55	115.7	2.37348
590	596.52	15.31	427.15	110.6	2.39140
600	607.02	16.28	434.78	105.8	2.40902
610	617.53	17.30	442.42	101.2	2.42644
620	628.07	18.36	450.09	96.92	2.44356
630	628.63	19.84	457.78	92.84	2.46048
640	649.22	20.64	465.50	88.99	2.47716
650	659.84	21.86	473.25	85.34	2.49364
660	670.47	23.13	481.01	81.89	2.50985
670	681.14	24.46	488.81	78.61	2.52589
680	691.82	25.85	496.62	75.50	2.54175
690	702.52	27.29	504.45	72.56	2.55731
700	713.27	28.80	512.33	69.76	2.57277
710	724.04	30.38	520.23	67.07	2.58810
720	734.82	32.02	528.14	64.53	2.60319
730	745.62	33.72	536.07	62.13	2.61803
740	756.44	35.50	544.02	59.82	2.63280
750	767.29	37.35	551.99	57.63	2.64737
760	778.18	39.27	560.01	55.54	2.66176
780	800.03	43.35	576.12	51.64	2.69013
800	821.95	47.75	572.30	48.08	2.71787
820	843.98	52.59	608.59	44.84	2.74504
840	866.08	57.60	624.95	41.85	2.77170
560	888.27	63.09	641.40	39.12	2.79783
880	910.58	68.98	657.95	36.61	2.82344
900	932.93	75.29	674.58	34.31	2.84856
920	955.38	82.05	691.28	32.18	2.87324
940	977.92	89.28	708.08	30.22	2.89748
960	1000.55	97.00	725.02	28.40	2.92128
980	1023.25	105.2	741.98	26.73	2.94468
1000	1046.04	114.0	758.94	25.17	2.96770

$T(K)$	$h(kJ/kg)$	p_r	$u(kJ/kg)$	v_r	$\phi(kJ/kg\ K)$
1020	1068.89	123.4	776.10	23.72	2.99034
1040	1091.85	133.3	793.36	22.39	3.01260
1060	1114.86	143.9	810.62	21.14	3.03449
1080	1137.89	155.2	827.88	19.98	3.05608
1100	1161.07	167.1	845.33	18.896	3.07732
1120	1184.28	179.7	862.79	17.886	3.09825
1140	1207.57	193.1	880.35	16.946	3.11883
1160	1230.92	207.2	897.91	16.064	3.13916
1180	1254.34	222.2	915.57	15.241	3.15916
1200	1277.79	238.0	933.33	14.470	3.17888
1220	1301.31	254.7	951.09	13.747	3.19334
1240	1324.93	272.3	968.95	13.069	3.21751
1260	1348.55	290.8	986.90	12.435	3.23638
1280	1372.24	310.4	1004.76	11.835	3.25510
1300	1395.97	330.9	1022.82	11.275	3.27345
1320	1419.76	352.5	1040.88	10.747	3.29160
1340	1443.60	375.3	1058.94	10.247	3.30959
1360	1467.49	399.1	1077.10	9.780	3.32724
1380	1491.44	424.2	1095.28	9.337	3.34474
1400	1515.42	450.5	1113.52	8.919	3.36200
1420	1539.44	478.0	1131.77	8.526	3.37901
1440	1563.51	506.9	1150.13	8.153	3.39586
1460	1587.63	537.1	1168.49	7.801	3.41247
1480	1611.79	568.8	1186.95	7.468	3.42892
1500	1635.97	601.9	1205.41	7.152	3.44516
1520	1660.23	636.5	1223.87	6.854	3.46120
1540	1684.51	672.8	1242.43	6.569	3.47712
1560	1708.83	710.5	1260.99	6.301	3.49276
1580	1733.17	750.0	1279.65	6.046	3.50829
1600	1757.57	791.2	1298.30	5.804	3.52364
1620	1782.00	834.1	1316.96	5.574	3.53879
1640	1806.46	878.9	1335.72	5.355	3.55381
1660	1830.96	925.6	1354.48	5.147	3.56867
1680	1855.50	974.2	1373.24	4.949	3.58335
1700	1880.1	1025	1392.7	4.761	3.5979
1750	1941.6	1161	1439.8	4.328	3.6336
1800	2003.3	1310	1487.2	3.944	3.6684
1850	2065.3	1475	1534.9	3.601	3.7023
1900	2127.4	1655	1682.6	3.295	3.7354

$T(K)$	$h(kJ/kg)$	p_r	$u(kJ/kg)$	v_r	$\phi(kJ/kg\ K)$
1950	2189.7	1852	1630.6	3.022	3.7677
2000	2252.1	2068	1678.7	2.776	3.7994
2050	2314.6	2303	1726.8	2.555	3.8303
2100	2377.4	2559	1775.3	2.356	3.8605
2150	2440.3	2837	1823.8	2.175	3.8901
2200	2503.2	3138	1872.4	2.012	3.9191
2250	2566.4	3464	1921.3	1.864	3.9474
2300	2629.68	3817	1994.25	1.729	4.78856
2350	2692.11	4197	2018.58	1.607	4.81586
2400	2756.54	4607	2088.71	1.495	4.84257
2450	2820.18	5048	2117.05	1.393	4.86883
2500	2886.61	5520	2166.42	1.300	4.89458
2550	2947.76	6030	2215.90	1.214	4.91986
2600	3011.69	6575	2265.51	1.135	4.94465
2650	3075.74	7159	2315.97	1.0623	4.96906
2700	3131.47	7782	2364.99	0.9957	4.99472
2750	3204.07	8448	2414.96	0.9342	5.01662
2800	3268.38	9159	2462.50	0.8104	5.03981
2850	3335.33	9916	2514.84	0.8248	5.06263
2900	3397.29	10722	2565.00	0.7762	5.08503
2950	3461.99	11579	2615.24	0.7312	5.10710
3000	3526.54	12490	2665.57	0.6893	5.12866
3050	3591.31	13458	2715.89	0.6404	5.15022
3100	3656.78	14483	2766.38	0.6143	5.17132
3150	3720.97	15569	2811.31	0.5806	5.92050
3200	3785.95	16720	2867.54	0.5493	5.21248
3250	3850.93	17939	2918.24	0.5120	5.232704
3300	3916.08	19224	2964.30	0.4926	5.22592
3350	3981.27	20582	3019.91	0.4671	5.27218
3400	4046.50	22016	3070.72	0.4432	5.29165
3450	4111.98	23295	3121.57	0.4250	5.31058
3500	4177.21	25123	3172.71	0.3998	5.32942

APPENDIX-II

GAS CONSTANTS AND SPECIFIC HEATS AT LOW PRESSURES AND 25°C

Gas	*M*	c_p *(kJ/kg-K)*	c_v *(kJ/kg-K)*	γ	*R* *(kJ/kg-K)*
Acetylene (C_2H_2)	26.036	1.6947	1.3753	1.232	0.3195
Air	28.97	1.0047	0.7176	1.4	0.287
Ammonia (NH_3)	17.032	2.089	1.5992	1.304	0.4882
Argon (A)	39.95	0.5208	0.3127	1.666	0.2081
Carbon dioxide (CO_2)	44.01	0.844	0.6552	1.288	0.1889
Corbon monoxide (CO)	28.01	1.0412	0.7444	1.399	0.2968
Chlorine (Cl_2)	70.914	0.4789	0.3617	1.324	0.1172
Ethane (C_2H_6)	30.068	1.7525	1.4761	1.187	0.2765
Ethylene (C_2H_4)	28.052	1.5297	1.2333	1.24	0.2764
Helium (He)	4.003	5.1954	3.1189	1.666	2.077
Hydrogen (H_2)	2.016	14.3136	10.190	1.4	4.125
Hydrazine (N_2H_4)	32.048	1.6453	1.3815	1.195	0.2594
Methane (CH_4)	16.043	2.1347	1.6164	1.321	0.5183
Neon (Ne)	20.183	1.0298	0.6179	1.666	0.4120
Nitrogen (N_2)	28.016	1.0399	0.7431	1.399	0.2968
Oxygen (O_2)	32	0.9185	0.6585	1.395	0.2598
Propane (C_3H_8)	44.094	1.6683	1.4799	1.127	0.1886
Sulphur dioxide (SO_2)	64.07	0.6225	0.4927	1.263	0.1298
Water vapour (H_2O)	18.016	1.8646	1.4033	1.329	0.4615
Xenon (Xe)	131.3	0.1582	0.0950	1.666	0.0633

APPENDIX-III
PHYSICAL PROPERTIES OF SELECTED FLUIDS

Temperature T (K)	*Density ρ (kg/m³)*	*Specific heat (c_p) (J/kg-K)*	*Thermal conductivity K (W/m-K)*	*Thermal diffusivity α × 10⁶ (m²/s)*	*Absolute viscosity μ × 10⁶ (N.s/m²)*	*Kinematic viscosity ν × 10⁶ (m²/s)*	*Prandtl number (Pr)*
Dry air at atmospheric pressure							
273	1.252	1011	0.0237	19.2	17.456	13.9	0.71
293	1.164	1012	0.0251	22.0	18.240	15.7	0.71
313	1.092	1014	0.0265	24.8	19.123	17.6	0.71
333	1.025	1017	0.0279	27.6	19.907	19.4	0.71
353	0.968	1019	0.0293	30.6	20.790	21.5	0.71
373	0.916	1022	0.0307	33.6	21.673	23.6	0.71
473	0.723	1035	0.0370	49.7	25.693	35.5	0.71
573	0.596	1047	0.0429	68.9	39.322	49.2	0.71
673	0.508	1059	0.0485	89.4	32.754	64.6	0.72
773	0.442	1076	0.0540	113.2	35.794	81.0	0.72
1273	0.268	1139	0.0762	240	48.445	181	0.74
Water at saturation pressure							
273	999.3	4226	0.558	0.131	1794	1.789	13.7
293	998.2	4182	0.597	0.143	993	1.006	7.0
313	992.2	4175	0.633	0.151	658	0.658	4.3
333	983.2	4181	0.658	0.159	472	0.478	3.00
353	971.8	4194	0.673	0.165	352	0.364	2.25
371	958.4	4211	0.682	0.169	278	0.294	1.75
393	963.4	4501	0.685	0.170	139	0.160	0.95
Refrigerant R12 (CCl_2F_2), saturated liquid							
223	1547	875.0	0.067	5.01	4.796	0.310	6.2
233	1519	884.7	0.069	5.14	4.238	0.279	5.4
243	1490	895.6	0.069	5.26	3.770	0.253	4.8
253	1461	907.3	0.071	5.39	3.433	0.235	4.4
263	1429	920.3	0.073	5.50	3.158	0.221	4.0
273	1397	934.5	0.073	5.57	2.990	0.214	3.8
283	1364	949.6	0.073	5.60	2.769	0.203	3.6
293	1330	965.9	0.073	5.60	2.633	0.198	3.5
303	1295	983.5	0.071	5.60	2.512	0.194	3.5
313	1257	1001.9	0.069	5.55	2.401	0.191	3.5
323	1216	1021.6	0.067	5.45	2.310	0.190	3.5
Unused engine oil, saturated liquid							
273	899.1	1796	0.147	911	3848	4280	471
293	888.2	1880	0.145	872	799	900	104
313	876.1	1964	0.144	834	210	240	28.7
333	864.1	2047	0.140	800	72.5	83.9	10.5
353	852.0	2131	0.138	769	32.0	37.5	4.90
373	840.0	2219	0.137	738	17.1	20.3	2.76
393	829.0	2307	0.135	710	10.3	12.4	1.75
413	816.9	2395	0.133	686	6.54	8.0	1.16
433	805.9	2483	0.132	663	4.51	5.6	0.84

APPENDIX-IV
HYDROCARBON FUELS

S. No.	Fuel group and general formula	Fuel	Molecular weight	Specific gravity	Bailing temp. at 1 atm, °C	Ignition temp. at 1 atm, °C	Constant pr. heating value kJ/kg		Critical compression ratio	A/F ratio	Octane rating research/Motor			
											ml TEL/Sal			
							Higher	Lower			0	3	0	3
1	2	3	4	5	6	7	8	9	10	11	12	13	14	15
1.	Paraffins	Methane, CH_4	16	0.3	– 162	732	55010	49538	12.6	17.2	120		120	
	C_nH_{2n+2}	Ethane, C_2H_6	30	0.37	– 89	504	51565	47148	12.4	16.1	115		99	
		Propane, C_3H_8	44	0.5	– 42	—	49974	45979	12.2	15.7	112		97	
		Butane, C_4H_8	58	0.579	– 0.6	431	49128	45343	5.5	15.5	94	104	90	104
		Hexane, C_6H_{14}	86	0.659	69	261	48312	44736	3.3	15.2	25	65	26	65
		Heptane, C_7H_{16}	100	0.684	98	247	60633	44560	3.0	15.2	0	44	0	47
		Triptane, C_7H_{16}	100	0.690	81	454	47947	44435	14.4	15.2	112		101	116
		Iso-octane, C_8H_{18}	114	0.692	99.4	447	47813	44347	7.3	15.1	100	116	100	116
		Dodecane, $C_{12}H_{26}$	170	0.749	216		47474	44116						
		Hexadecane, (Cetane) $C_{16}H_{34}$	226	0.773	287		47269	43957						
2.	Napthnes	Cyclo-hexane, C_6H_{12}	84	0.779	81	270	46582	43442	4.9	14.8	84	97	78	87
	C_nH_{2n}	Cyclo-heptane, C_7H_{14}	98	0.816	119	—	46708	43534	3.4	14.8	39	60	41	65
		Cyclo-octane, C_8H_{16}	112	0.841	151	—	46725	43551		14.8	71		58	
3.	Aromatics	Benzene, C_6H_6	78	0.879	80	592	41835	40143		13.3	—	—	115	
	C_nH_{2n-6}	Toluene, C_7H_8	92	0.867	111	568	42437	40528	15	13.5	120	120	109	113
		Xylene, C_8H_{10}	106	0.864	139	563	42877	40805	15.5	13.7	118	120	115	120
4.	Alcohol	Methanol, CH_3OH	32	0.792	—	—	22726	20105		6.4	107			
		Ethanol, C_2H_5OH	46	0.785	—	—	29726	26992		9.0		102		

APPENDIX-V
DATA ON FUEL PROPERTIES

Fuel	Formula (phase)	Molecular weight	Specific gravity	Heat of vaporization kJ/kg	Specific heat: Liquid kJ/kgK	Specific heat: Vapour c_p kJ/kg K	Higher heating value kJ/kg	Lower heating value kJ/kg	LHV of stoich. mixture kJ/kg	(A/F)	(F/A)	Fuel octane rating: RON	+ MON
Commercial petroleum fuels													
Gasoline	$C_nH_{1.87n}$ (1)	110	0.72–0.78	305	2.4	1.7	47300	44000	2830	14.6	0.0685	92.98	80-90
Light diesel	$C_nH_{1.8n}$ (1)	170	0.84–0.88	270	2.2	1.7	44800	42500	2740	14.5	0.0690	—	—
Heavy diesel	$C_nH_{1.7n}$ (1)	200	0.82–0.95	230	1.9	1.7	43800	41400	2760	14.4	0.0697	—	—
Natural gas	$C_nH_{3.8n}$ N_{01n}(g)	18	0.79	—	—	2	50000	45000	2900	14.5	0.069	—	—
Pure hydrocarbons													
Methane	CH_4(g)	16.04	(0.72)	509	0.63	2.2	55500	50000	2720	17.23	0.0580	120	120
Propane	C_3H_8(g)	44.10	0.51 (20)	426	2.5	1.6	50400	46400	2750	15.67	0.0638	112	97
Iso-octane	C_8H_{18}(I)	114.23	0.692	308	2.1	1.63	47800	44300	2750	15.13	0.0661	100	100
Cetane	$C_{16}H_{34}$(I)	226.44	0.773	358		1.6	47300	44000	2780	14.82	0.0675	—	—
Benzene	C_6H_6(I)	78.11	0.879	433	1.72	1.1	41900	40200	2820	13.27	0.0753	—	115
Toluene	C_7H_8(I)	92.14	0.867	412	1.68	1.1	42500	40600	2790	13.50	0.0741	120	109
Alcohols													
Methanol	CH_4O(I)	32.04	0.792	1103	2.6	1.72	22700	20000	2680	6.47	0.155	106	92
Ethanol	C_2H_6O(I)	46.07	0.785	840	2.5	1.93	29700	26900	2690	9.00	0.111	107	89
Other Fuels													
Carbon	C(s)	12.01	– 2	—	—	—	33800	33800	2700	11.51	0.0869		
Carbon monoxide	CO(g)	28.01	(1.25)	—	—	1.05	10100	10100	2910	2.467	0.405		
Hydrogen	H_2(g)	2.015	(0.0901)	—	—	1.44	142000	120000	3400	34.3	0.0292		

APPENDIX-VI

REQUIREMENTS FOR MOTOR GASOLINES (IS : 2796-1971)

S.No.	Characteristics	Requirements	
		83 Octane	*93 Octane*
(*i*)	Colour, visul	Orange	Red
(*ii*)	Copper-strip corrosion for 3 hours at 50°C	Not worse	No. 1
(*iii*)	Density at 150°C	Not limited but to be reported	
(*iv*)	Distillation :		
	(*a*) Initial boiling point	Not limited but to be reported	
	(*b*) Recovery up to 70°C, per cent by volume, *Min*	10	10
	(*c*) Recovery up to 125°C, per cent by volume, *Min*	50	50
	(*d*) Recovery up to 180°C, per cent by volume, *Min*	90	90
	(*e*) Final boiling point, *Max*	215°C	215°C
	(*f*) Residue, per cent by volume, *Max*	2	2
(*v*)	Octane number (Research method), *Min*	83	93
(*vi*)	Oxidation stability, in minutes, *Min*	360	360
(*vii*)	Residue on evaporation, mg/100 ml. *Max*	4.0	4.0
(*viii*)	Sulphur, total, per cent by weight, *Max*	0.25	0.20
(*ix*)	Lead content (as Pb), g/1, *Max*	0.56	0.80
(*x*)	Reid vapour pressure at 380°C, bar, *Max*	0.70	0.70

APPENDIX-VII

REQUIREMENT FOR DIESEL FUELS (IS : 1460-1974)

S.No.	Characteristics	Requirements	
		HSD	*LDO*
(*i*)	Acidity, inorganic	NIL	NIL
(*ii*)	Acidity, total mg of KOH/g (max)	0.50	—
(*iii*)	Ash, per cent by mass, max	0.01	0.02
(*iv*)	Carbon residue (Ramsbottom), per cent by mass (max)	0.20	1.50
(*v*)	Cetane number (min)	42	—
(*vi*)	Pour point (max)	6°C	12°C for winter 18°C for summer
(*vii*)	Copper strip corrosion for 3 hrs at 100°C	90	—
(*viii*)	Flash point :		
	(*a*) Abel, °C (min)	38	—
	(*b*) Pensky-Martens (closed), °C, (min)	—	66
(*ix*)	Kinematic viscosity, CS, at 38°C	20 to 7.5	2.5 to 15.7
(*x*)	Sediment, per cent by mass, (max)	0.05	0.10
(*xi*)	Total sulphur, per cent by mass, (max)	1.0	1.8
(*xii*)	Water content, per cent by volume, (max)	0.05	0.25
(*xiii*)	Total sediments, mg per 100 ml, (max)	1.0	—

APPENDIX-VIII
SPECIFICATIONS FOR DIESEL FUELS

Property		*BS 2869*	
		Class A1	*Class A2*
Viscosity, kinematic at 37.8°C (centistokes)	min.	1.5	1.5
Cetane number	max.	5.0	5.5
Carbon residue, Ramsbottom per cent by mass on 10 per cent residue	min.	50	45
Distillation, recovery at 350°C, per cent by volume	max.	0.2	0.2
Flashpoint, closed	min.	85	85.0
Pensky Martins °C	min.	56	56
Water content, per cent by volume	max.	0.05	0.05
Sediment, per cent	max.	0.01	0.01
Ash, per cent by mass	max.	0.01	0.01
Sulphur	max.	0.3	0.5
Copper corrosion test	max.	1	1
Cold filter plugging point (°C) max.	Summer (16 Mar./30 Sept.)	– 4	0
	Winter (1 Oct./15 March)	– 15	– 9

APPENDIX-IX
AVIATION TURBINE FUELS

Characteristics	*ASTM D1655*		*Mil-J-5624*		*Mil F-46005 Å*
Designation	Jet A	JP-1	JP-4 (up M = 2)	JP-5 M = 2 to 3	CITE-II
Flashpoint, °C (min-max)	43 – 46	43 (min)			
Freezing point, °C (max)	– 40	– 60	– 60	– 48	– 55
Gravity, API (min-max)	39 – 51	35 (max)	45 – 57	36 – 48	
Vapour pressure, Reid, icef/cm^2 cause (min-max)			0.14-0.21		0.21
Distillation, °C					
10% max	204	210		204	93
20% max			143		
50% max	232		188		163
90% max		254	243		
EP max	288	300		288	288
Heating value, lower, kcal/kg, min	10220	10167	10220	10167	
Sulphur, % by mass, max	0.3	0.2	0.4	0.4	0.4
Smoke point, mm (min)	25			20	
Aromatic, vol % max	20	20	25	25	25
Potential gum, mg/100 ml (max)	14	8	14	14	14

APPENDIX-X
SAE CLASSIFICATION OF LUBRICATING OILS

SAE viscosity number	*Viscosity units*	*Viscosity ranges*			
		18° C		*99°C*	
		Min.	*Max.*	*Min.*	*Max.*
5W	Centipoises	—	1200	—	—
	SUS	—	6000	—	—
10W	Centipoises	1200	2400	—	—
	SUS	6000*	12000	—	—
20W	Centipoises	2400	9600	—	—
	SUS	12000**	48000	—	—
20	Centistokes	—	—	5.7	9.6
	SUS	—	—	45	58
30	Centistokes	—	—	9.6	12.9
	SUS	—	—	58	70
40	Centistokes	—	—	12.9	16.8
	SUS	—	—	70	85
50	Centistokes	—	—	16.8	22.7
	SUS	—	—	85	110

* Minimum viscosity at – 18°C may bo waived provided viscosity at 99° C is below 40 SUS.

** Minimum viscosity at – 18°C may be waived provided viscosity at 99°C is not below 45 SUS.

APPENDIX-XI
MAJOR CLASSES OF ENGINE OIL ADDITIVES AND THEIR PRIMARY FUNCTIONS

Additive class	*Function*
Detergent	Control of high-temperature deposits. If overbased, also acts as effective acid neutralizer.
Dispersent	Control of low-temperature sludge and varnish deposits.
Anti-wear	Reduce wear and prevent scoring, galling and seizure.
Anti-rust	Reduce rusting by acid neutralization of formation by protective film.
V.I. improver	Increase V.I. of oil, thereby reducing sensitivity of oil viscosity to temperature.
Pour point depresent	Reduce pour point of oil by interfering with wax crystallization.
Anti-foam	Reduce oil foaming by causing collapse of bubbles due to air entrainment.
Anti-oxidant	Reduce oil oxidation to protect alloy bearings against corrosive attack.

APPENDIX-XII
TYPICAL COMPOSITIONS OF LPG

Constituents	*% by weight*		
	I	*II*	*III*
Ethane	1.8	4.4	1.7
Propane	93.1	64.0	34.1
Propylene	2.8	28.7	62.1
Butane	2.3	2.9	2.1
Total	100.0	100.0	100.0

APPENDIX-XIII
TYPICAL COMPOSITION OF NATURAL GAS

	Constituents	*% by volume*
1.	Methane, CH_4	80
2.	Ethane	7
3.	Propane	6
4.	Isobutane	1.5
5.	Butane	2.5
6.	Pentane plus vapours	3.0
Total		100.0

APPENDIX-XIV
GASEOUS FUELS

Type of gas	*Main Constituents*	*LHV = Kcal / m³ (Low heat value)*	*Theoretical air requirement (m³ air / m³ gas)*	*Ignition limits in % of air*	*Specific weights (kg / m³)*	*Methane number*
Hydrogen	H_2	2570	2.38	4 – 80	0.0889	0
Methane	CH_4	8550	9.50	5 – 15	0.7170	100.0
Ethane	C_3H_6	15370	16.90	3 – 14	1.3600	43.5
Propane	C_3H_8	22300	23.80	2.1 – 8.5	1.9600	35.0
Butane	C_4H_{10}	29500	32.00	1.5 – 8.5	2.6000	10.5
Blast-furnace gas	CO ; N_2 ; CO_2	700–950	0.74	12 – 70	1.2800	—
PRODUCER GAS						
Manufactured gas	CO ; H_2	100–1300	1.00	20 – 70	1.1000	33.0
Sewage gas (biogas)	CH_4 ; CO_2	5500	6.00	—	1.1000	130.0
Natural gas	CH_4 ; C_2H_6 ; C_3H_8 C_4H_{10} ; N_2	7000 – 9500	8 – 10.7	5 – 15	0.7600	90.0
Liquified gas (LPG)	C_3H_8 ; C_4H_{10}	23000 – 32000	—	1.5 – 20	—	10 – 35

1 Standard cubic metre (m³) equals 1 cubic metre of gas at 0°C and 760 mm of mercury.

APPENDIX-XV
GENERAL COMPOSITION OF BIOGAS PRODUCED FROM FARM WASTES

	Volume %
CH_4—Methane	54 to 70
CO_2—Carbon dioxide	27 to 45
H_2—Hydrogen	1 to 10
CO—Carbon monoxide	0 to 1
O_2—Oxygen	0 to 1
H_2S—Hydrogen sulphide	Traces

APPENDIX-XVI
PROPERTIES OF GOBAR GAS AS I.C. ENGINE FUEL

Calorific value	21500 kJ/m^3 19550 kJ/kg
Density	1.1 kg/m^3
Stoichiometric requirements for combustion (Air/Fuel Ratio)	6 m^3 Air/m^3 gas
Flammability limits	5 to 15% vol. in air-gas mixture
Knock resistance	130 methane number (Pure methane 100)
Ignition temperature	About 700°C

APPENDIX-XVII
TYPICAL VALVE TIMINGS FOR FOUR-STROKE SI ENGINES

Position	*Theoretical*	*Actual*	
		Low speed engine	*High speed engine*
Inlet valve opens (IVO) Inlet valve closes (IVC) Inlet valve is open for	TDC BDC 180°	10° b TDC 10° a BDC 200°	10° b TDC 60° a BDC 250°
Exhaust valve opens Exhaust valve closes Exhaust valve is open for	BDC TDC 180°	25° b BDC 5° a TDC 210°	55° b BDC 20° a TDC 255°
Valve overlap Spark.	Nil TDC	15° 15° b TDC	30° 30° b TDC

Note : Valve timing is different for different makes of engines.

b—before, a—after, TDC—Top dead centre, BDC—Bottom dead centre.

APPENDIX-XVIII
PORT TIMINGS FOR DIFFERENT TWO-STROKE ENGINES.

Engine type	*Exhaust*		*Lead angle deg.*	*Scavenge*		*Super-charge deg.*
	Opens	*Closes*		*Opens*	*Closes*	
1. Crankcase scavenging	60–76	60–67	10–20	45–60	45 – 60	0
2. Loop scavenging	60–76	60–76	10–20	45–60	45 – 60	0
3. Uniflow scavenging	80–90	38–59	20–40	38–55	38 – 55	0 – 70
4. Opposed piston	60–67	50–67	10–20	50–65	50 – 65	10 – 20

APPENDIX-XIX
APPLICATION OF I.C. ENGINES

Application	*Approx., engine power, kW*	*Predominant type SI or CI*	*Cycle*	*Cooling [A = Air W = water]*
1. Road vehicles				
(*i*) Mopeds, Scooters, Motor cycles	0.75 – 7	SI	2, 4	A
(*ii*) Small passenger car	15 – 75	SI	4	A, W
(*iii*) Large passenger car	75 – 200	SI	4	W
(*iv*) Light commercial	35 – 150	SI, CI	4	W
(*v*) Heavy commercial	120 – 300	CI	4	W
2. Railway locomotives	400 – 3000	CI	2, 4	W
3. Small aircraft				
(*i*) Helicopters	45 – 1500	SI	4	A
(*ii*) Airplanes	45 – 2700	SI	4	A
4. Marine				
(*i*) Motor boats	0.5 – 7.5	SI, CI	4	W
(*ii*) Ships	3500 – 22000	CI	2, 4	W
5. Off-road vehicles				
(*i*) Light vehicles (factory, airport etc.)	1.5 – 15	SI	2, 4	A, W
(*ii*) Agriculture	3 – 150	SI, CI	2, 4	A, W
(*iii*) Earth moving	40 – 750	CI	2, 4	W
(*iv*) Military	40 – 2000	CI	2, 4	A, W
6. Industrial, Stationary				
(*i*) Electric power	35 – 22000	CI	2, 4	W
(*ii*) Others	5 – 400	CI	2, 4	W
(*iii*) Gas Pipeline	750 – 5000	SI	2, 4	W
7. Home use-lawn movers	0.7 – 3	SI	2, 4	A

APPENDIX-XX
IGNITION SYSTEM CHARACTERISTICS

System (12 volt)	*Rise time* μ *sec*	*Arc duration* μ *sec*	*Energy mJ*	*Available voltage and drop off rpm*	
				kV	*rpm*
Conventional	80 – 200	1000 – 2000	20–60	20 – 25	2000
Transistor (TAC)	60 – 200	1000 – 3000	60–100	20 – 30	3000
Transistor (CD)*	1 – 100	5 – 300	5–100	15 – 30	8000
Magneto					
Low-tension	60+	500	20	18 – 25	6000
High-tension	50+	400	20	18 – 25	6000

* Eight cylinder engine, single coil.
+ At nominal speed (2500 rpm) and same capacitance load.

APPENDIX-XXI
COMPARATIVE IGNITION SYSTEM DATA OF SOME INDIAN AUTOMOBILES

		Hindustan Ambassador	*Fiat 1100*	*Jeep universal*	*Dodge / Fargo Model 89, M4 (Petrol type)*
1.	*Type of ignition*	Battery	Battery	Battery	Battery
2.	*Ignition coil*	Lucas Model LA 12	Prestolite	Prestolite 200713	Prestolite CAH 4001
3.	*Distributor*	Lucas DM 2	Lucas	Prestolite IAY-4401	IBR-4002
4.	*Control breaker gap*	0.36 – 0.40 mm	0.45 ± 0.03 mm	0.5 mm	0.457 – 0.559 mm
5.	*Type of automatic ignition advance*	Centrifugal and vacuum	Centrifugal	Centrifugal	Centrifugal and vacuum
6.	*Spark plug*	Mico Bosch HW 145 T 2 or KLG 50, 14mm	M14-1/225 14 mm 14 mm	Mico Bosch HW 145 T3 or KLG-TF-550,	Mico Bosch HW 145 T3 or KLG-TFS-50
7.	*Spark plug gap*	0.64 mm	0.5–0.6 mm	0.76 mm	0.7—10.81 mm

APPENDIX-XXII

RELATIVE HEAT LOSS DURING DIFFERENT PHASES OF THE CYCLE OF A SI ENGINE FOR A PARTICULAR DESIGN

Part of cylinder	*Fraction of total heat flow to coolant*	
	including piston friction	*not including piston friction*
Head and valve seat	0.5 to 0.55	0.57 to 0.63
Barrel	0.27 to 0.32	0.16 to 0.18
Exhaust port	0.17 to 0.22	0.20 to 0.25

APPENDIX-XXIII

RELATIVE HEAT LOSS TO COOLANT DURING DIFFERENT PHASES OF CYCLE

Phase	*Duration crankshaft degrees*	*Average relative area exposed*	*Estimated average temperature, K*	*Average temperature difference, °C*	*Heat loss*	
					Relative %	*% heat in fuel*
Inlet	180	0.7	320	– 110	– 2	– 0.3
Compression	150	0.7	470	35	1	0.2
Combustion	40	0.4	1600	1170	8	1.2
Expansion	125	0.7	2160	1720	40	6.0
Blow down	90	1.3	1820	1390	48	7.2
Exhaust	135	1.3	1150	720	5	0.7
				Total	100	15

APPENDIX-XXIV
EFFECT OF VARIABLES ON DELAY PERIOD

S.No.	*Increase in variable*	*Effect on delay period*	*Reason*
1.	Cetane number of fuel	Reduces	Reduces self-ignition temperature.
2.	Injection pressure	Reduces	Greater surface-volume ratio hence less physical delay.
3.	Injection advance angle	Increases	Pressures and temperatures lower when injection begins.
4.	Compression ratio	Reduces	Increases air temperature and pressure and reduces auto-ignition temperature.
5.	Intake temperature	Reduces	Increases air temperatures.
6.	Jacket water temperature	Reduces	Increases wall and hence air temperature.
7.	Fuel temperature	Reduces	Better vaporisation and increases chemical reaction.
8.	Intake pressure (supercharging)	Reduces	Increase in density reduces auto-ignition temperature.
9.	Speed	Reduces in milliseconds, increases in crank angle.	Less loss of heat more crank angle in a given time.
10.	Load (fuel-air-ratio)	Decreases	Operating temperature increases.
11.	Engine size	Little effect in milli seconds but crank angle decreases.	Low r.p.m.
12.	Type of combustion chamber	Lower for precombustion chamber	

APPENDIX-XXV
EFFECTS OF CHARACTERISTICS ON KNOCKING TENDENCY

	Characteristics	*Spark ignition engines*	*Compression ignition engines*
1.	Compression ratio	Low	High
2.	Inlet temperature	Low	High
3.	Inlet pressure	Low	High
4.	Ignition temp. of fuel	High	Low
5.	Time lag of fuel	Long	Short
6.	Speed	High	Low
7.	Cylinder size	Small	Large
8.	Combustion chamber wall temperature	Low	High

APPENDIX-XXVI
EXHAUST EMISSION STANDARDS FOR INDIAN VEHICLES

Refer table – 1 for exhaust emission standards for gasoline engines driven vehicles for two wheelers, three wheelers, passenger cars and Diesel engines driven vehicles of gross weight > 3.5 tonnes and also < 3.5 tonnes for Indian driving cycles.

Table-1

Category of vehicles and exhaust emissions	*Standards effective April 1991** **	*Standards effective April, 1996* ***	*Standards proposed April, 2000* ****
GASOLINE :			
Two Wheelers :			
(*a*) CO g/km	12 – 30	4.5	2.0
(*b*) HC g/km	8 – 12	—	—
(*c*) (HC + NO_x) g / km	—	3.6	1.5
Three Wheelers :			
(*a*) CO g / km	12 – 30	6.75	4.0
(*b*) HC g / km	8 – 12	—	—
(*c*) (HC + NO_x) g/km		5.40	1.5
Passenger Cars :			
(*a*) CO g/km	14.3 – 27.1	8.68 – 12.40	2.72
(*b*) HC g/km	2.0 – 2.9	—	—
(*c*) (HC + NO_x) g/km	—	3.00 – 4.36	0.97
DIESEL VEHICLES :			
A : Gross Vehicle Weight > 3.5 tonnes			
(*a*) CO g/kWh	14.0	11.2	4.5
(*b*) HC g/kWh	3.5	2.4	1.1
(*c*) NO_x g/kWh	18.0	14.4	8.0
(*d*) PM g/kWh	—	—	0.36
B : Gross Vehicle Weight < 3.5 tonnes			
(*a*) CO g/kWh	14.3 – 27.1	5.0 – 9.0	
(*b*) (HC + NO_x) g/kWh	2.7 – 6.9	2.0 – 4.0	

* For diesel vehicles, the standards were effective from April 1992.

** Based on reference mass of the vehicles and 'warm start' on Indian driving cycle.

*** Based on capacity of the engine and 'cold start' on Indian driving cycle.

**** Applicable to all categories of the engine and 'cold start' on Indian driving cycle.

Euro Norms at a Glance

The Euro norms for Petrol driven passenger cars, Diesel drawn passenger cars, diesel light duty vehicles and Diesel heavy duty vehicles are given below :

- **Petrol Driven Passenger Cars :**

	1991/92		*1996*	*1998*	*1996*	*2000*		*2005*
	INDIA	EURO-I	INDIA	INDIA	EURO-II	INDIA	EURO-III	EURO-IV
CO g/km	14.3–27.1	2.72	8.68–12.4	4.34–6.20	2.2	2.72	2.3	1.0
HC g/km	2.0–2.9	—	—	—	—	—	0.20	0.1
NO_x g/km	—	—	—	—	—	—	0.15	0.08
HC + NO_x g/kg	—	0.97	3.4 – 4.36	1.5–2.18	0.57	0.97		

Note : 1. Norms for Passenger cars fitted with catalytic converter only.

2. In case of Euro III and Euro IV COP = type approved forms.

- **Diesel Driven Passenger Cars :**

	1991/92		*1996*		*2000*	
	INDIA	EURO-I	INDIA	EURO-II	INDIA	EURO-III
CO g/km	14.0 (g/kWh)	2.72	5.0–9.0	1.00	2.72–6.90	0.6
HC g/km	3.5 (g/kWh)	–	–	—	—	—
NO_x g/km	18.0 (g/kWh)	—	—	—	—	—
HC + NO_xg/km	—	0.97 (IDI)	2.0–4.0	0.7	0.97–1.70	0.56
PM g/km	–	1.36 (DI)				
		0.14 (IDI)	—	0.08	0.14 – 0.25	0.05
		0.19 (DI)	—	—	—	—

Note : 1. In case of Euro II and Euro III type approved forms = COP norms.

2. In India there is also option for testing with engine dynamometer (in g/kWh) as these come under diesel light duty vehicles GVW < 3.5 tonnes.

3. In European norms passenger cars refers to passenger cars with seating capacity less than 6 and GVW less than 2.5 tonnes.

- **Diesel Light Duty Vehicles < 3.5 tonnes :**

	1991/92		*1996*		*2000*
	INDIA	EURO-I	INDIA	EURO-II	INDIA
CO	14.0 (g/kWh)	2.72 – 6.90[2] (g/km)	11.2[1] g/km or 5.0–9.0[3]	1.0–1.5	4.5[1] (g/kWh) or 2.75 – 6.90[3] (g/km)
HC	3.5[1] (g/kWh)	—	2.4[1] g/kWh or HC + NO_x norms	—	1.1 g/kWh or HC + NO_x norms
NO_x	18 (g/kWh)	—	14.4[1] g/kWh)	—	8.0[1] g/kWh
HC + NO_x	—	0.97–1.7[3] (g/km)	2.0 – 4.0[3] (g/km)	0.7–1.3[2]	0.97 – 1.70[3] (g/km)
PM	—	0.14–0.7 (g/km)	—		0.61 kWh or 0.14 – 0.25 (g/km[3])

1 = 13 mode 2 = EDC + EUDC 3 = Indian driving cycle.

- **Diesel Heavy Duty Vehicles > 3.5 tonnes :**

	1991/92		*1996*		*2000*	
	INDIA	EURO-I	INDIA	EURO-II	INDIA	EURO-III
CO g/kWh	14.0	4.5	11.2	4.00	4.5	2.1
HC g/kWh	3.5	1.10	2.4	1.10	1.10	0.66
NO_x g/kWh	18.0	8.00	14.4	7.00	8.00	5.0
PM > 85 kWg/kWh	—	0.36	—	0.15	0.36	0.1
PM < 85 kWg/kWh	—	0.61	—	0.15	0.61	0.1

APPENDIX-XXVII
COMPARISON OF THERMAL AND CATALYTIC CONVERTER

Oxidising systems	*Catalytic reactors*	*Thermal reactors*
Possible Advantages		
Reduces HC and CO	80–90%	80–90%
Reduces NO	No	No
Reduces aldehydes	50% or more	50% or more
Use same design for all vehicles	Hopefully	Hopefully
Long life	Up to 50000 miles	Up to 100000 miles
Possible Disadvantages		
Cost	High	High
Volume	High	Higher
Engine mounting required	No	Yes
Weight added	Some	Significant
Container durability problem	Yes	Yes
Potential over temp. problem	Yes	Yes
Raises engine compartment temperature	Depends upon location	Yes
Requires non-leaded fuel	Probably	No
Requires air injection	Some do	Some do
Lowers fuel economy	No	Probably yes (depends upon mixture requirement)
Decreases power	Depends upon back Pressure	Depends upon back Pressure
Loss of catalytic material due to attrition	Yes	No
May emit other toxic material	Yes	No

APPENDIX-XXVIII
EMISSION CHARACTERISTICS OF VARIOUS ENGINES

Emission	*Four-stroke turbocharged*	*Two-stroke air scavenged*	*Four-stroke, normally aspirated*	
			Medium speed	*High speed*
1. HC	Low	High	Low	High
2. NO_x	High	Medium	Low	Low
3. RCHO	Medium	Low	Low	High
4. Smoke	Low	Low	High	High

APPENDIX-XXIX
INFLUENCE OF OPERATIONAL MODES ON EMISSION LEVELS

Engine exhaust constituent	*Concentration values as measured in exhaust gas*			
	Idle	*Acceleration*	*Partial load*	*Full load*
Two-cycle engine				
1. HC, ppm C	250	500	350	550
2. NO_x, ppm	180	1200	1100	1250
3. RCHO, ppm	17.0	9.3	1.0	5.5
4. Smoke, Hartridge unit	4	44	4	10
5. Odour, diesel intensity truk	3.6	4.1	3.0	3.5
6. CO %	0.01	0.24	0.01	0.34
7. CO_2 %	0.83	5.42	3.79	5.29
Four cycle normally aspirated medium speed engine				
1. HC, ppm C	180	330	210	150
2. NO_x, ppm	330	920	590	780
3. RCHO, ppm	7.9	7.5	4.9	1.6
4. Smoke, Hartridge unit	4	44	4	10
5. Odour, diesel intensity (Truk)	3.6	4.1	3.4	3.5
6. CO %	0.02	0.08	0.04	0.26
7. CO_2%	2.56	3.40	5.33	6.68

APPENDIX-XXX
EXHAUST EMISSION FROM SELECTED ENGINES, mg/g FUEL

Engine	*CO*	*HC (as hexane)*	*NO_x*
1. SI engine, 50 kmph	241	6.1	16.3
2. SI engine, cold start	513	35	14.5
3. SI engine normal load	407	35	13
4. Regenerative gas turbine, 50 kmph	5.3	0.3	13.5
5. Gas turbine, cold start	5.0	1.1	—
6. Aircraft turbojet	3.3	0.4	5.5

APPENDIX-XXXI
VEHICLE EMISSION

• Pollutants Produced by Automobiles :

The incomplete and high temperature combustion process in automobile engines result in the formation of Carbon monoxide (CO), Hydrocarbons (HC) and Oxides of Nitrogen (NO_x) and are emitted into atmosphere. These gases are unpleasant to breathe, and are in many cases harmful and even dangerous to human beings, animals and plants. The harmful effects of these emission are given below :

Harmful action of pollutants

Pollutant	Major source in atmosphere	Major harmful actions	Results
CO	Automobiles 93%, power generation, etc. 7%	Hinders the exchange of oxygen in the blood and causes carbon monoxide poisoning. (Atmospheric CO at concentration of 30–40 ppm* numbs or paralyses the autonimic nervous system at 500 ppm or higher concentrations. It causes shortness of breath and headaches when body movements are attempted. At very high concentrations, it can cause death).	—
HC	Automobiles 57% Petroleum refining use of solvents, etc. 43%.	Irritates the linings of the respiratory organs.	A cause of photochemical smog**.
NO_x	Automobile 39%, factories, power generation, petroleum refining, etc. 61%	Irritates the eyes, nose and throat ; if the irritations is severe, it causes coughing, headaches and lung damage. Atmospheric NO_2 gives off an irritating odour at 3-5 ppm. Irritates the eyes and nose at 10-30 ppm. Causes coughing, headaches and dizziness at 30-50 ppm.	The major cause of photochemical smog**.
SO_2	*Automobiles (diesel) 1% * Factories, power generation, heating systems, etc. 99%.	Irritates respiratory system membranes and causes inflammation of the windpipe.	—

* ppm : Abbreviation for parts per million. Used as a unit to indicate concentration or content.
** Photochemical smog :
When HC and NO_x build up in the atmosphere and are struck by the sun's rays, a photochemical (light + chemical) reaction takes place which produces various types of chemical compounds and oxidants (chiefly O_3, which has strong oxidation properties, and results in the phenomenon known as *"smog"*). Photochemical smog obstructs the range of vision, irritates the eyes, and is alleged to be carcinogenic and to cause blight in forests.

• The Automotive Emission Control in India

In India, the air pollution control activities are governed by Ministry of Environment and Ministry of Surface Transport. The enforcement of vehicular emission is carried out by State Governments. The Emission Norms were notified by Ministry of Surface Transport to enforce in vehicles.

1. Motor cycle and three wheelers CO%–4.5%.
2. Motor cars–3.0%
3. Diesel engines–65 HSU (Heritage Smoke Unit)

At the production stage, the emission norms were set where the vehicle emission test was conducted (mass emission test).

The emission standard of US and Japan are more stringent as compared with EEC. In India the emission standard to be enforced to the manufacturer are adopted from EEC. The Supreme Court of India has directed in April 1999 that the Emission Standard which were scheduled to be enforced in the year 2005 should be preponed and be enforced in 1999. The norms to be adopted for year 2000 were called EURO-I and for year 2005 were known as EURO-II.

The basic difference of Emission of EURO II, EURO I are as under :

I. Passenger Cars	**gm/km**	**Euro-I Year 2000**	**Euro-II Year 2005**
Petrol Fuel	CO	2.30	1.00
	HC	0.20	0.10
	NO_x	0.15	0.08
Diesel Fuel			
	CO	0.64	0.50
	HC + HO_x	0.56	0.30
	NO_x	0.50	0.25
	PM	0.05	0.025

The EEC directions were implemented in European countries, EURO-II by (new models) Jan. 1, 1996 and all models Jan. 1, 1997, for Diesel Engines Oct. 1, 1999 (Direct Injection). However EURO-I was implemented in July 1992, itself.

II. For Heavy Duty vehicles the emission norms are as under :

	CO	NO_x	HC	PM
EURO-I	4.5	8.0	1.1	0.36–0.61
EURO-II	4.0	7.0	1.1	0.15–0.25

APPENDIX-XXXII

ASSEMBLY AND DISASSEMBLY OF PETROL AND DIESEL ENGINES OF AN AUTOMOBILE

A. *Engine mounting :*

- Automobile engines are mounted on the *frame.*
- According to the type of mounting used, it is attached to the frame at two, three or four points.
- At each mounting point, a flexible or elastic medium like rubber or spring is usually placed to insulate the automobile frame and body from vibration and noise.
- In a method of engine mounting, called *"floating power"*, only two rubber supports are used at the front and another at the rear of the engine. The front support is located above the rear support so that a line drawn through the centres of the front and rear supports passes through the centre of gravity of the engine.

B. *Engine removal :*

The procedure for removing the engine from the vehicle for *overhaul* is as follows :

1. Drain the cooling system by opening the drain cock.
2. Disconnect the battery at the positive terminal to avoid the possibility of short circuit.
3. Remove the air cleaner horn from the carburettor and disconnect the breather hose at the oil filler pipe.
4. Disconnect the carburettor choke and throttle controls by loosening the clamp bolts and set screws.
5. Disconnect the fuel tank from fuel pump line by unscrewing the connecting nut.
6. Plug the fuel line to prevent fuel leakage.
7. Remove the radiator stay bar.
8. Remove the upper and lower radiator hoses by loosening the hose clamps and slipping the clamps back on the hose.
9. Remove the four bolts from the fan hub and remove the fan hub and fan blades.
10. Remove the radiator attaching screws, and remove the radiator.
11. Remove the starting motor cables. Remove the starting motor.
12. Disconnect the wires from the generator/alternator. Disconnect the ignition primary wire at the ignition coil.
13. Disconnect oil pressure and temperature sending unit-wires at the units.
14. Disconnect the exhaust pipe at the exhaust manifold by removing the stud nuts.
15. Remove the two nuts and bolts from each front engine support. Disconnect the engine ground strap. Remove the engine supports. This will allow the engine to drop slightly and will permit access to the two top bolts on the flywheel housing.
16. Remove the rocker arm cover by removing the two holding nuts.
17. Remove two cylinder head bolts. Fit a suitable engine lifting bracket in place and retighten the cylinder head bolts previously removed. Attach the engine lifting bracket to a chain hoist or other lifting device. Take up all stacks.
18. Remove the bolts which attach the flywheel housing to the engine.
19. Pull the engine forward or roll the vehicle backwards until the clutch clears the flywheel housing. Lift the engine from the vehicle.

C. *Engine Disassembly :*

For disassembling the engine, it should be mounted in a suitable engine repair stand. If the stand is not available, disassembling should be done in a careful manner to prevent accident and damage to engine parts.

Engine disassembly is carried out in a sequence as follows, assuming that the engine is out of the vehicle, all the accessories have been removed and the oil has been drained :

1. Remove water pump.
2. Remove exhaust manifold
3. Remove oil filler tube
4. Remove water outlet fitting.
5. Remove thermostat
6. Remove crankshaft pulley
7. Remove oil pump
8. Remove crankcase ventilation valve

9. Remove rocker arm assembly
10. Remove cylinder head
11. Remove oil pan
12. Remove piston and connecting rod assembly
13. Remove timing gear cover
14. Remove front end plate
15. Remove flywheel housing
16. Remove clutch
17. Remove flywheel
18. Remove crankshaft
19. Remove exhaust valves and springs
20. Remove camshaft
21. Remove valve tappets.
22. Remove oil gallery plugs.

- *After disassembling all the parts, inspect them carefully, replace the defective parts and* **reassemble** *the engine.*

Index

A

B

C

D

E

F

G

T

U

V

W

12.7.5. Comparative Specifications of Some Vehicles

	Maruti Esteem	Fist Sienna		Ford Ikon		Hyundai Accent		Opel Corsa		Mitsubishi Lancer		Daewoo Cielo	Daewoo Naxia	Honda City		Opel Astra		Susuki Baleno
Specs	Petrol	Petrol	Diesel	Petrol	Diesel	Petrol	Diesel	Petrol (1.4 FLS)	Petrol (1.6 GLS)	Petrol	Diesel	Petrol	Petrol	Petrol 1.3	Petrol 1.5	Petrol	Diesel	Petrol
Engine	4-cyl in-line	4-cyl in-line	4-cyl in-line	4-cyl in-line SOHC	4-cyl in-line	4-cyl in-line	4-cyl in-line	4-cyl in-line	4-cyl in-line	4-cyl in-line SOHC	4-cyl in-line	4-cyl in-line	4-cyl in-line DOHC, 16-valve	4-cyl in-line SOHC	4-cyl in-line SOHC	4-cyl in-line	4-cyl in-line Turbo	4-cyl in-line SOHC 16-valve
Displacement	1298 cc	1242 cc	1698 cc	1597 cc	1753 cc	1495 cc	1527 cc	1389 cc	1598 cc	1468 cc	1998 cc	1498 cc	1498 cc	1343 cc	1493 cc	1598 cc	1700 cc	1590 cc
Bore × stroke	74 mm × 75.5 mm	70.8 mm × 78.86 mm	82.6 mm × 79.2 mm	NA	NA	NA	NA 82.0 mm	NA 93.0 mm	NA	75.5 mm × 81.5 mm	82.7 mm × 81.5 mm	76.5 mm × 81.5 mm	76.5 mm × 76.0 mm	75 mm × 84.5 mm	75 mm × 84.5 mm	80 mm × 79.5 mm	82.5 mm × 79.5 mm	75 mm × 90 mm
Maximum Power	65 bhp @ 6000 rpm	72 bhp @ 6000 rpm	645 bhp @ 5500 rpm	92 bhp @ 5600 rpm	58 bhp @ 4800 rpm	94 bhp @ 5500 rpm	57 bhp @ 5000 rpm	88 bhp @ 6000 rpm	92 bhp @ 5600 rpm	87 bhp @ 5500 rpm	68 bhp @ 3500 rpm	80 bhp @ 5600 rpm	92 bhp @ 5600 rpm	90 bhp @ 6400 rpm	100 bhp @ 6400 rpm	76 bhp @ 4500 rpm	68 bhp @ 5400 rpm	98 bhp @ 6000 rpm
Maximum Torque	10.1 kgm @ 4000 rpm	10.2 kgm @ 3000 rpm	12.2 kgm @ 2500 rpm	13.2 kgm @ 3400 rpm	11.02 kgm @ 2200 rpm	12.2 kgm @ 3500 rpm	9.5 kgm @ 2250 rpm	11.1 kgm @ 3200 rpm	12.8 kgm @ 3200 rpm	13.4 kgm @ 3300 rpm	12.5 kgm @ 3000 rpm	11.7 kgm @ 3200 rpm	13.5 kgm @ 3400 rpm	11.3 kgm @ 4700 rpm	13.0 kgm @ 4700 rpm	12.3 kgm @ 2800 rpm	13.4 kgm @ 2400 rpm	13.4 kgm @ 3000 rpm
Power to weight ratio	0.08	0.07	0.06	0.09	0.05	0.09	0.06	0.08	0.09	0.09	0.06	0.08	0.08	0.09	0.1*	0.08	0.06	0.1*
Torque to weight ratio	0.012	0.01	0.011	0.013	0.01	0.012	0.09	0.01	0.012	0.013	0.011	0.012	0.012	0.011	0.013	0.013	0.012	0.014*
Transmission	5-sp manual	5-sp manual	5-sp manual	5-sp manual	5-sp manual	5-sp manual	5-sp manual	5-sp manual	5-sp manual	5-sp manual	5-sp manual automatic	5-sp manual	5-sp manual	5-sp manual automatic	5-sp manual automatic	5-sp manual	5-sp manual	5-sp manual
Steering	Rack & pinion	Rack & pinion power assist	Rack & pinion power assist	Rack & pinion power assist	Rack & pinion power assist	Rack & pinion power assist	Rack & pinion power assist	Rack & pinion power assist	Rack & pinion power assist	Rack & pinion power assist	Rack & pinion power assist	Rack & pinion power assist	Rack & pinion power assist	Rack & pinion power assist	Rack & pinion power assist	Rack & pinion power assist	Rack & pinion power assist	Rack & pinion power assist
Turn circle dia	9.6 m	10.2 m	10.2 m	9.8 m	9.8 m	9.8 m	9.8 m	9.5 m	9.5 m	10.2 m	10.2 m	9.8 m	9.8 m	9.6 m	9.6 m	9.8 m	9.8 m	9.8 m
Front Brake	Disc	Disc	Disc	Disc	Disc	Disc	Disc	Disc	Disc	Disc	Disc	Disc	Disc	Disc	Disc	Disc	Disc	Disc
Rear Brake	Drum	Drum	Drum	Drum	Drum	Drum	Drum	Drum	Drum	Drum	Drum	Drum	Drum	Drum	Drum	Drum	Drum	Drum
Length	4080 mm	4100 mm	4290 mm	4140 mm	4140 mm	4250 mm	4250 mm	4056 mm	4056 mm	4290 mm	4290 mm	4482 mm	4482 mm	4225 mm	4225 mm	4239 mm	4239 mm	4225 mm
Width	1575 mm	1614 mm	1690 mm	1634 mm	1634 mm	1670 mm	1670 mm	1608 mm	1308 mm	1690 mm	1690 mm	1662 mm	1662 mm	1690 mm	1690 mm	1696 mm	1696 mm	1690 mm
Ground clearance	180 mm	140 mm	150 mm	NA	NA	172 mm	172 mm	NA	NA	150 mm	150 mm	165 mm	165 mm	170 mm	170 mm	136 mm	136 mm	170 mm
Wheel base	2365 mm	2360 mm	2500 mm	2486 mm	2486 mm	2440 mm	2440 mm	2443 mm	2443 mm	2500 mm	2500 mm	2520 mm	2520 mm	2500 mm	2500 mm	2517 mm	2517 mm	2480 mm
Kerb Weight	815 kg	1035 kg	1100 kg	917 kg	1077 kg	1006 kg	1006 kg	1060 kg	1070 kg	1010 kg	1095 kg	960 kg	1088 kg	980 kg	985 kg	960 kg	1140 kg	975 kg